# Molecular Breeding in Wheat, Maize and Sorghum

## Strategies for Improving Abiotic Stress Tolerance and Yield

# Molecular Breeding in Wheat, Maize and Sorghum

## Strategies for Improving Abiotic Stress Tolerance and Yield


*Edited by*

**Mohammad Anwar Hossain**
*Department of Genetics and Plant Breeding*
*Bangladesh Agricultural University*
*Mymensingh, Bangladesh*

**Mobashwer Alam**
*Centre for Horticultural Science*
*Queensland Alliance for Agriculture and Food Innovation*
*The University of Queensland*
*Brisbane, Australia*

**Saman Seneweera**
*National Institute of Fundamental Studies*
*Kandy, Sri Lanka*

**Sujay Rakshit**
*ICAR–Indian Institute of Maize Research*
*PAU Campus*
*Ludhiana, India*

**Robert Henry**
*Queensland Alliance for Agriculture and Food Innovation*
*The University of Queensland*
*Brisbane, Australia*



CABI

**CABI is a trading name of CAB International**

CABI
Nosworthy Way
Wallingford
Oxfordshire OX10 8DE
UK

Tel: +44 (0)1491 832111
Fax: +44 (0)1491 833508
E-mail: info@cabi.org
Website: www.cabi.org

CABI
WeWork
One Lincoln Street
24th Floor
Boston, MA 02111
USA

Tel: +1 (617)682-9015
E-mail: cabi-nao@cabi.org



A catalogue record for this book is available from the British Library, London, UK.

**Library of Congress Cataloging-in-Publication Data**

Names: Hossain, Mohammad Anwar, editor. | Alam, Mobashwer, editor. | Seneweera, Saman, editor. | Rakshit, Sujay, editor. | Henry, Robert J, editor.
Title: Molecular breeding in wheat, maize and sorghum : strategies for improving abiotic stress tolerance and yield / Mohammad Anwar Hossain, Mobashwer Alam, Saman Seneweera, Sujay Rakshit, Robert J Henry.
Description: Boston, MA : CAB International, [2021] | Includes bibliographical references and index. | Summary: "This book covers the latest understanding of molecular and genetic bases of abiotic stress tolerance and yield improvement of wheat, maize and sorghum, to develop strategies for improved stress tolerance and enhanced crop productivity"-- Provided by publisher.
Identifiers: LCCN 2021006801 (print) | LCCN 2021006802 (ebook) | ISBN 9781789245431 (hardback) | ISBN 9781789245448 (ebook) | ISBN 9781789245455 (epub)
Subjects: LCSH: Plant molecular genetics. | Wheat--Molecular genetics. | Corn--Molecular genetics. | Sorghum--Molecular genetics.
Classification: LCC QK981.4 .M633 2021 (print) | LCC QK981.4 (ebook) | DDC 572.8/2--dc23
LC record available at https://lccn.loc.gov/2021006801
LC ebook record available at https://lccn.loc.gov/2021006802

ISBN-13: 9781789245431 (hardback)
9781789245448 (ePDF)
9781789245455 (ePub)

DOI: 10.1079/9781789245431.0000

Commissioning Editor: Rebecca Stubbs
Editorial Assistant: Emma McCann
Production Editor: Tim Kapp

Typeset by SPi, Pondicherry, India
Printed and bound in the USA by Integrated Books International, Dulles, Virginia

# Contents

# About the Editors

**Dr Mohammad Anwar Hossain** is serving as a Professor in the Department of Genetics and Plant Breeding, Bangladesh Agricultural University (BAU), Mymensingh, Bangladesh. He received his BSc in Agriculture and MS in Genetics and Plant Breeding from BAU, Bangladesh. He also received an MS in Agriculture from Kagawa University, Japan in 2008 and a PhD in Abiotic Stress Physiology and Molecular Biology from Ehime University, Japan in 2011 through Monbukagakusho scholarship. As a Japan Society for the Promotion of Science (JSPS) postdoctoral researcher, he has worked on isolating low-phosphorus stress tolerance genes from rice at the University of Tokyo, Japan during the period 2015–2017. His current research programme focuses on understanding physiological, biochemical and molecular mechanisms underlying abiotic stresses in plants and the generation of stress-tolerant and nutrient-efficient plants through breeding and biotechnology. He has over 60 peer-reviewed publications and has edited 11 books (including this one) published by CRC Press, Springer, Elsevier, Wiley and CAB International.

**Dr Mobashwer Alam** is a plant geneticist and breeder, and a Research Fellow of the Queensland Alliance for Agriculture and Food Innovation (QAAFI), The University of Queensland. He achieved his doctoral degree from The University of Queensland. He was also a graduate of Bangladesh Agricultural University (MS and BSc). Dr Alam has multiple industry experience – as a Senior Plant Breeder (Grain Sorghum) at Nuseed Pty Ltd, Australia and an Assistant Manager (Seed Inspection, Sugacane) at Bangladesh Sugar and Food Industries Corporation, Bangladesh. Dr Alam was previously working as a Lecturer and Assistant Professor of the Department of Genetics and Plant Breeding at Patuakhali Science and Technology University, Bangladesh. Over the last 20 years, Dr Alam has been conducting multidisciplinary research, including plant breeding, quantitative genetics, genomics, plant physiology and crop modelling. He is currently involved in developing rapid breeding tools by using innovative technologies in selection and phenotyping.

**Professor Saman Seneweera** is a molecular plant physiologist whose main research aim is to improve crop yield and quality. His research broadly focuses on evaluating the impact of climate change on crops and pastures and developing adaptation strategies to mitigate climate stress. He uses cross-disciplinary science, spanning whole-plant biology, physiology, cell biology, molecular genetics and modelling to develop climate-insensitive crops. He is the Senior Professor and Director of the National Institute of Fundamental Studies (NIFS), Sri Lanka and an Honorary Professor at the University of Southern Queensland and University of Melbourne, Australia and Ruhuna University, Sri Lanka. Previously, he was the Discipline Leader of Plant Biology at the University of Southern Queensland and University of Melbourne. He has been playing a major role in building the Australian Grain Free Air $CO_2$ at the University of Melbourne and the Plant Biology Platform at the University of Southern Queensland. As Executive Director, he is now instrumental in establishing new strategic directions for Sri Lanka's premier research institute, the NIFS, for its broad expansion. He is a graduate of the University of Ruhuna (BSc Agri. Sci.), Sri Lanka and gained his PhD from the Western Sydney University, Australia. He is the recipient of many awards including the 'MSLE Research Excellence Award – 2011' from the University of Melbourne, a Japan Society for the Promotion of Science (JSPS) Fellowship in 2001 and a Science and Technology Fellow, Japan, 1999.

**Dr Sujay Rakshit** is Director of ICAR–Indian Institute of Maize Research, Ludhiana, Punjab (India). He gained his BSc (Ag.) Hons from Viswa-Bharati, Santiniketan in 1991. After completing an MSc in Genetics & Plant Breeding (1991–1993) from Banaras Hindu University and a PhD in Genetics (1993–1998) from the Indian Agricultural Research Institute (IARI), he started his scientific career at the Indian Institute of Pulses Research, Kanpur in 1996. In September 2000 he joined the erstwhile Directorate of Maize Research, New Delhi as well as a faculty member in the Division of Genetics, IARI. He did postdoctoral research on rice genomics at Iwate Biotechnology Research Center, Japan under a Japan Society for the Promotion of Science (JSPS) postdoctoral fellowship from 2003 to 2005. From November 2008 until March 2017 he worked as Principal Scientist (Plant Breeding) in the erstwhile Directorate of Sorghum Research. He has worked on genomics and molecular breeding of maize, sorghum, minor millets, rice and pulses. Since 24 March 2017 he has worked as Director of ICAR–Indian Institute of Maize Research, Ludhiana. He has 154 publications in journals of national and international repute and edited three books. He has three QPM hybrids, two baby hybrids, one normal maize and one forage sorghum variety to his credit besides 20 genetic stocks and over 700 DNA sequences. He is recipient of the following recognitions: ICAR Chaudhary Devilal Outstanding AICRP Award; Indian Society of Genetics & Plant Breeding (ISGPB) Fellow; NE

Borlaug Fellowship (US Department of Agriculture); JSPS Postdoctoral Fellowship for Foreign Researchers; BOYSCAST Fellowship; CSIR Scientist Travel Fellowship; Indian Science Congress Association Young Scientist Award; ICAR Jawaharlal Nehru Award; UNESCO Fellowship in Biotechnology; Best PhD Student of IARI Medal; IARI Merit Medal; Jawaharlal Nehru Memorial Award; Binani Gold Medal; and BHU Gold Medal, among others. He has visited Australia, Germany, the Netherlands, Singapore, Japan, Bangladesh, Malaysia and the USA. He has been President, Maize Technologists Association of India, New Delhi (2018–2020) and President, Agriculture & Forestry Section, 106th Indian Science Congress.

**Professor Robert Henry** conducts research on the development of new products from plants. His research targets improved understanding of the molecular basis of the quality of products derived from plants and genome analysis to capture novel genetic resources for diversification of food and energy crops. He is Professor of Innovation in Agriculture and Foundation Director of the Queensland Alliance for Agriculture and Food Innovation, an institute of The University of Queensland in partnership with the Queensland Government. He was previously Director of the Centre for Plant Conservation Genetics at Southern Cross University and Research Program Leader in the Queensland Agricultural Biotechnology Centre. He has been involved in establishing several Cooperative Research Centres in Australia and has contributed to the management of research funding by Rural Research and Development Corporations in Australia. He is a graduate of The University of Queensland (BSc (Hons)), Macquarie University (MSc (Hons)) and La Trobe University (PhD). He was awarded a higher doctorate (DSc) by The University of Queensland for his work on variation in plants, is a Fellow of the Royal Australian Chemical Institute, recipient of the Guthrie Medal for his contributions to cereal chemistry and a Fellow of the Australian Academy of Technological Sciences and Engineering.

# Contributors

**Andrea Acuna,** Crop Soil and Environmental Sciences, University of Arkansas, Fayetteville, Arkansas, USA.

**Mobashwer Alam,** Queensland Alliance for Agriculture and Food Innovation, The University of Queensland, 47 Mayers Road, Nambour, QLD 4560, Australia.

**Muhammad Amjad Ali,** Department of Plant Pathology, University of Agriculture, Faisalabad, Pakistan.

**John J. Amalraj,** Tamil Nadu Agricultural University, Coimbatore, India.

**Maria N. Arguello,** Horticulture and Crop Science, Ohio State University, Wooster, Ohio, USA.

**Arushi Arora,** Punjab Agricultural University, Ludhiana, India.

**Diana Ballesteros,** Crop Soil and Environmental Sciences, University of Arkansas, Fayetteville, Arkansas, USA.

**Bhoja R. Basnet,** International Maize and Wheat Improvement Center (CIMMYT), Carretera México-Veracruz Km. 45, El Batán, Texcoco, México, C.P. 56237.

**Salma Bhyan,** School of Science and Engineering, University of the Sunshine Coast, Sippy Downs, QLD 4556, Australia.

**Richard Boyles,** Plant and Environmental Sciences, Clemson University, Pee Dee Research and Extension Center, Florence, South Carolina, USA.

**Faïçal Brini,** Biotechnology and Plant Improvement Laboratory, Center of Biotechnology of Sfax (CBS), University of Sfax, 3018 Sfax, Tunisia.

**Yibo Cao,** State Key Laboratory of Plant Physiology and Biochemistry, College of Biological Sciences, China Agricultural University, Beijing 100094, People's Republic of China.

**S. Antony Ceasar,** Division of Plant Biotechnology, Entomology Research Institute, Loyola College, Chennai – 600034, India. Current address: Division of Plant Molecular Biology and Biotechnology, Department of Biosciences, Rajagiri College of Social Sciences, Kalamassery, Kochi – 683104, India.

**Nidhi Chakma,** Queensland Alliance for Agriculture and Food Innovation, The University of Queensland, 47 Mayers Road, Nambour, QLD 4560, Australia.

**Subhadra Chakrabarty,** Department of Plant Breeding, Justus Liebig University, Giessen, Germany.

**Moutoshi Chakraborty,** Institute of Biotechnology and Genetic Engineering (IBGE), Bangabandhu Sheikh Mujibur Rahman Agricultural University, Gazipur 1706, Bangladesh.

**Harmeet Singh Chawla,** Department of Plant Breeding, Justus Liebig University, Giessen, Germany.

**Dharampal Choudhary,** ICAR–Indian Institute of Maize Research, Ludhiana, Punjab, India.

**Mukesh Choudhary,** ICAR–Indian Institute of Maize Research, Ludhiana, Punjab, India.

**M.C. Dagla,** ICAR–Indian Institute of Maize Research, Ludhiana, Punjab, India.

**Mônica Furaste Danilevicz,** School of Biological Sciences and Institute of Agriculture, University of Western Australia, Crawley, WA 6009, Australia.

**Abhijit Kumar Das,** ICAR–Indian Institute of Maize Research, Ludhiana, Punjab, India.

**Haixiao Dong,** College of Plant Sciences, Jilin University, Changchun, People's Republic of China.

**Kate Dreher,** International Maize and Wheat Improvement Center (CIMMYT), Carretera México-Veracruz Km. 45, El Batán, Texcoco, México, C.P. 56237.

**Susanne Dreisigacker,** International Maize and Wheat Improvement Center (CIMMYT), Carretera México-Veracruz Km. 45, El Batán, Texcoco, México, C.P. 56237.

**David Edwards,** School of Biological Sciences and Institute of Agriculture, University of Western Australia, Crawley, WA 6009, Australia.

**Xiaorong Fan,** State Key Laboratory of Crop Genetics and Germplasm Enhancement, College of Resources and Environmental Sciences, Nanjing Agricultural University, Nanjing 210095, People's Republic of China; and Key Laboratory of Plant Nutrition and Fertilization in Low-Middle Reaches of the Yangtze River, Ministry of Agriculture, Nanjing Agricultural University, Nanjing 10095, People's Republic of China.

**Maria Laura Federico,** CONICET, Buenos Aires, Argentina; and INTA, EEA Famaillá, Ruta Prov. 301, Km 32, 4132 Famaillá, Tucumán, Argentina.

**Cassandria Tay Fernandez,** School of Biological Sciences and Institute of Agriculture, University of Western Australia, Crawley, WA 6009, Australia.

**Ertugrul Filiz,** Department of Plant and Animal Production, Cilimli Vocational School, Duzce University, Duzce, Turkey.

**Monika Garg,** National Agri-Food Biotechnology Institute, Sector 81, Mohali – 140306, Punjab, India.

**Claudia Teixeira Guimaraes,** Embrapa Maize and Sorghum, Road MG424 km 65, Sete Lagoas, MG 35701-970, Brazil.

**Karama Hamdi,** Biotechnology and Plant Improvement Laboratory, Center of Biotechnology of Sfax (CBS), University of Sfax, 3018 Sfax, Tunisia.

**K. Hariprasanna,** ICAR–Indian Institute of Millets Research, Rajendranagar, Hyderabad – 500030, India.

**Robert J. Henry,** Queensland Alliance for Agriculture and Food Innovation, The University of Queensland, Brisbane, QLD 4072, Australia.

**Firoz Hossain,** ICAR–Indian Agricultural Research Institute, New Delhi, India.

**Mohammad Anwar Hossain,** Department of Genetics and Plant Breeding, Bangladesh Agricultural University, Mymensingh – 2202, Bangladesh.

**Guanghui Hu,** Maize Institute, Heilongjiang Academy of Agricultural Sciences, Harbin, Heilongjiang, People's Republic of China.

**Zhaorong Hu,** State Key Laboratory for Agrobiotechnology, Key Laboratory of Crop Heterosis Utilization (MOE), Beijing Key Laboratory of Crop Genetic Improvement, China Agricultural University, Beijing 100193, People's Republic of China.

**B.S. Jat,** ICAR–Indian Institute of Maize Research, Ludhiana, Punjab, India.

**Caifu Jiang,** State Key Laboratory of Plant Physiology and Biochemistry, College of Biological Sciences, China Agricultural University, Beijing 100094, People's Republic of China.

**Faiz Ahmad Joyia,** Centre of Agricultural Biochemistry and Biotechnology (CABB), University of Agriculture, Faisalabad, Pakistan.

**Mehar C. Kamboj,** CCS–Haryana Agricultural University, Regional Research Station, Uchani, India.

**Chikkappa Karjagi,** ICAR–Indian Institute of Maize Research, Ludhiana, Punjab, India.

**Lovenpreet Kaur,** National Agri-Food Biotechnology Institute, Sector 81, Mohali – 140306, Punjab, India.

**Muhammad Sarwar Khan,** Centre of Agricultural Biochemistry and Biotechnology (CABB), University of Agriculture, Faisalabad, Pakistan.

**Emna Khanfir,** Biotechnology and Plant Improvement Laboratory, Center of Biotechnology of Sfax (CBS), University of Sfax, 3018 Sfax, Tunisia.

**Rajesh Khulbe,** ICAR–Vivekananda Parvatiya Krishi Anusandhan Sansthan, Almora, India.

**Bhupender Kumar,** ICAR–Indian Institute of Maize Research, Ludhiana, Punjab, India.
**Krishan Kumar,** ICAR–Indian Institute of Maize Research, Pusa Campus, New Delhi – 110012, India.
**Pardeep Kumar,** ICAR–Indian Institute of Maize Research, Ludhiana, Punjab, India.
**Pradeep Kumar,** ICAR–Indian Institute of Wheat and Barley Research (IIWBR), Karnal – 132001, Haryana, India.
**Santosh Kumar,** ICAR–Indian Institute of Maize Research, Ludhiana, Punjab, India.
**Fırat Kurt,** Department of Plant Production and Technologies, Faculty of Applied Sciences, Muş Alparslan University, Muş, Turkey.
**Zhao Li,** College of Plant Sciences, Jilin University, Changchun, People's Republic of China.
**Yun-Hua Liu,** Department of Soil and Crop Sciences, Texas A&M University, College Station, TX 77843, USA.
**Ningthaipuilu Longmei,** ICAR–Indian Institute of Maize Research, Ludhiana, Punjab, India.
**Damar López-Arredondo,** Institute of Genomics for Crop Abiotic Stress Tolerance, Department of Plant and Soil Science, Texas Tech University, Lubbock, TX 79409, USA; and State Key Laboratory of Crop Genetics and Germplasm Enhancement, College of Resources and Environmental Sciences, Nanjing Agricultural University, 210095 Nanjing, People's Republic of China.
**Jurandir Vieira de Magalhaes,** Embrapa Maize and Sorghum, Road MG424 km 65, Sete Lagoas, MG 35701-970, Brazil.
**T. Maharajan,** Division of Plant Biotechnology, Entomology Research Institute, Loyola College, Chennai – 600034, India.
**Jacob Marsh,** School of Biological Sciences and Institute of Agriculture, University of Western Australia, Crawley, WA 6009, Australia.
**R. Esten Mason,** Soil and Crop Sciences, Colorado State University, Fort Collins, Colorado, USA.
**Clémentine Mercé,** School of Biological Sciences and Institute of Agriculture, University of Western Australia, Crawley, WA 6009, Australia.
**Ghulam Mustafa,** Centre of Agricultural Biochemistry and Biotechnology (CABB), University of Agriculture, Faisalabad, Pakistan.
**Vignesh Muthusamy,** ICAR–Indian Agricultural Research Institute, New Delhi, India.
**Sudha K. Nair,** International Maize and Wheat Improvement Center (CIMMYT), ICRISAT Campus, Patancheru, Hyderabad – 502324, Telangana, India.
**Zhongfu Ni,** State Key Laboratory for Agrobiotechnology, Key Laboratory of Crop Heterosis Utilization (MOE), Beijing Key Laboratory of Crop Genetic Improvement, China Agricultural University, Beijing 100193, People's Republic of China.
**Kazutoshi Okuno,** Faculty of Life and Environmental Sciences, University of Tsukuba, Tennodai 1-1-1, Tsukuba, 305-8577 Japan. Current address: Kayada 2069-6, Yachiyo, Chiba, 276-0043 Japan.
**Geoffrey Onaga,** Cereals Research Programme, National Crops Resources Research Institute (NaCRRI), Namulonge, Uganda.
**Huiru Peng,** State Key Laboratory for Agrobiotechnology, Key Laboratory of Crop Heterosis Utilization (MOE), Beijing Key Laboratory of Crop Genetic Improvement, China Agricultural University, Beijing 100193, People's Republic of China.
**Jorge Fernando Pereira,** Embrapa Gado de Leite, Rua Eugênio do Nascimento 610, Juiz de Fora, MG, CEP 36038-330, Brazil.
**S.A.C.N. Perera,** Department of Agricultural Biology, Faculty of Agriculture, University of Peradeniya, Peradeniya, Sri Lanka.
**Bhukya Prakash,** ICAR–Directorate of Poultry Research, Hyderabad, India.
**Lydia Pramitha,** Tamil Nadu Agricultural University, Coimbatore, India.
**P. Rajendrakumar,** ICAR–Indian Institute of Millets Research, Rajendranagar, Hyderabad – 500030, India.
**Mantena V.L.N. Raju,** ICAR–Directorate of Poultry Research, Hyderabad, India.
**Sujay Rakshit,** ICAR–Indian Institute of Maize Research, PAU Campus, Ludhiana – 141004, India.

**Savaram V.R. Rao,** ICAR–Directorate of Poultry Research, Hyderabad, India.
**Ali Razzaq,** Centre of Agricultural Biochemistry and Biotechnology (CABB), University of Agriculture, Faisalabad, Pakistan.
**P. Sanjana Reddy,** ICAR–Indian Institute of Millets Research, Rajendranagar, Hyderabad – 500030, India.
**Trenton L. Roberts,** Crop Soil and Environmental Sciences, University of Arkansas, Fayetteville, Arkansas, USA.
**G. Victor Roch,** Division of Plant Biotechnology, Entomology Research Institute, Loyola College, Chennai – 600034, India.
**Umesh R. Rosyara,** International Maize and Wheat Improvement Center (CIMMYT), Carretera México-Veracruz Km. 45, El Batán, Texcoco, México, C.P. 56237.
**Pawan Saini,** CSB–Central Sericultural Research and Training Institute (CSR&TI), Pampore –192121, Jammu-Kashmir, India.
**Sindhu Sareen,** ICAR–Indian Institute of Wheat and Barley Research (IIWBR), Karnal – 132001, Haryana, India.
**Gajanan Saykhedkar,** International Maize and Wheat Improvement Center (CIMMYT), ICRISAT Campus, Patancheru, Hyderabad – 502324, Telangana, India.
**André Schaffasz,** Department of Plant Breeding, Justus Liebig University, Giessen, Germany.
**Armin Scheben,** Simons Center for Quantitative Biology, Cold Spring Harbor Laboratory, Cold Spring Harbor, NY, 11724, USA.
**Saman Seneweera,** National Institute of Fundamental Studies, Kandy, Sri Lanka.
**Natasha Sharma,** National Agri-Food Biotechnology Institute, Sector 81, Mohali – 140306, Punjab, India.
**Tariq Shehzad,** Syngenta Crop Protection, LLC, Research Triangle Park, North Carolina 27709, USA.
**Sonia Sheoran,** ICAR–Indian Institute of Wheat and Barley Research (IIWBR), Karnal – 132001, Haryana, India.
**Charan Singh,** ICAR–Indian Institute of Wheat and Barley Research (IIWBR), Karnal – 132001, Haryana, India.
**Ishwar Singh,** ICAR–Indian Institute of Maize Research, Pusa Campus, New Delhi – 110012, India.
**Prabha Singh,** ICAR–Indian Grassland and Fodder Research Institute, Jhansi – 284003, India.
**Vishal Singh,** ICAR–Indian Institute of Maize Research, Ludhiana, Punjab, India.
**Rod J. Snowdon,** Department of Plant Breeding, Justus Liebig University, Giessen, Germany.
**Nithya Subramanian,** Soil and Crop Sciences, Texas A&M University, College Station, Texas, USA.
**Qixin Sun,** State Key Laboratory for Agrobiotechnology, Key Laboratory of Crop Heterosis Utilization (MOE), Beijing Key Laboratory of Crop Genetic Improvement, China Agricultural University, Beijing 100193, People's Republic of China.
**Madhumal Thayil Vinayan,** International Maize and Wheat Improvement Center (CIMMYT), ICRISAT Campus, Patancheru, Hyderabad – 502324, Telangana, India.
**Yiqiao Wang,** State Key Laboratory of Plant Physiology and Biochemistry, College of Biological Sciences, China Agricultural University, Beijing 100094, People's Republic of China.
**Shabir Hussain Wani,** Mountain Research Centre for Field Crops, Sher-e-Kashmir University of Agricultural Sciences and Technology of Kashmir, Khudwani, Anantnag – 192101, Jammu and Kashmir, India.
**Steffen Windpassinger,** Department of Plant Breeding, Justus Liebig University, Giessen, Germany.
**Kerstin D. Wydra,** Erfurt University of Applied Sciences, Leipziger Str. 77, 90085 Erfurt, Germany.
**Yinping Jiao,** Institute of Genomics for Crop Abiotic Stress Tolerance, Department of Plant and Soil Science, Texas Tech University, Lubbock, TX 79409, USA.
**Mingming Xin,** State Key Laboratory for Agrobiotechnology, Key Laboratory of Crop Heterosis Utilization (MOE), Beijing Key Laboratory of Crop Genetic Improvement, China Agricultural University, Beijing 100193, People's Republic of China.

**Inès Yacoubi,** Biotechnology and Plant Improvement Laboratory, Center of Biotechnology of Sfax (CBS), University of Sfax, 3018 Sfax, Tunisia.

**Pranjal Yadava,** ICAR–Indian Institute of Maize Research, Pusa Campus, New Delhi – 110012, India.

**Yingyin Yao,** State Key Laboratory for Agrobiotechnology, Key Laboratory of Crop Heterosis Utilization (MOE), Beijing Key Laboratory of Crop Genetic Improvement, China Agricultural University, Beijing 100193, People's Republic of China.

**Kuohai Yu,** State Key Laboratory for Agrobiotechnology, Key Laboratory of Crop Heterosis Utilization (MOE), Beijing Key Laboratory of Crop Genetic Improvement, China Agricultural University, Beijing 100193, People's Republic of China.

**Yaping Yuan,** College of Plant Sciences, Jilin University, Changchun, People's Republic of China.

**Pervez Haider Zaidi,** International Maize and Wheat Improvement Center (CIMMYT), ICRISAT Campus, Patancheru, Hyderabad – 502324, Telangana, India.

**Hong-Bin Zhang,** Department of Soil and Crop Sciences, Texas A&M University, College Station, TX 77843, USA.

**Meiping Zhang,** College of Life Science, Jilin Agricultural University, Changchun, Jilin 130118, People's Republic of China. Current address: Department of Soil and Crop Sciences, Texas A&M University, College Station, TX 77843, USA.

**Zhiwu Zhang,** Department of Crop and Soil Sciences, Washington State University, Pullman, Washington, USA.

**Rajkumar U. Zunjare,** ICAR–Indian Agricultural Research Institute, New Delhi, India.

# Preface

Global food production must be increased by 70% to feed the ever-growing world population of 10 billion by 2050. Apart from rice, wheat, maize and sorghum are the three key cereals providing human health and nutrition for the majority of the world's population. Globally, wheat, maize and sorghum production are affected by various abiotic stresses causing significant yield losses. The situation is becoming worse due to climate change that may multiply the frequency and severity of such abiotic stresses. Sustainable wheat, maize and sorghum production delivering yields to meet ever-increasing demands and the development of biofortified wheat, maize and sorghum are major challenges for the scientific community, and will require the combined expertise of agronomists, farmers, breeders and molecular biologists. Recently, in response to global climate change, the increasing fragility of our natural resources and threats to food grain security across the globe, molecular breeding in wheat, maize and sorghum has attracted considerable interest from the scientific community. Many studies in various scientific disciplines dealing with different wheat, maize and sorghum species, in different environments, have focused on abiotic stress tolerance, grain and quality improvements, and wheat, maize and sorghum biofortification. Although significant progress has been made over the last few years, there is still a need to narrow down the yield gap between the most favourable environments and those under stress conditions. Strategies involving bridging the yield gap and increasing yield stability and adaptability to variable environmental conditions are of importance in assuring food security and sustainability in the future. Hence, there is an urgent need to improve our understanding of complex mechanisms regulating abiotic stress tolerance for developing modern wheat, maize and sorghum varieties that are more resilient to abiotic stresses as well as to increase the bioavailable concentrations of essential micronutrients.

Genetic improvement for abiotic stress tolerance is limited in conventional methods due to the complex nature of the trait, particularly due to the variability in intensity, frequency, duration and timing; and additionally, linkage drag of undesirable traits/genes with desirable traits. Transfer of favourable genes/alleles from diverse plant genetic resources limited by gene pool barriers gives molecular breeders a good option for developing new cultivars that can thrive in stressful environments. Over the last few decades, tremendous progress has been made in wheat, maize and sorghum genome analysis. The availability of the genome sequences along with other recent developments in sequencing and genotyping and genome editing technologies have resulted in considerable advancements in the area of wheat, maize and sorghum genomics. These in turn led to the development of DNA-based markers and resulted in the identification and fine mapping of quantitative trait loci associated with grain yield and yield-attributing traits, abiotic and biotic stress tolerances as well as grain quality traits in these crops. Tightly linked DNA markers and causal genes are used in marker-assisted selection in wheat, maize and sorghum breeding programmes and are able to shorten the time of variety development. Another use of DNA-based markers is overcoming the barrier of linkage drag, which refers to the presence of undesirable genes in the chromosomal region of the target gene that are difficult to avoid when using conventional breeding. Economic analysis has also shown the potential impacts of utilizing marker-assisted breeding to overcome the drawbacks of conventional breeding in wheat, maize and sorghum and ultimately reduce the cost of production and promote economic growth.

In this book, *Molecular Breeding in Wheat, Maize and Sorghum: Strategies for Improving Abiotic Stress Tolerance and Yield,* we present a collection of 29 chapters written by leading experts engaged with wheat, maize and sorghum molecular breeding. The chapters of this book aim to contribute the latest understandings of the molecular and genetic bases of abiotic stress tolerance, yield and quality improvement of wheat, maize and sorghum to develop strategies for improving abiotic stress tolerance that will lead to enhance productivity and better utilization of natural resources to ensure food security through modern breeding. We are extremely grateful to all the learned contributors and sincerely thank them for their contribution in compiling useful and updated information on different aspects of wheat, maize and sorghum breeding. We would like to extend our special thanks to Rebecca Stubbs, Commissioning Editor, CABI, UK, for initiating the book project and other staff of CABI for their generous help in completing this project on time. We strongly believe that the book will be a valuable resource for future environmental stress-related research, and can be considered a textbook for graduate students and teachers as well as a reference book for front-line wheat, maize and sorghum researchers around the globe.

Mymensingh, Bangladesh — Mohammad Anwar Hossain
Queensland, Australia — Mobashwer Alam
Kandy, Sri Lanka — Saman Seneweera
Punjab, India — Sujay Rakshit
Queensland, Australia — Robert Henry

# 1 Recent Understanding on Molecular Mechanisms of Plant Abiotic Stress Response and Tolerance

**Geoffrey Onaga[1] and Kerstin Wydra[2*]**
[1]*National Crops Resources Research Institute (NaCRRI), Namulonge, Uganda;*
[2]*Erfurt University of Applied Sciences, Erfurt, Germany*

## 1.1 Introduction

Abiotic stresses are growth- and yield-limiting non-biological factors that constrain crop production. The major abiotic stresses limiting crop productivity include drought, salinity, waterlogging, temperature extremes, nutrient imbalances, metal toxicities, ozone and UV-B irradiation. According to global estimates, over 90% of arable lands are prone to one or more of these stresses, and yield losses of up to 70% have been reported in major food crops (dos Reis *et al.*, 2012; Mantri *et al.*, 2012). While it is difficult to get accurate estimates of the effects of each abiotic stress on crop production, it is estimated that yield losses in agricultural crops are mostly caused by high temperature (40%), salinity (20%), drought (17%), low temperature (15%) and flooding (Ashraf *et al.*, 2008). The extent of some of these stresses has, however, increased in recent years. For instance, salinity in irrigated lands has increased by 37%, and more than 50% of arable land could be salt affected by the year 2050 (Munns and Tester, 2008; Qadir *et al.*, 2014). Changes in precipitation patterns have augmented the frequency and severity of drought stress (IPCC, 2018; Naumann *et al.*, 2018). The concentrations of greenhouse gases have increased and subsequently air and ocean temperatures have warmed (Raftery *et al.*, 2017). Heavy metals contamination of arable lands also increased in many parts of the world (Rehman *et al.*, 2018); and the frequency and severity of flooding/waterlogging are presently at alarming rates as a result of erratic weather patterns and sea level rise (Onaga and Wydra, 2016; Manik *et al.*, 2019). All these factors have significant influences on plant growth and crop yields and will be exacerbated by further direct and indirect impacts of climate change. Thus, improved understanding of the molecular, physiological and biochemical bases of plant stress response and tolerance is necessary to decipher promising, functionally relevant molecular mechanisms for accelerated development of abiotic stress-tolerant cultigens.

In response to abiotic stress challenge, plants have evolved intricate mechanisms allowing optimal perception and subsequent transduction of the stress signals. The first reaction to stress involves activation of extracellular and intracellular receptors/sensors localized at the cell membrane such as histidine kinases, phytochromes, receptor-like kinases and G-protein-coupled receptors. Second messengers such as reactive oxygen species (ROS), reactive nitrogen

*Corresponding author: Kerstin.Wydra@fh-erfurt.de

 *Molecular Breeding in Wheat, Maize and Sorghum: Strategies for Improving Abiotic Stress Tolerance and Yield* (eds M.A. Hossain *et al.*)
DOI: 10.1079/9781789245431.0001

species (RNS), reactive sulfur species (RSS), reactive carboxyl species (RCS), inositol phosphatase, $Ca^{2+}$ ions and abscisic acid (ABA) are subsequently triggered. Among the second messengers, ROS accumulation is depicted as the earliest signal in many plant abiotic stress responses which ultimately determine the cell fate. The response to these messengers can be either plastic or elastic (Cramer *et al.*, 2011), depending on cellular reactive species accumulation and oxidant–antioxidant balance. Plastic responses mainly occur through unrestrained accumulation of reactive species which cause oxidative damage to cellular components and structure, inhibiting vital cellular processes such as protein synthesis and energy metabolism. When stringently controlled, ROS accumulation, together with other secondary messengers – such as $H_2S$, methylglyoxal, NO and stored pools of $Ca^{2+}$– and stress receptors, form the signalling machinery that employs numerous stress-responsive downstream transducing molecules, including phosphorylation cascades, such as calcium-dependent protein kinase (CDPK), mitogen-activated protein kinase (MAPK) and dephosphorylation phosphatases (e.g. ABI1 and ABI2). Transcription factors (TFs) are activated or suppressed by protein kinases or phosphatases, respectively, and the modified TFs form a complex regulatory network, mediated by various molecules including co-regulators and cross-regulators such as G-box and W-box binders. Subsequently, TFs directly regulate the expression of stress-responsive genes by interacting with the specific *cis*-elements in their promoter regions. Genes expressed in plants exposed to abiotic stress include genes encoding metabolic and structural proteins as well as regulatory proteins. Abiotic stresses also lead to altered DNA methylation/demethylation, histone post-translational modifications (PTMs), remodelling of chromatin, small RNAs and long non-coding RNAs.

While new molecular pathways are yet to be discovered in most crop species, significant achievements have been made to discover some of the genes involved in the aforementioned molecular responses to stress in plants. This chapter provides an overview of the recent significant perspectives on molecules involved in response and tolerance to drought and salinity, the two major abiotic stresses affecting crop production, and highlights major molecular components identified in major cereals. For some general aspects, readers can refer to previous work on a related topic (Onaga and Wydra, 2016).

## 1.2 Molecular Mechanisms of Abiotic Stress Response and Tolerance

### 1.2.1 Drought stress

Plant response to drought is brought about by various mechanisms and depends on crop species, genotype, the age and stage of plant development, and the duration and severity of water loss. Under drought plants generally lose leaf water leading to decreased leaf osmotic potential, stomatal conductance, transpiration rates and photosynthesis (Farooq *et al.*, 2009). Drought stress can also impair mineral uptake by limiting nutrient supply through mineralization and reducing nutrient diffusion and mass flow in the soil (Samarah *et al.*, 2004). The consequence of these changes is a reduction in crop growth and yields, and sometimes increased susceptibility to biotic stresses. The reproductive phase of development is the most sensitive stage to drought stress in several crops and investigating drought effects during reproductive development is of great relevance due to its direct negative impact on crop yields.

The drought stress signal in plants is first perceived at the root level through receptors from the cell membrane/cell wall which convert extracellular stress signals into intracellular secondary messengers. Although several primary sensing mechanisms have been proposed, the true primary sensing receptors have not been clearly figured out due to the complexity of plant responses to drought stress. Several hypotheses have suggested either a redox imbalance or changes in the cell-wall integrity as a trigger to molecular responses to drought. Redox imbalance causes the pH of xylem sap to increase, triggering the loading and transportation of second messengers, such as ABA, throughout the plant. Several ABA transporters have been reported, including multidrug and toxin efflux transporters (MATEs), ATP-binding cassette (ABC), NITRATE TRANSPORTER 1/PEPTIDE TRANSPORTER FAMILY (NPF) and more recently

PLASMA MEMBRANE PROTEIN 1 (OsPM1) in rice (Takahashi *et al.*, 2020). Change in cell-wall integrity or hydraulic pressure is believed to be perceived by $Ca^{2+}$-permeable mechanosensitive channels (MCA1 and MCA2) (Yamanaka *et al.*, 2010). Recently, a member of non-selective cation channels, reduced hyperosmolality-induced [$Ca^{2+}$] increase 1 (OSCA1), which is one of the hyperosmolarity-gated calcium channels located at the plasma membrane, was suggested as the first potential osmosensor in plants (for review see Lamers *et al.*, 2020). The authors suggest that membrane tension or high extracellular osmotic potential experienced by plants exposed to drought is sensed by OSCA1. Activated OSCA1 then facilitates the influx of $Ca^{2+}$, triggering intracellular drought-stress signalling cascades. Similar to OSCA1, calcium-permeable stress-gated cation channel 1 (CSC1A) has been suggested as another hyperosmotic stress-induced $Ca^{2+}$ channel (Hou *et al.*, 2014). Both CSC1A and OSCA1 contain transmembrane domains, which are also present in the drought-responsive integral protein, Early Responsive to Dehydration 4 (ERD4). Moreover, the OSCA family members, ERD4, CSC1A and the transmembrane channel-like proteins (TMC) present in bacteria and yeast (Singh *et al.*, 2015) contain similar calcium-dependent channel (DUF221) domains. Thus, OSCAs form a functional complex together with ERD4, CSC1A and probably the key enzymes (such as phosphatidylinositol-4-kinase) that cause the influx of $Ca^{2+}$ into the cytoplasm. Several members of the *OSCA* gene family have been identified in *Arabidopsis* (e.g. *AtOSCA1.8*), rice (e.g. *OsOSCA1.4*), maize (e.g. *ZmOSCA4.1*) and wheat (e.g. *TaOSCA1.4*), seeming to have a conserved regulatory mechanism (Liu, Y. *et al.*, 2009; Ding *et al.*, 2019; Cao *et al.*, 2020). Micro RNAs (miRNAs) are thought to regulate *OSCA1* activity. miRNAs exert their functions through mRNA cleavage or translational repression of complementary target mRNAs, and they are crucial in regulating gene expression. Over 206 small regulatory RNAs have been identified in sorghum, 42 in wheat and 172 in maize (Kozomara and Griffiths-Jones, 2011). In maize, there is a significant negative correlation between the expression levels of miR5054 and *ZmOSCA2.4* under drought stress (Cao *et al.*, 2020). A depression of miR5054 potentially promotes expression of several other classes of *OSCAs*. Thus, miRNAs and their target genes could be useful for crop improvements to enhance drought tolerance. However, more studies are required to validate their mode of interaction and associated regulators.

The cytoplasmic pool of $Ca^{2+}$ is perceived by signal receptors such as Calmodulin (CaM)-like B proteins (CMLs), calcineurin B-like proteins (CBLs), CDPK and calmodulin-dependent protein kinase (CaMK) (Mittler *et al.*, 2006). CaM is conserved in all eukaryotes, whereas CMLs, CBLs and CDPKs are present in plants and some protists (Reddy and Reddy, 2004). Among these families, CBLs are one of the major sensor relays in plants responding to drought. All CBLs contain E-helix and F-helix (EF), which have a high affinity for $Ca^{2+}$ binding. CBL interacts with CBL-interacting protein kinases (CIPK), a family of sucrose non-fermenting-1 (SNF1)-related protein kinases (SnRKs), to form CBL–CIPK complexes, which are directed to a specific cellular target to phosphorylate downstream target proteins. CIPKs contain a protein–phosphatase interaction (PPI) domain capable of interacting with type 2C protein phosphatases (PP2Cs) to mediate the specific binding with CBLs. In plants, PYR/PYL/RCAR (pyrabactin resistance/pyrabactin resistance 1-like/regulatory component of ABA receptors), a ubiquitous ABA receptor, forms a complex with PP2C and SnRK2 kinase (PRY/PRL/RCARs–PP2Cs–SnRK2) in the cytosol (Soon *et al.*, 2012). When plants perceive drought stress signals, ABA binds to the PYR/PYL/RCAR-type receptor to form an activated signalling complex through inhibition of PP2Cs and release of SnRK2 (Fig. 1.1). SnRK2 is autoactivated when separated from PP2C and phosphorylates ABA-responsive elements-binding protein/ABA-responsive elements-binding factor (AREB/ABF) such as basic leucine zipper (bZIP) (AB15/ABF/ABREB), CBF4 and MYC/MYB proteins involved in the transcription of drought response genes encoding for proteins involved in water transport, osmotic balance, antioxidants and damage repair (Onaga and Wydra, 2016). Recently, SnRK2.6 has been shown to phosphorylate plasma-membrane intrinsic protein 2 (PIP2) (Grondin *et al.*, 2015), SNRK2-SUBSTRATE 1 (SNS1), FLOWERING BHLH 3 (FBH3)/AKS1 (Takahashi *et al.*, 2013), TATA BINDING PROTEIN-ASSOCIATED FACTOR 5 (TAF5), RBOHD/F, and modulate their activities

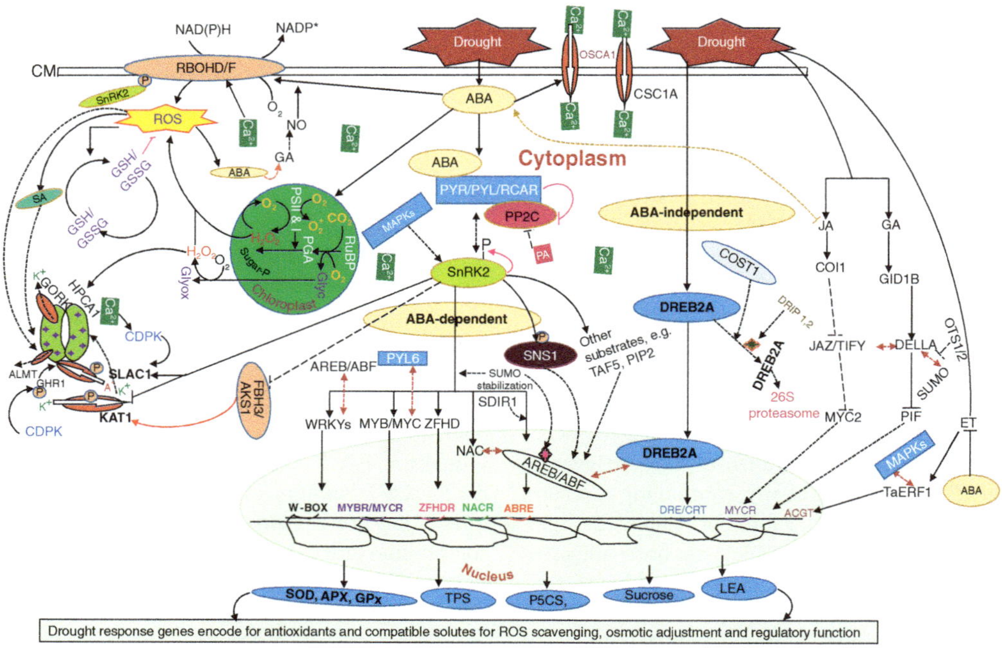


**Fig. 1.1.** The response to drought occurs via ABA-dependent and -independent signalling pathways. In the ABA-dependent pathway, ABA interacts with several molecules, including the plasma-membrane $Ca^{2+}$-permeable channels (e.g. OSCA1) and NADP(H) for ROS generation. ROS may promote further synthesis of ABA and other stress phytohormones like salicylic acid (SA), which stimulate the biosynthesis of glycine betaine (GB) and glutathione (GSH) to improve the tissue water status and protect cellular membranes (CM) from the effect of ROS. Gasotransmitters like NO also cooperate with GB to promote abiotic stress tolerance response. In chloroplast and peroxisomes, $H_2O_2$ can be produced in photosystem I and II (PSI, PSII) as well as during glycolysis and photorespiration. $H_2O_2$ mediates stomatal closure during drought by interacting with ABA up- and downstream targets. HPCA1 is a key component of $H_2O_2$ perception and signal modulation in the guard cells. Three main components of the ABA signalling pathway, PYR/PYL/RCAR5 ABA receptors, a PP2C negative regulator, PA, and a SnRK2 positive regulator, are shown. PA possibly binds PP2C to limit its activity and enhance ABA signalling during drought stress. SnRK2 phosphorylates several downstream factors, including TFs, ions channels (described in the text), SNRK2-SUBSTRATE 1 (SNS1), FLOWERING BHLH 3 (FBH3)/AKS1, TATA BINDING PROTEIN-ASSOCIATED FACTOR 5 (TAF5), RBOHD/F, and modulates their activities. FBH3/AKS1 is a transcriptional activator of KAT1. It is worth noting that PP2Cs dephosphorylate S-type anion channels in the absence of ABA causing stomatal opening. ROS possibly activate GHR1 and mediate the ABA activation of SLAC1. GHR1 is phosphorylated by CDPKs and SnRK2.6 (OST1) to activate SLAC1. SLAC1 activates guard-cell outward-rectifying $K^+$ (GORK) channels. The outward efflux of $K^+$ leads to loss of water from guard cells and stomatal closure. Inward-rectifying channels KAT1 ($K^+$ channel *Arabidopsis thaliana*1) facilitate inward flow of water into the guard cells, which leads to opening of the stomata (these channels are deactivated during drought stress (in the presence of ABA)). The key TF phosphorylated by SnRK2 is AREB/ABF. SnRK2 also possibly modulates the activity of MYB/MYCs, NACs, WRKYs and ZFHD TFs. MYC2 and NAC proteins are involved in JA signalling. In the ABA-independent pathway, DREB2A is the main regulator of drought response. Ubiquitin–proteasome pathway mediated by DREB2A-INTERACTING PROTEIN1 (DRIP1) negatively regulates drought tolerance by targeting DREB2A to 26S proteasome for degradation. DRIP1 is a ubiquitin E3 ligase harbouring a C3HC4-type RING domain. DRIPs regulate the levels of DREB2A under unstressed condition. SUMOylation probably stabilizes DREB2A and prevents its tagging for proteolysis under drought conditions. Other signalling pathways involved include the JA via MYC and gibberellic acid (GA) via the DELLA proteins. GA INSENSITIVE DWARF1 (GID1) tag DELLA proteins for degradation. DELLA proteins can be responsive to JA and could interact with the JA regulator, JAZ (jasmonic acid ZIM-domain protein),

*Continued*

in an ABA-dependent manner, suggesting the role of these proteins in drought response, although empirical evidence is still limited.

Apart from the ABA–SnRK2 pathway, an ABA-independent pathway exists in plants and is regulated by multiple families of TFs including the dehydration-responsive element/C-repeat (DRE/CRT) binding protein 2 and 1 (DREB2 and DREB1), NAC (NAM, ATAF and CUC) and zinc-finger homeodomain (HD) regulon. Both DREB1 and DREB2 belong to the plant-specific AP2 (APETALA2)/ERF (ethylene-responsive element-binding factor) family having an AP2/ERF DNA-binding motif and recognize DRE/CRT to activate downstream genes, including dehydration-responsive *late embryogenesis abundant* (*LEA*) genes, e.g. *RD29A*, which is induced by drought in *Arabidopsis*. In wheat, TaDREB promotes lignin formation under water stress (Egawa *et al.*, 2006) and in maize, transgenic plants of a DREB2 TF, *ZmDREBtv*, showed improved performance compared with the non-transgenic plants under drought conditions (Huynh *et al.*, 2019). The NAC acronym represents three proteins having a specific NAC domain: (i) NAM (no apical meristem); (ii) ATAF1/2 (*Arabidopsis thaliana* ACTIVATING FACTOR1/2); and (iii) CUC2 (cup-shaped cotyledon) in *Arabidopsis* (Aida *et al.*, 1997; Shao *et al.*, 2015). These TFs recognize NAC recognition site (NACR) to activate downstream genes (Fig. 1.1). NAC family members exist in maize (157 members) (Lu *et al.*, 2015), durum wheat (168 members) (Saidi *et al.*, 2017) and sorghum (131 members) (Sanjari *et al.*, 2019). The NAC domain proteins essentially regulate stress-inducible genes independent of ABA. Regardless, few of these proteins have been characterized in wheat, maize and sorghum, even though they are likely to play important roles in plant response to stress. In wheat, TtNAMB-2 and TaNAC69-1 participate in responses to drought stress (Tran *et al.*, 2004; Baloglu *et al.*, 2012). A NAC TF, SbSNAC1, from sorghum, also confers drought tolerance to transgenic *Arabidopsis*, whereas ZmNAC111 is associated with maize seedling drought tolerance (Mao *et al.*, 2015). At some point in the regulatory cycle, transcriptional repressors and enhancers interact with DRE/CRT and ABA-responsive element (ABRE) and hence initiate synergistic interactions. For instance, *Arabidopsis* NAC domain-containing protein 96 (ANAC096) interacts with ABFs to activate ABA-inducible drought-responsive genes (Xu *et al.*, 2013). TaNAC67 was demonstrated also to be inducible by ABA treatment (Mao *et al.*, 2014). Thus both ABA-dependent and -independent pathways appear to converge at some point with their interacting partners constituting a complex network. Regardless of the mode of interaction, these TFs activate expression of protective and regulatory proteins mediating physiological responses such as osmotic adjustments, reduced stomatal number and conductance, cuticular wax biosynthesis, early flowering, growth inhibition, increased root development, decreased leaf area, increased leaf thickness and leaf rolling, among others. The next section discusses the downstream targets of ABA-dependent and -independent genes in wheat, maize and sorghum and how these genes interact to bring about drought response and tolerance.

### *Downstream drought response targets that trigger tolerance in wheat, maize and sorghum*

It is well established that under drought stress, ABA production increases in both dicots and monocots as an important rapid response to minimize water loss. In wheat, endogenous ABA levels are mainly regulated by upstream genes encoding for a biosynthetic 9-*cis*-epoxycarotenoid dioxygenase (TaNCED) enzyme required for the cleavage of carotenoid precursors producing xanthoxin, which can subsequently be converted into ABA via ABA-aldehyde. Overexpression of *TaNCED* in *Arabidopsis* increases ABA and proline synthesis and limits water loss through

**Fig. 1.1.** Continued.

under drought. Below each TF are DNA-binding elements corresponding to DNA-binding domains of TFs (e.g. W-box, MYBR, etc.). The TFs and these elements fine-tune the activation of a defence response to drought, including increased expression of genes encoding for antioxidants such as superoxide dismutase (SOD), glutathione reductase (GR), guaiacol peroxidase (GPx) and ascorbate peroxidase (APX); compatible solutes and regulatory proteins. Dashed lines indicate possible but unconfirmed routes.

reduced stomatal conductance (Tong *et al.*, 2017). The first NCED-encoding gene *VIVPAROUS14* (*VIP14*) was discovered in maize *vp14* mutant (Tan *et al.*, 1997). Since then, NCED homologues have been identified in sorghum (e.g. SbNCED5) as well as other cereal crops (Ma *et al.*, 2019). Recently, Sato *et al.* (2018) showed that an upstream TF, NGATHA1, induces ABA biosynthesis by activating *NCED3* in *Arabidopsis* plants exposed to drought stress. Moreover, class II HOMEOBOX from *Arabidopsis thaliana* 1 and 3 (HAT1 and HAT3) and other TFs such as *Arabidopsis* NAC domain-containing protein 2 (ANAC2), 9-*cis*-epoxycarotenoid dioxygenase insensitive 1/Bodyguard 1 (BDG1) and WRKY57 have been reported to regulate ABA synthesis (Ali *et al.*, 2020). *NCED3* expression in leaves is also regulated by root-derived peptide CLE25 (CLAVATA3/EMBRYO-SURROUNDING REGION-RELATED 25) which transmits long-distance water-deficiency signals to the leaves for accumulation of ABA. *CLE25* expression signal in the root vascular tissues is translocated to the leaves where it binds to BARELY ANY MERISTEM (BAM) and BAM3 receptor-like protein kinases (RLKs) to induce ABA accumulation and stomatal closure in the leaves (Takahashi *et al.*, 2018). As mentioned earlier, conventionally ABA accumulation in plants is perceived by PYR/PYL/RCAR–PP2C–SnRK2 complex. In maize, *PYL* genes (e.g. *ZmPYL 8*, *9* and *12*) involved in the ABA pathway play a significant role in triggering drought resistance (He *et al.*, 2018). In wheat and sorghum, *TaPYLs* and *SbPYLs* are highly expressed in response to ABA treatments which suggests a similar role in the ABA pathway. Associated with PYL are the cytosolic SnRK2 kinases that phosphorylate downstream targets of ABA signalling. Studies have identified four, ten and 14 SnRK2 homologues in wheat, maize and sorghum, respectively (Huai *et al.*, 2008; Zhang *et al.*, 2016; Ma *et al.*, 2019). In wheat, *TaSnRK2.3*, *TaSnRK2.4* and *TaSnRK2.8* are the key genes required for drought stress signal transduction. When *TaSnRK2.3* and *TaSnRK2.8* are overexpressed in *Arabidopsis*, production of stress-related metabolites, such as proline, increases (Zhang *et al.*, 2010). In addition, increased transcript abundance of ABA biosynthetic genes (e.g. *ABA1* and *ABA2*), signalling components (e.g. *ABI3*, *ABI4*, *ABI5*) and ABA-independent genes (*CBF1*, *CBF2* and *CBF3*) is also observed in *TaSnRK2.8*-overexpressing lines. Expression and motif analysis indicates that the TaSnRK2 subclass III (*SRK2D/SnRK2.2*, *SRK2E/SnRK2.6/OST1* and *SRK2I/SnRK2.3*) is the most strongly induced by ABA and interacts with PP2C (*TaABI1*) to regulate ABA responses (Zhang *et al.*, 2010). Independent studies have identified SnRK2 homologues in sorghum, such as SbSAPK8 and SbSAPK9; as well as in maize, such as ZmSnRK2. Direct phosphorylation targets of activated SnRK2 in plants include slow anion channel 1 (SLAC1), SLAC1 homologue 3 (SLAH3) and inward-rectifying potassium channel 1 (KAT1) that control ion fluxes and guard-cell aperture. KAT1 is a transmembrane ion channel that facilitates cellular uptake of $K^+$ for an inward flow of water into the guard cells. In contrast, SLAC1 and SLAH3 facilitate the efflux of anions such as malate$^{2-}$, $Cl^-$ and $NO_3^+$ triggered by ABA accumulation, which leads to turgor loss and closure of stomatal apertures. SLAC1 is a nitrate sensor that has been evolved by both monocots and dicots. However, following the evolutionary split between dicots and monocots, SLAC1 from dicot species lost a distinct tandem amino acid motif for nitrate activation and did not develop a nitrate-dependent gating mechanism (Schäfer *et al.*, 2018). This change is partly reflected in the nitrate-dependent gating ability in monocots compared with dicots. The molecular structure of SLAC1-type anion channel is the recent important discovery that provides an opportunity to improve drought tolerance in cereals. For instance, the maize ZmSLAC1 and rice OsSLAC1 show a strong selectivity to nitrate over chloride (Sun *et al.*, 2016). Although SLAC1 is not fully studied in sorghum, it appears phylogenetically closer to the maize and rice SLAC1 than to the *Arabidopsis* AtSLAC1, which is a nitrate-activated chloride-permeable channel (Hedrich and Geiger, 2017; Müller *et al.*, 2017). It is important to further validate the structural and functional differences associated with SLAC1 in sorghum, wheat and maize and understand genotypic differences associated with stomatal closure during drought stress. This will provide additional information on their pivotal role in transpiration regulation during drought stress.

Apart from the SnRK2–SLAC1 pathway, other factors influencing stomatal closure in response to drought include complex physiological signals from ABA, ROS and $Ca^{2+}$ signalling pathways.

Increase in ABA is known to induce $Ca^{2+}$ currents by modulating $H_2O_2$, a major component of ROS in guard cells (reviewed in Takahashi *et al.*, 2020). Two plasma-membrane NADPH oxidases, respiratory burst oxidase homologues D and F (RbohD/F), play a crucial role in the production of $H_2O_2$ and have two EF-hand motifs for $Ca^{2+}$ binding (Kwak *et al.*, 2003). Moreover, RbohF is phosphorylated by SnRK2.6/OST1 (open stomata 1) to enhance $H_2O_2$ production (Sirichandra *et al.*, 2009). Accumulated $H_2O_2$ activates $Ca^{2+}$ channels to modulate stomatal closure. Recently, hydrogen peroxide-induced $Ca^{2+}$ increase 1 (HPCA) was reported as a sensor of $H_2O_2$ in guard-cell plasma membranes (Wu *et al.*, 2020). Mutants of *hpca1* impair stomatal closure modulated by ABA and $H_2O_2$. Thus, the extracellular domain of HPCA1 is a key component of $H_2O_2$ perception and signalling modulation in guard cells. The generation of $H_2O_2$ in response to ABA and jasmonic acid (JA) has also been reported to activate GUARD CELL HYDROGEN PEROXIDE-RESISTANT1 (GHR1). GHR1 is phosphorylated by CPKs/CDPKs and SnRK2.6 (OST1) to activate SLAC1. Activated SLAC1 causes plasma-membrane depolarization inducing guard-cell outward-rectifying $K^+$ (GORK) channels and subsequently $K^+$ efflux, which leads to turgor loss and closure of stomatal pores (reviewed in Ali *et al.*, 2020). Besides $Ca^{2+}$ and ROS oscillation, SnRK2 also interacts with other kinases, enzymes and TFs. As mentioned earlier, the major TF targets of SnRK2 activity include AREB/ABFs. The AREB/ABF–SnRK2 pathway is ubiquitous and functionally conserved in plants (for review see Fujita *et al.*, 2013). The *AREB* genes are highly expressed in response to drought in wheat (e.g. *TaAREB3* and *TaAREB5*), maize (e.g. *ZmbZIP4*) and sorghum (e.g. *Sb04g034190*). Their DNA-binding target genes are involved in encoding for the LEA proteins osmotin, proline and betaine. For instance, a wheat *bZIP* gene, *WLIP19*, is induced by drought and exogenous ABA application and increases transcriptional levels of cold-regulated genes (e.g. *COR15A* and *COR47*), dehydrins (*WDHN13*) and wheat response to ABA (*WRAB17*, *WRAB18*, *WRAB19*) (Kobayashi *et al.*, 2008). TaAREB3 also induces expression of *RD29A* and *RD29B* under stress conditions. Several other genes in the *LEA* family have been identified in wheat (e.g. *TaDHN18*, *TaDHN23* and *TaDHN31*) (Liu *et al.*, 2019), maize (e.g. *ZmLEA14tv*, *ZmLEA2*, *ZmRD20*, *ZmRD21*, *ZmRab18*, *ZmNHX3*, *ZmGEA6* and *ZmERD*) and sorghum (e.g. *SbLEA3*). Their protein products are rich in hydrophilic amino acids that maintain protein integrity and cell structure in plants exposed to stress. In modulating their downstream activity under osmotic stress, AREB/ABFs (AREB1/ABF2, AREB2/ABF4, ABF1 and ABF3) also form homo- or heterodimers (Yoshida *et al.*, 2010). Interestingly, these complex regulatory processes appear to function both upstream and downstream of AREB/ABFs. For instance, AREB/ABFs interact with NAC TFs (Jensen *et al.*, 2013). Moreover, the promoter region of stress-responsive *NAC* (*SNAC*) genes possesses ABRE sequences (Nakashima *et al.*, 2012), and *SNAC* genes such as *ANAC096* directly interact with ABF2 and ABF4 to activate ABA-dependent drought response genes (Xu *et al.*, 2013). In wheat, TaNAC67 was shown to increase expression of *RD29A* and *RD29B* genes, and in maize, ZmSNAC1 acts as a multiple stress-responsive TF in an ABA-dependent manner (Kimotho *et al.*, 2019). Moreover, seven *ZmNTLs* NAC TFs genes (*ZmNTL1*, *ZmNTL2*, *ZmNTL3*, *ZmNTL4*, *ZmNTL5*, *ZmNTL6* and *ZmNTL7*), were found to be strongly expressed in the stem and roots of plants exposed to ABA (Wang *et al.*, 2016). A RING finger E3 ligase, SDIR1, was also shown to act as a positive regulator of AREB/ABFs (Zhang *et al.*, 2007). AREB/ABFs also appear to interact with three WRKY TFs, ABO3/WRKY63, WRKY18 and WRKY40, upstream in response to ABA and drought stress (Ren *et al.*, 2010). In addition, WRKY63 binds to the promoter of *AREB1/ABF2* to positively regulate its expression. ABF3 also strongly interacts with JA and GA signalling pathways (Kumar *et al.*, 2019). On the other hand, MYC2 appears to be the interacting partner of ABF3 in the JA pathway because MYC2 interacts with ABA receptor PYL6, which provides a direct link between ABA and JA signalling pathways (Aleman *et al.*, 2016). The wheat MYB family member, TaMYB30-B, also induces the expression of ABA-dependent genes, including *RD29A* and *ERD1* (Zhang *et al.*, 2012). Moreover, RD22BP-1 and AtMYB2 have been shown to bind *cis*-elements in the *RD22* promoter and activate *RD22*, a gene in the ABA-dependent pathway. In maize, *ZmMYB30* is upregulated under drought, salt and ABA stresses, and positively modulates the expression of abiotic stress-induced genes in the ABA pathway (*ABF3, ATGolS2, AB15, DREB2A, 270 RD20, RD29B, RD29A* and *MYB2*) when exogenously expressed in *Arabidopsis* (Chen *et al.*, 2017). A maize homologue of NUCLEAR FACTOR Y (ZmNFYB2) was previously also shown to improve stomatal conductance and delayed onset of senescence in plants exposed to drought (Nelson *et al.*, 2007). NUCLEAR FACTOR Y (NF-Y) is a heterotrimeric TF with three distinct subunits, NF-YA, NF-YB and NF-YC, which binds to CCAAT box in the promoter region of target genes. In *Arabidopsis*, *NF-YA5* is upregulated in an ABA-dependent manner in leaves and roots of plants and modulates stomatal closure during drought stress (Li *et al.*, 2008). NF-Y subunit C (NF-YC) also interacts with ABF3, ABF4 and the nuclear zinc-finger protein CONSTANS (CO) to positively regulate a master floral gene, *SUPPRESSOR OF OVEREXPRESSION OF CONSTANS1 (SOC1)*, in response to drought (Weller and Ortega, 2015). Moreover, drought escape in *Arabidopsis* is regulated by ABA via CO and GIGANTEA (GI), which triggers the expression of the florigen genes *Flower Locus T (FT)*, *Twin Sister of FT (TSF)* and *SOC1* (Riboni *et al.*, 2013, 2016). It is relatively clear that drought escape is modulated by the relationship between CO and GI through ABA-dependent activation of *FT* genes. Thus, *NF-Y* genes crosstalk with other TFs and regulatory proteins modulated by ABA to ensure plant survival under drought stress. Phospholipase D (PLD) has also been reported to regulate stomatal closure in maize in an ABA-dependent manner (Ali *et al.*, 2020). $H_2O_2$ accumulation associated with ABA activation of OST1 and subsequently RbohD/F has been reported to enhance NO production through nitrate reductase 1 (NR1) activity. NO activates conversion of phospholipases C and D to phosphatidic acid (PA). PA binding decreases the phosphatase activity of ABI1, promoting ABA signalling (Ali *et al.*, 2020). Moreover, PA has been shown to enhance AtRbohD ROS generation causing the closure of stomata in response to ABA (Zhang *et al.*, 2009). Thus, PA acts as a positive regulator of ABA signalling leading to stomatal closure in plants exposed to drought stress. PLDs are also regulated by miR169 family during drought stress, most probably via the ABA signalling and/or an MAPK cascade (Shi *et al.*, 2017). Previous studies have implicated MAPK cascades in ABA-mediated responses (Jagodzik *et al.*, 2018). For instance, MAP3K18 apparently interacts with both SnRK2.6 kinase (Tajdel-Zielinska *et al.*, 2016) and PP2C phosphatase ABI1 (Mitula *et al.*, 2015). Specifically, ABA is implicated in regulating the MAPKKK17/18–MKK3–MPK1/MPK2/MPK7/MPK14 cascade (Li *et al.*, 2017), and a positive correlation between MAP3K18/MAP3K17 and the expression of ABA signal-transduction genes was shown in *Arabidopsis* (Danquah *et al.*, 2015). Several MAPK-related genes have been also shown to be expressed in wheat (e.g. *TaMAPK1*), maize (e.g. *ZmMAPK3*) and sorghum (e.g. *MAPK10* and *MPKKK7*) in response to drought, having functional association with ABA in regulating root hair growth and stomatal closure (Wang *et al.*, 2016). ABA also interacts with the ethylene signalling pathway in plants under water stress. In wheat, ABA-dependent TaMAPK1 physically interacts with ethylene-responsive factor 1 (TaERF1) in drought-stressed plants (Xu *et al.*, 2007). Moreover, plants overexpressing *TaERF1* were found to be highly sensitive to exogenous ABA treatment, resulting in rapid stomatal closure. Another wheat ethylene-response factor, TaERF3, contains both ACGT and ABRE motifs in its promoter and induces expression of ABA-dependent genes such as *RAB18* and *LEA3* (Rong *et al.*, 2014), confirming earlier reports implicating crosstalk between ethylene- and ABA-dependent pathways (Yu *et al.*, 2019).

Apart from ABA-dependent pathways, ABA-independent TFs (such as CBF/DREB1, DREB2, NAC) modulate drought stress response via drought-responsive element (DRE), CRT *cis*-acting elements, NAC recognition sequence (NACRS), MYBRS and W-box, among others. Well-known examples include *OsDREB2A* and *ZmDREB2, CBF2-1, CBF2-2, OsMYB3R2, MYB15* and *ANAC019* (Agarwal and Jha, 2010) and *OsWRKY08* (Song *et al.*, 2009). However, as mentioned earlier, evidence has shown that some ABA-independent genes possess ABRE in addition to DRE/CRT *cis*-element, causing their activation with and without ABA (Singh and Laxmi, 2015). AREB/ABF proteins physically interact with DREB1A/CBF3, DREB2A and DREB2C to modulate downstream gene expression (Lee *et al.*, 2010). In maize, ZmDREB1A induces both ABA-independent genes (e.g. *KIN1* and *KIN2*) and ABA-dependent genes (e.g. *RD17, ERD10* and *RD29A*) (Qin *et al.*, 2004). Thus, both signalling pathways hold great potential for a better understanding of how genes regulated by them determine drought response in plants. It can be argued that genotypes that integrate both signalling pathways cohesively might have stronger and more dynamic responses to drought stress.

During drought stress, plants' protein turnover mediated by autophagy also plays a critical role in signal transduction regulation (Stone, 2014; Yu and Xie, 2017), and a number of genes, termed autophagy related (ATG), are required for autophagosome formation. Interestingly, it has recently been revealed that a gene termed *constitutively stressed 1* (*COST1*) regulates constitutive activation of autophagy and leads to drought hypersensitivity and decreased autophagy in response to stress in *Arabidopsis* (Bao *et al.*, 2020). In addition, SUMO (small ubiquitin-like modifier) proteins have been found to be critical for plant response to stress (Srivastava *et al.*, 2017). In rice, knocking out SUMO protein (OTS1/2) increases drought tolerance by promoting a basic leucine zipper domain, OsbZIP23, SUMOylation. OsOTS1 interacts with OsbZIP23, a TF that regulates ABA and drought responses in rice. However, in OsOTS1 RNAi lines, OsbZIP23 is stabilized and has higher levels of SUMOylation leading to relatively higher transcription of drought tolerance genes (Srivastava *et al.*, 2017). It has also been demonstrated that SUMOylation stabilizes the DELLA proteins and OTS1/2 are the proteases that cleave SUMO from the DELLAs (Conti *et al.*, 2014). DELLA proteins integrate multiple hormone- and stress-related pathways (Colebrook *et al.*, 2014), including increased production of sugars, osmoprotectants, antioxidants and ROS scavengers. It would be interesting to determine if molecular mechanisms regulating autophagy and SUMOylation in *Arabidopsis* and rice apply in crop improvement against drought and other stresses. Taken together, drought stress response and tolerance in plants is complex, involving many different cellular and molecular components as well as metabolic pathways. The genes constituting these complex networks have been extensively studied but are not yet completely understood. Their cellular functions are depicted in Fig. 1.1, where SnRK is identified as a hub for both up and down signal integration. This review has also identified key genes involved in molecular regulation processes which have been reported as effective for improvement of drought tolerance in wheat, maize and sorghum (Table 1.1).

### 1.2.2 Salt stress

High salt levels in soil solution substantially impede plant productivity by affecting key biological processes, such as photosynthesis, energy metabolism, protein synthesis and lipid metabolism. The problem is more serious in the irrigated arid and semi-arid regions where there is high evapotranspiration and insufficient rainfall to leach the salts out of the plant root zones (Assaha *et al.*, 2017). There are two major plant divisions in response to salinity: the halophytes and the glycophytes, depending on their capacity to grow in saline conditions (Flowers *et al.*, 1977). Halophytes are equipped with specialized adaptation mechanisms such as the presence of salt glands and bladders, succulence, life cycle avoidance and salt-induced facultative metabolisms. Glycophytes, which encompass most cultivated crops, are mostly susceptible to salinity stress and yield losses of up to 80% can occur in saline conditions (electrical conductivity of the soil extract, ECe = 4–8 dS/m) (Panta *et al.*, 2014). However, this depends on the plant genus, species, cultivar within the species, duration of exposure and the stage of crop development at which stress occurs. For instance, in crop fields where salinity rises to about 10 dS/m, rice is severely affected, while wheat will produce some yield. Moreover, bread wheat is more tolerant than durum wheat, maize and sorghum (USDA-ARS, 2019). The within and outside-species genotypic variability in tolerance to saline conditions rationalizes the need to exploit these differences using molecular approaches.

Plants respond to soil salinity in two main phases. The first phase is osmotic, which arises from NaCl-induced reduction of soil water potential, restricting hydraulic conductance, water and solute uptake, causing stomatal closure and inhibition of cell expansion in the shoots (Munns and Tester, 2008). The second phase is ionic, arising from the build-up of toxic quantities of $Na^+$ and $Cl^-$ in the cells and tissues of the plants, which hinders the entry of important elements (such as $Ca^{2+}$ and $K^+$), enhances ROS production, interfering with cell redox and energy state, damaging cell membranes and subcellular organelles, slowing down cell signalling and causing premature

**Table 1.1.** Key genes involved in drought response and tolerance in wheat, maize and sorghum.

| Gene ID | Crop | Type/effect | Reference |
|---|---|---|---|
| *TaWRKY1* | Wheat | Transcription factor | He *et al.* (2016) |
| *TaWRKY33* | Wheat | Transcription factor | He *et al.* (2016) |
| *ZmDREB2A* | Maize | Transcription factor | Qin *et al.* (2007) |
| *TaRNAC1* | Wheat | Transcription factor | Chen *et al.* (2018) |
| *TaNAC29* | Wheat | Transcription factor | Xu *et al.* (2015) |
| *ZmbZIP4* | Maize | Transcription factor | Ma *et al.* (2018) |
| *Stay-green* (*stg2*) | Sorghum | QTL | Srinivas *et al.* (2008) |
| *Stg3* | Sorghum | QTL | Harris *et al.* (2006) |
| *Stg4* | Sorghum | QTL | Hayes *et al.* (2015) |
| *Stg5* | Sorghum | QTL | Hayes *et al.* (2015) |
| $qRDW_1$ | Sorghum | QTL | Mace *et al.* (2012) |
| *qTLA* | Sorghum | QTL | Mace *et al.* (2012) |
| *BLMC* | Sorghum | QTL | Burow *et al.* (2009) |
| *qRA* | Sorghum | QTL | Mace *et al.* (2012) |
| *TaMYB74* | Wheat | | Bi *et al.* (2016) |
| *TdSHN1* | Wheat | Cuticle biosynthesis-related gene | Djemal and Khoudi (2016) |
| *TaATT1* | Wheat | Cuticle biosynthesis-related gene | Bi *et al.* (2016) |
| *CER1* | | Cuticle formation | Djemal and Khoudi (2016) |
| *ZmFDL1/MYB94* | Wheat | Cuticle biosynthesis-related gene | La Rocca *et al.* (2015) |
| *ALDH22A1* | Maize | Aldehyde dehydrogenase | Huang *et al.* (2008) |
| *ASR1* | Maize | QTL | Jeanneau *et al.* (2002) |
| *NADP-ME* | Maize | NADP–malic enzyme | Laporte *et al.* (2002) |
| *PP2C* | Maize | PP2C | Liu, L. *et al.* (2009) |
| *RFP1* | Maize | RING finger protein | Liu *et al.* (2013) |
| *ZmACS6* | Maize | ACC synthase | Young *et al.* (2004) |
| *ZmCPK4* | Maize | Calcium-dependent protein kinase | Jiang *et al.* (2013) |
| *ZmDREB2.7* | Maize | DREB transcription factor | Liu *et al.* (2013) |
| *ABP9* | Maize | bZIP transcription factor | Zhang *et al.* (2011) |
| *CRT* | Wheat | Calreticulin, $Ca^{2+}$-binding protein | Jia *et al.* (2008) |
| *NAC2a* | Wheat | NAC transcription factor | Tang *et al.* (2012) |
| *PP2Ac-1* | Wheat | PP2A | Xu *et al.* (2007) |
| *Srg6* | Wheat | Stress response transcription factor | Tong *et al.* (2007) |
| *TaBTF3* | Wheat | Basic transcription factor 3 | Kang *et al.* (2013) |
| *WRKY19* | Wheat | WRKY-type transcription factor | Niu *et al.* (2012) |
| *WRKY2* | Wheat | WRKY-type transcription factor | Niu *et al.* (2012) |
| *TaWRKY44* | Wheat | WRKY-type transcription factor | Wang *et al.* (2015) |
| *ZmNF-YB2* | Maize | Nuclear factor Y transcription factor | Nelson *et al.* (2007) |
| *TaERF3* | Wheat | Transcription factor | Rong *et al.* (2014) |
| *TaRAP2.1* | Wheat | Transcription factor | Amalraj *et al.* (2016) |
| *TaER1* | Wheat | Transcription factor | Zheng *et al.* (2015) |
| *TAZFP34* | Wheat | Zinc-finger protein | Chang *et al.* (2016) |
| $qRT_6$ and $qRT_7$ | Sorghum | QTL | Li *et al.* (2014) |
| *Prf F* | Sorghum | Pre-flowering QTL | Kebede *et al.* (2001) |
| *HVA1* | Wheat | LEA protein | Sivamani *et al.* (2000) |
| *BADH* | Wheat | Betaine aldehyde dehydrogenase | Guo *et al.* (2000) |
| *TaNF-YB4* | Wheat | Nuclear factor Y transcription factor | Yadav *et al.* (2015) |
| *TaNAC69* | Wheat | More root biomass | Xue *et al.* (2011) |
| *CspA* and *CspB* | Wheat | High chlorophyll, proline and MDA | Yu *et al.* (2017) |
| *TaPEPKR2* | Wheat | Enhanced drought tolerance, higher root length | Zang *et al.* (2018) |
| *P5CS* | Wheat | Pyrroline-5 carboxylatesynthetase, proline biosynthesis | Vendruscolo *et al.* (2007) |
| *CIPK23* | Wheat | (CBL)-interacting protein kinase | Cui *et al.* (2018) |
| *TabZIP2* | Wheat | Increased single-seed weight | Luang *et al.* (2018) |
| *PEPC* | Maize | Proline biosynthesis | Qin *et al.* (2015) |

QTL, quantitative trait locus; ACC, aminocyclopropane-1-carboxylic acid; MDA, malondialdehyde.

senescence (Assaha *et al.*, 2017). Some plants are able to limit these counterproductive effects by limiting $Na^+$ uptake, enhancing $Na^+$ exclusion, adjusting cellular ionic balance (especially $Na^+$:$K^+$ ratio) and redistributing $Na^+$ in leaves (Ishikawa and Shabala, 2019). Both responses are governed by a multitude of physiological and molecular mechanisms that occur following conformational changes in membranes following depolarization or ligand binding in the presence of high concentration or entry of $Na^+$ in the root zone.

### *Molecular mechanisms of $Na^+$ entry and response in plants*

The most widely accepted model for $Na^+$ entry into the plant through the root system is the differential electrochemical gradient mediated by non-selective cation channels (NSCCs), although other transporters may also be involved (Assaha *et al.*, 2017). There are several classes of NSCC subdivided according to their voltage dependence or their responsiveness to ligands and physical stimuli (Demidchik and Maathuis, 2007; Kronzucker and Britto, 2011). These include constitutive NSCCs, amino acid-gated NSCCs (AAG-NSCCs), reactive oxygen species-activated NSCC (ROS-NSCCs), cyclic nucleotide-gated channels (CNGCs), glutamate receptors (GLRs) and mechanosensitive ion channels (MSCCs). Constitutive NSCCs are subdivided according to their response-membrane electrical potential change; these include depolarization-activated NSCCs (DA-NSCCs), hyperpolarization-activated NSCCs (HA-NSCCs) and voltage-insensitive NSCCs (VI-NSCCs). Although conclusive evidence is still scanty, cumulative research has demonstrated that VI-NSCCs are the main $Na^+$ entry route in plants, and this has been demonstrated in protoplasts of wheat root cortex, root cells of *Arabidopsis* and protoplasts of *Arabidopsis* roots (reviewed by Keisham *et al.*, 2018). VI-NSCCs are sensitive to $Ca^{2+}$ and are rapidly deactivated by cyclic adenosine monophosphate (cAMP) or cyclic guanosine monophosphate (cGMP) generated by turgor-sensitive membrane-located cyclases (Maathuis and Sanders, 2001). Thus, under salt stress, this is thought to be the main part of the ameliorative action of $Ca^{2+}$, cAMP or cGMP on salinity. Indeed, cellular cGMP and $Ca^{2+}$ elevation occurs simultaneously, and both cAMP and cGMP boost cytosolic $Ca^{2+}$ concentration and downregulate $Na^+$ influx and accumulation in *Arabidopsis* roots (Essah *et al.*, 2003; Isner and Maathuis, 2011). Other studies have shown that cAMP and cGMP are mainly perceived by CNGCs (non-selective) (Guo *et al.*, 2018). CNGCs are composed of hexa-transmembrane domains, calmodulin-binding domain (CAMB) and cyclic nucleotide-binding domain (CNBD). CNBD is the most conserved in both plants and animals and binds cAMP or cGMP for gating. On the contrary, a CAMB, CaM, appears to regulate CNBD affinity for cAMP or cGMP. CaM binds to the IQ motif of CNGCs to regulate downstream signalling (DeFalco *et al.*, 2016). Positive regulation has been recently substantiated, and it appears that the association is both $Ca^{2+}$-independent and -dependent due to polymorphism in the IQ peptide (Zhang *et al.*, 2019). To date, there have been limited studies able to prove the negative CaM regulation. More than ten CNGCs have been reported in *Arabidopsis* and rice genomes (Keisham *et al.*, 2018) and a few have been suggested to be negative regulators of salt tolerance. For instance, in rice, OsCNGC1 is suppressed in tolerant rice genotypes more than in salt-sensitive ones under salinity stress (Senadheera *et al.*, 2009), suggesting their negative role in salinity tolerance when highly expressed. *CNGC-like* genes are also present in the genomes of wheat, sorghum and maize. BLASTP searches on the plant genome databases has identified more than ten *CNGC* genes in both maize and sorghum (Saand *et al.*, 2015) and more than 20 in the wheat genome (Guo *et al.*, 2018). More studies are needed to shed light on the role of these CNGCs in wheat, maize and sorghum.

Other transmembrane receptors implicated in $Na^+$ response include GLRs and MSCC mechanosensors. GLRs are ligand-sensitive NSCCs gated by glutamate. Their role in $Na^+$ fluxes was demonstrated by Demidchik *et al.* (2004) in patch clamp experiments using *Arabidopsis* root protoplasts, and by Roy *et al.* (2008) in gene expression studies using *Xenopus laevis* oocytes. The positive effect of GLRs on salinity amelioration was shown also by Cheng *et al.* (2018) when they compared *Arabidopsis* mutants with the wild-type plants. *AtGLR3.4-1* and *AtGLR3.4-2* mutants were relatively sensitive to NaCl at seed germination and seedling stages. In addition, both mutants showed reduced expression of *salt overly sensitive* (*SOS3*, *SOS2* and *SOS1*) genes, indicating the positive effect of GLRs on crop salt tolerance. Unfortunately, no functional assays have been adequately carried out on wheat, maize and sorghum, so functional specificities of GLRs remain to be discovered in these cereals. MSCCs transduce mechanical stimuli such as osmotic pressure into the $Ca^{2+}$ influx. During salt stress, MSCCs are probably modulated by aquaporins (AQPs) and hydraulic effects on the cytoskeleton and exoskeleton (cell wall) (Ismail *et al.*, 2020). Among the AQPs, the plasma-membrane intrinsic proteins (PIPs) likely play a major role in $Na^+$ entry into the plant cells (Byrt *et al.*, 2017). This has been demonstrated in rice, where dopamine was shown to downregulate *PIP1;3* and reduce $Na^+$ uptake under saline conditions (Abdelkader *et al.*, 2012). A classic example of MSCCs are OSCAs (previously discussed; see drought stress response above). OSCAs induce $Ca^{2+}$ influx in response to osmotic pressure in the guard cells (Yuan *et al.*, 2014). This $Ca^{2+}$ influx probably plays a major role in gating AQPs to prevent further $Na^+$ influx under saline conditions. It has also been reported that salt-specific ionic stress disrupts cell-wall integrity, severing pectin crosslinking and displacing pectin-bound $Ca^{2+}$ with $Na^+$ (Feng *et al.*, 2018). Pectin is perceived by the extracellular domain of malectin-like RK, FERONIA (FER), and osmotic stress is detected once the co-receptor-glycosylphosphatidylinositol-anchored protein (GPI-AP) has been recruited, which activates cell-wall reinforcement during growth recovery (Feng *et al.*, 2018). More recently, Jiang *et al.* (2019) isolated a plasma-membrane-localized glucuronosyltransferase gene, *monocation-induced [$Ca^{2+}$] increases 1* (*moca1*), that is defective in salt-induced $Ca^{2+}$ spikes. MOCA1 transfers a negatively charged glucuronic acid to inositol phosphorylceramide (IPC) to form glycosyl-IPC (GIPC), which binds $Na^+$. *moca1* mutants induce elevated IPC:GIPC ratio on the plasma membrane (Lamers *et al.*, 2020). IPC is devoid of a negatively charged head, and *moca1* membranes show limited monovalent ion-binding capacity compared with the wild type. Thus, MOCA1-mediated GIPC–$Na^+$ interaction depolarizes the cell-surface potential, to facilitate $Ca^{2+}$ influx, and is a potential target for regulating $Na^+$ influx. The DA-NSCCs and OSCA1 are suggested to be the potential candidate channels for $Ca^{2+}$ influx (Ismail *et al.*, 2020). However, several questions remain unanswered regarding signal perception by DA-NSCCs, OSCA1 and other receptors with similar domains such as DORN1 (DOESN'T RESPOND TO NUCLEOTIDES), when $Na^+$ is perceived by MOCA1.

### *Plant adaptation strategies to internal $Na^+$ pools*

When excess $Na^+$ escapes to the cytosol it inhibits the $K^+$ outward rectifiers (KORs) such as GORK and $K^+$ uptake permease (KUP)/$K^+$ transporter (KT), which decreases their conductance and affects cellular homeostasis (Shabala *et al.*, 2006; Jayakannan *et al.*, 2013). Plants have to block or get rid of the excess $Na^+$ to maintain ion homeostasis mechanisms and avoid detrimental effects such as increased ROS accumulation and inhibition of plasma-membrane ATPase via membrane depolarization (Monetti *et al.*, 2014). Thus in addition to blocking $Na^+$ influx via $Ca^{2+}$-mediated deactivation of NSCCs and probably AQPs, getting rid of excessive cytosolic concentrations requires other downstream signalling components, including calmodulins (CaMs), CBLs, CDPKs and CBL–CIPKs (Møller *et al.*, 2009). CaMs and CBLs are small proteins that relay $Ca^{2+}$ signals to CDPKs. CBLs contain the Elongation Factor-hand (EF-hand) motif for perception and transfer of $Ca^{2+}$-dependent signals to CDPKs and CIPKs for propagation. It is well known that in plants exposed to saline conditions CBL4 or SOS3 is the CBL responding to upstream $Ca^{2+}$ signals (Fig. 1.2). $Ca^{2+}$ binding dimerizes SOS3 to activate CIPK24 (SOS2) and subsequent formation of CBL4/CIPK24 (SOS3/SOS2) protein kinase complex. SOS3-like calcium-binding protein 8 (SCaBP8) is phosphorylated by SOS2 to stabilize the SOS3/SOS2 complex for recruitment to the plasma membrane where it mediates SOS1 induction (Fig. 1.2). Apart from SOS3/SOS2 complex, PLD signalling pathway also interacts with SOS1 (Yu *et al.*, 2010). PLD derivative, PA, modulates mitogen-activated protein kinase 6 (MPK6), which phosphorylates SOS1 (Yu *et al.*, 2010). On the other hand, SOS2 interacts with vacuolar $Na^+/H^+$ antiporter (NHX) to sequester $Na^+$ into vacuoles in exchange for $H^+$ efflux from vacuoles or endosomes (Reguera *et al.*, 2015). NHX works in tandem with a voltage-independent $Ca^{2+}$-activated $K^+$-selective channel (VK channel) encoded by *TPK1* gene, sodium-permeable slow (SV) and

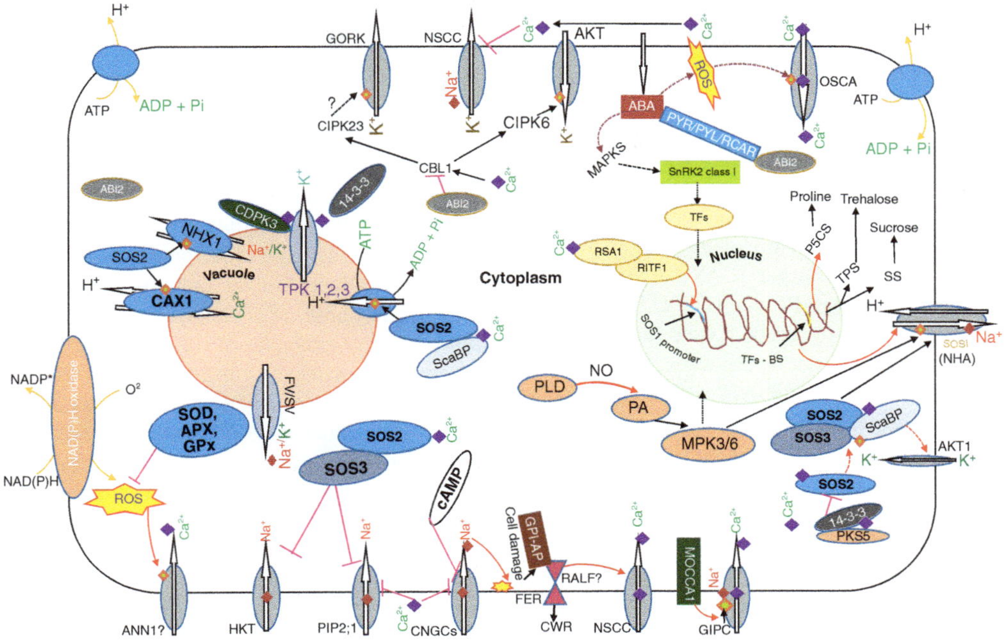


**Fig. 1.2.** Molecular regulation of salinity tolerance in plants. This illustration is prepared based on the findings from various articles cited throughout the manuscript and shows the main salt response and tolerance mechanisms such as $Na^+$ exclusion from the roots, compartmentalization in the vacuoles, cell-wall repair (CWR) and ROS remediation. The text provides the roles of the proteins indicated in Fig. 1.2. Question marks denote a response or proteins that might be related to tolerance. Proteins not clearly defined in the text, such as 14-3-3 and ABI2, are shown as negative regulators of SOS2. Also shown is that PKS5 modulates 14-3-3 at high cytosolic $Ca^{2+}$ to block inhibition of SOS2 by 14-3-3. GIPCs are salt stress sensors that directly bind $Na^+$ and trigger $Ca^{2+}$ influx via an unknown $Ca^{2+}$ channel, probably mediated by MOCCA1. This process is required for SOS activation. ROS generated by RbohD/F possibly activate Annexin 1 (ANN1)-mediated $Ca^{2+}$ signalling pathway. ANN1 can detect extracellular $Na^+$ and modulates ROS-activated $Ca^{2+}$ influx. The SOS2–SOS3 complex inhibits $Na^+$ uptake by HKTs. PIP2;1, CNGCs and GLRs could be blocked by exogenous $Ca^{2+}$, SOS complex or cAMPs. AKT is modulated by SCaBP8 for $K^+$ influx. Under salt stress FER possibly forms a complex with secreted peptides (RALFs) at the plasma membrane to trigger $Ca^{2+}$-mediated CWR. RALF and FER are possibly triggered by the perturbation of cell wall extensins (not shown). Salinity-induced ABA possibly activates SnRK2s either via the PYR/PYLs–PP2Cs or through MAPK cascades. The SnRK2s subclass I are of particular interest and could be activated via MAPKs. Both MPKs and SnRK2s could transduce signals to downstream TFs, including ABFs, ZIPs, MYBs, NACs, WRKYs and AP2/ERFs (not shown). These TFs modulate the induction of compatible solutes, antioxidants and repair modules. NHXs, CAX1, TPK1, and $H^+$-ATPase are involved in the regulation of ion homeostasis in the vacuole. The cation shunt $H^+$-pumping V-ATPase is provided by $K^+$-selective release via TPK1 channels. Slow (SV) and fast (FV) vacuolar channels represent $Na^+$ leak pathways into the cytosol. Their conductance is tightly regulated in tolerant genotypes to prevent intracellular ATP wastage in futile $Na^+$ cycling. PA is involved in activation of mitogen-activated protein kinase 6 (MPK6). MPK6 can directly phosphorylate SOS1. The $Ca^{2+}$-dependent kinase (CDPK3) and cytosolic $Ca^{2+}$ lead to activation of vacuolar two-pore $K^+$ channels (TPKs) and subsequent $K^+$ release from the vacuole. The binding of CDPK3 to TPKs at the vacuolar membrane could also be modulated by 14-3-3. Salt-induced $K^+$ loss from the cell is mediated by NSCCs and depolarization-activated outward-rectifying GORK $K^+$ channels. Similar to drought, elevated ABA inhibits ABI2 and induces phosphorylation of CBL, CIPK and CDPKs in $Ca^{2+}$-independent manner. Question marks indicate unconfirmed routes.

fast (FV) vacuolar channels, which release vacuolar $K^+$ for cytosolic homeostasis. Both SV and FV are a major pathway for $Na^+$ leakage into the cytosol. In spite of this, SV channels are a major vacuolar $Ca^{2+}$ channel and facilitate long-distance $Ca^{2+}$ signalling (Pottosin and Dobrovinskaya, 2018). Their relationship with NHX appears to be via the $Ca^{2+}$-dependent SOS3/SOS2 complex which modulates NHX, N-terminus of $H^+/Ca^{2+}$ exchanger (CAX), V-ATPase and pyrophosphatase (PPase) (Che-Othman *et al.*, 2017). In a more recent review (Shabala *et al.*, 2019), the interaction between SOS2, CAX1, plasma-membrane $H^+$-ATPases and tonoplast $H^+$-ATPases/$H^+$-PPases was suggested as a possible means to restore vacuolar $Ca^{2+}$ and $Na^+$ concentrations. Cellular $Na^+$ homeostasis is also regulated by high-affinity potassium transporter (HKT) proteins. The main function of HKTs is to unload $Na^+$ from the xylem transpiration stream and to load $Na^+$ into shoot phloem cells for shuttling to the roots, preventing excess $Na^+$ accumulation in the shoot (Jiang *et al.*, 2018).

Compared with *Arabidopsis*, cereals have multiple HKT isoforms. The first *HKT* gene, *TaHKT2;1*, was identified in wheat (Schachtman and Schroeder, 1994). Since then, several functional *HKT* genes have been identified in rice (e.g. *OsHKT1;5*), sorghum (e.g. *SbHKT1;4*), maize (e.g. *ZmHKT1;5*) and many other crops (Ren *et al.*, 2005; Wang *et al.*, 2015; Jiang *et al.*, 2018). In wheat, two genes, *Nax1* and *Nax2* (*TmHKT1;4* and *TmHKT1;4*), which extrude $Na^+$ from xylem in roots and shoots, have been reported to also regulate the expression of SOS1 (Assaha *et al.*, 2017). In *Arabidopsis*, SHORT ROOT IN SALT MEDIUM 1 (RSA1) interacting TF (RITF1) also modulates *SOS1* gene expression (Guan *et al.*, 2013). Thus, SOS proteins interact with other protein complexes or individual proteins to form functional modules that drive both subcellular and cellular signalling pathways. For more information on these transporters and other regulatory and osmoprotectant proteins, readers are referred to Onaga and Wydra (2016), Huang *et al.* (2020) and Ismail *et al.* (2020).

## 1.3 Conclusion

Plants' response to abiotic stress is limited to a few options, such as physiological avoidance or tolerance. The molecular mechanisms involved in the execution of either option are diverse and depend on the plants' ability to perceive the stress signals as well as on effective signal transmission through various pathways that evoke appropriate physiological and biochemical changes needed for adaptive response. This chapter has highlighted representative examples of the molecular mechanisms of plant response and tolerance to drought and salinity, the two major abiotic stresses affecting cereal production. Special emphasis has been placed on mechanisms that allow for interconnection of ABA, SnRK2, other ABA-dependent and -independent proteins for drought tolerance, and the network of $Ca^{2+}$ and SOS pathways for salinity tolerance. Emerging biological functions of $Ca^{2+}$, membrane channels and their interaction partners, and mechanisms such as the increasing role of lipids, autophagy and ubiquitination/SUMOylation, show promise in addressing abiotic stress challenges. Many of the TFs linked to these processes have also been identified and offer important targets for use in gene manipulation and regulation of abiotic stress response and tolerance. For drought, it is argued that genotypes that coherently integrate both ABA-dependent and -independent signalling pathways could have a stronger and more dynamic response to drought stress. In order to integrate both pathways in crops, suitable molecular tools are required. The question of the possible role of genomic selection and clustered regularly interspaced short palindromic repeats (CRISPR) genome editing in this regard remains lingering, especially when several gene networks are presented without a clear deployment strategy through use of these tools. A plausible option, for both drought and salinity, is to validate a core set of genes using both novel molecular tools and modern phenotyping technologies, to select those with high genetic effects and develop standardized protocols for their rapid integration and deployment in crops.

## References

Abdelkader, A., El-Khawas, S., Elsherif, N., Hassanein, R.A., Emam, M. and Hassan, R.E. (2012) Expression of aquaporin gene (Os PIP1-3) in salt-stressed rice (*Oryza sativa* L.) plants pre-treated with the neurotransmitter (dopamine). *Plant Omics* 5, 532–541.

Agarwal, P.K. and Jha, B. (2010) Transcription factors in plants and ABA dependent and independent abiotic stress signalling. *Biologia Plantarum* 54, 201–212. Available at: https://doi.org/10.1007/s10535-010-0038-7

Aida, M., Ishida, T., Fukaki, H., Fujisawa, H. and Tasaka, M. (1997) Genes involved in organ separation in *Arabidopsis*: an analysis of the cup-shaped cotyledon mutant. *The Plant Cell* 9, 841–857. Available at: https://doi.org/10.1105/tpc.9.6.841

Aleman, F., Yazaki, J., Lee, M., Takahashi, Y., Kim, A.Y. *et al.* (2016) An ABA increased interaction of the PYL6 ABA receptor with MYC2 transcription factor: a putative link of ABA and JA signaling. *Scientific Reports* 6, 28941.

Ali, S., Hayat, K., Iqbal, A. and Xie, L. (2020) Implications of abscisic acid in the drought stress tolerance of plants. *Agronomy* 10, 1323.

Amalraj, A., Luang, S., Kumar, M.Y., Sornaraj, P., Eini, O. *et al.* (2016) Change of function of the wheat stress-responsive transcriptional repressor *TaRAP2.1L* by repressor motif modification. *Plant Biotechnology Journal* 14(2), 820–832. Available at: https://doi.org/10.1111/pbi.12432

Ashraf, M., Athar, H.R., Harris, P.J.C. and Kwon, T.R. (2008) Some prospective strategies for improving crop salt tolerance. *Advances in Agronomy* 97, 45–110.

Assaha, D.V.M., Ueda, A., Saneoka, H., Al-Yahyai, R. and Yaish, M.W. (2017) The role of $Na^+$ and $K^+$ transporters in salt stress adaptation in glycophytes. *Frontiers in Physiology* 8, 509. Available at: https://doi.org/10.3389/fphys.2017.00509

Baloglu, M.C., Oz, M.T., Oktem, H.A. and Yucel, M. (2012) Expression analysis of *TaNAC69-1* and *TtNAMB-2*, wheat NAC family transcription factor genes under abiotic stress conditions in durum wheat (*Triticum turgidum*). *Plant Molecular Biology Reporter* 30(5), 1246–1252.

Bao, Y., Song, W.M., Wang, P., Yu, X., Li, B. *et al.* (2020) COST1 regulates autophagy to control plant drought tolerance. *Proceedings of the National Academy of Sciences USA* 117(13), 7482–7493. Available at: https://doi.org/10.1073/pnas.1918539117

Bi, H., Luang, S., Li, Y., Bazanova, N., Morran, S. *et al.* (2016) Identification and characterization of wheat drought-responsive MYB transcription factors involved in the regulation of cuticle biosynthesis. *Journal of Experimental Botany* 67, 5363–5380. Available at: https://doi.org/10.1093/jxb/erw298

Burow, G.B., Franks, C.D., Acosta-Martinez, V. and Xin, Z. (2009) Molecular mapping and characterization of BLMC, a locus for profue wax (bloom) and enhanced cuticular features of sorghum (*Sorghum bicolor* (L.) Moench.). *Theoretical and Applied Genetics* 118, 423–431.

Byrt, C.S., Zhao, M., Kourghi, M., Bose, J., Henderson, S. *et al.* (2017) Non-selective cation channel activity of aquaporin AtPIP2;1 regulated by $Ca^{2+}$ and pH: aquaporin ion channel. *Plant, Cell & Environment* 40, 802–815.

Cao, L., Zhang, P., Lu, X., Wang, G., Wang, Z. *et al.* (2020) Systematic analysis of the maize *OSCA* genes revealing *ZmOSCA* family members involved in osmotic stress and *ZmOSCA2.4* confers enhanced drought tolerance in transgenic *Arabidopsis*. *International Journal of Molecular Sciences* 21, 351. Available at: https://doi.org/10.3390/ijms21010351

Chang, H., Chen, D., Kam, J., Richardson, T., Drenth, J. *et al.* (2016) Abiotic stress upregulated TaZFP34 represses the expression of type-B response regulator and *SHY2* genes and enhances root to shoot ratio in wheat. *Plant Science* 252, 88–102. Available at: https://doi.org/10.1016/j.plantsci.2016.07.011

Che-Othman, H., Millar, A.H. and Taylor, N.L. (2017) Connecting salt stress signaling pathways with salinity-induced changes in mitochondrial metabolic processes in $C_3$ plants. *Plant, Cell & Environment* 40, 2875–2905.

Chen, D., Chai, S., McIntyre, C.L. and Xue, G.-P. (2018) Overexpression of a predominantly root-expressed NAC transcription factor in wheat roots enhances root length, biomass and drought tolerance. *Plant Cell Reports* 37, 225–237.

Chen, J., Nolan, T.M., Ye, H., Zhang, M., Tong, H. *et al.* (2017) *Arabidopsis* WRKY46, WRKY54, and WRKY70 transcription factors are involved in brassinosteroid-regulated plant growth and drought response. *The Plant Cell* 29, 425–1439. Available at: https://doi.org/10.1105/tpc.17.00364

Cheng, Y., Zhang, X., Sun, T., Tian, Q. and Zhang, W.H. (2018) Glutamate receptor homolog 3.4 is involved in regulation of seed germination under salt stress in *Arabidopsis*. *Plant & Cell Physiology* 59, 978–988. Available at: https://doi.org/10.1093/pcp/pcy034

Colebrook, E.H., Thomas, S.G., Phillips, A.L. and Hedden, P. (2014) The role of gibberellin signaling in plant responses to abiotic stress. *Journal of Experimental Biology* 217(Pt 1), 67–75. Available at: https://doi.org/10.1242/jeb.089938

Conti, L., Nelis, S., Zhang, C., Woodcock, A., Swarup, R. *et al.* (2014) Small ubiquitin-like modifier protein SUMO enables plants to control growth independently of the phytohormone gibberellin. *Developmental Cell* 28, 102–110.
Cramer, G.R., Urano, K., Delrot, S., Pezzotti, M. and Shinozaki, K. (2011) Effects of abiotic stress on plants: a systems biology perspective. *BMC Plant Biology* 11, 163. Available at: https://doi.org/10.1186/1471-2229-11-163
Cui, X.X., Du, Y.Y., Fu, J.J., Yu, T.T., Wang, C.C. *et al.* (2018) Wheat CBL-interacting protein kinase 23 positively regulates drought stress and ABA responses. *BMC Plant Biology* 18, 93.
Danquah, A., de Zélicourt, A., Boudsocq, M., Neubauer, J., Frey, N.F.D. *et al.* (2015) Identification and characterization of an ABA-activated MAP kinase cascade in *Arabidopsis thaliana*. *The Plant Journal* 82, 232–244. Available at: https://doi.org/10.1111/tpj.12808
DeFalco, T.A., Marshall, C.B., Munro, K., Kang, H.G., Moeder, W. *et al.* (2016) Multiple calmodulin-binding sites positively and negatively regulate *Arabidopsis* CYCLIC NUCLEOTIDE-GATED CHANNEL12. *The Plant Cell* 28(7), 1738–1751.
Demidchik, V. and Maathuis, F.J.M. (2007) Physiological roles of nonselective cation channels in plants: from salt stress to signaling and development. *New Phytologist* 175(3), 387–404. Available at: https://doi.org/10.1111/j.1469-8137.2007.02128.x
Demidchik, V., Adobea, P. and Tester, M.A. (2004) Glutamate activates sodium and calcium currents in the plasma membrane of *Arabidopsis* root cells. *Planta* 219, 167–175.
Ding, S., Feng, X., Du, H. and Wang, H. (2019) Genome-wide analysis of maize OSCA family members and their involvement in drought stress. *PeerJ* 7, e6765.
Djemal, R. and Khoudi, H. (2016) *TdSHN1*, a WIN1/SHN1-type transcription factor, imparts multiple abiotic stress tolerance in transgenic tobacco. *Environmental and Experimental Botany* 131, 89–100.
dos Reis, S.P., Lima, A.M. and de Souza, C.R.B. (2012) Recent molecular advances on downstream plant responses to abiotic stress. *International Journal of Molecular Sciences* 13, 8628–8647. Available at: https://doi.org/10.3390/ijms13078628
Egawa, C., Kobayashi, F., Ishibashi, M., Nakamura, T., Nakamura, C. and Takumi, S. (2006) Differential regulation of transcript accumulation and alternative splicing of a DREB2 homolog under abiotic stress conditions in common wheat. *Genes & Genetic Systems* 81, 77–91.
Essah, P.A., Davenport, R. and Tester, M. (2003) Sodium influx and accumulation in *Arabidopsis*. *Plant Physiology* 133(1), 307–318.
Farooq, M., Wahid, A., Kobayashi, N., Fujita, D. and Basra, S.M.A. (2009) Plant drought stress: effects, mechanisms and management. *Agronomy for Sustainable Development* 29, 185–212. Available at: https://doi.org/10.1051/agro:2008021
Feng, W., Kita, D., Peaucelle, A., Cartwright, H.N., Doan, V. *et al.* (2018) The FERONIA receptor kinase maintains cell-wall integrity during salt stress through $Ca^{2+}$ signaling. *Current Biology* 28(5), 666–675. Available at: https://doi.org/10.1016/j.cub.2018.01.023
Flowers, T.E.J., Troke, P. and Yeo, A.R. (1977) Mechanism of salt tolerance in halophytes. *Annual Review of Plant Physiology and Plant Molecular Biology* 28, 89–121.
Fujita, Y., Yoshida, T. and Yamaguchi-Shinozaki, K. (2013) Pivotal role of the AREB/ABF–SnRK2 pathway in ABRE-mediated transcription in response to osmotic stress in plants. *Physiologia Plantarum* 147(1), 15–27. Available at: https://doi.org/10.1111/j.1399-3054.2012.01635.x
Grondin, A., Rodrigues, O., Verdoucq, L., Merlot, S., Leonhardt, N. and Maurel, C. (2015) Aquaporins contribute to ABA-triggered stomatal closure through OST1-mediated phosphorylation. *The Plant Cell* 27(7), 945–954. Available at: https://doi.org/10.1105/tpc.15.00421
Guan, Q., Wu, J., Yue, X., Zhang, Y. and Zhu, J. (2013) A nuclear calcium-sensing pathway is critical for gene regulation and salt stress tolerance in Arabidopsis. *PLoS Genetics* 9(8), e1003755. Available at: https://doi.org/10.1371/journal.pgen.1003755
Guo, B.B., Zhang, Y.Y., Li, H.H., Du, L.L., Li, Y.Y. *et al.* (2000) Transformation of wheat with a gene encoding for the betaine aldehyde dehydrogenase (BADH). *Acta Botanica Sinica* 42, 279–283.
Guo, J., Islam, M.A., Lin, H., Ji, C., Duan, Y. *et al.* (2018) Genome-wide identification of cyclic nucleotide-gated ion channel gene family in wheat and functional analyses of *TaCNGC14* and *TaCNGC16*. *Frontiers in Plant Science* 9, 18. Available at: https://doi.org/10.3389/fpls.2018.00018
Harris, K., Subudhi, P., Borrell, A., Jordan, D., Rosenow, D. *et al.* (2006) Sorghum stay-green QTL individually reduce post-flowering drought induced leaf senescence. *Journal of Experimental Botany* 58, 327–338.
Hayes, C.M., Weers, B., Thakran, M., Burow, G., Xin, Z. *et al.* (2015) Discovery of a dhurrin QTL in *Sorghum bicolor*: co-localization of dhurrin biosynthesis and a novel stay-green QTL. *Crop Science* 56, 104–112.

He, G.-H., Xu, J.-Y., Wang, Y.-X., Liu, J.-M., Li, P.-S. *et al.* (2016) Drought-responsive WRKY transcription factor genes TaWRKY1 and TaWRKY33 from wheat confer drought and/or heat resistance in *Arabidopsis*. *BMC Plant Biology* 16, 116.

He, Z., Zhong, J., Sun, X., Wang, B., Terzaghi, W. and Dai, M. (2018) The maize ABA receptors ZmPYL8, 9, and 12 facilitate plant drought resistance. *Frontiers in Plant Science* 9, 422. Available at: https://doi.org/10.3389/fpls.2018.00422

Hedrich, R. and Geiger, D. (2017) Biology of SLAC1-type anion channels – from nutrient uptake to stomatal closure. *New Phytologist* 216, 46–61.

Hou, C.C., Tian, W., Kleist, T., He, K., Garcia, V. *et al.* (2014) DUF221 proteins are a family of osmosensitive calcium-permeable cation channels conserved across eukaryotes. *Cell Research* 24, 632–635. Available at: https://doi.org/10.1038/cr.2014.14

Huai, J., Wang, M., He, J., Zheng, J., Dong, Z. *et al.* (2008) Cloning and characterization of the SnRK2 gene family from *Zea mays*. *Plant Cell Reports* 27(12), 1861–1868.

Huang, L., Wu, D.-Z. and Zhang, G.-P. (2020) Advances in studies on ion transporters involved in salt tolerance and breeding crop cultivars with high salt tolerance. *Journal of Zhejiang University, Science B* 21(6), 426–441. Available at: https://doi.org/10.1631/jzus.b1900510

Huang, W., Ma, X., Wang, Q., Gao, Y., Xue, Y. *et al.* (2008) Significant improvement of stress tolerance in tobacco plants by overexpressing a stress-responsive *aldehyde dehydrogenase* gene from maize (*Zea mays*). *Plant Molecular Biology* 68, 451. Available at: https://doi.org/10.1007/s11103-008-9382-9

Huynh, T.T.H., Nguyen, T.L., Luu, H.L., Nguyen, H.H., Le, H.D. *et al.* (2019) Isolation and characterization of a DREB homolog gene from a local drought-tolerant maize cultivar. *Acta Biologica Cracoviensia s. Botanica* 62(2), 13–24.

IPCC (2018) Summary for policymakers. In: Masson-Delmotte, V., Zhai, P., Pörtner, H.-O., Roberts, D., Skea, J. *et al.* (eds) *Global Warming of 1.5°C. An IPCC Special Report on the impacts of global warming of 1.5°C above pre-industrial levels and related global greenhouse gas emission pathways, in the context of strengthening the global response to the threat of climate change, sustainable development, and efforts to eradicate poverty*. Intergovernmental Panel on Climate Change, Geneva, Switzerland, pp. 3–24.

Ishikawa, T. and Shabala, S. (2019) Control of xylem Na(+) loading and transport to the shoot in rice and barley as a determinant of differential salinity stress tolerance. *Physiologia Plantarum* 165(3), 619–631.

Ismail, A., El-Sharkawy, I. and Sherif, S. (2020) Salt stress signals on demand: cellular events in the right context. *International Journal of Molecular Sciences* 21(11), 3918. Available at: https://doi.org/10.3390/ijms21113918

Isner, J.-C. and Maathuis, F.J.M. (2011) Measurement of cellular cGMP in plant cells and tissues using the endogenous fluorescent reporter FlincG. *The Plant Journal* 65, 329–334.

Jagodzik, P., Tajdel-Zielinska, M., Ciesla, A., Marczak, M. and Ludwikow, A. (2018) Mitogen-activated protein kinase cascades in plant hormone signaling. *Frontiers in Plant Science* 9, 1387. Available at: https://doi.org/10.3389/fpls.2018.01387

Jayakannan, M., Bose, J., Babourina, O., Rengel, Z. and Shabala, S. (2013) Salicylic acid improves salinity tolerance in *Arabidopsis* by restoring membrane potential and preventing salt-induced $K^+$ loss via a GORK channel. *Journal of Experimental Botany* 64(8), 2255–2268. Available at: https://doi.org/10.1093/jxb/ert085

Jeanneau, M., Gerentes, D., Foueillassar, X., Zivy, M., Vidal, J., Toppan, A. and Perez, P. (2002) Improvement of drought tolerance in maize: towards the functional validation of the *Zm-Asr1* gene and increase of water use efficiency by over-expressing C4-PEPC. *Biochimie* 84(11), 1127–1135. Available at: https://doi.org/10.1016/s0300-9084(02)00024-x

Jensen, M.K., Lindemose, S., de Masi, F., Reimer, J.J., Nielsen, M. *et al.* (2013) ATAF1 transcription factor directly regulates abscisic acid biosynthetic gene *NCED3* in *Arabidopsis thaliana*. *FEBS Open Bio* 3, 21–7. Available at: https://doi.org/10.1016/j.fob.2013.07.006

Jia, X.Y., Xu, C.Y., Jing, R.L., Li, R.Z., Mao, X.-G., Wang, J.-P. and Chang, X.-P. (2008) Molecular cloning and characterization of wheat calreticulin (CRT) gene involved in drought-stressed responses. *Journal of Experimental Botany* 59, 739–751.

Jiang, S., Zhang, D., Wang, L., Pan, J., Liu, Y. *et al.* (2013) A maize calcium-dependent protein kinase gene, ZmCPK4, positively regulated abscisic acid signaling and enhanced drought stress tolerance in transgenic *Arabidopsis*. *Plant Physiology and Biochemistry* 71, 112–120. Available at: https://doi.org/10.1016/j.plaphy.2013.07.004

Jiang, Z., Song, G., Shan, X., Wei, Z., Liu, Y. *et al.* (2018) Association analysis and identification of ZmHKT1;5 variation with salt-stress tolerance. *Frontiers in Plant Science* 9, 1485. Available at: https://doi.org./10.3389/fpls.2018.01485

Jiang, Z., Zhou, X., Tao, M., Yuan, F., Liu, L. *et al.* (2019) Plant cell-surface GIPC sphingolipids sense salt to trigger $Ca^{2+}$ influx. *Nature* 572, 341–346. Available at: https://doi.org/10.1038/s41586-019-1449-z

Kang, G., Li, G., Ma, H., Wang, C. and Guo, T. (2013) Proteomic analysis on the leaves of *TaBTF3* gene virus-induced silenced wheat plants may reveal its regulatory mechanism. *Journal of Proteomics* 83, 130–143.

Kebede, H., Rosenow, D.T., Subudhi, P.K. and Nguyen, H.T. (2001) Quantitative trait loci influencing drought tolerance in grain sorghum (*Sorghum bicolor* L. Moench). *Theoretical and Applied Genetics* 103, 266–276.

Keisham, M., Mukherjee, S. and Bhatla, S.C. (2018) Mechanisms of sodium transport in plants – progresses and challenges. *International Journal of Molecular Sciences* 19(3), 647.

Kimotho, R., Hafiz, E. and Zhang, Z. (2019) Transcription factors involved in abiotic stress responses in maize (*Zea mays* L.) and their roles in enhanced productivity in the post genomics era. *PeerJ* 7(1), e7211. Available at: https://doi.org/10.7717/peerj.7211

Kobayashi, F., Maeta, E., Terashima, A., Kawaura, K., Ogihara, Y. and Takumi, S. (2008) Development of abiotic stress tolerance via bZIP-type transcription factor LIP19 in common wheat. *Journal of Experimental Botany* 59, 891–905. Available at: https://doi.org/10.1093/jxb/ern014

Kozomara, A. and Griffiths-Jones, S. (2011) miRBase: integrating microRNA annotation and deep-sequencing data. *Nucleic Acids Research* 39, D152–D157.

Kronzucker, H.J. and Britto, D.T. (2011) Sodium transport in plants: a critical review. *New Phytologist* 189, 54–81. Available at: https://doi.org/10.1111/j.1469-8137.2010.03540.x

Kwak, J.M., Mori, I.C., Pei, Z.M., Leonhardt, N., Torres, M.A. *et al.* (2003) NADPH oxidase *AtrbohD* and *AtrbohF* genes function in ROS-dependent ABA signaling in *Arabidopsis*. *The EMBO Journal* 22, 2623–2633.

Kumar, M., Kesawat, M.S., Ali, A., Lee, S.C., Gill, S.S. and Kim, A. (2019) Integration of abscisic acid signaling with other signaling pathways in plant stress responses and development. *Plants* 8(12), 592. Available at: https://doi.org/10.3390/plants8120592

Lamers, J., van der Meer, T. and Testerink, C. (2020) How plants sense and respond to stressful environments. *Plant Physiology* 182(4), 1624–1635. Available at: https://doi.org/10.1104/pp.19.01464

La Rocca, N., Manzotti, P.S., Cavaiuolo, M., Barbante, A., Dalla Vecchia, F. *et al.* (2015) The maize fused leaves1 (fdl1) gene controls organ separation in the embryo and seedling shoot and promotes coleoptile opening. *Journal of Experimental Botany* 66(19), 5753–5767. Available at: https://doi.org/10.1093/jxb/erv278

Laporte, M.M., Shen, B. and Tarczynski, M.C. (2002) Engineering for drought avoidance: expression of maize NADP-malic enzyme in tobacco results in altered stomatal function. *Journal of Experimental Botany* 53, 699–705.

Lee, S.-J., Kang, J.-Y., Park, H.-J., Kim, M.D., Bae, M.S., Choi, H.-I. and Kim, S.Y. (2010) DREB2C interacts with ABF2, a bZIP protein regulating abscisic acid-responsive gene expression, and its overexpression affects abscisic acid sensitivity. *Plant Physiology* 153, 716–727.

Li, R., Han, Y., Lv, P., Du, R. and Liu, G. (2014) Molecular mapping of the brace root traits in sorghum (*Sorghum bicolor* L. Moench). *Breeding Science* 64, 193–198.

Li, S., Castillo-González, C., Yu, B. and Zhang, X. (2017) The functions of plant small RNAs in development and in stress responses. *The Plant Journal* 90, 654–670.

Li, W.-X., Oono, Y., Zhu, J., He, X.-J., Wu, J.-M. *et al.* (2008) The *Arabidopsis* NF-YA5 transcription factor is regulated transcriptionally and posttranscriptionally to promote drought resistance. *The Plant Cell* 20(8), 2238–2251.

Liu, H., Xing, M., Yang, W., Mu, X., Wang, X. *et al.* (2019) Genome-wide identification of and functional insights into the late embryogenesis abundant (*LEA*) gene family in bread wheat (*Triticum aestivum*). *Scientific Reports* 9, 13375. Available at: https://doi.org/10.1038/s41598-019-49759-w

Liu, L., Hu, X., Song, J., Zong, X., Li, D. and Li, D. (2009a) Over-expression of a *Zea mays* L. protein phosphatase 2C gene (*ZmPP2C*) in *Arabidopsis thaliana* decreases tolerance to salt and drought. *Journal of Plant Physiology* 166(5), 531–542. Available at: https://doi.org/10.1016/j.jplph.2008.07.008

Liu, T.X., Zhang, L., Yuan, Z.L., Hu, X.L., Lu, M.H., Wan, W. and Wang, Y. (2013) Identification of proteins regulated by ABA in response to combined drought and heat stress in maize roots. *Acta Physiologiae Plantarum* 35, 501–513.

Liu, Y.H., Li, H.Y., Shi, Y.S., Song, Y.C., Wang, T.Y. and Li, Y. (2009b) A maize early responsive to dehydration gene, *ZmERD4*, provides enhanced drought and salt tolerance in *Arabidopsis*. *Plant Molecular Biology Reporter* 27, 542–548.

Lu, M., Sun, Q.-P., Zhang, D.-F., Wang, T.-Y. and Pan, J.-B. (2015) Identification of 7 stress-related NAC transcription factor members in maize (*Zea mays* L.) and characterization of the expression pattern of these genes. *Biochemical and Biophysical Research Communications* 462, 144–150.

Luang, S., Sornaraj, P., Bazanova, N., Jia, W., Eini, O. *et al.* (2018) The wheat TabZIP2 transcription factor is activated by the nutrient starvation-responsive SnRK3/CIPK protein kinase. *Plant Molecular Biology* 96, 543–561.

Ma, H., Liu, C., Li, Z., Ran, Q., Xie, G. *et al.* (2018) ZmbZIP4 contributes to stress resistance in maize by regulating ABA synthesis and root development. *Plant Physiology* 178, 753–770.

Ma, Q.B., Xia, Z.L., Cai, Z.D., Li, L., Cheng, Y.B., Liu, J. and Nian, H. (2019) *GmWRKY16* enhances drought and salt tolerance through an ABA-mediated pathway in *Arabidopsis thaliana*. *Frontiers in Plant Science* 9, 1979.

Mace, E.S., Singh, V., Van Oosterom, E.J., Hammer, G.L., Hunt, C.H. and Jordan, D.R. (2012) QTL for nodal root angle in sorghum (*Sorghum bicolor* L. Moench) co-locate with QTL for traits associated with drought adaptation. *Theoretical and Applied Genetics* 124(1), 97–109.

Manik, S.M.N., Pengilley, G., Dean, G., Field, B., Shabala, S. and Zhou, M. (2019) Soil and crop management practices to minimize the impact of waterlogging on crop productivity. *Frontiers in Plant Science* 10, 140. Available at: https://doi.org/10.3389/fpls.2019.00140

Mao, H., Wang, H., Liu, S., Li, Z., Yang, X. *et al.* (2015) A transposable element in a *NAC* gene is associated with drought tolerance in maize seedlings. *Nature Communications* 6, 8326. Available at: https://doi.org/10.1038/ncomms9326

Mao, X., Chen, S., Li, A., Zhai, C. and Jing, R. (2014) Novel NAC transcription factor TaNAC67 confers enhanced multi-abiotic stress tolerances in *Arabidopsis*. *PLoS ONE* 9(1), e84359. Available at: https://doi.org/10.1371/journal.pone.0084359

Mantri, N., Patade, V., Penna, S., Ford, R. and Pang, E. (2012) Abiotic stress responses in plants: present and future. In: Ahmad, P. and Prasad, M. (eds) *Abiotic Stress Responses in Plants*. Springer, New York, pp. 1–19.

Maathuis, F.J. and Sanders, D. (2001) Sodium uptake in *Arabidopsis* roots is regulated by cyclic nucleotides. *Plant Physiology* 127(4), 1617–1625.

Mittler, R., Kim, Y., Song, L., Coutu, J., Coutu, A. *et al.* (2006) Gain- and loss-of-function mutations in *Zat10* enhance the tolerance of plants to abiotic stress. *FEBS Letters* 580, 6537–6542.

Mitula, F., Tajdel-Zielinska, M., Ciesla, A., Kasprowicz-Maluski, A., Kulik, A. *et al.* (2015) *Arabidopsis* ABA-activated kinase MAPKKK18 is regulated by protein phosphatase 2C ABI1 and the ubiquitin–proteasome pathway. *Plant and Cell Physiology* 56(12), 2351–2367. Available at: https://doi.org/10.1093/pcp/pcv146

Møller, I.S., Gilliham, M., Jha, D., Mayo, G.M., Roy, S.J. *et al.* (2009) Shoot $Na^+$ exclusion and increased salinity tolerance engineered by cell type-specific alteration of $Na^+$ transport in *Arabidopsis*. *The Plant Cell* 21(7), 2163–2178.

Monetti, E., Kadono, T., Tran, D., Azzarello, E., Arbelet, B.D. *et al.* (2014) Deciphering early events involved in hyperosmotic stress-induced programmed cell death in tobacco BY-2 cells. *Journal of Experimental Botany* 65(5), 1361–1375. Available at: https://doi.org/10.1093/jxb/ert460

Müller, H.M., Schäfer, N., Bauer, H., Geiger, D., Lautner, S. *et al.* (2017) The desert plant *Phoenix dactylifera* closes stomata via nitrate-regulated SLAC1 anion channel. *New Phytologist* 216, 150–162.

Munns, R. and Tester, M. (2008) Mechanisms of salinity tolerance. *Annual Review of Plant Biology* 59, 651–681. Available at: https://doi.org/10.1146/annurev.arplant.59.032607.092911

Nakashima, K., Takasaki, H., Mizoi, J., Shinozaki, K. and Yamaguchi-Shinozaki, K. (2012) NAC transcription factors in plant abiotic stress responses. *Biochimica et Biophysica Acta* 1819, 97–103. Available at: https://doi.org/10.1016/j.bbagrm.2011.10.005

Naumann, G., Alfieri, L., Wyser, K., Mentaschi, L., Betts, R.A. *et al.* (2018) Global changes in drought conditions under different levels of warming. *Geophysical Research Letters* 45(7), 3285–3296. Available at: https://doi.org/10.1002/2017gl076521

Nelson, D.E., Repetti, P.P., Adams, T.R., Creelman, R.A., Wu, J. *et al.* (2007) Plant nuclear factor Y (NF-Y) B subunits confer drought tolerance and lead to improved corn yields on water-limited acres. *Proceedings of the National Academy of Sciences USA* 104, 16450–16455.
Niu, C.F., Wei, W., Zhou, Q.Y., Tian, A.G., Hao, Y.J. *et al.* (2012) Wheat *WRKY* genes *TaWRKY2* and *TaWRKY19* regulate abiotic stress tolerance in transgenic *Arabidopsis* plants. *Plant, Cell & Environment* 35, 1156–1170.
Onaga, G. and Wydra, K. (2016) Advances in plant tolerance to abiotic stresses. In: Abdurakhmonov, I.Y. (ed.) *Plant Genomics*. InTech, Rijeka, Croatia. Available at: https://doi.org/10.5772/64350
Panta, S., Flowers, T., Lane, P., Doyle, R.B., Haros, G. and Shabala, S. (2014) Halophyte agriculture: success stories. *Environmental and Experimental Botany* 107, 71–83.
Pottosin, I. and Dobrovinskaya, O. (2018) Two-pore cation (TPC) channel: not a shorthanded one. *Functional Plant Biology* 45, 83–92.
Qadir, M., Quillerou, E., Nangia, V., Murtaza, G., Singh, M. *et al.* (2014) Economics of salt-induced land degradation and restoration. *Natural Resources Forum* 38(4), 282–295.
Qin, F., Sakuma, Y., Li, J., Liu, Q., Li, Y.-Q., Shinozaki, K. and Yamaguchi-Shinozaki, K. (2004) Cloning and functional analysis of a novel DREB1/CBF transcription factor involved in cold-responsive gene expression in *Zea mays* L. *Plant and Cell Physiology* 45(8), 1042–1052. Available at: https://doi.org/10.1093/pcp/pch118
Qin, F., Kakimoto, M., Sakuma, Y., Maruyama, K., Osakabe, Y. *et al.* (2007) Regulation and functional analysis of ZmDREB2A in response to drought and heat stresses in *Zea mays* L. *The Plant Journal* 50, 54–69.
Qin, N., Xu, W., Hu, L., Li, Y., Wang, H. *et al.* (2015) Drought tolerance and proteomics studies of transgenic wheat containing the maize $C_4$ phosphoenolpyruvate carboxylase (PEPC) gene. *Protoplasma* 253, 1503–1512.
Raftery, A.E., Zimmer, A., Frierson, D.M.W., Startz, R. and Liu, P. (2017) Less than 2°C warming by 2100 unlikely. *Nature* 7, 637–643.
Reddy, V.S. and Reddy, A.S. (2004) Proteomics of calcium-signaling components in plants. *Phytochemistry* 65, 1745–1776. Available at: https://doi.org/10.1016/j.phytochem.2004.04.033
Reguera, M., Bassil, E., Tajima, H., Wimmer, M., Chanoca, A. *et al.* (2015) pH regulation by NHX-type antiporters is required for receptor-mediated protein trafficking to the vacuole in *Arabidopsis*. *The Plant Cell* 27, 1200–1217.
Rehman, K., Fatima, F., Waheed, I. and Akash, M.S.H. (2018) Prevalence of exposure of heavy metals and their impact on health consequences. *Journal of Cellular Biochemistry* 119(1), 157–184.
Ren, X., Chen, Z., Liu, Y., Zhang, H., Zhang, M. *et al.* (2010) ABO3, a WRKY transcription factor, mediates plant responses to abscisic acid and drought tolerance in *Arabidopsis*. *The Plant Journal* 63, 417–429.
Ren, Z.-H., Gao, J.-P., Li, L.-G., Cai, X.-L., Huang, W. *et al.* (2005) A rice quantitative trait locus for salt tolerance encodes a sodium transporter. *Nature Genetics* 37, 1141–1146.
Riboni, M., Galbiati, M., Tonelli, C. and Conti, L. (2013) *GIGANTEA* enables drought escape response via abscisic acid-dependent activation of the florigens and *SUPPRESSOR OF OVEREXPRESSION OF CONSTANS1*. *Plant Physiology* 162, 1706–1719. Available at: http://dx.doi.org/10.1104/pp.113.217729
Riboni, M., Robustelli Test, A., Galbiati, M., Tonelli, C. and Conti, L. (2016) ABA-dependent control of *GIGANTEA* signalling enables drought escape via up-regulation of *FLOWERING LOCUS T* in *Arabidopsis thaliana*. *Journal of Experimental Botany* 67, 6309–6322. Available at: http://dx.doi.org/10.1093/jxb/erw384
Rong, W., Qi, L., Wang, A., Ye, X., Du, L. and Liang, H. (2014) The ERF transcription factor TaERF3 promotes tolerance to salt and drought stresses in wheat. *Plant Biotechnology Journal* 12(4), 468–479. Available at: https://doi.org/10.1111/pbi.12153
Roy, S.J., Gilliham, M., Berger, B., Essah, P.A., Cheffings, C. *et al.* (2008) Investigating glutamate receptor-like gene co-expression in *Arabidopsis thaliana*. *Plant, Cell & Environment* 31, 861–871.
Saand, M.A., Xu, Y.-P., Munyampundu, J.-P., Li, W., Zhang, X.-R. and Cai, X.-Z. (2015) Phylogeny and evolution of plant cyclic nucleotide-gated ion channel (CNGC) gene family and functional analyses of tomato CNGCs. *DNA Research* 22(6), 471–483. Available at: https://doi.org/10.1093/dnares/dsv029
Saidi, M.N., Mergby, D. and Brini, F. (2017) Identification and expression analysis of the NAC transcription factor family in durum wheat (*Triticum turgidum* L. ssp. *durum*). *Plant Physiology and Biochemistry* 112, 117–128.

Samarah, N., Mullen, R. and Cianzio, S. (2004) Size distribution and mineral nutrients of soybean seeds in response to drought stress. *Journal of Plant Nutrition* 27, 815–835. Available at: https://doi.org/10.1081/PLN-120030673

Sanjari, S., Shirzadian-Khorramabad, R., Shobbar, Z.S. and Shahbazi, M. (2019) Systematic analysis of NAC transcription factors' gene family and identification of post-flowering drought stress responsive members in sorghum. *Plant Cell Reports* 38, 361–376. Available at: https://doi.org/10.1007/s00299-019-02371-8

Sato, H., Takasaki, H., Takahashi, F., Suzuki, T., Iuchi, S. *et al.* (2018) *Arabidopsis thaliana* NGA-THA1 transcription factor induces ABA biosynthesis by activating *NCED3* gene during dehydration stress. *Proceedings of the National Academy of Sciences USA* 115, E11178–E11187. Available at: https://doi.org/10.1073/pnas.1811491115

Schachtman, D.P. and Schroeder, J.I. (1994) Structure and transport mechanism of a high-affinity potassium uptake transporter from higher plants. *Nature* 370, 655–658. Available at: https://doi.org/10.1038/370655a0

Schäfer, N., Maierhofer, T., Herrmann, J., Jørgensen, M.E., Lind, C. *et al.* (2018) A tandem amino acid residue motif in guard cell SLAC1 anion channel of grasses allows for the control of stomatal aperture by nitrate. *Current Biology* 28, 1370–1379.e5.

Senadheera, P., Singh, R.K. and Maathuis, F.J. (2009) Differentially expressed membrane transporters in rice roots may contribute to cultivar dependent salt tolerance. *Journal of Experimental Botany* 60(9), 2553–2563. Available at: https://doi.org/10.1093/jxb/erp099

Shabala, S., Demidchik, V., Shabala, L., Cuin, T.A., Smith, S.J. *et al.* (2006) Extracellular $Ca^{2+}$ ameliorates NaCl-induced $K^+$ loss from *Arabidopsis* root and leaf cells by controlling plasma membrane $K^+$-permeable channels. *Plant Physiology* 141(4), 1653–1665. Available at: https://doi.org/10.1104/pp.106.082388

Shabala, S., Chen, G., Chen, Z.-H. and Pottosin, I.I. (2019) The energy cost of the tonoplast futile sodium leak. *New Phytologist* 225(3), 1105–1110.

Shao, H., Wang, H. and Tang, X. (2015) NAC transcription factors in plant multiple abiotic stress responses: progress and prospects. *Frontiers in Plant Science* 6, 902. Available at: https://doi.org/10.3389/fpls.2015.00902

Shi, J., Gao, H., Wang, H., Lafitte, H.R., Archibald, R.L. *et al.* (2017) ARGOS8 variants generated by CRISPR-Cas9 improve maize grain yield under field drought stress conditions. *Plant Biotechnology Journal* 15(2), 207–216. Available at: https://doi.org/10.1111/pbi.12603

Singh, A., Kushwaha, H.R., Soni, P., Gupta, H., Singla-Pareek, S.L. and Pareek, A. (2015) Tissue specific and abiotic stress regulated transcription of histidine kinases in plants is also influenced by diurnal rhythm. *Frontiers in Plant Science* 6, 711.

Singh, D. and Laxmi, A. (2015) Transcriptional regulation of drought response: a tortuous network of transcriptional factors. *Frontiers in Plant Science* 6, 895. Available at: https://doi.org/10.3389/fpls.2015.00895

Sirichandra, C., Gu, D., Hu, H.C., Davanture, M., Lee, S. *et al.* (2009) Phosphorylation of the *Arabidopsis* AtrbohF NADPH oxidase by OST1 protein kinase. *FEBS Letters* 583, 2982–2986.

Sivamani, E., Bahieldin, A., Wraith, J.M., Al-Niemi, T., Dyer, W.E., Ho, T.D. and Qu, R. (2000) Improved biomass productivity and water use efficiency under water deficit conditions in transgenic wheat constitutively expressing the barley *HVA1* gene. *Plant Science* 155(1), 1–9. Available at: https://doi.org/10.1016/s0168-9452(99)00247-2

Song, Y., Jing, S. and Yu, D. (2009) Overexpression of the stress-induced OsWRKY08 improves osmotic stress tolerance in *Arabidopsis*. *Chinese Science Bulletin* 54, 4671–4678.

Soon, F.-F., Ng, L.-M., Zhou, X.E., West, G.M., Kovach, K. *et al.* (2012) Molecular mimicry regulates ABA signaling by SnRK2 kinases and PP2C phosphatases. *Science* 335, 85–88.

Srinivas, G., Satish, K., Murali Mohan, S., Nagaraja Reddy, R., Madhusudhana, R. *et al.* (2008) Development of genic-microsatellite markers for sorghum staygreen QTL using a comparative genomic approach with rice. *Theoretical and Applied Genetics* 117, 283–296.

Srivastava, A.K., Zhang, C., Caine, R.S., Gray, J.E. and Sadanandom, A. (2017) Rice SUMO protease overly tolerant to salt 1 targets the transcription factor, OsbZIP23 to promote drought tolerance in rice. *The Plant Journal* 92(10), 1031–1043. Available at: https://doi.org/10.1111/tpj.13739

Stone, S.L. (2014) The role of ubiquitin and the 26S proteasome in plant abiotic stress signaling. *Frontiers in Plant Science* 5, 135. Available at: https://doi.org/10.3389/fpls.2014.00135

Sun, S.-J., Qi, G.-N., Gao, Q.-F., Wang, H.-Q., Yao, F.-Y. *et al.* (2016) Protein kinase OsSAPK8 functions as an essential activator of S-type anion channel OsSLAC1, which is nitrate-selective in rice. *Planta* 243, 489–500.

Tajdel-Zielinska, M., Mitula, F. and Ludwikow, A. (2016) Regulation of *Arabidopsis* MAPKKK18 by ABI1 and SnRK2, components of the ABA signaling pathway. *Plant Signaling & Behaviour* 11(4), e1139277. Available at: https://doi.org/10.1080/15592324.2016.1139277

Takahashi, F., Suzuki, T., Osakabe, Y., Betsuyaku, S., Kondo, Y. *et al.* (2018) A small peptide modulates stomatal control via abscisic acid in long-distance signalling. *Nature* 556(7700), 235–238. Available at: https://doi.org/10.1038/s41586-018-0009-2

Takahashi, F., Kuromori, T., Urano, K., Yamaguchi-Shinozaki, K. and Shinozaki, K. (2020) Drought stress responses and resistance in plants: from cellular responses to long-distance intercellular communication. *Frontiers in Plant Science* 11, 556972. Available at: https://doi.org/10.3389/fpls.2020.556972

Takahashi, Y., Ebisu, Y., Kinoshita, T., Doi, M., Okuma, E., Murata, Y. and Shimazaki, K. (2013) bHLH Transcription factors that facilitate $K^+$ uptake during stomatal opening are repressed by abscisic acid through phosphorylation. *Science Signaling* 6, ra48.

Tan, B.C., Schwartz, S.H., Zeevaart, J.A. and McCarty, D.R. (1997) Genetic control of abscisic acid biosynthesis in maize. *Proceedings of the National Academy of Sciences USA* 94, 12235–12240.

Tang, Y.M., Liu, M., Gao, S., Zhang, Z., Zhao, X. *et al.* (2012) Molecular characterization of novel *TaNAC* genes in wheat and overexpression of *TaNAC2a* confers drought tolerance in tobacco. *Physiologia Plantarum* 144(3), 210–224. Available at: https://doi.org/10.1111/j.1399-3054.2011.01539.x

Tong, S., Ni, Z., Peng, H., Dong, G. and Sun, Q. (2007) Ectopic overexpression of wheat *TaSrg6* gene confers water stress tolerance in *Arabidopsis*. *Plant Science* 172, 1079–1086.

Tong, S.M., Xi, H.X., Ai, K.J. and Hou, H.S. (2017) Overexpression of wheat *TaNCED* gene in *Arabidopsis* enhances tolerance to drought stress and delays seed germination. *Biologia Plantarum* 61, 64–72.

Tran, L.S.P., Nakashima, K., Sakuma, Y., Simpson, S.D., Fujita, Y. *et al.* (2004) Isolation and functional analysis of *Arabidopsis* stress-inducible NAC transcription factors that bind to a drought-responsive *cis*-element in the *early responsive to dehydration stress 1* promoter. *The Plant Cell* 16, 2481–2498. Available at: https://doi.org/10.1105/tpc.104.022699

USDA-ARS (2019) Agricultural Water Efficiency and Salinity Research Unit: Riverside, CA I Salt Tolerance Database I Fiber, Grain and Special Crops. US Department of Agriculture, Agricultural Research Service, Washington, DC. Available at: https://www.ars.usda.gov/pacific-west-area/riverside-ca/agricultural-water-efficiency-and-salinity-research-unit/docs/databases/fgs-crops/ (accessed 19 February 2021).

Vendruscolo, E.C.G., Schuster, I., Pileggi, M., Scapim, C.C., Molinari, H.B.C., Marur, C.C. and Vieira, L.G.E. (2007) Stress-induced synthesis of proline confers tolerance to water deficit in transgenic wheat. *Journal of Plant Physiology* 164, 1367–1376.

Wang, D., Yanchong, Y., Liu, Z., Shuo, L., Wang, Z. and Xiang, F. (2016) Membrane-bound NAC 1571 transcription factors in maize and their contribution to the oxidative stress response. *Plant Science* 250, 30–39. Available at: https://doi.org/10.1016/j.plantsci.2016.05.019

Wang, T.-T., Ren, Z.-J., Liu, Z.-Q., Feng, X., Guo, R.-Q. *et al.* (2014) SbHKT1;4, a member of the high-affinity potassium transporter gene family from *Sorghum bicolor*, functions to maintain optimal $Na^+/K^+$ balance under $Na^+$ stress. *Journal of Integrative Plants Biology* 56(3), 315–332.

Wang, X., Zeng, J., Li, Y., Rong, X., Sun, J. *et al.* (2015) Expression of *TaWRKY44*, a wheat *WRKY* gene, in transgenic tobacco confers multiple abiotic stress tolerances. *Frontiers in Plant Science* 6, 615. Available at: https://doi.org/10.3389/fpls.2015.00615

Weller, J.L. and Ortega, R. (2015) Genetic control of flowering time in legumes. *Frontiers in Plant Science* 6, 207. Available at: https://doi.org/10.3389/fpls.2015.00207

Wu, F., Chi, Y., Jiang, Z., Xu, Y., Xie, L. *et al.* (2020) Hydrogen peroxide sensor HPCA1 is an LRR receptor kinase in *Arabidopsis*. *Nature* 578(7796), 577–581. Available at: https://doi.org/10.1038/s41586-020-2032-3

Xu, Z., Wang, C., Xue, F., Zhang, H. and Ji, W. (2015) Wheat NAC transcription factor TaNAC29 is involved in response to salt stress. *Plant Physiology and Biochemistry* 96, 356–363.

Xu, Z.-S., Xia, L.-Q., Chen, M., Cheng, X.-G., Zhang, R.-Y. *et al.* (2007) Isolation and molecular characterization of the *Triticum aestivum* L. ethylene-responsive factor 1 (TaERF1) that increases multiple stress tolerance. *Plant Molecular Biology* 65, 719–732. Available at: https://doi.org/10.1007/s11103-007-9237-9

Xu, Z.Y., Kim, S.Y., Hyeon, D.Y., Kim, D.H., Dong, T. *et al.* (2013) The *Arabidopsis* NAC transcription factor ANAC096 cooperates with bZIP-type transcription factors in dehydration and osmotic stress responses. *The Plant Cell* 25, 4708–4724. Available at: https://doi.org/10.1105/tpc.113.119099

Xue, G.G., Way, H.H., Richardson, T., Drenth, J., Joyce, P.A. and McIntyre, C.L. (2011) Overexpression of TaNAC69 leads to enhanced transcript levels of stress up-regulated genes and dehydration tolerance in bread wheat. *Molecular Plant* 4, 697–712.

Yadav, D., Shavrukov, Y., Bazanova, N., Chirkova, L., Borisjuk, N. *et al.* (2015) Constitutive overexpression of the TaNF-YB4 gene in transgenic wheat significantly improves grain yield. *Journal of Experimental Botany* 66, 6635–6650.

Yamanaka, T., Nakagawa, Y., Mori, K., Nakano, M., Imamura, T. *et al.* (2010) MCA1 and MCA2 that mediate $Ca^{2+}$ uptake have distinct and overlapping roles in *Arabidopsis*. *Plant Physiology* 152, 1284–1296. Available at: https://doi.org/10.1104/pp.109.147371

Yoshida, T., Fujita, Y., Sayama, H., Kidokoro, S., Maruyama, K. *et al.* (2010) AREB1, AREB2, and ABF3 are master transcription factors that cooperatively regulate ABRE-dependent ABA signaling involved in drought stress tolerance and require ABA for full activation. *The Plant Journal* 61, 672–685.

Young, T.E., Meeley, R.B. and Gallie, D.R. (2004) ACC synthase expression regulates leaf performance and drought tolerance in maize. *The Plant Journal* 40, 813–825.

Yu, F. and Xie, Q. (2017) Non-26S proteasome endomembrane trafficking pathways in ABA signaling. *Trends in Plant Science* 22(11), 976–985. Available at: https://doi.org/10.1016/j.tplants.2017.08.009

Yu, L., Nie, J., Cao, C., Jin, Y., Yan, M. *et al.* (2010) Phosphatidic acid mediates salt stress response by regulation of MPK6 in *Arabidopsis thaliana*. *New Phytologist* 188(3), 762–763. Available at: https://doi.org/10.1111/j.1469-8137.2010.03422.x

Yu, T.-F., Xu, Z.-S., Guo, J.-K., Wang, Y.-X., Abernathy, B. *et al.* (2017) Improved drought tolerance in wheat plants over expressing a synthetic bacterial cold shock protein gene SeCspA. *Scientific Reports* 7(1), 44050. Available at: https://doi.org/10.1038/srep44050

Yu, Y., Wang, J., Li, S., Kakan, X., Zhou, Y. *et al.* (2019) Ascorbic acid integrates the antagonistic modulation of ethylene and abscisic acid in the accumulation of reactive oxygen species. *Plant Physiology* 179, 1861–1875.

Yuan, F., Yang, H., Xue, Y., Kong, D., Ye, R. *et al.* (2014) OSCA1 mediates osmotic-stress-evoked $Ca^{2+}$ increases vital for osmosensing in *Arabidopsis*. *Nature* 514, 367–371.

Zang, X., Geng, X., He, K., Wang, F., Tian, X. *et al.* (2018) Overexpression of the wheat (*Triticum aestivum* L.) TaPEPKR2 gene enhances heat and dehydration tolerance in both wheat and *Arabidopsis*. *Frontiers in Plant Science* 9, 1710.

Zhang, H., Mao, X., Wang, C. and Jing, R. (2010) Overexpression of a common wheat gene *TaSnRK2.8* enhances tolerance to drought, salt and low temperature in *Arabidopsis*. *PLoS One* 5(12), e16041. Available at: https://doi.org/10.1371/journal.pone.0016041

Zhang, H., Li, W., Mao, X., Jing, R. and Jia, H. (2016) Differential activation of the wheat SnRK2 family by abiotic stresses. *Frontiers in Plant Science* 7, 420.

Zhang, L., Zhao, G., Jia, J., Liu, X. and Kong, X. (2012) Molecular characterization of 60 isolated wheat MYB genes and analysis of their expression during abiotic stress. *Journal of Experimental Botany* 63, 203–214.

Zhang, X., Wang, L., Meng, H., Wen, H., Fan, Y. and Zhao, J. (2011) Maize ABP9 enhances tolerance to multiple stresses in transgenic Arabidopsis by modulating ABA signaling and cellular levels of reactive oxygen species. *Plant Molecular Biology* 75, 365–378.

Zhang, X., Liu, L., Chen, B., Qin, Z., Xiao, Y. *et al.* (2019) Progress in understanding the physiological and molecular responses of *Populus* to salt stress. *International Journal of Molecular Sciences* 20(6), 1312. Available at: https://doi.org/10.3390/ijms20061312

Zhang, Y., Yang, C., Li, Y., Zheng, N., Chen, H. *et al.* (2007) SDIR1 is a RING finger E3 ligase that positively regulates stress-responsive abscisic acid signaling in *Arabidopsis*. *The Plant Cell* 19(6), 1912–1929.

Zhang, Y., Zhu, H., Zhang, Q., Li, M., Yan, M. *et al.* (2009) Phospholipase Dα1 and phosphatidic acid regulate NADPH oxidase activity and production of reactive oxygen species in ABA-mediated stomatal closure in *Arabidopsis*. *The Plant Cell* 21, 2357–2377.

Zheng, J., Yang, Z., Madgwick, P.J., Carmo-Silva, E., Parry, M.A. and Hu, Y.-G. (2015) *TaER* expression is associated with transpiration efficiency traits and yield in bread wheat. *PLoS One* 10, e0128415. Available at: https://doi.org/10.1371/journal.pone.0128415

# 2 Breeding Strategies to Enhance Abiotic Stress Tolerance and Yield Improvement in Wheat, Maize and Sorghum

**Inès Yacoubi, Karama Hamdi and Faïçal Brini***
*Biotechnology and Plant Improvement Laboratory, Centre of Biotechnology of Sfax, University of Sfax, Sfax, Tunisia*

## 2.1 Introduction

Abiotic stress is the primary cause of crop losses worldwide, reducing average yields for most major crop plants by more than 50% (Bray *et al.*, 2000). The major abiotic stresses (high salinity, drought, cold and heat) negatively affect the survival, biomass production and yields of staple food crops by up to 70% (Kaur *et al.*, 2008; Thakur *et al.*, 2010). Under field conditions, crops are exposed to a combination of stresses occurring simultaneously (Sade *et al.*, 2018). Due to climate change abiotic and biotic stress combinations will be increased, which will severely affect crop growth (Pandey *et al.*, 2017). It is estimated that these changes will lead to a yield reduction of major crops such as maize, wheat and rice (Sade *et al.*, 2018). Tolerance to abiotic stresses is extremely complex. The tissue or organ affected by the stress, the level and the duration of the stress can all have a significant effect on the complexity of the response. Plants' molecular responses to abiotic stresses involve interactions and crosstalk with many molecular pathways. Systems biology and omics approaches have been used to elucidate some of the key regulatory pathways in plant responses to abiotic stress (Cramer *et al.*, 2011).

These advances have led to new evidence on stress tolerance which provides crop breeders with new tools for improvement to increase stress responses and in turn crop yields. The goal of plant breeding is to produce new improved varieties. Traits of concern to the breeders include disease resistance, abiotic stress tolerance and water-use efficiency (Pandey *et al.*, 2019). Hence, the main objectives of breeding programmes are to increase the yield with increases in stability ensuring the quality and nutritive value, and to produce types that suit the particular growing conditions and farming needs (Caligari, 2001). In breeding programmes, conventional breeding methods need to be complemented with molecular tools. The first stage of plant breeding has begun with domestication (Lin *et al.*, 2014). Over the past millennium, the approach was principally based on the introduction of variation through conventional crossing programmes within or between species in most cases. The major methods used in breeding programmes include pedigree selection, recurrent selection, backcrossing and induced mutagenesis (Oladosu *et al.*, 2019). The current genomic era, enabled by next-generation DNA sequencing technologies, offers new and

*Corresponding author: faical.brini@cbs.rnrt.tn

 *Molecular Breeding in Wheat, Maize and Sorghum: Strategies for Improving Abiotic Stress Tolerance and Yield* (eds M.A. Hossain *et al.*)
DOI: 10.1079/9781789245431.0002

powerful tools for targeted and precise selection (Lin *et al.*, 2014). The induction of breeding programmes of modern biotechnological approaches based on the use of molecular markers may contribute to the solution of these problems (Ni *et al.*, 2017). Molecular (DNA) markers have created a powerful and practicable tool to perform gene selection in plant breeding. The use of DNA markers in plant breeding is called marker-assisted selection (MAS). This technique and others, such as marker-assisted backcross breeding (MABB) and marker-assisted gene pyramiding, play a crucial role in the improvement of crop plants. Advancements in plant genetic engineering (genetic transformation and genome editing) have made it possible to transfer genes into crop plants from unrelated plants and even from non-plant organisms. These biotechnological approaches are a great option to improve crop plants with significant commercial properties like increased biotic stress resistance, abiotic stress tolerances, yield and quality grains (Fig. 2.1) (Pandey *et al.*, 2019). Since the remarkable advances in genomic tools, researchers at different international centres are collaborating to accelerate the breeding process through the application of DNA markers, genomics, proteomics, and laboratory-based assays for assessing resistance mechanisms and nutrient bioavailability (Gedil and Menkir, 2019). In recent years, marker/genomic-assisted selection using molecular markers and/or genotyping by sequencing have been used to increase the speed of traditional breeding approaches, providing breeders with material 2–3 years faster than with conventional breeding alone (Gilliham *et al.*, 2017). Molecular breeding provides great opportunities for plant breeders to evaluate germplasm, map genes and characterize complex traits. Moreover, it permits selection at seedling stage which reduces the breeding cycle of the crop (Pandey *et al.*, 2019).

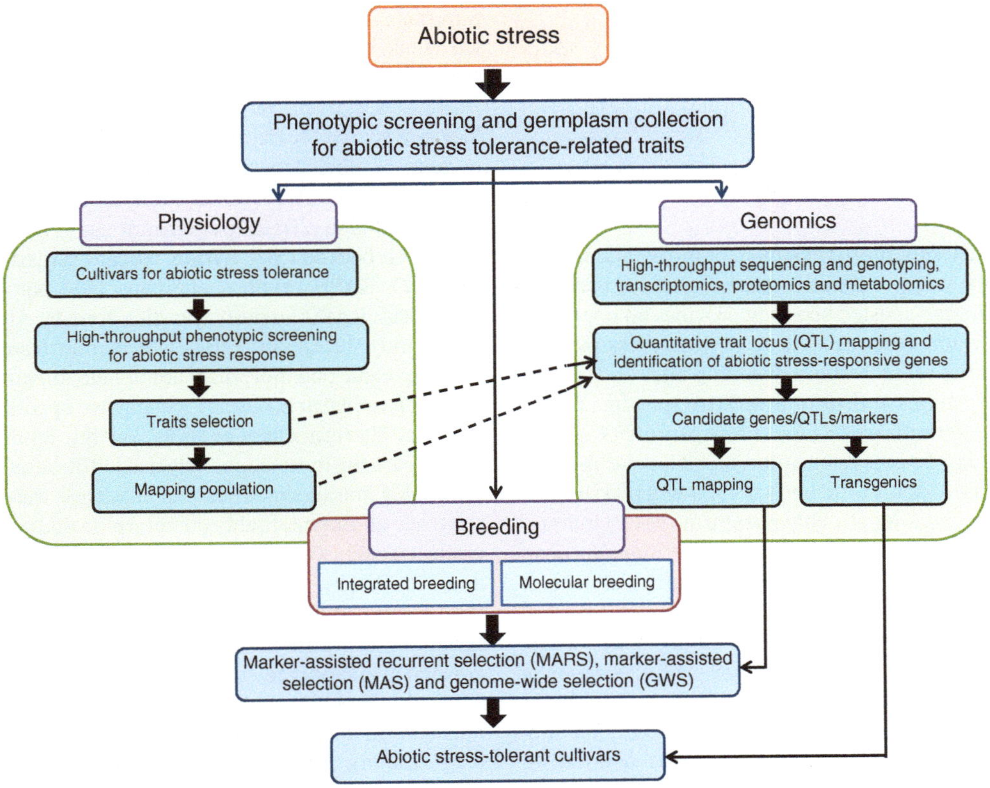


**Fig. 2.1.** A combined approach for integrating genomics, physiology and breeding techniques for developing improved crop cultivars with enhanced abiotic stress tolerance.

## 2.2 Molecular Breeding for Wheat Improvement

Wheat is the third most important food crop cultivated worldwide. It is a widely adapted crop possibly due to the complex nature of the plant's genome, which provides great plasticity to the crop (Acevedo *et al.*, 2002). The growth of the world's population requires at least a 70% increase in agricultural production by 2050 (Tester and Langridge, 2010; Pardey, 2011; FAO, 2019), which makes it necessary to increase wheat production capacity globally. However, the actual yield of modern varieties of wheat is largely limited by the influence of abiotic and biotic stresses (Iqbal *et al.*, 2007). Developing the DNA markers of economically valuable characters allowed carrying out a rapid assessment of plants for resistance to environmental stress factors, to identify varieties with a high genetic potential of productivity. A large number of genes and loci controlling the stability of various cereals to abiotic stresses, features and quality of grain yield were identified and mapped via DNA markers (Landjeva *et al.*, 2007; Zhang *et al.*, 2016).

### 2.2.1 Molecular markers in wheat

Currently, the development of molecular markers plays a pivotal role in marker-assisted breeding. Marker-assisted breeding provides an opportunity for wheat breeders to introgress/pyramid genes of interest into breeding lines and to identify genes and/or quantitative trait loci (QTLs) in germplasm to be used as parents (Vagndorf *et al.*, 2018). Application of molecular markers dates back to the early 1990s when restriction fragment length polymorphism (RFLP) markers were applied to wheat for gene mapping, varietal identification, characterization of wheat–rye recombinants and identification of homoeologous chromosome arms (Gupta *et al.*, 1999; Khan *et al.*, 2014). A set of 292 accessions of common wheat (*Triticum aestivum*) representing 21 germplasm pools based on geographical or breeding programme origins was assayed for RFLP diversity by Kim and Ward (2000). About 61% of all 233 scored bands were present in 75% or more of the accessions. They found that average genetic diversity for RFLPs and simple sequence repeats (SSRs) of *Triticum turgidum* L. ssp. *dicoccon* (Schrank) Thell. was higher than that observed in its closest domesticated relatives, namely *T. turgidum* L. ssp. *durum* (Desf.) Husn. and the hexaploid wheat *T. aestivum* L. ssp. *spelta*. Therefore, another category of PCR-based molecular markers, named randomly amplified polymorphic DNA (RAPD), has also been used. The RAPD method is rapid, with effectual outlay, and is proficient to execute analysis without the requirement for a preceding sequencing of the genome (Khan *et al.*, 2014). With establishment of RAPD as a prevailing marker, it has been used comprehensively for various purposes including estimation of genetic variability among wheat landraces (Autrique *et al.*, 1996; Pecetti *et al.*, 2001; Khan *et al.*, 2014). RAPD markers have been profitably used for recognition of cultivars (Hu and Quiros, 1991; Nybom, 1994; Rahman *et al.*, 2002; Khan *et al.*, 2014), estimation of genetic assets in wheat germplasm (Vierling and Nguyen, 1992; Nagaoka and Ogihara, 1997; Sun *et al.*, 1998; Mukhtar *et al.*, 2002; Mohapatra *et al.*, 2003), estimation of genetic polymorphism in wheat lines, species and cultivars (Joshi and Nguyen, 1993; Fahima *et al.*, 1999; Maric *et al.*, 2004; Irzykowski *et al.*, 2005) and to analyse phylogenetic relationships (Landry *et al.*, 1994). In 2012, RAPD analysis was conducted by Saleh (2012) to determine the genetic difference between five Syrian wheat cultivars using 21 RAPD primers, showing 18% polymorphism. RAPD analysis was also done by Al-Fares and Abu-Qaoud (2012) to demonstrate 95.5% genetic polymorphism among ten durum wheat (*T. turgidum* L.) varieties using five operon primers. Microsatellites or SSRs, on the other hand, were the most extensively used PCR-based molecular markers in wheat because they were relatively abundant, highly polymorphic and genome-specific (Xu and Crouch, 2008; Rasheed and Xia, 2019). Dograr *et al.* (2000) analysed the 16 winter-type durum wheat varieties using seven microsatellite markers and clearly distinguished the landrace selections, cultivars and advancing lines. They observed a polymorphism information content (PIC) value from 0.609 to 0.872 among the genotypes analysed. In 2013, Akfirat and Uncuoglu worked to differentiate seven winter wheat genotypes using 142 SSR primers and obtained a PIC value of 0.52.

Finally, Diversity Arrays Technology (DArT) is one of the ideal methods to determine and attain genetically diverse markers (Khan *et al.*, 2014). DArT was originally developed by Jaccoud *et al.* (2001). In wheat, DArT arrays have been developed by Akbari *et al.* (2006). This marker system is autonomous of sequences and capable of noticing a number of markers in a particular experimentation (Khan *et al.*, 2014). The genetic polymorphism of 50 Australian, 94 UK and 96 US wheat cultivars was calculated using DArT markers by White *et al.* (2008). At the whole-genome level, they found that Australian varieties have highest genetic diversity and UK ones have the least. A DArT marker system was developed by Jing *et al.* (2009) for assessing the genetic diversity of 16 *Triticum monococcum* L. ssp. *monococcum* genotypes of varied ecological origins and for the comparison of their relationships with hexaploid and tetraploid genomes (Khan *et al.*, 2014).

### 2.2.2 Functional wheat genomics

Functional genomics plays a crucial role in molecular breeding. In fact, it can be useful to develop MAS during wheat breeding. Functional markers are PCR-based molecular markers designed from sequence polymorphisms within functional genes; hence allelic variants can be diagnosed using functional markers (Liu *et al.*, 2012). This is contrary to neutral markers, because functional markers have strong associations with relevant phenotypes and are ideal molecular markers for MAS. Liu *et al.* (2012) documented 97 functional markers that detect 93 alleles at 30 loci in bread wheat. This number has increased during the last 5 years due to rapid advancements in wheat genomics. Currently, there are 157 functional markers documented for more than 100 loci underpinning adaptability, grain yield, disease resistance, end-use quality and tolerance to abiotic stresses. Functional markers are preferred for gene pyramiding and gene introgression and their use in genomic selection (GS) also enhances selection accuracy. In their work, Khalid *et al.* (2019) analysed a diversity panel consisting of advanced lines derived from synthetic hexaploid wheat for allelic variation at 87 functional genes or loci of breeding importance using 124 high-throughput, competitive allele-specific PCR (KASP) markers. Therefore, they recorded that the association analysis of functional genes with agronomic and biochemical traits under well-watered and water-limited conditions revealed that 21 marker–trait associations were consistently detected in both moisture conditions. Similarly, new technology such as plant genetic engineering has opened new avenues to modify crops and provided some solutions to solve specific needs. This technology can integrate foreign DNA into different plant cells to produce transgenic plants with new desirable traits. These biotechnological approaches are a great option to improve crop genotypes with significant commercial properties such as increased biotic or abiotic stress tolerances and for use as bioreactors producing proteins, edible vaccines and biodegradable plastics. Today, there has been further progress in enhancing the development of alternative marker-free systems technology as a research priority, to avoid the use of genes without any purpose after the transformation protocol as selectable and reporter marker genes. Typically, it is employed for the selection strategy that confers resistance to antibiotics and to herbicides (Pandey *et al.*, 2019).

### 2.2.3 Marker-assisted wheat breeding for improving quality traits

Wheat is grown in large parts of the world and is used for animal feed or for a wide range of products such as pasta, biscuits, cakes and bread (Vagndorf *et al.*, 2018). The end-use quality differs greatly between wheat cultivars and is influenced by several traits, such as grain hardness, grain protein content, gluten content and composition, and starch properties (Vagndorf *et al.*, 2018). Quality should therefore be an important focus in wheat breeding programmes. Thus, markers for wheat quality traits can be very useful as they enable screening of a high number of lines and can be used early in breeding programmes (Peña, 2002; Gale, 2005; Vagndorf *et al.*, 2018).

### *Grain hardness*

Grain hardness influences milling, flour and end-use properties of wheat. Flour from grain with hard endosperm texture has higher water absorption than flour from soft grain and is therefore preferred for bread-making. Grain hardness is primarily controlled by the *Hardness* locus on chromosome 5DS (Vagndorf *et al.*, 2018). This locus consists of three small genes: *Pina-D1*, *Pinb-D1* (*Puroindoline a/b*) and *grain softness protein-1* (*Gsp-1*). Wheat varieties with the wild-type alleles *Pina-D1a* and *Pinb-D1a* normally have soft grain, while deletions or other loss-of-function mutations in one or both *Pin* genes cause harder grain (Lillemo and Morris, 2000; Bhave and Morris, 2008).

### *Gluten*

The characteristic viscoelastic properties of wheat dough are due to a network of gluten proteins that is formed when flour is mixed with water. Thus, gluten is a major factor contributing to wheat quality (Vagndorf *et al.*, 2018). High grain protein content is typically associated with high quality, since roughly 80% of the grain protein is gluten (Shewry, 2009). However, both the amount and the composition of gluten affect wheat quality. Gluten consists of two types of proteins: polymeric glutenins and monomeric gliadins (Vagndorf *et al.*, 2018). Glutenins can be classified as low- or high-molecular-weight (LMW or HMW) subunits, while gliadins can be classified as α, β, γ or ω types (Payne, 1987; Barak *et al.*, 2015). The most important HMW glutenins, LMW glutenins and gliadins are encoded by the *Glu-1*, *Glu-3* and *Gli-1* loci, respectively. HMW glutenins generally have the largest impact on wheat quality. DNA markers have been developed to discriminate between different alleles of *Glu-1*, *Glu-3* and *Gli-1* loci (Liu *et al.*, 2012, 2014).

### *Wheat–rye translocation and falling number*

The wheat–rye translocation 1BL.1RS has been employed in many breeding programmes as it carries resistance genes against powdery mildew and rusts. Markers for the resistance genes can be used to test for the absence or presence of the translocation in wheat varieties (Mago *et al.*, 2002). Additionally, the 1BL.1RS translocation can have a negative effect on falling number. Falling number is an indirect measure of α-amylase enzyme activity (Vagndorf *et al.*, 2018). The α-amylases are encoded by the loci *α-Amy-1*, *α-Amy-2* and *α-Amy-3* located on the homoeologous chromosome groups 6, 7 and 5, respectively. High falling number reduces the risk of preharvest sprouting, which has a considerable negative impact on quality. Environmental conditions around the time of harvest influence falling number, but it is also influenced genetically. The b-allele of the *RhtD1* (*reduced height*) gene on chromosome 4D is correlated with increased falling number (Graybosch, 2001; Vagndorf *et al.*, 2018).

## 2.3 Molecular Breeding for Maize Improvement

Maize is one of the most important analysed among the plant genomes. Consequently, maize has been at the forefront in development and evaluation of an array of molecular markers for varied purposes in genetics and breeding (Prasanna *et al.*, 2010). Despite impressive progress made in the last few decades through conventional breeding, average maize yields remain low and demand is expected to increasingly exceed production in the coming years (Prasanna *et al.*, 2010). Molecular marker-assisted breeding is accelerating yield gains in the USA and elsewhere and offers tremendous potential for enhancing the productivity and value of Asian maize germplasm (Prasanna *et al.*, 2010). Molecular tools to enhance breeding efficiency and effectiveness have become integral to many maize research programmes worldwide (Prasanna *et al.*, 2010). This is now enabling the use of sophisticated biometric and modelling tools to predict genotypic value and conduct genotypic selection before evaluating phenotypes in large private breeding institutions (Prasanna *et al.*, 2010).

### 2.3.1 Molecular markers and mapping populations in maize

Advances in genomics have led to the identification of numerous DNA markers in maize during the last few decades, including thousands of mapped microsatellite or SSR markers, and, more recently, single-nucleotide polymorphisms (SNPs) and insertion/deletion (InDel) markers (Prasanna *et al.*, 2010). In addition to the SSRs and SNPs, a large number of genes controlling various aspects of plant development, biotic and abiotic stress resistances, quality characteristics, etc. have been cloned and characterized in maize, which are excellent assets for molecular marker-assisted breeding (Prasanna *et al.*, 2010).

At present, SSRs are the most widely used markers by maize researchers due to their availability in large numbers in the public domain (MaizeGDB; https://www.maizegdb.org, accessed 3 February 2021), simplicity and effectiveness. These PCR-based, genetically co-dominant markers are robust, reproducible, hyper-variable, abundant and uniformly dispersed in plant genomes (Powell *et al.*, 1996). While both SSRs and SNPs can be reliably applied on a large scale, SNPs are highly amenable for automation and therefore offer significant advantages for genetic and breeding purposes (Prasanna *et al.*, 2010). Compared with the genomes of other cultivated plant species, SNP frequency in maize is high, with one SNP being found every 28–124 bp (Tenaillon *et al.*, 2001; Ching *et al.*, 2002; Vroh Bi *et al.*, 2006; Prasanna *et al.*, 2010).

A database and resource for SNP discovery and trait dissection has been established for maize in which genotype, phenotype and polymorphism data can be accessed for diverse maize inbreds and populations (Zhao *et al.*, 2006a,b; panzea; https://www.panzea.org, accessed 3 February 2021). Several high-throughput genotyping platforms have been developed that allow rapid and simultaneous genotyping of up to a million SNP markers (Prasanna *et al.*, 2010). In addition to powerful marker systems, diverse mapping populations are available in maize as international maize genomic resources, including the intermated B73 × Mo17 (IBM) population (Lee *et al.*, 2002; Prasanna *et al.*, 2010) and the intermated recombinant inbred lines (IRILs) developed from the IBM population (Coe *et al.*, 2002; Cone *et al.*, 2002), etc. A genetic map of maize, ISU–IBM Map4, that integrates 2029 existing markers with 1329 new insertion/deletion polymorphism (IDP) markers has been developed using the IBM population (Fu *et al.*, 2006; Prasanna *et al.*, 2010).

The maize nested association mapping (NAM) population, comprising 5000 recombinant inbred lines (RILs) (200 RILs from each of 25 populations), is another important genetic resource developed in recent years. The NAM population is an approach for mapping genes underlying complex traits, in which the statistical power of QTL mapping is combined with the high (potentially gene-level) chromosomal resolution of association mapping (Yu *et al.*, 2008; Prasanna *et al.*, 2010). The RILs are 'nested' in the sense that they all share a common parent, but each population has a unique second parent. The common parental line used in all 25 families, B73, is the most important US maize breeding line. Descendants of B73 are widely deployed in US maize agricultural production, and the B73 genome has been sequenced (Schnable *et al.*, 2009). The remaining NAM parental inbred lines were chosen either on the basis of their agronomic importance in the USA or to capture as much of the genetic diversity present in maize as possible based on analysis of a worldwide collection using 94 SSRs (Liu *et al.*, 2003; Flint-Garcia *et al.*, 2005; Prasanna *et al.*, 2010).

Besides the well-demonstrated utility of molecular markers in genotype differentiation and analysis of genetic diversity in maize germplasm, application of DNA-based markers is also of considerable significance to tropical/subtropical maize production systems, for mapping and MAS for resistance to major biotic/abiotic stresses affecting production and productivity (Prasanna *et al.*, 2010; Gazal *et al.*, 2018).

### 2.3.2 Opportunities for enhancing the level and scope of molecular marker-assisted breeding in maize

Association mapping has been used in many crop species including maize, rice, wheat, barley, sorghum, sugarcane, soybean, potato and tomato, as well as trees such as eucalyptus, aspen and pine (Zhu *et al.*, 2008). There are many reports of successful association between DNA polymorphisms and qualitative traits in plants, but fewer reports for complex traits (Prasanna *et al.*, 2010). However, the genetic, genomic and statistical tools are now at hand to successfully apply association mapping for the dissection of complex traits in plants, which will harness the natural diversity in the crop-related gene pool to identify and use allelic variants for crop improvement (Yu *et al.*, 2006; Zhu *et al.*, 2008; Prasanna *et al.*, 2010). The use of molecular markers in crop improvement has been well established, but their application in routine crop breeding activities is still very limited, primarily due to inadequate access to high-throughput genotyping and phenotyping facilities (Prasanna *et al.*, 2010). There are excellent opportunities for undertaking high-throughput genotyping in maize (Prasanna *et al.*, 2010). Nearly a million maize SNPs are available in public databases (https://www.panzea.org, accessed 3 February 2021) and several high-throughput genotyping platforms have been developed for commercial use (Flint-Garcia *et al.*, 2005). These new technologies, and associated data handling and analysis tools, provide opportunities for the maize community to speed up research progress for large-scale diversity analysis, high-density linkage map construction, high-resolution QTL mapping, linkage disequilibrium (LD) analysis and genome-wide association studies (Prasanna *et al.*, 2010). Because the genomic sequence of maize is publicly available (Schnable *et al.*, 2009), resequencing of individual maize inbred lines can now give the entire genotype of that individual; that is to say, the allelic state of every SNP in the genome (Prasanna *et al.*, 2010). Expected advances in this technology should soon make it widely accessible, with SNPs available in every region of the genome and advancing enormously the possibilities for gene discovery and selection (Prasanna *et al.*, 2010).

## 2.4 Molecular Breeding for Sorghum Improvement

Sorghum is one of the world's major cereal crops and a dietary staple for more than 500 million people in sub-Saharan Africa and Asia (Mofokeng *et al.*, 2017). Sorghum is more nutritious than fine cereals and is the principal source of energy, protein, vitamins and minerals for millions of the poorest people in these regions. It grows well in harsh environments with minimum inputs, where other crops yield poorly. The area of sorghum production has declined globally over the past few decades, primarily due to susceptibility to biotic and abiotic factors and marginal economics. In this scenario, genetic enhancement of grain and fodder yield is a major challenge to sorghum breeders. Due to the advances in sequencing, genotyping, phenotyping, QTL mapping, genetic transformation and tissue culture technologies, we are beginning to visualize practical solutions for the genetic enhancement of sorghum through DNA marker-assisted breeding. At this point, it is essential to look back and critically review the advancements made to date so that we can formulate suitable strategies for the future sorghum improvement programmes.

### 2.4.1 Sorghum genetics and classification

Sorghum is a diploid ($2n = 2x = 20$) crop classified into two groups, the wild and the cultivated sorghums (Ayana *et al.*, 2000; Mofokeng *et al.*, 2017). The wild sorghums include *Sorghum halepense*, *Sorghum propinquum*, *Sorghum bicolor* ssp. *drummondii* and *S. bicolor* ssp. *verticilliflorum* (Mofokeng *et al.*, 2017). The cultivated sorghum has been classified into five major races, *bicolor*, *caudatum*, *durra*, *guinea* and *kafir*, and ten intermediate races based on panicle morphology (Mofokeng *et al.*, 2017). Due to the vast diversity in cultivated sorghum, breeders have developed an interest to exploit this diversity and to develop improved varieties in their programmes using various breeding methods and technologies. To develop cultivars it is important to characterize the germplasm available to identify potential candidate genotypes for further improvement.

### 2.4.2 Characterization and analysis of genetic diversity in sorghum

The germplasm can be characterized morphologically and genotypically using phenotypic descriptors, biochemical and molecular markers (Mofokeng *et al.*, 2017). The characterization of the genotypes gives descriptive information on the traits and aids in understanding the similarities and differences among genotypes (Mofokeng *et al.*, 2017).

#### *Morphological characterization*

Morphological or phenotypic descriptors are used to distinguish one accession from another (Mofokeng *et al.*, 2017). Most of the characterization and evaluation has been based on the recording of either or both qualitative and quantitative morphological characters (Mofokeng *et al.*, 2017). Geleta and Labuschagne (2005) characterized sorghum germplasm using qualitative traits that include leaf midrib colour, grain colour, glume colour, endosperm texture, pericarp colour, leaf trichomes, awns, pericarp thickness and panicle compactness. On the other hand, the quantitative traits are also useful for determination of genetic diversity among genotypes (Mofokeng *et al.*, 2017). The quantitative traits include plant height, days to maturity, leaf area, leaf width, leaf length, number of leaves, panicle length, grain yield per plant, grain size, 1000-grain weight, grain number per panicle, panicle width, number of primary branches per panicle and panicle weight (Punitha *et al.*, 2010; Mofokeng *et al.*, 2017). Morphological characterization is useful for the design of future breeding programmes, to identify duplicates, study patterns of genetic variation and establish the relationship between agronomic traits for direct and indirect selection (Mofokeng *et al.*, 2017).

Although the agronomical and morphological characterization provides useful information to breeders, the challenge is that they may be easily influenced by the environmental factors. This process/procedure may take a long time and the plants must be assessed during a fixed vegetative phase of the crop (Mofokeng *et al.*, 2017). Of late, breeders have resorted to the use of molecular marker systems because they are not subject to environmental influences, have fixed plant developmental stages and have the potential to give results within a short time (Mofokeng *et al.*, 2017).

#### *Characterization using molecular markers as new technologies*

The genetic improvement of sorghum has been made accessible through the use of easily assayable molecular genetics of DNA markers that enable accurate identification of genotype without the confounding effect of environment, thereby increasing heritability (Mofokeng *et al.*, 2017). MAS in sorghum reduces the length of time required for introgression of characters, unlike with the use of the pedigree breeding method (Mofokeng *et al.*, 2017). Also, the selection of progenies based on genetic values derived from molecular marker data substantially increases the rate of genetic gain, particularly if the number of cycles of evaluation or generations can be reduced (Meuwissen *et al.*, 2001; Mofokeng *et al.*, 2017). Various molecular markers that have been used in sorghum breeding include RAPD (Prakash *et al.*, 2006), amplified fragment length polymorphism (AFLP) (Wu *et al.*, 2006) and SSRs (Manzelli *et al.*, 2006), SNPs (Zeng *et al.*, 2011), microarrays (Buchanan *et al.*, 2005) and DArT (Mace *et al.*, 2009). The latter technique has been reported recently in several studies (Mofokeng *et al.*, 2017). These markers were mainly used for genetic diversity, cultivar identification, gene mapping and discovery, and gene pyramiding.

Molecular markers are useful for fingerprinting of collections for identification and germplasm management in plant breeding (Mofokeng *et al.*, 2017). They are also useful in a wide range of applications including genetic mapping and genome analysis (Li *et al.*, 2000), gene and QTL analysis (Blair and McCough, 1997) and in marker-assisted breeding (Weising *et al.*, 1998). For genotyping, SSRs were used in genetic diversity studies among elite sorghum inbred lines (Menz *et al.*, 2004; Mofokeng *et al.*, 2017), among germplasm collections from different geographic locations (Muraya *et al.*, 2011; Mofokeng *et al.*, 2017) and in the assessment of population genetic structure and relatedness within or among landraces (Folkertsma *et al.*, 2005). Wu and Huang (2006) used SSR for mapping the sorghum genome in comparison with the existing genetic linkage maps (Mofokeng *et al.*, 2017). In addition, SSRs can be used in conjunction with other molecular techniques (Geleta *et al.*, 2006; Mofokeng *et al.*, 2017).

## 2.5 Breeding Strategies to Enhance Abiotic Stress Tolerance in Crops

The recent progress in genomics permits to more efficiently assess and enhance diversity in germplasm collections, introgress valuable traits from new sources and identify genes that control key traits. MAS helps to reduce the impact of environment on breeder selection. Significant advances have been made in the development of *in vitro* selection methods. The broader use of traits from alien species and the manipulation of heterosis and polyploidy create new perspectives for improving yield potential and adaptation to abiotic stresses.

### 2.5.1 Use of forward genetics to elucidate mechanisms of abiotic stress tolerance

The complexity of plant stress adaptation makes breeding for abiotic stress tolerance, particularly for drought, complicated. Taking a reverse genetics approach based on candidate genes to increase the tolerance of crop plants is also fraught with danger. Therefore, a more powerful approach is to identify naturally occurring variation of abiotic stress tolerance in varieties, landraces and wild relatives of crops and to study the traits that contribute to tolerance. The genetic loci determining these traits can then be discovered by correlating trait values with the genetic variation in a large mapping population derived from the parents in whom the trait was originally identified to differ. Once the molecular bases of traits contributing to tolerance have been identified using this forward genetics approach, marker-assisted breeding and genetic modification (GM) technologies can be used to introduce these traits (genes) into current, high-yielding elite cultivars.

To undertake the above-mentioned process, it is necessary to have well-defined genetic materials and reliable and accurate quantitative phenotyping techniques (Salekdeh *et al.*, 2009; Berger *et al.*, 2010). While some components of abiotic stress tolerance can be relatively easily measured, such as the accumulation of $Na^+$ in leaf blades (Garthwaite *et al.*, 2005; Genc *et al.*, 2010; Shavrukov *et al.*, 2010), other traits are more difficult to measure, such as growth immediately after application of salt to obtain a measure of the 'osmotic' component of salinity stress (Reynolds and Tuberosa, 2008; Salekdeh *et al.*, 2009). For traits where it is more difficult to control the onset of the stress, such as drought tolerance, multiple measurements over time are also required both of the plant and of the environment in which it is growing. To identify the abiotic stress resistances' QTLs, a lot of work has been done by plant biologists but identified QTLs proved unstable across different environmental conditions due to their complex inheritance mechanism of abiotic stress tolerance. Most of the phenotypic variations in quantitative traits are due to few loci with great effect and less due to large loci with lower effect (Maki-Tanila and Hill, 2014). Recently, QTLs detected on the basis of high-density genetics maps have increased the understanding about the genetic control of abiotic stresses in cereals and other agricultural crops. This approach helps to discover the resistance genes and to understand the mechanism of plant adaptation under stress conditions (Table 2.1).

### 2.5.2 Phenomics technologies assist the genetic analysis of abiotic stress tolerance

Conventional methods for phenotyping plants are frequently laborious and destructive, and often involve the removal of plant biomass for analysis. Recent developments in high-throughput, non-destructive imaging technologies allow a researcher to obtain multiple images of the same plant at different time points and at different wavelengths, thereby offering new non-destructive methods for acquiring quantitative data on plant growth, health and water use under abiotic stress (Morison *et al.*, 2008; Jones *et al.*, 2009; Berger *et al.*, 2010). Some of these technologies have already been used to quantify traits related to drought, salt and heat tolerances in a number of crop plants (Rajendran *et al.*, 2009; Sirault *et al.*, 2009; Berger *et al.*, 2010). The majority of experiments, though, have focused on shoot traits. Despite the likely importance of tolerance mechanisms in roots, such as optimizing architecture to improve access to soil water, there have been few genetic studies of roots due to difficulties in phenotyping roots (Fleury *et al.*, 2010; Richards *et al.*, 2010; Zhu *et al.*, 2011).

**Table 2.1.** Important QTLs identified for abiotic stress tolerance in cereals.

| Crop | Abiotic stress/Plant trait | QTL/Gene | Chromosome no./Linkage group | Reference |
|---|---|---|---|---|
| Wheat | Cell-membrane stability | *qCMSa2AC* | 2A | Malik *et al.* (2015) |
| Wheat | Relative water content | *qCMSa2AC* | 2A | Malik *et al.* (2015) |
| Wheat | Net photosynthesis rate | *qPn2AC* | 2A | Malik *et al.* (2015) |
| Wheat | Root angle | *qRA.qgw-2A* | 2A1 | Christopher *et al.* (2013) |
| Wheat | Root angle | *qRA.qgw-3D* | 3D | Christopher *et al.* (2013) |
| | | *qRA.qgw-5D* | 5D | |
| Wheat | Number of roots | *qRN.qgw-1B* | 1B | Christopher *et al.* (2013) |
| Wheat | Root length | *qRl.ccsu-2B.1* | 2B | Bharti *et al.* (2014) |
| Wheat | Root dry weight | *qRdw.ccsu-2A.1* | 2A | Bharti *et al.* (2014) |
| | | *qRdw.ccsu-2A.2* | 2A | |
| Wheat | Na tolerance | *TaHKT1;5-D* | *Kna1* | Ren *et al.* (2005); James *et al.* (2006) |
| Wheat | $Na^+$ exclusion | *HKT1* | 2A | Genc *et al.* (2010) |
| Wheat | Na tolerance | *TaHKT1;5-D* | *Kna1* | James *et al.* (2006) |
| | Shoot Na accumulation | *TmHKT7-A2* | *Nax1* | |
| Wheat | Plasma membrane damage under high temperature | *qHtpmd.ksu.2B* | 2B | Talukder *et al.* (2014) |
| Wheat | Spikelet fertility under high temperature | *qHttmd.ksu-6A* | 6A | Talukder *et al.* (2014) |
| | | *qHttmd.ksu.1D* | 1D | |
| Maize | Arsenic concentration | *qLAC1* | 1 | Ding *et al.* (2011) |
| | | *qLAC5* | 5 | |
| | | *qLAC7* | 7 | |
| | | *qLAC8* | 8 | |
| | | *qLAC9a* | 9 | |
| Sorghum | Al tolerance | *SbMATE* | $Alt_{SB}$ | Magalhaes *et al.* (2004) |
| Sorghum | Leaf area | *qTLA3-8* | *SBI-08-II* | Mace *et al.* (2012) |

## 2.6 Conclusions

Plant genetic diversity and plant breeding are key elements in tackling climate change, and integration of plant breeding in climate change strategies is one of the best paths to sustainable food production by developing climate-smart crops: that is, abiotic and biotic stress-resistant crop varieties. Molecular breeding approaches, namely MAS, MABB, marker-assisted recurrent selection (MARS) and GS or genome-wide selection (GWS), offer opportunities for plant breeders to develop high-yielding crop cultivars with resilience to diseases precisely and in less time. For complex traits (mainly abiotic stresses) where multiple QTLs control the expression, new strategies like MARS and GS are employed to increase precision, reduce the cost of phenotyping and shorten the time frame to achieve disease resilience.

## Acknowledgement

This work was funded by grants provided by the Tunisian Ministry of Higher Education and Scientific Research-Tunisia.

## References

Acevedo, E., Silva, P. and Silva, H. (2002) Wheat growth and physiology. In: Curtis, B.C., Rajaram, S. and Gómez Macpherson, H. (eds) *Bread Wheat: Improvement and Production*. FAO Plant Production and Protection Paper no. 30. Food and Agriculture Organization of the United Nations, Rome, pp. 39–70.

Akbari, M., Wenzl, P., Caig, V., Carling, J., Xia, L. *et al.* (2006) Diversity arrays technology (DArT) for high-throughput profiling of the hexaploid wheat genome. *Theoretical and Applied Genetics* 113, 1409–1420.

Akfirat, F.S. and Uncuoglu, A.A. (2013) Genetic diversity of winter wheat (*Triticum aestivum* L.) revealed by SSR markers. *Biochemical Genetics* 51, 223–229.

Al-Fares, H. and Abu-Qaoud, H. (2012) Molecular characterization of genetic diversity in some durum wheat (*Triticum durum* Desf.) in Palestine. *African Journal of Biotechnology* 11, 12958–12963.

Autrique, E., Nachit, M.M., Monneveux, P., Tanksley, S.D. and Sorrels, M.E. (1996) Genetic diversity in durum wheat based on RFLPs, morphophysiological traits, and coefficient of parentage. *Crop Science* 36, 735–742.

Ayana, A., Bryngelsson, T. and Bekele, E. (2000) Geographic and altitudinal allozyme variation in sorghum (*Sorghum bicolor* (L.) Moench) landraces from Ethiopia and Eritrea. *Hereditas* 135, 1–12.

Barak, S., Mudgil, D. and Khatkar, B.S. (2015) Biochemical and functional properties of wheat Gliadins: a review. *Critical Reviews in Food Science and Nutrition* 55(3), 357–368.

Berger, B., Parent, B. and Tester, M. (2010) High-throughput shoot imaging to study drought responses. *Journal of Experimental Botany* 61, 3519–3528.

Bharti, S., Balyan, H.S. and Gupta, P. (2014) Quantitative trait loci analysis for some root traits in bread wheat (*Triticum aestivum* L.). *International Journal of Agriculture Science* 4, 214–221.

Bhave, M. and Morris, C.F. (2008) Molecular genetics of puroindolines and related genes: allelic diversity in wheat and other grasses. *Plant Molecular Biology* 66(3), 205–219.

Blair, M.W. and McCough, S.R. (1997) Microsatellites and sequence tagged site markers diagnostic for the bacterial blight resistance gene, *xa-5*. *Theoretical and Applied Genetics* 95, 174–184.

Bray, E.A., Bailey-Serres, J. and Weretilnyk, E. (2000) Responses to abiotic stresses. In: Gruissem, W., Buchannan, B. and Jones, R. (eds) *Biochemistry and Molecular Biology of Plants*. American Society of Plant Physiologists, Rockville, Maryland, pp. 1158–1249.

Buchanan, C.D., Lim, S., Salzman, R.A., Kagiampakis, I., Morishige, D.T. *et al.* (2005) *Sorghum bicolor*'s transcriptome response to dehydration, high salinity and ABA. *Plant Molecular Biology* 58, 699–720.

Caligari, P.D.S. (2001) Plant breeding and crop improvement. In: *Encyclopedia of Life Sciences*. Wiley, Chichester, UK, pp. 1–8.

Ching, A., Caldwell, K.S., Jung, M., Dolan, M., Smith, O.S. *et al.* (2002) SNP frequency, haplotype structure and linkage disequilibrium in elite maize inbred lines. *BMC Genetics* 3, 19.

Christopher, J., Christopher, M., Jennings, R., Jones, S., Fletcher, S. *et al.* (2013) QTL for root angle and number in a population developed from bread wheats (*Triticum aestivum*) with contrasting adaptation to water-limited environments. *Theoretical and Applied Genetics* 126, 1563–1574.

Coe, E., Cone, K., McMullen, M., Chen, S.S., Davis, G. *et al.* (2002) Access to the maize genome: an integrated physical and genetic map. *Plant Physiology* 128, 9–12.

Cone, K.C., McMullen, M.D., Bi, I.V., Davis, G.L., Yim, Y.-S. *et al.* (2002) Genetic, physical, and informatics resources for maize: on the road to an integrated map. *Plant Physiology* 130, 1598–1605.

Cramer, G.R., Urano, K., Delrot, S., Pezzotti, M. and Shinozaki, K. (2011) Effects of abiotic stress on plants: a systems biology perspective. *BMC Plant Biology* 11, 163.

Ding, D., Li, W., Song, G., Qi, H., Liu, J. and Tang, J. (2011) Identification of QTLs for arsenic accumulation in maize (*Zea mays* L.) using a RIL population. *PLoS One* 6, e25646.

Dograr, N., Akin-Yalin, S. and Akkaya, M.S. (2000) Discriminating durum wheat cultivars using highly polymorphic simple sequence repeat DNA markers. *Plant Breeding* 119, 360–362.

Fahima, T., Sun, G.L., Beharav, A., Krugman, T., Beiles, A. and Nevo, E. (1999) RAPD polymorphism of wild emmer wheat populations, *Triticum dicoccoides*, in Israel. *Theoretical and Applied Genetics* 98, 434–447.

FAO (2019) FAOSTAT. Crops. Rice, paddy. Food and Agriculture Organization of the United Nations, Rome. Available at: http://www.fao.org/faostat/en/#data/QC/visualize (accessed 4 February 2021).

Fleury, D., Jefferies, S., Kuchel, H. and Langridge, P. (2010) Genetic and genomic tools to improve drought tolerance in wheat. *Journal of Experimental Botany* 61, 3211–3222.

Flint-Garcia, S.A., Thuillet, A.C., Yu, J.M., Pressoir, G., Romero, S.M. *et al.* (2005) Maize association population: a high-resolution platform for quantitative trait locus dissection. *The Plant Journal* 44, 1054–1064.

Folkertsma, R.T., Rattunde, F.H., Chandra, S., Soma Raju, W. and Hash, C.T. (2005) The pattern of genetic diversity of Guinea-race *Sorghum bicolor* (L.) Moench landraces as revealed with SSR markers. *Theoretical and Applied Genetics* 111, 399–409.

Fu, Y., Wen, T.J., Ronin, Y.I., Chen, H.D., Guo, L. *et al.* (2006) Genetic dissection of intermated recombinant inbred lines using a new genetic map of maize. *Genetics* 174, 1671–1683.
Gale, K.R. (2005) Diagnostic DNA markers for quality traits in wheat. *Journal of Cereal Science* 41(2), 181–192.
Garthwaite, A.J., von Bothmer, R. and Colmer, T.D. (2005) Salt tolerance in wild *Hordeum* species is associated with restricted entry of $Na^+$ and $Cl^-$ into the shoots. *Journal of Experimental Botany* 56, 2365–2378.
Gazal, A., Dar, Z.A. and Lone, A.A. (2018) Molecular breeding for abiotic stresses in maize (*Zea mays* L.) 25. In: El-Esawi, M.A. (ed.) *Maize Germplasm: Characterization and Genetic Approaches for Crop Improvement*. IntechOpen, London. Available at: https://doi.org/10.5772/intechopen.71081
Gedil, M. and Menkir, A. (2019) An integrated molecular and conventional breeding scheme for enhancing genetic gain in maize in Africa. *Frontiers in Plant Science* 10, 1430.
Geleta, N. and Labuschagne, M.T. (2005) Qualitative traits variation in sorghum (*Sorghum bicolor* (L.) Moench) germplasm from eastern highlands of Ethiopia. *Biodiversity & Conservation* 14, 3055–3064.
Geleta, N., Labuschagne, M.T. and Viljoen, C.D. (2006) Genetic diversity analysis in sorghum germplasm as estimated by AFLP, SSR and morpho-agronomical markers. *Biodiversity & Conservation* 15, 3251–3265.
Genc, Y., Oldach, K., Verbyla, A., Lott, G., Hassan, M. *et al.* (2010) Sodium exclusion QTL associated with improved seedling growth in bread wheat under salinity stress. *Theoretical and Applied Genetics* 121, 877–894.
Gilliham, M., Able, J.A. and Roy, S.J. (2017) Translating knowledge about abiotic stress tolerance to breeding programmes. *The Plant Journal* 90, 898–917.
Graybosch, R.A. (2001) Uneasy unions: quality effects of rye chromatin transfers to wheat. *Journal of Cereal Science* 33, 316.
Gupta, P.K., Varshney, R.K., Sharma, P.C. and Ramesh, B. (1999) Molecular markers and their applications in wheat breeding. *Plant Breeding* 118, 369–390.
Hu, J. and Quiros, C.F. (1991) Identification of broccoli and cauliflower cultivars with RAPD markers. *Plant Cell Reports* 10, 505–511.
Iqbal, N., Firdissa, E., Khlestkina, K., Weidner, A., Röder, M.S. and Börner, A. (2007) The use of simple sequence repeat (SSR) markers to identify and map alien segments carrying genes for effective resistance to leaf rust in bread wheat. *Plant Genetic Resources* 5, 100–103.
Irzykowski, W., Soldatova, V., Gasich, E., Razgulaeva, N. and Jędryczka, M. (2005) RAPD analysis of *Sclerotinia sclerotiorum* from crucifers. *IOBC/WPRS Bulletins* 28(10), 69–82.
Jaccoud, D., Peng, K., Feinstein, D. and Kilian, A. (2001) Diversity arrays: a solid state technology for sequence information independent genotyping. *Nucleic Acids Research* 29(4), e25.
James, R.A., Davenport, R.J. and Munns, R. (2006) Physiological characterization of two genes for $Na^+$ exclusion in durum wheat, *Nax1* and *Nax2*. *Plant Physiology* 142, 1537–1547.
Jing, H.C., Bayon, C., Kanyuka, K., Berry, S., Wenzl, P. *et al.* (2009) DArT markers: Diversity analyses, genomes comparison, mapping and integration with SSR markers in *Triticum monococcum*. *BMC Genomics* 10(1), 458.
Jones, H.G., Serraj, R., Loveys, B.R., Xiong, L., Wheaton, A. and Price, A.H. (2009) Thermal infrared imaging of crop canopies for the remote diagnosis and quantification of plant responses to water stress in the field. *Functional Plant Biology* 36, 978–989.
Joshi, C.P. and Nguyen, H.T. (1993) RAPD (random amplified polymorphic DNA) analysis based intervarietal genetic relationships among hexaploid wheats. *Plant Science* 93, 95–103.
Kaur, G., Kumar, S., Nayyar, H. and Upadhyaya, H.D. (2008) Cold stress injury during the pod-filling phase in chickpea (*Cicer arietinum* L.): effects on quantitative and qualitative components of seeds. *Journal of Agronomy and Crop Science* 194, 457–464.
Khalid, M., Afzal, F., Gul, A., Amir, R., Ahmed, Z. *et al.* (2019) Molecular characterization of 87 functional genes in wheat diversity panel and their association with phenotypes under well-watered and water-limited conditions. *Frontiers in Plant Science* 10, 717.
Khan, M.K., Biotech, A., Shoudary, S., Hakki, E.E., Akkaya, M.S. and Thomas, G. (2014) From RFLP to DArT: molecular tools for wheat (*Triticum* spp.) diversity analysis. *Genetic Resources and Crop Evolution* 61(4), 1001–1032.
Kim, H.S. and Ward, R.W. (2000) Patterns of RFLP-based genetic diversity in germplasm pools of common wheat with different geographical or breeding program origins. *Euphytica* 115, 197–208.

Landjeva, S., Korzun, V. and Borner, A. (2007) Molecular markers: actual and potential contributions to wheat genome characterization and breeding, *Euphytica* 156, 271–296.

Landry, B.S., Li, R.Q., Cheung, W.Y. and Graner, R.L. (1994) Phylogeny analysis of 25 apple rootstocks using RAPD markers and tactical gene tagging. *Theoretical and Applied Genetics* 89, 847–852.

Lee, M., Sharopova, N., Beavis, W.D., Grant, D., Katt, M., Blair, D. and Hallauer, A. (2002) Expanding the genetic map of maize with the intermated B73 × Mo17 (IBM) population. *Plant Molecular Biology* 48, 453–461.

Li, Y.C., Fahima, T., Peng, J.H., Roder, M.S., Kirzhner, V.M., Beiles, A. and Nevo, E. (2000) Edaphic microsatellite DNA divergence in wild emmer wheat, *Triticum dicocoides*, at a microsite: Tagigha, Israel. *Theoretical and Applied Genetics* 101, 1029–1038.

Lillemo, M. and Morris, C.F. (2000) A leucine to proline mutation in puroindoline B is frequently present in hard wheats from northern Europe. *Theoretical and Applied Genetics* 100(7), 1100–1107.

Lin, T., Zhu, G., Zhang, J., Xu, X., Yu, Q. *et al.* (2014) Genomic analyses provide insights into the history of tomato breeding. *Nature Genetics* 46, 1220–1226.

Liu, K., Goodman, M.M., Muse, S., Smith, J.S., Buckler, E. and Doebley, J. (2003) Genetic structure and diversity among maize inbred lines as inferred from DNA microsatellites. *Genetics* 165, 2117–2128.

Liu, S., Rudd, J.C., Bai, G., Haley, S.D., Ibrahim, A.M.H. *et al.* (2014) Molecular markers linked to important genes in hard winter wheat. *Crop Science* 54(4), 1304.

Liu, Y., He, Z., Appels, R. and Xia, X. (2012) Functional markers in wheat: current status and future prospects. *Theoretical and Applied Genetics* 125(1), 1–10.

Mace, E.S., Rami, J.-F., Bouchet, S., Klein, P.E., Klein, R.R. *et al.* (2009) A consensus genetic map of sorghum that integrates multiple component maps and high-throughput diversity array technology (DArT) markers. *BMC Plant Biology* 9, 13.

Mace, E.S., Singh, V., van Oosterom, E.J., Hammer, G.L., Hunt, C.H. and Jordan, D.R. (2012) QTL for nodal root angle in sorghum (*Sorghum bicolor* L. Moench) co-locate with QTL for traits associated with drought adaptation. *Theoretical and Applied Genetics* 124, 97–109.

Magalhaes, J.V., Garvin, D.F., Wang, Y., Sorrells, M.E., Klein, P.E. *et al.* (2004) Comparative mapping of a major aluminum tolerance gene in sorghum and other species in the Poaceae. *Genetics* 167, 1905–1914.

Mago, R., Spielmeyer, W., Lawrence, G.J., Lagudah, E.S., Ellis, J.G. and Pryor, A. (2002) Identification and mapping of molecular markers linked to rust resistance genes located on chromosome 1RS of rye using wheat–rye translocation lines. *Theoretical and Applied Genetics* 104(8), 1317–1324.

Maki-Tanila, A. and Hill, W.G. (2014) Influence of gene interaction on complex trait variation with multilocus models. *Genetics* 198, 355–367.

Malik, S., Mehboob-ur-Rahman and Malik, T.A. (2015) Genetic mapping of potential QTLs associated with drought tolerance in wheat. *The Journal of Animal and Plant Science* 25(4), 1032–1040.

Manzelli, M., Pileri, L., Lacerenza, N., Benedettelli, S. and Vecchio, V. (2006) Genetic diversity assessment in Somali sorghum (*Sorghum bicolor* [L.] Moench) accessions using microsatellite markers. *Biodiversity & Conservation* 16, 1715–1730.

Maric, S., Bolaric, S., Martincic, J., Pejic, I. and Kozumplik, V. (2004) Genetic diversity of hexaploid wheat cultivars estimated by RAPD markers, morphological traits and coefficients of parentage. *Plant Breeding* 123, 366–369.

Menz, M., Klein, R., Unruh, N., Rooney, W., Klein, P. and Mullet, J. (2004) Genetic diversity of public inbreds of sorghum determined by mapped AFLP and SSR markers. *Crop Science* 44, 1236–1244.

Meuwissen, T.H.E., Hayes, B.J. and Goddard, M.E. (2001) Prediction of total genetic value using genome-wide dense marker maps. *Genetics* 157, 1819–1820.

Mofokeng, A.M., Shimelis, H. and Laing, M. (2017) Breeding strategies to improve sorghum quality. *Australian Journal of Crop Science* 11(02), 142–148.

Mohapatra, T., Krishanpal Singh, S.S., Swain, S.C., Sharma, R.K. and Singh, N.K. (2003) STMS based DNA fingerprints of the new plant type wheat lines. *Current Science* 84, 1125–1129.

Morison, J.I.L., Baker, N.R., Mullineaux, P.M. and Davies, W.J. (2008) Improving water use in crop production. *Philosophical Transactions of the Royal Society B: Biological Sciences* 363, 639–658.

Mukhtar, M.S., Rahman, M. and Zafar, Y. (2002) Assessment of diversity among wheat (*T. aestivum* L.) cultivars from a range of localities across Pakistan using random amplified polymorphic DNA analysis. *Euphytica* 128, 417–425.

Muraya, M.M., de Villiers, S., Parzies, H.K., Mutegi, E., Sagnard, F. *et al.* (2011) Genetic structure and diversity of wild sorghum populations (*Sorghum* spp.) from different eco-geographical regions of Kenya. *Theoretical and Applied Genetics* 123, 571–583.

Nagaoka, T. and Ogihara, Y. (1997) Applicability of inter-simple sequence repeat polymorphisms in wheat for use as DNA markers in comparison to RFLP and RAPD markers. *Theoretical and Applied Genetics* 94, 597–602.

Ni, F., Qi, J., Hao, Q., Lyu, B., Luo, M.C. *et al.* (2017) Wheat Ms2 encodes for an orphan protein that confers male sterility in grass species. *Nature Communications* 8, 15121.

Oladosu, Y., Rafii, M.Y., Samuel, C., Fatai, A., Magaji, U. *et al.* (2019) Drought resistance in rice from conventional to molecular breeding: a review. *International Journal of Molecular Sciences* 20, 3519.

Nybom, H. (1994) DNA fingerprinting – a useful tool in fruit breeding. *Euphytica* 77, 59–64.

Pandey, P., Irulappan, V., Bagavathiannan, M.V. and Senthil-Kumar, M. (2017) Impact of combined abiotic and biotic stresses on plant growth and avenues for crop improvement by exploiting physio-morphological traits. *Frontiers in Plant Science* 8, 537.

Pandey, K., Dangi, R., Prajapati, U., Kumar, S., Maurya, N.K. *et al.* (2019) Advance breeding and biotechnological approaches for crop improvement: a review. *International Journal of Chemical Studies* 7, 837–841.

Pardey, P.G. (2011) A strategic look at global wheat production, productivity and R and D developments. *Czech Journal of Genetics and Plant Breeding* 47, S9–S19.

Payne, P.I. (1987) Genetics of wheat storage proteins and the effect of allelic variation on breadmaking quality. *Annual Review of Plant Biology* 38, 141–153.

Pecetti, L., Doust, M.A., Calcagno, L., Raciti, C.N. and Boggini, G. (2001) Variation of morphological and agronomical traits, and protein composition in durum wheat germplasm from Eastern Europe. *Genetic Resources and Crop Evolution* 48, 609–620.

Peña, R.J. (2002) Wheat for bread and other foods. In: Curtis, B.C., Rajaram, S. and Gómez Macpherson, H. (eds) *Bread Wheat: Improvement and Production*. FAO Plant Production and Protection Paper no. 30. Food and Agriculture Organization of the United Nations, Rome, pp. 483–542.

Powell, W., Morgante, M., Andre, C., Hanafey, M., Vogel, J., Tingey, S. and Rafalski, A. (1996) The comparison of RFLP, RAPD, AFLP and SSR (microsatellite) markers for germplasm analysis. *Molecular Breeding* 2, 225–238.

Prakash, S.P.J., Biji, K.R., Gomez, S.M., Murthy, K.G. and Babu, R.C. (2006) Genetic diversity analysis of sorghum (*Sorghum bicolor* L. Moench) accessions using RAPD markers. *Indian Journal of Crop Science* 1, 109–112.

Prasanna, P.M., Pixley, K., Warburton, M.L. and Xie, C.X. (2010) Molecular marker-assisted breeding options for maize improvement in Asia. *Molecular Breeding* 26, 339–356.

Punitha, D., Ganesamurthy, K. and Rajarathinam, S. (2010) Metroglyph analysis of morphological variations in sorghum germplasm collections. *Electronic Journal of Plant Breeding* 1, 536–541.

Rahman, M., Hussain, D. and Zafar, Y. (2002) Estimation of genetic divergence among elite cotton cultivars–genotypes by DNA fingerprinting technology. *Crop Science* 42, 2137–2144.

Rajendran, K., Tester, M. and Roy, S.J. (2009) Quantifying the three main components of salinity tolerance in cereals. *Plant, Cell & Environment* 32, 237–249.

Rasheed, A. and Xia, X. (2019) From markers to genome-based breeding in wheat. *Theoretical and Applied Genetics* 132(3), 767–784.

Ren, Z.H., Gao, J.P., Li, L.G., Cai, X.L., Huang, W. *et al.* (2005) A rice quantitative trait locus for salt tolerance encodes a sodium transporter. *Nature Genetics* 37, 1141–1146.

Reynolds, M. and Tuberosa, R. (2008) Translational research impacting on crop productivity in drought-prone environments. *Current Opinion in Plant Biology* 11, 171–179.

Richards, R.A., Rebetzke, G.J., Watt, M., Condon, A.G., Spielmeyer, W. and Dolferus, R. (2010) Breeding for improved water productivity in temperate cereals: phenotyping, quantitative trait loci, markers and the selection environment. *Functional Plant Biology* 37, 85–97.

Sade, N., del Mar Rubio-Wilhelmi, M., Umnajkitikorn, K. and Blumwald, E. (2018) Stress-induced senescence and plant tolerance to abiotic stress. *Journal of Experimental Botany* 69, 845–853.

Saleh, B. (2012) Biochemical and genetic variation of some Syrian wheat varieties using NIR, RAPD and AFLPs techniques. *Journal of Plant Biology Research* 1, 1–11.

Salekdeh, G.H., Reynolds, M., Bennett, J. and Boyer, J. (2009) Conceptual framework for drought phenotyping during molecular breeding. *Trends in Plant Science* 14, 488–496.

Schnable, P.S., Ware, D., Fulton, R.S., Stein, J.C., Wei, F. *et al.* (2009) The B73 maize genome: complexity, diversity, and dynamics. *Science* 326, 1112–1115.

Shavrukov, Y., Gupta, N., Miyazaki, J., Baho, M., Chalmers, K. *et al.* (2010) *HvNax3* – a locus controlling shoot sodium exclusion derived from wild barley (*Hordeum vulgare* ssp. *spontaneum*). *Functional & Integrative Genomics* 10, 277–291.
Shewry, P.R. (2009) Wheat. *Journal of Experimental Botany* 60(6), 1537–1553.
Sirault, X.R.R., James, R.A. and Furbank, R.T. (2009) A new screening method for osmotic component of salinity tolerance in cereals using infrared thermography. *Functional Plant Biology* 36, 970–977.
Sun, Q., Ni, Z., Liu, Z., Gao, J. and Huang, T. (1998) Genetic relationships and diversity among Tibetan wheat, common wheat and European spelt wheat revealed by RAPD markers. *Euphytica* 99, 205–211.
Talukder, S.K., Babar, M.A., Vijayalakshmi, K., Poland, J., Prasad, P.V.V. *et al.* (2014) Mapping QTL for the traits associated with heat tolerance in wheat (*Triticum aestivum* L.). *BMC Genetics* 15, 97.
Tenaillon, M.I., Sawkins, M.C., Long, A.D., Gaut, R.L., Doebley, J.F. and Gaut, B.S. (2001) Patterns of DNA sequence polymorphism along chromosome 1 of maize (*Zea mays* ssp. *mays* L.). *Proceedings of the National Academy of Sciences USA* 98, 9161–9166.
Tester, M. and Langridge, P. (2010) Breeding technologies to increase crop production in a changing world. *Science* 327, 818–822.
Thakur, P., Kumar, S., Malik, J.A., Berger, J.D. and Nayyar, H. (2010) Cold stress effects on reproductive development in grain crops, an overview. *Environmental and Experimental Botany* 67, 429–443.
Vagndorf, N., Kristensen, P.S., Anderson, J.P., Jahoor, A. and Orabi, J. (2018) Marker-assisted breeding in wheat. In: Çiftçi, Y.Ö. (ed.) *Next Generation Plant Breeding*. IntechOpen, London. Available at: https://doi.org/10.5772/intechopen.74724
Vierling, R.A. and Nguyen, H.T. (1992) Use of RAPD markers to determine the genetic diversity of diploid, wheat genotypes. *Theoretical and Applied Genetics* 84, 835–838.
Vroh Bi, I., McMullen, M.D., Sanchez-Villeda, H., Schroeder, S., Gardiner, J. *et al.* (2006) Single nucleotide polymorphisms and insertion–deletions for genetic markers and anchoring the maize fingerprint contig physical map. *Crop Science* 46, 12–21.
Weising, K., Winter, P., Huttel, B. and Kahl, G. (1998) Microsatellite markers for molecular breeding. *Journal of Crop Production* 1, 113–143.
White, J., Law, J., MacKay, I., Chalmers, K., Smith, J., Kilian, A. and Powell, W. (2008) The genetic diversity of UK, US and Australian cultivars of *Triticum aestivum* measured by DArT markers and considered by genome. *Theoretical and Applied Genetics* 116, 439–453.
Wu, Y. and Huang, Y. (2006) An SSR genetic map of *Sorghum bicolor* (L.) Moench and its comparison to a published genetic map. *Genome* 50, 84–89.
Wu, Y., Huang, Y., Tauer, C.G. and Porter, D.R. (2006) Genetic diversity of sorghum accessions resistant to greenbugs as assessed with AFLP markers. *Genome* 49, 143–149.
Xu, Y. and Crouch, J.H. (2008) Marker-assisted selection in plant breeding: from publications to practice. *Crop Science* 48, 391–407.
Yu, J., Pressoir, G., Briggs, W.H., Vroh Bi, I., Yamasaki, M. *et al.* (2006) A unified mixed model method for association mapping that accounts for multiple levels of relatedness. *Nature Genetics* 38, 203–208.
Yu, J., Holland, J.B., McMullen, M.D. and Buckler, E.D. (2008) Genetic design and statistical power of nested association mapping in maize. *Genetics* 178, 539–551.
Zeng, L.Y., Guo, S.X., He, B., Sun, L.J., Peng, Y. *et al.* (2011) Genome-wide patterns of genetic variation in sweet and grain sorghum (*Sorghum bicolor*). *Genome Biology* 12(11), R114.
Zhang, P., Zhong, K., Shahid, M.Q. and Tong, H. (2016) Association analysis in rice: from application to utilization. *Frontiers in Plant Science* 7, 1202.
Zhao, M., Zhang, Z., Zhang, S., Li, W., Jeffers, D.P., Rong, T. and Pan, G. (2006a) Quantitative trait loci for resistance to banded leaf and sheath blight in maize. *Crop Science* 46, 1039–1045.
Zhao, W., Canaran, P., Jurkuta, R., Fulton, T., Glaubitz, J. *et al.* (2006b) Panzea: a database and resource for molecular and functional diversity in the maize genome. *Nucleic Acids Research* 34, D725–D757.
Zhu, C., Gore, M., Buckler, E.S. and Yu, J. (2008) Status and prospects of association mapping in plants. *Plant Genome* 1, 5–20.
Zhu, J., Ingram, P.A., Benfey, P.N. and Elich, T. (2011) From lab to field, new approaches to phenotyping root system architecture. *Current Opinion in Plant Biology* 14(3), 310–317.

# 3 Recent Advancement of Molecular Breeding for Improving Salinity Tolerance in Wheat

**Inès Yacoubi, Emna Khanfir, Karama Hamdi and Faïçal Brini***
*Biotechnology and Plant Improvement Laboratory, Centre of Biotechnology of Sfax, University of Sfax, Sfax, Tunisia*

## 3.1 Introduction

Wheat (*Triticum aestivum* L.) is unique among recently domesticated species; after originating in the Fertile Crescent, it spread to all parts of the world except Antarctica. Currently, wheat is among the three most important food crops including rice and maize. Wheat is cultivated on ~200 million hectares worldwide, providing one-fifth of the total caloric input of the global population (FAO, 2017). Estimates of the global population are 9–10 billion by 2050 (FAO, 2017), mostly living in the currently developing countries (Africa and South Asia) where wheat products are the most consumed staple food commodities. Low progress in yield improvement ranging from 0.8 to 1.0% annually will make it impossible to fulfil wheat production requirements by 2050. Among the key challenges in wheat production is elevating yield potential by increasing its ability to grow on marginal lands.

Soil salinity is a global problem that affects approximately 20% of irrigated land and reduces crop yields significantly (Qadir *et al.*, 2014). High soil salinity is one of the main agricultural challenges in the modern world (Rengasamy, 2016). Salt stress affects the growth and development of plants, thus reducing their yield (Arzani and Ashraf, 2016). Global cultivated lands cover approximately 1.5 billion hectares, and an estimated 32 million hectares are currently damaged by salinity. Irrigated lands, having the highest productivity, comprise just 230 million hectares, of which an estimated 20% have yields reduced by high soil salinity (Munns, 2005). However, little is known about the specific mechanisms through which wild wheat thrives in saline soil. Salinity reduces plant growth for two main reasons: (i) an osmotic effect due to the salt outside the root; and (ii) a toxic effect if the salt accumulates in leaves in excessive amounts (Munns, 2002).

Wheat is the most widely grown crop in the world. It is considered a moderately salt-tolerant and drought-tolerant crop, and, with barley, it is the preferred cereal in most arid and semi-arid agricultural regions. Much of the world's wheat is produced under irrigation; however, the area sown to rainfed wheat is very substantial and is expected to grow further as water for irrigation declines globally. As wheat is grown in many arid or semi-arid regions of the world, it is likely to incur salinity, whether caused by irrigation, land clearing or natural processes. This is because of large amounts of salt in the soil profile, deposited there by weathering of rocks or by

*Corresponding author: faical.brini@cbs.rnrt.tn

 *Molecular Breeding in Wheat, Maize and Sorghum: Strategies for Improving Abiotic Stress Tolerance and Yield* (eds M.A. Hossain *et al.*)
DOI: 10.1079/9781789245431.0003

deposition of oceanic salt by wind and rain. Bread wheat is moderately salt tolerant, compared with other cereals. In a field where salinity rises to 10 dS/m (about 100 mM NaCl), rice (*Oryza sativa* L.) will die before maturity, while bread wheat will produce a reduced yield. Even barley (*Hordeum vulgare* L.), the most tolerant cereal, dies after extended periods at salt concentrations higher than 250 mM NaCl (equivalent to 50% seawater). Durum wheat (*Triticum turgidum* L. ssp. *durum* (Desf.); tetraploid, AABB), which is used for making pasta and couscous, is less salt tolerant than bread wheat (Maas and Hoffman, 1977; Munns *et al.*, 2006). Bread wheat (*T. aestivum*), one of the most important staple crops globally, provides most of the calories for approximately 30% of the world's population (FAO, 2017). More than 800 million hectares (6%) of arable land is affected by salinity worldwide (Munns and Tester, 2008). Soil salinity is a major constraint on wheat grain yield (Lobell *et al.*, 2011). Therefore, understanding the mechanisms of response and adaptation to salt stress and then improving salinity tolerance of wheat are critical tasks for breeders and researchers alike. This chapter provides an outline of the mechanisms of wheat salinity tolerance and discusses the challenges of several breeding programmes that are in progress with regard to durum and bread wheats using the integration of trait-based and molecular selection for delivering improved wheat varieties adapted to saline conditions.

## 3.2 Key Mechanisms Controlling Salt Signalling Tolerance

The responses of plants, and especially wheat, to salinity are often complex and multifaceted, which makes experiments difficult to design and interpret. Therefore, plant response is different due to the complex nature of the stress, which leads to limitations in salinity tolerance breeding. For these reasons, identification of salinity tolerance mechanisms in wild wheat will facilitate the improvement of salinity tolerance in current, elite wheat varieties. However, little is known about the specific mechanisms by which plants can thrive in saline soils.

In the past decades, several wheat genetic stocks have been developed showing all three types of tolerance mechanism in plant salinity tolerance. These three interrelated concepts are classified into: (i) ion ($Na^+$ or $Cl^-$) exclusion; (ii) tolerance of tissue to accumulated $Na^+$ or $Cl^-$; and (iii) osmotic stress tolerance (Munns and Tester, 2008; Mujeeb-Kazi *et al.*, 2019). As described above, the exclusion of $Na^+$ from leaves is the main mechanism conferring salinity tolerance in wheat as it does not reach potentially toxic concentrations (Mujeeb-Kazi *et al.*, 2019). The ability to exclude $Na^+$ from the leaves is an important component for salinity tolerance and ensures that $Na^+$ is not accumulated in toxic concentrations. The limitation of $Na^+$ accumulation can be responsible for genotype differences in salinity tolerance in wheat. In fact, salt tolerance with both bread and durum wheat genotypes is associated with $Na^+$ exclusion and enhanced $K^+$:$Na^+$ ratio in the leaves. The osmotic effect of NaCl outside the roots causes growth reduction, at least in the early stages of the salinity response, with toxic effects of the salt accumulating in the older leaves (Munns *et al.*, 2011). Our understanding of the molecular basis for many of the transport processes, including the role of the high-affinity potassium transporter (HKT) family in $Na^+$ exclusion from leaves, is increasing (Munns and Tester, 2008). It has been shown that the long arm of chromosome 4D of *T. aestivum* contains a gene which influences the ability of wheat plants to discriminate between $Na^+$ and $K^+$. This discrimination most obviously affects transport from the roots to the shoots, in which less $Na^+$ and more $K^+$ accumulate in the plants (Gorham *et al.*, 1990). In bread wheat (*T. aestivum*), this is conferred by the *Kna1* locus on chromosome 4D which controls $K^+$ and $Na^+$ accumulation in the shoot. The candidate gene is *TaHKT1;5-D*, an $Na^+$ transporter which functions to remove $Na^+$ from the transpiration stream flowing to the leaves and increases the $K^+$:$Na^+$ ratio. This gene explains the relative salt tolerance of bread wheat compared with durum wheat (*T. turgidum* ssp. *durum*) lacking the D genome and homologues of *Kna1* on the A and B genomes (Mujeeb-Kazi *et al.*, 2019).

To improve the salt tolerance of durum wheat, many researchers have explored the natural diversity in shoot $Na^+$ exclusion within ancestral wheat germplasm. A genetic analysis, based on a population derived from a cross between a standard durum wheat genotype and a line containing introgressions from the A genome diploid ancestral wheat relative *Triticum monococcum* showing high $Na^+$ exclusion ability, revealed that two loci, *Nax1* and *Nax2*, were involved in excluding $Na^+$ ions (Munns *et al.*, 2000; Munns *et al.*, 2012). The study demonstrated the role of a gene in the *Nax2* locus, *TmHKT1;5-A*, which encodes an $Na^+$-selective transporter located on the plasma membrane of root cells surrounding the xylem vessels. This transporter is therefore ideally localized to withdraw $Na^+$ from the xylem and reduce the transport of $Na^+$ to the leaves. Field trials on saline soils demonstrated that *TmHKT1;5-A* presence significantly reduced leaf $Na^+$ concentration and increased durum wheat grain yield by 25% compared with near-isogenic lines without the *Nax2* locus (Munns *et al.*, 2012). Several HKT1-encoding genes, including *HKT1;1/2-like*, *HKT1;3-like*, *HKT1;4-like* and *HKT1;5-like*, have been identified and mapped to wheat homoeologous chromosome groups 2, 6, 2 and 4, respectively (Huang *et al.*, 2008). Among these, *Nax1* in chromosome arm 2AL co-segregated with sodium transporter gene *HKT1;4-A2*, which was shown to control $Na^+$ unloading from xylem in roots and sheaths and therefore was proposed as the functional candidate (Huang *et al.*, 2006). *Nax2* was mapped to the distal region of chromosome 5AL that is homoeologous to a region on chromosome 4DL containing *Kna1*. Based on synteny and phylogeny analysis with *Nax2*, *TmHKT1;5-A* was proposed to be the candidate of *Nax2* (Wang and Xia, 2018).

In barley, leaves of salt-tolerant crop cultivars tolerate high concentrations of $Na^+$ and $Cl^-$, which means that this species is able to grow productively in soils with high NaCl content in the soil solution. This capacity of organs to function while their tissues or cells contain high concentrations of $Na^+$ or $Cl^-$ is defined as 'tissue tolerance'. Basically, intracellular compartmentalization of $Na^+$ and $Cl^-$ is a key mechanism for tissue tolerance, where most of the ions are contained in vacuoles and the concentrations in the cytoplasm remain relatively low. This mechanism is also known as 'cellular tolerance' and is defined as the ability of a cell to compartmentalize $Na^+$ and $Cl^-$ in vacuoles at concentrations that would be toxic in the cytoplasm. Therefore, tissue tolerance is physiologically and genetically complex, and there is no evidence so far that any single gene or molecular marker is a quantitative predictor of this complex trait (Munns *et al.*, 2016; Mujeeb-Kazi *et al.*, 2019).

For a plant growing or surviving in saline soil, osmotic stress tolerance is essential. The osmotic effect on plant growth occurs instantly and causes reductions in growth rate. Consequently, growing leaves and shoot growth are most affected. Osmotic stress not only has an immediate effect on growth, but also has a greater effect on growth rates than the ionic stress that influences growth much later, especially at low to moderate salinity levels (Mujeeb-Kazi *et al.*, 2019). Cell osmotic pressure increases to match the increase in the osmotic pressure of the soil solution due to an increase in the solute content, not loss of water, so turgor and volume are maintained. This increase could be generated by the synthesis of organic solutes such as sugars that reach concentrations great enough to contribute a significant osmotic pressure, but this would be at the expense of growth as those solutes are no longer available for cell-wall and protein synthesis (Munns *et al.*, 2016). However, metabolic costs are needed to synthesize organic solutes for osmotic adjustment. The number of moles of ATP needed to use 1 mole of NaCl as an osmoticum is approximately 4 in root cells, and 7 in leaf cells, whereas the number required to synthesize an organic compound is a magnitude of higher order (Munns, 2002). Other physiological components such as the maintenance of plant water status, transpiration and transpiration-use efficiency, seed germination, production of antioxidants, early seedling growth and harvest index can affect salinity tolerance. However, these physiological components are still less studied, so further investigations are needed to understand the effects of salinity on these processes (Negrão *et al.*, 2017).

## 3.3 Conventional Breeding for Salt Tolerance in Wheat

Development of crop plants tolerant to salt stress is very important to meet the growing global demand for food. In the last decades conventional breeding has helped to improve yield potential and disease resistance in crops via knowledge of the physiological and biochemical phenomena related to a trait.

In a comprehensive review, Ashraf (1994) listed a few salt-tolerant lines/cultivars of different crops that had been developed through conventional breeding and some of them were tested under natural field conditions. For example, some lines/cultivars of alfalfa (*Medicago sativa* L.) such as AZ-Germ Salt II (Dobrenz *et al.*, 1989), AZ-90NDC-ST (Johnson *et al.*, 1991), AZ-97MEC and AZ-97MEC-ST (Al-Doss and Smith, 1998), ZS-9491 and ZS-9592 (Dobrenz, 1999) were tested under natural field conditions. Similarly, two salt-tolerant lines/cultivars of bread wheat, namely S24 (Ashraf and O'Leary, 1996) and KRL1-4 (Hollington, 2000), were evaluated on natural salt-affected soils. Besides, in India, Pakistan and Egypt, many experiments were conducted to increase salt tolerance in wheat using conventional breeding methods (Munns *et al.*, 2006). This work succeeded and many new wheat cultivars were created: the Indian KRL1-4, released by the Central Soil Salinity Research Institute (CSSRI) at Karnal; the Pakistani LU26S and SARC-1, released by the Saline Agriculture Research Cell (SARC) at Faisalabad; and the Egyptian Sakha 8, released by the Agricultural Research Centre at Giza. In India, almost all salt-tolerant wheat germplasm is derived from Kharchia 65 which is regarded as highly salt tolerant. CSSRI discharged KRL1-4 for saline areas, a cross between Kharchia 65 and WL711 (Hollington, 2000). KRL1-4 performed very well on saline soils in northern India, but not in Pakistan. This could be due to the heavier soils and greater problems of waterlogging (Hollington, 2000). Another derivative line of Kharchia 65 was created in the UK by S.A. Quarrie and A. Mahmood: a doubled haploid line, KTDH 19, from a cross of Kharchia 65 with a line identified with exceptional $Na^+$ exclusion, TW161. This derivative worked well in Spain (Hollington *et al.*, 1994), but in India and Pakistan it was late to mature and produced very low yield (Hollington, 2000).

The cultivar LU26S was developed in Pakistan and gave a high yield on the saline soils of Pakistan (Qureshi *et al.*, 1990). Also, LU26S was crossed with Kharchia and two salt-tolerant genotypes, S24 and S36, were selected (Ashraf and O'Leary, 1996). S24 presents a high salt tolerance, as high as Kharchia and SARC-1, possibly due to its low leaf $Na^+$ accumulation (Ashraf, 2002).

Other scientists tried conventional breeding for salt tolerance in wheat by using a large collection that was screened and grown in hydroponic or sand culture (Colmer *et al.*, 2005). A study by Jafari-Shabestari *et al.* (1995) evaluated 400 Iranian wheat varieties at one site in California over two seasons, irrigated with water at three salinity levels (1, 5 and 8 dS/m). They identified several accessions that were consistently high for grain yield in both low- and high-salinity treatments, but no cultivar was developed as a consequence (Jafari-Shabestari *et al.*, 1995). This could be due to the low correlation found between grain yield at high salinity and relative yield (yield in saline soil relative to non-saline soil), biomass or harvest index. In addition, Jafari-Shabestari *et al.* (1995) and Richards *et al.* (1987) noted a lack of correlation between relative yield and absolute yield that allowed them to conclude that the most efficient way to increase yields at high salinity was to select the highest-yielding lines at low salinity. Moreover, other researchers like Colmer *et al.* (2006) used wild relatives to improve salt tolerance of wheat by hybridizing *Aegilops tauschii* (DD) with durum wheat (AABB) to produce synthetic hexaploid wheat (Schachtman *et al.*, 1992; Mujeeb-Kazi and Diaz de Leon, 2002). They concluded that the variation in salt tolerance in the D genome influenced the salt tolerance of hexaploids too (Schachtman *et al.*, 1992). In fact, to introduce salt tolerance into bread wheat, Farooq *et al.* (1995) crossed *Aegilops cylindrica* (CCDD) with the Pakistani cultivars LU26 and Pak81. From the salinity tests conducted, the wheat lines WL1076 and WL41 were developed from the cross. These varieties were more salt and drought tolerant than their parents (Farooq *et al.*, 1995; Farooq, 2004).

Additionally, very salt-tolerant species which are the tall wheat grasses (E or J genomes), at low to moderate salinity and even up to seawater concentrations, behave like barley with a similar decline in biomass and more tolerance compared with bread wheat varieties. So, a cross was done with bread wheat and the E genome species *Lophopyrum elongatum* (syn. *Agropyron elongatum* and *Elytrigia elongata*) to improve salt tolerance of bread wheat (summarized by Colmer *et al.*, 2006). When testing the lines in the field, it was shown that the amphiploid had a higher salt tolerance but a lower yield than Chinese Spring, and that chromosome 3E had a major effect on salt tolerance. *Thinopyrum ponticum* (decaploid, E genome), a 'tall wheat grass', is very salt tolerant and is commonly used as a forage crop in saline lands. Thus, somatic hybridization techniques were used to transfer *T. ponticum* chromosomes into bread wheat, and field experiments were conducted with $F_4$ and $F_5$ generation lines grown in a soil with moderately high salinity levels (Chen *et al.*, 2004). The two hybrids produced a good yield whereas the parents died before maturity. Hence, the salt tolerance of *T. ponticum* appears to have been introduced into bread wheat, with the *T. ponticum* chromatin stably inherited (Chen *et al.*, 2004).

Wang *et al.* (2003) used *Thinopyrum junceum* (hexaploid, mixed E and J genomes, J1J1J2J-2EE) to produce recombinant lines of wheat that have segments of chromosome 5J. The data show that yield of these lines fluctuated with moderate salinity and yield was equal to or better than that of Kharchia 65.

Despite this progress, however, conventional breeding approaches as a whole are time-consuming and labour-intensive. Undesirable genes are often transferred in combination with desirable ones and reproductive barriers limit transfer of favourable alleles from interspecific and intergeneric sources (Ashraf and Akram, 2009).

## 3.4 Impact of High-Throughput Phenotyping and Genotyping in Salt Tolerance

Phenomics is a field of science at the junction of biology and informatics and solves the problems of rapid, accurate estimation of the plant phenotype; it was developed rapidly because of the need to analyse phenotypic characteristics in large-scale genetic and breeding experiments in plants. It is based on using the methods of computer image analysis and integration of biological data. Owing to automation, new approaches make it possible to considerably accelerate the process of estimating the characteristics of a phenotype, increase its accuracy and remove subjectivism (inherent to humans).

High-throughput field phenotyping represents a bottleneck in conventional breeding, marker-assisted selection (MAS) or genomic selection, where phenotyping is a key informant for establishing accuracy for statistical models (Desta and Ortiz, 2014). The main technologies of high-throughput plant phenotyping can be done in both controlled and field conditions, and it is based on image analysis, satisfying a majority of the goals formulated by scientists like enhancing nutrition efficiency, abiotic stress tolerance and crop quality traits (Hairmansis *et al.*, 2014). In addition, these technologies have been successfully employed in plant breeding and give good results (Araus and Cairns, 2014; Tardieu *et al.*, 2017). They also provide solutions for screening the plant model (*Arabidopsis thaliana*) and even a complex morpho-physiological trait in crop plants such as rice and soybean (De Diego *et al.*, 2017). Also, it successfully validates multivariate analysis of rosette growth in different salt concentrations and the interaction with varying nutritional composition of the growth medium (De Diego *et al.*, 2017). In this context, Berger *et al.* (2012) used image acquisition technology for assessing plant growth in response to salinity. With automated, high-throughput phenotyping facilities, such as The Plant Accelerator in Adelaide, Australia, and in facilities hosted by other partners of the International Plant Phenotyping Network (https://www.plant-phenotyping.org/, accessed 4 February 2021), it is possible to obtain a daily estimate of plant biomass from the start of an experiment, before salt imposition, through to the end. Importantly, these measurements also allow assessment of the shoot's ion-independent component of salt toxicity, which involves the inhibition of shoot growth from the moment of salt imposition (Berger *et al.*, 2012), before salt has had time to accumulate in the shoot and significantly affect the shoot's function.

To date, a plethora of imaging systems focus on shoot parameters (as recently reviewed by Fahlgren *et al.*, 2015). These systems appear to be particularly robust and reliable to quantify shoot growth under controlled environments. Some studies have reported root imaging under natural conditions and without destructive harvesting, such as imaging of roots using the Growth and Luminescence Observatory for Roots (GLO-Roots) system (Rellan-Alvarez *et al.*, 2015) or in a more artificial system using transparent growth media (such as gel or glass beads) (Courtois *et al.*, 2013; Topp *et al.*, 2013). It is believed that these technologies will provide an important platform for change in salinity research, especially when time resolution is incorporated to provide insights into the dynamic responses of plants to salinity. Non-destructive analyses, such as imaging, allow the monitoring of the same plant over multiple time points, such as before and after stress application, enabling the detection of small and dynamic differences in growth parameters.

## 3.5 Wheat Breeding Challenge: Integration of Physiological-Trait Breeding and Molecular Breeding

Wheat is moderately tolerant to salt stress and is known to be more sensitive than barley but more tolerant than rice, maize and sorghum. It has long been known that tetraploid wheat is less salt tolerant than bread wheat (Rawson *et al.*, 1988) and that a major factor behind this difference is that bread wheat is able to maintain a higher $K^+$:$Na^+$ concentration ratio in the leaves (Gorham *et al.*, 1990). Although plants' sensitivity to salinity is higher during the early seedling stage and reproductive stage, crops need to maintain functions at all stages of their life cycle to increase their ability to maintain yield under high salinity. For instance, yield can be reduced during the vegetative stage by affecting parameters such as tiller number per plant in cereals such as rice (Zeng and Shannon, 2000) or wheat (Maas *et al.*, 1994).

Breeding of salt-tolerant wheat varieties has been slow, despite its importance as a food crop and the considerable losses inflicted by salt stress (Ismail and Hories, 2017). Salinity stress impacts many aspects of a plant's physiology, making breeding for salt tolerance difficult. Traditional crop improvement programmes by hybridization have played a significant role in wheat improvement during past decades but have led to only modest progress in the development of salt-tolerant wheat varieties in several countries, including India, Pakistan, Australia and Egypt (Ismail and Hories, 2017; Mujeeb-Kazi *et al.*, 2019).

Wheat breeders have exploited superior traits that increase yield in target environments. This traditional wheat improvement was dependent on selection of traits without understanding the molecular mechanisms. Despite the slow progress of wheat genome sequencing and the development of functional and comparative genomics, wheat genetic improvement efforts undoubtedly did not fulfil the food requirements of the masses. More than these molecular–genetic approaches in terms of breeding, more salt-tolerant varieties are a way of ensuring sustainable production. It is more tractable to dissect the plant's response into traits that are hypothesized to be involved in the overall tolerance of the plant to salinity. However, it often can be difficult to identify the most important trait contributing to salinity tolerance in the given plant system and there are only a small number of examples where trait-based approaches have been successful in breeding for improved yield under salt stress.

Salinity tolerance is best measured by the ability of the plant to maintain yield in saline conditions relative to control conditions. Correlation studies can give an indication of which traits are the most important contributors to salinity tolerance. It is noteworthy that the lack of correlation does not necessarily mean that a particular trait does not play a part in salinity tolerance. A strong correlation between the $Na^+$ content in the third leaf and salt tolerance index is observed in many studies indicating that this trait may be associated with salinity tolerance (Negrão *et al.*, 2017).

Delivery of improved crop varieties adapted to saline conditions has been lagging for several reasons, including the huge knowledge gap in understanding the genetic basis of salinity tolerance in wheat and then applying the available knowledge to deliver salt-resilient crop varieties (Mujeeb-Kazi *et al.*, 2019). Trait-based breeding approaches, which often utilize molecular markers to improve selection efficiency, are starting to deliver new and significant gains (Fig. 3.1). To target the most important traits, it is important to know how they will influence yield. The challenge for breeders will be to efficiently integrate trait-based and molecular methods to increase yield in dry and saline environments where trait-based approaches have been successful in breeding for improved yield under drought. Knowledge of the target environment is very important in trait-based breeding as the benefit from a trait may be confined to specific environments. Salinity and drought have many traits in

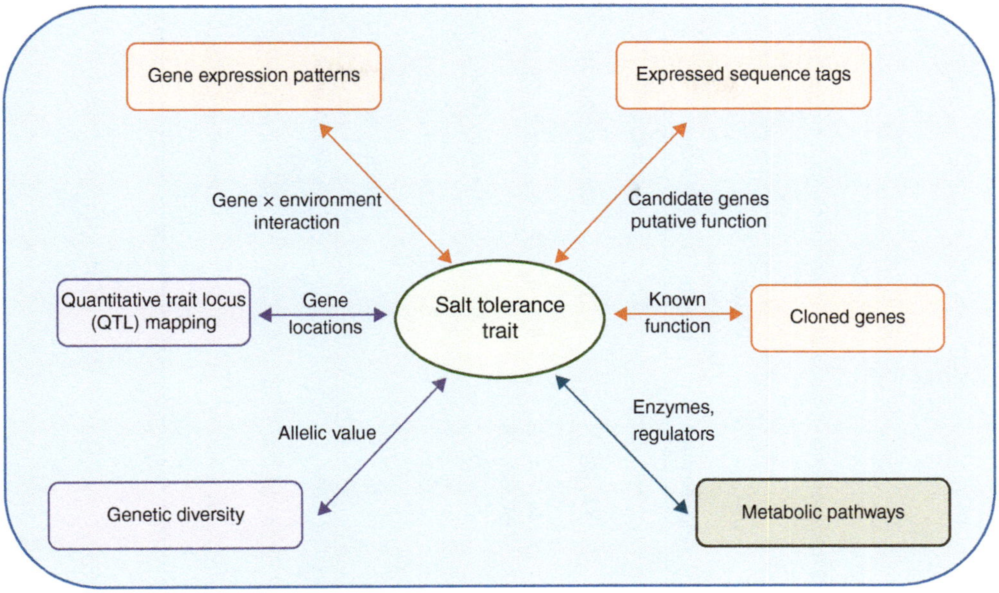


**Fig. 3.1.** Schematic diagram showing strategies and techniques involved in the production of salt-tolerant wheat cultivars.

common. Obvious examples are osmotic adjustment and the production of osmoprotectants including compounds and enzymes that detoxify reactive oxygen species (Munns and Richards, 2007). Less obvious traits are those for improved efficiency of water use, such as early vigour, long coleoptiles and transpiration efficiency, as described above. Early vigour maximizes growth when conditions are favourable early in the season or when water is more available and salt concentrations are lower. Transpiration efficiency optimizes water use when soil moisture is less available (Munns and Richards, 2007).

Two traits that are specific to salinity relate to the prevention of $Na^+$ toxicity in wheat: (i) $Na^+$ exclusion by roots and the associated high discrimination for $K^+$ over $Na^+$ in leaves; and (ii) tolerance of high internal $Na^+$ concentrations in leaves. To date there have been no releases of salt-tolerant wheat based on a particular trait. Successes have come from empirical selection, as described above. However, several breeding programmes are in progress in durum and bread wheats using trait-based selection (Table 3.1) (Munns and Richards, 2007). Early biomass growth is a good example of a trait that has a significant benefit in both dry and saline environments. A high transpiration efficiency, acknowledged as important in dry conditions, should also be important in saline soils. New selection methods using carbon isotope discrimination have resulted in the release of new wheat cultivars (Richards and Lukas, 2002) for dry conditions and should also be relevant for saline environments.

Dissecting the genetic basis of salinity tolerance mechanisms through a genetic characterization and/or transcriptomic approach would facilitate the selection of useful accessions as genetic sources for breeding salinity tolerance traits into commercial wheat.

It often can be difficult to identify which traits are the most important ones contributing to salinity tolerance in the given plant system. To ease this difficulty, we suggest the generation of graphs that show correlations between the proposed traits (e.g. leaf Na content) and a measure of salinity tolerance (e.g. salt tolerance index). Such correlations help to establish whether the measured traits are associated with each other (noting the limitation that a correlation cannot give definitive information on cause-and-effect relationships) and the correlation coefficient can give an indication of which traits are the most important contributors to salinity tolerance (for the analysed plant in the analysed environment) (Negrão *et al*., 2017).

**Table 3.1.** QTLs and genes mapped in populations of wheat selected for salinity tolerance studies. QTLs, gene location and wheat homoeologous groups are listed, along with osmotic and ionic stress tolerance.

| Process involved | Candidate gene or locus | Trait measured | Salt tolerance mechanism | Species | Chromosome | Wheat homoeologous groups | Reference |
|---|---|---|---|---|---|---|---|
| Shoot growth | | Tiller number | Osmotic stress and ionic tolerance | Bread wheat | 5A | 5 | Genc *et al.* (2010) |
| Shoot growth | | Dry matter production | Osmotic stress and ionic tolerance | Bread wheat | 1A, 3B | 1, 3 | Ma *et al.* (2007) |
| Shoot growth | | Dry matter production | Osmotic stress and ionic tolerance | Bread wheat | 2A, 4B, 5A, 5B, 6A, 6D, 7A | 2, 4, 5, 5, 6, 6, 7 | Genc *et al.* (2010) |
| Photosynthesis | *ERA1*, *PP2C*, *AAPK*, *PKS3* | Chlorophyll content | Ionic and osmotic stress tolerance | Bread wheat | 3D, 7A | 3, 7 | Ma *et al.* (2007) |
| Photosynthesis | *ERA1*, *PP2C*, *AAPK*, *PKS3* | Chlorophyll content | Ionic and osmotic stress tolerance | Bread wheat | 5B | 5 | Genc *et al.* (2010) |
| Shoot $Na^+$ accumulation | *HKT* (*Nax2*, *SKC1*), *SOS1* | Shoot $Na^+$ accumulation | Ionic tolerance | Durum wheat | 2A | 2 | Lindsay *et al.* (2004) |
| Shoot $Na^+$ accumulation | *HKT* (*Nax2*, *SKC1*), *SOS1* | Shoot $Na^+$ accumulation | Ionic tolerance | Bread wheat | 7A | 7 | Shavrukov *et al.* (2011) |
| Shoot $Na^+$ accumulation | *HKT* (*Nax2*, *SKC1*), *SOS1* | Shoot $Na^+$ accumulation | Ionic tolerance | Bread wheat | 2A, 2B, 7A | 2, 2, 7 | Genc *et al.* (2010) |
| Shoot $Na^+$ accumulation | *HKT* (*Nax2*, *SKC1*), *SOS1* | Shoot $Na^+$ accumulation | Ionic tolerance | Bread wheat | 2A, 6A | 2, 6 | Genc *et al.* (2013) |

Salt stress affects many aspects of plant growth such as biomass production, yield, photosynthesis and leaf metabolites (Chinnusamy *et al.*, 2006; Munns and Tester, 2008; Munns and Gilliham, 2015; Negrão *et al.*, 2017). Hence, there are many traits that could be recorded and analysed to accurately assess salinity tolerance of wheat. Total plant fresh mass, a measure of growth maintenance during salt stress, is the trait driving most of the variation across accessions. Growth maintenance has been widely acknowledged to be a good estimate of salinity tolerance (Genc *et al.*, 2007; Negrão *et al.*, 2017), especially at the seedling stage, since it is not possible to measure commercially relevant traits such as yield in young plants. Maintenance of growth, defined by an increase in mass, is one of the most important mechanisms contributing to salinity tolerance. According to Munns and Termaat (1986), leaf growth is more affected by salinity than root growth.

Molecular breeding, or MAS, refers to the technique of using DNA markers that are tightly linked to phenotypic traits to assist in a selection scheme for a particular breeding objective. Molecular markers have been extensively used for gene mapping and gene introgression through MAS to improve salinity tolerance in wheat. However, limited progress has been made in developing salt-tolerant cultivars via MAS approaches (Ashraf and Foolad, 2013). Arguably, the reasons for the limited progress in improving salt tolerance in wheat include: (i) the quantitative and physiologically complex nature of salt tolerance with an array of interacting mechanisms; and (ii) the inherent variability associated with environmental cues that impact salt tolerance and results from a combination of factors influencing ion exclusion, tissue tolerance to elevated levels of $Na^+$ and $Cl^-$, as well as osmotic adjustment. The complexity of salt tolerance is underscored by the fact that quantitative trait loci (QTLs) for salt tolerance have reportedly been found in at least 16 of the 21 wheat chromosomes. Among the different wheat species, it has been suggested that the D genome carries the gene *Kna1* that controls the relative concentrations of $K^+$ and $Na^+$ in shoots of plants grown in saline hydroponic culture (Wyn Jones, 1984; Gorham *et al.*, 1987, 1990).

## 3.6 Conclusion and Perspectives

Further improvement of salt tolerance in wheat and other important food crops will be critical to maintain production gains in lands that are progressively deteriorating, bringing marginal lands into production and keeping pace with increasing food demand. More diversity studies focused on screening are needed within the cultivated and wild relatives of both bread and durum wheats. This will help to identify additional loci and beneficial alleles associated with salt tolerance and $Na^+$ exclusion, as well as aid dissection of the salt tolerance phenotype into more 'defined/measurable' other important traits, which can then be combined in new genotypes using molecular breeding tools to improve salt tolerance of bread and durum wheats and deliver improved crop varieties adapted to saline conditions. The challenge for breeders will be to efficiently integrate trait-based/physiological and molecular methods to increase yield in dry and saline environments.

## Acknowledgement

This work was funded by grants provided by the Tunisian Ministry of Higher Education and Scientific Research-Tunisia.

## References

Al-Doss, A.A. and Smith, S. (1998) Registration of AZ-97MEC and AZ-97MEC-ST very nondormant alfalfa germplasm pools with increased shoot weight and differential response to saline irrigation. *Crop Science* 38(2), 568–568.

Araus, J.L. and Cairns, J.E. (2014) Field high-throughput phenotyping: the new crop breeding frontier. *Trends in Plant Science* 19(1), 52–61.

Arzani, A. and Ashraf, M. (2016) Smart engineering of genetic resources for enhanced salinity tolerance in crop plants. *Critical Reviews in Plant Sciences* 35(3), 146–189.

Ashraf, M. (1994) Genetic variation for salinity tolerance in spring wheat. *Heriditas* 120, 99–104.

Ashraf, M. (2002) Exploitation of genetic variation for improvement of salt tolerance in spring wheat. In: Ahmad, R. and Malik, K.A. (eds) *Prospects for Saline Agriculture*. Kluwer Academic Publishers, Dordrecht, the Netherlands, pp. 113–121.

Ashraf, M. and Akram, N.A. (2009) Improving salinity tolerance of plants through conventional breeding and genetic engineering: an analytical comparison. *Biotechnology Advances* 27(6), 744–752.

Ashraf, M. and Foolad, M.R. (2013) Crop breeding for salt tolerance in the era of molecular markers and marker-assisted selection. *Plant Breeding* 132(1), 10–20.

Ashraf, M. and O'Leary, J. (1996) Responses of some newly developed salt-tolerant genotypes of spring wheat to salt stress: 1. Yield components and ion distribution. *Journal of Agronomy and Crop Science* 176(2), 91–101.

Berger, B., de Regt, B. and Tester, M. (2012) Trait dissection of salinity tolerance with plant phenomics. In: Shabala, S. and Cuin, T. (eds) *Plant Salt Tolerance*. Methods in Molecular Biology (Methods and Protocols), Vol. 913. Humana Press, Totowa, New Jersey, pp. 399–413.

Chen, S.Y., Xia, G.M., Quan, T.Y., Xiang, F.N., Yan, J. and Chen, H.M. (2004) Introgression of salt-tolerance from somatic hybrids between common wheat and *Thinopyrum ponticum*. *Plant Science* 167, 773–779.

Chinnusamy, V., Zhu, J. and Zhu, J.-K. (2006) Salt stress signaling and mechanisms of plant salt tolerance. In: Setlow, J.K. (ed.) *Genetic Engineering*. Genetic Engineering: Principles and Methods, Vol. 27. Springer, Boston, Massachusetts, pp. 141–177.

Colmer, T.D., Munns, R. and Flowers, T.J. (2005) Improving salt tolerance of wheat and barley: future prospects. *Australian Journal of Experimental Agriculture* 45(11), 1425–1443.

Colmer, T.D., Flowers, T.J. and Munns, R. (2006) Use of wide crosses and wild relatives to improve salt tolerance of wheat. *Journal of Experimental Botany* 57, 1059–1078.

Courtois, B., Audebert, A., Dardou, A., Roques, S., Ghneim-Herrera, T. *et al.* (2013) Genome-wide association mapping of root traits in a japonica rice panel. *PLoS One* 8(11), e78037.

De Diego, N., Fürst, T., Humplík, J.F., Ugena, L., Podlešáková, K. and Spíchal, L. (2017) An automated method for high-throughput screening of *Arabidopsis* rosette growth in multi-well plates and its validation in stress conditions. *Frontiers in Plant Science* 8, 1702.

Desta, Z.A. and Ortiz, R. (2014) Genomic selection: genome-wide prediction in plant improvement. *Trends in Plant Science* 19(9), 592–601.

Dobrenz, A. (1999) *Salt-Tolerant Alfalfa*. Agripro Seeds, Inc., Shawnee Mission, Kansas, pp. 83–86.

Dobrenz, A.K., Robinson, D.L., Smith, S.E. and Poteet, D.C. (1989) Registration of AZ-GERM SALT-II nondormant alfalfa germplasm. *Crop Science* 29(2), 493.

Fahlgren, N., Gehan, M.A. and Baxter, I. (2015) Lights, camera, action: high-throughput plant phenotyping is ready for a close-up. *Current Opinion in Plant Biology* 24, 93–99.

FAO (2017) La production mondiale de céréales vers un nouveau record. Food and Agriculture Organization of the United Nations, Rome. Available at: http://www.fao.org (accessed 12 September 2017).

Farooq, S. (2004) Salt tolerance in *Aegilops* species: a success story from research and production to large-scale utilization of salt tolerant wheats. In: Taha, F.K., Ismail, S. and Jaradat, A. (eds) *Prospects of Saline Agriculture in the Arabian Peninsula*. Amherst Scientific Publishers, Amherst, Massachusetts, pp. 121–134.

Farooq, S., Asghar, M., Iqbal, N., Askari, E., Arif, M. and Shah, T.M. (1995) Production of salt-tolerant wheat germplasm through crossing cultivated wheat with *Aegilops cylindrica* – II. Field evaluation of salt-tolerant germplasm. *Cereal Research Communications* 23(3), 275–282.

Genc, Y., McDonald, G.K. and Tester, M. (2007) Reassessment of tissue $Na^+$ concentration as a criterion for salinity tolerance in bread wheat. *Plant, Cell & Environment* 30(11), 1486–1498.

Genc, Y., Oldach, K., Verbyla, A.P., Lott, G., Hassan, M. *et al.* (2010) Sodium exclusion QTL associated with improved seedling growth in bread wheat under salinity stress. *Theoretical and Applied Genetics* 121, 877–894.

Genc, Y., Oldach, K., Gogel, B., Wallwork, H., McDonald, G. and Smith, A. (2013) Quantitative trait loci for agronomic and physiological traits for a bread wheat population grown in environments with a range of salinity levels. *Molecular Breeding* 32, 39–59.

Gorham, J., Hardy, C., Wyn Jones, R.G., Joppa, L.R. and Law, C.N. (1987) Chromosomal location of a K/Na discrimination character in the D genome of wheat. *Theoretical and Applied Genetics* 74(5), 584–588.

Gorham, J., Jones, R.W. and Bristol, A. (1990) Partial characterization of the trait for enhanced $K^+$–$Na^+$ discrimination in the D genome of wheat. *Planta* 180(4), 590–597.

Hairmansis, A., Berger, B., Tester, M. and Roy, S.J. (2014) Image-based phenotyping for non-destructive screening of different salinity tolerance traits in rice. *Rice* 7(1), 1–10.
Hollington, P.A. (2000) Technological breakthroughs in screening/breeding wheat varieties for salt tolerance. In: Gupta, S.K., Sharma, S.K. and Tyagi, N.K. (eds) *Proceedings of the National Conference 'Salinity Management in Agriculture', December 1998*. Central Soil Salinity Research Institute, Karnal, India, pp. 273–289.
Hollington, P.A., Royo, A., Miller, T.E., Quarrie, S.A., Mahmood, A. and Aragues, R. (1994) The use of doubled haploid breeding techniques to develop wheat varieties for saline areas. In: *Proceedings of the 3rd Congress of the European Society of Agronomy*. Elsevier, Amsterdam, pp. 156–157.
Huang, S., Spielmeyer, W., Lagudah, E.S., James, R.A., Platten, J.D., Dennis, E.S. and Munns, R. (2006) A sodium transporter (HKT7) is a candidate for *Nax1*, a gene for salt tolerance in durum wheat. *Plant Physiology* 142(4), 1718–1727.
Huang, S., Spielmeyer, W., Lagudah, E.S. and Munns, R. (2008) Comparative mapping of HKT genes in wheat, barley and rice, key determinants of $Na^+$ transport and salt tolerance. *Journal of Experimental Botany* 59, 927–937.
Ismail, A.M. and Horie, T. (2017) Genomics, physiology, and molecular breeding approaches for improving salt tolerance. *Annual Review of Plant Biology* 68, 405–434.
Jafari-Shabestari, J., Corke, H., and Qualset, C.O. (1995) Field evaluation of tolerance to salinity stress in Iranian hexaploid wheat landrace accessions. *Genetic Resources and Crop Evaluation* 42, 147–156.
Johnson, D.W., Smith, S.E. and Dobrenz, A.K. (1991) Registration of AZ-90NDC-ST nondormant alfalfa germplasm with improved forage yield in saline environments. *Crop Science* 31(4), 1098–1099.
Lobell, D.B., Schlenker, W. and Costa-Roberts, J. (2011) Climate trends and global crop production since 1980. *Science* 333(6042), 616–620.
Lindsay, M.P., Lagudah, E.S., Hare, R.A. and Munns, R. (2004) A locus for sodium exclusion (*Nax1*), a trait for salt tolerance, mapped in durum wheat. *Functional Plant Biology* 31, 1105–1114.
Ma, I., Zhou, E., Huo, N., Zhow, R., Wang, G. and Jia, J. (2007) Genetic analysis of salt tolerance in a recombinant inbred population of wheat (*Triticum aestivum* L.). *Euphytica* 153, 109–117.
Maas, E.V. and Hoffman, G.J. (1977) Crop salt tolerance – current assessment. *Journal of the Irrigation and Drainage Division* 103(2), 115–134.
Maas, E.V., Lesch, S.M., François, E.L. and Grieve, C.M. (1994) Tiller development in salt-stressed wheat. *Crop Science* 34(6), 1594–1603.
Mujeeb-Kazi, A. and Diaz de Leon, J.L. (2002) Conventional and alien genetic diversity for salt tolerant wheats: focus on current status and new germplasm development. In: Ahmad, R. and Malik, K.A. (eds) *Prospects for Saline Agriculture*, Vol. 37. Kluwer Academic Publishers, Dordrecht, the Netherlands, pp. 69–82.
Mujeeb-Kazi, A., Munns, R., Rasheed, A., Ogbonnaya, F.C., Ali, N. *et al.* (2019) Breeding strategies for structuring salinity tolerance in wheat. *Advances in Agronomy* 155, 121–187.
Munns, R. (2002) Comparative physiology of salt and water stress. *Plant, Cell & Environment* 25(2), 239–250.
Munns, R. (2005) Genes and salt tolerance: bringing them together. *New Phytologist* 167(3), 645–663.
Munns, R. and Gilliham, M. (2015) Salinity tolerance of crops – what is the cost? *New Phytologist* 208(3), 668–673.
Munns, R. and Richards, R. (2007) Recent advances in breeding wheat for drought and salt stresses. In: Jenks, M.A., Hasegawa, P.M. and Mohan Jain, S. (eds) *Advances in Molecular Breeding Toward Drought and Salt Tolerant Crops*. Springer, Dordrecht, the Netherlands, pp. 565–585.
Munns, R. and Termaat, A. (1986) Whole-plant responses to salinity. *Australian Journal of Plant Physiology* 13(1), 143–160.
Munns, R. and Tester, M. (2008) Mechanisms of salinity tolerance. *Annual Review of Plant Biology* 59, 651–681.
Munns, R., Hare, R.A., James, R.A. and Rebetzke, G.J. (2000) Genetic variation for improving the salt tolerance of durum wheat. *Australian Journal of Agricultural Research* 51(1), 69–74.
Munns, R., James, R.A. and Läuchli, A. (2006) Approaches to increasing the salt tolerance of wheat and other cereals. *Journal of Experimental Botany* 57(5), 1025–1043.
Munns, R., James, R.A., Islam, A.K.M.R. and Colmer, T.D. (2011) *Hordeum marinum*–wheat amphiploids maintain higher leaf $K^+$:$Na^+$ and suffer less leaf injury than wheat parents in saline conditions. *Plant and Soil* 348(1–2), 365.
Munns, R., James, R.A., Xu, B., Athman, A., Conn, S.J. *et al.* (2012) Wheat grain yield on saline soils is improved by an ancestral $Na^+$ transporter gene. *Nature Biotechnology* 30(4), 360.

Munns, R., James, R.A., Gilliham, M., Flowers, T.J. and Colmer, T.D. (2016) Tissue tolerance: an essential but elusive trait for salt-tolerant crops. *Functional Plant Biology* 43(12), 1103–1113.
Negrão, S., Schmöckel, S. and Tester, M. (2017) Evaluating physiological responses of plants to salinity stress. *Annals of Botany* 119(1), 1–11.
Qadir, M., Quillérou, E., Nangia, V., Murtaza, G., Singh, M. *et al.* (2014) Economics of salt-induced land degradation and restoration. *Natural Resources Forum* 38(4), 282–295.
Qureshi, R., Rashid, A. and Ahmad, N. (1990) A procedure for quick screening of wheat cultivars for salt tolerance. In: El Bassam, N., Dambroth, M. and Loughman, B.C. (eds) *Genetic Aspects of Plant Mineral Nutrition*. Developments in Plant and Soil Sciences, Vol. 42. Springer, Dordrecht, the Netherlands, pp. 315–324.
Rawson, H.M., Richards, R.A. and Munns, R. (1988) An examination of selection criteria for salt tolerance in wheat, barley and triticale genotypes. *Australian Journal of Agricultural Research* 39, 759–772.
Rellán-Álvarez, R., Lobet, G., Lindner, H., Pradier, P.-L., Sebastian, J. *et al.* (2015) GLO-Roots: an imaging platform enabling multidimensional characterization of soil-grown root systems. *eLife* 4, e07597.
Rengasamy, P. (2016) Soil chemistry factors confounding crop salinity tolerance – a review. *Agronomy* 6(53), 1–11.
Richards, R.A. and Lukacs, Z. (2002) Seedling vigour in wheat – sources of variation for genetic and agronomic improvement. *Australian Journal of Agricultural Research* 53(1), 41–50.
Richards, R.A., Dennett, C.W., Qualset, C.O., Epstein, E., Norlyn, J.D. and Winslow, M.D. (1987) Variation in yield of grain and biomass in wheat, barley, and triticale in a salt-affected field. *Field Crops Research* 15, 277–287.
Schachtman, D.P., Lagudah, E.S. and Munns, R. (1992) The expression of salt tolerance from *Triticum tauschii* in hexaploid wheat. *Theoretical Applied Genetics* 84, 714–719.
Shavrukov, Y., Shamaya, N., Baho, M., Edwards, J., Ramsey, S. *et al.* (2011) Salinity tolerance and $Na^+$ exclusion in wheat: variability, genetics, mapping population and QTL analysis. *Czech Journal of Genetics and Plant Breeding* 47, 85–93.
Tardieu, F., Cabrera-Bosquet, L., Pridmore, T. and Bennett, M.J. (2017) Plant phenomics, from sensors to knowledge. *Current Biology* 27(15), R770–R783.
Topp, C.N., Iyer-Pascuzzi, A.S., Anderson, J.T., Lee, C.-R., Zurek, P.R. *et al.* (2013) 3D phenotyping and quantitative trait locus mapping identify core regions of the rice genome controlling root architecture. *Proceedings of the Natural Academic Sciences USA* 110(18), E1695–E1704.
Wang, M. and Xia, G. (2018) The landscape of molecular mechanisms for salt tolerance in wheat. *The Crop Journal* 6(1), 42–47.
Wang, R.R.-C., Li, X.-M. Hu, Z.-M., Zhang, J.-Y., Larson, S.R. *et al.* (2003) Development of salinity-tolerant wheat recombinant lines from a wheat disomic addition line carrying a *Thinopyrum junceum* chromosome. *International Journal of Plant Sciences* 164(1), 25–33.
Wyn Jones, R. (1984) Organic and inorganic solute contents as selection criteria for salt tolerance in the Triticeae. In: Staple, R.C. and Toennienssen, G.A. (eds) *Salinity Tolerance in Plants, Strategy for Crop Improvement*. Wiley, New York, pp. 189–203.
Zeng, L. and Shannon, M.C. (2000) Salinity effects on seedling growth and yield components of rice. *Crop Science* 40(4), 996–1003.

# 4 Genomics and Molecular Physiology for Improvement of Drought Tolerance in Wheat

**Sindhu Sareen[1]*, Pawan Saini[2], Charan Singh[1], Pradeep Kumar[1] and Sonia Sheoran[1]**
[1]*ICAR–Indian Institute of Wheat and Barley Research (IIWBR), Karnal, India;*
[2]*CSB–Central Sericultural Research and Training Institute (CSR&TI), Pampore, India*

## 4.1 Introduction

Global warming is responsible for frequent drought spells worldwide due to decreased precipitation and increased evaporation. The increase in temperature impacts water resources directly leading to drought incidence. Further, increase in air temperature, radiation stress, high levels of $CO_2$ and increase in the amount of greenhouse gases increase the frequency of drought and heat stress (Nguyen, 2002; Atlin *et al.*, 2017; Cassia *et al.*, 2018). The assumptions of climate change are becoming reality and it is predicted that there will be a 3–5°C increase in temperature and a 4–27% decrease in annual precipitation in different parts of world (Flato *et al.*, 2013). The frequency of global drought is projected to expand by beyond 20% by the end of this century, especially in regions like South America and Central and Western Europe (Prudhomme *et al.*, 2014). Similarly, water-stressed areas will increase all over the world by 2030, affecting 50 countries and about three billion people (Postel, 1999; Graham and Vance, 2003). Most crops are vulnerable to drought and suffer 50% or more losses in yield (Bray *et al.*, 2000). As about 80% of the global cereal cultivation is rainfed, the frequent occurrence of droughts will affect worldwide food security (Rosegrant *et al.*, 2002; Alexandratos and Bruinsma, 2012). Recurrent droughts and projections of future increases in dry spells and temperature have been reported from parts of Australia, Russia, China and Africa (Challinor *et al.*, 2010; Dronin and Kirilenko, 2011; Turner *et al.*, 2011; Hossain *et al.*, 2012; Spinoni *et al.*, 2014). The reduction in world wheat production of 5.5% since 1980 (Lobell *et al.*, 2011) has been attributed to recurring drought events (Li *et al.*, 2009; Mwadzingeni *et al.*, 2016a). In 2013, worldwide, approximately 65 million hectares of wheat production were affected by drought stress (FAO, 2013). On the other hand, bread wheat, which is the world's third most cultivated cereal crop covering >20% of area and providing 20% of calories globally (Godfray *et al.*, 2010), needs a 70% increase in production by 2050 to meet demand (Ray *et al.*, 2012).

*Corresponding author: sindhu.sareen@icar.gov.in


DOI: 10.1079/9781789245431.0004

## 4.2 Understanding the Complexity of Drought Tolerance in Wheat

The abiotic stresses, namely drought, salinity, excessive rainfall and high temperature, affect the development and growth of the plant negatively through metabolic and physiological changes. Drought is a complex phenomenon characterized by an intricate interplay of limited water availability, lower rainfall, reducing groundwater level and increasing temperature. Measuring and developing drought stress-tolerant crop plants by conventional breeding methods are difficult due to low heritability and complex environmental interactions (Asfaw *et al.*, 2012; Cabrera-Bosquet *et al.*, 2012; Hill *et al.*, 2013; Jha *et al.*, 2014). Plants respond to drought or water stress in a very complex fashion that includes various adaptation strategies; molecular, physiological and biochemical alterations. Water stress in plants is ameliorated by increasing water uptake and decreasing stomatal conductance (Jones, 1999). Drought is a complex trait; plant responses to drought are affected by various factors including growth conditions, physiology, genotype, developmental stage, drought severity and duration.

### 4.2.1 Morphological response

Morphological traits associated with drought tolerance in wheat affect stress tolerance under limited-moisture conditions and can be an indicator for plant adaption to water stress (Nouri-Ganbalani *et al.*, 2009; Anjum *et al.*, 2011). These traits have a major role in determining yield components and are also used in breeding programmes for enhancing grain yield of wheat varieties (Mollasadeghi *et al.*, 2011). Further, the green tissues above the flag leaf node are the main contributor to the synthesis and production of carbohydrates required to fill the grains (Kaul, 1974; Briggs and Aytenfisu, 1980). There are reports that the flag leaves contribute around 40% of the carbohydrates at the time of grain filling (Sharma *et al.*, 2003). The relationship of flag leaf with grain yield ranges from strong (Singh and Singh, 1992) to non-existent (McNeal and Berg, 1977) or dependent on the environment (Villegas *et al.*, 2007). The stress susceptibility index (SSI) used to quantify drought tolerance by Fischer and Maurer (1978) is based on grain yield losses under stressed versus non-stressed environments. This index has been used in bread wheat (Cedola *et al.*, 1994) and durum wheat (Mohammadi, 2012). In order to get varieties with high grain yield and stable performance, morphological traits are known to be useful tools to select for low SSI. Abdelali *et al.* (2019) reported that water regime significantly decreases grain yield per plant and that all morphological traits above flag leaf increase the grain yield per plant, days from sowing to anthesis and grain-filling period. In contrast, decrease in flag leaf length, area and peduncle length are observed and it has been reported that varieties with longer flag leaf and peduncle would have more drought resistance. Therefore, these traits merit attention in breeding programmes and can be used as selection criteria for higher grain yield under rainfed conditions.

Leaf waxiness and trichome density provide longer and lasting protection against water deficit stress (Richards *et al.*, 1986; Vogelmann, 1993; Grammatikopoulos and Manetas, 1994; Save *et al.*, 2000) by decreasing the amount of water lost from the plant to the environment. Most plants will close their stomata to reduce the transpiration rate under water deficit and drought stress conditions. The pores of stomata are responsible for the maximum loss of water (95–99%) from plants in well-watered conditions (Goodwin and Jenks, 2005). Leaf waxiness minimizes water loss from the leaf after stomatal closure, resulting in tolerance to drought and increasing water-use efficiency (WUE) (Richards *et al.*, 1986; Zhang *et al.*, 2007). Some genotypes roll their leaves under stress whereas some may not show leaf rolling (Izanloo *et al.*, 2008), categorized stress tolerance and susceptible genotypes, respectively. This process, mediated by the signalling molecule abscisic acid (ABA) produced in the roots, results in limiting transpiration and slowing production of reactive oxygen species (ROS) (Rizhsky *et al.*, 2002; Mahajan and Tuteja, 2005; Monneveux *et al.*, 2006). The processes can cause DNA nicking, amino acid and protein oxidation and lipid peroxidation (Reddy *et al.*, 2004), resulting in disruption of lipid bilayer structure and reducing the function of membrane-bound enzymes (Mahajan and

Tuteja, 2005; Miller *et al.*, 2010). Extended water deficit further causes changes in/loss of turgor and reduced cell growth (Steudle, 2000). However, plants experience a range of molecular changes which enable them to better cope with the short-term effects of water deficiency and allow more rapid response time. Osmotic adjustment (OA) has been observed in crops such as barley (González *et al.*, 2008), canola (Norouzi *et al.*, 2008) and maize (Hajlaoui *et al.*, 2010), and may contribute to photosynthetic and stomatal adjustment mechanisms. Many compounds like proline, polyols (mannitol and inositol) and quaternary ammonium compounds such as glycine betaine (Wang *et al.*, 2003; Yancey, 2005; Bartels and Phillips, 2010) are produced due to the OA. These OA compounds also surround the hydration shell around delicate proteins which further prevents their degradation under osmotic stress (Galinski, 1993).

### 4.2.2 Biochemical response

Drought stress is a well-known complex trait which hampers not only the physiological activities but also the biochemical processes in wheat and other crops. It is important to understand the drought-induced responses at various phenological phases of wheat which ultimately affect the grain yield. From the biochemical perspective, when drought occurs, it results in oxidative stress. The ROS are produced via metabolic processes such as respiration and photosynthesis through mitochondria, chloroplasts and peroxisomes during evolution (Saini *et al.*, 2018). Both these aerobic processes play a crucial role in plant growth and development. ROS cause the oxidation of nucleic acids (DNA, RNA), lipids and amino acids, membrane disintegration and enzyme inactivation, which finally leads to cell death (Bhattacharjee, 2005). Several factors, like genotype of the plant, phenology phase and duration, trigger many adverse effects during drought stress. In wheat, a higher content of malondialdehyde (MDA) triggered by $H_2O_2$ results in membrane disintegration, whereas a lower level of MDA provides drought tolerance (Zhang *et al.*, 2011). As a consequence of oxidative stress, activities of the Calvin cycle are hampered because of an increased amount of $H_2O_2$ (Hasanuzzaman *et al.*, 2013), which reduces reproductive growth and development. Reduced membrane stability is an indicator of the extent of lipid peroxidation caused by ROS (Sharma *et al.*, 2017) and also activates antioxidant enzymes: superoxide dismutase (SOD), catalase (CAT), peroxidase (POX), ascorbate peroxidase (APX), monodehydroascorbate reductase (MDHAR), dehydroascorbate reductase (DHAR), glutathione reductase (GR) and guaiacol peroxidase (GPX) (Ashraf and Foolad, 2007; Singh *et al.*, 2015). Levels of enzymatic and non-enzymatic antioxidant defence mechanisms play a major role during drought stress condition at different phenological stages of wheat. Drought-tolerant genotypes showed a higher level of enzymatic and non-enzymatic antioxidant defence systems and a lower level of oxidative damage factors (Hasheminasab *et al.*, 2012). Similarly, Nikolaeva *et al.* (2010) observed that GR and APX content increased in tolerant cultivars Ballada and Belchanka whereas no increase in GR and APX was observed after day 3 and day 5 of soil drought. In another study, Ourghi (a drought-tolerant wheat genotype) exhibited highest POX activity, phenolic content and lowest membrane damage, indicating greater stomatal closure than other genotypes (Outoukarte *et al.*, 2019). Under drought conditions, genes responsible for the activity of antioxidant enzymes have been observed in drought-tolerant wheat genotypes (Dudziak *et al.*, 2019). Similar trends of ROS activity along with OA through soluble sugars, proline and increased enzymatic and non-enzymatic antioxidant activities were observed at tillering and jointing stage during water stress. Complete recovery was observed for moderately stressed plants and no recovery for severely stressed plants (Abid *et al.*, 2018). This indicates the potential of plants during drought for instant recovery upon watering during vegetative stage for maintaining yield productivity. Under drought stress condition, both hexaploid (C306 and Hira) and tetraploid wheat (HW13 and A9-30-1) genotypes showed reduction in relative water content (RWC), chlorophyll and carotenoid content, nitrate reductase activity and membrane stability, whereas proline and ABA contents increased. From this it is clear that drought stress results not only because of one physiological parameter, but as a cumulative effect of all parameters/processes (Chandrasekar *et al.*, 2000).

### 4.2.3 Physiological response

Visible symptoms and attributes contributing to drought tolerance include leaf rolling, stay-green, stomatal closure, increased root length, WUE, epicuticular wax content, photochemical quenching, photoinhibition resistance, OA and membrane stability (Collins *et al.*, 2008; Khazaei *et al.*, 2013). The physiological responses of plants to water stress include leaf wilting, a reduction in leaf area, leaf abscission and the stimulation of root growth by directing nutrients to the underground parts of the plant. The reproductive stages are more sensitive to drought as the plant's resources are utilized to support root growth. Leaf rolling reduces the transpiration rate and canopy temperature (CT) and improves WUE (Townley-Smith and Hurd, 1979; Clarke, 1986). Maintaining higher RWC under these conditions supports normal growth and yield (Lazacano-Ferrat and Lovat, 1999; Costa Franca *et al.*, 2000; Beltrano *et al.*, 2006). The roots which are in direct contact with the soil feel immediate effects of reduced soil moisture content. Inhibition in the development of lateral roots and elongation of the main root are adaptive mechanisms to the moisture stress (Xiong *et al.*, 2006).

Drought has a higher impact on the photosynthetic apparatus, as it disrupts all major components which include electron transport, the carbon reduction cycle and stomatal control of the $CO_2$ supply, and also increases the accumulation of carbohydrates, peroxidative destruction of lipids and water imbalance. The limitation in $CO_2$ diffusion due to stomatal closure, reduced activity of photosynthetic enzymes and biochemical components and decreased photochemical efficiency of photosystem II (PS II) are the major effects on photosynthetic rate (Pandey and Shukla, 2015). The decreased photosynthetic activity in wheat under drought stress due to stomatal or non-stomatal mechanisms is on record (Ahmadi, 1998; Del Blanco *et al.*, 2000). As stomatal closure is one of the first responses to avoid water loss, $CO_2$ availability is limited leading to decrease in photosynthetic carbon assimilation and increase in photorespiration. Downregulation of Rubisco (ribulose-1,5-bisphosphate carboxylase-oxygenase) large subunit has been observed in drought-susceptible wheat genotypes (Bota *et al.*, 2004; Demirevska *et al.*, 2009). The probable reasons assigned to metabolic distortions of photosynthetic activity are imbalance between light capture and its utilization (Kingston-Smith and Foyer, 2000), decrease in Rubisco activity, loss of chloroplast membranes (Amirjani and Mahdiyeh, 2013), degradation of chloroplast structure and photosynthetic apparatus, chlorophyll photo-oxidation, destruction of chlorophyll substrate, inhibition of chlorophyll biosynthesis and increase of chlorophyllase activity (Kabiri *et al.*, 2014). Compared with stomatal limitations, the metabolic distortions are more complex (Rama *et al.*, 2014).

The decrease in chlorophyll content under drought stress is the result of pigment photo-oxidation and chlorophyll degradation. Both chlorophyll a and chlorophyll b are sensitive to soil moisture stress (Farooq *et al.*, 2009). Genotypes with high chlorophyll content gave higher seed yield under stress (Alaei, 2011). Positive correlations of grain yield in wheat with chlorophyll content, grain-filling period and the number of grains per spike were reported (Kilic and Yagbasanlar, 2010). The total chlorophyll at pre- and post-anthesis stages and stable photosynthetic rate were high in tolerant wheat cultivars (Dawood *et al.*, 2019). Water content, RWC, water loss rate, excised-leaf water retention (ELWR) and residual transpiration rate are some important characteristics that influence plant water relations. Leaf relative water content (LRWC), leaf water potential (LWP), stomatal conductance and rate of transpiration are influenced by leaf and CT. Drought and high temperature stress lead to higher vapour pressure deficits (VPD) and higher evapotranspiration. During reproductive and grain-filling phases, water is required for stem and peduncle elongation; spike and floret development; fertilization; grain growth and filling. Water moves through membranes facilitated by aquaporins. High temperature increases the hydraulic conductivity of membranes and plant tissues due to increased aquaporin activity, membrane fluidity and permeability (Martínez-Ballesta *et al.*, 2009). However, increased permeability of membranes may cause the developing organs to dehydrate. With increase in the rate of transpiration there is an increase in the pH of leaf sap, which promotes ABA accumulation leading to reduction in stomatal conductance. But increased cytokinin concentration in the xylem sap favours stomatal opening, decreasing the sensitivity of stomata towards ABA (Wilkinson and Davies, 2002). RWC is a measure of plant water status, reflecting the metabolic activity in tissues, and is used as the most meaningful index for dehydration tolerance. The change in water loss in terms of excised-leaf water loss may estimate the plant's leaf water relations, especially when comparing fully hydrated leaves with those under deficit irrigation, which gives an insight on cuticular thickness and cuticular transpiration (Clarke and McCaig, 1982). It reflects the balance between water supply to the leaf and transpiration rate. The drought-tolerant genotypes have less excised-leaf water loss and evapotranspiration water losses, and conserve their water content (Izanloo *et al.*, 2008). The enhanced ELWR, which is attributed to leaf rolling and reduction in exposed leaf area, could be a superior indirect selection criterion for drought tolerance leading to higher grain yield (Lonbani and Arzani, 2011). A significant positive correlation between RWC and grain yield under drought stress during the reproductive stages in wheat and barley is reported. Epicuticular waxes, the organic compounds of the cuticle, cover the plant surface to control the water loss through epidermal conductance (Ahmed *et al.*, 2012). These studies revealed that water stress results in damage to the oxygen-evolving complex of PS II (Scotnica *et al.*, 2000) and to the PS II reaction centres associated with the degradation of D1 protein (Murata *et al.*, 2007; Zlatev, 2009). When photosynthesis decreases, the light excitation energy is in excess, which causes photo-oxidative damage through accumulation of ROS and oxidative stress (Chaves *et al.*, 2009; Nishiyama *et al.*, 2011).

ABA production can affect drought adaptation through both dehydration avoidance and dehydration tolerance (Thompson *et al.*, 2007). ABA is known as a major chemical root-to-shoot stress signal (Schachtman and Goodger, 2018) which induces leaf expansion inhibition and stomatal closure. Moreover, ABA in wheat acts as a promotor for root growth which has a significant correlation with yield under drought stress (Xu *et al.*, 2013). It regulates tissue water content through stomatal oscillations and induces the expression of genes encoding proteins that control cellular dehydration tolerance (Zhu, 2002). Under drought conditions, ABA is synthesized in xylem and then transported to reproductive organs where it influences grain filling by modulating the expression of genes involved in carbohydrate metabolism and cell division. Accumulation of ABA in leaves and stem or root exudates increases with simultaneous reductions in leaf cytokinin content (Yang *et al.*, 2004). Reduced ethylene and 1-aminocyclopropane-1-carboxylic acid (ACC) concentrations and increased ABA concentration in developing wheat grains under mild drought increased the grain-filling rate. However, under severe drought, ethylene, ACC and ABA concentrations reached very high levels to maintain the grain-filling rate (Yang *et al.*, 2007).

### 4.2.4 Molecular response

Breeding for drought tolerance is complicated, largely due to the quantitative nature and significant genotype-by-environment interactions associated with drought tolerance. An alternative strategy, involving selection for less complex physiological traits that are associated with yield in water stress environments (leaf rolling, stay-green, canopy temperature depression (CTD), OA, water-soluble carbohydrates (WSC), deep roots and glaucousness), has been suggested as a more useful approach (Gupta *et al.*, 2017). Quantitative trait loci (QTLs) in large number have already been reported in wheat for several traits including coleoptile length, carbon isotope discrimination (CID) or $\Delta^{13}C$, WSC, root system, grain yield and related traits recorded under water stress (Gupta *et al.*, 2012).

Recent advancements in wheat genomics has deciphered many potential key genes, genomic regions and transcription regulators associated with morpho-physiological traits. Several approaches such as QTL mapping, genome-wide association mapping (GWAM) and marker-assisted backcross breeding (MABB) from wild gene pools are being used to identify new genes for genetic improvement of wheat under drought conditions (Mwadzingeni *et al.*, 2017). For drought and heat tolerance, meta-QTL (MQTL) analysis in wheat identified several QTLs for physiological traits, namely photosynthesis (17 MQTLs), soluble carbohydrates (24 MQTLs), water status (18 MQTLs), CID (nine MQTLs), CT (ten MQTLs), coleoptile vigour

(18 MQTLs) and stay-green (34 MQTLs) (Acuna-Galindo *et al.*, 2015). Several drought-responsive genes have been identified using RNA-seq technology (Liu *et al.*, 2015). Another RNA-seq study identified 309 differentially expressed genes in wheat that are involved in many critical processes including floral development, photosynthetic activity and stomatal development during reproductive stages in drought under field conditions (Ma *et al.*, 2017). A number of genomic regions/QTLs have been identified for yield-contributing traits, but a limited number of studies have focused on physiological traits such as leaf green area, leaf water content and WSC. Some of the efficient physiological traits associated with drought stress tolerance are examined next.

### *Photosynthesis process*

Photosynthesis consists of several components, including the photosystems and photosynthetic pigments, the electron transport system and $CO_2$ reduction pathways. It has been reported that yield progress is associated with increase in photosynthesis in terms of greater radiation-use efficiency (RUE) and increased stomatal conductance in wheat. In a recent association mapping study, common single-nucleotide polymorphism (SNP) markers have been identified for grain yield, biomass and RUE on chromosomes 5A and 7A using a 35K wheat breeders' array (Molero *et al.*, 2019). Under abiotic stress (heat, drought and salinity), photosynthetic efficiency and transpiration rates are decreased (Zandalinas *et al.*, 2016). Decrease in net photosynthetic rate is associated with stomatal closure, resulting in increased WUE (net $CO_2$ assimilation rate/transpiration) (Ruggiero *et al.*, 2017). Several QTLs for physiological traits such as ground cover (GC), chlorophyll content, normalized difference in vegetation index (NDVI) and CT have been identified using bi-parental populations under drought conditions in wheat (Pinto and Reynolds, 2015). In a recent study using high-density Illumina Select 90K SNPs, 24 QTLs for GC, NDVI and CTD were identified on 12 chromosomes, explaining 3.4–14.6% of the phenotypic variance. Out of these, five stable QTLs, namely *QGC-W.caas-7AL*, *QNDVI-S.caas-7AL*, *QGC-S.caas-3AS*, *QCTD-A.caas-5BS* and *QCTD-10.caas-5BS*, were identified in three environments for the physiological traits in the mapping population (Gao *et al.*, 2016). Glaucousness has been found responsible for providing tolerance against abiotic stresses, in particular to heat and drought stress. Previous studies in wheat reported two loci for *Wax production*, termed *W1* (chromosome 2B) and *W2* (chromosome 2D), and two *Inhibitor of wax* loci, *Iw1* (chromosome 2B) and *Iw2* (chromosome 2D), that inhibit glaucousness (Tsunewaki and Ebana, 1999).

### *Water-related traits*

OA maintains water absorption and cell turgor pressure of a plant under drought. It can be a potential selection criterion for drought tolerance in wheat (Cattivelli *et al.*, 2008). It has been reported that the wheat genotypes which contribute resources in the development of deep roots are able to extract the residual moisture during water stress conditions in the field (Lopes and Reynolds, 2010). A deep root system architecture (RSA) (i.e. seminal root number; nodal root number; primary root diameter, length and weight; adventitious seminal root weight) with higher root density increases the adaptability of plants under water stress conditions. Root length, density and root depth are the major components of RSA affecting water extraction in deep soils (Carvalho *et al.*, 2014). In a durum wheat study, the deep-rooted genotypes enhanced grain size by 16 to 35% under moisture stress and had grain yield advantage of 37 to 38% under low moisture stress when compared with the shallowest root types (El Hassouni *et al.*, 2018). In terminal drought, narrow root growth angle promoting deeper root growth is often associated with improved access to water and nutrients in deep soils. In a recent genome-wide association study (GWAS), Alahmad *et al.* (2019) identified seven marker–trait associations (MTAs) on chromosome 6A for seminal root angle (SRA) and this major quantitative trait locus (*qSRA-6A*) annotates loci related to gravitropism, polar growth and hormonal signalling.

CT is a trait which can be used as surrogate for the root development analysis. In wheat, genotypes with cooler temperature in the canopy showed 30% more yield associated with an increase of 40% in root dry weight at a depth of 60–120 cm (Lopes and Reynolds, 2010).

Pinto *et al.* (2010) identified five consistent QTLs on chromosomes 1Ba, 2Ba, 3Bb, 4Aa and 7Aa for cooler canopy explaining phenotypic variance from 7 to 14% under drought and heat stresses, respectively. The involvement of roots was inferred since cooler canopies are a result of higher transpiration rates which require sufficient access to water.

CID, or $\Delta^{13}C$, is a heritable trait and is under genetic control. $\Delta^{13}C$ in the kernel has been identified as a potential criterion for selecting high-WUE genotypes (Condon *et al.*, 2004). Only a few studies have been carried out to dissect the genetic control of this important trait. In an association mapping study, Mora *et al.* (2015) identified chromosomes 1A, 3A, 4A and 5A as being associated with the $\Delta^{13}C$ trait, which could be used as an indirect tool for increasing WUE in wheat.

Besides being a major contributor of carbon resources for grain yield, WSC also contributes to osmotic regulation as the osmolyte. WSC mobilization under well-watered conditions contributes 10–20% of grain yield and up to 30–50% of grain dry matter under drought stress during the grain-filling stage (Rebetzke *et al.*, 2008). GWAS detected associations of markers on chromosomes 1A, 1B, 1D, 2D, and 4A with WSC measured in water-limited conditions (Ovenden *et al.*, 2017). Two KASP (competitive allele-specific PCR) markers were also developed for the *TaSST-D1* and *TaSST-A1* genes underpinning WSC, out of which *TaSST-A1* showed good association with high WSC content in a synthetic derivatives (SYN-DER) diversity panel (Khalid *et al.*, 2019).

## 4.3 Breeding for Drought Tolerance in Wheat

Drought tolerance is a combinatorial trait that results from several genes influenced by various environmental elements (Kumar *et al.*, 2018). Selection for drought tolerance is hampered due to difficulty in precise phenotyping of associated potential traits under water-limited conditions (Reynold *et al.*, 2005). It is very difficult to screen a single trait index which indicates about both yield potential and drought tolerance (Chen *et al.*, 2012). Screening for identification of yield-contributing traits has been extensively done under drought tolerance in wheat (Mwadzingeni *et al.*, 2016b). Drought response is a morpho-physio-biochemical phenomenon wherein three mechanisms are well known: (i) drought escape; (ii) drought avoidance; and (iii) drought tolerance (Levitt, 1980). Crop plants use these mechanisms in combination to cope against water stress at particular phenological phases (Zhang *et al.*, 2012; Cheng *et al.*, 2016). These mechanisms are governed by many associated traits; morphological, physiological or biochemical (Tiwari *et al.*, 2017). During the drought escape situation, crop plants sense the occurrence of water stress and adjust the phenological phase to complete their life cycle early so that by the time water stress arrives, physiological activities will not be hampered that much. In the avoidance mechanism, crop plants attempt to avoid stress by reduction in leaf area, increased pubescence, leaf rolling, leaf reflectance, etc. Meanwhile, in the tolerance situation, the plant adjusts itself to withstand water stress through several adaptive features like maintaining RWC, deep root system, cell-membrane integrity, accumulation of osmolytes, presence of awns, assimilation of photosynthates and other associated traits. Thus, selection of the parental lines to breed for the development of improved high-yielding cultivars with drought tolerance should focus on the combination of different relevant traits rather than a single trait as selection criteria. Genetic improvement for drought tolerance of wheat can be done by introgression/incorporation of alleles with the existing allelic loci through conventional breeding, which could be further extended by the aid of molecular breeding tools. In wheat, development of high-yielding drought-adapted varieties is considered a key mitigation strategy against climate change. Further, the wild relatives and landraces offer an opportunity for utilization of untapped natural variation not only for drought, but also for other biotic and abiotic stresses through both conventional as well as genomics-assisted breeding (GAB) approaches.

### 4.3.1 Traditional breeding approaches

Concentrated efforts have been made towards the development of drought-tolerant wheat cultivars by the International Maize and Wheat

Improvement Center (CIMMYT), Mexico, through collaborative research. But the enhancement in yield is too low compared with that projected by the scientific community as needed to feed the global population by 2050. For achieving the required 70% increase in yield above that of the present scenario, the breeding material should be tested under water-limited conditions as the climate is changing at a rapid pace (Mwadzingeni *et al.*, 2016a). Identification of suitable genotypes for targeted growing environments and minimization of confounding effects of other abiotic stresses will be helpful for making effective selection for drought tolerance in wheat. For this, knowledge of phenotypic drought-associated traits is indispensable for a better understanding of drought tolerance mechanisms. Also, efficient utilization of wheat genetic resources is a prerequisite, as genetic resources are the key for unlocking the potential genetic variation left untapped due to the domestication bottleneck.

### *Utilization of wheat genetic resources for development of drought resistance*

The emergence of modern *Triticum aestivum* occurred due to agriculture. Thanks to nature for growing its ancestor (emmer) in an area with spontaneous occurrence of *Aegilops tauschii*, the interspecific hybridization that generated this species occurred (Dubcovsky and Dvorak, 2007). Domestication, centuries of cultivation and modern breeding have further restricted the genetic variability of several cultivated species, and wheat is not an exception (Reif *et al.*, 2005; Fu and Somers, 2009; Mir *et al.*, 2012; Voss-Fels *et al.*, 2015). It is important to remember that wheat was one of the first species to be domesticated and cultivated, further decreasing its variability due to constant selection cycles since then (Lev-Yadun *et al.*, 2000; Charmet, 2011; Riehl *et al.*, 2013). Among all crop species, wheat is probably the one in which most research has been invested regarding the use of wild and cultivated relatives as a source of variability for its improvement (Eduardo *et al.*, 2019). Another strategy in this field is the development of synthetic wheat, repeating the interspecific crosses that occurred in nature that led to the formation of hexaploid wheat (Liu *et al.*, 2006; Yang *et al.*, 2009). In this method, different accessions of the species *Triticum monococcum*, *Triticum turgidum* and *Ae. tauschii* can be used for the formation of new genetic constitutions of wheat, greatly increasing the genetic variability of the primary gene pool (Mujeeb-Kazi *et al.*, 2008). Numerous synthetic wheat germplasm pools have been developed by CIMMYT (Yang *et al.*, 2009). This illustrates an advantage that wheat possesses, as an allohexaploid, compared with diploid species. To produce new drought-resistant cultivars, plant breeders must identify germplasm with increased drought resistance. One opportunity is presented by the exploitation of wild germplasm of wheat. *Triticum dicoccoides* is an important source of drought-related genes and is highly suitable as a donor for improving drought resistance in cultivated wheat species (Peleg *et al.*, 2008; Peng *et al.*, 2013). Several drought tolerance genes of *T. dicoccoides* have been identified and characterized (Lucas *et al.*, 2011a,b; Kuzuoglu-Ozturk *et al.*, 2012). Interspecies variations are more profound than intraspecies variations. For example, *Ae. tauschii* is more drought tolerant than *Triticum* and wild emmer wheat (Ashraf *et al.*, 2009; Nevo and Chen, 2010). Wheat alien chromosome addition lines, created with individual chromosomes derived from wild germplasm, have been used to study the effect of individual chromosomes. By studying a set of wheat–rye disomic addition lines, Mohammadi *et al.* (2003) found that most of the genes controlling drought tolerance were located on chromosomes 7R, 5R and 3R. A wheat translocation line with an alien chromosome segment (7DL) from *Agropyron elongatum* was found to have improved water stress adaptation and higher root and shoot biomass compared with the control genotypes (Placido *et al.*, 2013). The presence of the 1RS translocation in spring cultivar 'Pavon' increased root biomass and was more tolerant to field environmental stresses than Pavon (Ehdaie *et al.*, 2003). These studies indicate that alien chromosome addition lines are promising sources of drought-related genes that could be used in wheat improvement programmes.

Prediction of grain yield in wheat could be done by measuring traits such as NDVI, CT, chlorophyll content, 1000-grain weight, spike length and number of grains per spike under stress (Babar *et al.*, 2006; Teal *et al.*, 2006; Jain *et al.*, 2014; Ramya *et al.*, 2015). There is genetic variation for drought tolerance in wheat cultivars (Rai *et al.*, 2018). The donor line HI1500 was selected on the basis of the presence of a positive allele for *Qchl/ct* associated with X*barc68-101*, *Qndvi* associated with X*gdm93* and *Qyield* associated with X*gwm165* (Jain *et al.*, 2014; Ramya *et al.*, 2015). A popular widely grown but drought-susceptible Indian bread wheat variety, HD2733, was improved by transferring QTLs linked to traits like chlorophyll content, CT, NDVI and grain yield under drought stress from a known drought-tolerant variety, HI1500, and 20 lines in $BC_2F_3$ and nine lines in $BC_1F_4$ (with an average yield of 265.39 g/plot) were selected on the basis of their superior performance against parental genotypes HD2733 (242.31 g/plot) and HI1500 (253.74 g/plot) (Rai *et al.*, 2018). Root dry weight, shoot dry weight, root length and RWC are suitable traits that can be used for screening of wheat lines for drought tolerance at early growth stage (Bansal *et al.*, 2016). Based on different morphological and physiological parameters and drought susceptibility index, IC333095, IC615005, Dharwad Dry and C306 were considered drought-tolerant genotypes (Bansal *et al.*, 2016). Plant height, grain number per spike, grain weight per spike, 1000-grain weight, grain plumpness, NDVI and CT could be recommended as indicators of drought tolerance improvement in spring wheat (Zhang *et al.*, 2019).

The membership function value of drought resistance (MFVD), which integrated the drought tolerance coefficients of different traits, could be used to evaluate drought resistance (Chen *et al.*, 2012). Through the MFVD analysis with the selected traits, *Ag. elongatum* 3E addition line was identified with super-high drought resistance; and 25 wheat addition lines were identified as high drought resistance germplasm (Liu *et al.*, 2015). These high drought-resistant wheat addition lines could be used in crossing with high-yield but drought-sensitive lines to generate lines being both high in drought resistance and yield. Two addition lines with *Aegilops peregrina* 4SV and *Ae. peregrina* 3UV chromosomes were identified with high drought resistance, and another eight lines were identified with relatively high drought resistance (Liu *et al.*, 2015). The MFVD analysis and drought resistance index (DI) under the two water regimes identified six SHW lines (SHW1, SHW3, SHW10, SHW16, SHW21 and SHW34) with higher drought resistance (Song *et al.*, 2017). The exploitation and utilization of drought-resistant germplasm is of great importance to meet the challenges of wheat production under changing climatic conditions. Previous works have indicated that plant height, number of effective tillers, grains per spike and grain weight are affected by water stress; similarly, biomass-related traits are significantly correlated with grain yield under water stress and have been considered as evaluation parameters of drought resistance (Duggan *et al.*, 2000; Blum, 2005; Nouri-Ganbalani *et al.*, 2009). Drought resistance is a quantitative character controlled by multiple genes (Shinozaki and Yamaguchi-Shinozaki, 2007). Large genetic variation exists in the primary gene pool of *Ae. tauschii*, which is an important resource for broadening the genetic diversity of common hexaploid wheat (Song *et al.*, 2017). The comprehensive evaluation of MFVD and DI showed six SHWs identified with high drought resistance (Song *et al.*, 2017).

### 4.3.2 Genomics-assisted breeding for drought tolerance in wheat

Genomics is the study of the structure and function of the entire genome of a living organism. The genome together with the environment influences the phenome of the crop plant. Therefore, selection based upon the phenome creates bias for selecting the desirable traits. But with the gene revolution, DNA-based molecular markers have become available for making selections at the gene level, which enhances the efficiency of selection of desirable genotypes. Among cereals, wheat is an important crop having a large number of genomic resources for different traits; morphological, physiological and biochemical features. GAB utilizes genomic resources with breeding practices for the development of superior cultivars with improvement in yield, quality and resistance to biotic and abiotic stresses. Assessment of correlations between genotype and phenotype for breed cultivars is the major objective of GAB. GAB includes genomics, transcriptomics, proteomics, metabolomics and ionomics approaches for identification of DNA markers associated with the trait of interest, which helps breeders to identify the phenotype of the genotype. But in wheat GAB has not contributed that much to drought tolerance. This may be due to the polygenic nature and complexity of drought tolerance. Attempts have been made by researchers to successfully utilize the variation present among the wheat genetic resources to develop improved cultivars using genomics approaches. However, the wheat genome sequence offers opportunities to wheat scientists to mine the sequences responsible for drought tolerance. The various GAB approaches such as marker-assisted breeding, genome-wide selection, association mapping, linkage mapping, transgenic technologies and genome editing provide an advantage for dissection of drought tolerance.

### *DNA markers associated with drought tolerance*

DNA-based molecular markers – such as restriction fragment length polymorphism (RFLP), amplified fragment length polymorphism (AFLP), randomly amplified polymorphic DNA (RAPD), simple sequence repeat (SSR), cleaved amplified polymorphic sequence (CAPS), sequence-characterized amplified region (SCAR), Diversity Arrays Technology (DArT) and SNP – serve as important tools for wheat breeders that enable better understanding of the complexity and dissection of the genes responsible for traits associated with drought tolerance using different kinds of marker-aided selection schemes – such as MABB, marker-assisted recurrent selection (MARS) and marker-assisted gene pyramiding (MAGP) – for the development of cultivars with drought tolerance. DNA markers are helpful in precise study of drought stress. Further, these markers could be employed for genetic tagging and mapping as DNA markers can track the presence of QTLs for drought tolerance in wheat. Also, they can be utilized for the development of drought-tolerant wheat lines. MAS is the most suitable approach for the development of drought tolerance via genetic linkage maps. RFLP markers were found to be associated with yield and other morpho-physiological traits such as LWP, CT, inhibition of chlorophyll synthesis and proline content in durum wheat for drought tolerance (Nachit *et al.*, 1993). QTLs for flag leaf senescence (FLS) have been detected and mapped on to chromosome 2D utilizing RFLP and SSR markers under normal as well as water-stressed conditions in doubled haploid (DH) populations of photoperiod-sensitive variety Beaver and photoperiod-insensitive variety Soissons (Verma *et al.*, 2004). In a DH population derived from the cross between Chinese Spring and SQ1 (a high ABA-expressing line), RFLP, AFLP and SSR markers were utilized for tagging of drought stress QTLs (Quarrie *et al.*, 2005). Twelve wheat genotypes were screened with RAPD and inter-simple sequence repeat (ISSR) markers under drought stress and polymorphic linked markers were identified that could be applied in MAS for drought tolerance in wheat (Barakat *et al.*, 2010). In another study, Milad *et al.* (2011) reported a correlation of *FLS* gene under water deficit condition with RAPD and ISSR markers. A DH population (CSDH) of 94 individuals was evaluated for additive and epistatic gene effects on total phenolic content, grain yield of the main stem, grain number per plant, 1000-grain weight and dry weight per plant for 8 years by drought combinations (moderate and severe drought stress). A total of 21 DArT markers exhibited significant additive effect and were located on chromosomes 1A, 1B, 2A, 2B, 2D, 3A, 3B, 3D, 4A and 4D. A subtle effect for total phenolic content on chromosome 4AL was noted (Czyczyło-Mysza *et al.*, 2019). Regarding drought-stress signalling, various markers have been documented. Molecular markers have also been found to be linked with transcription factors responsible for drought tolerance in wheat. Nowadays, sequence-based SNP markers have been found to be associated with transcription factors responsible for drought tolerance in wheat (Table 4.1).

### *Quantitative trait loci responsible for drought tolerance*

The locus governing the phenome of a quantitative trait in an environment is referred to as a quantitative trait locus. Drought is a complex trait as it is a combination of several morpho-physio-biochemical traits. For understanding of the genetic landscape of drought tolerance in wheat, intensive and integrative genetic, genomic and molecular research is compulsory for the assessment of genes and mechanisms underlying drought tolerance. The identification and introgression of QTLs is an appropriate approach

**Table 4.1.** DNA markers associated with drought signalling genes in wheat. (From Budak *et al*., 2015.)

| Marker | Chromosome location/Nearest marker | Drought signalling gene | Reference |
|---|---|---|---|
| SNP | 3A and 3BL | *DREB1* | Wei *et al*. (2009) |
| SNP S770 | *Xfbb117–Xmwg818* | | |
| SNP | 3D | | |
| RAPD | 3A | | Huseynova and Rustamova (2010) |
| SNP | *U16709.1* | *DREB1* | Mondini *et al*. (2012) |
| | *DQ323885.1* | *WRKY1* | |
| | *AF303376.1* | *HKT-1* | |
| | 3A | *DREB1A* | Edae *et al*. (2013) |
| | 3A, 3B, 3D | *ERA1-B* | |
| | – | *ERA1-D* | |
| | 6A | *1-FEH-A* | |
| | – | *1-FEH-B* | |
| | 5A | *TaSnRK2.8* | Zhang *et al*. (2013) |
| | – | *DREB2* and *DREB3* | Mondini *et al*. (2015) |

for improvement in yield of elite cultivars with drought tolerance (Qaseem *et al*., 2019) using MAS (Hamada *et al*., 2012). Dozens of QTLs associated with drought tolerance have been identified in wheat using bi-parental linkage mapping. For such type of mapping, evaluation of different mapping populations in different environments/seasons is a prerequisite. Different types of mapping populations, such as recombinant inbred lines (RILs), DH, chromosome substitution lines (CSSLs), backcross populations (BCs) and multi-parent advanced generation inter-cross (MAGIC), are available in wheat (Kalladan *et al*., 2013). Induction of drought during the reproductive phase severely affects the grain yield. In wheat, QTLs for drought tolerance in wheat have been identified through yield and yield-contributing traits under water stress conditions (Quarrie *et al*., 2006; Kirigwi *et al*., 2007; Salem *et al*., 2007; Yang *et al*., 2007; Maccaferri *et al*., 2008; Mathews *et al*., 2008; von Korff *et al*., 2008; McIntyre *et al*., 2009; Peleg *et al*., 2009; Pinto *et al*., 2010). For stay-green in wheat three QTLs (*QSg.bhu-1A*, *QSg.bhu-3B* and *QSg.bhu-7D*) were identified in a Chirya3/Sonalika RIL population and these QTLs explained 38.7% of the phenotypic variation (Kumar *et al*., 2010). Water deficit affects seed germination, vigour and seedling development and QTL mapping has been done for seed germination in wheat (Czyczyło-Mysza *et al*., 2014). Using a DH population of 96 individuals derived from Chinese Spring/SQ1 evaluated under drought conditions, three QTLs for stress susceptibility (SSI) on chromosomes 4B, 6B and 7A, two QTLs for mean productivity index (MP) on chromosomes 5A and 5B and one QTL each for stress tolerance index (STI), geometric mean productivity (GMP) and tolerance index (TOI) were detected. These QTLs accounted for 13–36% of the phenotypic variation (Dashti *et al*., 2007). Stomatal density and size of stomata play a crucial role in WUE during physiological processes like photosynthesis and transpiration which ultimately affect the grain yield. About 20 QTLs with additive effects and 19 QTLs having epistatic effects were detected under drought stress in a DH population of Hanxuan 10 (H10) and Lumai 14 (L14) which explained more than 10% of the phenotypic variation present on 5A and 2D chromosomes (Wang *et al*., 2016). In another study under irrigated and rainfed conditions of a DH population developed from Kukri/Excalibur at four different locations over 3 years, 32 QTLs for nine drought-responsive traits (germination percentage, days to anthesis, days to maturity, grain-filling duration, plant height, productive tillers per square metre, 1000-grain weight and grain yield per plot) were found (Gahlaut *et al*., 2017). Using composite interval mapping (CIM), ten QTLs were identified in water-limited condition for root length, shoot length and dry shoot weight in a RIL population derived from synthetic wheat (W7984)/Opata (SynOpRIL) (Khalid *et al*., 2018). Various QTLs with respect to agronomic, physiological and metabolic traits have been identified in wheat (Tables 4.2 and 4.3).

### *Genome-wide analysis of drought tolerance*

Genome-wide analysis includes GWAS and genomic selection (GS). The GWAS approach is a form of association mapping that is helpful in identification of allelic diversity of a particular trait, whereas GS is also a variant of MAS that is employed for the detection of minor QTLs with cumulative effect. Both these approaches utilize linkage disequilibrium (LD) and need appropriate statistical simulation models for precise identification of QTLs for the trait of interest. These approaches are advantageous over the bi-parental linkage mapping and MAS approaches (i.e. MABB and MARS) because of high resolution power, high precision and enhancement in genetic gain efficiency. In wheat also, these approaches showed potential for identification of QTLs for yield and yield attributes related with drought tolerance and for predicting genotype performance for breeding drought tolerance under water stress condition. The North American spring wheat diversity panel of 250 genotypes was scanned using 19,967 SNPs and identified 51 QTLs on ten chromosomes for drought resistance and yield attributes (Gizaw,

**Table 4.2.** QTLs identified for agronomic, physiological and metabolic traits in wheat using bi-parental linkage mapping. (From Sallam *et al.*, 2019.)

| Trait | Chromosome | Reference |
|---|---|---|
| **Agronomic traits** | | |
| Grain yield | 1B, 1D, 3B, 4A, 6D, 7D | Tahmasebi *et al.* (2017) |
| Grain weight per spike | 1B, 1D | Xu *et al.* (2017) |
| 1000-Grain weight | 1B, 1D, 2A, 2B, 3A, 3B, 4A, 4D, 6A, 6D, 7B, 7D | Peleg *et al.* (2009); Xu *et al.* (2017) |
| Grain number per square metre | 1B, 5A, 5B, 7D | Tahmasebi *et al.* (2017) |
| Grain number per spike | 1A, 2A, 2B, 3A, 6B | Peleg *et al.* (2009); Xu *et al.* (2017) |
| Harvest index | 1B, 2D, 4BS, 5A | Xu *et al.* (2017) |
| Spike number per plant | 1A, 2A, 2B, 2D, 4B, 5A, 7B | Xu *et al.* (2017) |
| Spikelet compactness | 1A, 1B, 2B, 5A, 5B, 6A, 6B, 7A | Peleg *et al.* (2009); Xu *et al.* (2017) |
| Spikelet number per spike | 1B, 1D, 2B, 3B, 4B, 5A, 6B, 7D | Peleg *et al.* (2009); Xu *et al.* (2017) |
| Sterile spikelet number per spike | 7A | Xu *et al.* (2017) |
| Fertile spikelet per spike | 2A | Xu *et al.* (2017) |
| Spike length | 2B, 7A, 7B | Xu *et al.* (2017) |
| Biomass | 1B | Xu *et al.* (2017) |
| Shoot biomass | 4B | Kadam *et al.* (2012) |
| Plant height | 1B, 4B, 7D | Peleg *et al.* (2009); Xu *et al.* (2017) |
| **Physiological traits** | | |
| Leaf area, growth rate, transpiration efficiency, WUE | 2A, 2D, 3A, 4B, 6A | Parent *et al.* (2015) |
| Carbon isotope ratio, osmotic potential, chlorophyll content, flag-leaf rolling index | 2B, 4A, 5A, 7B | Peleg and Blumwald (2011) |
| WSC | 1A, 1D, 2D, 4A, 6B, 7B, 7D | Yang *et al.* (2007) |
| Stomatal density index, aperture area, length; guard-cell area and length | 2B, 4AS, 5AS, 7AL, 7BL, 1BL, 4BS, 5BS, 7AS | Shahinnia *et al.* (2016) |
| Stomatal conductance, net photosynthetic rate | 5A, 6B | Xu *et al.* (2017) |
| Root length | 2D, 4B, 5D, 6B | Kadam *et al.* (2012) |
| Root biomass | 2D, 4BS | Kadam *et al.* (2012) |
| **Metabolic traits** | | |
| ABA | 1B, 2A, 3A, 4D, 5A, 6D, 7B | Iehisa *et al.* (2014) |
| Jasmonic acid, salicylic acid, ethylene | 6A | Castro *et al.* (2008) |

**Table 4.3.** Putative QTLs identified for drought tolerance in wheat. (From Mwadzingeni *et al.*, 2016a.)

| Chromosome | Trait associated with putative QTL | Mapping population | Reference |
|---|---|---|---|
| 2A | RWC, awn length, grain weight, coleoptile length, shoot length and extrusion length | Core Collection | Ahmad *et al.* (2014) |
| 1B, 4A, 4B, 7A, 7D | 1000-Grain weight | Core Collection | Nezhad *et al.* (2012) |
| 1A, 1D, 2B, 3A, 3B, 4B, 4D, 5B, 6A | Potential quantum efficiency of PS II, chlorophyll content, flag-leaf temperature, grain yield | RIL of C306/HUW206 | Kumar *et al.* (2012) |
| 1D, 2A, 2B, 2D, 3A, 4A, 4B, 5B, 5D, 6D, 7A, 7D | Root diameter, volume, surface area, crossings, forks and tips | Advanced backcross population of Devon/ Syn084 | Ibrahim *et al.* (2012) |
| 1D, 2B, 3A, 3B, 4A, 4B, 5A, 5B, 6A, 6D, 7A, 7B | Grain yield | RIL of Dharwar Dry/Sitta | Alexander *et al.* (2012) |
| 3BL | Grain yield | DH population of RAC875/Kukri | Bennett *et al.* (2012) |
| 1B, 2B, 3B, 5B, 7B, 7A | Grain weight, grain weight per spike, grain number per spike, spike per square metre, spike weight, spike harvest index and harvest index | RIL of Oste-Gata/ Massara-1 | Golabadi *et al.* (2011) |
| All chromosomes except 1D and 6A | Grain yield, number of grains per ear and chlorophyll a fluorescence | DH lines of Chinese Spring/SQ1 | Czyczyło-Mysza *et al.* (2011) |
| All chromosomes except 2A, 2D, 3D, 5D, 6D | Agronomic, phenological and physiological traits | RIL lines of Seri M82/ Babax | Pinto *et al.* (2010) |
| 1A, 3D, 7B | Stay-green | RILs of Chirya 3/Sonalika | Kumar *et al.* (2010) |
| 2B, 4A, 5A, 7B | Crop productivity, morpho-physiological and phenological traits | RILs of Langdon/G18-16 | Peleg *et al.* (2009) |
| 1B, 1D, 2B, 3A, 4A, 4B, 4D, 5A, 5B, 6A, 6B, 7A, 7B | Yield, anthesis and height | RILs of Seri M82/Babax | Mathews *et al.* (2008) |
| 6A | Coleoptiles, seedling vigour and plant height | RILs of Chuanmai 18/ Vigour 18 | Spielmeyer *et al.* (2007) |

2015). Sukumaran *et al.* (2018) phenotyped a panel of 208 durum wheat lines for grain yield, drought stress and heat stress, identifying 6211 DArT-seq SNPs. GWAS was performed and several common markers for stress indices were detected in drought and heat stress on chromosome 2B and 7A, respectively. A diverse panel of 94 bread wheat genotypes was phenotyped for grain yield and associated traits as well as arbuscular mycorrhizal fungus symbiosis under drought and watered conditions and subjected to GWAS utilizing 15511 SNP markers. Drought tolerance was found to be improved in the presence of mycorrhizae rather than its absence and chromosomes 3D and 7D have regions associated with mycorrhizae with drought stress tolerance (Lehnert *et al.*, 2018). GWAS conducted by Bhatta *et al.* (2018) revealed the presence of 90 novel MTAs on all three genomes (45 on A, 11 on B and 34 on D) linked with grain yield, yield-contributing traits and root traits under water stress condition. In another study, drought tolerance indices were assessed based upon grain yield-associated traits in 384 cultivars and advanced lines. Using GWAS, 175 SNPs showed association with at least one drought tolerance index and most of the associations were found on 4A chromosome, which revealed that 4A chromosome can play a crucial role in drought tolerance in wheat breeding programmes (Ballesta *et al.*, 2019). A total of 100 bread wheat genotypes were screened for yield and yield components and the same panel was genotyped using 15,600 SNP markers. Of these, 75 MTAs were identified and among these a further 37 MTA were obtained for drought stress condition. Most of the associations were found on B genome (Mathew *et al.*, 2019). An association panel of 320 diverse spring wheat accessions was genotyped with 9626 SNPs. The panel was phenotyped for yield and yield traits under watered and rainfed conditions and 19 MTAs were observed in rainfed condition (Gahlaut *et al.*, 2019). A consensus QTL locus which showed drought tolerance, designated as *QTL locus.4B.1*, further revealed the presence of two minor haplotypes associated with drought tolerance and plant stature in durum wheat accessions (Wang *et al.*, 2019). A set of 645 wheat landraces were evaluated for 16 seedling traits (root and shoot growth and water content under normal and water-limited conditions). The GWAS was performed using 52,118 DArT-seq markers. A total of 57 QTLs and 29 QTLs for seven and eight traits under normal and water-limited conditions were detected. Further, candidate genes under both conditions were identified and validated, which extends their scope for further utilization in genetic improvement programmes for drought tolerance in wheat (Lin *et al.*, 2019). Roots and shoots play an important role in the transportation of water from roots to other parts of the plant for utilization in various physiological processes like photosynthesis, transpiration, etc. A GWAS was performed among 323 wheat genotypes' root and shoot traits with a wheat 660K SNP array and 96 significantly associated loci were detected using a mixed linear model approach (Li *et al.*, 2019). SNP-assisted GWAS and haplotype-assisted GWAS were performed in a diverse panel of 240 wheat accessions comprising 171 SYN-DER and 69 modern cultivars and advanced lines, and 30 selection loci on the D genome that influence tolerance were identified (Afzal *et al.*, 2019). From 2018 onwards researchers have conducted GWAS routinely, which has rewarded them with several genomic regions associated with yield, yield-contributing traits and drought tolerance under drought condition across the world; this seems to be a main step towards development of improved cultivars with drought tolerance in the future. GS has emerged as an important molecular approach which predicts superior genotype based upon the breeding value estimated at genomic level, namely genomic estimated breeding value (GEBV). In wheat GS has been applied for various economic traits like grain yield, plant height, heading date (i.e. days to heading), lodging, preharvest sprouting (PHS), flour yield, flour protein, softness, and solvent retention capacity (SRC) for sucrose (SucSRC), water (H2O-SRC), lactic acid (LA-SRC) and sodium carbonate (NaCO-SRC) (Heffner *et al.*, 2011), as well as grain yield (Poland *et al.*, 2012; Rutkoski *et al.*, 2016), adult plant stem rust resistance (Rutkoski *et al.*, 2014), genotype-by-environment interaction and trait stability (Lopez-Cruz *et al.*, 2015; Huang *et al.*, 2016). In a study by Jafarzadeh *et al.* (2016), estimated GEBV for grain yield varied from 0.40 to 0.50 under drought stress condition. GS has not been utilized for drought tolerance improvement in wheat so far. However, initial results indicate an opportunity for wheat breeders to practise GS for drought tolerance for the development of superior wheat cultivars having better GEBV.

### *Transgenic technology for enhancing drought signalling*

Transgenic technology has emerged as an alternative approach to traditional breeding which involves the isolation, introduction, integration and expression of a foreign gene in a host plant (Saini *et al.*, 2020). It is very useful for the development of crop cultivars with enhancement of biotic and abiotic stress resistance/tolerance, quality and yield traits. *Bt* cotton with resistance against *Helicoverpa armigera* is the most remarkable product of this technology which proves its worthiness. In wheat, researchers have used this technology for the improvement of drought tolerance (Table 4.4).

**Table 4.4.** Genetic transformation of drought tolerance genes/transcription factors in wheat from other crops and wheat. (From Budak *et al.*, 2015; Khan *et al.*, 2019.)

| Source | Transgene | Improved traits | Reference |
|---|---|---|---|
| **From other crops** | | | |
| Barley | *HVA1* | ABA signalling, produces LEA3 for cell-membrane integrity, higher biomass production, WUE and RWC, drought and salt tolerance | Sivamani *et al.* (2000); Bahieldin *et al.* (2005) |
| Soybean | *GmDREB* | AIS, drought and salt tolerance | Shiqing *et al.* (2005) |
| | | AIS, twofold higher proline production, stay-green, SURV, drought tolerance | Wang *et al.* (2006) |
| Cotton | *GhDREB* | AIS, higher soluble sugars and chlorophyll production, improved drought, salt and freezing tolerance | Gao *et al.* (2009) |
| *Arabidopsis thaliana* | *DREB1A* | Higher SURV, WUE and yield under drought tolerance | Pierre *et al.* (2012) |
| | *AtWRKY30* | Higher biomass, photosynthesis, RWC, proline, soluble proteins, soluble sugars and antioxidant enzyme activities | El-Esawi *et al.* (2019) |
| | *AtHDG11* | More yield, higher proline content and photosynthesis, lower stomatal density, lower water loss rate, increased antioxidant enzymes CAT and SOD activity | Li *et al.* (2016) |
| | *OTS1* | Delayed senescence, higher RWC, photosynthesis and antioxidants | |
| | *DREB* | High proline content | Wang *et al.* (2006) |
| Rice | *SNAC1* | Activation of sucrose phosphate synthase, type 2C protein phosphatases and 1-phosphatidylinositol-3-phosphate-5-kinase genes for ABA signalling, high RWC, chlorophyll content and biomass, enhanced salinity and drought tolerance | Saad *et al.* (2013) |
| Alfalfa | *Alfalfa aldose reductase* (*MsALR*) | Antioxidant defence, ABA signalling, detoxification of aldehyde substrate, green biomass production, drought tolerance | Feher-Juhasz *et al.* (2014) |
| *Escherichia coli* | *Mannitol-1-phosphate dehydrogenase* (*mtlD*) | Improved biomass, mannitol accumulation | Abebe *et al.* (2003) |
| | *Beta encoding choline dehydrogenase* | Accumulation of glycine betaine | He *et al.* (2011) |
| | *ScCspA* | Higher proline, grain weight and grain yield, less reduction in chlorophyll, low MDA content | Yu *et al.* (2017) |
| | *CspA* and *CspB* | Lower water loss rate and MDA content, higher chlorophyll, proline and yield | |
| *Atriplex hortensis* | *BADH* | Higher BADH activity | Guo *et al.* (2000) |
| | | Accumulation of glycine betaine | Wang *et al.* (2010a) |
| | | Decreased PS II photoinhibition | Wang *et al.* (2010b) |

*Continued*

**Table 4.4.** Continued.

| Source | Transgene | Improved traits | Reference |
|---|---|---|---|
| *Vigna aconitifolia* | *P5CS* | Proline biosynthesis | Vendruscolo *et al.* (2007) |
| *Brachypodium distachyon* | *SBPase* | *SBPase* gene promoter fully drives the GUS expression | Alotaibi *et al.* (2019) |
| Sunflower | *HaHB4* | Increased yield and WUE | González *et al.* (2019) |
| Maize | *PEPC* | Higher proline, soluble sugars and WUE | Qin *et al.* (2015) |
| **From wheat** | | | |
| Xifengo20 (drought-tolerant wheat cultivar) | *TaWRKY2* | High survival rate, proline, soluble sugars and chlorophyll | Gao *et al.* (2018) |
| Xiaobaimai cultivar | *CIPK23* | High survival rate, increased osmolytes, induction of stomatal closure, enhanced ABA sensitivity | Cui *et al.* (2018) |
| *Triticum aestivum* | *P5cs* | Proline accumulation | Pavei *et al.* (2016) |
| TAM107 | *TaFER-5B* | Improved leaf Fe content and ROS, enhanced drought and temperature tolerance | Zhang *et al.* (2014) |
| | *TaPEPKR2* | Enhanced drought tolerance, higher root length | Zang *et al.* (2018) |
| Australian drought-tolerant genotype RAC875 | *TaSHN1* | Lower stomatal density and leaf water loss, improved recovery after severe drought | Bi *et al.* (2018) |
| | *TaNF-YB4* | More spikes | Yadav *et al.* (2015) |
| RAC785 | *DREB/CBF* gene *TaRAP2.1Lmut* | Enhanced ability to survive frost and drought | Amalraj *et al.* (2016) |
| *T. aestivum* | *TaNAC69* | More root biomass, longer roots | Xue *et al.* (2011) |
| RAC875 | *TabZIP2* | Fewer spikes and seeds, increased single-seed weight | Luang *et al.* (2018) |
| Langdon | *DREB* | Improved survival, slow growth, delayed flowering, less grain yield | Morran *et al.* (2011) |

LEA3, late embryogenesis abundant 3; AIS, ABA-independent signalling; SURV, survival and recovery under severe drought; BADH, betaine aldehyde dehydrogenase; SBPase, sedoheptulose-1,7-bisphosphatase; GUS, β-glucuronidase.

*Omics approaches for drought tolerance*

The recent advancement in genomics facilitates the rapid and accurate genome analysis in crop plants for the identification of gene(s) responsible for a specific effect. The RNA-seq and microarray approaches have been utilized for better understanding of the drought tolerance in wheat. The omics-based approaches such as transcriptomics, proteomics and metabolomics have revolutionized the functional genomics because these approaches are very helpful for determining the differential gene expression at different crop phenological stages during abiotic stress like drought, salinity, heat and cold. In wheat these functional genomics approaches have been used extensively for the identification of candidate genes for drought tolerance. A summary of candidate drought tolerance genes identified is presented in Table 4.5.

## 4.4 Conclusion and Future Perspectives

It is well known that drought is a complex, polygenic trait. It is not possible to measure drought

**Table 4.5.** Candidate genes identified through functional genomics in wheat.

| Transcriptomics/Proteomics/ Genetic manipulation studies | Gene regulation | Reference |
|---|---|---|
| Silicon application for drought tolerance enhancement | Upregulation of antioxidant, ascorbate–glutathione and phenylpropanoid pathway genes | Ma *et al.* (2016) |
| Overexpression of the wheat expansin gene *TaEXPA2* for improved drought tolerance | Overexpression in tobacco | Chen *et al.* (2016) |
| Dehydration and rehydration proteomic analysis | Induction of pathways related to carbohydrate and amino acid metabolism, antioxidants and defence, and ATP synthesis | Cheng *et al.* (2016) |
| Overexpression of *TaWRKY1* | Overexpression in tobacco | Ding *et al.* (2016) |
| *Aegilops longissima* substitution lines in Chinese Spring | Increased expression of ascorbate peroxidase, serpin-Z2B α-amylase genes under drought stress | Zhou *et al.* (2016) |
| Succinate dehydrogenase inhibitor (SHI) fungicide spray under drought stress | Cell-wall expansion, wax and defence genes | Ajigboye *et al.* (2017) |
| Pre-treatment of wheat seedlings with NaHS (sodium hydrosulfide) under drought | SOD, transport, CDPK, ABA, auxin, ribosome biogenesis | Li *et al.* (2017) |
| Durum wheat micro-RNA targets | Target genes of micro-RNAs under drought stress: ARFs, HD-Zip, SOD, ROS, HSPs | Liu, H. *et al.* (2017) |
| Drought response genes in developing wheat glumes | Enhanced expression of phenylpropanoid biosynthesis pathway genes in wheat glumes | Liu, C. *et al.* (2017) |
| Splice variation in wheat as an effect of drought | *HSFA1FD*, *HSFA6B*, HSP DnaJ alternatively spliced | Liu, Z. *et al.* (2017) |
| Wheat transcriptome changes under drought stress | *LTPL38* and α-*Amylase3* genes | Ma *et al.* (2017) |
| Response of He–Ne laser-pretreated wheat seedlings to drought stress | Altered expression of genes related to photosynthesis, nutrient uptake and transport | Qiu *et al.* (2017) |
| DEGs | Upregulation of osmoprotectants, ROS scavengers, biomass and energy | Hu *et al.* (2018) |
| DEGs | Upregulation of genes mostly involved in oxidation–reduction process, secondary metabolite biosynthesis, abiotic stress response, transferase activity and HSPs | Chaichi *et al.* (2019) |
| Response of wheat roots to drought | Gene regulatory networks showed 69 hub-genes integrating ABA-dependent and -independent pathways controlling sensing of drought, root growth, uptake regulation, purine metabolism, thiamine metabolism and antibiotics pathways, stomatal closure and senescence | Iquebal *et al.* (2019) |
| Drought-responsive DEGs | Genes enhancing upregulation of carotenoid biosynthesis pathway | Galvez *et al.* (2019) |

DEG, differentially expressed gene; CDPK, calcium-dependent protein kinase; ARF, auxin-response factor; HD-Zip, homeodomain-leucine zipper; HSP, heat-shock protein.

from a single trait index because of its combinatorial features. Traditional breeding involves utilization of genetic resources for significant gain in yield under diverse environments. Identification of suitable genotypes that can withstand water-limited conditions is a challenging task for wheat breeders. For a better understanding, precise phenotyping of drought tolerance-associated traits is necessary. Precise phenotyping helps breeders in selecting diverse genotypes which can be utilized for developing mapping populations such as RILs, DH, CSSLs and near-isogenic lines (NILs). These populations provide an opportunity to breeders for studying the nature of drought tolerance in wheat and further mapping the loci responsible for drought tolerance through bi-parental linkage mapping. Before wheat genome sequencing, it was very difficult to dissect drought tolerance genomic regions because of large genome size and repetitive sequences. But with the availability of sequencing approaches, a large number of genomic resources has become available which extend the scope of utilization of advanced genomics approaches such as GWAM and GS, MutMap$^+$, etc. A new genome editing approach, the clustered regularly interspaced short palindromic repeats (CRISPR)/CRISPR-associated protein 9 (Cas9) system, can also be utilized for enhancement of drought tolerance in wheat. Therefore, integration of genomic approaches with precise phenotyping is the need of the hour for improving drought tolerance in wheat.

## References

Abdelali, B., Hassan El, S., Mohamed El, Y. and Yahia, R. (2019) Morphological traits associated with drought stress tolerance in six Moroccan durum wheat varieties released between 1984 and 2007. *Journal of Crop Science and Biotechnology* 22, 345–353.

Abebe, T., Guenzi, A.A., Martin, B. and Cushman, J.C. (2003) Tolerance of mannitol-accumulating transgenic wheat to water stress and salinity. *Plant Physiology* 131, 1748–1755.

Abid, M., Ali, S., Qi, L.K., Zahoor, R., Tian, Z. *et al.* (2018) Physiological and biochemical changes during drought and recovery periods at tillering and jointing stages in wheat (*Triticum aestivum* L.). *Scientific Reports* 8, 4615.

Acuna-Galindo, M.A., Mason, R.E., Subramanian, N.K. and Hays, D.B. (2015) Meta-analysis of wheat qtl regions associated with adaptation to drought and heat stress. *Crop Science* 55, 477–492.

Afzal, F., Li, H., Gul, A., Subhani, A., Ali, A. *et al.* (2019) Genome-wide analyses reveal footprints of divergent selection and drought adaptive traits in synthetic-derived wheats. *G3: Genes, Genomes, Genetics* 9(6), 1957–1973.

Ahmad, M.Q., Khan, S.H., Khan, A.S., Kazi, A.M. and Basra, S. (2014) Identification of QTLs for drought tolerance traits on wheat chromosome 2A using association mapping. *International Journal of Agriculture and Biology* 16, 862–870.

Ahmadi, A.A. (1998) Effect of post-anthesis water stress on yield regulating processes in wheat (*Triticum aestivum* L.). PhD thesis, University of London, Ashford, UK.

Ahmed, M., Asif, M. and Goyal, A. (2012) Silicon the non-essential beneficial plant nutrient to enhanced drought tolerance in wheat. In: Goyal, A. (ed.) *Crop Plant*. IntechOpen, London. Available at: https://doi.org/10.5772/45647

Ajigboye, O.O., Lu, C., Murchie, E.H., Schlatter, C., Swart, G. and Ray, R.V. (2017) Altered gene expression by sedaxane increases PSII efficiency, photosynthesis and growth and improves tolerance to drought in wheat seedlings. *Pesticide Biochemistry and Physiology* 137, 49–61.

Alaei, Y. (2011) The effect of amino acids on leaf chlorophyll content in bread wheat genotypes under drought stress conditions. *Middle-East Journal of Science and Research* 10, 99–101.

Alahmad, S., El Hassouni, K., Bassi, F.M., Dinglasan, E., Youssef, C. *et al.* (2019) A major root architecture QTL responding to water limitation in durum wheat. *Frontiers in Plant Science* 10, 436.

Alexander, L.M., Kirigwi, F.M., Fritz, A.K. and Fellers, J.P. (2012) Mapping and quantitative trait loci analysis of drought tolerance in a spring wheat population using amplified fragment length polymorphism and Diversity Array Technology markers. *Crop Science* 52, 253–261.

Alexandratos, N. and Bruinsma J. (2012) *World Agriculture Towards 2030/2050: The 2012 Revision*. ESA Working Paper No. 12-03. Food and Agriculture Organization of the United Nations, Rome.

Alotaibi, S., Alyasi, H., El-Shehawi, A., Gaber, A., Hassan, M., Simkin, B.A.A. and Raines, C. (2019) Functional analysis of SBPase gene promoter in transgenic wheat under abiotic stresses. *Biotechnology* 18, 15–23.

Amalraj, A., Luang, S., Kumar, M.M., Sornaraj, P., Eini, O. *et al.* (2016) Change of function of the wheat stress-responsive transcriptional repressor Ta RAP 2.1 L by repressor motif modification. *Plant Biotechnology Journal* 14, 820–832.

Amirjani, M.R. and Mahdiyeh, M. (2013) Antioxidative and biochemical responses of wheat. *Journal of Agriculture and Biological Science* 8, 291–301.

Anjum, S.A., Xie, X., Wang, L., Saleem, M.F., Man, C. and Lei, W. (2011) Morphological, physiological and biochemical responses of plants to drought stress. *African Journal of Agriculture* 6, 2026–2032.

Asfaw, A., Blair, M.W. and Struik, P. (2012) Multienvironment quantitative trait loci analysis of photosynthate acquisition, accumulation, and remobilization traits in common bean under drought stress. *G3: Genes, Genomes, Genetics* 5, 579–595.

Ashraf, M. and Foolad, M.R. (2007) Roles of glycine betaine and proline in improving plant abiotic stress resistance. *Environmental and Experimental Botany* 59, 206–216.

Ashraf, M., Ozturk, M. and Athar, H.R. (2009) *Salinity and Water Stress: Improving Crop Efficiency*. Springer, New York.

Atlin, G.N., Cairns, J.E. and Das, B. (2017) Rapid breeding and varietal replacement are critical to adaptation of cropping systems in the developing world to climate change. *Global Food Security* 12, 31–37.

Babar, M.A., Reynolds, M.P., Van Ginkel, M., Klatt, A.R., Raun, W.R. and Stone, M.L. (2006) Spectral reflectance to estimate genetic variation for in-season biomass, leaf chlorophyll and canopy temperature in wheat. *Crop Science* 46, 1046–1057.

Bahieldin, A., Mahfouz, H.T., Eissa, H.F., Saleh, O.M., Ramadan, A.M. *et al.* (2005) Field evaluation of transgenic wheat plants stably expressing the HVA1 gene for drought tolerance. *Physiologia Plantarum* 123, 421–427. Available at: https://doi.org/10.1111/j.1399-3054.2005.00470.x

Ballesta, P., Mora, F. and Pozo, A.D. (2019) Association mapping of drought tolerance indices in wheat: QTL-rich regions on chromosome 4A. *Scientia Agricola* 77, e20180153.

Bansal, R., Pradheep, K., Kumari, J., Kumar, S., Yadav, M.C. *et al.* (2016) Physiological and biochemical evaluation for drought tolerance in wheat germplasm collected from arid western plains of India. *Indian Journal of Biochemistry & Biophysics* 53, 212–217.

Barakat, M.N., Al-Doss, A.A., Moustafa, K.A., Ahmed, E.I. and Elshafei, A.A. (2010) Morphological and molecular characterization of Saudi wheat genotypes under drought stress. *Journal of Food, Agriculture and Environment* 8, 220–228.

Bartels, D. and Phillips, J. (2010) Drought stress tolerance. In: Kempken, F. and Jung, C. (eds) *Genetic Modification of Plants*. Springer, Berlin/Heidelberg, Germany, pp. 139–157.

Beltrano, J., Ronco, M.G. and Arango, A.C. (2006) Soil drying and rewatering applied at three grain developmental stages affect differentially growth and grain protein deposition in wheat (*Triticum aestivum* L.). *Brazilian Journal of Plant Physiology* 18, 341–350.

Bennett, D., Reynolds, M., Mullan, D., Izanloo, A., Kuchel, H., Langridge, P. and Schnurbusch, T. (2012) Detection of two major grain yield QTLs in bread wheat (*Triticum aestivum* L.) under heat, drought and high yield potential environments. *Theoretical and Applied Genetics* 125, 1473–1485.

Bhatta, M., Morgounov, A., Belamkar, V. and Baenziger, P.S. (2018) Genome-wide association study reveals novel genomic regions for grain yield and yield-related traits in drought-stressed synthetic hexaploid wheat. *International Journal of Molecular Sciences* 19, 3011.

Bhattacharjee, S. (2005) Reactive oxygen species and oxidative burst: roles in stress, senescence and signal transduction in plants. *Current Science* 89, 1112–1121.

Bi, H., Shi, J., Kovalchuk, N., Luang, S., Bazanova, N. *et al.* (2018) Overexpression of the TaSHN1 transcription factor in bread wheat leads to leaf surface modifications, improved drought tolerance, and no yield penalty under controlled growth conditions. *Plant, Cell & Environment* 41(11), 2549–2566.

Blum, A. (2005) Drought resistance, water-use efficiency, and yield potential – are they compatible, dissonant, or mutually exclusive? *Australian Journal of Agricultural Research* 56, 1159–1168.

Bota, J., Flexas, J. and Medrano, H. (2004) Is photosynthesis limited by decreased Rubisco activity and RuBP content under progressive water stress? *New Phytopathology* 162, 671–681.

Bray, E.A., Bailey-Serres, J. and Weretilnyk, E. (2000) Responses to abiotic stresses. In: Gruissem, W., Buchannan, B. and Jones, R. (eds) *Biochemistry and Molecular Biology of Plants*. American Society of Plant Physiologists, Rockville, Maryland, pp. 1158–1249.

Briggs, K. and Aytenfisu, A. (1980) Relationships between morphological characters above the flag leaf node and grain yield in spring wheats. *Crop Science* 20, 350–354.

Budak, H., Hussain, B., Khan, Z., Ozturk, N.Z. and Ullah, N. (2015) From genetics to functional genomics: improvement in drought signaling and tolerance in wheat. *Frontiers in Plant Science* 6, 1012.

Cabrera-Bosquet, L., Crossa, J., Von Zitzewitz, J., Dolors Serret, M. and Araus, J.L. (2012) High-throughput phenotyping and genomic selection: the frontiers of crop breeding converge. *Journal of Integrated Plant Biology* 54, 312–320.

Carvalho, P., Azam-Ali, S. and Foulkes, M.J. (2014) Quantifying relationships between rooting traits and water uptake under drought in Mediterranean barley and durum wheat. *Journal of Integrated Plant Biology* 56, 455–469.

Cassia, R., Nocioni, M., Correa-Aragunde, N. and Lamattina, L. (2018) Climate change and the impact of greenhouse gases: $CO_2$ and NO, friends and foes of plant oxidative stress. *Frontiers in Plant Science* 9, 273.

Castro, A.M., Tacaliti, M.S., Gimenez, D., Tocho, E., Dobrovolskaya, O. *et al.* (2008) Mapping quantitative trait loci for growth responses to exogenously applied stress induced hormones in wheat. *Euphytica* 164, 719–727.

Cattivelli, L., Rizza, F., Badeck, F.W., Mazzucotelli, E., Mastrangelo, A.M. *et al.* (2008) Drought tolerance improvement in crop plants: an integrated view from breeding to genomics. *Field Crops Research* 105(1–2), 1–14.

Cedola, M.C., Scalfati, G., Soprano, M., Rascio, A. and Iannucci, A. (1994) Leaf morpho-physiological parameters as screening techniques for drought stress tolerance in *Triticum durum* Desf. *Journal of Genetics and Genome Research* 48, 229–236.

Chaichi, M., Sanjarian, F., Razavi, K. and Gonzalez-Hernandez, J.L. (2019) Analysis of transcriptional responses in root tissue of bread wheat landrace (*Triticum aestivum* L.) reveals drought avoidance mechanisms under water scarcity. *PLoS One* 14, e0212671.

Challinor, A.J., Simelton, E.S., Fraser, E.D., Hemming, D. and Collins, M. (2010) Increased crop failure due to climate change: assessing adaptation options using models and socioeconomic data for wheat in China. *Environmental Research Letters* 5, 034012.

Chandrasekar, V., Sairam, R.K. and Srivastava, G.C. (2000) Physiological and biochemical responses of hexaploid and tetraploid wheat to drought stress. *Journal of Agronomy and Crop Science* 185, 219–227.

Charmet, G. (2011) Wheat domestication: lessons for the future. *Comptes Rendus Biologies* 334, 212–220.

Chaves, M.M., Flexas, J. and Pinheiro, C. (2009) Photosynthesis under drought and salt stress-regulation mechanisms from whole plant to cell. *Annals of Botany* 103, 551–560.

Chen, X., Min, D., Yasir, T.A. and Hu, Y.G. (2012) Evaluation of 14 morphological, yield-related and physiological traits as indicators of drought tolerance in Chinese winter bread wheat revealed by analysis of the membership function value of drought tolerance (MFVD). *Field Crops Research* 137, 195–201.

Chen, Y., Han, Y., Zhang, M., Zhou, S., Kong, X. and Wang, W. (2016) Over expression of the wheat expansin gene TaEXPA2 improved seed production and drought tolerance in transgenic tobacco plants. *PLoS One* 11, e0153494.

Cheng, L.X., Wang, Y.P., He, Q., Li, H.J., Zhang, X.J. and Zhang, F. (2016) Comparative proteomics illustrates the complexity of drought resistance mechanisms in two wheat (*Triticum aestivum* L.) cultivars under dehydration and rehydration. *BMC Plant Biology* 16, 188–210.

Clarke, J.M. (1986) Effect of leaf rolling on leaf water loss in *Triticum* spp. *Canadian Journal of Plant Science* 66, 885–891.

Clarke, J.M. and McCaig, T.N. (1982) Evaluation of techniques for screening for drought resistance in wheat. *Crop Science* 22, 503.

Collins, N.C., Tardieu, F. and Tuberosa, R. (2008) Quantitative trait loci and crop performance under abiotic stress: where do we stand? *Plant Physiology* 147, 469–486.

Condon, A.G., Richards, R.A., Rebetzke, G.J. and Farquhar, G.D. (2004) Breeding for high water-use efficiency. *Journal of Experimental Botany* 55, 2447–2460.

Costa Franca, M.G., Pham-Thi, A.T., Pimentel, C., Pereyra Rossiello, R.O., Zuily-Fodil, Y. and Laffray, D. (2000) Differences in growth and water relations among *Phaseolus vulgaris* cultivars in response to induced drought stress. *Environmental and Experimental Botany* 43, 227–237.

Cui, X.X., Du, Y.Y., Fu, J.J., Yu, T.T., Wang, C.C. *et al.* (2018) Wheat CBL-interacting protein kinase 23 positively regulates drought stress and ABA responses. *BMC Plant Biology* 18, 93.

Czyczyło-Mysza, I., Marcińska, I., Skrzypek, E., Chrupek, M., Grzesiak, S. *et al.* (2011) Mapping QTLs for yield components and chlorophyll afluorescence parameters in wheat under three levels of water availability. *Plant Genetic Resources: Characterization and Utilization* 9, 291–295.
Czyczyło-Mysza, I., Marcinska, I., Skrzypek, E., Cyganek, K., Juzon, K. and Karbarz, M. (2014) QTL mapping for germination of seeds obtained from previous wheat generation under drought. *Open Life Sciences* 9, 374.
Czyczyło-Mysza, I.M., Cyganek, K., Dziurka, K., Quarrie, S., Skrzypek, E. *et al.* (2019) Genetic parameters and QTLs for total phenolic content and yield of wheat mapping population of CSDH lines under drought stress. *International Journal of Molecular Sciences* 20, 6064.
Dashti, H., Yazdi-Samadi, B., Ghannadha, M., Naghavi, M. and Quarri, S. (2007) QTL analysis for drought resistance in wheat using doubled haploid lines. *Integrated Journal of Agricultural Biology* 9, 98–102.
Dawood, M.F.A., Abeed, A.H.A. and Aldaby, E.E.S. (2019) Titanium dioxide nanoparticles model growth kinetic traits of some wheat cultivars under different water regimes. *Indian Journal of Plant Physiology* 24, 129–140.
Del Blanco, I.A., Rajaram, S., Kronstad, W.E. and Reynolds, M.P. (2000) Physiological performance of synthetic hexaploid wheat derived populations. *Crop Science* 40, 1257–1263.
Demirevska, K., Zasheva, D., Dimitrov, R., Simova-Stoilova, L., Stamenova, M. and Feller, U. (2009) Drought stress effects on Rubisco in wheat: changes in the Rubisco large subunit. *Acta Physiologiae Plantarum* 31, 1129–1138.
Ding, W., Fang, W., Shi, S., Zhao, Y., Li, X. and Xiao, K. (2016) Wheat WRKY type transcription factor gene TaWRKY1 is essential in mediating drought tolerance associated with an ABA-dependent pathway. *Plant Molecular Biology Reporter* 34, 1111–1126. Available at: https://doi.org/10.1007/s11105-016-0991-1
Dronin, N. and Kirilenko, A. (2011) Climate change, food stress, and security in Russia. *Regional Environmental Change* 11, 167–178.
Dubcovsky, J. and Dvorak, J. (2007) Genome plasticity a key factor in the success of polyploid wheat under domestication. *Science* 316, 1862–1866.
Dudziak, K., Zapalska, M., Börner, A., Szczerba, H., Kowalczyk, K. and Nowak, M. (2019) Analysis of wheat gene expression related to the oxidative stress response and signal transduction under short-term osmotic stress. *Scientific Reports* 9, 2743.
Duggan, B.L., Domitruk, D.R. and Fowler, D.B. (2000) Yield component variation in winter wheat grown under drought stress. *Canadian Journal of Plant Science* 80, 739–745.
Edae, E.A., Byrne, P.F., Manmathan, H., Haley, S.D., Moragues, M., Lopes, M.S. and Reynolds, M.P. (2013) Association mapping and nucleotide sequence variation in five drought tolerance candidate genes in spring wheat. *Plant Genome* 6, 1–13.
Eduardo, V., dos Santos, R.S., Busanello, C., Gustafson, P. and de Oliveira, A.C. (2019) Bread wheat: a role model for plant domestication and breeding. *Hereditas* 156, 1–11.
Ehdaie, B., Whitkus, R.W. and Waines, J.G. (2003) Root biomass, water-use efficiency, and performance of wheat–rye translocations of chromosomes 1 and 2 in spring bread wheat 'Pavon'. *Crop Science* 43, 710–717.
El-Esawi, M.A., Al-Ghamdi, A.A., Ali, H.M. and Ahmad, M. (2019) Overexpression of AtWRKY30 transcription factor enhances heat and drought stress tolerance in wheat (*Triticum aestivum* L.). *Genes* 10(2), 163.
El-Hassouni, K., Alahmad, S., Belkadi, B., Filali-Maltouf, A., Hickey, L. and Bassi, F. (2018) Root system architecture and its association with yield under different water regimes in durum wheat. *Crop Science* 58, 1–16.
FAO (2013) *FAO Statistical Yearbook 2013: World Food and Agriculture*. Food and Agriculture Organization of the United Nations, Rome. Available at: http://www.fao.org/3/i3107e/i3107e.pdf (accessed 9 February 2021).
Farooq, M., Wahid, A., Kobayashi, N., Fujita, D. and Basra, S.M.A. (2009) Plant drought stress: effects, mechanisms and management. *Agronomy for Sustainable Development* 29, 185–212.
Feher-Juhasz, E., Majer, P., Sass, L., Lantos, C., Csiszar, J. *et al.* (2014) Phenotyping shows improved physiological traits and seed yield of transgenic wheat plants expressing the alfalfa aldose reductase under permanent drought stress. *Acta Physiologiae Plantarum* 36, 663–673.
Fischer, R.A. and Maurer, R. (1978) Drought resistance in spring wheat cultivars. I. Grain yield responses. *Australian Journal of Agricultural Research* 29, 897–912.
Flato, G., Marotzke, J., Abiodun, B., Braconnot, P., Chou, S.C. *et al.* (2013) Evaluation of climate models. In: Stocker, T.F., Qin, D., Plattner, G.K., Tignor, M., Allen, S.K. *et al.* (eds) *Climate Change 2013: The*

*Physical Science Basis. Contribution of Working Group I to the Fifth Assessment Report of the Intergovernmental Panel on Climate Change*. Cambridge University Press, Cambridge, UK, pp. 741–882.

Fu, Y.B. and Somers, D.J. (2009) Genome-wide reduction of genetic diversity in wheat breeding. *Crop Science* 49, 161–168.

Gahlaut, V., Jaiswal, V., Tyagi, B.S., Singh, G., Sareen, S., Balyan, H.S. and Gupta, P.K. (2017) QTL mapping for nine drought-responsive agronomic traits in bread wheat under irrigated and rain-fed environments. *PLoS One* 12, e0182857.

Gahlaut, V., Jaiswal, V., Singh, S., Balyan, H.S. and Gupta, P.K. (2019) Multi-locus genome wide association mapping for yield and its contributing traits in hexaploid wheat under different water regimes. *Scientific Reports* 9, 19486.

Galinski, E.A. (1993) Compatible solutes of halophilic eubacteria: molecular principles, water–solute interaction, stress protection. *Cellular and Molecular Life Sciences* 49, 487–496.

Galvez, S., Mérida-García, R., Camino, C., Borill, P., Abrouk, M. *et al.* (2019) Hotspots in the genomic architecture of field drought responses in wheat as breeding targets. *Functional and Integrated Genomics* 19, 295.

Gao, F., Liu, J., Yang, L., Wu, X., Xiao, Y., Xia, X. and He, Z. (2016) Genome-wide linkage mapping of QTL for physiological traits in a Chinese wheat population using the 90K SNP array. *Euphytica* 209(3), 789–804.

Gao, H., Wang, Y., Xu, P. and Zhang, Z. (2018) Over expression of a WRKY transcription factor TaWRKY2 enhances drought stress tolerance in transgenic wheat. *Frontiers in Plant Science* 9, 997.

Gao, S.Q., Chen, M., Xia, L.Q., Xiu, H.J., Xu, Z.S. *et al.* (2009) A cotton (*Gossypium hirsutum*) DRE-binding transcription factor gene, GhDREB, confers enhanced tolerance to drought, high salt, and freezing stresses in transgenic wheat. *Plant Cell Reports* 28, 301–311.

Gizaw, S.A. (2015) Genome-wide association studies of drought resistance and yield potential in wheat (*Triticum aestivum* L.) using agronomic and remotely sensed traits. PhD dissertation, Washington State University, Pullman, Washington.

Godfray, H.C.J., Beddington, J.R., Crute, I.R., Haddad, L., Lawrence, D. *et al.* (2010) Food security: the challenge of feeding 9 billion people. *Science* 327(5967), 812–818.

Golabadi, M., Arzani, A., Maibody, S.M., Tabatabaei, B.S. and Mohammadi, S. (2011) Identification of microsatellite markers linked with yield components under drought stress at terminal growth stages in durum wheat. *Euphytica* 177, 207–221.

González, A., Martín, I. and Ayerbe, L. (2008) Yield and osmotic adjustment capacity of barley under terminal water-stress conditions. *Journal of Agronomy and Crop Science* 194, 81–91.

González, F.F., Capella, M., Ribichich, K.K., Curín, F., Giacomelli, J.J. *et al.* (2019) Field-grown transgenic wheat expressing the sunflower gene HaHB4 significantly out yields the wild type. *Journal of Experimental Botany* 70, 1669–1681.

Goodwin, M.S. and. Jenks, M.A. (2005) Plant cuticle function as a barrier to water loss. In: Jenks, M.A. and Hasegawa, P.M. (eds) *Plant Abiotic Stress*. Blackwell Publishing Ltd, Oxford, UK, pp. 14–36.

Graham, P.H. and Vance, C.P. (2003) Legumes: importance and constraints to greater use. *Plant Physiology* 131, 872–877.

Grammatikopoulos, G. and Manetas, Y. (1994) Direct absorption of water by hairy leaves of *Phlomis fruticosa* and its contribution to drought avoidance. *Canadian Journal of Botany* 72, 1805–1811.

Guo, B.B., Zhang, Y.Y., Li, H.H., Du, L.L., Li, Y.Y. *et al.* (2000) Transformation of wheat with a gene encoding for the betaine aldehyde dehydrogenase (BADH). *Acta Botanica Sinica* 42, 279–283.

Gupta, P.K., Balyan, H.S., Gahlaut, V. and Kulwal, P.L. (2012) Phenotyping, genetic dissection, and breeding for drought and heat tolerance in common wheat: status and prospects. *Plant Breeding Review* 36, 85–147.

Gupta, P.K., Balyan, H.S. and Gahlaut, V. (2017) QTL analysis for drought tolerance in wheat: present status and future possibilities. *Agronomy* 7, 2–21.

Hajlaoui, H., Ayeb, N.E., Garrec, J.P. and Denden, M. (2010) Differential effects of salt stress on osmotic adjustment and solutes allocation on the basis of root and leaf tissue senescence of two silage maize (*Zea mays* L.) varieties. *Industrial Crops and Products* 31, 122–130.

Hamada, A., Nitta, M., Nasuda, S., Kato, K., Fujita, M., Matsunaka, H. and Okumoto, Y. (2012) Novel QTLs for growth angle of seminal roots in wheat (*Triticum aestivum* L.). *Plant and Soil* 354, 395–405.

Hasanuzzaman, M., Nahar, K., Alam, M.M., Roychowdhury, R. and Fujita, M. (2013) Physiological, biochemical, and molecular mechanisms of heat stress tolerance in plants. *International Journal Molecular Sciences* 14, 9643–9684.

Hasheminasab, H., Assad, M.T., Ali, A. and Sahhafi, S.R. (2012) Influence of drought stress on oxidative damage and antioxidant defense systems in tolerant and susceptible wheat genotypes. *Journal of Agricultural Science* 4, 625–637.

He, C., Zhang, W., Gao, Q., Yang, A., Hu, X. and Zhang, J. (2011) Enhancement of drought resistance and biomass by increasing the amount of glycinebetaine in wheat seedlings. *Euphytica* 177, 151–167.

Heffner, E.L., Jannink, J.L. and Sorrells, M.E. (2011) Genomic selection accuracy using multifamily prediction models in a wheat breeding program. *The Plant Genome* 4(1), 65–75.

Hill, C.B., Taylor, J.D., Edwards, J., Mather, D., Bacic, A., Langridge, P. and Roessner, U. (2013) Whole-genome mapping of agronomic and metabolic traits to identify novel quantitative trait loci in bread wheat grown in a water-limited environment. *Plant Physiology* 162, 1266–1281.

Hossain, A., Da Silva, J.A.T., Lozovskaya, M.V. and Zvolinsky, V.P. (2012) High temperature combined with drought affect rainfed spring wheat and barley in south-eastern Russia: I. *Phenology and growth. Saudi Journal of Biological Sciences* 19, 473–487.

Hu, L., Xie, Y., Fan, S., Wang, Z., Wang, F. *et al.* (2018) Comparative analysis of root transcriptome profiles between drought-tolerant and susceptible wheat genotypes in response to water stress. *Plant Science* 272, 276–293.

Huang, M., Cabrera, A., Hoffstetter, A., Griffey, C., Van Sanford, D. *et al.* (2016) Genomic selection for wheat traits and trait stability. *Theoretical and Applied Genetics* 129, 1697–1710.

Huseynova, I. and Rustamova, S. (2010) Screening for drought stress tolerance in wheat genotypes using molecular markers. *Biological Science* 65, 132–139.

Ibrahim, S., Schubert, A., Pillen, K. and Leon, J. (2012) QTL analysis of drought tolerance for seedling root morphological traits in an advanced backcross population of spring wheat. *International Journal of Agricultural Sciences* 2, 619–629.

Iehisa, J.C., Matsuura, T., Mori, I.C. and Takumi, S. (2014) Identification of quantitative trait locus for abscisic acid responsiveness on chromosome 5A and association with dehydration tolerance in common wheat seedlings. *Journal of Plant Physiology* 171, 25–34.

Iquebal, M.A., Sharma, P., Jasrotia, R.S., Jaiswal, S., Kaur, A. *et al.* (2019) RNAseq analysis reveals drought-responsive molecular pathways with candidate genes and putative molecular markers in root tissue of wheat. *Scientific Reports* 9, 13917.

Izanloo, A., Condon, A.G., Langridge, P., Tester, M. and Schnurbusch, T. (2008) Different mechanisms of adaptation to cyclic water stress in two South Australian bread wheat cultivars. *Journal of Experimental Botany* 59, 3327–3346.

Jafarzadeh, J., Bonnett, D., Jannink, J.L., Akdemir, D., Dreisigacker, S. and Sorrells, M.E. (2016) Breeding value of primary synthetic wheat genotypes for grain yield. *PLoS One* 11, e0162860.

Jain, N., Singh, G.P., Singh, P.K., Ramya, P., Hari Krishna *et al.* (2014) Molecular approaches for wheat improvement under drought and heat stress. *Indian Journal of Genetics and Plant Breeding* 74, 578–583.

Jha, U.C., Chaturvedi, S.K., Bohra, A., Basu, P.S., Khan, M.S. and Barh, D. (2014) Abiotic stresses, constraints and improvement strategies in chickpea. *Plant Breeding* 133, 163–178.

Jones, H.G. (1999) Use of thermography for quantitative studies of spatial and temporal variation of stomatal conductance over leaf surfaces. *Plant, Cell & Environment* 22, 1043–1055.

Kabiri, R., Nasibi, F. and Farahbakhsh, H. (2014) Effect of exogenous salicylic acid on some physiological parameters and alleviation of drought stress in *Nigella sativa* plant under hydroponic culture. *Plant Protection* 50, 43–51.

Kadam, S., Singh, K., Shukla, S., Goel, S., Vikram, P. *et al.* (2012) Genomic associations for drought tolerance on the short arm of wheat chromosome 4B. *Functional and Integrated Genomics* 12, 447–464.

Kalladan, R., Worch, S., Rolletschek, H., Harshavardhan, V.T., Kuntze, L. *et al.* (2013) Identification of quantitative trait loci contributing to yield and seed quality parameters under terminal drought in barley advanced backcross lines. *Molecular Breeding* 32, 71–90.

Kaul, R. (1974) Potential net photosynthesis in flag leaves of severely drought-stressed wheat cultivars and its relationship to grain yield. *Canadian Journal of Plant Science* 54, 811–815.

Khalid, M., Gul, A., Amir, R., Ali, M., Afzal, F. *et al.* (2018) QTL mapping for seedling morphology under drought stress in wheat cross synthetic (W7984)/Opata. *Plant Genetic Resources: Characterization and Utilization* 16, 359–366.

Khalid, M., Afzal, F., Gul, A., Amir, R., Subhani, A. *et al.* (2019) Molecular characterization of 87 functional genes in wheat diversity panel and their association with phenotypes under well-watered and water-limited conditions. *Frontiers in Plant Science* 10, 717.

Khan, S., Anwar, S., Yu, S., Sun, M., Yang, Z. and Gao, Z.Q. (2019) Development of drought tolerant transgenic wheat: achievements and limitations. *International Journal of Molecular Sciences* 20, 3350.

Khazaei, H., Street, K., Bari, A., Mackay, M. and Stoddard, F.L. (2013) The FIGS (Focused Identification of Germplasm Strategy) approach identifies traits related to drought adaptation in *Vicia faba* genetic resources. *PLoS One* 8, e63107.

Kilic, H. and Yagbasanlar, T. (2010) The effect of drought stress on grain yield, yield components and some quality traits of durum wheat (*Triticum turgidum* ssp. durum). *Cultivars* 38, 164–170.

Kingston-Smith, A.H. and Foyer, C.H. (2000) Bundle sheath proteins are more sensitive to oxidative damage than those of the mesophyll in maize leaves exposed to paraquat or low temperatures. *Journal of Experimental Botany* 51, 123–130.

Kirigwi, F.M., Ginkel, M.V., Brown-Guedira, G., Gill, B.S., Paulsen, G.M. and Fritz, A.K (2007) Markers associated with a QTL for grain yield in wheat under drought. *Molecular Breeding* 20, 401–413.

Kumar, S., Sehgal, S.K., Kumar, U., Prasad, P.V., Joshi, A.K. and Gill, B.S. (2012) Genomic characterization of drought tolerance related traits in spring wheat. *Euphytica* 186, 265–276.

Kumar, S., Kumari, J., Bansal, R., Kuri, B.R., Upadhyay, D. *et al.* (2018) Multi-environmental evaluation of wheat genotypes for drought tolerance. *Indian Journal of Genetics and Plant Breeding* 78, 26–35.

Kumar, U., Joshi, A.K., Kumari, M., Paliwal, R., Kumar, S. and Röder, M.S. (2010) Identification of QTLs for stay green trait in wheat (*Triticum aestivum* L.) in the 'Chirya 3' × 'Sonalika' population. *Euphytica* 174, 437–445.

Kuzuoglu-Ozturk, D., Yalcinkaya, O.C., Akpinar, B.A., Mitou, G., Korkmaz, G., Gozuacik, D. and Budak, H. (2012) Autophagy-related gene, TdAtg8, in wild emmer wheat plays a role in drought and osmotic stress response. *Planta* 236, 1081–1092.

Lazacano-Ferrat, I. and Lovat, C.J. (1999) Relationship between relative water content, nitrogen pools, and growth of *Phaseolus vulgaris* L. and *P. acutifolius* A. Gray during water deficit. *Crop Science* 39, 467–475.

Le Roux, M.L., Kunert, K.J., van der Vyver, C., Cullis, C.A. and Botha, A.M. (2019) Expression of a small ubiquitin-like modifier protease increases drought tolerance in wheat (*Triticum aestivum* L.). *Frontiers in Plant Science* 10, 266.

Lehnert, H., Serfling, A., Friedt, W. and Ordon, F. (2018) Genome-wide association studies reveal genomic regions associated with the response of wheat (*Triticum aestivum* L.) to mycorrhizae under drought stress conditions. *Frontiers in Plant Science* 9, 1728.

Levitt, J. (1980) *Responses of Plants to Environmental Stresses*. Academic Press, New York, pp. 3642–3645.

Lev-Yadun, S., Gopher, A. and Abbo, S. (2000) The cradle of agriculture. *Science* 288, 1602–1603.

Li, H., Li, M., Wei, X., Zhang, X., Xue, R., Zhao, Y. and Zhao, H. (2017) Transcriptome analysis of drought-responsive genes regulated by hydrogen sulfide in wheat (*Triticum aestivum* L.) leaves. *Molecular Genetics and Genomics* 292, 1091–1110.

Li, L., Zheng, M., Deng, G., Liang, J., Zhang, H. *et al.* (2016) Over expression of AtHDG11 enhanced drought tolerance in wheat (*Triticum aestivum* L.). *Molecular Breeding* 36, 23.

Li, L., Peng, Z., Mao, X., Wang, J., Chang, X., Reynolds, M. and Jing, R. (2019) Genome-wide association study reveals genomic regions controlling root and shoot traits at late growth stages in wheat. *Annals of Botany* 124, 993–1006.

Li, Y., Ye, W., Wang, M. and Yan, X. (2009) Climate change and drought: a risk assessment of crop-yield impacts. *Climate Research* 39, 31–46.

Lin, Y., Yi, X., Tang, S., Chen, W., Wu, F. *et al.* (2019) Dissection of phenotypic and genetic variation of drought-related traits in diverse Chinese wheat landraces. *Plant Genome* 12, 190025.

Liu, C., Mi, H., Liu, H., Xie, S., Wu, Y. *et al.* (2017) Response to water deficit in glume of wheat: expression profiling by microarray analysis. *Euphytica* 213, 1–26.

Liu, C.Y., Yang, Z.Y. and Hu, Y.G. (2015) Drought resistance of wheat alien chromosome addition lines evaluated by membership function value based on multiple traits and drought resistance index of grain yield. *Field Crop Research* 179, 103–112.

Liu, H., Able, A.J., and Able, J.A. (2017) Water-deficit stress-responsive micro RNAs and their targets in four durum wheat genotypes. *Functional and Integrated Genomics* 17, 237–251.

Liu, S., Zhou, R., Dong, Y., Li, P. and Jia, J. (2006) Development, utilization of introgression lines using a synthetic wheat as donor. *Theoretical and Applied Genetics* 112, 1360–1373.

Liu, Z., Qin, J., Tian, X., Xu, S., Wang, Y. *et al.* (2017) Global profiling of alternative splicing landscape responsive to drought, heat and their combination in wheat (*Triticum asetivum* L.). *Plant Biotechnology Journal* 16, 714–726.

Lobell, D.B., Schlenker, W. and Costa-Roberts, J. (2011) Climate trends and global crop production since 1980. *Science* 333, 616–620.

Lonbani, M. and Arzani, A. (2011) Morpho-physiological traits associated with terminal drought-stress tolerance in triticale and wheat. *Agronomy Research* 9, 315–329.

Lopes, M.S. and Reynolds, M.P. (2010) Partitioning of assimilates to deeper roots is associated with cooler canopies and increased yield under drought in wheat. *Functional Plant Biology* 37, 147–156.

Lopez-Cruz, M., Crossa, J., Bonnett, D., Dreisigacker, S., Poland, J. *et al.* (2015) Increased prediction accuracy in wheat breeding trials using a marker–environment interaction genomic selection model. *G3: Genes, Genomes, Genetics* 5, 569–582.

Luang, S., Sornaraj, P., Bazanova, N., Jia, W., Eini, O. *et al.* (2018) The wheat TabZIP2 transcription factor is activated by the nutrient starvation-responsive SnRK3/CIPK protein kinase. *Plant Molecular Biology* 96, 543–561.

Lucas, S., Durmaz, E., Akpnar, B.A. and Budak, H. (2011a) The drought response displayed by a DRE-binding protein from *Triticum dicoccoides*. *Plant Physiology and Biochemistry* 49, 346–351.

Lucas, S., Dogan, E. and Budak, H. (2011b) TMPIT1 from wild emmer wheat: first characterisation of a stress-inducible integral membrane protein. *Gene* 483, 22–28.

Ma, D., Sun, D., Wang, C., Qin, H., Ding, H. *et al.* (2016) Silicon application alleviates drought stress in wheat through transcriptional regulation of multiple antioxidant defense pathways. *Journal of Plant Growth Regulation* 35, 1–10.

Ma, J., Li, R., Wang, H., Li, D., Wang, X. *et al.* (2017) Transcriptomics analyses reveal wheat responses to drought stress during reproductive stages under field conditions. *Frontiers in Plant Science* 8, 592.

Maccaferri, M., Sanguineti, M.C., Corneti, S., Ortega, J.L.A., Ben Salem, M. *et al.* (2008) Quantitative trait loci for grain yield and adaptation of durum wheat (*Triticum durum* Desf.) across a wide range of water availability. *Genetics* 178, 489–511.

McIntyre, C.L., Mathews, K.L., Rattey, A., Chapman, S.C., Drenth, J. *et al.* (2009) Molecular detection of genomic regions associated with grain yield and yield-related components in an elite bread wheat cross evaluated under irrigated and rainfed conditions. *Theoretical and Applied Genetics* 120, 527–541.

McNeal, F.H. and Berg, M.A. (1977) Flag leaf area in five spring wheat crosses and the relationship to grain yield. *Euphytica* 26, 739–744.

Mahajan, S. and Tuteja, N. (2005) Cold, salinity and drought stresses: an overview. *Archives Biochemistry and Biophysics* 444, 139–158.

Martínez-Ballesta, M.C., López-Pérez, L., Muries, B., Munoz Azcarate, O. and Carvajal, M. (2009) Climate change and plant water balance: the role of aquaporins – a review. In: Lichtfouse, E. (ed.) *Climate Change, Intercropping, Pest Control and Beneficial Microorganisms*. Sustainable Agriculture Reviews, Vol. 2. Springer, Dordrecht, the Netherlands, pp. 71–89.

Mathew, I., Shimelis, H., Shayanowako, A.I.T., Laing, M. and Chaplot, V. (2019) Genome-wide association study of drought tolerance and biomass allocation in wheat. *PLoS One* 14, e0225383. Available at: https://doi.org/10.1371/journal.pone.0225383

Mathews, K.L., Malosetti, M., Chapman, S., McIntyre, L., Reynolds, M., Shorter, R. and van Eeuwijk, F. (2008) Multi-environment QTL mixed models for drought stress adaptation in wheat. *Theoretical and Applied Genetics* 117, 1077–1091.

Milad, S.I., Wahba, L.E. and Barakat, M.N. (2011) Identification of RAPD and ISSR markers associated with flag leaf senescence under water-stressed conditions in wheat (*Triticum aestivum* L.). *Australian Journal of Crop Science* 5, 337–343.

Miller, G., Suzuki, N., Ciftci-Yilmaz, S. and Mittler, R. (2010) Reactive oxygen species homeostasis and signalling during drought and salinity stresses. *Plant, Cell & Environment* 33, 453–467.

Mir, R.R., Kumar, J., Balyan, H.S. and Gupta, P.K. (2012) Study of genetic diversity among Indian bread wheat (*Triticum aestivum* L.) cultivars released during last 100 years. *Genetic Resources and Crop Evolution* 59, 717–726.

Mohammadi, R. (2012) Genetic gain in grain yield and drought tolerance of durum wheat breeding lines under rainfed conditions in Iran. *Acta Agronomica Hungarica* 60, 417–432.

Mohammadi, R., Farshadfar, E., Aghaee-Sarbarzeh, M. and Sutka, J. (2003) Locating QTLs controlling drought tolerance criteria in rye using disomic addition lines. *Cereal Research Communications* 31, 257–264.

Molero, G., Joynson, R., Pinera-Chavez, F.J., Gardiner, L.J., Rivera-Amado, C. *et al.* (2019) Elucidating the genetic basis of biomass accumulation and radiation use efficiency in spring wheat and its role in yield potential. *Plant Biotechnology Journal* 17, 1276–1288.

Mollasadeghi, V., Shahryari, R., Imani, A.A. and Khayatnezhad, M. (2011) Factor analysis of wheat quantitative traits on yield under terminal drought. *American-Eurasian Journal of Agricultural & Environmental Sciences* 10, 157–159.

Mondini, L., Nachit, M., Porceddu, E. and Pagnotta, M.A. (2012) Identification of SNP mutations in DREB1, HKT1, and WRKY1 genes involved in drought and salt stress tolerance in durum wheat (*Triticum turgidum* L. var durum). *OMICS* 16, 178–187.

Mondini, L., Nachit, M.M. and Pagnotta, M.A. (2015) Allelic variants in durum wheat (*Triticum turgidum* L. var. durum) DREB genes conferring tolerance to abiotic stresses. *Molecular Genetics and Genomics* 290, 531–544.

Monneveux, P., Rekika, D., Acevedo, E. and Merah, O. (2006) Effect of drought on leaf gas exchange, carbon isotope discrimination, transpiration efficiency and productivity in field grown durum wheat genotypes. *Plant Science* 170, 867–872.

Mora, F., Castillo, D., Lado, B., Matus, I., Poland, J. *et al.* (2015) Genome-wide association mapping of agronomic traits and carbon isotope discrimination in a worldwide germplasm collection of spring wheat using SNP markers. *Molecular Breeding* 35, 69.

Morran, S., Eini, O., Pyvovarenko, T., Parent, B., Singh, R. *et al.* (2011) Improvement of stress tolerance of wheat and barley by modulation of expression of DREB/CBF factors. *Plant Biotechnology Journal* 9, 230–249.

Mujeeb-Kazi, A., Gul, A., Farooq, M., Rizwan, S. and Ahmad, I. (2008) Rebirth of synthetic hexaploids with global implications for wheat improvement. *Australian Journal of Agricultural Research* 59, 391–398.

Murata, N., Takahashi, S., Nishiyama, Y. and Allakhverdiev, S. (2007) Photoinhibition of photosystem II under environmental stress. *Biochimica et Biophysica Acta* 1767, 414–421.

Mwadzingeni, L., Shimelis, H., Dube, E., Laing, M.D. and Tsilo, T.J. (2016a) Breeding wheat for drought tolerance: progress and technologies. *Journal of Integrative Agriculture* 15, 935–943.

Mwadzingeni, L., Hussein, S., Samson, T., Toi, J. and Tsilo, T.J. (2016b) Screening of bread wheat genotypes for drought tolerance using phenotypic and proline analyses. *Frontiers in Plant Science* 7, 1276.

Mwadzingeni, L., Shimelis, H., Rees, D.J.G. and Tsilo, T.J. (2017) Genome-wide association analysis of agronomic traits in wheat under drought-stressed and non-stressed conditions. *PLoS One* 12, e0171692. Available at: https://doi.org/10.1371/journal.pone.0171692

Nachit, M.M., Baum, M., Autrique, E., Sorrells, M.E., Ali Dib, T. and Monneveux, P. (1993) Association of morphophysiological traits with RFLP markers in durum wheat. In: Monneveux, P. and Salem, M.B. (eds) *Tolérance à la Sécheresse des Céréales en Zone Méditerranéenne*. Diversité Génétique et Amélioration Variétale, Montpellier, France, pp. 159–171.

Nevo, E. and Chen, G.X. (2010) Drought and salt tolerances in wild relatives for wheat and barley improvement. *Plant, Cell & Environment* 33, 670–685.

Nezhad, K.Z., Weber, W., Roder, M., Sharma, S., Lohwasser, U. *et al.* (2012) QTL analysis for thousand-grain weight under terminal drought stress in bread wheat (*Triticum aestivum* L.). *Euphytica* 186, 127–138.

Nguyen, N. (2002) *Global Climate Changes and Rice Food Security*. Food and Agriculture Organization of the United Nations, Rome.

Nikolaeva, M.K., Maevskaya, S.N., Shugaev, A.G. and Bukhov, N.G. (2010) Effect of drought on chlorophyll content and antioxidant enzyme activities in leaves of three wheat cultivars varying in productivity. *Russian Journal of Plant Physiology* 57, 87–95.

Nishiyama, Y., Allakhverdiev, S.I. and Murata, N., (2011) Protein synthesis is the primary target of reactive oxygen species in the photoinhibition of photosystem II. *Physiologia Plantarum* 142, 35–46.

Norouzi, M., Toorchi, M., Hosseini Salekdeh, G., Mohammadi, S.A., Neyshabouri, M.R. and Aharizad, S. (2008) Effect of water deficit on growth, grain yield and osmotic adjustment in rapeseed. *Journal of Food, Agriculture and Environment* 6, 312–318.

Nouri-Ganbalani, A., Nouri-Ganbalani, G. and Hassanpanah, D. (2009) Effects of drought stress condition on the yield and yield components of advanced wheat genotypes in Ardabil, Iran. *Journal of Food, Agriculture and Environment* 7, 228–234.

Outoukarte, I., El Keroumi, A., Dihazi, A. and Naamani, K. (2019) Use of morpho-physiological parameters and biochemical markers to select drought tolerant genotypes of durum wheat. *Journal of Plant Stress Physiology* 5, 1–7.

Ovenden, B., Milgate, A., Wade, L.J., Rebetzke, G.J. and Holland, J.B. (2017) Genome-wide associations for water-soluble carbohydrate concentration and relative maturity in wheat using SNP and DArT marker arrays. *G3: Gene, Genome, Genetics* 7, 2821–2830.

Pandey, V. and Shukla, A. (2015) Acclimation and tolerance strategies of rice under drought stress. *Rice Science* 22, 147–161.

Parent, B., Shahinnia, F., Maphosa, L., Berger, B., Rabie, H. *et al.* (2015) Combining field performance with controlled environment plant imaging to identify the genetic control of growth and transpiration underlying yield response to water-deficit stress in wheat. *Journal of Experimental Botany* 66, 5481–5492.

Pavei, D., Gonçalves-Vidigal, M.M., Schuelter, A.A., Schuster, I., Vieira, E.E., Vendruscolo, E.E. and Poletine, J.J. (2016) Response to water stress in transgenic (p5cs gene) wheat plants (*Triticum aestivum* L.). *Australian Journal of Crop Science* 10, 776.

Peleg, Z. and Blumwald, E. (2011) Hormone balance and abiotic stress tolerance in crop plants. *Current Opinion in Plant Biology* 14, 290–295.

Peleg, Z., Saranga, Y., Krugman, T., Abbo, S., Nevo, E. and Fahima, T. (2008) Allelic diversity associated with aridity gradient in wild emmer wheat populations. *Plant, Cell & Environment* 31, 39–49.

Peleg, Z., Fahima, T., Krugman, T., Abbo, S., Yakir, D., Korol, A.B. and Saranga, Y. (2009) Genomic dissection of drought resistance in durum wheat × wild emmer wheat recombinant inbreed line population. *Plant, Cell & Environment* 32, 758–779.

Peng, J.H., Sun, D.F., Peng, Y.L. and Nevo, E. (2013) Gene discovery in *Triticum dicoccoides*, the direct progenitor of cultivated wheats. *Cereal Research Communications* 41, 1–22.

Pierre, C.S., Crossa, J.L., Bonnett, D., Yamaguchi-Shinozaki, K. and Reynolds, M.P. (2012) Phenotyping transgenic wheat for drought resistance. *Journal of Experimental Botany* 63, 1799–1808.

Pinto, R.S. and Reynolds, M.P. (2015) Common genetic basis for canopy temperature depression under heat and drought stress associated with optimized root distribution in bread wheat. *Theoretical and Applied Genetics* 128, 575–585.

Pinto, R.S., Reynolds, M.P., Mathews, K.L., McIntyre, C.L., Olivares-Villegas, J.J. and Chapman, S.C. (2010) Heat and drought adaptive QTL in a wheat population designed to minimize confounding agronomic effects. *Theoretical and Applied Genetics* 121, 1001–1021.

Placido, D.F., Campbell, M.T., Folsom, J.J., Cui, X.P., Kruger, G.R., Baenziger, P.S. and Walia, H. (2013) Introgression of novel traits from a wild wheat relative improves drought adaptation in wheat. *Plant Physiology* 161, 1806–1819.

Poland, J., Endelman, J., Dawson, J., Rutkoski, J., Wu, S. *et al.* (2012) Genomic selection in wheat breeding using genotyping by sequencing. *The Plant Genome* 5, 103–113.

Postel, S. (1999) Redesigning irrigated agriculture. In: Brown, L.R., Flavin, C. and French, H. (eds) *State of the World, 2000*. W.W. Norton and Co., New York, pp. 39–58.

Prudhomme, C., Giuntoli, I., Robinson, E.L., Clark, D.B., Arnell, N.W. *et al.* (2014) Hydrological droughts in the 21st century: hotspots and uncertainties from a global multi model ensemble experiment. *Proceedings of the National Academy of Sciences USA* 111, 3262–3267.

Qaseem, M.F., Qureshi, R., Shaheen, H. and Shafqat, N. (2019) Genome-wide association analyses for yield and yield-related traits in bread wheat (*Triticum aestivum* L.) under pre-anthesis combined heat and drought stress in field conditions. *PLoS One* 14, e0213407. Available at: https://doi.org/10.1371/journal.pone.0213407

Qin, N., Xu, W., Hu, L., Li, Y., Wang, H. *et al.* (2015) Drought tolerance and proteomics studies of transgenic wheat containing the maize $C_4$ phosphoenolpyruvate carboxylase (PEPC) gene. *Protoplasma* 253, 1503–1512.

Qiu, Z., Yuan, M., He, Y., Li, Y. and Zhang, L. (2017) Physiological and transcriptome analysis of He–Ne laser pretreated wheat seedlings in response to drought stress. *Scientific Reports* 7, 6108.

Quarrie, S.A., Steed, A., Calestani, C., Semikhodskii, A., Lebreton, C. *et al.* (2005) A high-density genetic map of hexaploid wheat (*Triticum aestivum* L.) from the cross Chinese Spring × SQ1 and its use to compare QTLs for grain yield across a range of environments. *Theoretical and Applied Genetics* 110, 865–880.

Quarrie, S.A., Quarrie, S.P., Radosevic, R., Rancic, D., Kaminska, A. *et al.* (2006) Dissecting a wheat QTL for yield present in a range of environments: from the QTL to candidate genes. *Journal of Experimental Botany* 57, 2627–2637.

Rai, N., Amasiddha, B., Sinha, N., Das, T.R., Patil, R. *et al.* (2018) Physiological and morphological evaluation of MABB derived lines under drought stress in bread wheat (*Triticum aestivum* L. em. Thell.). *Indian Journal of Genetics and Plant Breeding* 78, 417–425.

Rama, R., Nagaraja, R., Ragimasalawada, M., Sabbavarapu, M.M., Nadoor, S. and Patil, J.V. (2014) Detection and validation of stay-green QTL in post-rainy sorghum involving widely adapted cultivar, M35-1 and a popular stay-green genotype B35. *BMC Genomics* 15, 909.

Ramya, K.T., Jain, N., Ramya, P., Singh, P.K., Arora, A., Singh, G.P. and Prabhu, K.V. (2015) Genotypic variation for normalized difference vegetation index and its relationship with grain yield in wheat under terminal heat stress. *Indian Journal of Genetics and Plant Breeding* 75, 174–182.

Ray, D.K., Ramankutty, N., Mueller, N.D., West, P.C. and Foley, J.A. (2012) Recent patterns of crop yield growth and stagnation. *Nature Communications* 3, 1293.

Rebetzke, G.J., Condon, A.G., Farquhar, G.D., Appels, R. and Richards, R.A. (2008) Quantitative trait loci for carbon isotope discrimination are repeatable across environments and wheat mapping populations. *Theoretical and Applied Genetics* 118, 123–137.

Reddy, A.R., Chaitanya, K.V. and Vivekanandan, M., (2004) Drought-induced responses of photosynthesis and antioxidant metabolism in higher plants. *Journal of Plant Physiology* 161, 1189–1202.

Reif, J.C., Zhang, P., Dreisigacker, S., Warburton, M.L., van Ginkel, M. *et al.* (2005) Wheat genetic diversity trends during domestication and breeding. *Theoretical and Applied Genetics* 110, 859–864.

Reynolds, M.P., Mujeeb-Kazi, A. and Sawkins, M. (2005) Prospects for utilizing plant-adaptive mechanisms to improve wheat and other crops in drought- and salinity-prone environments. *Annals of Applied Biology* 146, 239–259.

Richards, R.A., Rawson, H.M. and Johnson, D.A. (1986) Glaucousness in wheat: its development and effect on water-use efficiency, gas exchange and photosynthetic tissue temperatures. *Australian Journal of Plant Physiology* 13, 465–473.

Riehl, S., Zeidi, M. and Conard, N.J. (2013) Emergence of agriculture in the foothills of the Zagros Mountains of Iran. *Science* 341, 65–67.

Rizhsky, L., Liang, H. and Mittler, R. (2002) The combined effect of drought stress and heat shock on gene expression in tobacco. *Plant Physiology* 130, 1143–1151.

Rosegrant, M., Cai, X., Cline, S. and Nakagawa, N. (2002) *The Role of Rainfed Agriculture in the Future of Global Food Production*. Environment and Production Technology Division Discussion Paper No. 90. International Food Policy Research Institute, Washington, DC.

Ruggiero, A., Punzo, P., Landi, S., Costa, A., VanOoosten, M., and Grillo, S. (2017) Improving plant water use efficiency through molecular genetics. *Horticulturae* 3, 31.

Rutkoski, J.E., Poland, J.A., Singh, R.P., Huerta-Espino, J., Bhavani, S. *et al.* (2014) Genomic selection for quantitative adult plant stem rust resistance in wheat. *The Plant Genome* 7, 1–10.

Rutkoski, J., Poland, J., Mondal, S., Autrique, E., Pérez, L.G. *et al.* (2016) Canopy temperature and vegetation indices from high-throughput phenotyping improve accuracy of pedigree and genomic selection for grain yield in wheat. *G3: Genes, Genomes, Genetics* 6, 2799–2808.

Saad, A.S.I., Li, X., Li, H.P., Huang, T., Gao, C.S. *et al.* (2013) A rice stress-responsive NAC gene enhances tolerance of transgenic wheat to drought and salt stresses. *Plant Science* 20, 33–40.

Saini, P., Gani, M., Kaur, J.J., Godara, L.C., Singh, C. *et al.* (2018) Reactive oxygen species (ROS): a way to stress survival in plants. In: Zargar, S.M. and Zargar, M.Y. (eds) *Abiotic Stress-Mediated Sensing and Signaling in Plants: An Omics Perspective*. Springer, Singapore, pp. 127–153.

Saini, P., Saini, P., Kaur, J.J., Francies, R.M., Gani, M. *et al.* (2020) Molecular approaches for harvesting natural diversity for crop improvement. In: Salgotra, R. and Zargar, S. (eds) *Rediscovery of Genetic and Genomic Resources for Future Food Security*. Springer, Singapore, pp. 67–169.

Salem, K.F.M., Roder, M.S. and Borner, A. (2007) Identification and mapping quantitative trait loci for stem reserve mobilisation in wheat (*Triticum aestivum* L.). *Cereal Research Communications* 35, 1367–1374.

Sallam, A., Alqudah, A.M., Dawood, M.F.A., Baenziger, P.S. and Börner, A. (2019) Drought stress tolerance in wheat and barley: advances in physiology, breeding and genetics research. *International Journal of Molecular Sciences* 20, 3137.

Save, R., Biel, C. and de Herralde, F. (2000) Leaf pubescence, water relations and chlorophyll fluorescence in two subspecies of *Lotus creticus* L. *Biologia Plantarum* 43, 239–244.

Schachtman, D.P. and Goodger, J.Q. (2018) Chemical root to shoot signaling under drought. *Trends in Plant Science* 13, 281–287.

Scotnica, J., Matouskova, M., Naus, J., Lazar, D. and Dvorak, L. (2000) Thermoluminescence of fluorescence study of changes in photosystem II: a 100- to 200-ns component between 4.2 and 300 K. *Proceedings of the National Academy of Sciences USA* 77, 5889–5893.

Shahinnia, F., Le Roy, J., Laborde, B., Sznajder, B., Kalambettu, P. *et al.* (2016) Genetic association of stomatal traits and yield in wheat grown in low rainfall environments. *BMC Plant Biology* 16, 150.

Sharma, P., Sareen, S., Saini, M. and Shefali (2017) Assessing genetic variation for heat stress tolerance in Indian bread wheat genotypes using morpho-physiological traits and molecular markers. *Plant Genetic Resources: Characterization and Utilization* 15, 539–547.

Sharma, S.N., Sain, R.S. and Sharma, R.K. (2003) The genetic control of flag leaf length in normal and late sown durum wheat. *Agricultural Sciences* 141, 323–233.

Shinozaki, K. and Yamaguchi-Shinozaki, K. (2007) Gene networks involved in drought stress response and tolerance. *Journal of Experimental Botany* 58, 221–227.

Shiqing, G., Huijun, X., Xianguo, C., Ming, C., Zhaoshi, X. *et al.* (2005) Improvement of wheat drought and salt tolerance by expression of a stress-inducible transcription factor GmDREB of soybean (*Glycine max*). *Chinese Science Bulletin* 50, 2714–2723.

Singh, A., Yadav, O.P., Gaikwad, K., Kumar, S. and Rai, R.D. (2015) Induced defense responses of contrasting bread wheat genotypes under differential salt stress imposition. *Indian Journal of Biochemistry and Biophysics* 52, 75–85.

Singh, D. and Singh, D. (1992) Effect of leaf-blade and awn on grain yield of rain-fed wheat (*Triticum aestivum* L.) at different stages of spike development. *Indian Journal of Agricultural Sciences* 62, 468–471.

Sivamani, E., Bahieldin, A., Wraith, J.M., Al-Niemi, T., Dyer, W.E. *et al.* (2000) Improved biomass productivity and water use efficiency under water deficit conditions in transgenic wheat constitutively expressing the barley *HVA1* gene. *Plant Science* 155, 1–9.

Song, Q., Liu, C., Bachir D.G., Chen L. and Hu, Y.G. (2017) Drought resistance of new synthetic hexaploid wheat accessions evaluated by multiple traits and antioxidant enzyme activity. *Field Crops Research* 210, 91–103.

Spielmeyer, W., Hyles, J., Joaquim, P., Azanza, F., Bonnett, D. *et al.* (2007) A QTL on chromosome 6A in bread wheat (*Triticum aestivum* L.) is associated with longer coleoptiles, greater seedling vigour and final plant height. *Theoretical and Applied Genetics* 115, 59–66.

Spinoni, J., Naumann, G., Carrao, H., Barbosa, P. and Vogt, J. (2014) World drought frequency, duration, and severity for 1951–2010. *International Journal of Climatology* 34, 2792–2804.

Steudle, E. (2000) Water uptake by plant roots: an integration of views. *Plant and Soil* 226(1), 45–56.

Sukumaran, S., Reynolds, M.P. and Sansaloni, C. (2018) Genome-wide association analyses identify QTL hotspots for yield and component traits in durum wheat grown under yield potential, drought, and heat stress environments. *Frontiers in Plant Science* 9, 81.

Tahmasebi, S., Heidari, B., Pakniyat, H. and McIntyre, C.L. (2017) Mapping QTLs associated with agronomic and physiological traits under terminal drought and heat stress conditions in wheat (*Triticum aestivum* L.). *Genome* 60, 26–45.

Teal, R.K., Tubana, B., Girma, K., Freeman, K.W., Arnall, D.B., Walsh, O. and Raun, W.R. (2006) In-season prediction of corn grain yield potential using normalized difference vegetation index. *Agronomy Journal* 98, 1488–1494.

Thompson, A.J., Andrews, J., Mulholland, B.J., McKee, J.M.T., Hilton, H.W., Black, C.R. and Taylor, I.B. (2007) Overproduction of abscisic acid in tomato increases transpiration efficiency and root hydraulic conductivity influences leaf expansion. *Plant Physiology* 143, 1905–1917.

Tiwari, V., Mamrutha, H.M., Sareen, S., Sheoran, S., Tiwari, R. *et al.* (2017) Managing abiotic stresses in wheat. In: Minhas, P., Rane, J. and Pasala R. (eds) *Abiotic Stress Management for Resilient Agriculture*. Springer, Singapore, pp. 313–337.

Townley-Smith, T.F. and Hurd, E.A. (1979) Testing and selecting for drought resistance in wheat. In: Mussell, H. and Staples, R.C. (eds) *Stress Physiology in Crop Plants*. Wiley, New York, pp. 447–464.

Tsunewaki, K. and Ebana, K. (1999) Production of near-isogenic lines of common wheat for glaucousness and genetic basis of this trait. *Genes & Genetic Systems* 74, 33–41.

Turner, N.C., Molyneux, N., Yang, S., Xiong, Y.C. and Siddique, K.H. (2011) Climate change in south-west Australia and north-west China: challenges and opportunities for crop production. *Crop & Pasture Science* 62, 445–456.

Vendruscolo, E.C.G., Schuster, I., Pileggi, M., Scapim, C.C., Molinari, H.B.C., Marur, C.C. and Vieira, L.G.E. (2007) Stress-induced synthesis of proline confers tolerance to water deficit in transgenic wheat. *Journal of Plant Physiology* 164, 1367–1376.

Verma, V., Foulkes, M.J., Worland, A.J., Sylvester-Bradley, R., Caligari, P.D.S. and Snape, J.W. (2004) Mapping quantitative trait loci for flag leaf senescence as a yield determinant in winter wheat under optimal and drought-stressed environments. *Euphytica* 135, 255–263.

Villegas, D., García del Moral, L.F., Rharrabti, Y., Martos, V. and Royo, C. (2007) Morphological traits above the flag leaf node as indicators of drought susceptibility index in durum wheat. *European Journal of Agronomy* 193, 103–116.

Vogelmann, T.C. (1993) Plant tissue optics. *Annual Review of Plant Physiology and Plant Molecular Biology* 44, 231–251.

von Korff, M., Grando, S., Del Greco, A., This, D., Baum, M. and Ceccarelli, S. (2008) Quantitative trait loci associated with adaptation to Mediterranean dryland conditions in barley. *Theoretical and Applied Genetics* 117, 653–669.

Voss-Fels, K., Frisch, M., Qian, L., Kontowski, S., Friedt, W., Gottwald, S. and Snowdon, R.J. (2015) Subgenomic diversity patterns caused by directional selection in bread wheat gene pools. *Plant Genome* 8(2), eplantgenome2015.03.0013.

Wang, G.P., Zhang, X., Li, F., Luo, Y. and Wang, W. (2010a) Overaccumulation of glycine betaine enhances tolerance to drought and heat stress in wheat leaves in the protection of photosynthesis. *Photosynthetica* 48, 117–126.

Wang, G.P., Hui, Z., Li, F., Zhao, M.-R., Zhang, J. and Wang, W. (2010b) Improvement of heat and drought photosynthetic tolerance in wheat by overaccumulation of glycinebetaine. *Plant Biotechnological Reports* 4, 213–222.

Wang, J.J., Yang, F.F., Chen, X.X., Liang, R.R., Zhang, L.L. *et al.* (2006) Induced expression of DREB transcriptional factor and study on its physiological effects of drought tolerance in transgenic wheat. *Acta Genetica Sinica* 33, 468–476.

Wang, S., Xu, S., Chao, S., Sun, Q., Liu, S. and Xia, G. (2019) A genome-wide association study of highly heritable agronomic traits in durum wheat. *Frontiers in Plant Science* 10, 919.

Wang, S.G., Jia, S.S., Sun, D.Z., Fan, H., Chang, X.P. and Jing, R.L. (2016) Mapping QTLs for stomatal density and size under drought stress in wheat (*Triticum aestivum* L.). *Journal of Integrative Agriculture* 15(9), 1955–1967.

Wang, W., Vinocur, B. and Altman, A. (2003) Plant responses to drought, salinity and extreme temperatures: towards genetic engineering for stress tolerance. *Planta* 218, 1–14.

Wei, B., Jing, R., Wang, C., Chen, J., Mao, X. *et al.* (2009) Dreb1 genes in wheat (*Triticum aestivum* L.): development of functional markers and gene mapping based on SNPs. *Molecular Breeding* 23, 13–22.

Wilkinson, S. and Davies, W.J. (2002) ABA-based chemical signalling: the co-ordination of responses to stress in plants. *Plant, Cell & Environment* 25, 195–210.

Xiong, L., Wang, R.G., Mao, G. and Koczan, J.M. (2006) Identification of drought tolerance determinants by genetic analysis of root response to drought stress and abscisic acid. *Plant Physiology* 142, 1065–1074.

Xu, W., Jia, L., Shi, W., Liang, J., Zhou, F., Li, Q. and Zhang, J. (2013) Abscisic acid accumulation modulates auxin transport in the root tip to enhance proton secretion for maintaining root growth under moderate water stress. *New Phytopathology* 197, 139–150.

Xu, Y.F., Li, S.S., Li, L.H., Ma, F.F., Fu, X.Y. *et al.* (2017) QTL mapping for yield and photosynthetic related traits under different water regimes in wheat. *Molecular Breeding* 37, 34.

Xue, G.G., Way, H.H., Richardson, T., Drenth, J., Joyce, P.A. and McIntyre, C.L. (2011) Over expression of TaNAC69 leads to enhanced transcript levels of stress up-regulated genes and dehydration tolerance in bread wheat. *Molecular Plant* 4, 697–712.

Yadav, D., Shavrukov, Y., Bazanova, N., Chirkova, L., Borisjuk, N. *et al.* (2015) Constitutive over expression of the TaNF-YB4 gene in transgenic wheat significantly improves grain yield. *Journal of Experimental Botany* 66, 6635–6650.

Yancey, P.H. (2005) Organic osmolytes as compatible, metabolic and counteracting cytoprotectants in high osmolarity and other stresses. *Journal of Experimental Biology* 208(15), 2819–2830.

Yang, D.L., Jing, R.L., Chang, X.P. and Li, W. (2007) Identification of quantitative trait loci and environmental interactions for accumulation and remobilization of water-soluble carbohydrates in wheat (*Triticum aestivum* L.) stems. *Genetics* 176, 571–584.

Yang, J., Zhang, J., Wang, Z., Xu, G. and Zhu, Q. (2004) Activities of key enzymes in sucrose-to-starch conversion in wheat grains subjected to water deficit during grain filling. *Plant Physiology* 135, 1621–1629.

Yang, W., Liu, D., Li, J., Zhang, L., Wei, H. *et al.* (2009) Synthetic hexaploid wheat and its utilization for wheat genetic improvement in China. *Journal of Genetics and Genomics* 36, 539–546.

Yu, T.T., Xu, Z.Z., Guo, J.J., Wang, Y.Y., Abernathy, B. *et al.* (2017) Improved drought tolerance in wheat plants over expressing a synthetic bacterial cold shock protein gene SeCspA. *Scientific Reports* 7, 44050.

Zandalinas, S.I., Rivero, R.M., Martinez, V., Gómez-Cadenas, A. and Arbona, V. (2016) Tolerance of citrus plants to the combination of high temperatures and drought is associated to the increase in transpiration modulated by a reduction in abscisic acid levels. *BMC Plant Biology* 16(1), 105.

Zang, X., Geng, X., He, K., Wang, F., Tian, X. *et al.* (2018) Overexpression of the wheat (*Triticum aestivum* L.) TaPEPKR2 gene enhances heat and dehydration tolerance in both wheat and *Arabidopsis*. *Frontiers in Plant Science* 9, 1710.

Zhang, H., Mao, X., Zhang, J., Chang, X. and Jing, R. (2013) Single nucleotide polymorphisms and association analysis of drought-resistance gene TaSnRK2.8 in common wheat. *Plant Physiology and Biochemistry* 70, 174–181.

Zhang, H., Xu, W., Wang, H., Hu, L., Li, Y. *et al.* (2014) Pyramiding expression of maize genes encoding phosphoenolpyruvate carboxylase (PEPC) and pyruvate orthophosphate dikinase (PPDK) synergistically improve the photosynthetic characteristics of transgenic wheat. *Protoplasma* 251, 1163–1173.

Zhang, J.Y., Broeckling, C.D., Sumner, L.W. and Wang, Z.Y. (2007) Heterologous expression of two *Medicago truncatula* putative ERF transcription factor genes, WXP1 and WXP2, in *Arabidopsis* led to increased leaf wax accumulation and improved drought tolerance, but differential response in freezing tolerance. *Plant Molecular Biology* 64, 265–278.

Zhang, Y., Wang, Z., Fan, Z., Li, J., Gao, X. *et al.* (2019) Phenotyping and evaluation of CIMMYT WPHYSGP nursery lines and local wheat varieties under two irrigation regimes. *Breeding Science* 69, 55–67.

Zhang, Y.J., Yang, J.S., Guo, S.J., Meng, J.J., Zhang, Y.L. *et al.* (2011) Over-expression of the *Arabidopsis* CBF1 gene improves resistance of tomato leaves to low temperature under low irradiance. *Plant Biology* 13, 362–367.

Zhang, Z., Liu, X., Wang, X., Zhou, M., Zhou, X., Ye, X. and Wei, X. (2012) An R2R3 MYB transcription factor in wheat, TaPIMP1, mediates host resistance to *Bipolaris sorokiniana* and drought stresses through regulation of defense- and stress-related genes. *New Phytopathology* 196, 1155–1170.

Zhou, J., Ma, C., Zhen, S., Cao, M., Zeller, F.J. *et al.* (2016) Identification of drought stress related proteins from 1Sl (1B) chromosome substitution line of wheat variety Chinese Spring. *Botanical Studies* 57, 20.

Zhu, J.K. (2002) Salt and drought stress signal transduction in plants. *Annual Review and Plant Biology* 53, 247–273.

Zlatev, Z. (2009) Drought-induced changes in chlorophyll fluorescence of young wheat plants. *Biotechnology and Biotechnological Equipment*, 438–441 Special Edition.

# 5 Molecular Breeding for Improving Heat Stress Tolerance in Wheat

**Kuohai Yu, Huiru Peng, Zhongfu Ni, Yingyin Yao, Zhaorong Hu, Mingming Xin and Qixin Sun***

*China Agricultural University, Beijing, People's Republic of China*

## 5.1 Introduction

Wheat (*Triticum aestivum* L., AABBDD) is one of the earliest domesticated food crops. It has been the basic staple food of civilizations in Europe, West Asia and North Africa for about 10,000 years (Eckardt, 2010). According to the latest data by the Food and Agriculture Organization of the United Nations (http://www.fao.org/, accessed 8 February 2021), wheat is the second most productive crop in the world after maize (FAO, 2020) and it is cultivated from 67°N in Scandinavia and Russia to 45°S in Argentina, including elevated regions in the tropics and subtropics (Shewry and Hey, 2015). As a temperate-season crop, wheat has as an optimal daytime growing temperature of 20 to 24°C during reproductive development (Farooq *et al.*, 2011). Recent studies have shown that wheat yields will decline by 4.1 to 6.4% for each increase of 1°C due to climate change while global surface temperature at the end of the 21st century will be 1.5°C higher than it is now, which will significantly intensify the growing demands on food supply (Liu *et al.*, 2016; IPCC, 2018).

To cope with climate change and ensure food security, scientists have studied the heat tolerance mechanism of model plants and crops, including wheat, for a long time. It is suggested that high temperature stress often results in an accumulation of reactive oxygen species (ROS), leading to oxidative stress, DNA damage, irreversible oxidation of proteins and lipids, and thylakoid membrane damage in cells (Mittler *et al.*, 2004; Suzuki *et al.*, 2012; Narayanan *et al.*, 2015). A previous study indicated that pre-flowering and flowering are the stages most sensitive to high temperature during wheat development (Cossani and Reynolds, 2012). Short-term high temperature in pre-flowering and flowering stages can reduce the number of grains per ear, which can be attributed to a reduction in the ability of pollen germination and the growth rate of the pollen tube (Feng *et al.*, 2014). In addition, high temperature not only causes damage to yield-related traits, but also causes deterioration of wheat quality. After thousands of years of evolution and domestication to protect itself from heat stress, wheat has evolved complex systems to enhance its tolerance to climate change. Therefore, understanding the complex response of wheat to heat stress, as well as potential new technologies, will shed light on important strategies for the development of new varieties to adapt to temperature changes (Ni *et al.*, 2018).

*Corresponding author: qxsun@cau.edu.cn

 *Molecular Breeding in Wheat, Maize and Sorghum: Strategies for Improving Abiotic Stress Tolerance and Yield* (eds M.A. Hossain *et al.*)



DOI: 10.1079/9781789245431.0005

## 5.2 Wheat Responses to Heat Stress

### 5.2.1 Morphological and growth responses

Heat stress can affect plant growth and development, leading to morphological and growth changes in wheat plants, hindering development processes and eventually resulting in loss of quantity and yield (Rahman *et al.*, 2009). Gupta *et al.* (2013) applied heat stress at 45°C for 2 h on 7-day-old wheat seedlings and revealed that heat stress reduced length and dry mass of shoot and root; it also decreased membrane stability index and reduced chlorophyll content by up to 18.52%. Heat stress at grain-filling stage can also reduce maximum quantum yield of photosystem II, grain weight and yield (Pradhan and Prasad, 2015). Besides, Lobell *et al.* (2012) detected a statistically significant acceleration of senescence from extreme heat stress, above and beyond the effects of increased average temperatures, indicating that heat stress presents an even greater challenge to wheat than implied by previous modelling studies. The response of wheat to heat stress at different developmental stages was also different. Liu *et al.* (2011) applied heat stress to wheat for a 3-day period at different times after flowering and found that heat treatment at 6–8 days after flowering had the greatest effect on starch accumulation. The contents of amylose and amylopectin decreased significantly in response to heat stress, and the decrease of amylopectin content was more serious than that of amylose (Liu *et al.*, 2011). It is reported that heat stress significantly decreased the expression level of starch biosynthesis-related genes, such as those encoding ADP-glucose pyrophosphorylase, starch synthase I, II and III, granule-bound starch synthase and starch-branching enzyme I and II, which may directly lead to the decrease of starch content in wheat after heat stress (Hurkman *et al.*, 2003). Besides, a systematic study of 101 cultivars with varied geographic origin showed that the duration and timing of heat stress significantly influenced 1000-kernel weight and reproductive tiller number, explaining 51.6% of its phenotypic variance (Balla *et al.*, 2019).

### 5.2.2 Cellular structure and physiological response

High temperature induces a series of physiological effects on plants, including cellular structure destruction, tissue dehydration, damage of photosystems, affecting respiration etc., to limit plant growth and development. Denaturation of proteins and increased levels of unsaturated fatty acids caused by heat shock disrupt water, ion and organic solute movement across membranes, leading to increased cell-membrane permeability and, in turn, inhibition of cellular function (Ni *et al.*, 2018). ROS accumulation associated with heat stress also damages membranes (Mohammed and Tarpley, 2009). It is reported that photosynthesis is quite sensitive to high temperature, whereas thylakoid membranes and photosystem II are the most heat-sensitive cell components (Ristic *et al.*, 2007). Heat stress can lead to a physical separation of the chlorophyll light-harvesting complex II from the photosystem II core complex and disruption of photosystem II-mediated electron transfer, and finally result in the decrease of photosynthetic rate under heat stress (Ristic *et al.*, 2008). Heat stress also accelerated leaf senescence during wheat maturity, causing loss of chloroplast integrity and rapid decline of photosystem II-mediated electron transport (Haque *et al.*, 2014). In addition, the increase of non-photorespiratory processes and the accumulation of ROS caused by heat stress may lead to the inactivation of chloroplast enzymes, which in turn reduces the net photosynthetic rate of leaves (Ainsworth and Ort, 2010). Therefore, the detoxification by antioxidant systems is of great significance to protect plant cells from heat stress injury.

## 5.3 Molecular-Genetic Bases of Heat Response in Wheat

### 5.3.1 Multi-omics research on heat stress response

The transcriptome of wheat has undergone great changes in response to heat stress, although its regulatory mechanisms are not fully understood. In a pioneering study, Qin *et al.* (2018) applied heat stress at 40°C on wheat seedlings, resulting in differential expression of about 10.7% of the transcriptome probe sets detected by microarray. Except for heat-shock protein (HSP) and heat-shock factor (HSF) encoding genes, those putative heat-responsive genes encoding proteins involved in phytohormone biosynthesis/signalling, calcium and sugar signalling pathways, ribosomal proteins, primary and secondary metabolisms were also up- or downregulated. Of these, 313 probes sets were differentially expressed in a heat-tolerant cultivar (TAM107) and a heat-sensitive cultivar (CS) after heat stress (Qin *et al.*, 2008). Kumar *et al.* (2015) identified 1525 differentially expressed transcripts under heat stress using RNA-seq analysis and recorded that metabolic processes, protein phosphorylation and oxidation–reduction processes were significantly influenced. Bread wheat is an allohexaploid containing three subgenomes (A, B and D) and exhibits stronger abiotic stress tolerance than its progenitors. Liu *et al.* (2018) reported that about 68.4% of wheat homoeologous genes are expressed differently under heat stress condition. Moreover, alternative splicing (AS) might also contribute to heat tolerance in wheat; besides, the patterns of a subset of genes were changed in response to heat stress and AS events occurred more in B subgenome than in A and D subgenomes. Comparison of genes regulated at AS and transcriptional levels showed that about 40% of heat stress-induced AS genes were subjected to transcriptional regulation, suggesting that both expression variation and AS changes play an important role in the heat stress response of bread wheat (Liu *et al.*, 2015, 2018).

Wheat responses to heat stress are not only at transcriptional level but also at translational level. The flag leaves of heat-tolerant wheat variety GW451 and heat-susceptible wheat variety WH147 were analysed by two-dimensional electrophoresis. It was found that there were more low-molecular-weight proteins in WH147 after heat treatment than before (Nandha *et al.*, 2018). Lu *et al.* (2017) carried out a daytime high temperature stress test on wheat at grain-filling stage. It was revealed that the expression levels of proteins related to chlorophyll synthesis, carbon fixation, protein turnover and redox regulation were significantly changed (Lu *et al.*, 2017). Zhang *et al.* (2017) used isobaric tags for relative and absolute quantitation (iTRAQ) methods to investigate the changes of protein expression profiles in the grain of wheat variety Jing411 under high stress condition, finding that 256 proteins were differentially expressed, among which 126 proteins were upregulated and 130 proteins were downregulated. Gene ontology enrichment analysis showed that these differentially expressed proteins were mainly involved in stimulus response, abiotic stress response and stress response, kinase activity and transferase activity (Zhang *et al.*, 2017).

Epigenetic modification contributes to the rapid and effective regulation of gene expression and ultimately improves plant stress tolerance. DNA methylation, microRNAs (miRNAs) and long non-coding RNAs (LncRNAs) have been proved to be important components of regulating heat tolerance in wheat. Gardiner *et al.* (2015) assayed 293,076 cytosine residues across the wheat genome in seedling samples at 12 and 27°C and found 0.1% of sites showing differential DNA methylation between the two temperatures. Although only minor differences in methylation patterns were observed between wheat at 12 and 27°C conditions, the authors confirmed that DNA methylation was associated with small changes in gene expression which have a temperature-dependent expression profile, including heat shock and stress-related genes (Gardiner *et al.*, 2015).

miRNAs are a class of 20–24 nt small RNAs which regulate gene expression by splicing targeted transcripts or by inhibiting translation and might play a role in regulating heat response in wheat (Xin *et al.*, 2010; Ragupathy *et al.*, 2016). Wang *et al.* (2012) found that wheat miR159 can bind to target genes *TaGAMYB1* and *TaGAMYB2* and directed their cleavage. Overexpression of wheat miR159 in rice leads to a heat-sensitive phenotype compared with wild type, indicating a contribution to the heat response (Wang *et al.*, 2012). In addition, Xin *et al.* (2011) found that LncRNA also participated in the heat response of wheat. LncRNA may act as a long transcript or by producing small interfering RNAs (siRNAs) or miRNAs. For example, *TahlnRNA27* may be the precursor of *micro2010*. Its expression level in heat-tolerant variety TAM107 was upregulated an hour after heat treatment. Interestingly, $H_3K_9$ acetylation may be involved in the heat stress response in wheat by regulating the expression of non-coding RNA (Xin *et al.*, 2011). Ravichandran *et al.* (2019) found that a total of

36 mature miRNAs were differentially expressed before and after heat stress in wheat, and parallel analysis of RNA ends (PARE) (degradome) sequencing validated that these responsive miRNAs targeted superoxide dismutases and an array of homeobox leucine-zipper proteins, F-box proteins and protein kinases (Ravichandran *et al.*, 2019).

### 5.3.2 Mapping quantitative trait loci related to heat tolerance

Heat tolerance is a quantitative trait controlled by many loci/genes (Bohnert *et al.*, 2006). Many efforts have been undertaken to elucidate the genetic basis of heat tolerance in wheat. Sun and Quick (1991) mapped heat tolerance-related genes on chromosomes 3A, 3B, 4A, 4B and 5A in durum wheat (Langdon) using chromosome substitution lines in terms of membrane thermal stability. Xu *et al.* (1996) then reported that chromosomes 3A, 3B and 3D were also associated with heat tolerance in wheat cultivar Hope.

With the rapid development of molecular marker technology, many heat tolerance-related quantitative trait loci (QTLs) have been identified on different chromosomes of wheat by using physiological and heat-sensitive indicators. For example, Pinto *et al.* (2010) located 16 QTLs closely related to high temperature using 167 recombinant inbred lines of Seri/Babax and took canopy temperature at the vegetative stage as a heat tolerance index, which accounted for 28% of phenotypic variation. They also found that the QTLs located at chromosomes 4A and 3B contributed to both heat and drought stress (Pinto *et al.*, 2010). In addition, using the number of grains per panicle and grain weight as heat sensitivity indicators, Mason *et al.* (2011) identified 14 QTLs on chromosomes 1B, 5A and 6D using 121 wheat recombinant inbred lines (Halberd × Karl92) that could explain 4.5 to 19.3% of phenotypic variation, respectively. Paliwal *et al.* (2012) identified four heat tolerance-related QTLs on chromosomes 2B, 7B and 7D with 1000-grain weight, filling time and heat-sensitive index of yield as indicators, which could explain 7.21–25.39% of phenotypic variation. By exploring Fv/Fm (maximum quantum efficiency of photosystem II) as a heat sensitivity indicator, Sharma *et al.* (2017) identified three important heat stress QTLs in wheat, two on chromosomes 3B and one on 1D, using 34,955 DArT-seq (Diversity Arrays Technology) and 27 SSR (simple sequence repeat) markers.

Although many heat tolerance-related QTLs have been mapped, no genes have been cloned, thus there is still a long way to go to understand the underlying molecular mechanisms in response to heat stress in wheat. However, with release of a wheat reference genome sequence, development of high-throughput sequencing and molecular marker technology, it is believed that the cloning of heat tolerance genes will be an important area for dissecting the genetic basis of wheat heat tolerance in the future.

### 5.3.3 Functional genes in response to heat stress

Heat shock rapidly changes the expression patterns of heat-related genes in plants (Bita and Gerats, 2013; Zhuang *et al.*, 2014). In recent years, many heat tolerance-related genes have been identified using reverse genetic methods including transcriptomics, proteomics and metabolomics, which help us to understand mechanisms at the molecular level and improve heat tolerance in wheat during breeding programmes.

*HSPs* are a set of stress genes induced rapidly under heat stress conditions. The proteins act as molecular chaperones in organisms to protect proteins from unpleasant folding in cells in response to heat stress (Vierling, 1991). Rampino *et al.* (2009) measured the acquired thermo-tolerance of different varieties of durum wheat in terms of cell-membrane stability and found that the accumulation of *HSP* transcripts was related to the duration of heat stress and significantly affected heat tolerance of durum wheat. *HSF* genes also play a central regulatory role in acquired thermo-tolerance. Bioinformatic and phylogenetic analyses identified 56 *HSF* members in wheat, which are classified into A, B and C classes. Upon heat stress, the transcript levels of A2- and A6-type *HSF* members were highly expressed, indicating an important regulatory role during heat stress (Xue *et al.*, 2014).

In addition to *HSP* and *HSF* genes, Zang *et al.* (2017) found a ferritin gene *TaFER* on chromosome 5B contributing to heat tolerance based on the transcriptome analysis. Moreover, *TaFER* transgenic plants showed improved heat tolerance in *Arabidopsis* associated to ROS accumulation (Zang *et al.*, 2017). In addition, Geng *et al.* (2018) found that upregulation and unconventional splicing of *TabZIP60* occurred in wheat seedlings in response to heat stress. Constitutive expression of the spliced form of *TabZIP60* (*TabZIP60s*) enhanced heat tolerance in *Arabidopsis*, but overexpression of the unspliced form (*TabZIP60u*) did not. RNA-seq analysis revealed that unconventional splicing of *TabZIP60* could contribute to heat tolerance in transgenic plants by modulating the expression of stress-related genes in the endoplasmic reticulum (Geng *et al.*, 2018).

As important signal molecules in cells, phytohormones are involved in the regulation of various physiological processes of plant growth and development, including heat tolerance (Ahammed *et al.*, 2016). Growing evidence shows that the plant hormone abscisic acid (ABA) is likely an important contributor in the regulation of heat tolerance in wheat. A wheat ABA-insensitive genetic variant had higher kernel weight and yield compared with its parental line, and it grew faster and lasted longer in leaf area under stressed conditions. Selecting an ABA-insensitive genotype may be an effective way to improve heat tolerance and drought resistance of wheat (Lu *et al.*, 1989). Exogenous ABA application improved grain yield by increasing the grain sink capacity and grain-filling rate through regulating endogenous hormone contents to promote endosperm cell division and photosynthetic accumulation under both normal and high temperature stress conditions (Ni *et al.*, 2018). In addition, ethylene is also reported to link with a yield penalty under heat stress that lowers ethylene content in the spike and was strongly associated with high grain yield (Valluru *et al.*, 2017). Moreover, Tian *et al.* (2020) reported that heat stress induces upregulation of the gene encoding 12-oxo-phytodienoic acid reductase (*TaOPR3*), which is involved in jasmonate biosynthesis. Wheat *TaOPR3* knockdown lines showed enhanced heat sensitivity, whereas overexpression lines exhibited improved heat tolerance (Tian *et al.*, 2020).

### 5.3.4 Improving heat tolerance of wheat by comprehensive strategies

High temperature adversely affects plant growth and causes severe crop yield losses worldwide, especially for chimonophilous wheat. Utilization of excellent genetic diversity is a priority for improvement of heat tolerance in wheat breeding programmes. More than 1200 wheat landraces collected from different heat-stressed areas were analysed; a highly significant correlation between chlorophyll content in leaves and 1000-grain weight was observed and a group of excellent germplasms was identified (Hede *et al.*, 1999). Based on the study of chlorophyll fluorescence, wheat heat tolerance has additive and cytoplasmic effects, and recurrent selection is an effective way to accumulate wheat heat tolerance-related genes (Moffatt *et al.*, 1990). Additionally, heat tolerance variations were observed in *Aegilops speltoides*, *Aegilops longissima* and *Aegilops searsii* (Pradhan *et al.*, 2012; Ni *et al.*, 2018). Only a few genetic materials in heat tolerance have been utilized due to the limitations of conventional breeding methods. However, marker-assisted selection (MAS) is an option for improving heat tolerance in wheat since more and more heat tolerance-related QTLs/genes have been reported. Besides, genetic modification can also be used to improve wheat heat tolerance. For example, overexpression of maize phosphoenolpyruvate carboxylase gene *ZmPEPC* in wheat enhanced photochemical and antioxidant enzyme activities, delayed chlorophyll degradation, altered contents of proline and other metabolites, and ultimately improved heat tolerance (Qi *et al.*, 2017). In addition, transgenic wheat with constitutive expression of maize *EFTu1* gene also exhibited enhanced heat tolerance (Fu *et al.*, 2008). It is worth noticing that the clustered regularly interspaced short palindromic repeats (CRISPR)/CRISPR-associated protein 9 (Cas9) system has been explored in wheat, which might be an efficient tool to improve heat tolerance in future studies.

## 5.4 Conclusion

Heat stress has become one of the limiting factors affecting wheat production. Genetic improvement of heat tolerance is a major scientific problem. However, researchers are still far from fully understanding the genetic and molecular basis because heat tolerance is a quantitative trait and controlled by many genes of minor effect, and the regulating network is also complicated. This chapter has focused on the recent studies of genetic and genomic research in response to heat stress in wheat, which needs further investigation in the future. However, it is believed that with the emphasis on genetic resource exploration and with better understanding of the molecular basis, heat tolerance will be improved during wheat breeding programmes in the future.

## References


Ahammed, G.J., Li, X., Zhou, J., Zhou, Y.H. and Yu, J.Q. (2016) Role of hormones in plant adaptation to heat stress. In: Ahammed, G.J. and Yu, J.Q. (eds) *Plant Hormones Under Challenging Environmental Factors*. Springer, Dordrecht, the Netherlands, pp. 1–21.

Ainsworth, E.A. and Ort, D.R. (2010) How do we improve crop production in a warming world? *Plant Physiology* 154, 526–530.

Balla, K., Karsai, I., Bonis, P., Kiss, T., Berki, Z. *et al.* (2019) Heat stress responses in a large set of winter wheat cultivars (*Triticum aestivum* L.) depend on the timing and duration of stress. *PLoS One* 14, e0222639.

Bita, C.E. and Gerats, T. (2013) Plant tolerance to high temperature in a changing environment: scientific fundamentals and production of heat stress-tolerant crops. *Frontiers in Plant Science* 4, 273.

Bohnert, H.J., Gong, Q., Li, P. and Ma, S. (2006) Unraveling abiotic stress tolerance mechanisms – getting genomics going. *Current Opinion in Plant Biology* 9, 180–188.

Cossani, C.M. and Reynolds, M.P. (2012) Physiological traits for improving heat tolerance in wheat. *Plant Physiology* 160, 1710–1718.

Eckardt, N.A. (2010) Evolution of domesticated bread wheat. *The Plant Cell* 22, 993.

FAO (2020) Crop prospects and food situation. In: FAO (ed.) *Quarterly Global Report No. 2, July 2020*. Food and Agriculture Organization of the United Nations, Rome, pp. 1–45.

Farooq, M., Bramley, H., Palta, J.A. and Siddique, K.H.M. (2011) Heat stress in wheat during reproductive and grain-filling phases. *Critical Reviews in Plant Sciences* 30, 491–507.

Feng, B., Liu, P., Li, G., Dong, S.T., Wang, F.H., Kong, L.A. and Zhang, J.W. (2014) Effect of heat stress on the photosynthetic characteristics in flag leaves at the grain-filling stage of different heat-resistant winter wheat varieties. *Journal of Agronomy and Crop Science* 200, 143–155.

Fu, J., Momčilović, I., Clemente, T.E., Nersesian, N., Trick, H.N. and Ristic, Z. (2008) Heterologous expression of a plastid EF-Tu reduces protein thermal aggregation and enhances $CO_2$ fixation in wheat (*Triticum aestivum*) following heat stress. *Plant Molecular Biology* 68, 277–288.

Gardiner, L.J., Quinton-Tulloch, M., Olohan, L., Price, J., Hall, N. and Hall, A. (2015) A genome-wide survey of DNA methylation in hexaploid wheat. *Genome Biology* 16, 273.

Geng, X., Zang, X., Li, H., Liu, Z., Zhao, A. *et al.* (2018) Unconventional splicing of wheat TabZIP60 confers heat tolerance in transgenic *Arabidopsis*. *Plant Science* 274, 252–260.

Gupta, N.K., Agarwal, S., Agarwal, V.P., Nathawat, N.S., Gupta, S. and Singh, G. (2013) Effect of short-term heat stress on growth, physiology and antioxidative defence system in wheat seedlings. *Acta Physiologiae Plantarum* 35, 1837–1842.

Haque, M.S., Kjaer, K.H., Rosenqvist, E., Sharma, D.K. and Ottosen, C.O. (2014) Heat stress and recovery of photosystem II efficiency in wheat (*Triticum aestivum* L.) cultivars acclimated to different growth temperatures. *Environmental and Experimental Botany* 99, 1–8.

Hede, A.R., Skovmand, B., Reynolds, M.P., Crossa, J., Vilhelmsen, A.L. and Stølen, O. (1999) Evaluating genetic diversity for heat tolerance traits in Mexican wheat landraces. *Genetic Resources and Crop Evolution* 46, 37–45.

Hurkman, W.J., McCue, K.F., Altenbach, S.B., Korn, A., Tanaka, C.K. *et al.* (2003) Effect of temperature on expression of genes encoding enzymes for starch biosynthesis in developing wheat endosperm. *Plant Science* 164, 873–881.

IPCC (2018) *Global Warming of 1.5°C. An IPCC Special Report on the Impacts of Global Warming of 1.5°C above Pre-Industrial Levels and Related Global Greenhouse Gas Emission Pathways, in the Context of Strengthening the Global Response to the Threat of Climate Change, Sustainable Development, and Efforts to Eradicate Poverty* (Masson-Delmotte, V., Zhai, P., Pörtner, H.O., Roberts, D., Skea, J. *et al.*, eds). Intergovernmental Panel on Climate Change, Geneva, Switzerland, pp. 1–26.

Kumar, R.R., Goswami, S., Sharma, S.K., Kala, Y.K., Rai, G.K. *et al.* (2015) Harnessing next generation sequencing in climate change: RNA-Seq analysis of heat stress-responsive genes in wheat (*Triticum aestivum* L.). *Omics* 19, 632–647.

Liu, B., Asseng, S., Müller, C., Ewert, F., Elliott, J. *et al.* (2016) Similar estimates of temperature impacts on global wheat yield by three independent methods. *Nature Climate Change* 6, 1130–1136.

Liu, P., Guo, W., Jiang, Z., Pu, H., Feng, C. *et al.* (2011) Effects of high temperature after anthesis on starch granules in grains of wheat (*Triticum aestivum* L.). *The Journal of Agricultural Science* 149, 159–169.

Liu, Z., Xin, M., Qin, J., Peng, H., Ni, Z., Yao, Y. and Sun, Q. (2015) Temporal transcriptome profiling reveals expression partitioning of homeologous genes contributing to heat and drought acclimation in wheat (*Triticum aestivum* L.). *BMC Plant Biology* 15, 152.

Liu, Z., Qin, J., Tian, X., Xu, S., Wang, Y. *et al.* (2018) Global profiling of alternative splicing landscape responsive to drought, heat and their combination in wheat (*Triticum aestivum* L.). *Plant Biotechnology Journal* 16, 714–726.

Lobell, D.B., Sibley, A. and Ortiz-Monasterio, J.I. (2012) Extreme heat effects on wheat senescence in India. *Nature Climate Change* 2, 186–189.

Lu, D.B., Sears, R.G. and Paulsen, G.M. (1989) Increasing stress resistance by *in vitro* selection for abscisic acid insensitivity in wheat. *Crop Science* 29, 939–943.

Lu, Y., Li, R., Wang, R., Wang, X., Zheng, W. *et al.* (2017) Comparative proteomic analysis of flag leaves reveals new insight into wheat heat adaptation. *Frontiers in Plant Science* 8, 1086.

Mason, E.R., Mondal, S., Beecher, F.W. and Hays, D.B. (2011) Genetic loci linking improved heat tolerance in wheat (*Triticum aestivum* L.) to lower leaf and spike temperatures under controlled conditions. *Euphytica* 180, 181–194.

Mittler, R., Vanderauwera, S., Gollery, M. and Van Breusegem, F. (2004) Reactive oxygen gene network of plants. *Trends in Plant Science* 9, 490–498.

Moffatt, J., Sears, R., Cox, T. and Paulsen, G. (1990) Wheat high temperature tolerance during reproductive growth: II. Genetic analysis of chlorophyll fluorescence. *Crop Science* 30, 886–889.

Mohammed, A.R. and Tarpley, L. (2009) Impact of high nighttime temperature on respiration, membrane stability, antioxidant capacity, and yield of rice plants. *Crop Science* 49, 313–322.

Nandha, A.K., Mehta, D.R., Tulsani, N.J., Umretiya, N. and Delvadiya, N. (2018) Proteomic analysis in wheat to study the effect of heat stress on flag leaf. *International Journal of Current Microbiology and Applied Sciences* 7, 3432–3439.

Narayanan, S., Prasad, P.V.V., Fritz, A.K., Boyle, D.L. and Gill, B.S. (2015) Impact of high night-time and high daytime temperature stress on winter wheat. *Journal of Agronomy and Crop Science* 201, 206–218.

Ni, Z., Li, H., Zhao, Y., Peng, H., Hu, Z., Xin, M. and Sun, Q. (2018) Genetic improvement of heat tolerance in wheat: recent progress in understanding the underlying molecular mechanisms. *The Crop Journal* 6, 32–41.

Paliwal, R., Roder, M.S., Kumar, U., Srivastava, J.P. and Joshi, A.K. (2012) QTL mapping of terminal heat tolerance in hexaploid wheat (*T. aestivum* L.). *Theoretical and Applied Genetics* 125, 561–575.

Pinto, R.S., Reynolds, M.P., Mathews, K.L., McIntyre, C.L., Olivares-Villegas, J.J. and Chapman, S.C. (2010) Heat and drought adaptive QTL in a wheat population designed to minimize confounding agronomic effects. *Theoretical and Applied Genetics* 121, 1001–1021.

Pradhan, G.P. and Prasad, P.V. (2015) Evaluation of wheat chromosome translocation lines for high temperature stress tolerance at grain filling stage. *PLoS One* 10, e0116620.

Pradhan, G.P., Prasad, P.V.V., Fritz, A.K., Kirkham, M.B. and Gill, B.S. (2012) High temperature tolerance in species and its potential transfer to wheat. *Crop Science* 52, 292–304.

Qi, X., Xu, W., Zhang, J., Guo, R., Zhao, M. *et al.* (2017) Physiological characteristics and metabolomics of transgenic wheat containing the maize $C_4$ phosphoenolpyruvate carboxylase (PEPC) gene under high temperature stress. *Protoplasma* 254, 1017–1030.

Qin, D., Wu, H., Peng, H., Yao, Y., Ni, Z. *et al.* (2008) Heat stress-responsive transcriptome analysis in heat susceptible and tolerant wheat (*Triticum aestivum* L.) by using wheat genome array. *BMC Genomics* 9, 432.

Ragupathy, R., Ravichandran, S., Mahdi, M.S., Huang, D., Reimer, E., Domaratzki, M. and Cloutier, S. (2016) Deep sequencing of wheat sRNA transcriptome reveals distinct temporal expression pattern of miRNAs in response to heat, light and UV. *Scientific Reports* 6, 39373.

Rahman, M., Chikushi, J., Yoshida, S. and Karim, A. (2009) Growth and yield components of wheat genotypes exposed to high temperature stress under control environment. *Bangladesh Journal of Agricultural Research* 34, 361–372.

Rampino, P., Mita, G., Pataleo, S., De Pascali, M., Di Fonzo, N. and Perrotta, C. (2009) Acquisition of thermotolerance and HSP gene expression in durum wheat (*Triticum durum* Desf.) cultivars. *Environmental and Experimental Botany* 66, 257–264.

Ravichandran, S., Ragupathy, R., Edwards, T., Domaratzki, M. and Cloutier, S. (2019) MicroRNA-guided regulation of heat stress response in wheat. *BMC Genomics* 20, 488–488.

Ristic, Z., Bukovnik, U. and Prasad, P.V.V. (2007) Correlation between heat stability of thylakoid membranes and loss of chlorophyll in winter wheat under heat stress. *Crop Science* 47, 2067–2073.

Ristic, Z., Bukovnik, U., Momcilovic, I., Fu, J. and Vara Prasad, P.V. (2008) Heat-induced accumulation of chloroplast protein synthesis elongation factor, EF-Tu, in winter wheat. *Plant Physiology* 165, 192–202.

Sharma, D.K., Torp, A.M., Rosenqvist, E., Ottosen, C.O. and Andersen, S.B. (2017) QTLs and potential candidate genes for heat stress tolerance identified from the mapping populations specifically segregating for F(v)/F(m) in wheat. *Frontiers in Plant Science* 8, 1668.

Shewry, P.R. and Hey, S.J. (2015) The contribution of wheat to human diet and health. *Food Energy Security* 4, 178–202.

Sun, Q.X. and Quick, J.S. (1991) Chromosomal locations of genes for heat tolerance in tetraploid wheat. *Cereal Research Communications* 19, 431–437.

Suzuki, N., Koussevitzky, S., Mittler, R. and Miller, G. (2012) ROS and redox signalling in the response of plants to abiotic stress. *Plant, Cell & Environment* 35, 259–270.

Tian, X., Wang, F., Zhao, Y., Lan, T., Yu, K. *et al.* (2020) Heat shock transcription factor A1b regulates heat tolerance in wheat and *Arabidopsis* through OPR3 and jasmonate signalling pathway. *Plant Biotechnology Journal* 18, 1109–1111.

Valluru, R., Reynolds, M.P., Davies, W.J. and Sukumaran, S. (2017) Phenotypic and genome-wide association analysis of spike ethylene in diverse wheat genotypes under heat stress. *New Phytologist* 214, 271–283.

Vierling, E. (1991) The roles of heat shock proteins in plants. *Annual Review of Plant Physiology and Plant Molecular Biology* 42, 579–620.

Wang, Y., Sun, F., Cao, H., Peng, H., Ni, Z., Sun, Q. and Yao, Y. (2012) TamiR159 directed wheat TaGAMYB cleavage and its involvement in anther development and heat response. *PLoS One* 7, e48445.

Xin, M., Wang, Y., Yao, Y., Xie, C., Peng, H., Ni, Z. and Sun, Q. (2010) Diverse set of microRNAs are responsive to powdery mildew infection and heat stress in wheat (*Triticum aestivum* L.). *BMC Plant Biology* 10, 123.

Xin, M., Wang, Y., Yao, Y., Song, N., Hu, Z. *et al.* (2011) Identification and characterization of wheat long non-protein coding RNAs responsive to powdery mildew infection and heat stress by using microarray analysis and SBS sequencing. *BMC Plant Biology* 11, 61.

Xu, R., Sun, Q. and Zhang, S. (1996) Chromosomal location of genes for heat tolerance as measured by membrane thermostability of common wheat cv. Hope. *Yi Chuan = Hereditas* 18, 1–3.

Xue, G.P., Sadat, S., Drenth, J. and McIntyre, C.L. (2014) The heat shock factor family from *Triticum aestivum* in response to heat and other major abiotic stresses and their role in regulation of heat shock protein genes. *Journal of Experimental Botany* 65, 539–557.

Zang, X., Geng, X., Wang, F., Liu, Z., Zhang, L. *et al.* (2017) Overexpression of wheat ferritin gene TaFER-5B enhances tolerance to heat stress and other abiotic stresses associated with the ROS scavenging. *BMC Plant Biology* 17, 14.

Zhang, Y., Pan, J., Huang, X., Guo, D., Lou, H. *et al.* (2017) Differential effects of a post-anthesis heat stress on wheat (*Triticum aestivum* L.) grain proteome determined by iTRAQ. *Scientific Reports* 7, 3468.

Zhuang, J., Zhang, J., Hou, X.L., Wang, F. and Xiong, A.S. (2014) Transcriptomic, proteomic, metabolomic and functional genomic approaches for the study of abiotic stress in vegetable crops. *Critical Reviews in Plant Sciences* 33, 225–237.

# 6 Molecular Breeding for Improving Waterlogging Tolerance in Wheat

R. Esten Mason[1*], Trenton L. Roberts[2], Richard Boyles[3], Andrea Acuna[2], Maria N. Arguello[4], Diana Ballesteros[2], and Nithya Subramanian[5]
[1]Colorado State University, Fort Collins, Colorado, USA; [2]University of Arkansas, Fayetteville, Arkansas, USA; [3]Clemson University, Florence, South Carolina, USA; [4]Ohio State University, Wooster, Ohio, USA; [5]Texas A&M University, College Station, Texas, USA

## 6.1 Introduction

Waterlogging negatively impacts 25% of worldwide area planted to wheat (*Triticum aestivum* L.) (Powell *et al.*, 2012) and as a non-aquatic species wheat is generally susceptible to yield loss or plant death from prolonged saturated soil conditions. Studies have shown genetic variation within some wheat germplasm pools and in wild relatives that could be exploited for increased tolerance to waterlogged soils, although molecular tools to do this are lacking (Boru *et al.*, 2001; McDonald *et al.*, 2001b; St Burgos *et al.*, 2001; Collaku and Harrison, 2002; Arguello *et al.*, 2016). Periodic flooding, drought and high temperature events already explain 35% or more of the variation in wheat yield globally and this is likely to increase with global climate change (Howden *et al.*, 2007). As such, more rapid development of waterlogging-tolerant cultivars is necessary to minimize losses in grain yield and maximize producer return.

Soils prone to waterlogging, such as the alluvial soils present in the southern span of the Mississippi River Delta region of the USA, have distinct physical characteristics, including restrictive soil layers known as fragipans which result in low infiltration rates. Soil waterlogging causes hypoxia followed by anoxia, which influences mineral nutrition and macro- and micronutrient accumulation resulting from chemical and biochemical reactions that occur when $O_2$ is depleted (Sairam *et al.*, 2008). Soil nutrient dynamic changes resulting from modification of soil pH and redox potential propitiate an inadequate nutritional status (Setter *et al.*, 2009). This often results in elemental toxicities for Mn, Fe (Shabala, 2011) and Al (Khabaz-Saberi *et al.*, 2012), in addition to possible deficiencies in N, P, K, Mg, Cu, Zn and others (Steffens *et al.*, 2005; Setter *et al.*, 2009).

Our current understanding of the genetic control of waterlogging tolerance in wheat is limited and a lack of molecular tools to complement or replace expensive and time-consuming phenotyping methods is a constraint to breeders. Reports of genetic loci associated with waterlogging tolerance are particularly limited when compared with other important crop species. In rice (*Oryza sativa* L.), for example, studies have been successful at identifying major quantitative trait loci (QTLs) for tolerance to hypoxia, with map-based cloning of several genes helping to confirm the importance of the ethylene signalling

*Corresponding author: esten.mason@colostate.edu


DOI: 10.1079/9781789245431.0006

pathway (Xu *et al.*, 2006; Hattori *et al.*, 2009). Studies in maize (*Zea mays* L.) have reported the fine mapping of QTLs involved in aerenchyma formation (Mano and Omori, 2009) and work in barley (*Hordeum vulgare* L.) has been successful in using a comparative mapping approach to confirm waterlogging tolerance QTLs across populations (Li *et al.*, 2008).

The success seen in other crops, despite the complexity of the stress itself and the tolerance mechanisms, shows the power of a combined phenotypic and molecular approach – and there is the potential for similar success in wheat. For example, enhanced elemental toxicity and deficiency tolerance are adaptive traits often associated with improved tolerance to soil waterlogging, although their genetic control is extremely understudied. Khabaz-Saberi *et al.* (2012) found wheat genotypes tolerant to high concentrations of $Al^{3+}$ had improved early root and shoot growth under waterlogging stress compared with susceptible genotypes. Increased tolerance to Al can be attributed to external mechanisms, such as Al-chelating organic acids that are exuded from the roots, and although major genes are known for dryland situations (Schnurbusch *et al.*, 2007; Cai *et al.*, 2008), current research is lacking to elucidate a similar mechanism to respond to waterlogging stress. In terms of molecular tools, genomic selection (GS) is an emerging form of marker-assisted selection (MAS) which aims to predict the performance (breeding values) of individuals using genome-wide marker data through training and validation of a prediction model (Meuwissen *et al.*, 2001). The training population consists of lines within the germplasm of a breeding programme having both phenotypic (i.e. response to waterlogging) and genotypic data and is used to estimate all marker effects simultaneously (Rutkoski *et al.*, 2011; Huang *et al.*, 2016). The genomic estimated breeding values (GEBVs) of individuals related to the training population can then be subsequently calculated using the respective GS model once the population is trained (Meuwissen *et al.*, 2001). Selection and advancement are carried out based at least in part on those lines predicted to have the highest GEBVs (Desta and Ortiz, 2014). While the pioneering GS studies focused on animal breeding (Meuwissen *et al.*, 2001; Hayes *et al.*, 2009), it has emerged as a valuable breeding tool for quantitative trait improvement in crops. In wheat, GS has been conducted for durable stem rust (*Puccinia graminis*) resistance (Rutkoski *et al.*, 2011), quality traits and grain yield (Huang *et al.*, 2016; Michel *et al.*, 2016), *Fusarium* head blight resistance (Arruda *et al.*, 2016) and other quantitative traits, but is limited in its use targeting tolerance to abiotic stresses including soil waterlogging.

In this chapter, the current understanding of the response of wheat to waterlogging stress, the genetic control of uptake and transport of macro- and micronutrients, and the QTLs and genes associated with tolerance mechanisms are summarized. Potential targets for molecular breeding through MAS and the potential for GS are discussed. This will provide a better understanding of the biology and genes underlying soil waterlogging tolerance, as well as clarity and direction for breeders for future molecular breeding targets to expedite cultivar development.

## 6.2 Soil Waterlogging Reduces Wheat Grain Yield

Crop production faces many constraints, including both biotic and abiotic factors (Spiertz, 2012). Waterlogging impacts 12% of agricultural land in the USA, 30% in France, 38% of the Indus Basin in Pakistan, as well as large areas in southern Australia and India (Boyer, 1982; Boru *et al.*, 2001; Kahlown and Azam, 2002; Jiang *et al.*, 2008). Soil waterlogging alters plant physiology and metabolism, resulting in reduced biomass but accelerated development, ultimately lowering grain yield and individual yield components. Waterlogging can impact wheat growth and development and grain yield if present at any growth stage and for any duration of time. Overall, mean grain yield losses of 7 to 45% have been reported (Belford, 1981; Cannell *et al.*, 1984; Musgrave and Ding, 1998; Collaku and Harrison, 2002; Arguello *et al.*, 2016). Cannell *et al.* (1984) reported reductions in grain yield and grain number per spike of 16 and 17%, respectively, when waterlogging occurred before emergence. Malik *et al.* (2001) reported shoot dry weight reductions between 67 and 72% in 3-week-old winter wheat plants subjected to waterlogging for 14 days, which also decreased seminal root growth threefold. Araki *et al.*

(2012a) reported reductions in 1000-kernel weight of 22 and 29% for plants subject to waterlogging stress during jointing and anthesis, respectively, the result of a shortened grain-filling duration and a decline in carbohydrate translocation. Robertson *et al.* (2009) observed a 50% reduction in tiller number after a 14-day waterlogging treatment and 71% lower shoot dry weight. Shao *et al.* (2013) reported a significant decrease in root length, root mass and root:shoot ratio when waterlogging occurred at the tillering or booting stage and an overall reduction in grain yield of 7.1 to 11.2%. Measurements of transpiration rate, stomatal conductance and net photosynthetic rate ($P_N$) were also affected, with $P_N$ decreased by 12 to 14% due to stomatal closure. The study concluded that a reduction in photosynthesis and thus grain yield could be explained by lower leaf turgor pressure and stomatal conductance due to decreased hydraulic conductivity, leading to a $CO_2$ deficit and a shortage of assimilate accumulation.

Collaku and Harrison (2002) reported mean grain yield reductions of 41% in US soft red winter wheat due to reductions in spike density and kernel number of 41 and 20%, respectively. A follow-up study showed kernel weight, chlorophyll content and tiller number to have moderately high heritability ranging from $H^2 = 0.31$ to 0.47, which was greater than for grain yield at $H^2 = 0.25$ (Collaku and Harrison, 2005). Similar results were reported in more modern soft red winter wheat germplasm evaluated in the field (Arguello *et al.*, 2016). While there are few reports of successful breeding strategies, Collaku and Harrison (2005) suggested that a selection index incorporating grain yield, kernel weight and tiller number could increase total grain yield by 17%.

## 6.3 Effect of Waterlogging on the Soil Environment

To understand the genetic response to soil waterlogging and identify potential targets for molecular breeding, it is important to understand the stress itself. There are two primary factors that inhibit plant growth when soils remain saturated for extended periods of time: (i) low ATP production resulting from a decrease in root respiration; and (ii) increase in trace element availability resulting in elemental toxicity or deficiency. In combination, these two abiotic factors can have large impacts on wheat yield resulting from poor root development, decreased stand and biomass, and senescence. An underdeveloped root system can also reduce tolerance to late-season heat and drought stress which require a vigorous root system for soil moisture exploitation.

Variation in site deposition has resulted in diverse and fertile soils in many crop production areas that are also lowly permeable and prone to soil waterlogging (Scott *et al.*, 1998; Snipes *et al.*, 2005). The biological shift in the soil environment resulting from prolonged saturation has a considerable impact on the chemistry of the soil. As $O_2$ depletion occurs there is a progression in the terminal electron acceptors utilized by soil microbes under hypoxic or anoxic conditions and will generally result in a net increase in $Fe^{2+}$, $Mn^{2+}$ and $Al^{3+}$, which are all highly available for plant uptake (Westerman, 1987; Stevenson and Cole, 1999). Phytotoxicity of Mn and Al have twofold effects as they negatively influence root growth and inhibit uptake of other essential elements such as Ca, Mg and K, causing deficiencies (Taylor and Foy, 1985; Carver and Ownby, 1995).

Soil texture is an important factor determining soil water content, water movement and water distribution through the root zone as well as the ability to cope with waterlogging stress (Evett, 2007). For example, in duplex soils water retention will start from the lower region and move towards the top of the soil profile. This results in seminal roots being exposed to hypoxia stress as soon as the soil water level rises, resulting in increased severity. In contrast, in sodic and heavy clay soils the waterlogging effect will move from the top to bottom of the soil profile, first affecting adventitious roots located in the upper profile (Setter and Waters, 2003). Cannell *et al.* (1984) reported that barley grown in a clay soil spent more time under anaerobic conditions compared with a sandy loam and thus needed a greater recovery time post waterlogging.

## 6.4 Effects of Anaerobic Conditions on Plant Metabolism

Waterlogging conditions alter the diffusion rate of $O_2$, reducing its availability by 320,000 times

(Armstrong, 1980; Lee *et al.*, 2006). A decrease in root $O_2$ availability leads to hypoxia under partial submergence or soil waterlogging and anoxia under complete. Under waterlogging, roots may be completely submerged (hypoxia) while the shoot remains above water level (Ahmed *et al.*, 2013). Hypoxia results in an energy crisis where the production of ATP can be reduced by 65 to 90% in affected tissues (Greenway and Gibbs, 2003). As oxygen is the main electron acceptor in the oxidative phosphorylation pathway (Dennis *et al.* 2000), a lack of it will inhibit photosynthesis, respiration and carbohydrate metabolism which are fundamental processes for energy production (Drew, 1997). Araki *et al.* (2012b) reported an increase in root respiration in wheat genotypes under waterlogging stress, which was related to a decrease in root growth and lower sugar metabolism in root cells. Lee *et al.* (2007) showed several ATP production processes including the enzymes NADH dehydrogenase and NADH ubiquinone oxidoreductase to be downregulated under hypoxic conditions in order to maintain ATP production and subserve aerenchyma formation. Alteration of ATP and $NAD^+$ cofactor limits coenzyme synthesis for the glycolysis cycle and redox reactions (Dennis *et al.*, 2000). Under normal conditions, plants will produce 2 moles of $CO_2$, 2 moles of $H_2O$ and 38 moles of ATP through the glucose pathway. In contrast, only 2 moles each of $CO_2$, $C_2H_5OH$ and ATP are produced under anaerobic conditions, reducing energy production by as much as 95% (Barrett-Lennard, 2003).

## 6.5 Plant Adaptations to Soil Waterlogging and Flooding

There are two classical responses to waterlogging or flooding in plants: (i) the low oxygen escape syndrome (LOES); and (ii) the quiescence strategy. The LOES strategy is an avoidance mechanism characterized by aerenchyma formation and shoot elongation in response to low $O_2$ following submergence (Bailey-Serres and Voesenek, 2008). The quiescence strategy involves the inhibition of growth through the downregulation of carbohydrate metabolism until re-emergence occurs (Xu *et al.*, 2006). Both the LOES and quiescence strategies have been well characterized in rice. Low $O_2$ triggers a stress pathway involving a gene family of ethylene-response factors (ERFs) that react to changes in redox, reactive oxygen species (ROS) and ATP supply (Limami *et al.*, 2014). In lowland rice, tolerance is characterized by a quiescence strategy in which the ERF *SUB1A* inhibits stem elongation and depletion of carbohydrates until flooding has subsided (Xu *et al.*, 2006). In deep-water rice the genetic haplotype is such that *SUB1A* is non-functional or slightly modified resulting in an LOES strategy where stem elongation promoted by gibberellic acid allows for escape from flooding if carbohydrate reserves are adequate (Hattori *et al.*, 2009). Internode elongation produces hollow stem structures that allow for adequate atmospheric gas exchange. Similar mechanisms in wheat have not been well characterized although early vigour may play a role in tolerance.

Under conditions of soil waterlogging wheat employs a combination of LOES and quiescence strategies. Root aerenchyma has been shown to form as early as 24 h post waterlogging, with well-defined aerenchyma present after 120 h (Jiang *et al.*, 2010). Depending on the mechanism by which it develops, aerenchyma can be schizogenous or lysigenous. Formation of lysigenous aerenchyma is found in wheat, barley, rice and maize (Evans, 2003) and can be triggered by abiotic stresses such as mineral deficiency, waterlogging and low $O_2$ (Drew and Lynch, 1980). In wheat, the development of aerenchyma tissue (Armstrong, 1980) and metabolic changes to modify energy production under low $O_2$ (Braendle and Crawford, 1987) are two of the principal adaptations involved in the waterlogging stress tolerance (Setter and Waters, 2003).

Root aerenchyma tissues allow for faster diffusion of $O_2$ for respiration and adequate production of ATP, resulting in less damage to cellular components. Both the phytohormone ethylene and $H_2O_2$ have been shown to regulate the programmed cell death involved in the development of aerenchyma and this adaptation is also known to play a role in tolerance to other abiotic factors such as salt and drought stresses. This is often accompanied by inhibition of seminal root growth and formation of lateral adventitious roots (Akhtar *et al.*, 1998; Malik *et al.*, 2001). Tolerant wheat varieties develop more adventitious roots containing more aerenchyma (Huang *et al.*, 1994b) and are more responsive to ethylene (Huang *et al.*, 1997). Boru *et al.* (2003) showed that tolerant spring wheats, Ducula, Prl/Sara and Vee/Myna, had greater root porosity compared with susceptible cultivars and this root porosity can lead to increased rooting depth in the waterlogged soil (Malik *et al.*, 2001). Ballesteros *et al.* (2015) found significant genotype-by-treatment interaction for root and shoot biomass production in response to waterlogging in a recombinant inbred line (RIL) population, indicating a possible adaptive mechanism. A successful quiescence strategy requires that normal growth and development be resumed following drainage of the soil and some variation for recovery of stomatal conductance and seminal root growth following waterlogging has been shown (Huang *et al.*, 1994b). Studies have shown carbon metabolism to be inhibited during waterlogging and to be preferentially allocated to root growth during recovery (Malik *et al.*, 2001). Collectively, both response strategies result in a stay-green phenotype and more biomass for subsequent yield production.

## 6.6 Genetic Control of Macronutrient Uptake and Transport

The solubility and availability of macronutrients become a concern with the changes in redox potential and soil pH that accompany soil waterlogging. There are six essential macronutrients in plants, including N, P, K and S, which are generally applied as commercial fertilizers to wheat, as well as Ca and Mg. For most macronutrients, there are established guidelines for critically low and optimal concentration ranges for growth and development (Table 6.1) and for the genes involved in nutrient uptake and transport (Table 6.2), although less is known about the genetic mechanisms of uptake to prevent potential nutrient deficiency or toxicity under waterlogging stress.

### 6.6.1 Nitrogen

N is the most thoroughly characterized nutrient in terms of the genetic control of uptake, transport and mobilization. Buchner and Hawkesford (2014) found complex expression patterns associated with 16 nitrate ($NO_3$-N) transporter genes in wheat that were homologous to those in *Arabidopsis thaliana* and responsible for uptake and distribution of $NO_3$-N within the plant. Tissue specificity and variation due to $NO_3$-N availability and senescence were observed. The source–sink relationship that drives remobilization of $NO_3$-N for grain fill is linked in large part to leaf senescence (Kong *et al.*, 2016). Both NAC-domain and WRKY transcription factors are upregulated during senescence and early grain-fill in wheat (Gregersen and Holm, 2007; Gregersen *et al.*, 2013) and are involved in the transition from early grain-fill to beginning of maturation in barley (Kohl *et al.*, 2015). The NAC transcription factor *Gpc-B1* was recently introduced into common wheat from ancestral wheat and transformed genotypes yielded higher grain protein content through accelerated leaf senescence and increased nutrient remobilization to the grain (Uauy *et al.*, 2006).

Low $O_2$ impacts both N uptake and reduction of $NO_3$-N to ammonium ($NH_4$-N) within the plant and waterlogging often results in N deficiency for wheat, which reduces shoot growth and tillering (Robertson *et al.*, 2009). Not only do prolonged waterlogged soil conditions influence the ability of wheat to uptake and assimilate N from the soil profile, they also significantly increase the N loss pathways by promoting leaching and denitrification, which are net losses from the soil system (Stevenson and Cole, 1999). While variation for shoot growth has been reported, the mechanism is not well understood (Huang *et al.*, 1994b). In tolerant species,

**Table 6.1.** Recommended macro- and micronutrient concentrations for wheat. (From Plank and Donohue, 2000).

| | Macronutrients (%) | | | | | | Micronutrients (ppm) | | | | |
|---|---|---|---|---|---|---|---|---|---|---|---|
| Statistic | N | P | K | Ca | Mg | S | B | Cu | Fe | Mn | Zn |
| Lower limit | 4.0 | 0.20 | 2.50 | 0.22 | 0.14 | 0.15 | 1.5 | 4.5 | 30.0 | 20 | 18 |
| Upper limit | 5.0 | 0.50 | 5.00 | 1.00 | 1.00 | 0.65 | 4.0 | 15.0 | 200.0 | 150 | 70 |
| Critically low | 3.0 | 0.15 | 2.00 | 0.15 | 0.10 | 0.10 | 1.0 | 3.0 | 25.0 | 15 | 15 |

**Table 6.2.** Important elements in wheat, their associated availability, function and known loci involved in uptake and transport.

| Element | Average soil availability | Key changes under hypoxia stress | Key interaction with other nutrients | Major function of nutrient | Known genes/QTLs | Reference |
|---|---|---|---|---|---|---|
| N | Low, commercially applied | Often deficient | Directly impacts absorption of P and S | Major component of chlorophyll and amino acids | *NPF* gene family | Buchner and Hawkesford (2014) |
| P | Low, commercially applied | Availability increases | High P can result in deficiencies in Fe and Zn | Component of DNA and ATP | *PHT1* gene family | Teng *et al*. (2017) |
| K | Generally low, commercially applied | Root K decreases and inhibits root growth | Impacts N uptake when deficient, antagonistic to absorption of Ca and Mg | $K^+$ involved in nearly all plant processes, but essential for plant water regulation | | Fageria (2001) |
| S | Medium, lower availability in sandy soils | | | Critical for S-containing amino acids, important for nitrate reductase | | Fageria (2001) |
| Ca | Medium, decreases in low-pH soils | | With Mg and P neutralizes organic acids | Enzyme activation, neutralizes organic acids and cell-wall structure | | Fageria (2001) |
| Mg | Medium, decreases in low-pH soils | Often deficient | Carrier of P in the plant and activates kinase enzymes, decreases uptake of K and Ca | Essential component of chlorophyll | | Fageria (2001) |
| Na | Low unless sodic or salt-affected soil | | Will replace K when K levels are low in the soil, but will not substitute for K in metabolic processes | Na/K signalling processes | *Nax1* | James *et al*. (2011) |

*Continued*

**Table6.2.** Continued.

| Element | Average soil availability | Key changes under hypoxia stress | Key interaction with other nutrients | Major function of nutrient | Known genes/QTLs | Reference |
|---|---|---|---|---|---|---|
| Al | High, increases when pH < 5 | Availability of $Al^{3+}$ can result in toxicity | Deficiency of mineral nutrients and decreased water uptake due to inhibition of root elongation | Impairs DNA synthesis and root growth | *Alt1* | Kinraide (1988); Delhaize *et al.* (1993a); Rout *et al.* (2001); Ma *et al.* (2005) |
| B | Medium | | Positive interactions with Ca | Cell-wall biosynthesis, reproductive growth | *Bo1* | Schnurbusch *et al.* (2007) |
| Cl | Medium | | | Photosynthesis, stomatal regulation | | Fageria (2001) |
| Cu | Low | | | Enzyme activation, photosynthesis, respiration | 4BL/5RL | Graham *et al.* (1987) |
| Fe | High | Availability of $Fe^{2+}$ can result in toxicity | In excess inhibits root growth and reduces uptake of N, P, K, Mg and other nutrients | Chlorophyll development, abiotic stress regulation | *NAM-B1* | Uauy *et al.* (2006) |
| Mn | Medium | Availability of $Mn^{2+}$ can result in toxicity | | Enzyme activation, photosynthesis, respiration | *CAX2* | Hirschi *et al.* (2000); Delhaize *et al.* (2003) |
| Mo | Low | | Catalyses nitrate reductase for available N | Frost tolerance, N metabolism | *Cbf14* | Al-Issawi *et al.* (2013) |
| Zn | Medium | Often deficient | Can reduce effects of B toxicity | Heat tolerance, membrane integrity, protection against ROS | *NAM-B1* | Torun *et al.* (2000); Hacisalihoglu and Kochian (2003); Cakmak (2008); Peck and McDonald (2010) |

the enzyme nitrate reductase is more active compared with nitrite reductase, which can alleviate stress by regenerating $NAD^+$ from NADH (Bailey-Serres and Voesenek, 2008). The production of NO by nitrate reductase is required for the formation of ethylene-induced lysigenous root aerenchyma in response to hypoxia (Wany *et al.*, 2017). It has also been shown that the application of exogenous N will remedy some negative effects of soil waterlogging in wheat and barley (Wu *et al.*, 2014; Simpson *et al.*, 2016).

### 6.6.2 Phosphorus

P is a limiting factor for barley grain yield under waterlogging stress (Ylivainio *et al.*, 2018) and in production systems where wheat follows flooded rice (Saleque *et al.*, 2006). Inorganic phosphate is taken up by the root hairs and low P can trigger changes in root anatomy, similar to the response to soil waterlogging and flooding. P uptake and *in vivo* remobilization are mediated by the *PHT1* gene family (Teng *et al.*, 2017), of which there are 14 in wheat. The expression of some *PHT1* genes (*TaPHT1.1/1.9*, *1.2* and *1.10*) is root specific and is involved in P uptake during flowering (Teng *et al.*, 2017). Wheat lines over-expressing *PHR1-A1* had higher grain number per spike and grain yield in field trials. The *PHT1* family is known to be upregulated when P is limiting and during salt stress, although generally independent of changes in root anatomy (Li *et al.*, 2019; Roch *et al.*, 2019). The *PHT1* genes may also be developmentally or growth stage specific (Grun *et al.*, 2018). Low P triggers aerenchyma formation and root growth in rice plants grown in solution (Lu *et al.*, 1999). As stated by Roch *et al.* (2019), very little is known about *PHT1* genes in wheat compared with other species including rice, although given that P can be limiting during waterlogging stress, they may play a key role in tolerance.

### 6.6.3 Potassium

Plants deficient in K often have decreased leaf number and size, which has negative effects on yield performance (Pettigrew, 2008). Most K uptake in wheat occurs during the rapid vegetative growth phase (commonly referred to as 'green-up'). K is often deficient in wheat–rice cropping systems (Panaullah *et al.*, 2006) and a decrease in K can be seen in wheat seedlings even after short periods of hypoxia. The role of K in tolerance to hypoxia may be attributed to maintaining K along with sugars in the roots for membrane integrity and subsequent translocation to the shoot (Greenway *et al.*, 1992; Colmer and Greenway, 2011). Imbalance of K is often associated with salt stress due to lower K:Na. Leakage of K from roots occurs as early as within 24 h of anoxia and increases significantly after 48 to 72 h. Genotypes differ in their ability recover root K concentrations which correlates with resumption of seminal root growth following anoxia (Goggin and Colmer, 2007). Cuin *et al.* (2009) reported a decrease in total shoot K under salt stress but an increase in shoot sap K concentration, which may contribute to osmotic adjustment. Waterlogging also decreases the concentration of K and lowers K:Na in wheat leaves (Saqib *et al.*, 2004b). In their review, Rengel and Damon (2008) highlight previous studies demonstrating that genotypic differences exist for K uptake and utilization. The genetic basis of K-use efficiency appears to be highly quantitative according to a QTL mapping study by Kong *et al.* (2013), who identified 29 QTLs for K-use efficiency under natural field conditions.

## 6.7 Genetic Control of Micronutrient and Plant Non-Essential Element Uptake and Transport

Under waterlogging stress, micronutrient concentrations can reach toxic levels in the plant, negatively influencing wheat growth and yield, and resulting in plant death (Colmer and Voesenek, 2009; Shabala, 2011; Khabaz-Saberi *et al.*, 2012). Ding and Musgrave (1995) evaluated the elemental concentration of 11 elements, including B, Ca, Cu, K, Mg, Na, S, Zn, Fe, Mn and P, in 14 wheat genotypes during a 3-year experiment. Root samples were evaluated using inductively coupled plasma spectroscopy (ICP) and significant increases of Fe, Mn and P were observed along with grain yield reductions of between 28 to 49%. Khabaz-Saberi *et al.* (2005) also reported an increase in soil Fe and Na concentrations up to tenfold while evaluating six different wheat genotypes under waterlogging conditions on acidic soil.

### 6.7.1 Aluminium

Al is present in all soils and in concentrations up to 300,000 mg Al/kg, although much of this is not available to the plant at soil pH above 5.0 (Lindsay, 1979). The concentration of Al increases in low-pH or acidic soils and can result in toxicity (Rout *et al.*, 2001). Al can be found in almost all plants but is not a plant essential element and most often will have net negative effects associated with toxicity if soil conditions such as low soil pH favour high Al availability in the soil solution (Havlin *et al.*, 2014). The primary effects of Al toxicity are root related and result in severely stunted, shallow root systems (Taylor, 1989). In severe cases, Al toxicity will disrupt DNA synthesis and alter cell-membrane potential (Wallace and Anderson, 1984; Kinraide, 1988). There are several mechanisms used by plants for protection from Al toxicity. First is the capacity of the plant to either eliminate the toxins generated by Al or tolerate Al (Taylor, 1991; Tice *et al.*, 1992). Second is the ability to prevent the entrance of Al to the symplasm (exclusion) by immobilization of Al or production of chelating ligands (Taylor, 1991; Delhaize *et al.*, 1993b). The third mechanism is to defend cell-wall and membrane function and structure from lesions through variations in the apoplastic pathway (Tice *et al.*, 1992). Wheat has been shown to mostly use an exclusion mechanism to avoid toxicity from Al rather than internal tolerance mechanisms (Carver and Ownby, 1995). For example, Al-tolerant wheat genotypes excreted five to ten times more malic acid from roots than susceptible wheat plants (Delhaize *et al.*, 1993b). Malic acid is an organic acid which chelates Al and protects the root apex from high levels of bioavailable Al in the soil (Ma *et al.*, 2001). A major QTL for Al tolerance has been shown to localize to chromosome 4DL and an Al-activated malate transporter gene (*ALTM1*) was identified in this same region (Ma *et al.*, 2005). Gene-based markers for *ALTM1* have since been widely used for MAS of Al tolerance, with tolerant genotypes having greater root growth in low-pH and high-Al soil conditions.

### 6.7.2 Iron

Similar to Al, flooded soils can accumulate high concentrations of $Fe^{2+}$ due to oxido-reduction reactions that occur (Gotoh and Patrick, 1974). Fe-tolerance mechanisms have been classified into three strategies: (i) exclusion/avoidance; (ii) inclusion/avoidance; and (iii) inclusion/tolerance (Becker and Asch, 2005). Exclusion/avoidance is the mechanism by which the plant does not take up $Fe^{2+}$ in order to prevent $Fe^{2+}$ injuries in the shoot tissue. Inclusion/avoidance implies uptake of $Fe^{2+}$ by the root and then immobilization of Fe in the leaf apoplast. Inclusion/tolerance is when the plant is resistant to high amounts of $Fe^{2+}$ in the leaf cells, possibly through enzymatic detoxification (Becker and Asch, 2005). Khabaz-Saberi *et al.* (2010) evaluated the effects of different Fe concentrations in waterlogged acid soils on wheat cultivars, including a tolerant wheat genotype, Siete Cerros, and an intolerant wheat genotype, BH1146, finding that the Siete Cerros cultivar used the exclusion mechanism to overcome $Fe^{2+}$ stress.

### 6.7.3 Sodium

Salinity stress occurs in both waterlogged and acidic soils (Setter *et al.*, 2009). Similar to Al, Na is not considered a plant essential element, but is found in all plant species due to its presence in soil and the plant's inability to exclude or differentiate it from K (Havlin *et al.*, 2014). Barrett-Lennard *et al.* (1999) reported increases in $Na^{+}$ and $Cl^{-}$ of 360 to 650% and 110 to 170%, respectively, in waterlogged wheat shoots. Waterlogging increases the intensity of salt stress by decreasing the ratio K:Na in wheat leaves (Saqib *et al.*, 2004a) and the formation of aerenchyma is associated with improved tolerance to both stresses (Saqib *et al.*, 2005). Cuin *et al.* (2009) found Na exclusion from the shoot to be correlated with salinity tolerance, and this occurred more often in hexaploid wheat compared with tetraploid durum species (*Triticum turgidum*).

A transporter of pyruvate, a precursor to abscisic acid, isolated from the tolerant wheat line Shanrong 3 was recently shown to increase salinity tolerance in wheat (Zhao *et al.*, 2016).

The NAC transcription factor *TaNAC29*, possibly located on chromosome 2BS, increases salt tolerance likely through antioxidant-induced activity (Huang *et al.*, 2015). A similar function was found for the ERF transcription factor *TaERF3* (Rong *et al.*, 2014). Genc *et al.* (2010) reported 40 QTLs associated with salinity tolerance under dryland conditions, including a cluster of QTLs on chromosome 2AL, possibly homologous to the *Nax1* locus identified in *T. turgidum* (Huang *et al.*, 2006). James *et al.* (2011) used interspecific hybridization to transfer *Nax1* and *Nax2* from *T. turgidum* to bread wheat. The combination of *Nax1* and *Nax2* decreased leaf $Na^+$ concentration by 60% under a salinity treatment. When salinity was combined with hypoxia to simulate waterlogging, *Nax1* maintained its ability to exclude Na and increase K:Na, while *Nax2* did not. The tolerance mechanism of *Nax1* appears to be the storage of Na in the leaf sheath as opposed to the leaf blade (James *et al.*, 2006).

### 6.7.4 Manganese

Mn toxicity is a global constraint to crop production in waterlogged soils as it interferes with the formation and function of Mg enzymes essential to plant chloroplasts and photosynthesis (Foy, 1984; Socha and Guerinot, 2014). As a result, it is characterized by a marked decrease in photosynthesis which may be attributed to the disruption of the activity of rubisco, an enzyme involved in the first step of carbon fixation (Houtz *et al.*, 1988). Unlike Al, the dominant mechanisms to deal with Mn appear to be controlled internally after plant uptake and transport to the shoot (Horst, 1988; Carver and Ownby, 1995). Plants tolerant to $Mn^{2+}$ are able to maintain low concentrations in the cytoplasm by accumulating $Mn^{2+}$ into the vacuole. In addition, plants use active transport to mobilize $Mn^{2+}$ from cytoplasm out of the cell or into the vacuole, while $Na^+$ is taken up by the root through facilitated diffusion (Quiquampoix *et al.*, 1993). Additional tolerance mechanisms include intra- and extracellular chelation, translocation (Kumar *et al.*, 1995), storage of the excess $Mn^{2+}$ in the epidermis (Memon *et al.*, 1981) and the formation of glandular trichomes (Blamey *et al.*, 1986), hair-like structures derived from the epidermis that secrete secondary metabolites (Levin, 1973). The presence of excess $Mn^{2+}$ in wheat reduced vegetative biomass by 54%, due to a reduction in both tiller number and plant height, and a reduction in root dry biomass by 25% (Ohki, 1984). Several $Mn^{2+}$-tolerant wheat cultivars were found to have higher photosynthetic rates, respiration, and higher levels of chlorophyll a and chlorophyll b compared with a susceptible cultivar (Macfie and Taylor, 1992).

While the specific mechanisms of $Mn^{2+}$ toxicity tolerance are not well studied in wheat, metal transporters have been proposed as candidates in *A. thaliana* (Kochian *et al.*, 2004) The expression of the antiporter *CAX2* (*calcium exchange 2*) was identified as a key factor involved in modifying vacuolar exchange activity to increase ability to uptake $Ca^{2+}$ and $Mn^{2+}$ (Hirschi *et al.*, 2000). Delhaize *et al.* (2003) reported an increase in $Mn^{2+}$ tolerance in transgenic *A. thaliana* plants due to expression of the gene *ShMTP1* which increased $Mn^{2+}$ sequestration allowing for high concentrations of $Mn^{2+}$ without impacting plant development. A clear relationship has been shown between a reduction in soil redox potential and increases in Fe and Mn solubility in Indian and western Australian waterlogged soils (Yaduvanshi *et al.*, 2012).

### 6.7.5 Other micronutrients

The QTLs and genes associated with elemental toxicities have been reported for B, Al and Cu (Ma *et al.*, 2005; Balint *et al.*, 2007; Schnurbusch *et al.*, 2007; Cai *et al.*, 2008), although not under waterlogging stress. Lines containing the tolerant *Bo1* allele on chromosome 7BL identified by Schnurbusch *et al.* (2007) show low shoot B accumulation which is correlated with maintenance of biomass and yield. Compared with other elements, tolerance to Cu appears to be more quantitative, with QTLs having been identified on 14 of the 21 wheat chromosomes (Ganeva *et al.*, 2003; Balint *et al.*, 2007, 2009). Soriano and Alvaro (2019) performed a meta-analysis of 634 root-related QTLs of which a 6A region was associated with a heavy metal transport/detoxification gene model. The mechanism(s) associated with tolerance to Zn deficiency remain(s) unclear, although evidence points towards not just efficient uptake and utilization, but also the ability to maintain activity of Zn-requiring enzymes under low Zn concentrations (Hacisalihoglu and Kochian, 2003). High levels of intrinsic Zn at the seedling stage have been correlated with higher grain yield under waterlogging (Faran *et al.*, 2019). Zn availability in the soil is primarily controlled by soil pH, with redox having almost no impact on Zn availability.

## 6.8 Other Quantitative Trait Loci and Genes Associated with Waterlogging Tolerance

Molecular markers and candidate genes allow plant breeders to be more efficient by combining the process of conventional phenotypic selection with MAS for trait improvement (Gupta *et al.*, 1999). Information on the genetic control of waterlogging tolerance in wheat is limited compared with other abiotic stress. St Burgos *et al.* (2001) identified five and ten QTLs for seedling survival and seedling growth under waterlogging, respectively, in a wheat × spelt (*Triticum spelta*) population. Of these 15 QTLs, 11 were contributed by the spelt parent, Olberkulmer. Interestingly, QTLs on chromosomes 3B and 5A co-localized for both seedling survival and coleoptile electrolytic leakage. Ballesteros *et al.* (2015) evaluated a RIL population for biomass and root traits under both field and greenhouse waterlogging experiments and identified ten genomic regions. Of these, eight were constitutively expressed under both waterlogging and non-waterlogging scenarios, and two regions on chromosomes 1B and 6D were found to be adaptive to waterlogging stress only. In addition, a QTL on chromosome 1D explained up to 24% of the phenotypic variation for chlorophyll content. Yu and Chen (2013) reported 46 QTLs for shoot and root biomass distributed throughout the wheat genome, including 20 favourable alleles from synthetic hexaploid lines.

The most successful genetic studies of flooding tolerance come from rice where QTL studies have subsequently led to the map-based cloning of several agronomically important genes. For example, the *Sub1* locus was originally identified as a QTL on chromosome 9 (Xu and Mackill, 1996) and was eventually shown through map-based cloning to be an *ERF* gene (Xu *et al.*, 2006). *ERF* genes are known to play a role in wheat waterlogging tolerance as well. Wei *et al.* (2019) observed differential expression for *TaERFVII.1* between the tolerant genotype Nonglin46 and the susceptible wheat Yangmai16. Constitutive expression of *TaERFVII.1* resulted in increased grain weight per plant, higher leaf chlorophyll and higher expression of other waterlogging-responsive genes. Pretreatment of wheat seedlings with the ethylene precursor 1-aminocyclopropanecarboxylic acid (ACC) also enhances tolerance to hypoxia (Yamauchi *et al.*, 2014). In barley, QTLs have been mapped for leaf chlorosis, plant survival and plant biomass under waterlogging and many of these QTLs have been shown to be conserved across populations and studies (Li *et al.*, 2008; Zhou, 2011). Boru *et al.* (2001) estimated area under the chlorosis curve in spring wheat to be controlled by four genes with additive genetic effects and concluded that two genes provided adequate tolerance to waterlogging, although no molecular markers were identified in that study. While the information for wheat QTLs associated with waterlogging stress is limited, these studies do provide potential targets for MAS and future research to better understand these QTL regions.

## 6.9 Sources of Genetic Variation: Bread Wheat

Significant genetic variation is key to both improvements through breeding and for molecular genetic studies of a quantitative trait. While little genetic variation appears to exist for waterlogging tolerance in *T. aestivum* at the time of germination (St Burgos *et al.*, 2001), variability starting at the early seedling stage has been demonstrated (Table 6.3). This includes studies on spring wheat germplasm from Mexico, India and Australia (Boru *et al.*, 2001; Villareal *et al.*, 2001; Setter *et al.*, 2009), European winter wheat (Dickin *et al.*, 2009) and soft winter wheat from eastern US breeding programmes (Collaku and Harrison, 2002; Ballesteros *et al.*, 2015; Arguello *et al.*, 2016). Wheat seedlings exposed immediately to anoxic conditions (complete submergence) died within 24 h (Waters *et al.*,

**Table 6.3.** Waterlogging-tolerant germplasm and mechanisms in wheat.

| Genotype | Accession no. | Wheat class | Tolerance mechanism | Reference |
|---|---|---|---|---|
| SARC 1 | | Spring | Resumption of root elongation and K recovery | Goggin and Colmer (2007) |
| Ducula 4 | | Spring | Resumption of root elongation and K recovery | Boru *et al.* (2001, 2003) |
| HD2329 | | Spring | Resumption of root elongation and K recovery | Boru *et al.* (2003) |
| Terral LA422 | | Soft winter | Yield maintenance | Collaku and Harrison (2002) |
| Jaypee | | Soft winter | Yield component maintenance, root and shoot biomass | Collaku and Harrison (2002); Ballesteros *et al.* (2015) |
| Savannah | | Soft winter | Root aerenchyma | Huang *et al.* (1994b) |
| USG 3555 | | Soft winter | Yield maintenance | Arguello *et al.* (2016) |
| Pioneer Brand 26R22 | | Soft winter | High yield under waterlogging | Arguello *et al.* (2016) |
| AR01167-3-1 | | Soft winter | High yield under waterlogging | Arguello *et al.* (2016) |
| Magnolia | | Soft winter | High yield under waterlogging | Arguello *et al.* (2016) |
| Sarc-6 | | Spring | Na exclusion and high K:Na via root aerenchyma | Saqib *et al.* (2005) |
| CIGM86.953 | 613278 | Spring synthetic hexaploid | Reduced chlorosis | Villareal *et al.* (2001) |
| CIGM90.863-SH64 | 613279 | Spring synthetic hexaploid | Reduced chlorosis | Villareal *et al.* (2001) |
| CIGM89.567-SH54 | 613280 | Spring synthetic hexaploid | Reduced chlorosis | Villareal *et al.* (2001) |
| CIGM92.1723-SH82 | 613281 | Spring synthetic hexaploid | Reduced chlorosis | Villareal *et al.* (2001) |
| Nonglin46 | | Spring | Ethylene signalling pathway via the ERF *TaERFVII.1* | Wei *et al.* (2019) |
| Westonia | | Spring | Al tolerance | Setter *et al.* (2009) |
| KRL19 | | Spring | Al tolerance | Setter *et al.* (2009) |

1991), but tolerance improves with a hypoxic pretreatment followed by anoxia (Greenway *et al.*, 1992). Genotypes SARC 1, Ducula 4 and HD2329 were able to resume seminal root growth and had faster recovery of root K concentration following anoxia (Goggin and Colmer, 2007). Collaku and Harrison (2002) found a range in yield loss from 15 to 60% for a set of 15 cultivars and were able to identify genotypes such as Terral LA422 that exhibited both waterlogging tolerance in terms of yield maintenance and high yield under non-waterlogging conditions. Tiller number and grain number were the phenotypic characters most affected by waterlogging and identified as breeding targets. In the same study, other cultivars such as Jaypee showed small reductions in yield components under waterlogging and were in the high-yielding group under non-waterlogging conditions. Jaypee has also shown higher root and shoot biomass, root length and seedling height under waterlogging stress compared with a susceptible line (Ballesteros *et al.*, 2015). The cultivar Savannah was identified as waterlogging tolerant compared with a susceptible genotype due to the formation of root aerenchyma (Huang *et al.*, 1994a). Wheat cultivars Pioneer Brand 26R22, AR01167-3-1, Magnolia and USG 3555 had significantly higher grain yield than other genotypes under field waterlogging stress, with USG 3555 also showing a non-significant grain yield reduction (Arguello *et al.*, 2016). Research in Australia and India has shown elemental toxicities to be the most limiting factor in a waterlogged environment and genetic variations for Mn, Fe, Al, B and Na tolerance have been reported in spring wheat cultivars adapted to these regions (Setter *et al.*, 2009).

## 6.10 Sources of Genetic Variation: Wheat Relative Species

Wheat relatives and diploid species have continued to be an important source of alleles in breeding for resistance to biotic and abiotic stresses. For waterlogging stress tolerance, Davies and Hillman (1988) reported the hexaploid species *Triticum macha* to be more flood tolerant than *T. aestivum* and *T. spelta* based on vegetative growth, higher inflorescence number and higher grain weight. Spring-type synthetic hexaploid wheat derived from hybridizing durum (*T. turgidum*) with the D-progenitor species (*Aegilops tauschii*) has shown reduced leaf chlorosis (Villareal *et al.*, 2001) and increased root porosity (Boru *et al.*, 2003). Yu and Chen (2013) also found synthetic wheat lines W7984 and SHW-L1 to be a valuable source of waterlogging-tolerant alleles. In a *T. aestivum* × *T. spelta* population, the spelt parent Olberkulmer had more vigorous coleoptile growth and better survival compared with the bread wheat parent (St Burgos *et al.*, 2001). The waterlogging-tolerant haplophyte species *Hordeum marinum* Huds., commonly known as sea barley grass, is a wheat wild relative adapted to salt marshes and a potential source of both salinity and waterlogging tolerance alleles (McDonald *et al.*, 2001a). Amphiploid lines containing the genomes of both *H. marinum* and *T. aestivum* show improved waterlogging and salinity tolerance attributed to increased wheat porosity and Na exclusion compared with hexaploid wheat (Malik *et al.*, 2009, 2011). *Thinopyrum elongatum*, a species also from a salt marsh habitat, has been suggested as a potential source of waterlogging and salt tolerance alleles for hexaploid wheat but attempts to introgress this tolerance mechanism into wheat have not been successful thus far (McDonald *et al.*, 2001b). Amphiploids containing the 4E chromosome from *T. elongatum* showed improved root growth under waterlogging and were similar in their response to tetrasomic lines carrying multiple 4B and 4D chromosomes (Taeb *et al.*, 1993). The authors attributed the tolerance to increased doses of group 4 chromosomes that would likely carry beneficial alleles. This same dosage effect was not seen for the group 2 chromosomes and as such they might carry a novel resistance gene not present in *T. aestivum*. It should be noted that *Nax1*, which improves both salt and waterlogging tolerance, was first identified on chromosome 2A in *T. turgidum* (James *et al.*, 2006).

## 6.11 Use of Physiology to Understand Waterlogging and Other Complex Traits

Increasing the yield potential and stress tolerance of wheat is vital to global food security and the development of new tools for plant breeders to use for genetic improvement can aid in this process (Araus *et al.*, 2008). Physiological-based breeding tools are a complementary tool to make genetic gains in yield and stress tolerance through introgression of novel alleles that may not be incorporated via phenotypic selection alone (Reynolds *et al.*, 2009). Examples in other crops and stresses include the use of canopy temperature (CT) measurements as a diagnostic marker for plant water status (Saint Pierre *et al.*, 2010; Mason and Singh, 2014) and the use of carbon isotope discrimination to identify drought-tolerant cultivars (Rebetzke *et al.*, 2008). The use of spectral reflectance measurements, such as the normalized difference vegetative index (NDVI) or water indices, either ground based or via unmanned aerial systems (UAS), to assess plant N status, plant biomass or predict yield is another example of the utility of these tools. The normalized water index-3 has shown nearly perfect correlation with yield ($r = 0.95$) under severe drought stress as a predictive measure of CT and plant water potential (Gutierrez *et al.*, 2010). A similar utility was observed by Arguello *et al.* (2016), where NDVI was predictive of both grain yield ($R^2 = 0.77$) and biomass ($R^2 = 0.64$) under waterlogging stress but not in the control. Musgrave and Ding (1998) showed photosynthesis under waterlogging to be predictive of yield ($R^2 = 0.61$) in eight wheat genotypes varying for waterlogging tolerance and reductions in photosynthesis have been shown to be delayed in tolerant wheat cultivars (Zheng *et al.*, 2009). Similar results have been observed for measurements of chlorophyll, such as the use of a SPAD meter, with tolerant genotypes showing delayed onset of chlorosis (Boru *et al.*, 2001; Zheng *et al.*, 2009). Moderate correlations between chlorophyll and yield have also been shown in field waterlogging experiments (Collaku and Harrison, 2002; Ballesteros *et al.*, 2015). Additionally, NDVI has been shown to be a sensitive index for estimating biomass of flooded grass canopies but has limitations based on water depth (Beget and Di Bella, 2007).

## 6.12 Genomic Selection versus Marker-Assisted Selection for Genetic Improvement

Genome-wide selection or GS is an emerging approach in plant breeding programmes which relies on genome-wide molecular markers (Heffner *et al.*, 2009). The GS approach was proposed by Meuwissen *et al.* (2001) and subsequently gained popularity in animal breeding programmes (Luan *et al.*, 2009; Daetwyler *et al.*, 2010; Hayes and Goddard, 2010; Meuwissen and Goddard, 2010). The GS approach has great potential in plant breeding programmes, where it allows breeders to select plants based on GEBVs instead of observed phenotypic performance (Nakaya and Isobe, 2012). In GS, the effect of all markers is estimated simultaneously based on phenotypic and marker data from a training population. Marker effects are then used to predict GEBVs of individuals for which only genotypic data are available. The main frameworks of GS and MAS are similar, where both methods consist of training and breeding phases (Nakaya and Isobe, 2012). In the training phase, QTLs are identified in MAS, whereas GEBVs are predicted in GS in a training population. In the breeding phase, favourable individuals are selected based on the markers in MAS and the GEBVs in GS.

## 6.13 Brief Genomic Selection Methodology Overview

Genomic selection utilizes two types of populations: (i) a training population; and (ii) a breeding population (Meuwissen *et al.*, 2001; Heffner *et al.*, 2009). A training population consists of lines most relevant to the breeding programme. The training population is genotyped with a large number of markers and phenotyped for traits of interest to train the GS prediction model. In the prediction model, the genome-wide markers are considered as random effects and all marker effects on the phenotype are estimated simultaneously in a single model. GS assumes one or more markers are in linkage disequilibrium (LD) with each QTL affecting the trait, allowing for the capture of variance for all QTLs. The prediction model is used to account for the total additive genetic variance to estimate breeding value of individuals based on the sum of all marker effects in the training population. Using the model, GEBVs are calculated for individuals in the breeding population with only genotypic information and used in selecting individuals for advancement.

## 6.14 Statistical Methods to Predict Genomic Estimated Breeding Values

Statistical approaches for predicting GEBVs differ based on their assumptions and treatment of marker effects and include parametric, semi-parametric and non-parametric approaches (Meuwissen *et al.*, 2001; Heffner *et al.*, 2009; Lorenz *et al.*, 2011). The parametric methods include ridge regression–best linear unbiased prediction (RR-BLUP) and Bayesian methods which use a shrinkage procedure (Heffner *et al.*, 2009). In RR-BLUP all marker effects are estimated simultaneously assuming a normal (Gaussian) distribution with an equal variance (Meuwissen *et al.*, 2001). The RR-BLUP approach treats marker effects as random and shrinks equally all QTL effects towards zero, thus is capable of capturing minor effects using a large number of markers (minor QTLs). As RR-BLUP calculates GEBVs by contracting all marker effects equally, it can lead to an underestimation of large-effect QTLs. To overcome this limitation, Meuwissen *et al.* (2001) proposed using marker-specific shrinkage effects (Bayesian methods). The Bayesian approach assumes the variances of the marker effects are different and is estimated by using a prior distribution for these variances (Meuwissen *et al.*, 2001). Different types of prior distribution have been proposed, including Bayes A (Bayesian shrinkage regression), Bayes B (Bayesian variable selection) and Bayes $C_{\pi}$, which combines all markers with non-zero effects and estimates a common variance for them (Habier *et al.*, 2011). The Bayesian least absolute shrinkage and selection operator (LASSO) has an exponential prior on the marker variances, giving a double-exponential distribution for the marker effects resulting in less and more shrinkage on large- and small-effect markers, respectively (Park and Casella, 2008).

The semi-parametric and non-parametric methods used in GS include kernel regression and reproducing kernel Hilbert spaces (RKHS), and random forest (RF) regression methods. Kernel regression is based on genetic distance, which regresses marker effects to a smoothing parameter to control the distribution of the QTL effects (Gianola *et al.*, 2006). RF (Breiman, 2001) is a machine learning method which captures non-linear relationships between phenotypes and marker genotypes by building a non-linear prediction model. Semi- and non-parametric methods can capture non-additive effects which may improve predictions (Dudley and Johnson, 2009).

## 6.15 Genomic Selection for Quantitative Traits in Wheat and Other Crops

The expected genetic gain from selection can be expressed as $R = ir\sigma_A$, where $R$ is the response to selection, $i$ is the selection intensity, $r$ is the selection accuracy and $\sigma_A$ is the standard deviation of breeding values (Falconer and Mackay, 1996). Correlation between the predicted breeding value (GEBV) and the true breeding (phenotypic) value determines the accuracy ($r$) of GS models (Goddard and Hayes, 2007). Accuracy, $r$, is analogous to the square root of the narrow-sense heritability ($h$). In empirical studies, GS had greater accuracy than traditional MAS in bi-parental crosses (Lorenzana and Bernardo, 2009), multiple bi-parental crosses (Massman *et al.*, 2013), nested association mapping populations (Guo *et al.*, 2012), genome-wide association mapping–MAS in a multi-family wheat population (Heffner *et al.*, 2011a,b) and phenotypic selection using pedigree information (Crossa *et al.*, 2010, 2011). Overall, moderate to high accuracies of the GEBVs have been reported for quantitative traits, implying GS is a potential method for plant breeding with reduction in the breeding cycle and high genetic gain per time compared with phenotypic selection and MAS.

## 6.16 Genomic Selection for Abiotic Stress Tolerance Traits

Plant responses to abiotic stress are complex and rarely controlled by a single gene – making them suitable for a GS approach. Leplat *et al.* (2016) used chlorophyll fluorescence as a proxy for determining Mn-deficiency tolerance in a winter barley population of 248 genotypes. A G-BLUP model (analogous to RR-BLUP) incorporating both phenotypes and genotypes outperformed phenotypes alone, with prediction accuracies based on different cross-validation schemes ranging from $r$ = 0.21 to 0.73 depending on environment. Ovenden *et al.* (2018) reported the predictability for water-soluble carbohydrate concentration to range from $r$ = 0.474 to 0.535 for the well-watered experiments and from $r$ = 0.445 to 0.481 under drought stress, under a cross-validation approach within an individual experiment. Across experiments the mean predictability was higher for well-watered ($r$ = 0.502) compared with water deficit ($r$ = 0.455). The authors did not find GEBVs to be an accurate predictor of phenotype given that the residual genetic variance that could not be explained was higher than the additive genetic variance explained by the models. Incorporating a crop growth model for N improved the prediction of grain number only slightly under variable N-stress scenarios (Ly *et al.*, 2017). The model,

APSIM-Wheat, which simulated winter wheat growth based on previous studies at variable N levels and incorporated environment variables and their interactions, was able to capture additional genotype by environment (G × E) interactions compared with an additive effects model only. The overall benefit of incorporating G × E effects was likely limited due to the very high heritability for grain number (0.96) and a lack of genotype × N treatment interaction.

## 6.17 Genomic Selection for Waterlogging Stress Tolerance

Given its quantitative inheritance, GS is a suitable approach for breeding improved waterlogging tolerance and training populations need to be developed that capture the breadth of genetic variation available. Tolerance to elemental toxicity and deficiency is an important adaptive trait improving waterlogging tolerance in wheat, although to date there are no reports on GS for waterlogging tolerance in wheat. To explore the use of GS to predict nutrient uptake under waterlogging stress, a training population consisting of 240 inbred lines of soft red winter wheat was evaluated in controlled field waterlogging experiments. Elemental concentrations from total plant extracts were determined on wheat shoots by ICP–atomic emission spectroscopy (AES) for N, P, K, Ca, Mg, Na, K, Fe, Mn, Zn, Cu, B, S and Al. The training population was genotyped using the Illumina 9K iSelect assay for wheat (Cavanagh *et al.*, 2013) with the final marker data set after filtering consisting of 5287 single-nucleotide polymorphisms (SNPs). The GEBVs for the training population and the marker effects for 3450 SNP loci were calculated using phenotypic data, kinship information from TASSEL and genome-wide marker data using an RR-BLUP model. When waterlogging stress was imposed at the late tillering stage, shoot tissue concentrations of P, Ca, S, Na, Fe, Mn, Al and B increased by 6 to 1077% (Fig. 6.1). Prediction accuracy for GS ($r_{GS}$) was assessed through cross-validation of the GS model for 500 cycles. Prediction accuracies increased with training population size and were moderate for grain yield ($r_{GS}$ = 0.39), Na concentration ($r_{GS}$ = 0.47) and Mn concentration ($r_{GS}$ = 0.37), but lower for Al concentration ($r_{GS}$ = 0.31) (Fig. 6.2). These accuracies are in the general range of previous reports for quantitative traits. Additional phenotypic and marker data, in particular for the D genome where marker coverage is low, should be a target for future analyses. In addition, the inclusion of large-effect loci as fixed effects in the prediction model, for example *Alt1* for Al, should be explored in future studies.

## 6.18 Doubled Haploids for Winter Wheat

The long life cycle (8–12 months) of winter wheat compared with other crops is a constraint to breeders and results in low genetic gain per unit time. However, the process of inbred line development

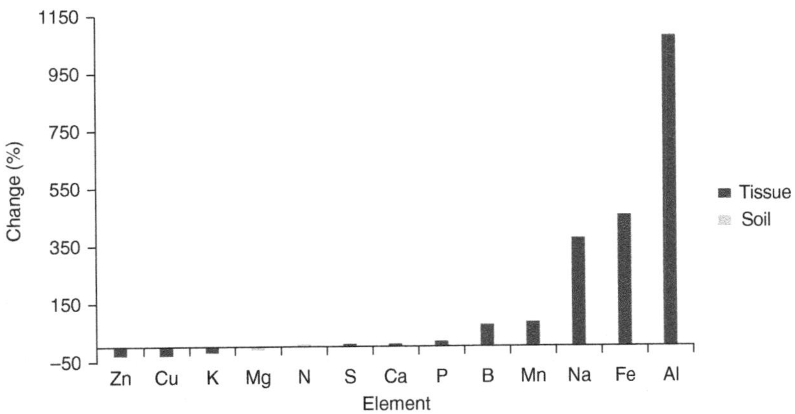


**Fig. 6.1.** Change (%) in shoot tissue and soil elemental concentrations for 28 wheat genotypes grown under soil waterlogging stress.

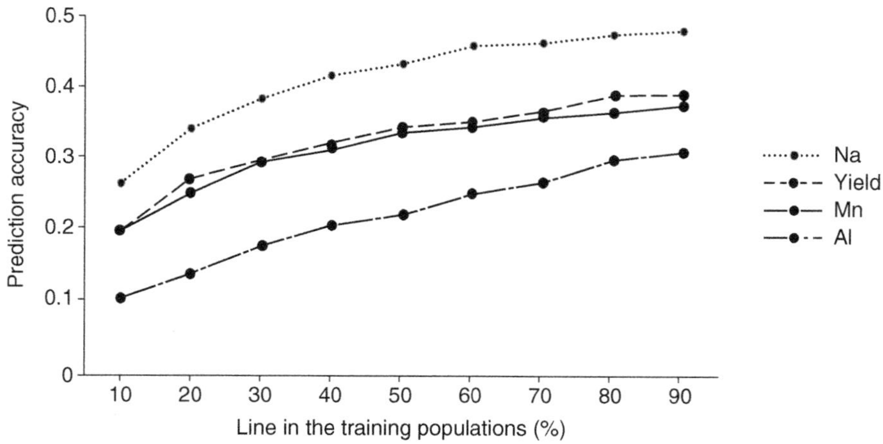


**Fig. 6.2.** Impact of training population size ($n$ = 240) on prediction accuracy for grain yield and element uptake in wheat under waterlogging stress.

can be expedited through the use of doubled haploid (DH) technology. The creation of pure wheat lines by way of interspecific crosses was first reported in the early 1970s (Ouyang *et al.*, 1973) and is now common practice in wheat breeding programmes in Europe, Australia and Canada. The wheat by maize technique involves pollination of the wheat flower with maize pollen resulting in fertilization of the wheat ovule followed by chemical treatment to duplicate chromosomes of the haploid plant and produce DH lines (Laurie and Bennett, 1986; Laurie and Bennett, 1988; Tadesse *et al.*, 2013; Niu *et al.*, 2014). This has been proposed as a breeding scheme in winter wheat which combines the use of GS at an accuracy of $r = 0.5$ with a single seed descent approach which can result in a twofold increase in genetic gain, with the breeding cycle to advanced testing reduced from 7 to 5 years (Heffner *et al.*, 2010). The use of DH technology in addition to GS can reduce this time to 4 years or less.

## 6.19 Putting It All Together: A Pipeline for Genetic Improvement of Waterlogging Tolerance in Wheat

To expedite genetic gain for waterlogging tolerance in wheat, a rapid breeding pipeline within a GS framework is necessary (Fig. 6.3). The first step is to characterize potential elemental and nutrient deficiencies and toxicities for a given target environment through soil and tissue analysis in wheat cropping areas prone to waterlogging. Second is the establishment of reliable field or greenhouse screening nurseries for general tolerance to hypoxia and for target elemental toxicities and deficiencies. Third is the characterization of a training population that segregates for waterlogging tolerance traits (and alleles) and also encompass the scope of genetic diversity within a breeding programme for other traits (i.e. grain yield), in order to train prediction models. Fourth is the development, GEBV prediction and phenotyping of new breeding lines. The use of DHs or other methods of rapid generation will expedite this effort. As new lines are advanced, retraining of prediction models with data collected on advanced breeding lines will be necessary to maintain relatedness between the training and breeding populations. In terms of research and understanding the biology of waterlogging tolerance, a more thorough characterization of elemental transporters and the role they play in exclusion or uptake of micro and macro elements under waterlogging stress is needed. A better understanding of the genetic control of adaptive mechanisms including the production of hollow stems, such as those seen in rice, and genetic variation for the formation of root aerenchyma is also warranted. With global climate change, it is likely that soil waterlogging will continue to grow in terms of its impact on wheat production and a proactive approach is necessary to circumvent the negative effects of this abiotic stress.

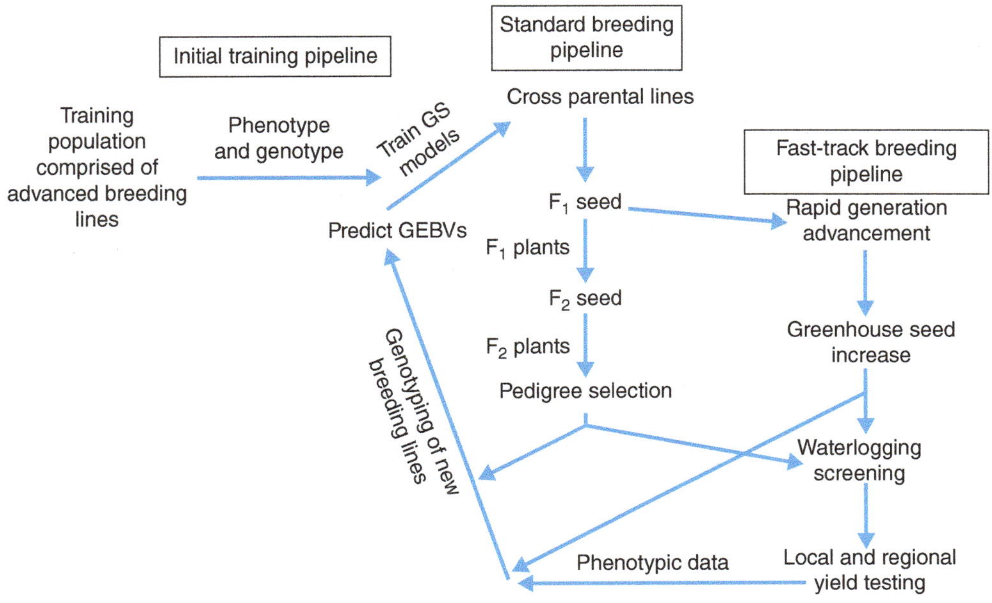


**Fig. 6.3.** Standard and fast-track breeding pipelines within a GS framework for development of waterlogging-tolerant wheat varieties and germplasm.

## References


Ahmed, F., Rafii, M.Y., Ismail, M.R., Juraimi, A.S., Rahim, H.A., Asfaliza, R. and Latif, M.A. (2013) Waterlogging tolerance of crops: breeding, mechanism of tolerance, molecular approaches, and future prospects. *BioMed Research International* 2013, 963525.

Akhtar, J., Gorham, J., Qureshi, R.H. and Aslam, M. (1998) Does tolerance of wheat to salinity and hypoxia correlate with root dehydrogenase activities or aerenchyma formation? *Plant and Soil* 201, 275–284.

Al-Issawi, M., Rihan, H.Z., Woldie, W.A., Burchett, S. and Fuller, M.P. (2013) Exogenous application of molybdenum affects the expression of CBF14 and the development of frost tolerance in wheat. *Plant Physiology and Biochemistry* 63, 77–81.

Araki, H., Hamada, A., Hossain, M.A. and Takahashi, T. (2012a) Waterlogging at jointing and/or after anthesis in wheat induces early leaf senescence and impairs grain filling. *Field Crops Research* 137, 27–36.

Araki, H., Hossain, M.A. and Takahashi, T. (2012b) Waterlogging and hypoxia have permanent effects on wheat root growth and respiration. *Journal of Agronomy and Crop Science* 198, 264–275.

Araus, J.L., Slafer, G.A., Royo, C. and Serret, M.D. (2008) Breeding for yield potential and stress adaptation in cereals. *Critical Reviews in Plant Sciences* 27, 377–412.

Arguello, M.N., Mason, R.E., Roberts, T.L., Subramanian, N., Acuña, A. *et al.* (2016) Performance of soft red winter wheat subjected to field soil waterlogging: grain yield and yield components. *Field Crops Research* 194, 57–64.

Armstrong, W. (1980) Aeration in higher plants. *Advances in Botanical Research* 7, 225–332.

Arruda, M., Lipka, A., Brown, P., Krill, A., Thurber, C. *et al.* (2016) Comparing genomic selection and marker-assisted selection for *Fusarium* head blight resistance in wheat (*Triticum aestivum*). *Molecular Breeding* 36, 1–11.

Bailey-Serres, J. and Voesenek, L.A.C.J. (2008) Flooding stress: acclimations and genetic diversity. *Annual Review of Plant Biology* 59, 313–339.

Balint, A.F., Roder, M.S., Hell, R., Galiba, G. and Borner, A. (2007) Mapping of QTLs affecting copper tolerance and the Cu, Fe, Mn and Zn contents in the shoots of wheat seedlings. *Biologia Plantarum* 51, 129–134.

Balint, A.F., Szira, F., Roder, M.S., Galiba, G. and Borner, A. (2009) Mapping of loci affecting copper tolerance in wheat – the possible impact of the vernalization gene Vrn-A1. *Environmental and Experimental Botany* 65, 369–375.

Ballesteros, D.C., Mason, R.E., Addison, C.K., Acuna, M.A., Arguello, M.N. *et al.* (2015) Tolerance of wheat to vegetative stage soil waterlogging is conditioned by both constitutive and adaptive QTL. *Euphytica* 201, 329–343.
Barrett-Lennard, E.G. (2003) The interaction between waterlogging and salinity in higher plants: causes, consequences and implications. *Plant and Soil* 253, 35–54.
Barrett-Lennard, E.G., van Ratingen, P. and Mathie, M.H. (1999) The developing pattern of damage in wheat (*Triticum aestivum* L.) due to the combined stresses of salinity and hypoxia: experiments under controlled conditions suggest a methodology for plant selection. *Australian Journal of Agricultural Research* 50, 129–136.
Becker, M. and Asch, F. (2005) Iron toxicity in rice – conditions and management concepts. *Journal of Plant Nutrition and Soil Science* 168, 558–573.
Beget, M.E. and Di Bella, C.M. (2007) Flooding: the effect of water depth on the spectral response of grass canopies. *Journal of Hydrology* 335, 285–294.
Belford, R.K. (1981) Response of winter wheat to prolonged waterlogging under outdoor conditions. *The Journal of Agricultural Science* 97, 557–568.
Blamey, F., Joyce, D., Edwards, D. and Asher, C. (1986) Role of trichomes in sunflower tolerance to manganese toxicity. *Plant and Soil* 91, 171–180.
Boru, G., van Ginkel, M., Kronstad, W.E. and Boersma, L. (2001) Expression and inheritance of tolerance to waterlogging stress in wheat. *Euphytica* 117, 91–98.
Boru, G., van Ginkel, M., Trethowan, R.M., Boersma, L. and Kronstad, W.E. (2003) Oxygen use from solution by wheat genotypes differing in tolerance to waterlogging. *Euphytica* 132, 151–158.
Boyer, J.S. (1982) Plant productivity and environment. *Science* 218, 443–448.
Braendle, R. and Crawford, R.M.M. (1987) Rhizome anoxia tolerance and habitat specialization in wetland plants. In: Crawford, R.M.M. (ed.) *Plant Life in Aquatic and Amphibious Habitats*. Blackwell Scientific Publications, Oxford, UK, pp. 397–410.
Breiman, L. (2001) Random forests. *Machine Learning* 45, 5–32.
Buchner, P. and Hawkesford, M.J. (2014) Complex phylogeny and gene expression patterns of members of the NITRATE TRANSPORTER 1/PEPTIDE TRANSPORTER family (NPF) in wheat. *Journal of Experimental Botany* 65, 5697–5710.
Cai, S.B., Bai, G.H. and Zhang, D.D. (2008) Quantitative trait loci for aluminum resistance in Chinese wheat landrace FSW. *Theoretical and Applied Genetics* 117, 49–56.
Cakmak, I. (2008) Enrichment of cereal grains with zinc: agronomic or genetic biofortification? *Plant and Soil* 302, 1–17.
Cannell, R.Q., Belford, R.K., Gales, K., Thomson, R.J. and Webster, C.P. (1984) Effects of waterlogging and drought on winter-wheat and winter barley grown on a clay and a sandy loam soil. 1. Crop growth and yield. *Plant and Soil* 80, 53–66.
Carver, B.F. and Ownby, J.D. (1995) Acid soil tolerance in wheat. *Advances in Agronomy* 54, 117–173.
Cavanagh, C., Chao, S., Wang, S., Huang, B., Stephen, S. *et al.* (2013) Genome-wide comparative diversity uncovers multiple targets of selection for improvement in hexaploid wheat landraces and cultivars. *Proceedings of the National Academy of Sciences USA* 110, 8057–8062.
Collaku, A. and Harrison, S.A. (2002) Losses in wheat due to waterlogging. *Crop Science* 42, 444–450.
Collaku, A. and Harrison, S.A. (2005) Heritability of waterlogging tolerance in wheat. *Crop Science* 45, 722–727.
Colmer, T.D. and Greenway, H. (2011) Ion transport in seminal and adventitious roots of cereals during $O_2$ deficiency. *Journal of Experimental Botany* 62, 39–57.
Colmer, T.D. and Voesenek, L. (2009) Flooding tolerance: suites of plant traits in variable environments. *Functional Plant Biology* 36, 665–681.
Crossa, J., de los Campos, G., Perez, P., Gianola, D., Burgueno, J. *et al.* (2010) Prediction of genetic values of quantitative traits in plant breeding using pedigree and molecular markers. *Genetics* 186, 713–724.
Crossa, J., Pérez, P., de los Campos, G., Mahuku, G., Dreisigacker, S. and Magorokosho, C. (2011) Genomic selection and prediction in plant breeding. *Journal of Crop Improvement* 25, 239–261.
Cuin, T.A., Tian, Y., Betts, S.A., Chalmandrier, R. and Shabala, S. (2009) Ionic relations and osmotic adjustment in durum and bread wheat under saline conditions. *Functional Plant Biology* 36, 1110–1119.
Daetwyler, H.D., Pong-Wong, R., Villanueva, B. and Woolliams, J.A. (2010) The impact of genetic architecture on genome-wide evaluation methods. *Genetics* 185, 1021–1031.
Davies, M.S. and Hillman, G.C. (1988) Effects of soil flooding on growth and grain yield of populations of tetraploid and hexaploid species of wheat. *Annals of Botany* 62, 597–604.

Delhaize, E., Craig, S., Beaton, C.D., Bennet, R.J., Jagadish, V.C. and Randall, P.J. (1993a) Aluminum tolerance in wheat (*Triticum aestivum* L.) (I. Uptake and distribution of aluminum in root apices). *Plant Physiology* 103, 685–693.

Delhaize, E., Ryan, P.R. and Randall, P.J. (1993b) Aluminum tolerance in wheat (*Triticum aestivum* L.). (II. Aluminum-stimulated excretion of malic acid from root apices). *Plant Physiology* 103, 695–702.

Delhaize, E., Kataoka, T., Hebb, D.M., White, R.G. and Ryan, P.R. (2003) Genes encoding proteins of the cation diffusion facilitator family that confer manganese tolerance. *The Plant Cell* 15, 1131–1142.

Dennis, E.S., Dolferus, R., Ellis, M., Rahman, M., Wu, Y. *et al.* (2000) Molecular strategies for improving waterlogging tolerance in plants. *Journal of Experimental Botany* 51, 89–97.

Desta, Z.A. and Ortiz, R. (2014) Genomic selection: genome-wide prediction in plant improvement. *Trends in Plant Science* 19, 592–601.

Dickin, E., Bennett, S. and Wright, D. (2009) Growth and yield responses of UK wheat cultivars to winter waterlogging. *Journal of Agricultural Science* 147, 127–140.

Ding, N. and Musgrave, M.E. (1995) Relationship between mineral coating on roots and yield performance of wheat under waterlogging stress. *Journal of Experimental Botany* 46, 939–945.

Drew, M.C. (1997) Oxygen deficiency and root metabolism: injury and acclimation under hypoxia and anoxia. *Annual Review of Plant Physiology and Plant Molecular Biology* 48, 223–250.

Drew, M.C. and Lynch, J. (1980) Soil anaerobiosis, microorganisms, and root function. *Annual Review of Phytopathology* 18, 37–66.

Dudley, J.W. and Johnson, G.R. (2009) Epistatic models improve prediction of performance in corn. *Crop Science* 49, 763–770.

Evans, D.E. (2003) Aerenchyma formation. *New Phytologist* 161, 35–49.

Evett, S.R. (2007) Soil water and monitoring technology. In: Lascano, R.J. and Sojka, R.E. (eds) *Irrigation of Agricultural Crops*, 2nd edn. Agronomy Monographs, Vol. 30. American Society of Agronomy, Crop Science Society of America and Soil Science Society of America, Madison, Wisconsin, pp. 25–84.

Fageria, V.D. (2001) Nutrient interactions in crop plants. *Journal of Plant Nutrition* 24, 1269–1290.

Falconer, D.S. and Mackay, T.F.C. (1996) *Introduction to Quantitative Genetics*, 4th edn. Longmans Green, Harlow, UK.

Faran, M., Farooq, M., Rehman, A., Nawaz, A., Saleem, M.K., Ali, N. and Siddique, K.H.M. (2019) High intrinsic seed Zn concentration improves abiotic stress tolerance in wheat. *Plant and Soil* 437, 195–213.

Foy, C.D. (1984) Physiological effects of hydrogen, aluminum, and manganese toxicities in acid soil. In: Adams, F. (ed.) *Soil Acidity and Liming*, Vol. 12, 2nd edn. American Society of Agronomy, Crop Science Society of America and Soil Science Society of America, Madison, Wisconsin, pp. 57–97.

Ganeva, G., Landjeva, S. and Merakchijska, M. (2003) Effects of chromosome substitutions on copper toxicity tolerance in wheat seedlings. *Biologia Plantarum* 47, 621–623.

Genc, Y., Oldach, K., Verbyla, A.P., Lott, G., Hassan, M. *et al.* (2010) Sodium exclusion QTL associated with improved seedling growth in bread wheat under salinity stress. *Theoretical and Applied Genetics* 121, 877–894.

Gianola, D., Fernando, R.L. and Stella, A. (2006) Genomic-assisted prediction of genetic value with semiparametric procedures. *Genetics* 173, 1761–1776.

Goddard, M.E. and Hayes, B.J. (2007) Genomic selection. *Journal of Animal Breeding and Genetics* 124, 323–330.

Goggin, D.E. and Colmer, T.D. (2007) Wheat genotypes show contrasting abilities to recover from anoxia in spite of similar anoxic carbohydrate metabolism. *Journal of Plant Physiology* 164, 1605–1611.

Gotoh, S. and Patrick, W.H. (1974) Transformation of iron in a waterlogged soil as influenced by redox potential and pH. *Soil Science Society of America Journal* 38, 66–71.

Graham, R.D., Ascher, J.S., Ellis, P.A.E. and Shepherd, K.W. (1987) Transfer to wheat of the copper efficiency factor carried on rye chromosome arm 5RL. *Plant and Soil* 99, 107–114.

Greenway, H. and Gibbs, J. (2003) Mechanisms of anoxia tolerance in plants. II. Energy requirements for maintenance and energy distribution to essential processes. *Functional Plant Biology* 30, 999–1036.

Greenway, H., Waters, I. and Newsome, J. (1992) Effects of anoxia on uptake and loss of solutes in roots of wheat. *Australian Journal of Plant Physiology* 19, 233–247.

Gregersen, P.L. and Holm, P.B. (2007) Transcriptome analysis of senescence in the flag leaf of wheat (*Triticum aestivum* L.). *Plant Biotechnology Journal* 5, 192–206.

Gregersen, P.L., Culetic, A., Boschian, L. and Krupinska, K. (2013) Plant senescence and crop productivity. *Plant Molecular Biology* 82, 603–622.

Grun, A., Buchner, P., Broadley, M.R. and Hawkesford, M.J. (2018) Identification and expression profiling of Pht1 phosphate transporters in wheat in controlled environments and in the field. *Plant Biology* 20, 374–389.
Guo, Z., Tucker, D.M., Lu, J., Kishore, V. and Gay, G. (2012) Evaluation of genome-wide selection efficiency in maize nested association mapping populations. *Theoretical and Applied Genetics* 124, 261–275.
Gupta, P.K., Varshney, R.K., Sharma, P.C. and Ramesh, B. (1999) Molecular markers and their applications in wheat breeding. *Plant Breeding* 118, 369–390.
Gutierrez, M., Reynolds, M.P. and Klatt, A.R. (2010) Association of water spectral indices with plant and soil water relations in contrasting wheat genotypes. *Journal of Experimental Botany* 61, 3291–3303.
Habier, D., Fernando, R.L., Kizilkaya, K. and Garrick, D.J. (2011) Extension of the Bayesian alphabet for genomic selection. *BMC Bioinformatics* 12, 186.
Hacisalihoglu, G. and Kochian, L.V. (2003) How do some plants tolerate low levels of soil zinc? Mechanisms of zinc efficiency in crop plants. *New Phytologist* 159, 341–350.
Hattori, Y., Nagai, K., Furukawa, S., Song, X.J., Kawano, R. *et al.* (2009) The ethylene response factors SNORKEL1 and SNORKEL2 allow rice to adapt to deep water. *Nature* 460, 1026–1030.
Havlin, J.L., Tisdale, S.L., Nelson, W.L. and Beaton, J.D. (2014) *Soil Fertility and Nutrient Management: An Introduction to Nutrient Management*, 8th edn. Pearson, Upper Saddle River, New Jersey.
Hayes, B. and Goddard, M. (2010) Genome-wide association and genomic selection in animal breeding. *Genome* 53, 876–883.
Hayes, B.J., Bowman, P.J., Chamberlain, A.C., Verbyla, K. and Goddard, M.E. (2009) Accuracy of genomic breeding values in multi-breed dairy cattle populations. *Genetics Selection Evolution* 41, 51.
Heffner, E.L., Sorrells, M.E. and Jannink, J.L. (2009) Genomic selection for crop improvement. *Crop Science* 49, 1–12.
Heffner, E.L., Lorenz, A.J., Jannink, J.-L. and Sorrells, M.E. (2010) Plant breeding with genomic selection: gain per unit time and cost all rights reserved. *Crop Science* 50, 1681–1690.
Heffner, E.L., Jannink, J.L., Iwata, H., Souza, E. and Sorrells, M.E. (2011a) Genomic selection accuracy for grain quality traits in biparental wheat populations. *Crop Science* 51, 2597–2606.
Heffner, E.L., Jannink, J.L. and Sorrells, M.E. (2011b) Genomic selection accuracy using multifamily prediction models in a wheat breeding program. *Plant Genome* 4, 65–75.
Hirschi, K.D., Korenkov, V.D., Wilganowski, N.L. and Wagner, G.J. (2000) Expression of *Arabidopsis* CAX2 in tobacco: altered metal accumulation and increased manganese tolerance. *Plant Physiology* 124, 125–134.
Horst, W. (1988) The physiology of manganese toxicity. In: Graham, R.D., Hannam, R.J. and Uren, N.C. (eds) *Manganese in Soils and Plants*. Developments in Plant and Soil Sciences, Vol. 33. Springer, Dordrecht, the Netherlands, pp. 175–188.
Houtz, R.L., Nable, R.O., Cheniae, G.M. (1988) Evidence for effects on the *in vivo* activity of ribulose-bisphosphate carboxylase/oxygenase during development of Mn toxicity in tobacco. *Plant Physiology* 86, 1143–1149.
Howden, S.M., Soussana, J.F., Tubiello, F.N., Chhetri, N., Dunlop, M. and Meinke, H. (2007) Adapting agriculture to climate change. *Proceedings of the National Academy of Sciences USA* 104, 19691–19696.
Huang, B., Johnson, J.W., Nesmith, S. and Bridges, D.C. (1994a) Growth, physiological and anatomical responses of two wheat genotypes to waterlogging and nutrient supply. *Journal of Experimental Botany* 45, 193–202.
Huang, B., Johnson, J.W., Nesmith, D.S. and Bridges, D.C. (1994b) Root and shoot growth of wheat genotypes in response to hypoxia and subsequent resumption of aeration. *Crop Science* 34, 1538–1544.
Huang, B., Johnson, J.W., Box, J.E. and Nesmith, D.S. (1997) Root characteristics and hormone activity of wheat in response to hypoxia and ethylene. *Crop Science* 37, 812–818.
Huang, M., Cabrera, A., Hoffstetter, A., Griffey, C., Van Sanford, D. *et al.* (2016) Genomic selection for wheat traits and trait stability. *Theoretical and Applied Genetics* 129, 1697–1710.
Huang, Q., Wang, Y., Li, B., Chang, J., Chen, M., Li, K., Yang, G. and He, G. (2015) TaNAC29, a NAC transcription factor from wheat, enhances salt and drought tolerance in transgenic *Arabidopsis*. *BMC Plant Biology* 15, 268.
Huang, S.B., Spielmeyer, W., Lagudah, E.S., James, R.A., Platten, J.D., Dennis, E.S. and Munns, R. (2006) A sodium transporter (HKT7) is a candidate for Nax1, a gene for salt tolerance in durum wheat. *Plant Physiology* 142, 1718–1727.

James, R.A., Davenport, R.J. and Munns, R. (2006) Physiological characterization of two genes for $Na^+$ exclusion in durum wheat, Nax1 and Nax2. *Plant Physiology* 142, 1537–1547.

James, R.A., Blake, C., Byrt, C.S. and Munns, R. (2011) Major genes for $Na^+$ exclusion, Nax1 and Nax2 (wheat HKT1;4 and HKT1;5), decrease $Na^+$ accumulation in bread wheat leaves under saline and waterlogged conditions. *Journal of Experimental Botany* 62, 2939–2947.

Jiang, D., Fan, X.M., Dai, T.B. and Cao, W.X. (2008) Nitrogen fertiliser rate and post-anthesis waterlogging effects on carbohydrate and nitrogen dynamics in wheat. *Plant and Soil* 304, 301–314.

Jiang, Z., Song, X.F., Zhou, Z.Q., Wang, L.K., Li, J.W., Deng, X.Y. and Fan, H.Y. (2010) Aerenchyma formation: programmed cell death in adventitious roots of winter wheat (*Triticum aestivum*) under waterlogging. *Functional Plant Biology* 37, 748–755.

Kahlown, M.A. and Azam, M. (2002) Individual and combined effect of waterlogging and salinity on crop yields in the Indus basin. *Irrigation and Drainage* 51, 329–338.

Khabaz-Saberi, H., Setter, T.L. and Waters, I. (2005) Waterlogging induces high to toxic concentrations of iron, aluminum, and manganese in wheat varieties on acidic soil. *Journal of Plant Nutrition* 29, 899–911.

Khabaz-Saberi, H., Rengel, Z., Wilson, R. and Setter, T. (2010) Variation for tolerance to high concentration of ferrous iron ($Fe^{2+}$) in Australian hexaploid wheat. *Euphytica* 172, 275–283.

Khabaz-Saberi, H., Barker, S.J. and Rengel, Z. (2012) Tolerance to ion toxicities enhances wheat (*Triticum aestivum* L.) grain yield in waterlogged acidic soils. *Plant and Soil* 354, 371–381.

Kinraide, T.B. (1988) Proton extrusion by wheat roots exhibiting severe aluminum toxicity symptoms. *Plant Physiology* 88, 418–423.

Kochian, L.V., Hoekenga, O.A. and Piñeros, M.A. (2004) How do crop plants tolerate acid soils? Mechanisms of aluminum tolerance and phosphorous efficiency. *Annual Review of Plant Biology* 55, 459–493.

Kohl, S., Hollmann, J., Erban, A., Kopka, J., Riewe, D., Weschke, W. and Weber, H. (2015) Metabolic and transcriptional transitions in barley glumes reveal a role as transitory resource buffers during endosperm filling. *Journal of Experimental Botany* 66, 1397–1411.

Kong, F.M., Guo, Y., Liang, X., Wu, C.H., Wang, Y.Y., Zhao, Y. and Li, S.S. (2013) Potassium (K) effects and QTL mapping for K efficiency traits at seedling and adult stages in wheat. *Plant and Soil* 373, 877–892.

Kong, L.A., Xie, Y., Hu, L., Feng, B. and Li, S.D. (2016) Remobilization of vegetative nitrogen to developing grain in wheat (*Triticum aestivum* L.). *Field Crops Research* 196, 134–144.

Kumar, P.B.A.N., Dushenkov, V., Motto, H. and Raskin, I. (1995) Phytoextraction: the use of plants to remove heavy metals from soils. *Environmental Science & Technology* 29, 1232–1238.

Laurie, D.A. and Bennett, M.D. (1986) Wheat × maize hybridization. *Canadian Journal of Genetics and Cytology* 28, 313–316.

Laurie, D.A. and Bennett, M.D. (1988) The production of haploid wheat plants from wheat × maize crosses. *Theoretical and Applied Genetics* 76, 393–397.

Lee, T.G., Jang, C.S., Kim, J.Y., Kim, D.S., Park, J.H., Kim, D.Y. and Seo, Y.W. (2006) A Myb transcription factor (TaMyb1) from wheat roots is expressed during hypoxia: roles in response to the oxygen concentration in root environment and abiotic stresses. *Physiologia Plantarum* 129, 375–385.

Lee, T.G., Jang, C.S., Kim, J.Y., Seong, R.C., Kim, I.G., Kim, D.S. and Seo, Y.W. (2007) Expressed sequence tags from wheat roots under hypoxia. *Russian Journal of Plant Physiology* 54, 659–668.

Leplat, F., Jensen, J. and Madsen, P. (2016) Genomic prediction of manganese efficiency in winter barley. *Plant Genome* 9, 13.

Levin, D.A. (1973) The role of trichomes in plant defense. *The Quarterly Review of Biology* 48, 3–15.

Li, H., Vaillancourt, R., Mendham, N. and Zhou, M. (2008) Comparative mapping of quantitative trait loci associated with waterlogging tolerance in barley (*Hordeum vulgare* L.). *BMC Genomics* 9, 401.

Li, Y., Wang, X., Zhang, H., Wang, S.L., Ye, X.S. *et al.* (2019) Molecular identification of the phosphate transporter family 1 (PHT1) genes and their expression profiles in response to phosphorus deprivation and other abiotic stresses in *Brassica napus*. *PLoS One* 14, 23.

Limami, A.M., Diab, H. and Lothier, J. (2014) Nitrogen metabolism in plants under low oxygen stress. *Planta* 239, 531–541.

Lindsay, W.L. (1979) *Chemical Equilibria in Soils*. Wiley, New York.

Lorenz, A.J., Chao, S.M., Asoro, F.G., Heffner, E.L., Hayashi, T. *et al.* (2011) Genomic selection in plant breeding: knowledge and prospects. In: Sparks, D.L. (ed) *Advances in Agronomy*, Vol. 110. Academic Press, San Diego, California, pp. 77–123.

Lorenzana, R.E. and Bernardo, R. (2009) Accuracy of genotypic value predictions for marker-based selection in biparental plant populations. *Theoretical and Applied Genetics* 120, 151–161.

Lu, Y., Wassmann, R., Neue, H.U. and Huang, C. (1999) Impact of phosphorus supply on root exudation, aerenchyma formation and methane emission of rice plants. *Biogeochemistry* 47, 203–218.

Luan, T., Woolliams, J.A., Lien, S., Kent, M., Svendsen, M. and Meuwissen, T.H.E. (2009) The accuracy of genomic selection in Norwegian red cattle assessed by cross-validation. *Genetics* 183, 1119–1126.

Ly, D., Chenu, K., Gauffreteau, A., Rincent, R., Huet, S. *et al.* (2017) Nitrogen nutrition index predicted by a crop model improves the genomic prediction of grain number for a bread wheat core collection. *Field Crops Research* 214, 331–340.

Ma, H.X., Bai, G.H., Carver, B.F. and Zhou, L.L. (2005) Molecular mapping of a quantitative trait locus for aluminum tolerance in wheat cultivar Atlas 66. *Theoretical and Applied Genetics* 112, 51–57.

Ma, J.F., Ryan, P.R. and Delhaize, E. (2001) Aluminium tolerance in plants and the complexing role of organic acids. *Trends in Plant Science* 6, 273–278.

McDonald, M.P., Galwey, N.W. and Colmer, T.D. (2001a) Waterlogging tolerance in the tribe Triticeae: the adventitious roots of *Critesion* marinum have a relatively high porosity and a barrier to radial oxygen loss. *Plant, Cell & Environment* 24, 585–596.

McDonald, M.P., Galwey, N.W., Ellneskog-Staam, P. and Colmer, T.D. (2001b) Evaluation of *Lophopyrum elongatum* as a source of genetic diversity to increase the waterlogging tolerance of hexaploid wheat (*Triticum aestivum*). *New Phytologist* 151, 369–380.

Macfie, S.M. and Taylor, G.J. (1992) The effects of excess manganese on photosynthetic rate and concentration of chlorophyll in *Triticum aestivum* grown in solution culture. *Physiologia Plantarum* 85, 467–475.

Malik, A., Colmer, T.D., Lambers, H. and Schortemeyer, M. (2001) Changes in physiological and morphological traits of roots and shoots of wheat in response to different depths of waterlogging. *Australian Journal of Plant Physiology* 28, 1121–1131.

Malik, A.I., English, J.P. and Colmer, T.D. (2009) Tolerance of *Hordeum marinum* accessions to $O_2$ deficiency, salinity and these stresses combined. *Annals of Botany* 103, 237–248.

Malik, A.I., Islam, A.K.M.R. and Colmer, T.D. (2011) Transfer of the barrier to radial oxygen loss in roots of *Hordeum marinum* to wheat (*Triticum aestivum*): evaluation of four *H. marinum*–wheat amphiploids. *New Phytologist* 190, 499–508.

Mano, Y. and Omori, F. (2009) High-density linkage map around the root aerenchyma locus Qaer1.06 in the backcross populations of maize Mi29 × teosinte '*Zea nicaraguensis*'. *Breeding Science* 59, 427–433.

Mason, R. and Singh, R. (2014) Considerations when deploying canopy temperature to select high yielding wheat breeding lines under drought and heat stress. *Agronomy* 4, 191.

Massman, J.M., Jung, H.J.G. and Bernardo, R. (2013) Genomewide selection versus marker-assisted recurrent selection to improve grain yield and stover-quality traits for cellulosic ethanol in maize. *Crop Science* 53, 58–66.

Memon, A.R., Chino, M., Hidaka, H., Hara, K. and Yatazawa, M. (1981) Manganese toxicity in field grown tea plants and the microdistribution of manganese in the leaf tissues as revealed by electron probe X-ray micrography. *Soil Science and Plant Nutrition* 27, 317–328.

Meuwissen, T. and Goddard, M. (2010) Accurate prediction of genetic values for complex traits by whole-genome resequencing. *Genetics* 185, 623–631.

Meuwissen, T.H.E., Hayes, B.J. and Goddard, M.E. (2001) Prediction of total genetic value using genome-wide dense marker maps. *Genetics* 157, 1819–1829.

Michel, S., Ametz, C., Gungor, H., Epure, D., Grausgruber, H., Löschenberger, F. and Buerstmayr, H. (2016) Genomic selection across multiple breeding cycles in applied bread wheat breeding. *Theoretical and Applied Genetics* 129, 1179–1189.

Musgrave, M.E. and Ding, N. (1998) Evaluating wheat cultivars for waterlogging tolerance. *Crop Science* 38, 90–97.

Nakaya, A. and Isobe, S.N. (2012) Will genomic selection be a practical method for plant breeding? *Annals of Botany* 110, 1303–1316.

Niu, Z.X., Jiang, A.X., Abu Hammad, W., Oladzadabbasabadi, A., Xu, S.S., Mergoum, M. and Elias, E.M. (2014) Review of doubled haploid production in durum and common wheat through wheat × maize hybridization. *Plant Breeding* 133, 313–320.

Ohki, K. (1984) Manganese deficiency and toxicity effects on growth, development, and nutrient composition in wheat. *Agronomy Journal* 76, 213–218.

Ouyang, T., Hu, H., Chuang, C. and Tseng, C. (1973) Induction of pollen plants from anthers of *Triticum aestivum* L. cultured in vitro. *Scientia Sinica* 16, 79–95.

Ovenden, B., Milgate, A., Wade, L.J., Rebetzke, G.J. and Holland, J.B. (2018) Accounting for genotype-by-environment interactions and residual genetic variation in genomic selection for water-soluble carbohydrate concentration in wheat. *G3: Genes, Genomes, Genetics* 8, 1909–1919.

Panaullah, G.M., Timsina, J., Saleque, M.A., Ishaque, M., Pathan, A. *et al.* (2006) Nutrient uptake and apparent balances for rice–wheat sequences. III. Potassium. *Journal of Plant Nutrition* 29, 173–187.

Park, T. and Casella, G. (2008) The Bayesian lasso. *Journal of the American Statistical Association* 103, 681–686.

Peck, A.W. and McDonald, G.K. (2010) Adequate zinc nutrition alleviates the adverse effects of heat stress in bread wheat. *Plant and Soil* 337, 355–374.

Pettigrew, W.T. (2008) Potassium influences on yield and quality production for maize, wheat, soybean and cotton. *Physiologia Plantarum* 133, 670–681.

Phukan, U.J., Mishra, S. and Shukla, R.K. (2016) Waterlogging and submergence stress: affects and acclimation. *Critical Reviews in Biotechnology* 36, 956–966.

Plank, C.O. and Donohue, S.J. (2000) Small grain (barley, oats, rye, wheat). In: Campbell, C.R. (ed.) *Reference Sufficiency Ranges for Plant Analysis in the Southern Region of the United States*. Southern Cooperative Series Bulletin No. 394. North Carolina Department of Agriculture and Consumer Services Agronomic Division, Rayleigh, North Carolina, pp. 29–31.

Powell, N., Ji, X., Ravash, R., Edlington, J. and Dolferus, R. (2012) Yield stability for cereals in a changing climate. *Functional Plant Biology* 39, 539–552.

Quiquampoix, H., Loughman, B.C. and Ratcliffe, R.G. (1993) A $^{31}$P-NMR study of the uptake and compartmentation of manganese by maize roots. *Journal of Experimental Botany* 44, 1819–1827.

Rebetzke, G., Condon, A., Farquhar, G., Appels, R. and Richards, R. (2008) Quantitative trait loci for carbon isotope discrimination are repeatable across environments and wheat mapping populations. *Theoretical and Applied Genetics* 118, 123–137.

Rengel, Z. and Damon, P.M. (2008) Crops and genotypes differ in efficiency of potassium uptake and use. *Physiologia Plantarum* 133, 624–636.

Reynolds, M., Manes, Y., Izanloo, A. and Langridge, P. (2009) Phenotyping approaches for physiological breeding and gene discovery in wheat. *Annals of Applied Biology* 155, 309–320.

Robertson, D., Zhang, H.P., Palta, J.A., Colmer, T. and Turner, N.C. (2009) Waterlogging affects the growth, development of tillers, and yield of wheat through a severe, but transient, N deficiency. *Crop & Pasture Science* 60, 578–586.

Roch, G.V., Maharajan, T., Ceasar, S.A. and Ignacimuthu, S. (2019) The role of PHT1 family transporters in the acquisition and redistribution of phosphorus in plants. *Critical Reviews in Plant Sciences* 38, 171–198.

Rong, W., Qi, L., Wang, A.Y., Ye, X.G., Du, L.P. *et al.* (2014) The ERF transcription factor TaERF3 promotes tolerance to salt and drought stresses in wheat. *Plant Biotechnology Journal* 12, 468–479.

Rout, G.R., Samantaray, S. and Das, P. (2001) Aluminium toxicity in plants: a review. *Agronomie* 21, 3–21.

Rutkoski, J.E., Heffner, E.L. and Sorrells, M.E. (2011) Genomic selection for durable stem rust resistance in wheat. *Euphytica* 179, 161–173.

Saint Pierre, C., Crossa, J., Manes, Y. and Reynolds, M.P. (2010) Gene action of canopy temperature in bread wheat under diverse environments. *Theoretical and Applied Genetics* 120, 1107–1117.

Sairam, R.K., Kumutha, D., Ezhilmathi, K., Deshmukh, P.S. and Srivastava, G.C. (2008) Physiology and biochemistry of waterlogging tolerance in plants. *Biologia Plantarum* 52, 401–412.

Saleque, M.A., Timsina, J., Panaullah, G.M., Ishaque, M., Pathan, A. *et al.* (2006) Nutrient uptake and apparent balances for rice–wheat sequences. II. Phosphorus. *Journal of Plant Nutrition* 29, 157–172.

Saqib, M., Akhtar, J. and Qureshi, R.H. (2004a) Pot study on wheat growth in saline and waterlogged compacted soil. I. Grain yield and yield components. *Soil & Tillage Research* 77, 169–177.

Saqib, M., Akhtar, J. and Qureshi, R.H. (2004b) Pot study on wheat growth in saline and waterlogged compacted soil. II. Root growth and leaf ionic relations. *Soil & Tillage Research* 77, 179–187.

Saqib, M., Akhtar, J. and Qureshi, R.H. (2005) $Na^+$ exclusion and salt resistance of wheat (*Triticum aestivum*) in saline-waterlogged conditions are improved by the development of adventitious nodal roots and cortical root aerenchyma. *Plant Science* 169, 125–130.

Schnurbusch, T., Collins, N.C., Eastwood, R.F., Sutton, T., Jefferies, S.P. and Langridge, P. (2007) Fine mapping and targeted SNP survey using rice–wheat gene colinearity in the region of the bo1 boron toxicity tolerance locus of bread wheat. *Theoretical and Applied Genetics* 115, 451–461.

Scott, D., Ferguson, J.A., Hanson, L., Fugitt, T. and Smith, E. (1998) *Agricultural Water Management in the Mississippi Delta Region of Arkansas*. Research Bulletin No. 959. Arkansas Agriculture Experiment Station, University of Arkansas, Fayetteville, Arkansas.

Setter, T.L. and Waters, I. (2003) Review of prospects for germplasm improvement for waterlogging tolerance in wheat, barley and oats. *Plant and Soil* 253, 1–34.

Setter, T.L., Waters, I., Sharma, S.K., Singh, K.N., Kulshreshtha, N. *et al.* (2009) Review of wheat improvement for waterlogging tolerance in Australia and India: the importance of anaerobiosis and element toxicities associated with different soils. *Annals of Botany* 103, 221–235.

Shabala, S. (2011) Physiological and cellular aspects of phytotoxicity tolerance in plants: the role of membrane transporters and implications for crop breeding for waterlogging tolerance. *New Phytologist* 190, 289–298.

Shao, G.C., Lan, J.J., Yu, S.E., Liu, N., Guo, R.Q. and She, D.L. (2013) Photosynthesis and growth of winter wheat in response to waterlogging at different growth stages. *Photosynthetica* 51, 429–437.

Simpson, N.L., Brennan, R.F. and Anderson, W.K. (2016) Grain yield increases in wheat and barley to nitrogen applied after transient waterlogging in the high rainfall cropping zone of western Australia. *Journal of Plant Nutrition* 39, 974–992.

Snipes, C.E., Nichols, S.P., Poston, D.H., Walker, T.W., Evans, L.P. and Robinson, H.R. (2005) *Current Agricultural Practices of the Mississippi Delta, Mississippi*. Bulletin No. 1143. Agricultural & Forestry Experiment Station, Mississippi State University, Mississippi State, Mississippi.

Socha, A.L. and Guerinot, M.L. (2014) Mn-euvering manganese: the role of transporter gene family members in manganese uptake and mobilization in plants. *Frontiers in Plant Science* 5, 106.

Soriano, J.M. and Alvaro, F. (2019) Discovering consensus genomic regions in wheat for root-related traits by QTL meta-analysis. *Scientific Reports* 9, 14.

Spiertz, H. (2012) Avenues to meet food security. The role of agronomy on solving complexity in food production and resource use. *European Journal of Agronomy* 43, 1–8.

St Burgos, M., Messmer, M.M., Stamp, P. and Schmid, J.E. (2001) Flooding tolerance of spelt (*Triticum spelta* L.) compared to wheat (*Triticum aestivum* L.) – a physiological and genetic approach. *Euphytica* 122, 287–295.

Steffens, D., Hutsch, B., Eschholz, T., Lošák, T., Schubert, S. and Liebig, J. (2005) Water logging may inhibit plant growth primarily by nutrient deficiency rather than nutrient toxicity. *Plant, Soil and Environment* 51, 545–552.

Stevenson, F.J. and Cole, M.A. (1999) *Cycles of Soil: Carbon, Nitrogen, Phosphorus, Sulfur, Micronutrients*, 2nd edn. Wiley, New York.

Tadesse, W., Tawkaz, S., Inagaki, M.N., Picard, E. and Baum, M. (2013) *Methods and Applications of Doubled Haploid Technology in Wheat Breeding – A Technical Manual*. International Center for Agricultural Research in the Dry Areas (ICARDA), Beirut.

Taeb, M., Koebner, R.M.D. and Forster, B.P. (1993) Genetic variation for waterlogging tolerance in the Triticeae and the chromosomal location of genes conferring waterlogging tolerance in *Thinopyrum elongatum*. *Genome* 36, 825–830.

Taylor, G.J. (1989) Multiple metal stress in *Triticum aestivum* – differentiation between additive, multiplicative, antagonostic, and synergistic effects. *Canadian Journal of Botany* 67, 2272–2276.

Taylor, G.J. (1991) Current views of the aluminum stress response the physiological basis of tolerance. *Current Topics in Plant Biochemistry and Physiology* 10, 57–93.

Taylor, G.J. and Foy, C.D. (1985) Mechanisms of aluminum tolerance in *Triticum aestivum* L. (wheat). II. Differential pH induced by spring cultivars in nutrient solutions. *American Journal of Botany* 72, 702–706.

Teng, W., Zhao, Y.Y., Zhao, X.Q., He, X., Ma, W.Y. *et al.* (2017) Genome-wide identification, characterization, and expression analysis of PHT1 phosphate transporters in wheat. *Frontiers in Plant Science* 8, 14.

Tice, K.R., Parker, D.R. and DeMason, D.A. (1992) Operationally defined apoplastic and symplastic aluminum fractions in root tips of aluminum-intoxicated wheat. *Plant Physiology* 100, 309–318.

Torun, B., Bozbay, G., Gultekin, I., Braun, H.J., Ekiz, H. and Cakmak, I. (2000) Differences in shoot growth and zinc concentration of 164 bread wheat genotypes in a zinc-deficient calcareous soil. *Journal of Plant Nutrition* 23, 1251–1265.

Uauy, C., Distelfeld, A., Fahima, T., Blechl, A. and Dubcovsky, J. (2006) A NAC gene regulating senescence improves grain protein, zinc, and iron content in wheat. *Science* 314, 1298–1301.

Villareal, R.L., Sayre, K., Banuelos, O. and Mujeeb-Kazi, A. (2001) Registration of four synthetic hexaploid wheat (*Triticum turgidum/Aegilops tauschii*) germplasm lines tolerant to waterlogging. *Crop Science* 41, 274–274.

Wallace, S.U. and Anderson, I.C. (1984) Aluminum toxicity and DNA synthesis in wheat roots. *Agronomy Journal* 76, 5–8.

Wany, A., Kumari, A. and Gupta, K.J. (2017) Nitric oxide is essential for the development of aerenchyma in wheat roots under hypoxic stress. *Plant, Cell & Environment* 40, 3002–3017.

Waters, I., Morrell, S., Greenway, H. and Colmer, T.D. (1991) Effect of anoxia on wheat seedlings 2. Influence of $O_2$ supply prior to anoxia on tolerance to anoxia, alcoholic fermentation, and sugar levels. *Journal of Experimental Botany* 42, 1437–1447.

Wei, X.N., Xu, H.J., Rong, W., Ye, X.G. and Zhang, Z.Y. (2019) Constitutive expression of a stabilized transcription factor group VII ethylene response factor enhances waterlogging tolerance in wheat without penalizing grain yield. *Plant, Cell & Environment* 42, 1471–1485.

Westerman, R.L. (1987) Soil reaction: acidity, alkalinity and salinity. In: Heyne, E.G. (ed.) *Wheat and Wheat Improvement*, 2nd edn. Agronomy Monographs, Vol. 13. American Society of Agronomy, Crop Science Society of America and Soil Science Society of America, Madison, Wisconsin, pp. 340–344.

Wu, J.-D., Li, J.-C., Wei, F.-Z., Wang, C.-Y., Zhang, Y. and Sun, G. (2014) Effects of nitrogen spraying on the post-anthesis stage of winter wheat under waterlogging stress. *Acta Physiologiae Plantarum* 36, 207–216.

Xu, K. and Mackill, D.J. (1996) A major locus for submergence tolerance mapped on rice chromosome 9. *Molecular Breeding* 2, 219–224.

Xu, K., Xu, X., Fukao, T., Canlas, P., Maghirang-Rodriguez, R. *et al.* (2006) Sub1A is an ethylene-response-factor-like gene that confers submergence tolerance to rice. *Nature* 442, 705–708.

Yaduvanshi, N.P.S., Setter, T.L., Sharma, S.K., Singh, K.N. and Kulshreshtha, N. (2012) Influence of waterlogging on yield of wheat (*Triticum aestivum*), redox potentials, and concentrations of microelements in different soils in India and Australia. *Soil Research* 50, 489–499.

Yamauchi, T., Watanabe, K., Fukazawa, A., Mori, H., Abe, F. *et al.* (2014) Ethylene and reactive oxygen species are involved in root aerenchyma formation and adaptation of wheat seedlings to oxygen-deficient conditions. *Journal of Experimental Botany* 65, 261–273.

Ylivainio, K., Jauhiainen, L., Uusitalo, R. and Turtola, E. (2018) Waterlogging severely retards P use efficiency of spring barley (*Hordeum vulgare*). *Journal of Agronomy and Crop Science* 204, 74–85.

Yu, M. and Chen, G.-Y. (2013) Conditional QTL mapping for waterlogging tolerance in two RILs populations of wheat. *SpringerPlus* 2, 245.

Zhao, Y., Ai, X.H., Wang, M.C., Xiao, L.T. and Xia, G.M. (2016) A putative pyruvate transporter TaBASS2 positively regulates salinity tolerance in wheat via modulation of ABI4 expression. *BMC Plant Biology* 16, 12.

Zheng, C.F., Jiang, D., Liu, F.L., Dai, T.B., Jing, Q. and Cao, W.X. (2009) Effects of salt and waterlogging stresses and their combination on leaf photosynthesis, chloroplast ATP synthesis, and antioxidant capacity in wheat. *Plant Science* 176, 575–582.

Zhou, M. (2011) Accurate phenotyping reveals better QTL for waterlogging tolerance in barley. *Plant Breeding* 130, 203–208.

# 7 Molecular Breeding for Improving Aluminium Resistance in Wheat

**Jorge Fernando Pereira***
*Embrapa Gado de Leite, Juiz de Fora, MG, Brazil*

## 7.1 Introduction

Wheat is one of the most cultivated crops worldwide. Although it can be used in industrial applications and animal feed, it is mainly used as a human food and represents a substantial part of the diet of several billions of people (Shewry, 2009). Depending on the gluten content, wheat flour can be used to produce pastries, breads, cookies, cakes and other products. Around 95% of the wheat cultivated worldwide is bread wheat (*Triticum aestivum*), which has a hexaploid genome (AABBDD). Most of the remaining 5% is durum wheat (*Triticum turgidum* spp. *durum*), which is a tetraploid species (AABB) lacking the D genome. Wheat is adapted to a wide range of temperate environments (Shewry, 2009), but improving its adaptation to tropical regions is required (Pereira *et al.*, 2019). This is especially important because the mean wheat yield globally is lower than required to meet the demand expected by 2050 (Ray *et al.*, 2013). One way to improve wheat yield is to increase productivity under Al stress, which plants face when grown in acid soils and subsoils, because both bread and durum wheats are impacted by Al toxicity.

Al is the third most abundant element and the most common metal in the Earth's crust, even though it plays no essential role in any biological system (Exley, 2009). It is largely harmless to plants when they grow under neutral or slightly acid soils. Under these conditions, most Al is unavailable to plants because it occurs in minerals and is complexed with oxides, silicates and other compounds. However, when the soil pH becomes more acidic, some of these minerals are solubilized and Al ions are released into the soil solution. Depending on the soil pH, a number of Al species occur, but $Al(H_2O)_6^{3+}$, or simply $Al^{3+}$, becomes prevalent below pH 5.0. The trivalent cation $Al^{3+}$ is highly toxic to plants, and it injures several sites in the plant cell (Zheng and Yang, 2005). The root apex is the primary site for Al toxicity (Ryan *et al.*, 1993) and the most evident effect is the inhibition of root growth. A reduced root system can explore a lower fraction of soil, decreasing the likelihood of finding water and nutrients. Additionally, it may provide weaker anchor for the plant, making it potentially more prone to lodging, and could affect root–microbe interactions. Even in hydroponics, when water is constantly available, Al-sensitive plants might suffer disturbed stomatal conductance and transpiration, which could lead to lower water status in leaves (Silva *et al.*, 2018b; Gavassi *et al.*, 2020). Thus, Al toxicity affects plant development. Acid soils pose additional nutritional challenges to plants because they are often leached

*Corresponding author: jorge.pereira@embrapa.br



DOI: 10.1079/9781789245431.0007

of basic cations (such as $K^+$, $Ca^{2+}$ and $Mg^{2+}$) and have reduced phosphate availability. Thus, plants suffering from Al toxicity are more likely to experience nutrient deficiencies. Nearly half of the world's cultivated or potentially cultivated lands show acidity (von Uexküll and Mutert, 1995); therefore, acidic soils are one of the major stresses for plant productivity worldwide.

Agronomic strategies are available for managing soil acidity and Al toxicity. These include the application of soil ameliorants such as gypsum ($CaSO_4 \cdot 2H_2O$), which returns the basic cations that are commonly leached from acid soils, or lime ($CaCO_3$) that has the added benefit of raising the pH. Although these management practices allow deeper root growth, they will not immediately overcome the problem because these compounds are expensive and/or not readily available to small farmers in many countries, and lime especially moves slowly down the soil profile. As a result, the most sustainable way to cope with Al toxicity is the amelioration of soil pH in combination with the use of Al-resistant genotypes. Additionally, greater Al resistance allows plants to better cope with drought stress (Yang *et al.*, 2013), to increase yield in acid soil under elevated $CO_2$ (Dong *et al.*, 2019) and potentially to better desorb phosphate from minerals in the soil (Kochian *et al.*, 2004). Thus, breeding wheat for greater Al resistance is important. Wheat has genetic variability for that trait (see Fig. 7.1 below) and understanding the mechanisms underlying this variability could be useful to breed new cultivars.

In this context, this chapter aims at describing the main physiological mechanisms associated with Al resistance in wheat and how the research about these mechanisms has evolved to its current status. Practical aspects of phenotyping and using the molecular basis to increase Al resistance, which can be easily introduced in breeding programmes, are detailed. This chapter discusses the reliability of methods to screen root growth under Al stress, the allelic variation of genes associated with the main Al resistance mechanism in wheat, the quantitative trait loci (QTLs) and genomic regions that might contain minor Al tolerance genes, the use of wheat wild relatives, the uncertainties of developing transgenic wheat for greater Al resistance and the development of Al-resistant lines of durum wheat.

## 7.2 Screening Wheat Root Growth under Aluminium Stress

The most common approach to phenotype Al resistance/tolerance is the evaluation of root development. Most of the studies employing this approach have used hydroponics (or nutrient solution) because it provides many advantages (Shavrukov *et al.*, 2012), including isolation of the Al stress, greater control of environmental conditions, easy observation of the roots and is not destructive. Various measurements can be performed and used as indicators of Al resistance/tolerance, such as relative root length, relative root weight, root regrowth and haematoxylin staining (Polle *et al.*, 1978; Taylor and Foy, 1985; Delhaize *et al.*, 1993a; Tang *et al.*, 2002; Cai *et al.*, 2008; Raman *et al.*, 2008; Pereira *et al.*, 2010).

Haematoxylin is a dye that forms a complex with Al resulting in a coloured stain (Polle *et al.*, 1978). The intensity of the haematoxylin stain is lighter in Al-resistant/tolerant genotypes and darker in Al-sensitive plants because the Al-sensitive plants take up more Al (Polle *et al.*, 1978; Delhaize *et al.*, 1993a). There is close agreement in screening Al tolerance based on haematoxylin staining and root length when seedlings are grown in nutrient solution after haematoxylin staining (Delhaize *et al.*, 1993a). Nevertheless, an indicator of Al resistance largely used is relative root length (RRL), also called relative root growth (RRG) (Ryan *et al.*, 1995; Papernik *et al.*, 2001; Tang *et al.*, 2002; Raman *et al.*, 2005; Guo *et al.*, 2007b; Zhou *et al.*, 2007a; Raman *et al.*, 2008; Pereira *et al.*, 2010; Aguilera *et al.*, 2016; Pereira, 2018). RRL is assessed by growing plants in conditions with and without Al stress (Fig. 7.1) and it is a better indicator of Al resistance/tolerance in wheat than relative root dry weight (Rengel and Jurkic, 1992). Comparison of RRL between studies is not feasible, because the level of Al stress in the solution varies based on the amount of Al added, pH and nutrients. Thus, when accessing Al resistance, the use of controls (preferably both Al-resistant and Al-sensitive genotypes) is recommended.

The downside of hydroponics is that the level of Al resistance is not always highly correlated with results obtained from field trials. Some studies have found a good relationship between the length of the roots in hydroponics and Al resistance/tolerance in the field as, for instance, for 43

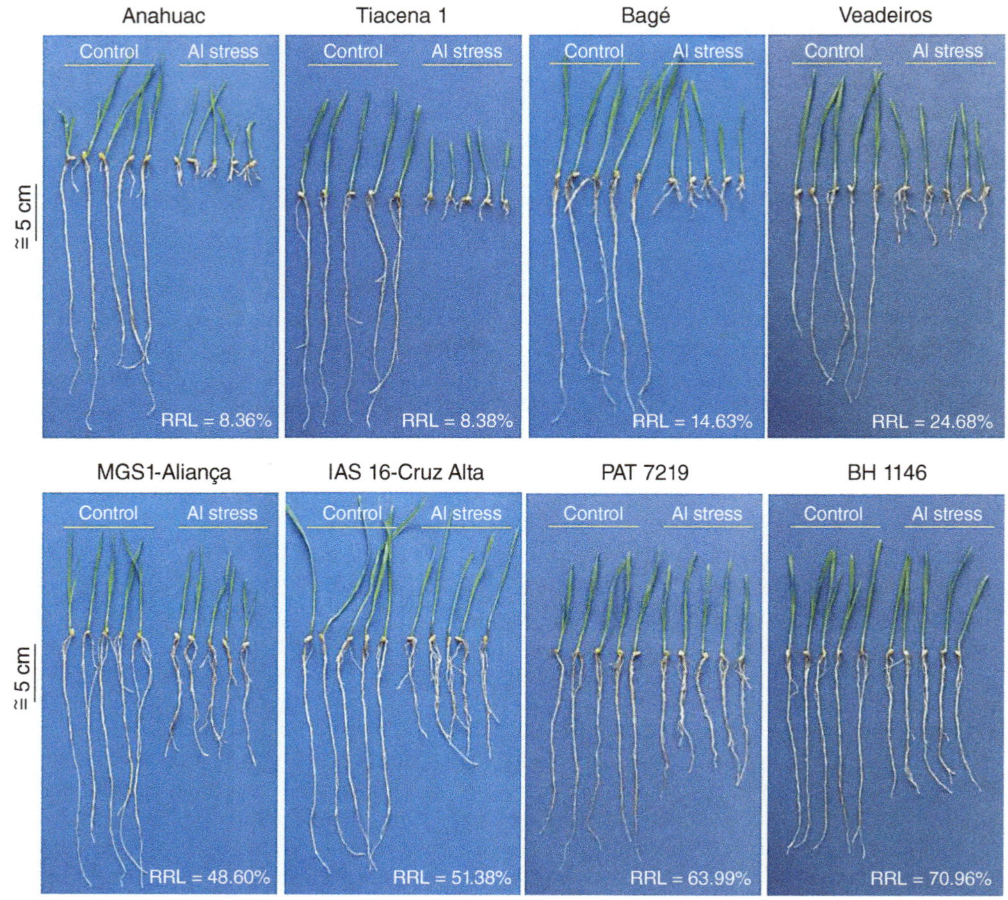


**Fig. 7.1.** Natural variability of Al resistance in wheat accessed by relative root length (RRL) in short-term soil experiments. These experiments require acid soil, from the area where the wheat plants are targeted to be cultivated, and limed soil, which is the same acid soil in which pH is increased and the soluble Al is reduced to near 0%. Wheat plants are grown under limed soil (control) and acid soil (Al stress). After 6 days of growth, roots are carefully removed, the length of the longest root is measured, and the mean root length and standard error (SE) of the plants in each condition are calculated. The RRL for each genotype is calculated as 100 × (mean root length under Al stress/mean root length in the control condition) and the errors associated with deriving the RRL are estimated as $SE_{RRL} = RRL\ [(SE_x/x)^2 + (SE_y/y)^2]^{1/2}$, where $x$ represents the mean root length in the control condition and $y$ is the mean root length under the Al treatment. The RRL of 8.36% (as obtained for cultivar Anahuac) indicates that the mean root length of plants grown under Al stress is 8.36% of the root length of the same genotype when grown without Al stress. The higher the RRL, the greater the ability of the plant to resist the Al toxicity. When the RRL is compared with the RRL of Al-resistant (such as BH 1146) and Al-sensitive (such as Anahuac) genotypes, the levels of Al resistance can be determined as, in this example, Al-sensitive (Tiacena 1), moderately sensitive (Bagé and Veadeiros), moderately resistant (MGS1-Aliança and IAS 16-Cruz Alta) and Al-resistant (PAT 7219). The data used for this figure were published by Pereira (2018).

wheat genotypes where these parameters were reported as highly correlated ($r = 0.71–0.85$, $P < 0.01$) (Baier *et al.*, 1995). However, when 338 wheat genotypes were evaluated in both hydroponics and field conditions, only an intermediate correlation ($r = 0.56$, $P < 0.001$) was observed (Aguilera *et al.*, 2016). Additionally, non-consistent responses were obtained when Al-tolerant and Al-sensitive wheat cultivars and their progenies were evaluated in soil and nutrient solution experiments (Bona *et al.*, 1994). This indicates that the nutrient solution can be useful to screen for Al resistance, but the best genotypes found in hydroponics will not necessarily perform well under acidic soil conditions in the field. For example, among the 338 wheat genotypes evaluated by Aguilera *et al.* (2016), 106 were considered Al-resistant in nutrient solution, but only eight remained classified as resistant in field trials. The reasons for the low correlation between hydroponic and soil screens that have been raised previously for barley (Ferreira *et al.*, 2018) include: (i) Al toxicity is virtually the only stress limiting root growth in the nutrient solution but a mix of several stresses is found in acid soil; (ii) the pH of the nutrient solution with or without Al stress is the same, while the soil pH may vary even between short distances; (iii) the level of Al toxicity might be different between the nutrient solution and the soil; and (iv) the availability of other nutrients (such as P and K) is different between hydroponics and soil.

One alternative to bypass the low correlation between Al resistance in hydroponics and field may be the use of short-term soil experiments (Pereira, 2018). In these experiments, a 92% correlation was found between the RRL of 20 wheat cultivars and the classification index of Al toxicity in the field proposed by Aguilera *et al.* (2016). Thus, a short-term soil experiment could be used to establish a more realistic level of Al stress in hydroponics. Several levels of Al stress can be tested in nutrient solution and the level that correlates most with the RRL of the short-term soil experiment can be selected for future experiments. In this way, the Al toxicity in nutrient solution could be adjusted to represent the amount of Al stress present in the soil from the region where the wheat is targeted to grow. This adjustment is required both to screen the wheat root growth under Al stress and to allow a fast, efficient and reliable method for breeding wheat for greater Al resistance.

## 7.3 Efflux of Organic Acid Anions as the Major Mechanisms of Aluminium Resistance in Wheat

Various physiological mechanisms can explain the variability of root growth when plants are cultivated under Al stress. Broadly, the mechanisms that prevent Al entering the plant cell (exclusion mechanisms) are usually classified as resistance mechanisms. Those that detoxify Al once it enters the plant cell (non-exclusion mechanisms) are considered tolerance mechanisms. Although these terms (resistance and tolerance) have been previously used interchangeably, many research groups now use more specific terms for clarity (Brunner and Sperisen, 2013; Kochian *et al.*, 2015; Pereira *et al.*, 2015; Ryan, 2018). The most commonly studied mechanism of Al resistance in plants relies on the exudation of low-molecular-weight organic acids by the roots. Most organic acids at the near-neutral pH of the cytoplasm occur as fully dissociated anions, and they are probably transported outside the cell as organic anions (OAs) (Ryan *et al.*, 2001). The OAs are thought to chelate Al cations, resulting in a lower availability of toxic $Al^{3+}$, which consequently reduces the level of Al toxicity in the environment. Studies showing that OAs can help plants cope with Al stress include the greater root growth, nutrient content and yield in maize when citrate was added to the nutrient solution containing Al (Bartlett and Riego, 1972) and other subsequent reports as the greater root growth of cotton growing under Al stress when different OAs were added to nutrient solution (Hue *et al.*, 1986). Years later, the ability of plants to exude OAs by the root apex was investigated (Kitagawa *et al.*, 1986; Miyasaka *et al.*, 1991). In those reports, the Al-stimulated efflux of malate was greater in roots of an Al-resistant wheat cultivar in comparison with an Al-sensitive cultivar (Kitagawa *et al.*, 1986) and an Al-tolerant cultivar of snapbean exuded ten times more citrate than an Al-sensitive cultivar when Al was added in a hydroponic system (Miyasaka *et al.*, 1991).

In wheat, Al-resistant genotypes can exude OAs (malate and/or citrate) from the root apex, which is the main mechanism of Al resistance (Fig. 7.2). The malate efflux by roots of wheat growing under Al stress was first detected by Kitagawa *et al.* (1986) but detailed and compelling data about malate as the most excreted OA stimulated by Al in wheat roots were reported when near-isogenic lines (NILs) contrasting for Al resistance were evaluated in nutrient solution (Delhaize *et al.*, 1993b). The exudation of malate was ten times greater in the Al-resistant

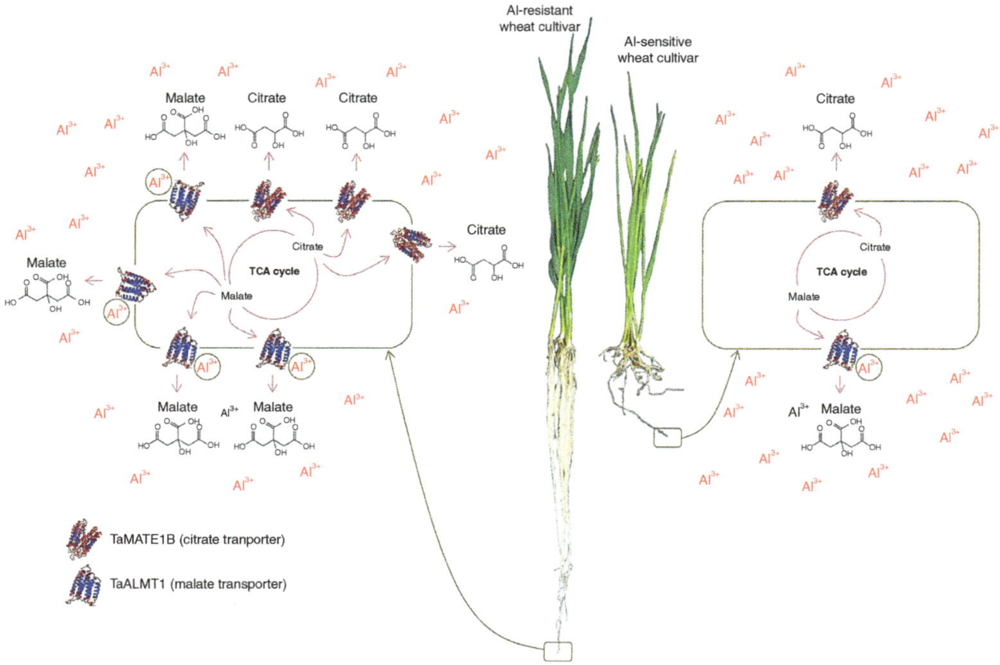


**Fig. 7.2.** Al resistance mechanism in wheat based on the malate and citrate efflux by the root apex. The efflux of both organic acids is mostly detected in the root apex and is dependent on malate (TaALMT1) and citrate (TaMATE1B) transporters located on the cell membrane. The malate and citrate exuded are thought to chelate toxic Al ions (especially the trivalent cation $Al^{3+}$), whose resulting complexes are less toxic than $Al^{3+}$. Thus, the amount of malate and citrate exuded is positively correlated with the root growth under Al stress. The three-dimensional structures of the ALMT and MATE transporters used in the figure were obtained by submitting the sequences of *TaALMT1* (GenBank DQ072260) and *HvAACT1* (GenBank KX278713) to the SWISS-MODEL (Biasini *et al.*, 2014).

genotypes, and 35-fold more malate was detected from the first 3–5 mm of the root apex than from the mature root (Delhaize *et al.*, 1993b). The malate efflux by the root apex was specifically induced by $Al^{3+}$ and not by other trivalent cations ($La^{3+}$ and $Fe^{3+}$). Although these earlier studies were performed on NILs and their parental genotypes, subsequent experiments confirmed the importance of malate exudation for Al tolerance in wheat. A high correlation ($r^2 = 0.84$) was detected between the root length of 36 wheat cultivars and the amount of malate exuded by their root apices (Ryan *et al.*, 1995). Furthermore, a significant correlation between root growth under Al toxicity and malate efflux by the root apex has been reported for several wheat genotypes with the exception of some wheat genotypes from Japanese origin (Papernik *et al.*, 2001; Raman *et al.*, 2005, 2008; Sasaki *et al.*, 2006). Consequently, Al-stimulated malate efflux is considered the main contributor to a general mechanism of Al resistance in wheat.

Although Delhaize *et al.* (1993b) detected other OAs exuded by the wheat root apex, they were exuded in relatively small quantities or were relatively poor chelators of Al (Delhaize *et al.*, 1993b). However, a study that aimed to determine how the parent genomes of triticale (× Triticosecale Wittmack), a synthetic self-pollinated hybrid derived from wheat and rye, contributed to Al resistance provided evidence that some wheat genotypes also secrete citrate. In that study, triticale obtained from Al-resistant wheat genotype showed higher citrate efflux than triticale originated from Al-sensitive wheat (Stass *et al.*, 2008). The citrate efflux was closely related to the performance of the wheat parents, with significant differences found between the Al-resistant and Al-sensitive genotypes for citrate exudation (Stass *et al.*, 2008). Soon after, 26 wheat genotypes originating from several countries were tested for citrate efflux. Among those, the root tips of four Brazilian wheat cultivars showed constitutive (not Al activated) citrate efflux (Ryan *et al.*, 2009). For the first time, citrate efflux was also implicated in Al resistance in wheat.

## 7.4 Organic Acid Transporters in Wheat

Early evidence raised the possibility that the synthesis of OAs could be the major component for the Al tolerance mechanism based on OA efflux. For instance, an increased concentration of OA was found in sorghum roots treated with Al (Cambraia *et al.*, 1983). The overexpression of genes coding for enzymes of the tricarboxylic acid (TCA) cycle such as citrate synthase, which produces citrate from oxaloacetate, and malate dehydrogenase, which causes oxidation of oxaloacetate to form malate, have been reported as a strategy to increase Al tolerance (de la Fuente *et al.*, 1997; Tesfaye *et al.*, 2001). In those reports, the overexpression of a bacterial citrate synthase gene in tobacco and papaya improved Al tolerance by increasing root growth under Al stress (de la Fuente *et al.*, 1997), and the overexpression of a malate dehydrogenase gene in alfalfa resulted in greater root growth at higher Al concentrations (50 and 100 μM Al) (Tesfaye *et al.*, 2001). However, the strategy of increasing the expression of enzymes of the TCA cycle does not appear to be easily replicated because another study about the overexpression of bacterial citrate synthase in tobacco failed to increase accumulation of citrate in roots or increase Al-activated efflux of citrate from roots (Delhaize *et al.*, 2001). Similarly, internal levels or the ability of the triticale root cells to produce malate and citrate do not regulate the Al-induced efflux of OA in triticale (Hayes and Ma, 2003). In wheat, the differential Al tolerance is not correlated with the internal content of organic acids either in shoots or roots (Foy *et al.*, 1990). Studies have shown that Al stress does not change the internal concentration of malate in excised root apices even after 6 h of exposure to Al (Ryan *et al.*, 1995). However, these excised root apices can exude malate three times more after 6 h than just after excision (Ryan *et al.*, 1995). Thus, the ability of the cell to synthesize OA does not seem to be the limiting step for the Al resistance mechanism based on the efflux of OAs by the root apex (Ryan *et al.*, 2001).

Since plants are able to produce more OA when demanded, the limiting step for the efflux of OA by the root apex is not their synthesis but their transport from cytosol to outside the cell. The exudation of OAs requires specific transporters. In wheat, a large inward current was triggered by the addition of $AlCl_3$ in a culture of protoplasts (plant cells without the cell wall) isolated from the root apex of an Al-resistant cultivar (Ryan *et al.*, 1997). As long as $Al^{3+}$ was present, the current remained active and was sensitive to niflumic acid, a compound that inhibits Al-activated efflux of malate from wheat roots (Ryan *et al.*, 1997, 2001). The channel was activated by $Al^{3+}$ and was detected in protoplasts from the root tip but not from the mature roots. In addition, a multistate channel with single-channel conductance of between 27 and 66 pS was detected (Ryan *et al.*, 1997). A follow-up study demonstrated that the channel was permeable to malate$^{2-}$ anions and that channels obtained from an Al-tolerant NIL had greater frequency of responses and larger current density in response to $Al^{3+}$ (Zhang *et al.*, 2001). These results indicated that Al-resistant wheat genotypes have channels in the root apex that are involved in malate efflux (Fig. 7.2).

Insights about the citrate transporter responsible for citrate efflux by the wheat root tip were reported years later (Ryan *et al.*, 2009; Tovkach *et al.*, 2013). The wheat citrate transporter is located on the plasma membrane and, when overexpressed in oocytes, inward and outward currents were detected if 10 mM citrate was added, but no current was observed in control oocytes. This means the citrate transporter facilitates the efflux of citrate, or other anions, at negative membrane potentials (Tovkach *et al.*, 2013). When transferred to transgenic rice and tobacco, the wheat citrate transporter generated citrate efflux in these species up to four times more than in the non-transgenic controls (Tovkach *et al.*, 2013). These results indicated that citrate efflux by wheat root apices requires a citrate transporter (Fig. 7.2). Interestingly, the amount of citrate secreted by the wheat root apex is around tenfold less than the amount of malate exuded (Ryan *et al.*, 2009, 2014). Another important difference between the malate and citrate exudation by the wheat root apex is that $Al^{3+}$ must be present in the environment to activate the malate efflux, whereas the citrate efflux is constitutive.

## 7.5 Genes Coding for Organic Acid Transporters in Wheat

The importance of chromosome 4D for Al resistance of hexaploid wheat was described nearly half a century ago when Slootmaker (1974) argued that the D genome carried one or more genes associated with tolerance to high soil acidity. Later, Al tolerance genes were detected in several chromosome arms including 4DL (Aniol and Gustafson, 1984). The chromosome arm 4DL was also the location of a single dominant locus controlling Al tolerance named *Alt2* (Luo and Dvořák, 1996) and a restriction fragment length polymorphism (RFLP) marker that explained 85% of the phenotypic variation in Al tolerance in a locus designated as $Alt_{BH}$ (Riede and Anderson, 1996). Others have also reported the major contribution of chromosome arm 4DL for greater Al tolerance and malate efflux in wheat (Milla and Gustafson, 2001; Papernik *et al.*, 2001).

For the isolation of the Al resistance gene from the chromosome arm 4DL, NILs were used to access the differential gene expression in roots grown under Al stress (Sasaki *et al.*, 2004). One gene, named *ALMT1* (*aluminium-activated malate transporter*), was detected as more expressed in the roots of the Al-resistant NIL. Several characteristics of *ALMT1* strongly supported its role as a gene coding for a malate transporter in wheat associated with Al resistance: (i) Al resistance and higher *ALMT1* expression co-segregated in $F_2$ population whose parents contrasted for Al resistance; (ii) *ALMT1* coded for a membrane protein that showed higher expression in root apices of the Al-resistant than the Al-sensitive NIL; (iii) when *ALMT1* cDNA and malate were injected in *Xenopus laevis* oocytes, an inward current was activated by the addition of $AlCl_3$; (iv) malate efflux from the roots was activated by $Al^{3+}$ in transgenic rice overexpressing *ALMT1*; and (v) transgenic tobacco cells overexpressing *ALMT1* exhibited Al-activated malate efflux as well as lower Al accumulation and greater regrowth under Al stress (Sasaki *et al.*, 2004). The *ALMT1* gene isolated from wheat was the founding member of the ALMT family since ALMT1 did not belong to any existing protein family (Sasaki *et al.*, 2004). Subsequently, genes homologous to *ALMT1* have been isolated from several plant species and shown to have a number of functions in various physiological processes including Al resistance (Sharma *et al.*, 2016). Interestingly, ALMT transporters from various species (*Arabidopsis*, barley, rice and wheat) have been described as permeable to γ-aminobutyric acid (GABA), an endogenous plant signalling molecule (Ramesh *et al.*, 2018). This might explain why malate efflux is negatively correlated with endogenous GABA concentrations in wheat root apices (Ramesh *et al.*, 2015, 2018). However, the impact of these interesting findings in the Al resistance of wheat and other species remains to be clarified.

Since several genes from the ALMT family have been isolated, the wheat *ALMT1* gene was renamed *TaALMT1*, in which the first two letters indicate the species (*Triticum aestivum*). The *TaALMT1* gene has six exons and five introns (Fig. 7.3). From the first ATG to the stop codon, the sequences vary from 3951 to 3968 bp due to intronic sequences (Raman *et al.*, 2005), but the TaALMT1 protein is 459 amino acids long in several wheat genotypes (Sasaki *et al.*, 2004; Raman *et al.*, 2005). Variability has been detected in exons and introns of *TaALMT1*, but these differences are not associated with Al resistance (Sasaki *et al.*, 2004, 2006; Raman *et al.*, 2005, 2008). Instead, Al resistance in wheat is correlated with polymorphisms in the *TaALMT1* promoter region (Sasaki *et al.*, 2006; Raman *et al.*, 2008). Seven alleles have been detected, and these alleles exhibit different blocks of sequences (named as A to D) in the *TaALMT1* upstream region with each block having a particular number and fashion of repeats (Sasaki *et al.*, 2006; Raman *et al.*, 2008; Pereira *et al.*, 2015) (Fig. 7.3). Except in lines of Japanese origin, the general pattern is that greater number of repeats in the *TaALMT1* promoter (as in alleles Types V and VI) is associated with greater levels of *TaALMT1* expression (Sasaki *et al.*, 2006). The levels of *TaALMT1* expression are strongly correlated with the root growth under Al stress when evaluated among 13 wheat genotypes ($r^2 = 0.86$) and among 18 non-Japanese lines ($r^2 = 0.94$) (Raman *et al.*, 2005; Sasaki *et al.*, 2006). Thus, the polymorphisms at the *TaALMT1* promoter are associated with higher gene expression that is correlated with Al resistance.

Although the expression of *TaALMT1* is constitutive (Sasaki *et al.*, 2004; Raman *et al.*, 2005), the efflux of malate is not constant, as $Al^{3+}$ must be in the environment to activate the malate transporter. This mechanism prevents wheat root cells from exuding malate in conditions that do not require chelation of $Al^{3+}$. In fact, *TaALMT1* was unable to facilitate the efflux of malate from root apices of an Al-resistant wheat genotype when the plants were grown under alkaline pH (Silva *et al.*, 2018a), a situation where toxic Al anions are not detected. Surprisingly, *TaALMT1* activity was induced by several anions at alkaline pH in the absence of $Al^{3+}$ when expressed in transgenic tobacco and *X. laevis* oocytes (Ramesh *et al.*, 2015). Recently, greater root growth and higher malate efflux were detected in an Al-resistant wheat line with high *TaALMT1* expression after growing for 5 weeks at alkaline pH (Kamran *et al.*, 2020). As the importance of *TaALMT1* in alkaline pH seems to be debatable, new experiments are required to shine greater light on this matter.

The gene coding for the citrate transporter in wheat belongs to a different family from ALMT, which is named as MATE (*multidrug and toxic compound extrusion*). MATE transporters were first identified to facilitate the efflux of a variety of secondary compounds in prokaryotic cells (Moriyama *et al.*, 2008; Takanashi *et al.*, 2014). Proteins from the MATE family usually have 12 transmembrane helices, and phylogenetic analyses have revealed 14 small subfamilies with the MATE proteins from plants belonging to the subfamily 2B (Moriyama *et al.*, 2008). MATE proteins in plants are not exclusively linked to the transport of citrate but are responsible for transporting a number of compounds (such as plant hormones and secondary metabolites) associated with various physiological functions (Takanashi *et al.*, 2014). Although some ALMT and MATE proteins are responsible for the efflux of OAs by the root apex in several plant species, this seems to be a case of convergent evolution, in which protein families without a common ancestor exhibit similar functions (Delhaize *et al.*, 2007; Liu *et al.*, 2009; Ryan and Delhaize, 2010).

The first members of the MATE family reported to have a role in plant Al resistance were the *SbMATE* in sorghum and *HvAACT1* (also called *HvMATE*) in barley (Furukawa *et al.*, 2007; Magalhaes *et al.*, 2007; Wang *et al.*, 2007). Subsequently, MATE transporters associated with citrate efflux and Al resistance were identified in *Arabidopsis* and rye (Liu *et al.*, 2009; Silva-Navas *et al.*, 2012). The identification of the *MATE* gene associated with Al resistance in wheat took advantage of the *HvAACT1* gene from barley. That gene was used to identify seven homologous expressed sequence tags (ESTs) in wheat, and primers were designed based on a sequence assembled from the wheat ESTs (Tovkach *et al.*, 2013). The primers detected a single cDNA fragment from root apices of the wheat cultivar Carazinho (Al-resistant genotype that shows citrate efflux), whereas no product was obtained from root apices of Egret (Al-sensitive wheat cultivar showing no citrate efflux). The gene, named as *TaMATE1B*, codes for a citrate transporter that is constitutively expressed in roots of the Al-resistant wheat cultivar (Tovkach *et al.*, 2013). Additional analysis revealed that *TaMATE1B* has homologues on chromosome 4A and chromosome 4D named *TaMATE1A* and *TaMATE1D*, respectively. However, the homologues on chromosomes 4A and 4D have low expression in roots of Al-resistant and Al-sensitive wheat cultivars (Carazinho and Egret), which excluded their role as responsible for the constitutive citrate efflux in wheat (Tovkach *et al.*, 2013). Thus, *TaMATE1B* is the gene associated with citrate efflux by the wheat root apices. The *TaMATE1B* coding region is 3933 bp long with 12 introns (Fig. 7.3) and the TaMATE1B protein has 554 amino acid residues with seven to 11 predicted transmembrane domains. When the TaMATE1B protein was fused with green fluorescent protein and transiently expressed in leek (*Allium ampeloprasum*), the chimeric protein was detected only at the cell periphery indicating that TaMATE1B is located on the plasma membrane (Tovkach *et al.*, 2013).

When sequences of *TaMATE1B* were compared, only two bases within the 11th intron were detected as polymorphic between wheat cultivars contrasting for citrate efflux (Carazinho and Egret). The main polymorphism between

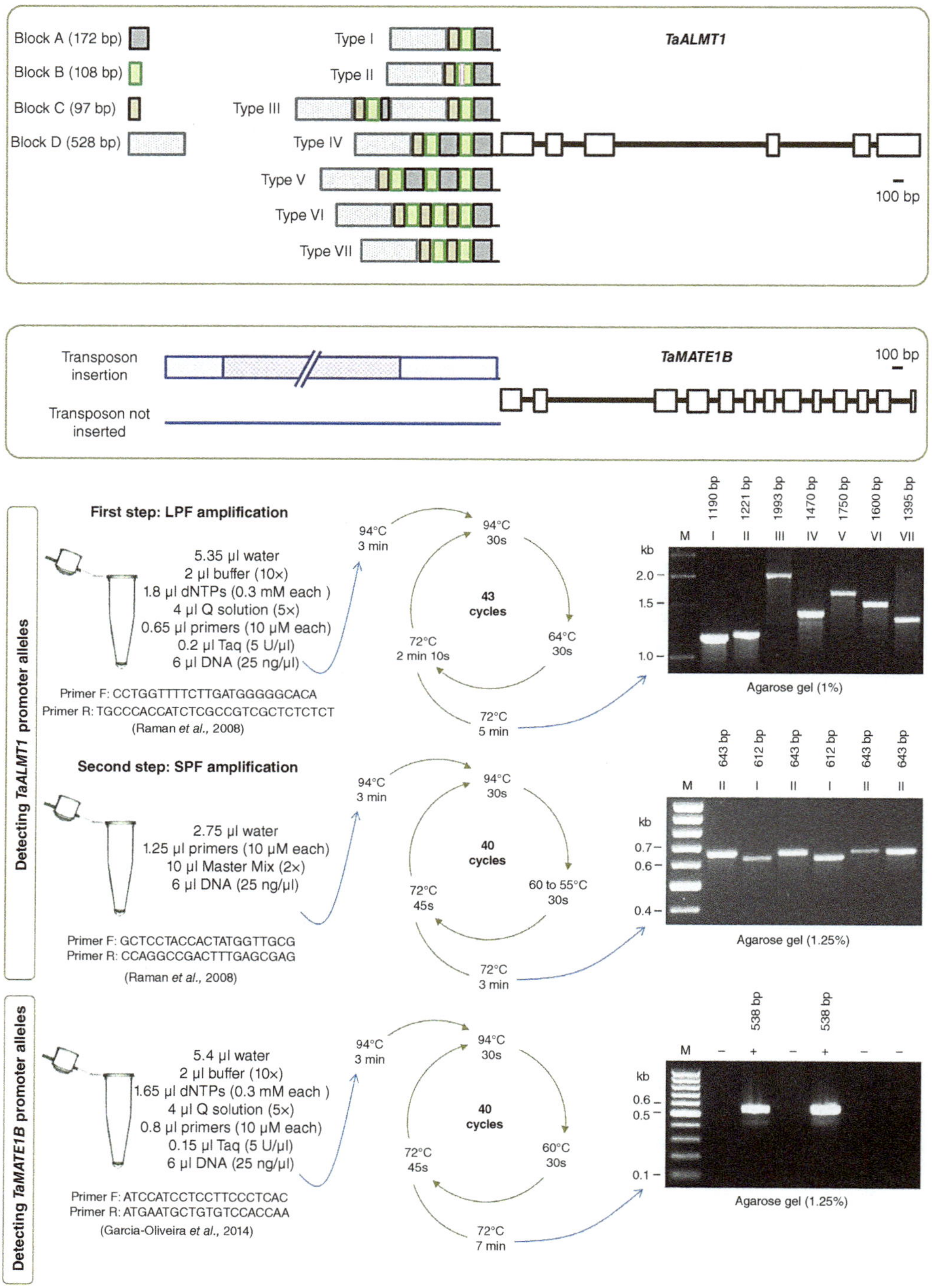


**Fig. 7.3.** *TaALMT1* and *TaMATE1B* promoter alleles and PCR conditions for their detection. So far, *TaALMT1* has seven promoter types with different repeats of four blocks named as A, B, C and D (note that block B in Type II allele has a 31 bp insertion and block A in Type III allele is 75 bp shorter). *TaMATE1B* has two alleles differing by the insertion of a 11.1 kb transposon-like element located 25 bp upstream of the start codon (Tovkach *et al.*, 2013). The purple rectangle represents the ~3.9 kb sequence that is

*Continued*

*TaMATE1B* from these cultivars is a long insertion (11.1 kb) located 25 bp upstream from the ATG start codon (Tovkach *et al.*, 2013). This long insertion seems to be a hybrid repetitive element where part of the sequence (approximately 3.9 kb) is homologous to a *Sukkula* retrotransposon. The evidence that *TaMATE1B* is a citrate transporter associated with Al resistance in wheat includes: (i) the superior allele (with the insertion in the promoter region) co-segregated with Al resistance in a population derived from Carazinho and Egret; (ii) the superior allele was detected in other wheat cultivars that exhibit citrate efflux; (iii) transgenic expression of *TaMATE1B* resulted in greater citrate efflux from whole tobacco seedlings and from the excised root apices of rice; (iv) expression of *TaMATE1B* cRNA in oocytes of *X. laevis* showed novel inward (negative) and outward (positive) currents when citrate was added; and (v) expression of *TaMATE1B* in Al-resistant cultivar (Carazinho) was significantly higher (50-fold greater) than the expression in cultivar Egret (Al-sensitive) when the apical 10 mm region of the root was analysed (Tovkach *et al.*, 2013).

A second member of the MATE family, named *TaMATE2*, was isolated in wheat due its similarity with other MATE genes from rye (Garcia-Oliveira *et al.*, 2018). *TaMATE2* homoeologues were located at chromosomes 1A, 1B and 1D (referred to as *TaMATE2-A*, *TaMATE2-B* and *TaMATE2-D*), and all three homoeologues were expressed in root apices and shoot tissues of Al-resistant and Al-sensitive genotypes under Al stress. However, higher expression was detected in root apices of the Al-resistant cultivar (Garcia-Oliveira *et al.*, 2018). *TaMATE2* might also encode a citrate transporter, but it remains unclear if it has a role in wheat Al resistance. The action of two citrate transporters in alleviating Al toxicity is not unprecedented, because it has been reported for other plant species (Liu *et al.*, 2018; Wang *et al.*, 2019). In soybean, two citrate transporters appear to mediate citrate efflux and contribute to the external detoxification of Al (Wang *et al.*, 2019), while in rice bean, the citrate efflux under Al stress is biphasic, with an early phase of a low citrate efflux (associated with *VuMATE2*) followed by a later phase of an increased secretion (associated with *VuMATE1*) (Liu *et al.*, 2018).

## 7.6 PCR to Detect *TaALMT1* and *TaMATE1B* Promoter Alleles

For both *TaALMT1* and *TaMATE1B*, polymorphisms in the promoter regions are associated with increased gene expression, greater OA efflux and improved Al resistance in wheat. Thus, detection of those polymorphisms is important to characterize germplasm and could be a great asset for wheat breeding programmes. Several groups have used this information to evaluate wheat germplasm (Sasaki *et al.*, 2006; Raman *et al.*, 2008; Han *et al.*, 2013; Garcia-Oliveira *et al.*, 2014; Pereira *et al.*, 2015; Aguilera *et al.*, 2016; Pereira, 2018). *TaALMT1* promoter Types V and VI (here designated as superior alleles due to their typical association with greater malate efflux and Al resistance) are the most common alleles detected in Brazilian wheat genotypes (Pereira *et al.*, 2015). These superior alleles are not always frequently detected in wheat germplasm from other countries, as *TaALMT1* promoter Type VI was not observed and the Type V was less frequently identified in a

**Fig. 7.3.** Continued.

homologous to a *Sukkula* retrotransposon. The *TaALMT1* and *TaMATE1B* coding regions are based on sequences from ET8 (DQ072260) and Carazinho (KC152459), respectively, while the promoter alleles are based on previous reports (Sasaki *et al.*, 2006; Raman *et al.*, 2008; Tovkach *et al.*, 2013; Pereira *et al.*, 2015). *TaALMT1* promoter alleles are first detected with primers LPF (Long Promoter Fragment), which allows the amplification of all seven alleles. When Types I and II are detected, a second step is required using primers SPF (Short Promoter Fragment), in which the 31 bp difference in size between these alleles is easier to discriminate. For SPF amplification, the annealing temperature starts at 60°C and falls 1°C every two cycles until reaching 55°C. The alleles for *TaMATE1B* are detected as presence (indicated in the gel as '+') or absence (indicated as '–') of the transposon insertion in the promoter (Garcia-Oliveira *et al.*, 2014). The primers amplify a 538 bp fragment containing part of the insertion (the first 123 bp before the ATG) and part of the *TaMATE1B* coding region (the first 415 bp).

collection of cultivars and landraces of Chinese wheat (Han *et al.*, 2013). *TaALMT1* promoter Types III and VII are detected only in a small number of genotypes (Raman *et al.*, 2008; Han *et al.*, 2013; Pereira *et al.*, 2015). For *TaMATE1B* promoter alleles, the transposon-like insertion (here designated as a superior allele due to its association with citrate efflux) is not frequently detected in large collections of wheat genotypes (Pereira *et al.*, 2015; Aguilera *et al.*, 2016) but seems to be widespread among Portuguese wheat cultivars (Garcia-Oliveira *et al.*, 2014).

Amplifying *TaALMT1* promoter alleles might not be a trivial endeavour because most of them are detected as reasonably large PCR products (up to 1993 bp). Thus, sharing details about the PCR methodology might help some research groups. The conditions that have been used to screen *TaALMT1* and *TaMATE1B* alleles are detailed in Fig. 7.3. The PCR markers can be used for marker-assisted breeding, a strategy that allowed successful introgression of the *TaALMT1* promoter Type V (superior allele) into a high-yielding Al-sensitive cultivar in only three backcrossing generations (Soto-Cerda *et al.*, 2015). Introgressing superior *TaALMT1* alleles is imperative to achieve greater Al resistance in wheat because this gene (alone or with the genomic region harbouring it) has been shown to have a great impact on Al resistance of wheat, for example the chromosome arm 4DL accounted for 85% of the phenotypic variation for Al resistance (Riede and Anderson, 1996). Additionally, *TaALMT1* explained 92% of the phenotypic variation for Al resistance (Raman *et al.*, 2005) and 69% of the phenotypic variation in a collection of 338 wheat genotypes evaluated under acidic conditions in the field (Aguilera *et al.*, 2016). However, the impact of the superior *TaMATE1B* allele for significant root growth under Al stress deserves a more thorough analysis (see next section).

## 7.7 The Importance of Citrate Efflux for Greater Aluminium Resistance in Bread Wheat

Citrate is theorized to provide higher Al tolerance than malate because the stability constants for the complex Al–citrate are significantly greater than for the Al–malate complex (Jones, 1998). In fact, citrate was shown to better chelate $Al^{3+}$ than malate (Delhaize *et al.*, 1993b; Ryan *et al.*, 1995, 2001). However, wheat genotypes with the allele associated with citrate efflux (superior *TaMATE1B* allele) have not necessarily demonstrated a significantly better Al resistance (Ryan *et al.*, 2009; Zhou *et al.*, 2013; Pereira *et al.*, 2015; Aguilera *et al.*, 2016, 2019; Han *et al.*, 2016; Dong *et al.*, 2018, 2019; Pereira, 2018). Thus, the impact of *TaMATE1B* for significant Al resistance in wheat needs to be better explored.

One strategy used to access the importance of *TaMATE1B* in bread wheat was to evaluate backcrossed germplasm where *TaMATE1B* was combined with superior or inferior *TaALMT1* alleles (Han *et al.*, 2016). In this work, two wheat cultivars with inferior *TaMATE1B* allele were used: EGA-Burke (which contains superior *TaALMT1* allele) and Egret (which has an inferior *TaALMT1* allele). After backcrosses with Carazinho (which exhibits the superior *TaMATE1B* allele), NILs for EGA-Burke and Egret were obtained, and these NILs contrasted for the *TaMATE1B* allele. When evaluated under Al stress in hydroponics, the sensitive cultivar (Egret) with the superior *TaMATE1B* allele had significant root growth up to 10 μM $AlCl_3$, evidencing the impact of the superior *TaMATE1B* allele when combined with the inferior *TaALMT1* allele. The superior *TaMATE1B* allele resulted in almost threefold greater root length in 10 μM $AlCl_3$ than the inferior *TaMATE1B* allele. However, when the superior *TaMATE1B* allele was combined with the superior *TaALMT1* allele, difference in root length was observed only at higher levels of Al stress (40 and 60 μM $AlCl_3$). Although the difference was significant, the root growth was only about 20% greater in 60 μM $AlCl_3$. When the same lines were analysed in soil culture, significant differences were found only for the Al-sensitive cultivar carrying the superior *TaMATE1B* allele (Han *et al.*, 2016). Thus, the superior *TaMATE1B* allele seems to be more important for wheat genotypes carrying inferior *TaALMT1* alleles. A similar behaviour was discussed by Pereira (2018), since the superior *TaMATE1B* allele seemed important for genotypes having the *TaALMT1* allele IV (inferior allele) when evaluated in short-term soil experiments.

However, by using a new index for selecting Al-resistant wheat genotypes, the superior *TaMATE1B* allele appeared to benefit wheat grown on acidic soil independent of the *TaALMT1* alleles (Aguilera *et al.*, 2019). More than 86% of the wheat genotypes with the *TaMATE1B* superior allele were classified as Al-resistant or moderately resistant in the field or hydroponics (Aguilera *et al.*, 2016). In this same study, 29 out of 32 genotypes containing both *TaMATE1B* and *TaALMT1* superior alleles were Al-resistant or moderately resistant in the field. Brazilian wheat cultivars internationally recognized as important Al resistance sources (i.e. IAC 5-Maringá, Toropi and Trintecinco) have superior *TaALMT1* and *TaMATE1B* alleles (Aguilera *et al.*, 2016). Additionally, the cultivars with the best two root lengths (cultivars MGS1-Aliança and IAC 21-Iguaçú), among 33 Brazilian wheat cultivars evaluated in a short-term soil experiment, both had the superior *TaALMT1* and *TaMATE1B* alleles (Pereira *et al.*, 2015). An additive effect of the superior alleles of *TaALMT1* and *TaMATE1B* on the root length of EGA-Burke (Al-resistant genotype) was observed in a very acid soil (pH 4.12) after 24 days of growth (Dong *et al.*, 2018); however, no additive effect was observed for grain yield (Dong *et al.*, 2019). Alternatively, in an Al-sensitive wheat genotype (Egret) containing either the *TaMATE1B* superior allele or the *TaALMT1* superior allele, *TaMATE1B* was as effective as *TaALMT1* in increasing grain yield, although *TaALMT1* improved shoot and root biomass better (Dong *et al.*, 2019). Thus, the combination of these superior alleles is interesting because they may show greater malate and citrate efflux by the root apex.

Similar results were obtained when another citrate transporter (*HvAACT1*) was transgenically expressed in the wheat cultivar (Fielder) which contains a superior *TaALMT1* allele (Zhou *et al.*, 2013). Although only one genetically modified line was analysed, called *Ta*(Fielder):T2_8, that line achieved greater root growth under Al stress in hydroponics than conventional Al-resistant wheat cultivars (Carazinho and Fielder) (Zhou *et al.*, 2013). However, that line seems to have multiple *HvAACT1* inserts (Zhou *et al.*, 2013), which may have exaggerated its root growth. Nevertheless, the efflux of citrate appears to improve Al resistance even if the wheat genotype presents greater malate efflux.

On the other hand, an increase in the allelic frequency of the *TaMATE1B* superior allele was not detected when *TaMATE1B* allelic variation was evaluated in genotypes obtained over 90 years of breeding (from 1922 to 2013) (Aguilera *et al.*, 2016). This indicated that, unlike for the increased use of superior *TaALMT1* alleles, the *TaMATE1B* superior allele had little selection pressure and consequently low adaptive power (Aguilera *et al.*, 2016). Although both *TaMATE1B* and *TaALMT1* alleles were highly correlated ($r = 0.71$, $P < 0.001$) with Al resistance in field conditions, that correlation was similar to the correlation of the *TaALMT1* alleles alone (0.69, $P < 0.001$), meaning that the *TaMATE1B* alleles were responsible for a smaller proportion of the Al resistance (Aguilera *et al.*, 2016). Most of the genotypes carrying the *TaMATE1B* superior allele did not outperform the genotypes without the insertion but with the same *TaALMT1* allele (Pereira *et al.*, 2015). Although wheat lines with the superior *TaMATE1B* allele demonstrated improved Al resistance in hydroponic culture, the level of resistance was considerably lower than observed for lines that possessed only *TaALMT1* (Ryan *et al.*, 2009). This means *TaMATE1B* contributes only a small percentage of the Al resistance in wheat.

Even though the citrate efflux is secondary in comparison with malate efflux for conferring greater Al resistance in wheat, pyramiding superior alleles for both *TaALMT1* and *TaMATE1B* is likely to be an important strategy to achieve better root growth under Al toxicity. No adverse effects have been reported for wheat cultivars carrying superior alleles for both these genes.

## 7.8 The Search for Other Genes Associated with Root Growth of Wheat under Aluminium Stress

Although the efflux of OAs (malate and citrate) by the root apex is the major mechanism for Al resistance in wheat, other genes with minor effect (minor genes) contribute a small fraction of the phenotypic variation for Al tolerance in wheat. Here, the term *tolerance* is used because the minor genes could be associated with detoxification of Al once it enters the plant cell (non-exclusion mechanisms). The Al tolerance mechanisms in plants include chelation of toxic Al in the cytosol (either by organic acids and phenolic substances), accommodation of the toxic Al into less-sensitive cellular compartments (e.g. cell wall of mature leaves and vacuoles) and activation of metabolic pathways to deal with the toxic effects of Al (e.g. enzymes of the oxidative stress response) (Brunner and Sperisen, 2013). Some genes, proteins and QTLs associated with Al tolerance/resistance have been identified in wheat through various studies, such as gene expression analysis (Sasaki *et al.*, 2004; Guo *et al.*, 2007a; Salvador-Moreno *et al.*, 2018), proteomics (Oh *et al.*, 2014; Yang *et al.*, 2018), response to oxidative stress (Darkó *et al.*, 2004; Nasr *et al.*, 2011; Xu *et al.*, 2012; Aggarwal *et al.*, 2015), QTL detection by bi-parental mapping (Riede and Anderson, 1996; Papernik *et al.*, 2001; Ma *et al.*, 2005, 2006; Raman *et al.*, 2005; Guo *et al.*, 2007b; Zhou *et al.*, 2007b; Cai *et al.*, 2008; Navakode *et al.*, 2009; Ryan *et al.*, 2009; Dai *et al.*, 2013; Liu *et al.*, 2015; Farokhzadeh *et al.*, 2019, 2020) and QTL detection by association mapping (Raman *et al.*, 2010; Navakode *et al.*, 2014; Froese and Carter, 2016; Emebiri *et al.*, 2020).

One of the studies employing gene expression analysis in wheat under Al stress ended up isolating *TaALMT1*, the most important gene for Al resistance in wheat (Sasaki *et al.*, 2004). In another study investigating gene expression under Al stress in wheat, NILs obtained by crossing Atlas 66 (Al-resistant) and Chisholm (Al-sensitive) were used (Guo *et al.*, 2007a). Root apices from the lines, named as Chisholm-S, and Chisholm-T, received the Al treatment. A total of 1065 putatively unique genes were identified with most of them either unclassified or no hits. Among all these genes, 57 were differentially expressed between the NILs for at least one time point of Al exposure. *TaALMT1* was found to be more abundantly expressed in both Al-stressed and unstressed root tips of Chisholm-T than in Chisholm-S. Other genes that showed abundant expression in Chisholm-T included those for ent-kaurenoic acid oxidase-1, β-glucosidase, lectin, histidine kinase and phosphoenolpyruvate carboxylase (Guo *et al.*, 2007a). Another experiment employing NILs to evaluate gene expression in wheat under Al stress used lines OK91G106, named Century-T (Al-resistant), and OK91G108, named Century-S (Al-sensitive), which share 96.9% genetic similarity between them. This study also used the cultivars Atlas66 (Al-resistant) and Bounty (Al-sensitive) (Houde and Diallo, 2008). Total RNA was isolated from the root tips of plants that grew for 24 h under Al stress and the expression analysis was performed using the Affymetrix GeneChip® Wheat Genome Array. The number of genes differentially expressed in all four cultivars was 263 genes, with 83 of them being highly significantly regulated. Among the genes associated with the Al response were those for pyruvate dehydrogenase, alternative oxidase, galactonolactone oxidase, ABC transporter, ascorbate oxido-reducatase, *TaALMT1*, glutathione *S*-transferase, germin/oxalate oxidase, fructose 1,6-bisphosphatase, cysteine-rich proteins, cytochrome P450 monooxygenase, cellulose synthase, zinc-finger transcription factor, disease resistance response protein and F-box containing domain protein (Houde and Diallo, 2008). The most recent study about gene expression triggered by Al toxicity used the cultivar Chinese Spring to investigate candidate genes associated with Al resistance (Salvador-Moreno *et al.*, 2018). Four-day-old plants grown in nutrient solution without Al stress were transferred to nutrient solution with $AlCl_3$ for 24 h. RNA was extracted from root tips and allowed to hybridize in the Agilent 4×44K wheat gene expression microarray. Fifty-nine genes from wheat showed a tenfold increase in expression under Al treatment while 77 exhibited downregulation. Among the upregulated genes detected in wheat were those for proteases, phosphoenolpyruvate carboxykinases, protease inhibitors, and genes of the oxidative stress response (Salvador-Moreno *et al.*, 2018). Other genes that were reported as expressed under Al stress in wheat include five genes called *wali1* to *wali5* (Snowden and Gardner, 1993; Richards *et al.*, 1994); genes encoding 1,3-β-glucanase and cytoskeletal fimbrin-like (Cruz-Ortega *et al.*, 1997); peroxidase, cysteine proteinase, phenylalanine-ammonia lyase and oxalate oxidase (Hamel *et al.*, 1998); and a multidrug-resistance (MDR) protein member of the ATP-binding cassette (ABC) protein superfamily (Sasaki *et al.*, 2002).

Proteomics has also been employed to study the response to Al stress in wheat. Oh *et al.* (2014) analysed the proteome of seedling roots from the wheat cultivar Keumkang grown under Al stress and control conditions. After separation by two-dimensional electrophoresis and peptide identification, 19 protein spots were found as upregulated and 28 as downregulated. The 19 upregulated proteins were β-amylase,

*S*-adenosylmethionine synthetase, oxalate oxidase 2 precursor, malate dehydrogenase, cysteine synthase, intracellular chloride channel, triosephosphate isomerase, ascorbate peroxidase, ribulose bisphosphate carboxylase small chain, quinone reductase 2, phosphoglycerate kinase, chloroplast precursor, methionine synthase 2 enzyme, elongation factor 1-γ2, succinyl-CoA ligase α2 subunit, annexin p35, UDP-D-glucuronate decarboxylase, β-D-glucan exohydrolase and fructose-bisphosphate aldolase (Oh *et al.*, 2014). Another study used the isobaric tags for relative and absolute quantitation (iTRAQ) technique to find 7041 proteins from roots of ET8, an Al-resistant wheat line, which was grown under Al stress and control treatments (Yang *et al.*, 2018). Among these proteins, 97 were differentially expressed with 47 upregulated and 50 downregulated. When the authors focused in proteins associated with root cell-wall components, nine proteins were selected and five were identified as plasma-membrane $H^+$-ATPase, peroxidase, glycosyltransferase, lipoxygenase and 14-3-3 protein (Yang *et al.*, 2018). The proteins identified in these proteomic studies need to be further studied to evaluate their putative role in Al tolerance of wheat.

An interesting point about gene expression and proteomic studies is the typical use of resistant/tolerant genotypes versus sensitive materials. However, genotypes that are not Al-resistant may also yield interesting results. For instance, Pereira (2018) reported that two genotypes (BRS 208 and Veadeiros) with inferior alleles for both *TaALMT1* and *TaMATE1B* presented the greatest RRL (33.4 and 24.7%, respectively) among six cultivars with the same inferior alleles. Although these cultivars were classified as Al-sensitive and moderately sensitive (Aguilera *et al.*, 2016), characterizing the genes and/or proteins associated with this minor Al tolerance could still be useful to learn more about mechanisms of Al tolerance in wheat. Other, Al-resistant genotypes that do not show significant malate efflux by the roots, such as the ones found by Raman *et al.* (2010), could also reveal important Al tolerance genes in wheat.

Other proteins that have a minor role in Al tolerance of wheat are the enzymes of the antioxidant system. This system is involved in response to Al stress in plants (Yamamoto *et al.*, 2003), although the production of reactive oxygen species (ROS) in wheat seedlings seems to be activated primarily by low pH exposure rather than by the Al stress (Babourina *et al.*, 2006). Nevertheless, differential responses of the antioxidant system in wheat plants contrasting for Al resistance/tolerance have been reported. For instance, roots of Al-sensitive and Al-tolerant variants of the wheat cultivar Mv Palma were evaluated for ROS formation and activity of antioxidant enzymes under Al stress (Darkó *et al.*, 2004). Three enzymes (ascorbate peroxidase, glutathione reductase and catalase) were detected in higher amounts or exclusively detected in the Al-tolerant lines that were shown to form fewer superoxides and peroxides (Darkó *et al.*, 2004). When two wheat genotypes contrasting for Al resistance, Raj-3077 (Al-resistant) and Raj-4120 (Al-sensitive), were evaluated for four antioxidant enzymes after being treated with various levels of Al stress, all enzymes were induced in the roots of the Al-resistant genotypes (Aggarwal *et al.*, 2015). In addition, the $H_2O_2$ content was lower in the roots of the Al-resistant genotype that also exhibited increased *TaALMT1* expression and higher malate exudation. Thus, mechanisms for Al resistance (malate efflux) and Al tolerance (antioxidation system) can act together in the response to Al stress. However, when the wheat cultivars Darab (Al-sensitive) and Maroon (Al-tolerant) were exposed to Al stress, no significant difference in the activity of ascorbate peroxidase, catalase, glutathione reductase and superoxide dismutase was found in the Al-tolerant genotype (Nasr *et al.*, 2011). In fact, an increased catalase activity was observed at the higher Al concentration only for the Al-sensitive genotype (Nasr *et al.*, 2011). The Al-sensitive genotype Yangmai-5 also showed higher activity of superoxide dismutase and peroxidase under Al exposure than the Al-tolerant genotype Jian-864, although the Al-tolerant genotype presented higher activity of other antioxidant enzymes and greater antioxidant capacity (Xu *et al.*, 2012). Pretreatment of wheat seedlings by $H_2O_2$ has been shown to improve the antioxidant defence capacity and consequently prevent ROS accumulation (Xu *et al.*, 2011). The pretreatment enhanced Al acclimatization during subsequent Al exposure, which was greater in an Al-sensitive genotype than in an Al-tolerant genotype (Xu *et al.*, 2011). Thus, breeding programmes can access the antioxidant capacity of wheat genotypes to breed a more effective antioxidant system that helps wheat plants better cope with Al stress.

Another strategy used to identify minor genes or genomic regions associated with Al tolerance in wheat is QTL mapping. QTL mapping can identify loci harbouring genes that contribute to variation in a complex trait (Alonso-Blanco and Koornneef, 2000). Recombination frequency is crucial to locate QTLs and two main types of recombination can be used: (i) recombination events that occurred in a mapping population (bi-parental mapping); and (ii) historical recombination events in a random mating population (association mapping) (Mackay *et al.*, 2009). Both strategies have been used to map QTLs for Al resistance in wheat. The importance of these types of investigation relies on narrowing the genomic regions associated with a specific trait, which is particularly interesting for plants with big genomes such as the 17 Gb wheat genome (Brenchley *et al.*, 2012). This type of information was important for mapping the loci $Alt_{SB}$ and *Alp*, which resulted in the isolation of the genes *SbMATE* and *HvMATE* associated with citrate efflux and Al resistance in sorghum and barley, respectively (Magalhaes *et al.*, 2004, 2007; Wang *et al.*, 2007).

QTLs detected by bi-parental mapping revealed several genomic regions associated with Al tolerance in wheat (Table 7.1). A QTL was commonly detected on chromosome arm 4DL by different studies that used various mapping populations or ditelosomic lines (Riede and Anderson, 1996; Papernik *et al.*, 2001; Ma *et al.*, 2005, 2006; Raman *et al.*, 2005; Guo *et al.*, 2007b; Zhou *et al.*, 2007b; Cai *et al.*, 2008; Navakode *et al.*, 2009; Dai *et al.*, 2013; Liu *et al.*, 2015; Farokhzadeh *et al.*, 2019). This QTL is probably associated with *TaALMT1* as a major gene for Al resistance in wheat. When focusing on Al-resistant wheat cultivars but contrasting for the ability to exude citrate, a major QTL on chromosome arm 4BL was detected (Ryan *et al.*, 2009). This QTL harbours the *TaMATE1B* gene. Several other QTLs that are not associated with *TaALMT1* and *TaMATE1B* have been found on chromosomes 5AS and 7AS (Papernik *et al.*, 2001); 5AS and 2DL (Ma *et al.*, 2006); 3BL (Zhou *et al.*, 2007b); 2A and 3BL (Cai *et al.*, 2008); 3B (Navakode *et al.*, 2009); 3BL (Dai *et al.*, 2013); and 2DL and 7BL (Liu *et al.*, 2015). More recently, several QTLs for above-ground traits (including yield) have been detected throughout the wheat genome when a wheat population was evaluated in field trials under Al stress and normal conditions (Farokhzadeh *et al.*, 2019, 2020). Nevertheless, the genes underlying these QTLs outside chromosomes 4BL and 4DL have not been isolated yet.

Different markers and populations have been used in association mapping studies to detect genomic regions associated with Al tolerance in wheat (Raman *et al.*, 2010; Navakode *et al.*, 2014; Froese and Carter, 2016; Emebiri *et al.*, 2020). Using genotyping by Diverse Arrays Technology (DArT) markers, association with Al tolerance was found in 15 out of 21 wheat chromosomes (Raman *et al.*, 2010) and on chromosomes 1AL, 1DL, 3BL and 6AS (Navakode *et al.*, 2014). The 9K and 90K SNP (single-nucleotide polymorphism) platforms (Cavanagh *et al.*, 2013; Wang *et al.*, 2014) were employed to genotype two diverse populations of wheat leading to the identification of 55 loci associated with Al tolerance (Froese and Carter, 2016). DArT and SNP markers were used in 300 accessions of synthetic hexaploid wheat and significant association with Al tolerance was found in 24 loci distributed across 11 wheat chromosomes (Emebiri *et al.*, 2020). These association mapping studies identified regions that are in agreement with the QTLs identified by bi-parental mapping and also found new loci and candidate genes associated with Al tolerance. However, depending on the population or the type of markers used, some expected associations were not detected. For instance, SNP markers within chromosome 4D (where the *TaALMT1* gene is located) were not associated with the Al tolerance in one population (Froese and Carter, 2016), and the chromosomes associated with Al tolerance do not harbour the *TaALMT1* and *TaMATE1B* genes (Navakode *et al.*, 2014). In the same way, no statistically significant association was detected with loci mapped to chromosome 4B where *TaMATE1B* gene is located (Emebiri *et al.*, 2020). These results indicate that new association studies are important to validate QTLs previously detected and strategies should be employed to isolate the genes underlying these QTLs.

## 7.9 Exploiting Wheat Wild Relatives

Wild relatives are the closely related species or ancestors of a particular crop. Breeders have

**Table 7.1.** Quantitative trait loci (QTLs) for aluminium resistance/tolerance identified in wheat.

| Mapping methodology | Traits evaluated | QTLs and their contribution to Al resistance | Reference |
|---|---|---|---|
| RILs derived from BH 1146 (Al-resistant) and Anahuac (Al-sensitive) genotyped with 83 RFLP markers | Al tolerance in nutrient solution | One RFLP marker was located 1.1 cM from a major gene on chromosome arm 4DL, which explained 85% of the phenotypic variation for Al tolerance | Riede and Anderson (1996) |
| 27 ditelosomic lines of Chinese Spring (moderately Al-resistant) | Relative root growth | Lack of chromosome arms 4DL, 5AS and 7AS resulted in decreased Al tolerance. The loss of 4DL resulted in the largest decrease in Al tolerance, the largest increase in Al accumulation and the largest decrease in Al-induced malate exudation | Papernik *et al*. (2001) |
| 118 RILs obtained from Atlas 66 (Al-tolerant) and Century (Al-sensitive) genotyped with 131 polymorphic SSR markers | Haematoxylin staining and root growth after Al stress | One QTL was detected on chromosome arm 4DL, where *TaALMT1* is found. This major QTL accounted for nearly 50% of the phenotypic variation for Al tolerance and had high LOD score (about 21) | Ma *et al*. (2005) |
| Five DH populations derived from Cranbrook (Al-sensitive) and Halberd (Al-tolerant); Sunco (Al-sensitive) and Tasman (Al-tolerant); Diamondbird (Al-tolerant) and Janz (Al-sensitive); Currawong (Al-tolerant) and CD87 (Al-sensitive); and Spica (Al-sensitive) and Maringa (Al-tolerant), whose linkage maps were available or constructed by using nine polymorphic SSR markers from chromosome 4D | Root growth and haematoxylin staining | DH populations derived from different crosses revealed a single major QTL for Al tolerance on chromosome 4D. For the population derived from Sunco and Tasman, the QTL had a LOD score of 62.35 and accounted for 80% of the variation. The population obtained from Cranbrook and Halberd showed the same QTL with a LOD score of 44.54 and explained 93% of the variation. In the population derived from Diamondbird and Janz, *TaALMT1* explained 92% of the total phenotypic variance for Al tolerance with a LOD score of 83.7 | Raman *et al*. (2005) |
| 90 RILs derived from Annong 8455 (Al-sensitive) and CS-SM3DS7A (line where chromosome 7A from Chinese Spring was substituted by the same chromosome of Sumai 3) mapped with 381 AFLP markers and 168 SSR markers, and analysis of 31 ditelosomic lines derived from Chinese Spring (moderately Al-resistant) | Root elongation, referred to as stress root growth (SRG), and root resistance index (RRI) | Three QTLs that enhanced root growth under Al stress were detected: (i) *Qalt.pser-4D* (chromosome arm 4DL), which explained 18.4% of phenotypic variance and had LOD scores of 2.97 for SRG and 2.58 for RRI; (ii) *Qalt.pser-5A* (chromosome arm 5AS), which explained 10.5% of phenotypic variance with LOD scores of 1.85 for SRG and 1.80 for RRI; and (iii) *Qalt.pser-2D* (chromosome arm 2DL), which explained 12.8% of phenotypic variance with LOD scores of 2.1 for SRG and 1.92 for RRI. Ditelosomic lines of Chinese Spring with lower root growth indicated Al resistance genes on chromosomes 5AS, 7AS, 2DL and 4DL | Ma *et al*. (2006) |

*Continued*

**Table 7.1.** Continued.

| Mapping methodology | Traits evaluated | QTLs and their contribution to Al resistance | Reference |
|---|---|---|---|
| Six NILs derived from Atlas 66 (Al-tolerant) and Chisholm and Century (Al-sensitive cultivars) genotyped with 200 AFLP (82 polymorphic) and 88 SSR markers (one polymorphic between parents and NILs) | Relative root growth (RRG) and haematoxylin stain score (HSS) | Al tolerance in Al-tolerant NILs was shown to be controlled by at least one common gene/QTL, most likely conferred by the major QTL on 4DL that harbours *TaALMT1*. Other QTLs with minor effects might be responsible for Al tolerance in Atlas 66 | Guo *et al*. (2007b) |
| 192 RILs derived from Atlas 66 (Al-tolerant) and a Chisholm (Al-sensitive) mapped with 50 polymorphic SSR markers | Haematoxylin stain score (HSS) and net root growth (NRG) | Two QTLs for Al resistance were detected in Atlas 66: (i) one major QTL on chromosome 4D (that co-segregated with *TaALMT1*); and (ii) a minor QTL located on chromosome arm 3BL. Together, they explained about 58% of the phenotypic variation for HSS and 50% for NRG | Zhou *et al*. (2007b) |
| 199 RILs derived from Chinese landrace FSW (Al-resistant) and Chinese line ND35 (Al-sensitive) genotyped with 116 polymorphic SSR markers | Net root growth (NRG) and haematoxylin staining score (HSS) | Three QTLs explaining up to 81.9% of the phenotypic variation for HSS and 78.3% of the variation for NRG were detected. Mean LOD scores were: 15.9 for HSS and 13.7 for NRG on QTL *Qalt.pser-4DL*; 12.9 for HSS and 12.6 for NRG on *Qalt.pser-3BL*; and 3.3 for HSS and 2.1 for NRG on *Qalt.pser-2A* | Cai *et al*. (2008) |
| 57 DH populations derived from Chinese Spring (moderately Al-resistant) and CS′ (‘Synthetic’ 3B), where chromosome 3B of Chinese Spring was substituted, genotyped with 14 polymorphic SSR markers on chromosome 3B, and use of substitution and introgression lines of Chinese Spring | Haematoxylin staining and root tolerance index (RTI) | The introgression lines revealed a major QTL on chromosome arm 4DL with LOD score 6.69, explaining about 31% of the phenotypic variation for RTI. The DH population revealed a major Al tolerance QTL on chromosome 3B (*QaltCS.ipk-3B*), which explained 49% of the phenotypic variation | Navakode *et al*. (2009) |
| 67 $F_2$ individuals derived from Carazinho (Al-resistant with citrate efflux) and EGA-Burke (Al-resistant without citrate efflux) genotyped with 676 polymorphic DArT markers and six polymorphic SSR markers | Citrate efflux and net root growth | Based on data from DArT markers, a major QTL, *Qcec-4BL*, was detected on chromosome 4BL which explained more than 50% of the phenotypic variation for citrate efflux. The SSR markers revealed a significant QTL also on chromosome 4BL, which explained 51% of the phenotypic variation. The QTL explained 56% of the phenotypic variation for net root growth | Ryan *et al*. (2009) |

| | | | |
|---|---|---|---|
| Association mapping in 1057 accessions of common wheat, originating from various parts of the world, genotyped with 178 DArT markers | Haematoxylin staining, relative root elongation and malate efflux | At least 16 genomic regions located on chromosomes 1A, 1B, 2A, 2B, 2D, 3A, 3B, 4A, 4B, 4D, 5B, 6A, 6B, 7A and 7B were significantly associated with Al resistance via GLM. When using a more stringent analysis (MLM), only three loci on chromosomes 3B, 4D and 5B were identified. The *TaALMT1* locus explained up to 95% of the variation in Al resistance | Raman *et al.* (2010) |
| 217 RILs derived from Chinese landrace FSW (Al-resistant) cultivar Wheaton (Al-sensitive) genotyped with 35 polymorphic SSR markers | Net root growth (NRG) and haematoxylin staining score (HSS) | Two QTLs accounting for about 74.9% of the phenotypic variation for HSS and 72.1% for NRG. QTL on chromosome 4DL showed LOD score of 64.86 for HSS and 57.80 for NRG, and on chromosome 3BL, the LOD scores were 6.72 for HSS and 7.80 for NRG | Dai *et al.* (2013) |
| Association mapping in a core collection of 96 winter wheat accessions, originating from several countries, genotyped with 525 DArT markers | Haematoxylin staining | Significant or highly significant marker–trait associations were identified on chromosomes 1AL, 1DL, 3BL and 6AS by MLM. All associations except for chromosome 1AL were also detected by GLM. The major locus on chromosome 4DL (harbouring *TaALMT1*) was not detected | Navakode *et al.* (2014) |
| 144 RILs derived from Jagger (acidic soil-tolerant) and 2174 (moderately tolerant to acidic soil) genotyped with 9K SNP markers and 400 SSR markers | Visual ratings of acidic soil tolerance under field conditions | Three QTLs for acidic soil tolerance were detected. *QAlmt.osu-4D* (chromosome 4DL) was the major QTL, explaining up to 38.5% of the total phenotypic variation (LOD score from 4.6 to 13.3). The other two QTLs were: *QAlmt.osu-2D* (on chromosome 2DL), observed in Jagger, explaining up to 32% of the total phenotypic variation (LOD score from 1.2 to 7.0); and *QAlmt.osu-7B* (chromosome 7BL), detected in 2174, explaining up to 25.1% of the total phenotypic variation (LOD score from 2.4 to 6.1) | Liu *et al.* (2015) |
| Association mapping in two diverse populations of winter wheat (Population A with 459 accessions of soft white wheat and Population B with 401 accessions of soft white and hard red) genotyped using 90K (Population A) or 9K (Population B) SNP arrays | Field phenotyping where plants from both populations were rated on a 1 to 5 scale (where a score of 1 indicated tolerance and 5 indicated sensitivity); Population B was also phenotyped for root length in hydroponics | A total of 55 loci were detected and significant association with Al tolerance, in both field and hydroponics, was identified on all chromosomes except 1D, 5D and 6D. Fifteen loci, located on chromosomes 1A, 1B, 2A, 2B, 3B, 4A, 5A, 5B, 6A, 6B and 7A, were equivalent between the two populations. Loci with superior stability across distinct environments were detected in chromosomes 1A, 3B, 4D, 6A, 6B and 7A. Region of chromosome 4DL, where *TaALMT1* resides, was not detected in Population B | Froese and Carter (2016) |

*Continued*

**Table 7.1.** Continued.

| Mapping methodology | Traits evaluated | QTLs and their contribution to Al resistance | Reference |
|---|---|---|---|
| 167 RILs derived from SeriM82 (moderately tolerant to drought and environmental stresses) and Babax (tolerant to drought and environmental stresses) genotyped with 477 DNA markers (SSR, AFLP and DArT) | Measurements of flag leaf, canopy temperature and its degradation at grain-filling stage, RWC, cell-membrane stability, WSC concentration, Al amount and grain yield | Single-locus QTL analysis revealed 48 QTLs (32 suggestive and 16 putative), which explained 4.57 to 11.29% of the phenotypic variation with a LOD score of 2.06 to 4.40 for all traits. The QTLs were detected throughout the wheat genome (52.08, 29.17 and 18.75% in the A, B and D genomes, respectively). The strongest QTL associated with Al concentration was detected on linkage group 7A | Farokhzadeh *et al.* (2019) |
| 167 RILs derived from SeriM82 (moderately tolerant to drought and environmental stresses) and Babax (tolerant to drought and environmental stresses) genotyped with 477 DNA markers (SSR, AFLP and DArT) | Vegetative growth period, grain-filling period, grain yield, biomass yield, straw yield and harvest index | Single-locus QTL analysis detected nine putative and 31 suggestive QTLs, which explained from 4.57 to 14.21% of the phenotypic variation with a LOD score of 2.01 to 5.72. Major or stable Al tolerance QTLs were detected on linkage groups 1A, 1B, 1D-a, 2A-b, 2A-d, 2B, 2D, 4A, 4B, 6A-a, 6B, 7A and 7D-a | Farokhzadeh *et al.* (2020) |
| Association mapping in a collection of 300 synthetic hexaploid wheat accessions, obtained from crosses between *Aegilops tauschii* and durum parents, with 438 DArT markers and 6223 SNPs | Haematoxylin staining | Statistically significant association with Al tolerance was detected for 24 loci located to chromosomes 1B, 1D, 2A, 2B, 4A, 4D, 5A, 5B, 6A, 6D and 7A. Genomes A and B harboured 71% of the QTLs identified. The phenotypic variation for Al tolerance varied from 3 to 17% for the individual markers. Association with *TaALMT1* was detected but not with *TaMATE1B* | Emebiri *et al.* (2020) |

RIL, recombinant inbred line; DH, doubled haploid; SSR, simple sequence repeat; AFPL, amplified fragment length polymorphism; NIL, near-isogenic lines; RWC, relative water content; WSC, water-soluble carbohydrate; LOD, logarithm of the odds; GLM, general linear model; MLM, mixed linear model.

exploited a number of these wild relatives to isolate or introgress important genes for several traits in bread wheat (Hajjar and Hodgkin, 2007). Advantages of using wild relatives include reversing breeding bottlenecks and helping identify stress tolerance genes and associated regulatory regions (Yumurtaci, 2015).

When focusing on Al tolerance, a number of wheat wild relatives have been evaluated, but they usually do not outperform modern Al-resistant hexaploid wheat cultivars. For instance, only 22 out of 523 accessions showed some level of Al tolerance among 25 *Triticum* species evaluated by Berzonsky and Kimber (1986). However, the tolerance levels found in the 22 accessions were not as high as the Al resistance of Atlas 66, an Al-resistant hexaploid wheat cultivar. Additionally, none of the 731 accessions of six *Triticum* species studied by Ryan *et al.* (2010) displayed Al resistance. In the same study, five out of 29 accessions of *Aegilops tauschii* had moderate levels of Al resistance which were comparable to ET8, an Al-resistant hexaploid wheat genotype. Interestingly, all the 29 accessions plus 360 other genotypes of *Ae. tauschii* lack the promoter repeats commonly found in Al-resistant hexaploid wheat (Ryan *et al.*, 2010). *Aegilops uniaristata* is another wheat wild relative that exhibits some level of Al tolerance (Berzonsky and Kimber, 1986). This species was used to obtain three substitution lines where the chromosome 3N from *Ae. uniaristata* replaced wheat chromosomes 3A, 3B or 3D of Chinese Spring, a hexaploid wheat cultivar showing moderate Al resistance (Miller *et al.*, 1997). When evaluated in hydroponics under Al stress, all lines carrying the chromosome 3N achieved larger increase in yield in lower Al concentrations than the Chinese Spring parent and BH 1146, an Al-resistant cultivar. However, at higher Al concentrations, the lines carrying the chromosome 3N were as affected as BH 1146 (Miller *et al.*, 1997).

The wild species *Leymus racemosus* that belongs to tribe *Triticeae* (*Poaceae*) was also targeted for studies about Al tolerance (Mohammed *et al.*, 2013). Thirteen addition lines, two substitution lines of wheat (Chinese Spring) and *L. racemosus*, along with 15 addition lines from 11 wheat wild relatives, were used to search for new genetic resources to improve wheat tolerance to Al. None of the 15 addition lines from 11 wheat wild relatives achieved greater root growth under Al stress than Chinese Spring. However, three wheat–*L. racemosus* addition lines (A, E and O) showed better Al tolerance than Chinese Spring at 50 μM Al when grown for 48 h, but in Al concentrations up to 200 μM, only one line (E) had greater Al tolerance than Chinese Spring (Mohammed *et al.*, 2013). When evaluated for 5 days at 10 μM Al, the A and E lines achieved greater root growth than Chinese Spring. In long-term experiments (15 days), the two addition lines also showed better growth parameters than Chinese Spring. Thus, the addition of *L. racemosus* chromosomes A and E improved Al tolerance in hexaploid wheat. The introgressed chromosomes did not affect the Al concentration in the root apices and the *TaALMT1* expression, which was comparable among the addition lines and Chinese Spring. When cell-membrane integrity was evaluated, the addition line E had the lowest reduction for that trait. Thus, the improved Al tolerance was attributed to improved cell-membrane integrity (Mohammed *et al.*, 2013). Although the wheat–*L. racemosus* addition lines were not compared with the wheat cultivars that exhibited greater Al resistance than Chinese Spring, the greater cell-membrane integrity of the addition line E should be further investigated. It would also be interesting to study the mechanisms of Al tolerance different from OA efflux detected in *Andropogon virginicus*, a wild species of *Poaceae* (Ezaki *et al.*, 2013). Pyramiding malate and citrate efflux with mechanisms different from OA efflux could eventually improve even further the wheat root growth under Al stress.

## 7.10 Transgenic Approach to Increase Aluminium Resistance in Hexaploid Wheat

Transgenesis to increase root growth of wheat under Al stress could explore a variety of genes (Ryan *et al.*, 2011). However, only a few attempts have been reported with all of them based on the overexpression of genes coding for OA transporters (Magalhaes *et al.*, 2007; Pereira *et al.*, 2010; Zhou *et al.*, 2013).

Wheat transformation with the *SbMATE* gene under control of the maize *ubiquitin* (*ubi*) promoter was performed, along with several other experiments, to establish the role of *SbMATE* in coding for a citrate transporter in sorghum associated with Al resistance (Magalhaes *et al.*, 2007). Four transgenic $T_1$ lines, originated from Bobwhite (an Al-sensitive wheat genotype widely used as target for wheat transformation), were evaluated in nutrient solution under low Al stress (5 μM $AlCl_3$). The transgenic lines had greater root length under Al stress compared with the non-transformed wild-type Bobwhite (Magalhaes *et al.*, 2007). The transgenic line 6001D achieved an average RRL near twofold greater than the non-transformed control. However, as cited in Pereira *et al.* (2010), the increased Al resistance of those lines was not maintained in subsequent generations.

Overexpression of the wheat gene coding for malate transporter, *TaALMT1*, was used to obtain genetically modified (GM) wheat lines with greater Al resistance (Pereira *et al.*, 2010). The Al-sensitive genotype Bobwhite (line SH 98026) was transformed with a construct in which *TaALMT1* was under the control of the *ubi* promoter. Nine $T_2$ homozygous transgenic lines were accessed for root growth under Al stress, malate efflux and *TaALMT1* expression, and compared with Bobwhite, ET8 (an Al-resistant genotype) and null segregant lines. Malate efflux from the root apices of all $T_2$ lines was significantly greater than that of Bobwhite and one null segregant line, while most lines showed similar malate efflux to the Al-resistant genotype. All transgenic lines exhibited greater root growth under Al stress on nutrient solution when compared with the Al-sensitive wild type (Bobwhite) and the null segregant lines. However, when compared to the Al-resistant genotype (ET8), only some lines were more Al-resistant at the higher $Al^{3+}$ concentrations (40 and 60 μM $AlCl_3$) (Pereira *et al.*, 2010). Plants were also compared in two soil experiments. In the first one, plants were grown for 4 days, and the three tested $T_2$ lines performed better than Bobwhite while the line T2_20A.8 exhibited greater root length than ET8. In the second soil experiment, plants were evaluated after 31 days and the transgenic line (T2_4.4) and Al-resistant control (ET8) showed larger root systems than Bobwhite, with T2_4.4 achieving greater root dry weight than both the Al-resistant and Al-sensitive controls (Pereira *et al.*, 2010). In these experiments, the $T_2$ line exhibiting the greatest root length under 20 μM $AlCl_3$ (T2_5.4) had an average RRL nearly eightfold greater than the non-transformed control. This increase is significantly higher than the near twofold increment in root growth of $T_1$ wheat lines expressing *SbMATE* (Magalhaes *et al.*, 2007).

Similar to the previous reports, the *HvAACT1* gene that codes for a citrate transporter in barley was overexpressed in wheat (Zhou *et al.*, 2013). *HvAAACT1* was engineered to be under control of the *ubi* promoter and Bobwhite (line SH 98026) was also used as receptor genotype. Additionally, a second wheat genotype, the Al-resistant cultivar Fielder, was used for transformation (Zhou *et al.*, 2013). All $T_2$ lines presented higher *HvAACT1* expression than the controls (Bobwhite, Fielder, Carazinho and a null segregant line), and most lines had citrate efflux greater than their respective wild types (Bobwhite and Fielder) and the null segregants. The best $T_2$ transgenic line derived from Bobwhite exhibited about 70% increase in root length compared with non-transformed Bobwhite when the plants were grown at low level of Al stress (4 μM $AlCl_3$). For the only $T_2$ line obtained from Fielder, the *HvAACT1* overexpression allowed a near 20% increase in root length compared with non-transformed Fielder growing at high level of Al stress (30 μM $AlCl_3$) (Zhou *et al.*, 2013). The root growth of that line was even greater than of the Al-resistant wheat cultivar Carazinho (Zhou *et al.*, 2013). However, that line was not evaluated in soil. When the root growth of the transgenic $T_2$ lines derived from Bobwhite was evaluated under soil conditions, the plants did not outperform Carazinho. Their root growth was clearly greater than the non-transformed Bobwhite and one null segregant, but Carazinho showed the best root growth under soil conditions. When comparing the results obtained from the GM lines derived from Bobwhite, the increment in root growth was poorer than the twofold increase observed by overexpressing *SbMATE* (Magalhaes *et al.*, 2007) and much lower than the eightfold improvement by overexpressing *TaALMT1* (Pereira *et al.*, 2010). This indicates that, as similarly observed for the naturally occurred *TaALMT1* and *TaMATE1B* genes, the overexpression of *TaALMT1* seems to be much more important for improving root growth under Al stress than overexpressing citrate transporters.

The low number of genes used so far to obtain GM wheat for greater Al resistance is possibly related to: (i) the difficulty in transforming wheat, although some modern protocols are reported to be more efficient (Ishida *et al.*, 2015; Hayta *et al.*, 2019); and (ii) the questionable acceptance of GM wheat. Progress to obtain commercial cultivars of transgenic wheat is slow, with no GM wheat of any nature approved yet for commercial cultivation (Araus *et al.*, 2019). Since other plants (such as *Arabidopsis*, rice and tobacco) are routinely easier to transform, most of the experiments to establish the role of a specific gene in Al resistance/tolerance (such as those encoding OA transporters) have been performed with those species. Nevertheless, more research on this topic is necessary to clarify the importance of the transgenic approach to increase Al resistance in wheat. There are still a number of questions regarding the use of that approach, such as: (i) the performance of the GM plants under field conditions; (ii) the impact of gene overexpression on productivity and quality of the wheat; (iii) the consequences of combining the overexpression of genes for higher OA synthesis (i.e. citrate synthase and malate dehydrogenase) with greater OA efflux (i.e. *TaMATE1B* and *TaALMT1*); (iv) the acceptance of the GM wheat plants by farmers, industry and consumers; and (v) the impact of root-specific promoters instead of constitutive promoters. This last question needs to be investigated because it would prevent the expression of the gene in parts of the plant that are not relevant to Al resistance. For instance, interesting results may be obtained by evaluating GM wheat plants where the *TaMATE1B* gene is under control of a superior *TaALMT1* promoter (as Types V or VI). That strategy has been previously used in *Arabidopsis*, in which the *AtALMT1* promoter conferred a significantly higher level of gene expression than the *AtMATE* promoter (Liu *et al.*, 2012). When the *AtMATE* gene was placed under the control of the *AtALMT1* promoter, the Al resistance significantly increased and the carbon-use efficiency for Al resistance was enhanced (Liu *et al.*, 2012).

## 7.11 Improving Aluminium Resistance of Durum Wheat

Durum wheat (*T. turgidum* spp. *durum*) is known for its poor growth under Al stress compared with bread wheat (*T. aestivum*) (Bona *et al.*, 1992, 1995). In fact, no Al-tolerant or moderately tolerant genotype has been detected in durum wheat (Bona *et al.*, 1992). None of the 631 accessions of durum wheat evaluated in nutrient solution containing 10 μM $AlCl_3$ exhibited Al resistance (Ryan *et al.*, 2010), and all durum wheat genotypes evaluated by Wayima *et al.* (2019) were Al-sensitive. Interestingly, some studies have detected Al tolerance in durum lines, but after using a marker specific to the D genome, the lines were subsequently found to be bread wheat that likely contaminated the durum grain stocks (Foy, 1996; Han *et al.*, 2014; Wayima *et al.*, 2019). The lower Al resistance of durum wheat might have restricted its use in agriculture (Ryan *et al.*, 2010). Because of the superior rheological properties of durum wheat pasta doughs and the cooking quality of durum wheat pasta (Autran and Feillet, 1985), production of this wheat is important in many regions of the world. Breeding greater Al resistance in durum wheat is important to expand its production to regions where soil/subsoil acidity is a problem.

In contrast to bread wheat that contains a hexaploid genome (AABBDD), durum wheat has a tetraploid genome (AABB). This means that the major gene responsible for Al resistance in hexaploid wheat, *TaALMT1* located on the chromosome 4D (D genome), is absent from the durum wheat genome. Thus, Al sensitivity of durum wheat might be essentially attributed to the lack of *TaALMT1*. Even *TaMATE1B* may not 'naturally' contribute to Al resistance in durum wheat, otherwise the research performed so far would have detected some level of Al resistance. The superior *TaMATE1B* allele in hexaploid wheat, which is associated with a transposon insertion upstream from its coding region, is probably a relatively recent evolutionary event (Pereira and Ryan, 2019) that does not appear to have occurred during the evolution of durum wheat. Wayima *et al.* (2019) argued that the hybridization of the D genome with the A and B genomes during the evolution of hexaploid wheat may have resulted in genome instability that could have activated transposable elements. Thus, the current evidence indicates that durum wheat germplasm lacks the transposon insertion upstream of the *TaMATE1B* gene that leads to citrate efflux by the root tip.

Nevertheless, lines of durum wheat have been developed where the chromosome 4D of Chinese Spring (bread wheat) replaced its homeologue (chromosome 4B) (Joppa and Williams, 1988; Luo and Dvořák, 1996). These lines were more Al-resistant than the durum wheat cultivar Langdon (Luo and Dvořák, 1996) or had a performance under Al stress comparable to Al-resistant hexaploid wheat (Wayima *et al.*, 2019). The disomic substitution lines developed by Luo and Dvořák (1996) were used to introgress the chromosome arm 4DL into Jandaroi, a modern durum wheat cultivar from Australia (Han *et al.*, 2014). Al-resistant lines with *TaALMT1* in a homozygous state and Al-sensitive sister lines without *TaALMT1* were obtained, and all resistant lines showed greater root growth than Jandaroi and the sensitive lines up to 60 μM $AlCl_3$ (Han *et al.*, 2014). All resistant lines were able to exude malate and express *TaALMT1* in a slightly lower fashion than some controls. Measurements of root growth in unlimed soil revealed that the resistant lines had longer roots than the sensitive lines and Jandaroi, but the Al-resistant bread wheat, ET8 and Chinese Spring, showed the greatest root length in the acid soil (Han *et al.*, 2014). Although the durum wheat lines with the chromosome 4DL exhibited increased Al resistance, most lines grown under non-acid soil in a glasshouse had lower grain weight than the sensitive lines (Han *et al.*, 2014). Nevertheless, the importance of incorporating *TaALMT1* for greater Al resistance in durum wheat was evidenced.

To reduce the size of the chromosome 4D introduced into durum wheat and lower the chance of detrimental effects on grain weight, a new line of durum wheat was obtained (Han *et al.*, 2016). For that, the disomic substitution lines derived from cultivar Langdon were crossed with the durum wheat cultivar Capelli that had a mutation in the pairing homoeologous gene (*ph1c*). That mutation allows non-homologous chromosomes to pair at a high frequency. The new durum line, named SF *TaALMT1* line, has only a small translocation of chromosome 4D (harbouring the *TaALMT1* gene) into chromosome 4B of durum wheat. That small introgression containing *TaALMT1* was able to increase Al resistance in durum wheat in hydroponics to a similar level as the disomic substitution lines previously developed by Han *et al.* (2014). Both lines were consistently more Al-resistant than Jandaroi (Han *et al.*, 2016). However, when root growth was measured in soil, the SF *TaALMT1* line showed lower growth of both thick and fine roots than the previous developed line (Han *et al.*, 2016). In this same work, the *TaMATE1B* gene was introgressed in durum wheat by using a co-dominant marker. The line containing *TaMATE1B*, which did not have *TaALMT1*, achieved greater total root growth in acid soil than the durum line containing *TaALMT1*. This result contrasts with what happens in bread wheat, in which *TaALMT1* is responsible for most of the root growth under Al stress (Raman *et al.*, 2005; Aguilera *et al.*, 2016, 2019; Han *et al.*, 2016). It is not clear why *TaMATE1B* conferred a stronger Al resistance in durum wheat than *TaALMT1*. Nevertheless, the separate introgressions of *TaALMT1* and *TaMATE1B* increased the Al resistance in durum wheat, and the introgression of the small fragment containing *TaALMT1*, as in the SF *TaALMT1* line, appears to have avoided the smaller grains that were observed when the full 4DL arm was introgressed (Han *et al.*, 2016).

When the durum wheat line with the superior *TaMATE1B* allele was compared with a null sister line not having that superior allele, *TaMATE1B* allowed 76% greater total root growth under subsoil acidity when reaching anthesis (Pooniya *et al.*, 2020). The citrate efflux conditioned by the superior *TaMATE1B* allele also resulted in vertical root growth that reached 1 m in comparison to only 25 cm of the null sister line. However, at final harvest, the plants did not present significant differences for grain yield, shoot biomass, number of spikes and grains per plant (Pooniya *et al.*, 2020). An interesting point is that the experiments were performed in rhizoboxes under glasshouse conditions and the soil was watered twice a week, keeping field capacity close to 90%. These conditions are very different from the field, and the greater root growth shown by the durum wheat line having the *TaMATE1B* superior allele will probably be beneficial for grain yield under field conditions. Ongoing experiments are being performed to verify that hypothesis (E. Delhaize, Canberra, 2019, personal communication).

## 7.12 Conclusions and Outlook

The importance of citrate efflux for root growth in acidic soils of hexaploid wheat that already has the superior *TaALMT1* allele is still a topic of debate. However, the evidence to date is that citrate exudation by the root tip is not detrimental for plant development and yield. Thus, when targeting molecular breeding of hexaploid wheat for greater Al resistance, pyramiding superior alleles of *TaALMT1* and *TaMATE1B* should be considered. These alleles can be easily detected by PCR, which allows to screen their introgression by crossing and backcrossing. When considering durum wheat, both the introgression of *TaALMT1* or *TaMATE1B* increases Al resistance. In contrast to bread wheat, *TaMATE1B* improves root growth of durum wheat in acid soil with higher levels than *TaALMT1*. The greater importance of *TaMATE1B* for durum wheat remains to be explained. The breeding of durum wheat is currently facing the task of pyramiding *TaALMT1* and *TaMATE1B*, and a greater Al resistance is expected by combining both these OA transporter genes in durum wheat. When considering genetic modification for greater Al resistance in wheat, several questions still need to be answered. The uncertain acceptance of GM wheat is an important factor that may change if, among other things, GM wheat with superior Al resistance also shows significantly increased yield under field conditions.

Although Al resistance mediated by OA efflux by wheat roots is the most studied mechanism in wheat, the characterization of minor genes that are associated with Al tolerance may contribute to greater wheat root growth under Al stress. Progress towards this goal may use the better accuracy and lower price of sequencing and SNP detection allied with new bioinformatics tools that can facilitate the analysis of the wheat transcriptome and QTL mapping. Eventually, any minor gene associated with Al tolerance should be combined with superior *TaALMT1* and *TaMATE1B* alleles. Nevertheless, breeding for greater root growth under Al stress in wheat requires a coherent method to screen root growth. Efforts to adjust the level of Al toxicity in nutrient solution are valid if the phenotyping aims to not only access Al resistance among lines but also to improve root growth under acidic soil/subsoil in the field. When assertive phenotyping is combined with the detection of superior alleles for *TaALMT1* and *TaMATE1B* and use of potential minor genes, wheat breeding programmes can obtain cultivars with better growth under Al stress.

## References


Aggarwal, A., Ezaki, B. and Tripathi, B.N. (2015) Two detoxification mechanisms by external malate detoxification and anti-peroxidation enzymes cooperatively confer aluminum tolerance in the roots of wheat (*Triticum aestivum* L.). *Environmental and Experimental Botany* 120, 43–54.

Aguilera, J.G., Minozzo, J.A., Barichello, D., Fogaça, C.M., Silva Júnior, J.P., Consoli, L. and Pereira, J.F. (2016) Alleles of organic acid transporter genes are highly correlated with wheat resistance to acidic soil in field conditions. *Theoretical and Applied Genetics* 129, 1317–1331.

Aguilera, J.G., Teodoro, P.E., Silva Júnior, J.P., Pereira, J.F., Zuffo, A.M. and Consoli, L. (2019) Selection of aluminum-resistant wheat genotypes using multienvironment and multivariate indices. *Agronomy Journal* 111, 2804–2810.

Alonso-Blanco, C. and Koornneef, M. (2000) Naturally occurring variation in *Arabidopsis*: an under-exploited resource for plant genetics. *Trends in Plant Science* 5, 22–29.

Aniol, A. and Gustafson, J.P. (1984) Chromosome location of genes controlling aluminum tolerance in wheat, rye, and triticale. *Canadian Journal of Genetics and Cytology* 26, 701–705.

Araus, J.L., Serret, M.D. and Lopes, M.S. (2019) Transgenic solutions to increase yield and stability in wheat: shining hope or flash in the pan? *Journal of Experimental Botany* 70, 1419–1424.

Autran, J.C. and Feillet, P. (1985) Genetic and technological basis of protein quality for durum wheat in pasta. In: *Agriculture: Protein Evaluation in Cereals and Legumes*. Report EUR 10404. Commission of the European Communities, Thessaloniki, Greece, pp. 59–71.

Babourina, O., Ozturk, L., Cakmak, I. and Rengel, Z. (2006) Reactive oxygen species production in wheat roots is not linked with changes in $H^+$ fluxes during acidic and aluminium stresses. *Plant Signaling & Behavior* 1, 70–75.

Baier, A.C., Somers, D.J. and Gusiafson, J.P. (1995) Aluminium tolerance in wheat: correlating hydroponic evaluations with field and soil performances. *Plant Breeding* 114, 291–296.

Bartlett, R.J. and Riego, D.C. (1972) Effect of chelation on the toxicity of aluminum. *Plant and Soil* 37, 419–423.

Berzonsky, W.A. and Kimber, G. (1986) Tolerance of *Triticum* species to aluminum. *Plant Breeding* 97, 275–278.

Biasini, M., Bienert, S., Waterhouse, A., Arnold, K., Studer, G. *et al.* (2014) SWISS-MODEL: modelling protein tertiary and quaternary structure using evolutionary information. *Nucleic Acids Research* 42, W252–W258.

Bona, L., Wright, R.J. and Baligar, V.C. (1992) Acid soil tolerance of *Triticum aestivum* L. and *Triticum durum* Desf. wheat genotypes. *Cereal Research Communications* 20, 95–101.

Bona, L., Carver, B.F., Wright, R.J. and Baligar, V.C. (1994) Aluminum tolerance of segregating wheat populations in acidic soil and nutrient solutions. *Communications in Soil Science and Plant Analysis* 25, 327–339.

Bona, L., Baligar, V.C. and Wright, R.J. (1995) Soil acidity effects on agribotanical traits of durum and common wheat. In: Date, R.A., Grundon, N.J., Rayment, G.E. and Probert, M.E. (eds) *Plant–Soil Interactions at Low pH: Principles and Management*. Developments in Plant and Soil Sciences, Vol. 64. Springer, Dordrecht, the Netherlands, pp. 425–428.

Brenchley, R., Spannagl, M., Pfeifer, M., Barker, G.L., D'Amore, R. *et al.* (2012) Analysis of the bread wheat genome using whole-genome shotgun sequencing. *Nature* 491, 705–710.

Brunner, I. and Sperisen, C. (2013) Aluminum exclusion and aluminum tolerance in woody plants. *Frontiers in Plant Science* 4, 172.

Cai, S., Bai, G.H. and Zhang, D. (2008) Quantitative trait loci for aluminum resistance in Chinese wheat landrace FSW. *Theoretical and Applied Genetics* 117, 49–56.

Cambraia, J., Galvani, F.R., Estevão, M.M. and Sant'Anna, R. (1983) Effects of aluminum on organic acid, sugar and amino acid composition of the root system of sorghum (*Sorghum bicolor* L. Moench). *Journal of Plant Nutrition* 6, 313–322.

Cavanagh, C.R., Chao, S., Wang, S., Huang, B.E., Stephen, S. *et al.* (2013) Genome-wide comparative diversity uncovers multiple targets of selection for improvement in hexaploid wheat landraces and cultivars. *Proceedings of the National Academy of Sciences USA* 110, 8057–8062.

Cruz-Ortega, R., Cushman, J.C. and Ownby, J.D. (1997) cDNA clones encoding 1,3-β-glucanase and a fimbrin-like cytoskeletal protein are induced by Al toxicity in wheat roots. *Plant Physiology* 114, 1453–1460.

Dai, J., Bai, G., Zhang, D. and Hong, D. (2013) Validation of quantitative trait loci for aluminum tolerance in Chinese wheat landrace FSW. *Euphytica* 192, 171–179.

Darkó, É., Ambrus, H., Stefanovits-Bányai, É., Fodor, J., Bakos, F. and Barnabás, B. (2004) Aluminium toxicity, Al tolerance and oxidative stress in an Al-sensitive wheat genotype and in Al-tolerant lines developed by *in vitro* microspore selection. *Plant Science* 166, 583–591.

de la Fuente, J.M., Ramírez-Rodríguez, V., Cabrera-Ponce, J.L. and Herrera-Estrella, L. (1997) Aluminum tolerance in transgenic plants by alteration of citrate synthesis. *Science* 276, 1566–1568.

Delhaize, E., Craig, S., Beaton, C.D., Bennet, R.J., Jagadish, V.C. and Randall, P.J. (1993a) Aluminum tolerance in wheat (*Triticum aestivum* L.) I. Uptake and distribution of aluminum in root apices. *Plant Physiology* 103, 685–693.

Delhaize, E., Ryan, P.R. and Randall, P.J. (1993b) Aluminum tolerance in wheat (*Triticum aestivum* L.) II. Aluminum-stimulated excretion of malic acid from root apices. *Plant Physiology* 103, 695–702.

Delhaize, E., Hebb, D.M. and Ryan, P.R. (2001) Expression of a *Pseudomonas aeruginosa* citrate synthase gene in tobacco is not associated with either enhanced citrate accumulation or efflux. *Plant Physiology* 125, 2059–2067.

Delhaize, E., Gruber, B.D. and Ryan, P.R. (2007) The roles of organic anion permeases in aluminium resistance and mineral nutrition. *FEBS Letters* 581, 2255–2262.

Dong, J., Hunt, J., Delhaize, E. and Tang, C. (2018) The impact of elevated $CO_2$ on acid-soil tolerance of hexaploid wheat (*Triticum aestivum* L.) genotypes varying in organic anion efflux. *Plant and Soil* 428, 401–413.

Dong, J., Grylls, S., Hunt, J., Armstrong, R., Delhaize, E. and Tang, C. (2019) Elevated $CO_2$ (free-air $CO_2$ enrichment) increases grain yield of aluminium-resistant but not aluminium-sensitive wheat (*Triticum aestivum*) grown in an acid soil. *Annals of Botany* 123, 461–468.

Emebiri, L.C., Raman, H. and Ogbonnaya, F.C. (2020) Synthetic hexaploid wheat as a source of novel genetic loci for aluminium tolerance. *Euphytica* 216, 135.

Exley, C. (2009) Darwin, natural selection and the biological essentiality of aluminium and silicon. *Trends in Biochemical Sciences* 34, 589–593.
Ezaki, B., Jayaram, K., Higashi, A. and Takahashi, K. (2013) A combination of five mechanisms confers a high tolerance for aluminum to a wild species of Poaceae, *Andropogon virginicus* L. *Environmental and Experimental Botany* 93, 35–44.
Farokhzadeh, S., Fakheri, B.A., Nezhad, N.M., Tahmasebi, S. and Mirsoleimani, A. (2019) Mapping QTLs of flag leaf morphological and physiological traits related to aluminum tolerance in wheat (*Triticum aestivum* L.). *Physiology and Molecular Biology of Plants* 25, 975–990.
Farokhzadeh, S., Fakheri, B.A., Nezhad, N.M., Tahmasebi, S., Mirsoleimani, A. and McIntyre, C.L. (2020) Genetic control of some plant growth characteristics of bread wheat (*Triticum aestivum* L.) under aluminum stress. *Genes and Genomics* 42, 245–261.
Ferreira, J.R., Minella, E., Delatorre, C.A., Delhaize, E., Ryan, P.R. and Pereira, J.F. (2018) Conventional and transgenic strategies to enhance the acid soil tolerance of barley. *Molecular Breeding* 38, 12.
Foy, C.D. (1996) Tolerance of durum wheat lines to an acid, aluminum-toxic subsoil. *Journal of Plant Nutrition* 19, 1381–1394.
Foy, C.D., Lee, E.H., Coradetti, C.A. and Taylor, G.J. (1990) Organic acids related to differential aluminum tolerance in wheat (*Triticum aestivum*) cultivars. In: Beusichem, M.L. (ed.) *Proceedings of XI International Plant Nutrition Colloquium, Wageningen, the Netherlands, 29 July–5 August 1989*. Kluwer Academic Publishers, Dordrecht, the Netherlands, pp. 301–329.
Froese, P.S. and Carter, A.H. (2016) Single nucleotide polymorphisms in the wheat genome associated with tolerance of acidic soils and aluminum toxicity. *Crop Science* 56, 1662–1677.
Furukawa, J., Yamaji, N., Wang, H., Mitani, N., Murata, Y. *et al.* (2007) An aluminum-activated citrate transporter in barley. *Plant and Cell Physiology* 48, 1081–1091.
Garcia-Oliveira, A.L., Martins-Lopes, P., Tolrá, R., Poschenrieder, C., Tarquis, M. *et al.* (2014) Molecular characterization of the citrate transporter gene *TaMATE1* and expression analysis of upstream genes involved in organic acid transport under Al stress in bread wheat (*Triticum aestivum*). *Physiologia Plantarum* 152, 441–452.
Garcia-Oliveira, A.L., Benito, C., Guedes-Pinto, H. and Martins-Lopes, P. (2018) Molecular cloning of *TaMATE2* homoeologues potentially related to aluminium tolerance in bread wheat (*Triticum aestivum* L.). *Plant Biology* 20, 817–824.
Gavassi, M.A., Dodd, I.C., Puértolas, J., Silva, G.S., Carvalho, R.F. and Habermann, G. (2020) Aluminum-induced stomatal closure is related to low root hydraulic conductance and high ABA accumulation. *Environmental and Experimental Botany* 179, 104233.
Guo, P., Bai, G., Carver, B., Li, R., Bernardo, A. and Baum, M. (2007a) Transcriptional analysis between two wheat near-isogenic lines contrasting in aluminum tolerance under aluminum stress. *Molecular Genetics and Genomics* 277, 1–12.
Guo, P.G., Bai, G.H., Li, R.H., Brett, C. and Michael, B. (2007b) Molecular characterization of Atlas 66-derived wheat near-isogenic lines contrasting in aluminum (Al) tolerance. *Agricultural Sciences in China* 6, 522–528.
Hajjar, R. and Hodgkin, T. (2007) The use of wild relatives in crop improvement: a survey of developments over the last 20 years. *Euphytica* 156, 1–13.
Hamel, F., Breton, C. and Houde, M. (1998) Isolation and characterization of wheat aluminum-regulated genes: possible involvement of aluminum as a pathogenesis response elicitor. *Planta* 205, 531–538.
Han, C., Dai, S.F., Liu, D.C., Pu, Z.J., Wei, Y.M. *et al.* (2013) *TaALMT1* promoter sequence compositions, acid tolerance, and Al tolerance in wheat cultivars and landraces from Sichuan in China. *Genetics and Molecular Research* 12, 5602–5616.
Han, C., Ryan, P.R., Yan, Z. and Delhaize, E. (2014) Introgression of a 4D chromosomal fragment into durum wheat confers aluminium tolerance. *Annals of Botany* 114, 135–144.
Han, C., Zhang, P., Ryan, P.R., Rathjen, T.M., Yan, Z. and Delhaize, E. (2016) Introgression of genes from bread wheat enhances the aluminium tolerance of durum wheat. *Theoretical and Applied Genetics* 129, 729–739.
Hayes, J.E. and Ma, J.F. (2003) Al-induced efflux of organic acid anions is poorly associated with internal organic acid metabolism in triticale roots. *Journal of Experimental Botany* 54, 1753–1759.
Hayta, S., Smedley, M.A., Demir, S.U., Blundell, R., Hinchliffe, A. *et al.* (2019) An efficient and reproducible *Agrobacterium*-mediated transformation method for hexaploid wheat (*Triticum aestivum* L.). *Plant Methods* 15, 121.

Houde, M. and Diallo, A.O. (2008) Identification of genes and pathways associated with aluminum stress and tolerance using transcriptome profiling of wheat near-isogenic lines. *BMC Genomics* 9, 400.
Hue, N.V., Craddock, G.R. and Adams, F. (1986) Effect of organic acids on aluminum toxicity in subsoils. *Soil Science Society of America Journal* 50, 28–34.
Ishida, Y., Tsunashima, M., Hiei, Y. and Komari, T. (2015) Wheat (*Triticum aestivum* L.) transformation using immature embryos. In: Wang, K. (ed.) *Agrobacterium Protocols*. Methods in Molecular Biology, Vol. 1223. Springer, New York, pp. 189–198.
Jones, D.L. (1998) Organic acids in the rhizosphere – a critical review. *Plant and Soil* 205, 25–44.
Joppa, L.R. and Williams, N.D. (1988) Langdon durum disomic substitution lines and aneuploid analysis in tetraploid wheat. *Genome* 30, 222–228.
Kamran, M., Ramesh, S.A., Gilliham, M., Tyerman, S.D. and Bose, J. (2020) Role of *TaALMT1* malate–GABA transporter in alkaline pH tolerance of wheat. *Plant, Cell & Environment* 43, 2443–2459.
Kitagawa, T., Morishita, T., Tachibana, Y., Namai, H. and Ohta, Y. (1986) Differential aluminum resistance of wheat varieties and organic acid secretion. *Japanese Journal of Soil Science and Plant Nutrition* 57, 352–358.
Kochian, L.V., Hoekenga, O.A. and Pineros, M.A. (2004) How do crop plants tolerate acid soils? Mechanisms of aluminum tolerance and phosphorous efficiency. *Annual Review of Plant Biology* 55, 459–493.
Kochian, L.V., Piñeros, M.A., Liu, J. and Magalhaes, J.V. (2015) Plant adaptation to acid soils: the molecular basis for crop aluminum resistance. *Annual Review of Plant Biology* 66, 571–598.
Liu, J., Magalhaes, J.V., Shaff, J. and Kochian, L.V. (2009) Aluminum-activated citrate and malate transporters from the MATE and ALMT families function independently to confer *Arabidopsis* aluminum tolerance. *The Plant Journal* 57, 389–399.
Liu, J., Luo, X., Shaff, J., Liang, C., Jia, X. *et al.* (2012) A promoter-swap strategy between the *AtALMT* and *AtMATE* genes increased *Arabidopsis* aluminum resistance and improved carbon-use efficiency for aluminum resistance. *The Plant Journal* 71, 327–337.
Liu, M., Yu, M., Li, G., Carver, B.F. and Yan, L. (2015) Genetic characterization of aluminum tolerance in winter wheat. *Molecular Breeding* 35, 205.
Liu, M.Y., Lou, H.Q., Chen, W.W., Piñeros, M.A., Xu, J.M. *et al.* (2018) Two citrate transporters coordinately regulate citrate secretion from rice bean root tip under aluminum stress. *Plant, Cell & Environment* 41, 809–822.
Luo, M.C. and Dvořák, J. (1996) Molecular mapping of an aluminum tolerance locus on chromosome 4D of Chinese Spring wheat. *Euphytica* 91, 31–35.
Ma, H.X., Bai, G.H., Carver, B.F. and Zhou, L.L. (2005) Molecular mapping of a quantitative trait locus for aluminum tolerance in wheat cultivar Atlas 66. *Theoretical and Applied Genetics* 112, 51–57.
Ma, H.X., Bai, G.H. and Lu, W.Z. (2006) Quantitative trait loci for aluminum resistance in wheat cultivar Chinese Spring. *Plant and Soil* 283, 239–249.
Mackay, T.F., Stone, E.A. and Ayroles, J.F. (2009) The genetics of quantitative traits: challenges and prospects. *Nature Reviews Genetics* 10, 565–577.
Magalhaes, J.V., Garvin, D.F., Wang, Y., Sorrells, M.E., Klein, P.E. *et al.* (2004) Comparative mapping of a major aluminum tolerance gene in sorghum and other species in the Poaceae. *Genetics* 167, 1905–1914.
Magalhaes, J.V., Liu, J., Guimaraes, C.T., Lana, U.G., Alves, V.M., *et al.* (2007) A gene in the multidrug and toxic compound extrusion (MATE) family confers aluminum tolerance in sorghum. *Nature Genetics* 39, 1156–1161.
Milla, M.R. and Gustafson, J.P. (2001) Genetic and physical characterization of chromosome 4DL in wheat. *Genome* 44, 883–892.
Miller, T.E., Iqbal, N., Reader, S.M., Mahmood, A., Cant, K.A. and King, I.P. (1997) A cytogenetic approach to the improvement of aluminium tolerance in wheat. *New Phytologist* 137, 93–98.
Miyasaka, S.C., Buta, J.G., Howell, R.K. and Foy, C.D. (1991) Mechanism of aluminum tolerance in snapbeans: root exudation of citric acid. *Plant Physiology* 96, 737–743.
Mohammed, Y.S.A., Eltayeb, A.E. and Tsujimoto, H. (2013) Enhancement of aluminum tolerance in wheat by addition of chromosomes from the wild relative *Leymus racemosus*. *Breeding Science* 63, 407–416.
Moriyama, Y., Hiasa, M., Matsumoto, T. and Omote, H. (2008) Multidrug and toxic compound extrusion (MATE)-type proteins as anchor transporters for the excretion of metabolic waste products and xenobiotics. *Xenobiotica* 38, 1107–1118.

Nasr, N., Carapetian, J., Heidari, R., Rezaei, S.A., Abbaspour, N. *et al.* (2011) The effect of aluminium on enzyme activities in two wheat cultivars. *African Journal of Biotechnology* 10, 3354–3364.

Navakode, S., Weidner, A., Lohwasser, U., Röder, M.S. and Börner, A. (2009) Molecular mapping of quantitative trait loci (QTLs) controlling aluminium tolerance in bread wheat. *Euphytica* 166, 283–290.

Navakode, S., Neumann, K., Kobiljski, B., Lohwasser, U. and Börner, A. (2014) Genome wide association mapping to identify aluminium tolerance loci in bread wheat. *Euphytica* 198, 401–411.

Oh, M.W., Roy, S.K., Kamal, A.H.M., Cho, K., Cho, S.W. *et al.* (2014) Proteome analysis of roots of wheat seedlings under aluminum stress. *Molecular Biology Reports* 41, 671–681.

Papernik, L.A., Bethea, A.S., Singleton, T.E., Magalhaes, J.V., Garvin, D.F. and Kochian, L.V. (2001) Physiological basis of reduced Al tolerance in ditelosomic lines of Chinese Spring wheat. *Planta* 212, 829–834.

Pereira, J.F. (2018) Initial root length in wheat is highly correlated with acid soil tolerance in the field. *Scientia Agricola* 75, 79–83.

Pereira, J.F. and Ryan, P.R. (2019) The role of transposable elements in the evolution of aluminium resistance in plants. *Journal of Experimental Botany* 70, 41–54.

Pereira, J.F., Zhou, G., Delhaize, E., Richardson, T., Zhou, M. and Ryan, P.R. (2010) Engineering greater aluminium resistance in wheat by over-expressing *TaALMT1*. *Annals of Botany* 106, 205–214.

Pereira, J.F., Barichello, D., Ferreira, J.R., Aguilera, J.G., Consoli, L. *et al.* (2015) *TaALMT1* and *TaMATE1B* allelic variability in a collection of Brazilian wheat and its association with root growth on acidic soil. *Molecular Breeding* 35, 169.

Pereira, J.F., Cunha, G.R.D. and Moresco, E.R. (2019) Improved drought tolerance in wheat is required to unlock the production potential of the Brazilian Cerrado. *Crop Breeding and Applied Biotechnology* 19, 217–225.

Polle, E., Konzak, C.F. and Kattrick, J.A. (1978) Visual detection of aluminum tolerance levels in wheat by hematoxylin staining of seedling roots. *Crop Science* 18, 823–827.

Pooniya, V., Palta, J.A., Chen, Y., Delhaize, E. and Siddique, K.H. (2020) Impact of the *TaMATE1B* gene on above and below-ground growth of durum wheat grown on an acid and $Al^{3+}$-toxic soil. *Plant and Soil* 447, 73–84.

Raman, H., Zhang, K., Cakir, M., Appels, R., Garvin, D.F. *et al.* (2005) Molecular characterization and mapping of *ALMT1*, the aluminium-tolerance gene of bread wheat (*Triticum aestivum* L.). *Genome* 48, 781–791.

Raman, H., Ryan, P.R., Raman, R., Stodart, B.J., Zhang, K. *et al.* (2008) Analysis of *TaALMT1* traces the transmission of aluminum resistance in cultivated common wheat (*Triticum aestivum* L.). *Theoretical and Applied Genetics* 116, 343–354.

Raman, H., Stodart, B., Ryan, P.R., Delhaize, E., Emebiri, L. *et al.* (2010) Genome-wide association analyses of common wheat (*Triticum aestivum* L.) germplasm identifies multiple loci for aluminium resistance. *Genome* 53, 957–966.

Ramesh, S.A., Tyerman, S.D., Xu, B., Bose, J., Kaur, S. *et al.* (2015) GABA signalling modulates plant growth by directly regulating the activity of plant-specific anion transporters. *Nature Communications* 6, 7879.

Ramesh, S.A., Kamran, M., Sullivan, W., Chirkova, L., Okamoto, M. *et al.* (2018) Aluminum-activated malate transporters can facilitate GABA transport. *The Plant Cell* 30, 1147–1164.

Ray, D.K., Mueller, N.D., West, P.C. and Foley, J.A. (2013) Yield trends are insufficient to double global crop production by 2050. *PLoS One* 8, e66428.

Rengel, Z. and Jurkic, V. (1992) Genotypic differences in wheat Al tolerance. *Euphytica* 62, 111–117.

Richards, K.D., Snowden, K.C. and Gardner, R.C. (1994) *wali6* and *wali7*: genes induced by aluminum in wheat (*Triticum aestivum* L.) roots. *Plant Physiology* 105, 1455–1456.

Riede, C.R. and Anderson, J.A. (1996) Linkage of RFLP markers to an aluminum tolerance gene in wheat. *Crop Science* 36, 905–909.

Ryan, P.R. (2018) Assessing the role of genetics for improving the yield of Australia's major grain crops on acid soils. *Crop and Pasture Science* 69, 242–264.

Ryan, P.R. and Delhaize, E. (2010) The convergent evolution of aluminium resistance in plants exploits a convenient currency. *Functional Plant Biology* 37, 275–284.

Ryan, P.R., DiTomaso, J.M. and Kochian, L.V. (1993) Aluminum toxicity in roots: investigation of spatial sensitivity and the role of the root cap. *Journal of Experimental Botany* 44, 437–446.

Ryan, P.R., Delhaize, E. and Randall, P.J. (1995) Malate efflux from root apices and tolerance to aluminium are highly correlated in wheat. *Functional Plant Biology* 22, 531–536.

Ryan, P.R., Skerrett, M., Findlay, G.P., Delhaize, E. and Tyerman, S.D. (1997) Aluminum activates an anion channel in the apical cells of wheat roots. *Proceedings of the National Academy of Sciences USA* 94, 6547–6552.
Ryan, P.R., Delhaize, E. and Jones, D.L. (2001) Function and mechanism of organic anion exudation from plant roots. *Annual Review of Plant Biology* 52, 527–560.
Ryan, P.R., Raman, H., Gupta, S., Horst, W.J. and Delhaize, E. (2009) A second mechanism for aluminum resistance in wheat relies on the constitutive efflux of citrate from roots. *Plant Physiology* 149, 340–351.
Ryan, P.R., Raman, H., Gupta, S., Sasaki, T., Yamamoto, Y. and Delhaize, E. (2010) The multiple origins of aluminium resistance in hexaploid wheat include *Aegilops tauschii* and more recent *cis* mutations to *TaALMT1*. *The Plant Journal* 64, 446–455.
Ryan, P.R., Tyerman, S.D., Sasaki, T., Furuichi, T., Yamamoto, Y. *et al.* (2011) The identification of aluminium-resistance genes provides opportunities for enhancing crop production on acid soils. *Journal of Experimental Botany* 62, 9–20.
Ryan, P.R., James, R.A., Weligama, C., Delhaize, E., Rattey, A. *et al.* (2014) Can citrate efflux from roots improve phosphorus uptake by plants? Testing the hypothesis with near-isogenic lines of wheat. *Physiologia Plantarum* 151, 230–242.
Salvador-Moreno, N., Ryan, P.R., Holguín, I., Delhaize, E., Benito, C. and Gallego, F.J. (2018) Transcriptional profiling of wheat and wheat–rye addition lines to identify candidate genes for aluminum tolerance. *Biologia Plantarum* 62, 741–749.
Sasaki, T., Ezaki, B. and Matsumoto, H. (2002) A gene encoding multidrug resistance (MDR)-like protein is induced by aluminum and inhibitors of calcium flux in wheat. *Plant and Cell Physiology* 43, 177–185.
Sasaki, T., Yamamoto, Y., Ezaki, B., Katsuhara, M., Ahn, S.J. *et al.* (2004) A wheat gene encoding an aluminum-activated malate transporter. *The Plant Journal* 37, 645–653.
Sasaki, T., Ryan, P.R., Delhaize, E., Hebb, D.M., Ogihara, Y. *et al.* (2006) Sequence upstream of the wheat (*Triticum aestivum* L.) *ALMT1* gene and its relationship to aluminum resistance. *Plant and Cell Physiology* 47, 1343–1354.
Sharma, T., Dreyer, I., Kochian, L. and Piñeros, M.A. (2016) The ALMT family of organic acid transporters in plants and their involvement in detoxification and nutrient security. *Frontiers in Plant Science* 7, 1488.
Shavrukov, Y., Genc, Y. and Hayes, J. (2012) The use of hydroponics in abiotic stress tolerance research. In: Asao, T. (ed.) *Hydroponics: A Standard Methodology for Plant Biological Researches*. InTech, Rijeka, Croatia, pp. 39–66.
Shewry, P.R. (2009) Wheat. *Journal of Experimental Botany* 60, 1537–1553.
Silva, C.M., Zhang, C., Habermann, G., Delhaize, E. and Ryan, P.R. (2018a) Does the major aluminium-resistance gene in wheat, *TaALMT1*, also confer tolerance to alkaline soils? *Plant and Soil* 424, 451–462.
Silva, G.S., Gavassi, M.A., Nogueira, M.A. and Habermann, G. (2018b) Aluminum prevents stomatal conductance from responding to vapor pressure deficit in *Citrus limonia*. *Environmental and Experimental Botany* 155, 662–671.
Silva-Navas, J., Benito, C., Téllez-Robledo, B., El-Moneim, D.A. and Gallego, F.J. (2012) The *ScAACT1* gene at the *Qalt5* locus as a candidate for increased aluminum tolerance in rye (*Secale cereale* L.). *Molecular Breeding* 30, 845–856.
Slootmaker, L.A.J. (1974) Tolerance to high soil acidity in wheat related species, rye and triticale. *Euphytica* 23, 505–513.
Snowden, K.C. and Gardner, R.C. (1993) Five genes induced by aluminum in wheat (*Triticum aestivum* L.) roots. *Plant Physiology* 103, 855–861.
Soto-Cerda, B.J., Inostroza-Blancheteau, C., Mathias, M., Penaloza, E., Zuñiga, J. *et al.* (2015) Marker-assisted breeding for *TaALMT1*, a major gene conferring aluminium tolerance to wheat. *Biologia Plantarum* 59, 83–91.
Stass, A., Smit, I., Eticha, D., Oettler, G. and Horst, W.J. (2008) The significance of organic-anion exudation for the aluminum resistance of primary triticale derived from wheat and rye parents differing in aluminum resistance. *Journal of Plant Nutrition and Soil Science* 171, 634–642.
Takanashi, K., Shitan, N. and Yazaki, K. (2014) The multidrug and toxic compound extrusion (MATE) family in plants. *Plant Biotechnology* 31, 417–430.
Tang, Y., Garvin, D.F., Kochian, L.V., Sorrells, M.E. and Carver, B.F. (2002) Physiological genetics of aluminum tolerance in the wheat cultivar Atlas 66. *Crop Science* 42, 1541–1546.

Taylor, G.J. and Foy, C.D. (1985) Mechanisms of aluminum tolerance in *Triticum aestivum* L. (wheat). I. Differential pH induced by winter cultivars in nutrient solutions. *American Journal of Botany* 72, 695–701.

Tesfaye, M., Temple, S.J., Allan, D.L., Vance, C.P. and Samac, D.A. (2001) Overexpression of malate dehydrogenase in transgenic alfalfa enhances organic acid synthesis and confers tolerance to aluminum. *Plant Physiology* 127, 1836–1844.

Tovkach, A., Ryan, P.R., Richardson, A.E., Lewis, D.C., Rathjen, T.M. *et al.* (2013) Transposon-mediated alteration of *TaMATE1B* expression in wheat confers constitutive citrate efflux from root apices. *Plant Physiology* 161, 880–892.

von Uexküll, H.R. and Mutert, E. (1995) Global extent, development and economic impact of acid soils. *Plant and Soil* 171, 1–15.

Wang, H., Ji, F., Zhang, Y., Hou, J., Liu, W., Huang, J. and Liang, W. (2019) Interactions between hydrogen sulphide and nitric oxide regulate two soybean citrate transporters during the alleviation of aluminium toxicity. *Plant, Cell & Environment* 42, 2340–2356.

Wang, J., Raman, H., Zhou, M., Ryan, P.R., Delhaize, E. *et al.* (2007) High-resolution mapping of the *Alp* locus and identification of a candidate gene *HvMATE* controlling aluminium tolerance in barley (*Hordeum vulgare* L.). *Theoretical and Applied Genetics* 115, 265–276.

Wang, S., Wong, D., Forrest, K., Allen, A., Chao, S. *et al.* (2014) Characterization of polyploid wheat genomic diversity using a high-density 90 000 single nucleotide polymorphism array. *Plant Biotechnology Journal* 12, 787–796.

Wayima, E.F., Ligaba-Osena, A., Dagne, K., Tesfaye, K., Machuka, E.M. *et al.* (2019) Screening of diverse Ethiopian durum wheat accessions for aluminum tolerance. *Agronomy* 9, 440.

Xu, F.J., Jin, C.W., Liu, W.J., Zhang, Y.S. and Lin, X.Y. (2011) Pretreatment with $H_2O_2$ alleviates aluminum-induced oxidative stress in wheat seedlings. *Journal of Integrative Plant Biology* 53, 44–53.

Xu, F.J., Li, G., Jin, C.W., Liu, W.J., Zhang, S.S. *et al.* (2012) Aluminum-induced changes in reactive oxygen species accumulation, lipid peroxidation and antioxidant capacity in wheat root tips. *Biologia Plantarum* 56, 89–96.

Yamamoto, Y., Kobayashi, Y., Devi, S.R., Rikiishi, S. and Matsumoto, H. (2003) Oxidative stress triggered by aluminum in plant roots. *Plant and Soil* 255, 239–243.

Yang, Y., Ma, L., Zeng, H., Chen, L.Y., Zheng, Y. *et al.* (2018) iTRAQ-based proteomics screen for potential regulators of wheat (*Triticum aestivum* L.) root cell wall component response to Al stress. *Gene* 675, 301–311.

Yang, Z.B., Rao, I.M. and Horst, W.J. (2013) Interaction of aluminium and drought stress on root growth and crop yield on acid soils. *Plant and Soil* 372, 3–25.

Yumurtaci, A. (2015) Utilization of wild relatives of wheat, barley, maize and oat in developing abiotic and biotic stress tolerant new varieties. *Emirates Journal of Food and Agriculture* 27, 1–23.

Zhang, W.H., Ryan, P.R. and Tyerman, S.D. (2001) Malate-permeable channels and cation channels activated by aluminum in the apical cells of wheat roots. *Plant Physiology* 125, 1459–1472.

Zheng, S.J. and Yang, J.L. (2005) Target sites of aluminum phytotoxicity. *Biologia Plantarum* 49, 321–331.

Zhou, G., Delhaize, E., Zhou, M. and Ryan, P.R. (2013) The barley *MATE* gene, *HvAACT1*, increases citrate efflux and $Al^{3+}$ tolerance when expressed in wheat and barley. *Annals of Botany* 112, 603–612.

Zhou, L.L., Bai, G.H., Carver, B. and Zhang, D.D. (2007a) Identification of new sources of aluminum resistance in wheat. *Plant and Soil* 297, 105–118.

Zhou, L.L., Bai, G.H., Ma, H.X. and Carver, B.F. (2007b) Quantitative trait loci for aluminum resistance in wheat. *Molecular Breeding* 19, 153–161.

# 8 Molecular Breeding for Enhancing Iron and Zinc Content in Wheat Grains

**Lovenpreet Kaur, Natasha Sharma and Monika Garg***
*National Agri-Food Biotechnology Institute, Mohali, India*

## 8.1 Introduction

Wheat is the most commonly grown crop around the world and is a major source of energy, nutrients and dietary fibre. Its consumption is high in temperate countries, where it is a major staple food crop. However, the concentration or bioavailability of Fe and Zn are low in wheat products and this leads to micronutrient deficiencies in the countries where it is consumed as a main food by the population. Two major reasons contributing to low contents of bioavailable Fe and Zn in wheat grains are: (i) the fewer amounts of these minerals in wheat grains; and (ii) the low bioavailability of these nutrients due to the occurrence of phytates in wheat bran. Although wheat varieties with high micronutrient contents have been developed with various breeding approaches, this has not worked for all micronutrients. Merging breeding approaches with genetic engineering techniques, such as quantitative trait locus (QTL) analysis, marker-assisted breeding, gene cloning and gene transformation from wild wheat relatives, is significant for developing micronutrient-rich wheat cultivars.

## 8.2 Importance of Iron and Zinc in Daily Life

Micronutrient malnutrition is one of the most important challenges the world is facing today. Key micronutrients lacking in diets globally are Fe, Zn, iodine, Se and vitamin A (Ghandilyan *et al.*, 2006). Dietary micronutrient deficiency can severely affect human health, increase health-care expenditures, and also reduce the economic progress of countries. Fe and Zn are the two most vital micronutrients necessary for normal human growth and development. Fe deficiency is the most ubiquitous and widespread across the globe (WHO, 2017). It causes Fe-deficiency anaemia (50% of cases) throughout the world (Fig. 8.1) and is associated with poor pregnancy outcomes, impaired cognitive development, lowered immunity and increased tiredness (WHO, 2017). Zn is a cofactor for many enzymes and regulatory proteins in the body and also plays a vital role in DNA and RNA synthesis along with gene expression (Garcia-Oliveira *et al.*, 2018). Zn deficiency is related with stunted growth and neurobehavioural difficulties in children below 5 years of age and affects around 155 million children worldwide (Nriagu *et al.*, 2008; WHO, 2013). Zn deficiency is also reported to cause anaemia as Fe absorption in the intestine is controlled by Zn (Chang *et al.*, 2010; Graham *et al.*, 2012). Deficiency of Zn and Fe in the diet affects more than two billion people around the world (White and Broadley, 2009; WHO, 2012), mainly pregnant women and children below

*Corresponding author: mkajgarg@gmail.com

 *Molecular Breeding in Wheat, Maize and Sorghum: Strategies for Improving Abiotic Stress Tolerance and Yield* (eds M.A. Hossain *et al.*)

DOI: 10.1079/9781789245431.0008

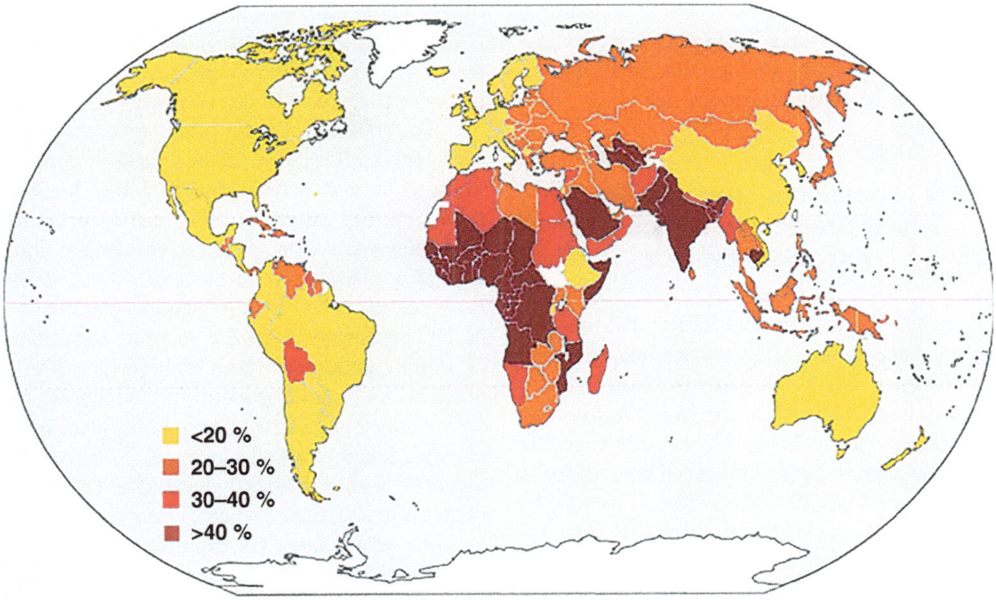


**Fig. 8.1.** Anaemia prevalence in different parts of the world, showing that African and Asian countries have high prevalence of anaemia.

5 years old. Fe can bind to proteins or S as Fe–S cofactors or as haem (Balk and Schaedler, 2014). The bioavailability of haem Fe is very high due to its stability and uptake by a specific transporter in the small intestine (Knutson, 2017). Zn binds to proteins in the form of Zn-finger structures.

There are two different strategies to address these challenges of increasing Fe and Zn contents in plant sources. These are: (i) fortification, i.e. adding chemical forms of minerals to the diet; and (ii) biofortification, i.e. improving crops to produce more minerals. Many countries have made it mandatory to fortify wheat, maize and rice flours with Fe. However, Zn fortification of cereal flours is mainly on a voluntary basis in several countries (Brown *et al.*, 2010). In the UK, milled white flour is fortified with inorganic Fe forms (16.5 mg/kg) according to The Flour and Bread Regulations 1998 (http://www.legislation.gov.uk/uksi/1998/141/contents/made, accessed 10 February 2021). However, this strategy is most difficult to implement in developing nations where milling is carried out at domestic or small-scale level. Therefore, biofortification is required in developing nations which lack the proper infrastructure and capacity to implement fortification programmes.

## 8.3 Wheat: The Most Important Staple Crop

Wheat (*Triticum aestivum* L., $2n = 6x = 42$, AABBDD) is the world's most important cereal crop, which contributes about 30% of the total cereal consumption in the world (FAO, 2003). It provides approximately 20% of dietary energy and proteins to human populations consuming it (Braun *et al.*, 2010). Wheat provides substantial amounts of mineral elements which are beneficial for human health. Fe and Zn are micronutrients that are recognized by the World Health Organization as limiting in wheat (Ortiz-Monasterio *et al.*, 2007). Although wheat is poor in micronutrients Fe and Zn, it is dominant in the diet of most resource-scarce populations in developing regions (Graham *et al.*, 2001). Therefore, it is necessary to develop wheat cultivars with improved micronutrient concentrations to lessen malnutrition among people consuming primarily cereals. Hence, biofortification of wheat is a promising strategy to ameliorate Fe and Zn deficiencies in developing countries.

## 8.4 Location of Iron and Zinc in Wheat Grains and Difficulties Associated with It

Fe is mainly found in the aleurone and the embryo of wheat grains (Fig. 8.2). Upon seed germination, the embryo grows into a young seedling and this requires several enzymes using Fe and Zn as cofactors (Bastow *et al.*, 2018). Plants store Fe as the protein ferritin or in bodies derived from vacuoles (Connorton *et al.*, 2017a). Ferritin is considered the most bioavailable form of stored Fe in plant tissues and is mainly present in amyloplasts in the endosperm (Balmer *et al.*, 2006). In wheat grains, Fe and Zn are mostly present in the aleurone layer which is lost during milling. Furthermore, Fe in these tissues is bound to phytate and polyphenols, which reduces the bioavailability of Fe to humans (Borg *et al.*, 2009; Hurrell *et al.*, 2010; Neal *et al.* 2013; De Brier *et al.*, 2016). This presents a serious challenge to human nutrition as mineral–phytate complexes are insoluble. Thus, in spite of good Fe and Zn contents in wheat grains, the tissue localization and speciation (chelation, binding to protein particles or antinutrients) ultimately determines their bioavailability. Fe–nicotianamine complexes have been found in wheat flour extracts (Eagling *et al.*, 2014a). Studies based on Caco-2 cells (Eagling *et al.*, 2014b) and murine models (Lee *et al.*, 2009) have proved that nicotianamine increases the bioavailability of Fe and Zn.

The presence of minerals in bread and other cereal products is estimated by their total content in the grain and by their bioavailability. Conventional milling practices of wheat by grinding seeds between stones produces wholemeal flour having a mixture of all parts of wheat seeds. However, the roller milling introduced at the end of 19$^{th}$ century enabled the accurate separation of starchy endosperm from embryo (germ) and outer layers (including aleurone), which are together called bran. This increased the availability of white bread at affordable prices. However, removal of bran significantly lowers the Fe and Zn contents in the refined wheat flour in comparison to whole wheat flour. For example, white wheat flour contains 6.7 mg Fe/kg and 8.4 mg Zn/kg, whereas whole wheat flour contains comparatively higher Fe and Zn contents: 28.2 mg Fe/kg and 28.6 mg Zn/kg (Tang *et al.*, 2008). As a result, various campaigns and public health-care advertisements have focused on increasing the consumption of whole-grain wheat products.

## 8.5 Transport of Iron and Zinc from Roots to Seed

The accumulation of Fe and Zn in wheat involves several processes, including mobilization and

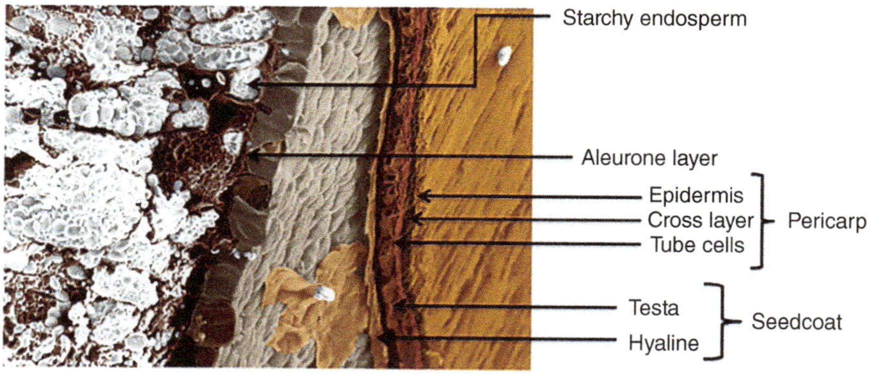


**Fig. 8.2.** Scanning electron micrograph of wheat seed showing different tissue layers. Fe and Zn are mainly localized in the living cells of the aleurone layer between the seedcoat and starchy endosperm. (Adapted from Kumar *et al.*, 2017.)

uptake of minerals from the soil, absorption by plant roots from the plant rhizosphere, translocation of minerals from the roots, mineral remobilization from plant vegetative tissues and their accumulation in bioavailable forms in seeds. Each of these processes is controlled by several genes (Bouis and Welch, 2010). Plant uptake of Fe and Zn from the soil occurs via two processes: (i) the first strategy involves the direct uptake of $Fe^{2+}$ and $Zn^{2+}$ ions from soil by ZRT/IRT-like proteins (ZIPs); and (ii) the second strategy involves the secretion of phytosiderophores which are capable of chelating Fe cations from the soil. The soluble Fe–siderophore complexes are then taken up by yellow stripe-like (YSL) transporters (Fig. 8.3) (Sperotto *et al.*, 2012). The Fe-chelation strategy is common in monocots like wheat. The chelated Fe cations reach the developing grain via vascular tissues (xylem and phloem). The organic acids citrate and malate enable Fe transport in the xylem, whereas nicotianamine enables Fe and Zn transport intracellularly as well as in the phloem (Connorton *et al.*, 2017a). Different classes of proteins from multiple gene families are involved in Fe and Zn transport. These are ZIP, YSL, MFS (major facilitator superfamily transporter), MTP (metal tolerance protein), HMA (heavy metal ATPase), FPN (ferroportin), NRAMP (natural resistance-associated macrophage protein) and VIT (vacuolar iron transporter). Epidermal cells of the plant roots absorb free $Zn^{2+}$ ions as well as phytosiderophore-bound Fe and Zn from the soil and move them to the pericycle through the apoplast and symplast, which results in their accumulation in basal shoots or leaf tissues. The radial movement of Fe and Zn ions through the root occurs with the help of metal chelators such as nicotianamine (Rellan-Alvarez *et al.*, 2010; Deinlein *et al.*, 2012). The vacuolar accumulation of Zn eventually affects the transport of Zn to the shoots (Morel *et al.*, 2009; Haydon *et al.*, 2012). In the xylem, Zn is transported in the form of a chelated complex with organic acids such as citrate or as a free ion (Lu *et al.*, 2013), whereas Fe is transported only in the chelated form with citrate (Rellan-Alvarez *et al.*, 2010). ZIP, YSL and MTP protein families are involved in the translocation of Fe and Zn complexes across the plasma membrane from xylem to phloem in plant roots or basal shoots or their remobilization from leaves during grain filling. In wheat, accumulation of nutrients in grains occurs via phloem (Zee and O'Brien, 1970). In phloem, Fe and Zn complexed with nicotianamine are transported and accumulated in protein storage vacuoles (PSVs) in the aleurone layer or as ferritin in amyloplasts (Borg *et al.*, 2009; Tauris *et al.*, 2009).

## 8.6 Biofortification of Wheat for Essential Micronutrients

Enhancing micronutrient content in biofortified wheat has appeared as a key solution to the problem of hidden hunger arising due to micronutrient deficiency in the diet of people from developing and poor nations (Pfeiffer and McClafferty, 2007; Chattha *et al.*, 2017). Although biofortified cereals and other foods do not have as high levels of nutrients as those in dietary supplements or artificially fortified food products, they can increase the dietary intake of micronutrients among people who consume them daily (Bouis *et al.*, 2011). The goal of biofortification is to reach poor rural people whose diets are deficient in micronutrients and who have inadequate access to supplements, markets and health-care facilities. The supplementation of Fe and Zn to the soil has its own set of limitations such as its high cost, labour intensiveness, continuous demand for inputs, environmental concerns, nutrient solubility, nutrient mobility in soil, food toxicity when used in excess, and, above all, it may or may not enhance the nutrient content in the edible portion of crops (Garg *et al.*, 2018). To guarantee global nutritional and food security, it becomes necessary to develop and release suitable biofortified wheat varieties which have global acceptance as well. The successful biofortified crop should enhance the micronutrient status of human populations, should be easily available to farmers, and should be high yielding and disease resistant. In addition, its awareness and acceptability among farmers and consumers are of utmost importance (Bouis and Welch, 2010). Biofortification should involve various genetic, breeding and agronomic strategies to produce crops with higher micronutrient levels. Also, efforts should be made to reduce the levels of antinutrients in food crops (Bouis, 2003). Unlike mineral supplementation programmes, the effort,

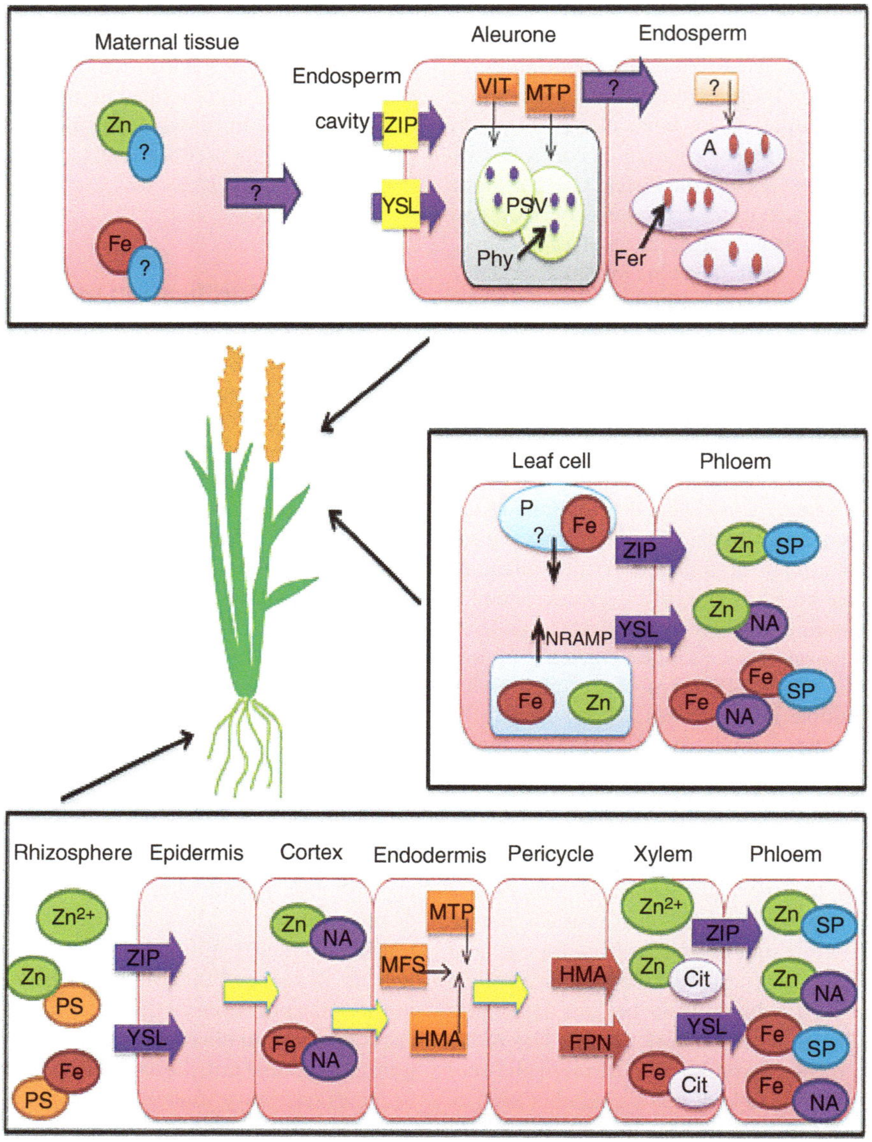


**Fig. 8.3.** Fe and Zn uptake and translocation pathway to the grain in wheat. ZIP, YSL, MFS, MTP, HMA, FPN, NRAMP and VIT are putative classes of transport proteins. Phytosiderophore-bound Fe and Zn along with free $Zn^{2+}$ are absorbed by the root epidermal cells from the soil and moved to the pericycle by the apoplast and symplast, accumulated into xylem and transferred to root phloem, basal shoot or leaf tissues. From leaf cell plastids (P) and vacuoles (V), Fe and Zn are transferred to phloem. In the aleurone layer, Fe and Zn are sequestered in protein storage vacuoles (PSVs) bound to phytate (Phy), and a small quantity of these nutrients enters the endosperm and is stored as ferritin (Fer) in amyloplasts (A). ZIP, ZRT/IRT-like protein; YSL, yellow stripe-like transporter; MFS, major facilitator superfamily transporter; MTP, metal tolerance protein; HMA, heavy metal ATPase; FPN, ferroportin; NRAMP, natural resistance-associated macrophage protein; VIT, vacuolar iron transporter; NA, nicotianamine; PS, phytosiderophore; Cit, citrate; SP, small proteins; ?, unknown transporters. (Adapted from Borrill *et al.*, 2014.)

time and cost for producing biofortified crops is a one-time investment and people can take advantage of these crops over several decades and generations. In addition, using biofertilizers to enhance crop productivity also involves very high monetary and labour costs. On the other hand, biofortification through breeding is an economical and sustainable way to combat micronutrient malnutrition among the masses (White and Broadley, 2005; Graham *et al.*, 2007).

Classical breeding approaches of biofortification aim to exploit the genetic variation among crops and their wild relatives for mineral contents, and also involve marker-assisted breeding (Grusak, 2002). Advancements in genome sequencing have reduced the time requirement and helped in increasing the gene libraries. It has also made it cheaper to focus on genes fundamental to QTLs or to identify these genes straightforwardly using genome-wide association studies. More than 10 years of studies on Fe homeostasis genes and their expression in crops have provided sufficient understanding about developing various strategies for crop biofortification.

In comparison to the total number of released wheat varieties, the number of biofortified varieties for Fe and Zn is quite small; moreover, only a few known cultivars are released for high Fe content (Table 8.1). Still, continuous efforts are being made across the globe to identify and develop superior breeding lines.

## 8.7 Breeding Strategies to Increase Bioavailable Forms of Iron and Zinc

Biofortification through breeding techniques is the most recognized method of biofortification. It provides a sustainable, economical and environmentally friendly substitute to transgenic and agronomic strategies of biofortification. Appropriate genotypic discrepancy in the traits of

**Table 8.1.** List of wheat varieties biofortified for various micronutrients.

| Variety | Biofortified for | Year of release | Institute | Reference |
|---|---|---|---|---|
| HI 8627 (Malav Kirti) | Carotene | 2005 | Indian Agricultural Research Institute (IARI), India | IARI (2019) |
| HD 2932 (Pusa Wheat 111) | Zn | 2007 | IARI, India | IARI (2019) |
| BHU 1, Akshai (BHU 3), BHU 5, HU 6, BHU 17, BHU 18 | Zn | 2014 | International Center for Tropical Agriculture (CIAT), Colombia; International Maize and Wheat Improvement Center (CIMMYT), Mexico; HarvestPlus, USA | Velu *et al.* (2015); Singh and Govindan (2017) |
| Abhay (ZincShakth) | Zn | 2015 | Nirmal Seeds, India; HarvestPlus | Velu *et al.* (2015, 2018) |
| Zincol | Zn | 2015 | CIMMYT; National Agricultural Research Center, Pakistan | Singh and Govindan (2017) |
| NABIMG-9, NABIMG-10, NABIMG-11 | Anthocyanin | 2016 | National Agri-Food Biotechnology Institute, India | Garg *et al.* (2016) |
| Zinc-Shakti (Chitra) | Zn | 2016 | HarvestPlus | Singh and Govindan (2017) |
| HPBW-01 (PBW 1 Zn) | Fe and Zn | 2017 | Punjab Agricultural University, India | Singh and Govindan (2017); Yadava *et al.* (2017) |
| WB02 | Fe and Zn | 2017 | Indian Institute of Wheat and Barley Research (IIWBR), India | Singh and Govindan (2017); Yadava *et al.* (2017) |
| BARI Gom 33 | Zn | 2017 | Bangladesh Agricultural Research Institute (BARI) collaborated with CIMMYT | CIMMYT (2019) |

interest is essential for conventional breeding programmes to be achievable. Breeding techniques exploit this variation among species to enhance the mineral and vitamin contents in crops. In the conventional plant breeding approach, donor lines having high nutrient content are crossed with recipient lines having desired agronomic traits for several generations. This is done until the target of desired nutrient content and desired agronomic traits is achieved in the generations. However, the major limitation of conventional breeding programmes is the presence of limited genetic variation among different species. In many cases, this drawback is overcome by crossing wheat with distantly related species and moving the desired character slowly into the commercial varieties. In addition, mutagenesis can be used to get the desired character in the genome of the species.

Several programmes involving conventional breeding techniques have been introduced by various international organizations. The Healthgrain Project (2005–2010) was carried out in the European Union and involved 15 countries and a budget of £10 million. Its main aim was to develop high-quality food crops with high nutrient content (Fardart, 2010; Tighe *et al.*, 2010; Lafiandra *et al.*, 2014). In addition, the HarvestPlus programme involving conventional breeding programmes among cereal and vegetable crops has been launched by CGIAR along with the International Center for Tropical Agriculture (CIAT) and the International Food Policy Research Institute. Figure 8.4 illustrates different breeding strategies to enhance Fe and Zn contents in wheat using different genetic approaches.

## 8.8 Genetic Variation

In breeding, to understand genetic variation is the key step. The present wheat cultivars have

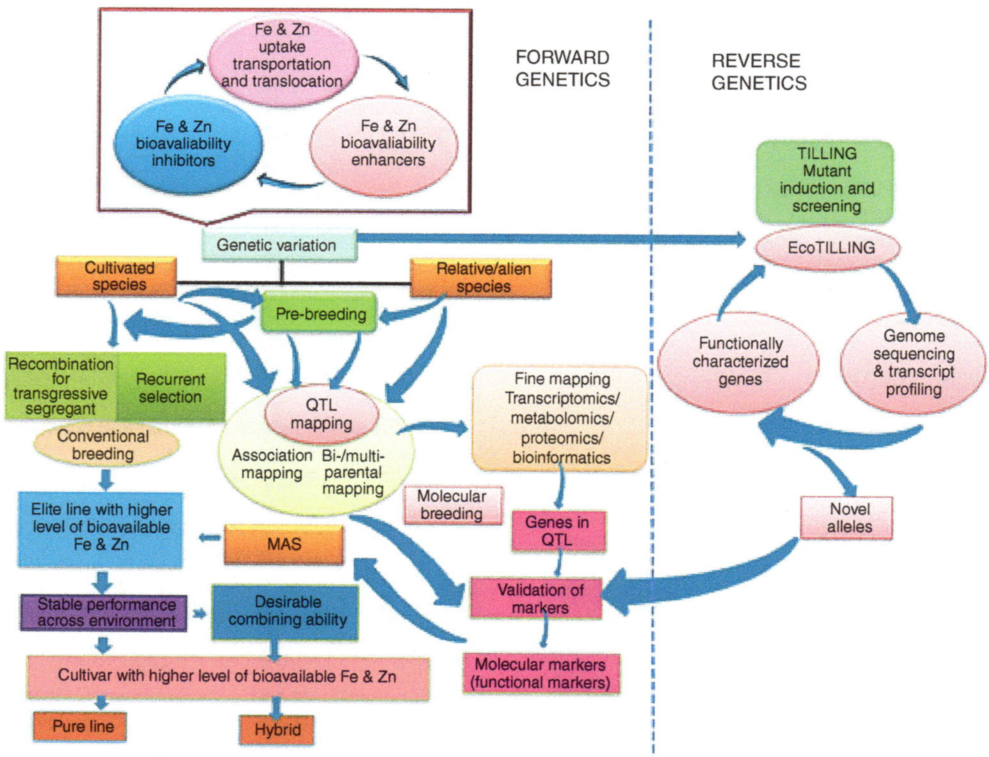


**Fig. 8.4.** Schematic representation of different breeding strategies using various genetic approaches for Fe and Zn improvement in wheat. TILLING, targeting induced local lesions in genomes; EcoTILLING, ecotype TILLING. (Adapted from Garcia-Oliveira *et al.*, 2018.)

less variation for Zn and Fe contents. Wide variation in grain Fe and Zn concentrations has been noticed in wild species and landraces of wheat (Monasterio and Graham, 2000; Ortiz-Monasterio *et al.*, 2007; Velu *et al.*, 2011a). The available genetic variation from different wild species and landraces can be utilized in the biofortification breeding programme for the development of nutrient-enriched wheat germplasms with comparative yield potential and stress tolerance (Velu *et al.*, 2014). Wheat–rye chromosome addition lines and their rye parents have been recorded for their capacity to improve Zn uptake from soils and build up tissue use efficiency (Cakmak *et al.*, 1998; Schlegel *et al.*, 1998; Erenoglu *et al.*, 1999). In triticale and rye, Zn efficiency is very high, and the genes behind this can be used to develop wheat cultivars with improved Zn efficiency. Synthetic hexaploid wheats (*Triticum turgidum* × *Aegilops tauschii* and *T. turgidum* × *Triticum monococcum*), developed by transfer of the DD genome from *Ae. tauschii* ($2n = 2x = 14$, DD) or the AA genome from *T. monococcum* ($2n = 2x = 14$, AA) to tetraploid wheat (*T. turgidum*, $2n = 4x = 28$, AABB), have been recorded to have a clear effect on total Zn content (Cakmak *et al.*, 1999).

## 8.9 Genotype × Environment Interaction

Apart from genetic variation, precision phenotyping plays an important role in Fe and Zn biofortification through breeding. There are different soil and environmental factors such as pH, temperature, radiation, precipitation, organic matter and soil texture which have the potential to influence concentration, solubility and absorption of micronutrients by plant roots (Tisdale and Nelson, 1975; Cakmak, 2008a; Joshi *et al.*, 2010). The other environmental factors, such as increasing atmospheric $CO_2$, are likely to lead to a further drop in Fe content in wheat (Myers *et al.*, 2014). Soil composition is also an issue for breeding to increase Zn concentration in wheat (Trethowan *et al.*, 2007). Therefore, increase in uptake efficiency or mobilization to the grain through breeding has less impact on grain Zn concentration due to Zn availability in the soil (Ortiz-Monasterio *et al.*, 2007, 2011). Genotype and location interactions have significant impact on Zn and Fe contents in both wild and improved wheat cultivars (Oury *et al.*, 2006; Ortiz-Monasterio *et al.*, 2007; Trethowan *et al.*, 2007; Gomez-Becerra *et al.*, 2010a). The International Maize and Wheat Improvement Center (CIMMYT), Mexico, developed high-Zn wheat lines which when evaluated in India at multiple locations acknowledged that wheat grain Zn concentrations were highly unstable (Joshi *et al.*, 2010) as the growth of the elite lines differed with location. The greater gene × environment (G × E) interaction for Zn concentration has another reason for its quantitative inheritance in wheat (Trethowan *et al.*, 2007). Contrarily, a recent study conducted in South Asia to test biofortified wheat lines at multiple locations revealed high heritability and high genetic interaction between locations for grain Zn, proposing that G × E may not be a serious concern in breeding high-Zn wheat genotypes (Velu *et al.*, 2012). Understanding the connection between micronutrients and different parameters like grain yield, plant height, grain size and end-use quality parameters would facilitate the selection of mineral-dense progenies through breeding with desired phenological and consumer-preferred traits. According to former studies, grain Zn and Fe are positively correlated in wheat (Morgounov *et al.*, 2007; Genc *et al.*, 2009; Peleg *et al.*, 2009; Zhang *et al.*, 2010; Gomez-Becerra *et al.*, 2010a; Velu *et al.*, 2011a,b, 2012), suggesting that the alleles for both Zn and Fe deposition in the grain co-segregate or are pleiotropic, and therefore Zn and Fe can be improved simultaneously. Rawat *et al.* (2009a,b) have shown in the flag leaves of *Aegilops* species that Zn and Fe are positively correlated. In many studies, there is no negative linkage of grain Zn and Fe with grain yield (Graham *et al.*, 1999; Welch and Graham, 2004; Velu *et al.*, 2012). On the contrary, some reports showed a slightly negative association between Zn and grain yield in wheat (Morgounov *et al.*, 2007; Peleg *et al.*, 2009; Zhao *et al.*, 2009; Gomez-Becerra *et al.*, 2010b).

## 8.10 Quantitative Trait Locus Mapping

Past evidence has shown that grain Zn and Fe are quantitatively inherited traits in wheat (Trethowan *et al.*, 2005, 2007). Therefore, identifying the QTLs that regulate the accumulation of high mineral nutrient concentration in wheat grains and the yield-related agronomic traits would allow breeders to develop biofortified cultivars more efficiently by using closely linked molecular markers to screen and select the most favourable genotypes. Many QTLs have been identified in diverse germplasm and advanced breeding lines for Fe and Zn; some selected studies are presented in Table 8.2. QTL analysis provides an effective means of resolving quantitative traits into single components to study their relative impacts on a specific trait. Such QTL analyses for micronutrient concentration have been conducted on wheat (Shi *et al.*, 2008; Genc *et al.*, 2009) and related species (Peleg *et al.*, 2009; Tiwari *et al.*, 2009). In recent years, many QTLs of micronutrient concentration in wheat grain have been mapped using recombinant inbred lines or doubled haploid populations (Distelfeld *et al.*, 2007; Ozkan *et al.*, 2007; Shi *et al.*, 2008; Peleg *et al.*, 2009; Tiwari *et al.*, 2009). Epistatic and QTL × environment (Q × E) interactions are important genetic components. Most quantitative traits are greatly affected by either one of them or both. The detection of Q × E interaction was greatly affected by the accuracy of phenotyping in a particular environment (Xu and Crouch, 2008). Epistasis and Q × E analyses have been conducted in wheat (Yang *et al.*, 2007; Zhang *et al.*, 2008, 2009) and other plant species. However, differences in mapping results among various environments have resulted in unreliable indications of the significance of Q × E interactions in some cases (Jansen *et al.*, 1995).

A major QTL *Gpc-B1* (transcription factor *NAM-B1*) from wild emmer wheat (*T. turgidum* ssp. *dicoccoides*) was mapped on chromosome arm 6BS (Joppa *et al.*, 1997). It was cloned and confirmed to be effective in improving Fe and Zn by 18 and 12%, respectively (Uauy *et al.*, 2006; Distelfeld *et al.*, 2007). A molecular marker *Xuhw89* was tightly linked to the *Gpc-B1* locus with a 0.1 cM genetic distance (Distelfeld *et al.*, 2006). A QTL mapping study organized on the diploid A genome of wheat showed identification of two QTLs for grain Fe on chromosomes 2A and 7A, and one QTL for grain Zn on chromosome 7A (Tiwari *et al.*, 2009). A major QTL has been found on chromosome 5B in *T. monococcum* that is linked with high grain content of Fe, Zn, Co and Mn (Ozkan *et al.*, 2007). In addition, Singh *et al.* (2010) also identified QTLs for Fe and Zn, which are *QFe.pau-2A* and *QFe.pau-7A* for Fe and *QZn.pau-7A* for Zn, from *Aegilops kotschyi* and *Aegilops peregrine*. Furthermore, Genc *et al.* (2009) reported four QTLs for Zn (on chromosomes 3D, 4B, 6B and 7A) and a single QTL for Fe in a doubled haploid population. Recently, a new study on Chinese winter wheat gave QTLs for Fe and Zn accumulation after evaluation under two different environments (Xu *et al.*, 2012).

## 8.11 Novel Experiments using Sequence Data Resources

The low cost of sequencing and the data generated from it are quite helpful in finding new alternatives for exploring and finding gene functions associated with regulating Fe and Zn content in grains. During the last 5 years, data on the genome sequencing of wheat in the public domain have increased substantially. Identifying and distinguishing homeologous genome sequences in wheat genome-specific contigs have been made possible through the EnsemblPlants database. In addition, physical maps of bacterial artificial chromosome (BAC) libraries have been constructed for reference (Paux *et al.*, 2008). Moreover, accessing the genome sequences through whole-genome shotgun sequencing, RNA-seq and exome capture (Saintenac *et al.*, 2011; Trick *et al.*, 2012; Winfield *et al.*, 2012) has been used to develop genome-specific markers for genetic mapping and identifying single-nucleotide polymorphisms in wheat (Wilkinson *et al.*, 2012; Allen *et al.*, 2013).

In a study by Cantu *et al.* (2011), RNA-seq was used to find differentially expressed genes in lines having decreased NAM genes expression. RNA-seq was used to identify genes related to transporter proteins, hormone-regulated genes and transcription factors. This study produced data related to early senescence stages and nutrient remobilization.

RNA-seq data (Duan *et al.*, 2012) and homoeologue-specific gene models could deliver enhanced resolution to transcriptome studies (Krasileva *et al.*, 2013). In addition, knowledge from one particular species could be used to identify candidate genes in other species. Borrill *et al.* (2014) studied sequences of the NRAMPs in rice to identify wheat homologues. Genes

**Table 8.2.** List of QTLs identified for Fe and Zn in wheat.

| Trait | Cross/Parents | QTL | Reference |
|---|---|---|---|
| Fe and Zn | *Triticum dicoccoides* | *GPC-B1* | Joppa *et al.* (1997); Uauy *et al.* (2006); Distelfeld *et al.* (2007) |
| Fe and Zn | *T. dicoccoides* | *TtNAM-B1* | Distelfeld *et al.* (2007) |
| Fe | RIL (*Triticum boeoticum* × *Triticum monococcum*) | *QFe.pau-7A*<br>*QFe.pau-2A* | Tiwari *et al.* (2009) |
| Fe and Zn | RIL (Xiaoyan × 54 Jing 411) | *QZn-5A*<br>*QFe-5A2*<br>*QGpc-5A1*<br>*QGpc-6A* | Xu *et al.* (2012) |
| Zn | RIL (PBW343 × KenyaSwara) | *QGzncpk.cimmyt-1BS*<br>*QGzncpk.cimmyt-2Bc*<br>*QGzncpk.cimmyt-3AL* | Hao *et al.* (2014) |
| Fe and Zn | RIL (*Triticum spelta* (H+26 (PI348449)) × *Triticum aestivum* cv. HUW 234) | *QZn.bhu-2B*<br>*QZn.bhu-6A*<br>*QFe.bhu-3B* | Srinivasa *et al.* (2014) |
| Fe and Zn | DH (Berkut × Krichauff) hexaploid (Adana99 × 70711) | *QGfe.ada-2B*<br>*QGfe.ada-2B*<br>*QGZn.ada-2B*<br>*QGfe.ada-2B*<br>*QFe.bhu-2B* | Tiwari *et al.* (2016); Velu *et al.* (2016) |
| Fe | Tetraploid (Saricanak98 × MM5/4) | *QGfe.sar-5B* | Velu *et al.* (2016) |
| Zn | Tetraploid (Saricanak98 × MM5/4) | *Qzneff.sar-6A*<br>*Qzneff.sar-6B* | Velu *et al.* (2016) |
| Zn | DH (Berkut × Krichauff) | *QZn.bhu-1B*<br>*QZn.bhu-2* | Tiwari *et al.* (2016); Velu *et al.* (2016) |
| Zn | Tetraploid (Saricanak98 × MM5/4) | *QGzn.sar-1B*<br>*QGzn.sar-6B*<br>*QGZn.sar-1B* | Velu *et al.* (2016) |
| Zn | Hexaploid (Adana99 × 70711) | *QGzn.ada-6B*<br>*QGzn.ada-1D*<br>*QGzn.ada-7B* | Velu *et al.* (2016) |
| Fe and Zn | RIL (synthetic hexaploid wheat × *T. spelta*) | *QGZn.cimmyt-7B_1P2*<br>*QGFe.cimmyt-4A_P2*<br>*QGZn.cimmyt-7B_1P2*<br>*QGZn.cimmyt-7B_1P1* | Crespo-Herrera *et al.* (2017) |
| Fe and Zn | *Triticum dicoccon* (PI94624)/ *Aegilops squarrosa* [409] × BCN | *QGFe.iari-2A*<br>*QGFe.iari-5A*<br>*QGFe.iari-7A*<br>*QGFe.iari-7B*<br>*QGZn.iari-2A*<br>*QGZn.iari-4A*<br>*QGZn.iari-5A*<br>*QGZn.iari-7A*<br>*QGZn.iari-7B* | Krishnappa *et al.* (2017) |

RIL, recombinant inbred line; DH, doubled haploid.

involved in Fe and Zn transport in wheat could be cloned and characterized in mutant yeast strains and could be eventually expressed in transgenic wheat plants to enhance the nutrient content in grains.

## 8.12 Marker-Assisted Breeding

Nowadays, breeders have mainly focused on marker-assisted selection (MAS) rather than conventional breeding for generating improved varieties. There are various molecular markers used in MAS that are highly associated with important traits. Therefore, molecular markers prove as tools used to spot the presence of a desired character in crossing, which greatly enhances the efficiency of selection. With MAS phenotypic selection, it would be easy and less time consuming and would also save costs for the selection of desired traits (Koide *et al.*, 2009). This method proves to be a more accurate tool in introducing novel cultivars and enabling breeding research in crops based on the genotype rather than on the phenotype. After finding the link between molecular markers and genes of interest, the development of new cultivars with specified traits could be made at an early level (Zhu *et al.*, 2012). Pyramiding of linked genes into a single line or cultivar is one of the common applications of MAS. In addition to MAS, marker-assisted backcrossing (MABC) also utilizes markers, but here markers are used to select target loci which reduce the length of the donor segment holding a target locus and stimulate the recovery of the recurrent parent (RP) genome during backcrossing (Charcosset, 1997; Hospital, 2001; Hasan *et al.*, 2015). MABC has the main purpose of transferring the desired character or targeted gene into the recipient along with recovering the RP characters and/or genes. One successful example of MAS is the transfer of *Gpc-B1* locus (Uauy *et al.*, 2006). The presence of *Gpc-B1* locus in near-isogenic lines increased Fe and Zn grain concentrations (Distelfeld *et al.*, 2007). This gene is being widely used in breeding programmes over several continents for marker-assisted breeding (Kumar *et al.*, 2011; Randhawa *et al.*, 2013; Tabbita *et al.*, 2013).

## 8.13 Classical Breeding

Utilizing this variation, HarvestPlus has worked extensively to increase the Fe and Zn contents in wheat. Breeding for Fe and Zn was decided by expected bioavailability percentage, daily intake of wheat per capita, type of food preparation and estimated average requirements. The preliminary breeding target is to increase Fe and Zn levels by 25 and 10 mg/kg, respectively.

There are several varieties of wheat having 4–10 ppm higher Zn content which have been released by HarvestPlus. Six varieties (BHU 1, BHU 3, BHU 5, BHU 6, BHU 17 and BHU 18) were released in India (2014) and one variety in Pakistan (2015) (Zincol). Recently, HPBW-01 (PBW 1 Zn) and WB02 varieties with high Fe and Zn contents have been released by Punjab Agricultural University and the Indian Institute of Wheat and Barley Research (IIWBR), respectively (Garg *et al.*, 2018). Several scientists have observed the use of plant breeding to increase Fe and Zn contents in wheat (Cakmak *et al.*, 1999; Monasterio and Graham, 2000; Cakmak *et al.*, 2004; Welch *et al.*, 2005). In some countries, high-Zn wheat varieties developed by CIMMYT are also being released after testing by national programmes (Velu *et al.*, 2012; Baloch *et al.*, 2015). Some of these varieties are Zinc-Shakti, HPBW-01 and Ankur Shiva, released in India by both public and private partners, and one variety named Nohely-F2018 has also been released in Mexico. Interestingly, BARI Gom 33 (Kachu/Solala) released in Bangladesh during 2017 showed 7–8 mg Zn/kg advantage and offered resistance against wheat blast.

There has been successful improvement in the concentration of Fe in grains, with the access of chromosomal regions in modern wheat from wild *Aegilops* species (Neelam *et al.*, 2011). In addition, to overcoming the genetic variability for low Fe and Zn levels in wheat, more than 180 lines of *Aegilops* have been analysed. These different lines can be used for fixing the Fe and Zn contents in wheat grain. Amphiploids (Tiwari *et al.*, 2010) and partial amphiploids (Rawat *et al.*, 2009b) produced by crossing *Ae. kotschyi* accessions with wheat lines appeared to have higher grain Fe and Zn contents. This exploration for higher grain Fe and Zn contents has been carried out in different countries but the significant exploration is from India (158 lines and accessions) as compared with other countries. Addition of chromosome pairs $1S^l$ (Wang *et al.*, 2011), $2S^l$ (Wang *et al.*, 2011; Kumari *et al.*, 2012) and $7S^l$ (Wang *et al.*, 2011) of *Aegilops longissima* into wheat showed higher Fe and Zn contents in wheat grain. The substitution of wheat 4B chromosome with the 3 Mb chromosome of *Aegilops biuncialis* led to increased Fe and Zn contents (Farkas *et al.*, 2014). Interspecific hybrids of *Ae. longissima* with *T. turgidum* and *Ae. kotschyi* produced after crossing with tetraploid and hexaploid genotypes also showed higher levels of Fe and Zn in wheat grain (Tiwari *et al.*, 2008; Sheikh *et al.*, 2018).

Another important criterion for enhancing grain Fe and Zn contents is the lowering of antinutrients such as phytic acid. Significant genetic variation in the phytate content of different wheat species has been detected in different studies (Erdal *et al.*, 2002; Welch *et al.*, 2005; White and Broadley, 2009) and wheat genotypes and mutants with low phytic acid contents will be desirable (Guttieri *et al.*, 2004).

## 8.14 Transgenic Strategies to Increase Bioavailable Forms of Iron and Zinc

The transgenic approach involves the exploitation of unlimited genetic resources for the transfer of desirable genes from one species to another using molecular biology and genetic engineering techniques. This approach is independent of the evolutionary history and taxonomic status of donors and recipients, and, hence, exogenous genes from bacteria and other organisms can also be introduced into plants to get the desirable characteristics (Aken and Dotty, 2009). Different phenomena including mineral uptake, transport and grain deposition in wheat are critically important for understanding and identifying the bottlenecks and target genes in molecular and genetic studies. Researchers have also tried to address the challenges of Fe and Zn in wheat by this strategy. The most common genes targeted by transgenic strategies for Fe and Zn enhancement across the plant kingdom are those encoding ferritin and nicotianamine synthase (NAS) (Garg *et al.*, 2018).

In one of the transgenic studies in wheat, Borg *et al.* (2012) cloned and carried out endosperm-targeted overexpression of wheat ferritin genes (*TaFer1-A*), finding a 1.5- to 1.9-fold increase in grain Fe content in comparison to normal wheat. Similarly, improvement of 1.1- to 1.6-fold in the Fe content of wheat grain by soybean ferritin expression cassette has also been reported (Sui *et al.*, 2012). Singh *et al.* (2017) showed that overexpression of *OsNAS2* gene in wheat resulted in the production of grain Fe content up to 93.1 μg/g. Similarly, Beasley *et al.* (2019) also carried out cloning and overexpression of *OsNAS2* in wheat and reported 80 μg Fe/g in grains under field conditions. Connorton *et al.* (2017b) cloned and expressed vacuolar Fe transporter (*TaVIT*) in wheat and observed double Fe concentration in wheat flour after gene expression. The silencing of wheat *ABCC13* transporter could be helpful to decrease the phytic acid content in wheat (Bhati *et al.*, 2016).

Researchers have tried to increase Fe and Zn levels in the starchy endosperm also. The *NAS* expression in wheat led to an increase in the concentration of Fe and Zn in grains (Masuda *et al.*, 2009; Zheng *et al.*, 2010; Johnson *et al.*, 2011; Singh *et al.*, 2017). As nicotianamine serves as a chelator for Fe and Zn in their ionic forms, increasing its level would therefore also increase Fe and Zn contents in wheat. In contrast, overexpression of metal transporter genes redirects minerals into the starchy endosperm which leads to increase of a single mineral, due to the high specificity of metal transporters, unless several genes are overexpressed at the same time. For example, expression of *TaVIT2* using a starchy endosperm-specific promoter doubled the Fe content of the white flour fraction (Connorton *et al.*, 2017b). The increase in Fe in transgenic lines is not followed by P increase, showing accumulation of Fe without being linked with phytic acid. More recently, studies have shown that preventing Fe storage in the vacuoles, while at the same time overexpressing the Fe-storage protein ferritin especially in the starchy endosperm cells, greatly increases Fe (Wu *et al.*, 2018). Hence, it is now approved that transgenesis can be used to increase the contents of bioavailable Fe and Zn in the starchy endosperm of cereals several-fold by redirecting mineral transport and/or providing a sink to sequester Fe.

## 8.15 Agronomic Strategies for Iron and Zinc Improvement in Wheat

In agronomic biofortification, micronutrients are applied to the soil and/or directly to the plant leaves. This method results in increased grain yield of wheat, due to which an increase was recorded in the use of Zn fertilizer from 0 in 1994 to 400,000 tonnes per annum over 10–15 years in Turkey (Cakmak, 2008a). In many countries like Turkey, application of Zn fertilizers has increased grain Zn concentration, and this ultimately has contributed to human nutrition and health, especially in rural regions where 50% of energy intake comes from wheat on a daily basis (Cakmak *et al.*, 2008a). The improvement in Zn concentration of wheat grain with application of Zn-containing fertilizers depends on Zn deficiency of soil (Cakmak, 2008a; Zou *et al.*, 2012). After testing various forms of Zn fertilizer, Zn as $ZnSO_4$ gives more promising results compared with other Zn forms (Zhang *et al.*, 2012). Several studies have recorded improvement in grain Zn concentration after application of Zn fertilizers to the soil (Hao *et al.*, 2003; Cakmak *et al.*, 2010). Another advantage of Zn application is protection from soilborne pathogens (Cakmak, 2012). Among the sources of Fe for agronomic fortification, $FeSO_4$ is the most important. The combination of N fertilizers with Zn and Fe has been reported to further increase the yield and uptake of these elements in the seed (Kutman *et al.*, 2011; Pascoalino *et al.*, 2018; Liu *et al.*, 2019).

Alternatively, foliar application of Zn on the plants resulted in better grain yield and Zn concentration. Micro elements applied as foliar spray prove to be more beneficial than soil spray. However, application rates are minor as compared with soil application, but it could be achieved easily and the crop reacts to nutrient application immediately (Zayed *et al.*, 2011). Normally, Zn fertilizer solution for foliar spray contains 2–5 g $ZnSO_4 \cdot 7H_2O$ per litre. According to this, the amount of Zn used for foliar spray is about 1 kg/ha or less, which counts as five times less than used for soil application and is also considered as safe for the ecosystem (Cakmak *et al.*, 2010; Boonchuay *et al.*, 2013; Ram *et al.*, 2016). Foliar application of Zn during grain developmental stages results in increased grain Zn concentration as the leaf epidermis absorbs Zn and transports it via phloem (Haslett *et al.*, 2001) to other parts and translocates it into developing wheat grains (Haslett *et al.*, 2001; Erenoglu *et al.*, 2011). Furthermore, foliar application of Zn also increased Zn content in starchy endosperm. In addition, Fe concentration in wheat grains is also elevated by foliar Zn application (Fang *et al.*, 2008; Habib, 2009; Aref, 2010; Zeidan *et al.*, 2010). Increase of 20–70% in Fe concentration in grains of bread wheat has been recorded (Shukla and Warsi, 2000; Habib, 2009; Zeidan *et al.*, 2010; Zhang *et al.*, 2010).

However, such agronomic practices are less effective for Fe, except if combined with increased N fertilization (Aciksoz *et al.*, 2011) which may not be economically or environmentally acceptable. According to published data, the foliar application of Zn fertilizer is economical because its cost of application is less than the benefits earned through it (Harris *et al.*, 2007; Shivay *et al.*, 2008; Manzeke *et al.*, 2014; Joy *et al.*, 2016). Residual effects of Zn fertilization can be seen for up to 10 years because a small amount of the Zn applied to soil is taken up in a single season by annual crops. Hence, there is no need to undertake Zn fertilization every year (Brennan, 2001; Alloway, 2008; Cakmak, 2008b; Singh, 2008).

## 8.16 Challenges in Wheat Iron and Zinc Biofortification

The exact mechanisms of transportation and regulation of Fe and Zn are different among different species in monocots. So, there is no particular mechanism of action in all plants. Zn transportation in wheat is dependent upon two major factors: (i) the root–shoot barrier; and (ii) grain filling. However, in rice excess of Zn is stored in both roots as well as shoots and hence these barriers are reduced (Jiang *et al.*, 2008; Stomph *et al.*, 2009). In addition, Zn is loaded from xylem to grains without transferring to phloem and this further reduces the barriers in rice (Zee, 1971), as shown in Fig. 8.5.

Furthermore, in rice, both secretion of phytosiderophores and direct uptake of Fe from the soil take place (Bughio *et al.*, 2002; Inoue *et al.*, 2009; Nozoye *et al.*, 2011), whereas wheat absorbs Fe only via phytosiderophores (Romheld and Marschner, 1986; Zaharieva and Romheld, 2000; Murata *et al.*, 2006; Walker and Connolly, 2008). Other challenges include a lack of knowledge about the specificity of transporters and the control of flux through different pathways. Further research is required on the role of source and sink in metal deposition and limits of manipulating the particular metal on the total metal content of the grain.

Also, Cd is a strong competitor of Zn that is toxic and is hazardous to human health (Welch, 1999; Reeves and Chaney, 2008; Cakmak, 2009).

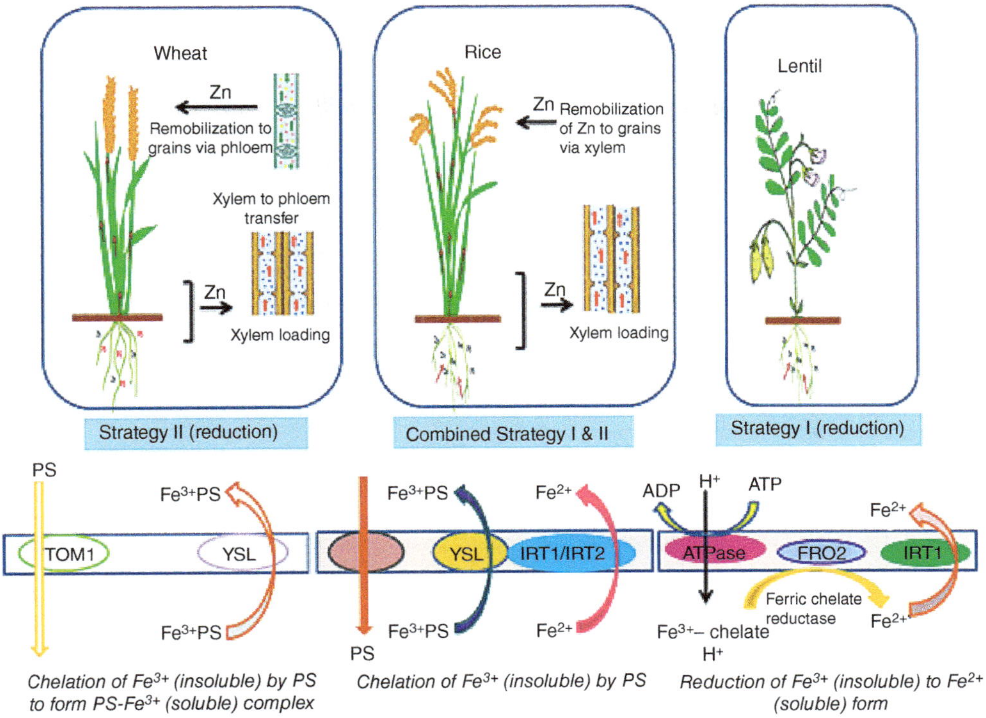


**Fig. 8.5.** Different strategies operating in monocot and dicot plants for Fe transport from root to shoot (bottom) and different mechanisms of Zn filling in grain of these plants (top). PS, phytosiderophore; TOM1, transporter of mugineic acid; YSL, yellow stripe-like transporter; IRT1/IRT2, iron transporter 1/iron transporter 2; FRO2, ferric reductase oxidase 2.

Hence, Zn biofortification in wheat through breeding or transgenic approaches might increase the risk of Cd accumulation in plants. Furthermore, Zn is a heavy metal and at high concentrations, it poses several toxic effects on plants such as inhibitory effects on the root uptake and retardation of root-to-shoot translocation (Welch, 1999; Cakmak *et al.*, 2000; Jiao *et al.*, 2004).

## 8.17 Conclusions and Future Directions

Use of biofortified wheat varieties with increased contents of Fe and Zn and other vital nutrients is the new face of modern agriculture. It will help in combatting hidden hunger and micronutrient deficiencies, especially among people of the poor and developing world. Biofortification requires one-time investment but is better than conventional agronomic techniques which have budgetary and legal constraints. Hence, biofortification is a cheap, sustainable and environmentally friendly approach to increase micronutrient contents in plants. Usually increasing crop yield dilutes the mineral content in grains but applying biofortification strategies will help in preventing this micronutrient dilution. This will require the joint efforts of scientists, breeders, farmers and consumers, however, to get sustainable results. The new wheat varieties enriched in Fe and Zn contents will have the potential to improve the health and lives of millions of people across the world.


## Acknowledgement

The authors are thankful to the National Agri-Food Biotechnology Institute, India for providing support to this work.

## References


Aciksoz, S.B., Yazici, A., Ozturk, L. and Cakmak, I. (2011) Biofortification of wheat with iron through soil and foliar application of nitrogen and iron fertilizers. *Plant and Soil* 349(1–2), 215–225. Available at: https://doi.org/10.1007/s11104-011-0863-2

Aken, B.V. and Doty, S.L. (2009) Transgenic plants and associated bacteria for phytoremediation of chlorinated compounds. *Biotechnology and Genetic Engineering Reviews* 26, 43–64. Available at: https://doi.org/10.5661/bger-26-43

Allen, A.M., Barker, G.L.A., Wilkinson, P., Burridge, A., Winfield, M. *et al.* (2013) Discovery and development of exome-based, co-dominant single nucleotide polymorphism markers in hexaploid wheat (*Triticum aestivum* L.). *Plant Biotechnology Journal* 11, 279–295. Available at: https://doi.org/10.1111/pbi.12009

Alloway, B.J. (2008) *Zinc in Soils and Crop Nutrition*, 2nd edn. International Zinc Association and International Fertilizer Association, Brussels and Paris.

Aref, F. (2010) Zinc and boron fertilization on concentration and uptake of iron and manganese in the corn grain. *Journal of American Science* 6, 236–242.

Balk, J. and Schaedler, T.A. (2014) Iron cofactor assembly in plants. *Annual Review of Plant Biology* 65, 125–153. Available at: https://doi.org/10.1146/annurev-arplant-050213-035759

Balmer, Y., Vensel, W.H., Dupont, F.M., Buchanan, B.B. and Hurkman, W.J. (2006) Proteome of amyloplasts isolated from developing wheat endosperm presents evidence of broad metabolic capability. *Journal of Experimental Botany* 57, 1591–1602. Available at: https://doi.org/10.1093/jxb/erj156

Baloch, Q.B., Makhdum, M.I., Mujahid, M.Y. and Noreen, S.N. (2015) Biofortification: high zinc wheat programme – the potential agricultural options for alleviating malnutrition in Pakistan. *International Journal of Food and Allied Sciences* 1, 36–39. Available at: https://doi.org/10.21620/ijfaas.2015136-39

Bastow, E.L., Garcia de la Torre, V.S., Maclean, A.E., Green, R.T., Merlot, S., Thomine, S. and Balk, J. (2018) Vacuolar iron stores gated by NRAMP3 and NRAMP4 are the primary source of iron in germinating seeds. *Plant Physiology* 177, 1267–1276. Available at: https://doi.org/10.1104/pp.18.00478

Beasley, J.T., Julien, P., Bonneau, J.T., Sánchez-Palacios, L.T., Moreno-Moyano, D.L. *et al.* (2019) Metabolic engineering of bread wheat improves grain iron concentration and bioavailability. *Plant Biotechnology Journal* 17, 1514–1526. Available at: https://doi.org/10.1111/pbi.13074

Bhati, K.K., Alok, A., Kumar, A., Kaur, J., Tiwari, S. and Pandey, A.K. (2016) Silencing of *ABCC13* transporter in wheat reveals its involvement in grain development, phytic acid accumulation and lateral root formation. *Journal of Experimental Botany* 67(14), 4379–4389. Available at: https://doi.org/10.1093/jxb/erw224

Boonchuay, P., Cakmak, I., Rerkasem, B. and Prom-U-Thai, C. (2013) Effect of different foliar zinc application at different growth stages on seed zinc concentration and its impact on seedling vigor in rice. *Soil Science and Plant Nutrition* 59, 180–188. Available at: https://doi.org/10.1080/00380768.2013.763382

Borg, S., Brinch-Pedersen, H., Tauris, B. and Holm, P. (2009) Iron transport, deposition and bioavailability in the wheat and barley grain. *Plant and Soil* 325, 15–24. Available at: https://doi.org/10.1007/s11104-009-0046-6

Borg, S., Brinch-Pedersen, H., Tauris, B., Madsen, L.H., Darbani, B., Noeparvar, S. and Holm, P.B. (2012) Wheat ferritins: improving the iron content of the wheat grain. *Journal of Cereal Science* 56, 204–213. Available at: http://dx.doi.org/10.1016/j.jcs.2012.03.005

Borrill, P., Connorton, J., Balk, J., Miller, T., Sanders, D. and Uauy, C. (2014) Biofortification of wheat grain with iron and zinc: integrating novel genomic resources and knowledge from model crops. *Frontiers in Plant Science* 5, 53. Available at: https://doi.org/10.3389/fpls.2014.00053

Bouis, H.E. (2003) Micronutrient fortification of plants through plant breeding: can it improve nutrition in man at low cost? *Proceedings of the Nutrition Society* 62, 403–411. Available at: https://doi.org/10.1079/pns2003262

Bouis, H.E. and Welch, R.M. (2010) Biofortification – a sustainable agricultural strategy for reducing micronutrient malnutrition in the global South. *Crop Science* 50, S20–S32.

Bouis, H.E., Hotz, C., McClafferty, B., Meenakshi, J.V. and Pfeiffer, W.H. (2011) Biofortification: a new tool to reduce micronutrient malnutrition. *Food and Nutrition Bulletin* 32(1_suppl1), 31S–40S. Available at: https://doi.org/10.1177/15648265110321S105

Braun, H.J., Atlin, G. and Payne, T. (2010) Multi-location testing as a tool to identify plant response to global climate change. In: Reynolds, M.P. (ed.) *Climate Change and Crop Production*. CAB International, Wallingford, UK, pp. 115–138. Available at: http://dx.doi.org/10.1079/9781845936334.0115

Brennan, R.F. (2001) Residual value of zinc fertiliser for production of wheat. *Australian Journal of Experimental Agriculture* 41, 541–547. Available at: https://doi.org/10.1071/EA00139

Brown, K.H., Hambidge, K.M., Ranum, P. and Zinc Fortification Working Group (2010) Zinc fortification of cereal flours: current recommendations and research needs. *Food and Nutrition Bulletin* 31(1_suppl1), S62–S74. Available at: https://doi.org/10.1177/15648265100311S106

Bughio, N., Yamaguchi, H., Nishizawa, N.K., Nakanishi, H. and Mori, S. (2002) Cloning an iron-regulated metal transporter from rice. *Journal of Experimental Botany* 53, 1677–1682. Available at: https://doi.org/10.1093/jxb/erf004

Cakmak, I. (2008a) Enrichment of cereal grains with zinc: agronomic or genetic biofortification. *Plant and Soil* 302, 1–17. Available at: https://doi.org/10.1007/s11104-007-9466-3

Cakmak, I. (2008b) Zinc deficiency in wheat in Turkey. In: Alloway, B.J. (ed.) *Micronutrient Deficiencies in Global Crop Production*. Springer, Dordrecht, the Netherlands pp. 181–200.

Cakmak, I. (2009) Enrichment of fertilizers with zinc: an excellent investment for humanity and crop production in India. *Journal of Trace Elements in Medicine and Biology* 23, 281–289. Available at: https://doi.org/10.1016/j.jtemb.2009.05.002

Cakmak, I. (2012) HarvestPlus zinc fertilizer Project: HarvestZinc. *Better Crops* 96, 17–19.

Cakmak, I., Torun, B., Erenoglu, B., Ozturk, L., Marschner, H. *et al.* (1998) Morphological and physiological differences in the response of cereals to zinc deficiency. *Euphytica* 100, 349–357.

Cakmak, I., Cakmak, O., Eker, S., Ozdemir, A., Watanabe, N. and Braun, H.J. (1999) Expression of high zinc efficiency of *Aegilops tauschii* and *Triticum monococcum* in synthetic hexaploid wheats. *Plant and Soil* 215, 203–209.

Cakmak, I., Ozkan, H., Braun, H.J., Welch, R.M. and Romheld, V. (2000) Zinc and iron concentrations in seeds of wild, primitive and modern wheats. *Food and Nutrition Bulletin* 21, 401–403.

Cakmak, I., Torun, A., Millet, E., Feldman, M., Fahima, T. *et al.* (2004) *Triticum dicoccoides*: an important genetic resource for increasing zinc and iron concentration in modern cultivated wheat. *Soil Science and Plant Nutrition* 50, 1047–1054. Available at: https://doi.org/10.1080/00380768.2004.10408573

Cakmak, I., Pfeiffer, W.H. and McClafferty, B. (2010) Biofortification of durum wheat with zinc and iron. *Cereal Chemistry* 87, 10–20. Available at: https://doi.org/10.1094/CCHEM-87-1-0010

Cantu, D., Pearce, S.P., Distelfeld, A., Christiansen, M.W., Uauy, C. *et al.* (2011) Effect of the down-regulation of the *High Grain Protein Content* (*GPC*) genes on the wheat transcriptome during monocarpic senescence. *BMC Genomics* 12, 492. Available at: https://doi.org/10.1186/1471-2164-12-492

Chang, S., El Arifeen, S., Bari, S., Wahed, M.A. *et al.* (2010) Supplementing iron and zinc: double blind, randomized evaluation of separate or combined delivery. *European Journal of Clinical Nutrition* 64, 153–160. Available at: https://doi.org/10.1038/ejcn.2009.127

Charcosset, A. (1997) Marker-assisted introgression of quantitative trait loci. *Genetics* 147, 1469–1485.

Chattha, M.U., Hassan, M.U., Khan, I., Chattha, M.B., Mahmood, A. *et al.* (2017) Biofortification of wheat cultivars to combat zinc deficiency. *Frontiers in Plant Science* 8, 281. Available at: https://doi.org/10.3389/fpls.2017.00281

CIMMYT (2019) The case for rushing farmer access to BARI Gom 33. International Maize and Wheat Improvement Center, Mexico City. Available at: https://www.cimmyt.org/news/the-case-for-rushing-farmer-access-to-bari-gom-33/ (accessed 2 December 2019).

Connorton, J.M., Balk, J. and Rodriguez-Celma, J. (2017a) Iron homoestasis in plants – a brief overview. *Metallomics* 9, 813–823. Available at: https://doi.org/10.1039/C7MT00136C

Connorton, J.M., Jones, E.R., Rodríguez-Ramiro, I., Fairweather-Tait, S., Uauy, C. and Balk, J. (2017b) Wheat vacuolar iron transporter TaVIT2 transports Fe and Mn and is effective for biofortification. *Plant Physiology* 174, 2434–2444. Available at: https://doi.org/10.1104/pp.17.00672

Crespo-Herrera, L.A., Govindan, V., Stangoulis, J., Hao, Y. and Singh, R.P. (2017) QTL mapping of grain Zn and Fe concentrations in two hexaploid wheat RIL populations with ample transgressive segregation. *Frontiers in Plant Science* 8, 1–12. Available at: https://doi.org/10.3389/fpls.2017.01800

De Brier, N., Gomand, S.V., Donner, E., Paterson, D., Smolders, E., Delcour, J.A. and Lombi, E. (2016) Element distribution and iron speciation in mature wheat grains (*Triticum aestivum* L.) using synchrotron X-ray fluorescence near-edge structure (XANES) imaging. *Plant, Cell & Environment* 39, 1835–1847. Available at: https://doi.org/10.1111/pce.12749

Deinlein, U., Weber, M., Schmidt, H., Rensch, S., Trampczynska, A. *et al.* (2012) Elevated nicotianamine levels in *Arabidopsis halleri* roots play a key role in zinc hyper accumulation. *The Plant Cell* 24, 708–723. Available at: https://doi.org/10.1105/tpc.111.095000

Distelfeld, A., Uauy, C., Fahima, T. and Dubcovsky, J. (2006) Physical map of the wheat high-grain protein content gene *Gpc-B1* and development of a high-throughput molecular marker. *New Phytologist* 169, 753–763. Available at: https://doi.org/10.1111/j.1469-8137.2005.01627.x

Distelfeld, A., Cakmak, I., Peleg, Z., Ozturk, L., Yazici, A.M. *et al.* (2007) Multiple QTL-effects of wheat *Gpc-B1* locus on grain protein and micronutrient concentrations. *Physiologia Plantarum* 129, 635–643. Available at: https://doi.org/10.1111/j.1399-3054.2006.00841.x

Duan, J., Xia, C., Zhao, G., Jia, J. and Kong, X. (2012) Optimizing *de novo* common wheat transcriptome assembly using short-read RNA-Seq data. *BMC Genomics* 13, 392. Available at: https://doi.org/10.1186/1471-2164-13-392

Eagling, T., Neal, A.L., McGrath, S.P., Fairweather-Tait, S., Shewry, P.R. and Zhao, F.J. (2014a) Distribution and speciation of iron and zinc in grain of two wheat genotypes. *Journal of Agricultural and Food Chemistry* 62, 708–716. Available at: https://doi.org/10.1021/jf403331p

Eagling, T., Wawer, A.A., Shewry, P.R., Zhao, F.J. and Fairweather-Tait, S.J. (2014b) Iron bioavailability in two commercial cultivars of wheat: comparison between wholegrain and white flour and the effects of nicotianamine and 2′-deoxymugineic acid on iron uptake into Caco-2 cells. *Journal of Agricultural and Food and Chemistry* 62, 10320–10325. Available at: https://doi.org/10.1021/jf5026295

Erdal, I., Yilmaz, A., Taban, S., Eker, S., Torun, B. and Cakmak, I. (2002) Phytic acid and phosphorus concentrations in seeds of wheat cultivars grown with and without zinc fertilization. *Journal of Plant Nutrition* 25, 113–127.

Erenoglu, B., Cakmak, I., Romheld, V., Derici, R. and Rengel, Z. (1999) Uptake of zinc by rye, bread wheat and durum wheat cultivars differing in zinc efficiency. *Plant and Soil* 209, 245–252.

Erenoglu, E.B., Kutman, U.B., Ceylan, Y., Yildiz, B. and Cakmak, I. (2011) Improved nitrogen nutrition enhances root uptake, root-to-shoot translocation and remobilization of zinc ($^{65}Zn$) in wheat. *New Phytologist* 189, 438–448. Available at: https://doi.org/10.1111/j.1469-8137.2010.03488.x

Fang, Y., Wang, L., Xin, Z., Zhao, L., An, X. and Hu, Q. (2008) Effect of foliar application of zinc, selenium, and iron fertilizers on nutrients concentration and yield of rice grain in China. *Journal of Agricultural and Food Chemistry* 56, 2079–2084. Available at: https://doi.org/10.1021/jf800150z

FAO (2003) *Food Outlook,* November 2010. Global Information and Early Warning System on Food and Agriculture. Food and Agriculture Organization of the United Nations, Rome. Available at: http://www.fao.org/3/al993e/al993e00.pdf (accessed 11 November 2019).

Fardart, A. (2010) New hypotheses for the health-protective mechanisms of whole-grain cereals: what is beyond fibre? *Nutrition Research Reviews* 23(1), 65–134. Available at: https://doi.org/10.1017/S0954422410000041

Farkas, A., Molnár, I., Dulai, S., Rapi, S., Oldal, V. *et al.* (2014) Increased micronutrient content (Zn, Mn) in the $3M^b$(4B) wheat–*Aegilops biuncialis* substitution and $3M^b$.4BS translocation identified by GISH and FISH. *Genome* 57, 61–67. Available at: http://dx.doi.org/10.1139/gen-2013-0204

Garcia-Oliveira, A.L., Chander, S., Ortiz, R., Menkir, A. and Gedil, M. (2018) Genetic basis and breeding perspectives of grain iron and zinc enrichment in cereals. *Frontiers in Plant Science* 9, 937. Available at: https://doi.org/10.3389/fpls.2018.00937

Garg, M., Chawla, M., Chunduri, V., Kumar, R., Sharma, S. *et al.* (2016) Transfer of grain colors to elite wheat cultivars and their characterization. *Journal of Cereal Science* 71, 138–144. Available at: https://doi.org/10.1016/j.jcs.2016.08.004

Garg, M., Sharma, N., Sharma, S., Kapoor, P., Kumar, A., Chunduri, V. and Arora, P. (2018) Biofortified crops generated by breeding, agronomy and transgenic approaches are improving lives of millions of people around the world. *Frontiers in Nutrition* 5, 1–33. Available at: https://dx.doi.org/10.3389/fnut.2018.00012

Genc, Y., Verbyla, A.P., Torun, A.A., Cakmak, I., Willsmore, K., Wallwork, H. and McDonald, G.K. (2009) Quantitative trait loci analysis of zinc efficiency and grain zinc concentration in wheat using whole genome average interval mapping. *Plant and Soil* 314, 49. Available at: http://dx.doi.org/10.1007/s11104-008-9704-3

Ghandilyan, A., Vreugdenhil, D. and Aarts, M.G.M. (2006) Progress in the genetic understanding of plant iron and zinc nutrition. *Physiologia Plantarum* 126, 407–417. Available at: https://doi.org/10.1111/j.1399-3054.2006.00646.x

Gomez-Becerra, H.F., Erdem, H., Yazici, A., Tutus, Y., Torun, B., Ozturk, L. and Cakmak, I. (2010a) Grain concentrations of protein and mineral nutrients in a large collection of spelt wheat grown under different environments. *Journal of Cereal Science* 52, 342–349.

Gomez-Becerra, H.F., Yazici, A., Ozturk, L., Budak, H., Peleg, Z. *et al.* (2010b) Genetic variation and environmental stability of grain mineral nutrient concentrations in *Triticum dicoccoides* under five environments. *Euphytica* 171, 39–52. Available at: https://doi.org/10.1007/s10681-009-9987-3

Graham, R.D., Senadhira, D., Beebe, S., Iglesias, C. and Monasterio, I. (1999) Breeding for micronutrient density in edible portions of staple food crops conventional approaches. *Field Crops Research* 60, 57–80.

Graham, R.D., Welch, R.M. and Bouis, H.E. (2001) Addressing micronutrient malnutrition through enhancing the nutritional quality of staple foods: principles, perspectives and knowledge gaps. *Advances in Agronomy* 70, 77–142. Available at: https://doi.org/10.1016/S0065-2113(01)70004-1

Graham, R.D., Welch, R.M., Saunders, D.A., Ortiz-Monasterio, I., Bouis, H.E. *et al.* (2007) Nutritious subsistence food systems. *Advances in Agronomy* 92, 1–74.

Graham, R.D., Knez, M. and Welch, R.M. (2012) How much nutritional iron deficiency in humans globally is due to an underlying zinc deficiency? *Advances in Agronomy* 115, 1–40. Available at: https://doi.org/10.1016/B978-0-12-394276-0.00001-9

Grusak, M. (2002) Enhancing mineral content in plant food products. *Journal of the American College of Nutrition* 21(sup3), 178S–183S. Available at: https://doi.org/10.1080/07315724.2002.10719263

Guttieri, M., Bowen, D., Dorsch, J.A., Raboy, V. and Souza, E. (2004) Identification and characterization of a low phytic acid wheat. *Crop Science* 44, 418–424.

Habib, M. (2009) Effect of foliar application of Zn and Fe on wheat yield and quality. *African Journal of Biotechnology* 8, 6795–6798. Available at: https://academicjournals.org/article/article1380103701_Habib.pdf

Hao, M.D., Wei, X.R. and Dang, T.H. (2003) Effect of long-term applying zinc fertilizer on wheat yield and zinc absorption by wheat in dryland. *Plant Nutrition and Fertilizing Science* 9, 377–380.

Hao, Y., Velu, G., Peña, R.J., Singh, S. and Singh, R.P. (2014) Genetic loci associated with high grain zinc concentration and pleiotropic effect on kernel weight in wheat (*Triticum aestivum* L.). *Molecular Breeding* 34, 1893–1902.

Harris, D., Rashid, A., Miraj, G., Arif, M. and Shah, H. (2007) On-farm seed priming with zinc sulphate solution – a cost-effective way to increase the maize yields of resource-poor farmers. *Field Crops Research* 102, 119–127.

Hasan, M.M., Rafii, M.Y., Ismail, M.R., Mahmood, M., Rahim, H.A. *et al.* (2015) Marker-assisted backcrossing: a useful method for rice improvement. *Biotechnology & Biotechnological Equipment* 29, 237–254. Available at: https://doi.org/10.1080/13102818.2014.995920

Haslett, B.S., Reid, R.J. and Rengel, Z. (2001) Zinc mobility in wheat: uptake and distribution of zinc applied to leaves or roots. *Annals of Botany* 87(3), 379–386. Available at: https://doi.org/10.1006/anbo.2000.1349

Haydon, M.J., Kawachi, M., Wirtz, M., Hillmer, S., Hell, R. and Kramer, U. (2012) Vacuolar nicotianamine has critical and distinct roles under iron deficiency and for zinc sequestration in *Arabidopsis*. *The Plant Cell* 24, 724–737. Available at: https://doi.org/10.1105/tpc.111.095042

Hospital, F. (2001) Size of donor chromosome segments around introgressed loci and reduction of linkage drag in marker-assisted backcross programs. *Genetics* 158, 1363–1379. Available at: https://www.ncbi.nlm.nih.gov/pmc/articles/PMC1461714/

Hurrell, R., Ranum, P., de Pee, S., Biebinger, R., Hulthen, L., Johnson, Q. and Lynch, S. (2010) Revised recommendations for iron fortification of wheat flour and an evaluation of the expected impact of current national wheat flour fortification programs. *Food and Nutrition Bulletin* 31(1 Suppl.), S7–S21.

IARI (2019) Annual Report 2019. Indian Agricultural Research Institute, New Delhi. Available at: https://www.iari.res.in/files/Publication/annual_report/IARIAnnualReportRevised2019_20_30092020.pdf (accessed 15 February 2021).

Inoue, H., Kobayashi, T., Nozoye, T., Takahashi, M., Kakei, Y. *et al.* (2009) Rice OsYSL15 is an iron-regulated iron(III)-deoxymugineic acid transporter expressed in the roots and is essential for iron uptake in early growth of the seedlings. *Journal of Biological Chemistry* 284, 3470–3479. Available at: https://doi.org/10.1074/jbc.M806042200

Jansen, R.C., Van Ooijen, J.W., Stam, P., Lister, C. and Dean, C. (1995) Genotype-by-environment interaction in genetic mapping of multiple quantitative trait loci. *Theoretical and Applied Genetics* 91, 33–37. Available at: https://doi.org/10.1007/BF00220855

Jiang, W., Struik, P.C., Van, K.H., Zhao, M., Jin, L.N. and Stomph, T.J. (2008) Does increased zinc uptake enhance grain zinc mass concentration in rice? *Annals of Applied Biology* 153, 135–147. Available at: https://doi.org/10.1111/j.1744-7348.2008.00243.x

Jiao, Y., Grant, C.A. and Bailey, L.D. (2004) Effects of phosphorus and zinc fertilizer on cadmium uptake and distribution in flax and durum wheat. *Journal of the Science of Food and Agriculture* 84, 777–785.

Johnson, A.A., Kyriacou, B., Callahan, D.L., Carruthers, L., Stangoulis, J., Lombi, E. and Tester, M. (2011) Constitutive overexpression of the *OsNAS* gene family reveals single-gene strategies for effective iron-and zinc-biofortification of rice endosperm. *PLoS One* 6(9), 24476. Available at: https://doi.org/10.1371/journal.pone.0024476

Joppa, L.R., Du, C.H., Hart, G.E. and Hareland, G.A. (1997) Mapping gene(s) for grain protein in tetraploid wheat (*Triticum turgidum* L.) using a population of recombinant inbred chromosome lines. *Crop Science* 37, 1586–1589.

Joshi, A.K., Crossa, I., Arun, B., Chand, R., Trethowan, R., Vargas, M. and Ortiz-Monasterio, I. (2010) Genotype × environment interaction for zinc and iron concentration of wheat grain in eastern Gangetic plains of India. *Field Crops Research* 116, 268–277. Available at: https://doi.org/10.1016/j.fcr.2010.01.004

Joy, E.J.M., Ahmad, W., Zia, M.H., Kumssa, D.B., Young, S.D. *et al.* (2016) Valuing increased zinc (Zn) fertilizer-use in Pakistan. *Plant and Soil* 411, 139–150.

Knutson, M.D. (2017) Iron transport proteins: gateways of cellular and systemic iron homeostasis. *Journal of Biological Chemistry* 292(31), 12735–12743. Available at: https://doi.org/10.1074/jbc.R117.786632

Koide, Y., Kobayashi, N., Xu, D. and Fukuta, Y. (2009) Resistance genes and selection DNA markers for blast disease in rice (*Oryza sativa* L.). *Japan Agricultural Research Quarterly* 43, 255–280. Available at: https://doi.org/10.6090/jarq.43.255

Krasileva, K., Buffalo, V., Bailey, P., Pearce, S., Ayling, S. *et al.* (2013) Separating homeologs by phasing in the tetraploid wheat transcriptome. *Genome Biology* 14, R66. Available at: https://doi.org/10.1186/gb-2013-14-6-r66

Krishnappa, G., Singh, A.M., Chaudhary, S., Ahlawat, A.K., Singh, S.K. *et al.* (2017) Molecular mapping of the grain iron and zinc concentration, protein content and thousand kernel weight in wheat (*Triticum aestivum* L.). *PLoS One* 12, e0174972. Available at: https://doi.org/10.1371/journal.pone.0174972

Kumar, J., Jaiswal, V., Kumar, A., Kumar, N., Mir, R.R. *et al.* (2011) Introgression of a major gene for high grain protein content in some Indian bread wheat cultivars. *Field Crops Research* 123, 226–233. Available at: https://doi.org/10.1016/j.fcr.2011.05.013

Kumar, A., Sen, A., Upadhyay, P.K. and Singh, R.K. (2017) Effect of zinc, iron and manganese levels on quality, micro and macro nutrients content of rice and their relationship with yield. *Communications in Soil Science and Plant Analysis* 48(13), 1539–1551. Available at: https://doi.org/10.1080/00103624.2017.1373799

Kumari, N., Rawat, N., Tiwari, V., Prasad, R., Tripathi, S., Randhawa, G. and Dhaliwal, H. (2012) Evaluation and identification of wheat–*Aegilops* addition lines controlling high grain iron and zinc concentration and mugineic acid production. *Cereal Research Communications* 40, 53–61. Available at: https://doi.org/10.1556/CRC.40.2012.1.7

Kutman, U.B., Yildiz, B. and Cakmak, I. (2011) Improved nitrogen status enhances zinc and iron concentrations both in the whole grain and the endosperm fraction of wheat. *Journal of Cereal Science* 53, 118–125. Available at: https://doi.org/10.1016/j.jcs.2010.10.006

Lafiandra, D., Riccardi, G. and Shewry, P.R. (2014) Improving cereal grain carbohydrates for diet and health. *Journal of Cereal Science* 59, 312–326. Available at: https://doi.org/10.1016/j.jcs.2014.01.001

Lee, S., Jeon, U.S., Kim, Y.K., Persson, D.P., Husted, S. *et al.* (2009) Iron fortification of rice seeds through activation of the *nicotianamine synthase* gene. *Proceedings of the National Academy of Sciences USA* 106, 22014–22019. Available at: https://doi.org/10.1073/pnas.0910950106

Liu, D.Y., Liu, Y.M., Zhang, W., Chen, X.P. and Zou, C.Q. (2019) Zinc uptake, translocation, and remobilization in winter wheat as affected by soil application of Zn fertilizer. *Frontiers in Plant Science* 10, 426. Available at: https://doi.org/10.3389/fpls.2019.00426

Lu, L., Tian, S., Zhang, J., Yang, X., Labavitch, J.M. *et al.* (2013) Efficient xylem transport and phloem remobilization of Zn in the hyperaccumulator plant species *Sedum alfredii*. *New Phytologist* 198, 721–731. Available at: https://doi.org/10.1111/nph.12168

Manzeke, G.M., Mtambanengwe, F., Nezomba, H. and Mapfumo, P. (2014) Zinc fertilization influence on maize productivity and grain nutritional quality under integrated soil fertility management in Zimbabwe. *Field Crops Research* 166, 128–136.

Masuda, H., Usuda, K., Kobayashi, T., Ishimaru, Y., Kakei, Y. *et al.* (2009) Overexpression of the barley nicotianamine synthase gene *HvNAS1* increases iron and zinc concentrations in rice grains. *Rice* 2(4), 155–166. Available at: https://doi.org/10.1007/s12284-009-9031-1

Monasterio, I. and Graham, R.D. (2000) Breeding for trace minerals in wheat. *Food and Nutrition Bulletin* 21(4), 392–396. Available at: https://doi.org/10.1177/156482650002100409

Morel, M., Crouzet, J., Gravot, A., Auroy, P., Leonhardt, N., Vavasseur, A. and Richaud, P. (2009) AtHMA3, a $P_{1B}$-ATPase allowing Cd/Zn/Co/Pb vacuolar storage in *Arabidopsis*. *Plant Physiology* 149, 894–904. Available at: https://doi.org/10.1104/pp.108.130294

Morgounov, A., Gomez-Becerra, H.F., Abugalieva, A., Dzhunusova, M., Yessimbekova, M. *et al.* (2007) Iron and zinc grain density in common wheat grown in Central Asia. *Euphytica* 155, 193–203.

Murata, Y., Ma, J.F., Yamaji, N., Ueno, D., Nomoto, K. and Iwashita, T. (2006) A specific transporter for iron(III)–phytosiderophore in barley roots. *The Plant Journal* 46, 563–572. Available at: https://doi.org/10.1111/j.1365-313X.2006.02714.x

Myers, S.S., Zanobetti, A., Kloog, I., Huybers, P., Leakey, A.D. *et al.* (2014) Increasing $CO_2$ threatens human nutrition. *Nature* 510(7503), 139–142. Available at: https://doi.org/10.1038/nature13179

Neal, A.L., Geraki, K., Borg, S., Quinn, P., Mosselmans, J.F., Brinch-Pedersen, H. and Shewry, P.R. (2013) Iron and zinc complexation in wild-type and ferritin-expressing wheat grain: implications for mineral transport into developing grain. *Journal of Biological Inorganic Chemistry* 18, 557–570. Available at: https://doi.org/10.1007/s00775-013-1000-x

Neelam, K., Rawat, N., Tiwari, V.K., Kumar, S., Chhuneja, P. *et al.* (2011) Introgression of group 4 and 7 chromosomes of *Ae. peregrina* in wheat enhances grain iron and zinc density. *Molecular Breeding* 28, 623–634.

Nozoye, T., Nagasaka, S., Kobayashi, T., Takahashi, M., Sato, Y. *et al.* (2011) Phytosiderophore efflux transporters are crucial for iron acquisition in graminaceous plants. *Journal of Biological Chemistry* 286(7), 5446–5454. Available at: https://doi.org/10.1074/jbc.M110.180026

Nriagu, J., Afeiche, M., Linder, A., Arowolo, T., Ana, G. *et al.* (2008) Lead poisoning associated with malaria in children of urban areas of Nigeria. *International Journal of Hygiene and Environmental Health* 211, 591–605. Available at: https://doi.org/10.1016/j.ijheh.2008.05.001

Ortiz-Monasterio, J.I., Palacios-Rojas, N., Meng, E., Pixley, K., Trethowan, R. and Pena, R.J. (2007) Enhancing the mineral and vitamin content of wheat and maize through plant breeding. *Journal of Cereal Science* 46, 293–307. Available at: https://doi.org/10.1016/j.jcs.2007.06.005

Ortiz-Monasterio, I., Trethowan, R., Holm, P.B., Cakmak, I., Borg, S., Tauris, B.E.B. and Brinch-Pedersen, H. (2011) Breeding, transformation, and physiological strategies for the development of wheat with high zinc and iron grain concentration. In: Bonjean, A.P., Angus, W.J. and Van Ginkel, M. (eds) *The World Wheat Book, A History of Wheat Breeding*, Vol. 2. Lavoisier, Paris, pp. 951–977.

Oury, F.X., Leenhardt, F., Remesy, C., Chanliaud, E., Duperrier, B., Balfouriera, F. and Charmet, G. (2006) Genetic variability and stability of grain magnesium, zinc and iron concentration in bread wheat. *European Journal of Agronomy* 25, 177–185. Available at: https://doi.org/10.1016/j.eja.2006.04.011

Ozkan, H., Brandolini, A., Torun, A., Altintas, S., Eker, S. *et al.* (2007) Natural variation and identification of microelements content in seeds of einkorn wheat (*Triticum monococcum*). In: Buck, H.T., Nisi, J.E. and Salomon, N. (eds) *Wheat Production in Stressed Environments*. Springer, Berlin, Germany, pp. 455–462.

Pascoalino, J.A.L., Thompson, J.A., Wright, G., Franco, F.A., Scheeren, P.L. *et al.* (2018) Grain zinc concentrations differ among Brazilian wheat genotypes and respond to zinc and nitrogen supply. *PLoS One* 13(7), e0199464. Available at: https://doi.org/10.1371/journal.pone.0199464

Paux, E., Sourdille, P., Salse, J., Saintenac, C., Choulet, F. *et al.* (2008) A physical map of the 1-gigabase bread wheat chromosome 3B. *Science* 322(5898), 101–104. Available at: https://doi.org/10.1126/science.1161847

Peleg, Z., Cakmak, I., Ozturk, L., Yazici, A., Jun, Y. *et al.* (2009) Quantitative trait loci conferring grain mineral nutrient concentrations in durum wheat × wild emmer wheat RIL population. *Theoretical and Applied Genetics* 119, 353–369. Available at: https://doi.org/10.1007/s00122-009-1044-z

Pfeiffer, W.H. and McClafferty, B. (2007) HarvestPlus: breeding crops for better nutrition. *Crop Science* 47(S3), S88–S105. Available at: https://doi.org/10.2135/cropsci2007.09.0020IPBS

Ram, H., Rashid, A., Zhang, W., Duarte, A.P., Phattarakul, N. *et al.* (2016) Biofortification of wheat, rice and common bean by applying foliar zinc fertilizer along with pesticides in seven countries. *Plant and Soil* 403, 389–401. Available at: https://doi.org/10.1007/s11104-016-2815-3

Randhawa, H.S., Asif, M., Pozniak, C., Clarke, J.M., Graf, R.J. *et al.* (2013) Application of molecular markers to wheat breeding in Canada. *Plant Breeding* 132, 458–471. Available at: https://doi.org/10.1111/pbr.12057

Rawat, N., Tiwari, V.K., Neelam, K., Randhawa, G.S., Singh, K., Chhuneja, P. and Dhaliwal, H.S. (2009a) Development and characterization of *Triticum aestivum–Aegilops kotschyi* amphiploids with high grain iron and zinc. *Plant Genetic Resources* 7, 271–280. Available at: https://doi.org/10.1017/S1479262109356592

Rawat, N., Tiwari, V.K., Singh, N., Randhawa, G.S., Singh, K., Chhuneja, P. and Dhaliwal, H.S. (2009b) Evaluation and utilization of *Aegilops* and wild *Triticum* species for enhancing iron and zinc content in wheat. *Genetic Resources and Crop Evolution* 56, 53. Available at: https://doi.org/10.1007/s10722-008-9344-8

Reeves, P.G. and Chaney, R.L. (2008) Bioavailability as an issue in risk assessment and management of food cadmium: a review. *Science of the Total Environment* 398, 13–19. Available at: https://doi.org/10.1016/j.scitotenv.2008.03.009

Rellan-Alvarez, R., Giner-Martinez-Sierra, J., Orduna, J., Orera, I., Rodriguez-Castrillon, J.A. *et al.* (2010) Identification of a tri-iron(III), tri-citrate complex in the xylem sap of iron-deficient tomato resupplied with iron: new insights into plant iron long-distance transport. *Plant & Cell Physiology* 51, 91–102. Available at: https://doi.org/10.1093/pcp/pcp170

Römheld, V. and Marschner, H. (1986) Evidence for a specific uptake system for iron phytosiderophores in roots of grasses. *Plant Physiology* 80, 175–180. Available at: https://doi.org/10.1104/pp.80.1.175

Saintenac, C., Jiang, D. and Akhunov, E.D. (2011) Targeted analysis of nucleotide and copy number variation by exon capture in allotetraploid wheat genome. *Genome Biology* 12(9), R88. Available at: https://doi.org/10.1186/gb-2011-12-9-r88

Schlegel, R., Cakmak, I., Torun, B., Eker, S., Tolay, I. *et al.* (1998) Screening for zinc efficiency among wheat relatives and their utilisation for alien gene transfer. *Euphytica* 100, 281–286. Available at: https://doi.org/10.1023/A:1018376827876

Sheikh, I., Sharma, P., Verma, S.K., Kumar, S., Kumar, N. *et al.* (2018) Development of intron targeted amplified polymorphic markers of metal homeostasis genes for monitoring their introgression from *Aegilops* species to wheat. *Molecular Breeding* 38, 47. Available at: https://doi.org/10.1007/s11032-018-0809-y

Shi, R.L., Li, H.W., Tong, Y.P., Jing, R.L., Zhang, F.S. and Zou, C.Q. (2008) Identification of quantitative trait locus of zinc and phosphorus density in wheat (*Triticum aestivum* L.) grain. *Plant and Soil* 306, 95–104. Available at: https://doi.org/10.1007/s11104-007-9483-2

Shivay, Y.S., Kumar, D. and Prasad, R. (2008) Effect of zinc-enriched urea on productivity, zinc uptake and efficiency of an aromatic rice–wheat cropping system. *Nutrient Cycling in Agroecosystems* 81, 229–243. Available at: https://doi.org/10.1007/s10705-007-9159-6

Shukla, S.K. and Warsi, A.S. (2000) Effect of sulphur and micronutrients on growth, nutrient content and yield of wheat (*Triticum aestivum* L). *Indian Journal of Agricultural Research* 34, 203–205.

Singh, K., Chhuneja, P., Tiwari, V.K., Rawat, N., Neelam, K. *et al.* (2010) Mapping of QTL for grain iron and zinc content in diploid A genome wheat and validation of these loci in U and S genomes. In: *Proceedings of the Plant and Animal Genomes XVIII Conference, San Diego, California, 9–13 January 2010*, pp. 9–13.

Singh, M.V. (2008) Micronutrient deficiencies in crops and soils in India. In: Brown, A. (ed.) *Micronutrient Deficiencies in Global Crop Production.* Springer, New York, pp. 93–125.

Singh, R. and Govindan, V. (2017) Zinc-biofortified wheat: harnessing genetic diversity for improved nutritional quality. *Science Brief: Biofortification* No. 1, May 2017. CIMMYT, HarvestPlus and Global Crop Diversity Trust, Bonn, Germany.

Singh, S.P., Vogel-Mikus, K., Arcon, I., Vavpetic, P., Jeromel, L. *et al.* (2013) Pattern of iron distribution in maternal and filial tissues in wheat grains with contrasting levels of iron. *Journal of Experimental Botany* 64(11), 3249–3260. Available at: https://doi.org/10.1093/jxb/ert160

Singh, S.P., Keller, B., Gruissem, W. and Bhullar, N.K. (2017) Rice *NICOTIANAMINE SYNTHASE 2* expression improves dietary iron and zinc levels in wheat. *Theoretical and Applied Genetics* 130, 283–292. Available at: https://doi.org/10.1007/s00122-016-2808-x

Sperotto, R.A., Ricachenevsky, F.K., Waldow, V.D. and Fett, J.P. (2012) Iron biofortification in rice: it's a long way to the top. *Plant Science* 190, 24–39. Available at: https://doi.org/10.1016/j.plantsci.2012.03.004

Srinivasa, J., Arun, B. Mishra, V.K., Singh, G.P., Velu, G. *et al.* (2014) Zinc and iron concentration QTL mapped in a *Triticum spelta* × *T. aestivum* cross. *Theoretical and Applied Genetics* 127, 1643–1651. Available at: https://doi.org/10.1007/s00122-014-2327-6

Stomph, T.J., Wen, J. and Struik, P.C. (2009) Zinc biofortification of cereals: rice differs from wheat and barley. *Trends in Plant Science* 14, 123–124. Available at: https://doi.org/10.1016/j.tplants.2009.01.001

Sui, X., Zhao, Y., Wang, S., Duan, X., Xu, L. and Liang, R. (2012) Improvement Fe content of wheat (*Triticum aestivum*) grain by soybean ferritin expression cassette without vector backbone sequence. *Journal of Agricultural Biotechnology* 20, 766–773.

Tabbita, F., Lewis, S., Vouilloz, J.P., Ortega, M.A., Kade, M., Abbate, P.E. and Barneix, A.J. (2013) Effects of the *Gpc-B1* locus on high grain protein content introgressed into Argentinean wheat germplasm. *Plant Breeding* 132, 48–52. Available at: https://doi.org/10.1111/pbr.12011

Tang, J., Zou, C., He, Z., Shi, R., Ortiz-Monasterio, I., Qu, Y. and Zhang, Y. (2008) Mineral element distributions in milling fractions of Chinese wheats. *Journal of Cereal Science* 48, 821–828.

Tauris, B., Borg, S., Gregersen, P.L. and Holm, P.B. (2009) A roadmap for zinc trafficking in the developing barley grain based on laser capture microdissection and gene expression profiling. *Journal of Experimental Botany* 60, 1333–1347. Available at: https://doi.org/10.1093/jxb/erp023

Tighe, P., Duthie, G., Vaughan, N., Brittenden, J., Simpson, W.G. *et al.* (2010) Effect of increased consumption of whole-grain foods on blood pressure and other cardiovascular risk markers in healthy middle-aged persons: a randomized controlled trial. *American Journal of Clinical Nutrition* 92(4), 733–740. Available at: https://doi.org/10.3945/ajcn.2010.29417

Tisdale, S.L. and Nelson, W.L. (1975) *Soil Fertility and Fertilizer*, 3rd edn. Macmillan, New York.

Tiwari, C., Wallwork, H., Arun, B., Mishra, V.K., Velu, G. *et al.* (2016) Molecular mapping of quantitative trait loci for zinc, iron and protein content in the grains of hexaploid wheat. *Euphytica* 207, 563–570.

Tiwari, V.K., Rawat, N., Neelam, K., Randhawa, G.S., Singh, K., Chhuneja, P. and Dhaliwal, H.S. (2008) Development of *Triticum turgidum* subsp. durum *Aegilops longissima* amphiploids with high iron and zinc content through unreduced gamete formation in $F_1$ hybrids. *Genome* 51, 757–766. Available at: https://doi.org/10.1139/G08-057

Tiwari, V.K., Rawat, N., Chhuneja, P., Neelam, K., Aggarwal, R. *et al.* (2009) Mapping of quantitative trait loci for grain iron and zinc concentration in diploid A genome wheat. *Journal of Heredity* 100, 771–776.

Tiwari, V.K., Rawat, N., Neelam, K., Kumar, S., Randhawa, G.S. and Dhaliwal, H.S. (2010) Substitutions of 2S and 7U chromosomes of *Aegilops kotschyi* in wheat enhance grain iron and zinc concentration. *Theoretical Applied Genetics* 121, 259–269. Available at: https://doi.org/10.1007/s00122-010-1307-8

Trethowan, R.M., Reynolds, M.P., Sayre, K.D. and Ortiz-Monasterio, I. (2005) Adapting wheat cultivars to resource conserving farming practices and human nutritional needs. *Annals of Applied Biology* 146, 404–413. Available at: https://doi.org/10.1111/j.1744-7348.2005.040137.x

Trethowan, R.M., Reynolds, M.P., Ortiz-Monasterio, I. and Ortiz, R. (2007) The genetic basis of the green revolution in wheat production. *Plant Breeding Reviews* 28, 39–58. Available at: https://doi.org/10.1002/9780470168028.ch2

Trick, M., Adamski, N.M., Mugford, S.G., Jiang, C.C., Febrer, M. and Uauy, C. (2012) Combining SNP discovery from next-generation sequencing data with bulked segregant analysis (BSA) to fine-map genes in polyploid wheat. *BMC Plant Biology* 12, 14. Available at: https://doi.org/10.1186/1471-2229-12-14

Uauy, C., Distelfeld, A., Fahima, T., Blechl, A. and Dubcovsky, J. (2006) A NAC gene regulating senescence improves grain protein, zinc, and iron content in wheat. *Science* 314, 1298–1301. Available at: https://doi.org/10.1126/science.1133649

Velu, G., Ortiz-Monasterio, I., Singh, R.P. and Payne, T. (2011a) Variation for grain micronutrients concentration in wheat core-collection accessions of diverse origin. *Asian Journal of Crop Science* 3, 43–48. Available at: https://doi.org/10.3923/ajcs.2011.43.48

Velu, G., Singh, R.P., Huerta-Espino, J. and Peña, R.J. (2011b) Breeding for enhanced zinc and iron concentration in CIMMYT spring wheat germplasm. *Czech Journal of Genetics and Plant Breeding* 47, S174–S177. Available at: https://doi.org/10.17221/3275-CJGPB

Velu, G., Singh, R.P., Huerta-Espino, J., Peña-Bautista, R.J., Arun, B. *et al.* (2012) Performance of biofortified spring wheat genotypes in target environments for grain zinc and iron concentrations. *Field Crops Research* 137, 261–267. Available at: https://doi.org/10.1016/j.fcr.2012.07.018

Velu, G., Ortiz-Monasterio, I., Cakmak, I., Hao, Y. and Singh, R.P. (2014) Biofortification strategies to increase grain zinc and iron concentrations in wheat. *Journal of Cereal Science* 59, 365–372. Available at: https://doi.org/10.1016/j.jcs.2013.09.001

Velu, G., Singh, R., Balasubramaniam, A., Mishra, V.K., Chand, R. *et al.* (2015) Reaching out to farmers with high zinc wheat varieties through public–private partnerships: an experience from eastern-Gangetic plains of India. *Advances in Food Technology and Nutritional Sciences* 1(3), 73–75. Available at: https://doi.org/10.17140/AFTNSOJ-1-112

Velu, G., Guzman, C., Mondal, S., Autrique, J.E., Huerta, J. and Singh, R.P. (2016) Effect of drought and elevated temperature on grain zinc and iron concentrations in CIMMYT spring wheat. *Journal of Cereal Science* 69, 182–186. Available at: https://doi.org/10.1016/j.jcs.2016.03.006

Velu, G., Singh, R.P., Crespo-Herrera, L., Juliana, P., Dreisigacker, S. *et al.* (2018) Genetic dissection of grain zinc concentration in spring wheat for mainstreaming biofortification in CIMMYT wheat breeding. *Scientific Reports* 8, 13526. Available at: https://doi.org/10.1038/s41598-018-31951-z

Walker, E.L. and Connolly, E.L. (2008) Time to pump iron: iron-deficiency-signaling mechanisms of higher plants. *Current Opinion in Plant Biology* 11, 530–535. Available at: https://doi.org/10.1016/j.pbi.2008.06.013

Wang, K., Gao, L., Wang, S., Zhang, Y., Li, X., Zhang, M. *et al.* (2011) Phylogenetic relationship of a new class of LMW-GS genes in the M genome of *Aegilops comosa*. *Theoretical and Applied Genetics* 122(7), 411–425.

Welch, R.M. (1999) Importance of seed mineral nutrient reserves in crop growth and development. In: Rengel, Z. (ed.) *Mineral Nutrition of Crops: Fundamental Mechanisms and Implications*. Food Products Press, New York, pp. 205–226.

Welch, R.M. and Graham, R.D. (2004) Breeding for micronutrients in staple food crops from a human nutrition perspective. *Journal of Experimental Botany* 55, 353–364. Available at: https://doi.org/10.1093/jxb/erh064

Welch, R.M., House, W.A., Ortiz-Monasterio, I. and Cheng, Z. (2005) Potential for improving bioavailable zinc in wheat grain (*Triticum* species) through plant breeding. *Journal of Agricultural and Food Chemistry* 53, 2176–2180. Available at: https://doi.org/10.1021/jf040238x

White, P.J. and Broadley, M.R. (2005) Biofortifying crops with essential mineral elements. *Trends in Plant Science* 10, 586–593. Available at: https://doi.org/10.1016/j.tplants.2005.10.001

White, P.J. and Broadley, M.R. (2009) Biofortification of crops with seven mineral elements often lacking in human diets – iron, zinc, copper, calcium, magnesium, selenium and iodine. *New Phytologist* 182, 49–84.

WHO (2012) World Health Statistics 2012. World Health Organization, Geneva, Switzerland. Available at: https://www.who.int/gho/publications/world_health_statistics/EN_WHS2012_Full.pdf (accessed 20 November 2019).

WHO (2013) World Health Report 2013: Research for Universal Health Coverage. World Health Organization, Geneva, Switzerland. Available at: https://apps.who.int/iris/bitstream/handle/10665/85761/9789240690837_eng.pdf?sequence=2 (accessed 5 December 2019).

WHO (2017) The World Health Statistics 2017: Monitoring Health for the SDGs. World Health Organization, Geneva, Switzerland. Available at: https://www.who.int/gho/publications/world_health_statistics/2017/EN_WHS2017_TOC.pdf?ua=1 (accessed 2 December 2019).

Wilkinson, P., Winfield, M., Barker, G., Allen, A., Burridge, A., Coghill, J. and Edwards, K.J. (2012) CerealsDB 2.0: an integrated resource for plant breeders and scientists. *BMC Bioinformatics* 13, 219. Available at: https://doi.org/10.1186/1471-2105-13-219

Winfield, M.O., Wilkinson, P.A., Allen, A.M., Barker, G.L.A., Coghill, J.A. *et al.* (2012) Targeted re-sequencing of the allohexaploid wheat exome. *Plant Biotechnology Journal* 10, 733–742. Available at: https://doi.org/10.1111/j.1467-7652.2012.00713.x

Wu, T.Y., Gruissem, W. and Bhullar, N.K. (2018) Targeting intracellular transport combined with efficient uptake and storage significantly increases grain iron and zinc levels in rice. *Plant Biotechnology Journal* 17, 9–20. Available at: https://doi.org/10.1111/pbi.12943

Xu, Y.B. and Crouch, J.H. (2008) Marker-assisted selection in plant breeding: from publications to practice. *Crop Science* 48, 391–407.

Xu, Y.F., An, D.G., Liu, D.C., Zhang, A.M., Xu, H.X. and Li, B. (2012) Molecular mapping of QTLs for grain zinc, iron and protein concentration of wheat across two environments. *Field Crops Research* 138, 57–62.

Yadava, D.K., Choudhury, P.R., Hossain, F. and Kumar, D. (2017) *Biofortified Varieties: Sustainable Way to Alleviate Malnutrition*. Indian Council of Agricultural Research, New Delhi.

Yang, D.L., Jing, R.L., Chang, X.P. and Li, W. (2007) Identification of quantitative trait loci and environmental interactions for accumulation and remobilization of water-soluble carbohydrates in wheat (*Tiiticum aestivum* L.) stems. *Genetics* 176, 571–584.

Zaharieva, T. and Römheld, V. (2000) Specific $Fe^{2+}$ uptake system in strategy I plants inducible under Fe deficiency. *Journal of Plant Nutrition* 23, 1733–1744. Available at: https://doi.org/10.1080/01904160009382137

Zayed, B.A., Salem, A.K.M. and Sharkawy, H.M. (2011) Effect of different micronutrient treatments on rice (*Oryza sativa* L.) growth and yield under saline soil conditions. *World Journal of Agricultural Sciences* 7, 179–184.

Zee, S.Y. (1971) Vascular tissue and transfer cell distribution in the rice spikelet. *Australian Journal of Biological Sciences* 25, 411–414.

Zee, S.Y. and O'Brien, T.P. (1970) A special type of tracheary element associated with xylem-discontinuity in floral axis of wheat. *Australian Journal of Biological Sciences* 23, 783–791.

Zeidan, M.S., Mohamed, M.F. and Hamouda, H.A. (2010) Effect of foliar fertilization of Fe, Mn and Zn on wheat yield and quality in low sandy soils fertility. *World Journal of Agricultural Sciences* 6, 696–699.

Zhang, K.P., Tian, J.C., Zhao, L. and Wang, S.S. (2008) Mapping QTLs with epistatic effects and QTL × environment interactions for plant height using a doubled haploid population in cultivated wheat. *Journal of Genetics and Genomics* 35, 119–127. Available at: https://doi.org/10.1016/S1673-8527(08)60017-X

Zhang, K.P., Tian, J.C., Zhao, L., Liu, B. and Chen, G.F. (2009) Detection of quantitative trait loci for heading date based on the doubled haploid progeny of two elite Chinese wheat cultivars. *Genetica* 135, 257–265. Available at: http://doi.org/10.1007/s10709-008-9274-6

Zhang, Y., Song, Q., Jan, Y., Tang, J., Zhao, R. *et al.* (2010) Mineral element concentrations in grains of Chinese wheat cultivars. *Euphytica* 174, 303–313.

Zhang, Y.Q., Sun, Y.X., Ye, Y.L., Karim, M.R., Xue, Y.F. *et al.* (2012) Zinc biofortification of wheat through fertilizer applications in different locations of China. *Field Crops Research* 125, 1–7. Available at: https://doi.org/10.1016/j.fcr.2011.08.003

Zhao, F.J., Su, Y.H., Dunham, S.J., Rakszegi, M., Bedo, Z., McGrath, S.P. and Shewry, P.R. (2009) Variation in mineral micronutrient concentrations in grain of wheat lines of diverse origin. *Journal of Cereal Science* 49, 290–295. Available at: https://doi.org/10.1016/j.jcs.2008.11.007

Zheng, L., Cheng, Z., Ai, C., Jiang, X., Bei, X. *et al.* (2010). Nicotianamine, a novel enhancer of rice iron bioavailability to humans. *PLoS One* 5(4), 10190. Available at: https://doi.org/10.1371/journal.pone.0010190

Zhu, X., Chen, S., Yang, J., Zhou, S., Zeng, L. *et al.* (2012) The identification of *Pi50*(t), a new member of the rice blast resistance *Pi2*/*Pi9* multigene family. *Theoretical and Applied Genetics* 124, 1295–1304. Available at: https://doi.org/10.1007/s00122-012-1787-9

Zou, C.Q., Zhang, Y.Q., Rashid, A., Ram, H., Savasli, E. *et al.* (2012) Biofortification of wheat with zinc through zinc fertilization in seven countries. *Plant and Soil* 361, 119–130. Available at: https://doi.org/10.1007/s11104-012-1369-2

# 9 Recent Advancements of Molecular Breeding and Functional Genomics for Improving Nitrogen-, Phosphorus- and Potassium-Use Efficiencies in Wheat

**T. Maharajan[1], G. Victor Roch[1] and S. Antony Ceasar[1,2]***
[1]*Loyola College, Chennai, India;* [2]*Rajagiri College of Social Sciences, Kochi, India*

## 9.1 Introduction

Among the cereals, wheat (*Triticum aestivum*) is the major food source for humans. It was the first ever domesticated food cereal crop and for 8000 years it has been the main food of the major civilizations of Europe, West Asia and North Africa (Fuller, 2007; Giraldo *et al.*, 2019). It belongs to grass family *Poaceae* of tribe *Triticeae* (Sramkova *et al.*, 2009). Globally, four wheat species, namely bread wheat or common wheat (*T. aestivum*; $2n = 42$, hexaploid, AABBDD), durum wheat (*Triticum durum*; $2n = 28$, tetraploid, AABB), spelt wheat (*Triticum spelta*; $2n = 42$, hexaploid, AABBDD) and dwarf Polish wheat (*Triticum polonicum*; $2n = 4x = 28$, tetraploid, AABB), are available (Vasil, 2007; Sramkova *et al.*, 2009). Of them, common wheat is the wheat species largely (95%) cultivated worldwide and used for human consumption, and the other three species cover the remaining 5% of the global wheat production (Shewry, 2009). As wheat supplies >20% of the daily protein and energy consumed globally, it is considered a principal food crop for 35% of the human population (Vasil, 2007; Shewry, 2009). Wheat-derived components are also valuable raw materials for varied food-based industries and animal feed (Shewry, 2009). Wheat contains high amounts of carbohydrate, proteins, fibres, lipids, vitamins, minerals and phytochemicals (Shewry and Hey, 2015). Besides, it has agronomic adaptability, easy grain storage and easy conversion of grain into flour for making many different foods. Therefore, it serves as a staple food crop in developed and developing regions and countries like Western Europe, West Asia, North America, Nigeria, South Africa, China, India and Mexico (Shewry and Hey, 2015). Presently, wheat is the most widely grown crop in the world, on more than 218 million hectares, and its world trade is greater than for all other crops combined (Giraldo *et al.*, 2019). Also, wheat production ranks second worldwide compared with the other major cereals.

Wheat is the second most important food crop after rice (*Oryza sativa*) as 80 million farmers depend on wheat for their livelihoods in the developing world (Giraldo *et al.*, 2019). Low-quality wheat grain is used by industry to make adhesives, paper additives and several other products including alcohol (Giraldo *et al.*, 2019). In the 'green revolution' era, numerous dwarf wheat varieties were introduced which showed very

*Corresponding author: antony_sm2003@yahoo.co.in

 *Molecular Breeding in Wheat, Maize and Sorghum: Strategies for Improving Abiotic Stress Tolerance and Yield* (eds M.A. Hossain *et al.*)
DOI: 10.1079/9781789245431.0009

good response to fertilizer application. This wheat response emphasized the significant usage of different fertilizers for high wheat production. Remarkable research work in the past has confirmed the relationship between the use of fertilizers with macro- (N, P and K) and micronutrients and higher yield. N, P and K fertilizers were applied extensively to improve the shelf-life, growth, yield and resistance to stresses in wheat and other plants (Khan *et al.*, 2015). The global wheat consumption is increasing even in countries with unfavourable climates for wheat production. It is an important crop for ensuring food security. According to estimates by the Food and Agriculture Organization of the United Nations, by 2050 the world will require an additional 198 million tonnes of wheat to meet the future global food demand; wheat production needs to be augmented by 77% in the developing countries (Sharma *et al.*, 2015). Such a significant increase in wheat yield is possible only by integrated approaches including marker-assisted breeding (MAB) methods, employing useful genes from any plants, animals or microorganisms to attain the dramatic yields, eliminating productivity losses caused by pests, pathogens and weeds (Oerke *et al.*, 1994), minimizing losses due to abiotic factors (nutrient, drought, salinity, etc.) and post-harvest spoilage during storage. This increase must focus on improving productivity on land that is already under cultivation rather than bringing new land into use by destruction of forests, grasslands, etc. (Vasil, 2003). Added to that, attaining higher yields with improved nutritional quality must also be the focus of the plant breeders. In the context of growing food demand worldwide, plant breeders, plant physiologists, biotechnologists and ecophysiologists work to investigate various new ways to improve the acquisition and use efficiencies of N, P and K in wheat.

### 9.1.1 The importance and implications of nitrogen

N is the most essential among all the nutrients used by plants for their growth, development and yield (Barker and Bryson, 2016). N is required by plants in the greatest amount, accounting for 1.5–2.0% of plant dry matter and about 16% of total plant protein (Frink *et al.*, 1999; Chen *et al.*, 2003; Alvarez *et al.*, 2012). N is also the essential component of all proteins and enzymes and is involved in many vital metabolic processes of energy transformation (Street and Kidder, 1997). Sufficient N in plants is one of the major key factors for plant metabolic systems (Ahanger and Agarwal, 2017) and crop production (Nadeem *et al.*, 2014). As N is an essential constituent of chlorophylls, it is associated closely with the photosynthetic process (Nursuaidah *et al.*, 2014). N also improves plant nutritional make-up including protein content, crop quality, forage crops, leaf, fruit and seed production (Wang *et al.*, 2003; Pettigrew, 2008; Marschner and Marschner, 2011). It also plays a very important role in the N signalling connected to the phytohormones abscisic acid (ABA), indole-3-acetic acid (IAA) and cytokinins in many plant species (Kiba *et al.*, 2011; Pavlíková *et al.*, 2012). Particularly, the metabolism and translocation of cytokinins are regulated by N status in plants (Sakakibara *et al.*, 2006; Pavlíková *et al.*, 2012). In plants, sufficient N increases the height, dry matter, protein content and yield (Bahmanyar and Ranjbar, 2008). In natural ecosystems and agriculture, N occurs in many forms such as nitrate, nitrite, ammonium, amino acids, etc. and plants take up N directly from fertilizers, air, water, rain and other available sources (Bernhard, 2010; Khajuria and Kanae, 2013). In low-N (LN) soils, optimum N fertilization has almost doubled crop production (Shaviv, 2001). Globally, ~120 million tonnes of N are used as fertilizer every year (Salim and Raza, 2020). Of this, 90% of N fertilizer is consumed by wheat, maize (*Zea mays*) and rice. N application in LN-input soils increases dry matter accumulation, tiller number (TN) and number of green leaves, plant height (PH), straw yield, number of grains per spike (GPS) and grain yield (GY) and enhances the cumulative uptake of N-P-K (Chaturvedi, 2006). Overuse of fertilizers also causes serious environmental problems. On the other hand, N deficiency is common among all plants and agronomic crops including wheat. N deficiency minimizes the biomass production and GY in wheat (Zhai and Li, 2001; Berry *et al.*, 2002; Zhai and Li, 2005). Furthermore, LN during the tillering period of wheat reduces the production of new tillers and other tillers up to three leaves will stop growing and die without producing a spike (Jeuffroy and Bouchard, 1999).

### 9.1.2 The importance and implications of phosphorus

P is one of the most indispensable macronutrients required by plants and animals for their growth, development, metabolism and other regulatory processes. P plays a key role in many vital processes of plants including seed germination, flowering and fruit formation. P is an important component of cell membranes as phospholipids, and it does a multitude of functions in energy transfer, photosynthesis, many aspects of metabolism, intracellular signalling and gene replication and expression (Baker *et al.*, 2015; Maharajan *et al.*, 2018; Roch *et al.*, 2019). After N, P is considered the next most essential nutrient contributing about 0.2% of dry weight (Theodorou and Plaxton, 1993). Nearly 97% of the absorbed P by humans is deposited in the skeleton together with Ca (Penido and Alon, 2012). P is a main structural constituent of vital biomolecules including ATP, NADPH, nucleic acids, phospholipids and sugar-phosphates involved in the metabolism of proteins and lipids, transpiration, photosynthesis and secondary metabolism (Poirier and Bucher, 2002; Baker *et al.*, 2015; Maharajan *et al.*, 2018). P constitutes up to 0.5% of cell dry weight (Nussaume *et al.*, 2011) and it affects the physiochemical and biological properties of the soil (Richardson *et al.*, 2009).

In soil, P is available in organic form existing in plant residues, microbial tissues and manures and inorganic form existing in many soils that is not readily accessible to plants due to complexation with cations like $Fe^{2+}$, $Al^{3+}$ and $Ca^{2+}$ (Baker *et al.*, 2015; Maharajan *et al.*, 2018). Globally, about 70% of arable land is P deficient, with P concentration of ~1 μmol/l against the optimum plant P requirement of ~30 μmol/l (Adhya *et al.*, 2015). Around 5.7 billion hectares of agricultural land worldwide lack P (Batjes, 1997; Heuer *et al.*, 2009). Plants absorb inorganic phosphate (Pi) from soil solution in monovalent forms ($H_2PO_4^-$) which tend to change according to the soil pH (Maharajan *et al.*, 2018; Roch *et al.*, 2019). P deficiency adversely affects the growth and yield of several crop plants including rice (Wissuwa and Ae, 2001), maize (Plenet *et al.*, 2000), wheat (Lázaro *et al.*, 2010), sorghum (*Sorghum bicolor*) (Camacho *et al.*, 2002), common bean (*Phaseolus vulgaris*) (Bonser *et al.*, 1996), soybean (*Glycine max*) (Mahamood *et al.*, 2009) and small millets (Ceasar *et al.*, 2014; Maharajan *et al.*, 2019). Pi-deficient soil reduces food production, and hence to improve crop yield farmers are forced to apply Pi fertilizers, which are very expensive. Pi fertilizers are obtained from mining rock phosphate (RP), the major source of Pi fertilizers. Annually, 50 million tonnes of Pi fertilizer are required for crop production worldwide and this non-renewable RP is projected to be exhausted in the next 100 years (Sattari *et al.*, 2012; Ceasar, 2018). Intense agricultural practices have led to a depleted Pi level in various soils which calls for a collective effort from biotechnologists, agronomists and plant breeders to develop newer crop plants with high P-acquisition efficiency (PAE) and P-use efficiency (PUE) to acclimatize under low-P (LP) soils. PAE is defined as the ability of plants to uptake Pi from the rhizosphere, and PUE is the efficiency of allocation/mobilization of P within the plant for sustaining biomass production (van de Wiel *et al.*, 2016). Among the essential N-P-K nutrients, P is the most crucial nutrient for wheat growth. P level in the soil varies from area to area and most wheat-cultivatable soil suffers from P deficiency (Ahmad *et al.*, 1992) which makes the application of P to the soil imperative for sustainable wheat production (Alam *et al.*, 2003). Pi fertilizer application enhances wheat crop growth and GPS due to an increased rate of photosynthesis (Reuter *et al.*, 1995). Although the fertilizer should be crop- and site-specific, factors like application method and application time of Pi fertilizer also affect the fertilizer's efficiency (Ahmad *et al.*, 1992). The Pi-fertilized plants show improved growth with enhanced agronomic traits and grain production (Iqbal *et al.*, 2003). P deficiency is a yield-reducing factor in wheat as well; for example, P deficiency reduces the TN and leaf area (LA) by producing smaller and fewer number of leaves (Sato and Jiang, 1996). P deficiency in wheat also causes early maturation (Hussain *et al.*, 2008), reduced root growth and its spreading pattern (Lynch and Brown, 2001). In general, P deficiency limits biological yield, PH, TN, P efficiency (PE), harvest index (HI) and overall agronomic production (Ali *et al.*, 2014).

### 9.1.3 The importance and implications of potassium

K is another important macronutrient for plant growth, contributing ~10% of the plant total dry weight (TDW) (Adams and Shin, 2014). K also regulates various biochemical and physiological mechanisms and directly influences photosynthesis due to its role in ATP formation (Maathuis, 2009; Wang, M. *et al.*, 2013). K has an essential role in enzyme activation (Hussain *et al.*, 1997), plant growth (Bukhsh *et al.*, 2011), grain production in cereals (Ahmad *et al.*, 2009) and for sustainable crop production (Anser Ali and Hussain, 2012) in agriculture. It is an osmoregulator which regulates the movement of $CO_2$ and water through stomata (Gorcek and Erdal, 2015). K is an extremely mobile macronutrient in plants that is abundantly present in the vascular region of cells and young parts of the plant. It plays a major role as a cationic inorganic element and therefore plants cannot survive without its presence (Mengel, 2007). It is also linked to many physiological processes that help in water relations, assimilation and transportation (Pettigrew, 2008; Zlatev and Lidon, 2012). Both transcription and translation are not possible without sufficient K supply to plants (Wakeel *et al.*, 2011) as K is frequently associated with the binding of transfer RNA (tRNA) to ribosomes. K also plays a vital role in minimizing the heavy-metal stress from Cd, Cr and Al (Shaviv, 2001). In cereals like wheat, K regulates the reaction of photosynthesis (Dibb and Thompson, 1985). K uptake in turn regulates the uptake of other nutrients; for example, plants require a large amount of K to increase the uptake of N. Combined application of N and K has a tremendous effect on wheat growth, GY and TN (Bundy and Andraski, 2004). As N enhances wheat vegetative growth, K increases the root growth (Deng *et al.*, 2004). Most K is stored in wheat straws and grains (Saifullah *et al.*, 2002). In low-K (LK) soils, commercial fertilizer is generally used as a supplement to fulfil the K requirement of the crop. Globally, 15 kg/ha is the average use of K, while its application rate is 0.8 kg/ha (Wakeel and Magen, 2017). Of the applied K, about 4 kg K/ha is leached down with rainwater, another 4 kg K/ha is removed after harvest and some amount is fixed in soil (Salim and Raza, 2020). Continuous cultivation of the same crop leads to low level of K and it gets lower day after day. The available K in soil is not sufficient to fulfil the demand of the crop. Until now K has received very little attention and/or is ignored. Since plants require a fair quantity of K for their optimum growth and yield, K deficiency in soil decreases the production of wheat (McCauley *et al.*, 2009). K deficiency severely affects the time of grain filling and decreases the seed size in wheat (Saifullah *et al.*, 2002). K deficiency also reduces the spike growth and grain number per spike (GNPS) (Bly and Woodard, 2003; Tripathi *et al.*, 2003), the quantity of spikes in wheat (Tao *et al.*, 2006) and the uptake of P and N in wheat (Saifullah *et al.*, 2002).

Most soils in the world lack the essential macronutrients N, P and K which affect the production of crops, including wheat. In the context of global food security, wheat plays a key role both in developed and developing countries. To increase yields from soils deficient in these triple nutrients, farmers are forced to apply fertilizers which are not only harmful to the environment but also the sources of these fertilizers like RP are getting depleted. In order to overcome this crisis and to enhance the wheat production, the need of the hour is to improve the acquisition and use efficiency of N, P and K which calls for in-depth understanding and applications of plant breeding and functional genomics studies. This chapter attempts to gather all the necessary information on the influence of the three essential macronutrients N, P and K, as well as on quantitative trait locus (QTL) and functional genomics studies related to these three nutrients, for developing wheat varieties with improved N-use efficiency (NUE), PUE and K-use efficiency (KUE) for utilization in sustainable agriculture and ensuring food security in developing countries like those in Asia and Africa.

## 9.2 Application of Molecular Breeding for Improving Nutrient-Use Efficiency in Wheat

Over the past four decades, global wheat yield has improved remarkably due to the green revolution and post green revolution through high-input-responsive varieties along with the application of more chemical fertilizers and pesticides and ample water supply. This approach now seems to be exhausted amidst our hope to raise productivity, primarily due to low N, P and K nutrient retrieval from the fertilized soils (Ladha *et al.*, 2005; Xu *et al.*, 2015). Over the years, new wheat genotypes to maximize nutrient absorption were not developed. On the other hand, efficient farming management practices like tillage, seeding, weeding, pest control, irrigation and harvesting together escalated the overall production costs (Balafoutis *et al.*, 2017). Considering the high production costs and progressively unpredictable sources for fertilizers, the focus of future agricultural systems must be on improving yield productivity with low input of fertilizers and irrigation water. Nutrient-use efficiency (NtUE) in wheat can be understood through two step approaches: (i) decoding the processes affecting wheat under nutrient-deficiency stress and identifying the various responses of wheat and its genotypes to cope with the low-nutrient-input stress (Bilal *et al.*, 2018); and (ii) exploiting genetic variability (both natural and induced) through innovative molecular breeding programmes and functional genomics to improve the wheat crop efficiency under low-nutrient-input soils (Bilal *et al.*, 2018). And breeding wheat cultivars with improved NtUE is a prerequisite for lowering the overall production costs. NtUE improvement in wheat is an indispensable prerequisite for expanding its production into marginal lands with low nutrient availability. Wheat cultivars with improved NtUE also protect the environment by reducing fertilizer application, the rate of nutrient application loss to ecosystems and input costs. On the other hand, improved NtUE wheat cultivars will enhance wheat yield and maintain soil and groundwater quality. In this context, the identification and development of wheat varieties with superior GY with high N-recovery efficiency under low input conditions is a top breeding priority (Vinod and Heuer, 2012). Despite the existence of significant genotypic differences in NtUE in wheat, genetic selection for this trait has not been carried out systematically due to complexity involved in the overall phenotype and its evaluation and the non-availability of genetic tools to use (Singh *et al.*, 1998; Han *et al.*, 2015; van Bueren and Struik, 2017). Plants in general are more efficient in the absorption and use of nutrients in controlled experimental environments. However, both genetic and physiological traits change drastically in the field due to plants' complex interactions with environmental variables. This calls for a systematic breeding programme to develop wheat cultivars with high nutrient-acquisition efficiency (NtAE) and NtUE traits for enhanced nutrient absorption, transport, use and mobilization for augmented wheat production (Baligar *et al.*, 2001). The recent use of high-throughput markers offers ease and high precision in this area of research to improve wheat yield under low-nutrient-input soils (Chen *et al.*, 2014; Thomson *et al.*, 2017) where N-recovery efficiency for farmer-managed fields under rainfed conditions is 20–30%, for irrigated conditions 30–55% (Roberts, 2008; Liu *et al.*, 2016) and although it reaches up to 57% in experimental plots, rarely exceeds 50% for wheat. To broaden the genetic basis of wheat and to enrich existing varieties, high-resolution research on the wild species of wheat and its native landraces becomes vital for exploiting the untapped pool of useful QTLs and genes (Kole *et al.*, 2015; Stein *et al.*, 2018). Genetic selection and plant breeding techniques facilitate the development of wheat varieties resistant to pests, diseases and adverse environmental conditions like drought, submergence and salinity. Nevertheless, proper genetic selection tactics are crucial for improving NtUE in wheat.

Molecular linkage genetic maps, QTLs and characterizing these QTLs to their map positions in the wheat genome, their phenotypic effects and interactions with other QTLs and loci have facilitated exploring genetic loci associated with drought, salinity, NtUE, disease and insect resistance in crop plants (Gahlaut *et al.*, 2017; Gupta *et al.*, 2017; Ren *et al.*, 2017, 2018). The recent fast progress in genome sequencing technologies and MAB approaches have led to a change in breeding methods (Vinod and Heuer, 2012). Association mapping is widely used in plant breeding to identify genes and QTLs determining quantitatively inherited variation based on a diverse set of fixed lines. QTLs/genes can be identified through historical phenotypic data and eventually gene functions, underused alleles and allele combinations which are useful for crop improvement (Flint-Garcia *et al.*, 2005; Ersoz *et al.*, 2009). Genome-wide association mapping is based on the strength of linkage disequilibrium across a diverse population apart from identifying the relationships between markers and traits of agronomic and evolutionary interest (Clark *et al.*, 2007; Zhao *et al.*, 2007). Unveiling the genetic base of agronomic, physiological and morphological traits in wheat is essential for developing new and improved wheat varieties. This genetic information can be used to select parental lines efficiently for hybridization. Unlimited access to the enormous online wealth of genomic and plant breeding resources, including high-quality genome sequences (International Wheat Genome Sequencing Consortium, 2014), dense single-nucleotide polymorphism (SNP) maps (Chao *et al.*, 2009; van Poecke *et al.*, 2013), extensive germplasm collections and public databases of genomic information (Varshney *et al.*, 2005; Perez-de-Castro *et al.*, 2012), is very useful for plant breeders and researchers.

## 9.3 Identification of Quantitative Trait Loci for Improving Nitrogen-Use Efficiency in Wheat

N is the most important macronutrient for wheat growth in natural ecosystems and the green revolution mainly focused on the development of high-yielding modern wheat varieties with N fertilization. The ability of the plant to absorb and use N under various environmental and ecological conditions depends primarily on the genetic make-up and physiological components of the plant (Baligar *et al.*, 2001). NUE is mostly dependent on interactions and the optimum use of N, water availability, light intensity, disease pressure and genotype. N-efficient wheat genotypes will assist in identification of more efficient N-management strategies in wheat production (Cormier *et al.*, 2016). In this context, NUE improvement of wheat varieties is of great concern. All the identified QTLs associated with NUE-related agro-morphological traits in wheat are mentioned below.

Seventeen QTLs for N uptake (NUP) were detected in 120 double haploid lines (DHLs) of winter wheat derived from Hanxuan 10 and Lumai 14 genotypes using 395 markers (132 amplified fragment length polymorphisms (AFLPs), 257 simple sequence repeats (SSRs) and six expressed sequence tags (ESTs)) (Table 9.1) (An *et al.*, 2006). Of the 17 NUP QTLs, nine QTL were detected under LN conditions (LNC) and eight QTLs were detected under high-N (HN) conditions (HNC) in the field.

In the same study, four QTLs were detected for shoot dry weight (SDW), root dry weight (RDW) and TN via hydroponic culture supplied with all sufficient nutrients. In total, 96 DHLs were derived from a cross between Chinese Spring and SQ1 and they were used to identify more QTLs for GY components such as ears per plant (EPP), grains per ear (GPE) and 1000-grain weight using 596 restriction fragment length polymorphism (RFLP), AFLP and SSR markers under LNC (Quarrie *et al.*, 2005). Similarly, many QTLs for kernel-related traits such as protein content in grain (PCG), wet gluten content (WGC), dough tractility (DT), test weight (TW), water absorption (ABS), zeleny sedimentation value (ZEL), kernel hardness (KH), kernel length (KL), kernel width (KW), kernel diameter ratio (KDR) and 1000-grain weight were identified in 188 recombinant inbred lines (RILs) derived from Kenong 9204 × Jing 411 using 1127 SSR markers under both LNC and HNC (Table 9.1) (Cui *et al.*, 2016). For kernel-related traits, 17 QTLs each for KW, total kernel weight (TKW), KL and KDR, nine QTLs each for PCG, WGC and DT, eight QTLs each for TW and ABS and six QTLs each for ZEL and KH were detected. All the identified QTL traits responded well to LN stress. The same 188 RILs were used to detect 22 and 12 QTLs for GY and yield difference between the value (YDDV), respectively, under LNC using 174 genomic simple sequence repeat (g-SSR), 42 EST-derived microsatellite (e-SSR), 289 Diversity Arrays Technology (DArT), 15 sequence-tagged sites (STS), 27 sequence-related amplified polymorphism (SRAP) and three inter-simple sequence repeat (ISSR) markers (Cui *et al.*, 2014).

Two bread wheat cultivars, Arche and Recital, were used to develop 241 DHLs and they were used to detect 104 QTLs for physiological traits and 44 QTLs for agronomic traits under LN stress using 197 SSR markers (Fontaine *et al.*, 2009). Both physiological and agronomic QTLs were located on chromosomes 2A, 2B, 2D, 5A, 5B and 5D. Similarly, 131 RILs derived from a cross of Chuan 35050 and Shannong 483 were used to detect 87 QTLs for six morphological traits such as number of axial roots, maximal root length (MRL), shoot height (SH), SDW,

**Table 9.1.** List of QTLs identified for various traits of N, P and K nutrient-use efficiency in wheat.

| Study no. | Population type | Total no. of RILs/DHLs | Crosses | Trait name[a] | Type of markers used | Total no. of markers | No. of QTLs | LOD[b] score | Reference |
|---|---|---|---|---|---|---|---|---|---|
| **Nitrogen** | | | | | | | | | |
| 1 | DHL | 96 | Chinese Spring × SQ1 | EPP, GPE and 1000-grain weight | RFLP, AFLP and SSR | 595 | 50 | 2.0 | Quarrie *et al.* (2005) |
| 2 | DHL | 120 | Hanxuan 10 × Lumai | SDW, RDW, TN and NUP | AFLP, SSR and EST | 395 | 17 | 2.08–7.25 | An *et al.* (2006) |
| 3 | DHL | 241 | Arche × Recital | LRL, PRL, LRN, TRL, SRL, RUE, RDM, TDM, NUR, NTA, LA, TDM, ADM, RDM and NTA | SSR | 183 | 32 | 2.61–5.13 | Laperche *et al.* (2006) |
| 3 | DHL | 241 | Arche × Recital | FLAA, CFLL, CNRFLL, DTH, FLLDW, FLLS, GDHDM, GDHPR, PCG, GNPS, GSDM, GSPR, FLLNC, PROT, QPG and 1000-grain weight | SSR | 197 | 148 | 2.50 | Fontaine *et al.* (2009) |
| 4 | RIL | 131 | Chuan 35050 × Shannong 483 | No. of axial roots, MRL, SH, SDW, RDW, TDW, RNC, SNC, TNC, SNUtE, RNUtE and TNUtE root | SSR | 380 | 87 | 3.0–10.2 | Guo *et al.* (2012) |
| 5 | RIL | 142 | Xiaoyan 54 × Jing 411 | MRL, PRL, LRL SRN and TRL | SSR | 470 | 16 | 2.10 –29.89 | Ren *et al.* (2012) |
| 6 | RIL | 131 | Chuan 35050 × Shannong 483 | RFW, SFW, TFW, RSFW, RRFW, SDW, TDW, RSDW and RRDW | DArT, SSR, EST-SSR, and other molecular and biochemical markers | 719 | 113 | 2.5–33.0 | Sun *et al.* (2013) |
| 7 | DHL | 188 | Kenong 9204 × Jing 411 | GY and YDDV | g-SSR, e-SSR, DArT, STS, SRAP and ISSR | 550 | 34 | 2.01 –10.24 | Cui *et al.* (2014) |
| 8 | RIL | 182 | Xiaoyan 54 × Jing 411 | NHI, NUP, $NUtE_{DM}$, $NUtE_{GY}$, StNUP | SSR, EST-SSR | 541 | 21 | – | Xu *et al.* (2014) |
| 9 | RIL | 188 | Kenong 9204 × Jing 411 | PCG, WGC, DT, TW, ABS, ZEL, KH, KL, KW, KDR and 1000-grain weight | SSR | 1,127 | 109 | 2.07 –18.23 | Cui *et al.* (2016) |
| 10 | RIL | 142 | Xiaoyan 54 × Jing 411 | MRL and RDW | SSR | 470 | 5 | 2.7–6.5 | Ren *et al.* (2017) |
| 11 | NAM | | | RDW, SDW and TDW | SNP | 2,050 | 67 | >2.5 | Ren *et al.* (2018) |

| | | | | | | | | | |
|---|---|---|---|---|---|---|---|---|---|
| **Phosphorus** | | | | | | | | | |
| 1 | RIL | 114 | W7984 × Opata85 | WPUE and SPUE | RFLP | 918 | 5 | 2.05–3.35 | Weidong *et al.* (2001) |
| 2 | DHL | 92 | Lovrin 10 × Chinese Spring | TN, SDW, SPU and PUE | SSR | 253 | 20 | 3.3–17.9 | Su *et al.* (2006) |
| 3 | DHL | 119 | Hanxuan 10 × Lumai 14 | PC and P content | SSR | 395 | 10 | 2.05–5.74 | Shi *et al.* (2008) |
| 4 | DHL | 120 | Hanxuan 10 × Lumai 14 | SDW, TN, PUP, SPUtE, BY, GY, EN, GNE, TGW and 1000-grain weight | | | 195 | 2.01 –12.98 | Su *et al.* (2009) |
| 5 | RIL | 142 | Xiaoyan 54 × Jing 411 | MRL and RDW | SSR | 470 | 5 | 2.5–5.4 | Ren *et al.* (2017) |
| 6 | RIL | 138 | SHW-L1 × Chuanmai 32 | Root diameter | SSR, DArT and SSR | 121,222 | 16 | 2.04–5.36 | Wu *et al.* (2017) |
| 7 | RIL | 181 | TN 18 × LM 6 | GNPS, GPC, GPUtE, GY, PE, PH, PUPE and PUtE | DArT, SNP and SSR | 10,739 | 178 | 3.00 –31.55 | Yuan *et al.* (2017) |
| **Potassium** | | | | | | | | | |
| 1 | RIL | 152 | Langdon × wild emmer | K concentration | SSR and DArT | 690 | 8 | 3.2–16.7 | Peleg *et al.* (2009) |
| 2 | RIL | 131 | Chuan 35050 × Shannong 483 | RDW, SDW, RSDW, RRDW, TDW, RKUE, SKUE, TKU, RKC, SKC, RSKC/RKC, TKC, GY, StW, SN, GNPS, 1000-grain weight GKC, StKC, GKUE, StKUE | DArTs, SSR, EST-SSR and other molecular and biochemical loci | 719 | 138 | 3.00–9.66 | Kong *et al.* (2013) |
| 3 | DHL | 168 | Huapei 3 × Yumai 57 | TRL, TRS, TRV, RDW, SDW, RAD, RSDW, RRDW, CHL, POD, RA, SOD, SKC, RKC and MDA | SSR, EST and ISSR | 322 | 48 | 3.01 –13.76 | Zhao *et al.* (2014) |
| 4 | RIL | 131 | Chuan 35050 × Shannong 483 | RKCE, SKCE, TKCE, RKC, SKC, TKC, RSKC, RKUE, SKUE and TKUE | SSR | 719 | 127 | 3.10–8.12 | Gong *et al.* (2015) |

*Continued*

**Table 9.1.** Continued.

| Study no. | Population type | Total no. of RILs/DHLs | Crosses | Trait name[a] | Type of markers used | Total no. of markers | No. of QTLs | LOD[b] score | Reference |
|---|---|---|---|---|---|---|---|---|---|
| 5 | RIL | 184 | Tainong 18 × Linmai 6 | RDW, SDW, TDW, RSDW, RKC, SKC, TKC, RKCE, SKCE, TKCE, SKUE, RKUE, TKUE, RCaCE, SCaCE, TCaCE, RCaC, SCaC, TCaC, SCaUE, RCaUE, TCaUE, RMgCE, SMgCE, TMgCE, SMgC, RMgC, TMgC, SMgUE, TMgUE, RMgUE, PH, SN, GNPS, 1000-grain weight, GWP, TAW, HI, GKCE, StKCE, GKC, StKC, AKC, KHI, GKUE, StKUE, AKUE, GCaCE, StCaCE, ACaCE, GCaC, StCaC, ACaC, CaHI, GCaUE, StCaUE, ACaUE, GMgCE, StMgCE, GMgC, AMgC, MgHI, GMgUE and StMgUE | SSR | 10,739 | 306 | 24.56 | Shen *et al.* (2019) |

[a]ABS, water absorption; ACaC, above-ground Ca content; ACaCE, above-ground Ca concentration; ACaUE, above-ground Ca-use efficiency; ADM, aerial dry matter; AKC, above-ground K content; AKUE, above-ground K-use efficiency; AMgC, above-ground Mg content; BY, biomass yield; CaHI, Ca-harvest index; CFLL, carbon in flag leaf lamina; CHL, chlorophyll concentration; CNRFLL, C:N ratio in flag leaf lamina; DT, dough tractility; DTH, heading date; EN, ear number; EPP, ears per plant; FLLA, flag leaf lamina area; FLLDW, flag leaf lamina dry weight; FLLNC, flag leaf lamina N content; FLLS, flag leaf lamina senescence; GCaC, grain Ca content; GCaCE, grain Ca concentration; GCaUE, grain Ca-use efficiency; GDHDM, glutamate dehydrogenase activity expressed per dry matter; GDHPR, glutamate dehydrogenase activity expressed per protein; GKC, grain K content; GKCE, grain K concentration; GKUE, grain K-use efficiency; GMgC, grain Mg content; GMgCE, grain Mg concentration; GMgUE, grain Mg-use efficiency; GNE, grain number per ear; GNPS, grain number per spike; GPC, grain P concentration; GPE, grains per ear, GPUtE, grain P-utilization efficiency; GSDM, glutamine synthetase activity expressed per dry matter; GSPR, glutamine synthetase activity expressed per protein; GWP, grain weight per plant; GY, grain yield; HI, harvest index; KDR, kernel diameter ratio; KH, kernel hardness; KHI, K-harvest index; KL, kernel length; KW, kernel width; LA, leaf area; LRL, lateral root length; LRN, lateral root number; MDA, malondialdehyde; MgHI, Mg-harvest index; MRL, maximal root length; NHI, N-harvest index; NTA, N total amount; NUP, N uptake; $NUtE_{DM}$, N-utilization efficiency for above-ground dry matter; $NUtE_{GY}$, N-utilization efficiency for grain yield; NUR, N-specific uptake rate; PC, P concentration; PCG, protein content in grain; PE, P efficiency; POD, peroxidase; PH, plant height; PRL, primary root length; PROT, protein content of the flag leaf; PUE, P-use efficiency; PUP, P uptake; PUPE, P uptake-use efficiency; PUtE, P-utilization efficiency; QPG, quantity of protein per grain; RA, root activity; RAD, root average diameter; RCaC, root Ca content; RCaCE, root Ca concentration; RCaUE, root Ca-use efficiency; RDM, root dry matter; RDW, root dry weight; RFW, root fresh weight; RKC, root K content; RKCE, root K concentration; RKUE, root K-use efficiency; RMgC, root Mg content; RMgCE, root Mg concentration; RMgUE, root Mg-use efficiency; RNC, root N content; RNUtE, root N-utilization efficiency; RRDW, ratio of root dry weight; RRFW; ratio of root fresh weight; RSDW, ratio of shoot dry weight; RSFW, ratio of shoot fresh weight; RSKC, ratio of shoot K content; RUE, radiation-use efficiency; SCaC, shoot Ca content; SCaCE, shoot Ca concentration; SCaUE, shoot Ca-use efficiency; SDW, shoot dry weight; SFW, shoot fresh weight; SH, shoot height; SKC, shoot K content; SKCE, shoot K concentration; SKUE, shoot K-use efficiency; SMgC, shoot Mg content; SMgCE, shoot Mg concentration; SMgUE, shoot Mg-use efficiency; SN, spike number; SNC, shoot N content; SNUtE, shoot N-utilization efficiency; SOD, superoxide dismutase; SPU, shoot P uptake; SPUE, shoot P-use efficiency; SPUtE, shoot P-utilization efficiency; SRL, specific root length; SRN, seminal root tip number; StCaC, straw Ca content; StCaCE, straw Ca concentration; StCaUE, straw Ca-use efficiency; StKC, straw K content; StKCE, straw K concentration; StKUE, straw K-use efficiency; StMgCE, straw Mg concentration; StMgUE, straw Mg-use efficiency; StNUP, straw N uptake; StW, straw weight, TAW, total above-ground weight; TCaC, total Ca content; TCaCE, total Ca concentration; TCaUE, total Ca-use efficiency; TDM, total dry matter; TDW, total dry weight; TFW, total fresh weight; TGW, total grain weight; TKC, total K content; TKCE, total K concentration; TKUE, total K-use efficiency; TMgC, total Mg content; TMgCE, total Mg concentration; TMgUE, total Mg-use efficiency; TN, tiller number; TNC, total N content; TNUtE, total N-utilization efficiency; TRL, total root length; TRS, total root surface area; TRV, total root volume; TW, test weight; WGC, wet gluten content; WPUE, whole-plant K-use efficiency; YDDV, yield difference between the value; ZEL, zeleny sedimentation value.

[b]Logarithm of the odds.

RDW and TDW which were located on 13 chromosomes (1A, 1B, 1D, 2B, 3A, 3B, 4A, 4B, 5D, 6A, 6B, 7A and 7B) using 380 SSR markers (Guo *et al.*, 2012). Employing the same 131 RILs and markers, a total of 144 QTLs were identified for root N content (RNC), shoot N content (SNC) and total N content (TNC) and another 144 QTLs were found for to be associated with root N-utilization efficiency (RNUtE), shoot N-utilization efficiency (SNUtE) and total N-utilization efficiency (TNUtE) (Table 9.1) (Guo *et al.*, 2012). Using 183 SSR markers, 32 QTLs were detected in 241 DHLs derived from the cross between Arche and Recital under LNC (Laperche *et al.*, 2006). Out of 32 QTLs, six were specific for root architectural traits such as lateral root length (LRL), primary root length (PRL), lateral roots number (LRN), total root length (TRL) and specific root length (SRL) and QTLs were located on chromosomes 4B, 5A, 5D and 7B; six QTLs were for C:N efficiencies such as radiation-use efficiency (RUE), root dry matter (RDM), total dry matter (TDM), N-specific uptake rate (NUR) and N total amount (NTA) and were located on chromosomes 1B, 2B, 6A, 6B, 7A, 7B and 7D; and the remaining 20 QTLs were for integrative traits such as LA, TDM, aerial dry matter (ADM), RDM and NTA and were located on chromosomes 1A, 1B, 2B, 4B, 5A, 5B and 6B (Laperche *et al.*, 2006).

A high-resolution study using 470 SSR markers helped to identify many root-related QTLs; two for MRL, three for PRL, six for LRL and seminal root tip number (SRN) and five for TRL. They were detected in 142 RILs of wheat derived from Xiaoyan 54 and Jing 411 and located on chromosomes 2B, 7B, 1A, 6A, 6D, 3A, 3D and 4B (Ren *et al.*, 2012). Similarly, the same set of RILs were used to detect three and two QTLs for RDW and MRL, respectively, under LNC in hydrophobic culture and they were located on chromosomes 2B, 4B, 4D, 5A and 6A (Ren *et al.*, 2017). The same group recently identified 67 QTLs for RDW, SDW and TDW under LNC and HNC at the seedling stage using as many as 2059 SNP markers via nested association mapping (NAM) (Ren *et al.*, 2018). In NAM methods, RIL populations derived from crosses between a single female parent Yanzhan 1 and four different male parents (Hussar, Cayazheda 29, Yunnan and Yutian) were used to detect 67 QTLs for RDW, SDW and TDW under LNC and HNC (Ren *et al.*, 2018). Furthermore, 131 RILs derived from the cross between Chuan 35050 and Shannong 483 were used for the identification of 18 QTLs for root fresh weight (RFW), 15 for shoot fresh weight (SFW), 16 for total fresh weight (TFW), 22 for ratio of SFW (RSFW) and ratio of RFW (RRFW), 18 for SDW, 14 for TDW and 25 for ratio of SDW (RSDW) and ratio of RDW (RRDW) through 361 DArT, 170 SSR, 100 EST-SSR, and 88 other molecular and biochemical markers (Table 9.1) (Sun *et al.*, 2013). Identified QTLs were mapped on chromosomes 1A, 1D, 2B, 2D, 4A, 4B, 5B, 5D, 6B, 7A and 7B. A set of 182 RILs derived from a cross between wheat cultivars Xiaoyan 54 and Jing 411 were used to detect one QTL for N-harvest index (NHI), two for NUP, five for N-utilization efficiency in above-ground dry matter ($NUtE_{DM}$), nine for NUtE for GY ($NUtE_{GY}$) and four QTLs for straw NUP (StNUP) using 523 SSR and 18 EST-SSR markers (Xu *et al.*, 2014). Among the factors affecting plants under LN soil, responses of plants to cope with N deficiency in soil and QTLs to improve NUE in wheat would help not only to optimize N inputs in agriculture but also to increase wheat yield under LN-input soils.

## 9.4 Identification and Characterization of Genes Involved in Nitrogen Transport

Plant N transporters were first characterized in the model plant *Arabidopsis thaliana* (Tsay *et al.*, 1993). N transporter gene families in plants are largely divided into low-affinity (*nitrate transporter 1 (NRT1)/peptide transporter (PTR)*) and high-affinity (*NRT2*) (Tsay *et al.*, 2007; Fan *et al.*, 2017). *T. aestivum NRT 2;3* (*TaNRT2;3*) was highly expressed in wheat roots under LNC and HNC and it was not expressed in shoots under both N conditions (Zhao *et al.*, 2004). Therefore, it was suggested that TaNRT2;3 might be a high-affinity N transporter in wheat as it was not confirmed with its $K_M$ (Michaelis constant) value. Similarly, TaNRT2;1 was also considered a high-affinity transporter (Yin *et al.*, 2007) ($K_M = 50\mu M$), its expression level was 30- to 33-fold higher in roots at 1 and 4 h, respectively, under LNC and its expression was not found in leaves. The expression patterns of low-affinity N transporters such as *TaNPF2;1, 2;2, TaNPF4;1, TaNPF6;1, 6;2, 6;3, 6;4, 6;5, 6;6* and *TaNPF7;1* and *7;2* were analysed in roots and shoots of 3-week-old hydroponically grown wheat plants (Buchner and Hawkesford, 2014) (Fig. 9.1). Among these low-affinity N transporters, *TaNPF6;1, 6;2, 6;3, 6;4* and *TaNPF7;1* were highly expressed in roots and *TaNPF2;1, 2;2, TaNPF4;1, TaNPF6;3, 6;4, 6;6* and *TaNPF7;1* had higher expression in shoots under LNC than under HNC. Expression patterns of two putative low-affinity (*TaNRT1;1* and *1;2*) and three putative high-affinity (*TaNRT2;1, 2;2* and *2;3*) N transporters were identified under LNC and HNC in the roots of wheat colonized by four species of arbuscular mycorrhiza fungi (AMF), namely *Acaulospora delicata, Claroideoglomus etunicatum, Funneliformis mosseae* and *Rhizophagus intraradices* (Duan *et al.*, 2015). Under LNC, all five genes (*TaNRT1;1, 1;2, TaNRT2;1, 2;2* and *2;3*) were expressed in AMF-colonized roots but their expression levels were higher in roots colonized by *A. delicata* and *R. intraradices* compared with the other two AMF species. Contrastingly, under HNC, the expression of *TaNRT1;1, TaNRT2;1* and *2;2* was highly induced by *F. mosseae* and *R. intraradices* and *TaNRT1;2* and *TaNRT2;3* expression levels were high in roots inoculated by *C. etunicatum* and *F. mosseae* and *R. intraradices*. The same group also analysed five *NRT* genes (*TaNRT1;1, 1;2, TaNRT2;1, 2;2* and *2;3*) in the roots of two winter wheat genotypes, XY107 (efficient N up-taker) and XY6 (inefficient N up-taker), in response to drought with contrasting N supplies (LNC and HNC) during the vegetative and reproductive stages (Duan *et al.*, 2016). At vegetative and reproductive stages, the expression of *TaNRT1;1* and *1;2* was highly induced by drought stress in the root of XY6 genotype under LNC compared with the XY6 genotype. Contrastingly, *TaNRT2;1, 2;2* and *2;3* were highly induced by drought stress in the root of XY107 under LNC compared with XY6 genotype. The expression levels of eight *NRT1* family genes (*TaNRT1;1* to *1;8*) and five *NRT2* family genes (*TaNRT2;1* to *2;5*) were analysed in the root of Yumai 34 genotype under LNC (Guo, T. *et al.*, 2014). The transcripts of *TaNRT1;1* and *TaNRT2;3* were highly expressed at 2 and 4 days; *TaNRT1;3* at 2, 4 and 6 days; *TaNRT1;4, 1;7* and *1;8* and *TaNRT2;1* and *2;2* at 2 days under LNC.

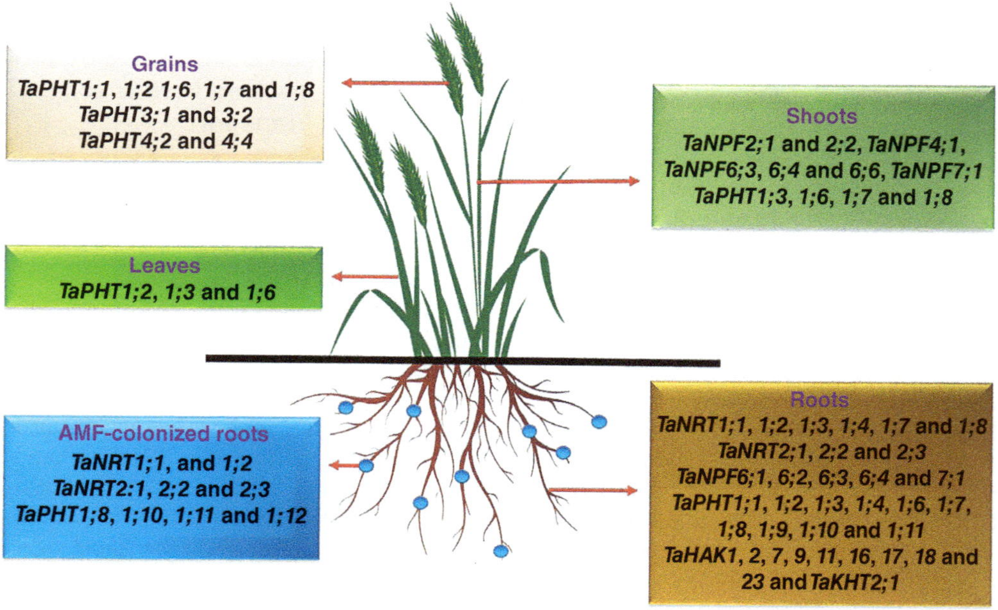


**Fig. 9.1.** N, P and K family transporters identified in direct and AMF-mediated (indirect) uptake, export and remobilization of nutrients in wheat. N, P and K transporters identified in wheat shoots, leaves, grains, roots and AMF-colonized roots are listed in rectangular boxes in different colours. The AMF-colonized roots are indicated by blue circular structures in the root zone.

## 9.5 Identification of Quantitative Trait Loci for Improving Phosphorus-Use Efficiency of Wheat

Under LP conditions (LPC), two and three QTLs were detected for whole-plant PUE (WPUE) and shoot PUE (SPUE), respectively, in 114 RILs developed from a cross between W7984 and Opata85 using 918 RFLP markers (Table 9.1) (Weidong *et al.*, 2001). The QTLs for SPUE were located on chromosomes 2D, 3B and 6D and for WPUE on chromosome 2D. Similarly, under HP conditions (HPC), two QTLs for SPUE located on chromosomes 1B and 5A and three QTLs for WPUE on chromosomes 2B, 5A and 7A were identified in the same study (Weidong *et al.*, 2001). Totally, 20 QTLs associated with panicle number and dry weight matter were identified in 92 DHLs derived from Lovrin 10 and Chinese Spring under LPC using 253 SSR markers (Su *et al.*, 2006). Among them, six QTLs were found for TN and they were located on chromosomes 1A, 2D, 3A, 4B and 5D, and four QTLs for SDW were located on chromosomes 5A, 5D and 7B. Six QTLs for shoot P uptake (SPU) were located on chromosomes 2A, 4A, 4B, 5A and 5D and four QTLs for PUE were located on chromosomes 5A, 6B and 7A under LPC (Su *et al.*, 2006). Four QTLs for P concentration (PC) and six QTLs for P content were identified in 119 DHLs developed from a cross between Hanxuan 10 and Lumai 14 by 395 SSR markers (Table 9.1) (Shi *et al.*, 2008). The PC QTLs were located on chromosomes 3A, 3B, 4A and 4D and the P content QTLs were located on 2A, 2B, 2D, 3A, 3D and 4A. The parents Hanxuan 10 and Lumai 14 were used to develop 120 DHLs which were used to detect 195 QTLs for SDW, TN, P uptake (PUP), shoot P-utilization efficiency (SPUtE), biomass yield (BY), GY, ear number (EN), grain number per ear (GNE), total grain weight (TGW) and 1000-grain weight which were identified at seedling and maturity stages under LPC and HPC (Su *et al.*, 2009). Among the 195 QTLs, 63 QTLs were detected at the seedling stage and 132 QTLs were detected at maturity stage and they were distributed on different loci of 21 chromosomes (Su *et al.*, 2009). Remarkably, three RILs were developed from crosses among four Chinese genotypes, namely Weimai 8 × Luohan 2, Weimai 8 × Yannong 19 and Weimai 8 × Jimai 20, which comprised 179, 175 and 172 lines, respectively. Among these genotypes, Weimai 8 was more P-utilization efficient than Yannong 19, Luohan 2 and Jimai 20 (Zhang and Wang, 2015). These three related RILs were used to identify 43 QTLs for 12 traits such as PH, MRL, numbers of axial roots, SDW, stem and leaf dry weight (SLDW), RDW, P-utilization efficiency (PUtE) of stem and leaf (SLPUtE), root PUtE (RPUtE), SPUtE, stem and leaf PC (SLPC), root PC (RPC) and shoot PC (SPC) under LPC using 576 DArT and 496 EST-SSR markers (Table 9.1) (Zhang and Wang, 2015). Among the 43 QTLs, two QTLs for PH were detected on chromosomes 3B and 5A, six for RDW, four for RL, one for RPC, four for SDW, four for SLDW, five for SLPC, five for SLPUtE, two for SPC and seven for SPU (Zhang and Wang, 2015). Yuan *et al.* (2017) developed 184 RILs derived from the cross between TN 18 and LM 6 grown under hydroponic culture trials and field trials. In total, 178 QTLs were identified for several traits including GNPS, grain PC (GPC), grain PUtE (GPUtE), GY, PE, PH, P-uptake efficiency (PUPE) and PUtE using 5548 DArT, 5085 SNP and 106 SSR or EST-SSR markers (Table 9.1). One and four QTLs for RDW and MRL, respectively, were detected on chromosomes 2B, 4B, 5A, 6B and 7B in 142 RILs of wheat derived from Xiaoyan 54 and Jing 411 using 470 SSR markers (Ren *et al.*, 2017). Furthermore, seven and nine QTLs for root diameter were identified in 138 RILs derived from SHW-L1 and Chuanmai 32 under LPC and HPC, respectively, by 120,370 SNP, 733 DArT and 119 SSR markers (Wu *et al.*, 2017), and the seven and nine QTLs were located on chromosomes 1A, 2A, 2B, 2D, 3B, 3D, 4A and 7D. Unlike rice and maize, the study of QTLs is in its juvenile stage in wheat although it is one of the major cereal crops. When we compare wheat RILs with those of other cereals, only very few wheat RILs have been deployed in research without much diversity for mining QTLs for PUE in wheat. Furthermore, the identified QTLs in wheat ought to be further mapped and validated to develop new wheat varieties with enhanced PUE in LP soils worldwide.

## 9.6 Identification and Functional Characterization of Genes Involved in Phosphorus Transport

P is an important nutrient which is acquired from soil as Pi and redistributed within the plant by several specific transporters. Five types of phosphate transporter family, namely PHT1, PHT2, PHT3, PHT4 and PHT5/vacuolar phosphate transporter1 (VPT1), have previously been reported in many plants. Each transporter family is located in different organs and plasma membranes of plants, like PHT1 in plasma membranes, PHT2 in plastids (Versaw and Harrison, 2002; Rausch *et al.*, 2004), PHT3 in mitochondria (Rausch and Bucher, 2002), PHT4 in plastid envelope or the Golgi apparatus (Guo *et al.*, 2008) and PHT5 in tonoplasts/vacuoles (Liu, T.Y. *et al.*, 2016), and each plays a different role in Pi transport. Among these, PHT1 transporter proteins, which belong to the major facilitator superfamily (MFS) transporters, are the important ones that are involved in Pi uptake from root, Pi transport from root to shoot via xylem, remobilization of Pi from older leaves to younger leaves and maintenance of Pi homeostasis in plants (Baker *et al.*, 2015; Roch *et al.*, 2019). To date, *PHT1* genes were identified and functionally characterized in *Arabidopsis* (Muchhal *et al.*, 1996; Mudge *et al.*, 2002), rice (Paszkowski *et al.*, 2002), maize (Liu, F. *et al.*, 2016), wheat (Grün *et al.*, 2018), sorghum (Walder *et al.*, 2015), barley (Rae *et al.*, 2003; Sisaphaithong *et al.*, 2012) and foxtail millet (Ceasar *et al.*, 2014) and other millets (Maharajan *et al.*, 2019). This section discusses the identification and characterization of *PHT* genes in wheat in detail.

The first full-length *PHT1* gene sequence of *T. aestivum TaPHT1;2*, *1;3* and *1;6* was analysed by Davies *et al.* (2002). The expression pattern of these three individual PHT1 family transporters was analysed in roots and leaves of four wheat varieties (Xiaoyan 54, Lovrin 10, 81(85)-5-3-3-3 (line), Jing 411) under LPC and HPC. The *TaPHT1;2* gene was highly induced in roots under LPC in all the genotypes. *TaPHT1;3* was highly expressed only in the root of Lorvin 10 genotype under LPC when compared with the other three genotypes. In leaves, *TaPHT1;2*, *1;3* and *1;6* were expressed in Xiaoyan 54, Lovrin 10 and Jing 411 under both LPC and HPC but their expression level was higher only under LPC in these three genotypes (Davies *et al.*, 2002). Yeast complement analysis revealed that *TaPHT1;2* encodes a high-affinity Pi transporter with an apparent $K_M$ of 23.6 μM (Guo, C. *et al.*, 2014). Overexpression of *TaPHT1;2* significantly increases plant dry weight and Pi acquisition. These results suggest that TaPHT1;2 acts as a high-affinity Pi transporter and facilitates Pi uptake under LP soils (Zeng *et al.*, 2002; Guo, C. *et al.*, 2014).

Later, the expression pattern of *PHT1* genes was studied in the AMF-colonized roots of wheat (Glassop *et al.*, 2005; Sisaphaithong *et al.*, 2012). Four *PHT1* genes, namely *TaPHT1;8*, *1;10*, *1;11* and *1;12*, were identified in wheat roots colonized by *Glomus* spp., *Scutellospora calospora* and *R. irregularis* and the higher expression of these four *PHT1* genes in wheat was confirmed by RT-PCR (Fig. 9.1) (Glassop *et al.*, 2005; Sisaphaithong *et al.*, 2012). The expression pattern of *TaPHT1;4* was studied in leaves and roots of Shixin 828 genotype under LPC and HPC (Liu *et al.*, 2013). *TaPHT1;4* transcript was detected in the root and was highly induced under LPC in that study. Also, yeast complementation analysis in LPC confirmed that TaPHT1;4 ($K_M$ = 35.3 μM) is a high-affinity transporter and plays a critical role in wheat Pi acquisition under Pi deprivation (Liu *et al.*, 2013). Miao *et al.* (2009) characterized the promoter of *TaPHT1;2* in roots of P-efficient (Xiaoyan 54) and P-inefficient genotypes (Jing 411) under both LPC and HPC. The promotor of *TaPHT1;2* was more strongly activated in the roots of P-efficient genotypes than in those of P-inefficient genotypes irrespective of Pi condition. Another study showed that the expression levels of *TaPHT1;1*, *1;2*, *1;9* and *1;10* were higher in root at flowering stage under LPC compared with HPC in a field experiment (Teng *et al.*, 2013). The same group also identified 12 *PHT1* genes in the whole genome of wheat grown in the field and hydroponically and analysed the expression pattern of these genes in shoot and root tissues of Xiaoyan 54 genotype under both low and high Pi conditions (Teng *et al.*, 2017). Those authors reported that *TaPHT1;1*, *1;2*, *1;8* and *1;10* were predominantly expressed in roots and *TaPHT1;6* and *1;7* in shoots under both Pi conditions, but these genes were highly induced by LP in the hydroponic culture (Fig. 9.1). Similarly, *TaPHT1;3*, *1;6*, *1;7* and *1;8* were expressed in root tissues and *TaPHT1;3*, *1;6* and *1;8* were expressed in shoot tissues under LPC and HPC. The study suggests that *TaPHT1;1*, *1;2*, *1;3*, *1;6*, *1;7*, *1;8* and *1;10* contributed to Pi uptake and *TaPHT1;6*, *1;7* and *1;8* were involved in Pi translocation from old leaves to younger leaves (Teng *et al.*, 2017).

After the advent of whole-genome sequencing in wheat, 14 putative members of the *PHT1* family genes were identified in Hereward cultivar under LPC and HPC in both hydroponics and field conditions (Grün *et al.*, 2018). The 14 *PHT1* genes were located on chromosomes 1 (*TaPHT1;14*), 2 (*TaPHT1;2*, *1;9* and *1;13*), 4 (*TaPHT1;1*, *1;2*, *1;5* and *1;7*), 5 (*TaPHT1;3*, *1;4* and *1;6*), 6 (*TaPHT1;8*) and 7 (*TaPHT1;10*) (Grün *et al.*, 2018). Gene expression analysis showed that *TaPHT1;2* gene was induced greater than five-fold more than all other *PHT1* genes in roots of hydroponically grown 0-day-old seedlings under HPC. *TaPHT1;1* and *1;2* genes were upregulated in roots after 3 days of Pi starvation and the expression levels were further increased to 27- to 50-fold higher in roots after 6 days of Pi starvation when compared with day 0. Similarly, the expression levels of *TaPHT1;8* and *1;10* were more than threefold higher in roots on day 6 under LPC. The expression level of *TaPHT1;6* and *1;11* was eight- to 15-fold higher in roots on day 12 compared with day 0 under LPC. The expression of *TaPHT1;1*, *1;2* and *1;8* was found in ears, glumes, grains and rachis. *TaPHT1* also showed different levels of expression in various tissues during the developmental stages of wheat. For example, in roots, the expression level of *TaPHT1;1*, *1;2* and *1;8* increased at the beginning of stem elongation, decreased at booting and anthesis, and again increased during ripening. Expression of *TaPHT1;5* in root was the highest at tillering and milk ripening, the lowest at elongation and at an intermediate level in ear tissues. Similarly, the expression of *TaPHT1;6* increased in root at tillering, stem elongation, anthesis and ear, particularly in the rachis. Notably, *TaPHT1;6* was expressed only in the ear and particularly in the rachis, whereas transcript of *TaPHT1;7* was found to be very low in all tissues except in ear tissues at late booting stage under Pi starvation. Expression of *TaPHT1;10* and *TaPHT1;11* was much higher in roots throughout the wheat developmental stages than in shoot tissues and was the highest at maturity but decreased in the grain (Grün *et al.*, 2018).

Recently, Shukla *et al.* (2016) identified 13 *PHT1*, one *PHT2*, three *PHT3* and four *PHT4* family genes in root, leaf, stem, flag leaf, rachis, glume, aleurone, endosperm and embryo of C306 genotype of wheat during grain-filling stage. Notably, the expression patterns of *TaPHT1;1*, *1;2* and *1;4* were higher in aleurone, *TaPHT3;1* in embryo and *TaPHT4;2* in endosperm of C306 genotype under LPC in 28-day-old seedlings. *TaPHT2;1* was upregulated in the leaves of wheat under LPC (Guo *et al.*, 2013; Aziz *et al.*, 2014). Knockout of *TaPHT2;1* reduced the Pi concentration in the chloroplast under LPC and HPC (Guo *et al.*, 2013). This result suggests that *TaPHT2;1* has an important role in the mediation of Pi translocation from the cytosol to the chloroplast under LPC and HPC. Similarly, *TaPHT3;1* in embryo and rachis, *TaPHT3;2* in the aleurone and *TaPHT4;2* and *4;4* in endosperm were involved in wheat grain development and Pi allocation within grains (Shukla *et al.*, 2016). *PHT1* expression is directly proportional to PUE in wheat (Aziz *et al.*, 2014), where the highly P-efficient wheat cultivar Chinese 80–55 has a higher Pi acquisition and accumulates higher Pi concentrations in all organs than the less-efficient cultivar Machete. This correlated with differential organ-specific expression of Pi transporters *TaPHT1.4*, *1;6* and *1;10* (Aziz *et al.*, 2014). When comparing research on *PHT* family genes in wheat with that on *A. thaliana* and rice, very little research has been carried out even with regard to the important *PHT1* genes. Similarly, minimally studied *PHT2*, *PHT3* and unexplored *PHT4* and *PHT5* in wheat call for a systematic functional and molecular study on these genes, their tissue-specific expression, affinity, AMF association and their functional characterization, which will aid development of new varieties of food security-important wheat with high PUE and other nutritional efficiencies.

## 9.7 Identification of Quantitative Trait Loci for Improving Potassium-Use Efficiency in Wheat

Eight QTLs for K concentration in grain were identified in 152 RILs derived from a cross between durum wheat Langdon and wild emmer using 197 SSR and 493 DArT markers under field condition (Peleg *et al.*, 2009). Similarly, as many as 138 QTLs for biomass weight, yield and KUE-related traits were detected on 32 chromosomes at seedling and adult stage of 131 RILs derived from the cross of Chuan 35050 and Shannong 483 by 361 DArTs, 170 SSRs, 100 EST-SSRs, and 88 other molecular and biochemical loci (Kong *et al.*, 2013). Of these 138 QTLs, 50 were found to be associated with RDW, SDW, ratio of RDW and SDW (RSDW/RDW), TDW, root KUE (RKUE), shoot KUE (SKUE) and total KUE (TKUE), and 37 QTLs for root K content (RKC), shoot K content (SKC), ratio of RKC and

SKC (RSKC/RKC) and total K content (TKC), 33 QTLs for GY, straw weight (StW), spike number (SN), GNPS and 1000-grain weight and 18 QTLs for grain K content (GKC), straw K content (StKC), grain KUE (GKUE) and straw KUE (StKUE) were identified under LK conditions (LKC) (Table 9.1) (Kong *et al.*, 2013). The same 131 RILs were used to detect 87 QTLs for six morphological traits such as number of axial roots, MRL, SH, RDW, SDW and TDW using 380 SSR markers (Guo *et al.*, 2012). In the same study, 11, 20 and 23 QTLs for RKC, SKC and TKC, respectively, and 23, 15 and 13 QTLs for RKUE, SKUE and TKUE, respectively, were identified using the same 380 SSR markers under LKC (Guo *et al.*, 2012).

Later, at the seedling stage under LKC, 168 DHLs derived from a cross between Huapei 3 and Yumai 57 were used to investigate 29 QTLs for morphological and 19 QTLs for physiological traits in root and shoots, respectively, using 284 SSR, 37 EST and one ISSR markers (Zhao *et al.*, 2014). Of the 29 morphological QTLs, five for TRL, three for total root surface area (TRS), five for total root volume (TRV), and four for each of RDW, SDW, root average diameter (RAD), RRDW and RSDW were detected under LKC and they were distributed on chromosomes 1A, 1B, 1D, 2A, 3A, 3B, 4B, 5B, 6A, 6B and 7B. Similarly, under LKC, among 19 physiological trait QTLs, one for chlorophyll concentration (CHL), two for peroxidase (POD), three for each of root activity (RA), superoxide dismutase (SOD), SKC and RKC, and four for malondialdehyde (MDA) were detected that were located on chromosomes 1B, 1D, 3A, 4A, 4B, 4D, 5B, 7B and 7D (Table 9.1) (Zhao *et al.*, 2014). Furthermore, 131 RILs derived from a Chuan 35050 × Shannong 483 cross were used to identify 127 QTLs detected on 20 chromosomes associated with RKC, SKC, total K content (TKC), root K concentration (RKCE), shoot K concentration (SKCE), total K concentration (TKCE), root KUE (RKUE), shoot KUE (SKUE), ratio of SKC (RSKC), ratio of RKC (RRKC) and total K-utilization efficiency (TKUtE) (Gong *et al.*, 2015). Of these 127 QTLs, 35, 55 and 37 were found for the RKCE, SKCE and TKCE, four for the RKC, SKC, TKC and RSKC, and three for the RKUE, SKUE and TKUE, respectively. More recently, 217 QTLs for seedling traits and 89 QTLs for adult traits were detected at the seedling stage in a hydroponic culture and at the mature stage in a field trail, respectively, in 184 RILs derived from a cross between Tainong 18 and Linmai 6 under low- and high-K conditions (Shen *et al.*, 2019). Among 217 QTLs in a greenhouse trail, 28 QTLs for four biomass weight (RDW, SDW, TDW, RSDW and RRDW), 52 for K efficiency-related traits (RKC, SKC, TKC, RKCE, SKCE, TKCE, SKUE, RKUE and TKUE), 67 for Ca efficiency-related traits (root Ca concentration (RCaCE), shoot Ca concentration (SCaCE), total Ca concentration (TCaCE), root Ca content (RCaC), shoot Ca content (SCaC), total Ca content (TCaC), Ca-use efficiency (CaUE) of shoot (SCaUE), root CaUE (RCaUE), total CaUE (TCaUE)) and 70 for Mg efficiency-related traits (root Mg concentration (RMgCE), shoot Mg concentration (SMgCE), total Mg concentration (TMgCE), shoot Mg content (SMgC), root Mg content (RMgC), total Mg content (TMgC), Mg-use efficiency (MgUE) of shoot (SMgUE), total MgUE (TMgUE), root MgUE (RMgUE)) have been detected (Table 9.1). Similarly, among 89 adult QTLs in a field trial, 21 QTLs for yield traits (PH, SN, GNPS, 1000-grain weight, grain weight per plant (GWP), total above-ground weight (TAW) and HI), 22 QTLs for K efficiency-related traits (such as grain K concentration (GKCE), straw K concentration (StKCE), grain K content (GKC), straw K content (StKC), above-ground K content (AKC), K-harvest index (KHI), grain KUE (GKUE), straw KUE (StKUE), above-ground KUE (AKUE)), 35 QTLs for Ca efficiency-related traits (grain Ca concentration (GCaCE), straw Ca concentration (StCaCE), above-ground Ca concentration (ACaCE), grain Ca content (GCaC), straw Ca content (StCaC), above-ground Ca content (ACaC), Ca-harvest index (CaHI), grain CaUE (GCaUE), straw CaUE (StCaUE), above-ground CaUE (ACaUE)) and 11 QTL for Mg efficiency-related traits (grain Mg concentration (GMgCE), straw Mg concentration (StMgCE), grain Mg content (GMgC), above-ground Mg content (AMgC), Mg-harvest index (MgHI), grain MgUE (GMgUE) and straw MgUE (StMgUE)) have been reported. Under LKC, the majority of traits related to K content and KC were decreased but KUE-related traits were increased at the seedling and maturity stages in both trials. Increases in Ca and Mg contents and decreases in Ca- and Mg-use efficiencies under LKC in both trials may facilitate further understanding of the effects of K deficiency on

K, Ca and Mg nutrition at the phenotypic and QTL levels.

## 9.8 Identification and Functional Characterization of Potassium Transporters

Plants have various K transporters that have been grouped into four families such as $K^+$ transporter (KT)/high-affinity $K^+$ transporter (HAK)/ $K^+$ uptake transporter (KUP), transport of $K^+$ (Trk)/high-affinity $K^+$ transporters (HKT), $K^+$ efflux anti-porter (KEA) and cation/hydrogen exchanger (CHX). All these K transporters play a vital role in plant growth and development as well as K uptake and transport. Among the four families, HAK/KUP/KT is the largest K transporter family and it is further subdivided into four groups: clusters I, II, III and IV. The first K transporter, *T. aestivum TaHKT1*, was identified in root of wheat by Schachtman and Schroeder (1994). TaHKT1 was reported to be a high-affinity transporter ($K_M$ = 10–40 μM) based on yeast complementation assay. *TaKHT2;1* transporter was found to be a $Na^+/K^+$ symporter when expressed in *Xenopus* oocytes and downregulation of its expression *in planta* reduces root $Na^+$ accumulation and improves growth in saline conditions. Recently, the expression patterns of 56 *HAK* family transporter genes were identified in wheat root at various time periods (0, 1, 3, 6, 9, 12 and 24 h) under LKC (Cheng *et al.*, 2018). Among 56 *HAK* family genes, *TaHAK1, 2, 7, 9, 11, 16, 17, 18* and *23* were expressed in roots at all time periods under LKC. Notably, the transporters *TaHAK9, 11, 16* and *17* had higher expression at 3 h, *TaHAK1, 12* and *23* at 12 h and *TaHAK2* and *7* at 9 h than at other time periods. TaHAK1 transporter was functionally characterized and found to be localized in plasma membrane involved in the transport of K from root to shoot. Upon an imposition of K deficiency treatment, *TaHAK2, 9, 11, 16* and *17* were upregulated for a short period and then rapidly downregulated, suggesting that they might be involved in LK responses in wheat. Similarly, K deficiency treatment resulted in continuous upregulation of *TaHAK1, 7, 18* and *23*, suggesting that they might function in the K uptake and maintain normal growth of the plant under LKC. In wheat, the study of K transporters is at an early stage stage when compared with P and N transporter studies. Further study on K transporters' structure, function, affinity and other molecular studies will throw more light on the complex function of K transporters in K uptake, transport, remobilization and utilization in wheat and other plants. Apart from P and N transporter studies in wheat, further high-resolution study on various aspects of K transporters in wheat and future development of new wheat varieties with high acquisition and use efficiencies for K will surely aid in enhancing wheat production globally, which is intrinsically connected to food security of the poor in the developing world.

## 9.9 Improving Nitrogen-, Phosphorus- and Potassium-Use Efficiencies Through Transgenic Approach

Development of biotechnological tools helps us to characterize the function of key genes in plants. Some important transporter and transcription factors related to N and P transport in wheat are overexpressed in wheat transgenic modification. Transcription factors of CCAAT box-binding were conserved among all eukaryotes, which are called nuclear factor Y (NF-Y) in plants (Mantovani, 1999). The NF-Y transcription factors are heterotrimers and composed of three different protein subunits: NFYA, NFYB and NFYC. The NF-Y family members play a vital role in photosynthesis, nodule development, control of flowering and seed development in plants (Yamamoto *et al.*, 2009; Zanetti *et al.*, 2010; Stephenson *et al.*, 2011; Yan *et al.*, 2011). Similarly, several studies reported that the NF-Y family members enhance plant growth and development under several abiotic stresses such as drought (Nelson *et al.*, 2007; Ni *et al.*, 2013), salinity (Zhao *et al.*, 2009; Leyva-González *et al.*, 2012), cold (Leyva-González *et al.*, 2012; Shi *et al.*, 2014) and low-nutrient stresses (N, P and other nutrients) (Qu *et al.*, 2015). For example, overexpression of *TaNFYA-B1* in wheat stimulated root development, upregulated the expression of N and P transporters, and increased N and P uptake and GY under LNC and LPC (Qu *et al.*, 2015). This study revealed that NF-Y transcription factors are involved in root development and N- and P-use management in wheat. Furthermore, NAC transcription factors family genes were also involved in many developmental and physiological processes of plants, such as embryo and shoot meristem development, lateral root formation, auxin signalling, defence, abiotic stress responses and senescence (Olsen *et al.*, 2005; Nakashima *et al.*, 2012; Nuruzzaman *et al.*, 2012). In transgenic wheat, overexpression of *TaNAC-S* delayed leaf senescence and increased GY and grain N concentration (Zhao *et al.*, 2015). Similarly, overexpression of *TaNAC2-5A* in transgenic wheat increased GY, grain N concentration, NHI and root growth under LNC (He *et al.*, 2015). This clearly reveals that the NAC transcription factors play a crucial role in uptake of N from the soil. Recently, wheat miRNA transcription factors such as *TamiR444a* and *TamiR2275*, when overexpressed in tobacco, improved plant growth and biomass along with increasing N concentration and content, NUP, photosynthetic parameters and antioxidant enzymatic activities under LNC (Gao *et al.*, 2016; Qiao *et al.*, 2018). This report suggests that wheat miRNA transcription factors may help to improve NUE in wheat under LNC.

The *PHT1* genes were also overexpressed through transgenic modification in wheat. Until now, only two *PHT1* genes were used for overexpression analysis in wheat. Under LPC, overexpression of *TaPHT1;2* increased the plant dry mass accumulation, total P content and photosynthetic efficiencies in transgenic wheat (Guo, C. *et al.*, 2014). Similarly, overexpression of *TaPHT1;4* also improved the root and leaf growth and increased uptake of P under LPC (Liu *et al.*, 2013). By contrast, downregulation of *TaPHT1;4* reduced plant growth, fresh and dry weight of shoot and root, and Pi acquisition in the roots and leaves under LPC (Liu *et al.*, 2013). In total, 12 *PHT1* genes were identified in wheat. Hence, the overexpression study of other *PHT1* genes of wheat would help in understanding the complex mechanism of plant Pi uptake, translocation and homeostasis. MYB–coiled-coil (MYB-CC) type transcription factor of *phosphate starvation response 1* (*PHR1*) regulates the expression of *Pi starvation-responsive/induced* (*PS-R/PSI*) genes in plants (Rubio *et al.*, 2001). *TaPHR1* and *TaPHR3* overexpressed in transgenic wheat upregulated other *PSI* genes, stimulated lateral root branching and increased PUP, GY and GNPS under LPC (Wang, J. *et al.*, 2013; Zheng *et al.*, 2020). *Phosphate2* (*PHO2*) plays a key role in Pi starvation-signalling in plants (Huang *et al.*, 2013). Overexpression of *TaPHO2* increased PUP and GY under LPC (Ouyang *et al.*, 2016). Transcription factors of basic helix-loop-helix (bHLH) are involved in various biological processes in plants including flower induction, trichome and root hair development, organelle differentiation, biosynthesis metabolisms of flavonoids, isoquinoline alkaloids and anthocyanins, and nodule vascular patterning (Godiard *et al.*, 2011). Yang *et al.* (2016) overexpressed *TabHLH1* in transgenic wheat which increased plant growth, biomass, Pi and N acquisition under LNC and LPC. Notably, until now, K transporter genes have not been overexpressed in wheat. Therefore, researchers need to focus on overexpression of the genes already identified as related to KUE in wheat, which may help to better understand their function. Also, few of the N- and P-related genes overexpressed in wheat compared with rice and other crops.

## 9.10 Impact of Climate Change on Wheat Nitrogen, Phosphorus and Potassium Management

Natural resources such as light, carbon, water and nutrients are essential for the normal growth and yield of plants (Vitousek *et al.*, 1997). These natural resources are severely affected by climate change. For example, the intensity and frequency of thunderstorms, tropical storms, hurricanes, cyclones, tornadoes, hailstorms, ice storms, dust storms and erosion of beaches during storms are the effects of climate change that severely affect the management of N, P, K and other nutrients (Moore and Allard, 2011). Importantly, the changes in climatic conditions along with geographical gradients greatly affect soil nutrient cycling. Therefore, climate change intensively affects agriculture that feeds 6 billion people globally; furthermore, in Asia and the Pacific, more than two-thirds of the working population depends on agriculture for their livelihoods. The increases in temperature and atmospheric $CO_2$ severely affect soil microbial physiology, altering the availability of soil P in terrestrial ecosystems (Delgado-Baquerizo *et al.*, 2013). Similarly, global mean temperature and precipitation are projected to show variation over the 21st century where the Mediterranean region will be drier while the northern hemisphere latitude region will be wetter, although a general global trend would be towards dryness (Cook *et al.*, 2014). Apart from the elevated temperature and $CO_2$, other climate-change-driven effects such as agricultural runoff, drought, ozone depletion and water pollution have been intrinsically linked to human activity (Vitousek *et al.*, 1997). The global mean temperature has increased by >1.5°C over the last 100 years due to both natural and human causes (Karmakar *et al.*, 2016). In the 21st century it will rise by another 2–3°C, contributed mainly by intensified agriculture (Compant *et al.*, 2010). Importantly, P is affected by temperature, rainfall, drought and $CO_2$ concentration (Short *et al.*, 2016). Temperature at its optimum level is very vital for nutrient uptake, plant growth and development. For example, optimum temperature for vegetative growth of wheat is 20–30°C and optimum temperature for yield is 15°C (Tuteja and Gill, 2013). Low soil temperature reduces root growth, diffusion of P and the subsequent lowering of energy leading to decreased kinetic energy, fluidity of protoplasm, solubility of certain solids and reaction velocities (Amstrong, 1999). Extremely low temperature will also decrease soil-bound enzyme activities to a large extent (Tuteja and Gill, 2013). Low soil temperatures also limit the plants with slower metabolic processes, poor growth and development. Soil temperature below 25°C reduced translocation of Pi from the shoots to roots, Pi uptake by the roots, or both, having a great negative impact on root growth, root extension, Pi uptake, yield and growth of plants with and without Pi fertilizer (Ketcheson, 1957). Elevated temperature (34–40°C) also decreased biomass, GPS and yields in winter wheat (Ju *et al.*, 2010; Tuteja and Gill, 2013). These findings clearly indicate that low and high temperatures suppress the nutrient uptake, translocation, soil microbial activity, kinetic energy, root and shoot growth and overall yield of plants. Furthermore, elevated $CO_2$ modifies the root development, biomass allocation and nutrient uptake including N, P and K in wheat during vegetative growth. For example, elevated $CO_2$ reduced the N, P and K contents in roots and acquisition of N, P and K in wheat under drought and moisture conditions (Van Vuuren *et al.*, 1997). Similarly, elevated $CO_2$ reduced the leaf N concentration in wheat under LNC (Sinclair *et al.*, 2000). The change in the climate adversely affects not only the soil chemistry and properties but also the growth and yield of plants. At this juncture, knowledge about the impact of climate change on nutrient management and especially the management of N, P and K is very important to ensure global food security. A high-resolution study of various climatic factors and their cumulative effects on N, P and K availability, uptake and mobilization, and scientific and eco-friendly N, P and K mitigation and recovery methods, is needed to understand the complex N, P and K management in the changing climate and improve it to ensure food security, particularly in the developing world.

## 9.11 Conclusions and Outlook

In conclusion, the deficiency of the essential macronutrients N, P and K in soils is a major problem for global wheat production. Low N, P and K have caused severe yield losses in major crops including wheat. Developing and improving wheat varieties for enhanced wheat production under low N, P and K soils worldwide is a major challenge not only for farmers but also for plant breeders and scientists. The balanced management of N, P and K nutrients is indispensable to maintain soil fertility, as well as uptake and transport of these nutrients from soil to grain, to have higher yield with nutrient qualities. Improving the use efficiencies of N, P and K in wheat is an important component of agronomic and economic aspects apart from the well-discussed environmental issues. To improve the internal NUE, PUE and KUE in wheat, the identification, mapping and validation of QTLs will play an important role in marker-assisted selection and MAB. An array of QTLs for N-P-K identified so far will help to develop wheat varieties with better NtUE which are suitable for poor farmers. To decode this complex nutrient efficiency problem in wheat, functional genomics studies cannot be ignored. Apart from the identification of N, P and K transporters, various studies have been carried out on their characterization, tissue-specific expression, structure, function, transgenic studies, etc. in order to understand these transporters for improving crop efficiency of major cereals like wheat. Further study on the post-transcriptional and post-translational regulation of N, P and K transporters needs to be fine-tuned to exploit them in higher wheat production where the global population is ever-increasing and food demand is on a continuous rise. The recent clustered regularly interspaced short palindromic repeats (CRISPR)/CRISPR-associated protein 9 (Cas9) gene-editing tool is projected to play a vital role in crop improvement. The expression of functionally characterized N, P and K genes may be edited or silenced by CRISPR/Cas9 in the future to improve the wheat yield under low N, P and K soils. Furthermore, a combined holistic research with next-generation sequencing (NGS), genotyping by sequencing (GBS), high-density and SNP linkage maps, genome-wide association studies (GWAS) and functional genomics will give more insights for identifying wheat cultivars with higher yield under these nutrient-deficient soils. The crosstalk among N, P, K and other nutrients in the soil, enhancing their mutual uptake and utilization in wheat and other major cereals, should also be paid more attention in our tactics as the lion's share of the population in developing countries is still facing the basic problem of malnutrition. This chapter will offer a new opportunity for creative partnerships among plant breeders, geneticists, physiologists, crop physiologists and soil scientists and others to work collectively towards improved wheat production all over the world to ensure global food security.

## Acknowledgement

The research work in the authors' laboratory is funded by the Department of Biotechnology, Government of India (under grant BT/PR21321/GET/119/76/2016).

## References

Adams, E. and Shin, R. (2014) Transport, signaling, and homeostasis of potassium and sodium in plants. *Journal of Integrative Plant Biology* 56, 231–249.

Adhya, T.K., Kumar, N., Reddy, G., Podile, A.R., Bee, H. and Samantaray, B. (2015) Microbial mobilization of soil phosphorus and sustainable P management in agricultural soils. *Current Science* 108, 1280–1287.

Ahanger, M.A. and Agarwal, R. (2017) Salinity stress induced alterations in antioxidant metabolism and nitrogen assimilation in wheat (*Triticum aestivum* L.) as influenced by potassium supplementation. *Plant Physiology and Biochemistry* 115, 449–460.

Ahmad, M., Riaz, A., Ishaque, M. and Malik, A. (2009) Response of maize hybrids to varying potassium application in Pakistan. *Pakistan Journal of Agricultural Sciences* 46, 179–184.

Ahmad, N., Saleem, M.T. and Twyford, I. (1992) Phosphorus research in Pakistan – a review. In: *Proceeding of Symposium, On the role of phosphorus in crop production*. National Fertilizer Development Centre (NFDC), Islamabad, pp. 59–92.

Alam, S.M., Shah, S.A. and Akhter, M. (2003) Varietal differences in wheat yield and phosphorus use efficiency as influenced by method of phosphorus application. *Songklanakarin Journal of Science Technology* 25, 175–181.

Ali, M.S., Sutradhar, A., Edano, M.L., Edwards, J.T. and Girma, K. (2014) Response of winter wheat grain yield and phosphorus uptake to foliar phosphite fertilization. *International Journal of Agronomy* 2014, 1–9.

Alvarez, J.M., Vidal, E.A. and Gutiérrez, R.A. (2012) Integration of local and systemic signaling pathways for plant N responses. *Current Opinion in Plant Biology* 15, 185–191.

Amstrong, D. (1999) Important factors affecting crop response to phosphorus. *Better Crops* 83, 16–19.

An, D., Su, J., Liu, Q., Zhu, Y., Tong, Y. *et al.* (2006) Mapping QTLs for nitrogen uptake in relation to the early growth of wheat (*Triticum aestivum* L.). *Plant and Soil* 284, 73–84.

Anser Ali, M. and Hussain, S. (2012) Nutritional and physiological significance of potassium application in maize hybrid crop production. *Pakistan Journal of Nutrition* 11, 187–202.

Aziz, T., Finnegan, P.M., Lambers, H. and Jost, R. (2014) Organ-specific phosphorus-allocation patterns and transcript profiles linked to phosphorus efficiency in two contrasting wheat genotypes. *Plant, Cell & Environment* 37, 943–960.

Bahmanyar, M. and Ranjbar, G. (2008) The role of potassium in improving growth indices and increasing. *Journal of Applied Sciences* 8, 1280–1285.

Baker, A., Ceasar, S.A., Palmer, A.J., Paterson, J.B., Qi, W., Muench, S.P. and Baldwin, S.A. (2015) Replace, reuse, recycle: improving the sustainable use of phosphorus by plants. *Journal of Experimental Botany* 66, 3523–3540.

Balafoutis, A., Beck, B., Fountas, S., Vangeyte, J., Wal, T.V.D. *et al.* (2017) Precision agriculture technologies positively contributing to GHG emissions mitigation, farm productivity and economics. *Sustainability* 9, 1339.

Baligar, V., Fageria, N. and He, Z. (2001) Nutrient use efficiency in plants. *Communications in Soil Science and Plant Analysis* 32, 921–950.

Barker, A.V. and Bryson, G.M. (2016) Nitrogen. In: Barker, A.V. and Pilbeam, D.J. (eds) *Handbook of Plant Nutrition*. CRC Press, Boca Raton, Florida, pp. 37–66.

Batjes, N. (1997) A world dataset of derived soil properties by FAO–UNESCO soil unit for global modelling. *Soil Use and Management* 13, 9–16.

Bernhard, A. (2010) The nitrogen cycle: processes, players, and human impact. *Nature Education Knowledge* 2, 12–20.

Berry, P., Sylvester-Bradley, R., Philipps, L., Hatch, D.J., Cuttle, S.P., Rayns, F. and Gosling, P. (2002) Is the productivity of organic farms restricted by the supply of available nitrogen? *Soil Use and Management* 18, 248–255.

Bilal, H.M., Aziz, T., Maqsood, M.A., Farooq, M. and Yan, G. (2018) Categorization of wheat genotypes for phosphorus efficiency. *PLoS One* 13, 10.

Bly, A.G. and Woodard, H.J. (2003) Foliar nitrogen application timing influence on grain yield and protein concentration of hard red winter and spring wheat. *Agronomy Journal* 95, 335–338.

Bonser, A.M., Lynch, J. and Snapp, S. (1996) Effect of phosphorus deficiency on growth angle of basal roots in *Phaseolus vulgaris*. *New Phytologist* 132, 281–288.

Buchner, P. and Hawkesford, M.J. (2014) Complex phylogeny and gene expression patterns of members of the nitrate transporter 1/peptide transporter family (NPF) in wheat. *Journal of Experimental Botany* 65, 5697–5710.

Bukhsh, M.A., Ahmad, R., Iqbal, J., Rehman, A., Hussain, S. and Ishaque, M. (2011) Potassium application reduces bareness in different maize hybrids under crowding stress conditions. *Pakistan Journal of Agricultural Science* 48, 41–48.

Bundy, L.G. and Andraski, T.W. (2004) Diagnostic tests for site-specific nitrogen recommendations for winter wheat. *Agronomy Journal* 96, 608–614.

Camacho, R., Malavolta, E., Guerrero-Alves, J. and Camacho, T. (2002) Vegetative growth of grain sorghum in response to phosphorus nutrition. *Scientia Agricola* 59, 771–776.

Ceasar, S.A. (2018) Feeding world population amidst depleting phosphate reserves: the role of biotechnological interventions. *The Open Biotechnology Journal* 12, 51–55.

Ceasar, S.A., Hodge, A., Baker, A. and Baldwin, S.A. (2014) Phosphate concentration and arbuscular mycorrhizal colonisation influence the growth, yield and expression of twelve PHT1 family phosphate transporters in foxtail millet (*Setaria italica*). *PLoS One* 9, e108459.

Chao, S., Zhang, W., Akhunov, E., Sherman, J., Ma, Y., Luo, M.C. and Dubcovsky, J. (2009) Analysis of gene-derived SNP marker polymorphism in US wheat (*Triticum aestivum* L.) cultivars. *Molecular Breeding* 23, 23–33.

Chaturvedi, I. (2006) Effects of different nitrogen levels on growth, yield and nutrient uptake of wheat (*Triticum aestivum* L.). *International Journal of Agricultural Sciences* 2, 372–374.

Chen, Q., Yi, K., Huang, G., Wang, X., Liu, F., Wu, Y. and Wu, P. (2003) Cloning and expression pattern analysis of nitrogen-starvation-induced genes in rice. *Acta Botanica Sinica* 45, 974–980.

Chen, H., Xie, W., He, H., Yu, H., Chen, W. *et al.* (2014) A high-density SNP genotyping array for rice biology and molecular breeding. *Molecular Plant* 7, 541–553.

Cheng, X., Liu, X., Mao, W., Zhang, X., Chen, S., Zhan, K. and Xu, H. (2018) Genome-wide identification and analysis of HAK/KUP/KT potassium transporters gene family in wheat (*Triticum aestivum* L.). *International Journal of Molecular Sciences* 19, 3969.

Clark, R.M., Schweikert, G., Toomajian, C., Ossowski, S., Zeller, G. *et al.* (2007) Common sequence polymorphisms shaping genetic diversity in *Arabidopsis thaliana*. *Science* 317, 338–342.

Compant, S., Van Der Heijden, M.G. and Sessitsch, A. (2010) Climate change effects on beneficial plant–microorganism interactions. *FEMS Microbiology Ecology* 73, 197–214.

Cook, B.I., Smerdon, J.E., Seager, R. and Coats, S. (2014) Global warming and 21st century drying. *Climate Dynamics* 43, 2607–2627.

Cormier, F., Foulkes, J., Hirel, B., Gouache, D., Moënne-Loccoz, Y. and Le Gouis, J. (2016) Breeding for increased nitrogen-use efficiency: a review for wheat (*Triticum aestivum* L.). *Plant Breeding* 135, 255–278.

Cui, F., Fan, X., Zhao, C., Zhang, W., Chen, M., Ji, J. and Li, J. (2014) A novel genetic map of wheat: utility for mapping QTL for yield under different nitrogen treatments. *BMC Genetics* 15, 57.

Cui, F., Fan, X., Chen, M., Zhang, N., Zhao, C. *et al.* (2016) QTL detection for wheat kernel size and quality and the responses of these traits to low nitrogen stress. *Theoretical and Applied Genetics* 129, 469–484.
Davies, T., Ying, J., Xu, Q., Li, Z., Li, J. and Gordon-Weeks, R. (2002) Expression analysis of putative high-affinity phosphate transporters in Chinese winter wheats. *Plant, Cell & Environment* 25, 1325–1339.
Delgado-Baquerizo, M., Maestre, F.T., Gallardo, A., Bowker, M.A., Wallenstein, M.D. *et al.* (2013) Decoupling of soil nutrient cycles as a function of aridity in global drylands. *Nature* 502, 672–676.
Deng, L.W.,Tan, X.Q., Li, J., Chen, Z., Tang, Y.Z. and Zhu, H. (2004) Treatment and reuse of piggery wastewater by composting process of straw. *Transactions of the Chinese Society of Agricultural Engineering* 20, 255–259.
Dibb, D.W. and Thompson, W.R. (1985) Interaction of potassium with other nutrients. In: Munson, R.D. (ed.) *Potassium in Agriculture*. American Society of Agronomy, Crop Science Society of America and Soil Science Society of America, Madison, Wisconsin, pp. 515–533.
Duan, J., Tian, H., Drijber, R.A. and Gao, Y. (2015) Systemic and local regulation of phosphate and nitrogen transporter genes by arbuscular mycorrhizal fungi in roots of winter wheat (*Triticum aestivum* L.). *Plant Physiology and Biochemistry* 96, 199–208.
Duan, J., Tian, H. and Gao, Y. (2016) Expression of nitrogen transporter genes in roots of winter wheat (*Triticum aestivum* L.) in response to soil drought with contrasting nitrogen supplies. *Crop and Pasture Science* 67, 128–136.
Ersoz, E.S., Yu, J. and Buckler, E.S. (2009) Applications of linkage disequilibrium and association mapping in maize. In: Kriz, A.L. and Larkins, B.A. (eds) *Molecular Genetic Approaches to Maize Improvement*. Springer, New York, pp. 173–195.
Fan, X., Naz, M., Fan, X., Xuan, W., Miller, A.J. and Xu, G. (2017) Plant nitrate transporters: from gene function to application. *Journal of Experimental Botany* 68, 2463–2475.
Flint-Garcia, S.A., Thuillet, A.C., Yu, J., Pressoir, G., Romero, S.M. *et al.* (2005) Maize association population: a high-resolution platform for quantitative trait locus dissection. *The Plant Journal* 44, 1054–1064.
Fontaine, J.X., Ravel, C., Pageau, K., Heumez, E., Dubois, F., Hirel, B. and Le Gouis, J. (2009) A quantitative genetic study for elucidating the contribution of glutamine synthetase, glutamate dehydrogenase and other nitrogen-related physiological traits to the agronomic performance of common wheat. *Theoretical and Applied Genetics* 119, 645–662.
Frink, C.R., Waggoner, P.E. and Ausubel, J.H. (1999) Nitrogen fertilizer: retrospect and prospect. *Proceedings of the National Academy of Sciences USA* 96, 1175–1180.
Fuller, D.Q. (2007) Contrasting patterns in crop domestication and domestication rates: recent archaeobotanical insights from the old world. *Annals of Botany* 100, 903–924.
Gahlaut, V., Jaiswal, V., Tyagi, B.S., Singh, G., Sareen, S., Balyan, H.S. and Gupta, P.K. (2017) QTL mapping for nine drought-responsive agronomic traits in bread wheat under irrigated and rain-fed environments. *PLoS One* 12, e0182857.
Gao, S., Guo, C., Zhang, Y., Zhang, F., Du, X., Gu, J. and Xiao, K. (2016) Wheat microRNA member TaMIR444a is nitrogen deprivation-responsive and involves plant adaptation to the nitrogen-starvation stress. *Plant Molecular Biology Reporter* 34, 931–946.
Giraldo, P., Benavente, E., Manzano-Agugliaro, F. and Gimenez, E. (2019) Worldwide research trends on wheat and barley: a bibliometric comparative analysis. *Agronomy* 9, 352.I.
Glassop, D., Smith, S.E. and Smith, F.W. (2005) Cereal phosphate transporters associated with the mycorrhizal pathway of phosphate uptake into roots. *Planta* 222, 688–698.
Godiard, L., Lepage, A., Moreau, S., Laporte, D., Verdenaud, M., Timmers, T. and Gamas, P. (2011) MtbHLH1, a bHLH transcription factor involved in *Medicago truncatula* nodule vascular patterning and nodule to plant metabolic exchanges. *New Phytologist* 191, 391–404.
Gong, X.P., Liang, X., Guo, Y., Wu, C.H., Zhao, Y. *et al.* (2015) Quantitative trait locus mapping for potassium use efficiency traits at the seedling stage in wheat under different nitrogen and phosphorus treatments. *Crop Science* 55, 2690–2700.
Gorcek, Z. and Erdal, S. (2015) Lipoic acid mitigates oxidative stress and recovers metabolic distortions in salt-stressed wheat seedlings by modulating ion homeostasis, the osmo-regulator level and antioxidant system. *Journal of the Science of Food and Agriculture* 95, 2811–2817.
Grün, A., Buchner, P., Broadley, M. and Hawkesford, M. (2018) Identification and expression profiling of Pht1 phosphate transporters in wheat in controlled environments and in the field. *Plant Biology* 20, 374–389.
Guo, B., Jin, Y., Wussler, C., Blancaflor, E., Motes, C. and Versaw, W.K. (2008) Functional analysis of the *Arabidopsis* PHT4 family of intracellular phosphate transporters. *New Phytologist* 177, 889–898.

Guo, C., Zhao, X., Liu, X., Zhang, L., Gu, J. *et al.* (2013) Function of wheat phosphate transporter gene TaPHT2; 1 in Pi translocation and plant growth regulation under replete and limited Pi supply conditions. *Planta* 237, 1163–1178.

Guo, C., Guo, L., Li, X., Gu, J., Zhao, M. *et al.* (2014a) TaPT2, a high-affinity phosphate transporter gene in wheat (*Triticum aestivum* L.), is crucial in plant Pi uptake under phosphorus deprivation. *Acta Physiologiae Plantarum* 36, 1373–1384.

Guo, T., Xuan, H., Yang, Y., Wang, L., Wei, L., Wang, Y. and Kang, G. (2014b) Transcription analysis of genes encoding the wheat root transporter NRT1 and NRT2 families during nitrogen starvation. *Journal of Plant Growth Regulation* 33, 837–848.

Guo, Y., Kong, F.M., Xu, Y.F., Zhao, Y., Liang, X. *et al.* (2012) QTL mapping for seedling traits in wheat grown under varying concentrations of N, P and K nutrients. *Theoretical and Applied Genetics* 124, 851–865.

Gupta, P.K., Balyan, H.S. and Gahlaut, V. (2017) QTL analysis for drought tolerance in wheat: present status and future possibilities. *Agronomy* 7, 5.

Han, M., Okamoto, M., Beatty, P.H., Rothstein, S.J. and Good, A.G. (2015) The genetics of nitrogen use efficiency in crop plants. *Annual Review of Genetics* 49, 269–289.

He, X., Qu, B., Li, W., Zhao, X., Teng, W., Ma, W. and Tong, Y. (2015) The nitrate-inducible NAC transcription factor TaNAC2-5A controls nitrate response and increases wheat yield. *Plant Physiology* 169, 1991–2005.

Heuer, S., Lu, X., Chin, J.H., Tanaka, J.P., Kanamori, H. *et al.* (2009) Comparative sequence analyses of the major quantitative trait locus phosphorus uptake 1 (Pup1) reveal a complex genetic structure. *Plant Biotechnology Journal* 7, 456–471.

Huang, T.K., Han, C.L., Lin, S.I., Chen, Y.J., Tsai, Y.C. *et al.* (2013) Identification of downstream components of ubiquitin-conjugating enzyme PHOSPHATE2 by quantitative membrane proteomics in *Arabidopsis* roots. *The Plant Cell* 25, 4044–4060.

Hussain, I., Balko, L., Russell, W., Belger, E., Buttrey, S. *et al.* (1997) Soil fertility and fertilizers: elements required in plant nutrition. *Pakistan Journal of Biological Sciences* 2, 723–728.

Hussain, N., Khan, M.B. and Ahmad, R. (2008) Influence of phosphorus application and sowing time on performance of wheat in calcareous soils. *International Journal of Agriculture & Biology* 10, 399–404.

International Wheat Genome Sequencing Consortium (2014) A chromosome-based draft sequence of the hexaploid bread wheat (*Triticum aestivum*) genome. *Science* 345, 1251788.

Iqbal, Z., Latif, A., Ali, S. and Iqbal, M.M. (2003) Effect of fertigated phosphorus on P use efficiency and yield of wheat and maize. *Songklanakarin Journal of Science Technology* 25, 697–702.

Jeuffroy, M.H. and Bouchard, C. (1999) Intensity and duration of nitrogen deficiency on wheat grain number. *Crop Science* 39, 1385–1393.

Ju, Z., Hu, C., Zhang, Y. and Chen, S. (2010) Effects of temperature rising on soil hydrothermal properties, winter wheat growth and yield. In: Darnhofer, I. and Grötzer, M. (eds) *Proceedings of the 9th European IFSA (International Farming Systems Association) Symposium, Vienna, 4–7 July 2010*. Universität für Bodenkultur, Vienna, pp. 1307–1316.

Karmakar, R., Das, I., Dutta, D. and Rakshit, A. (2016) Potential effects of climate change on soil properties: a review. *Science International* 4, 51–73.

Ketcheson, J. (1957) Some effects of soil temperature on phosphorus requirements of young corn plants in the greenhouse. *Canadian Journal of Soil Science* 37, 41–47.

Khajuria, A. and Kanae, S. (2013) Potential and use of nitrate in agricultural purposes. *Journal of Water Resource and Protection* 5, 529–533.

Khan, A., Khan, S., Khan, M.A., Qamar, Z. and Waqas, M. (2015) The uptake and bioaccumulation of heavy metals by food plants, their effects on plants nutrients, and associated health risk: a review. *Environmental Science and Pollution Research* 22, 13772–13799.

Kiba, T., Kudo, T., Kojima, M. and Sakakibara, H. (2011) Hormonal control of nitrogen acquisition: roles of auxin, abscisic acid, and cytokinin. *Journal of Experimental Botany* 62, 1399–1409.

Kole, C., Muthamilarasan, M., Henry, R., Edwards, D., Sharma, R. *et al.* (2015) Application of genomics-assisted breeding for generation of climate resilient crops: progress and prospects. *Frontiers in Plant Science* 6, 563.

Kong, F.M., Guo, Y., Liang, X., Wu, C.H., Wang, Y.Y., Zhao, Y. and Li, S.S. (2013) Potassium (K) effects and QTL mapping for K efficiency traits at seedling and adult stages in wheat. *Plant and Soil* 373, 877–892.

Ladha, J.K., Pathak, H., Krupnik, T.J., Six, J. and van Kessel, C. (2005) Efficiency of fertilizer nitrogen in cereal production: retrospects and prospects. *Advances in Agronomy* 87, 85–156.

Laperche, A., Devienne-Barret, F., Maury, O., Le Gouis, J. and Ney, B. (2006) A simplified conceptual model of carbon/nitrogen functioning for QTL analysis of winter wheat adaptation to nitrogen deficiency. *Theoretical and Applied Genetics* 113, 1131–1146.

Lázaro, L., Abbate, P., Cogliatti, D. and Andrade, F. (2010) Relationship between yield, growth and spike weight in wheat under phosphorus deficiency and shading. *The Journal of Agricultural Science* 148, 83–93.

Leyva-González, M.A., Ibarra-Laclette, E., Cruz-Ramírez, A. and Herrera-Estrella, L. (2012) Functional and transcriptome analysis reveals an acclimatization strategy for abiotic stress tolerance mediated by *Arabidopsis* NF-YA family members. *PLoS One* 7, e48138.

Liu, F., Xu, Y., Jiang, H., Jiang, C., Du, Y., Gong, C. and Cheng, B. (2016a) Systematic identification, evolution and expression analysis of the *Zea mays* PHT1 gene family reveals several new members involved in root colonization by arbuscular mycorrhizal fungi. *International Journal of Molecular Sciences* 17, 930.

Liu, T.Y., Huang, T.K., Yang, S.Y., Hong, Y.T., Huang, S.M. *et al.* (2016b) Identification of plant vacuolar transporters mediating phosphate storage. *Nature Communications* 7, 11095.

Liu, X., Zhao, X., Zhang, L., Lu, W., Li, X. and Xiao, K. (2013) TaPht1;4, a high-affinity phosphate transporter gene in wheat (*Triticum aestivum*), plays an important role in plant phosphate acquisition under phosphorus deprivation. *Functional Plant Biology* 40, 329–341.

Liu, Z., Zhu, C., Jiang, Y., Tian, Y., Yu, J. *et al.* (2016c) Association mapping and genetic dissection of nitrogen use efficiency-related traits in rice (*Oryza sativa* L.). *Functional and Integrative Genomics* 16, 323–333.

Lynch, J.P. and Brown, K.M. (2001) Topsoil foraging – an architectural adaptation of plants to low phosphorus availability. *Plant and Soil* 237, 225–237.

Maathuis, F.J. (2009) Physiological functions of mineral macronutrients. *Current Opinion in Plant Biology* 12, 250–258.

Mahamood, J., Abayomi, Y. and Aduloju, M. (2009) Comparative growth and grain yield responses of soybean genotypes to phosphorous fertilizer application. *African Journal of Biotechnology* 8, 1030–1036.

Maharajan, T., Ceasar, S.A., Ajeesh Krishna, T.P., Ramakrishnan, M., Duraipandiyan, V., Naif Abdulla, A.D. and Ignacimuthu, S. (2018) Utilization of molecular markers for improving the phosphorus efficiency in crop plants. *Plant Breeding* 137, 10–26.

Maharajan, T., Ceasar, S.A., Krishna, T.P.A. and Ignacimuthu, S. (2019) Phosphate supply influenced the growth, yield and expression of PHT1 family phosphate transporters in seven millets. *Planta* 250, 1433–1448.

Mantovani, R. (1999) The molecular biology of the CCAAT-binding factor NF-Y. *Gene* 239, 15–27.

Marschner, H. and Marschner, P. (2011) *Marschner's Mineral Nutrition of Higher Plants*, 3rd edn. Academic Press, London/Waltham, Massachusetts.

McCauley, A., Jones, C. and Jacobsen, J. (2009) Plant nutrient functions and deficiency and toxicity symptoms. *Nutrient Management Module* 9, 1–16.

Mengel, K. (2007) Potassium. In: Barker, A.V. and Pilbeam, D.J. (eds) *Handbook of Plant Nutrition*. CRC Press, Boca Raton, Florida, pp. 91–120.

Miao, J., Sun, J., Liu, D., Li, B., Zhang, A., Li, Z. and Tong, Y. (2009) Characterization of the promoter of phosphate transporter TaPHT1.2 differentially expressed in wheat varieties. *Journal of Genetics and Genomics* 36, 455–466.

Moore, B. and Allard, G. (2011) *Abiotic Disturbances and Their Influence on Forest Health: A Review*. Forest Health & Biosecurity Working Paper No. FBS/35E. Food and Agriculture Organization of the United Nations, Rome.

Muchhal, U.S., Pardo, J.M. and Raghothama, K.G. (1996) Phosphate transporters from the higher plant *Arabidopsis thaliana*. *Proceedings of the National Academy of Sciences USA* 93, 10519–10523.

Mudge, S.R., Rae, A.L., Diatloff, E. and Smith, F.W. (2002) Expression analysis suggests novel roles for members of the Pht1 family of phosphate transporters in *Arabidopsis*. *The Plant Journal* 31, 341–353.

Nadeem, S.M., Ahmad, M., Zahir, Z.A., Javaid, A. and Ashraf, M. (2014) The role of mycorrhizae and plant growth promoting rhizobacteria (PGPR) in improving crop productivity under stressful environments. *Biotechnology Advances* 32, 429–448.

Nakashima, K., Takasaki, H., Mizoi, J., Shinozaki, K. and Yamaguchi-Shinozaki, K. (2012) NAC transcription factors in plant abiotic stress responses. *Biochimica et Biophysica Acta* 1819, 97–103.

Nelson, D.E., Repetti, P.P., Adams, T.R., Creelman, R.A., Wu, J., Warner, D.C. and Hinchey, B.S. (2007) Plant nuclear factor Y (NF-Y) B subunits confer drought tolerance and lead to improved corn yields on water-limited acres. *Proceedings of the National Academy of Sciences USA* 104, 16450–16455.

Ni, Z., Hu, Z., Jiang, Q. and Zhang, H. (2013) GmNFYA3, a target gene of miR169, is a positive regulator of plant tolerance to drought stress. *Plant Molecular Biology* 82, 113–129.

Nursuaidah, H., Motior, M., Nazia, A. and Islam, M. (2014) Growth and photosynthetic responses of long bean (*Vigna unguiculata*) and mung bean (*Vigna radiata*) response to fertilization. *Journal of Animal and Plant Sciences* 24, 573–578.

Nuruzzaman, M., Sharoni, A.M., Satoh, K., Moumeni, A., Venuprasad, R. *et al.* (2012) Comprehensive gene expression analysis of the NAC gene family under normal growth conditions, hormone treatment, and drought stress conditions in rice using near-isogenic lines (NILs) generated from crossing Aday Selection (drought tolerant) and IR64. *Molecular Genetics and Genomics* 287, 389–410.

Nussaume, L., Kanno, S., Javot, H., Marin, E., Nakanishi, T.M. and Thibaud, M.C. (2011) Phosphate import in plants: focus on the PHT1 transporters. *Frontiers in Plant Science* 2, 83.

Oerke, E., Dehne, H., Schonbeck, F. and Weber, A. (1994) *Crop Production and Crop Protection: Estimated Losses in Major Food and Cash Crops*. Elsevier Science BV, Amsterdam.

Olsen, A.N., Ernst, H.A., Leggio, L.L. and Skriver, K. (2005) NAC transcription factors: structurally distinct, functionally diverse. *Trends in Plant Science* 10, 79–87.

Ouyang, X., Hong, X., Zhao, X., Zhang, W., He, X. *et al.* (2016) Knock out of the PHOSPHATE 2 gene TaPHO2-A1 improves phosphorus uptake and grain yield under low phosphorus conditions in common wheat. *Scientific Reports* 6, 29850.

Pavlíková, D., Neuberg, M., Zizkova, E., Motyka, V. and Pavlik, M. (2012) Interactions between nitrogen nutrition and phytohormone levels in *Festulolium* plants. *Plant, Soil and Environment* 58, 367–372.

Paszkowski, U., Kroken, S., Roux, C. and Briggs, S.P. (2002) Rice phosphate transporters include an evolutionarily divergent gene specifically activated in arbuscular mycorrhizal symbiosis. *Proceedings of the National Academy of Sciences USA* 99, 13324–13329.

Peleg, Z., Cakmak, I., Ozturk, L., Yazici, A., Jun, Y. *et al.* (2009) Quantitative trait loci conferring grain mineral nutrient concentrations in durum wheat × wild emmer wheat RIL population. *Theoretical and Applied Genetics* 119, 353–369.

Penido, M.G.M. and Alon, U.S. (2012) Phosphate homeostasis and its role in bone health. *Pediatric Nephrology* 27, 2039–2048.

Perez-de-Castro, A.M., Vilanova, S., Cañizares, J., Pascual, L.M., Blanca, J.J. *et al.* (2012) Application of genomic tools in plant breeding. *Current Genomics* 13, 179–195.

Pettigrew, W.T. (2008) Potassium influences on yield and quality production for maize, wheat, soybean and cotton. *Physiologia Plantarum* 133, 670–681.

Plenet, D., Mollier, A. and Pellerin, S. (2000) Growth analysis of maize field crops under phosphorus deficiency: II. Radiation-use efficiency, biomass accumulation and yield components. *Plant and Soil* 224, 259–272.

Poirier, Y. and Bucher, M. (2002) Phosphate transport and homeostasis in *Arabidopsis*. In: Somerville, C.R. (ed.) *The Arabidopsis Book*. American Society of Plant Biologists, Rockville, Maryland, pp. 1–35.

Qiao, Q., Wang, X., Yang, M., Zhao, Y., Gu, J. and Xiao, K. (2018) Wheat miRNA member TaMIR2275 involves plant nitrogen starvation adaptation via enhancement of the N acquisition-associated process. *Acta Physiologiae Plantarum* 40, 183.

Qu, B., He, X., Wang, J., Zhao, Y., Teng, W., Shao, A. and Li, Z. (2015) A wheat CCAAT box-binding transcription factor increases the grain yield of wheat with less fertilizer input. *Plant Physiology* 167, 411–423.

Quarrie, S., Steed, A., Calestani, C., Semikhodskii, A., Lebreton, C. *et al.* (2005) A high-density genetic map of hexaploid wheat (*Triticum aestivum* L.) from the cross Chinese Spring × SQ1 and its use to compare QTLs for grain yield across a range of environments. *Theoretical and Applied Genetics* 110, 865–880.

Rae, A.L., Cybinski, D.H., Jarmey, J.M. and Smith, F.W. (2003) Characterization of two phosphate transporters from barley; evidence for diverse function and kinetic properties among members of the Pht1 family. *Plant Molecular Biology* 53, 27–36.

Rausch, C. and Bucher, M. (2002) Molecular mechanisms of phosphate transport in plants. *Planta* 216, 23–37.

Rausch, C., Zimmermann, P., Amrhein, N. and Bucher, M. (2004) Expression analysis suggests novel roles for the plastidic phosphate transporter Pht2;1 in auto- and heterotrophic tissues in potato and *Arabidopsis*. *The Plant Journal* 39, 13–28.

Ren, Y., He, X., Liu, D., Li, J., Zhao, X. *et al.* (2012) Major quantitative trait loci for seminal root morphology of wheat seedlings. *Molecular Breeding* 30, 139–148.

Ren, Y., Qian, Y., Xu, Y., Zou, C., Liu, D. *et al.* (2017) Characterization of QTLs for root traits of wheat grown under different nitrogen and phosphorus supply levels. *Frontiers in Plant Science* 8, 2096.

Ren, Y., Xu, Y., Teng, W., Li, B. and Lin, T. (2018) QTLs for seedling traits under salinity stress in hexaploid wheat. *Ciencia Rural Santa Maria* 48, e20170446.

Reuter, D., Dyson, C., Elliott, D., Lewis, D. and Rudd, C. (1995) An appraisal of soil phosphorus testing data for crops and pastures in South Australia. *Australian Journal of Experimental Agriculture* 35, 979–995.

Richardson, A.E., Hocking, P.J., Simpson, R.J. and George, T.S. (2009) Plant mechanisms to optimise access to soil phosphorus. *Crop and Pasture Science* 60, 124–143.

Roberts, T.L. (2008) Improving nutrient use efficiency. *Turkish Journal of Agriculture and Forestry* 32, 177–182.
Roch, V.G., Maharajan, T., Ceasar, S.A. and Ignacimuthu, S. (2019) The role of PHT1 family transporters in the acquisition and redistribution of phosphorus in plants. *Critical Reviews in Plant Sciences* 38, 171–198.
Rubio, V., Linhares, F., Solano, R., Martín, A.C., Iglesias, J., Leyva, A. and Paz-Ares, J. (2001) A conserved MYB transcription factor involved in phosphate starvation signaling both in vascular plants and in unicellular algae. *Genes & Development* 15, 2122–2133.
Saifullah, A., Ranjha, M., Yaseen, M. and Akhtar, M. (2002) Response of wheat to potassium fertilization under field conditions. *Pakistan Journal of Agricultural Sciences* 39, 269–272.
Sakakibara, H., Takei, K. and Hirose, N. (2006) Interactions between nitrogen and cytokinin in the regulation of metabolism and development. *Trends in Plant Science* 11, 440–448.
Salim, N. and Raza, A. (2020) Nutrient use efficiency (NUE) for sustainable wheat production: a review. *Journal of Plant Nutrition* 43, 297–315.
Sato, K. and Jiang, H.Y. (1996) Gram-negative bacterial flora on the root surface of wheat (*Triticum aestivum*) grown under different soil conditions. *Biology and Fertility of Soils* 23, 273–281.
Sattari, S.Z., Bouwman, A.F., Giller, K.E. and van Ittersum, M.K. (2012) Residual soil phosphorus as the missing piece in the global phosphorus crisis puzzle. *Proceedings of the National Academy of Sciences USA* 109, 6348–6353.
Schachtman, D.P. and Schroeder, J.I. (1994) Structure and transport mechanism of a high-affinity potassium uptake transporter from higher plants. *Nature* 370, 655–658.
Sharma, I., Tyagi, B., Singh, G., Venkatesh, K. and Gupta. O. (2015) Enhancing wheat production – a global perspective. *Indian Journal of Agricultural Sciences* 85, 3–13.
Shaviv, A. (2001) Advances in controlled-release fertilizers. *Advances in Agronomy* 71, 1–49.
Shen, X., Yuan, Y., Zhang, H., Guo, Y., Zhao, Y., Li, S. and Kong, F. (2019) The hot QTL locations for potassium, calcium, and magnesium nutrition and agronomic traits at seedling and maturity stages of wheat under different potassium treatments. *Genes* 10, 607.
Shewry, P.R. (2009) Wheat. *Journal of Experimental Botany* 60, 1537–1553.
Shewry, P.R. and Hey, S.J. (2015) The contribution of wheat to human diet and health. *Food and Energy Security* 4, 178–202.
Shi, H., Ye, T., Zhong, B., Liu, X., Jin, R. and Chan, Z. (2014) AtHAP5A modulates freezing stress resistance in *Arabidopsis* through binding to CCAAT motif of AtXTH21. *New Phytologist* 203, 554–567.
Shi, R., Li, H., Tong, Y., Jing, R., Zhang, F. and Zou, C. (2008) Identification of quantitative trait locus of zinc and phosphorus density in wheat (*Triticum aestivum* L.) grain. *Plant and Soil* 306, 95–104.
Short, F.T., Kosten, S., Morgan, P.A., Malone, S. and Moore, G.E. (2016) Impacts of climate change on submerged and emergent wetland plants. *Aquatic Botany* 135, 3–17.
Shukla, V., Kaur, M., Aggarwal, S., Bhati, K.K., Kaur, J., Mantri, S. and Pandey, A.K. (2016) Tissue specific transcript profiling of wheat phosphate transporter genes and its association with phosphate allocation in grains. *Scientific Reports* 6, 39293.
Sinclair, T.R., Pinter, P.J. Jr, Kimball, B.A., Adamsen, F.J., LaMorte, R.L. *et al.* (2000) Leaf nitrogen concentration of wheat subjected to elevated [$CO_2$] and either water or N deficits. *Agriculture, Ecosystems & Environment* 79, 53–60.
Singh, U., Ladha, J., Castillo, E., Punzalan, G., Tirol-Padre, A. and Duqueza, M. (1998) Genotypic variation in nitrogen use efficiency in medium- and long-duration rice. *Field Crops Research* 58, 35–53.
Sisaphaithong, T., Kondo, D., Matsunaga, H., Kobae, Y. and Hata, S. (2012) Expression of plant genes for arbuscular mycorrhiza-inducible phosphate transporters and fungal vesicle formation in sorghum, barley, and wheat roots. *Bioscience, Biotechnology, and Biochemistry* 76, 2364–2367.
Sramkova, Z., Gregova, E. and Sturdik, E. (2009) Genetic improvement of wheat – a review. *Nova Biotech* 9, 27–51.
Stein, J.C., Yu, Y., Copetti, D., Zwickl, D.J., Zhang, L. *et al.* (2018) Genomes of 13 domesticated and wild rice relatives highlight genetic conservation, turnover and innovation across the genus *Oryza*. *Nature Genetics* 50, 285–296.
Stephenson, T.J., McIntyre, C.L., Collet, C. and Xue, G.P. (2011) TaNF-YB3 is involved in the regulation of photosynthesis genes in *Triticum aestivum*. *Functional and Integrative Genomics* 11, 327–340.
Street, J. and Kidder, G. (1997) *Soils and Plant Nutrition*. Fact Sheet No. SL8. University of Florida, Institute of Food and Agriculture Sciences, Corporative Extension Service, Gainesville, Florida.
Su, J., Xiao, Y., Li, M., Liu, Q., Li, B. *et al.* (2006) Mapping QTLs for phosphorus-deficiency tolerance at wheat seedling stage. *Plant and Soil* 281, 25–36.
Su, J.Y., Zheng, Q., Li, H.W., Li, B., Jing, R.L., Tong, Y.P. and Li, Z.S. (2009) Detection of QTLs for phosphorus use efficiency in relation to agronomic performance of wheat grown under phosphorus sufficient and limited conditions. *Plant Science* 176, 824–836.

Sun, J., Guo, Y., Zhang, G., Gao, M., Zhang, G. *et al.* (2013) QTL mapping for seedling traits under different nitrogen forms in wheat. *Euphytica* 191, 317–331.

Tao, Y., Zhang, S., Jian, W., Yuan, C. and Shan, X. (2006) Effects of oxalate and phosphate on the release of arsenic from contaminated soils and arsenic accumulation in wheat. *Chemosphere* 65, 1281–1287.

Teng, W., Deng, Y., Chen, X.P., Xu, X.F., Chen, R.Y. *et al.* (2013) Characterization of root response to phosphorus supply from morphology to gene analysis in field-grown wheat. *Journal of Experimental Botany* 64, 1403–1411.

Teng, W., Zhao, Y.Y., Zhao, X.Q., He, X., Ma, W.Y. *et al.* (2017) Genome-wide identification, characterization, and expression analysis of PHT1 phosphate transporters in wheat. *Frontiers in Plant Science* 8, 543.

Theodorou, M.E. and Plaxton, W.C. (1993) Metabolic adaptations of plant respiration to nutritional phosphate deprivation. *Plant Physiology* 101, 339–344.

Thomson, M.J., Singh, N., Dwiyanti, M.S., Wang, D.R., Wright, M.H. *et al.* (2017) Large-scale deployment of a rice 6 K SNP array for genetics and breeding applications. *Rice* 10, 40.

Tripathi, S.C., Sayre, K., Kaul, J. and Narang, R. (2003) Growth and morphology of spring wheat (*Triticum aestivum* L.) culms and their association with lodging: effects of genotypes, N levels and ethephon. *Field Crops Research* 84, 271–290.

Tsay, Y.F., Schroeder, J.I., Feldmann, K.A. and Crawford, N.M. (1993) The herbicide sensitivity gene CHL1 of *Arabidopsis* encodes a nitrate-inducible nitrate transporter. *Cell* 72, 705–713.

Tsay, Y.F., Chiu, C.C., Tsai, C.B., Ho, C.H. and Hsu, P.K. (2007) Nitrate transporters and peptide transporters. *FEBS Letters* 581, 2290–2300.

Tuteja, N. and Gill, S.S. (2013) *Climate Change and Plant Abiotic Stress Tolerance*. Wiley, New York.

van Bueren, E.T.L. and Struik, P.C. (2017) Diverse concepts of breeding for nitrogen use efficiency: a review. *Agronomy for Sustainable Development* 37, 50.

van de Wiel, C.C., van der Linden, C.G. and Scholten, O.E. (2016) Improving phosphorus use efficiency in agriculture: opportunities for breeding. *Euphytica* 207, 1–22.

van Poecke, R.M., Maccaferri, M., Tang, J., Truong, H.T., Janssen, A. *et al.* (2013) Sequence-based SNP genotyping in durum wheat. *Plant Biotechnology Journal* 11, 809–817.

Van Vuuren, M.M., Robinson, D., Fitter, A.H., Chasalow, S.D., Williamson, L. and Raven, J.A. (1997) Effects of elevated atmospheric $CO_2$ and soil water availability on root biomass, root length, and N, P and K uptake by wheat. *New Phytologist* 135, 455–465.

Varshney, R.K., Graner, A. and Sorrells, M.E. (2005) Genomics-assisted breeding for crop improvement. *Trends in Plant Science* 10, 621–630.

Vasil, I.K. (2003) The science and politics of plant biotechnology – a personal perspective. *Nature Biotechnology* 21, 849–851.

Vasil, I.K. (2007) Molecular genetic improvement of cereals: transgenic wheat (*Triticum aestivum* L.). *Plant Cell Reports* 26, 1133–1154.

Versaw, W.K. and Harrison, M.J. (2002) A chloroplast phosphate transporter, PHT2;1, influences allocation of phosphate within the plant and phosphate-starvation responses. *The Plant Cell* 14, 1751–1766.

Vinod, K. and Heuer, S. (2012) Approaches towards nitrogen- and phosphorus-efficient rice. *AoB Plants* 2012, pls028.

Vitousek, P.M., Mooney, H.A., Lubchenco, J. and Melillo, J.M. (1997) Human domination of Earth's ecosystems. *Science* 277, 494–499.

Wakeel, A. and Magen, H. (2017) Potash use for sustainable crop production in Pakistan: a review. *International Journal of Agriculture & Biology* 19, 381–390.

Wakeel, A., Farooq, M., Qadir, M. and Schubert, S. (2011) Potassium substitution by sodium in plants. *Critical Reviews in Plant Sciences* 30, 401–413.

Walder, F., Brule, D., Koegel, S., Wiemken, A., Boller, T. and Courty, P.E. (2015) Plant phosphorus acquisition in a common mycorrhizal network: regulation of phosphate transporter genes of the Pht1 family in sorghum and flax. *New Phytologist* 205, 1632–1645.

Wang, J., Sun, J., Miao, J., Guo, J., Shi, Z. *et al.* (2013a) A phosphate starvation response regulator Ta-PHR1 is involved in phosphate signalling and increases grain yield in wheat. *Annals of Botany* 111, 1139–1153.

Wang, M., Zheng, Q., Shen, Q. and Guo, S. (2013b) The critical role of potassium in plant stress response. *International Journal of Molecular Sciences* 14, 7370–7390.

Wang, W., Vinocur, B. and Altman, A. (2003) Plant responses to drought, salinity and extreme temperatures: towards genetic engineering for stress tolerance. *Planta* 218, 1–14.

Weidong, C., Jizeng, J. and Jiyun, J. (2001) Identification and interaction analysis of QTL for phosphorus use efficiency in wheat seedlings. In: Horst, W.J., Schenk, M.K., Bürkert, A., Claassen, N., Flessa, H.

*et al.* (eds) *Plant Nutrition*. Developments in Plant and Soil Sciences, Vol. 92. Springer, Dordrecht, Netherlands, pp. 76–77.

Wissuwa, M. and Ae, N. (2001) Further characterization of two QTLs that increase phosphorus uptake of rice (*Oryza sativa* L.) under phosphorus deficiency. *Plant and Soil* 237, 275–286.

Wu, F., Yang, X., Wang, Z., Deng, M., Ma, J., Chen, G. and Liu, Y. (2017) Identification of major quantitative trait loci for root diameter in synthetic hexaploid wheat under phosphorus-deficient conditions. *Journal of Applied Genetics* 58, 437–447.

Xu, X., Liu, X., He, P., Johnston, A.M., Zhao, S., Qiu, S. and Zhou, W. (2015) Yield gap, indigenous nutrient supply and nutrient use efficiency for maize in China. *PLoS One* 10, e0140767.

Xu, Y., Wang, R., Tong, Y., Zhao, H., Xie, Q. *et al.* (2014) Mapping QTLs for yield and nitrogen-related traits in wheat: influence of nitrogen and phosphorus fertilization on QTL expression. *Theoretical and Applied Genetics* 127, 59–72.

Yamamoto, A., Kagaya, Y., Toyoshima, R., Kagaya, M., Takeda, S. and Hattori, T. (2009) *Arabidopsis* NF-YB subunits LEC1 and LEC1-LIKE activate transcription by interacting with seed-specific ABRE-binding factors. *The Plant Journal* 58, 843–856.

Yan, W.H., Wang, P., Chen, H.X., Zhou, H.J., Li, Q.P., Wang, C.R. and Zhang, Q.F. (2011) A major QTL, Ghd8, plays pleiotropic roles in regulating grain productivity, plant height, and heading date in rice. *Molecular Plant* 4, 319–330.

Yang, T., Hao, L., Yao, S., Zhao, Y., Lu, W. and Xiao, K. (2016) TabHLH1, a bHLH-type transcription factor gene in wheat, improves plant tolerance to Pi and N deprivation via regulation of nutrient transporter gene transcription and ROS homeostasis. *Plant Physiology and Biochemistry* 104, 99–113.

Yin, L.P., Li, P., Wen, B., Taylor, D. and Berry, J.O. (2007) Characterization and expression of a high-affinity nitrate system transporter gene (TaNRT2.1) from wheat roots, and its evolutionary relationship to other NTR2 genes. *Plant Science* 172, 621–631.

Yuan, Y., Gao, M., Zhang, M., Zheng, H., Zhou, X. *et al.* (2017) QTL mapping for phosphorus efficiency and morphological traits at seedling and maturity stages in wheat. *Frontiers in Plant Science* 8, 614.

Zanetti, M.E., Blanco, F.A., Beker, M.P., Battaglia, M. and Aguilar, O.M. (2010) AC subunit of the plant nuclear factor NF-Y required for rhizobial infection and nodule development affects partner selection in the common bean–*Rhizobium etli* symbiosis. *The Plant Cell* 22, 4142–4157.

Zeng, Y.J., Ying, J., Liu, J.Z., Sun, J.H., Li, B., Xiao, H.S. and Li, Z.S. (2002) Function analysis of a wheat phosphate transporter in yeast mutant. *Acta Genetica Sinica* 29, 1017–1020.

Zhai, B. and Li, S. (2001) Effects of nitrogen stress and complement on the yield and its component in winter wheat. *Journal of Northwest Sci-Tech University of Agriculture and Forestry* 39, 53–56.

Zhai, B. and Li, S. (2005) Response to nitrogen deficiency and compensation on growth and yield of winter wheat. *Plant Nutrition and Fertilizer Science* 11, 308–313.

Zhang, H. and Wang, H. (2015) QTL mapping for traits related to P-deficient tolerance using three related RIL populations in wheat. *Euphytica* 203, 505–520.

Zheng, X., Liu, C., Qiao, L., Zhao, J., Han, R. *et al.* (2020) The MYB transcription factor TaPHR3-A1 is involved in phosphate signaling and governs yield-related traits in bread wheat *(Triticum aestivum* L.). *Journal of Experimental Botany*, 1–47.

Zhao, B., Ge, L., Liang, R., Li, W., Ruan, K., Lin, H. and Jin, Y. (2009) Members of miR-169 family are induced by high salinity and transiently inhibit the NF-YA transcription factor. *BMC Molecular Biology* 10, 29.

Zhao, D., Derkx, A.P., Liu, D.C., Buchner, P. and Hawkesford, M.J. (2015) Overexpression of a NAC transcription factor delays leaf senescence and increases grain nitrogen concentration in wheat. *Plant Biology* 17, 904–913.

Zhao, F.M., Zhu, H.T., Ding, X.H., Zeng, R.Z., Zhang, Z.M. *et al.* (2007) Detection of QTLs for important agronomic traits and analysis of their stabilities using SSSLs in rice. *Agricultural Sciences in China* 6, 769–778.

Zhao, X.Q., Li, Y.J., Liu, J.Z., Li, B., Liu, Q.Y. *et al.* (2004) Isolation and expression analysis of a high-affinity nitrate transporter TaNRT2.3 from roots of wheat. *Acta Botanica Sinica* 46, 347–354.

Zhao, Y., Li, X., Zhang, S., Wang, J., Yang, X. *et al.* (2014) Mapping QTLs for potassium-deficiency tolerance at the seedling stage in wheat (*Triticum aestivum* L.). *Euphytica* 198, 185–198.

Zlatev, Z. and Lidon, F.C. (2012) An overview on drought induced changes in plant growth, water relations and photosynthesis. *Emirates Journal of Food and Agriculture* 24, 57–72.

# 10 Molecular Breeding for Improving Yield in Wheat: Recent Advances and Future Perspectives

**S.A.C.N. Perera[1]* and Saman Seneweera[2]**
[1]*University of Peradeniya, Peradeniya, Sri Lanka;*
[2]*National Institute of Fundamental Studies, Kandy, Sri Lanka*

## 10.1 Introduction

Wheat is the staple diet for many, playing an important role in human nutrition by providing 20% of energy and protein intakes globally (FAO, 2017). It is also a widely adapted crop which can be cultivated in different regions of the world, at altitudes ranging from sea level to 4570 m above sea level and at latitudes in the range of 30–60°N and 27–40°S, providing year-round harvest (Nuttonson, 1955). Total wheat production in the world in 2017 was around 772 million tonnes with China, India, Russia, the USA and France being the leading producers (FAO, 2017). The demand for wheat is increasing annually, and most of the countries in Asia and sub-Saharan Africa are currently net importers of wheat. Accordingly, a total volume of 184 million tonnes of wheat was traded internationally in 2016, amounting to a worth of US$36 billion (Shiferaw *et al.*, 2013).

Wheat yields have recorded a net increase over the years. For example, in 1961, the world wheat production was 222 million tonnes, averaging a productivity level of 1.2 tonnes per hectare, compared with 772 million tonnes in 2017. Over the years, an average annual grain yield increment of 1% has been recorded, but currently the demand has been calculated to increase by 1.7% per annum, creating a deficit between supply and demand (Dixon *et al.*, 2009). Moreover, it has been estimated that grain yield will have to be doubled from that of 2005 to meet demand by the year 2050 (Ray *et al.*, 2013). To meet this demand, a considerable increase in grain yield, which presently lies at 3 t/ha, will be required (FAO, 2017).

Increasing wheat yield has been a primary objective in global wheat breeding programmes. The net grain yield increment can be via directly targeting the yield character itself or indirectly by focusing on traits such as resistance/tolerance to biotic (pests, diseases and weeds) and abiotic (drought, salinity, effects of changing climate) stresses. Out of the different objectives of wheat breeding, this chapter focuses on the direct increment of wheat yields via genetic improvement of the crop.

## 10.2 Wheat Genetics and Genetic Resources

Wheat is recorded to have originated 8000–10,000 years ago and has made a significant contribution to human civilization (Dubcovsky

*Corresponding author: chandrikaperera2003@yahoo.com

 *Molecular Breeding in Wheat, Maize and Sorghum: Strategies for Improving Abiotic Stress Tolerance and Yield* (eds M.A. Hossain *et al.*)
DOI: 10.1079/9781789245431.0010

and Dvorak, 2007; Brenchley *et al.*, 2012). Genetically, wheat is a polyploid with cultivated species comprising of diploids ($2n = 2x = 14$, AA), tetraploids ($2n = 4x = 28$, BBAA) and hexaploids ($2n = 6x = 42$, BBAADD) (Devos *et al.*, 2009). Out of these the tetraploids and hexaploids are formed by duplications of different genomes. Currently, there are two widely cultivated wheat species commonly referred to as durum wheat and bread (common) wheat. Common wheat, *Triticum aestivum*, is an allohexaploid comprising of three genomes A, B and D with a large genome of 5.1 Gbp (Doležel *et al.*, 1989) and more than 124,000 gene loci annotated by the International Wheat Genome Sequencing Consortium in 2014. It is reported to have evolved from hybridization between the tetraploid wheat species *Triticum turgidum* ($2n = 4x = 28$, BBAA) and the diploid progenitor *Aegilops tauschii* var. *strangulata* ($2n = 2x = 14$, DD) (Feldman, 2001; Gill and Friebe, 2001). Durum wheat (*Triticum turgidum* L. ssp. *durum*) is an allotetraploid comprising of A and B genomes. Tetraploid and hexaploid wheat performs similar to diploids in meiosis by pairing only between the homologous chromosomes due to the suppression activity of *Ph1* and *Ph2* genes (Riley and Chapman, 1958; Sears, 1976). Of these two, bread wheat is the most common comprising over 95% of the total wheat production. Spring wheat accounts for 65% of the total bread wheat production while the remaining 35% is of the winter wheat type (Braun *et al.*, 2010). Being polyploids, cultivated wheat is characterized by high levels of phenotypic buffering. This situation has been highly advantageous in the genetic improvement process as it compensates for the deletions and disruptions that occur in wheat, providing fine gene dosage effects and facilitating the use of aneuploidy techniques in breeding (Dubcovsky and Dvorak, 2007).

The genetic resources in a crop are included in three gene pools as primary, secondary and tertiary based on the constitution of their genomes, evolutionary pathways and the ability for cross-pollination. Landraces, first domesticated types and the wild relatives are identified in the primary gene pool of wheat (Feuillet *et al.*, 2007; Qi *et al.*, 2007), while the polyploid *Triticum* and other species (such as *Aegilops*) that share one or more homologous genomes represent the secondary gene pool (Feuillet *et al.*, 2007). Wheat tertiary gene pool consists of most of the perennial diploids or polyploids which do not have common genomes of the cultivated species. Of the three gene pools, cultivated wheat can be crossed with members in the primary gene pool and also with the secondary gene pool but with some difficulties. But the cultivated wheat cannot be hybridized by traditional methods with members in the tertiary gene pool (Ogbonnaya, 2013).

Wheat germplasm collection and conservation have been undertaken in different countries to provide material for breeding, especially to incorporate resistance to biotic and abiotic stresses. Only about 10% of the conserved germplasm has already been used in wheat breeding (Chapman, 1986). Low levels of characterization and evaluation for specific traits, unwillingness of breeders to disturb the already developed elite cultivars and long time required for incorporation of traits from poorly defined germplasm are some reasons for the usage of low levels of conserved germplasm in wheat breeding. However, the most important fact is that there is conserved and unutilized germplasm to be used in future wheat breeding for yield, both in a conventional manner and also by advanced molecular breeding methods (Maxted *et al.*, 2016).

## 10.3 Yield Components and Harvest Index of Wheat

Crop yield is a complex character governed by several interrelated factors. The genotype of the plant, its environment and complex interactions between these two factors determine the yield. The yield of any given genotype is also complex and accounted for by several to many alleles governing different yield components. Division of the total yield among its simpler yield components is the best approach in understanding and determining the total yield. In wheat, the total yield is divided into two broad aspects, namely the number of grains per unit land area and the average grain weight. Further, the total wheat yield is determined by the number of plants per unit land area and the numbers of spikes per plant, spikelets per spike and grains per spikelet. These traits are directly associated with crop yield potential.

Understanding the physiological basis of crop yield and yield components is of paramount importance in the genetic manipulation of wheat for higher yields (Araus *et al.* 2004; de Oliveira Silva *et al.* 2020). While yield is a highly complex character, the yield components are easier to work with and manipulate in breeding programmes (Slafer, 2003; Philipp *et al.*, 2018). The knowledge on the scientific basis of yield components can be directly used in parental selection or further related analysis in the process of genetic improvement for yield. However, most of the yield components are negatively related to each other (Slafer, 2003) and this is a major concern in use of the yield component approach in breeding for a net increase in total grain yield in wheat (Fischer, 1996; Philipp *et al.*, 2018).

Harvest index of wheat is the proportion of dry matter of the plant that is stored in the grain; accordingly, an increase in harvest index will result in the increase of grain yield (Qin *et al.*, 2013). During the process of genetic improvement of wheat, the harvest index has been increased from 0.35 in 1951 to 0.50 by 1995–2013 (Wiesmeier *et al.*, 2014). A maximum harvest index of 0.62 has been predicted for modern winter wheat cultivars (Austin, 1993) and considering the already achieved levels of harvest index, further yield increment should focus on the increase in total dry matter of the wheat plant (Parry *et al.*, 2010). On the contrary, there are suggestions that the breeding programmes for yield improvement should focus on the harvest index rather than the total dry matter because high levels of total dry matter exceeding 18,000 kg/ha have already been recorded (Parry *et al.*, 2010). Considering both views, it is also suggested that the wheat yield has plateaued (Yu *et al.*, 2017). Accordingly, it will be essential to break the barriers arresting the wheat yield increment to meet the expected demand on grain yield by the year 2050 (Tadesse *et al.*, 2019) and this would require improvements in both yield components and harvest index, which is the greatest challenge for wheat breeders in the 21st century.

## 10.4 Conventional Approaches in Breeding Wheat for Yield

Scientific wheat breeding was initiated in the 19th century with the selection and crossing of different wheat landraces and cultivars, representing diverse origins, to develop new varieties with agriculturally desirable traits. The fundamental objective at these initial stages of wheat improvement was the increase of wheat yields. In Italy, in the early 20th century, a Japanese wheat variety named Akakokomugi was crossed with Italian wheat breeding lines to produce some highly desirable cultivars (Salvi *et al.*, 2013). These cultivars, for example Ardito and Mentana, showed insensitivity to photoperiod and reduced height, which were very desirable traits that played a crucial role in increasing wheat yield potential. Later, it was revealed that these lines possessed the alleles *Ppd-D1* governing photoperiod insensitivity and *Rht8c* governing the short length of the straw. These wheat lines produced in Italy laid the foundation for the improved wheat cultivars in the Mediterranean region, Soviet Russia, China and even in a few countries of the South American continent (Yang and Smale, 1996).

A Japanese landrace, Shiro Daruma, which possessed genes responsible for dwarfism, was used for breeding the wheat variety Norin10. In the 1950s this line was used by Dr Norman Borlaug to produce the semi-dwarf ideotype in the 'green revolution' (Borlaug, 2008), marking a milestone in the improvement of cereal yields in the world. The varieties of the green revolution were resistant to lodging and highly responsive to inputs. The semi-dwarf trait of these cultivars was governed by *Rht1* and *Rht2*, which are the genes insensitive to gibberellic acid-controlled growth resulting in the semi-dwarf stature of plants. Photoperiod insensitivity, governed by the three genes *Ppd1*, *Ppd2* and *Ppd3*, was the second dominant trait that characterized these cultivars (Hedden, 2003). The evaluation of these cultivars was carried out in a shuttle breeding programme in two different environments to maximize the advantage of photoperiod insensitivity for expanded cultivation of wheat across different environments. Ultimately, these two traits governed by dominant genes were responsible for the yield increment in green revolution wheat cultivars by improving the harvest index and facilitating the cultivation in larger extents compared with traditional cultivars.

Over the years, the green revolution has been instrumental in the increased yield of wheat and other cereals facilitating to feed the ever-increasing human population. Since the beginning of the green revolution, until the application of molecular markers, wheat breeding depended on the introgression of various desirable traits into cultivated varieties through conventional means by selecting, crossing and backcrossing procedures. The international wheat breeding programmes were led by the International Maize and Wheat Improvement Center (CIMMYT) and the International Center for Agriculture Research in the Dry Areas (ICARDA), and such conventional breeding programmes were successful in the improvement of grain yield to very high levels. However, the conventional breeding programmes are highly time-consuming and currently the yield levels have plateaued, not keeping up with the increase in demand for grain yield. Under such circumstances, the molecular approaches that have been developing over the last three decades together with the high-throughput phenotyping platforms are perceived to break the barriers in the genetic improvement of wheat in increasing the grain yields to expected levels.

## 10.5 Wheat Molecular Breeding

Molecular breeding methods apply DNA fingerprinting and genotyping techniques in the identification of the allelic make-up of the plant to predict the phenotype. The novel molecular marker technologies – marker-assisted selection (MAS), genome-wide studies, targeting induced local lesions in genomes (TILLING), gene silencing and genomic selection (GS), etc. – provide more gene-targeted approaches in breeding crops, in addition to expediting the process of genetic improvement compared with conventional plant breeding techniques. Over the last three decades, molecular breeding approaches have been progressively used in breeding wheat for higher yields. These efforts have considerably enhanced the cisgenic genetic improvement process in transferring the desirable alleles and allelic combinations by maximized germplasm and progeny screening and selection.

In conventional breeding, scientists focus on the phenotype or the manifestation of the genotype. A major difference of molecular breeding is the ability to detect the gene and the allelic effects for breeding to be more targeted and gene specific. Accordingly, molecular breeding techniques have assisted in classifying wheat yield-related genes (Table 10.1), elucidating that many are individuals of multigenic families rather than single genes (Nadolska-Orczyk *et al.*, 2017).

In addition, the genes and the specific alleles rendering different phenotypes in wheat have been extensively studied (Table 10.2) and utilized in breeding via the molecular breeding techniques.

The widely adopted molecular techniques to genetically improve yields, the resources that are needed and their limitations are discussed in the following sections.

### 10.5.1 Molecular markers

It is essential to have specific resources for the molecular breeding of a crop to be efficient and successful. The availability of polymorphic molecular markers is one such primary essentiality. A variety of molecular marker types have so far been used in wheat breeding. Beginning from first-generation marker resources, restriction fragment length polymorphism (RFLP), random amplified polymorphic DNA (RAPD), amplified fragment length polymorphism (AFLP) and simple sequence repeat (SSR) markers have been extensively used in genome mapping, quantitative trait locus (QTL) analysis and MAS in wheat. Out of these primary marker systems SSRs have been the preferred choice of

**Table 10.1.** Categories of yield-related genes in wheat.

| Gene/transcription factor category | Governed yield trait |
|---|---|
| Transcription factors governing spike development | Grain number |
| Genes governing metabolism or signalling growth regulators cytokinin, gibberellin and brassinosteroids | Plant architecture resulting in grain yield and stem hardiness |
| Genes governing cell division and cell proliferation | Grain size |
| Floral regulators | Inflorescence architecture governing seed number |
| Genes involved in carbohydrate metabolism | Plant architecture and grain yield |

**Table 10.2.** Some of the yield-determining genes in wheat and their orthologues in other cereals.

| Gene | Known orthologues (crop) | Reference |
|---|---|---|
| *TaCKX6-D1* | *OsCKX2* (rice) | Nadolska-Orczyk *et al.* (2017) |
| *TaTEF-7A* | *OsTEF1* (rice) | Nadolska-Orczyk *et al.* (2017) |
| *TaGW2* (*6A*, *6B*) | *OsGW2* (rice) | Nadolska-Orczyk *et al.* (2017) |
| *TaTGW6-A1* (*4A*) | *OsTGW6* (rice) | Hu *et al.* (2016) |
| *TaGS5-3A-T* | *OsGS5* (rice) | Nadolska-Orczyk *et al.* (2017) |
| *TaSus1* (*7A*), *TaSus2* (*2A*, *2B*, *2D*) | *OsSUS* (rice); *ZmSUS1* (maize) | Jiang *et al.* (2011) |
| *TaNAC2-5A* | *HvNAC* (barley) | Christiansen *et al.* (2016) |

many in MAS in wheat (Gupta and Varshney, 2000; Zhang *et al.*, 2015; Su *et al.*, 2018) while the low-throughput and practically difficult RFLP and RAPD markers are rarely used nowadays. The high-throughput molecular marker system of today, namely single-nucleotide polymorphism (SNP), has also been developed and utilized in QTL mapping in wheat (Rostoks *et al.*, 2005; Su *et al.*, 2018; Xin *et al.*, 2020; Yang *et al.*, 2020).

Development of molecular markers for wheat has been comparatively challenging due to its polyploid nature, including the subgenomes and the larger genome size. This constraint has been eased now with the elucidation of the whole-genome sequence of wheat in 2012 (Brenchley *et al.*, 2012). Molecular breeding research, in the beginning, relied on linked markers which were located in proximity to a gene of interest but not directly within the gene itself. This was a handicap to harness the potential of molecular breeding. Later, functional or gene-specific markers, which are present within the polymorphic sites of genes, have been identified to be used in wheat breeding (Liu *et al.*, 2012). Among the functional markers derived from expressed sequence tags (EST), EST-SNPs (Rafalski, 2002; Wang *et al.*, 2017) and EST-SSRs (Varshney *et al.*, 2007; Yang *et al.*, 2016) have been prominently used in wheat. By now there is a large collection of molecular markers, deposited in publicly available sequence databases, to be effectively used in wheat breeding.

### 10.5.2 High-throughput genotyping

Genotyping at the beginning of the molecular era was limited to simultaneous screening of only a few loci. However, the genotyping facilities underwent rapid improvements giving rise to high-throughput genotyping assays. High-throughput genotyping technologies utilize SNPs in detecting genetic loci. The technological advances in detecting SNPs at large scale, along with the methods for next-generation sequencing (NGS) that provided a method of discovery of SNPs in data sets, were instrumental in facilitating high-throughput genotyping. The simultaneous development of bioinformatics tools enabled the management of huge data sets and the extraction of important genetic information that resulted from high-throughput genotyping assays. The early experiments using high-throughput genotyping relied on the fixed array SNP platforms using microarrays and subsequently flexible SNP platforms were developed. The developments in high-throughput genotyping technologies were coupled with the involvement of large-scale commercial investment by life science companies that mediated to develop sophisticated sequencing and genotyping platforms. The related developments combined the advances in different disciplines including computer science, bioengineering, automation and nanotechnology (Thompson, 2014). The high-throughput genotyping facilities are expensive to set up, but the end result has been the decrease in the cost per sample resulting from the high throughput of the experiment and accordingly wheat breeding has also been benefited by high-throughput genotyping techniques (Rimbert *et al.*, 2018). High-throughput genotyping will be essential in future, in tagging the array of multiple genes governing wheat yields.

### 10.5.3 High-throughput phenotyping (phenomics)

The precision and the success of gene identification in molecular biology depend on the accurate phenotyping of germplasm, progeny and breeding lines. The efforts on molecular breeding of wheat for higher yields were crucially dependent on the phenotypic data of yield parameters and other related agronomic traits. Further, there was a need for the phenotyping facilities to keep pace with the genotyping, resulting in the need for high-throughput phenotyping to generate better-quality data free from manual errors (Araus *et al.*, 2018). Generally, plants are grown in densely populated plots in the field, which affects the kinds of traits that can be measured effectively at the plot level compared with those at the single plant level. The data on parameters such as canopy size, light penetration, water use, chlorophyll fluorescence, whole plant architecture, etc. combined with genomics can be used to develop the high-yielding plant ideotype and precision models.

Recording the biological features in changing environments across a multitude of spatial and temporal dimensions has been a daunting task, limiting breeding research to very-low-throughput phenotyping efforts. However, novel high-throughput phenotyping platforms have evolved during the last two decades facilitating the combining of phenomics with other 'omics'-based approaches and leading the path towards precision breeding that is highly essential in improving a complex trait such as yield. The developments in sensor technology coupled with computer vision tools, ranging from charge-coupled devices (CCDs), complementary metal–oxide–semiconductor (CMOS) technologies to three-dimensional shoot imaging approaches – have revolutionized the plant phenotyping technologies (Pieruschka and Schurr, 2019). These developments have led to the installation of high-throughput facilities, from greenhouses to extensive high-throughput field-phenotyping (HTFP) facilities, which use technologies such as magnetic resonance imaging (MRI), positron emission tomography (PET), computer tomography (CT) or even a combination of them. As of today, the high-throughput phenotyping facilities use satellite imaging, unmanned aerial vehicles (UAVs) and ground-based vehicles and sensors for proximal phenotyping, providing plant breeders with technologically advanced options for large-scale precision phenotyping (Chawade *et al.*, 2019).

The feasibility of extensive phenotyping for important traits leads to the novel technique referred to as 'phenomics', which comes under the current 'omics' approaches. With the development of remote sensing technologies, high-throughput phenotyping became a practical feasibility allowing the recording of phenotypic data in a large population within a very short time period. Accordingly, the phenotyping technique of spectral imagery has been used to record yield, biomass and other yield-related traits in breeding wheat for higher yields (Reynolds *et al.*, 2015). In another study, a robotic phenotyping platform was used to measure the canopy height of a wheat population (Lyra *et al.*, 2020). Genetic dissection of lodging (Singh *et al.*, 2019) and early selection for desirable traits (Hu *et al.*, 2020) are some recent examples of the use of high-throughput phenotyping in wheat breeding for higher yields.

## 10.6 Quantitative Trait Locus/Genome Mapping and Marker-Assisted Selection

Gene/QTL mapping is the location of genes/QTLs to different chromosomes and specific places along chromosomes. Location of genes/QTLs is essential for subsequent MAS and breeding. The availability of populations such as $F_2$s, backcrosses, recombinant inbred lines (RILs) and doubled haploid lines (DHLs) that are segregating for yield and related traits and a high-density marker coverage of the genome are essential requirements for mapping genes for yield in wheat. The mapped genes/QTLs for yield are expected to enhance breeding programmes by defining more gene-targeted selection criteria and identifying desirable parents in a cross (Austin, 1993).

The first-ever genome map of wheat was developed in 1989 using RFLP markers (Chao *et al.*, 1989). Following this, many QTL maps of wheat have been developed using AFLP, SSR, Diverse Arrays Technology (DArT), SNP and other functional markers. There have been several genome maps specifically concentrating on the mapping of yield-related traits: length and width of the kernel, spike number and length, fertile and sterile grain numbers per spike, and the weight of 1000 grains (Li *et al.*, 2007; Ramya *et al.*, 2010; Sun *et al.*, 2010; Carter *et al.*, 2011; Deng *et al.*, 2011; Heidari *et al.*, 2011; Cui *et al.*, 2012; Liu *et al.*, 2012).

Despite the achievements in mapping yield-related QTLs in wheat, there has been a deficit in actually transferring such identified genes/QTLs from the mapping populations to a cultivated variety followed by successful cultivation of that variety by wheat farmers (Slafer, 2003). The reasons for this drawback would be the intrinsic complications: the complexity of the trait involving many QTLs governing yields (Stuber *et al.*, 1999), the low resolution of yield-related QTLs, the negative correlation of yield components negating the net effect of identified QTLs and the more significant gene × environment (G × E) interaction. The grain yield, the final product of any crop, is determined by the genotypic potential (G), the environmental effect (E) and the G × E interaction (Yan and Kang, 2002). In the presence of G × E interaction, the phenotyping experiments for tagging QTLs should carefully be designed to capture the effects of genotype. Under such circumstances, the selection of genotypes based solely on the mean yield is inadequate as described by Sharifi *et al.* (2017).

Mapping genes/QTLs subsequently allows MAS in crops. MAS is based on the linkage of a trait to molecular markers facilitating selection via the linked or functional markers, without being entirely dependent on the phenotypic expression. MAS possess the advantages of revealing the true genetic effects independent of the environment and the facility for detection of the phenotype at any growth stage without waiting for the ultimate expression of the trait.

It has been recorded that more than half of the wheat varieties that are recommended have been developed with the utilization of molecular markers at some stage of the breeding programme (Varshney *et al.*, 2007) and the percentage has risen to almost 100% by now. The time taken for the development of such varieties has been recorded to be about half that taken with the methods utilizing conventional breeding strategies alone. MAS has also been used in wheat in pyramiding genes mainly for pathogen resistance (Anderson *et al.*, 2007), with records for yield traits being scarce. Therefore, it is highly important to consider the efficiency of the approach used compared with the phenotypic selection, the appropriateness of the marker and the contribution of the gene to the phenotype in using MAS to improve such a complex and QTL-governed trait as yield.

## 10.7 Marker-Assisted Backcrossing

Marker-assisted backcross breeding (MABB) involves the use of molecular markers to screen for the individuals having the favourable alleles in backcross breeding. MABB can be used to select the background for donor alleles and the foreground for the recurrent parent alleles (Young and Tanksley, 1989). MABB is highly effective in selecting individuals compared with conventional methods. MABB assists in reducing the number of backcross generations and limits the number of individuals to be maintained and advanced to subsequent generations, minimizing the costs and time involved in backcrossing.

The use of gene-based markers enhances the efficiency of MABB and in the absence of such markers, it is important to use markers which are tightly linked to the QTL of concern. Marker-assisted backcrossing (MABC) can also be used to introgress multiple QTLs, or gene pyramiding, which assists in developing desirable phenotypes which are stable and durable. However, in gene pyramiding, the population size should match with the number of QTLs to be introgressed and the number of QTLs that are to be simultaneously introgressed should be kept at a manageable level for success of MABC in a lesser number of generations. For example, with MABC, genes such as qualitative and quantitative are rapidly transferred from wild progenitors to advanced breeding lines, simultaneously pyramiding them into a single line. MABC has been applied in wheat to introgress QTLs for high concentrations of Fe and Zn in the grain (Hao *et al.*, 2014; Srinivasa *et al.*, 2014). Identification of yield-related QTLs for introgression has been challenging so far, yet it remains a task for the future in improving wheat for higher yields.

## 10.8 Genomics in Wheat Breeding

Genomics is the study about the entire genome in a single assay. The availability of a large volume of genetic information on multiple traits and alleles has facilitated the genome breeding approaches targeting to improve multiple alleles concurrently (Langridge, 2005). With the advancement of molecular breeding, a wide array of genomic resources, such as high-density genome and QTL maps, a variety of genetic stocks, physical maps, including bacterial artificial chromosome (BAC) libraries and the whole-genome sequence information, became available in primarily important crops including wheat. The whole-genome breeding of these crops is being assisted with transcriptomics, metabolomics and proteomics, which are identified as 'omics approaches' in molecular biology, and wheat breeders have started utilizing the whole-genome breeding strategies to improve multiple traits including yield (Fig. 10.1).

### 10.8.1 Association/linkage disequilibrium mapping in wheat

Association mapping is a technique based on linkage disequilibrium (LD) in tagging QTLs. Association mapping can be applied to scan the entire genome, which is referred to as genome-wide-association studies (GWAS). One major advantage of GWAS is the ability to use the natural populations such as landraces and germplasm collections in mapping QTLs, avoiding the necessity of segregating populations which are needed for linkage mapping. Compared with the segregating populations, the natural populations have undergone many cycles of recombination, resulting in higher resolution in mapped QTLs.

LD is defined as the non-random association of markers. In association studies, the extent of LD around a specific locus defines the map resolution and the number of markers needed for the whole-genome scan (Rafalski and Morgante, 2004). In such genome scans, the genes with even small effects can be detected in a sufficiently large population. For example, the associations of genes accounting for about 8% of the variation can be detected in a population of 500 while a population of 100 detects only the associations of over 15% (Varshney *et al.*, 2007).

LD mapping has been applied in wheat to tag QTLs responsible for several traits, some of which are related to yield. For example, LD analysis has revealed 62 loci governing kernel size and milling quality on chromosomes 2D, 5A and 5B (Breseghello and Sorrells, 2005). Besides, GWAS studies on landraces and wild relatives of wheat have identified QTLs associated with morphological traits in normal irrigated, heat and

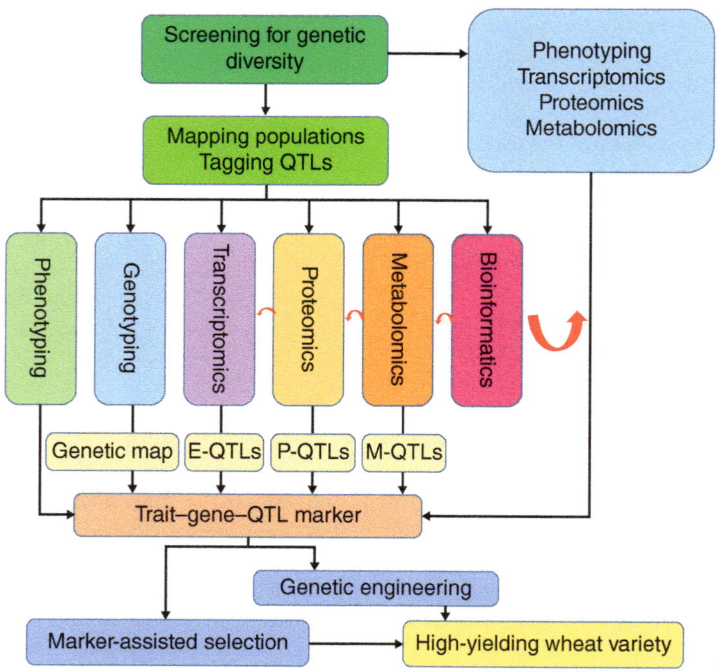


**Fig. 10.1.** Integration of different 'omics' approaches in developing high-yielding wheat varieties.

drought environments and with disease resistance (Liu *et al.*, 2014; Kertho *et al.*, 2015; Sukumaran *et al.*, 2017).

### 10.8.2 Comparative genomics in wheat breeding

Genomes of plants are highly diverse in their size, ploidy levels, number of chromosomes and base sequences. Despite the millions of years of separate genome evolution creating this diversity, a significant collinearity has been observed especially among the members of the families *Poaceae* (Ahn *et al.*, 1993; Devos, 2005), *Brassicaceae* (Lagercrantz, 1998), etc. Such syntenic regions in the genome have facilitated comparative mapping and QTL identification across species. This molecular approach is referred to as comparative or synteny mapping, and it allows the transfer of genetic information from the species having abundant sequence and genetic information (including model species) to species lacking such information.

Both rice and wheat are included in the same family *Poaceae*, and they differ in plant architecture related to the inflorescence. Rice is considered the model species for grasses due to its smaller genome size and the availability of genome sequence information much earlier than for wheat. Accordingly, prior to the elucidation of the genome sequence, wheat benefited by comparative mapping with rice. A comprehensive analysis of ESTs revealed chromosomal rearrangements in wheat compared with rice, with non-consistent physical locations in the non-conserved regions (La Rota and Sorrells, 2004). In spite of the disturbed micro-synteny between rice and wheat genomes, gene isolation has been facilitated in wheat via map-based cloning approaches (Stein and Graner, 2004), in addition to the usage of molecular marker resources derived from syntenic regions across rice and wheat (Khlestkina *et al.*, 2004). Hanif *et al.* (2016) and several other research teams have reported the tagging of yield QTLs of wheat through comparative mapping with rice. Accordingly, even after availability of the whole-genome sequence of wheat, the genetic information derived for rice in morphology and yield-related traits can be examined for cross-transferability avoiding the duplication of complex experiments in increasing wheat yields.

### 10.8.3 Genomic selection

GS is a novel marker-assisted breeding technique adopted to train a prediction model using genome-wide markers and phenotypes from a reference population. The prediction model thus developed can be used for the prediction of breeding values based on the marker data of the genome. GS is a more effective selection method, especially for polygenic traits such as yield (Bernardo and Yu, 2007). In addition, the rate of genetic gain in populations can be increased by using the GS technique. The use of GS in breeding wheat has been shown to be promising (Heffner *et al.*, 2011). Several studies have revealed moderate to high prediction levels in breeding for grain quality (Velu *et al.*, 2016), disease resistance (Ornella *et al.*, 2012; Arruda *et al.*, 2015) and importantly for grain yield (Crossa *et al.*, 2010; Dawson *et al.*, 2013). Sukumaran *et al.* (2017) reported the development of prediction models and testing them on 287 advanced elite spring wheat lines phenotyped for grain yield, grain number and 1000-grain weight in 18 different environments in different countries. The results indicated a high potential for utilizing GS in marker-assisted breeding. Accordingly, once the GS models are validated, it will be possible to use such models in combination with MAS to improve even a complex trait such as yield.

### 10.8.4 Ideotype breeding and systems modelling

Genomic approaches in wheat breeding can facilitate the development of an ideotype of wheat. A genetic ideotype is an ideal plant having a majority of desirable alleles delivering the potential for economically positive traits resulting in high yields. An ideotype can be developed by 'designing' chromosomes to contain the intended positive alleles, and this can be achieved by effective employment of genome breeding techniques. Advanced technologies of marker screening and association genetic studies facilitate the building of key chromosome blocks and haplotype structures for ideotype breeding.

In ideotype breeding, it is important to determine the allelic combinations that would make up a high-yielding ideotype of wheat. Combining desirable alleles at multiple loci into an adapted high-yielding cultivar of wheat would result in a wheat ideotype. Kuchel *et al.* (2003) reported the enrichment of a high-yielding wheat cultivar with alleles for correcting multiple deficiencies. Christy *et al.* (2020), in a study using 200 wheat genotypes, reported the identification of optimum allelic combinations of the photoperiod (*Ppd*) and vernalization (*Vrn*) genes for the determination of an allelic-based phenological model for the production of a defined ideotype.

Plant systems modelling is a mathematical method to model for quantitative representation, integration and simulation of biological parameters in a plant. Systems models can assist MAS and QTL mapping, as demonstrated by Gu *et al.* (2014) in analysing the impact of genetic variation in the rate of leaf photosynthesis via QTLs on crop biomass production. Systems models do possess a huge potential in directing crop improvement for higher biomass leading to higher grain yields. Systems models in combination with genomics and phenomics are referred to as the three pillars in breeding for high-yielding photosynthetically efficient crops (Cheng *et al.*, 2019). Accordingly, ideotype breeding combined with these three pillars is envisaged to be a strategy for the future in the development of high-yielding wheat cultivars enhanced with other desirable traits.

## 10.9 Other Molecular Technologies and Next-Generation Approaches in Wheat Breeding

Cisgenesis is defined as the transfer of genes within the compatible species of the same genus using molecular biological approaches. The molecular approach of cisgenesis (Schouten *et al.*, 2006) is similar to that of classical breeding, but the molecular approach of cisgenesis possesses the advantages of comparatively rapid transfer of required genes and the lesser constraints related to linkage drag (Mondal *et al.*, 2016). In addition, the transfer of genes with limited levels of allelic variations is also facilitated through this method. There are some examples of the use of cisgenesis in cereals, with transfer of the high-molecular-weight glutenin subunit *1Dy10* gene responsible for superior bread-making quality from hexaploid bread wheat to tetraploid durum wheat (Gadaleta *et al.*, 2008) being an example of the technology's use in wheat. The possibility of the transfer of the selected genes via gene cassettes is being explored (Ellis *et al.*, 2014), but the limitation is the need for genome-editing technologies, which are still being experimented, to be practical realities. However, the rapid development of molecular technologies is expected to provide practical approaches for yield enhancement in wheat.

Among the recent molecular biological tools, genome editing is considered to be a powerful tool in the genetic improvement for yield via targeted mutagenesis, gene editing and gene stacking. Already, transcription activator-like effector nucleases (TALEN) and clustered regularly interspaced short palindromic repeats (CRISPR)/CRISPR-associated protein 9 (Cas9) gene editing techniques have been applied in wheat (Wang *et al.*, 2014). In addition, RNAi-based gene silencing technologies have been utilized in wheat in the genetic analysis including functional gene analysis (Fu *et al.*, 2007) and increasing grain hardiness (Gasparis *et al.*, 2013), giving evidence for future potential in breeding for increased yields. Furthermore, mutant induction using TILLING strategies is also expected to be a future technology in polyploid species such as wheat in breeding for higher yields (Nadolska-Orczyk *et al.*, 2017).

## 10.10 Future Perspectives and Conclusions

Molecular breeding approaches are considered to be the best option for overcoming the challenges posed by yield stagnation in crops, including wheat. Wheat possesses a complex genome and yield is a super complex trait to improve genetically. A deficiency of knowledge in the physiology, cell biology and biochemistry of the individual traits of yield components and the lack of progress in the introgression of elusive yield-related QTLs have limited the progress in breeding for yield using both conventional tools after the green revolution and even at the initial stages of the molecular era. However, the molecular breeding approaches that have evolved over the last three decades provide gene-targeted high-throughput options to breed wheat for higher yields. The availability of rich genetic diversity of wheat, including members of different ploidy levels and wild relatives, is the most important natural resource that can be used in breeding wheat for increased yields. The molecular approaches for detection of yield-responsible transcripts and development of novel mutants can be used for determining the important functional genes responsible for yield. Modern molecular techniques, along with the availability of the whole-genome sequence of wheat, have been proven to be highly efficient in mining wheat germplasm for allele-specific desirable traits. The identification of physiological, biochemical and morphological parameters defining a high-yielding wheat ideotype is fundamental for breeding to achieve such a genotype. The molecular technologies such as QTL analysis, GWAS, MABB and GS, coupled with advanced statistical analysis methods, can be utilized for locating genes making the ideotype. The use of high-throughput genotyping and phenotyping platforms and other 'omics' approaches, sequencing facilities and bioinformatics tools can greatly enhance and expedite this process. The judicious and targeted use of molecular breeding tools will be instrumental in breaking the yield plateau in wheat for the second major revolution in increasing wheat yields.

## References


Ahn, S., Anderson, J.A., Sorrells, M.E. and Tanksley, S.D. (1993) Homeologous relationships of rice, wheat and maize chromosomes. *Molecular Genetics and Genomes* 241, 483–490.

Anderson, J., Chao, S. and Liu, S. (2007) Breeding using a major QTL for Fusarium head blight resistance in wheat. *Crop Science* 47(S3), S112–S119. Available at: https://doi.org/10.2135/cropsci2007.04.0006IPBS

Araus, J.L., Slafer, G.A., Reynolds, M.P. and Royo, C. (2004) Physiology of yield and adaptation in wheat and barley breeding. In: Nguyen, H.T. and Blum, A. (eds) *Physiology and Biotechnology Integration for Plant Breeding*. Marcel Dekker, New York, pp. 1–49.

Araus, J.L., Kefauver, S.C., Zaman-Allah, M., Olsen, M.S. and Cairns, J.E. (2018) Translating high-throughput phenotyping into genetic gain. *Trends in Plant Science* 23, 451–466.

Arruda, M.P., Brown, P.J., Lipka, A.E., Krill, A.M., Thurber, C. and Kolb, F.L. (2015) Genomic selection for predicting *Fusarium* head blight resistance in a wheat breeding programme. *The Plant Genome* 8(3), plantgenome2015.01.0003. Available at: https://doi.org/10.3835/plantgenome2015.01.0003

Austin, R.B. (1993) Augmenting yield-based selection. In: Hayward, M.D., Bosemark, N.O. and Romagosa, I. (eds) *Plant Breeding: Principles and Prospects*. Springer, Dordrecht, the Netherlands, pp. 391–405.

Bernardo, R. and Yu, J. (2007) Prospects for genome-wide selection for quantitative traits in maize. *Crop Science* 47, 1082–1090. Available at: https://doi.org/10.2135/cropsci2006.11.0690

Borlaug, N.E. (2008) Feeding a world of 10 billion people: our 21st century challenge. In: Scanes, C.G. and Miranowski, J.A. (eds) *Perspectives in World Food and Agriculture 2004*. Iowa State Press, Ames, Iowa, pp. 32–56.

Braun, H.J., Atlin, G. and Payne, T. (2010) Multi-location testing as a tool to identify plant response to global climate change. In: Reynolds, M.P. (ed) *Climate Change and Crop Production*. CAB International, Wallingford, UK, pp. 115–138.

Brenchley, R., Spannagl, M., Pfeifer, M., Barker, G.L., D'Amore, R.M. *et al.* (2012) Analysis of the bread wheat genome using whole-genome shotgun sequencing. *Nature* 491(7426), 705–710.

Breseghello, F. and Sorrells, M.E. (2005) Association mapping of kernel size and milling quality in wheat. *Genetics* 172, 1165–1177.

Carter, A.H., Garland-Campbell, K. and Kidwell, K.K. (2011) Genetic mapping of quantitative trait loci associated with important agronomic traits in the spring wheat (*Triticum aestivum* L.) 'Louise' × 'Penawawa'. *Crop Science* 51, 84–95. Available at: https://doi.org/10.2135/cropsci2010.03.0185

Chao, S., Sharp, P.J., Worland, A.J., Warham, E.J., Koebner, R.M. and Gale, M.D. (1989) RFLP-based genetic maps of wheat homoeologous group 7 chromosomes. *Theoretical and Applied Genetics* 78(4), 495–504. Available at: https://doi.org/10.1007/bf00290833

Chawade, A., van Ham, J., Blomquist, H., Bagge, O., Alexandersson, E. and Ortiz, R. (2019) High-throughput field-phenotyping tools for plant breeding and precision agriculture. *Agronomy* 9, 258. Available at: https://doi.org/10.3390/agronomy9050258

Cheng, Y., Wang, Y., Han, Y., Li, D., Zhang, Z. *et al.* (2019) The stimulatory effects of nanochitin whisker on carbon and nitrogen metabolism and on the enhancement of grain yield and crude protein of winter wheat. *Molecules* 24, 1752. Available at: https://doi.org/10.3390/molecules24091752

Chapman, C.G.D. (1986) The role of genetic resources in wheat breeding. *Plant Genetic Resources Newsletter* 65, 2–5.

Christiansen, M.W., Matthewman, C., Podzimska-Sroka, D., O'Shea, C., Lindemose, S. *et al.* (2016) Barley plants over-expressing the NAC transcription factor gene *HvNAC005* show stunting and delay in development combined with early senescence. *Journal of Experimental Botany* 67, 5259–5273. Available at: https://doi.org/10.1093/jxb/erw286

Christy, B., Riffkin, P., Richards, R., Partington, D., Acuna, T.B. *et al.* (2020) An allelic based phenological model to predict phasic development of wheat (*Triticum aestivum* L.). *Field Crops Research* 249, 107722. Available at: https://doi.org/10.1016/j.fcr.2020.107722

Crossa, J., de los Campos, G., Pérez, P., Gianola, D., Burgueño, J. *et al.* (2010) Prediction of genetic values of quantitative traits in plant breeding using pedigree and molecular markers. *Genetics* 186(2), 713–724. Available at: https://doi.org/10.1534/genetics.110.118521

Cui, F., Ding, A.M., Li, J., Zhao, C.H., Wang, L., *et al.* (2012) QTL detection of seven spike-related traits and their genetic correlations in wheat using two related RIL populations. *Euphytica* 186, 177–192. Available at: https://doi.org/10.1007/s10681-011-0550-7

Dawson, J.C., Endelmana, J.B., Heslot, N., Crossa, J., Poland, J. *et al.* (2013) The use of unbalanced historical data for genomic selection in an international wheat breeding program. *Field Crops Research* 154, 12–22. Available at: https://doi.org/10.1016/j.fcr.2013.07.020

Deng, S.M., Wu, X.R., Wu, Y.Y., Zhou, R.H., Wang, H.G., Jia, J.Z. and Liu, S.B. (2011) Characterization and precise mapping of a QTL increasing spike number with pleiotropic effects in wheat. *Theoretical and Applied Genetics* 122, 281–289. Available at: https://doi.org/10.1007/s00122-010-1443-1

de Oliveira Silva, A., Slafer, G.A., Fritz, A.K. and Lollato, R.P. (2020) Physiological basis of genotypic response to management in dryland wheat. *Frontiers in Plant Science* 10, 1644. Available at: https://doi.org/10.3389/fpls.2019.01644

Devos, K.M. (2005) Updating the 'crop circle'. *Current Opinion in Plant Biology* 8, 155–162.

Devos, K.M., Doležel, J. and Feuillet, C. (2009) Genome organization and comparative genomics. In: Carver, B.F. (ed.) *Wheat Science and Trade*. Wiley-Blackwell, Ames, Iowa, pp. 327–367. Available at: https://doi.org/10.1002/9780813818832

Dixon, J., Braun, H.J. and Crouch, J. (2009) Transitioning wheat research to serve the future needs of the developing world. In: Dixon, J., Braun, H.J. and Kosina, P. (eds) *Wheat Facts and Futures*. CIMMYT, Mexico City, pp. 1–19.

Doležel, J., Binarová, P. and Lucretti, S. (1989) Analysis of nuclear DNA content in plant cells by flow cytometry. *Biologia Plantarum* 31(2), 113–120. Available at: https://doi.org/10.1007/BF02907241

Dubcovsky, J. and Dvorak, J. (2007) Genome plasticity a key factor in the success of polyploid wheat under domestication. *Science* 316(5833), 1862–1866.

Ellis, J.G., Lagudah, E.S., Spielmeyer, W. and Dodds, P.N. (2014) The past, present and future of breeding rust resistant wheat. *Frontiers in Plant Science* 5, 641. Available at: https://doi.org/10.3389/fpls.2014.00641

FAO (2017) FAOSTAT Database. Food and Agriculture Organization of the United Nations, Rome. Available at: http://faostat.fao.org/ (accessed 18 September 2019).

Feldman, M. (2001) Origin of cultivated wheat. In: Bonjean, A.P. and Angus, W.J. (eds) *The World Wheat Book: A History of Wheat Breeding*. Intercept Ltd, London, pp. 3–56.

Feuillet, C., Langridge, P. and Waugh, R. (2007) Cereal breeding takes a walk on the wild side. *Trends in Genetics* 24, 24–32.

Fischer, R.A. (1996) Wheat physiology at CIMMYT and raising the yield plateau. In: Reynolds, M.P., Rajaram, S. and McNab, A. (eds) *Increasing Yield Potential in Wheat: Breaking the Barriers*. CIMMYT, Mexico City, pp. 195–203.

Fu, D., Uauy, C., Blechl, A. and Dubcovsky, J. (2007) RNA interference for wheat functional gene analysis. *Transgenic Research* 16, 689–701. Available at: https://doi.org/10.1007/s11248-007-9150-7

Gadaleta, A., Giancaspro, A., Blechl, A.E. and Blanco, A. (2008) A transgenic durum wheat line that is free of marker genes and expresses *1Dy10*. *Journal of Cereal Science* 48, 439–445. Available at: https://doi.org/10.1016/j.jcs.2007.11.005

Gasparis, S., Orczyk, W. and Nadolska-Orczyk, A. (2013) *Sina* and *Sinb* genes in triticale do not determine grain hardness contrary to their orthologs *Pina* and *Pinb* in wheat. *BMC Plant Biology* 13(1), 190. Available at: https://doi.org/10.1186/1471-2229-13-190

Gill, B.S. and Friebe, B. (2001) Cytogenetics, phylogeny and evolution of cultivated wheats. In: Bonjean, A.P. and Angus, W.J. (eds) *The World Wheat Book: A History of Wheat Breeding*. Intercept Ltd, London, pp. 57–72.

Gu, J., Yin, X., Stomph, T.-J. and Struik, P.C. (2014) Can exploiting natural genetic variation in leaf photosynthesis contribute to increasing rice productivity? A simulation analysis. *Plant, Cell & Environment* 37, 22–34.

Gupta, P.K. and Varshney, R.K. (2000). The development and use of microsatellite markers for genetic analysis and plant breeding with emphasis on bread wheat. *Euphytica* 113, 163–185.

Hanif, M., Gao, F., Liu, J., Wen, W., Zhang, Y. *et al.* (2016) *TaTGW6*-A1, an ortholog of rice *TGW6*, is associated with grain weight and yield in bread wheat. *Molecular Breeding* 36, 1. Available at: https://doi.org/10.1007/s11032-015-0425-z

Hao, Y., Velu, G., Peña, R.J., Singh, S. and Singh, R.P. (2014) Genetic loci associated with high grain zinc concentration and pleiotropic effect on kernel weight in wheat (*Triticum aestivum* L.). *Molecular Breeding* 34, 1893–1902. Available at: https://doi.org/10.1007/s11032-014-0147-7

Hedden, P. (2003) The genes of the green revolution. *Trends in Genetics* 19, 5–9.

Heffner, E.L., Jannink, J., Iwata H., Souza, E. and Sorrells, M.E. (2011) Genomic selection accuracy for grain quality traits in biparental wheat populations. *Crop Science* 51(6), 2597–2606. Available at: https://doi.org/10.2135/cropsci2011.05.0253

Heidari, B., Sayed-Tabatabaei, B.E., Saeidi, G., Kearsey, M. and Suenaga, K. (2011) Mapping QTL for grain yield, yield components, and spike features in a doubled haploid population of bread wheat. *Genome* 54(6), 517–527.

Hu, M.-J., Zhang, H.-P., Cao, J.-J., Zhu, X.-F., Wang, S.-X. *et al.* (2016) Characterization of an IAA-glucose hydrolase gene *TaTGW6* associated with grain weight in common wheat (*Triticum aestivum* L.). *Molecular Breeding* 36, 25. Available at: https://doi.org/10.1007/s11032-016-0449-z

Hu, Y., Knapp, S. and Schmidhalter, U. (2020) Advancing high-throughput phenotyping of wheat in early selection cycles. *Remote Sensing* 12, 574. Available at: https://doi.org/10.3390/rs12030574

Jiang, Q., Hou, J., Hao, C., Wang, L., Ge, H., Dong, Y. and Zhang, X. (2011) The wheat (*T. aestivum*) sucrose synthase 2 gene (*TaSus2*) active in endosperm development is associated with yield traits. *Functional & Integrative Genomics* 11(1), 49–61. Available at: https://doi.org/10.1007/s10142-010-0188-x

Kertho, A., Mamidi, S., Bonman, J.M., McClean, P.E. and Acevedo, M. (2015) Genome-wide association mapping for resistance to leaf and stripe rust in winter-habit hexaploid wheat landraces. *PLoS One* 10, e0129580. Available at: https://doi.org/10.1371/journal.pone.0129580

Khlestkina, E.K., Than, M.H.M., Pestsova, E.G., Röder, M.S., Malyshev, S.V., Korzun, V. and Börner, A. (2004) Mapping of 99 microsatellite loci in rye (*Secale cereale* L.) including 39 expressed sequence tags. *Theoretical and Applied Genetics* 109, 725–732.

Kuchel, H., Wraner, P., Fox, R.L., Chalmers, K., Howes, N., Langridge, P. and Jefferies, S.P. (2003) Whole genome based marker assisted selection strategies in wheat breeding. In: Pogna, N.E., Romano, M., Pogna, E.A. and Galterio, G. (eds) *Proceedings of the Tenth International Wheat Genetics Symposium, Paestum, Italy, 1–6 September 2003*. Istituto Sperimentale per la Cerealicoltura, Rome, pp. 144–147.

Lagercrantz, U. (1998) Comparative mapping between *Arabidopsis thaliana* and *Brassica nigra* indicates that *Brassica* genomes have evolved through extensive genome replication accompanied by chromosome fusions and frequent rearrangements. *Genetics* 150, 1217–1228.

Langridge, P. (2005) Molecular breeding of wheat and barley. In: Tuberosa, R., Phillips, R.L. and Gale, M. (eds) *In the Wake of Double Helix: From the Green Revolution to the Gene Revolution.* Avenue Media, Bologna, Italy, pp. 279–286.

La Rota, C.M. and Sorrells, M.E. (2004) Comparative DNA sequence analysis of mapped wheat ESTs reveals complexity of genome relationships between rice and wheat. *Functional & Integrative Genomics* 4, 34–46.

Li, S.S., Jia, J.Z., Wei, X.Y., Zhang, X.C., Li, L.Z. *et al.* (2007) An intervarietal genetic map and QTL analysis for yield traits in wheat. *Molecular Breeding* 20, 167–178. Available at: https://doi.org/10.1007/s11032-007-9080-3

Liu, G., Jia, L., Lu, L., Qin, D., Zhang, J. and Guan, P. (2014) Mapping QTLs of yield-related traits using RIL population derived from common wheat and Tibetan semi-wild wheat. *Theoretical and Applied Genetics* 127, 2415–2432. Available at: https://doi.org/10.1007/s00122-014-2387-7

Liu, Y., He, Z., Appels, R. and Xia, X. (2012) Functional markers in wheat: current status and future prospects. *Theoretical and Applied Genetics* 125, 1–10. Available at: https://doi.org/10.1007/s00122-012-1829-3

Lyra, D.H., Virlet, N., Sadeghi-Tehran, P., Hassall, K.L., Luzie, U. *et al.* (2020) Functional QTL mapping and genomic prediction of canopy height in wheat measured using a robotic field phenotyping platform. *Journal of Experimental Botany* 71(6), 1885–1898. Available at: https://doi.org/10.1093/jxb/erz545

Maxted, N.A., Amri, N.P., Castaneda-Alvarez, S., Dias, M.E., Dulloo, H. *et al.* (2016) Joining up the dots: a systematic perspective of crop wild relatives conservation and use. In: Maxted, N., Ehsan, M. and Ford Lloyd, B.V. (eds) *Crop Genepool Use: Capturing Wild Relative and Landrace Diversity for Crop Improvement*. CAB International, Wallingford, UK, pp 87–124.

Mondal, S., Rutkoski, J.E, Velu, G., Singh, P.K., Crespo-Herrera, L.A. *et al.* (2016) Harnessing diversity in wheat to enhance grain yield, climate resilience, disease and insect pest resistance and nutrition through conventional and modern breeding approaches. *Frontiers in Plant Science* 6(7), 991. Available at: https://doi.org/10.3389/fpls.2016.00991

Nadolska-Orczyk, A., Rajchel, I.K., Orczyk, W. and Gasparis, S. (2017) Major genes determining yield-related traits in wheat and barley. *Theoretical and Applied Genetics* 130, 1081–1098. Available at: https://doi.org/10.1007/s00122-017-2880-x

Nuttonson, M.Y. (1955) *Wheat–Climatic Relationships and the Use of Phenology in Ascertaining the Thermal and Photothermal Requirements of Wheat*. American Institute of Crop Ecology, Silver Spring, Maryland.

Ogbonnaya, F.C., Abdalla, O., Mujeeb-Kazi, A., Kazi, A.G., Gosnian, N. and Lagudah, E.S. (2013) Synthetic hexaploids: harnessing species of the primary gene pool for wheat improvement. *Plant Breeding Review* 37, 35–122.

Ornella, L., Singh, S., Perez, P., Burgueño, J., Singh, R. *et al.* (2012) Genomic prediction of genetic values for resistance to wheat rust. *The Plant Genome* 5(3), 136–148. Available at: https://doi.org/10.3835/plantgenome2012.07.0017

Parry, M.A.J., Reynolds, M.P., Salvucci, M.E., Raines, C.A., Andralojc, P.J. *et al.* (2010) Raising yield potential of wheat: II. Increasing photosynthetic capacity and efficiency. *Journal of Experimental Botany* 62(2), 453–467. Available at: https://doi.org/10.1093/jxb/erq304

Philipp, N., Weichert, H., Bohra, U., Weschke, W., Schulthess, A.W. *et al.* (2018) Grain number and grain yield distribution along the spike remain stable despite breeding for high yield in winter wheat. *PLoS One* 13(10), e0205452. Available at: https://doi.org/10.1371/journal.pone.0205452

Pieruschka, R. and Schurr, U. (2019) Plant phenotyping: past, present, and future. *Plant Phenomics* 2019, 7507131. Available at: https://doi.org/10.34133/2019/7507131

Qi, L., Friebe, B., Zhang, P. and Gill, B.S. (2007) Homoeologous recombination, chromosome engineering and crop improvement. *Chromosome Research* 15, 3–19.

Qin, X.L., Weiner, J., Qi, L., Xiong, Y.C. and Li, F.M. (2013) Allometric analysis of the effects of density on reproductive allocation and harvest index in 6 varieties of wheat (*Triticum*). *Field Crops Research* 144, 162–166. Available at: https://doi.org/10.1016/j.fcr.2012.12.011

Rafalski, J.A. (2002) Applications of single nucleotide polymorphisms in crop genetics. *Current Opinion in Plant Biology* 5, 94–100.

Rafalski, A. and Morgante, M. (2004) Corn and humans: recombination and linkage disequilibrium in two genomes of similar size. *Trends in Genetics* 20, 103–111.

Ramya, P., Chaubal, A., Kulkarni, K., Gupta, L., Kadoo, N. and Dhaliwal, H.S. (2010) QTL mapping of 1000-kernel weight, kernel length, and kernel width in bread wheat (*Triticum aestivum* L.). *Journal of Applied Genetics* 51, 421–429. Available at: https://doi.org/10.1007/BF03208872

Ray, D.K., Mueller, N.D., West, P.C. and Foley, J.A. (2013) Yield trends are insufficient to double global crop production by 2050. *PLoS One* 8(6), e66428. Available at: https://doi.org/10.1371/journal.pone.0066428

Reynolds, M., Tattaris, M., Cossani, C.M., Ellis, M., Yamaguchi-Shinozaki, K. and Pierre, K.C. (2015) Exploring genetic resources to increase adaptation of wheat to climate change. In: Ogihara, Y., Takumi, S. and Handa, H. (eds) *Advances in Wheat Genetics: From Genome to Field.* Springer, Tokyo, pp. 355–368. Available at: https://doi.org/10.1007/978-4-431-55675-6_41

Riley, R. and Chapman, V. (1958) Genetic control of the cytologically diploid behaviour of hexaploid wheat. *Nature* 182, 713–715.

Rimbert, H., Darrier, B., Navarro, J., Kitt, J., Choulet, F. *et al.* (2018) High throughput SNP discovery and genotyping in hexaploid wheat. *PLoS ONE* 13(1), e0186329. Available at: https://doi.org/10.1371/journal.pone.0186329

Rostoks, N., Borevitz, J.O., Headley, P.E., Russel, J., Mudie, S. *et al.* (2005) Single-feature polymorphism discovery in the barley transcriptome. *Genome Biology* 6, R54.

Salvi, S., Porfiri, O. and Ceccarelli, S. (2013) Nazareno Strampelli, the 'Prophet' of the green revolution. *Journal of Agricultural Science* 151, 1–5. Available at: https://doi.org/10.1017/S0021859612000214

Sears, E.R. (1976) Genetic control of chromosome pairing in wheat. *Annual Review of Genetics* 10, 31–51.

Sharifi, P., Hashem, A., Rahman, E., Ali, M. and Abouzar, A. (2017) Evaluation of genotype × environment interaction in rice based on AMMI model in Iran. *Rice Science* 24, 173–180.

Shiferaw, B., Smale, M., Braun, H.J., Duveiller, E., Reynolds, M. and Muricho, G. (2013) Crops that feed the world 10. Past successes and future challenges to the role played by wheat in global food security. *Food Security* 5, 291–317.

Schouten, H.J., Krens, F.A. and Jacobsen, E. (2006) Cisgenic plants are similar to traditionally bred plants: international regulations for genetically modified organisms should be altered to exempt cisgenesis. *EMBO Reports* 7(8), 750–753. Available at: https://doi.org/10.1038/sj.embor.7400769

Singh, D., Wang, X., Kumar, U., Gao, L., Noor, M. *et al.* (2019) High-throughput phenotyping enabled genetic dissection of crop lodging in wheat. *Frontiers in Plant Science* 10, 394. Available at: https://doi.org/10.3389/fpls.2019.00394

Slafer, G.A. (2003) Genetic basis of yield as viewed from a crop physiologist's perspective. *Annals of Applied Biology* 142, 117–128.

Srinivasa, J., Arun, B., Mishra, V.K., Singh, G.P., Velu, G. *et al.* (2014). Zinc and iron concentration QTL mapped in a *Triticum spelta* × *T. aestivum* cross. *Theoretical and Applied Genetics* 127, 1643–1651. Available at: https://doi.org/10.1007/s00122-014-2327-6

Stein, N. and Graner, A. (2004) Map-based gene isolation in cereal genomes. In: Gupta, P.K. and Varshney, R.K. (eds) *Cereal Genomics*. Kluwer Academic Publishers, Dordrecht, the Netherlands, pp. 331–360.

Stuber, C.W., Polacco, M.L. and Senior, M. (1999) Synergy of empirical breeding, marker-assisted selection, and genomics to increase crop yield potential. *Crop Science* 39, 1571–1583.

Su, Q., Zhang, X., Zhang, W., Zhang, N., Song, L. *et al.* (2018) QTL detection for kernel size and weight in bread wheat (*Triticum aestivum* L.) using a high-density SNP and SSR-based linkage map. *Frontiers in Plant Science* 9, 1484. Available at: https://doi.org/10.3389/fpls.2018.01484

Sukumaran, S., Crossa, J., Jarquin, D., Lopes, M. and Reynolds, M.P. (2017) Genomic prediction with pedigree and genotype × environment Interaction in spring wheat grown in South and West Asia, North Africa, and Mexico. *G3: Genes, Genomes Genetics* 7, 481–495. Available at: https://doi.org/10.1534/g3.116.036251

Sun, X.C., Marza, F., Ma, H.X., Carver, B.F. and Bai, G.H. (2010) Mapping quantitative trait loci for quality factors in an inter-class cross of US and Chinese wheat. *Theoretical and Applied Genetics* 120, 1041–1051. Available at: https://doi.org/10.1007/s00122-009-1232-x

Tadesse, W., Garcia, M.S., Asefa, S.G., Amri, A., Bishaw, Z., Ogbonnaya, F.C. and Baum, M. (2019) Genetic gains in wheat breeding and its role in feeding the world. *Crop Breeding Genetics and Genomes* 1, e190005. Available at: https://doi.org/10.20900/cbgg20190005

Thompson, M.J. (2014) High-throughput SNP genotyping to accelerate crop improvement. *Plant Breeding and Biotechnology* 2, 195–212. Available at: https://doi.org/10.9787/PBB.2014.2.3.195

Varshney, R.K., Langridge, P. and Graner, A. (2007) Application of genomics to molecular breeding of wheat and barley. *Advances in Genetics* 58, 121–155. Available at: https://doi.org/10.1016/S0065-2660(06)58005-8

Velu, G., Crossa, J., Singh, R.P., Hao, Y., Dreisigacker, S. *et al.* (2016) Genomic prediction for grain zinc and iron concentrations in spring wheat. *Theoretical and Applied Genetics* 129, 1595–1605. Available at: https://doi.org/10.1007/s00122-016-2726-y

Wang, J., Li, R., Mao, X. and Jing, R. (2017) Functional analysis and marker development of *TaCRT-D* gene in common wheat (*Triticum aestivum* L.). *Frontiers in Plant Science* 8, 1557. Available at: https://doi.org/10.3389/fpls.2017.01557

Wang, Y., Cheng, X., Shan, Q., Zhang, Y., Liu, J., Gao, C. and Qiu, J.L. (2014) Simultaneous editing of three homoeoalleles in hexaploid bread wheat confers heritable resistance to powdery mildew. *Nature Biotechnology* 32, 947–951. Available at: https://doi.org/10.1038/nbt.2969

Wiesmeier, M., Hübner, R., Dechow, R., Maier, H., Spörlein, P. *et al.* (2014) Estimation of past and recent carbon input by crops into agricultural soils of southeast Germany. *European Journal of Agronomy* 61, 10–23. Available at: https://doi.org/10.1016/j.eja.2014.08.001

Xin, F., Zhu, T., Wei, S., Han, Y., Zhao, Y. *et al.* (2020) QTL mapping of kernel traits and validation of a major QTL for kernel length-width ratio using SNP and bulked segregant analysis in wheat. *Scientific Reports* 10, 25. Available at: https://doi.org/10.1038/s41598-019-56979-7

Yan, W. and Kang, M.S. (2002) *Genotype-by-environment interaction. GGE Biplot Analysis: A Graphical Tool for Breeders, Geneticists and Agronomists*. CRC Press, Boca Raton, Florida, pp. 1–10.
Yang, L., Zhao, D., Meng, Z., Xu, K., Yan, J. *et al.* (2020) QTL mapping for grain yield-related traits in bread wheat via SNP-based selective genotyping. *Theoretical and Applied Genetics* 133, 857–872. Available at: https://doi.org/10.1007/s00122-019-03511-0
Yang, N. and Smale, M. (1996) *Indicators of Wheat Genetic Diversity and Germplasm Use in the People's Republic of China*. NRG Paper Series 96-04. CIMMYT, Mexico City.
Yang, Z.J., Peng, Z.S. and Yang, H. (2016) Identification of novel and useful EST-SSR markers from *de novo* transcriptome sequence of wheat (*Triticum aestivum* L.). *Genetics and Molecular Research* 15(1), gmr.15017509. Available at: https://doi.org/10.4238/gmr.15017509
Young, N.D. and Tanksley, S.D. (1989) RFLP analysis of the size of chromosomal segments retained around the *Tm-2* locus of tomato during backcross breeding. *Theoretical and Applied Genetics* 77(3), 353–359. Available at: https://doi.org/10.1007/BF00305828
Yu, Q.Y., Wu, W.B., You, L.Z., Zhu, T.J., Vliet, J.V. *et al.* (2017) Assessing the harvested area gap in China. *Agricultural Systems* 153, 212–220. Available at: https://doi.org/10.1016/j.agsy.2017.02.003
Zhang, G., Wang, Y., Guo, Y., Zhao, Y., Kong, F. and Li, S. (2015) Characterization and mapping of QTLs on chromosome 2D for grain size and yield traits using a mutant line induced by EMS in wheat. *The Crop Journal* 3, 135–144. Available at: https://doi.org/10.1016/j.cj.2014.11.002

# 11 Tools for Transforming Wheat Breeding: Genomic Selection, Rapid Generation Advance and Database-Based Decision Support

**Umesh R. Rosyara*, Kate Dreher, Bhoja R. Basnet and Susanne Dreisigacker***
*International Maize and Wheat Improvement Center (CIMMYT), Texcoco, Mexico*

## 11.1 Introduction

Plant breeding is the science and art of developing new crop varieties (Fehr, 1991; Sleper and Poehlman, 1995). The art component includes the breeder's imagination of a crop ideotype together with his/her visual skills, while various principles from a variety of sciences are used to improve the overall genetic potential of a crop. With an increased understanding of crop genetics, plant breeders can keep track of how genomic regions and cultivar are inherited or combined in a new line. In addition, genetic architecture governing economically important traits has been elucidated. In the last two decades, several factors have especially allowed this increased depth of genomic information: (i) rapid evolution of new sequencing and genotyping technologies has led to better marker platforms with enhanced depth and density across at reduced costs; (ii) phenotyping, sequencing and genotyping methods are increasingly fully automated; and (iii) the production of large data sets, 'big data', has been rising along with the use of prediction and machine learning methods. Beside advances in genomics, there have been remarkable developments in other areas, such as: (i) shortening breeding cycles via rapid generation advance (RGA) techniques; (ii) high-throughput phenotyping; and (iii) full or partial automation or mechanization of breeding trials. Clearly, the next-generation technologies have increased implications in the current breeding methodology in any crop, including wheat. The fundamental basis of increased genetic gains, however, remains the same as of today for quantitative traits: the breeder's equation by Lush (1937) (Fig. 11.1).

Rate of genetic gain ($\Delta_g$) is driven as a function of the heritability of a trait (as a measure of accuracy), selection intensity, genetic variability and the cycle length (Fig. 11.1). The rate of genetic gain can therefore be increased when:

**1.** *The heritability is improved* – either through direct selection of a trait per se via excellent trial management or using indirect selection such as marker-aided selection or genomic selection (GS).
**2.** *Stronger selection intensity is applied without reducing desirable genetic variability* – this can be achieved by identifying individuals via precision genotyping and phenotyping.

*Corresponding authors: to rosyaraur@gmail.com; s.dreisigacker@cgiar.org


DOI: 10.1079/9781789245431.0011

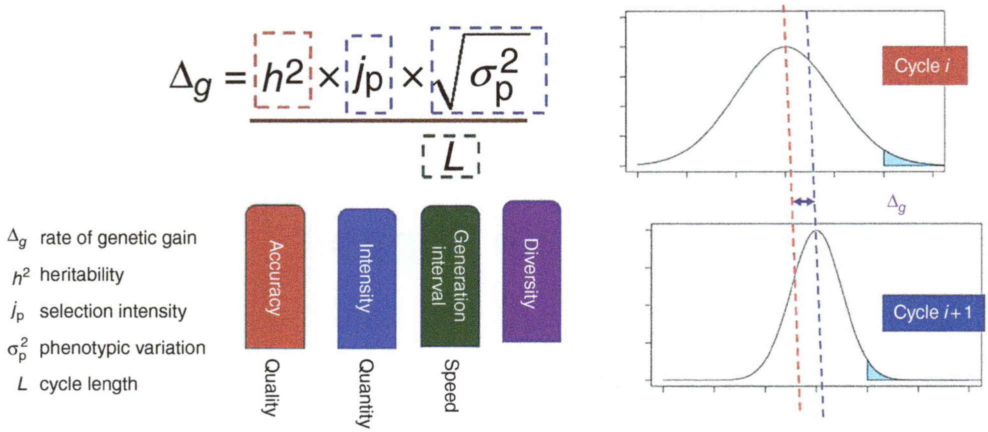


**Fig. 11.1.** The breeder's equation: rate of genetic gain and factors affecting genetic gain – heritability, selection intensity, generation interval and diversity, accuracy of selection can be defined as the correlation between estimated ($\hat{u}$) and true breeding value ($u$), $h = r(u, \hat{u})$.

**3.** *The generation/cycle interval is shortened* – via techniques such as speed breeding, RGA or doubled haploid (DH) production.
**4.** *The genetic variability is increased by growing larger populations* – without compromising the quality of breeding trial management.

Rate of genetic gain should also be considered in terms of funds spent, i.e. rate of genetic gain per unit funds spent ($\Delta_{g/M}$), where $M$ is the total spend. Thus, rates of genetic gain are also tied with overall management of breeding programmes in terms of funding, time and human resource management. Modern breeding programmes draw on high-throughput genotyping/phenotyping, effective trial management and overall operational excellence, data collection, data storage and analytics, and decision support. When a large amount of data can be collected, stored, analysed and provided in meaningful ways to breeders, it can complement the breeders' required memory/experience or artistic skills. Most futuristic thinking includes complete automation of the breeding process, where breeding is completely done by robots/artificial intelligence (i.e. 'robo-breeders'). This chapter discusses some of the strategies that have started and will further transform wheat breeding in the next few years.

## 11.2 Genomic Selection

In contrast to conventional marker-assisted selection (MAS), GS involves a larger number of genome-wide markers, irrespective of linkage or association with quantitative trait loci (QTL) or genes of interest (Bernardo 1994; Meuwissen *et al.*, 2001). The first step in GS is model development where one or more statistical models are utilized and tested using a training population which is both genotyped and phenotyped (Fig. 11.2). Once a model is developed and validated, it is used to calculate genomic estimated breeding values (GEBVs) in an application population, from which individual lines are selected for crossing or generation advance.

In recent years, GS has gained popularity in plant and animal breeding (e.g. Bernardo and Yu, 2007; Van Raden 2008; Crossa *et al.*, 2010, 2017; Lehermeier *et al.*, 2013). In practice, the main purpose of applying GS is to predict the genetic merit of lines ahead of time so that candidate individuals can be selected in an earlier breeding generation(s) and with higher accuracy over phenotype-based selection with reduced cost.

Breeding values in GS are predicted using statistical equations. The basic genomic prediction model can be represented by the following mixed-model equation:

$$y = Zu + \varepsilon \qquad (11.1)$$

where $\boldsymbol{y}$ is a vector of phenotypes consisting of phenotypic value or of adjusted means (e.g. best linear unbiased predictor (BLUP), best linear unbiased estimator (BLUE)), $\boldsymbol{u}$ is a vector of random genetic values, $\boldsymbol{\varepsilon}$ is the vector of residuals and $\mathbf{Z}$ is

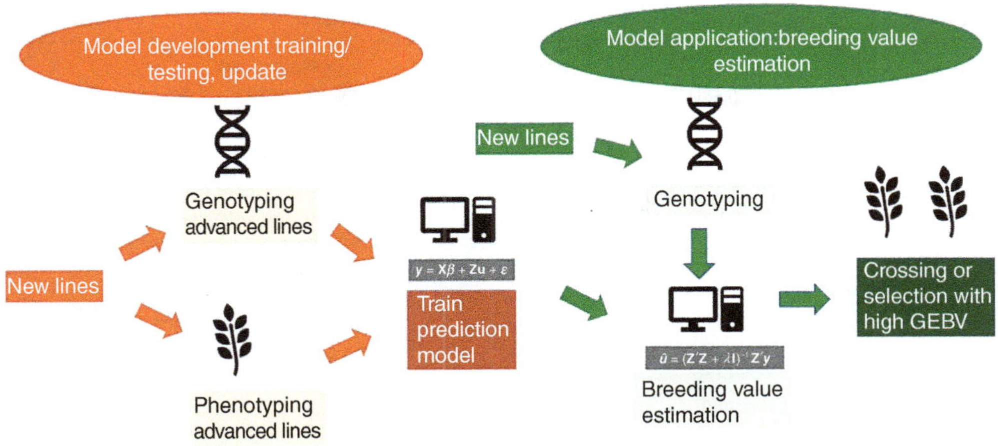


**Fig. 11.2.** Example GS cycle: model development, update and utilization.

the design matrix based on marker scores. As genome-wide prediction uses a large number of markers, it creates a $p > n$ issue; that is, the number of markers/parameters ($p$) is greater than the number of samples ($n$). Therefore, shrinkage needs to be integrated into the prediction model. With Gaussian methods several prior distributions can be fitted to deal with this issue (Pérez and de los Campos, 2014). The breeding value with shrinkage, $\lambda$, included would be: $\mathrm{BLUP}[u] = \hat{u} = (\mathbf{Z}'\mathbf{Z} + \lambda \mathbf{I})^{-1}\mathbf{Z}'y$. This equation can be extended to additional fixed-effect covariates and replacing the marker matrix with a marker-based relatedness matrix, a method also known as GBLUP:

$$y = \mathbf{X}\boldsymbol{\beta} + \mathbf{Z}u + \boldsymbol{\varepsilon} \tag{11.2}$$

where $\boldsymbol{\beta}$ is a vector of fixed effects and **X** and **Z** are design matrices. The $\boldsymbol{u}$ was assumed to follow a Gaussian distribution $\boldsymbol{u} \sim N(0, \mathbf{A}\sigma_g^2)$, where **A** is the genomic relationship matrix and $\sigma_g^2$ is the additive genetic variance. The residuals $\boldsymbol{\varepsilon}$ was assumed to follow a Gaussian normal distribution $\boldsymbol{\varepsilon} \sim \mathrm{N}(0, \mathbf{I}\sigma_g^2)$, where **I** is the identity matrix. With group effects used as fixed effect $\mathbf{X} \in \mathbf{G}$, where **G** is the matrix of group relationship $\{0,1\}$ for one or more groups, the additive relationship matrix ($\mathbf{A_M}$) can be calculated as $\mathbf{A}_\mathrm{M} = \mathbf{M}\mathbf{M}^\mathrm{T}$, where $\mathbf{M} \in \{1, 0, -1\}$ depending upon whether marker is homozygous reference or heterozygous or homozygous alternate alleles. The model can be extended to fit a Gaussian kernel-based relationship matrix $\mathbf{K_M}$, $K_{ij} = \exp[-(D_{ij} / \theta)]$, where $D_{ij}$ is the Euclidean distance normalized to interval $[0,1]$ and $\theta$ is the scale parameter (Rosyara *et al.*, 2016). The Gaussian kernel model can indirectly account for epistatic effects (Sehgal *et al.*, 2020). The model performance is usually evaluated based on accuracy, i.e. the correlation between actual and predicted breeding values. There are several factors influencing prediction accuracy: the prediction model, training population size, trait heritability, genetic relationship between training and validation sets, population structure and training population composition.

Studies on GS have been growing in wheat during the last decade (Rutkoski *et al.*, 2010, 2012, 2014; Poland *et al.*, 2012; Crossa *et al.*, 2014; Juliana *et al.*, 2017, 2018, 2019; Sehgal *et al.*, 2020) and different traits show different levels of prediction accuracy (Table 11.1). The International Maize and Wheat Improvement Center (CIMMYT) Wheat Program started to explore GS more aggressively as a breeding tool in 2010. Since then, CIMMYT has made significant contributions developing and testing various new genome-wide prediction models in CIMMYT wheat data sets. GS became operational in the CIMMYT spring bread wheat programme in 2013 and today is routinely applied at the first yield trial stage to improve selection accuracy in a large set of lines with limited phenotypic information.

Implementing GS in a breeding programme should often justify its advantages over the standard breeding approaches. A routine way of applying GS in animal breeding is to predict the breeding value of potential parents for mating,

**Table 11.1.** Genomic prediction accuracy for example traits in wheat.

| Trait | Prediction accuracy | Reference |
|---|---|---|
| Grain yield | 0.35–0.40 (within cycle)<br>0.05–0.15 (between cycles) | Juliana *et al.* (2018); Sehgal *et al.* (2020) |
| 1000-Kernel weight | 0.45–0.81 | Battenfield *et al.* (2016) |
| Hardiness | −0.02–0.82 | Battenfield *et al.* (2016) |
| Grain protein | 0.32–0.95 | Battenfield *et al.* (2016) |
| Mixogram performance | 0.41–0.85 | Battenfield *et al.* (2016) |
| Leaf rust (seedling) | 0.31–0.74 | Juliana *et al.* (2017) |
| Leaf rust (adult) | 0.12–0.56 | Juliana *et al.* (2017) |
| Stem rust (adult) | 0.31–0.65 | Juliana *et al.* (2017) |
| Yellow rust (seedling) | 0.70–0.78 | Juliana *et al.* (2017) |
| Yellow rust (adult) | 0.34–0.71 | Juliana *et al.* (2017) |
| Fusarium head blight (severity) | 0.34–0.64 | Rutkoski *et al.* (2012) |
| Fusarium head blight (deoxynivalenol content) | 0.18–0.55 | Rutkoski *et al.* (2012) |

which in plant breeding could be translated to an 'elite line crossing model', namely crossing two lines with high GEBV with the assumption that they will produce a population with transgressive segregants. An alternative model is cross-prediction: the 'best combiner model' where essentially a breeding value is assigned to crosses rather individuals (Lado *et al.*, 2017). In some cases, the objective of GS can also simply be to reduce phenotyping cost via selecting lines that are likely to succeed or discarding lines that are likely to fail (Beyene *et al.*, 2019). The cost of genotyping ($C_G$) can be compared with the cost of phenotyping ($C_P$) and how both relate to increased genetic gains estimated. The relative efficiency of genomic selection ($GS_{eff}$) can be represented by the equation:

$$GS_{eff} = \frac{(i_{GS} r_{M,G}) / L_{GS}}{(i_{PS} r_{P,G}) / L_{PS}} \quad (11.3)$$

where $i_{GS}$ is the selection intensity when using GS, $r_{M,G}$ is the accuracy of genomic selection, $L_{GS}$ is the cycle length using GS, $i_{PS}$ is the selection intensity using phenotypic selection, $r_{P,G}$ is the accuracy of phenotypic selection (i.e. heritability) and $L_{PS}$ is the cycle length using phenotypic selection. Assuming that selection intensity and cycle length are fixed, the advantage of GS over phenotypic (PS) would be simply the ratio of accuracy between the two: $r_{M,G}/r_{P,G}$. Let us assume $r_{P,G} = 0.25$ and $r_{M,G} = 0.50$, the cost of phenotyping and genotyping is equally \$15 per line and the cycle length is 7 years, the genetic gain can potentially be doubled applying GS. When $r_{M,G}$ is high, selection intensity ($i_{GS}$) can be increased, which will further increase $GS_{eff}$. Another potential point of intervention is to decrease cycle length ($L_{GS}$), by combining RGA with GS (Voss-Fels *et al.*, 2019), as discussed in the following sections.

## 11.3 Rapid Generation Advance

In a self-pollinated crop such as wheat, a breeding cycle involves a cross followed by various generations of selfing to generate genetically stable lines for performing yield tests. The cycle usually requires five or six generations of selfing (Fig. 11.3). In many environments where only a single season can be planted each year, this step thus requires 5–6 years. Alternatively, when shuttle breeding or winter nurseries are possible, two, sometimes up to three, seasons per year can be planted. This reduces the selfing generations to 3–4 years. The process can be further accelerated by RGA methods such as speed breeding.

The RGA method has already been targeted by breeders for some time in several crops to speed up the breeding process (Depauw and Clarke 1976; Heu *et al.*, 1982; Carandang *et al.*, 2006; Tanio *et al.*, 2006; Gaur *et al.*, 2007; Bhattarai *et al.*, 2009; Ishigaki, 2010; Wang *et al.*, 2011; Rizal *et al.*, 2014). One common technology for RGA is the production of DHs (Niu *et al.*, 2014; Patial *et al.*, 2019). However, DHs

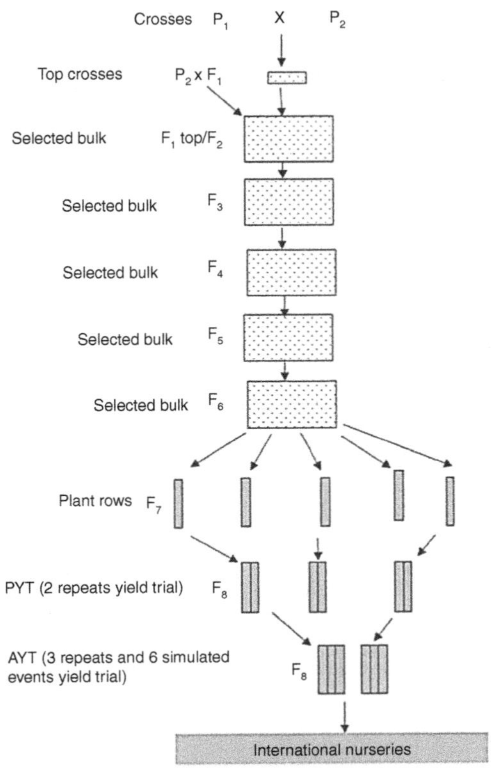


| Model 1 | Model 2 | Model 3 |
|---|---|---|
| Year 1 | Year 1, season 1 | Year 1, season 1 |
| Year 2 | Year 1, season 2 | Year 1, season 2 |
| Year 3 | Year 2, season 1 | Year 2, season 1 |
| Year 4 | Year 2, season 2 | Year 2, season 2 |
| Year 5 | Year 3, season 1 | Year 2, season 3 |
| Year 6 | Year 3, season 2 | Year 3, season 1 |
| Year 7 | Year 4, season 1 | Year 3, season 2 |
| Year 8 | Year 4, season 2 | Year 3, season 3 |
| Year 9 | Year 5, season 1 | Year 4, season 1 |
| Year 10 | Year 6, season 1 | Year 5, season 1 |
| Year 11 | Year 7, season 1 | Year 6, season 1 |

**Fig. 11.3.** Proposed spring wheat breeding schemes: one season per year (model 1); shuttle breeding or an off season for generation advance (model 2); and speed breeding with three seasons per year (model 3), in which a maximum of six seasons is possible in wheat but may be a logistical challenge in some circumstances. PYT, preliminary yield trial; AYT, advanced yield trial.

have the disadvantage of being expensive to deploy in a larger breeding programme. An alternative to DH production is speed breeding (Ghosh *et al.*, 2018; Watson *et al.*, 2018), which includes extending the daily exposure of plants to light along with early seed harvest. Speed breeding has been reported to achieve as much as six generations per year in spring bread wheat (*Triticum aestivum*), durum wheat (*Triticum durum*) and other crops. Thus, individual plants can be advanced up to the $F_6$ generation within 1 year. The downside of using speed breeding, however, is that phenotypic characterization is almost impossible and can only partly be compensated with the screening for major genes or QTL with linked molecular markers. Voss-Fels *et al.* (2019) suggested to combine GS with speed breeding in wheat. At different institutions, several deployment options must still be simulated and tested for different traits to validate this combined approach for transforming wheat breeding. The CIMMYT wheat programme will test such a scheme characterized by a rapid turnover of generations as bulks and early generation GS utilizing the earliest possible phenotypic and genotypic data to select parents for the subsequent breeding cycle. Combining the accelerated breeding process with data-intensive techniques such as GS requires robust data management.

## 11.4 Breeding Data Management

Rate of genetic gain per unit of investment in a breeding programme depends upon several activities, such as the level of mechanization or automation, human resources, the agronomy applied, reliable data generation and data management. Effective management of data (collection, storage, analysis, visualization) becomes increasingly important in line with the development of a robust breeding decision-support system. Data management, ideally, should include all types of data: phenotypic, genotypic, climate data, etc. With the implementation of GS, high-throughput phenotyping/phenomics and any other new technologies, efficient data management becomes more relevant. Developments in the area of computation technology make this more automated, faster, as well as cheaper.

In a breeding programme, the most routine and obvious step is the collection of phenotypic data. Instead of using traditional field books, a more efficient way to gather data today is by hand-held devices paired with custom mobile apps. For example, Field Book (Rife and Poland, 2014) and KDSmart (http://www.kddart.org/kdsmart.html, accessed 15 February 2010) are used at CIMMYT, which can automatically load phenotypic data into a database. Ideally such hand-held devices have barcode scanners. Sample tracking is an additional key element for both automated genotyping and phenotyping. Sample tracking for genotyping includes tracing samples from field to the laboratory, to an eventual genotyping service provider and back to the field. There are several laboratory management systems (LIMS) in use by different molecular laboratories. CIMMYT has a custom-built sample tracking system that is a DNA sample tracker (https://github.com/itucimmyt/Sample_tracking_system_CE, accessed 15 February 2021).

### 11.4.1 Databases and associated breeding information management systems

A database is a collection of information that is organized in tables and stored on a computer system. This information can be updated or modified as required with the help of a database management system (DBMS), which is a software for creating and managing data in the database. A DBMS interface is a user interface which allows for the ability to input queries to a database without using the query language itself. The DBMS interface provides users and programmers a defined process for data retrieval, management, updating and creation. Some of the popular DBMSs are SolarWinds Database Performance Analyzer, Oracle RDBMS, IBM DB2, Microsoft® SQL Server, SAP Sybase ASE, MySQL, Microsoft Access, Informix, SQLite, PostgreSQL, AmazonRDS, MongoDB, Redis, OrientDB, phpMyAdmin, SQL Developer, Seqel PRO, Hadoop HDFS, Cloudera, MariaDB and Informix Dynamic Server.

Generic databases can be divided into two major types: (i) relational or sequence databases; and (ii) non-relational or non-sequence databases. A relational database is a digital database used on the relational model of data (Codd, 1970) and the software used to maintain such a database is called a relational database management system (RDBMDS). The RDBMDS uses schema, which are templates used to dictate the structure of data that will be stored within the database. The relational model organizes the data into one or more tables (or 'relations') of columns and rows. Each row has a unique key that identifies each row. Rows are called records or tuples while columns are called attributes. Many of the known RDBMDSs have an option for using Structured Query Language (SQL) for querying and maintaining; for example, Oracle (https://www.oracle.com/database/, accessed 15 February 2021), PostgreSQL (https://www.postgresql.org/, accessed 15 February 2021) and MySQL (https://www.mysql.com/, accessed 15 February 2021). Relational databases follow the strict schema, enabling data to be predictable and easily assessable. Use of strict schema means each new entry must have a different component that should fit to a preformed template. In contrast to relational databases, the non-relational databases are more forgiving in their structure and form.

Although the conventional RDBMDSs are designed to build and store, manage and analyse large-scale data, they face a performance problem when it comes to a large matrix of data like that of a genomic experiment meant to perform GS (Nti-Addae *et al.*, 2019). To deal with large data sets in the genomic context, non-relational databases are sometimes chosen including MonetDB (Cijvat *et al.*, 2015), NoSQL systems (Guimaraes *et al.*, 2015) and MongoDB, a document-based NoSQL database, used in Gigwa (Sempéré *et al.*, 2016), for example. Nti-Addae *et al.* (2019) found that the Hierarchical Data Format (HDF5) is the most efficient way to store high-volume and complex data, when they performed benchmarking against relational and non-relational databases. Desired features of DBMSs include menu-based interfaces for web clients or browsing (browsing interface), forms-based interfaces (e.g. users can fill out form entries to insert new data), graphical user interfaces (which display schema in which the user can specify a query by manipulating a diagram), natural language interfaces, and speech input and output. Online or web database applications are very convenient in breeding contexts where multiple users can access and manage data remotely via the Internet. The online databases are a popular type to serve genomic information to the community. Several plant genomics databases are available online, including the Genome Database for Rosaceae (Jung *et al.*, 2008), CottonGen (https://www.cottongen.org/, accessed 21 February 2021), GrainGenes (https://wheat.pw.usda.gov/GG3/, accessed 15 February 2021), SoyaBase (https://www.soybase.org/, accessed 15 February 2021), MaizeGDB (https://maizegdb.org/, accessed 15 February 2021) and TAIR (https://www.arabidopsis.org/, accessed 15 February 2021).

### 11.4.2 Databases and connection to breeding software

Usually, breeding programmes are interested in connecting applications such as R programming (https://www.r-project.org/about.html, accessed 15 February 2021) to a database to enable the analysis and visualization of data. Several packages on CRAN (Comprehensive R Archive Network) provide (or relate to) interfaces between databases and R; for example, MySQL (dbConnect, TSMySQL), Oracle (RODM, ROracle), PostgreSQL (RpgSQL, RPostgreSQL, TSPostgreSQL), SQLite (shSQLite, RSQLite, RSQLite.extfuns, TSSQLite), Cassandra (RCassandra), MongoDB (RMongo, rmongodb), fame (fame), TSfame (TSfame) and HDF5 (h5r, hdf5). In similar way, several analytical and visualization modules (software) written in other programming languages such as Java or Python can be connected to databases to create a complete breeding decision system. In this context, the application programming interface (API), which defines the interaction between software intermediaries, defines the kinds of calls or requests that can be made, how to make them and the data format to be used. To set a standard for applications and facilitate interoperability among breeding applications, a consortium of plant breeding community members has contributed to the development of the public plant Breeding Application Programming Interface (BrAPI) (Selby *et al.*, 2019; https://brapi.org/, accessed 15 February 2021).

### 11.4.3 Comprehensive breeding systems; custom-built breeding databases and software

As many breeding programmes have limited database development capabilities, strategies are implemented to facilitate the adoption of comprehensive breeding software by the development of 'pre-designed' configurable database systems. Many components must be interconnected when designing a breeding data management system (Figs 11.4 and 11.5). Databases should relate to the breeding practices and to the analytical tools to make breeding decisions. A modern comprehensive breeding system (CBS) should facilitate the input of data, manage the data and provide results to breeders when needed; thus many processes must be designed and developed at the back end of the system.

One of first attempts to create a database in wheat at CIMMYT was the International Wheat Information System (IWIS). IWIS was developed to create a seamless system that starts from germplasm conservation to breeding, testing and release (Fox and Skovmand, 1996). IWIS includes a Wheat Pedigree Management System (WPMS), a Wheat Data Management System (WDMS) for phenotypic and environmental data generated in field trials, and a gene bank management module for wheat gene bank management including passport and characterization data. IWIS holds a suite of tools that are used by CIMMYT breeders to manage pedigrees, design crossing blocks and field books, print field tags and envelopes, and manage phenotypic data collected through the International Wheat Improvement Network (IWIN). The system was

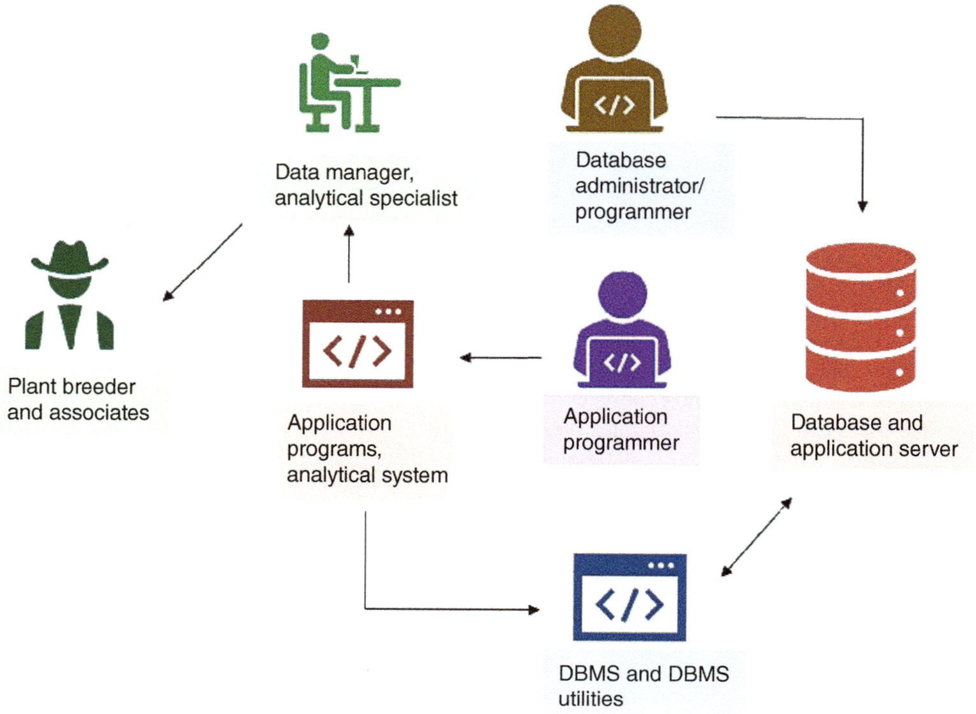


**Fig. 11.4.** Components of a breeding database system.

first built in 1974 on a VAX platform (IWIS1) and migrated to Windows NT (New Technology with a 32-bit operating system) in 2012 (IWIS2). IWIS is still being used but does not meet the data management requirements of current breeding programmes.

Many private and public breeding programmes are using different CBSs. Some of them are designed for a specific breeding programme, while others are available commercially or publicly for multiple breeding programmes. Some commercially or publicly available CBSs are the Breeding Management System (BMS) (https://integrated-breeding.net/, accessed 15 February 2021), PRISM (https://www.teamcssi.com/prism.html, accessed 15 February 2021), Breeding4Results (B4R) (https://b4r.irri.org/, accessed 21 February 2021), Agrobase (https://www.agronomix.com/, accessed 15 February 2021) and Genovix® (https://www.agronomix.com/, accessed 15 February 2021). The B4R is a Prostgres database. The Agrobase generation II is a relational database and has plant breeding software utilizing Microsoft SQL Server or SQL Server Express associated that can be run over Citrix®, the cloud or VPN connections. Genovix also builds on a Microsoft SQL Server. Besides these breeding data management systems, some specialized data management systems such as Germinate have evolved that are meant to manage gene bank data. Germinate is a generic database, developed in MySQL (Shaw *et al.*, 2017).

The CBSs are designed to house a database that is connected to a suite of applications that work together to collect, store and analyse research data (Fig. 11.5). Individual CBSs differ by crop and in features, application attached or connected. The objective of any CBS is to help manage breeding data across all phases of the crop improvement cycle, from programme planning to decision making. In addition to data storage and retrieval, the desired features include trial management, nursery and seed inventory, pedigree management, genotypic data management, statistical analysis, visualizations and decision support.

With the increase in 'big data' and more data-driven breeding decisions, it becomes fundamental to breeding programmes to set up

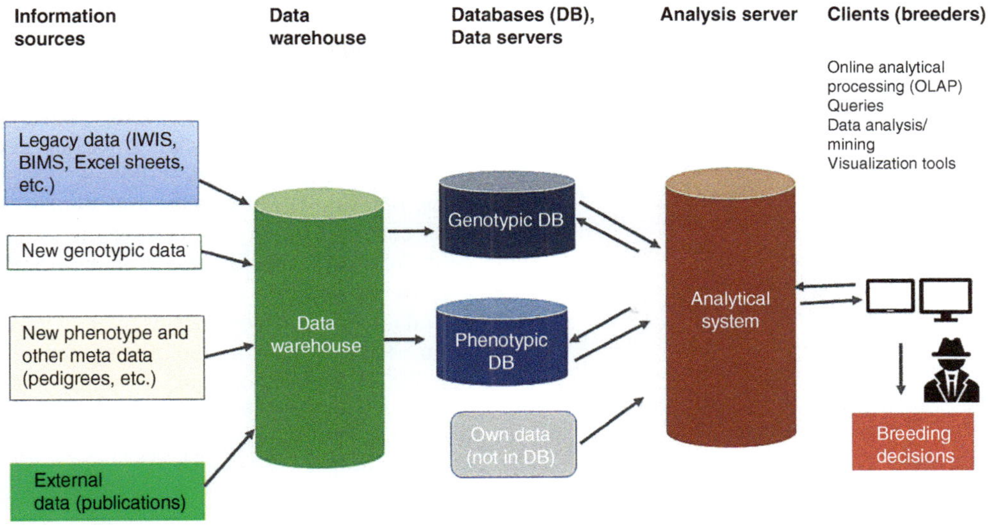


**Fig. 11.5.** Example of the computational infrastructure and its connectivity needed to make wheat breeding decisions.

a CBS. The components of a CBS consist of databases, an analytical system and a user interface (UI) (Fig. 11.5). In a usual cycle, data can come from several sources that can be collected and pre-processed in a data warehouse before loading to the databases (phenotypic, genotypic and combined). The stored data can be connected to an analytical system which will deliver required information to the UI. The ideal UI for the breeder or data analyst must have some desirable features such as informative with required data and plots, fast with an easy understandable structure, clicks over typing, connected to hand-held devices, synchronizing functions from the field and best fit of current breeding practices.

### 11.4.4 Genomic Open-source Breeding Informatics Initiative

The Genomic Open-source Breeding Informatics Initiative (GOBii) (www.gobii.org, accessed 15 February 2021) is the first large-scale, public-sector effort to systematically apply low- and high-density genotypic information for breeding staple crops in the developing world and to deliver increased rates of genetic gain. This helps public-sector actors to upgrade their breeding systems to make them comparable with those of private breeding institutions in developed countries. CIMMYT's Global Wheat Program is a key partner in fulfilling the GOBii project mission. GOBii has been working on three fronts: (i) genotypic database development; (ii) development of analysis tools; and (iii) integration of new and existing tools and databases. The objective is to develop products that can communicate with each other, thus helping users to make informed breeding decisions. Thus, the vision of GOBii was to provide an integrated solution when using genomic data in breeding programmes. GOBii has been actively collaborating with the James Hutton Institute (JHI) to improve the Flapjack software, adding analysis features and the visualization for GOBii data outputs. With GOBii support, JHI has extended features of marker-assisted backcrossing (MABC), $F_1$ and line-pedigree verification in its recent release/pre-release versions. GOBii is also collaborating with Diverse Arrays Technology (DArT)™ to integrate quality control (QC) protocols into the GOBii database. The GOBii has also developed a GS pipeline implemented in Galaxy (www.gobii.org, accessed 15 February 2021).

As the amount of genotypic data has grown substantially in the last decade, management of these data requires a more scalable database system that can be connected to tools required to make breeding decisions. For this reason, the GOBii project has developed a modern genomic data management (GOBii-GDM) system (Nti-Addae *et al.*, 2019). The GOBii-GDM is a scalable and responsive genomic data management system, with a BrAPI-enabled web-service layer, integrated genotyping data QC, integrated web-based file management, and a Marker Tools Portal to access all GOBii-GDM tools and key breeding tools developed in Galaxy and Flapjack (https://ics.hutton.ac.uk/flapjack/, accessed 15 February 2021). The GOBii-GDM includes a database, imputation systems and decision-support tools for plant breeders. More tools are under development under GOBii and other projects. The ultimate goal is to develop a connected analytical ecosystem including genotypic database, phenotypic database and analytical tools.

## 11.5 Conclusion

Modern breeding approaches/tools, which have become increasingly available, have the potential to increase breeding efficiency and rate of genetic gain while reducing overall costs. Breeding is an industry and should include both innovation and management to succeed. For a successful implementation of novel approaches such as GS and RGA combined with data management systems, careful validation, applied logistics solutions, capacity building and the allocation of sufficient resources are required. It is important to manage the related change processes well, for the breeding programmes to succeed.

## References


Battenfield, S.D., Guzman, C., Gaynor, R.C., Singh, R.P., Peña, R.J. and Dreisigacker, S. (2016) Applying genomic selection for prediction of processing and end-use quality traits in CIMMYT spring bread wheat breeding program. *The Plant Genome* 9(2), plantgenome2016.01.0005.

Bernardo, R. (1994) Prediction of maize single-cross performance using RFLPs and information from related hybrids. *Crop Science* 34, 20–25.

Bernardo, R. and Yu, J. (2007) Prospects for genome-wide selection for quantitative traits in maize, *Crop Science* 47, 1082–1090.

Beyene Y., Gowda, M., Olsen, M., Robbins, K.R., Pérez-Rodríguez, P. *et al.* (2019) Empirical comparison of tropical maize hybrids selected through genomic and phenotypic selections. *Frontiers in Plant Science* 10, 1502.

Bhattarai, S.P., De La Pena, R.C., Midmore, D.J. and Palchamy, K. (2009) *In vitro* culture of immature seed for rapid generation advancement in tomato. *Euphytica* 167, 23–30.

Carandang, F.M., Shanmugasundaram, S. and Carpena, A.L. (2006) Rapid generation advancement in soybeans using immature seeds. *Philippine Journal of Crop Science* 31, 53–59.

Cijvat, R., Manegold, S., Kersten, M., Klau, G.W., Schönhuth, A., Marschall, T. and Zhang, Y. (2015) Genome sequence analysis with MonetDB. *Datenbank-Spektrum* 15, 185–191.

Codd, E.F. (1970) A relational model of data for large shared data banks. *Communications of the ACM* 13, 377–387.

Crossa, J., de los Campos, G., Pérez, P., Gianola, D., Burgueño, J. *et al.* (2010) Prediction of genetic values of quantitative traits in plant breeding using pedigree and molecular markers. *Genetics* 186, 713–724.

Crossa, J., Pérez, P., Hickey, J., Burgueño, J., Ornella, L. *et al.* (2014) Genomic prediction in CIMMYT maize and wheat breeding programs. *Heredity* 112, 48–60.

Crossa, J., Pérez-Rodríguez, P., Cuevas, J., Montesinos-López, O., Jarquín, D. *et al.* (2017) Genomic selection in plant breeding: methods, models, and perspectives. *Trends in Plant Science* 22, 961–975.

Depauw, R.M. and Clarke, J.M. (1976) Acceleration of generation advancement in spring wheat. *Euphytica* 25, 415–418.

Fehr, W.R. (1991) *Principles of Cultivar Development: Theory and Technique*. Macmillan, New York.

Fox, P.N. and Skovmand, B. (1996) The International Wheat Information System at work in breeding. In: Cooper, M. and Hammer, G.L. (eds) *Plant Adaption and Crop Improvement*. CAB International, Wallingford, UK, pp. 317–326.

Gaur, P.M., Srinivasan, S., Gowda, C.L.L. and Rao, B.V. (2007) Rapid generation advancement in chickpea. *Journal of SAT Agricultural Research* 3, 1–3.

Ghosh, S., Watson, A., Gonzalez-Navarro, O.E., Ramirez-Gonzalez, R.H., Yanes, L. *et al.* (2018) Speed breeding in growth chambers and glasshouses for crop breeding and model plant research. *Nature Protocols* 13, 2944–2963.

Guimaraes, V., Hondo, F., Almeida, R., Vera, H., Holanda, M. *et al.* (2015) A study of genomic data provenance in NoSQL document-oriented database systems. In: *Proceedings of the 2015 IEEE International Conference on Bioinformatics and Biomedicine (BIBM)*. IEEE, Washington, DC, pp. 1525-1531.

Heu, M.H., Chung, G.S., Kim, D.K., Sasaki, T. and Vergara, B.S. (1982) Rapid generation advance in breeding rice for low temperature tolerance. In: *Proceedings of the International Rice Research Conference, Los Baños, Philippines, 19–23 April 1982.*

Ishigaki, Y. (2010) Establishment of cultivation technique with rapid generation advancement of *Cyclamen persicum* by sowing seeds right after picking seeds. *Bulletin of Gifu Prefecture Research Institute of Agricultural Science in the Hill and Mountain Area* 6, 7–12.

Juliana, P., Singh, R.P., Singh, P.K., Crossa, J., Huerta-Espino, J. *et al.* (2017) Genomic and pedigree-based prediction for leaf, stem, and stripe rust resistance in wheat. *Theoretical and Applied Genetics* 130, 1415–1430.

Juliana, P., Singh, R.P., Poland, J., Mondal, S., Crossa, J. *et al.* (2018) Prospects and challenges of applied genomic selection – a new paradigm in breeding for grain yield in bread wheat. *Plant Genome* 11, 180017.

Juliana, P., Poland, J., Huerta-Espino, J., Shrestha, S., Crossa, J. *et al.* (2019) Improving grain yield, stress resilience and quality of bread wheat using large-scale genomics. *Nature Genetics* 51, 1530–1539.

Jung, S., Staton, M., Lee, T., Blenda, A., Svancara, R., Abbott, A. and Main, D. (2008) GDR (Genome Database for Rosaceae): integrated web-database for Rosaceae genomics and genetics data. *Nucleic Acids Research* 36, D1034–D1040.

Lado, B., Battenfield, S., Guzmán, C., Quincke, M., Singh, R.P. *et al.* (2017) Strategies for selecting crosses using genomic prediction in two wheat breeding programs. *Plant Genome* 10(2), plantgenome2016.12.0128.

Lehermeier, C., Wimmer, V., Albrecht, T., Auinger, H.J., Gianolaet, D. *et al.* (2013) Sensitivity to prior specification in Bayesian genome-based prediction models. *Statistical Applications in Genetics and Molecular Biology* 12, 1–17.

Lush, J.L. (1937) *Animal Breeding Plans*. Iowa State Press, Ames, Iowa.

Meuwissen, T.H.E., Hayes, B.J. and Goddard, M.E. (2001) Prediction of total genetic value using genome-wide dense marker maps. *Genetics* 157, 1819–1829.

Niu, Z., Jiang, A., Abu Hammad, W., Oladzadabbasabadi, A., Xu, S.S., Mergoum, M. and Elias, E.M. (2014) Review of doubled haploid production in durum and common wheat through wheat × maize hybridization. *Plant Breeding* 133, 313–320.

Nti-Addae, Y., Matthews, D., Jun Ulat, V., Syed, R., Sempéré, G. *et al.* (2019) Benchmarking database systems for genomic selection implementation. *Database* 19, 1-10.

Patial, M., Pal, D., Thakur, A., Bana, R.S. and Patial, S. (2019) Doubled haploidy techniques in wheat (*Triticum aestivum* L.): an overview. *Proceedings of the National Academy of Sciences India, Section B: Biological Sciences* 89, 27–41.

Pérez, P. and de los Campos, G. (2014) Genome-wide regression and prediction with the BGLR statistical package. *Genetics* 198, 483–495.

Poland, J., Endelman, J., Dawson, J., Rutkoski, J., Wu, S. *et al.* (2012) Genomic selection in wheat breeding using genotyping-by-sequencing. *Plant Genome* 5, 103–113.

Rife, T.W. and Poland, J.A. (2014) Field Book: an open-source application for field data collection on android. *Crop Science* 54, 1624.

Rizal, G., Karki, S., Alcasid, M., Montecillo, F., Acebron, K. *et al.* (2014) Shortening the breeding cycle of sorghum, a model crop for research. *Crop Science* 54, 520–529.

Rosyara, U.R., De Jong, W.S., Douches, D.S. and Endelman, J.B. (2016) Software for genome-wide association studies in autopolyploids and its application to potato. *Plant Genome* 9, 2.

Rutkoski, J.E., Hefner, E.L. and Sorrells, M.E. (2010) Genomic selection for durable stem rust resistance in wheat. *Euphytica* 179, 161–173.

Rutkoski, J., Benson, J., Jia, Y., Brown-Guedira, G., Jannink, J.L. and Sorrells, M. (2012) Evaluation of genomic prediction methods for fusarium head blight resistance in wheat. *Plant Genome* 5, 51.

Rutkoski, J.E., Poland, J.A., Singh, R.P., Huerta-Espino, J., Bhavani, S. *et al.* (2014) Genomic selection for quantitative adult plant stem rust resistance in wheat. *Plant Genome* 7, 3.

Sehgal, D., Rosyara, U., Mondal, S., Singh, R., Poland, J. and Dreisigacker, S. (2020) Incorporating genome-wide association mapping results into genomic prediction models for grain yield and yield stability in CIMMYT spring bread wheat. *Frontiers in Plant Science* 11, 197.

Selby, P., Abbeloos, R., Backlund, J.E., Salido, M.B., Bauchet, G. *et al.* (2019), BrAPI – an application programming interface for plant breeding applications. *Bioinformatics* 35, 4147–4155.

Sempéré, G., Philippe, F., Dereeper, A., Ruiz, M., Gautier, S. and Larmande, P. (2016) Gigwa – genotype investigator for genome-wide analyses. *Gigascience* 5, 25.

Shaw, P.D., Raubach, S., Hearne, S.J., Dreher, K., Bryan, G. *et al.* (2017) Germinate 3: development of a common platform to support the distribution of experimental data on crop wild relatives. *Crop Science* 57, 1259–1273.

Sleper, D.A. and Poehlman, J.M. (1995) *Breeding Field Crops*. Iowa State University Press, Ames, Iowa.

Tanio, M., Kato, K., Ishikawa, N., Tabiki, T., Nishio, Z. *et al.* (2006) Effect of shuttle breeding with rapid generation advancement on heading traits of Japanese wheat. *Breeding Science* 56, 311–320.

Van Raden, P.M. (2008) Efficient methods to compute genomic predictions. *Journal of Dairy Science* 91, 4414–4423.

Voss-Fels, K.P., Herzog, E., Dreisigacker, S., Sukumaran, S., Watson, A. *et al.* (2019) 'SpeedGS' to accelerate genetic gain in spring wheat. In: Miedaner, T. and Korzun, V. (eds) *Applications of Genetic and Genomic Research in Cereals*. Woodhead Publishing Series in Food Science, Technology and Nutrition. Woodhead Publishing, Duxford, UK, pp. 303–332.

Wang, X.F., Wang, Y.X., Zhang, G.Y. and Ma, Z.Y. (2011) An integrated breeding technology for accelerating generation advancement and trait introgression in cotton. *Plant Breeding* 130, 569–573.

Watson, A., Ghosh, S., Williams, M., Cuddy, W.S., Simmonds, J. *et al.* (2018) Speed breeding is a powerful tool to accelerate crop research and breeding. *Nature Plants* 4, 23–29.

# 12 CRISPR-Mediated Gene Editing in Wheat for Abiotic Stress Tolerance

**David Edwards[1] and Armin Scheben[2*]**
[1]*University of Western Australia, Crawley, Australia;* [2]*Cold Spring Harbor Laboratory, Cold Spring Harbor, New York, USA*

## 12.1 Introduction

Wheat is the most widely grown crop worldwide, covering over 200 million hectares in a wide range of soils and climates. Wheat includes many species of the genus *Triticum*, with hexaploid bread wheat (*Triticum aestivum*) making up ~90% of all current wheat production (FAO, 2018). Despite the popularity of wheat, its productivity lags substantially behind other cereal crops such as maize and rice. A common cause of reduced wheat productivity is abiotic stress (Boyer, 1982; Abhinandan *et al.*, 2018). In its diverse environments, wheat is exposed to abiotic stresses including drought, heat, salinity and metal toxicity. These abiotic stresses can lead to severe yield penalties, particularly by reducing grain number and grain weight. Drought and heat stress are among the most important stresses in wheat and these stresses are likely to become more prevalent in the future due to climate change. Each 1°C of warming has been predicted to reduce global wheat production by 6% while also increasing the variability of wheat yields in different regions (Asseng *et al.*, 2015). Generating abiotic stress-tolerant wheat varieties is therefore important for ensuring food security.

Genetic improvement of wheat traits, including the introduction of the 'green revolution' semi-dwarfing genes *Rht1* and *Rht2* (Peng *et al.*, 1999), is a key contributor to the continued success of this crop. Although irrigation and other agronomic practices also have an important role to play in reducing yield loss through abiotic stress, genetic improvement is a sustainable approach that does not rely on additional resources. To provide the genetic variation from which improved traits can be selected, breeders have mostly used conventional approaches such as recombination between different wheat varieties and chemical mutagenesis using ethyl methanesulfonate (Dong *et al.*, 2009; Uauy *et al.*, 2009). Introgression of DNA from more distant relatives has also been used, particularly for introducing biotic stress resistance genes (Paull *et al.*, 1994). The genetic basis of abiotic stress tolerance in crops such as wheat has been extensively studied (Dwivedi *et al.*, 2017). Harnessing genetic variation to improve abiotic stress tolerance, however, is complicated because often many loci are each making small phenotypic contributions to the tolerance trait (Mickelbart *et al.*, 2015). Nevertheless, conventional breeding can improve wheat abiotic stress tolerance despite trait complexity. For example, Voss-Fels *et al.* (2019) found that modern elite wheat cultivars outperform older cultivars under low-rainfall conditions due to improved water-use efficiency.

*Corresponding author: scheben@cshl.edu

 *Molecular Breeding in Wheat, Maize and Sorghum: Strategies for Improving Abiotic Stress Tolerance and Yield* (eds M.A. Hossain *et al.*)
DOI: 10.1079/9781789245431.0012

Although improving quantitative abiotic stress tolerance traits is tractable with conventional breeding, it can take large populations and many generations to combine multiple favourable alleles in a suitable genetic background. Further, local recombination rates are suppressed in many regions in the wheat genome (Choulet *et al.*, 2014), potentially hindering breeding efforts. In some cases, recombination and chemical mutagenesis may therefore be too time-consuming to rapidly accelerate genetic gains in wheat.

In recent years, the clustered regularly interspaced short palindromic repeats (CRISPR)/CRISPR-associated protein 9 (Cas9) genome editing system has been efficiently applied to rapidly improve wheat and other crops (Belhaj *et al.*, 2015; Scheben *et al.*, 2017). Genome editing has attracted enormous interest from crop scientists in the public and private sectors and it is expected to have a major impact on agriculture (Scheben and Edwards, 2017; Gao, 2018). A large portion of wheat genome editing has focused on traits such as pathogen resistance, male sterility for hybrid breeding, herbicide resistance and low gluten content because these traits are often controlled by few loci and offer substantial rapid yield gains or commercial benefits. Genome editing technology and transformation protocols are also making the editing of abiotic stress genes easier, but the detection of target genes and useful alleles remains challenging. In some cases, these challenges have been addressed; for instance, wheat genome editing experiments have been carried out on genes related to drought and metal toxicity. These experiments provide insight into how abiotic stress tolerance in wheat can be quickly improved using genome editing. In this chapter, the relevant advances in genome editing technology and how they will enable improvement of abiotic stress tolerance in wheat are highlighted.

## 12.2 Advances in Genome Editing Technology

### 12.2.1 The wheat genome as a target for genome editing

Hexaploid bread wheat has three homeologous subgenomes (A, B, D) and a total repeat content over 85% in a genome of ~16 Gb (Appels *et al.*, 2018). This complex genome structure means that in many cases three homeoalleles of a gene need to be edited to generate a target agronomic trait (Shan *et al.*, 2014; Wang *et al.*, 2014). Disrupting a single gene may not lead to a phenotype because effects are buffered by redundant homeologues. Depending on the amount of sequence divergence between homeologues, individual guide RNAs may need to be designed for each one. Additionally, if homeologues have subfunctionalized and only a single homeologue needs to be targeted, a unique guide for that homeologue needs to be developed, which can make it more difficult to find unique sites to use for targeting. Polyploid crops like bread wheat therefore require increased efficiency in order to simultaneously target multiple alleles (Howells *et al.*, 2018). As in other monocots, delivery of the genome editing machinery in wheat can also be challenging, depending on the genotype being edited. These features can make it more challenging to edit wheat than diploid dicotyledonous plants such as tomato or *Arabidopsis*.

Despite the technical challenges in wheat genome editing, wheat was one of the plant species first modified by CRISPR/Cas9 (Shan *et al.*, 2013). However, many wheat genotypes, including the reference variety Chinese Spring, are recalcitrant to transformation. Varieties such as the spring wheat Fielder allow efficient transformation (Ishida *et al.*, 2013), but the lack of a high-quality reference genome and an elite background for these non-reference genotypes prevents them from being ideal models. Recent work has addressed this challenge in efficient editing by showing that haploid induction allows one-step editing in wheat by using Cas9 delivered by maize pollen (Kelliher *et al.*, 2019). Together with other innovations that streamline crop transformation, such as overexpression of the morphogenetic regulators *BABYBOOM* (*Bbm*) and *WUSCHEL2* (*WS2*) (Lowe *et al.*, 2016), haploid induction is making wheat genome editing less laborious. These efforts are an important step towards allowing genome editing experiments directly in high-yielding commercial cultivars. In addition, more targeted editing of individual bases using base editors has now also been demonstrated in wheat (Zong *et al.*, 2017, 2018; Li *et al.*, 2018). Furthermore, the amount of high-quality data available to help design and guide genome editing experiments is increasing rapidly. The publication of the latest wheat reference genome and annotation (Appels *et al*., 2018), together with the first wheat pangenome (Montenegro *et al*., 2017), further facilitate genome editing efforts, particularly the design of guide RNAs. Genomic databases for crops are also expanding the availability of data on wheat variation and known quantitative trait loci (QTLs) (Blake *et al*., 2016; Scheben *et al*., 2018; Tello-Ruiz *et al*., 2018).

### 12.2.2 Disrupting and replacing functional sequences in wheat

Genome editing can introduce site-specific mutations. The most common genome editing system, CRISPR/Cas9, uses a guide RNA with a ~20 bp sequence to target the Cas9 protein to a complementary sequence in the genome (Jinek *et al*., 2012). The Cas9 protein induces a double-strand break at the target site, allowing introduction of mutations arising during non-homologous end joining (NHEJ). The same process can be used to knock-in a sequence by providing a donor DNA template that can be integrated via the homology-directed repair (HDR) pathway. One of the simplest applications of genome editing in wheat crop improvement is to disrupt a gene by introducing insertions–deletions (InDels) in its coding region or promotor. This can prevent a protein being produced or reduce the functionality or amount of the protein. Disrupting a gene with CRISPR/Cas in this way is relatively simple, because after the Cas protein induces a double-strand break at a specific site the cell's own error-prone NHEJ repair pathway generates InDels at the target location. Often multiple guide RNAs will be used to target different regions of the coding sequences and ensure a gene is fully disrupted. This method of disrupting genes is a much more common type of edit than gene knock-in. Of the 22 published wheat genome editing experiments provided by a non-exhaustive table in a recent review (see Table 4 in Kumar *et al*., 2019), 19 involved gene knockouts and the remainder involved InDels. The polyploidy of the wheat genome also makes it particularly amenable for gene disruption experiments that aim to modulate gene dosage. In other genomes, disruption of certain genes may be lethal, but in wheat the gene copies on the other subgenomes can buffer potentially deleterious effects of a gene disruption. Testing and combining mutations in different homeologues of a wheat gene could therefore allow subtle manipulation of dosage-dependent traits.

Although disrupting genes and modulating their dosage can bestow agronomically important traits, some traits may be best improved via nucleotide-specific mutations. Many agronomic traits are the result of point mutations in the coding regions of genes, a prominent example being the semi-dwarf trait associated with gain-of-function point mutations in the genes *Rht1* and *Rht2* (Pearce *et al*., 2011). Point mutations and more complex nucleotide-specific mutations can be achieved with genome editing by providing a DNA template that is inserted at the location of the Cas-induced double-strand break. The DNA template can contain an arbitrary sequence, although the sequence length may be limited to several kilobases. This approach can use the low-error HDR pathway (Gil-Humanes *et al*., 2017) or even the NHEJ pathway (Li *et al*., 2016) to integrate the DNA template. For instance, Gil-Humanes *et al*. (2017) used viral delivery vectors to help integrate green fluorescent protein sequence into the three homoeoalleles of the wheat *ubiquitin* gene. These authors achieved an editing efficiency of about 1%. Although such successful examples of gene targeting in wheat exist, the low efficiency complicates broader application. Newer CRISPR/Cas methods demonstrated in bacteria use transposons to allow efficient insertion of larger sequences up to 10 kb (Klompe *et al*., 2019; Strecker *et al*., 2019). Deploying this approach in plants could simplify the current inefficient approaches. The larger size of potential sequence insertions means that blocks of genes could potentially be inserted, for example from a QTL hotspot.

Precise base editing can also be used to generate point mutations, without requiring DNA template or induction of double-strand breaks (Komor *et al*., 2016). By fusing a deaminase enzyme to a Cas protein, it is possible to induce four types of transition point mutations (C→T, G→A, A→G and T→C). In wheat, base editors relying on cytidine deaminase (Zong *et al*., 2017) and adenosine deaminase (Li *et al*., 2018) have shown high editing efficiencies. Zong *et al*. (2018) were able to edit blue fluorescent protein to produce green fluorescent protein via a point mutation. Finally, prime editing overcomes the limitations of base editing by enabling the introduction of all 12 base-to-base conversions as well as InDels without requiring double-strand breaks (Anzalone *et al.*, 2019). In prime editing, the guide RNA is replaced by a prime editing guide RNA that includes an RNA template with a sequence to be introduced into the genome. A catalytically impaired Cas9 protein, that only induces a nick in a single strand, is fused with a reverse transcriptase. The target region is nicked, and the RNA is reverse transcribed to DNA and integrated into the genome. The first prime editing optimized for wheat was recently published (Lin *et al.*, 2020). These innovative approaches have not yet been widely tested in plants and some technical challenges remain to be overcome. Base editors, for instance, can still generate off-target mutations within the deamination window spanning several bases (Zong *et al.*, 2017). Nevertheless, base editing and prime editing are now expanding the wheat genome editing toolkit, so that modifying practically any base in the genome will soon be possible. This flexibility will be important when tackling complex quantitative traits like abiotic stress tolerance.

### 12.2.3 Editing efficiency and regeneration

Two key components of genome editing are the delivery of the editing machinery (Cas and guide RNA transgenes) and the generation of edited plants not bearing transgenes. Delivery in wheat is mostly carried out using biolistic transformation (Gil-Humanes *et al.*, 2017) and *Agrobacterium*-mediated transformation (Howells *et al.*, 2018), although leaf infiltration with *Agrobacterium* was initially challenging. Delivery without integration of transgenes has been achieved in wheat using ribonucleoprotein delivery (Liang *et al.*, 2017) and pollen haploid induction (Kelliher *et al.*, 2019). The efficiency of these methods is dependent on tissue and genotype. When competent genotypes and tissues have been identified, plants need to be regenerated from transformed cells. This has been particularly difficult in monocots including wheat. Lowe *et al.* (2016) were able to dramatically improve transformation efficiencies in maize by overexpressing *Bbm* and *WS2* genes after *Agrobacterium*-mediated transformation. In wheat, the regeneration-related wheat gene *TaCB1* has been reported to have similar effects (Wang *et al.*, 2019). A pending patent also describes how a GRF-GIF chimera boosts transformation (UC Davis, 2020), without causing the deleterious phenotype of overexpressed *Bbm*.

An emerging technology in transformation and regeneration in wheat editing is haploid induction using pollen (Kelliher *et al.*, 2019). By using the pollen of a maize line transformed with a CRISPR/Cas cassette, it was possible to fertilize cytoplasmic-male-sterile wheat spikes. Genome editing occurs in the wheat genome and during fertilization of the wheat ovule, the maize chromosome is eliminated. The efficiency of haploid induction genome editing in wheat is low, but the one-step genome editing process saves valuable time in the breeding process. Time is saved with haploid induction editing and ribonucleoprotein delivery, because no backcrossing is required to excise the CRISPR/Cas transgenes from the edited genome. Avoiding an intermediate step with integrated transgenes is also practical from a commercial point of view, because transgenes can trigger governmental regulation during the research and development phase of a wheat variety. In addition, farmers and grain millers can lose entire export markets if their product is classified as genetically modified. The increasing availability of efficient transgene-free editing approaches will help make wheat genome editing routine in small and large laboratories.

## 12.3 Applications of Genome Editing to Improve Abiotic Stress Tolerance

### 12.3.1 Drought

Drought is a major abiotic stressor of wheat, casing substantial yield loss (Araus *et al.*, 2008; Budak *et al.*, 2015), although increasing the drought tolerance of crops is challenging because it is a polygenic trait involving complex signalling networks (Kuzuoglu-Ozturk *et al.*, 2012; Hu and Xiong, 2014). However, progress is being made and fifteen genes associated with increased yield and drought survival under field conditions are summarized in a review of the genetic mechanisms of abiotic stress tolerance in crops (Mickelbart *et al.*, 2015). Experiments have highlighted several genes that have potential as genome editing targets in breeding programmes. In maize, increased constitutive expression of the gene *ARGOS8*, a negative regulator of ethylene response, improved yield under field drought stress conditions (Shi *et al.*, 2016). To generate the allelic variant with increased expression, CRISPR/Cas9 was used to insert the native promotor of the maize gene *GOS2* into the upstream region of *ARGOS8* via HDR. *ARGOS* genes are also highly conserved in wheat and its wild relatives, showing induction under drought conditions (Zhao *et al.*, 2017). The potential of these genes as targets to improve wheat drought tolerance was further supported in Zhao *et al.* (2017) by overexpression of *TaARGOS-D* in *Arabidopsis*, which was associated with improved drought tolerance.

The wheat genes *TaSG1* (Tian *et al.*, 2013), *TaDREB2* (Egawa *et al.*, 2006) and *TaERF3* (Rong *et al.*, 2014) have also been implicated in improved drought tolerance. A mutation in *TaSG1* induced by ethyl methanesulfonate mutagenesis led to a stay-green phenotype with enhanced drought tolerance in wheat (Tian *et al.*, 2013). Additional studies showed that the *TaSG1* mutation is associated with altered cytokinin metabolism (Wang *et al.*, 2016). In wheat and barley, overexpression of the regulatory genes *DREB2* and *DREB3* increased drought tolerance (Lata and Prasad, 2011; Morran *et al.*, 2011; Shavrukov *et al.*, 2016). Finally, the transcription factor *TaERF3* increases drought tolerance in wheat (Rong *et al.*, 2014), likely through promoting root elongation and root hair development (Cheng *et al.*, 2016). Expression of *TaDREB2* and *TaERF* showed upregulation in response to shock drought stress in the wheat reference variety Chinese Spring (Kim *et al.*, 2018). Kim *et al.* (2018) also demonstrated genome editing of *TaDREB2* and *TaERF3* in wheat; however, they were only able to edit the B and D subgenomes due to guide RNA mismatches on the A subgenome. Additionally, the InDels introduced into the target genes are unlike to improve drought tolerance, which is more likely to be achieved via insertion of a strong constitutive promotor. Despite these challenges, technical advances in CRISPR/Cas technology and further functional understanding of drought response will pave the way to genome-edited wheat varieties with improved drought tolerance.

### 12.3.2 Metal toxicity

Metals occur in agricultural soils naturally and via industrial contamination, posing a potential threat to crop productivity. Although the effects of various metals on crops differ, metal toxicity resulting in reduced yield is common in crops including wheat (Athar and Ahmad, 2002). Additionally, metals such as As and Cd can be accumulated in crops and taken up by humans, causing health risks (Clemens, 2019). Increasing the tolerance of wheat to metals while reducing accumulation of toxic metals is therefore an important breeding goal in wheat.

Al toxicity limits wheat growth in acidic soils due to prevalence of $Al^{3+}$ cations that inhibit root elongation (Delhaize and Ryan, 1995; Rout *et al.*, 2001). Many plants have developed $Al^{3+}$ tolerance via efflux of organic acids such as malate and citrate from root apices (Ma, 2000). These organic acids can detoxify the $Al^{3+}$ by chelating it. Substantial phenotypic variation for Al tolerance has been attributed to *TaALMT1* (Raman *et al.*, 2005), which encodes an $Al^{3+}$-activated anion channel permeable to malate (Sasaki *et al.*, 2004). By using a conventional transgenic approach to overexpress *TaALMT1* with a strong maize ubiquitin promotor, wheat Al tolerance in acidic soils could be increased (Pereira *et al.*, 2010). A non-transgenic alternative would be to use genome editing to insert optimized native wheat promotors to drive *TaALMT1* overexpression. In barley, lines with the sorghum transgene *SbMATE* also displayed increased Al tolerance, although the transgene *TaALMT1* delivered a greater increase in tolerance (Zhou *et al.*, 2014). By comparing similar genes encoding transporters in different grasses, it may be possible to apply genome editing to optimize the expression and function of genes such as *TaALMT1* and thus further increase Al tolerance.

In rice, genome editing has been applied to develop lines with low accumulation of Cd (Tang *et al.*, 2017) and As (Wang *et al.*, 2017) by knocking out the metal transporter *OsNramp5* and overexpression of the transcription factor *OsARM1*, respectively. In wheat, *Nramp5* is less highly expressed and the protein has a lower Cd transport activity than in rice (Sui *et al.*, 2018), and overall wheat accumulates substantially less As (Su *et al.*, 2010) and Cd (Sui *et al.*, 2018) than rice. Nevertheless, minimizing Cd accumulation in wheat is an important breeding goal (Rizwan *et al.*, 2016). The durum wheat *Cdu1* locus on chromosome 5B explains >80% of the phenotypic variation in Cd concentration in grain (Wiebe *et al.*, 2010). Discovery of a candidate gene was difficult, until the recent assembly of the durum wheat genome allowed the detection of the *TdHMA3-B1a* allele for low Cd accumulation in the QTL region (Maccaferri *et al.*, 2019). This underlines the importance of reference genomes for discovery of genome editing targets. Additional candidate wheat genes have been implicated in tolerance of Cd toxicity, including the transmembrane gene *TaTM20* (Kim *et al.*, 2008). There are also transporter genes such as *TaABCC3* (Bhati *et al.*, 2015) that may help reduce the toxicity of various heavy metals. The discovery of further genes, particularly transporters, involved in metal toxicity and accumulation will provide targets for validation using genome editing. Although gene knockouts may provide agronomic benefits in some cases, engineering of the expression and protein structure of candidate genes will yield the greatest benefits in enhancing tolerance to metal toxicity and reducing negative human health impacts of heavy metal uptake.

## 12.4 Crop Wild Relatives are a Source of Variation for Breeding

Although breeding is often considered a genetic bottleneck, modern wheat breeding has not necessarily narrowed the available genetic diversity (van de Wouw *et al.*, 2010; Voss-Fels *et al.*, 2019). Nevertheless, harnessing novel alleles from wheat relatives plays an important role in improving traits. There are over 500 species in the tribe *Triticeae*, with different levels of relatedness to wheat. By seeking genetic diversity outside the confines of wheat, breeders can take advantage of millions of years of evolution in different genetic backgrounds and environments. For instance, drought tolerance loci detected by a genome-wide association study in the wheat progenitor *Aegilops tauschii* (Qin *et al.*, 2016) can provide breeding targets for wheat. Indeed, substantial genetic diversity in wheat is already the product of introgressions from wild species (Feuillet *et al.*, 2008). An example of an introgression from a wild wheat relative is the introduction of the *Sr22* rust resistance gene derived from *Triticum boeoticum* (Paull *et al.*, 1994). However, this gene also illustrates the difficulties of introgression breeding, because the introgressed sequence carrying *Sr22* also carried linked loci with undesirable effects on time to maturity. This 'linkage drag' is among the reasons why introgression breeding with wild species is laborious and has not been widely used for complex traits such as abiotic stress tolerance.

Genome editing opens up new avenues for rapidly introducing alleles from wild relatives into wheat, without the complications of recombination such as linkage drag. Although a wheat pangenome was recently assembled providing novel gene sequences and variants in bread wheat (Montenegro *et al.*, 2017), genomic data for most non-crop *Triticeae* remain scarce. However, increased interest in plant pangenomics is now driving major sequencing projects in different crops and their relatives (Golicz *et al.*, 2016, 2020; Khan *et al.*, 2020). As the cost of high-quality genome assembly in *Triticeae* (Monat *et al.*, 2019) and resequencing decreases further, genomic data will become available to uncover new alleles for abiotic stress tolerance. Furthermore, wild plants are not just sources of alleles to be introduced into existing crops, they can also be domesticated into entirely new crops. Recent publications have emphasized the potential for crop domestication using genome editing (Osterberg *et al.*, 2017; Zsögön *et al.*, 2017). Conceptually, this involves the editing of key domestication genes that confer the most important traits for agricultural cultivation such as flowering time and seed quantity and quality. Rather than transferring complex, poorly understood abiotic stress tolerance traits from wild plants into crops, it is possible to transfer well-understood domestication genes into wild plants to make crops with novel genetic backgrounds. New approaches for genome editing of wild plants hold much promise in giving breeders access to a different range of traits missing from existing crops (Shan *et al.*, 2020). Domestication of wild species was recently demonstrated in tomato (Zsögön *et al.*, 2018) and the orphan crop groundcherry (Lemmon *et al.*, 2018). Wild emmer wheat (*Triticum dicoccoides*), a progenitor of bread wheat, is one example of a future candidate for domestication as it may contain valuable adaptations to abiotic stress that are not found in the wheat gene pool.

## 12.5 Conclusion and Perspective

To develop hardier elite wheat varieties, breeders can now apply efficient genome editing to increase tolerance to drought, metal toxicity and other abiotic stresses. Fuelled by intense interest in biomedical applications, the genome editing toolbox is still rapidly expanding, making it ever easier to introduce any mutation into any region of the wheat genome. A limiting factor remains the lack of knowledge of many of the pathways underlying complex abiotic stress tolerance traits. Understanding how wild wheat relatives have adapted to their environments offers a practical approach to identify new candidate genes for abiotic stress tolerance. With more applied knowledge on *Triticeae* genomes, new breeding approaches can be deployed with a greater focus on targeted changes in traits. Cycles of breeding by recombination can be accelerated with cycles of genome editing to optimize wheat for different environments across the world.

## References


Abhinandan, K., Skori, L., Stanic, M., Hickerson, N.M.N., Jamshed, M. and Samuel, M.A. (2018) Abiotic stress signaling in wheat – an inclusive overview of hormonal interactions during abiotic stress responses in wheat. *Frontiers in Plant Science* 9, 734. Available at: https://doi.org/10.3389/fpls.2018.00734

Anzalone, A.V., Randolph, P.B., Davis, J.R., Sousa, A.A., Koblan, L.W. *et al.* (2019) Search-and-replace genome editing without double-strand breaks or donor DNA. *Nature* 576, 149–157. Available at: https://doi.org/10.1038/s41586-019-1711-4

Appels, R., Eversole, K., Feuillet, C., Keller, B., Rogers, J. *et al.* (2018) Shifting the limits in wheat research and breeding using a fully annotated reference genome. *Science* 361, eaar7191. Available at: https://doi.org/10.1126/science.aar7191

Araus, J.L., Slafer, G.A., Royo, C. and Serret, M.D. (2008) Breeding for yield potential and stress adaptation in cereals. *Critical Reviews in Plant Sciences* 27, 377–412. Available at: https://doi.org/10.1080/07352680802467736

Asseng, S., Ewert, F., Martre, P., Rotter, R.P., Lobell, D.B. *et al.* (2015) Rising temperatures reduce global wheat production. *Nature Climate Change* 5, 143–147. Available at: https://doi.org/10.1038/Nclimate2470

Athar, R. and Ahmad, M. (2002) Heavy metal toxicity: effect on plant growth and metal uptake by wheat, and on free living *Azotobacter*. *Water, Air, and Soil Pollution* 138, 165–180. Available at: https://doi.org/10.1023/A:1015594815016

Belhaj, K., Chaparro-Garcia, A., Kamoun, S., Patron, N.J. and Nekrasov, V. (2015) Editing plant genomes with CRISPR/Cas9. *Current Opinion in Biotechnology* 32, 76–84. Available at: https://doi.org/10.1016/j.copbio.2014.11.007

Bhati, K.K., Sharma, S., Aggarwal, S., Kaur, M., Shukla, V. *et al.* (2015) Genome-wide identification and expression characterization of ABCC-MRP transporters in hexaploid wheat. *Frontiers in Plant Science* 6, 448. Available at: https://doi.org/10.3389/fpls.2015.00488

Blake, V.C., Birkett, C., Matthews, D.E., Hane, D.L., Bradbury, P. and Jannink, J.L. (2016) The Triticeae Toolbox: combining phenotype and genotype data to advance small-grains breeding. *Plant Genome* 9, plantgenome2014.12.0099. Available at: https://doi.org/10.3835/plantgenome2014.12.0099

Boyer, J.S. (1982) Plant productivity and environment. *Science* 218, 443–448. Available at: https://doi.org/10.1126/science.218.4571.443

Budak, H., Hussain, B., Khan, Z., Ozturk, N.Z. and Ullah, N. (2015) From genetics to functional genomics: improvement in drought signaling and tolerance in wheat. *Frontiers in Plant Science* 6, 1012. Available at: https://doi.org/10.3389/fpls.2015.01012

Cheng, S.F., Zhou, D.X. and Zhao, Y. (2016) *WUSCHEL*-related homeobox gene *WOX11* increases rice drought resistance by controlling root hair formation and root system development. *Plant Signaling & Behavior* 11, e1130198. Available at: https://doi.org/10.1080/15592324.2015.1130198

Choulet, F., Alberti, A., Theil, S., Glover, N., Barbe, V. *et al.* (2014) Structural and functional partitioning of bread wheat chromosome 3B. *Science* 345, 1249721. Available at: https://doi.org/10.1126/science.1249721

Clemens, S. (2019) Safer food through plant science: reducing toxic element accumulation in crops. *Journal of Experimental Botany* 70, 5537–5557. Available at: https://doi.org/10.1093/jxb/erz366

Delhaize, E. and Ryan, P.R. (1995) Aluminum toxicity and tolerance in plants. *Plant Physiology* 107, 315–321. Available at: https://doi.org/10.1104/pp.107.2.315

Dong, C.M., Dalton-Morgan, J., Vincent, K. and Sharp, P. (2009) A modified TILLING method for wheat breeding. *Plant Genome* 2, 39–47. Available at: https://doi.org/10.3835/plantgenome2008.10.0012

Dwivedi, S.L., Scheben, A., Edwards, D., Spillane, C. and Ortiz, R. (2017) Assessing and exploiting functional diversity in germplasm pools to enhance abiotic stress adaptation and yield in cereals and food legumes. *Frontiers in Plant Science* 8, 1461. Available at: https://doi.org/10.3389/fpls.2017.01461

Egawa, C., Kobayashi, F., Ishibashi, M., Nakamura, T., Nakamura, C. and Takumi, S. (2006) Differential regulation of transcript accumulation and alternative splicing of a *DREB2* homolog under abiotic stress conditions in common wheat. *Genes & Genetic Systems* 81, 77–91. Available at: https://doi.org/10.1266/ggs.81.77

FAO (2018) FAOSTAT, Compare Data. Food and Agriculture Organization of the United Nations, Rome. Available at: http://www.fao.org/faostat/en/#compare (accessed 8 August 2018).

Feuillet, C., Langridge, P. and Waugh, R. (2008) Cereal breeding takes a walk on the wild side. *Trends in Genetics* 24, 24–32. Available at: https://doi.org/10.1016/j.tig.2007.11.001

Gao, C. (2018) The future of CRISPR technologies in agriculture. *Nature Reviews Molecular Cell Biology* 5, 275–276. Available at: https://doi.org/10.1038/nrm.2018.2

Gil-Humanes, J., Wang, Y., Liang, Z., Shan, Q., Ozuna, C.V. *et al.* (2017) High-efficiency gene targeting in hexaploid wheat using DNA replicons and CRISPR/Cas9. *The Plant Journal* 89, 1251–1262. Available at: https://doi.org/10.1111/tpj.13446

Golicz, A.A., Batley, J. and Edwards, D. (2016) Towards plant pangenomics. *Plant Biotechnology Journal* 14, 1099–1105. Available at: https://doi.org/10.1111/pbi.12499

Golicz, A.A., Bayer, P.E., Bhalla, P.L., Batley, J. and Edwards, D. (2020) Pangenomics comes of age: from bacteria to plant and animal applications. *Trends in Genetics* 36, 132–145. Available at: https://doi.org/10.1016/j.tig.2019.11.006

Howells, R.M., Craze, M., Bowden, S. and Wallington, E.J. (2018) Efficient generation of stable, heritable gene edits in wheat using CRISPR/Cas9. *BMC Plant Biology* 18, 215. Available at: https://doi.org/10.1186/s12870-018-1433-z

Hu, H.H. and Xiong, L.Z. (2014) Genetic engineering and breeding of drought-resistant crops. *Annual Review of Plant Biology* 65, 715–741. Available at: https://doi.org/10.1146/annurev-arplant-050213-040000

Ishida, Y., Hiei, Y. and Komari, T. (2013) High efficiency wheat transformation mediated by *Agrobacterium tumefaciens*. *In Vitro Cellular & Developmental Biology – Animal* 49, S24–S25.

Jinek, M., Chylinski, K., Fonfara, I., Hauer, M., Doudna, J.A. and Charpentier, E. (2012) A programmable dual-RNA – guided DNA endonuclease in adaptive bacterial immunity. *Science* 337, 816–821. Available at: https://doi.org/10.1126/science.1225829

Kelliher, T., Starr, D., Su, X., Tang, G., Chen, Z. *et al.* (2019) One-step genome editing of elite crop germplasm during haploid induction. *Nature Biotechnology* 37, 287–292. Available at: https://doi.org/10.1038/s41587-019-0038-x

Khan, A.W., Garg, V., Roorkiwal, M., Golicz, A.A., Edwards, D. and Varshney, R.K. (2020) Super-pangenome by integrating the wild side of a species for accelerated crop improvement. *Trends in Plant Science* 25, 148–158. Available at: https://doi.org/10.1016/j.tplants.2019.10.012

Kim, D., Alptekin, B. and Budak, H. (2018) CRISPR/Cas9 genome editing in wheat. *Functional & Integrative Genomics* 18, 31–41. Available at: https://doi.org/10.1007/s10142-017-0572-x

Kim, Y.Y., Kim, D.Y., Shim, D., Song, W.Y., Lee, J. *et al.* (2008) Expression of the novel wheat gene *TM20* confers enhanced cadmium tolerance to bakers' yeast. *Journal of Biological Chemistry* 283, 15893–15902. Available at: https://doi.org/10.1074/jbc.M708947200

Klompe, S.E., Vo, P.L.H., Halpin-Healy, T.S. and Sternberg, S.H. (2019) Transposon-encoded CRISPR-Cas systems direct RNA-guided DNA integration. *Nature* 571, 219–225. Available at: https://doi.org/10.1038/s41586-019-1323-z

Komor, A.C., Kim, Y.B., Packer, M.S., Zuris, J.A. and Liu, D.R. (2016) Programmable editing of a target base in genomic DNA without double-stranded DNA cleavage. *Nature* 533, 420–424. Available at: https://doi.org/10.1038/nature17946

Kumar, R., Kaur, A., Pandey, A., Mamrutha, H.M. and Singh, G.P. (2019) CRISPR-based genome editing in wheat: a comprehensive review and future prospects. *Molecular Biology Reports* 46, 3557–3569. Available at: https://doi.org/10.1007/s11033-019-04761-3

Kuzuoglu-Ozturk, D., Yalcinkaya, O.C., Akpinar, B.A., Mitou, G., Korkmaz, G., Gozuacik, D. and Budak, H. (2012) Autophagy-related gene, *TdAtg8*, in wild emmer wheat plays a role in drought and osmotic stress response. *Planta* 236, 1081–1092. Available at: https://doi.org/10.1007/s00425-012-1657-3

Lata, C. and Prasad, M. (2011) Role of DREBs in regulation of abiotic stress responses in plants. *Journal of Experimental Botany* 62, 4731–4748. Available at: https://doi.org/10.1093/jxb/err210

Lemmon, Z.H., Reem, N.T., Dalrymple, J., Soyk, S., Swartwood, K.E. *et al.* (2018) Rapid improvement of domestication traits in an orphan crop by genome editing. *Nature Plants* 4, 766–770. Available at: https://doi.org/10.1038/s41477-018-0259-x

Li, C., Zong, Y., Wang, Y.P., Jin, S., Zhang, D.B. *et al.* (2018) Expanded base editing in rice and wheat using a Cas9-adenosine deaminase fusion. *Genome Biology* 19, 59. Available at: https://doi.org/10.1186/s13059-018-1443-z

Li, J., Meng, X., Zong, Y., Chen, K., Zhang, H. *et al.* (2016) Gene replacements and insertions in rice by intron targeting using CRISPR-Cas9. *Nature Plants* 2, 16139. Available at: https://doi.org/10.1038/nplants.2016.139

Liang, Z., Chen, K.L., Li, T.D., Zhang, Y., Wang, Y.P. *et al.* (2017) Efficient DNA-free genome editing of bread wheat using CRISPR/Cas9 ribonucleoprotein complexes. *Nature Communications* 8, 14261. Available at: https://doi.org/10.1038/ncomms14261

Lin, Q., Zong, Y., Xue, C., Wang, S., Jin, S. *et al.* (2020) Prime genome editing in rice and wheat. *Nature Biotechnology* 38, 582–585. Available at: https://doi.org/10.1038/s41587-020-0455-x

Lowe, K., Wu, E., Wang, N., Hoerster, G., Hastings, C. *et al.* (2016) Morphogenic regulators *baby boom* and *wuschel* improve monocot transformation. *The Plant Cell* 28, 1998–2015. Available at: https://doi.org/10.1105/tpc.16.00124

Ma, J.F. (2000) Role of organic acids in detoxification of aluminum in higher plants. *Plant and Cell Physiology* 41, 383–390. Available at: https://doi.org/10.1093/pcp/41.4.383

Maccaferri, M., Harris, N.S., Twardziok, S.O., Pasam, R.K., Gundlach, H. *et al.* (2019) Durum wheat genome highlights past domestication signatures and future improvement targets. *Nature Genetics* 51, 885–895. Available at: https://doi.org/10.1038/s41588-019-0381-3

Mickelbart, M.V., Hasegawa, P.M. and Bailey-Serres, J. (2015) Genetic mechanisms of abiotic stress tolerance that translate to crop yield stability. *Nature Reviews Genetics* 16, 237–251. Available at: https://doi.org/10.1038/nrg3901

Monat, C., Padmarasu, S., Lux, T., Wicker, T., Gundlach, H. *et al.* (2019) TRITEX: chromosome-scale sequence assembly of Triticeae genomes with open-source tools. *Genome Biology* 20, 284. Available at: https://doi.org/10.1186/s13059-019-1899-5

Montenegro, J.D., Golicz, A.A., Bayer, P.E., Hurgobin, B., Lee, H. *et al.* (2017) The pangenome of hexaploid bread wheat. *The Plant Journal* 90, 1007–1013. Available at: https://doi.org/10.1111/tpj.13515

Morran, S., Eini, O., Pyvovarenko, T., Parent, B., Singh, R. *et al.* (2011) Improvement of stress tolerance of wheat and barley by modulation of expression of DREB/CBF factors. *Plant Biotechnology Journal* 9, 230–249. Available at: https://doi.org/10.1111/j.1467-7652.2010.00547.x

Osterberg, J.T., Xiang, W., Olsen, L.I., Edenbrandt, A.K., Vedel, S.E. *et al.* (2017) Accelerating the domestication of new crops: feasibility and approaches. *Trends in Plant Science* 22, 373–384. Available at: https://doi.org/10.1016/j.tplants.2017.01.004

Paull, J.G., Pallotta, M.A., Langridge, P. and The, T.T. (1994) RFLP markers associated with *Sr22* and recombination between chromosome 7A of bread wheat and the diploid species *Triticum boeoticum*. *Theoretical and Applied Genetics* 89, 1039–1045. Available at: https://doi.org/10.1007/Bf00224536

Pearce, S., Saville, R., Vaughan, S.P., Chandler, P.M., Wilhelm, E.P. *et al.* (2011) Molecular characterization of *Rht-1* dwarfing genes in hexaploid wheat. *Plant Physiology* 157, 1820–1831. Available at: https://doi.org/10.1104/pp.111.183657

Peng, J.R., Richards, D.E., Hartley, N.M., Murphy, G.P., Devos, K.M. *et al.* (1999) 'Green revolution' genes encode mutant gibberellin response modulators. *Nature* 400, 256–261. Available at: https://doi.org/10.1038/22307

Pereira, J.F., Zhou, G.F., Delhaize, E., Richardson, T., Zhou, M.X. and Ryan, P.R. (2010) Engineering greater aluminium resistance in wheat by over-expressing *TaALMT1*. *Annals of Botany* 106, 205–214. Available at: https://doi.org/10.1093/aob/mcq058

Qin, P., Lin, Y., Hu, Y.D., Liu, K., Mao, S.S. *et al.* (2016) Genome-wide association study of drought-related resistance traits in *Aegilops tauschii*. *Genetics and Molecular Biology* 39, 398–407. Available at: https://doi.org/10.1590/1678-4685-Gmb-2015-0232

Raman, H., Zhang, K.R., Cakir, M., Appels, R., Garvin, D.F. *et al.* (2005) Molecular characterization and mapping of *ALMT1*, the aluminium-tolerance gene of bread wheat (*Triticum aestivum* L.). *Genome* 48, 781–791. Available at: https://doi.org/10.1139/g05-054

Rizwan, M., Ali, S., Abbas, T., Zia-ur-Rehman, M., Hannan, F. *et al.* (2016) Cadmium minimization in wheat: a critical review. *Ecotoxicology and Environmental Safety* 130, 43–53. Available at: https://doi.org/10.1016/j.ecoenv.2016.04.001

Rong, W., Qi, L., Wang, A.Y., Ye, X.G., Du, L.P. *et al.* (2014) The ERF transcription factor *TaERF3* promotes tolerance to salt and drought stresses in wheat. *Plant Biotechnology Journal* 12, 468–479. Available at: https://doi.org/10.1111/pbi.12153

Rout, G.R., Samantaray, S. and Das, P. (2001) Aluminium toxicity in plants: a review. *Agronomie* 21, 3–21. Available at: https://doi.org/10.1051/agro:2001105

Sasaki, T., Yamamoto, Y., Ezaki, B., Katsuhara, M., Ryan, P.R., Delhaize, E. and Matsumoto, H. (2004) A wheat gene encoding an aluminum-activated malate transporter. *The Plant Journal* 37, 645–653. Available at: https://doi.org/10.1111/j.1365-313X.2003.01991.x

Scheben, A. and Edwards, D. (2017) Genome editors take on crops. *Science* 355, 1122–1123. Available at: https://doi.org/10.1126/science.aal4680

Scheben, A., Wolter, F., Batley, J., Puchta, H. and Edwards, D. (2017) Towards CRISPR/Cas crops – bringing together genomics and genome editing. *New Phytologist* 216, 682–698. Available at: https://doi.org/10.1111/nph.14702

Scheben, A., Verpaalen, B., Lawley, C.T., Chan, K.C.-K., Bayer, P., Batley, J. and Edwards, D. (2018) CropSNPdb: a database of SNP array data for *Brassica* crops and hexaploid bread wheat. *The Plant Journal* 98, 142–152. Available at: https://doi.org/10.1111/tpj.14194

Shan, S.C., Soltis, P.S., Soltis, D.E. and Yang, B. (2020) Considerations in adapting CRISPR/Cas9 in nongenetic model plant systems. *Applications in Plant Sciences* 8, e11314. Available at: https://doi.org/10.1002/aps3.11314

Shan, Q.W., Wang, Y.P., Li, J., Zhang, Y., Chen, K.L. *et al.* (2013) Targeted genome modification of crop plants using a CRISPR-Cas system. *Nature Biotechnology* 31, 686–688. Available at: https://doi.org/10.1038/nbt.2650

Shan, Q.W., Wang, Y.P., Li, J. and Gao, C.X. (2014) Genome editing in rice and wheat using the CRISPR/Cas system. *Nature Protocols* 9, 2395–2410. Available at: https://doi.org/10.1038/nprot.2014.157

Shavrukov, Y., Baho, M., Lopato, S. and Langridge, P. (2016) The *TaDREB3* transgene transferred by conventional crossings to different genetic backgrounds of bread wheat improves drought tolerance. *Plant Biotechnology Journal* 14, 313–322. Available at: https://doi.org/10.1111/pbi.12385

Shi, J., Gao, H., Wang, H., Lafitte, H.R., Archibald, R.L. *et al.* (2016) *ARGOS8* variants generated by CRISPR-Cas9 improve maize grain yield under field drought stress conditions. *Plant Biotechnology Journal* 15, 207–216. Available at: https://doi.org/10.1111/pbi.12603

Strecker, J., Ladha, A., Gardner, Z., Schmid-Burgk, J.L., Makarova, K.S., Koonin, E.V. and Zhang, F. (2019) RNA-guided DNA insertion with CRISPR-associated transposases. *Science* 365, 48–53. Available at: https://doi.org/10.1126/science.aax9181

Su, Y.H., McGrath, S.P. and Zhao, F.J. (2010) Rice is more efficient in arsenite uptake and translocation than wheat and barley. *Plant and Soil* 328, 27–34. Available at: https://doi.org/10.1007/s11104-009-0074-2

Sui, F.Q., Chang, J.D., Tang, Z., Liu, W.J., Huang, X.Y. and Zhao, F.J. (2018) *Nramp5* expression and functionality likely explain higher cadmium uptake in rice than in wheat and maize. *Plant and Soil* 433, 377–389. Available at: https://doi.org/10.1007/s11104-018-3849-5

Tang, L., Mao, B.G., Li, Y.K., Lv, Q.M., Zhang, L.P. *et al.* (2017) Knockout of *OsNramp5* using the CRISPR/Cas9 system produces low Cd-accumulating *indica* rice without compromising yield. *Scientific Reports* 7, 14438. Available at: https://doi.org/10.1038/s41598-017-14832-9

Tello-Ruiz, M.K., Naithani, S., Stein, J.C., Gupta, P., Campbell, M. *et al.* (2018) Gramene 2018: unifying comparative genomics and pathway resources for plant research. *Nucleic Acids Research* 46, D1181–D1189. Available at: https://doi.org/10.1093/nar/gkx1111

Tian, F.X., Gong, J.F., Zhang, J., Zhang, M., Wang, G.K., Li, A.X. and Wang, W. (2013) Enhanced stability of thylakoid membrane proteins and antioxidant competence contribute to drought stress resistance in the *tasg1* wheat stay-green mutant. *Journal of Experimental Botany* 64, 1509–1520. Available at: https://doi.org/10.1093/jxb/ert004

Uauy, C., Paraiso, F., Colasuonno, P., Tran, R.K., Tsai, H. *et al.* (2009) A modified TILLING approach to detect induced mutations in tetraploid and hexaploid wheat. *BMC Plant Biology* 9, 115. Available at: https://doi.org/10.1186/1471-2229-9-115

UC Davis (2020) Improved plant regeneration method using GRFs, GIFs or chimeric GRF-GIF proteins. University of California, Davis, California. Available at: https://techtransfer.universityofcalifornia.edu/NCD/31641.html (accessed 5 March 2020).

van de Wouw, M., van Hintum, T., Kik, C., van Treuren, R. and Visser, B. (2010) Genetic diversity trends in twentieth century crop cultivars: a meta analysis. *Theoretical and Applied Genetics* 120, 1241–1252. Available at: https://doi.org/10.1007/s00122-009-1252-6

Voss-Fels, K.P., Stahl, A., Wittkop, B., Lichthardt, C., Nagler, S. *et al.* (2019) Breeding improves wheat productivity under contrasting agrochemical input levels. *Nature Plants* 5, 706–714. Available at: https://doi.org/10.1038/s41477-019-0445-5

Wang, F.Z., Chen, M.X., Yu, L.J., Xie, L.J., Yuan, L.B. *et al.* (2017) *OsARM1*, an R2R3 MYB transcription factor, is involved in regulation of the response to arsenic stress in rice. *Frontiers in Plant Science* 8, 1868. Available at: https://doi.org/10.3389/fpls.2017.01868

Wang, K., Gong, Q. and Ye, X.G. (2019) Recent developments and applications of genetic transformation and genome editing technologies in wheat. *Theoretical and Applied Genetics* 133, 1603–1622. Available at: https://doi.org/10.1007/s00122-019-03464-4

Wang, W.Q., Hao, Q.Q., Tian, F.X., Li, Q.X. and Wang, W. (2016) The stay-green phenotype of wheat mutant *tasg1* is associated with altered cytokinin metabolism. *Plant Cell Reports* 35, 585–599. Available at: https://doi.org/10.1007/s00299-015-1905-7

Wang, Y.P., Cheng, X., Shan, Q.W., Zhang, Y., Liu, J.X., Gao, C.X. and Qiu, J.L. (2014) Simultaneous editing of three homoeoalleles in hexaploid bread wheat confers heritable resistance to powdery mildew. *Nature Biotechnology* 32, 947–951. Available at: https://doi.org/10.1038/nbt.2969

Wiebe, K., Harris, N.S., Faris, J.D., Clarke, J.M., Knox, R.E., Taylor, G.J. and Pozniak, C.J. (2010) Targeted mapping of *Cdu1*, a major locus regulating grain cadmium concentration in durum wheat (*Triticum turgidum* L. var *durum*). *Theoretical and Applied Genetics* 121, 1047–1058. Available at: https://doi.org/10.1007/s00122-010-1370-1

Zhao, Y., Tian, X.J., Li, Y.Y., Zhang, L.Y., Guan, P.F. *et al.* (2017) Molecular and functional characterization of wheat *ARGOS* genes influencing plant growth and stress tolerance. *Frontiers in Plant Science* 8, 170. Available at: https://doi.org/10.3389/fpls.2017.00170

Zhou, G.F., Pereira, J.F., Delhaize, E., Zhou, M.X., Magalhaes, J.V. and Ryan, P.R. (2014) Enhancing the aluminium tolerance of barley by expressing the citrate transporter genes *SbMATE* and *FRD3*. *Journal of Experimental Botany* 65, 2381–2390. Available at: https://doi.org/10.1093/jxb/eru121

Zong, Y., Wang, Y., Li, C., Zhang, R., Chen, K. *et al.* (2017) Precise base editing in rice, wheat and maize with a Cas9-cytidine deaminase fusion. *Nature Biotechnology* 35, 438–440. Available at: https://doi.org/10.1038/nbt.3811

Zong, Y., Song, Q.N., Li, C., Jin, S., Zhang, D.B. *et al.* (2018) Efficient C-to-T base editing in plants using a fusion of nCas9 and human APOBEC3A. *Nature Biotechnology* 36, 950–953. Available at: https://doi.org/10.1038/nbt.4261

Zsögön, A., Cermak, T., Voytas, D. and Peres, L.E.P. (2017) Genome editing as a tool to achieve the crop ideotype and *de novo* domestication of wild relatives: case study in tomato. *Plant Science* 256, 120–130. Available at: https://doi.org/10.1016/j.plantsci.2016.12.012

Zsögön, A., Cermak, T., Naves, E.R., Notini, M.M., Edel, K.H. *et al.* (2018) *De novo* domestication of wild tomato using genome editing. *Nature Biotechnology* 36, 1211–1216. Available at: https://doi.org/10.1038/nbt.4272

# 13 Application of Pangenomics for Wheat Molecular Breeding

**Cassandria Tay Fernandez, Jacob Marsh, Mônica Furaste Danilevicz, Clémentine Mercé and David Edwards***
*University of Western Australia, Crawley, Australia*

## 13.1 Introduction

### 13.1.1 What is a pangenome?

Wheat breeding has been supported by genetic and genomic technologies through the application of molecular genetic markers. These markers can be tightly linked to phenotypic traits which makes them valuable as proxies for phenotyping, targeting agronomically important traits for breeding. To properly utilize molecular breeding techniques, associated genetic markers need to be identified; however, many approaches for discovery and analysis of single-nucleotide polymorphism (SNPs) use a reference genome assembly and this may lead to bias due to the extent of gene presence/absence variation in plant species (Golicz *et al.*, 2016a; Hurgobin and Edwards, 2017).

A pangenome consists of the full repertoire of genes from all sequenced variations of the same species, genus or a larger group. These genes are divided by their presence and/or absence in the species: genes that belong in the 'core genome' are genes that are present in all individuals, while genes that are present only in some individuals are placed in the 'accessory/variable genome' (also known as the 'dispensable genome'). Among genotyping-by-sequencing and genotyping arrays, using pangenomes is an effective way to capture crop diversity. Pangenomes capture a larger scope of the diversity of a species than a single reference genome due to the range of individuals used for its construction (Golicz *et al.*, 2016b). In conjunction with genotyping-by-sequencing and genotyping arrays, pangenomes have been used to capture the scope of genomic diversity in a species and have so far been successfully used as a reference for gene functional studies and evolutionary analyses (Scheben *et al.*, 2016). For example, a gene function study using the tomato pangenome identified variation in the *TomLoxC* promoter which contributes to tomato flavour and had been selected against during domestication (Gao *et al.*, 2019; Zhang *et al.*, 2019). In the same vein, the construction of a pigeon pea pangenome found 225 SNPs associated with nine agronomically important traits in the crop, one of which being a gene associated negatively with seed weight distribution (Zhao *et al.*, 2020).

One approach to constructing a pangenome involves producing multiple, whole-genome *de novo* assemblies which are compared with one another to identify genomic differences (Li *et al.*, 2014; Gordon *et al.*, 2017; Zhao *et al.*,

*Corresponding author: dave.edwards@uwa.edu.au

 *Molecular Breeding in Wheat, Maize and Sorghum: Strategies for Improving Abiotic Stress Tolerance and Yield* (eds M.A. Hossain *et al.*)
DOI: 10.1079/9781789245431.0013

2018). While this can place the majority of genes in a chromosomal location, it requires very high levels of sequencing coverage. In addition, sequencing assembly errors or variations in assembly and annotation quality can lead to a misinterpretation of true genomic differences (Schatz *et al.*, 2014; Hurgobin and Edwards, 2017). In contrast, the iterative mapping and assembly approach uses a single reference genome as a template and sequentially adds non-redundant sequences from other genotypes (Montenegro *et al.*, 2017; Hübner *et al.*, 2019; Zhang *et al.*, 2019). Presence/absence variants (PAVs) are then called by remapping reads back to the improved pangenome reference. The iterative mapping and assembly approach is considerably less expensive and time-consuming than the *de novo* assembly and annotation approach and allows for the pooling of large numbers of individuals with relatively low sequencing coverage, producing presence/absence variation information across large populations and capturing rare genes that are found only in few individuals. This has been applied in several crops, including *Brassica oleracea* (Golicz *et al.*, 2016b), sesame (Yu *et al.*, 2019) and wheat (Montenegro *et al.*, 2017). However, unlike the *de novo* assembly and comparison approach, it cannot position the majority of genes on the genome. The *de novo* assembly and the iterative assembly methods are highly complementary and a combination of a small number of high-quality *de novo* assemblies together with low-coverage (>10×) data from a large number of individuals allows for a highly detailed assessment of gene PAVs, the capture of rare genes and the physical placement of variable genes.

Pangenome information can be applied to existing germplasm resources to support crop improvement through the selection of variable genes (Edwards *et al.*, 2012; Dwivedi *et al.*, 2017; Varshney *et al.*, 2019). It can also be used to compare related species to identify novel genes or orthologous genes with a shared history. Pangenomes benefit from the availability of novel data for gene discovery and gene prioritization and could reduce the problem with widespread redundancy in research by acting as its own specific repository that is easily accessible (Scheben and Edwards, 2018a). Thanks to the reducing cost of high-throughput DNA sequencing, these detailed and informative references are more easily produced, have already been used to accelerate the study and production of crops and can continue to be a useful molecular breeding tool for wheat (Abberton *et al.*, 2016; Danilevicz *et al.*, 2020).

### 13.1.2 The first wheat pangenome

There have been rapid developments in wheat genomics in recent years with the reduced cost of DNA sequencing together with approaches for whole-genome assembly that are applicable to these large and complex genomes. In 2011, isolated chromosome-arm sequencing reduced some of the assembly complexity of wheat by reducing the amount of represented repetitive regions and also by eliminating the potential confusion brought by the presence of homeologous chromosomes, leading to the assembly of approximately 40% of the 7DS chromosome arm, including all of its known genes, placing them in order based on synteny with related grass species (Berkman *et al.*, 2011). The following year, low-copy and unique regions of the chromosome arm 7BS were also assembled, defining the 7BS/4AL translocation at the gene level (Berkman *et al.*, 2012). The subsequent year saw the publication of all group 7 chromsome arms, giving highlights into wheat genome evolution and domestication (Berkman *et al.*, 2013). This approach was subsequently adopted by the International Wheat Genome Sequencing Consortium (IWGSC) to assemble all chromosome arms of wheat, providing the first reference for this large and complex genome (International Wheat Genome Sequencing Consortium, 2014). The same year, the largest wheat chromosome, 3B, was assembled through a bacterial artificial chromosome (BAC) approach which resulted in 1036 contigs, representing 82% of the chromosome (Choulet *et al.*, 2014). In addition, 7A, 7B and 7D wheat microRNAs (Deng *et al.*, 2014) and SNPs (Lai *et al.*, 2015) were identified and characterized which showed their evolutionary roles in polyploidization. In 2017, the assemblies of two modern wheat ancestors, wild emmer (Avni *et al.*, 2017) and *Aegilops tauschii* (Zhao *et al.*, 2017), were released, followed by a near-complete assembly of domesticated wheat (Clavijo *et al.*, 2017). Finally, in 2018, the IWGSC published a fully annotated reference of the Chinese Spring cultivar: IWGSC RefSeq assembly v1.0 (International Wheat Genome Sequencing Consortium, 2018). This first version has since been improved using optical maps together with PacBio SMRT sequencing. The updated IWGSC RefSeq v2.0 was released in 2019 and is available online (https://wheat-urgi.versailles.inra.fr/Seq-Repository/Assemblies, accessed 16 February 2021).

The first wheat pangenome was published in 2017 and applied the iterative assembly approach and a chromosome-arm-based reassembly of Chinese Spring (Montenegro *et al.*, 2017). This pangenome was produced from 18 cultivars and revealed the extent of presence/absence variation in this species for the first time. This study predicted a total of 128,656 genes of which 82,725 genes (64.3%) were identified as core (a genome browser of this assembly is available at http://appliedbioinformatics.com.au/cgi-bin/gb2/gbrowse/WheatPan/ and data are available at http://wheatgenome.info/wheat_genome_databases.php, both accessed 16 February 2021). Analysis of gene ontology revealed that variable genes are enriched for terms including stress response, reflecting results from previous studies in rice (Yao *et al.*, 2015) and *B. oleracea* (Golicz *et al.*, 2016b). Analysis of the gene content in the 18 cultivars, many of which are closely related to current breeding lines, demonstrated that the Chinese Spring line has a significantly different gene content from current modern varieties. There were 245 genes that were present only in the Chinese Spring genome, while 12,150 genes were found in every one of the 18 recent cultivars but were absent from Chinese Spring.

### 13.1.3 Resequencing additional varieties

An international collaboration has been established to expand on this first wheat pangenome. Established in 2016, the 10+ Wheat Genomes Project (http://www.10wheatgenomes.com, accessed 16 February 2021) currently includes 18 members from eight countries, and the cultivars were selected to represent diverse agroecological zones (Europe, Australia, Canada, Japan, USA). The project will take a *de novo* assembly approach, with 12 cultivars sequenced to date (April 2020; Table 13.1), although currently the data are available only to project partners.

### 13.1.4 Scope for future sequencing approaches

#### *Bi-directional variant graphs for pangenome assembly*

As sequencing technology advances, faster and more accurate methods of constructing pangenomes will be developed. One of the newer methods of assembling a pangenome involves the usage of bi-directed variation graphs (VGs). VGs utilize sequencing reads supported by linear genome references to produce a graphical representation of the variable regions (Garrison *et al.*, 2018; Rakocevic *et al.*, 2019). VG tools are fast and robust assemblers which avoid assembly bias towards a reference genome. They enable a population-scale representation of the genome that allows for the continuous addition of novel sequencing data (Eggertsson *et al.*, 2017; Jackman *et al.*, 2017; Rakocevic *et al.*, 2019). Unlike other pangenome approaches that have to realign and reassemble new sequences, new data

**Table 13.1.** Cultivars used in the two wheat pangenome projects.

| wheatgenome.info (Montenegro *et al.*, 2017) | 10+ Wheat Genomes Project |
|---|---|
| AB-1 | Mace |
| Alsen | Lancer |
| BX-1 | CDC Landmark |
| CH7 | Julius |
| Drysdale | Norin61 |
| Excalibur | ArinaLrFor |
| Gladius | Jagger |
| H45 | Cadenza |
| Kukri | Paragon |
| OpataM85 | Kronos |
| Pastor | Robigus |
| RAC | Claire |
| Volcani | |
| W7984 | |
| Westonia | |
| Wyalkatchem | |
| Xi-1 | |
| Yp-1 | |

can be continuously added to VG-constructed pangenomes reducing the time it takes for genomic analyses to incorporate the latest pangenomic data.

Pangenomes separate core genes from variants (into the core genome and accessory/variable genome), making it easier to identify variants for further study. In a VG assembly, PAVs can be studied by looking at the variant locations which are more widely represented by the set of graph paths than in typical pangenome assemblies. Nevertheless, iterative/*de novo* pangenomes are useful tools and can be combined with VG pangenomes or used alone for further exploration into identifying important variation and function of wheat genes.

Presently, Variation Graph and Graph Genome are two toolkits that are able to manipulate human-sized genomic data. These programs are also able to perform variant calling in these genomes and are more accurate in calling SNPs when compared with established methods that utilize linear analysis tools (Garrison *et al.*, 2018; Rakocevic *et al.*, 2019).

VG applications in wheat and other plant genomes have been limited by topological constraints and scalability, as it requires a large amount of computational power and storage. As of 2020, there have been few tools available to perform the same analyses that iterative and *de novo* pangenome assemblies can undertake, such as common downstream analyses and visualization of the graph genomes. The potential for VG to become the standard of pangenome assembly exists but improvements need to be made to VG tools in order to make them viable for wheat pangenomic studies and analyses.

### *Long-read sequencing and genome mapping*

Until recently, most plant genome sequencing was performed using short-read sequencing technology from Illumina, which has a very low error rate (<1%) and is inexpensive, even for the large quantities of data required for wheat genome sequencing. However, the short length of the reads limits the ability to assemble repetitive regions, leading to collapsed repeats and fragmented assemblies. More recently, third-generation sequencing technologies offered by Pacific Bioscience (Pacbio) and Oxford Nanopore Technology (ONT) are capable of generating long and ultra-long reads (up to 2.3 Mb per read) (Payne *et al.*, 2019) to assemble unresolved, complex and repetitive regions of the genome. A recent study using bread wheat employed long-read sequencing associated with other mapping methods to incorporate long tandem repeats to pseudomolecules (Kapustová *et al.*, 2019).

Long-read sequencing can span hundreds of kilobases, bridging some of the gapped regions, offering a less ambiguous *de novo* genome assembly and the characterization of structural variants (SVs). However, these long reads have a higher error rate. Recently, new bioinformatics tools have aimed to address this limitation; for example, Nanopolish (https://github.com/jts/nanopolish, accessed 16 February 2021) can improve base calling accuracy to 99.77% in the human genome sequenced using ONT (Koren *et al.*, 2019). Even though the wheat genome is much more complex than the human genome, the improvement of bioinformatics tools for long-read sequence reads could improve the assembly of repetitive regions. In addition, it is possible to use a hybrid approach in which Illumina short reads are mapped into the long-read assembly to polish it and increase base calling confidence (Koren *et al.*, 2012; Yuan *et al.*, 2017; Mahmoud *et al.*, 2019). The use of the hybrid approach has improved the assembly quality of bread wheat, enabling the identification of complete families of agronomic genes and evidence of chromosomal translocation (Clavijo *et al.*, 2017).

Additional technologies such as optical mapping, the sequencing of BACs and chromosome conformation capture (Hi-C) have also been employed in conjunction with short- and/or long-read sequencing to decrease genome gaps and improve scaffolding in plants (Deschamps *et al.*, 2018; Keeble-Gagnère *et al.*, 2018). Optical mapping uses linkage information based on the physical location of restriction endonuclease sites to assist the ordering scaffolds and identification of assembly errors or structural variation (Zhou and Schwartz, 2004). Similarly, Hi-C estimates the physical location of sequence read pairs by measuring their contact frequency in the genome, providing linkage information to assist in building chromosome-level assemblies (Lieberman-Aiden *et al.*, 2009; Ghurye *et al.*, 2019). In wild emmer wheat, optical mapping enabled the study of wild emmer evolution and domestication as well as a putative mutation in genes controlling shattering, which is one of the key domestication traits in cereals (Avni *et al.*, 2017; Zhu *et al.*, 2019). Various combinations of these scaffolding methods have been observed to significantly enhance the quality of the *de novo* assembly of bread wheat, durum wheat and the wild emmer genome, allowing for a reliable comparison of variable regions between different lines (Staňková *et al.*, 2016; Avni *et al.*, 2017; Keeble-Gagnère *et al.*, 2018; Salina *et al.*, 2018; Kapustová *et al.*, 2019; Maccaferri *et al.*, 2019; Zhu *et al.*, 2019).

#### *Tools for pangenome analysis*

Analysing pangenome variability can be a daunting task as the majority of genome analysis tools were developed for linear reference genomes. However, novel resources are being developed to improve variant detection and characterization in wheat pangenomes (Borrill *et al.*, 2019). Pretzel is one of the new platforms developed to incorporate past research into high-quality genome and pangenome assemblies, it enables the visualization of genetic maps and physical genome sequence (https://github.com/plantinformatics/pretzel, accessed 16 February 2021). Similarly, the Wcpheat@URGI portal (https://wheat-urgi.versailles.inra.fr/, accessed 16 February 2021) aims to integrate physical maps, gene expression, QTLs, SNPs and other genetic and phenomic data from several collaborative projects on wheat (Alaux *et al.*, 2018). The Wcpheat@URGI platform also offers dedicated tools to explore different aspects of the data, allowing for investigation of specific genes and meta-analysis across data sets (Alaux *et al.*, 2018). Some earlier platforms also provide wheat networks based on gene orthology, such as WheatNet (https://www.inetbio.org/wheatnet, accessed 16 February 2021) and Knetminer (http://knetminer.rothamsted.ac.uk, accessed 16 February 2021), that also offers a RNA-Seq co-expression network (Lee *et al.*, 2017). With the growing amount of wheat information, databases such as WheatGenome.info (Lai *et al.*, 2012) and the International Wheat Information System (WheatIS) are working towards collating all data involving wheat to make the available resources more accessible and accelerate the production of new crop lines (Scheben *et al.*, 2018; Scheben and Edwards, 2018a).

A growing number of publications describe using deep neural networks (DNNs) to address the current challenges in genome assembly and functional characterization (Wang *et al.*, 2020). DNNs have been applied in a wide range of applications, developing models for prediction, classification and clustering in several fields of research (LeCun *et al.*, 2015). DNNs are a part of machine learning, they employ multiple functions to autonomously detect underlying patterns in large data sets (LeCun *et al.*, 2015). Several models for the prediction of genomic features using DNNs have been published to determine regulatory regions in the genome (Li *et al.*, 2018), predict association of non-coding mutations to disease (Zhou *et al.*, 2019) and to build a chromatin accessibility map (Kelley *et al.*, 2016). A detailed review of the use of DNNs for genomic analysis is available from Zou *et al.* (2019) and Wang *et al.* (2020) who focus on plant genomics. Moreover, some DNN models can be downloaded for testing and reuse from Kipoi (https://kipoi.org, accessed 16 February 2021), a repository for sharing trained predictive models in genomics (Avsec *et al.*, 2019).

## 13.2 Application

### 13.2.1 Identification of presence/absence variants associated with stress

SNPs and SVs, such as copy number variations (CNVs) and PAVs, have all been studied to understand the genetic constituents of wheat diversity. A pangenome can be used to define PAVs which can then be associated with agronomic traits such as yield and environmental stress tolerance and subsequently used to enhance abiotic stress adaptation (Dolatabadian *et al.*, 2017; Anderson *et al.*, 2020). In other crops such as *Brassica napus*, a comparative pangenomic analysis identified novel resistance gene analogues that were driven by SNPs and PAVs that could not be identified from a single reference alone (Hurgobin *et al.*, 2018; Dolatabadian *et al.*, 2020).

A combination of phenotyping and linkage analysis of SNP, PAV and simple sequence repeat (SSR) markers was used to identify genomic regions on chromosomes 3B and 1D associated with heat stress (Sharma *et al.*, 2017). CNVs responsible for cold tolerance such as FR-2 have also been identified, all of which contribute to the overall fitness of the plants (Knox *et al.*, 2010). PAVs have also been associated with biotic stresses and contribute to resistance to disease. PAV and SNP markers associated with adult plant stripe rust resistance (*YrBai*) in Baidatou were identified to control the resistance to the Chinese prevalent *Pst* (*Puccinia striiformis* f.sp. *tritici*) race CYR33 in adult plants and can possibly be used to integrate resistance in other cultivars (Li *et al.*, 2018). More than 80 leaf rust (*Puccinia triticina*) resistance genes have been identified, including *Lr10* which is a PAV that, when overexpressed in transgenic plants, confers enhanced leaf rust resistance (Feuillet *et al.*, 2003). Other PAVs have been associated with stripe rust resistance (*Yr36*) (Fu *et al.*, 2009), stem rust resistance (*Sr22* and *Sr24*) (Mago *et al.*, 2005; Periyannan *et al.*, 2011) and tan spot/*Stagonospora nodorum* blotch resistance (*Tsn1*) (Liu *et al.*, 2006).

### 13.2.2 Identification of presence/absence variants associated with yield

Yield is genetically influenced by many factors. Several PAVs have been found to influence the yield of wheat in a number of different ways. The heading date of wheat crops was found to be influenced by a PAV region of photoperiod-insensitive alleles (Nishida *et al.*, 2013) and can possibly be used to manipulate the growth of wheat crops. A genome-wide association study of durum wheat identified a PAV marker that corresponds to late embryogenesis abundant (LEA) glycoproteins, located in the roots and the seeds of wheat, responding to abiotic stresses (Gao and Lan, 2016; Roselló *et al.*, 2019) and could possibly be used to enhance abiotic resistance in wheat and other crops. The same study identified a PAV marker associated with a basic leucine zipper (bZIP) transcription factor family protein which was observed to affect root growth in other studies and has marked it for potential use in improving new varieties (Zhang *et al.*, 2017; Roselló *et al.*, 2019).

Genome-wide QTL mapping and candidate gene analyses have aided in the discovery of yield-related candidate genes and PAVs (Nigro *et al.*, 2019). Four candidate genes associated with grain yield were identified in soft red winter wheat, one of which is an orthologue of the rice aberrant panicle organization protein (APO1) and one an orthologue of *Arabidopsis thaliana* that affects the number of grains per square metre, which can possibly be exploited for commercial use (Ward *et al.*, 2019).

### 13.2.3 Potential to bring variable genes in from related species

The average level of genetic diversity in modern wheat cultivars has dropped almost threefold compared with the AA and BB genomes of wild emmer and an over fourfold decrease occurs in the DD subgenome compared with *Ae. tauschii* (Tanno and Willcox, 2006). The reduction of diversity suggests that key traits of agronomic importance in modern hexaploid wheat such as biotic/abiotic stress resistance and local adaptation may have been lost during domestication and breeding (Golicz *et al.*, 2020). Historical gene flow into domesticated hexaploid wheat from its wild progenitors has recently been shown to have introduced novel variants which reduced the deleterious mutation burden and carried adaptive traits (He *et al.*, 2019).

The wild relatives of crops (CWRs) contain a reservoir of genetic traits and variability. Characterizing the genomes of wheat CWRs and mapping the evolutionary trends that have taken place in specific genomic regions will assist breeders who wish to exploit this natural variation. The effectiveness of utilizing close wild-type ancestors and multiple cultivated accessions in order to uncover functional allele differences has also been shown in other important crops such as soybean (Valliyodan *et al.*, 2019) and *B. napus* (Bayer *et al.*, 2017).

Pangenomes need not be restricted to discrete taxonomic groups; incorporating CWRs from the diploid and tetraploid genera into a single wheat pangenome would support the comparison of these genomes at the species level. A pangenome of the representing diverse wheat genera would enable the classification of core (present in all individuals) and dispensable (absent in some individuals) genes, as well as taxon-specific genes (present in all individuals of one cladistic group, absent in all others). For wheat, Liu *et al.* (2016) sequenced one chromosome from a single accession and compared it with the Chinese Spring reference, finding large SVs up to 159.3 Mb in size linked to adaptation. Tulpova *et al.* (2019) generated a physical map of 7DS to facilitate gene mapping and cloning. In 2020, 7DL of the reference pangenome of wheat and 7DL of *Ae. tauschii* were compared and identified domestication signatures associated with yield and grain quality (Feng *et al.*, 2020). These highlight the potential for identifying novel adaptive SVs present within the dispensable genome of wheat.

Studies involving interspecific hybridization between bread wheat and related species highlight the high degree of cross-compatibility between hexaploid wheat and its wild relatives, as well as the potential for crop improvement through introgression of adaptive traits (Devi *et al.*, 2019). Introgression through recombination is a stochastic process that may not produce progeny with the desired traits. Furthermore, hybridization often transfers additional undesirable regions linked to QTLs and regions of interest – a phenomenon known as linkage drag, which has been shown to have had a significant detrimental effect on root growth in European bread wheat (Voss-Fels *et al.*, 2017).

Advancements in gene editing technology such as with the clustered regularly interspaced short palindromic repeats (CRISPR)/CRISPR-associated protein 9 (Cas9) system offer a promising alternative for the targeted introduction of novel adaptive alleles from related species as well as to introduce neutral variants to counteract recent deleterious mutations in crop populations (Scheben *et al.*, 2017; Scheben and Edwards, 2018b). Continued improvements in genome editing promise to increase target specificity and reduce off-target effects as recent research has focused on engineering modified Cas9 cassettes (Jaganathan *et al.*, 2018) and selecting optimal guide RNAs for use in hexaploid wheat (Scheben and Edwards, 2017; Arndell *et al.*, 2019). A pangenome provides a complete genomic profile to map the evolutionary history of genes of interest in the wild population and to predict potential off-target interactions and target specificity in the recipient crop population.

## 13.3 Conclusion and Future Perspective

Pangenomes can be used by both researchers and breeders alike to develop elite wheat cultivars through the discovery and integration of genetic variations associated with agronomically beneficial traits. By providing a reference that accommodates for variation in individuals, variants whose presence and/or absence control abiotic stress resistance and yield can be identified. This tool has only become more informative as more wheat varieties are sequenced, new sequencing approaches such as long-read sequencing and genome mapping are utilized, and tools for pangenomic analysis are developed. With pangenomics, variable genes from wild wheat relatives and related species can be used to optimize wheat molecular breeding and develop improved varieties tailored for the changing global environment.

## References


Abberton, M., Batley, J., Bentley, A., Bryant, J., Cai, H. *et al.* (2016) Global agricultural intensification during climate change: a role for genomics. *Plant Biotechnology Journal* 14, 1095–1098.

Alaux, M., Rogers, J., Letellier, T., Flores, R., Alfama, F. *et al.* (2018) Linking the International Wheat Genome Sequencing Consortium bread wheat reference genome sequence to wheat genetic and phenomic data. *Genome Biology* 19, 111.

Anderson, R., Bayer, P.E. and Edwards, D. (2020) Climate change and the need for agricultural adaptation. *Current Opinion in Plant Biology* 56, 197–202.

Arndell, T., Sharma, N., Langridge, P., Baumann, U., Watson-Haigh, N.S. and Whitford, R. (2019) gRNA validation for wheat genome editing with the CRISPR-Cas9 system. *BMC Biotechnology* 19, 71.

Avni, R., Nave, M., Barad, O., Baruch, K., Twardziok, S.O. *et al.* (2017) Wild emmer genome architecture and diversity elucidate wheat evolution and domestication. *Science* 357, 93–97.

Avsec, Ž., Kreuzhuber, R., Israeli, J., Xu, N., Cheng, J. *et al.* (2019) The Kipoi repository accelerates community exchange and reuse of predictive models for genomics. *Nature Biotechnology* 37, 592–600.

Bayer, P.E., Hurgobin, B., Golicz, A.A., Chan, C.K., Yuan, Y. *et al.* (2017) Assembly and comparison of two closely related *Brassica napus* genomes. *Plant Biotechnology Journal* 15, 1602–1610.

Berkman, P.J., Skarshewski, A., Lorenc, M.T., Lai, K., Duran, C. *et al.* (2011) Sequencing and assembly of low copy and genic regions of isolated *Triticum aestivum* chromosome arm 7DS. *Plant Biotechnology Journal* 9, 768–775.

Berkman, P.J., Skarshewski, A., Manoli, S., Lorenc, M.T., Stiller, J. *et al.* (2012) Sequencing wheat chromosome arm 7BS delimits the 7BS/4AL translocation and reveals homoeologous gene conservation. *Theoretical and Applied Genetics* 124, 423–432.

Berkman, P.J., Visendi, P., Lee, H.C., Stiller, J., Manoli, S. *et al.* (2013) Dispersion and domestication shaped the genome of bread wheat. *Plant Biotechnology Journal* 11, 564–751.

Borrill, P., Harrington, S.A. and Uauy, C. (2019) Applying the latest advances in genomics and phenomics for trait discovery in polyploid wheat. *The Plant Journal* 97, 56–72.

Choulet, F., Alberti, A., Theil, S., Glover, N., Barbe, V. *et al.* (2014) Structural and functional partitioning of bread wheat chromosome 3B. *Science* 345, 1249721.

Clavijo, B.J., Venturini, L., Schudoma, C., Accinelli, G.G., Kaithakottil, G. *et al.* (2017) An improved assembly and annotation of the allohexaploid wheat genome identifies complete families of agronomic genes and provides genomic evidence for chromosomal translocations. *Genome Research* 27, 885–896.

Danilevicz, M.F., Tay Fernandez, C.G., Marsh, J.I., Bayer, P.E. and Edwards, D. (2020) Plant pangenomics: approaches, applications and advancements. *Current Opinion in Plant Biology* 54, 18–25.

Deng, P., Nie, X., Wang, L., Cui, L., Liu, P. *et al.* (2014) Computational identification and comparative analysis of miRNAs in wheat group 7 chromosomes. *Plant Molecular Biology Reporter* 32, 487–500.

Deschamps, S., Zhang, Y., Llaca, V., Ye, L., Sanyal, A. *et al.* (2018) A chromosome-scale assembly of the sorghum genome using nanopore sequencing and optical mapping. *Nature Communications* 9, 4844.

Devi, U., Grewal, S., Yang, C.-Y., Hubbart-Edwards, S., Scholefield, D. *et al.* (2019) Development and characterisation of interspecific hybrid lines with genome-wide introgressions from *Triticum timopheevii* in a hexaploid wheat background. *BMC Plant Biology* 19, 183.

Dolatabadian, A., Patel, D.A., Edwards, D. and Batley, J. (2017) Copy number variation and disease resistance in plants. *Theoretical and Applied Genetics* 130, 2479–2490.

Dolatabadian, A., Bayer, P.E., Tirnaz, S., Hurgobin, B., Edwards, D. and Batley, J. (2020) Characterization of disease resistance genes in the *Brassica napus* pangenome reveals significant structural variation. *Plant Biotechnology Journal* 18, 969–982.

Dwivedi, S.L., Scheben, A., Edwards, D., Spillane, C. and Ortiz, R. (2017) Assessing and exploiting functional diversity in germplasm pools to enhance abiotic stress adaptation and yield in cereals and food legumes. *Frontiers in Plant Science* 8, 1461.

Edwards, D., Wilcox, S., Barrero, R.A., Fleury, D., Cavanagh, C.R. *et al.* (2012) Bread matters: a national initiative to profile the genetic diversity of Australian wheat. *Plant Biotechnology Journal* 10, 703–708.

Eggertsson, H.P., Jonsson, H., Kristmundsdottir, S., Hjartarson, E., Kehr, B. *et al.* (2017) Graphtyper enables population-scale genotyping using pangenome graphs. *Nature Genetics* 49, 1654–1660.

Feng, K., Cui, L., Wang, L., Shan, D., Tong, W. *et al.* (2020) The improved assembly of 7DL chromosome provides insight into the structure and evolution of bread wheat. *Plant Biotechnology Journal* 18, 732–742.

Feuillet, C., Travella, S., Stein, N., Albar, L., Nublat, A. and Keller, B. (2003) Map-based isolation of the leaf rust disease resistance gene *Lr10* from the hexaploid wheat (*Triticum aestivum* L.) genome. *Proceedings of the National Academy of Sciences USA* 100, 15253–15258.

Fu, D., Uauy, C., Distelfeld, A., Blechl, A., Epstein, L. *et al.* (2009) A kinase-START gene confers temperature-dependent resistance to wheat stripe rust. *Science* 323, 1357–1360.

Gao, J. and Lan, T. (2016) Functional characterization of the late embryogenesis abundant (LEA) protein gene family from *Pinus tabuliformis* (Pinaceae) in *Escherichia coli*. *Scientific Reports* 6, 19467–19467.

Gao, L., Gonda, I., Sun, H., Ma, Q., Bao, K. *et al.* (2019) The tomato pan-genome uncovers new genes and a rare allele regulating fruit flavor. *Nature Genetics* 51, 1044–1051.

Garrison, E., Sirén, J., Novak, A.M., Hickey, G., Eizenga, J.M. *et al.* (2018) Variation graph toolkit improves read mapping by representing genetic variation in the reference. *Nature Biotechnology* 36, 875–879.

Ghurye, J., Rhie, A., Walenz, B.P., Schmitt, A., Selvaraj, S. *et al.* (2019) Integrating Hi-C links with assembly graphs for chromosome-scale assembly. *PLoS Computational Biology* 15, e1007273.

Golicz, A.A., Batley, J. and Edwards, D. (2016a) Towards plant pangenomics. *Plant Biotechnology Journal* 14, 1099–1105.

Golicz, A.A., Bayer, P.E., Barker, G.C., Edger, P.P., Kim, H. *et al.* (2016b) The pangenome of an agronomically important crop plant *Brassica oleracea*. *Nature Communications* 7, 13390.
Golicz, A.A., Bayer, P.E., Bhalla, P.L., Batley, J. and Edwards, D. (2020) Pangenomics comes of age: from bacteria to plant and animal applications. *Trends in Genetics* 36, 132–145.
Gordon, S.P., Contreras-Moreira, B., Woods, D.P., Des Marais, D.L., Burgess, D. *et al.* (2017) Extensive gene content variation in the *Brachypodium distachyon* pan-genome correlates with population structure. *Nature Communications* 8, 2184.
He, F., Pasam, R., Shi, F., Kant, S., Keeble-Gagnere, G. *et al.* (2019) Exome sequencing highlights the role of wild-relative introgression in shaping the adaptive landscape of the wheat genome. *Nature Genetics* 51, 896–904.
Hübner, S., Bercovich, N., Todesco, M., Mandel, J.R., Odenheimer, J. *et al.* (2019) Sunflower pan-genome analysis shows that hybridization altered gene content and disease resistance. *Nature Plants* 5, 54–62.
Hurgobin, B. and Edwards, D. (2017) SNP discovery using a pangenome: has the single reference approach become obsolete? *Biology* 6, 21.
Hurgobin, B., Golicz, A.A., Bayer, P.E., Chan, C.K., Tirnaz, S. *et al.* (2018) Homoeologous exchange is a major cause of gene presence/absence variation in the amphidiploid *Brassica napus*. *Plant Biotechnology Journal* 16, 1265–1274.
International Wheat Genome Sequencing Consortium (2014) A chromosome-based draft sequence of the hexaploid bread wheat (*Triticum aestivum*) genome. *Science* 345, 1251788.
International Wheat Genome Sequencing Consortium (2018) Shifting the limits in wheat research and breeding using a fully annotated reference genome. *Science* 361, eaar7191.
Jackman, S.D., Vandervalk, B.P., Mohamadi, H., Chu, J., Yeo, S. *et al.* (2017) ABySS 2.0: resource-efficient assembly of large genomes using a Bloom filter. *Genome Research* 27, 768–777.
Jaganathan, D., Ramasamy, K., Sellamuthu, G., Jayabalan, S. and Venkataraman, G. (2018) CRISPR for crop improvement: an update review. *Frontiers in Plant Science* 9, 985–985.
Kapustová, V., Tulpová, Z., Toegelová, H., Novák, P., Macas, J. *et al.* (2019) The dark matter of large cereal genomes: long tandem repeats. *International Journal of Molecular Sciences* 20, 2483.
Keeble-Gagnère, G., Rigault, P., Tibbits, J., Pasam, R., Hayden, M. *et al.* (2018) Optical and physical mapping with local finishing enables megabase-scale resolution of agronomically important regions in the wheat genome. *Genome Biology* 19, 112.
Kelley, D.R., Snoek, J. and Rinn, J.L. (2016) Basset: learning the regulatory code of the accessible genome with deep convolutional neural networks. *Genome Research* 26, 990–999.
Knox, A.K., Dhillon, T., Cheng, H., Tondelli, A., Pecchioni, N. and Stockinger, E.J. (2010) CBF gene copy number variation at Frost Resistance-2 is associated with levels of freezing tolerance in temperate-climate cereals. *Theoretical and Applied Genetics* 121, 21–35.
Koren, S., Schatz, M.C., Walenz, B.P., Martin, J., Howard, J.T. *et al.* (2012) Hybrid error correction and *de novo* assembly of single-molecule sequencing reads. *Nature Biotechnology* 30, 693–700.
Koren, S., Phillippy, A.M., Simpson, J.T., Loman, N.J. and Loose, M. (2019) Reply to 'Errors in long-read assemblies can critically affect protein prediction'. *Nature Biotechnology* 37, 127–128.
Lai, K., Berkman, P.J., Lorenc, M.T., Duran, C., Smits, L. *et al.* (2012) WheatGenome.info: an integrated database and portal for wheat genome information. *Plant Cell Physiology* 53, e2.
Lai, K., Lorenc, M.T., Lee, H.C., Berkman, P.J., Bayer, P.E. *et al.* (2015) Identification and characterization of more than 4 million intervarietal SNPs across the group 7 chromosomes of bread wheat. *Plant Biotechnology Journal* 13, 97–104.
Lecun, Y., Bengio, Y. and Hinton, G. (2015) Deep learning. *Nature* 521, 436–444.
Lee, T., Hwang, S., Kim, C.Y., Shim, H., Kim, H. *et al.* (2017) WheatNet: a genome-scale functional network for hexaploid bread wheat, *Triticum aestivum*. *Molecular Plant* 10, 1133–1136.
Li, Y., Shi, W. and Wasserman, W.W. (2018) Genome-wide prediction of *cis*-regulatory regions using supervised deep learning methods. *BMC Bioinformatics* 19, 202.
Li, Y.-H., Zhou, G., Ma, J., Jiang, W., Jin, L.-G. *et al.* (2014) *De novo* assembly of soybean wild relatives for pan-genome analysis of diversity and agronomic traits. *Nature Biotechnology* 32, 1045–1052.
Lieberman-Aiden, E., Van Berkum, N.L., Williams, L., Imakaev, M., Ragoczy, T. *et al.* (2009) Comprehensive mapping of long-range interactions reveals folding principles of the human genome. *Science* 326, 289–293.
Liu, M., Stiller, J., Holušová, K., Vrána, J., Liu, D., Doležel, J. and Liu, C. (2016) Chromosome-specific sequencing reveals an extensive dispensable genome component in wheat. *Scientific Reports* 6, 36398.

Liu, Z., Friesen, T.L., Ling, H., Meinhardt, S.W., Oliver, R.P., Rasmussen, J.B. and Faris, J.D. (2006) The *Tsn1–ToxA* interaction in the wheat–*Stagonospora nodorum* pathosystem parallels that of the wheat-tan spot system. *Genome* 49, 1265–1273.

Maccaferri, M., Harris, N.S., Twardziok, S.O., Pasam, R.K., Gundlach, H. *et al.* (2019) Durum wheat genome highlights past domestication signatures and future improvement targets. *Nature Genetics* 51, 885–895.

Mago, R., Bariana, H.S., Dundas, I.S., Spielmeyer, W., Lawrence, G.J., Pryor, A.J. and Ellis, J.G. (2005) Development of PCR markers for the selection of wheat stem rust resistance genes *Sr24* and *Sr26* in diverse wheat germplasm. *Theoretical and Applied Genetics* 111, 496–504.

Mahmoud, M., Zywicki, M., Twardowski, T. and Karlowski, W.M. (2019) Efficiency of PacBio long read correction by 2nd generation Illumina sequencing. *Genomics* 111, 43–49.

Montenegro, J.D., Golicz, A.A., Bayer, P.E., Hurgobin, B., Lee, H. *et al.* (2017) The pangenome of hexaploid bread wheat. *The Plant Journal* 90, 1007–1013.

Nigro, D., Gadaleta, A., Mangini, G., Colasuonno, P., Marcotuli, I. *et al.* (2019) Candidate genes and genome-wide association study of grain protein content and protein deviation in durum wheat. *Planta* 249, 1157–1175.

Nishida, H., Yoshida, T., Kawakami, K., Fujita, M., Long, B. *et al.* (2013) Structural variation in the 5′ upstream region of photoperiod-insensitive alleles *Ppd-A1a* and *Ppd-B1a* identified in hexaploid wheat (*Triticum aestivum* L.), and their effect on heading time. *Molecular Breeding* 31, 27–37.

Payne, A., Holmes, N., Rakyan, V. and Loose, M. (2019) BulkVis: a graphical viewer for Oxford nanopore bulk FAST5 files. *Bioinformatics* 35, 2193–2198.

Periyannan, S.K., Bansal, U.K., Bariana, H.S., Pumphrey, M. and Lagudah, E.S. (2011) A robust molecular marker for the detection of shortened introgressed segment carrying the stem rust resistance gene *Sr22* in common wheat. *Theoretical and Applied Genetics* 122, 1–7.

Rakocevic, G., Semenyuk, V., Lee, W.-P., Spencer, J., Browning, J. *et al.* (2019) Fast and accurate genomic analyses using genome graphs. *Nature Genetics* 51, 354–362.

Roselló, M., Royo, C., Sanchez-Garcia, M. and Soriano, J. (2019) Genetic dissection of the seminal root system architecture in Mediterranean durum wheat landraces by genome-wide association study. *Agronomy* 9, 364.

Salina, E.A., Nesterov, M.A., Frenkel, Z., Kiseleva, A.A., Timonova, E.M. *et al.* (2018) Features of the organization of bread wheat chromosome 5BS based on physical mapping. *BMC Genomics* 19, 80.

Schatz, M.C., Maron, L.G., Stein, J.C., Hernandez Wences, A., Gurtowski, J. *et al.* (2014) Whole genome *de novo* assemblies of three divergent strains of rice, *Oryza sativa*, document novel gene space of *aus* and *indica*. *Genome Biology* 15, 506.

Scheben, A. and Edwards, D. (2017) Genome editors take on crops. *Science* 355, 1122–1123.

Scheben, A. and Edwards, D. (2018a) Bottlenecks for genome-edited crops on the road from lab to farm. *Genome Biology* 19, 178.

Scheben, A. and Edwards, D. (2018b) Towards a more predictable plant breeding pipeline with CRISPR/Cas-induced allelic series to optimize quantitative and qualitative traits. *Current Opinion in Plant Biology* 45, 218–225.

Scheben, A., Yuan, Y. and Edwards, D. (2016) Advances in genomics for adapting crops to climate change. *Current Plant Biology* 6, 2–10.

Scheben, A., Wolter, F., Batley, J., Puchta, H. and Edwards, D. (2017) Towards CRISPR/Cas crops – bringing together genomics and genome editing. *New Phytologist* 216, 682–698.

Scheben, A., Chan, C.-K.K., Mansueto, L., Mauleon, R., Larmande, P. *et al.* (2018) Progress in single-access information systems for wheat and rice crop improvement. *Briefings in Bioinformatics* 20, 565–571.

Sharma, D.K., Torp, A.M., Rosenqvist, E., Ottosen, C.-O. and Andersen, S.B. (2017) QTLs and potential candidate genes for heat stress tolerance identified from the mapping populations specifically segregating for Fv/Fm in wheat. *Frontiers in Plant Science* 8, 1668.

Staňková, H., Hastie, A.R., Chan, S., Vrána, J., Tulpová, Z. *et al.* (2016) BioNano genome mapping of individual chromosomes supports physical mapping and sequence assembly in complex plant genomes. *Plant Biotechnology Journal* 14, 1523–1531.

Tanno, K. and Willcox, G. (2006) How fast was wild wheat domesticated? *Science* 311, 1886.

Tulpova, Z., Luo, M.C., Toegelova, H., Visendi, P., Hayashi, S. *et al.* (2019) Integrated physical map of bread wheat chromosome arm 7DS to facilitate gene cloning and comparative studies. *New Biotechnology* 48, 12–19.

Valliyodan, B., Cannon, S.B., Bayer, P.E., Shu, S., Brown, A.V. *et al.* (2019) Construction and comparison of three reference-quality genome assemblies for soybean. *The Plant Journal* 100, 1066–1082.

Varshney, R.K., Thudi, M., Roorkiwal, M., He, W., Upadhyaya, H.D. *et al.* (2019) Resequencing of 429 chickpea accessions from 45 countries provides insights into genome diversity, domestication and agronomic traits. *Nature Genetics* 51, 857–864.

Voss-Fels, K.P., Qian, L., Parra-Londono, S., Uptmoor, R., Frisch, M. *et al.* (2017) Linkage drag constrains the roots of modern wheat. *Plant, Cell & Environment* 40, 717–725.

Wang, H., Cimen, E., Singh, N. and Buckler, E. (2020) Deep learning for plant genomics and crop improvement. *Current Opinion in Plant Biology* 54, 34–41.

Ward, B.P., Brown-Guedira, G., Kolb, F.L., Van Sanford, D.A., Tyagi, P., Sneller, C.H. and Griffey, C.A. (2019) Genome-wide association studies for yield-related traits in soft red winter wheat grown in Virginia. *PLoS One* 14, e0208217.

Yao, W., Li, G., Zhao, H., Wang, G., Lian, X. and Xie, W. (2015) Exploring the rice dispensable genome using a metagenome-like assembly strategy. *Genome Biology* 16, 187.

Yu, J., Golicz, A.A., Lu, K., Dossa, K., Zhang, Y. *et al.* (2019) Insight into the evolution and functional characteristics of the pan-genome assembly from sesame landraces and modern cultivars. *Plant Biotechnology Journal* 17, 881–892.

Yuan, Y., Bayer, P.E., Batley, J. and Edwards, D. (2017) Improvements in genomic technologies: application to crop genomics. *Trends in Biotechnology* 35, 547–558.

Zhang, B., Zhu, W., Diao, S., Wu, X., Lu, J., Ding, C. and Su, X. (2019) The poplar pangenome provides insights into the evolutionary history of the genus. *Communications Biology* 2, 215.

Zhang, W., Chen, S., Abate, Z., Nirmala, J., Rouse, M.N. and Dubcovsky, J. (2017) Identification and characterization of *Sr13*, a tetraploid wheat gene that confers resistance to the Ug99 stem rust race group. *Proceedings of the National Academy of Sciences USA* 114, E9483–E9492.

Zhao, G., Zou, C., Li, K., Wang, K., Li, T. *et al.* (2017) The *Aegilops tauschii* genome reveals multiple impacts of transposons. *Nature Plants* 3, 946–955.

Zhao, J., Bayer, P.E., Ruperao, P., Saxena, R.K., Khan, A.W. *et al.* (2020) Trait associations in the pangenome of pigeon pea (*Cajanus cajan*). *Plant Biotechnology Journal* 18, 1946–1954.

Zhao, Q., Feng, Q., Lu, H., Li, Y., Wang, A. *et al.* (2018) Pan-genome analysis highlights the extent of genomic variation in cultivated and wild rice. *Nature Genetics* 50, 278–284.

Zhou, J., Park, C.Y., Theesfeld, C.L., Wong, A.K., Yuan, Y. *et al.* (2019) Whole-genome deep-learning analysis identifies contribution of noncoding mutations to autism risk. *Nature Genetics* 51, 973–980.

Zhou, S. and Schwartz, D.C. (2004) The optical mapping of microbial genomes. *ASM News* 70, 323–330.

Zhu, T., Wang, L., Rodriguez, J.C., Deal, K.R., Avni, R. *et al.* (2019) Improved genome sequence of wild emmer wheat Zavitan with the aid of optical maps. *G3: Genes, Genomes, Genetics* 9, 619–624.

Zou, J., Huss, M., Abid, A., Mohammadi, P., Torkamani, A. and Telenti, A. (2019) A primer on deep learning in genomics. *Nature Genetics* 51, 12–18.

# 14 Recent Advancement of Molecular Understanding for Combating Salinity Stress in Maize

**Yibo Cao, Yiqiao Wang and Caifu Jiang***
*China Agricultural University, Beijing, People's Republic of China*

## 14.1 Introduction

Maize is a glycophytic crop that is moderately sensitive to soil salinity. Meanwhile, the soil salinization of global maize farmlands is constantly increasing (Hanks *et al.*, 1978; Schnable, 2015). Hence, soil salinization is becoming a serious problem for world maize production and there is an urgent need for breeding salt-tolerant maize varieties. Under salt environment, maize plants take up excessive amounts of cations and anions (e.g. $Na^+$ and $Cl^-$), which then impair numerous metabolic processes and lead to ion toxicity (Munns and Tester, 2008; Yang and Guo, 2018). In addition, high concentrations of soil salts decrease the external osmotic potential, making it difficult for maize roots to take up water and resulting in physiological drought (Munns and Tester, 2008). Therefore, the molecular breeding of salt-tolerant maize varieties principally includes, but is not limited to, improvement of the tolerances to major harmful cations (mainly $Na^+$) and anions (mainly $Cl^-$) and osmotic stress.

Existing knowledge has shown that natural maize inbred lines show large variation in salt sensitivity (Fig. 14.1) (Fortmeier and Schubert, 1995; Zhao *et al.*, 2010; Gao *et al.*, 2016; Zhang *et al.*, 2018; Zhang, M. *et al.*, 2019), which is attributable to variations in tolerances to $Na^+$ and $Cl^-$ toxicity, and/or osmotic stress, and is a quantitative trait (Zhao *et al.*, 2010; Gao *et al.*, 2016). Recent studies have identified several quantitative trait loci (QTLs) that underlie the natural variations of $Na^+$ and $K^+$ homeostasis and salt tolerance in maize (Table 14.1) (Zhang *et al.*, 2018; Cao *et al.*, 2019, 2020; Zhang, M. *et al.*, 2019) and have shown that the natural maize population confers large variations in osmoregulatory substances (e.g. glycine betaine) (Saneoka *et al.*, 1995; Yang *et al.*, 1995; Niu *et al.*, 2007), suggesting that further studies towards identification and application of the salt tolerance-associated genetic variations could provide a promising route of breeding salt-tolerant maize varieties. Moreover, dozens of genes associated with maize salt tolerance have been identified (Table 14.1), providing further genetic resources for improving maize salt tolerance by marker-assisted breeding or bioengineering strategies. The following sections review recent progress in the fundamental understanding of maize salt tolerance and the advancement of molecular breeding for salt-tolerant maize.

*Corresponding author: cfjiang@cau.edu.cn


DOI: 10.1079/9781789245431.0014

**Fig. 14.1.** The appearance of selected maize inbred lines grown under control and salinity (100 mM NaCl) conditions. Bar = 10 cm.

## 14.2 Tolerance to Salinity-Induced Osmotic Stress in Maize

Osmotic stress is one of the major stresses imposed upon maize plants by high concentrations of salts, which makes it difficult for maize roots to take up water from the soil solution (Muhammad *et al.*, 2015). Accordingly, maize has evolved a series of osmotic stress-responsive signalling pathways and metabolic rearrangements to cope with the salinity-induced low osmotic potential and water deficiency (Liu *et al.*, 2010; Hu *et al.*, 2015; Zhang *et al.*, 2016; Mittal *et al.*, 2017; Qi *et al.*, 2018).

### 14.2.1 Perception and signalling of salinity-induced osmotic stress in maize

The maize plants grown under salinity condition confer a decreased rate of leaf growth, which is partially ascribed to the osmotic stress caused by the salts surrounding the root system (Muhammad *et al.*, 2015). It has been suggested that osmotic stress starts immediately following the onset of salinity treatment, which is sensed by an osmoreceptor and then converted into second signalling messengers, mostly the increase of the concentration of cytoplasmic $Ca^{2+}$. The featured $Ca^{2+}$ increase is recognized by numerous $Ca^{2+}$-binding proteins. This subsequently activates/inactivates the downstream responses, enabling the adaption to the osmotic stress (Yuan *et al.*, 2014; Zhang *et al.*, 2016; Mittal *et al.*, 2017). The *Arabidopsis AtOSCA1* (*reduced hyperosmolality-induced [$Ca^{2+}$] increase 1*) encodes an osmoreceptor that mediates the increase of the cytoplasmic $Ca^{2+}$ under osmotic stress treatment and is essential for the tolerance to osmotic stress (Yuan *et al.*, 2014). Although the osmosensor in maize is yet to be known, the maize orthologue of AtOSCA1 (GRMZM2G456000) may act as an osmosensor. In addition, Liu *et al.* (2010) showed under osmotic stress treatment that the cytoplasmic free $Ca^{2+}$ concentration in root tip protoplasts of maize increased significantly. This supports the perspective that $Ca^{2+}$ acts as a second messenger in maize osmotic response, which may act downstream of the osmosensor. Other studies in maize indicated that the transcript levels of some genes (e.g. *ZmCBLs* and *ZmCDPKs*) encoding $Ca^{2+}$-binding proteins like CBLs (calcineurin B-like proteins) and CDPKs ($Ca^{2+}$-dependent protein kinases) are induced by salinity and osmotic stress conditions. These genes may be involved in recognizing $Ca^{2+}$ signalling and regulating downstream responses (Zhang *et al.*, 2016; Mittal *et al.*, 2017). Moreover, many studies showed that reactive oxygen species (ROS) act as an important second

**Table 14.1.** List of genes likely associate with salt tolerance in maize.

| Gene name | Gene ID | Annotation | Cloning method | Genetic evidence | Reference |
|---|---|---|---|---|---|
| *ZmSOS1* | *GRMZM2G098494* | $Na^+/H^+$ antiporter | QTL | – | Luo *et al.* (2017) |
| *ZmSOS3* | *GRMZM2G007555* | CBL protein | QTL | – | Luo *et al.* (2017) |
| *ZmNSA1* | *GRMZM2G000397* | EF-hand family protein | GWAS | Maize | Cao *et al.* (2020) |
| *ZmCPK11* | *GRMZM2G047486* | CDPK 11 | – | *Arabidopsis* | Borkiewicz *et al.* (2019) |
| *ZmHKT1* | *GRMZM2G047616* | HKT1 family transporter | QTL | Maize | Zhang *et al.* (2018) |
| *ZmHKT2* | *GRMZM2G135674* | HKT2 family transporter | QTL | Maize | Cao *et al.* (2019) |
| *ZmHAK4* | *GRMZM2G425999* | HAK family transporter | GWAS | Maize | Zhang, M. *et al.* (2019) |
| *ZmSAG4* | *GRMZM2G077295* | COP II complex SEC23 protein | GWAS | *Arabidopsis* | Luo *et al.* (2019) |
| *ZmSAG6* | *GRMZM2G106056* | Double-strand break repair protein MRE11 | GWAS | *Arabidopsis* | Luo *et al.* (2019) |
| *ZmDi19-1* | *GRMZM2G014066* | Drought-induced protein 19 family | – | *Arabidopsis* | Zhang, X. *et al.* (2019) |
| *ZmMYB3R* | *GRMZM2G081919* | MYB-transcription factor | – | *Arabidopsis* | Wu *et al.* (2019) |
| *ZmSCE1e* | *GRMZM2G038851* | A putative SUMO-conjugating enzyme | – | *Arabidopsis* | Wang *et al.* (2019) |
| *ZmGPDH1* | *GRMZM2G155348* | Glycerol-3-phosphate dehydrogenase | – | *Arabidopsis* | Zhao *et al.* (2019) |
| *ZmNPF6.4* | *GRMZM2G086496* | NPF family transporter | – | Maize | Wen *et al.* (2017) |
| *ZmALMT1* | *DQ358745* | ALMT family transporter | – | Maize | Pineros *et al.* (2008) |
| *ZmALMT2* | *GRMZM5G858653* | ALMT family transporter | – | Maize | Ligaba *et al.* (2012) |
| *ZmCLC-d* | *GRMZM2G397836* | CLC family protein | – | *Arabidopsis* | Wang, S. *et al.* (2015) |

SUMO, small ubiquitin-related modifier.

messenger mediating the osmotic response in *Arabidopsis*, which act together with $Ca^{2+}$ to mediate long-distance transduction of osmotic signals (Yuan *et al.*, 2014; Kurusu *et al.*, 2015; AbdElgawad *et al.*, 2016; Martiniere *et al.*, 2019). It is possible that such a ROS-mediated osmotic response is conserved in maize as well, but it is yet to be confirmed. Taking this into consideration it may be said that the exact mechanism of how a maize crop senses salinity-induced osmotic stress and subsequently converts it into second signalling messengers, along with the mechanism of decoding the second messengers to activate/inactivate the downstream responses, remain largely unknown. This happens to be the major bottleneck restricting the breeding of maize varieties with increased tolerance to salinity-induced osmotic stress.

### 14.2.2 Abscisic acid-mediated response to salinity-induced osmotic stress in maize

The stress phytohormone abscisic acid (ABA) plays important roles in plants' response to environmental stresses, especially in the responses to osmotic stress and the regulation of stomatal movement (Yoshida *et al.*, 2014; Munemasa *et al.*, 2015; Sah *et al.*, 2016). In *Arabidopsis*, it has been shown that osmotic stress increases the biosynthesis of ABA, which binds to receptors and activates downstream signalling pathways (e.g. the ABA–PYL (pyrabactin resistance-like)–OST1 (open stomata 1) –SLAC1 (slow anion channel-associated 1) pathway) (Furihata *et al.*, 2006; Fujii *et al.*, 2009; Fujii and Zhu, 2009; Fujita *et al.*, 2009; Yoshida *et al.*, 2014) regulating the transcript levels of many osmotic responsive genes. This promotes the tolerance to salinity-induced osmotic stress. In maize, studies are increasingly demonstrating that ABA signalling confers the transcriptional regulation of numerous osmotic responsive genes. For example, while osmotic stress treatment increases ABA biosynthesis in maize (Bahrun *et al.*, 2002; Jia *et al.*, 2002), the overexpression of *AtLOS5* (a key regulator of ABA biosynthesis) in maize increases the ABA level, promotes water uptake and maintains a better water status under salinity conditions (Zhang *et al.*, 2016). ABA treatment increases the transcript levels of maize *BADH* (*Betaine Aldehyde Dehydrogenase*) promoting the synthesis of the osmoregulator glycine betaine (Zhang *et al.*, 2012). The proteomic analysis of the phosphoproteins in maize *vp5* mutant (deficient in ABA biosynthesis) revealed that the activation of many transcription factors, enzymes and transporters is associated with the ABA signalling pathway under osmotic stress treatment (Hu *et al.*, 2015). These observations together suggest that the ABA-dependent signalling pathways play important roles in the response to salinity-induced osmotic stress in maize. However, large gaps remain in our knowledge of ABA-mediated tolerance to salinity-induced osmotic stress.

In *Arabidopsis*, the understanding of ABA-mediated regulation of stomatal movement and its responses to salt stress has progressed greatly in recent years. A likely model is that, under normal condition, PP2Cs (type 2C protein phosphatases) bind and dephosphorylate OST1 protein kinase, a member of the SnRKs (sucrose non-fermenting-1 related kinases) family, repressing the kinase activity of OST1. This keeps stomata open. Under salinity condition, the ABA concentration in guard cells increases, ABA binds to the PYR/PYL/RCAR (pyrabactin resistance/pyrabactin resistance 1-like/regulatory component of ABA receptors), the receptors then bind and inhibit the activity of PP2Cs, thereby releasing the kinase activity of OST1. The active OST1 then phosphorylates and activates the SLAC1 anion channel, which in turn promotes anion efflux and stomatal closure (Geiger *et al.*, 2009; Chen *et al.*, 2010; Maierhofer *et al.*, 2014; Wege *et al.*, 2014). The maize plants grown under salt conditions confer reduced rates of water loss from the stomata, leading to a higher leaf temperature (Fig. 14.2a). As previous studies have shown that ABA promotes stomatal closure and salinity stress increases ABA synthesis in maize (Jossier *et al.*, 2010; Munemasa *et al.*, 2015), hence ABA signalling may promote maize stomatal closure under salinity conditions. This enables a better adaptation to salinity-induced water deficiency. The *Arabidopsis* SLAC1 anion transporter plays important roles in regulating stomatal movement by mediating anions (e.g. $Cl^-$ and $NO_3^-$) efflux in guard cells (Geiger *et al.*, 2009; Chen *et al.*, 2010; Brandt *et al.*, 2012). Similarly, the maize SLAC1 orthologue (ZmSLAC1) also confers the regulation of stomatal movement by mediating $NO_3^-$ efflux in guard cells, and OST1 can phosphorylate and activate ZmSLAC1-mediated $NO_3^-$ efflux (Qi *et al.*, 2018) (Fig. 14.2b). These observations suggest that the OST1–ZmSLAC1–$NO_3^-$ pathway may confer the regulation of stomatal movement in response to salinity-induced osmotic stress in maize, which may act at the downstream of ABA perception.

### 14.2.3 Accumulations of osmoregulatory substances promote salt tolerance in maize

In order to cope with the low osmotic potential caused by high concentration of soil salts, plants increase the concentrations of numerous organic or inorganic osmoticums in cytoplasm, which then reduces the difference in osmotic pressure outside and inside the plasma membrane, thus facilitating water uptake, the maintenance of cell turgor and the adaption to salinity-induced osmotic stress (Munns and Tester, 2008; Yang and Guo, 2018; Nadeem *et al.*, 2019). In general, $Cl^-$ and $K^+$ are labelled as the main inorganic osmoticums, which are 'cheap' osmoregulatory substances compared with the organic osmotic solutes synthesized via various energy-costing processes (Sanders, 1981). However, the organic substances also play essential roles (Rodriguez *et al.*, 1997). Existing studies have showed that salinity stress increases the contents of almost all known organic osmoticums, such as alanine, glutamate, asparagine, glycine betaine and sucrose (Gavaghan *et al.*, 2011), with some of them having been shown to confer osmotic tolerance in maize.

Glycine betaine is one of the most important inorganic osmoregulatory substances (Munns

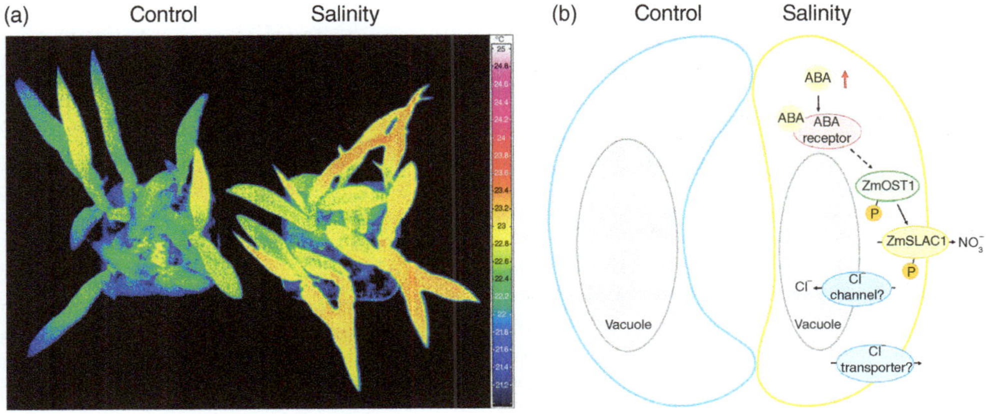


**Fig. 14.2.** ABA-mediated regulation of stomatal movement under salinity condition in maize. (a) False-colour infrared image of maize plants grown under control and salinity (100 mM NaCl) conditions. (b) A proposed model of the regulation of stomatal movement under salinity condition in maize.

and Tester, 2008; Nadeem *et al.*, 2019). The biosynthesis of glycine betaine in maize is induced by various environmental stresses (e.g. salinity, extreme temperature and drought stress) (Brunk *et al.*, 1989). Intriguingly, some maize inbred lines are deficient in glycine betaine (Saneoka *et al.*, 1995; Yang *et al.*, 1995). With these inbred lines, Yang *et al.* (1995) generated betaine-containing (*Bet1/Bet1*) and betaine-deficient (*bet1/bet1*) near-isogenic lines (NILs) and showed that the *Bet1/Bet1* plants confer greater salt tolerance than *bet1/bet1* plants, in terms of dry matter accumulation, leaf area expansion rate, leaf water content and carbon assimilation rate (Saneoka *et al.*, 1995). This indicates that glycine betaine is an important osmoticum promoting maize growth and development under salinity conditions. Therefore, future efforts towards identifying the genetic variation underlying the natural variations of the glycine betaine content are valuable. As studies have shown that there were some insertions and deletions in the exon of *BADH* of some cereal crops (e.g. rice) and these natural variations lead to premature termination of the protein translation (Niu *et al.*, 2007), it will be valuable to determine whether the natural variations of glycine betaine contents in maize are associated with *BADH*, especially for the purpose of breeding salt-tolerant maize by increasing the content of glycine betaine.

Free amino acids are important inorganic osmoregulatory substances and salinity stress significantly increases the contents of numerous free amino acids in maize (Chyzhykova and Palladina, 2006), suggesting that the accumulation of free amino acids may promote salinity tolerance. In agreement with this perspective, Voetberg and Sharp (1991) showed that the osmotic stress induced an up to tenfold increase of proline content in the root growth zone of maize seedling (Voetberg and Sharp, 1991), and the *Arabidopsis* plants overexpressing *ZmDi19-1* conferred higher proline contents and were more tolerant to salinity stress (Zhang, X. *et al.*, 2019). This further supports the notion that the accumulation of proline promotes maize salt tolerance.

As the accumulations of osmoregulatory substances promote maize salt tolerance, a possible strategy for breeding salt-tolerant maize is increasing the concentrations of osmoticums. Chimenti *et al.* (2006) screened 20 inbred lines grown under water-deficiency conditions, identified the lines with the highest and lowest osmolytes, then crossed them and obtained high-osmolytes (average of 0.47 MPa) and low-osmolytes (average of 0.06 MPa) populations. The follow-up study showed that the high-osmolytes population exhibited higher leaf area duration and grain yields under drought conditions (Chimenti *et al.*, 2006), suggesting that increasing the contents of osmoticums provides a strategy for breeding drought-tolerant maize. Expectantly, such a programme may facilitate the breeding of maize varieties tolerant to salinity-induced osmotic stress.

## 14.3 Sodium Homeostasis under Salinity Stress in Maize

### 14.3.1 The maintenance of $Na^+$ homeostasis is essential for maize salt tolerance

$Na^+$ is the most abundant soluble cation in the salinity farmlands (Munns and Tester, 2008). Under salinity conditions, plants take up an excessive amount of $Na^+$, which leads to ion toxicity by competing with $K^+$ and causing deregulation of numerous important physiological processes (Munns and Tester, 2008; Park *et al.*, 2016). Therefore, the maintenance of low $Na^+$ concentration and $Na^+$:$K^+$ ratio in the cytoplasm is essential for plant salt tolerance, especially for the glycophytic species (Zhu, 2003; Deinlein *et al.*, 2014; Yang and Guo, 2018). It has been shown that, under salinity conditions, the uptake and delivery of $Na^+$ from the soil to shoot involve the following consecutive steps. First, $Na^+$ enters root epidermal cells via non-selective cation channels and probably also via HKTs (high-affinity $K^+$ transporters) (e.g. AtHKT1) (Tester and Davenport, 2003; Munns and Tester, 2008). Second, $Na^+$ moves across the endodermis by apoplastic and symplastic pathways and then enters the stele. Finally, $Na^+$ is loaded from the stelar cells into the xylem vessels and is delivered to the shoot via the transpiration stream (Munns and Tester, 2008). Meanwhile, there are mechanisms that restrict the amount of $Na^+$ delivered to the shoot and these mechanisms are mainly conferred by $Na^+$-selective transporters. For instance, the plasma-membrane NHXs ($Na^+$/$H^+$ antiporters) (e.g. AtSOS1) act at the soil–cell interface to pump $Na^+$ back into the soil (Shi *et al.*, 2002; Zhu, 2002; Yang and Guo, 2018). The HKT1 family transporters act at the cell–xylem interface to retrieve $Na^+$ from the xylem vessels, thus reducing root-to-shoot $Na^+$ delivery (Moller *et al.*, 2009). The $Na^+$ transporters located on tonoplasts mediate cytoplasm-to-vacuole compartmentalization of $Na^+$ (Apse *et al.*, 2003). These mechanisms act together to maintain the cytoplasmic $Na^+$ concentration at non-toxic level, thereby promoting salt tolerance (Zhu, 2002; Munns and Tester, 2008; Yang and Guo, 2018). The maintenance of $Na^+$ homeostasis is essential for maize salt tolerance as well (Hajibagheri *et al.*, 1989; Schubert and Läuchli, 1990; de Azevedo *et al.*, 2004; Akram *et al.*, 2007). Notably, recent studies are increasingly showing that the variations in salt tolerance among natural maize inbred lines are substantially attributable to the differences in $Na^+$ contents (Zhang *et al.*, 2018; Zhang, M. *et al.*, 2019; Cao *et al.*, 2020), suggesting that the identification and application of the genetic variants promoting $Na^+$ homeostasis could provide a feasible strategy of improving maize salt tolerance. The following sections introduce current understanding of the mechanisms regulating $Na^+$ homeostasis under salinity conditions in maize.

### 14.3.2 Regulation of $Na^+$ homeostasis at soil–cell interface in maize

The regulation of $Na^+$ homeostasis at the soil–cell interface includes two distinct processes: (i) soil-to-cell $Na^+$ influx; and (ii) cell-to-soil $Na^+$ efflux (Tester and Davenport, 2003). It has been suggested that, under conditions with high concentration of $Na^+$, the soil-to-cell $Na^+$ influx is a passive process (Munns and Tester, 2008). Research has shown that the soil-to-cell $Na^+$ influx in maize is sensitive to $Ca^{2+}$ treatment, supporting the notion that the $Na^+$ influx is dependent (at least partially) upon non-selective cation channels (NSCCs) (see Fig. 14.3 below) (Roberts and Tester, 1997; Tyerman and Skerrett, 1999). In addition, while the studies in rice showed that the OsHKT2;1 can mediate $Na^+$ influx into root under $K^+$ starvation (Horie *et al.*, 2007), its maize orthologue (ZmHKT2) showed undetectable $Na^+$-transporting activity (Cao *et al.*, 2019). This suggests the functional divergence of HKT2 family transporters. Moreover, as numerous $K^+$-selective channels can mediate $Na^+$ influx (e.g. AtHAK5) (Li *et al.*, 2018), it remains unexclusive that some maize $K^+$ transporters can mediate $Na^+$ influx into root under salinity conditions.

Most of the $Na^+$ that enters root cells is transported back to soil by plasma-membrane $Na^+$/$H^+$ antiporters (Tester and Davenport, 2003). In *Arabidopsis*, the $Na^+$/$H^+$ antiporter SOS1 is the major transporter mediating cell-to-soil efflux of $Na^+$. The activity of SOS1 is dependent upon the $H^+$ gradient across the plasma membrane and is regulated by additional factors (Shi *et al.*, 2000;

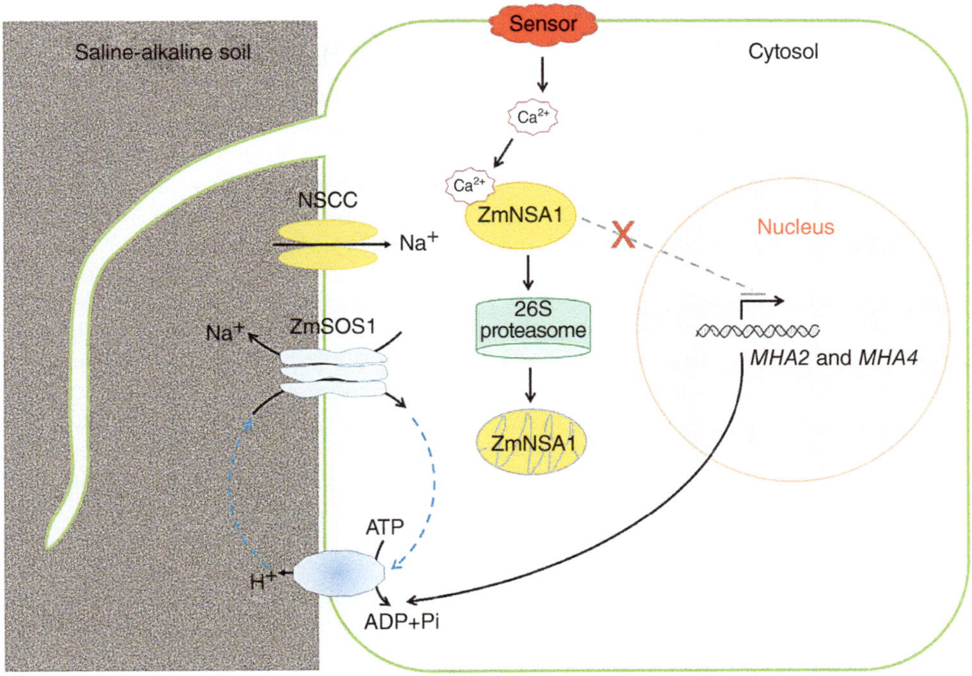


**Fig. 14.3.** An updated model of the maintenance of $Na^+$ homeostasis at the soil–cell interface.

Yang and Guo, 2018). A simple model is that, under salinity treatment, the concentration of cytoplasmic $Ca^{2+}$ increases. The change of $Ca^{2+}$ concentration is decoded by the $Ca^{2+}$-binding protein SOS3, SOS3 then binds to and activates SOS2, the active SOS3/SOS2 complex then phosphorylates and activates the $Na^+/H^+$ antiporter SOS1 (Halfter *et al.*, 2000; Qiu *et al.*, 2002; Yang *et al.*, 2019). As the SOS pathway has been identified in many other species (e.g. rice and barley), this suggests that the SOS pathway is an important and conserved mechanism mediating $Na^+$ homeostasis under salinity conditions (Martínez-Atienza *et al.*, 2007).

The orthologues of major components of the SOS pathway have been identified in maize, including one AtSOS1 orthologue, 43 CIPKs (CBL-interacting protein kinases) and 11 CBLs (Chen *et al.*, 2011). While Shi and others showed that the overexpression of *AtSOS1* improves salt tolerance in *Arabidopsis* (Shi *et al.*, 2000, 2003), Selvakumar *et al.* (2018) observed that mycorrhizal colonization promotes $Na^+$ homeostasis likely by increasing the transcript levels of *ZmSOS1* (*Zea mays salt overly sensitive 1*) in maize. This supports the perspective that ZmSOS1 mediates $Na^+$ homeostasis and salt tolerance in maize. In addition, Luo *et al.* (2017) showed that natural variation of the SOS pathway may be associated with the variation of salt tolerance between maize inbred lines PH6WC and PH4CV. In that study, the authors used plant height of salt-grown maize plant (SPH) and plant height-based salt tolerance index (ratio of plant height between saline and control fields, PHI) as the phenotypes to map salt tolerance QTLs and observed a major QTL for both SPH and PHI on chromosome 1. The study further suggested that the candidate genes of the QTL are *GRMZM2G007555* and *GRMZM2G098494*, which are the orthologues of AtSOS3 and AtSOS1, respectively. These results together suggest that the SOS pathway plays an important role in the regulation of $Na^+$ homeostasis under salinity condition in maize, with the potential for favourable variations of SOS pathway factors to provide tools for improving maize salt tolerance.

The $Na^+$-transporting activity of the SOS1 $Na^+/H^+$ antiporter is dependent upon the $H^+$ gradient across the plasma membrane (membrane potential) (Yang and Guo, 2018; Yang *et al.*, 2019). Under saline-alkaline stress treatment, the high-pH environment surrounding the roots reduces the membrane potential and thus attenuates the function of SOS1 (Yang and Guo, 2018). The maize plants grown under saline-alkaline ($NaHCO_3$) condition confer higher concentrations of tissue $Na^+$ compared with those grown under saline (NaCl) condition (Guo *et al.*, 2017; Cao *et al.*, 2020), which is consistent with the notion that high pH mars the $Na^+$ homeostasis and boosts the $Na^+$ toxicity. A recent study by Cao *et al.* (2020) identified an EF-hand $Ca^{2+}$-binding protein-coding gene *ZmNSA1* (*$Na^+$ Content under Saline-Alkaline Condition*), which confers natural variation of shoot $Na^+$ content under saline-alkaline ($NaHCO_3$) condition, by genome-wide association study (GWAS) analysis in maize. The loss-of-function mutant *ZmNSA1$^{UFMu}$* shows lower shoot $Na^+$ content and is more tolerant to saline-alkaline stress compared with the wild types, while the *ZmNSA1*-overexpressing plants are hypersensitive to saline-alkaline stress. The study observed that the functional variation of ZmNSA1 is ascribed to a 4 bp deletion located in the 3′-untranslated region (3′-UTR) of *ZmNSA1*, which decreases the abundance of ZmNSA1 protein by reducing the translation efficiency of *ZmNSA1* mRNA. The saline-alkaline treatment has insignificant effect on the transcript levels of *ZmNSA1* but triggers the degradation of ZmNSA1 protein. ZmNSA1 enables the identification of a mechanism regulating $Na^+$ homeostasis under saline-alkaline condition: namely, under saline-alkaline treatment, the concentration of cytosolic $Ca^{2+}$ increases, $Ca^{2+}$ binds to ZmNSA1 and triggers its degradation via the 26S proteasome pathway, then increases the transcript levels of maize plasma-membrane $H^+$-ATPase (*MHA2* and *MHA4*), thereby promoting root $H^+$ efflux and enhancing root-to-soil $Na^+$ efflux mediated by SOS1 $Na^+/H^+$ antiporter (Fig. 14.3) (Cao *et al.*, 2020). ZmNSA1 provides a target for breeding saline-alkaline tolerance in maize varieties, by either gene editing or marker-assisted breeding.

### 14.3.3 Regulation of root-to-shoot Na⁺ delivery at the cell–xylem vessel interface in maize

Similar to the observations in other glycophytic species, the shoot $Na^+$ exclusion is essential for the salt tolerance of maize (Zhu, 2002; Munns *et al.*, 2012; Zhang *et al.*, 2018; Zhang, M. *et al.*, 2019). Current knowledge has shown that $Na^+$ retrieval from the transpiration stream at the cell–xylem vessel interface is an important regulatory process determining shoot $Na^+$ exclusion (Berthomieu *et al.*, 2003; Ren *et al.*, 2005; Zhang *et al.*, 2018; Zhang, M. *et al.*, 2019). It has been shown that the HKT1 family of $Na^+$-selective transporters promotes shoot $Na^+$ exclusion by retrieving $Na^+$ from the xylem vessels (Ren *et al.*, 2005; Sunarpi *et al.*, 2005; Davenport *et al.*, 2007; Munns *et al.*, 2012) and the variations of the *HKT1* locus confer the natural variation in shoot $Na^+$ exclusion in numerous species, including *Arabidopsis*, rice, wheat, maize and others (Ren *et al.*, 2005; Munns *et al.*, 2012). Recently, Zhang *et al.* (2018) showed that the maize HKT1 family transporter (ZmHKT1) confers the variation in shoot $Na^+$ exclusion between the inbred lines Zheng58 and Chang7-2. In that study, the authors observed that Chang7-2 is hypersensitive to salt stress due to the excessive accumulation of $Na^+$ in the shoot tissue and identified *ZmNC1* (*$Na^+$ Content 1*) as a major QTL regulating shoot $Na^+$ exclusion, which encodes the HKT1 family transporter ZmHKT1 (GRMZM2G047616). A 390 bp LTR/Gypsy retrotransposon insertion results in frameshifting and truncation of ZmHKT1, which then leads to excessive shoot $Na^+$ accumulation and salt hypersensitivity. The plants lacking ZmHKT1 accumulate more $Na^+$ in the shoot and are more sensitive to salt stress compared with the wild types. As *ZmHKT1* is preferentially expressed in the root stele (including the parenchyma cells) and encodes a plasma membrane-localized $Na^+$-selective transporter, the authors suggested that ZmHKT1 promotes leaf $Na^+$ exclusion and salt tolerance by withdrawing $Na^+$ from the xylem vessels (see Fig. 14.4 below). As the favourable *HKT1* alleles in wheat have been shown to increase yield, the favourable allele of maize *ZmHKT1* may provide a tool for breeding salt-tolerant maize varieties by marker-assisted breeding (Zhang *et al.*, 2018).

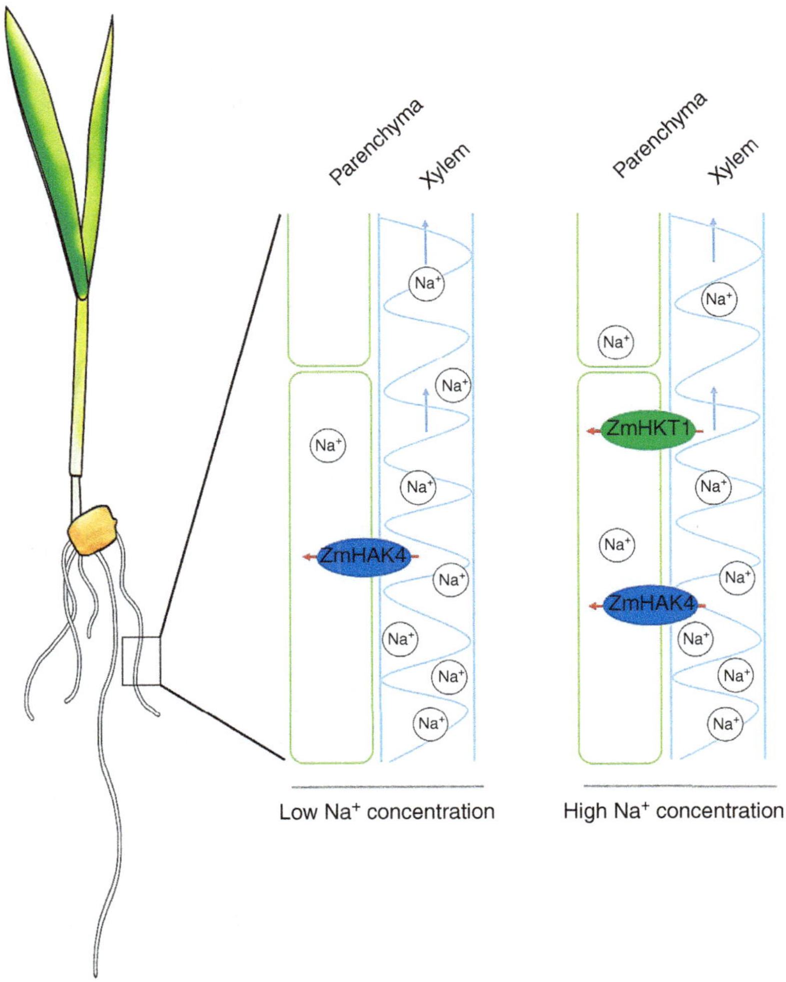


**Fig. 14.4.** A proposed model of ZmHKT1- and ZmHAK4-mediated retrieval of xylem $Na^+$.

The HAK (high-affinity $K^+$ transporter)/KUP ($K^+$ uptake transporter)/KT ($K^+$ transporter) families of transporters were first observed in fungi as a $K^+$ transporter and then their homologues have been identified from almost all tested organisms (Li *et al.*, 2018). The studies in *Arabidopsis* and rice have shown that HAK/KUP/KT transporters are involved in root $K^+$ acquisition and root-to-shoot $K^+$ translocation (Rubio *et al.*, 2008). In contrast, a recent study in maize showed that the HAK/KUP/KT family transporter ZmHAK4 is an $Na^+$-selective transporter regulating the root-to-shoot translocation of $Na^+$ (Zhang, M. *et al.*, 2019). In that study, the authors identified *ZmNC2* (*$Na^+$ Content 2*), which encodes the HAK family transporter ZmHAK4, as a major QTL regulating shoot $Na^+$ content under salinity condition by GWAS analysis. A natural *ZmHAK4*-deficient allele, containing a 12,586 bp insertion, shows decreased *ZmHAK4* transcript level and increased shoot $Na^+$ content. The mutants lacking *ZmHAK4* confer greater shoot $Na^+$ content and are hypersensitive to salt stress compared with the wild types. Given *ZmHAK4* is preferentially expressed in root stele and encodes an $Na^+$-selective transporter mediating inward $Na^+$ transport, the authors proposed that ZmHAK4 promotes shoot $Na^+$ exclusion by retrieving $Na^+$ from the xylem vessels at the cell–xylem vessel interface, which is in the same way as ZmHKT1 (Zhang, M. *et al.*, 2019) (Fig. 14.4). In addition, they showed that lack of *ZmHAK4* leads to defect of shoot $Na^+$ exclusion under conditions with $Na^+$ concentration ranging from submillimolar to over 100 mM, but *ZmHKT1* mutants show such a defect only under high $Na^+$ concentration (e.g. over 100 mM) (Fig. 14.4). This suggests that ZmHAK4 and ZmHKT1 likely confer distinct roles in retrieving $Na^+$ from the xylem vessel. Moreover, the authors showed that orthologues of ZmHAK4 were identified in wheat and rice and many other species, suggesting that ZmHAK4 and its orthologues may identify a novel evolutionarily conserved salt tolerance mechanism (Zhang, M. *et al.*, 2019).

Taken together, two categories of $Na^+$-selective transporters (ZmNC1/ZmHKT1 and ZmNC2/ZmHAK4) act at the cell–xylem vessel interface to promote shoot $Na^+$ exclusion and salt tolerance. Future breeding programmes could combine the favourable alleles of *ZmNC1/ZmHKT1* and *ZmNC2/ZmHAK4* in high-yielding maize backgrounds to develop commercial salt-tolerant maize varieties.

### 14.3.4 $Na^+$ translocation within and between tissues in maize

The $Na^+$ translocation within and between tissues to compartmentalize it into tissues that are less sensitive to $Na^+$ toxicity provides additional strategies of avoiding $Na^+$ toxicity. For example, the wheat salt tolerance QTL *Nax1* promotes $Na^+$ loading into the xylem in leaf sheaths, thus keeping low $Na^+$ concentration in leaf blades (James *et al.*, 2006). The rice *OsHKT1;1* expression is predominantly detected in the vicinity of the xylem and phloem in leaves, and the $Na^+$ content in both xylem sap and phloem sap are increased in the mutant under salt conditions (Wang, R. *et al.*, 2015; Ismail and Horie, 2017), suggesting that OsHKT1;1 may contribute to both $Na^+$ retrieval from the xylem and loading into the phloem for $Na^+$ recirculation. The *Arabidopsis* AtHKT1;1 can load $Na^+$ into the phloem then promote shoot-to-root recirculation of $Na^+$ (Berthomieu *et al.*, 2003). In maize, it has been shown that the plants tend to compartmentalize $Na^+$ into stems and leaf sheaths, thus maintaining low $Na^+$ accumulation in the leaf blade which is more sensitive to salt stress (Isla and Aragues, 2010). However, the mechanisms that underlie $Na^+$ compartmentalization within and between tissues in maize remain largely unknown.

### 14.3.5 $Na^+$ compartmentalization into tonoplast in maize

Existing knowledge has indicated that compartmentalization of $Na^+$ into vacuoles is important for plant cells to maintain low cytoplasmic $Na^+$ (Yang and Guo, 2018). In *Arabidopsis*, the $Na^+/H^+$ antiporter NHX1 located in the tonoplast membrane can mediate $Na^+$ compartmentalization into the vacuole, and its activity is dependent upon the $H^+$ gradient outside and inside the tonoplast (Apse *et al.*, 2003; Yang and Guo, 2018). There are six *NHX* family genes (*ZmNHX1–6*) in maize, with the transcript level of some *ZmNHXs* upregulated by salt stress (Zorb *et al.*, 2005). It is therefore possible that the $Na^+$ compartmentalization into vacuoles is mediated by the NHX family $Na^+/H^+$ antiporter in maize. Nevertheless, such a perspective is yet to be confirmed by future study.

## 14.4 Potassium Homeostasis under Salinity Conditions in Maize

$K^+$ is an essential macro element that is important for numerous physiological processes and adaptation to a broad range of environmental constraints (Shabala and Pottosin, 2014). Previous studies in *Arabidopsis* have shown that $K^+$ concentration in the cytoplasm of root cells decreases rapidly under salt stress treatment and that high cytosolic $K^+$ promotes salt tolerance (Shabala *et al.*, 2006), indicating that the maintenance of $K^+$ homeostasis is important for plant salt tolerance. Salinity stress decreases $K^+$ uptake and increases $K^+$ efflux mainly by inducing membrane depolarization, which impairs the activity of the $K^+$-transporting channels (e.g. AKT1) and causes $K^+$ efflux via $K^+$-selective outward-rectifying $K^+$ currents (KORC) (Shabala *et al.*, 2006; Shabala and Pottosin, 2014). Under salinity conditions, the factors mediating xylem loading of $K^+$ (e.g. AtKUP, OsHAK1, OsHAK5 and OsAKT1) are essential for the root-to-shoot $K^+$ translocation, shoot $K^+$ supply and salt tolerance in *Arabidopsis* and rice (Very and Sentenac, 2003; Wang and Wu, 2017). Similarly, the maize hybrids (Pioneer 32B33 and Pioneer 30Y87) with higher $K^+$ content and $K^+$:$Na^+$ ratio are more tolerant to salt stress in terms of yield of biomass (Akram *et al.*, 2007), indicating that the $K^+$ homeostasis is important for maize salt tolerance.

Class II HKTs (HKT2) act as $K^+$-selective transporters or $Na^+/K^+$ symporters (Munns and Tester, 2008), with their ion preferences being partially dependent upon the concentrations of $Na^+$ and $K^+$ in the environment (Horie *et al.*, 2001; Yao *et al.*, 2010). The maize genome encodes only one HKT2 family transporter. Recently, Cao *et al.* (2019) showed that *ZmHKT2* is a major QTL responsible for the variation in shoot $K^+$ content between the inbred line W22 and the progenitor grass teosinte. *ZmHKT2* encodes a $K^+$-selective transporter, which reduces shoot $K^+$ content likely by removing $K^+$ from xylem flow. The plants lacking *ZmHKT2* confer higher $K^+$ concentration in xylem sap and shoot tissues compared with those of the wild types. A non-synonymous variation (SNP389) accounts for the functional divergence of *ZmHKT2*, with the maize (W22) allele showing a decreased $K^+$-transporting activity, thereby resulting in increased shoot $K^+$ content and salt tolerance. *ZmHKT2* provides a potential gene target for improving maize salt tolerance by promoting $K^+$ homeostasis (Cao *et al.*, 2019).

## 14.5 Regulation of Chloride Homeostasis in Maize

### 14.5.1 Maize is sensitive to high concentration of $Cl^-$

$Cl^-$ is a nutritional element essential for numerous physiological processes like stomatal movement, maintenance of cell turgor and photosynthesis (Felle, 1994; Rivalta *et al.*, 2011; Wege *et al.*, 2014; Franco-Navarro *et al.*, 2016). $Cl^-$ is also the most abundant anion in salinized farmlands, excess of which is harmful to most plants. Therefore, plants have evolved multiple strategies to avoid $Cl^-$ toxicity (Li *et al.*, 2016a). First, root-to-soil $Cl^-$ efflux is essential for $Cl^-$ homeostasis under high $Cl^-$ conditions. For example, previous studies in *Populus euphratica* showed that $Cl^-$ efflux in the root of salt-tolerant lines is significantly higher than that in salt-sensitive lines (Sun *et al.*, 2009). Second, $Cl^-$ can be compartmentalized into vacuoles which then reduces the $Cl^-$ concentration in the cytoplasm. Third, studies in citrus and grapevine showed that the $Cl^-$ concentration in the shoot of salt-tolerant varieties is lower than in salt-sensitive lines, suggesting that reducing root-to-shoot $Cl^-$ translocation promotes shoot $Cl^-$ homeostasis and salt tolerance (Storey and Walker, 1999; Storey *et al.*, 2003). Moreover, studies conducted in barley indicated that $Cl^-$ is preferentially accumulated in the leaf epidermis cells rather than the mesophyll cells that are more sensitive to $Cl^-$ toxicity (Huang and Van Steveninck, 1989; Fricke *et al.*, 1996). This indicates that $Cl^-$ compartmentalization into $Cl^-$-tolerant tissues promotes tolerance to high $Cl^-$ stress. Existing knowledge indicates that these $Cl^-$ homeostasis mechanisms are substantially conferred by transporters with $Cl^-$-transporting activities. For example, several *Arabidopsis* $Cl^-$ transporters such as CLCs (chloride channels), ALMTs (aluminium-activated malate transporters) and DTX (detoxification efflux carrier)/MATEs (multidrug and toxic compound extrusion transporters), located in the tonoplast of guard cells, mediate $Cl^-$ flux from the cytoplasm to vacuoles (Pineros *et al.*, 2008; Jossier *et al.*, 2010; Ligaba *et al.*, 2012; Nguyen *et al.*, 2016; Zhang *et al.*, 2017). The $Cl^-$ transporters (e.g. some ALMT family members) reducing $Cl^-$ loading into the xylem vessels prevent root-to-shoot translocation of $Cl^-$ (Li *et al.*, 2016b).

Maize is sensitive to high concentration of $Cl^-$, hence understanding the mechanisms by which maize copes with high $Cl^-$ stress is important, essentially for the purpose of improving maize salt tolerance. In recent years, the maize orthologues of numerous $Cl^-$ transporters have been identified, including nitrate transporter/peptide transporter protein (NPF), slow-type anion channel-associated homologues (SLAHs), CLCs, ALMTs, cation/chloride cotransporters (CCCs) and MATEs (Li *et al.*, 2017; Zhang *et al.*, 2017). The following section reviews the molecular understanding of the function of these $Cl^-$ transporters.

### 14.5.2 Current understanding of $Cl^-$ transporters in maize

*Peptide transporter proteins (NPFs)*

The *Arabidopsis* $Cl^-$ transporter AtNPF2.4 mediates the loading of $Cl^-$ into xylem (Li *et al.*, 2016b). The $Cl^-$ contents in the shoot tissues of *AtNPF2.4* mutants are significantly lower than those in the wild types under high $Cl^-$ conditions. *AtNPF2.4* mainly expresses in the vascular bundle of primary and lateral roots, and the transcript levels of *AtNPF2.4* decrease dramatically under salt stress. In contrast, AtNPF2.5 promotes $Cl^-$ efflux in the root under high $Cl^-$ conditions and its transcript level is induced significantly by salt stress, with plants overexpressing *AtNPF2.5* being hypersensitive to high $Cl^-$ stress due to excessive accumulation of $Cl^-$ (Li *et al.*, 2016a). In maize, the functions of ZmNPF6.4 and ZmNPF6.6 have been characterized. ZmNPF6.6 is a high-affinity $NO_3^-$ transporter which is also capable of transporting $Cl^-$ in the low-affinity range. ZmNPF6.6 can transport $Cl^-$ when the $NO_3^-$ concentration is low or absent, and this transport capacity is strongly inhibited by $NO_3^-$, possibly due to $NO_3^-$ occupation of the substrate-binding site (Wen *et al.*, 2017). Instead, ZmNPF6.4 is a high-affinity $Cl^-$-selective transporter and $NO_3^-$ cannot inhibit its $Cl^-$-transporting activity. Similar to the discoveries in *Arabidopsis*, the differences in the affinities to $Cl^-$ or $NO_3^-$ between ZmNPF6.4 and ZmNPF6.6 are associated with a single determinative amino acid. ZmNPF6.4 contains a Tyr370 residue which makes it a high-affinity $Cl^-$-selective transporter. Mutation of the NPF6.4 Tyr370 to His (Y370H) changes its selectivity from $Cl^-$ to $NO_3^-$. These results together suggest that many maize NPF family transporters mediate $Cl^-$ transport with different affinities. Future studies demonstrating the physiological roles of ZmNPF6.4, ZmNPF6.6 and other maize NPF family transporters will be valuable for the understanding of maize response to high $Cl^-$ stress (Wen *et al.*, 2017).

*Slow-type anion channel-associated homologues (SLAHs)*

In *Arabidopsis*, *AtSLAH1* and *AtSLAH3* encode SLAH family transporters regulating $Cl^-$ translocation from root to shoot. The $Cl^-$ concentrations in the shoots of *slah1* and *slah3* mutants are significantly lower than in the wild types under salinity conditions. AtSLAH1 is a $Cl^-$ transporter mediating the loading of $Cl^-$ into xylem. AtSLAH3 is an $NO_3^-$ transporter, but it can interact with SLAH1 to form a heterotrimer and then increase the $Cl^-$-transporting activity of AtSLAH1 (Cubero-Font *et al.*, 2016; Qiu *et al.*, 2016). *Arabidopsis* SLAC1 is a transporter with high permeability to $NO_3^-$ and low permeability to $Cl^-$ which plays an important role in regulating the stomatal movement (Geiger *et al.*, 2009; Brandt *et al.*, 2012). The maize orthologue of AtSLAC1 has been identified, which is permeable to $NO_3^-$ but not $Cl^-$. The plants lacking *ZmSLAC1* show a defect of stomatal closure under salt stress and ABA treatment (Qi *et al.*, 2018). Still, whether other maize SLAH family transporters are capable of transporting $Cl^-$, and their roles in responses to high $Cl^-$ stress, remain unknown.

*Chloride channels (CLCs)*

Previous studies have shown that CLC family transporters show different patterns of subcellular localization. For example, AtCLCa, AtCLCb, AtCLCc and AtCLCg are located in tonoplast, AtCLCd and AtCLCf are located in Golgi, AtCLCe is located in chloroplast membranes. The functions of CLC family transporters have been associated with turgor maintenance, stomatal movement, nutrient transport and metal tolerance, with CLCa, CLCc and CLCg able to regulate stomatal movement by mediating $Cl^-$ compartmentalization into vacuoles (Hechenberger *et al.*, 1996; Grattan and Grieve, 1999; Diedhiou and Golldack, 2006; Marmagne *et al.*, 2007; Jossier *et al.*, 2010; Nguyen *et al.*, 2016). ZmCLCd is the orthologue of AtCLCd (Yang *et al.*, 2011). The *Arabidopsis* plants overexpressing *ZmCLCd* are more tolerant to salinity stress and accumulate less $Cl^-$ in root and shoot tissue compared with the wild types (Wang, S. *et al.*, 2015), suggesting that ZmCLCd may confer $Cl^-$ homeostasis and salt tolerance in maize. However, the precise salt tolerance roles of ZmCLCd and other maize CLC family transporters are largely unknown.

*Aluminium-activated malate transporters (ALMTs)*

In *Arabidopsis*, ALMT family transporters are located in either the plasma membrane or tonoplast, capable of transporting organic and inorganic anions (Hoekenga *et al.*, 2006; Kovermann *et al.*, 2007; Meyer *et al.*, 2010, 2011; De Angeli *et al.*, 2013). Previous studies have shown that ALMT family transporters are involved in numerous biological processes. For example, AtALMT1 and TaALMT1 can bind $Al^{3+}$ then activate the malate efflux, thereby promoting $Al^{3+}$ tolerance (Sasaki *et al.*, 2004; Hoekenga *et al.*, 2006). *AtALMT9* mainly expresses in the vasculature, its transcript level can be induced by NaCl treatment, and the $Cl^-$ content in the shoots of *almt9* mutants is lower than that in the wild types under salt condition (Baetz *et al.*, 2016). ALMT family transporters have also been shown to be involved in the regulation of stomatal movement, anion homeostasis, fruit quality and seed development (Meyer *et al.*, 2011; De Angeli *et al.*, 2013; Sharma *et al.*, 2016). The maize *ZmALMT1* mainly expresses in the maturation zone of the root. ZmALMT1 is located in the plasma membrane and is capable of transporting malate, citrate, $Cl^-$, $NO_3^-$ and $SO_4^{2-}$ (Pineros *et al.*, 2008). ZmALMT2 is characterized as an anion (e.g. $NO_3^-$ and $Cl^-$) transporter that is capable of transporting malate and $NO_3^-$ under high $Al^{3+}$ condition (Krill *et al.*, 2010; Ligaba *et al.*, 2012). Given that many maize ALMT proteins are capable of transporting $Cl^-$, it is therefore possible that these transporters may be involved in the maintenance of $Cl^-$ homeostasis under high $Cl^-$ stress conditions; nevertheless, such a perspective is yet to be confirmed by future study.

### *Multidrug and toxic compound extrusion (MATE) family transporters*

Similar to AMLT family transporters, MATE family transporters are able to transport malate and citrate in response to Al stress and drought stress (Omote *et al.*, 2006; Marinova *et al.*, 2007; Zhang *et al.*, 2014, 2017). Meanwhile, many MATE family transporters are also permeable to $Cl^-$. For example, the *Arabidopsis* MATE family transporters AtDTX33 and AtDTX35 are located in the tonoplast of root hairs and guard cells, regulating cell turgor under drought condition by mediating $Cl^-$ efflux (Zhang *et al.*, 2017). The maize genome encodes 49 MATE family transporters, they show large differences in tissue specificity of expression (Zhu *et al.*, 2016). Previous studies have identified *ZmMATE1* and *ZmMATE2* as the candidate genes of Al tolerance QTLs (Maron *et al.*, 2010; Guimaraes *et al.*, 2014), they show citrate- and malate-transporting activities, but with no detectable $Cl^-$-transporting activity (Maron *et al.*, 2010). However, the roles of maize MATE family transporters in the regulation of $Cl^-$ homeostasis under high $Cl^-$ conditions remain unknown.

### *Cation/chloride cotransporters (CCCs)*

The CCC family proteins are cation/$Cl^-$ cotransporters. According to the kinds of coupled cations, CCC family proteins are divided into three categories: (i) $Na^+$–$Cl^-$ cotransporter (NCC); (ii) $K^+$–$Cl^-$ cotransporter (KCC); and (iii) $Na^+$–$K^+$–$Cl^-$ cotransporter (NKCC) (Russell, 2000). The *Arabidopsis AtCCC* encodes an NKCC. The plants lacking *AtCCC* show higher shoot $Cl^-$ content and lower root $Cl^-$ content compared with those of the wild types, suggesting that AtCCC is involved in the regulation of $Cl^-$ translocation from the root to shoot (Colmenero-Flores *et al.*, 2007). Given that the roles of maize CCCs are unclear, future efforts unravelling the function of maize CCC family proteins will facilitate the breeding of salt-tolerant maize.

## 14.6 Other Understandings of Maize Salt Tolerance

In addition to the salt tolerance genes and mechanisms described above, many other genes that may be associated with maize salt tolerance have been identified. For example, Luo *et al.* (2019) identified 57 loci significantly associated with salt tolerance in terms of survival rate by GWAS analysis, with 47 candidate genes. The authors also showed that *Arabidopsis* plants overexpressing some of the candidate genes (e.g. *SAG4/GRMZM2G077295* and *SAG6/GRMZM2G106056*) are more tolerant to salinity stress. Xie *et al.* (2019) identified eight candidate genes associated with the changes of plant height and weight under salt stress by GWAS analysis, and showed that the candidate genes (*GRMZM2G158568*, *LOC103631163*, *LOC100280148* and *LOC103651154*) exhibit significant differences in transcript levels between the salt-resistant and salt-sensitive inbred lines.

Moreover, the GWAS analysis of six shoot and root growth-related traits also identified some candidate genes that may be associated with maize salt tolerance (Sandhu *et al.*, 2020). Although these GWAS analyses identified a large number of genes that may be associated with maize salt tolerance, the functions of these genes are yet to be confirmed and the exact mechanisms remain to be elucidated by further study.

## 14.7 Future Issues for Breeding Salt-Tolerant Maize

Great progress has been made towards the molecular understanding of maize salt tolerance in recent years. Nevertheless, the breeding of commercial salt-tolerant maize hybrids remains challenging. Deciphering the following issues may facilitate the molecular breeding for combating salinity stress in maize.

**1.** Natural maize inbred lines show large variations in tolerance to high concentrations of $Na^+$ and $Cl^-$, and salt-induced osmotic stress, indicating that the identification and application of salt-tolerant genetic variations could provide a promising route of developing salt-tolerant maize varieties. Hence, future studies towards the systemic identification and functional study of salt tolerance QTLs are of great significance.

**2.** It remains largely unknown how maize senses $Na^+$, converts it into second signalling messengers (e.g. $Ca^{2+}$ and ROS), decodes the messengers and then activates/inactivates the downstream responses. Future efforts towards identifying the key signalling components of maize salt response are essential for building up the salt response network of maize.

**3.** As decades of effort have achieved significant progress in the understanding of plant salt tolerance, especially in the model species *Arabidopsis*, future reverse genetic studies determining the roles of the maize orthologues of the key factors promoting $Na^+$, $Cl^-$ and osmotic tolerances are valuable.

**4.** Existing studies of maize response to salinity stress have mainly focused on the salt tolerance aspects, with the other agronomic traits (e.g. tolerance to other stresses and yield) barely being evaluated. As it has been shown that the salt-tolerant crops often show severe yield penalties, it is essential to take other agronomic traits into consideration when developing salt tolerance in maize.

**5.** Most commercial maize varieties are hybrids and current studies of maize salt tolerance were mostly conducted using the inbred lines. Future works towards understanding of the hybrid vigour for maize salt tolerance will be valuable, especially for the purpose of developing commercial maize varieties.

## References


AbdElgawad, H., Zinta, G., Hegab, M.M., Pandey, R., Asard, H. and Abuelsoud, W. (2016) High salinity induces different oxidative stress and antioxidant responses in maize seedlings organs. *Frontiers in Plant Science* 7, 276.

Akram, M., Malik, M.A., Ashraf, M.Y., Saleem, M.F. and Hussain, M. (2007) Competitive seedling growth and $K^+/Na^+$ ratio in different maize (*Zea mays L.*) hybrids under salinity stress. *Pakistan Journal of Botany* 39, 2553–2563.

Apse, M.P., Sottosanto, J.B. and Blumwald, E. (2003) Vacuolar cation/$H^+$ exchange, ion homeostasis, and leaf development are altered in a T-DNA insertional mutant of AtNHX1, the *Arabidopsis* vacuolar $Na^+/H^+$ antiporter. *The Plant Journal* 36, 229–239.

Baetz, U., Eisenach, C., Tohge, T., Martinoia, E. and De Angeli, A. (2016) Vacuolar chloride fluxes impact ion content and distribution during early salinity stress. *Plant Physiology* 172, 1167–1181.

Bahrun, A., Jensen, C.R., Asch, F. and Mogensen, V.O. (2002) Drought-induced changes in xylem pH, ionic composition, and ABA concentration act as early signals in field-grown maize (*Zea mays L.*). *Journal of Experimental Botany* 53, 251–263.

Berthomieu, P., Conejero, G., Nublat, A., Brackenbury, W.J., Lambert, C. *et al.* (2003) Functional analysis of AtHKT1 in *Arabidopsis* shows that $Na^+$ recirculation by the phloem is crucial for salt tolerance. *The EMBO Journal* 22, 2004–2014.

Borkiewicz, L., Polkowska-Kowalczyk, L., Ciesla, J., Sowinski, P., Jonczyk, M. *et al.* (2019) Expression of maize calcium-dependent protein kinase (ZmCPK11) improves salt tolerance in transgenic *Arabidopsis* plants by regulating sodium and potassium homeostasis and stabilizing photosystem II. *Physiologia Plantarum* 168, 38–57.

Brandt, B., Brodsky, D.E., Xue, S., Negi, J., Iba, K. *et al.* (2012) Reconstitution of abscisic acid activation of SLAC1 anion channel by CPK6 and OST1 kinases and branched ABI1 PP2C phosphatase action. *Proceedings of the National Academy of Sciences USA* 109, 10593–10598.

Brunk, D.G., Rich, P.J. and Rhodes, D. (1989) Genotypic variation for glycine betaine among public inbreds of maize. *Plant Physiology* 91, 1122–1125.

Cao, Y., Liang, X., Yin, P., Zhang, M. and Jiang, C. (2019) A domestication-associated reduction in $K^+$-preferring HKT transporter activity underlies maize shoot $K^+$ accumulation and salt tolerance. *New Phytologist* 222, 301–317.

Cao, Y., Zhang, M., Liang, X., Li, F., Shi, Y., Yang, X. and Jiang, C. (2020) Natural variation of an EF-hand $Ca^{2+}$-binding-protein coding gene confers saline-alkaline tolerance in maize. *Nature Communications* 11, 186.

Chen, X., Gu, Z., Xin, D., Hao, L., Liu, C. *et al.* (2011) Identification and characterization of putative CIPK genes in maize. *Journal of Genetics and Genomics* 38, 77–87.

Chen, Y.H., Hu, L., Punta, M., Bruni, R., Hillerich, B. *et al.* (2010) Homologue structure of the SLAC1 anion channel for closing stomata in leaves. *Nature* 467, 1074–1080.

Chimenti, C.A., Marcantonio, M. and Hall, A.J. (2006) Divergent selection for osmotic adjustment results in improved drought tolerance in maize (*Zea mays L.*) in both early growth and flowering phases. *Field Crops Research* 95, 305–315.

Chyzhykova, O.A. and Palladina, T.O. (2006) The role of amino acids and sugars in supporting of osmotic homeostasis in maize seedlings under salinization conditions and treatment with synthetic growth regulators. *Ukrains'kyi Biokhimichnyi Zhurnal* 78, 124–129.

Colmenero-Flores, J.M., Martinez, G., Gamba, G., Vazquez, N., Iglesias, D.J., Brumos, J. and Talon, M. (2007) Identification and functional characterization of cation–chloride cotransporters in plants. *The Plant Journal* 50, 278–292.

Cubero-Font, P., Maierhofer, T., Jaslan, J., Rosales, M.A., Espartero, J. *et al.* (2016) Silent S-type anion channel subunit SLAH1 gates SLAH3 open for chloride root-to-shoot translocation. *Current Biology* 26, 2213–2220.

Davenport, R.J., Munoz-Mayor, A., Jha, D., Essah, P.A., Rus, A. and Tester, M. (2007) The $Na^+$ transporter AtHKT1;1 controls retrieval of $Na^+$ from the xylem in *Arabidopsis*. *Plant, Cell & Environment* 30, 497–507.

De Angeli, A., Zhang, J., Meyer, S. and Martinoia, E. (2013) AtALMT9 is a malate-activated vacuolar chloride channel required for stomatal opening in *Arabidopsis*. *Nature Communications* 4, 1804.

de Azevedo Neto, A.D., Prisco, J.T., Enéas-Filho, J., de Lacerda, C.F., Silva, J.V., da Costa, P.H.A. and Gomes-Filho, E. (2004) Effects of salt stress on plant growth, stomatal response and solute accumulation of different maize genotypes. *Brazilian Journal of Plant Physiology* 16, 31–38.

Deinlein, U., Stephan, A.B., Horie, T., Luo, W., Xu, G. and Schroeder, J.I. (2014) Plant salt-tolerance mechanisms. *Trends in Plant Science* 19, 371–379.

Diedhiou, C. and Golldack, D. (2006) Salt-dependent regulation of chloride channel transcripts in rice. *Plant Science* 170, 793–800.

Felle, H.H. (1994) The $H^+/Cl^-$ symporter in root-hair cells of *Sinapis alba* (an electrophysiological study using ion-selective microelectrodes). *Plant Physiology* 106, 1131–1136.

Fortmeier, R. and Schubert, S. (1995) Salt tolerance of maize (*Zea mays L.*): the role of sodium exclusion. *Plant, Cell & Environment* 18, 1041–1047.

Franco-Navarro, J.D., Brumos, J., Rosales, M.A., Cubero-Font, P., Talon, M. and Colmenero-Flores, J.M. (2016) Chloride regulates leaf cell size and water relations in tobacco plants. *Journal of Experimental Botany* 67, 873–891.

Fricke, W., Leigh, R.A. and Tomos, A.D. (1996) The intercellular distribution of vacuolar solutes in the epidermis and mesophyll of barley leaves changes in response to NaCl. *Journal of Experimental Botany* 47, 1413–1426.

Fujii, H. and Zhu, J.K. (2009) *Arabidopsis* mutant deficient in 3 abscisic acid-activated protein kinases reveals critical roles in growth, reproduction, and stress. *Proceedings of the National Academy of Sciences USA* 106, 8380–8385.

Fujii, H., Chinnusamy, V., Rodrigues, A., Rubio, S., Antoni, R. *et al.* (2009) *In vitro* reconstitution of an abscisic acid signalling pathway. *Nature* 462, 660–664.

Fujita, Y., Nakashima, K., Yoshida, T., Katagiri, T., Kidokoro, S. *et al.* (2009) Three SnRK2 protein kinases are the main positive regulators of abscisic acid signaling in response to water stress in *Arabidopsis*. *Plant and Cell Physiology* 50, 2123–2132.

Furihata, T., Maruyama, K., Fujita, Y., Umezawa, T., Yoshida, R., Shinozaki, K. and Yamaguchi-Shinozaki, K. (2006) Abscisic acid-dependent multisite phosphorylation regulates the activity of a transcription activator AREB1. *Proceedings of the National Academy of Sciences USA* 103, 1988–1993.

Gao, Y., Lu, Y., Wu, M., Liang, E., Li, Y. *et al.* (2016) Ability to remove $Na^+$ and retain $K^+$ correlates with salt tolerance in two maize inbred lines seedlings. *Frontiers in Plant Science* 7, 1716.

Gavaghan, C.L., Li, J.V., Hadfield, S.T., Hole, S., Nicholson, J.K. *et al.* (2011) Application of NMR-based metabolomics to the investigation of salt stress in maize (*Zea mays*). *Phytochemical Analysis* 22, 214–224.

Geiger, D., Scherzer, S., Mumm, P., Stange, A., Marten, I. *et al.* (2009) Activity of guard cell anion channel SLAC1 is controlled by drought-stress signaling kinase-phosphatase pair. *Proceedings of the National Academy of Sciences USA* 106, 21425–21430.

Grattan, S.R. and Grieve, C.M. (1999) Salinity mineral nutrient relations in horticultural crops. *Scientia Horticulturae* 78, 127–157.

Guimaraes, C.T., Simoes, C.C., Pastina, M.M., Maron, L.G., Magalhaes, J.V. *et al.* (2014) Genetic dissection of Al tolerance QTLs in the maize genome by high density SNP scan. *BMC Genomics* 15, 153.

Guo, R., Shi, L., Yan, C., Zhong, X., Gu, F. *et al.* (2017) Ionomic and metabolic responses to neutral salt or alkaline salt stresses in maize (*Zea mays L.*) seedlings. *BMC Plant Biology* 17, 41.

Hajibagheri, M.A., Yeo, A.R., Flowers, T.J. and Collins, J.C. (1989) Salinity resistance in *Zea mays*: fluxes of potassium, sodium and chloride, cytoplasmic concentrations and microsomal membrane lipids. *Plant, Cell & Environment* 12, 753–757.

Halfter, U., Ishitani, M. and Zhu, J.K. (2000) The *Arabidopsis* SOS2 protein kinase physically interacts with and is activated by the calcium-binding protein SOS3. *Proceedings of the National Academy of Sciences USA* 97, 3735–3740.

Hanks, R.J., Ashcroft, G.L., Rasmussen, V.P. and Wilson, G.D. (1978) Corn production as influenced by irrigation and salinity – Utah studies. *Irrigation Science* 1, 47–59.

Hechenberger, M., Schwappach, B., Fischer, W.N., Frommer, W.B., Jentsch, T.J. and Steinmeyer, K. (1996) A family of putative chloride channels from *Arabidopsis* and functional complementation of a yeast strain with a *CLC* gene disruption. *Journal of Biological Chemistry* 271, 33632–33638.

Hoekenga, O.A., Maron, L.G., Pineros, M.A., Cancado, G.M., Shaff, J. *et al.* (2006) *AtALMT1*, which encodes a malate transporter, is identified as one of several genes critical for aluminum tolerance in *Arabidopsis*. *Proceedings of the National Academy of Sciences USA* 103, 9738–9743.

Horie, T., Yoshida, K., Nakayama, H., Yamada, K., Oiki, S. and Shinmyo, A. (2001) Two types of HKT transporters with different properties of $Na^+$ and $K^+$ transport in *Oryza sativa*. *The Plant Journal* 27, 129–138.

Horie, T., Costa, A., Kim, T.H., Han, M.J., Horie, R. *et al.* (2007) Rice OsHKT2;1 transporter mediates large $Na^+$ influx component into $K^+$-starved roots for growth. *The EMBO Journal* 26, 3003–3014.

Hu, X., Li, N., Wu, L., Li, C., Li, C. *et al.* (2015) Quantitative iTRAQ-based proteomic analysis of phosphoproteins and ABA-regulated phosphoproteins in maize leaves under osmotic stress. *Scientific Reports* 5, 15626.

Huang, C.X. and Van Steveninck, R.F. (1989) Maintenance of low $Cl^-$ concentrations in mesophyll cells of leaf blades of barley seedlings exposed to salt stress. *Plant Physiology* 90, 1440–1443.

Isla, R. and Aragues, R. (2010) Yield and plant ion concentrations in maize (*Zea mays L.*) subject to diurnal and nocturnal saline sprinkler irrigations. *Field Crops Research* 116, 175–183.

Ismail, A.M. and Horie, T. (2017) Genomics, physiology, and molecular breeding approaches for improving salt tolerance. *Annual Review of Plant Biology* 68, 405–434.

James, R.A., Davenport, R.J. and Munns, R. (2006) Physiological characterization of two genes for $Na^+$ exclusion in durum wheat, *Nax1* and *Nax2*. *Plant Physiology* 142, 1537–1547.

Jia, W., Wang, Y., Zhang, S. and Zhang, J. (2002) Salt-stress-induced ABA accumulation is more sensitively triggered in roots than in shoots. *Journal of Experimental Botany* 53, 2201–2206.

Jossier, M., Kroniewicz, L., Dalmas, F., Le Thiec, D., Ephritikhine, G. *et al.* (2010) The *Arabidopsis* vacuolar anion transporter, AtCLCc, is involved in the regulation of stomatal movements and contributes to salt tolerance. *The Plant Journal* 64, 563–576.

Kovermann, P., Meyer, S., Hortensteiner, S., Picco, C., Scholz-Starke, J. *et al.* (2007) The *Arabidopsis* vacuolar malate channel is a member of the ALMT family. *The Plant Journal* 52, 1169–1180.

Krill, A.M., Kirst, M., Kochian, L.V., Buckler, E.S. and Hoekenga, O.A. (2010) Association and linkage analysis of aluminum tolerance genes in maize. *PLoS One* 5, e9958.

Kurusu, T., Kuchitsu, K., and Tada, Y. (2015) Plant signaling networks involving $Ca^{2+}$ and Rboh/Nox-mediated ROS production under salinity stress. *Frontiers in Plant Science* 6, 427.

Li, B., Qiu, J., Jayakannan, M., Xu, B., Li, Y. *et al.* (2016a) AtNPF2.5 modulates chloride ($Cl^-$) efflux from roots of *Arabidopsis thaliana*. *Frontiers in Plant Science* 7, 2013.

Li, B., Byrt, C., Qiu, J., Baumann, U., Hrmova, M. *et al.* (2016b) Identification of a stelar-localized transport protein that facilitates root-to-shoot transfer of chloride in *Arabidopsis*. *Plant Physiology* 170, 1014–1029.

Li, B., Tester, M. and Gilliham, M. (2017) Chloride on the move. *Trends in Plant Science* 22, 236–248.

Li, W., Xu, G., Alli, A. and Yu, L. (2018) Plant HAK/KUP/KT $K^+$ transporters: function and regulation. *Seminars in Cell and Development Biology* 74, 133–141.

Ligaba, A., Maron, L., Shaff, J., Kochian, L. and Pineros, M. (2012) Maize ZmALMT2 is a root anion transporter that mediates constitutive root malate efflux. *Plant, Cell & Environment* 35, 1185–1200.

Liu, Z., Ma, Z., Guo, X., Shao, H., Cui, Q. and Song, W. (2010) Changes of cytosolic $Ca^{2+}$ fluorescence intensity and plasma membrane calcium channels of maize root tip cells under osmotic stress. *Plant Physiology and Biochemistry* 48, 860–865.

Luo, M., Zhao, Y., Zhang, R., Xing, J., Duan, M. *et al.* (2017) Mapping of a major QTL for salt tolerance of mature field-grown maize plants based on SNP markers. *BMC Plant Biology* 17, 140.

Luo, X., Wang, B., Gao, S., Zhang, F., Terzaghi, W. and Dai, M. (2019) Genome-wide association study dissects the genetic bases of salt tolerance in maize seedlings. *Journal of Integrative Plant Biology* 61, 658–674.

Maierhofer, T., Diekmann, M., Offenborn, J.N., Lind, C., Bauer, H. *et al.* (2014) Site- and kinase-specific phosphorylation-mediated activation of SLAC1, a guard cell anion channel stimulated by abscisic acid. *Science Signaling* 7, ra86.

Marinova, K., Pourcel, L., Weder, B., Schwarz, M., Barron, D. *et al.* (2007) The *Arabidopsis* MATE transporter TT12 acts as a vacuolar flavonoid/$H^+$-antiporter active in proanthocyanidin-accumulating cells of the seed coat. *The Plant Cell* 19, 2023–2038.

Marmagne, A., Vinauger-Douard, M., Monachello, D., de Longevialle, A.F., Charon, C. *et al.* (2007) Two members of the *Arabidopsis* CLC (chloride channel) family, AtCLCe and AtCLCf, are associated with thylakoid and Golgi membranes, respectively. *Journal of Experimental Botany* 58, 3385–3393.

Maron, L.G., Pineros, M.A., Guimaraes, C.T., Magalhaes, J.V., Pleiman, J.K. *et al.* (2010) Two functionally distinct members of the MATE (multi-drug and toxic compound extrusion) family of transporters potentially underlie two major aluminum tolerance QTLs in maize. *The Plant Journal* 61, 728–740.

Martínez-Atienza, J., Jiang, X., Garciadeblas, B., Mendoza, I., Zhu, J.K., Pardo, J.M. and Quintero, F.J. (2007) Conservation of the salt overly sensitive pathway in rice. *Plant Physiology* 143, 1001–1012.

Martiniere, A., Fiche, J.B., Smokvarska, M., Mari, S., Alcon, C. *et al.* (2019) Osmotic stress activates two reactive oxygen species pathways with distinct effects on protein nanodomains and diffusion. *Plant Physiology* 179, 1581–1593.

Meyer, S., Mumm, P., Imes, D., Endler, A., Weder, B. *et al.* (2010) AtALMT12 represents an R-type anion channel required for stomatal movement in *Arabidopsis* guard cells. *The Plant Journal* 63, 1054–1062.

Meyer, S., Scholz-Starke, J., De Angeli, A., Kovermann, P., Burla, B., Gambale, F. and Martinoia, E. (2011) Malate transport by the vacuolar AtALMT6 channel in guard cells is subject to multiple regulation. *The Plant Journal* 67, 247–257.

Mittal, S., Mallikarjuna, M.G., Rao, A.R., Jain, P.A., Dash, P.K. and Thirunavukkarasu, N. (2017) Comparative analysis of CDPK family in maize, *Arabidopsis*, rice, and sorghum revealed potential targets for drought tolerance improvement. *Frontiers in Chemistry* 5, 115.

Moller, I.S., Gilliham, M., Jha, D., Mayo, G.M., Roy, S.J. *et al.* (2009) Shoot $Na^+$ exclusion and increased salinity tolerance engineered by cell type-specific alteration of $Na^+$ transport in *Arabidopsis*. *The Plant Cell* 21, 2163–2178.

Muhammad, F., Mubshar, H., Abdul, W. and Kadambot, H.M. (2015) Salt stress in maize: effects, resistance mechanisms, and management. *Agronomy for Sustainable Development* 35, 461–481.

Munemasa, S., Hauser, F., Park, J., Waadt, R., Brandt, B. and Schroeder, J.I. (2015) Mechanisms of abscisic acid-mediated control of stomatal aperture. *Current Opinion in Plant Biology* 28, 154–162.

Munns, R. and Tester, M. (2008) Mechanisms of salinity tolerance. *Annual Review of Plant Biology* 59, 651–681.

Munns, R., James, R.A., Xu, B., Athman, A., Conn, S.J. *et al.* (2012) Wheat grain yield on saline soils is improved by an ancestral $Na^+$ transporter gene. *Nature Biotechnology* 30, 360–364.

Nadeem, M., Li, J., Yahya, M., Wang, M., Ali, A. *et al.* (2019) Grain legumes and fear of salt stress: focus on mechanisms and management strategies. *International Journal of Molecular Sciences* 20, 799.

Nguyen, C.T., Agorio, A., Jossier, M., Depre, S., Thomine, S. and Filleur, S. (2016) Characterization of the chloride channel-like, AtCLCg, involved in chloride tolerance in *Arabidopsis thaliana*. *Plant and Cell Physiology* 57, 764–775.

Niu, X., Zheng, W., Lu, B.R., Ren, G., Huang, W. *et al.* (2007) An unusual posttranscriptional processing in two *betaine aldehyde dehydrogenase* loci of cereal crops directed by short, direct repeats in response to stress conditions. *Plant Physiology* 143, 1929–1942.

Omote, H., Hiasa, M., Matsumoto, T., Otsuka, M. and Moriyama, Y. (2006) The MATE proteins as fundamental transporters of metabolic and xenobiotic organic cations. *Trends in Pharmacological Sciences* 27, 587–593.

Park, H.J., Kim, W.Y. and Yun, D.J. (2016) A new insight of salt stress signaling in plant. *Molecules and Cells* 39, 447–459.

Pineros, M.A., Cancado, G.M., Maron, L.G., Lyi, S.M., Menossi, M. and Kochian, L.V. (2008) Not all ALMT1-type transporters mediate aluminum-activated organic acid responses: the case of ZmALMT1 – an anion-selective transporter. *The Plant Journal* 53, 352–367.

Qi, G.N., Yao, F.Y., Ren, H.M., Sun, S.J., Tan, Y.Q. *et al.* (2018) The S-type anion channel ZmSLAC1 plays essential roles in stomatal closure by mediating nitrate efflux in maize. *Plant and Cell Physiology* 59, 614–623.

Qiu, J., Henderson, S.W., Tester, M., Roy, S.J. and Gilliham, M. (2016) SLAH1, a homologue of the slow type anion channel SLAC1, modulates shoot $Cl^-$ accumulation and salt tolerance in *Arabidopsis thaliana*. *Journal of Experimental Botany* 67, 4495–4505.

Qiu, Q.S., Guo, Y., Dietrich, M.A., Schumaker, K.S. and Zhu, J.K. (2002) Regulation of SOS1, a plasma membrane $Na^+/H^+$ exchanger in *Arabidopsis thaliana*, by SOS2 and SOS3. *Proceedings of the National Academy of Sciences USA* 99, 8436–8441.

Ren, Z.H., Gao, J.P., Li, L.G., Cai, X.L., Huang, W. *et al.* (2005) A rice quantitative trait locus for salt tolerance encodes a sodium transporter. *Nature Genetics* 37, 1141–1146.

Rivalta, I., Amin, M., Luber, S., Vassiliev, S., Pokhrel, R. *et al.* (2011) Structural-functional role of chloride in photosystem II. *Biochemistry* 50, 6312–6315.

Roberts, S.K. and Tester, M. (1997) A patch clamp study of $Na^+$ transport in maize roots. *Journal of Experimental Botany* 48, 431–440.

Rodriguez, H.G., Roberts, J., Jordan, W.R. and Drew, M.C. (1997) Growth, water relations, and accumulation of organic and inorganic solutes in roots of maize seedlings during salt stress. *Plant Physiology* 113, 881–893.

Rubio, F., Nieves-Cordones, M., Aleman, F. and Martinez, V. (2008) Relative contribution of AtHAK5 and AtAKT1 to $K^+$ uptake in the high-affinity range of concentrations. *Physiologia Plantarum* 134, 598–608.

Russell, J.M. (2000) Sodium–potassium–chloride cotransport. *Physiological Reviews* 80, 211–276.

Sah, S.K., Reddy, K.R. and Li, J. (2016) Abscisic acid and abiotic stress tolerance in crop plants. *Frontiers in Plant Science* 7, 571.

Sanders, D. (1981) Physiological control of chloride transport in *Chara corallina*: II. The role of chloride as a vacuolar osmoticum. *Plant Physiology* 68, 401–406.

Sandhu, D., Pudussery, M.V., Kumar, R., Pallete, A., Markley, P., Bridges, W.C. and Sekhon, R.S. (2020) Characterization of natural genetic variation identifies multiple genes involved in salt tolerance in maize. *Functional & Integrative Genomics* 20, 261–275.

Saneoka, H., Nagasaka, C., Hahn, D.T., Yang, W.J., Premachandra, G.S., Joly, R.J. and Rhodes, D. (1995) Salt tolerance of glycinebetaine-deficient and -containing maize lines. *Plant Physiology* 107, 631–638.

Sasaki, T., Yamamoto, Y., Ezaki, B., Katsuhara, M., Ahn, S.J. *et al.* (2004) A wheat gene encoding an aluminum-activated malate transporter. *The Plant Journal* 37, 645–653.

Schnable, J.C. (2015) Genome evolution in maize: from genomes back to genes. *Annual Review of Plant Biology* 66, 329–343.

Schubert, S. and Läuchli, A. (1990) Sodium exclusion mechanism at the root surface of 2 maize cultivars. *Plant and Soil* 123, 205–209.

Selvakumar, G., Shagol, C.C., Kim, K., Han, S. and Sa, T. (2018) Spore associated bacteria regulates maize root $K^+/Na^+$ ion homeostasis to promote salinity tolerance during arbuscular mycorrhizal symbiosis. *BMC Plant Biology* 18, 109.

Shabala, S. and Pottosin, I. (2014) Regulation of potassium transport in plants under hostile conditions: implications for abiotic and biotic stress tolerance. *Physiologia Plantarum* 151, 257–279.

Shabala, S., Demidchik, V., Shabala, L., Cuin, T.A., Smith, S.J. *et al.* (2006) Extracellular $Ca^{2+}$ ameliorates NaCl-induced $K^+$ loss from *Arabidopsis* root and leaf cells by controlling plasma membrane $K^+$-permeable channels. *Plant Physiology* 141, 1653–1665.

Sharma, T., Dreyer, I., Kochian, L. and Pineros, M.A. (2016) The ALMT family of organic acid transporters in plants and their involvement in detoxification and nutrient security. *Frontiers in Plant Science* 7, 1488.

Shi, H., Ishitani, M., Kim, C. and Zhu, J.K. (2000) The *Arabidopsis thaliana* salt tolerance gene *SOS1* encodes a putative $Na^+/H^+$ antiporter. *Proceedings of the National Academy of Sciences USA* 97, 6896–6901.

Shi, H., Quintero, F.J., Pardo, J.M. and Zhu, J.K. (2002) The putative plasma membrane $Na^+/H^+$ antiporter SOS1 controls long-distance $Na^+$ transport in plants. *The Plant Cell* 14, 465–477.

Shi, H., Lee, B.H., Wu, S.J. and Zhu, J.K. (2003) Overexpression of a plasma membrane $Na^+/H^+$ antiporter gene improves salt tolerance in *Arabidopsis thaliana*. *Nature Biotechnology* 21, 81–85.

Storey, R. and Walker, R.R. (1999) Citrus and salinity. *Scientia Horticulturae* 78, 39–81.

Storey, R., Schachtman, D.P. and Thomas, M.R. (2003) Root structure and cellular chloride, sodium and potassium distribution in salinized grapevines. *Plant, Cell & Environment* 26, 789–800.

Sun, J., Chen, S., Dai, S., Wang, R., Li, N. *et al.* (2009) NaCl-induced alternations of cellular and tissue ion fluxes in roots of salt-resistant and salt-sensitive poplar species. *Plant Physiology* 149, 1141–1153.

Sunarpi, Horie, T., Motoda, J., Kubo, M., Yang, H. *et al.* (2005) Enhanced salt tolerance mediated by AtHKT1 transporter-induced Na unloading from xylem vessels to xylem parenchyma cells. *The Plant Journal* 44, 928–938.

Tester, M. and Davenport, R. (2003) $Na^+$ tolerance and $Na^+$ transport in higher plants. *Annals of Botany* 91, 503–527.

Tyerman, S.D. and Skerrett, M. (1999) Root ion channels and salinity. *Scientiae Horticulturae* 78, 175–235.

Very, A.A. and Sentenac, H. (2003) Molecular mechanisms and regulation of $K^+$ transport in higher plants. *Annual Review of Plant Biology* 54, 575–603.

Voetberg, G.S. and Sharp, R.E. (1991) Growth of the maize primary root at low water potentials: III. Role of increased proline deposition in osmotic adjustment. *Plant Physiology* 96, 1125–1130.

Wang, H., Wang, M. and Xia, Z. (2019) Overexpression of a maize SUMO conjugating enzyme gene (*ZmSCE1e*) increases SUMOylation levels and enhances salt and drought tolerance in transgenic tobacco. *Plant Science* 281, 113–121.

Wang, R., Jing, W., Xiao, L., Jin, Y., Shen, L. and Zhang, W. (2015a) The rice high-affinity potassium transporter1;1 is involved in salt tolerance and regulated by an MYB-type transcription factor. *Plant Physiology* 168, 1076–1090.

Wang, S., Su, S.Z., Wu, Y., Li, S.P., Shan, X.H., Liu, H.K., Wang, S. and Yuan, Y.P. (2015b) Overexpression of maize chloride channel gene ZmCLC-d in *Arabidopsis thaliana* improved its stress resistance. *Biologia Plantarum* 59, 55–64.

Wang, Y. and Wu, W.H. (2017) Regulation of potassium transport and signaling in plants. *Current Opinion in Plant Biology* 39, 123–128.

Wege, S., De Angeli, A., Droillard, M.J., Kroniewicz, L., Merlot, S. *et al.* (2014) Phosphorylation of the vacuolar anion exchanger AtCLCa is required for the stomatal response to abscisic acid. *Science Signaling* 7, ra65.

Wen, Z., Tyerman, S.D., Dechorgnat, J., Ovchinnikova, E., Dhugga, K.S. and Kaiser, B.N. (2017) Maize NPF6 proteins are homologs of *Arabidopsis* CHL1 that are selective for both nitrate and chloride. *The Plant Cell* 29, 2581–2596.

Wu, J., Jiang, Y., Liang, Y., Chen, L., Chen, W. and Cheng, B. (2019) Expression of the maize MYB transcription factor ZmMYB3R enhances drought and salt stress tolerance in transgenic plants. *Plant Physiology and Biochemistry* 137, 179–188.

Xie, Y., Feng, Y., Chen, Q., Zhao, F., Zhou, S. *et al.* (2019) Genome-wide association analysis of salt tolerance QTLs with SNP markers in maize (*Zea mays L.*). *Genes & Genomics* 41, 1135–1145.

Yang, G., Zou, H.D., Wu, Y., Liu, H.K. and Yuan, Y.P. (2011) Identification and characterisation of candidate genes involved in chilling responses in maize (*Zea mays L.*). *Plant Cell Tissue and Organ Culture* 106, 127–141.

Yang, W.J., Nadolska-Orczyk, A., Wood, K.V., Hahn, D.T., Rich, P.J. *et al.* (1995) Near-isogenic lines of maize differing for glycinebetaine. *Plant Physiology* 107, 621–630.

Yang, Y. and Guo, Y. (2018) Elucidating the molecular mechanisms mediating plant salt-stress responses. *New Phytologist* 217, 523–539.

Yang, Z., Wang, C., Xue, Y., Liu, X., Chen, S. *et al.* (2019) Calcium-activated 14-3-3 proteins as a molecular switch in salt stress tolerance. *Nature Communications* 10, 1199.

Yao, X., Horie, T., Xue, S., Leung, H.Y., Katsuhara, M. *et al.* (2010) Differential sodium and potassium transport selectivities of the rice OsHKT2;1 and OsHKT2;2 transporters in plant cells. *Plant Physiology* 152, 341–355.

Yoshida, T., Mogami, J. and Yamaguchi-Shinozaki, K. (2014) ABA-dependent and ABA-independent signaling in response to osmotic stress in plants. *Current Opinion in Plant Biology* 21, 133–139.

Yuan, F., Yang, H., Xue, Y., Kong, D., Ye, R. *et al.* (2014) OSCA1 mediates osmotic-stress-evoked $Ca^{2+}$ increases vital for osmosensing in *Arabidopsis*. *Nature* 514, 367–371.

Zhang, F., Li, L., Jiao, Z., Chen, Y., Liu, H. *et al.* (2016) Characterization of the calcineurin B-like (CBL) gene family in maize and functional analysis of ZmCBL9 under abscisic acid and abiotic stress treatments. *Plant Science* 253, 118–129.

Zhang, H., Zhu, H., Pan, Y., Yu, Y., Luan, S. and Li, L. (2014) A DTX/MATE-type transporter facilitates abscisic acid efflux and modulates ABA sensitivity and drought tolerance in *Arabidopsis*. *Molecular Plant* 7, 1522–1532.

Zhang, H., Zhao, F.G., Tang, R.J., Yu, Y., Song, J. *et al.* (2017) Two tonoplast MATE proteins function as turgor-regulating chloride channels in *Arabidopsis*. *Proceedings of the National Academy of Sciences USA* 114, E2036–E2045.

Zhang, L., Gao, M., Hu, J., Zhang, X., Wang, K. and Ashraf, M. (2012) Modulation role of abscisic acid (ABA) on growth, water relations and glycinebetaine metabolism in two maize (*Zea mays L.*) cultivars under drought stress. *International Journal of Molecular Sciences* 13, 3189–3202.

Zhang, M., Cao, Y., Wang, Z., Wang, Z.Q., Shi, J. *et al.* (2018) A retrotransposon in an HKT1 family sodium transporter causes variation of leaf $Na^+$ exclusion and salt tolerance in maize. *New Phytologist* 217, 1161–1176.

Zhang, M., Liang, X., Wang, L., Cao, Y., Song, W., Shi, J., Lai, J. and Jiang, C. (2019) A HAK family $Na^+$ transporter confers natural variation of salt tolerance in maize. *Nature Plants* 5, 1297–1308.

Zhang, X., Cai, H., Lu, M., Wei, Q., Xu, L. *et al.* (2019) A maize stress-responsive Di19 transcription factor, ZmDi19-1, confers enhanced tolerance to salt in transgenic *Arabidopsis*. *Plant Cell Reports* 38, 1563–1578.

Zhao, K.F., Song, J., Fan, H., Zhou, S. and Zhao, M. (2010) Growth response to ionic and osmotic stress of NaCl in salt-tolerant and salt-sensitive maize. *Journal of Integrative Plant Biology* 52, 468–475.

Zhao, Y., Liu, M., He, L., Li, X., Wang, F. *et al.* (2019) A cytosolic $NAD^+$-dependent GPDH from maize (ZmGPDH1) is involved in conferring salt and osmotic stress tolerance. *BMC Plant Biology* 19, 16.

Zhu, H., Wu, J., Jiang, Y., Jin, J., Zhou, W. *et al.* (2016) Genomewide analysis of MATE-type gene family in maize reveals microsynteny and their expression patterns under aluminum treatment. *Journal of Genetics and Genomics* 95, 691–704.

Zhu, J.K. (2002) Salt and drought stress signal transduction in plants. *Annual Review of Plant Biology* 53, 247–273.

Zhu, J.K. (2003) Regulation of ion homeostasis under salt stress. *Current Opinion in Plant Biology* 6, 441–445.

Zorb, C., Noll, A., Karl, S., Leib, K., Yan, F. and Schubert, S. (2005) Molecular characterization of $Na^+/H^+$ antiporters (ZmNHX) of maize (*Zea mays L.*) and their expression under salt stress. *Journal of Plant Physiology* 162, 55–66.

# 15 Isolation of Genes/Quantitative Trait Loci for Drought Stress Tolerance in Maize

**Pardeep Kumar[1], Mukesh Choudhary[1], B.S. Jat[1], M.C. Dagla[1], Vishal Singh[1], Abhijit Kumar Das[1], Santosh Kumar[1], Ningthaipuilu Longmei[1], Robert J. Henry[2] and Shabir Hussain Wani[3]***

*[1]ICAR–Indian Institute of Maize Research, Ludhiana, India; [2]University of Queensland, Brisbane, Australia; [3]Sher-e-Kashmir University of Agricultural Sciences and Technology of Kashmir, Khudwani, India*

## 15.1 Introduction

Increasing population in the 21st century makes fulfilling global food demand a difficult task for scientists within the context of the challenges associated with the deteriorating quality of cultivated land and depleting water tables along with increasingly unreliable weather patterns due to climate change. Considering the present situation, it seems it will be difficult to achieve the projected target of increasing food production by 70% by 2050 (Wani and Sah, 2014). Although good progress has been achieved for major crops in terms of genetic gains for yield, vulnerability to climate variability has emerged as a serious concern in sustaining the higher yields that have been attained as well as enhancing yields further. Abiotic stresses are a major yield constraint factor with estimated losses due to drought, waterlogging and heat being serious concerns for productivity in different crops (Qin *et al.*, 2011; Bailey-Serres *et al.*, 2012; Gosal and Wani, 2018; Ahmad *et al.*, 2019). One of the reasons for stalled yields under limited water may be that higher sowing density leads to more competition for water and nutrients. In addition, the confounding effects of various stresses like drought with heat make it difficult to cope with multiple stresses together (Maleki *et al.*, 2019). Therefore, the concern of food security in this period of environmental change has driven plant researchers in different crops to shift to breeding for climate-resilient genotypes (Kole *et al.*, 2015; Pandita *et al.*, 2019). Although maize, being a hardy crop, has an inherent potential to perform well under suboptimal environments with minimum crop management, maize productivity is truly influenced by different stresses such as drought, heat and waterlogging (Ahuja *et al.*, 2010).

Maize, being a rainfed crop, is highly prone to be exposed to drought as evident from wide fluctuations in maize yields over the years in rainfed environments. For example, nearly 80% of rainfed maize grown in South and South-East Asia has an average yield of half or less than

*Corresponding author: shabirhussainwani@gmail.com, shabirhwani@skuastkashmir.ac.in


DOI: 10.1079/9781789245431.0015

half that of irrigated maize. Further, the area of rainfed maize is almost six times that of irrigated maize and is increasing at 1.8% per year (Edmeades *et al.*, 2006). Reductions in the irrigated area in major maize-growing areas are due to the reducing moisture level in the soil that induces water scarcity and limits maize yields worldwide (Cooper *et al.*, 2014). Development of drought tolerance in maize would be a major breakthrough allowing introduction of hybrids into drought-affected areas. However, drought is a complex phenomenon as it influences numerous processes associated with plant development and improvement, including osmotic modification, cell reinforcement capacities and photosynthetic rate (Cramer *et al.*, 2011; Wani *et al.*, 2018b, 2020; Yadav *et al.*, 2020). These processes are controlled and regulated by various proteins and their differential expression in tolerant cultivars reflects their role in biochemical pathways and phases of crop growth and development.

Drought stress affects seed germination in maize and also hampers growth at the seedling stage. There are different factors like water absorption, imbibition and activity of various metabolic enzymes which reduce seed germination under drought conditions. After successful germination in drought-prone condition there is significant growth reduction at the plumule and radicle growth stage which results in stunting of seedling vigour (Gharoobi *et al.*, 2012). The growth of the seedling under drought is measured by the elongation rate of roots and shoots and is significantly reduced under stress conditions. However, reduction in root elongation is less than that of shoot elongation under drought stress in maize (Khodarahmpour, 2011).

Conventional breeding has been exploited at its fullest for the production of high-yielding maize hybrids but has shown little success in breeding climate-resilient varieties owing to the multigenic nature of stress tolerance. Although wild relatives have been found to exhibit tolerance against different abiotic stresses, their use has been limited due to associated problems like linkage drag (Choudhary *et al.*, 2017). Molecular breeding has evolved to overcome such problems and provided the base to develop climate-smart cultivars. In the context of molecular mapping, several quantitative trait loci (QTLs) for drought, heat and waterlogging have been mapped in maize. In addition to molecular mapping, genomic selection (GS), genome-wide association studies (GWAS) and doubled haploid (DH) technologies are boosting maize stress breeding (Azhar *et al.*, 2020). After the development of molecular markers, the field of 'omics' emerged at a rapid pace to facilitate the identification of abiotic stress-associated QTLs in maize through characterization of abiotic stress-related genes at the transcriptional and protein levels (Cushman and Bohnert, 2000; Cattivelli *et al.*, 2002; Coraggio and Tuberosa, 2004; Tuberosa and Salvi, 2004; Ganie *et al.*, 2020). Cheaper high-throughput sequencing platforms along with transcriptomics and proteomics approaches helped to reveal the molecular basis of drought tolerance-related gene action. The recently recognized but most important approach known as phenomics provided a better platform for precise phenotyping of complex traits. Precise and accurate phenotyping upon which the true genetic associations can be built is the most important part of any breeding programme. Molecular mapping without accurate phenotyping can be considered 'a gun without a bullet'. In a nutshell, it can be considered that the efficiency of drought stress breeding programmes in maize warrants the integration of high-throughput phenotyping for stress-associated target traits, genomics-assisted breeding, omics approaches and agronomic management via adoption of conservation agricultural practices. This chapter focuses on target traits for drought stress, progress in mapping for drought tolerance-associated genes/QTLs identification and expression studies, introgression strategies followed by the possibilities of integrating the concept of speed breeding in maize drought breeding programmes for better utilization of wild relatives.

## 15.2 Target Traits for Drought Tolerance

Target traits for drought stress screening are mainly morphological and physiological, together known as morpho-physiological traits, which can further be categorized into: (i) drought-constitutive traits, low yield reduction under immediate drought condition; and (ii) drought-responsive traits, high yield reduction under severe drought (Blum, 2006; Tuberosa, 2012). Compared with drought-responsive traits, breeding for drought tolerance has had great success by targeting constitutive traits which function by avoiding the drought (Blum, 2006; Blum, 2011). Drought tolerance can be explored by targeting traits which exhibit a vast amount of variability, high heritability and positive correlation with yield under moisture stress conditions. The measurement of such traits should be precise, rapid, cost-effective and non-destructive. However, the target traits should not compensate for better yields under normal or optimum conditions (Monneveux and Ribaut, 2006).

There are several secondary morphological traits that serve as useful criteria for drought tolerance in maize. Being complex in nature, drought stress affects the crop at various stages over space and time and hence the crop exhibits complex physiological responses which are often unpredictable. The significant impact of drought stress in maize is delay in silking, which results in an enhanced anthesis–silking interval (ASI). The enhanced ASI results in failure of zygote formation due to missed nicking of fertilization and hence results in severe yield losses (Sari-Gorla *et al*., 1999). When drought stress occurs around 7–10 days before flowering, cob growth tends to get slower relative to tassel growth and hence results in delayed silking compared with anthesis, which ultimately widens the gap between pollen shedding and silking (Edmeades, 2013). Other typical symptoms of drought stress are converting the green colour of the plant into green-grey and rolling of leaves on the lower plant followed by the upper leaves of the plant. Furthermore, closure of stomata hampers the photosynthesis drastically and hence hampers growth (Choudhary *et al*., 2019). The delayed senescence or stay-green trait is another important trait that helps to continue assimilate synthesis by protecting the photosynthetic machinery (Borrell *et al*., 2001).

Drought stress during the grain-filling period also affects grain development in areas prone to moisture stress at harvest time. However, in such conditions, high levels of accumulation of assimilates in maize stem contributes to yield (Jurgens *et al*., 1978). Several hormones play a key role in combating abiotic stresses particularly drought stress (Wani and Kumar, 2015; Kumar *et al*., 2016; Wani *et al*., 2016a; Dar *et al*., 2017; Soliman *et al*., 2018). For example, the role of both abscisic acid (ABA) and ethylene is to interact with other growth regulators which pass the signal from root to aerial tissues through reactive oxygen species and induce drought tolerance (Ribaut *et al*., 2009). Root structure is an important trait to explore for drought stress tolerance as it helps to extract the accessible groundwater from far down in the soil (Robertson *et al*., 1993). Heat-tolerant maize cultivars have better root structure that reacts to drought by diverting development and dry tissue accumulation away from the shoot to the root (Hsiao and Xu, 2000).

Osmotic modification is another significant tolerance component that includes the dynamic accumulation of solutes in the cell (Ribaut *et al*., 2009). The accumulation of solutes in the cell helps in maintaining the desired water potential and hence constraining the loss of turgor to avoid the adverse effects that may arise from shrinkage of the cell. In addition, under prolonged stress, osmotic solutes help in the adjustment of different macromolecular structures. Finally, stomata act in response to moisture stress and restrict their opening to restrict water loss. Maize is quite sensitive to both excess moisture and limited moisture and hence is considered an optimum water-requiring crop (Tardieu and Simonneau, 1998). Mild drought stress can be tolerated in maize, but severe drought stress needs to be addressed by developing drought-tolerant cultivars. The initial step of any drought stress breeding programme starts with the screening of cultivars for identification of tolerant cultivars, which in turn relies upon choosing target environments and precise phenotyping of target traits.

## 15.3 Target Environment for Evaluation

Choice of appropriate selection strategy and selection environment is critical to attain high genetic gains in drought stress breeding programmes. The most common strategy is to select high-yielding cultivars (under optimal conditions) and test them for drought stress at multiple locations (Magorokosho *et al*., 2003). However, in field trials, although moisture stress can be applied by withholding irrigation at a targeted stage, confounding effects imposed by non-controllable heat stress can lead to erroneous results. Hence, controlled condition chambers (managed temperature and humidity) are a better place to screen the genotypes for drought stress. Moreover, in maize, being a cross-pollinated crop, hybrids give higher yield compared with traditional varieties under drought, which proves that heterosis may play a role in stress tolerance (Bruce *et al.*, 2002). Hence, breeders should be cautious as lines derived from tolerant hybrids may exhibit only partial drought stress tolerance.

Precise design of the experiment and statistical analysis, field homogeneity in terms of soil conditions and fertility will together increase the reliability of genotypic evaluations (Ribaut *et al.*, 2009). Multi-location testing is key to estimating the stability of cultivars via assessing genotype × environment interactions. It is critical for drought tolerance breeding where plant growth vigour is highly affected by environmental variation. Several versions of elite lowland tropical populations have been developed as drought-tolerant populations (Bänziger *et al.*, 2005; Monneveux *et al.*, 2005).

## 15.4 Phenomics for Drought Stress Screening

In the last decades there have been several advances in different areas of modern computational biology such as bioinformatics in the genomic revolution. However, the practical utility of data generated by these techniques in the last few years has not met expectations (Xu and Crouch, 2008; Passioura, 2010). The reasons for this could include the lack of precise phenotyping for large sets of germplasm as conventional phenotyping approaches have been time-consuming, destructive and expensive. Stage-specific data recording is prone to biases when dealing with large germplasm collections. Hence, recording of precise and reproducible phenotypic data for complex drought tolerance-associated traits in large germplasm sets was a cumbersome job, even in the initial evolutionary stage of phenomics (Mir *et al.*, 2012). Consequently, more efforts were put into automation of phenomics and now a large number of high-throughput platforms are available to generate quick and precise phenotyping data; these employ advanced technologies such as infrared cameras that have the ability to scan temperature profiles and transpiration, hyperspectral imaging, computational tomography and fluorescent microscopy to assess photosynthesis and photosynthetic rates, three-dimensional live imaging cameras to capture growth responses at micro levels, and magnetic resonance imaging for studying root or leaf physiology (Finkel, 2009; Gupta *et al.*, 2012). Several software programs such as WinRhizo and Automatic Root Image Analysis (ARIA) have been developed to collect large amounts of data about root architectural traits such as root length, root number and root width by digital image (Cobb *et al.*, 2013; Pace *et al.*, 2014). These phenotyping tools will assist drought breeding programmes in maize by generating precise phenotyping data to give better genotypic–phenotypic associations, leading to better utilization of available whole-genome sequencing data.

## 15.5 Genomics-Assisted Breeding Approach: Boon for Drought Tolerance

Genetic variability created by inter-crossing of potential parents acts as the base upon which effective selections can be made to identify an ideal ideotype that can perform better under moisture stress (Bänziger *et al.*, 2000). Hence, existence of adequate genetic variation, high heritability and high selection intensity has been suggested to assist in better selection (Falconer, 1989). Drought tolerance in maize, being complex in nature, can be improved by targeting multiple traits rather than single trait selection. Instead, a selection indices approach has been suggested as a better approach for improvement of complex traits (Beissinger *et al.*, 2018). Morpho-physiological traits responsible for preventing moisture loss, increased water-use efficiency and better yields can be targeted for pyramiding (Subbarao *et al.*, 2005). Besides that, early vigour, quick establishment, better root architecture, stomatal conductance and leaf waxiness have been suggested as important traits for moisture stress tolerance (Parry *et al.*, 2005). Hence it is important to develop water-use-efficient and better-yielding cultivars that are capable of maintaining optimal physiological activities like accumulation of osmolytes, production of antioxidants, stress-responsive proteins and transcription factors (Bänziger *et al.*, 2000; Gosal *et al.*, 2009; Wani *et al.*, 2013; Joshi *et al.*, 2016; Dutta *et al.*, 2018). Traditional breeding approaches based on multi-location and multi-season yield testing help to identify drought stress-tolerant stable cultivars (Babu *et al.*, 2003). Yield and yield-attributing traits having a strong positive correlation with grain yield are preferred traits under both moisture stress and optimal conditions (Edmeades *et al.*, 2001).

However, traditional breeding approaches have had limited success in improving drought tolerance for maize due to the long time required for developing varieties and the complex nature of the stress. Hence, molecular breeding serves as a better approach for achieving rapid genetic gains in drought stress breeding programmes (Whitford *et al.*, 2010; Mickelbart *et al.*, 2015; Telem *et al.*, 2016). The extensive use of genetic markers in drought tolerance breeding programmes can be witnessed through various studies of QTL mapping/gene isolation. Identification of genomic regions governing drought tolerance is the foundation step in maize to execute the QTLian breeding, i.e. identification and introgression of QTLs (Choudhary *et al.*, 2019). The initial attempt at QTL mapping for drought tolerance in maize was executed using Polj17 (drought resistant) and F-2 (drought sensitive) to generate an $F_2$ population (Lebreton *et al.*, 1995). In that study, QTLs for ABA content and stomata conductance were mapped (using restriction fragment length polymorphism (RFLP) markers) at the same position, revealing the regulation of the latter by the former. Since then, a number of studies have been recorded using QTL mapping in maize for important morpho-physiological traits under drought stress condition. A compiled list of drought tolerance mapped QTLs in maize has been provided in Table 15.1. In different studies of QTL mapping, many QTLs were identified for morphological traits like male flowering, female flowering, ASI, yield and cob number (Agrama and Moussa, 1996; Ribaut *et al.*, 1996, 1997; Sari-Gorla *et al.*, 1999). Later, considering the importance of roots and root-related traits in imparting tolerance to drought, different QTLs for root architecture and root-associated traits along with yield traits were identified by different researchers: one QTL for root traits (Landi *et al.*, 2010), 22 QTLs for root-associated traits such as root density, root dry weight, sugar concentration and leaf ABA content through composite interval mapping in $F_{2:3}$ population (Rahman *et al.*, 2011). Similarly, Trachsel *et al.* (2016) identified a total of 17 QTLs for stomatal conductance, leaf water content, ASI and grain yield. Besides QTL mapping, association mapping is another important approach for better resolution of QTLs as it utilizes the historical recombination events in natural populations as an association panel. Setter *et al.* (2010) conducted association mapping to identify single-nucleotide polymorphisms (SNPs) related to genes involved in carbohydrate and ABA metabolite accumulation during drought stress. An aldehyde oxidase gene was found to regulate silk ABA concentration under drought stress. Later, SNP-based genome-wide association mapping was conducted using 5000 inbred lines as an association panel (Li *et al.*, 2016). The study revealed significant associations of SNPs with drought tolerance-associated candidate genes. However, most of the QTLs identified to date are of minor effect except for a few major QTLs. The identified major QTLs need to be stable QTLs for better performance under optimal and drought stress conditions (Wani *et al.*, 2018a). This is to ensure that the stress tolerance QTLs do not behave adversely under optimum moisture conditions, which is unacceptable in any stress breeding programme.

Meta-QTL (mQTL) analysis helped to identify consensus QTLs for grain yield under optimal and stress conditions using the information from previous mapping studies (Goffinet and Gerber, 2000; Li *et al.*, 2011; Swamy *et al.*, 2011). The mQTL study conducted using three populations and several environments revealed seven genomic regions for grain yield and one genomic region for ASI. Among these, six mQTLs for grain yield mapped on chromosomes 1, 4, 5 and 10 under moisture stress and optimal environments and hence were classified as stable QTLs. However, care should be taken when identifying the stable QTLs under drought stress only (but not under optimal moisture), because such QTLs become a waste if they exhibit unfavourable effects in favourable conditions, i.e. in the

**Table 15.1.** List of mapped QTLs for drought tolerance in maize.

| Study no. | Traits | No. of QTLs | Mapping population[a] (cross) | Marker type[b] | Environment | Chromosome | PVE[c] (%) | Mapping method[d] | Reference |
|---|---|---|---|---|---|---|---|---|---|
| 1. | Male flowering, female flowering, ASI | 23 | $F_{2:3}$ (Ac7643S5 × Ac7729/TZSRWS5) | RFLP | Field | 1, 2, 4, 5, 6, 8, 9, 10 | 4.4–15.1 | CIM | Ribaut *et al.* (1996) |
| 2. | Yield, ASI, ears per plant | 11 | $F_{2:3}$ (SD34 × SD35) | RFLP | Field | 1, 3, 5, 6, 8 | 9.4–49.6 | IM | Agrama and Moussa (1996) |
| 3. | Grain yield, expression of ear number, expression of kernel number | 7 | $F_{2:3}$ (Ac7643S5 × Ac7729/TZSRWS5) | RFLP | Field | 1, 4, 6, 9, 10 | 4.6–12.9 | CIM | Ribaut *et al.* (1997) |
| 4. | Leaf ABA concentration, yield | 17 | $F_4$ (Os420 × IABO78) | RFLP | Field | 1, 2, 3, 4, 6, 7, 9, 10 | – | CIM | Sanguineti *et al.* (1999) |
| 5. | Male flowering time, female flowering time, ASI, plant height | 17 | RILs (B73 × H99) | RFLP, SSR and AFLP | Field | 1, 2, 4, 5, 7, 8, 9 | 4.4–54.7 | IM | Sari-Gorla *et al.* (1999) |
| 6. | Grain yield, drought tolerance index | 17 | $F_{2:3}$ (Lo964 × Lo1016) | RFLP, SSR and AFLP | Field | 1, 2, 4, 6, 8, 10 | 9.5–47.2 | PLABQTL | Tuberosa *et al.* (2002) |
| 7. | Grain yield and associated traits | 20 | $F_{2:3}$ (X178 × B73) | SSR | Field | 1, 2, 3, 5, 7, 8, 9 | 4.31–31.28 | CIM | Xiao *et al.* (2005) |
| 8. | Leaf elongation rate in correspondence with ASI | 5 | RILs (Ac7643 × Ac7729/TZSR) | RFLP | Greenhouse | 1, 2, 5, 8 | 7.5–13.7 | CIM | Welcker *et al.* (2007) |
| 9. | Plant height, flower time, yield and yield components | 21 | RILs (5003 × P138) | SSR | Field | 1, 4, 5, 6, 8 | 1.68–13.30 | CIM | Guo *et al.* (2008) |
| 10. | Plant height, ear height, ASI, grain yield per plant, ear number per plant, 100-kernel weight, kernel number | 44 | $F_{2:3}$ (Ac7643S5 × Ac7729/TZSRWS5) | RFLP | Field | 1, 2, 3, 4, 5, 7, 8, 9, 10 | 4.32–9.56 | CIM | Prasanna *et al.* (2009) |
| 11. | Male flowering, ASI, grain yield, kernel number, 100-kernel weight, plant height | 81 | RIL (CML444 × SC-Malawi) | RFLP, SSR | Field | 1, 3, 8 | 0.1–24.1 | CIM | Messmer *et al.* (2009) |
| 12. | Root traits, yield | 1 | NIL (Lo964 × Lo1016) | SSR | Field | 1 | – | mQTL | Landi *et al.* (2010) |
| 13. | Leaf chlorophyll, senescence, root electrical capacitance | 17 | RIL (CML444 × SC-Malawi) | SSR | Field | 1, 2, 4, 5, 6, 10 | 4.4–22.9 | CIM | Messmer *et al.* (2011) |
| 14. | Root architecture-associated traits, sugar concentration, biomass, relative water content, leaf ABA | 22 | $F_{2:3}$ (DTP79 × B73) | RFLP | Greenhouse | 1, 3, 5, 6, 7, 9 | 0.2–52.2 | CIM | Rahman *et al.* (2011) |

| | | | | | | | | | |
|---|---|---|---|---|---|---|---|---|---|
| 15. | ASI, plant height, grain yield, ear height, ear setting | 25 | $F_{2:3}$ (D5 × 7924) | SSR | Rain shelter | 1, 2, 3, 4, 6, 8, 9, 10 | 5.39–15.64 | CIM | Zhu *et al.* (2011) |
| 16. | Leaf temperature, seedling dry matter | 9 | RILs (Zong3 × 87-1) | SSR | Greenhouse | 1, 2, 9, 10 | 6.2–11.0 | CIM | Liu *et al.* (2011) |
| 17. | Grain yield and associated traits, visually scored drought score, relative water content, osmotic potential, relative sugar content | 64 | $F_{2:3}$ (DTP79 × B73) | RFLP, SSR and AFLP | Field | 1, 2, 3, 4, 5, 7, 8, 10 | 0.1–45.41 | CIM | Nikolić *et al.* (2012) |
| 18. | Grain yield per plant, yield components | 45 | $F_{2:3}$ (B73 × DTP79) | SSR | Field | 1, 2, 3, 4, 5, 6, 7, 8, 10 | 0.1–28.86 | CIM | Nikolić *et al.* (2013) |
| 19. | Grain yield, ASI | 83 and 62 | RILs (CML44 × MALAWI) $F_{2:3}$ (CML440 × CML504) $F_{2:3}$ (CML444 × CML441) | SNP | Field | 1, 2, 3, 4, 5, 6, 7, 8, 10 | 1.7–17.8 | CIM | Almeida *et al.* (2013) |
| 20. | Root-associated traits | 15 | RIL (3 populations) | SNP | Greenhouse | 1, 2, 3, 4, 5, 7, 8, 9 | – | CIM | Burton *et al.* (2014) |
| 21. | Stomatal conductance, leaf relative water content, ASI, grain yield | 17 | $BC_1F_{2:3}$ (DTPWC9F1 × LPSC7F64) | SNP | Field | 1, 2, 3, 4, 6, 7, 9, 10 | – | CIM | Trachsel *et al.* (2016) |
| 22. | Grain yield, ear length, kernel number per row, ear weight, 100-kernel weight | 169 | NAM (11 parents) | SNP | Field | 1, 3, 10 | 23.7–66.3 | Joint linkage analysis | Li *et al.* (2016) |

[a]RIL, recombinant inbred line; NIL, near-isogenic line; NAM, nested association mapping.
[b]SSR, simple sequence repeat; AFLP, amplified fragment length polymorphism.
[c]PVE, phenotypic variation explained.
[d]CIM, composite interval mapping; IM, interval mapping; PLABQTL, (QTL analysis in PLant Breeding And Biology) software package.

absence of drought stress. Hence, due care should be taken about the performance of drought stress tolerance QTLs in optimal moisture environments.

Molecular breeding involves marker-assisted backcrossing (MABC), marker-assisted recurrent selection (MARS) and marker-assisted gene pyramiding. Marker-assisted selection is a method for indirect selection for targets for which direct selection is ineffective and hence speeds up the selection process (Collard and Mackill, 2008). The integration of major genes or QTLs into commonly adapted cultivars was accomplished by marker-assisted backcross breeding. Marker-assisted pyramiding is the process of combining two or more genes in the background of a common elite cultivar to impart the stress resilience but is used mainly for major QTLs (Gazal *et al.*, 2016). Similarly, MABC is used for improving the defects of otherwise elite cultivars by introgression of a major QTL. However, for complex traits such as drought stress tolerance, being governed by a large number of minor QTLs, MARS, GS and GWAS are more effective approaches for achieving better genetic gains. Accurate phenotyping and cheaper sequencing platforms act as important elements of these approaches, mainly GS (Bhat *et al.*, 2016). In GS costs are reduced by avoiding repeated phenotyping, a costly affair. However, besides precise phenotyping, selection of a representative training population with subsequent addition of diverse lines over time is a crucial factor to achieve success in GS (Bhat *et al.*, 2016). The superiority of GS over traditional breeding can be demonstrated by the comparatively higher yield gains (two- to fourfold) in drought stress environments (Beyene *et al.*, 2015). Molecular breeding strategies thus give plant breeders adequate opportunities to produce high-yielding, drought-tolerant cultivars. However, molecular breeding cannot work in isolation from conventional breeding; instead, most breeding approaches complement each other. The identified QTLs need to be verified for confirming their functionality. Hence, backcross-derived lines differing for the parental alleles at a major QTL (*root-ABA1*) were developed and screened in maize (Tuberosa *et al.*, 1998; Sanguineti *et al.*, 1999). The study revealed the strong and reliable effect of a major QTL on ABA concentration in leaf, which was later verified in subsequent studies under various water regimes (Giuliani *et al.*, 2005; Landi *et al.*, 2005, 2007). Association mapping is another important and more effective approach for identification of QTLs with better resolution and is based on historical recombinations in natural populations, hence also has the statistical power to detect rare variants (Mackay and Powell, 2007). Setter *et al.* (2010) identified SNPs related to an aldehyde oxidase gene involved in carbohydrate and ABA metabolite accumulation during moisture stress through association mapping and found the association of SNP with silk ABA concentration. Li *et al.* (2016) reported that 52 of 354 genes demonstrated substantial differential expression in well-watered and water-stressed conditions in B73 inbred line. The integrated use of all the technologies like focused morphological, physiological and biochemical traits, conventional breeding to develop the mapping populations and molecular techniques (MABC, MARS, GS, GWAS and transgenic technology) leads to the development of drought-tolerant cultivars as summarized in Fig. 15.1.

## 15.6 Transgenic Technology

Molecular breeding has its own limitations in the form of mostly minor genes available for drought tolerance. Hence, considering the limitations of a lack of a major source of variation for drought tolerance, genome editing technologies and transformation techniques offer a better solution for the identification, isolation and transfer of genes regulating drought tolerance (Pathak *et al.*, 2014; Maazou *et al.*, 2016; Sah *et al.*, 2016; Wani *et al.*, 2016b, 2017; Atif *et al.*, 2019; Kumar *et al.*, 2020). Although a lot of research papers are available on gene identification for drought tolerance, few have been revalidated through reverse genetics and transcriptomics. A maize transgenic (containing *betA* gene from *Escherichia coli*) having a regulatory action on choline dehydrogenase (involved in glycine betaine synthesis) exhibited seedling drought tolerance and high economic yield after drought due to a higher level of glycine betaine accumulation (Quan *et al.*, 2004). A gene from tobacco that encoded mitogen-activated protein kinase kinase kinase (MAPKKK) was found to activate an oxidative signal cascade and induce

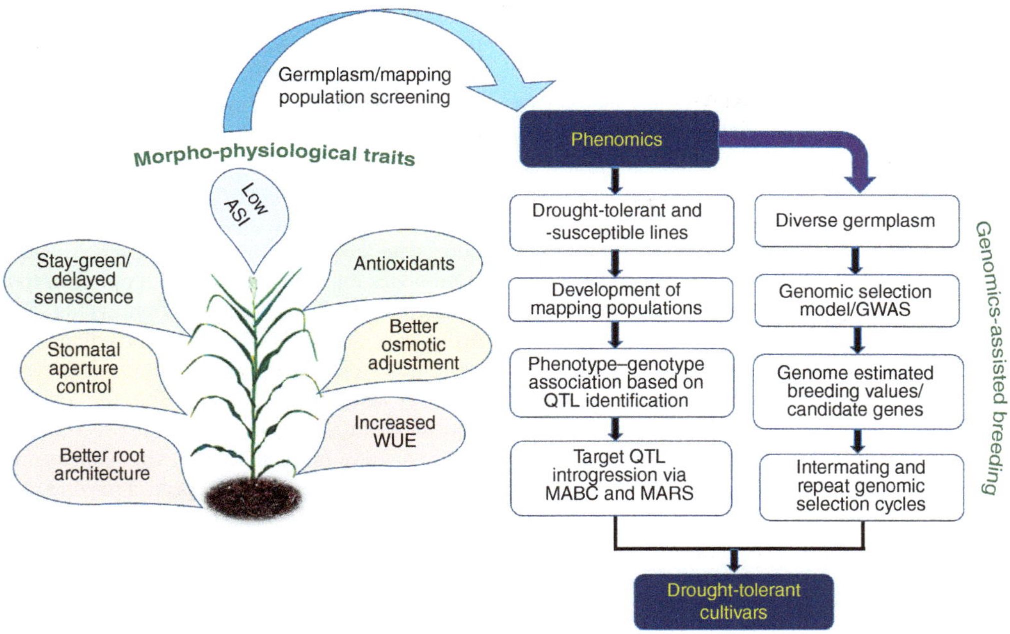


**Fig. 15.1.** Target traits and genomics-assisted breeding approaches for improvement of drought stress tolerance. Precise phenotyping via utilization of phenomics along with genomics-assisted breeding approaches is crucial for achieving higher genetic gains in drought stress breeding programmes. WUE, water-use efficiency.

tolerance for different abiotic stresses in transgenic maize. The transgenic maize had a significantly higher rate of photosynthesis than the non-transgenic maize under drought conditions. This indicated that MAPKKK controlled a mechanism that prevented the damage to photosynthesis machinery from dehydration (Shou *et al*., 2004). The genes *CspA* from *E. coli* and *CspB* from *Bacillus subtilis* were shown to induce drought tolerance in maize (Castiglioni *et al*., 2008). In another study, transgenic maize containing *betA* (encoding choline dehydrogenase from *E. coli*) and *TsVP* (encoding vacuolar $H^+$-pyrophosphatase (V-$H^+$-PPase) from *Thellungiella halophila*) genes exhibited higher glycine betaine content and $H^+$-PPase activity under drought stress and hence displayed higher relative water content, greater solute accumulation and lower cell damage (Wei *et al*., 2011). Liu *et al*. (2015) cloned the *ZmSDD1* gene and investigated its role in drought tolerance, finding that transgenic maize with overexpression of *ZmSDD1* had a better capacity to resist drought conditions. In recent work, the *PYRABACTIN RESISTANCE 1* (*PYR1*)/*PYR1-like protein* (*PYL*)/*regulatory components of the ABA receptor* (*RCARs*), which encode for the ABA receptors, were found to play pivotal roles in ABA signalling and hence in imparting drought tolerance (He *et al*., 2018; Kumar *et al*. 2020). A late embryogenesis abundant (LEA) protein is involved in drought and salt stresses. A putative gene *ZmLEA14tv* from maize cultivar Tevang 1 was isolated and expressed in plants. The transgenic maize with *ZmLEA14tv* in drought condition showed enhanced seed germination ability and twice the seedlings survived in transgenic plants, which indicated enhanced drought tolerance (Minh *et al*., 2019).

## 15.7 Conclusion

Improving drought stress tolerance in any crop is a relatively difficult task owing to its complex nature. Hence, the best strategy is to carry out precise and accurate screening under controlled conditions followed by the validation in field conditions. A selection index combining related traits such as ASI with grain yield is also an efficient approach for genetic improvement of complex traits compared with selection for yield alone. Although showing some success for the development of drought-tolerant cultivars, conventional breeding approaches are slow. Hence, integration of conventional breeding approaches with molecular markers/genomics evolved at a rapid pace to give birth to the field of genomics-assisted breeding (Singh *et al.*, 2020). Genetic mapping studies helped to identify a large number of minor genes through QTL mapping and genome-wide association but only a few major genes conferring drought tolerance in maize have been identified. However, the potential of identified major genes for introgression also remains unexplored as evident from the limited introgression studies. This could be due to lack of tightly linked markers or lack of stability of QTLs under optimum moisture conditions. Hence, fine mapping and expression studies can be explored to precisely locate the exact genomic position of drought tolerance genes and to dissect the physiological mechanism and signalling pathways governing their expression. Further genomics-assisted breeding tools such as GS offer an opportunity to speed up maize drought improvement programmes by helping to introgress major genes and minor genes. Genome sequencing has also further supplemented rapid mutation detection and gene discovery. Omics approaches too have emerged as a powerful tool to support high-throughput phenotyping and expression studies in drought breeding programmes. Although limited success has been achieved using molecular breeding, the integration of existing and emerging approaches will surely achieve more success in the near future. Furthermore, the use of speed breeding approaches will surely benefit drought stress maize breeding programmes by harnessing beneficial alleles from wild relatives.

## References


Agrama, H.A.S. and Moussa, M.E. (1996) Mapping QTLs in breeding for drought tolerance in maize. *Euphytica* 91, 89–97.

Ahmad, B., Zaid, A., Sadiq, Y., Bashir, S. and Wani, S.H. (2019) Role of selective exogenous elicitors in plant responses to abiotic stress tolerance. In: Hasanuzzaman, M., Hakeem, K., Nahar, K. and Alharby, H. (eds) *Plant Abiotic Stress Tolerance*. Springer, Cham, Switzerland, pp. 273–290.

Ahuja, I., de Vos, R.C., and Bones, A.M. (2010) Plant molecular stress responses face climate change. *Trends in Plant Science* 15, 664–674.

Almeida, G.D., Makumbi, D., Magorokosho, C., Nair, S., Borém, A. *et al.* (2013) QTL mapping in three tropical maize populations reveals a set of constitutive and adaptive genomic regions for drought tolerance. *Theoretical and Applied Genetics* 126, 583–600.

Atif, R.M., Shahid, L., Waqas, M., Ali, B., Rashid, M.A.R. *et al.* (2019) Insights on calcium-dependent protein kinases (CPKs) signaling for abiotic stress tolerance in plants. *International Journal of Molecular Sciences* 20(21), 5298.

Azhar, M.T., Wani, S.H., Chaudhary, M.T., Jameel, T., Kaur, P. and Du, X. (2020) Heat tolerance in cotton: morphological, physiological, and genetic perspectives. In: Wani, S.H. and Kumar, V. (eds) *Heat Stress Tolerance in Plants: Physiological, Molecular and Genetic Perspectives*. Wiley, Chichester, UK, pp. 1–22.

Babu, R.C., Nguyen, B.D., Chamarerk, V., Shanmugasundaram, P., Chezhian, P. *et al.* (2003) Genetic analysis of drought resistance in rice by molecular markers: association between secondary traits and field performance. *Crop Science* 43, 1457–1469.

Bailey-Serres, J., Lee, S.C. and Brinton, E. (2012) Waterproofing crops: effective flooding survival strategies. *Plant Physiology* 160, 1698–1709.

Bänziger, M., Edmeades, G.O., Beck, D. and Bellon, M. (2000) *Breeding for Drought and Nitrogen Stress Tolerance in Maize: From Theory to Practice*. CIMMYT, Mexico City.

Bänziger, M., Setimela, P.S., Hodson, D. and Vivek, B. (2005) Breeding for improved abiotic stress tolerance in maize adapted to Southern Africa. *Agricultural Water Management* 80, 212–224.

Beissinger, T., Kruppa, J., Cavero, D., Ha, N.T., Erbe, M. and Simianer, H. (2018) A simple test identifies selection on complex traits. *Genetics* 209(1), 321–333.

Beyene, Y., Semagn, K., Mugo, S., Tarekegne, A., Babu, R. *et al.* (2015) Genetic gains in grain yield through genomic selection in eight bi-parental maize populations under drought stress. *Crop Science* 55(1), 154–163.

Bhat, J.A., Ali, S., Salgotra, R.K., Mir, Z.A., Dutta, S., Jadon, V. and Singh, G.P. (2016) Genomic selection in the era of next generation sequencing for complex traits in plant breeding. *Frontiers in Genetics* 7, 221.

Blum, A. (2006) Drought adaptation in cereal crops: a prologue. In: Ribaut, J.M. (ed.) *Drought Adaptation in Cereals*. The Haworth Press, Inc., Binghamton, New York, pp. 3–15.

Blum, A. (2011) *Plant Breeding for Water-Limited Environments*. Springer, London.

Borrell, A., Hammer, G. and Van Oosterom, E.R.I.K. (2001) Stay-green: a consequence of the balance between supply and demand for nitrogen during grain filling. *Annals of Applied Biology* 138, 91–95.

Bruce, W.B., Edmeades, G.O. and Barker, T.C. (2002) Molecular and physiological approaches to maize improvement for drought tolerance. *Journal of Experimental Botany* 53, 13–25.

Burton, A.L., Johnson, J.M., Foerster, J.M., Hirsch, C.N., Buell, C.R. *et al.* (2014) QTL mapping and phenotypic variation for root architectural traits in maize (*Zea mays* L.). *Theoretical and Applied Genetics* 127, 2293–2311.

Castiglioni, P., Warne, R.D., Bensen, R.J., Anstrom, D.C., Harrison, J. *et al.* (2008) Bacterial RNA chaperones confer abiotic stress tolerance in plants and improved grain yield in maize under water-limited conditions. *Plant Physiology* 147, 446–455.

Cattivelli, L., Baldi, P., Crosatti, C., Di Fonzo, N., Faccioli, P. *et al.* (2002) Chromosome regions and stress-related sequences involved in resistance to abiotic stress in Triticeae. *Plant Molecular Biology* 48, 649–665.

Choudhary, M., Singh, V., Muthusamy, V. and Wani, S.H. (2017) Harnessing crop wild relatives for crop improvement. *International Journal of Life Sciences* 6(2), 73–85.

Choudhary, M., Wani, S.H., Kumar, P., Bagaria, P.K., Rakshit, S., Roorkiwal, M. and Varshney, R.K. (2019) QTLian breeding for climate resilience in cereals: progress and prospects. *Functional & Integrative Genomics* 19, 685–701.

Cobb, J.N., DeClerck, G., Greenberg, A., Clark, R. and McCouch, S. (2013) Next-generation phenotyping: requirements and strategies for enhancing our understanding of genotype–phenotype relationships and its relevance to crop improvement. *Theoretical and Applied Genetics* 126, 867–887. Available at: https://doi.org/10.1007/s00122-013-2066-0

Collard, B.C. and Mackill, D.J. (2008) Marker-assisted selection: an approach for precision plant breeding in the twenty-first century. *Philosophical Transactions of the Royal Society B: Biological Sciences* 363(1491), 557–572.

Cooper, M., Gho, C., Leafgren, R., Tang, T. and Messina, C. (2014) Breeding drought-tolerant maize hybrids for the US corn-belt: discovery to product. *Journal of Experimental Botany* 65, 6191–6204.

Coraggio, I. and Tuberosa, R. (2004) Improving crops tolerance to abiotic stresses. In: Christou, P. and Klee, H. (eds) *Handbook of Plant Biotechnology*. Wiley, Chichester, UK, pp. 413–468.

Cramer, G.R., Urano, K., Delrot, S., Pezzotti, M. and Shinozaki, K. (2011) Effects of abiotic stress on plants: a systems biology perspective. *BMC Plant Biology* 11(1), 163.

Cushman, J.C. and Bohnert, H.J. (2000) Genomic approaches to plant stress tolerance. *Current Opinion in Plant Biology* 3, 117–124.

Dar, N.A., Amin, I., Wani, W., Wani, S.A., Shikari, A.B., Wani, S.H. and Masoodi, K.Z. (2017) Abscisic acid: a key regulator of abiotic stress tolerance in plants. *Plant Gene* 11, 106–111.

Dutta, T., Neelapu, N.R., Wani, S.H. and Challa, S. (2018) Compatible solute engineering of crop plants for improved tolerance toward abiotic stresses. In: Wani, S.H. (ed.) *Biochemical, Physiological and Molecular Avenues for Combating Abiotic Stress Tolerance in Plants*. Academic Press, London, pp. 221–254.

Edmeades, G.O. (2013) *Progress in Achieving and Delivering Drought Tolerance in Maize: An Update*. ISAAA, Ithaca, New York.

Edmeades, G.O., Cooper, M., Lafitte, R., Zinselmeier, C., Ribaut, J.M. *et al.* (2001) Abiotic stresses and staple crops. In: Nösberger, J., Geiger, H.H. and Struik, P.C. (eds) *Crop Science: Progress and Prospects. Third International Crop Science Congress*. CAB International, Wallingford, UK, pp. 137–154.

Edmeades, G.O., Bänziger, M., Campos, H. and Schussler, J.R. (2006) Improving tolerance to abiotic stresses in staple crops: a random or planned process? In: Lamkey, K.R. and Lee, M. (eds) *Plant Breeding: The Arnel R. Hallauer International Symposium*. Blackwell Publishing, Ames, Iowa, pp. 293–309.

Falconer, D.S. (1989) *Introduction to Quantitative Genetics*, 3rd edn. Longman, London.

Finkel, E. (2009) With 'phenomics,' plant scientists hope to shift breeding into overdrive. *Science* 325, 380–381. Available at: https://doi.org/10.1126/science.325_380

Ganie, S.A., Ahammed, G.J. and Wani, S.H. (2020) Vascular plant one zinc-finger (VOZ) transcription factors: novel regulators of abiotic stress tolerance in rice (*Oryza sativa* L.). *Genetic Resources and Crop Evolution* 67, 799–807.
Gazal, A., Dar, Z.A., Wani, S.H., Lone, A.A., Shikari, A.B., Ali, G. and Abidi, I. (2016) Molecular breeding for enhancing resilience against biotic and abiotic stress in major cereals. *SABRAO Journal of Breeding & Genetics* 48(1), 1–32.
Gharoobi, B., Ghorbani, M. and Nezhad, M.G. (2012) Effects of different levels of osmotic potential on germination percentage and germination rate of barley, corn and canola. *Iran Journal of Plant Physiology* 2(2), 413–417.
Giuliani, S., Sanguineti, M.C., Tuberosa, R., Bellotti, M., Salvi, S. and Landi, P. (2005) *Root-ABA1*, a major constitutive QTL, affects maize root architecture and leaf ABA concentration at different water regimes. *Journal of Experimental Botany* 56, 3061–3070.
Goffinet, B. and Gerber, S. (2000) Quantitative trait loci: a meta-analysis. *Genetics* 155(1), 463–473.
Gosal, S.S. and Wani, S.H. (2018) Plant genetic transformation and transgenic crops: methods and applications. In: Gosal, S.S. and Wani, S.H. (eds) *Biotechnologies of Crop Improvement*, Vol. 2. Springer, Cham, Switzerland, pp. 1–23.
Gosal, S.S., Wani, S.H. and Kang, M.S. (2009) Biotechnology and drought tolerance. *Journal of Crop Improvement* 23(1), 19–54.
Guo, J.F., Su, G.Q., Zhang, J.P. and Wang, G.Y. (2008) Genetic analysis and QTL mapping of maize yield and associate agronomic traits under semiarid land condition. *African Journal of Biotechnology* 7, 1829–1838.
Gupta, P.K., Balyan, H.S., Gahlaut, V. and Kulwal, P. (2012) Phenotyping, genetic dissection, and breeding for drought and heat tolerance in common wheat: status and prospects. *Plant Breeding Reviews* 36, 85–168.
He, Z., Zhong, J., Sun, X., Wang, B., Terzaghi, W. and Dai, M. (2018) The maize ABA receptors ZmPYL8, 9, and 12 facilitate plant drought resistance. *Frontiers in Plant Science* 9, 422.
Hsiao, T.C. and Xu, L.K. (2000) Sensitivity of growth of roots versus leaves to water stress: biophysical analysis and relation to water transport. *Journal of Experimental Botany* 51, 1595–1616.
Joshi, R., Wani, S.H., Singh, B., Bohra, A., Dar, Z.A. *et al.* (2016) Transcription factors and plants response to drought stress: current understanding and future directions. *Frontiers in Plant Science* 7, 1029.
Jurgens, S.K., Johnson, R.R. and Boyer, J.S. (1978) Dry matter production and translocation in maize subjected to drought during grain fill. *Agronomy Journal* 70, 678–682.
Khodarahmpour, Z. (2011) Effect of drought stress induced by polyethylene glycol (PEG) on germination indices in corn (*Zea mays* L.) hybrids. *African Journal of Biotechnology* 10(79), 18222–18227.
Kole, C., Muthamilarasan, M., Henry, R., Edwards, D., Sharma, R. *et al.* (2015) Application of genomics-assisted breeding for generation of climate resilient crops: progress and prospects. *Frontiers in Plant Science* 6, 563.
Kumar, K., Gambhir, G., Dass, A., Tripathi, A.K., Singh, A. *et al.* (2020) Genetically modified crops: current status and future prospects. *Planta* 251, 91.
Kumar, V., Sah, S.K., Khare, T., Shriram, V. and Wani, S.H. (2016) Engineering phytohormones for abiotic stress tolerance in crop plants. In: Ahammed, G. and Yu, J.Q. (eds) *Plant Hormones Under Challenging Environmental Factors*. Springer, Dordrecht, the Netherlands, pp. 247–266.
Landi, P., Sanguineti, M.C., Salvi, S., Giuliani, S., Bellotti, M. *et al.* (2005) Validation and characterization of a major QTL affecting leaf ABA concentration in maize. *Molecular Breeding* 15, 291–303.
Landi, P., Sanguineti, M.C., Liu, C., Li, Y., Wang, T.Y. *et al.* (2007) *Root-ABA1* QTL affects root lodging, grain yield, and other agronomic traits in maize grown under well-watered and water-stressed conditions. *Journal of Experimental Botany* 58, 319–326.
Landi, P., Giuliani, S., Salvi, S., Ferri, M., Tuberosa, R. and Sanguineti, M.C. (2010) Characterization of *root-yield-1.06*, a major constitutive QTL for root and agronomic traits in maize across water regimes. *Journal of Experimental Botany* 61, 3553–3562.
Lebreton, C., Lazic-Jancic, V., Steed, A., Pekic, S. and Quarrie, S.A. (1995) Identification of QTL for drought responses in maize and their use in testing causal relationships between traits. *Journal of Experimental Botany* 46, 853–865.
Li, C., Sun, B., Li, Y., Liu, C., Wu, X. *et al.* (2016) Numerous genetic loci identified for drought tolerance in the maize nested association mapping populations. *BMC Genomics* 17(1), 894.

Li, Y., Yang, M., Dong, Y., Wang, Q., Zhou, Y. *et al.* (2011) Three main genetic regions for grain development revealed through QTL detection and meta-analysis in maize. *Molecular Breeding* 30, 195–211. Available at: https://doi.org/10.1007/s11032-011-9610-x

Liu, Y., Subhash, C., Yan, J., Song, C., Zhao, J. and Li, J. (2011) Maize leaf temperature responses to drought: thermal imaging and quantitative trait loci (QTL) mapping. *Environmental and Experimental Botany* 71(2), 158–165.

Liu, Y., Qin, L., Han, L., Xiang, Y. and Zhao, D. (2015) Overexpression of maize SDD1 (*ZmSDD1*) improves drought resistance in *Zea mays* L. by reducing stomatal density. *Plant Cell, Tissue and Organ Culture* 122, 147–159.

Maazou, A.R., Tu, J., Qiu, J. and Liu, Z. (2016) Breeding for drought tolerance in maize (*Zea mays* L.). *American Journal of Plant Science* 7(14), 1858–1870.

Mackay, I. and Powell, W. (2007) Methods for linkage disequilibrium mapping in crops. *Trends in Plant Science* 12, 57–63.

Magorokosho, C., Pixley, K.V. and Tongoona, P. (2003) Selection for drought tolerance in two tropical maize populations. *African Crop Science Journal* 11, 151–161.

Maleki, M., Ghorbanpour, M., Nikabadi, S. and Wani, S.H. (2019) *In vitro* screening of crop plants for abiotic stress tolerance. In: Wani, S.H. (ed.) *Recent Approaches in Omics for Plant Resilience to Climate Change*. Springer, Cham, Switzerland, pp. 75–91.

Messmer, R., Fracheboud. Y., Bänziger, M., Vargas, M., Stamp, P. and Ribaut, J.M. (2009) Drought stress and tropical maize: QTL-by-environment interactions and stability of QTLs across environments for yield components and secondary traits. *Theoretical and Applied Genetics* 119, 913–930.

Messmer, R., Fracheboud, Y., Bänziger, M., Stamp, P. and Ribaut, J.M. (2011) Drought stress and tropical maize: QTLs for leaf greenness, plant senescence, and root capacitance. *Field Crops Research* 124, 93–103.

Mickelbart, M.V., Hasegawa, P.M. and Bailey-Serres, J. (2015) Genetic mechanisms of abiotic stress tolerance that translate to crop yield stability. *Nature Reviews Genetics* 16, 237–251.

Minh, B.M., Linh, N.T., Hanh, H.H., Hien, L.T.T., Thang, N.X., Hai, V.N. and Hue, H.T.T. (2019) A *LEA* gene from a Vietnamese maize landrace can enhance the drought tolerance of transgenic maize and tobacco. *Agronomy* 9, 62.

Mir, R.R., Zaman-Allah, M., Sreenivasulu, N., Trethowan, R. and Varshney, R.K. (2012) Integrated genomics, physiology and breeding approaches for improving drought tolerance in crops. *Theoretical and Applied Genetics* 125, 625–645.

Monneveux, P. and Ribaut, J.M. (2006) Secondary traits for drought tolerance improvement in cereals. In: Ribaut, J.M. (ed.) *Drought Adaptation in Cereals*. The Haworth Press, Inc., Binghamton, New York, pp. 97–143.

Monneveux, P., Sanchez, C., Beck, D. and Edmeades, G.O. (2005) Drought tolerance improvement in tropical maize source populations. *Crop Science* 46, 180–191.

Nikolić, A., Ignjatović-Micić, D., Dodig, D., Anđelković, V. and Lazić-Jančić, V. (2012) Identification of QTLs for yield and drought-related traits in maize: assessment of their causal relationships. *Biotechnology & Biotechnological Equipment* 26(3), 2952–2960.

Nikolić, A., Andelković, V., Dodig, D., Mladenović-Drinić, S., Kravić, N. and Ignjatović-Micić, D. (2013) Identification of QTL-s for drought tolerance in maize, II: yield and yield components. *Genetika* 45(2), 341–350.

Pace, J., Lee, N., Naik, H.S., Ganapathysubramanian, B. and Lübberstedt, T. (2014) Analysis of maize (*Zea mays* L.) seedling roots with the high-throughput image analysis tool *ARIA* (Automatic Root Image Analysis). *PLoS One* 9(9), e108255.

Pandita, D. and Wani, S.H. (2019) MicroRNA as a tool for mitigating abiotic stress in rice (*Oryza sativa* L.). In: Wani, S.H. (ed.) *Recent Approaches in Omics for Plant Resilience to Climate Change*. Springer, Cham, Switzerland, pp. 109–133.

Parry, M.A.J., Flexas, J. and Medrano, H. (2005) Prospects for crop production under drought: research priorities and future directions. *Annals of Applied Biology* 147, 211–226.

Passioura, J.B. (2010) Scaling up: the essence of effective agricultural research. *Functional Plant Biology* 37, 585–591. Available at: https://doi.org/10.1071/FP10106

Pathak, M.R., Teixeira da Silva, J.A. and Wani, S.H. (2014) Polyamines in response to abiotic stress tolerance through transgenic approaches. *GM Crops & Food* 5(2), 87–96.

Prasanna, B.M., Beiki, A.H., Sekhar, J.C., Srinivas, A. and Ribaut, J.M. (2009) Mapping QTLs for component traits influencing drought stress tolerance of maize (*Zea mays* L) in India. *Journal of Plant Biochemistry and Biotechnology* 18(2), 151–160.
Qin, F., Shinozaki, K., and Yamaguchi-Shinozaki, K. (2011) Achievements and challenges in understanding plant abiotic stress responses and tolerance. *Plant & Cell Physiology* 52(9), 1569–1582.
Quan, R., Shang, M., Zhang, H., Zhao, Y. and Zhang, J. (2004) Engineering of enhanced glycine betaine synthesis improves drought tolerance in maize. *Plant Biotechnology Journal* 2, 477–486.
Rahman, H., Pekic, S., Lazic-Jancic, V., Quarri, S.A., Shah, S.M.A., Pervez, A. and Shah, M.M. (2011) Molecular mapping of quantitative trait loci for drought tolerance in maize plants. *Genetics Molecular Research* 10(2), 889–901.
Ribaut, J.M., Hoisington, D.A., Deutsch, J.A., Jiang, C., González-de-León, D. (1996) Identification of quantitative trait loci under drought conditions in tropical maize. 1. Flowering parameters and the anthesis-silking interval. *Theoretical and Applied Genetics* 92, 905–914.
Ribaut, J.M., Jiang, C., Gonzalez-de-Leon, D., Edmeades, G.O. and Hoisington, D.A. (1997) Identification of quantitative trait loci under drought conditions in tropical maize. II: Yield components and marker selection strategies. *Theoretical and Applied Genetics* 94, 887–896.
Ribaut, J.M., Betran, J., Monneveux, P. and Setter, T. (2009) Drought tolerance in maize. In: Bennetzen, J. and Hake, S. (eds) *Handbook of Maize: Its Biology*. Springer, New York, pp. 311–344.
Robertson, M.J., Fukai, S., Ludlow, M.M. and Hammer, G.L. (1993) Water extraction by grain sorghum in a sub-humid environment. II. Extraction in relation to root growth. *Field Crops Research* 33, 99–112.
Sah, S.K., Kaur, G. and Wani, S.H. (2016) Metabolic engineering of compatible solute trehalose for abiotic stress tolerance in plants. In: Iqbal, N., Nazar R. and Khan, N.A. (eds) *Osmolytes and Plants Acclimation to Changing Environment: Emerging Omics Technologies*. Springer, New Delhi, pp. 83–96.
Sanguineti, M.C., Tuberosa, R., Landi, P., Salvi, S., Maccaferri, M., Casarini, E. and Conti, S. (1999) QTL analysis of drought-related traits and grain yield in relation to genetic variation for leaf abscisic acid concentration in field-grown maize. *Journal of Experimental Botany* 50, 1289–1297.
Sari-Gorla, M., Krajewski, P., Di-Fonzo, N., Villa, M. and Frova, C. (1999) Genetic analysis of drought tolerance in maize by molecular markers. II. Plant height and flowering. *Theoretical and Applied Genetics* 99, 289–295.
Setter, T.L., Yan, J., Warburton, M., Ribaut, J.M., Xu, Y. *et al.* (2010) Genetic association mapping identifies single nucleotide polymorphisms in genes that affect abscisic acid levels in maize floral tissues during drought. *Journal of Experimental Botany* 62, 701–716.
Shou, H., Bordallo, P. and Wang, K. (2004) Expression of the nicotiana protein kinase (NPK1) enhanced drought tolerance in transgenic maize. *Journal of Experimental Botany* 55, 1013–1019.
Singh, L., Sharma, D., Parmar, N., Singh, K.H., Jain, R. *et al.* (2020) Genetic diversity studies in Indian mustard (*Brassica juncea* L. Czern and Coss) using molecular markers. In: Wani, S.H., Thakur, A. and Jeshima Khan, Y. (eds) *Brassica Improvement*. Springer, Cham, Switzerland, pp. 215–244.
Soliman, S., El-Keblawy, A., Mosa, K.A., Helmy, M. and Wani, S.H. (2018) Understanding the phytohormones biosynthetic pathways for developing engineered environmental stress-tolerant crops. In: Gosal, S.S. and Wani, S.H. (eds) *Biotechnologies of Crop Improvement*, Vol. 2. Springer, Cham, Switzerland, pp. 417–450.
Subbarao, G.V., Ito, O., Serraj, R., Crouch, J.J., Tobita, S. *et al.* (2005) Physiological perspectives on improving crop adaptation to drought – justification for a systematic component-based approach. In: Pessarakli, M. (ed.) *Handbook of Photosynthesis*, 2nd edn. Marcel and Dekker, New York, pp. 577–594.
Swamy, B.M., Vikram, P., Dixit, S., Ahmed, H.U. and Kumar, A. (2011) Meta-analysis of grain yield QTL identified during agricultural drought in grasses showed consensus. *BMC Genomics* 12, 319. Available at: https://doi.org/10.1186/1471-2164-12-319
Tardieu, F.T. and Simonneau, T. (1998) Variability among species of stomatal control under fluctuating soil water status and evaporative demand: modeling isohydric and anisohydric behaviors. *Journal of Experimental Botany* 49, 419–432.
Telem, R.S., Wani, S.H., Singh, N.B., Sadhukhan, R. and Mandal, N. (2016) Single nucleotide polymorphism (SNP) marker for abiotic stress tolerance in crop plants. In: Al-Khayri, J., Jain, S. and Johnson, D. (eds) *Advances in Plant Breeding Strategies: Agronomic, Abiotic and Biotic Stress Traits*. Springer, Cham, Switzerland, pp. 327–343.

Trachsel, S., Dapeng, S., Felix, M., San, V., Hongjian, Z. *et al.* (2016) Identification of QTL for early vigor and stay-green conferring tolerance to drought in two connected advanced backcross populations in tropical maize (*Zea mays* L.). *PloS One* 11(3), e0149636.

Tuberosa, R. (2012) Phenotyping for drought tolerance of crops in the genomics era. *Frontiers in Physiology* 3, 347.

Tuberosa, R. and Salvi, S. (2004) QTLs and genes for tolerance to abiotic stress in cereals. In: Gupta, P.K. and Varshney, R. (eds) *Cereal Genomics*. Kluwer, Dordrecht, the Netherlands, pp. 253–315.

Tuberosa, R., Sanguineti, M.C., Landi, P., Salvi, S., Casarini, E. and Conti, S. (1998) RFLP mapping of quantitative trait loci controlling abscisic acid concentration in leaves of drought-stressed maize (*Zea mays* L.). *Theoretical and Applied Genetics* 97, 744–755.

Tuberosa, R., Sanguineti, M.C., Landi, P., Giuliani, M.M., Salvi, S. and Conti, S. (2002) Identification of QTLs for root characteristics in maize grown in hydroponics and analysis of their overlap with QTLs for grain yield in the field at two water regimes. *Plant Molecular Biology* 48, 697–712.

Wani, S.H. and Kumar, V. (2015) Plant stress tolerance: engineering ABA: a potent phytohormone. *Transcriptomics* 3(2), 1000113.

Wani, S.H. and Sah, S.K. (2014) Biotechnology and abiotic stress tolerance in rice. *Journal of Rice Research* 2, e105.

Wani, S.H., Singh, N.B., Haribhushan, A. and Mir, J.I. (2013) Compatible solute engineering in plants for abiotic stress tolerance – role of glycine betaine. *Current Genomics* 14(3), 157–165.

Wani, S.H., Kumar, V., Shriram, V. and Sah, S.K. (2016a) Phytohormones and their metabolic engineering for abiotic stress tolerance in crop plants. *The Crop Journal* 4(3), 162–176.

Wani, S.H., Sah, S.K., Hossain, M.A., Kumar, V. and Balachandran, S.M. (2016b) Transgenic approaches for abiotic stress tolerance in crop plants. In: Al-Khayri, J., Jain, S. and Johnson, D. (eds) *Advances in Plant Breeding Strategies: Agronomic, Abiotic and Biotic Stress Traits*. Springer, Cham, Switzerland, pp. 345–396.

Wani, S.H., Dutta, T., Neelapu, N.R.R. and Surekha, C. (2017) Transgenic approaches to enhance salt and drought tolerance in plants. *Plant Gene* 11, 219–231.

Wani, S.H., Choudhary, M., Kumar, P., Akram, N.A., Surekha, C., Ahmad, P. and Gosal, S.S. (2018a) Marker-assisted breeding for abiotic stress tolerance in crop plants. In: Gosal, S.S. and Wani, S.H. (eds) *Biotechnologies of Crop Improvement*, Vol. 3. Springer, Cham, Switzerland, pp. 1–23.

Wani, S.H., Tripathi, P., Zaid, A., Challa, G.S., Kumar, A. *et al.* (2018b) Transcriptional regulation of osmotic stress tolerance in wheat (*Triticum aestivum* L.). *Plant Molecular Biology* 97(6), 469–487.

Wani, S.H., Kumar, V., Khare, T., Guddimalli, R., Parveda, M., Solymosi, K., Suprasanna, P. and Kishor, P.K. (2020) Engineering salinity tolerance in plants: progress and prospects. *Planta* 251, 76.

Wei, A., He, C., Li, B., Li, N. and Zhang, J. (2011) The pyramid of transgenes *TsVP* and *BetA* effectively enhances the drought tolerance of maize plants. *Plant Biotechnology Journal* 9, 216–229.

Welcker, C., Boussuge, B., Bencivenni, C., Ribaut, J.M. and Tardieu, F. (2007) Are source and sink strengths genetically linked in maize plants subjected to water deficit? A QTL study of the responses of leaf growth and of anthesis-silking interval to water deficit. *Journal of Experimental Botany* 58, 339–349.

Whitford, R., Gilbert, M. and Langridge, P. (2010) Biotechnology in agriculture. In: Reynolds, M.P. (ed.) *Climate Change and Crop Production*. CABI Series in Climate Change, Vol. 1. CAB International, Wallingford, UK, pp. 219–244.

Xiao, Y.N., Li, X.H., George, M.L., Li, M.S., Zhang, S.H. and Zheng, Y.L. (2005) Quantitative trait locus analysis of drought tolerance and yield in maize in China. *Plant Molecular Biology Reporter* 23, 155–165.

Xu, Y.B. and Crouch, J.H. (2008) Marker-assisted selection in plant breeding: from publications to practice. *Crop Science* 48, 391–407.

Yadav, A., Singh, J., Ranjan, K., Kumar, P., Khanna, S. *et al.* (2020) Heat shock proteins: master players for heat-stress tolerance in plants during climate change. In: Wani, S.H. and Kumar, V. (eds) *Heat Stress Tolerance in Plants: Physiological, Molecular and Genetic Perspectives*. Wiley, Chichester, UK, pp. 198–211.

Zhu, J.J., Wang, X.P., Sun, C.X., Zhu, X.M., Li, M. *et al.* (2011) Mapping of QTL associated with drought tolerance in a semi-automobile rain shelter in maize (*Zea mays* L.). *Agricultural Sciences in China* 10(7), 987–996.

# 16 The Genetic Architecture and Breeding Towards Cold Tolerance in Maize: Review

**Haixiao Dong[1], Zhao Li[1], Guanghui Hu[2], Yaping Yuan[1] and Zhiwu Zhang[3]***
*[1]Jilin University, Changchun, People's Republic of China; [2]Heilongjiang Academy of Agricultural Sciences, Harbin, People's Republic of China; [3]Washington State University, Pullman, Washington, USA*

## 16.1 Introduction

Maize (*Zea mays* L.) originated from tropical regions, so it is inherently sensitive to low temperatures (Sobkowiak *et al.*, 2014; Zhao *et al.*, 2016; Di Fenza *et al.*, 2017). Over years, the planting area of maize has been expanding due to the increasing demand for maize as an industry commodity and food. As there are still cooler regions with higher latitude or elevation, cold tolerance is still the bottleneck for the expansion. Additionally, cold stress frequently happens in the spring, impairing the maize product and its benefits.

The relevant research was reviewed for cold stress in 2005 (Marocco *et al.*, 2005) and for maize growth and development in 2014 (Sánchez *et al.*, 2014). This chapter reviews the global adaptation of maize, the effect of cold stress, existing cold-tolerant or cold-sensitive maize varieties or mutants, research on linkage analysis, and genome-wide association studies (GWAS) and gene expression profiling in maize cold response. In addition, the potential usage of genomic selection (GS) to accelerate the breeding process is explored. The objectives are to integrate knowledge for the benefit of geneticists to understand the genetic architecture of cold tolerance and for breeders to select 'hyper-tolerant' maize varieties adapted to broader and changeable environments.

## 16.2 Maize Cold Tolerance and Global Adaptation

Originating in Mexico under a tropical climate, maize prefers warm weather for its growth. Unlike heat stress, which happens only during some extremely hot and dry summers, cold stress is more common and can negatively impact maize throughout its entire life cycle. When cold stress occurs during the sowing-to-emergence period, the germination process can be delayed and/or the germination rate can be reduced. Cold stress occurs most often during the seedling stage in the early spring, when cold temperatures are accompanied by strong winds. Leaves suffering from this stress will wilt, droop or even die. Long periods of low temperatures will often cause anthocyanin accumulation and reduced chlorophyll content. At flowering time, low temperatures can prevent anthers from shedding pollen and thus affect seed setting. During the milking

*Corresponding author: zhiwu.zhang@wsu.edu

 *Molecular Breeding in Wheat, Maize and Sorghum: Strategies for Improving Abiotic Stress Tolerance and Yield* (eds M.A. Hossain *et al.*)

DOI: 10.1079/9781789245431.0016

stage, cold weather can terminate grain filling, resulting in small seeds.

Nevertheless, through domestication and artificial selection, maize has adapted to temperate climates. For example, maize cultivation has expanded to cooler, higher-latitude and mountainous regions. The optimal temperature for maize production is about 30–35°C; the minimum temperature is about 10–12°C. The temperature limitations and optimums for each of the various developmental phases of maize growth are summarized by Sánchez *et al.* (2014). Breeding programmes are continuing the efforts to improve the adaptability.

## 16.3 Maize Germplasm for Cold Tolerance

Both cold-tolerant and cold-sensitive maize varieties are valuable to diversify germplasm for cold tolerance improvement and to explore the mechanisms for low temperature response. When these germplasms are exposed to low temperatures, reduction of biomass production is expected to be the ultimate criterion to differentiate tolerant and sensitive lines. To shorten the time of evaluation and simplify the protocol, however, different criteria have been proposed. For example, Hu *et al.* (2017) conducted a cold tolerance evaluation of the maize 282 diverse panel using germination tests under normal and chilling conditions. Bhosale *et al.* (2007) evaluated the chilling tolerance of five European *flint* maize, five European *dent* maize inbred lines and their 25 factorial crosses, based on chlorophyll concentration, plant height and fresh weight measurements. Schapendonk *et al.* (1989) used chlorophyll fluorescence measurements to screen cold tolerance in four maize populations that were developed by divergent mass selection for contrasting resistance to chlorosis.

The differences among the evaluation methods resulted in different performance aspects of maize germplasm being measured (Table 16.1). The classification of tolerant versus sensitive also varied among the different criteria. Some maize lines were defined as cold-sensitive in multiple studies including Mo17, G50, CM 109 and Penjalinan; whereas other lines were defined as cold-tolerant in multiple studies consistently, including Huang C, S68911, KW 1074 and Z7. On the other hand, some lines were classified differently across studies due to various measurements or evaluation criteria. For example, the B73 line was classified as cold-tolerant in some older studies (Stewart *et al.*, 1990; Chen *et al.*, 2013) but as cold-sensitive in more recent studies (Bilska-Kos *et al.*, 2018).

## 16.4 Maize Mutants for Cold Tolerance

Natural and artificial mutants with altered phenotypes or physiological responses provide good resources for breeding and for studies using forward and reverse genetics. Researchers have applied various methods, including ethyl methanesulfonate (EMS), *Mu* insertion and irradiation, to create tens of thousands of maize mutants, including the ones listed by libraries/databases that are publicly available (Table 16.2).

Most of the time, EMS-induced mutation triggers C→T changes, resulting in C/G to T/A substitutions, which is an advantage over other methods that often cause large structural variations. Another advantage of EMS is that the mutations are randomly distributed throughout the genome (Greene *et al.*, 2003). Traditionally, map-based cloning for maize EMS mutants is time- and labour-consuming. Now, new gene mapping strategies like SHOREmap (Schneeberger *et al.*, 2009), MutMap (Abe *et al.*, 2012) and BSR-Seq (Liu *et al.*, 2012) have been developed. These methods combine whole-genome/transcriptome sequencing and bulked segregant analysis (BSA), greatly simplifying the gene mapping process. Creating a new EMS-induced library or using the currently available libraries to test those mutant lines in response to different abiotic stresses can be beneficial for evaluating their tolerance. Those mutants found to be tolerant or sensitive to the target abiotic stressor could then be gene mapped using a BSA strategy.

Many mutants showing tolerance or sensitivity to cold stress have been studied in *Arabidopsis*, rice and some other species. Recently, researchers have used mutant libraries to discover novel genes important to chilling stress. In *Arabidopsis*, Wang *et al.* (2016) screened for chilling tolerance in 11,000 T-DNA insertion mutant

**Table 16.1.** Performance and mechanism of cold-tolerant or -sensitive maize germplasm.

| Tolerant | Sensitive | Performance and/or mechanism | Reference |
|---|---|---|---|
| CFD04_349 | CFD04_332 | Sensitive line had weak growth and aerial biomass. Tolerant line had higher chlorophyll content, glucose-6-phosphate dehydrogenase activity and sucrose:starch ratio | Duran Garzon *et al.* (2020) |
| M54 | 753F | Tolerant line had protection on PSII and accumulated secondary metabolites. Genes related to photosystem structure and regulating electron transport were less affected while genes related to secondary metabolism and unsaturated fatty acid synthesis were upregulated | Li *et al.* (2019) |
| S68911 (Zm-T) | B73 (Zm-S) | Sensitive line demonstrated inhibition of net $CO_2$ assimilation and a clear decrease in F′v/F′m, Fv/Fm and ϕPSII | Bilska-Kos *et al.* (2018) |
| 220 and P9-10 | Y1518 and PH4CV | Tolerant lines had better germination | Li *et al.* (2018) |
| CO (439,438, 450, 435 and 445) | CO (437, 436, and 440) | Tolerant lines had better germination | Farooqi *et al.* (2016) |
| KW 1074 | CM 109 | Sensitive line demonstrated the reduction of aquaporins (PIP2;3) | Bilska-Kos *et al.* (2016) |
| KW 1074 | CM 109 | Sensitive line exhibited reduction of pectin content and PME activity | Bilska-Kos *et al.* (2017) |
| S68911 | S50676 and S160 | Tolerant line had less damage to the photosynthetic apparatus | Sobkowiak *et al.* (2016) |
| ETH-DH7 | ETH-DL3 | Genes encoding membrane/cell-wall proteins were induced | Sobkowiak *et al.* (2014) |
| Fengdan 3 | Zhengdan 958 | RuBPCase and PEPCase showed different temperature sensitivity | Chen *et al.* (2013) |
| B73 | Mo17 | *ZmCIPKs* had different expression patterns among the two lines | Chen *et al.* (2011) |
| KW 1074 | CM 109 | Inhibition of photosynthesis was reversible in tolerant line but not in sensitive line | Bilska and Sowinski (2010) |
| Huang C | Mo17 | Tolerant line had higher growth rate of mesocotyl, coleoptile, CAT and POD activities, Pro content in root, and lower plasma-membrane permeability | Gao *et al.* (2006) |
| KW 1074 | CM 109 | Sensitive line had more responses on chloroplast ultrastructure and photosynthetic efficiency | Sowinski *et al.* (2005) |
| Z7 | Penjalinan | Root hydraulic conductance (Lo) declined in both lines. Recovery exhibited only in the tolerant line | Aroca *et al.* (2005) |
| Z7 | Penjalinan | Tolerant line showed less damage in dark-adapted leaves | Ribas-Carbo *et al.* (2000) |
| CM 7 and Co 151 | S 215 and EP 1 | Sensitive lines had higher galactolipase activity in aged chloroplasts | Kaniuga *et al.* (1998) |
| CO 328 | CO 316 | SOD had increased activity in tolerant line | Pinhero *et al.* (1997) |
| B73 and B49 | G50 and G84 | Sensitive lines had slower shoot growth rate | Stewart *et al.* (1990) |

PSII, photosystem II; F′v/F′m, effective photochemical efficiency; Fv/Fm, maximal photochemical efficiency of photosystem II; ϕPSII, quantum yield of photosystem II; PIP2;3, plasma-membrane intrinsic protein 2;3; PME, pectin methylesterase; RuBPCase, ribulose-1,5-bisphosphate carboxylase; PEPCase, phosphoenolpyruvate carboxylase; *ZmCIPKs*, genes coding for calcineurin B-like protein (CBL)-interacting protein kinases in *Zea mays*; CAT, catalase; POD, peroxidase; Pro, proline; SOD, superoxide dismutase.

**Table 16.2.** Maize mutant library.

| Library name | Type | Size | Link |
|---|---|---|---|
| Photosynthetic Mutant Library | *Mu* transposon | ~2100 | http://pml.uoregon.edu/photosyntheticml.html (accessed 22 February 2021) |
| Maize EMS induced Mutant Database (MEMD) | EMS | 4264 | http://www.elabcaas.cn/memd/ (accessed 22 February 2021) |
| Maize Genetics Cooperation Stock Center | EMS; UniformMu transposon; RescueMu transposon; *Ac/Ds/Ds-GFP* | Over 100,000 individually pedigreed samples | http://maizecoop.cropsci.uiuc.edu/ (accessed 22 February 2021) https://www.maizegdb.org/data_center/phenotype (accessed 22 February 2021) |

lines and identified 54 associated with the disruption of 49 genes that display drastic chilling sensitivity. Gao *et al.* (2017) screened more than 3700 homozygous SALK T-DNA insertion lines under chilling conditions and identified 41 lines that exhibited visible phenotypes. In rice, Jiang *et al.* (2007) generated around 20,000 *Ds* insertion rice lines and subjected several thousand lines to abiotic stresses, including chilling. These causal genes are potentially useful for improving cold tolerance in maize, although very few of the maize mutants were reported or evaluated to be associated with cold tolerance (Table 16.3).

## 16.5 Linkage Analysis

Linkage is the tendency for genes of interest and the genetic markers to be inherited together because they are located nearby on the same chromosome. Linkage analysis is the way of mapping the genes of interest using genetic markers as reference. The gene locations are estimated at the positions where the probability is maximized across all possible locations. Locations are represented by genetic distance with units of centimorgan, equivalent to 1% of the recombination rate. Through domestication, maize accumulated considerable genetic diversity for resistance to low temperatures, providing the foundation to map genes controlling cold tolerance. For example, European *flint* and tropical plateau germplasm materials have a stronger resistance to low temperatures than *dent* germplasm resources (Soldati *et al.*, 1999).

Studies have shown that cold resistance is a complex quantitative trait controlled by multiple genes, with different genetic mechanisms operating in different plant development periods. Further studies are necessary to understand the specific genetic mechanisms underlying different cold resistance traits across different germplasms. Linkage analysis has been used to identify quantitative trait loci (QTLs) to understand the genetic basis of maize resistance to low temperature. Many QTLs are related to seedling activity and basic physiological processes such as photosynthesis. Although the molecular mechanisms of plant adaptation to low temperature are still not fully understood and rarely reported (Hund *et al.*, 2004), three general mechanisms have been discovered including photosynthetic structure, cell-wall protection and developmental processes. The discovery of these mechanisms took many years and is based on field data and transcription levels of different maize inbred lines that were sensitive to low temperature (Sobkowiak *et al.*, 2014).

Although a substantial number of QTLs have been reported to be associated with cold stress, only a small number are consistent across studies, indicating that different genetic backgrounds have a strong impact on cold tolerance (Huang *et al.*, 2013). By using different populations, multiple QTLs related to the photosynthetic system and nearly 200 QTLs related to maize cold tolerance have been located. Jompuk *et al.* (2005) used 226 $F_{2:3}$ families derived from ETH-DH7 and ETH-DL3, and identified 29 QTLs related to chlorophyll fluorescence parameters, leaf greenness, leaf area, crown dry weight and

**Table 16.3.** Summary of maize mutants associated with cold tolerance.

| Mutant name | Mutant feature | Gene encodes | Reference |
|---|---|---|---|
| *ZmSiR* knock-down | *ZmSiR* knock-down maize plants were more susceptible to cold or oxidative stress than wild type | Sulfite reductase (SiR) | Xia *et al.* (2018) |
| *Adh1⁻Adh2⁻* doubly null mutant | *Adh1⁻Adh2⁻* doubly null seedlings had lower survival rate than non-doubly null maize seedlings (Silverado $F_1$) at 2°C | Alcohol dehydrogenase (ADH, EC 1.1.1.1) | Peters and Frenkel (2004) |
| *Dwarf-5* | *Dwarf-5* maize seedlings had a higher growth-constraining temperature (Pe) than tall segregates | – | Stoddart and Lloyd (1986) |
| Neuffer 3:116-1<br>Coe 7843.2-3 (wh)<br>97:1072-7 (var)<br>E8A-1<br>Coe 9744 × 31 (w/t) | Displayed low temperature-induced virescence | – | Alberte *et al.* (1974) |
| M11 mutant | Chlorophyll accumulation was sensitive to low temperature | – | Millerd and McWilliam (1968); Millerd *et al.* (1969) |

crown N content. With the same population, 19 QTLs were found to be related to chlorophyll fluorescence parameters, $CO_2$ exchange rate, leaf greenness, crown dry matter content and crown N content traits. Additionally, 27 QTLs were related to $C_4$ circulating enzyme activity, antioxidant content and chlorophyll.

Using 168 $F_{2:3}$ family lines derived from Lo964 × Lo1016, 60 QTLs were located with the traits such as root length, root weight, root diameter, root number, germination rate, germination index, leaf area, plant dry weight and operating efficiency of photosystem II (PhiPSII) (Hund *et al.*, 2004) and seven new QTLs were identified for leaf area (Hund *et al.*, 2005). Fracheboud *et al.* (2002) mapped 233 recombinant inbred lines (RILs), derived from Ac7643 × Ac7729, under conditions of low temperature (15°C) for QTLs related to chlorophyll fluorescence, gas exchange, photosynthetic pigment and other traits. They found a total of 25 QTLs, of which eight genetic regions were significantly correlated with five traits associated with photosynthesis organs. Four QTLs were located on chromosome 1 (~146 cM), chromosome 2 (~138 cM), chromosome 3 (~70 cM) and chromosome 9 (~62 cM). These QTLs explained about 28% of the phenotypic variation. Additionally, a gene associated with functional chloroplast development under cold stress was located near the QTL on chromosome 3 (~70 cM).

At present, many studies have been conducted on the location of cold tolerance QTLs in maize bud and seedling stages. By using 243 RIL families derived from B73 × Mo17, Hu *et al.* (2017) located six QTLs on chromosomes 4, 5, 6, 7 and 9 under different degrees of low temperature stress. Each of these six QTLs explained between 3.39 and 11.29% of the phenotypic variation in response to cold. Li *et al.* (2018) constructed three $F_{2:3}$ families with cold-tolerant inbred lines 220 and P9-10 and cold-sensitive materials Y1518 and PH4CV. They identified 43 QTLs associated with germination rate that explained between 0.62 and 39.44% of the phenotypic variation. Seven of the 43 QTLs explained more than 10% of the phenotypic variation and 16 QTLs associated with higher resistance were from cold-tolerant lines. With an isolated maize population derived from cold-resistant inbred EP42 and cold-sensitive inbred A661, Rodríguez *et al.* (2013) performed QTL mapping under conditions of 15°C. They detected ten QTLs using photosystem II electron transport and found three significant genetic regions on chromosomes 2, 4 and 8 related to seedling growth and development during low temperature stress.

## 16.6 Genome-Wide Association Studies

With reduced genotyping cost, the density of genetic markers can be high enough to guarantee that a gene of interest has the same linkage phase with at least one genetic marker without recombination. In such case, it is not necessary to estimate the recombination rate between the genes and their nearest markers. Instead, GWAS can be employed for gene mapping. There are two potential reasons that cause a gene of interest to be in linkage disequilibrium with a marker. The first reason is that they are in complete linkage, i.e. they are nearby each other on the same chromosome. The second reason is that the gene of interest is not in linkage with the marker and they may even be on different chromosomes. The common causes include selection, non-random mating and random drift. Therefore, population selection is critical for GWAS.

A collection of 282 core inbred lines from the USA has been used to map genes of several traits, including cold tolerance, using germination traits under cold stress of 8°C (Hu *et al.*, 2017). The genotype data from a GBS (genotyping-by-sequencing) data set and maize 50K array and their integration were used. A total of 17 loci were detected, with seven single-nucleotide polymorphisms (SNPs) located in candidate genes. Four SNPs were located within 366 kb or less of a candidate gene and the 18 candidate genes appeared to be involved in the regulation of maize cold tolerance.

Huang *et al.* (2013) used 125 maize inbred lines to conduct GWAS on ten traits during seedling germination, finding 43 SNPs related to maize cold tolerance. Of these 43 SNPs, 31 were related to 40 candidate genes. Strigens *et al.* (2013) performed GWAS with 375 maize inbred lines under different experimental conditions to evaluate seedling and chlorophyll fluorescence parameters. They identified 19 SNPs associated with cold resistance, which explained 5.7–52.5% of the phenotypic variation in early growth stage performance, chlorophyll fluorescence, and the candidate genes' function on ethylene conduction, synthesis of brassinolides and lignin.

Yan *et al.* (2017) identified 32 significant loci and 36 genes related to stress resistance using 338 lines. Among these 36 genes, ten were induced by low temperature stress. Revilla *et al.* (2016) used 598 maize inbred lines to conduct GWAS with 49,585 SNPs for evaluating days to germination, relative traits of chlorophyll content and the quantum efficiency of photosystem II. They identified 275 associated loci.

## 16.7 Gene Expression Profiling

With microarray and sequencing technologies, the activities of thousands of genes, or even genes of an entire genome, can be measured simultaneously. These profiles can reveal the responses to low temperatures as plants initiate a series of transcriptional regulation processes (Chinnusamy *et al.*, 2007; Ding *et al.*, 2019). Many genes were identified with transcriptional response to cold conditions using different approaches (Table 16.4), including genes encoding transcription factors and functional enzymes in different biological processes.

Using microarray analysis in multiple maize varieties, Sobkowiak *et al.* (2014) found that photosynthesis-related genes were preferentially repressed by cold stress. In contrast, they found that numerous genes related to basic biological activity, including transcription, gene expression regulation, protein phosphorylation and cell-wall organization, were induced by cold stress (Sobkowiak *et al.*, 2014). More importantly, 20 genes encoding membrane/cell-wall proteins were exclusively induced in the cold-tolerant ETH-DH7 line (Sobkowiak *et al.*, 2014). Di Fenza *et al.* (2017) studied the response of the primary roots at seedling emergence to cold stress using microarray analysis. They found that some RNA-binding proteins and pathogenesis-related proteins have the same expression pattern under cold growing conditions.

With the development of next-generation sequencing, RNA-seq analysis has also been adopted to explore gene expression in maize under cold stress. Shan *et al.* (2013) identified that in maize variety Zheng58, many genes including *C-repeat binding factor (CBF)/dehydration-responsive element* (*DREB*), *mitogen-activated protein kinase* (*MAPK*) and *NAC* (*NAM*, *ATAF* and *CUC*) were upregulated, while gibberellic acid (GA) metabolism-related gene (*GA20ox*)

**Table 16.4.** Genes with transcriptional response to low temperature stress identified in maize.

| Gene ID | Gene name | Function | Reference |
|---|---|---|---|
| *GRMZM2G051256* | *ZmMYB-IF35* | Alleviates PS II photoinhibition, scavenges ROS, reduces ion leakage | Meng and Sui (2019) |
| *GRMZM2G006745* | *ZmDREB2A* | Transcription regulation | Nguyen *et al.* (2009) |
| *GRMZM2G479760* | *ZmbZIP4* | Transcription regulation, positively regulates number of stress response genes and some ABA synthesis-related genes | Ma *et al.* (2018) |
| *GRMZM2G124037* | *ZmDREB1A* | Transcription regulation | Qin *et al.* (2004); Hu *et al.* (2011); Shan *et al.* (2013) |
| *GRMZM2G040030* | *ZmCOR413* | Downstream of *ZmDREB1A* | Hu *et al.* (2011) |
| *GRMZM2G064701* | | | Zhao *et al.* (2019) |
| *GRMZM2G128971* | | | Waters *et al.* (2017) |
| *GRMZM2G179120* | *ZmFAD2.1, ZmFAD7, ZmSLD1* | Possible participation in the regulation of fatty acid desaturation | Sobkowiak *et al.* (2014) |
| *GRMZM2G119258* | *ZmUSP* | Response to stress | Lu *et al.* (2017) |
| *GRMZM2G028568* | – | Signal transduction, homologue of rice cell-wall-associated receptor-like cytoplasmic kinase | Anderson *et al.* (1994) |
| *GRMZM2G173534* | *ZmmICE1* | Transactivation activities, contains a highly conserved bHLH domain and C-terminal region of ICE-like proteins | Lu *et al.* (2017) |
| *GRMZM2G079348* | *Zmcat3* | Encodes maize catalase, isozyme 3 | Shan *et al.* (2013) |
| *GRMZM2G459663* | – | $Ca^{2+}$ binding, ICE binding | Shan *et al.* (2013) |

PSII, photosystem II; ROS, reactive oxygen species; ABA, abscisic acid; bHLH, basic helix-loop-helix; ICE, interleukin 1β-converting enzyme.

and *ent-kaurene synthase* were downregulated. Therefore, under cold stress, maize seedlings may reduce GA level and thereby reduce growth rate to cope with abiotic stress. Transcriptomic responses were analysed between two maize inbred lines with contrasting low temperature sensitivity using RNA-seq (Zhao *et al.*, 2016). The study found that genes related to the G-protein-coupled receptor protein signalling pathway, binding and catalytic activity were significantly upregulated.

## 16.8 Genomic Selection

Maize yield was increased from 2.17 t/ha in the 1940s to 11.43 t/ha in the 2010s. Molecular breeding holds the promise to boost the yield increase and resistance to low temperature stress (Revilla *et al.*, 2005). Although the discovered QTLs have not been used in practice for breeding to improve cold tolerance (Leipner *et al.*, 2008), the genome approach may eventually lead to the development of superior varieties for cold tolerance. The genome approach is based on dense genetic markers covering the whole genome regardless of whether they have significant effects or not. This approach was first introduced through maize breeding by Dr Rex Bernardo in 1994 using restriction fragment length polymorphism (RFLP) markers to define kinship among individuals (Bernardo, 1994).

As best linear unbiased prediction (BLUP) was used to evaluate individuals' genetic potential, the genomic approach gained its popular name of genome BLUP, or gBLUP, after the introduction of Bayesian methods (Meuwissen *et al.*, 2001) and of gBLUP into the MTDFREMO package (Zhang *et al.*, 2007), a widely used software for animal breeding using pedigree. In the broader sense, the genome approach was named as genomic selection, or as genomic prediction in human genetics, regardless of using marker-based kinship to predict individuals' genetic potential directly or evaluating all marker effects and adding them together for the prediction.

Many software packages have been developed to implement GS, including rrBLUP (Endelman, 2011), GAPIT (Lipka *et al.*, 2012) and BGLR (Pérez and de los Campos, 2014). rrBLUP and BGLR implemented both methods of using marker-based kinship to predict individuals' genetic potential directly and of evaluating all marker effects and adding them together for the prediction. Both ridge regression and gBLUP methods were implemented in rrBLUP. BGLR implements a large collection of Bayesian regression models, including BayesA (Meuwissen *et al.*, 2001), BayesB (Habier *et al.*, 2011), BayesC (Habier *et al.*, 2011), Bayesian least absolute shrinkage and selection operator (LASSO) (BL) (Park and Casella, 2008), Bayesian ridge regression (BRR) and Bayesian reproducing kernel Hilbert spaces regression (RKHS) (Wahba, 1990). GAPIT was designed for both GWAS and GS. In addition to gBLUP, GAPIT implemented compressed BLUP and SUPER BLUP, which are superior to other methods for traits with low heritability and for traits controlled by a few genes, respectively.

Over 100 GS studies have been conducted in maize. Various populations have been newly created or directly used in GS, including inbred line populations like natural association panels (Zhang, H. *et al.*, 2019), the US nested mapping association (NAM) population (Zhang, H. *et al.*, 2019), the European *dent* and *flint* panel (Allier *et al.*, 2020) and doubled haploid (DH) lines (Sitonik *et al.*, 2019). There are also studies utilizing hybrid populations, including single-cross hybrids from multiple inbred lines following incomplete diallel mating design (Vidotti *et al.*, 2019) or test crosses of multiple inbred lines to a common parent (Galic *et al.*, 2019). These studies covered a diversity of important agronomic traits including grain yield, plant height, flowering time, etc. (Alves *et al.*, 2019; Lyra *et al.*, 2019; Millet *et al.*, 2019; Zhang, H. *et al.*, 2019; Allier *et al.*, 2020). GS studies about abiotic/biotic stress were focused on drought tolerance (Cerrudo *et al.*, 2018) and disease resistance (Cooper *et al.*, 2019; Vidotti *et al.*, 2019). However, no GS study has been reported about maize cold tolerance yet.

Considering the success of GS in other traits especially drought tolerance, it should also be useful to predict the cold tolerance ability. Selecting inbred lines or hybrids predicted to be tolerant or discarding those with likely hyper-sensitivity would improve breeding efficiency.

## 16.9 Perspective

Maize yield may always stay on top in the prioritization of breeding objectives; however, working on its component traits such cold tolerance could be more effective to increase yield indirectly, especially for cooler areas and vulnerable environments where chilling frequently takes place. Technical advances, such as sequencing, gene editing, gene mapping through GWAS and GS, make the option of improving cold tolerance more viable than ever. These efforts benefit the harvesting of diversity of existing maize germplasm and of mutants existing or created in the future towards the development of maize production.

## Acknowledgement

This project was partially supported by the US Department of Agriculture, National Institute of Food and Agriculture (Hatch Project 1014919, Award numbers 2016-68004-24770 and 2018-70005-28792), the National Natural Science Foundation of China (31901434), the Washington Grain Commission (Endowment and Award numbers 126593 and 134574) and the National Key Research and Development Program of China (2016YFD0101202, 2018YFD0300200 and 2016YFD0101203). The authors thank Linda R. Klein for helpful comments and editing the manuscript.

## References

Abe, A., Kosugi, S., Yoshida, K., Natsume, S., Takagi, H. *et al.* (2012) Genome sequencing reveals agronomically important loci in rice using MutMap. *Nature Biotechnology* 30(2), 174–178. Available at: https://doi.org/10.1038/nbt.2095

Alberte, R.S., Hesketh, J.D., Hofstra, G., Thornber, J.P., Naylor, A.W *et al.* (1974) Composition and activity of the photosynthetic apparatus in temperature-sensitive mutants of higher plants. *Proceedings of the National Academy of Sciences USA* 71(6), 2414–2418.

Allier, A., Teyssedre, S., Lehermeier, C., Charcosset, A. and Moreau, L. (2020) Genomic prediction with a maize collaborative panel: identification of genetic resources to enrich elite breeding programs. *Theoretical and Applied Genetics* 133(1), 201–215. Available at: https://doi.org/10.1007/s00122-019-03451-9

Alves, F.C., Granato, I.S.C., Galli, G., Lyra, D.H., Fritsche-Neto, R. and de los Campos, G. (2019) Bayesian analysis and prediction of hybrid performance. *Plant Methods* 15, 14. Available at: https://doi.org/10.1186/s13007-019-0388-x

Anderson, M.D., Prasad, T.K., Martin, B.A. and Stewart, C.R. (1994) Differential gene expression in chilling-acclimated maize seedlings and evidence for the involvement of abscisic acid in chilling tolerance. *Plant Physiology* 105(1), 331–339. Available at: https://doi.org/10.1104/pp.105.1.331

Aroca, R., Amodeo, G., Fernandez-Illescas, S., Herman, E.M., Chaumont, F. and Chrispeels, M.J. (2005) The role of aquaporins and membrane damage in chilling and hydrogen peroxide induced changes in the hydraulic conductance of maize roots. *Plant Physiology* 137(1), 341–353. Available at: https://doi.org/10.1104/pp.104.051045

Bernardo, R. (1994) Prediction of maize single-cross performance using RFLPs and information from related hybrids. *Crop Science* 34(1), 20–25. Available at: https://doi.org/10.2135/cropsci1994.0011183X003400010003x

Bhosale, S.U., Rymen, B., Beemster, G.T.S., Melchinger, A.E. and Reif, J.C. (2007) Chilling tolerance of central European maize lines and their factorial crosses. *Annals of Botany* 100(6), 1315–1321. Available at: https://doi.org/10.1093/aob/mcm215

Bilska, A. and Sowinski, P. (2010) Closure of plasmodesmata in maize (*Zea mays*) at low temperature: a new mechanism for inhibition of photosynthesis. *Annals of Botany* 106(5), 675–686. Available at: https://doi.org/10.1093/aob/mcq169

Bilska-Kos, A., Szczepanik, J. and Sowinski, P. (2016) Cold induced changes in the water balance affect immunocytolocalization pattern of one of the aquaporins in the vascular system in the leaves of maize (*Zea mays* L.). *Journal of Plant Physiology* 205, 75–79. Available at: https://doi.org/10.1016/j.jplph.2016.08.006

Bilska-Kos, A., Solecka, D., Dziewulska, A., Ochodzki, P., Jonczyk, M., Bilski, H. and Sowinski, P. (2017) Low temperature caused modifications in the arrangement of cell wall pectins due to changes of osmotic potential of cells of maize leaves (*Zea mays* L.). *Protoplasma* 254(2), 713–724. Available at: https://doi.org/10.1007/s00709-016-0982-y

Bilska-Kos, A., Panek, P., Szulc-Glaz, A., Ochodzki, P., Cislo, A. and Zebrowski, J. (2018) Chilling-induced physiological, anatomical and biochemical responses in the leaves of *Miscanthus* × *giganteus* and maize (*Zea mays* L.). *Journal of Plant Physiology* 228, 178–188. Available at: https://doi.org/10.1016/j.jplph.2018.05.012

Cerrudo, D., Cao, S., Yuan, Y., Martinez, C., Suarez, E.A. *et al.* (2018) Genomic selection outperforms marker assisted selection for grain yield and physiological traits in a maize doubled haploid population across water treatments. *Frontiers in Plant Science* 9, 366. Available at: https://doi.org/10.3389/fpls.2018.00366

Chen, C., Dong, Z., Gao, J., Xu, T., Jiao, L., Lu, L. and Zhang, F. (2013) [Effects of different accumulated temperature on photosynthetic performances of spring maize varieties during grain-filling period]. *Ying Yong Sheng Tai Xue Bao = The Journal of Applied Ecology* 24(6), 1593–1600.

Chen, X., Gu, Z., Xin, D., Hao, L., Liu, C. *et al.* (2011) Identification and characterization of putative *CIPK* genes in maize. *Journal of Genetics and Genomics = Yi Chuan Xue Bao* 38(2), 77–87. Available at: https://doi.org/10.1016/j.jcg.2011.01.005

Chinnusamy, V., Zhu, J. and Zhu, J.K. (2007) Cold stress regulation of gene expression in plants. *Trends in Plant Science* 12(10), 444–451. Available at: https://doi.org/10.1016/j.tplants.2007.07.002

Cooper, J.S., Rice, B.R., Shenstone, E.M., Lipka, A.E. and Jamann, T.M. (2019) Genome-wide analysis and prediction of resistance to Goss's wilt in maize. *The Plant Genome* 12(2), 180045. Available at: https://doi.org/10.3835/plantgenome2018.06.0045

Di Fenza, M., Hogg, B., Grant, J. and Barth, S. (2017) Transcriptomic response of maize primary roots to low temperatures at seedling emergence. *PeerJ* 5, e2839. Available at: https://doi.org/10.7717/peerj.2839

Ding, Y., Shi, Y. and Yang, S. (2019) Advances and challenges in uncovering cold tolerance regulatory mechanisms in plants. *The New Phytologist* 222(4), 1690–1704. Available at: https://doi.org/10.1111/nph.15696

Duran Garzon, C., Lequart, M., Rautengarten, C., Bassard, S., Sellier-Richard, H. *et al.* (2020) Regulation of carbon metabolism in two maize sister lines contrasted for chilling tolerance. *Journal of Experimental Botany* 71(1), 356–369. Available at: https://doi.org/10.1093/jxb/erz421

Endelman, J.B. (2011) Ridge regression and other kernels for genomic selection in the R package rrBLUP. *The Plant Genome* 4(3), 250–255. Available at: https://doi.org/10.3835/plantgenome2011.08.0024

Farooqi, M.Q.U., Sa, K.J., Hong, T.K. and Lee, J.K. (2016) Bulk segregant analysis (BSA) for improving cold stress resistance in maize using SSR markers. *Genetics and Molecular Research* 15(4), gmr15049326. Available at: https://doi.org/10.4238/gmr15049326

Fracheboud, Y., Ribaut, J.M., Vargas, M., Messmer, R. and Stamp, P. (2002) Identification of quantitative trait loci for cold-tolerance of photosynthesis in maize (*Zea mays* L.). *Journal of Experimental Botany* 53(376), 1967–1977. Available at: https://doi.org/10.1093/jxb/erf040

Galic, V., Franic, M., Jambrovic, A., Ledencan, T., Brkic, A., Zdunic, Z. and Simic, D. (2019) Genetic correlations between photosynthetic and yield performance in maize are different under two heat scenarios during flowering. *Frontiers in Plant Science* 10, 566. Available at: https://doi.org/10.3389/fpls.2019.00566

Gao, C., Hu, J., Zheng, Y. and Zhang, S. (2006) [Antioxidant enzyme activities and proline content in maize seedling and their relationships to cold endurance]. *Ying Yong Sheng Tai Xue Bao = The Journal of Applied Ecology* 17(6), 1045–1050.

Gao, J., Wallis, J.G., Jewell, J.B. and Browse, J. (2017) Trimethylguanosine synthase1 (TGS1) is essential for chilling tolerance. *Plant Physiology* 174(3), 1713–1727. Available at: https://doi.org/10.1104/pp.17.00340

Greene, E.A., Codomo, C.A., Taylor, N.E., Henikoff, J.G., Till, B.J. *et al.* (2003) Spectrum of chemically induced mutations from a large-scale reverse-genetic screen in *Arabidopsis*. *Genetics* 164(2), 731–740.

Habier, D., Fernando, R.L., Kizilkaya, K. and Garrick, D.J. (2011) Extension of the Bayesian alphabet for genomic selection. *BMC Bioinformatics* 12(1), 186. Available at: https://doi.org/10.1186/1471-2105-12-186

Hu, G., Li, Z., Lu, Y., Li, C., Gong, S. *et al.* (2017) Genome-wide association study identified multiple genetic loci on chilling resistance during germination in maize. *Scientific Reports* 7(1), 10840. Available at: https://doi.org/10.1038/s41598-017-11318-6

Hu, Y., Zhang, L., Zhao, L., Li, J., He, S. *et al.* (2011) Trichostatin A selectively suppresses the cold-induced transcription of the *ZmDREB1* gene in maize. *PloS One* 6(7), e22132. Available at: https://doi.org/10.1371/journal.pone.0022132

Huang, J., Zhang, J., Li, W., Hu, W., Duan, L. *et al.* (2013) Genome-wide association analysis of ten chilling tolerance indices at the germination and seedling stages in maize. *Journal of Integrative Plant Biology* 55(8), 735–744. Available at: https://doi.org/10.1111/jipb.12051

Hund, A., Fracheboud, Y., Soldati, A., Frascaroli, E., Salvi, S. and Stamp, P. (2004) QTL controlling root and shoot traits of maize seedlings under cold stress. *Theoretical and Applied Genetics* 109(3), 618–629. Available at: https://doi.org/10.1007/s00122-004-1665-1

Hund, A., Frascaroli, E., Leipner, J., Jompuk, C., Stamp, P. and Fracheboud, Y. (2005) Cold tolerance of the photosynthetic apparatus: pleiotropic relationship between photosynthetic performance and specific leaf area of maize seedlings. *Molecular Breeding* 16(4), 321–331. Available at: https://doi.org/10.1007/s11032-005-1642-7

Jiang, S., Bachmann, D., La, H., Ma, Z., Venkatesh, P.N. *et al.* (2007) *Ds* insertion mutagenesis as an efficient tool to produce diverse variations for rice breeding. *Plant Molecular Biology* 65(4), 385–402. Available at: https://doi.org/10.1007/s11103-007-9233-0

Jompuk, C., Fracheboud, Y., Stamp, P. and Leipner, J. (2005) Mapping of quantitative trait loci associated with chilling tolerance in maize (*Zea mays* L.) seedlings grown under field conditions. *Journal of Experimental Botany* 56(414), 1153–1163. Available at: https://doi.org/10.1093/jxb/eri108

Kaniuga, Z., Saczynska, V. and Miskiewicz, E. (1998) Galactolipase activity but not the level of high-melting-point phosphatidylglycerol is related to chilling tolerance in differentially sensitive *Zea mays* inbred lines. *Plant Cell Reports* 17(11), 897–901. Available at: https://doi.org/10.1007/s002990050505

Leipner, J., Jompuk, C., Camp, K.H., Stamp, P. and Fracheboud, Y. (2008) QTL studies reveal little relevance of chilling-related seedling traits for yield in maize. *Theoretical and Applied Genetics* 116(4), 555–562. Available at: https://doi.org/10.1007/s00122-007-0690-2

Li, M., Sui, N., Lin, L., Yang, Z. and Zhang, Y. (2019) Transcriptomic profiling revealed genes involved in response to cold stress in maize. *Functional Plant Biology* 46(9), 830–844. Available at: https://doi.org/10.1071/FP19065

Li, X., Wang, G., Fu, J., Li, L., Jia, G. *et al.* (2018) QTL mapping in three connected populations reveals a set of consensus genomic regions for low temperature germination ability in *Zea mays* L. *Frontiers in Plant Science* 9, 65. Available at: https://doi.org/10.3389/fpls.2018.00065

Lipka, A.E., Tian, F., Wang, Q., Peiffer, J., Li, M. *et al.* (2012) GAPIT: genome association and prediction integrated tool. *Bioinformatics* 28(18), 2397–2399. Available at: https://doi.org/10.1093/bioinformatics/bts444

Liu, S., Yeh, C.T., Tang, H.M., Nettleton, D. and Schnable, P.S. (2012) Gene mapping via bulked segregant RNA-Seq (BSR-Seq). *PloS One* 7(5), e36406. Available at: https://doi.org/10.1371/journal.pone.0036406

Lu, X., Yang, L., Yu, M., Lai, J., Wang, C. *et al.* (2017) A novel *Zea mays* ssp. *mexicana* L. MYC-type ICE-like transcription factor gene *ZmmICE1*, enhances freezing tolerance in transgenic *Arabidopsis thaliana*. *Plant Physiology and Biochemistry* 113, 78–88. Available at: https://doi.org/10.1016/j.plaphy.2017.02.002

Lyra, D.H., Galli, G., Alves, F.C., Granato, I.S.C., Vidotti, M.S. *et al.* (2019) Modeling copy number variation in the genomic prediction of maize hybrids. *Theoretical and Applied Genetics* 132(1), 273–288. Available at: https://doi.org/10.1007/s00122-018-3215-2

Ma, H., Liu, C., Li, Z., Ran, Q., Xie, G. *et al.* (2018) ZmbZIP4 contributes to stress resistance in maize by regulating ABA synthesis and root development. *Plant Physiology* 178(2), 753–770. Available at: https://doi.org/10.1104/pp.18.00436

Marocco, A., Lorenzoni, C. and Fracheboud, Y. (2005) Chilling stress in maize. *Maydica* 50(3–4), 571–580.

Meng, C. and Sui, N. (2019) Overexpression of maize *MYB-IF35* increases chilling tolerance in *Arabidopsis*. *Plant Physiology and Biochemistry* 135, 167–173. Available at: https://doi.org/10.1016/j.plaphy.2018.11.038

Meuwissen, T.H., Hayes, B.J. and Goddard, M.E. (2001) Prediction of total genetic value using genome-wide dense marker maps. *Genetics* 157(4), 1819–1829. Available at: http://www.ncbi.nlm.nih.gov/entrez/query.fcgi?cmd=Retrieve&db=PubMed&dopt=Citation&list_uids=11290733

Millerd, A. and McWilliam, J.R. (1968) Studies on a maize mutant sensitive to low temperature I. Influence of temperature and light on the production of chloroplast pigments. *Plant Physiology* 43(12), 1967–1972. Available at: https://doi.org/10.1104/pp.43.12.1967

Millerd, A., Goodchild, D.J. and Spencer, D. (1969) Studies on a maize mutant sensitive to low temperature II. Chloroplast structure, development, and physiology. *Plant Physiology* 44(4), 567–583. Available at: https://doi.org/10.1104/pp.44.4.567

Millet, E.J., Kruijer, W., Coupel-Ledru, A., Alvarez Prado, S., Cabrera-Bosquet, L. *et al.* (2019) Genomic prediction of maize yield across European environmental conditions. *Nature Genetics* 51(6), 952–956. Available at: https://doi.org/10.1038/s41588-019-0414-y

Nguyen, H.T., Leipner, J., Stamp, P. and Guerra-Peraza, O. (2009) Low temperature stress in maize (*Zea mays* L.) induces genes involved in photosynthesis and signal transduction as studied by suppression subtractive hybridization. *Plant Physiology and Biochemistry* 47(2), 116–122. Available at: https://doi.org/10.1016/j.plaphy.2008.10.010

Park, T. and Casella, G. (2008) The Bayesian Lasso. *Journal of the American Statistical Association* 103(482), 681–686. Available at: https://doi.org/10.1198/016214508000000337

Pérez, P. and de los Campos, G. (2014) Genome-wide regression and prediction with the BGLR statistical package. *Genetics* 198(2), 483–495. Available at: https://doi.org/10.1534/genetics.114.164442

Peters, J.S. and Frenkel, C. (2004) Relationship between alcohol dehydrogenase activity and low-temperature in two maize genotypes, Silverado $F_1$ and *Adh1⁻Adh2⁻* doubly null. *Plant Physiology and Biochemistry* 42(10), 841–846. Available at: https://doi.org/10.1016/j.plaphy.2004.10.004

Pinhero, R.G., Rao, M.V., Paliyath, G., Murr, D.P. and Fletcher, R.A. (1997) Changes in activities of antioxidant enzymes and their relationship to genetic and paclobutrazol-induced chilling tolerance of maize seedlings. *Plant Physiology* 114(2), 695–704. Available at: https://doi.org/10.1104/pp.114.2.695

Qin, F., Sakuma, Y., Li, J., Liu, Q., Li, Y.Q., Shinozaki, K. and Yamaguchi-Shinozaki, K. (2004) Cloning and functional analysis of a novel DREB1/CBF transcription factor involved in cold-responsive gene expression in *Zea mays* L. *Plant & Cell Physiology* 45(8), 1042–1052. Available at: https://doi.org/10.1093/pcp/pch118

Revilla, P., Butrón, A., Cartea, M.E., Malvar, R.A. and Ordás, A. (2005) Breeding for cold tolerance. In: Ashraf, M. and Harris, P.J.C. (eds) *Abiotic Stresses: Plant Resistance Through Breeding and Molecular Approaches*. The Haworth Press, Inc., New York, pp. 301–398.

Revilla, P., Rodriguez, V.M., Ordas, A., Rincent, R., Charcosset, A. *et al.* (2016) Association mapping for cold tolerance in two large maize inbred panels. *BMC Plant Biology* 16(1), 127. Available at: https://doi.org/10.1186/s12870-016-0816-2

Ribas-Carbo, M., Aroca, R., Gonzalez-Meler, M.A., Irigoyen, J.J. and Sanchez-Diaz, M. (2000) The electron partitioning between the cytochrome and alternative respiratory pathways during chilling recovery in two cultivars of maize differing in chilling sensitivity. *Plant Physiology* 122(1), 199–204. Available at: https://doi.org/10.1104/pp.122.1.199.

Rodríguez, V.M., Velasco, P., Garrido, J.L., Revilla, P., Ordás, A. and Butrón, A. (2013) Genetic regulation of cold-induced albinism in the maize inbred line A661. *Journal of Experimental Botany* 64(12), 3657–3667. Available at: https://doi.org/10.1093/jxb/ert189

Sánchez, B., Rasmussen, A. and Porter, J.R. (2014) Temperatures and the growth and development of maize and rice: a review. *Global Change Biology* 20(2), 408–417. Available at: https://doi.org/10.1111/gcb.12389

Schapendonk, A.H., Dolstra, O. and van Kooten, O. (1989) The use of chlorophyll fluorescence as a screening method for cold tolerance in maize. *Photosynthesis Research* 20(3), 235–247. Available at: https://doi.org/10.1007/BF00034067

Schneeberger, K., Ossowski, S., Lanz, C., Juul, T., Petersen, A.H. *et al.* (2009) SHOREmap: simultaneous mapping and mutation identification by deep sequencing. *Nature Methods* 6(8), 550–551. Available at: https://doi.org/10.1038/nmeth0809-550

Shan, X., Li, Y., Jiang, Y., Jiang, Z., Hao, W. and Yuan, Y. (2013) Transcriptome profile analysis of maize seedlings in response to high-salinity, drought and cold stresses by deep sequencing. *Plant Molecular Biology Reporter* 31(December), 1485–1491. Available at: https://doi.org/10.1007/s11105-013-0622-z

Sitonik, C., Suresh, L.M., Beyene, Y., Olsen, M.S., Makumbi, D. *et al.* (2019) Genetic architecture of maize chlorotic mottle virus and maize lethal necrosis through GWAS, linkage analysis and genomic prediction in tropical maize germplasm. *Theoretical and Applied Genetics* 132(8), 2381–2399. Available at: https://doi.org/10.1007/s00122-019-03360-x

Sobkowiak, A., Jonczyk, M., Jarochowska, E., Biecek, P., Trzcinska-Danielewicz, J. *et al.* (2014) Genome-wide transcriptomic analysis of response to low temperature reveals candidate genes determining divergent cold-sensitivity of maize inbred lines. *Plant Molecular Biology* 85(3), 317–331. Available at: https://doi.org/10.1007/s11103-014-0187-8

Sobkowiak, A., Jonczyk, M., Adamczyk, J., Szczepanik, J., Solecka *et al.* (2016) Molecular foundations of chilling-tolerance of modern maize. *BMC Genomics* 17, 125. Available at: https://doi.org/10.1186/s12864-016-2453-4

Soldati, A., Stehli, A. and Stamp, P. (1999) Temperature adaptation of tropical highland maize (*Zea mays* L.) during early growth and in controlled conditions. *European Journal of Agronomy* 10(2), 111–117. Available at: https://doi.org/10.1016/S1161-0301(98)00057-4

Sowinski, P., Rudzinska-Langwald, A., Adamczyk, J., Kubica, I. and Fronk, J. (2005) Recovery of maize seedling growth, development and photosynthetic efficiency after initial growth at low temperature. *Journal of Plant Physiology* 162(1), 67–80. Available at: https://doi.org/10.1016/j.jplph.2004.03.006

Stewart, C.R., Martin, B.A., Reding, L. and Cerwick, S. (1990) Seedling growth, mitochondrial characteristics, and alternative respiratory capacity of corn genotypes differing in cold tolerance. *Plant Physiology* 92(3), 761–766. Available at: https://doi.org/10.1104/pp.92.3.761

Stoddart, J.L. and Lloyd, E.J. (1986) Modification by gibberellin of the growth–temperature relationship in mutant and normal genotypes of several cereals. *Planta* 167(3), 364–368. Available at: https://doi.org/10.1007/BF00391340

Strigens, A., Freitag, N.M., Gilbert, X., Grieder, C., Riedelsheimer, C. *et al.* (2013) Association mapping for chilling tolerance in elite flint and dent maize inbred lines evaluated in growth chamber and field experiments. *Plant, Cell & Environment* 36(10), 1871–1887. Available at: https://doi.org/10.1111/pce.12096

Vidotti, M.S., Matias, F.I., Alves, F.C., Perez-Rodriguez, P., Beltran, G.A. *et al.* (2019) Maize responsiveness to *Azospirillum brasilense*: insights into genetic control, heterosis and genomic prediction. *PloS One* 14(6), e0217571. Available at: https://doi.org/10.1371/journal.pone.0217571

Wahba, G. (1990) *Spline Models for Observational Data.* Society for Industrial and Applied Mathematics, Philadelphia, Pennsylvania. Available at: https://doi.org/10.1137/1.9781611970128

Wang, S., Bai, G., Wang, S., Yang, L., Yang, F. *et al.* (2016) Chloroplast RNA-binding protein RBD1 promotes chilling tolerance through 23S rRNA processing in *Arabidopsis*. *PLoS Genetics* 12(5), e1006027. Available at: https://doi.org/10.1371/journal.pgen.1006027

Waters, A.J., Makarevitch, I., Noshay, J., Burghardt, L.T., Hirsch, C.N., Hirsch, C.D. and Springer, N.M. (2017) Natural variation for gene expression responses to abiotic stress in maize. *The Plant Journal* 89(4), 706–717. Available at: https://doi.org/10.1111/tpj.13414

Xia, Z., Wang, M. and Xu, Z. (2018) The maize sulfite reductase is involved in cold and oxidative stress responses. *Frontiers in Plant Science* 9, 1680. Available at: https://doi.org/10.3389/fpls.2018.01680

Yan, J., Wu, Y., Li, W., Qin, X., Wang, Y. and Yue, B. (2017) Genetic mapping with testcrossing associations and $F_{2:3}$ populations reveals the importance of heterosis in chilling tolerance at maize seedling stage. *Scientific Reports* 7(1), 3232. Available at: https://doi.org/10.1038/s41598-017-03585-0

Zhang, H., Yin, L., Wang, M., Yuan, X. and Liu, X. (2019) Factors affecting the accuracy of genomic selection for agricultural economic traits in maize, cattle, and pig populations. *Frontiers in Genetics* 10, 189. Available at: https://doi.org/10.3389/fgene.2019.00189

Zhang, Z., Todhunter, R.J., Buckler, E.S. and Van Vleck, L.D. (2007) Technical note: use of marker-based relationships with multiple-trait derivative-free restricted maximal likelihood. *Journal of Animal Science* 85(4), 881–885. Available at: https://doi.org/10.2527/jas.2006-656

Zhao, L., Hu, G., Liu, X., Zhou, Y., Li, Y. *et al.* (2016) Transcriptome sequencing identified genes and gene ontologies associated with early freezing tolerance in maize. *Frontiers in Plant Science* 7, 1477. Available at: https://doi.org/10.3389/fpls.2016.01477

Zhao, X., Wei, J., He, L., Zhang, Y., Zhao, Y. *et al.* (2019) Identification of fatty acid desaturases in maize and their differential responses to low and high temperature. *Genes* 10(6), 445. Available at: https://doi.org/10.3390/genes10060445

# 17 Physiological and Molecular Mechanisms Underlying Excess Moisture Stress Tolerance in Maize: Molecular Breeding Opportunities to Increase Yield Potential

**Sudha K. Nair*, Pervez Haider Zaidi, Madhumal Thayil Vinayan and Gajanan Saykhedkar**
*International Maize and Wheat Improvement Center (CIMMYT), Hyderabad, India*

## 17.1 Introduction

In the Asian tropics, more than 80% of the maize-growing area is rainfed and highly prone to various climatic extremes/variabilities. Erratic/uneven distribution pattern of monsoon rains often causes intermittent drought, coupled with heat and/or waterlogging, at different crop growth stage(s) within the same season. This is a major factor responsible for the realized low productivity of rainfed maize. Studies have warned that the Asian tropics will experience an increasing frequency of extreme weather conditions with high variability beyond the current capacity to cope (Lobell *et al.*, 2011; Cairns *et al.*, 2012). Impacts of climate change are evident in the form of shifting crop seasons due to significant inter-annual variation in rainfall, coupled with increased frequency of extreme weather events causing severe drought and/or waterlogging.

Waterlogging is a major problem for maize production in several maize agroecologies in South and South-East Asia where rainfall is erratic and intense and soil drainage capacity is poor. Rainfed maize is grown during the monsoon season in the Asian tropics and occasionally temporary waterlogging results in anaerobic conditions, even in well-drained fields. These waterlogged soils adversely affect various crop growth stages, overall plant stand and final grain yield. Flood and waterlogging frequently affect more than 18% of the total maize production area in South and South-East Asia, causing production losses of 25–30% annually (Zaidi *et al.*, 2010; Cairns *et al.*, 2012). Moreover, the increasing demand for maize in Asia is rapidly transforming cropping systems in the region from rice monoculture to more profitable rice–maize systems. Maize production in the rice–maize system frequently faces the problem of early-stage excessive soil moisture, as the soils of paddy fields are often saturated due to late monsoon rains.

When the amount of water in the soil exceeds field capacity (the upper limit of plant-available water), it results in waterlogging or excessive soil moisture. Excess moisture in the root zone in the form of free water fills the air spaces in the soil, resulting in poor soil aeration and eventually

*Corresponding author: sudha.nair@cgiar.org

 *Molecular Breeding in Wheat, Maize and Sorghum: Strategies for Improving Abiotic Stress Tolerance and Yield* (eds M.A. Hossain *et al.*)
DOI: 10.1079/9781789245431.0017

hindering plant growth and development. Waterlogging causes major changes in the soil's physical and chemical (and thereby biological) properties, which results in multiple stresses, including oxygen stress caused by low/no $O_2$ in the root zone, nutrient imbalance and biotic stress due to anaerobic conditions. Most crop plants, including maize, are obligate aerobes and therefore require well-drained soil for optimal growth, development and productivity. Unlike wetland crops such as rice, there is no aeration system in maize plants for gaseous exchange between above-ground plant parts and roots. Therefore, maize roots suffer with progressive decline in $O_2$ in the rhizosphere. $O_2$ supply in most soils is depleted within 48 h of waterlogging (Fausey *et al.*, 1985) and therefore maize roots cannot perform critical, life-sustaining functions such as nutrient and water uptake, which eventually affects various plant functions and causes severe plant injuries and yield losses.

In the Asian tropics, waterlogging may affect rainfed maize crops at different crop stages, starting from pre-emergence until harvest.

- *Pre-emergence stage waterlogging*. This is a common problem in tropical maize, especially if the maize crop is planted in the following situations:
  - at the onset of the monsoon rains and if the rains continue for a few days after planting;
  - in a poorly drained field that is excessively irrigated after planting;
  - in low-lying areas/patches in a poorly levelled field; and
  - under highly saturated field conditions immediately after rice harvest (rice–maize cropping system in Asia) or in *char* or *diara* land (areas located near riverbanks that are vacated after the river water recedes after the rainy season).
- *Early seedling stage waterlogging:*
  - heavy rainfall early in the season coupled with poor drainage;
  - excessive irrigation coupled with poor drainage; and
  - undulating field with low-lying patches preventing the runoff of excess water.
- *Vegetative growth stage waterlogging:*
  - heavy and concentrated mid-season rainfall coupled with poor drainage;
  - excessive irrigation coupled with poor drainage; and
  - poorly levelled field with low-lying patches where water stagnates after rain or irrigation.
- *Flowering/grain-filling stage waterlogging:*
  - late-planted spring-season maize crops in eastern India and Bangladesh (and similar ecologies elsewhere) that are exposed to early monsoon rains in low-lying areas.

Understanding the impact of excess moisture on maize plants at various growth stages, and studying the phenological, physiological and molecular responses of tolerant maize genotypes towards adaptation to excess moisture stress, could help define ways in which this trait could be improved through targeted breeding.

## 17.2 Impact of Excess Moisture Stress on Maize Plants

Being a non-wetland crop of tropical origin, maize is highly susceptible to waterlogging at almost all the crop stages, especially before tassel emergence (Zaidi *et al.*, 2004; Kuang *et al.*, 2012; Ren *et al.*, 2014). Studies on the responses to excessive moisture showed a wide range of changes in maize plants at the molecular, biochemical, physiological, anatomical and morphological levels, and has been reviewed extensively (Kennedy *et al.*, 1992; Perata and Alpi, 1993; Ricard *et al.*, 1994; Drew, 1997). Excessive moisture condition in the rhizosphere severely affects various plant functions and causes stress injury.

### 17.2.1 Water deficit

An early response of plants to waterlogging is stomatal closure and reduced water flux from roots to shoots. Waterlogging reduces $K^+$ levels in the guard cells resulting in stomatal closure and hence slower water flux from root to shoot. Decrease in water uptake was also found to be associated with root decay and wilting. The first symptom of excess moisture injury is wilting of leaves. Reduced stomatal conductance is among the earliest responses to excess moisture in maize, followed by leaf yellowing, inhibition of root growth, alteration in root and shoot morphology, leaf senescence and brace root development from above-ground nodes (Rathore *et al.*, 1997; Zaidi and Singh, 2001, 2002; Zaidi *et al.*, 2003; Kaur *et al.*, 2019). Under inundated conditions, maize root respires anaerobically and therefore very little energy is available (only 2 moles of ATP per 1 mole of glucose versus 38 moles of ATP produced during aerobic respiration). Consequently, due to energy starvation, the root cell membrane is unable to maintain structural integrity; the membrane becomes porous, resulting in the leaching of mineral nutrients and organic substances from root tissues (Table 17.1). Leaching-induced changes in osmotic gradient in the root cortex result in inhibition of the radial movement of water from root hairs across the cortex into the xylem. Consequently, the water supply to above-ground plant parts is reduced and the plant suffers water deficit stress.

**Table 17.1.** Effect of 6 days of waterlogging on the leakage of mineral ions from the detached leaves of pea. (From Jackson and Kowalewska, 1983.)

| | Leakage (μg/ml/leaflet) | |
|---|---|---|
| Mineral ion | Control | Waterlogging |
| $K^+$ | 0.18 | 41.63 |
| $Ca^{2+}$ | 0.73 | 9.28 |
| $HPO_3^{2-}$ / $PO_3^{4-}$ | 0.0 | 8.38 |
| $Mg^{2+}$ | 0.0 | 6.70 |
| $NH_4^+$ | 0.0 | 0.39 |
| $NO_3^-$ | 0.0 | 0.08 |

### 17.2.2 Nutrient imbalance

Soil waterlogging causes mineral nutrient imbalances in plants due to its effects on nutrient availability in the soil, uptake and transport by roots, and distribution within the plant. There are three main aspects of plant nutrition that are affected by excessive moisture:

- reduced soil conditions change the availability of important mineral nutrients in the soil;
- anaerobiosis reduces ATP availability, resulting in less energy available for nutrient uptake by the roots; and
- reduced transport of water reduces nutrient transport in the leaves.

Prolonged waterlogging significantly affects nutrient homeostasis in maize roots, which results in deficiency of important nutrients and toxicity of some other nutrients (Table 17.2). Under anoxia condition, $NO^-_3$ is reduced to $NH^+_4$ and $SO^{2-}_4$ is converted to $H_2S$, and both become unavailable to maize plants. In addition, being a highly soluble form, $NO^-_3$ leaches down rapidly with stagnating water conditions. Availability of P may increase or decrease depending upon soil pH during waterlogging. In waterlogged soil, anaerobic microbes start using soil oxidized compost as electron acceptors after the consumption of the dissolved $O_2$. In reduction reactions, pH is modified and increases the availability of some elements (Ponnamperuma, 1972); for example, $Fe^{3+}$ and $Mn^{4+}$ are reduced to $Fe^{2+}$ and $Mn^{2+}$, and thereby concentrations of Fe and Mn increase in the soil (Silva *et al.*, 2003). The highly soluble $Fe^{2+}$ and $Mn^{2+}$ cause severe toxicity in maize plants (Rathore *et al.*, 1996).

**Table 17.2.** Waterlogging-induced changes in tissue ion concentrations in tropical maize (*Zea mays* L.).

| | Element | | | | | | | | | | | |
|---|---|---|---|---|---|---|---|---|---|---|---|---|
| Tissue | N | P | K | Ca | Mg | Fe | Mn | Cu | Zn | Mo | B | Reference |
| Ear leaf | ↓ | ↓ | ↓ | | | ↑ | ↑ | ↓ | ↓ | ↑ | ↓ | Lal and Taylor (1970) |
| Shoot | ↓ | ↓ | ↓ | ↓ | ↓ | | | | | | | Sharpiro *et al.* (1956) |
| Roots | ↑ | ↑ | ↑ | | ↑ | | | | | | | Singh and Ghildyal (1980) |
| Leaves | ↓ | ↑ | ↓ | ↓ | | ↑ | ↑ | ↓ | ↓ | | ↓ | Rathore *et al.* (1997) |
| Whole plant | ↓ | ↑ | ↓ | → | → | | | | | | | Devitt and Francis (1972) |

↓, decrease; ↑, increase; →, no change.

### 17.2.3 Plant functions

Waterlogging affects several plant functions like photosynthesis, root respiration, hormonal balances, biochemical changes and plant growth and development changes in general. Sharp decline in aerobic respiration in root tissues is one the key effects on maize plants due to soil waterlogging. Lack of $O_2$ triggers anaerobic respiration resulting in energy starvation and the end products are toxic substances such as ethanol, lactate, malate, alanine, etc. (Fig. 17.1) which can cause severe plant injuries. Increased anaerobic respiration causes rapid depletion of carbohydrate in the roots, resulting in 'carbohydrate starvation' (Setter *et al.*, 1987). Activity of the enzyme $NAD^+$-alcohol dehydrogenase (ADH) increases exponentially in the tolerant maize genotypes, whereas in susceptible genotypes ADH activity declines rapidly (Fig. 17.2). Liao and Lin (1995) suggested that ADH activity was positively correlated with the magnitude of excess moisture injury and species with higher ethanol production were less tolerant. Ethanol accumulation might have a 'self-poisoning' role in flood-intolerant genotypes. According to the metabolic theory of Crawford (1967), flooding tolerance is achieved by minimization of ethanol production and is associated with rerouting from ethanol fermentation to malate production. Although shoot tissues are usually not exposed to waterlogging conditions, photosynthesis is affected due to: (i) reduced stomatal conductance (Zaidi *et al.*, 2003; Else *et al.*, 2009); (ii) reduced chlorophyll content and active leaf area due to enhanced senescence (Rathore *et al.*, 1997; Zaidi and Singh, 2002; Ren *et al.*, 2016; Kaur *et al.*, 2019); (iii) reduction in activity of key photosynthetic enzymes (Bradford, 1982; Liao and Lin, 1994; Ren *et al.*, 2016); and (iv) reduced translocation of photosynthates due to interference in phloem translocation and reduced sink size (Krishnamoorthy, 1993).

Waterlogging causes significant imbalance of plant hormones. The levels of hormones synthesized in the roots (cytokinins and gibberellins) decrease sharply, while those synthesized in shoots (auxins, ethylene and abscisic acid (ABA)) accumulate in the leaves and stems (Reid and Bradford, 1984). Several morphological and anatomical responses are observed among plants exposed to excess moisture, such as increased surface rooting, adventitious rooting from above-ground nodes, formation of aerenchyma in cortical region of maize roots, rapid stomatal closure and epinastic bending (Zaidi *et al.*, 2003) mediated by phytohormones such as ethylene

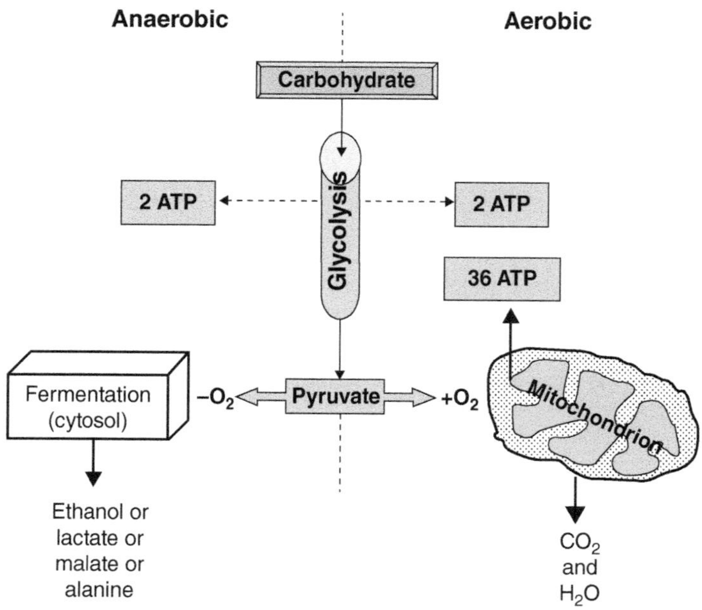


**Fig. 17.1.** Energy generation through aerobic and anaerobic respiration in plants. (From Zaidi *et al.*, 2005.)

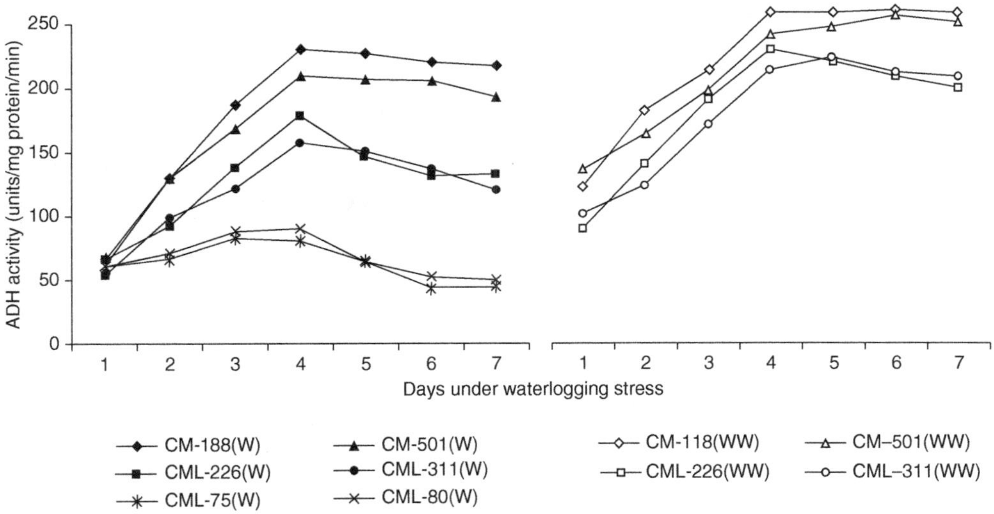


**Fig. 17.2.** NAD+-alcohol dehydrogenase (ADH) activity in new developed adventitious roots of different maize genotypes under excess soil moisture stress at knee-high stage. (From Zaidi *et al.*, 2003.)

and ABA (Jackson *et al.*, 1993). Jackson (1989, 1990) suggested that ethylene is the principal mediator for promoting the development of aerenchyma in maize roots as well as other plants (Fig. 17.3). Auxin and gibberellins are prerequisites for ethylene action and have a triggering rather than a regulatory function. Anaerobic metabolism at an accelerated rate in waterlogged roots is important as it supplies enough energy for root survival (Mohanty *et al.*, 1993) by mobilizing starch reserves, induced by α-amylase activity (Perata *et al.*, 1992). Excessive soil moisture significantly affects the activity of nitrate reductase and glutamine synthetase, two key enzymes in N metabolism (Reggiani *et al.*, 1988).

Waterlogging during the vegetative stages severely inhibits maize plant growth and development, including plant height, leaf area and total biomass (Rathore *et al.*, 1997; Zaidi *et al.*, 2003; Ren *et al.*, 2014; Shin *et al.*, 2016; Panozzo *et al.*, 2019). If the crop is exposed to waterlogging during the vegetative stage, the apical meristem is severely damaged, resulting in irreversible effects on the growth and development of growing maize plants. The impact of the excessive moisture stress is also apparent in reproductive behaviour. Female flowering is comparatively more susceptible to this stress than male flowering (Zaidi and Singh, 2001; Shah *et al.*, 2012). Delayed silking results in a long anthesis–silking interval (ASI) and severe barrenness with exposure to the stress at two-leaf (V2) and seven-leaf (V7) growth stages (Table 17.3). However, waterlogging at later growth stages did not contribute to the increase in ASI.

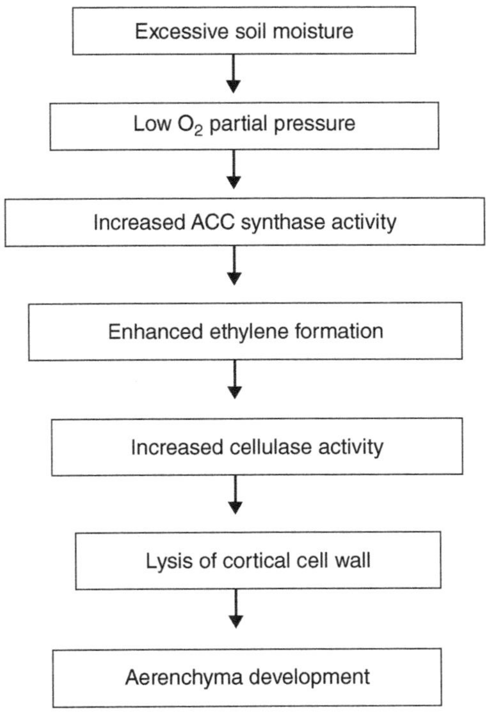


**Fig. 17.3.** Schematic representation of aerenchyma development in root cortex under excessive soil moisture stress. ACC, aminocyclopropane-1-carboxylic acid. (From Jackson *et al.*, 1993.)

**Table 17.3.** Means of the traits observed in maize genotypes under normal and excess moisture stress imposed for 10 days at different growth stages.

| | | Excess moisture at different growth stages | | | |
|---|---|---|---|---|---|
| Trait | Normal moisture | V2 stage | V7 stage | VT stage | R1 stage |
| Plant mortality (%) | 1.3 | 87.8** | 70.3** | 32.6** | 13.6* |
| Plant height (cm) | 121.3 | 45.8** | 79.6** | 116.6$^{NS}$ | 122.3$^{NS}$ |
| Leaf area (dm$^{2}$/plant) | 286.3 | 92.3** | 169.7** | 269.2$^{NS}$ | 279.5$^{NS}$ |
| Dry weight (g/plant) | 64.9 | 14.2** | 33.6** | 53.7$^{NS}$ | 61.5$^{NS}$ |
| ASI (days) | 3.3 | 25.3** | 10.7** | 3.7$^{NS}$ | 3.2$^{NS}$ |
| Yield (t/ha) | 2.46 | 0.49** | 1.01** | 2.03* | 2.27$^{NS}$ |

V2, two-leaf stage; V7, seven-leaf stage; VT, tassel emergence stage; R1, silk emergence stage.
*$P < 0.05$; **$P < 0.01$; NS, non-significant difference.

## 17.3 Phenological Adaptations and Physiological Mechanisms Leading to Excess Moisture Stress Tolerance in Maize

Tolerant genotypes of non-wetland crop species, including maize, respond to waterlogging through specific morpho-physiological and biochemical adaptation mechanisms, which might be useful criteria for the selection of genotypes with increased tolerance to waterlogging stress (Crawford and Braendle, 1996; Zaidi *et al.*, 2007a). Surface rooting and adventitious (brace) root formation are the most prominent morphological adaptations observed in tolerant genotypes. Under excessive moisture conditions, root geotropism has been observed in maize genotypes (Fig. 17.4a). However, if the standing water depth is deeper, then the advantage of surface rooting is significantly minimized. Within 2–3 days of exposure to excessive moisture, nodal roots (Fig. 17.4b) are initiated from both below- and above-ground nodes (Rathore *et al.*, 1996). In tolerant genotypes, the primary root system was found to be completely replaced by newly developed adventitious roots below the ground surface (Zaidi *et al.*, 2002). Newly emerged adventitious roots in maize genotypes under waterlogging conditions were found to have large air spaces in the cortical region (Rathore *et al.*, 1996; Zaidi *et al.*, 2007a), which helps in increasing the supply of $O_2$, nutrients and water, and improves anchorage where severe damage of primary roots under excess moisture has occurred. A major anatomical acclimatization in maize to waterlogging stress is through aerenchyma formation in the cortical region of roots, especially in waterlogging-induced adventitious roots (Fig. 17.5). Aerenchyma development in maize roots can be stimulated within 12–24 h of waterlogging associated with the shortage of $O_2$. Unlike its ancestor teosinte, root aerenchyma is not a constitutive trait in maize. It is a stress-induced trait developed under waterlogging or with other low $O_2$-related stresses.

Apart from morphological and anatomical changes, various metabolic adaptations have also been observed in maize and other non-wetland crop plants. In waterlogging-tolerant genotypes, the tricarboxylic acid (TCA) cycle switches off under anaerobic conditions leading to increased phosphofructokinase (PFK) activity, which results in an enhanced rate of glycolysis, known as the 'pasture effect'. The high rate of pasture effect under anaerobic conditions helps cells to maintain sufficient energy supply, provided carbohydrate supply is sustained in the tissues. Another important metabolic adaptation in respiration is through the regulatory induction of ADH under anoxia, increasing alcoholic fermentation as a way to alleviate the adverse effect of anoxia on energy production (Ap Rees *et al.*, 1987). To negate the phytotoxicity of ethanol, the end product of alcoholic fermentation, a low ethanol concentration is maintained in tolerant lines by the high excretion of ethanol out of root tissues into the surrounding water (Zaidi *et al.*, 2007a). Anaerobic treatment drastically alters the profile of total protein synthesis in maize seedlings. During anaerobiosis, there was an immediate repression of ongoing protein synthesis and it was observed that

**Fig. 17.4.** Morphological adaptation in maize under excessive moisture: surface roots in susceptible genotype (a) and brace roots in tolerant genotype (b), showing adventitious rooting from the nodes present above the ground surface (c).

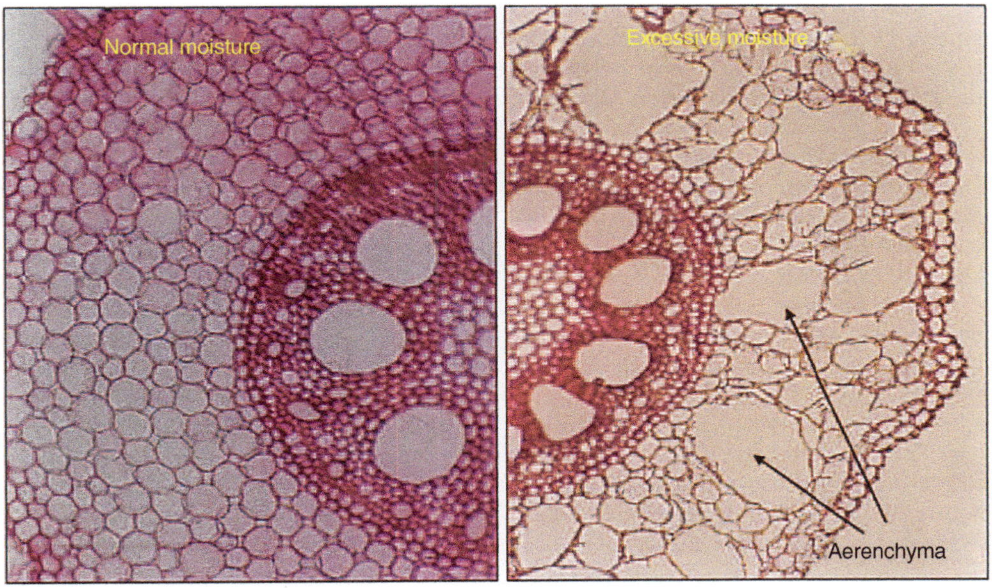


**Fig. 17.5.** Anatomical adaptation under excessive moisture in maize through development of aerenchymatous spaces in cortical region of roots.

20 new anoxia-induced proteins (ANPs), which accounted for >70% of the total translation, were selectively synthesized (Sachs *et al.*, 1996). Most of these ANPs are identified as enzymes of glycolysis or sugar-phosphate metabolism (Subbaiah and Sachs, 2003).

## 17.4 Molecular Signature of Excess Moisture Stress Tolerance

Interaction of plants with the growing conditions triggers molecular changes at the subcellular level, which further translate into tissue-specific fine-tuning that supports survival in a given environment. Reliable molecular signatures can provide insights into the plant's subcellular tolerance mechanisms to different environmental stresses. This, in turn, helps plant technologists understand stress adaptation mechanisms and utilize them to tailor more sturdy crops. Excess moisture stress causes improper aeration in the rhizosphere and exchange of gases is highly affected in the area surrounding the plant root. $O_2$ diffuses approximately 10,000 times slower in water than in air, and the flow of $O_2$ into waterlogged soil is around 320,000 times less (Watanabe *et al.*, 2013). Roots of the plants are usually in contact with $O_2$ at a partial pressure equivalent to the gaseous atmosphere. In waterlogged soil, there could be two situations of reduction in $O_2$ level: (i) anoxia, where there is complete lack of $O_2$ as a result of long-term flooding; and (ii) hypoxia, where there is a suboptimal level of $O_2$ due to short-term flooding. The latter is the most common form of excess moisture stress witnessed in maize-growing environments. Molecular responses of plants in these excess moisture conditions are mainly due to reduced $O_2$ levels.

A class of ethylene-response transcription factors known as the group VII ethylene-response factors (ERF-VIIs) regulate the expression of a large number of genes involved in adaptive responses to flooding and low $O_2$ (Voesenek and Bailey-Serres, 2015). Rice is a model plant, where a lot of research has been done to study molecular mechanisms leading to excess moisture stress tolerance. In rice, the *SUB1* locus on chromosome 9 was found to confer up to 69% of phenotypic variation in submergence tolerance (Xu and Mackill, 1996). Plants with the favourable *SUB1* allele were capable of surviving 2 weeks or longer of complete inundation (Fukao *et al.*, 2006). This locus was cloned, and the multi-genic locus included three genes of ERF-VIIs, designated as *SUB1A*, *SUB1B* and *SUB1C* (Xu *et al.*, 2006), with *SUB1A* found to be sufficient for submergence tolerance. Studies have shown that the favourable allele of *SUB1A* (*SUB1A-1*) functions in a hierarchical hormonal pathway involving ethylene, ABA and gibberellic acid that regulates growth elongation during submergence (Bailey-Serres *et al.*, 2012). It has been shown that *SUB1A* also aids recovery from submergence, as evident from upregulation of mRNAs encoding antioxidant enzymes during submergence (Jung *et al.*, 2010) and less oxidative damage upon de-submergence (Fukao *et al.*, 2011). Other rice *ERF-VII* genes, *SNORKEL1* and *SNORKEL2*, were also found, specifically in deep-water rice (Hattori *et al.*, 2009). These tandemly repeated ERF-VIIs are responsible for enhanced internode elongation under flooded conditions, enabling plants to outgrow floodwaters. Some deep-water rice cultivars can increase their height by 25 cm/day (Vergara *et al.*, 1976). In *Arabidopsis*, five *ERF-VII* genes, *HRE1*, *HRE2*, *RAP2.2*, *RAP2.3* and *RAP2.12*, were recognized as key regulators for flooding and low $O_2$ tolerance (Fukao *et al.*, 2019). There seems to be differing mechanisms involving *ERF-VII* genes in different systems (Fukao *et al.*, 2019). All of the *Arabidopsis* ERF-VIIs are substrates of the N-end rule pathway related to their stability, regulated by the N-end rule of proteolysis (NERP). However, some rice ERF-VIIs such as SUB1A and SUB1C are not degraded via this pathway (Gibbs *et al.*, 2011).

Plant survival under excess moisture requires tolerance to multiple stresses such as submergence, reoxygenation and dehydration (Fukao *et al.*, 2019). Many ERF-VIIs themselves are involved in the adaptation to these processes, by activating the detoxification of reactive oxygen species (ROS) and ABA responsiveness (Fukao *et al.*, 2011). Jasmonate is also found to be a pivotal hormone in activation of ROS and detoxification systems under reoxygenation in *Arabidopsis* (Yuan *et al.*, 2017). The biological roles of NO in plants undergoing hypoxia and de-submergence also have been elucidated to fit well with a cellular signalling process, where ROS are recognized as an integral part (Foyer *et al.*, 2017). Plant cells have evolved efficient mechanisms to sense $O_2$ levels and to transduce those concentrations into a molecular response suited to the different flooding and hypoxic scenarios, aimed at detecting a cellular energy crisis. Studies have presented strong evidence for the role of $Ca^{2+}$ signalling as a strong, fast and flexible primary signal of energy stress (Igamberdiev and Hill, 2018). Downstream signal transduction events may not be unique to hypoxic stress, but to any stress condition that limits energy availability, thus creating an adjustable response to multiple stresses (Fukao *et al.*, 2019).

Through transcriptional profiling of a set of submergence-tolerant and -sensitive lines in maize, it was observed that sensitive genotypes displayed a considerably higher induction of ROS marker genes *Alternative Oxidase 1a* (*AOX1a*), *WRKY6* and *CYP81D8* in response to submergence and submergence-tolerant genotypes maintained lower ROS levels (Campbell *et al.*, 2015). This suggested that the lower survival and poor recovery of the sensitive lines could partly be due to higher ROS levels and downstream transcriptome changes that accompany oxidative stresses. Panozzo *et al.* (2019) also observed that *AOX1a* was an important gene in the responses of genotypes to excess moisture conditions. Submergence-tolerant maize lines exhibited higher expression of hypoxia-associated energy homeostasis genes (Campbell *et al.*, 2015). They also observed that 18 *ERFs* were upregulated in tolerant and sensitive genotypes, suggesting that ERFs are an integral component of the early submergence response in maize also.

The molecular changes happening at cellular level under excess moisture stress also lead to recognizable anatomical and morphological changes in tolerant plants, aiding their survival, as discussed in Section 17.3. Plants without constitutive aerenchyma in the roots, like maize, form cortical lysigenous aerenchyma within 12–24 h of waterlogging through a programmed cell death process that involves ethylene, $Ca^{2+}$ and ROS (Steffens *et al.*, 2012). Rajhi *et al.* (2011) profiled the transcriptomes of the cortex and stele of waterlogged seedling roots of maize. Their study identified genes expressed preferentially in the cortex in a waterlogging- or ethylene-dependent manner that are associated with transcription, $Ca^{2+}$ signalling, ROS regulation and cell expansion. Evans (2004) outlined the five stages in molecular signalling following hypoxia to aerenchyma formation in maize, leading from hypoxia-induced ethylene signalling to cell lysis. Promotion of growth of adventitious root primordium was also found to be initiated by ethylene- and ROS-dependent signalling, through a mechanical force on the overlying epidermis. This cell-to-cell mechano-signalling led to localized cell death of the tightly attached epidermal cells through ethylene signalling and ROS production, aiding the emergence of adventitious roots (Steffens *et al.*, 2012).

Micro RNAs (miRNAs) have been found to have an active role in response to multiple stresses, by modulating their target mRNAs. Under excess moisture conditions, miRNAs have been found to regulate plant responses through morphological adaptation, management of energy supply, control of flowering and oxidative stress response (Fukao *et al.*, 2019). In maize roots, miR159, miR166, miR167, miR164 and miR172 were found to have differential expression towards modulation in root development, leading to adventitious root formation (Zhang *et al.*, 2008; Liu *et al.*, 2012; Zhai *et al.*, 2013). In maize, submergence was found to induce miR399, predictably downregulating a gene coding for a granule-bound starch synthase (Zhang *et al.*, 2008) that could be a possible mechanism to manage energy supply. Similarly, miR395 and miR474, regulating targets involved in the management of fundamental processes under energy stress, were found to have differential action during short-term and long-term hypoxia treatments (Zhang *et al.*, 2008). miR408, miR528 and miR397, which target mRNAs producing activation of oxidation/reduction enzymes containing cupredoxin domains, were found to be differentially regulated under hypoxia in maize (Liu *et al.*, 2012) and *Arabidopsis* (Licausi *et al.*, 2011), reflecting a possible activation of oxidative stress response.

## 17.5 Genetic Studies on Excess Moisture Stress Tolerance in Maize

It is imperative to understand the genetic architecture of traits to devise effective breeding strategies. Stress tolerance traits are generally studied on grain yield under stress, or on component traits exhibited in the form of morphological or physiological changes exhibited under stress. Component traits are comparatively easier to study, considering the possibility to achieve high heritability in phenotyping and reduced dimensionality. Polygenic inheritance of waterlogging stress tolerance with partial dominance of tolerance over susceptibility was inferred from a study involving tropical maize lines exposed to water logging at V7 to eight-leaf (V8) stage for 7 days (Zaidi *et al.*, 2007b, 2010). In diallel and line cross tester studies, Zaidi *et al.* (2010) observed a predominance of additive variance over non-additive variance controlling grain yield under waterlogging. In another study, under similar phenotyping conditions, Zaidi *et al.* (2007b) observed that the contribution of mid- and best-parent heterosis increased under waterlogging for grain yield along with component traits, compared with optimal soil moisture conditions. That study further suggested that per se performance of lines was a relatively more important factor in determining hybrid performance under excess moisture stress, while under optimum soil moisture conditions mid-parent heterosis was relatively more important than per se performance of mid-parent. Significant phenotypic correlation was observed between hybrid and mid-parent yields under excess moisture stress ($r = 0.66$), suggesting that the performance of hybrid progenies under excessive moisture conditions could be predicted and improved to some extent on the basis of their inbred parents that have been systematically selected and improved for excess moisture stress.

Genetic mapping is another way of understanding the genetic factors underlying a trait, by studying recombination in controlled or uncontrolled crosses. With the advent of molecular markers, linkage mapping has emerged as a commonly used method to achieve this understanding and to utilize this information in directed crop breeding. Linkage mapping or quantitative trait locus (QTL) mapping is carried out using two (or a few) genetically diverse parents which are crossed in various ways to produce a progeny with maximum genetic variation for a particular trait to elucidate linkage between specific markers and the genetic loci controlling the trait. Considering its major limitation that only allelic diversity that segregates between the parents of the particular population can be studied, genome-wide association study (GWAS) in a panel of breeding-relevant diverse lines came into practice.

After the first GWAS reported in maize a decade ago (Belo *et al.*, 2008), there has been a large number of publications on GWAS in maize for traits ranging from nutritional quality to abiotic and biotic stress tolerance and grain yield (Xiao *et al.*, 2017). It must be emphasized that QTL mapping and GWAS are complementary strategies towards genetic mapping and when effectively combined, can overcome the limitations of each other (Korte and Farlow, 2013). Based on this understanding, huge resources for joint linkage and association mapping have been created in temperate maize (Yu *et al.*, 2008; Dell'Acqua *et al.*, 2015). Several association mapping panels have been assembled at the International Maize and Wheat Improvement Center (CIMMYT), like the DTMA (Drought Tolerant Maize for Africa) panel, IMAS (Improved Maize for African Soils) panel, CAAM (CIMMYT Asia Association Mapping) panel and HTAM (Heat Tolerant Maize for Asia) panel, which have been used in GWAS of many traits relevant to the tropics (Nair *et al.*, 2015; Suwarno *et al.*, 2015; Zaidi *et al.*, 2016; Cao *et al.*, 2017; Gowda *et al.*, 2018; Hindu *et al.*, 2018; Rashid *et al.*, 2018). Similarly, hundreds of articles have been published on QTL mapping for various traits relevant to tropical maize, but there are very few studies where both these approaches are judiciously employed to discover and validate trait-associated markers.

Three mapping populations developed to study waterlogging tolerance in teosinte (*Zea nicaraguensis*) identified seven QTLs for constitutive aerenchyma under waterlogging, accounting for between 6 and 25% of phenotypic variance (Mano *et al.*, 2016). The QTL with the largest effect was identified on chromosomal bin 1.05–1.07 in all the populations studied. Construction of a high-density linkage map aided in developing fine-mapping populations of this QTL on chromosome 1 (*Qaer1.06*) (Mano *et al.*, 2009). In addition, for adventitious root development, three QTLs were identified with low effect size of 0.04–0.06 in one of the three populations. Further, one QTL with large effect size of 0.13–0.42 was identified for tolerance to reducing conditions. There were efforts eventually to pyramid these teosinte QTLs into parents of successful varieties. A similar study involving another subspecies of teosinte (*Zea luxurians*) identified a different set of QTLs for constitutive aerenchyma formation, thereby indicating the possibility of pyramiding multiple genomic regions from the different teosinte subspecies into cultivated maize (Mano *et al.*, 2008). Mano *et al.* (2005) mapped QTLs for adventitious root formation under waterlogged conditions on chromosomes 4 and 8 using a maize × teosinte (*Zea huehuetenangensis*) cross, in which the favourable alleles were contributed by teosinte. Flooding tolerance has also been studied in Brazilian germplasm and Anjos e Silva *et al.* (2005) identified three loci corresponding to glutamine synthetase (on chromosome 5), zein (on chromosome 4) and triosephosphate isomerase (on chromosome 3) which together explained up to 30% of phenotypic variance for shoot and root dry matter under flooded conditions. Qiu *et al.* (2007) mapped QTLs in Chinese germplasm for various secondary traits like plant height, root length, shoot dry weight, root dry weight and total dry weight under waterlogging stress at V2 stage under glasshouse conditions. They identified QTLs for all these traits, with a major QTL identified on chromosome 9 for shoot dry weight and root dry weight. Table 17.4 provides a compilation of the QTLs reported for several excess moisture tolerance-associated traits.

At CIMMYT, Zaidi *et al.* (2015) studied a recombinant inbred line (RIL) population, formed from waterlogging-tolerant and elite susceptible lines, for grain yield and secondary traits under waterlogging stress in lines per se and test crosses at V7–V8 stage. The study was designed in this way as the per se performance of lines as well as the performance of test crosses have practical implications in breeding. Based on the per se performance, a total of six QTLs on chromosomes 1, 2, 3, 5, 7 and 10 were identified, and one QTL was identified on chromosome 5 using the RIL test cross data set for grain yield under waterlogging stress, together explaining around 40% of phenotypic variance. QTLs on chromosome 1, 3 and 5 were contributed by the waterlogging-tolerant parent, while the rest by the elite but waterlogging-susceptible parent. The additive effect sizes of the QTLs ranged from 0.1 to 0.6 t/ha. Three QTLs were detected for number of nodes with brace root on chromosomes 7, 8 and 10, together explaining close to 15% of phenotypic variance.

The grain yield QTL on chromosome 1 and brace root QTL on chromosome 7 co-located with previously identified constitutive QTLs for aerenchyma formation (Mano *et al.*, 2010) that were contributed by the teosinte accession, *Z. nicaraguensis*. A total of ten QTLs were detected across all the chromosomes for the waterlogging component traits, namely root lodging, stem lodging, brace roots and chlorophyll content in lines per se. The detected QTLs accounted for 4–14% of phenotypic variance. For secondary traits studied in test crosses, one QTL was detected each for ASI, brace roots and mortality percentage, ranging from 3 to 12% of phenotypic variance. Additionally, eight bi-parental populations were developed by crossing two common parents with eight different waterlogging-tolerant lines to derive *c.*700 $F_{2:3}$ families. They were evaluated for per se performance under waterlogging and a total of 39 QTLs across seven bi-parental populations were found for grain yield and component traits. Nine QTLs for grain yield were identified explaining phenotypic variance of 1–20%, and 30 QTLs for component traits explained phenotypic variance ranging from 1 to 24%. The physical coordinates flanking these QTLs were compared against the QTL regions that were detected in the RIL families of WLT × WLS population, which revealed around 65% overlap indicating consensus regions that were identified across populations (D.D. Van, 2021, unpublished results).

Apart from QTL mapping studies carried out in controlled crosses, association mapping was also conducted in diverse germplasm for tolerance to waterlogging stress. Zhang *et al.* (2012) conducted GWAS followed by linkage mapping for validating the associations identified from the study. They identified major associations with single-nucleotide polymorphisms (SNPs) located within genes *GRMZM2G012046*, *GRMZM2G009808*, *GRMZM2G137108* and *GRMZM2G369629*, on chromosomes 5, 6 and 9. The linkage mapping study in an advanced backcross population validated the associations identified on chromosomes 5 and 9. Apart from these, many associations identified at lower significance threshold were also found to be located in QTL regions reported in other published studies.

Yu *et al.* (2019) utilized the current understanding of the molecular basis for waterlogging tolerance and conducted a candidate gene–association mapping analysis. In a waterlogging trial at two-leaf stage of maize seedlings in greenhouse conditions, they identified and characterized 19 *ERFVIIs* of maize (*ZmERFVIIs*) and associated the genetic variations of each *ZmERFVII* gene with waterlogging tolerance, evaluated in terms of survival rate under long-term stress at the seedling stage, in a diverse population of maize consisting of 368 inbred lines from global germplasm. A key candidate gene, *ZmEREB180*, was found to be strongly associated with survival rate under multiple environments. Variations in the 5′-untranslated region (5′-UTR) of *ZmEREB180* affected its expression in different varieties and showed significant correlation with survival rate. They conducted functional analysis of

**Table 17.4.** QTLs mapped for excess moisture tolerance traits in maize.

| Trait | Chromosome bin | Flanking/nearest marker | PVE (%) | Wild/cultivated accessions used in mapping population | Reference |
|---|---|---|---|---|---|
| Adventitious root formation | 8.05 | *umc1777–bnlg240* | 25.0 | *Zea huehuetenangensis* | Mano *et al.* (2005) |
| | 4.07 | *umc1869–bnlg1189* | 9.0 | | |
| | 8.03 | *phi115–phi014* | 10.0 | | |
| Shoot dry matter | 5.06 | *Phi085* | 16.0 | Cultivated | Anjos e Silva *et al.* (2005) |
| | 4.04 | *Phi074* | 15.0 | | |
| | 3.04 | *Phi029* | 14.0 | | |
| Root dry matter | 4.04 | *Phi074* | 11.0 | | |
| | 3.04 | *Phi029* | 11.0 | | |
| Plant height | 1.07 | *bnlg1556* | 13.5 | Cultivated | Qiu *et al.* (2007) |
| | 1.08 | *bnlg1643* | 6.0 | | |
| | 4.06 | *umc2027* | 4.7 | | |
| | 7.03 | *umc1567* | 4.9 | | |
| Shoot dry weight | 4.06 | *umc1299* | 4.2 | | |
| Root length | 7.03 | *umc1567* | 7.4 | | |
| Total dry weight | 4.06 | *umc1299* | 4.6 | | |
| Plant height | 2.05 | *nc131* | 4.2 | | |
| | 6.04 | *umc1918* | 5.8 | | |
| Shoot dry weight | 3.04 | *umc1347* | 3.9 | | |
| | 4.05 | *phi026* | 5.6 | | |
| | 9.04 | *umc1519* | 37.3 | | |
| Root length | 7.02 | *bnlg657* | 6.3 | | |
| Root dry weight | 6.04 | *umc1918* | 4.2 | | |
| | 9.04 | *umc1519* | 26.3 | | |
| Total dry weight | 3.04 | *umc1347* | 4.1 | | |
| | 4.05 | *phi026* | 5.5 | | |
| | 9.04 | *umc1519* | 33.3 | | |
| Plant height | 10.03 | *umc2067* | 7.1 | | |
| | 10.04 | *umc1053* | 5.2 | | |
| Shoot dry weight | 4.05 | *nc005* | 5.1 | | |
| | 6.04 | *umc1918* | 5.4 | | |
| Root length | 7.02 | *bnlg657* | 4.0 | | |
| Root dry weight | 1.04 | *bnlg1811* | 4.4 | | |
| Total dry weight | 9.04 | *umc1519* | 31.7 | | |
| Root aerenchyma formation | 2.06–2.07 | *umc1763–umc2402* | 9.1 | *Zea luxurians* | Mano *et al.* (2008) |
| | 2.05–2.06 | *bnlg1887–umc1763* | 9.1 | | |
| | 3.09–3.1 | *umc1136–umc2048* | 6.3 | | |
| | 5.05–5.06 | *mmc0282–umc1019* | 4.2 | | |
| | 9.07–9.08 | *bnlg1191–bnlg1129* | 4.5 | | |
| | 10.04–10.05 | *umc2003–bnlg1074* | 3.7 | | |
| Root aerenchyma formation | 1.06 | *umc2396* | 17.0 | *Zea nicaraguensis* | Mano *et al.* (2010) |
| | 7.01 | *umc1159* | 12.0 | | |
| | 1.07 | *bnlg1556* | 8.0 | | |
| Grain yield (t/ha) | 1.08–1.09 | *PZA03301.2–PZA01921.20* | 5.0 | Cultivated | Zaidi *et al.* (2015) |
| | 3.06–3.06 | *PZA02212.1–PZA02654.3* | 4.2 | | |

*Continued*

**Table 17.4.** Continued.

| Trait | Chromosome bin | Flanking/nearest marker | PVE (%) | Wild/cultivated accessions used in mapping population | Reference |
|---|---|---|---|---|---|
| | 5.04–5.04 | *PZA02164.16–PZA01796.1* | 8.0 | | |
| | 7.02–7.02 | *PHM4353.31–PZA02612.1* | 6.1 | | |
| | 10.03–10.03 | *PZA01677.1–PZA02941.7* | 3.6 | | |
| Root lodging (%) | 1.07–1.08 | *PZA02014.3–PZA03001.15* | 3.8 | | |
| | 3.07–3.08 | *PZA03458.1–PHM13742.5* | 6.3 | | |
| | 10.04–10.04 | *PZA03713.1–PZA03196.1* | 6.1 | | |
| Stem lodging (%) | 3.06–3.06 | *PZA02402.1–PZA02212.1* | 8.6 | | |
| | 4.09–4.09 | *PZA00636.7–PZA00521.3* | 9.2 | | |
| | 7.03–7.04 | *PHM9162.135–PHM1912.23* | 6.4 | | |
| | 8.01–8.03 | *PZA01601.1–PZA01186.1* | 7.1 | | |
| Brace root (no.) | 7.03–7.04 | *PHM9162.135–PHM1912.23* | 4.8 | | |
| | 8.06–8.06 | *PZA03698.1–PZA00838.2* | 4.2 | | |
| Chlorophyll (SPAD) | 2.0–2.02 | *PZA00365.2–PHM12952.13* | 13.6 | | |
| Grain yield_test cross (t/ha) | 5.05–5.05 | *PZA02383.1–PZA02209.2* | 3.4 | | |
| ASI_test cross (days) | 3.07–3.08 | *PZA03154.4–PHM15964.16* | 6.8 | | |
| Brace root_test cross (no.) | 7.01–7.02 | *PHM2691.31–PHM4353.31* | 2.6 | | |
| Plant mortality_test cross (%) | 5.02–5.03 | *PZA01371.1–PHM5484.22* | 12.3 | | |
| Leaf senescence | 6 (bin: NA) | NA | 22.0 | Cultivated | Campbell *et al.* (2015) |
| Constitutive aerenchyma | 1.02–1.03 | *bnlg1007–bnlg1484* | 11.0 | *Z. nicaraguensis* | Mano *et al.* (2016) |
| | 1.05 | *umc1297–IDP5918* | 17.0–25.0 | | |
| | 1.06–1.07 | *umc1128–bnlg1347* | 8.0–17.0 | | |
| | 1.11 | *umc1862* | 12.0 | | |
| | 5.09 | *umc1153* | 6.0 | | |
| | 7.01 | *umc1159* | 12.0 | | |
| | 8.05 | *umc1712–bnlg162* | 7.0 | | |
| Adventitious root development | 3.04 | *bnlg1113* | 6.0 | | |
| | 7.04 | *dupssr13* | 4.0 | | |
| | 8.03 | *umc2075* | 4.0 | | |
| Tolerance to reducing conditions | 4.07–4.11 | NA | 13.0–42.0 | | |

PVE, proportion of phenotypic variance explained by the QTL; SPAD, Soil Plant Analysis Development (chlorophyll meter); NA, information not available.

*ZmEREB180* and showed that this gene played a positive role in maize waterlogging tolerance by promoting the formation of adventitious roots and coordinating ROS homeostasis. In another study under similar waterlogging stress conditions, Yu *et al.* (2018) conducted GWAS in the same panel and identified a total of 110 associations spanning 16 genomic regions, of which 14 co-localized with previously detected waterlogging tolerance-related QTLs. Single locus associations explained 3–11% of the phenotypic variance observed. They resequenced and studied the expression profile of the gene *GRMZM2G110141* on chromosome 6, which had the most significantly associated SNP signals, and found variations in the 5′-UTR of the gene leading to leaf injury responses under waterlogging.

At CIMMYT, association mapping for waterlogging tolerance at V7–V8 stage was conducted in two association mapping panels: (i) DTMA panel, primarily formed by lines with adaptation to sub-Saharan Africa and Latin America; and (ii) CAAM panel, a panel of 419 Asia-adapted lines. DTMA panel lines were evaluated for secondary traits associated with waterlogging tolerance, whereas CAAM panel lines were test crossed with testers of two heterotic groups and phenotyped for grain yield under waterlogging stress and its component traits. For grain yield under stress, 21 significant associations in the test crosses with HG-A tester (CML451) and 18 significant associations in the test crosses with HG-B tester (CL02450) were identified. Eighteen associations were validated based on the linkage mapping studies done on RILs and multiple $F_{2:3}$ families, mentioned above, to develop KASP (competitive allele-specific PCR) assays for targeted introgression of valuable alleles tagged to specific waterlogging-tolerant donors identified in the panel (Table 17.5) (S.K. Nair, 2021 unpublished results).

Studies undertaken to date on the genetic analysis of waterlogging tolerance have shown that this is a polygenic trait, mostly explained by low- to moderate-effect QTLs. Waterlogging tolerance in teosinte has shown some large-effect QTLs, especially affecting morphological modifications associated with the tolerance mechanism, like the constitutive aerenchyma formation and adventitious root development. Utilizing a wild relative of maize as a trait source in breeding programmes would require long pre-breeding

**Table 17.5.** Synopsis of GWAS and QTL mapping studies for grain yield and component traits for waterlogging tolerance in maize at CIMMYT.

| Trait | No. of significant associations | $R^2$ (%) range |
|---|---|---|
| GWAS line per se: DTMA_AM | | |
| Reduction in root weight (%) | 19 | 2.6–9.3 |
| Reduction in shoot weight (%) | 18 | 3.5–12.3 |
| Reduction in total weight (%) | 21 | 3.4–13.4 |
| Differential emergence | 34 | 2.8–12.4 |
| Days to differential emergence | 28 | 2.6–9.5 |
| GWAS test crosses: CAAM_AM | | |
| GY_WL_CML451 | 21 | 2.5–8.6 |
| GY_WL_CL02450 | 18 | 2.4–7.9 |
| | No. of QTLs | |
| $F_{2:3}$–QTL | | |
| Grain yield | 9 | 1.1–20.4 |
| Brace root | 6 | 1.8–23.7 |
| Surface root | 6 | 0.1–5.1 |
| ASI_WL | 10 | 1.3–17.7 |
| % Mortality | 8 | 6.2–56.7 |
| RIL_QTL | | |
| Grain yield | 6 | 3.4–6.1 |
| Root lodging (%) | 3 | 3.8–6.3 |
| Stem lodging (%) | 4 | 5.2–13.0 |
| Brace root (no.) | 3 | 2.6–4.8 |

AM, association mapping; GY, grain yield; WL, waterlogging.

processes, to remove the unwanted traits of teosinte. It is encouraging to note that some genetic mapping studies in cultivated germplasm have identified QTLs in genomic regions previously reported for constitutive root aerenchyma and adventitious root formation in teosinte. QTLs identified for waterlogging tolerance in cultivated germplasm in maize are much smaller in size and would require pyramiding of multiple QTLs from multiple sources to have a discernible effect on the trait, as expressed in terms of grain yield under stress.

## 17.6 Molecular Breeding for Excess Moisture Tolerance

The duration of excess moisture stress, the stage of cultivation and other environmental factors during this stress often determine the magnitude of yield loss in the maize crop (Zaidi *et al.*, 2012). Zaidi *et al.* (2004) reported V2 and V7 stages of the crop as the most susceptible to excess moisture, particularly in medium to heavy textured soils. Nielsen (2019) further suggested that high temperatures could exacerbate the effect of excess moisture stress leading to plant mortality particularly during the vegetative stage of the maize crop. Developing hybrids that withstand excess moisture stress should be a major focus of maize breeding programmes targeting products to regions prone to this stress. Substantial variability has been recorded among the maize germplasm evaluated under excess moisture stress (Zaidi *et al.*, 2004, 2010) that could be effectively utilized in breeding programmes. Designing a breeding pipeline often entails an in-depth understanding of the target population of environments (TPE), along with a detailed genetic analysis of the trait of interest and the effect of the stresses in combination with other factors unique to the TPE. Breeders often consider performance under high-yielding environment as a major criterion of selection; however, products from these pipelines are often unsuited for low-input stress-prone environments. Banziger *et al.* (2006) suggested that with adequate weighing of performance of entries in managed stress-prone environments in the breeding pipeline, the yield levels of entries could be marginally increased simultaneously in high- and even in severely stressed low-yielding environments. This strategy has been used for the development of several potential hybrids with high yield under optimal conditions along with resilience to excess moisture stress within Asia (Das *et al.*, 2019).

One of the major bottlenecks in breeding for excess moisture stress is the low heritability generally observed for grain yield under the stress. This presents a significant hindrance to the gains that could be achieved in the breeding programme. In addition, grain yield under excess moisture stress is reported to be controlled by both additive and non-additive gene action. Hence breeding pipelines need to be designed to exploit these effects for making significant progress to excess moisture tolerance. The use of secondary traits could greatly improve the selection response, particularly under stress conditions. Several successes have been reported for gains observed when selections are done primarily on secondary traits like ASI for drought tolerance in maize (Edmeades *et al.*, 1999; Monneveux *et al.*, 2006). An ideal secondary trait is one that could be effectively used as a proxy for grain yield under the stress, should be highly heritable and show high genetic variation in the working germplasm, and should be stable and economical to score (Lafitte *et al.*, 2003). For waterlogging tolerance, brace root has been identified as one of the most effective surrogates for excess moisture stress, due to higher heritability, good genetic variability and ease of observation (Zaidi *et al.*, 2007c, 2015).

Initial studies for waterlogging stress tolerance in maize have identified several lines within the working germplasm, although at low frequency, having tolerance to excess moisture stress (Zaidi *et al.*, 2004, 2007c, 2010; Mano *et al.*, 2006; Amin *et al.*, 2014; Kaur *et al.*, 2019). A multi-parent population breeding approach, to increase favourable allele frequency through recurrent selection, is being followed within CIMMYT dedicated towards simultaneous improvement of drought and waterlogging tolerance. Given the complexity of the trait, there is a need for evaluating a large number of genotypes at multiple locations to identify suitable recombinants, which often limits the gains that could be attained. Hence investing in high-throughput methodologies along with appropriate screening and selection techniques could greatly assist breeders with making substantial progress.

Molecular breeding, utilizing the findings from genetic and molecular studies on the trait, is found to be an efficient breeding strategy in enhancing genetic gains. This is especially true in the case of traits like excess moisture stress where managed phenotyping is cumbersome and repeatability is generally low. In this context, wherever possible, marker-based selection methods like targeted introgression and genomic selection are extremely beneficial. Targeted introgression follows particular markers linked to traits, which allows them to be tracked with reasonable precision and hence tailors the product with the required trait, but keeps the background intact. Trait-linked markers that are discovered and deployed by way of marker-assisted selection (MAS) require prior knowledge of the precise location and effect size of QTLs and germplasm/environment specificity. This strategy also requires that the genetic architecture of the trait supports such interventions by way of having large-effect genes/QTLs.

Genomic prediction is another form of MAS where genome-wide markers are used to estimate the breeding value of individuals (Meuwissen *et al.*, 2001). This concept, which was originally developed in dairy cattle, found its way to crop plants, especially in crops like maize. While the public-sector maize breeding programmes in the tropics have been slow in venturing into genomic prediction strategy, the commercial seed sector has been routinely practising genomic prediction in its breeding pipelines with success. Genomic prediction brought a paradigm shift in the way plant breeding is done, shifting the unit of selection from individual lines to alleles (Lorenz and Nice, 2017). A number of factors like heritability, trait architecture, marker density, training population size and relationship between the training and prediction populations are critical to the accuracy of the predicted breeding value (Combs and Bernardo, 2013). Genomic prediction will be highly beneficial and cost-efficient in driving genetic gains in the breeding programmes when medium- to high-density genotyping costs and turnaround times decrease sufficiently to replace resource-intensive, field-based precision phenotyping at least partially.

There are not many published reports on using molecular markers in breeding for excess moisture stress tolerance in maize, understandably, due to the polygenic nature of the trait and most of the detected QTLs not being significant enough to provide discernible effect differences under actual field conditions under stress. Rice has the most well-documented effort for improving submergence tolerance in terms of MAS using *SUB1A*, a submergence QTL with about 69% effect in phenotypic variation (Xu and Mackill, 1996). Using marker-assisted backcrossing, a small genomic region containing *SUB1A* from a landrace FR13A was introgressed into modern high-yielding varieties in diverse genetic backgrounds, such as Swarna, Samba Mahsuri, IR64, Thadokkam 1 (TDK1), CR1009 and BR11 (Septiningsih *et al.*, 2009). Multiple evaluations of submergence tolerance in the greenhouse and farmers' fields confirmed that all converted lines exhibit significantly greater tolerance to complete submergence as compared with their original parents (Sarkar *et al.*, 2009; Septiningsih *et al.*, 2009; Singh *et al.*, 2009).

In maize, Mano *et al.* (2016) used teosinte, as a flooding-tolerant donor, as it is tolerant to flooding in its low-lying and frequently flooded natural growing habitat. Teosinte possesses unique morphological and physiological characteristics related to flooding tolerance, including the capacity to form constitutive aerenchyma (Mano *et al.*, 2006, 2007), tolerance to flooding under reducing soil conditions (Mano and Omori, 2013), a barrier to radial oxygen loss (Abiko *et al.*, 2012) and the ability to form adventitious roots at the soil surface during flooding (Mano *et al.*, 2009). Mano *et al.* (2016) developed near-isogenic lines (NILs) separately for these component traits possessing one or more QTLs for the given trait, using teosinte as the donor parent. These NILs were later used for pyramiding effective flooding tolerance-related QTLs. To our knowledge, the field performance of the final variety incorporating multiple pyramided component traits is not yet in the public domain.

Based on the various genetic studies conducted on waterlogging in cultivated maize germplasm, both by GWAS and QTL mapping, CIMMYT developed backcrosses of elite lines in the two heterotic groups with waterlogging-tolerant parents, (DT/LN/EM-46-3-1 × CML311-2-1-3)-B-F239-1-1-1-BB and Saracura. The strategy was to incorporate waterlogging tolerance in elite lines using inter-crosses of backcrosses developed using multiple waterlogging tolerance donors. The backcrosses were selected using markers validated in genetic analyses, and favourable families carrying the trait markers were selected. Doubled haploids were developed for fixing favourable combinations and test crossed with corresponding heterotic group testers. The test crosses thus developed are in the process of evaluation under waterlogging conditions (S.K. Nair, 2021, unpublished results).

Rapid-cycle genomic selection (RCGS) is considered an efficient strategy in population improvement, offering better prediction accuracy of breeding and genetic values by incorporating all markers compared with using a subset of markers significantly associated with the QTL, namely marker-assisted recurrent selection (MARS) (Zhang *et al.*, 2017). Massman *et al.* (2013) reported that RCGS had a superior response for stover yield and grain yield indices that were 14–50% higher than those of MARS. Beyene *et al.* (2015) evaluated realized genetic gains in grain yield from RCGS in eight bi-parental maize lines; they found that the average gain from RCGS per cycle across eight populations was 0.086 tons/ha[1] and that hybrids derived from cycle 3 produced 7.3% (0.176 tons/ha) higher grain yield than those developed through the conventional pedigree breeding method. RCGS in bi-parental populations offered the advantage of significant time efficiency over conventional breeding methods, as up to three cycles of RCGS can be conducted within a year. Vivek *et al.* (2017) conducted RCGS in bi-parental populations and found that top crosses from cycle 2 of RCGS produced 4–43% higher grain yield than top crosses from cycle 2 of the same populations formed based on phenotypic selection. Zhang *et al.* (2017) conducted RCGS in a multi-parental population involving 18 parents with an objective of increasing the favourable allele frequency for drought stress tolerance in the population and further deriving improved inbred lines. Results from that study indicated that realized grain yield from cycle 1 to cycle 4 reached 0.225 tons/ha per cycle, which is equivalent to 0.100 tons/ha per year over a 4.5-year breeding period from the initial cross to the last cycle.

RCGS was carried out for simultaneous improvement of source germplasm belonging to the two heterotic groups for waterlogging stress and drought stress tolerance in Asia. Four waterlogging tolerance donors, four drought tolerance donors and two elite lines were included in population development for RCGS in each of the heterotic groups. As part of the RCGS strategy, $S_2$ test crosses formed after $F_1$ inter-crosses of parental crosses carried out in a half-diallel design were phenotyped in multiple stress environments. These families were also genotyped using 300 genome-wide markers and marker effects were estimated. After an initial cycle of phenotype-based selection with an intensity of 5–10%, two cycles were raised based on intermating of plants selected based on genomic estimated breeding value (GEBV). Genetic gain studies from test crosses of various cycles indicated an increase in genetic gain under waterlogging stress of 38 kg/ha per year in HG-A population and 113 kg/ha per year in HG-B population with two cycles of genomic selection (Das *et al.*, 2020). Genomic selection within bi-parental populations in the breeding pipeline has been proven to be a resource- and a time-efficient method towards enhancing genetic gains in stress tolerance breeding programmes (Zhang *et al.*, 2015; Beyene *et al.*, 2019). At CIMMYT, the waterlogging tolerance breeding programme has initiated this strategy, where 50% of the lines entering Stage 1 test crosses from a full-sib family (forming the training set of the genomic selection model) are used to predict the performance of the other 50% of the lines (tester/validation set) based on their GEBVs. These selected lines can enter Stage 2 test crossing without actual test crossing and Stage 1 field-based evaluations. In general, genomic selection will be a strategy to look forward to for improvement of polygenic traits in changing times where genotyping technologies are becoming more accessible and breeder oriented.

## 17.7 Conclusion

Excess moisture tolerance in plants is a well-researched topic, with most of the morphological and anatomical adaptations, hormonal and metabolite pathways, and molecular mechanisms elegantly elucidated. Maize also adds to the repertoire of crops where molecular signals leading to various tolerance mechanisms leading to excess moisture tolerance have been studied, although no gene has been cloned for this trait in maize. Genetic studies to understand the hereditary basis of the trait and possible selection targets, in terms of linked markers, are not many. This could be due to the difficulty in precision phenotyping for the trait and lack of excess moisture-tolerant germplasm in many breeding programmes. Also, most of the studies deal with the trait in terms of studying morphological adaptations, rather than grain yield under stressed growing conditions, and hence lack any proven impact in breeding programmes. The polygenic nature of the trait, with no large-effect QTLs identified for the trait in the cultivated germplasm, makes it tougher to undertake MAS. Pyramiding of QTLs from specific donors has been attempted with good success. RCGS, which is a form of recurrent selection using genome-wide markers, has been successful in rapid population improvement to derive improved waterlogging-tolerant lines. Genomic selection has the capacity to remodel breeding strategies with a positive impact on cost, time and efficiency towards achieving enhanced genetic gains under excess moisture stress.

## Endnote

[1] Conversion factor from tons to tonnes = 0.9.

## References


Abiko, T., Kotula, L., Shiono, K., Malik, A.I., Colmer, T.D. and Nakazono, M. (2012) Enhanced formation of aerenchyma and induction of a barrier to radial oxygen loss in adventitious roots of *Zea nicaraguensis* contribute to its waterlogging tolerance as compared with maize (*Zea mays* ssp. *mays*). *Plant, Cell & Environment* 35, 1618–1630.

Amin, M., Amiruzzaman, M., Ahmed, A. and Ali, M. (2014) Combining ability study in waterlogged tolerant maize (*Zea mays* L.). *Bangladesh Journal of Agricultural Research* 39, 283–291.

Anjos e Silva, S.D. dos, Sereno, M.J.C. de M., Lemons e Silva, C.F., Oliveira, A.C. de and Barbosa Neto, J.F. (2005) Genetic parameters and QTL for tolerance to flooded soils in maize. *Crop Breeding and Applied Biotechnology* 5, 287–293.

Ap Rees, T., Jenkin, L.E.T., Smith, A.M. and Wilson, P.M. (1987) The metabolism of flood tolerant plant. In: Crawford, R.M.M. (ed.) *Plant Life in Aquatic and Amphibious Habitat*. Blackwell Scientific, Oxford, UK, pp. 227–238.

Bailey-Serres, J., Lee, S.C. and Brinton, E. (2012) Waterproofing crops: effective flooding survival strategies. *Plant Physiology* 160, 1698–1709.

Banziger, M., Setimela, P.S., Hodson, D. and Vivek, B. (2006) Breeding for improved abiotic stress tolerance in maize adapted to southern Africa. *Agricultural Water Management* 80, 212–224.

Belo, A., Zheng, P., Luck, S., Shen, B., Meyer, D.J. *et al.* (2008) Whole genome scan detects an allelic variant of *fad2* associated with increased oleic acid levels in maize. *Molecular Genetics and Genomics* 279, 1–10.

Beyene, Y., Semagn, K., Mugo, S., Tarekegne, A., Babu, R. *et al.* (2015) Genetic gains in grain yield through genomic selection in eight bi-parental maize populations under drought stress. *Crop Science* 55, 154–163.

Beyene, Y., Gowda, M., Olsen, M., Robbins, K.R., Perez-Rodriguez, P. *et al.* (2019) Empirical comparison of tropical maize hybrids selected through genomic and phenotypic selections. *Frontiers in Plant Science* 10, 1502.

Bradford, K.J. (1982) Regulation of shoot response to root stress by ethylene, abscisic acid and cytokinins. In: Wareing, P.F. (ed.) *Plant Growth Substances 1982*. Academic Press, New York and London, pp. 599–608.

Cairns, J.E., Sonder, K., Zaidi, P.H., Verhulst, N., Mahuku, G. *et al.* (2012) Maize production in a changing climate: impacts, adaptation, and mitigation strategies. *Advances in Agronomy* 114, 1–58.

Campbell, M.T., Proctor, C.A., Dou, Y., Schmitz, A.J., Phansak, P. *et al.* (2015) Genetic and molecular characterization of submergence response identifies *Subtol6* as a major submergence tolerance locus in maize. *PLoS One* 10, e0120385.

Cao, S., Loladze, A., Yuan, Y., Wu, Y., Zhang, A. *et al.* (2017) Genome-wide analysis of tar spot complex resistance in maize using genotyping-by-sequencing SNPs and whole-genome prediction. *Plant Genome* 10, plantgenome2016.10.0099.

Combs, E. and Bernardo, R. (2013) Accuracy of genomewide selection for different traits with constant population size, heritability, and number of markers. *Plant Genome* 6, 30.
Crawford, R.M.M. (1967) Alcohol dehydrogenase activity in relation to flooding of roots. *Journal of Experimental Botany* 18, 454–464.
Crawford, R.M.M. and Braendle, R. (1996) Oxygen deprivation stress in changing environment. *Journal of Experimental Botany* 47, 145–159.
Das, R.R., Vinayan, M.T., Seetharam, K., Vishwanadh, S., Salahuddin, A. *et al.* (2019) One-size doesn't fit to all: maize for various stress-prone agro ecologies in Asian tropics. Poster presented at *13th Asian Maize Conference and Expert Consultation on 'Maize for Food, Feed, Nutrition and Environmental Security', Ludhiana, India, 8–10 October 2019*.
Das, R.R., Vinayan, M.T., Patel, M.B., Phagna, R.K., Singh, S.B. *et al.* (2020) Genetic gains with rapid-cycle genomic selection for combined drought and waterlogging tolerance in tropical maize (*Zea mays* L.). *Plant Genome* 13(3), e20035.
Dell'Acqua, M., Gatti, D.M., Pea, G., Cattonaro, F., Coppens, F. *et al.* (2015) Genetic properties of the MAGIC maize population: a new platform for high definition QTL mapping in *Zea mays*. *Genome Biology* 16, 167.
Devitt, A.C. and Francis, C.M. (1972) The effect of waterlogging on the mineral nutrient content in *Trifolium subterraneum*. *Australian Journal of Experimental Agriculture and Animal Husbandry* 12, 614–617.
Drew, M.C. (1997) Oxygen deficiency and root metabolism: injury and acclimation under hypoxia and anoxia. *Annual Review of Plant Physiology and Plant Molecular Biology* 48, 223–250.
Edmeades, G., Chapman, S. and Lafitte, H. (1999) Selection improves drought tolerance in tropical maize populations: I. Gains in biomass, grain yield, and harvest index. *Crop Science* 39, 1306–1315.
Else, M.A., Janowiak, F., Atkinson, C.J. and Jackson, M.B. (2009) Root signals and stomatal closure in relation to photosynthesis, chlorophyll a fluorescence and adventitious rooting of flooded tomato plants. *Annals of Botany* 103, 313–323.
Evans, D.E. (2004) Aerenchyma formation. *New Phytologist* 161, 35–49.
Fausey, N.R., VanToai, T.T. and McDonald, M.B. (1985) Response of ten maize cultivars to flooding. *Transactions of the American Society of Agricultural and Biological Engineers* 28, 1794–1797.
Foyer, C.H., Ruban, A.V. and Noctor, G. (2017) Viewing oxidative stress through the lens of oxidative signalling rather than damage. *Biochemical Journal* 474, 877–883.
Fukao, T., Xu, K., Ronald, P.C. and Bailey-Serres, J. (2006) A variable cluster of ethylene response factor-like genes regulates metabolic and developmental acclimation responses to submergence in rice. *The Plant Cell* 18, 2021–2034.
Fukao, T., Yeung, E. and Bailey-Serres, J. (2011) The submergence tolerance regulator SUB1A mediates crosstalk between submergence and drought tolerance in rice. *The Plant Cell* 23, 412–427.
Fukao, T., Barrera-Figueroa, B.E., Juntawong, P. and Pena-Castro, J.M. (2019) Submergence and waterlogging stress in plants: a review highlighting research opportunities and understudied aspects. *Frontiers in Plant Science* 10, 340.
Gibbs, D.J., Lee, S.C., Isa, N.M., Gramuglia, S., Fukao, T. *et al.* (2011) Homeostatic response to hypoxia is regulated by the N-end rule pathway in plants. *Nature* 479, 415–418.
Gowda, M., Beyene, Y., Makumbi, D., Semagn, K., Olsen, M.S. *et al.* (2018) Discovery and validation of genomic regions associated with resistance to maize lethal necrosis in four biparental populations. *Molecular Breeding* 38, 66.
Hattori, Y., Nagai, K., Furukawa, S., Song, X.J., Kawano, R. *et al.* (2009) The ethylene response factors *SNORKEL1* and *SNORKEL2* allow rice to adapt to deep water. *Nature* 460, 1026–1030.
Hindu, V., Palacios-Rojas, N., Babu, R., Suwarno, W.B., Rashid, Z. *et al.* (2018) Identification and validation of genomic regions influencing kernel zinc and iron in maize. *Theoretical and Applied Genetics* 131, 1443–1457.
Igamberdiev, A.U. and Hill, R.D. (2018) Elevation of cytosolic $Ca^{2+}$ in response to energy deficiency in plants: the general mechanism of adaptation to low oxygen stress. *Biochemical Journal* 475, 1411–1425.
Jackson, M.B. (1989) Regulation of aerenchyma formation in roots and shoots by oxygen and ethylene. In: Osborne, D.J. and Jackson, M.B. (eds) *Separation of Plants: Physiology, Biochemistry and Molecular Biology*. Springer, Heidelberg, Germany, pp. 262–274.
Jackson, M.B. (1990) Hormones and developmental changes in plants subjected to submergence and soil waterlogging. *Aquatic Botany* 38, 49–72.

Jackson, M.B. and Kowalewska, A.K.B. (1983) Positive and negative massages from root induce foliar desiccation and stomatal closure in flooded pea plants. *Journal of Experimental Botany* 34, 493–506.

Jackson, M.B., Brailsford, R.W. and Else, M.A. (1993) Hormone and plant adaptation to poor aeration: a review. In: Kuo, C.G. (ed.) *Adaptation of Food Crops to Temperature and Water Stress*. AVRDC Publications, Taipei, ROC, pp. 231–243.

Jung, K.H., Seo, Y.S., Walia, H., Cao, P., Fukao, T. *et al.* (2010) The submergence tolerance regulator *Sub1A* mediates stress-responsive expression of *AP2/ERF* transcription factors. *Plant Physiology* 152, 1674–1692.

Kaur, G., Zurweller, B., Motavalli, P.P. and Nelson, K.A. (2019) Screening corn hybrids for soil waterlogging tolerance at an early growth stage. *Agriculture* 9, 33.

Kennedy, R.A., Rumpho, M.E. and Fox, T.C. (1992) Anaerobic metabolism in plants. *Plant Physiology* 100, 1–6.

Korte, A. and Farlow, A. (2013) The advantages and limitations of trait analysis with GWAS: a review. *Plant Methods* 9, 29.

Krishnamoorthy, H.N. (1993) *Waterlogging. Physiology of Plant Growth and Development*. ATMA Ram & Sons, New Delhi, pp. 467–488.

Kuang, W., Xianjiang, Y., Xiuqing, C. and Yafeng, X. (2012) Experimental study on water production function for waterlogging stress on corn. *Procedia Engineering* 28, 598–603.

Lafitte, H.R., Blum, A. and Atlin, G. (2003) Using secondary traits to help identify drought-tolerant genotypes. In: Fischer, K.S., Lafitte, R.H., Fukai, S., Atlin, G. and Hardy, B. (eds) *Breeding Rice for Drought-prone Environments*. International Rice Research Institute, Los Baños, Philippines, pp. 37–48.

Lal, R. and Taylor, G.S. (1970) Drainage and nutrient effects in a field lysimeter study. II. Mineral uptake in maize. *Soil Science Society of America Journal* 34, 245–248.

Liao, C.T. and Lin, C.H. (1994) Effect of flooding stress on photosynthetic activities of *Momordica charantia*. *Plant Physiology and Biochemistry* 32, 1–5.

Liao, C.T. and Lin, C.H. (1995) Effect of flood stress on morphology and aerobic metabolism of *Momordica charantia*. *Environmental and Experimental Botany* 35, 105–113.

Licausi, F., Weits, D.A., Pant, B.D., Scheible, W.R., Geigenberger, P. and van Dongen, J.T. (2011) Hypoxia responsive gene expression is mediated by various subsets of transcription factors and miRNAs that are determined by the actual oxygen availability. *New Phytologist* 190, 442–456.

Liu, Z., Kumari, S., Zhang, L., Zheng, Y. and Ware, D. (2012) Characterization of miRNAs in response to short-term waterlogging in three inbred lines of *Zea mays*. *PLoS One* 7, e39786.

Lobell, D.B., Schlenker, W. and Costa-Roberts, J. (2011) Climate trends and global crop production since 1980. *Science* 333, 616–620.

Lorenz, A. and Nice, L. (2017) Training population design and resource allocation for genomic selection in plant breeding. In: Varshney, R., Roorkiwal, M. and Sorrells, M. (eds) *Genomic Selection for Crop Improvement*. Springer, Cham, Switzerland, pp. 7–22.

Mano, Y. and Omori, F. (2013) Flooding tolerance in interspecific introgression lines containing chromosome segments from teosinte (*Zea nicaraguensis*) in maize (*Zea mays* subsp. *mays*). *Annals of Botany* 112, 1125–1139.

Mano, Y., Muraki, M., Fujimori, M., Takamizo, T. and Kindiger, B. (2005) Identification of QTL controlling adventitious root formation during flooding conditions in teosinte (*Zea mays* ssp. *huehuetenangensis*) seedlings. *Euphytica* 142, 33–42.

Mano, Y., Omori, F., Takamizo, T., Kindiger, B., Bird, R.M. and Loaisiga, C.H. (2006) Variation for root aerenchyma formation in flooded and non-flooded maize and teosinte seedlings. *Plant and Soil* 281, 269–279.

Mano, Y., Omori, F., Takamizo, T., Kindiger, B., Bird, R.M., Loaisiga, C.H. and Takahashi, H. (2007) QTL mapping of root aerenchyma formation in seedlings of a maize × rare teosinte '*Zea nicaraguensis*' cross. *Plant and Soil* 295, 103–113.

Mano, Y., Omori, F., Kindiger, B. and Takahashi, H. (2008) A linkage map of maize × teosinte *Zea luxurians* and identification of QTLs controlling root aerenchyma formation. *Molecular Breeding* 21, 327–337.

Mano, Y., Omori, F., Loaisiga, C.H. and Bird, R.M. (2009) QTL mapping of above-ground adventitious roots during flooding in maize × teosinte '*Zea nicaraguensis*' backcross population. *Plant Root* 3, 3–9.

Mano, Y., Omori, F. and Takeda, K. (2010) Construction of intraspecific linkage maps, detection of a chromosome inversion, and mapping of QTL for constitutive root aerenchyma formation in the teosinte *Zea nicaraguensis*. *Molecular Breeding* 29, 137–146.

Mano, Y., Omori, F., Tamaki, H., Mitsuhashi, S. and Takahashi, W. (2016) DNA marker-assisted selection approach for developing flooding-tolerant maize. *Japan Agricultural Research Quarterly* 50, 173–182.
Massman, J., Jung, G. and Bernardo, R. (2013) Genomewide selection versus marker-assisted recurrent selection to improve grain yield and stover-quality traits for cellulosic ethanol in maize. *Crop Science* 53, 58–66.
Meuwissen, T.H.E., Hayes, B.J. and Goddard, M.E. (2001) Prediction of total genetic value using genome-wide dense marker maps. *Genetics* 157, 1819–1829.
Mohanty, B., Wilson, P.M. and ap Rees, T. (1993) Effect of anoxia on growth and carbohydrate metabolism in suspension cultures of soybean and rice. *Phytochemistry* 34, 75–82.
Monneveux, P., Sanchez, C., Beck, D. and Edmeades, G. (2006) Drought tolerance improvement in tropical maize source populations. *Crop Science* 46, 180–191.
Nair, S.K., Babu, R., Magorokosho, C., Mahuku, G., Semagn, K. *et al.* (2015) Fine mapping of *Msv1*, a major QTL for resistance to maize streak virus leads to development of production markers for breeding pipelines. *Theoretical and Applied Genetics* 128, 1839–1854.
Nielsen, R. (2019) Effects of flooding or ponding on corn prior to tasseling. *Corny News Network*, May 2019. Department of Agronomy, Purdue University, West Lafayette, Indiana. Available at: https://www.agry.purdue.edu/ext/corn/news/timeless/pondingyoungcorn.html (accessed 23 February 2021).
Panozzo, A., Dal Cortivo, C., Ferrari, M., Vicelli, B., Varotto, S. and Vamerali, T. (2019) Morphological changes and expressions of *AOX1A*, *CYP81D8*, and putative *PFP* genes in a large set of commercial maize hybrids under extreme waterlogging. *Frontiers in Plant Science* 10, 62.
Perata, P. and Alpi, A. (1993) Plant responses to anaerobiosis. *Plant Science* 93, 1–17.
Perata, P., Pozueta-Romero, J., Akazawa, T. and Yamaguchi, J. (1992) Effect of anoxia on starch breakdown in rice and wheat seeds. *Planta* 188, 611–618.
Ponnamperuma, F.N. (1972) The chemistry of submerged soils. *Advances in Agronomy* 24, 29–96.
Qiu, F., Zheng, Y., Zhang, Z. and Xu, S. (2007) Mapping of QTL associated with waterlogging tolerance during the seedling stage in maize. *Annals of Botany* 99, 1067–1081.
Rajhi, I., Yamauchi, T., Takahashi, H., Nishiuchi, S., Shiono, K. *et al.* (2011) Identification of genes expressed in maize root cortical cells during lysigenous aerenchyma formation using laser microdissection and microarray analyses. *New Phytologist* 190, 351–368.
Rashid, Z., Singh, P.K., Vemuri, H., Zaidi, P.H., Prasanna, B.M. and Nair, S.K. (2018) Genome-wide association study in Asia-adapted tropical maize reveals novel and explored genomic regions for sorghum downy mildew resistance. *Scientific Reports* 8, 366.
Rathore, T.R., Warsi, M.Z.K., Lothrop, J.E. and Singh N.N. (1996) Production of maize under excess soil moisture (waterlogging) conditions. In: Singh, N.N., Mauria, S., Lothrop, J.E., De Leon, C. and Baldos, D.P. (eds) *Proceedings of the Sixth Asian Regional Maize Workshop, New Delhi, 30 October–3 November 1995*. ICAR and CIMMYT, New Delhi, pp. 56–63.
Rathore, T.R., Warsi, M.Z.K., Zaidi, P.H. and Singh N.N. (1997) Waterlogging problem for maize production in Asian region. *TAMNET News Letter* 4, 13–14.
Reggiani, R., Cantu, C.A., Brambilla, I. and Bertani, A. (1988) Accumulation and inter-conversion of amino acids in rice roots under anoxia. *Plant and Cell Physiology* 29, 981–987.
Reid, D.M. and Bradford, K.J. (1984) Effects of flooding on hormone relations. In: Kozlowski, T.T. (ed.) *Flooding and Plant Growth*. Academic Press, Orlando, Florida, pp. 195–219.
Ren, B., Zhang, J., Li, X., Fan, X., Dong, S., Liu, P. and Zhao, B. (2014) Effects of waterlogging on the yield and growth of summer maize under field conditions. *Canadian Journal of Plant Science* 94, 23–31.
Ren, B., Zhang, J., Dong, S., Liu, P., and Zhao, B. (2016) Effects of waterlogging on leaf mesophyll cell ultrastructure and photosynthetic characteristics of summer maize. *PLoS One* 11, e0161424.
Ricard, B., Couee, I., Raymond, P., Salgio, P., Saint-Ges, V. and Pradet, A. (1994) Plant metabolism under hypoxia and anoxia. *Plant Physiology and Biochemistry* 32, 1–10.
Sachs, M.M., Subbaiah, C.C. and Saab, I.N. (1996) Anaerobic gene expression and flooding tolerance in maize. *Journal of Experimental Botany* 47, 1–15.
Sarkar, R.K., Panda, D., Reddy, J.N., Patnaik, S.S.C., Mackill, D.J. and Ismail, A.M. (2009) Performance of submergence tolerant rice (*Oryza sativa*) genotypes carrying the *Sub1* quantitative trait locus under stressed and non-stressed natural field conditions. *Indian Journal of Agricultural Sciences* 79, 876–883.
Septiningsih, E.M., Pamplona, A.M., Sanchez, D.L., Neeraja, C.N., Vergara, G.V. *et al.* (2009) Development of submergence-tolerant rice cultivars: the *Sub1* locus and beyond. *Annals of Botany* 103, 151–160.

Setter, T.L., Kupkanchanakul, T., Kupkanchanakul, K., Bhekasut, P., Weingweera, A. and Greenway, H. (1987) Concentrations of $CO_2$ and $O_2$ in floodwater and in internodal lacunae of floating rice growing at 1–2 meter water depths. *Plant, Cell & Environment* 10, 767–776.

Shah, N.A., Srivastava, J.P., Teixeira da Silva, J.A. and Shahi, J.P. (2012) Morphological and yield responses of maize (*Zea mays* L.) genotypes subjected to root zone excess soil moisture stress. *Plant Stress* 6, 59–72.

Sharpiro, R.E., Taylor, G.S. and Volk, G.W. (1956) Soil oxygen content and oxygen uptake by maize. *Soil Science Society of America Journal* 20, 193–197.

Shin, S., Kim, S.G., Jung, S.G., Kim, C.G., Son, B.Y. *et al.* (2016) Evaluation of waterlogging tolerance with the degree of foliar senescence at early vegetative stage of maize (*Zea mays* L.). *Journal of Crop Science and Biotechnology* 19, 393–400.

Silva-Cardenas, R.I., Ricard, B., Saglio, P. and Hill, R.D. (2003) Hemoglobin and hypoxic acclimation in maize root tips. *Russian Journal of Plant Physiology* 50, 821–826.

Singh, R. and Ghildyal, B.P. (1980) Soil submergence effects on nutrient uptake, growth, and yield of five maize cultivars. *Agronomy Journal* 72, 737–741.

Singh, S., Mackill, D.J. and Ismail, A.M. (2009) Responses of *SUB1* rice introgression lines to submergence in the field: yield and grain quality. *Field Crops Research* 113, 12–23.

Steffens, B., Kovalev, A., Gorb, S.N. and Sauter, M. (2012) Emerging roots alter epidermal cell fate through mechanical and reactive oxygen species signaling. *The Plant Cell* 24, 3296–3306.

Subbaiah, C. and Sach, M.M. (2003) Molecular and cellular adaptations of maize to flooding stress. *Annals of Botany* 90, 119–127.

Suwarno, W., Pixley, K., Palacios-Rojas, N., Kaeppler, S. and Babu, R. (2015) Genome-wide association analysis reveals new targets for carotenoid biofortification in maize. *Theoretical and Applied Genetics* 128, 851–864.

Vergara, B., Jackson, B. and De Datta, S. (1976) Deep water rice and its response to deep water stress. In: *Climate and Rice*. International Rice Research Institute, Los Baños, Philippines, pp. 301–319.

Vivek, B.S., Krishna, G.K., Vengadessan, V., Babu, R., Zaidi, P.H. *et al.* (2017) Use of genomic estimated breeding values results in rapid genetic gains for drought tolerance in maize. *Plant Genome* 10, plantgenome2016.07.0070.

Voesenek, L.A. and Bailey-Serres, J. (2015) Flood adaptive traits and processes: an overview. *New Phytologist* 206, 57–73.

Watanabe, K., Nishiuchi, S., Kulichikhin, K. and Nakazono, M. (2013) Does suberin accumulation in plant roots contribute to waterlogging tolerance? *Frontiers in Plant Science* 4, 178.

Xiao, Y., Liu, H., Wu, L., Warburton, M. and Yan, J. (2017) Genome-wide association studies in maize: praise and stargaze. *Molecular Plant* 10, 359–374.

Xu, K. and Mackill, D.J. (1996) A major locus for submergence tolerance mapped on rice chromosome 9. *Molecular Breeding* 2, 219–224.

Xu, K., Xu, X., Fukao, T., Canlas, P., Maghirang-Rodriguez, R. *et al.* (2006) *Sub1A* is an ethylene-response-factor-like gene that confers submergence tolerance to rice. *Nature* 442, 705–708.

Yu, F., Liang, K., Zhang, Z., Du, D., Zhang, X. *et al.* (2018) Dissecting the genetic architecture of waterlogging stress-related traits uncovers a key waterlogging tolerance gene in maize. *Theoretical and Applied Genetics* 131, 2299–2310.

Yu, F., Liang, K., Fang, T., Zhao, H., Han, X., Cai, M. and Qiu, F. (2019) A group VII ethylene response factor gene, *ZmEREB180*, coordinates waterlogging tolerance in maize seedlings. *Plant Biotechnology Journal* 17, 2286–2298.

Yu, J., Holland, J.B., McMullen, M.D. and Buckler, E.S. (2008) Genetic design and statistical power of nested association mapping in maize. *Genetics* 178(1), 539–551.

Yuan, L.B., Dai, Y.S., Xie, L.J., Yu, L.J., Zhou, Y. *et al.* (2017) Jasmonate regulates plant responses to postsubmergence reoxygenation through transcriptional activation of antioxidant synthesis. *Plant Physiology* 173, 1864–1880.

Zaidi, P.H. and Singh, N.N. (2001) Effect of waterlogging on growth, biochemical compositions and reproduction in maize. *Journal of Plant Biology* 28, 61–69.

Zaidi, P.H. and Singh, N.N. (2002) Identification of morpho-physiological traits for excess soil moisture tolerance in maize. In: Bora, K.K., Singh, K. and Kumar, A. (eds) *Stress and Environmental Physiology*. Scientific Publishers, Jodhpur, India, pp. 172–183.

Zaidi, P.H., Rafique, S. and Singh, N.N. (2003) Response of maize (*Zea mays* L.) genotypes to excess moisture stress: morpho-physiological effects and basis of tolerance. *European Journal of Agronomy* 19, 383–399.
Zaidi, P.H., Rafique, S., Rai, P.K., Singh, N.N. and Srinivasan, G. (2004) Tolerance to excess moisture in maize (*Zea mays* L.): susceptible crop stages and identification of tolerant genotypes. *Field Crops Research* 90, 189–202.
Zaidi, P.H., Mani Selvan, P., Rizvi, R., Chauhan, S., Singh, R.P. *et al.* (2005) Advances in excessive moisture (water-logging) tolerance in tropical maize. In: Zaidi, P.H. and Singh, N.N. (eds) *Stresses on Maize in Tropics*. Directorate of Maize Research (ICAR), New Delhi, pp. 100–134.
Zaidi, P.H., Maniselvan, P., Yadav, P., Singh, A.K., Rizvi, R. *et al.* (2007a) Stress-adaptive changes in tropical maize (*Zea mays* l.) under excessive soil moisture stress. *Maydica* 52, 159–173.
Zaidi, P.H., Maniselvan, P., Sultana, R., Srivastava, A., Singh, A.K. *et al.* (2007b) Association between line *per se* and hybrid performance under excessive soil moisture stress in tropical maize (*Zea mays* L.) *Field Crops Research* 101, 117–126.
Zaidi, P.H., Maniselvan, P., Sultana, R., Yadav, M., Singh, R.P. *et al.* (2007c) Importance of secondary traits in improvement of maize (*Zea mays* L.) for enhancing tolerance to excess soil moisture stress. *Cereal Research Communications* 35, 1427–1435.
Zaidi, P.H., Maniselvan, P., Srivastava, A., Poonam, Y. and Singh, R. (2010) Genetic analysis of waterlogging tolerance in tropical maize. *Maydica* 55, 17–26.
Zaidi, P.H., Rashid, Z., Vinayan, M.T. and Babu, T.A. (2012) Pre-germination anaerobic stress tolerance in tropical maize (*Zea mays* L.). *Australian Journal of Crop Science* 6, 1703–1711.
Zaidi, P.H., Rashid, Z., Vinayan, M.T., Almeida, G.D., Phagna, R.K. and Babu, R. (2015) QTL mapping of agronomic waterlogging tolerance using recombinant inbred lines derived from tropical maize (*Zea mays* L.) germplasm. *PLoS One* 10, e0124350.
Zaidi, P.H., Seetharam, K., Krishna, G., Krishnamurthy, L., Gajanan, S. *et al.* (2016) Genomic regions associated with root traits under drought stress in tropical maize (*Zea mays* L.). *PLoS One* 11, e0164340.
Zhai, J., Zhao, Y., Simon, S.A., Huang, S., Petsch, K. *et al.* (2013) Plant microRNAs display differential 3′ truncation and tailing modifications that are ARGONAUTE1 dependent and conserved across species. *The Plant Cell* 25, 2417–2428.
Zhang, X., Tang, B., Yu, F., Li, L., Wang, M. *et al.* (2012) Identification of major QTL for waterlogging tolerance using genome-wide association and linkage mapping of maize seedlings. *Plant Molecular Biology Reporter* 31, 594–606.
Zhang, X., Perez-Rodriguez, P., Semagn, K., Beyene, Y., Babu, R. *et al.* (2015) Genomic prediction in biparental tropical maize populations in water-stressed and well-watered environments using low-density and GBS SNPs. *Heredity* 114, 291–299.
Zhang, X., Perez-Rodriguez, P., Burgueno, J., Olsen, M., Buckler, E. *et al.* (2017) Rapid cycling genomic selection in a multiparental tropical maize population. *G3: Genes, Genomics, Genetics* 7, 2315–2326.
Zhang, Z., Wei, L., Zou, X., Tao, Y., Liu, Z. and Zheng, Y. (2008) Submergence-responsive microRNAs are potentially involved in the regulation of morphological and metabolic adaptations in maize root cells. *Annals of Botany* 102, 509–519.

# 18 Recent Molecular Breeding Advances for Improving Aluminium Tolerance in Maize and Sorghum

**Claudia Teixeira Guimaraes* and Jurandir Vieira de Magalhaes**
*Embrapa Maize and Sorghum, Sete Lagoas, Brazil*

## 18.1 Introduction

Maize is the most produced cereal in the world, whereas sorghum is a crop of great importance for global agriculture that is used for food, feed and bioenergy production. Acidic soils occupy a large portion of the world's surface, including tropical and subtropical areas that are suitable for agriculture. Al is one of the major limiting factors for grain yield on low-pH soils. Although liming can be used to neutralize $Al^{3+}$ in superficial soil layers, neutralization of $Al^{3+}$ in the subsoil is difficult and expensive. In order to cope with this problem, soil amendments should be combined with genetic tolerance to Al toxicity for achieving sustainable yields on acid soils.

Citrate transporters belonging to the multidrug and toxic compound extrusion (MATE) family of membrane transporters in sorghum and maize, *SbMATE* and *ZmMATE1*, respectively, play a major role in Al tolerance. However, these MATE members show regulatory differences, as well as peculiarities in their genetic effect and mode of action. These aspects, which are discussed in this chapter, have to be considered to design successful breeding programmes in order to achieve maximum Al tolerance and, consequently, to improve grain and biomass production in regions of the world with Al toxicity.

## 18.2 Effects of Aluminium Toxicity in Plants

Al toxicity is a major constraint for agricultural production on acidic soils, which comprise over 30% of arable lands on Earth (von Uexküll and Mutert, 1995). At soil pH below 5, $Al^{3+}$ damages plant root systems, inhibiting root growth due to changes in cell division and elongation (Kochian *et al.*, 2004). Al toxicity also reduces lateral root growth and induces root thickening (Foy *et al.*, 1978). Roots intoxicated by Al are unable to explore deeper soil layers, limiting water and nutrient uptake (Foy *et al.*, 1978). Hence, Al-sensitive cultivars become more prone to drought stress and mineral nutrient deficiency, which results in yield losses.

## 18.3 Genetic Control of Aluminium Tolerance in Sorghum and Maize

Plants possess two major mechanisms to cope with Al toxicity: (i) Al exclusion mechanisms

*Corresponding author email: claudia.guimaraes@embrapa.br

 *Molecular Breeding in Wheat, Maize and Sorghum: Strategies for Improving Abiotic Stress Tolerance and Yield* (eds M.A. Hossain *et al.*)
DOI: 10.1079/9781789245431.0018

that prevent Al from entering the root apex; and (ii) internal mechanisms that minimize the toxic effects of internal Al, either by detoxification or by translocation to organelles. The Al exclusion mechanism mediated by root-apical citrate exudation takes place both in sorghum (Magalhaes *et al.*, 2007) and maize (Maron *et al.*, 2010), in addition to other species. Via this mechanism, citrate in the rhizosphere forms non-toxic complexes with $Al^{3+}$, excluding it from the root cells (Kochian *et al.*, 2004).

Several Al tolerance genes have been isolated and characterized in different plant species (Delhaize *et al.*, 2012; Kochian *et al.*, 2015; Zhang *et al.*, 2019), with a few of them being identified in maize and sorghum (Table 18.1).

### 18.3.1 Sorghum

In sorghum, a single locus, $Alt_{SB}$, located on chromosome 3, explained a large proportion of the variation for Al tolerance (Magalhaes *et al.*, 2004). The underlying gene, *SbMATE*, encodes an Al-activated citrate transporter of the MATE family (Magalhaes *et al.*, 2007). *SbMATE* is induced by Al in the root apex of Al-tolerant sorghum lines and variation in *SbMATE* expression was associated with allelic effects on citrate exudation and thereby Al tolerance (Magalhaes *et al.*, 2007).

Most sorghum accessions are sensitive to Al. In fact, 80% of sorghum accessions representing a wide genetic variability in the species were characterized as highly sensitive to Al toxicity (Caniato *et al.*, 2007, 2011). Al tolerance in sorghum is likely to have a relatively recent origin and Al-tolerant accessions belong primarily to the *guinea* and the *caudatum* races (Caniato *et al.*, 2011).

The extensive variability of Al tolerance in near-isogenic lines (NILs) derived from parents carrying different functional alleles of *SbMATE* was linked with *SbMATE* expression, which confirmed the importance of factors acting in *cis* to control *SbMATE* expression and hence Al tolerance in sorghum (Melo *et al.*, 2013). However, incomplete transfer of Al tolerance to NILs was associated with a reduction in *SbMATE* expression, which indicated the involvement of regulatory loci acting in *trans* (Melo *et al.*, 2013). *WRKY* and *zinc finger-DHHC* transcription factors, designated *SbWRKY1* and *SbZNF1*, were isolated by positional cloning and found to influence genetic background effects on *SbMATE* expression, which leads to incomplete transfer of Al tolerance from parents to NILs (Melo *et al.*, 2019). WRKY proteins are a large family of transcription factors that share the conserved amino acid signature WRKYGQK followed by a zinc finger-like motif at the N-terminal end (Eulgem *et al.*, 2000). SbZNF1 is a DHHC-like *S*-acyl transferase zinc finger that was also implicated in the transactivation of *SbMATE* (Melo *et al.*, 2019). Variation in *SbMATE* expression likely results from changes in tandemly repeated *cis* sequences flanking a MITE (miniature inverted repeat transposable element) in the promoter region of *SbMATE*, which recruits SbWRKY1 and SbZNF1 to modulate *SbMATE* expression in a dose-dependent manner (Melo *et al.*, 2019).

Four *sensitive to proton rhizotoxicity 1* (*STOP1*)-like genes (*SbSTOP1a, SbSTOP1b, SbSTOP1c* and *SbSTOP1d*) with different expression profiles were identified in response to Al in sweet sorghum (Huang *et al.*, 2018). All SbSTOP1 proteins contain four putative $Cys_2His_2$ zinc-finger domains and SbSTOP1d has the highest sequence identity with OsART1 (*Oryza sativa* Al resistance transcription factor 1). SbSTOP1d was found to transcriptionally regulate both *SbSTAR2* and *SbMATE* (Huang *et al.*, 2018). *SbSTAR2* is a sorghum homologue of *OsSTAR2* (*Oryza sativa sensitive to Al rhizotoxicity 2*), which encodes the

**Table 18.1.** Al tolerance genes in sorghum and maize.

| Gene | Plant species | Function | Reference |
|---|---|---|---|
| *SbMATE* | Sorghum | Citrate transport | Magalhaes *et al.* (2007) |
| *SbSTAR2* | Sorghum | Homologue to *OsSTAR2* | Huang *et al.* (2018) |
| *SbSTOP1* | Sorghum | Regulation of *SbSTAR2* expression | Huang *et al.* (2018) |
| *SbZNF1* | Sorghum | Regulation of *SbMATE* expression | Melo *et al.* (2019) |
| *SbWRKY1* | Sorghum | Regulation of *SbMATE* expression | Melo *et al.* (2019) |
| *ZmMATE1* | Maize | Citrate transport | Maron *et al.* (2010) |

transmembrane domain of a bacterial-type ATP-binding cassette (ABC) (Huang *et al.*, 2009). In rice, *OsSTAR2* is transcriptionally regulated by OsART1 (Yamaji *et al.*, 2009; Tsutsui *et al.*, 2011). Together, OsSTAR2 and OsSTAR1 constitute an ABC transporter that has been suggested to mediate the efflux of UDP-glucose into the cell wall, which presumably alters the cell-wall composition, limiting Al accumulation and reducing Al toxicity in rice (Huang *et al.*, 2009). These findings suggest the existence of another Al tolerance mechanism in sorghum based on *SbSTAR2* (Huang *et al.*, 2018).

### 18.3.2 Maize

Genetic variability for Al tolerance in maize has been detected in experiments conducted both in the field and in hydroponics, where Al inhibition of root growth is typically used to quantify Al tolerance (Lutz *et al.*, 1971; Guimaraes *et al.*, 2014). The inheritance of Al tolerance assessed in hydroponics is affected by the genetic background. For example, a single dominant locus primarily controlled Al tolerance in maize populations derived from crossing Al-sensitive and Al-tolerant lines (Rhue *et al.*, 1978; Garcia and Silva, 1979). In contrast, polygenes acting additively explained most of the genetic variation for Al tolerance in bi-parental populations, with a minor contribution of dominance and epistatic effects (Magnavaca *et al.*, 1987; Sawazaki and Furlani, 1987; Prioli *et al.*, 2000).

Three to five quantitative trait loci (QTLs) explaining from 5 to 30% of the phenotypic variance for Al tolerance were detected in different mapping studies in maize (Table 18.2) (Sibov *et al.*, 1999; Ninamango-Cárdenas *et al.*, 2003; Conceição *et al.*, 2009; Guimaraes *et al.*, 2014; Matonyei *et al.*, 2020). These results indicate a predominantly oligogenic control of maize Al tolerance.

The maize Al tolerance QTL, *qALT6*, located at bin 6.00 on chromosome 6, explained 30% of

**Table 18.2.** Al tolerance QTLs in maize.

| QTL | Chromosome position (bin) | Closest marker | $R^2$ | Parents | Population | Trait | Reference |
|---|---|---|---|---|---|---|---|
| *Alm2* | 6.00 | *csu70* | 7.67* | Cat-100-6 × S1587-17 | 56 $F_{2:3}$ families | Net root growth | Sibov *et al.* (1999) |
| *Alm1* | 10.01 | *umc130* | 24.20* | | | | |
| *QTL1* | 2.06 | *umc139* | 10.9 | Cateto Al237 × L53 | 168 $F_{3:4}$ families | Net seminal root length | Ninamango-Cárdenas *et al.* (2003) |
| *QTL2* | 6.00 | *bnlg238* | 5.30 | | | | |
| *QTL3* | 6.05 | *mmc241* | 15.60 | | | | |
| *QTL4* | 8.04 | *csu155* | 7.40 | | | | |
| *QTL5* | 8.07 | *bnlg1828* | 8.60 | | | | |
| | 4.03 | *umc1550* | 10.00 | Diallel between L20, L08 and L06 (Al-sensitive) × L10 and L09 (Al-tolerant) | 96 $F_2$ from each of 10 hybrids | Root regrowth | Conceição *et al.* (2009) |
| | 6.05 | *bnlg1154* | 9.00 | | | | |
| | 8.05 | *umc1202* | 7.00 | | | | |
| | 10.01 | *umc1318* | 15.00 | | | | |
| *qALT2* | 2.08 | *S2_212940514* | 15.47 | Cateto Al237 × L53 | 118 RILs | Relative net root growth | Guimaraes *et al.* (2014) |
| *qALT3* | 3.06 | *S3_187460236* | 27.51 | | | | |
| *qALT5* | 5.03 | *S5_30301926* | 17.56 | | | | |
| *qALT6* | 6.00 | *ZmMATE1* | 30.54 | | | | |
| *qALT8* | 8.03 | *S8_22681622* | 19.86 | | | | |
| *qALT1.09* | 1.09 | *PZA00356_8* | 5.50 | 203B-14 × SCH3 | 180 $F_{2:3}$ families | Relative net root growth | Matonyei *et al.* (2020) |
| *qALT5.03* | 5.03 | *PZA001870_20* | 5.40 | | | | |
| *qALT8.05* | 8.05 | *PHM10525_9* | 0.50 | | | | |
| *qALT9.01* | 9.01 | *PHM229_15* | 5.50 | | | | |
| *qALT10.02* | 10.02 | *PZB01301_5* | 4.80 | | | | |

*$P < 0.05$ ($F$-test).

Al tolerance in a recombinant inbred line (RIL) population derived from Cateto Al237 (Al-tolerant) × L53 (Al-sensitive) (Guimaraes *et al.*, 2014). QTLs on chromosome 6, in similar position to *qALT6*, were detected by Ninamango-Cárdenas *et al.* (2003) and Sibov *et al.* (1999). Sibov *et al.* (1999) designated the Al tolerance locus on chromosome 6 *Alm1*, which acted in semi-dominant fashion in a segregating population derived from Cat-100-6 × S1587-17, an Al-sensitive somaclonal variant of Cat-100-6 (Moon *et al.*, 1997). The Al-tolerant parents that originated the mapping populations where *qALT6* and *Alm1* were mapped were Cateto Al237 (Ninamango-Cárdenas *et al.*, 2003; Guimaraes *et al.*, 2014) and Cat-100-6 (Moon *et al.*, 1997; Sibov *et al.*, 1999), respectively. Interestingly, both lines belong to *cateto* maize races, which were classified as 'ancient commercial races' according to Paterniani and Goodman (1977). *Cateto* maize races were cultivated by native South American Indians in pre-Colombian times and were subsequently cultivated for commercial purposes (Paterniani and Goodman, 1977). In the early 1980s, *cateto* accessions from Brazil were found to be highly Al-tolerant based on nutrient solution screenings (Magnavaca *et al.*, 1987; Sawazaki and Furlani, 1987) and were then widely used both for physiological and genetic studies of Al tolerance in maize.

The major Al tolerance QTL on chromosome 6, *qALT6*, co-localized with *ZmMATE1*, which encodes a citrate transporter of the MATE family (Maron *et al.*, 2010) homologous to sorghum *SbMATE* (Magalhaes *et al.*, 2007). *ZmMATE1* is induced by Al and is expressed at higher levels in the root apex of Cateto Al237 compared with L53 (Al-sensitive) (Maron *et al.*, 2010). Higher *ZmMATE1* expression in Cateto Al237 was associated with copy number variation (CNV) for *ZmMATE1*. Cateto Al237 was found to harbor three copies in tandem of *ZmMATE1*, whereas a single copy is found in the Al-sensitive line, L53 (Maron *et al.*, 2013). Two other lines were found to possess a similar three-copy CNV, Cat-100-6 and Il677a, a sweet corn from Bolivia, among 278 genetically diverse maize lines evaluated (Maron *et al.*, 2013). Thus, the superior allele of *ZmMATE1* is rare in maize and seems to be primarily originated in South American regions with acidic soils (Maron *et al.*, 2013).

Maize landraces and inbred lines from Kenya have also been shown to be highly Al-tolerant (Gudu *et al.*, 2001; Ouma *et al.*, 2013; Matonyei *et al.*, 2020; Table 18.3). However, these Al-tolerant maize genotypes have lower *ZmMATE1* expression compared with Cateto Al237 (Matonyei *et al.*, 2014, 2017, 2020). Five QTLs and four epistatic interactions explained ~51% of the phenotypic variance for Al tolerance in an $F_{2:3}$ population derived from a highly tolerant Kenyan line, 203B-14, and no QTL overlapping with *ZmMATE1* was found using this population (Matonyei *et al.*, 2020). Thus, evidence from QTL mapping and expression analyses suggested that *ZmMATE1* does not play a significant role in Al tolerance in 203B-14, whereas other candidate genes such as *ZmNrat1*, *ZmMATE3*, *ZmWRKY* and *ZmART1* may possibly contribute to Al tolerance in 203B-14 (Matonyei *et al.*, 2020).

## 18.4 Molecular Tools for Marker-Assisted Breeding

### 18.4.1 Sorghum

Allele-specific markers for the *SbMATE* gene have been developed based on polymorphisms within the $Alt_{SB}$ locus, where *SbMATE* is located (Caniato *et al.*, 2014). These polymorphisms were associated with Al tolerance and were converted into the KASP (competitive allele-specific PCR) and ARMS (amplification-refractory mutation system) marker systems (Hufnagel *et al.*, 2018). These markers, along with DNA-pooling strategies, offer a wide range of possibilities for allele mining approaches aimed at identifying superior alleles of *SbMATE* in different sorghum germplasm.

SbWRKY1 and SbZNF1 modulate the expression of *SbMATE* and consequently sorghum Al tolerance. The allele-specific markers designed to differentiate the alleles from contrasting parental lines (Melo *et al.*, 2019) can also be used to select superior alleles of these transcription factors to maximize *SbMATE* expression in marker-assisted breeding.

The effect of *SbMATE* was validated in a field site with high Al saturation and a single 'Al-tolerant' allele of *SbMATE* was found to increase grain yield by approximately 600 kg/ha,

**Table 18.3.** Maize genotypes possessing high Al tolerance.

| Identification | Genotype | Origin | RTi[a] | RNRG[b] | Reference |
|---|---|---|---|---|---|
| 5A | Landrace | Nandi, Kenya | 0.9 | | Gudu *et al.* (2001) |
| 1X1 | Landrace | Vihiga, Kenya | 1.1 | | Gudu *et al.* (2001) |
| 4D | Landrace | Butere-Mumias, Kenya | 1.0 | | Gudu *et al.* (2001) |
| 203B | Landrace | Muranga, Kenya | 1.0 | 107 | Gudu *et al.* (2001) |
| 6D | Landrace | Nandi, Kenya | 0.9 | | Gudu *et al.* (2001) |
| 2B4 | Landrace | Vihiga, Kenya | 0.9 | | Gudu *et al.* (2001) |
| 401 | Landrace | Kilifi, Kenya | 0.9 | | Gudu *et al.* (2001) |
| 2A1 | Landrace | Vihiga, Kenya | 0.9 | | Gudu *et al.* (2001) |
| 4C3 | Landrace | Butere-Mumia, Kenya | 0.9 | | Gudu *et al.* (2001) |
| 203B-14 | Line | Kenya | | 105 | Ouma *et al.* (2013); Matonyei *et al.* (2020) |
| CON5 | Landrace | Kenya | | 104 | Ouma *et al.* (2013) |
| CatAl237/67 × L3-5 | Hybrid | Brazil | | 103 | Ouma *et al.* (2013) |
| S396-15-1 | Line | Kenya | | 101 | Ouma *et al.* (2013) |
| 203B-39 | Line | Kenya | | 100 | Ouma *et al.* (2013) |
| 203B-2 | Line | Kenya | | 98 | Ouma *et al.* (2013) |
| Cateto Al237 | Line | Brazil | | 97 | Guimaraes *et al.* (2014) |

[a]Root tolerance index.
[b]Relative net root growth (%).

both in inbred lines and in hybrids (Carvalho *et al.*, 2016). The activity of *SbMATE* does not cause yield penalty in the absence of Al toxicity. Additionally, *SbMATE* explained 16% of the genetic variation for grain yield in sorghum genotypes grown in a soil with low P availability in western Africa (Leiser *et al.*, 2014). These results suggest a pleiotropic effect of *SbMATE* on tolerance to different abiotic stresses, such as Al toxicity and P deficiency, which in general coexist on acid, tropical soils.

### 18.4.2 Maize

An allele-specific marker based on the KASP system was also developed for *ZmMATE1*-based marker-assisted selection (Barros *et al.*, 2016). The underlying polymorphism is located ~3 kb upstream of the *ZmMATE1* start codon and allelic variation at the marker locus was associated with variation in *ZmMATE* expression (Barros *et al.*, 2016). Marker-assisted introgression of *ZmMATE1* from Cateto Al237 into the Al-sensitive background of L53 led to a twofold increase in Al tolerance (Guimaraes *et al.*, 2014) and improved Al tolerance in *flint* and *dent* tropical elite lines.

## 18.5 Molecular Breeding Strategies to Improve Aluminium Tolerance in Sorghum and Maize

Knowledge of the genetic complexity of Al tolerance is essential when predicting the effectiveness of molecular breeding strategies to produce cultivars adapted to acid soils with Al toxicity. The molecular tools currently available for Al tolerance in sorghum and maize can be used for marker-assisted introgression approaches of major Al tolerance genes in both species. As the superior alleles of both *ZmMATE1* and *SbMATE* are rare and likely originated in specific regions of the world, it is expected that the target alleles are absent in elite germplasm that has not been previously selected for Al tolerance. In the case where superior alleles of *ZmMATE1* and *SbMATE* are found within the breeding germplasm, a locally adapted source would be the preferred choice for marker-assisted introgression approaches. Otherwise, the sorghum accessions SC283 and SC566 are important sources of Al tolerance controlled by *SbMATE*. In maize, Cateto Al237 is the most studied source of the *ZmMATE1*-based Al tolerance, whereas two lines identified by Maron *et al.* (2013) and NILs introgressed with *ZmMATE1* by Guimaraes *et al.* (2014) can also be used.

## 18.6 Conclusions and Remarks

As shown in this chapter, target genes with major effects and molecular tools are available for marker-assisted breeding for improving Al tolerance both in sorghum and maize. However, wide adaptation to acid soils should be sought by pyramiding genes controlling different traits such as drought tolerance, P acquisition, resistance to diseases and other stresses commonly found in each agroecological environment.

## References


Barros, B.A., Mitre, L.M., Pinto, M.O., Magalhães, J.V., Guimarães, L.J.M. *et al.* (2016) Marcador alelo-específico associado com níveis de expressão do gene *ZmMATE1* em milho. In: Paes, M.C.D., Menezes, C.B. and Karam, D. (eds) *Proceedings of the Conference 31° Congresso Nacional de Milho e Sorgo, Bento Gonçalves, Brazil, 25–29 September 2016*. Associação Brasileira de Milho e Sorgo, Sete Lagoas, Brazil, pp. 127–130.

Caniato, F.F., Guimaraes, C.T., Schaffert, R.E., Alves, V.M.C., Kochian, L.V. *et al.* (2007) Genetic diversity for aluminum tolerance in sorghum. *Theoretical and Applied Genetics* 114, 863–876.

Caniato, F.F., Guimaraes, C.T., Hamblin, M., Billot, C., Rami, J.-F. *et al.* (2011) The relationship between population structure and aluminum tolerance in cultivated sorghum. *PLoS One* 6, e20830.

Caniato, F.F., Hamblin, M.T., Guimaraes, C.T., Zhang, Z., Schaffert, R.E. *et al.* (2014) Association mapping provides insights into the origin and the fine structure of the sorghum aluminum tolerance locus, $Alt_{SB}$. *PLoS One* 9, e87438.

Carvalho, G. Jr, Schaffert, R.E., Malosetti, M., Viana, J.H.M., Menezes, C.B. *et al.* (2016) Back to acid soil fields: the citrate transporter SbMATE is a major asset for sustainable grain yield for sorghum cultivated on acid soils. *G3: Genes, Genomes, Genetics* 6, 476–484.

Conceição, L.D.H.C.S., Tessele, C. and Barbosa Neto, J.F. (2009) Diallel analysis and mapping of aluminum tolerance in corn inbred lines. *Maydica* 54, 55–61.

Delhaize, E., Ma, J.F. and Ryan, P.R. (2012) Transcriptional regulation of aluminium tolerance genes. *Trends in Plant Science* 17, 341–348.

Eulgem, T., Rushton, P.J., Robatzek, S. and Somssich, I.E. (2000) The WRKY superfamily of plant transcription factors. *Trends in Plant Science Reviews* 5, 1360–1385.

Foy, C.D., Chaney, R.L. and White, M.C. (1978) The physiology of metal toxicity in plants. *Annual Review of Plant Physiology* 29, 511–566.

Garcia, O. Jr and Silva, W.J. (1979) Análise qenética da tolerância ao alumínio em milho. *Ciência e Cultura* 31, 585.

Gudu, S., Maina, S.M., Onkware, A.O., Ombakho, G. and Ligeyo, D.O. (2001) Screening of Kenyan maize germplasm for tolerance to low pH and aluminium for use in acid soils of Kenya. In: Friesen, D.K. and Palmer, A.F.E. (eds) *Proceedings of the Seventh Eastern Southern Africa Region Maize Conference, Nairobi, 5–11 February 2001*. International Maize and Wheat Improvement Center, Nairobi, pp. 216–221.

Guimaraes, C.T., Simões, C.C., Pastina, M.M., Maron, L.G., Magalhães, J.V. *et al.* (2014) Genetic dissection of Al tolerance QTLs in the maize genome by high density SNP scan. *BMC Genomics* 15, 153.

Huang, C.F., Yamaji, N., Mitani, N., Yano, M., Nagamura, Y. *et al.* (2009) A bacterial-type ABC transporter is involved in aluminum tolerance in rice. *The Plant Cell* 21, 655–667.

Huang, S., Gao, J., You, J., Liang, Y., Guan, K. *et al.* (2018) Identification of STOP1-like proteins associated with aluminum tolerance in sweet sorghum (*Sorghum bicolor* L.). *Frontiers in Plant Science* 9, 258.

Hufnagel, B., Guimaraes, C.T., Craft, E.J., Shaf, J.E., Schafert, R.E. *et al.* (2018) Exploiting sorghum genetic diversity for enhanced aluminum tolerance: allele mining based on the $Alt_{SB}$ locus. *Scientific Reports* 8, 10094.

Kochian, L.V., Hoekenga, O.A. and Piñeros, M.A. (2004) How do crop plants tolerate acid soils? Mechanisms of aluminum tolerance and phosphorous efficiency. *Annual Review of Plant Physiology and Plant Molecular Biology* 55, 459–493.

Kochian, L.V., Pineros, M.A., Liu, J. and Magalhaes, J.V. (2015) Plant adaptation to acid soils: the molecular basis for crop aluminum resistance. *Annual Review of Plant Biology* 66, 571–598.

Leiser, W.L., Rattunde, H.F.W., Weltzien, E., Cisse, N., Abdou, M. *et al.* (2014) Two in one sweep: aluminum tolerance and grain yield in P-limited soils are associated to the same genomic region in West African sorghum. *BMC Plant Biology* 14, 206.

Lutz, J.A. Jr, Hawkins, G.W. and Genter, C.F. (1971) Differential response of corn inbreds and single crosses to certain properties of an acid soil. *Agronomy Journal* 63, 803–805.

Magalhaes, J.V., Garvin, D.F., Wang, Y., Sorrells, M.E., Klein, P.E. *et al.* (2004) Comparative mapping of a major aluminum tolerance gene in sorghum and other species in the Poaceae. *Genetics* 167, 1905–1914.

Magalhaes, J.V., Liu, J., Guimarães, C.T., Lana, U.G.P., Alves, V.M.C. *et al.* (2007) A gene in the multidrug and toxic compound extrusion (MATE) family confers aluminum tolerance in sorghum. *Nature Genetics* 39, 1156–1161.

Magnavaca, R., Gardner, C.O. and Clark, R.B. (1987) Inheritance of aluminum tolerance in maize. In: Gabelman, H.W. and Loughman, B.C. (eds) *Genetic Aspects of Plant Mineral Nutrition*. Martinus Nijhoff Publishers, Dordrecht, the Netherlands, pp. 201–212.

Maron, L.G., Piñeros, M.A., Guimarães, C.T., Magalhaes, J.V., Pleiman, J.K. *et al.* (2010) Two functionally distinct members of the MATE (multi-drug and toxic compound extrusion) family of transporters potentially underlie two major aluminum tolerance QTLs in maize. *The Plant Journal* 61, 728–740.

Maron, L.G., Guimarães, C.T., Kirst, M., Albert, P.S., Birchler, J.A. *et al.* (2013) Aluminum tolerance in maize is associated with higher MATE1 gene copy number. *Proceedings of the National Academy of Sciences USA* 110, 5241–5246.

Matonyei, T.K., Cheprot, R., Liu, J., Piñeros, M.A., Shaff, J.E. *et al.* (2014) Physiological and molecular analysis of aluminum tolerance in selected Kenyan maize lines. *Plant and Soil* 377, 357–367.

Matonyei, T.K., Sirmah, P.K., Sitienei, A., Ouma, E.O., Ligeyo, D.O. *et al.* (2017) The expression of *ZmMATE1* gene at seminal root tip does not explain aluminum toxicity tolerance in a Kenyan maize breeding line. *International Journal of Scientific Research and Innovative Technology* 4, 45–59.

Matonyei, T.K., Barros, B.A., Guimarães, R.G.N., Ouma, E.O., Cheprot, R.K. *et al.* (2020) Aluminum tolerance mechanisms in Kenyan maize germplasm are independent from the citrate transporter ZmMATE1. *Scientific Reports* 10, 7320.

Melo, J.O., Lana, U.G.P., Pineros, M.A., Alves, V.M.C., Guimaraes, C.T. *et al.* (2013) Incomplete transfer of accessory loci influencing *SbMATE* expression underlies genetic background effects for aluminum tolerance in sorghum. *The Plant Journal* 73, 276–288.

Melo, J.O., Martins, L.G.C., Barros, B.A., Pimenta, M.R., Lana, U.G.P. *et al.* (2019) Repeat variants for the SbMATE transporter protect sorghum roots from aluminum toxicity by transcriptional interplay in *cis* and *trans*. *Proceedings of the National Academy of Sciences USA* 116, 313–318.

Moon, D.H., Ottoboni, L.M.M., Souza, A.P., Sibov, S.T., Gaspar, M. *et al.* (1997) Somaclonal-variation-induced aluminum-sensitive mutant from an aluminum-inbred maize tolerant line. *Plant Cell Reports* 16, 686–691.

Ninamango-Cárdenas, F.E., Guimarães, C.T., Martins, P.R., Parentoni, S.N., Carneiro, N.P. *et al.* (2003) Mapping QTLs for aluminum tolerance in maize. *Euphytica* 130, 223–232.

Ouma, E., Ligeyo D., Matonyei, T., Agalo, J., Were, B. *et al.* (2013) Enhancing maize grain yield in acid soils of western Kenya using aluminium tolerant germplasm. *Journal of Agricultural Science and Technology* 3, 33–46.

Paterniani, E. and Goodman, M.M. (1977) *Races of Maize in Brazil and Adjacent Areas*. International Maize and Wheat Improvement Center, Mexico City.

Prioli, A.J., Scapim, C.A., Prati, R.M., Prioli, S.M.A.P., Bravo, J.P. *et al.* (2000) Genetic analysis of means and variances of aluminum tolerance in maize. *Acta Scientiarum* 22, 869–875.

Rhue, R.D., Grogan, C.O., Stockmeyer, E.W. and Everett, H.L. (1978) Genetic control of aluminum tolerance in corn. *Crop Science* 18, 1063–1067.

Sawazaki, E. and Furlani, P.R. (1987) Tolerância ao alumínio em milho Cateto. *Bragantia* 46, 269–278.

Sibov, S.T., Gaspar, M.J., Ottoboni, L.M.M., Arruda, P. and Souza, A.P. (1999) Two genes controlling aluminum tolerance in maize: genetic and molecular mapping analyses. *Genome* 42, 475–482.

Tsutsui, T., Yamaji, N. and Ma, J.F. (2011) Identification of a *cis*-acting element of ART1, a $C_2H_2$-type zinc-finger transcription factor for aluminum tolerance in rice. *Plant Physiology* 156, 925–931.

von Uexküll, H.R. and Mutert, E. (1995) Global extent, development and economic impact of acid soils. *Plant and Soil* 171, 1–15.

Yamaji, N., Huang, C.F., Nagao, S., Yano, M., Sato, Y. *et al.* (2009) A zinc finger transcription factor ART1 regulates multiple genes implicated in aluminum tolerance in rice. *The Plant Cell* 21, 3339–3349.

Zhang, X., Yan, L., Jingjing, H. and Jixing, X. (2019) Molecular mechanisms for coping with Al toxicity in plants. *International Journal of Molecular Science* 20, 1551.

# 19 Physiological and Molecular Interventions for Improving Nitrogen-Use Efficiency in Maize

**Ishwar Singh[1], Krishan Kumar[1], Prabha Singh[2], Pranjal Yadava[1] and Sujay Rakshit[3]***

*[1]ICAR–Indian Institute of Maize Research, New Delhi, India; [2]ICAR–Indian Grassland and Fodder Research Institute, Jhansi, India; [3]ICAR–Indian Institute of Maize Research, Ludhiana, India*

## 19.1 Introduction

N is an essential element for plant vegetative as well as reproductive growth. It is an important constituent in most biomolecules such as DNA, RNA, proteins and enzymes. Owing to the critical role of N in crop development and hence yield, nitrogenous fertilizers are usually utilized by farmers in the field. Although, up to some extent, these fertilizers are very helpful for increasing crop yield, excessive use of fertilizers is harmful for the environment and the ecosystem, on one hand, and increases the cost of cultivation, on the other. It was reported that the nitrogen-use efficiency (NUE) of cereal crops, including *Zea mays* (maize), does not exceed 40–50% and the remaining 50–60% of N either leaches into the soil or evaporates into the environment (Raun and Johnson, 1999; Tilman *et al.*, 2002; Galloway *et al.*, 2014). There are many adverse effects of excess nitrogenous fertilizers, such as:

**1.** Sometimes excessively applied nitrogenous fertilizer promotes the growth of non-native plants which compete with and suppress the growth of native plants, thereby creating an ecological imbalance in that particular area.
**2.** Too much N also causes the depletion of other important nutrients, such as Ca, P, Mg, etc. When these nutrients deplete from the soil, other toxic nutrients like Al could accumulate.
**3.** Excess $NO_3^-$ from soil leaches into underground waters and drinking of $NO_3^-$-contaminated water causes various diseases like blue baby syndrome or methaemoglobinaemia (Majumdar, 2003). Other than this, when this N enters the food chain through vegetables and fruits, it may cause thyroid disorders, cancers and neural tube defects (Ward *et al.*, 2018).
**4.** The excess N from fertilizer application may lead to the formation of $N_2O$ in the atmosphere, which is a potent greenhouse gas resulting in depletion of the atmospheric ozone layer, thereby causing global warming (Wuebbles, 2009; Galloway *et al.*, 2014).
**5.** Nitrogenous fertilizers are very popular among farmers but manufacturing of nitrogenous fertilizer is an energy-intensive process because natural gas is used as both reactant and heat source for the reaction. This also adds to the cost of fertilizer, which ultimately may create a

*Corresponding author: s.rakshit@icar.gov.in


DOI: 10.1079/9781789245431.0019

burden on the farmer's income as well as the environment (Smith, 2002).

Maize is one of the major cereal crops with wide adaptability under diverse agroclimatic conditions. Among the cereals, maize has the highest genetic yield potential. Being a $C_4$ crop, water- and C-use efficiency of this crop are higher than other cereals like wheat and rice. It is also a day-neutral crop and can be grown in any season. This crop has wide utilities as food, fodder and feed as well as industrial uses. Owing to its myriad uses and the ever-increasing global population, demand for maize is increasing continuously. Like high-yielding crop varieties, single-cross maize hybrids required high levels of N for optimum production. Thus, to meet the increased global demand and obtain high yield, as well as in view of the prevalence of N deficiency in most soils globally, farmers excessively use N-based chemical fertilizers. As mentioned above, plants can utilize only half of the applied N from fertilizer. Therefore, development of novel strategies and maize genotypes which can utilize externally applied N more efficiently would be of prime importance to reduce the cost of cultivation, environmental pollution and achieve sustainable agriculture.

**Table 19.1.** General information about nitrogen.

| Name | Nitrogen |
|---|---|
| Discovered by | Daniel Rutherford in 1772 |
| Availability in air | 78–79% by weight |
| Classified as | Non-metallic/macro element |
| Colour | Colourless |
| Present as | $N_2$, $NO_2$, $NH_3$, $NH_4^+$, $NO_3^-$, C–$NH_2$ |
| Plant uses | $NH_4^+$, $NO^-$, C–$NH_2$ |
| Losses by | Leaching, volatilization, crop removal, erosion, etc. |
| Deficiency symptom | Yellowing of older leaves (V-shaped yellowing) |
| Excess symptom | Extra green colour to leaves and succulent growth |
| Contribution in agriculture | Promotes yield and quality |

## 19.2 Importance of Nitrogen in Plant Growth and Development

For efficient plant growth and production, balanced nutrition is essential. Proper nutrient management plays a key role in increased food grain production and quality. In this regard, N occupies a central role in plant metabolism as it is an essential component of chlorophyll, proteins, enzymes, nucleic acids, etc. (Table 19.1). N is the fourth most important component of any living organism after C, H and O. Up to 4% of plant dry matter is N. Chlorophyll is the molecule which captures light energy and converts it into carbohydrate by the process of photosynthesis. Being part of the chlorophyll, N directly affects photosynthetic capacity of the plant and ultimately yield. Low N affects chlorophyll content resulting in reduced flow of available photosynthates to the growing regions, affecting overall plant growth. Protein is the building block of any living entity and DNA contains all its genetic information, which is a key molecule for plant survival. N affects crop yield by increasing leaf area, canopy structure, net assimilation rate and photosynthetic capacity of the plant. For achieving higher yields, N fertilizer application is unavoidable and indispensable as it also affects grain weight and quality. Maximum leaf area along with total leaf biomass are major determinants of crop yield. A constituent of chlorophyll, N imparts dark green colour to leaves and promotes vegetative growth of other plant parts. In maize, zein is the main storage protein in the endosperm and globulin in the embryo. These play an important role in seed germination and initial plant vigour (Moose and Below, 2009). During early stage development, N promotes growth of roots which in turn absorb other mineral nutrients like P, K, etc. Root branching pattern influences N uptake from soil as the root is the point of entry of N into the plant. Absorbed N is assimilated into amino acids in the root or leaf tissues (Moose and Below, 2009). Maximum N uptake and assimilation take place during early growth stages. However, N uptake continues briefly after flowering. Kernel number and size among yield-attributing traits are influenced maximally by N application in maize (John and Schmitt, 2008). Starch synthesis in the endosperm of maize seeds is greatly influenced by N.

Being a macronutrient, N plays myriad roles in cellular physiology. Low N availability leads to poor growth in plants (Hull and Liu, 2005), chlorosis, necrosis of leaves and disorders in many physiological and biochemical characteristics

(Bray *et al.*, 2000). Deficiency of N in maize causes reduced growth, chlorosis in leaves (changing of the green colour of leaves into yellow), especially V-shaped chlorosis of lower leaves (Chun, 2005). Deficiency appears first in the lower leaves because N is mobile in the plant and under deficient condition, it moves to the younger leaves (Soltabayeva *et al.*, 2018). The anthesis–silking interval (ASI) is most sensitive to any stress, including N stress. ASI beyond 5 days drastically affects grain yield (Elings *et al.*, 1996). Recently, Yadava *et al.* (2020) studied the effect of N starvation in maize and observed reduced growth along with reductions in leaf chlorophyll content, total soluble proteins and total biomass accumulation.

Excess nitrogenous fertilizer is also harmful for plants as it imparts dark green colour in leaves, succulent vegetative growth over reproductive growth and less fruit quantity with poorer quality. Plants take up N from soil in the form of $NO_3^-$ in aerobic soils and $NH_4^+$ in flooded wetland or acidic soils. $NO_3^-$ is the most common absorbable form rather than $NH_4^+$, which is less available in soil. Soil and climatic condition play an important role in N uptake and utilization.

## 19.3 What is Nitrogen-Use Efficiency and How to Manage It?

An essential nutrient for plant growth and development, N is unavailable in its most prevalent form as atmospheric $N_2$. Plants instead depend upon combined, or fixed, forms of N such as $NH_4^+$ and $NO_3^-$, also competing with microbes for the limited N that is available in soil. Thus, N is a major limiting nutrient for most agricultural ecosystems. Therefore, for restoring soil fertility and increasing crop yield, inorganic nitrogenous fertilizer application has become the most common practice. The nitrogenous fertilizers contain relatively soluble forms of N, which makes the absorption and assimilation easy from soil (Reetz *et al.*, 2015). However, plants cannot utilize the complete exogenous N applied in the form of chemical fertilizers.

The term ‘nitrogen-use efficiency’ refers to grain yield per unit of supplied N (by soil and fertilizers) (Moll *et al.*, 1987). It has been estimated that since 1961, total N applied has increased by a factor of 4.4 and total protein production has increased only by a factor of 3.1; thus in actual terms NUE has declined from 66 to 46% (Lassaletta *et al.*, 2016). NUE is made up of two components: (i) nitrogen-uptake efficiency (NUpE), which is the efficiency of absorption/uptake of per unit of supplied N; and (ii) nitrogen-utilization efficiency (NUtE), which is the efficiency of assimilation and remobilization of plant N to ultimately produce grain. In general, NUE of applied N includes recovery efficiency (RE), physiological efficiency (PE) and agronomic efficiency (AE) (Huggins and Pan, 2003). The RE of applied N reflects the efficiency of above-ground N uptake per unit of N applied. PE is the efficiency by which N in above-ground plant parts is converted to grain, and AE is grain yield per unit of N applied (Hirel *et al.*, 2007). Large genetic variability has been observed in maize for both uptake efficiency and utilization efficiency of N (Han *et al.*, 2015). High-yielding varieties of maize are not necessarily the best ones, when N supply is reduced, because most of the breeding programmes are carried out under non-limiting N conditions. In intensive agriculture systems, where enormous amounts of nitrogenous fertilizer are applied, most of it is lost by leaching and runoff into the surface water or groundwater and poses a serious threat to human health (Majumdar, 2003; Ward *et al.*, 2018).

Initially, it was reported that NUpE is more important for better yield, because a significant positive correlation was established between use efficiency and accumulation of N (Sinclari and Vadez, 2002). Eventually other findings reported that under N-limited condition, N utilization affects grain yield in a significant way. Moreover, NUtE encourages more N uptake and accumulation; for example, a plant which has faster growth rate at early developmental stage will have a more structured root system which in turn will promote more N uptake. Therefore, both N uptake and N utilization are dependent on one another and their contribution for plant growth and yield cannot be described in isolation. NUE can be increased by improving either NUpE or NUtE or both; however, due to variable soil and climatic conditions, it is difficult to quantify the actual amount of N available to or absorbed by plants.

Variation in the NUpE affects all phases of maize growth including the development, activity and senescence of leaves and the initiation, growth and composition of ovules (Uhart and Andrade, 1995). In maize, primary N assimilation and mobilization occur simultaneously in source leaves and contribute almost equally to the supply of N to the developing ear (Hirel *et al.*, 2007). It is therefore likely that the supply of N assimilates, their efficient translocation and their conversion during kernel formation contribute to the overall plant NUE (Canas *et al.*, 2009).

Low soil N is a major maize production constraint worldwide, particularly in the developing countries. The global demand for N fertilizers was 105 million tonnes in 2017 and is predicted to reach 111.5 million tonnes by 2022 (FAO, 2019). Maize is an N-responsive crop with higher demand for N fertilizers. NUE of maize varies from 25 to 50% (Tilman *et al.*, 2002). This suggests that wide cultivation of maize is associated with significant contamination of the environment due to N loss. Hence, high NUE should be a major breeding emphasis. Significant interaction of genotype and N fertilizer application has been reported in maize by a number of studies (Tollenaar *et al.*, 1997; Presterl *et al.*, 2002; Gallais and Coque, 2005; Haegele *et al.*, 2013; Mastrodomenico *et al.*, 2018). This interaction may be due to differential uptake of N applied or differential utilization of N. The former is important under conditions of high N input, while the latter is relevant under conditions of low N input (Gallais and Hirel, 2004). The complex nature of soil, microorganism and plant interaction makes selection of genotypes for high NUE difficult (Gallais and Coque, 2005). Screening a large number of maize genotypes, first under hydroponic conditions followed by evaluation under field conditions could be a better strategy for identifying N-use-efficient maize genotypes.

## 19.4 Traits Influencing Nitrogen-Uptake Efficiency

NUpE is regulated by plant demand and affected by various factors (Ismande and Touraine, 1994) as described below.

### 19.4.1 Root system architecture

Roots, being the underground part of a plant, help in nutrient uptake and absorption. N is a mobile element in soil as well as in plants. In soil it moves by mass flow. Under N-sufficient conditions with adequate water supply, plant nutrient uptake remains unaffected by root morphology and size. However, under water scarcity, length and spatial distribution of roots play an important role for nutrient accumulation (Sinclair and Vadez, 2002). It has been reported that N-efficient genotypes allocate relatively more N to roots at early growth stages of the plant (Niu *et al.*, 2007). Root senescence is also an important feature for nutrient uptake. It has been reported that major N uptake for grain happens post anthesis. Therefore, a plant staying green at later maturation stage to promote root growth is a desirable feature to enhance NUpE (Chun *et al.*, 2005; Yadava *et al.*, 2019).

### 19.4.2 Root nitrogen transporter system

In primary roots, the apical section is involved in $NO_3^-$ sensing and signalling, and the basal section in $NO_3^-$ acquisition. However, root tips exhibit higher capacity to absorb N and this capacity is dependent on plant age (Sorgonà *et al.*, 2011). N is absorbed by specific transporters located on root cell membranes, which may have high or low affinity. The high- and low- affinity nitrate transporters (NRTs) play a major role in $NO_3^-$ transport when soil N concentration is low and high, respectively. In higher plants, nitrate transporters are classified mainly into two families, namely: (i) the nitrate transporter 1/peptide transporter (NRT1/PTR) family (NPF); and (ii) the NRT2 family, also called the major facilitator superfamily (Léran *et al.*, 2014). The availability of the maize genome sequence has improved our understanding of how N uptake is carried out under both N-sufficient and N-deficient conditions. Garnett *et al.* (2013) reported that $NO_3^-$ uptake capacity in maize plants after 40 days of leaf emergence is correlated with *ZmNRT2.1* and *ZmNRT2.2* genes encoding two high-affinity transporters. Yu *et al.* (2014) reported that expression of *ZmNRT2.1* and *ZmNRT2.2* was enhanced and inhibited in the roots of maize plants under high and low $NO_3^-$, respectively. Recently, it has been found that two maize NPF6 transporters, ZmNPF6.4 and ZmNPF6.6, are permeable to both $NO_3^-$ and $Cl^-$ and the latter transporter has selectivity for substrate when roots are exposed to high $NO_3^-$ (Wen *et al.*, 2017). Different *NRTs* exhibit differential expression patterns during the plant life cycle and under different levels of $NO_3^-$ concentration (Garnett *et al.*, 2013). For instance, increased expression level of many high-affinity *NRT2s* has been reported in *Arabidopsis*, maize and *Triticum aestivum* (wheat) under N deficiency, which in turn increased $NO_3^-$ uptake (Okamoto *et al.*, 2003; Garnett *et al.*, 2013, 2015; Buchner and Hawkesford, 2014).

In soil, the concentration of $NH^+_4$ is much lower than that of $NO_3^-$ and plants also prefer $NO_3^-$ over $NH^+_4$. In higher plants, $NH^+_4$ is taken up by transporters of the ammonium transporter/methylammonium permease/Rhesus (AMT/MEP/Rh) family, located on root cell membranes, that show different cellular distribution and substrate affinities (Yuan *et al.*, 2007). Two ammonium transporters, ZmAMT1.1 and ZmAMT1.3, located in the rhizodermis have been reported as a high-affinity transport system.

### 19.4.3 Interaction with microorganisms

Arbuscular mycorrhizal fungi (AMF) are symbiont fungi which enter the root cortical cells and form arbuscules. This symbiotic association has been shown to improve nutrient uptake in plants. Therefore, for improving NUE, developing varieties with efficient symbiosis with AMF could be one of the strategies (Verzeaux *et al.*, 2017). Another possible approach could be to exploit the potential of various N-fixing bacteria such as diazotrophs which have the ability to colonize the roots of cereals (Parnell *et al.*, 2016). These N-fixing bacteria do not form root nodules as rhizobium, but they colonize the root surface and sometimes enter root tissue and provide sufficient amount of N to the plant. They are sold as biofertilizers or phytostimulators (Kuan *et al.*, 2016). Some other plant growth-promoting bacteria (PGPB) have also been reported to release hormones for root development, thus increasing nutrient acquisition (Cassán and Diaz-Zorita, 2016).

## 19.5 Traits Influencing Nitrogen-Utilization Efficiency

### 19.5.1 Nitrate assimilation

After the uptake of $NO_3^-$, it is generally transported to shoots for further assimilation. In the cytosol of leaf mesophyll cells, this $NO_3^-$ is converted into $NO_2^-$ by the enzyme nitrate reductase (NR; EC 1.6.6.1). Further, the enzyme nitrite reductase (NiR; EC 1.7.7.1) catalyses the reduction of $NO_2^-$ into $NH_4^+$ in the plastids of these cells. Wang and Loussaert (2015) have shown that overexpression of *yeast nitrate transporter* (*YNT1*) in maize driven by root-preferred promoter led to yield improvement in the transgenic *YNT1*-overexpressing lines. Subsequently, irrespective of their origin, $NH_4^+$ ions (both synthesized via $NO_3^-$ reduction and absorbed) are incorporated into the amino acid (organic form) glutamine and glutamate via assimilation primarily by glutamine synthetase (GS; EC 6.3.1.2) and sometimes by ferredoxin-dependent glutamate synthase (Fd-GOGAT; EC 1.4.7.1), also known as the GS/GOGAT cycle (Krapp, 2015). The generated glutamine and glutamate are used as amino group donors for most of the other amino acids, which are further incorporated into proteins and nucleic acids directly or transported through the phloem stream providing organic N to developing organs. In both $C_3$ and $C_4$ plants, various isoenzymes of GS and GOGAT are present which are located in different cellular compartments. Luo *et al.* (2015) reported that the gene encoding GS1 isoenzymes coincides with a quantitative trait locus (QTL) for low N tolerance. That study confirmed that GS plays an important role in NUtE. It has also been reported that some variability exists in long-distance transport of amino acids in phloem (Yesbergenova-Cuny *et al.*, 2016), especially in the concentrations of amino acids and their translocation in phloem tissue for improving grain yield. Although amino acid translocation to developing sink organs and utilization play an important role, most of the studies have focused on vegetative organs such as root and shoot. The developing ear and tassel also play an important role in improving NUtE (Seebauer *et al.*, 2004; Liao *et al.*, 2012; Pan *et al.*, 2015).

### 19.5.2 Canopy photosynthesis per unit of nitrogen

It is well-established fact that foliar N content is the major determinant for leaf photosynthesis. Being a $C_4$ plant, maize has a much higher NUtE than most of the $C_3$ cereal plants. The best strategy for efficient N use is to absorb as well as utilize the whole N for grain yield. It has been reported that if N concentration in vegetative organs is the same, N-efficient maize hybrids still have greater photosynthetic rate at kernel-filling stage resulting in a higher number of kernels per ear (Chen *et al.*, 2006). One possible explanation for this behaviour might be that there is strong positive feedback on photosynthetic rate from the sink requirement in N-efficient cultivars. So, it can be concluded that for higher NUtE, it is important to develop more active reproductive sink units.

## 19.6 Identification and Use of Quantitative Trait Loci Related to Nitrogen-Use Efficiency

A QTL is a locus (section of DNA) that correlates with variation of a quantitative trait in the phenotype of a population of organisms. QTLs are mapped by identifying which molecular markers correlate with an observed trait. As NUE is defined as the production of grain yield per unit of N from soil and fertilizers, NUE must be a multi-gene/QTL-controlled trait. Numerous studies suggest that NUE and its related physiological traits such as N accumulation and re-translocation are mainly controlled by additive gene effects (Pollmer *et al.*, 1979; Below, 1997; Chen *et al.*, 2003). Therefore, identifying QTLs linked to NUE is a promising way for genetic improvement of NUE. The first maize NUE trait-related QTLs were reported by Agrama *et al.* (1999). In most of the studies in which QTL mapping was performed using recombinant inbred lines (RILs) (Table 19.2), the size of the characterized chromosomal regions was generally between 5 and 30 cM depending on the size of the population of RILs and on the measured agronomic and phenotypic traits (Gallais and Hirel, 2004; Jansen *et al.*, 2015). Agrama *et al.* (1999) found significant differences in QTLs controlling grain yield and its components under high and low N supply, and the contribution of these QTLs to the phenotypes is between 11.8% (grain weight) and 42.1% (yield). Later studies have shown coincidences between QTLs and traits related to NUE which are controlled by specific genes involved in N uptake, metabolism and remobilization. Bertin and Gallais (2001) found QTLs related to NUE at high rather than low N supply, possibly because plant response to N is higher at high N inputs. They further reported that on chromosome 1 the region near to *bnlg1643* locus may be involved in grain yield determination under optimal N fertility, probably through efficient plant N uptake. Moreover, at the end of chromosome 6, locus *umc1653* plays a role in the adaptation to N-stress conditions through efficient grain filling (Coque and Gallais, 2006). Further, QTLs for root architecture and GS activity may be important for grain yield both under N-sufficient and N-deficient conditions. Analysing coincidences between QTLs for agronomic and physiological traits and key genes of N uptake and metabolism could be useful to identify genes/QTLs involved in the variations in NUE. By this means, one QTL for leaf GS1 enzyme activity has been shown to be coincident with a QTL for yield, one QTL for 1000-kernel weight was coincident with the *Gln3* (*gln1-4*) locus and two QTLs for 1000-kernel weight and yield were coincident with the *Gln4* (*gln1-3*) locus (Hirel *et al.*, 2001; Gallais and Hirel, 2004). Based on this information, *gln1-3* and *gln1-4* mutants, as well as the *gln1-3*/*gln1-4* double mutant, were isolated and analysed. The *gln1-4* phenotype displayed reduced kernel size and *gln1-3* reduced kernel number, with both phenotypes displayed in the *gln1-3*/*gln1-4* double mutant. All three mutants have not produced any kernels when grown under the N-limiting conditions (Hirel *et al.*, 2007). Moreover, when *gln1-3* gene was over-expressed constitutively in the leaves, kernel number increased by 30%, indicating that *gln1-3* plays a major role in kernel yield (Martin *et al.*, 2006). Based on these results, the *GS* locus on chromosome 5 appeared to be a good candidate gene which can, at least partially, explain the variation in NUE (Gallais and Hirel, 2004). These studies highlighted the possibility of increasing maize yield by optimizing N metabolic traits. Considering the importance of root traits in

**Table 19.2.** List of cloned/fine-mapped and important QTLs for NUE and associated traits in maize.

| Study no. | QTLs/loci mapped[a] | Mapping population | Cross | Genotyping markers[b] | Chromosome or bin no. | Reference |
|---|---|---|---|---|---|---|
| 1. | Two joint QTLs having additive effects across environments for GY per plant and NUE, as well as for biomass and N harvest index | RIL | B100 × LP2 | SSR, SNP | 1, 9 | Mandolino *et al.* (2018) |
| 2. | Five stable, low-N stress-specific QTLs | RIL | 178 × K12 | SSR | 2.07, 2.03/2.04, 4.00/4.01, 5.02/5.03, 8.05/8.06 | He *et al.* (2018) |
| 3. | Five important QTL clusters in which QTLs for NUE and root system architecture-related traits coincided | RIL | Ye478 × Wu312 | SSR | 1.04, 2.04, 3.04, 3.05/3.06, 6.07/6.08 | Li *et al.* (2015) |
| 4. | A major QTL for AARL under low N co-localizes with QTLs previously reported for GY and N uptake | RIL | Z3 × 87-1 | SSR | 1 | Liu *et al.* (2008) |
| 5. | Eight QTL clusters in which QTLs for root architecture and traits related to N uptake, N remobilization and GY yield coincide positively | RILs | F-2 × Io | RFLP, SSR | 2.6, 3.7, 4.2, 4.6, 5.1, 5.4, 5.5, 5.6 | Coque *et al.* (2008) |
| 6. | Two QTL clusters having QTLs for N remobilization and leaf senescence | | | | 6.4, 7.4 | |
| 7. | Genomic locus playing role in adaptation to N-stress conditions through efficient grain filling | RILs | F-2 × Io | SSR | 6 | Coque and Gallais (2006) |
| 8. | One locus in which QTLs for CY, GS and NR enzyme activity, and $NO_3^-$ content coincide | RILs | F-2 × Io | RFLP | 5 (*gln4* locus) | Hirel *et al.* (2001) |
| 9. | Six QTLs involved in expression of GY components under low-N stress | $F_{2:3}$ | B73 × G79 | RFLP | 1, 2, 7, 9, 10 | Agrama *et al.* (1999) |

[a]GY, grain yield; AARL, average axial root length.
[b]SSR, simple sequence repeat; SNP, single-nucleotide polymorphism; RFLP, restriction fragment length polymorphism.

NUE, various groups are working on QTLs related to maize and the most important QTLs were detected on chromosome bins 1.03, 1.06, 1.08, 2.03, 2.04, 7.02, 8.06 and 10.04 (Tuberosa *et al.*, 2003) In a meta-analysis study, Luo *et al.* (2015) identified 21 consensus QTLs (cQTLs) strongly induced for low-N tolerance. He *et al.* (2018) identified five stable QTLs specific to low-N stress (*qPH5b*, *qLL4a*, *qSPAD-BEL2b*, *qSPADBEL8b* and *qGLN2a*) (Table 19.2). Out of these five, four QTLs have been reported for the first time and it is expected that candidate genes of these low-N stress-specific QTLs may be induced by low-N stress, which in turn might play a pivotal role in maize adaptation to N deficiency. Recently, map-based cloning was successfully utilized in tobacco to identify homologous genes involved in NUE (Edwards *et al.*, 2017). This approach has not been successful in maize so far. However, the recent approach of balanced multi-parental (MAGIC) populations (Dell'Acqua *et al.*, 2015) and ultra-high-density maps (Liu *et al.*, 2015; Su *et al.*, 2017) could provide powerful tools leading to higher power and definition in QTL mapping for complex traits such as NUE.

## 19.7 Identification of Nitrogen-Responsive Genes

A detailed understanding of NUE is necessary for optimizing fertilizer input without compromising crop yield. To understand and improve NUE, two major approaches can be adopted: (i) study the response of crop plants to N stress to identify the major genes and metabolic/biological processes affected by it; and (ii) utilization of natural or induced genetic variation in NUE. From microarray and other gene expression studies, various genes involved in C assimilation, N and C metabolism, $NO_3^-$ assimilation (*NR* and *GS*) and $NO_3^-$ transport (*NRT1* and *NRT2*) were found differentially regulated under conditions of varied N application (Gutiérrez *et al.*, 2007; Amiour *et al.*, 2012; Schlüter *et al.*, 2012; Plett *et al.*, 2016; Jiang *et al.*, 2018; Yadava *et al.*, 2020). By transcriptomics and transgenic approaches, the roles of a set of genes (encoding transcription factors, nitrogen transporters and kinases) related to the regulation of N assimilation have also been investigated. Transcriptomics has played a key role in understanding NUE in plant systems, including *Arabidopsis*, rice, wheat and maize. Overexpression of *OsENOD93-1* (*early nodulin 93*) has been shown to lead to accumulation of higher concentrations of total amino acids and total N in roots (Sun *et al.*, 2014). Heterotrimeric G-proteins that regulate NUE were reported in rice (Kurai *et al.*, 2011; Sun *et al.*, 2014). Similar work has been carried out in *Arabidopsis* with Dof1 (DNA-binding with one finger) (Table 19.3). In maize, by integrating meta-analysis and large-scale gene expression data, Luo *et al.* (2015) mined 30 candidate low-N stress tolerance genes and a further 12 most important maize orthologues were identified by *in silico* analyses of genes with known functions in NUE in model plants. Apart from genes, micro RNAs (miRNAs) responsive to $NO_3^-$ stress have also been identified (Fisher *et al.*, 2013). In *Arabidopsis*, miR393/*AFB3* has been shown as a unique N-responsive module controlling root system architecture in response to external and internal N availability (Vidal *et al.*, 2010). The repression of miR528a/b and miR169i/j/k in maize roots under $NO_3^-$ stress suggested their role in integrating $NO_3^-$ signals into root developmental changes (Trevisan *et al.*, 2012). Although significant progress has been made in the identification and characterization of genes and miRNAs playing critical roles in N uptake, translocation and homeostasis in *Arabidopsis* and rice, only a few studies are available in maize pertaining to the same. The potential candidate genes/miRNAs playing a pivotal role in NUE in *Arabidopsis* and rice (Table 19.3) or their maize orthologues might be utilized and explored for improving NUE in maize.

## 19.8 Nitrogen Signalling and Transduction for Improving Nitrogen-Use Efficiency

Signal transduction for any physiological process involves sensing and processing of stimuli. In the case of NUE in maize, Dof1 plays an important role. Dof1 is a transcription factor, a member of the DNA binding with one finger (Dof) family. It is unique to plants and facilitates expression of a range of genes associated with organic acid metabolism (Yanagisawa, 2004). The overexpression of *Dof1* in *Arabidopsis* under the control of a maize *pyruvate phosphate dikinase* (*PPDK*) promoter resulted in increased concentration of amino acids, especially glutamine, under low-N stress conditions, which in turn helped the transgenic plants to tolerate N stress (Yanagisawa, 2004). However, Cavalar *et al.* (2007) reported that Dof1 does not play a major role in the control of N or C metabolism in maize. Recently, Peña *et al.* (2017) have shown that

**Table 19.3.** List of key genes validated for improving N uptake and utilization in various crop plants that may be potentially useful for improving NUE in maize.

| Gene name; source | Gene function | Crop engineered | Overexpression/silencing phenotype | Reference |
|---|---|---|---|---|
| Alanine aminotransferase (*AlaAT*); barley (*Hordeum vulgare*) | Involved in both C and N metabolism; catalyses the reversible reaction (by transfer of an amino group) converting alanine and 2-oxoglutarate to glutamate and pyruvate, and vice versa | *Brassica napus* spp. *oleifera*<br>*O. sativa* | Overexpression using root-specific promoter resulted in significant increase in root biomass and NUE in transgenic lines | Good *et al.* (2007); Shrawat *et al.* (2008) |
| *AlaAT*; from barley, mouse (both cytoplasmic and mitochondrial isoforms) and *Pyrococcus furiosus* | | *Arabidopsis* | Overexpression of different variants of *AlaAT* conferred diverse NUE phenotypes under different external conditions | McAllister and Good (2015) |
| *Dof1*; maize (*Zea mays*) | Regulates expression of genes encoding enzymes for C-skeleton production (e.g. upregulation of PEPC) and thereby modulates C/N network | *Arabidopsis*<br>*Solanum tuberosum* | Overexpression resulted in enhanced N and amino acid contents (i.e. N assimilation) and increased plant growth under low-N conditions | Yanagisawa (2004) |
| | | *O. sativa* | Overexpression resulted in increased C and N assimilation under low-N conditions | Kurai *et al.* (2011) |
| | | *Triticum aestivum*<br>*Sorghum bicolor* | Constitutive overexpression resulted in reduction in photosynthesis, plant height and biomass in transgenic lines, while tissue-specific expression under *rbcS1* (Rubisco subunit 1) promoter resulted in increased biomass and yield | Peña *et al.* (2017) |
| *NRT2.3b*; rice (*Oryza sativa*) | $NO_3^-$ uptake, enhances the pH-buffering capacity of the plant | *O. sativa* | Constitutive overexpression resulted in improved plant growth, yield and NUE by 40% in transgenic lines | Fan *et al.* (2016) |
| *NPF6.5* (*NRT1.1B*); *O. sativa* | $NO_3^-$ uptake and subsequent root-to-shoot transportation, $NO_3^-$ signalling | *O. sativa* | Overexpression significantly improved NUE and yield | Hu *et al.* (2015) |

*Continued*

**Table 19.3.** Continued.

| Gene name; source | Gene function | Crop engineered | Overexpression/silencing phenotype | Reference |
|---|---|---|---|---|
| *NRT2.1*; *O. sativa* | Interacts with OsNAR2.1 and plays a role in $NO_3^-$ transport | *O. sativa* | Constitutive overexpression (using *ubiquitin* promoter) resulted in decreased NUE, while overexpression under native promoter resulted in increased NUE | Chen *et al.* (2016) |
| *NLP6* and *NLP8*; *Z. mays* | $NO_3^-$ signalling and metabolism, induction of $NO_3^-$ -responsive genes | *Arabidopsis* | Overexpression restored $NO_3^-$ signalling and assimilation in *nlp7* mutant, increased biomass and yield | Cao *et al.* (2017) |
| *ENOD93-1*; *Oryza sativa* | Encodes the early nodulin 93 protein and expressed at high levels in roots, especially at the panicle emergence stage, localized in mitochondria; exact role not known | *O. sativa* | Overexpression resulted in higher accumulation of total N and amino acids in roots, increased shoot dry biomass and seed yield under N-limiting condition | Sun *et al.* (2014) |

PEPC, phosphoenolpyruvate carboxylase.

overexpression of the maize *Dof1* in wheat led to an improvement in growth and productivity. These studies in wheat and *Arabidopsis* led to the conclusion that Dof1 transcription factor is an interesting candidate for increasing nutrient-use efficiency and yield potential of cereals. Other than *Dof1*, some regulatory genes such as *NLP7*, *PHR1* and protein kinase *AtCIP8* are key elements in regulating $NO_3^-$ response in *Arabidopsis* (Castaings *et al.*, 2009). Nine *NLP* genes were identified in maize; out of them *ZmNLP6* and *ZmNLP8* regulate $NO_3^-$ signalling in *Arabidopsis* and were able to increase plant biomass and yield when overexpressed in the model species (Cao *et al.*, 2017). In spite of studies related to identification of NUE-related QTLs and genes, the molecular mechanism governing NUE in cereals, including maize, is not fully understood. Therefore, more work is required in this field for better understanding of $NO_3^-$ signalling so that this knowledge can be used for genetic manipulation and transformation for better yield results.

## 19.9 Conclusion

NUE is the key to sustainable agriculture in the coming future. Over the past few years, remarkable progress has been achieved in identifying molecular targets for improvement of NUE in various crops including maize. Various important genes having crucial roles in root development, N assimilation, N uptake and utilization, N signalling and sensing can be used for improving NUE. As discussed above, NUE is a complex phenomenon. So, neither conventional breeding nor the transgenic approach alone is sufficient to increase the NUE. Soil N status and the plant's inherent nature towards N uptake and utilization are also important. For improving NUE in any field crop, a holistic approach which includes soil management, agronomic practices, conventional breeding as well as transgenic approaches with specific transgenes will help in maintaining crop yield with minimum N input and will ensure environmentally friendly crop production.

Roots play a key role in N uptake. In this regard, root response studies should be undertaken more systematically. Roots being an underground system are inherently difficulty to phenotype, but several high-throughput screening techniques may help breeders and physiologists effectively. Selection of genotypes recording efficient growth under low N is indicative of better root architecture. Hence, this should be a general strategy for selection of efficient genotypes. Transcriptome analyses have identified a large number of genetic elements influencing NUE. These can be used as biomarkers for better NUE. However, these need further validation in a wider number of genotypes.

Intensive research on molecular and genetic aspects of NUE has led to the identification of many new genes, QTLs and alleles that could be deployed to develop new genotypes. The future direction of the research efforts should be towards understanding the interaction of NUE-related genes with cellular small RNA flux and perturbing the system performance through metabolic engineering and genome editing techniques. It is expected that these efforts would ultimately lead to commercialization of new improved maize hybrids with high NUE in the near future.

## References

Agrama, H.A.S., Zakaria, A.G., Said, F.B. and Tuinstra, M. (1999) Identification of quantitative trait loci for nitrogen use efficiency in maize. *Molecular Breeding* 5(2), 187–195.

Amiour, N., Imbaud, S., Clément, G., Agier, N., Zivy, M. *et al.* (2012) The use of metabolomics integrated with transcriptomic and proteomic studies for identifying key steps involved in the control of nitrogen metabolism in crops such as maize. *Journal of Experimental Botany* 63, 5017–5033.

Below, F.E. (1997) Growth and productivity of maize under nitrogen stress. In: Edmeades, G.O., Bänziger, M., Mickelson, H.R. and Peña-Valdivia, C.B. (eds) *Developing Drought- and Low N-tolerant Maize.* CIMMYT Mexico City, pp. 235–240.

Bertin, P. and Gallais, A. (2001) Physiological and genetic basis of nitrogen use efficiency. II. QTL detection and coincidences. *Maydica* 46, 53–68.

Bray, E.A., Bailey-Serres, J. and Weretilnyk, E. (2000) Responses to abiotic stresses. In: Buchanan, B.B., Gruissem, W. and Jones, R.L. (eds) *Biochemistry and Molecular Biology of Plants*. American Society of Plant Physiologists, Rockville, pp. 1158–1203.

Buchner, P. and Hawkesford, M.J. (2014) Complex phylogeny and gene expression patterns of members of the NITRATE TRANSPORTER 1/PEPTIDE TRANSPORTER family (NPF) in wheat. *Journal of Experimental Botany* 65, 5697–5710.

Canas, R.A., Quillere, I., Christ, A. and Hirel, B. (2009) Nitrogen metabolism in the developing ear of maize (*Zea mays*): analysis of two lines contrasting in their mode of nitrogen management. *New Phytologist* 184, 340–352.

Cao, H., Qi, S., Sun, M., Li, Z., Yang, Y., Crawford, N.M. and Yong, W. (2017) Overexpression of the maize *ZmNLP6* and *ZmNLP8* can complement the *Arabidopsis* nitrate regulatory mutant *nlp7* by restoring nitrate signalling and assimilation. *Frontiers in Plant Science* 8, 1703.

Cassán, F. and Diaz-Zorita, M. (2016) *Azospirillum* sp. in current agriculture: from the laboratory to the field. *Soil Biology and Biochemistry* 103, 117–130.

Castaings, L., Camargo, A., Pocholle, D., Gaudon, V., Texier, Y. *et al.* (2009) The nodule inception like protein 7 modulates nitrate sensing and metabolism in *Arabidopsis*. *The Plant Journal* 57(3), 426–435.

Cavalar, M., Phlippen, Y., Kreuzaler, F. and Peterhaensel, C. (2007) A drastic reduction in DOF1 transcript levels does not affect $C_4$ specific gene expression in maize. *Journal of Plant Physiology* 164, 1665–1674.

Chen, F., Mi, G., Chun, L., Liu, J., Wang, Y. and Zhang, F. (2003) Combination ability analysis of traits related to nitrogen use efficiency in maize. *Scientia Agricultura Sinica* 36, 134–139.

Chen, F.J., Chun, L., Bao, J., Zhang, F. and Mi, G. (2006) Vegetative growth and photosynthetic characteristics of maize hybrids differing in nitrogen use efficiency. *Journal of Maize Sciences* 14(6), 127–130.

Chen, J.G., Zhang, Y., Tan, Y., Zhang, M., Zhu, L., Xu, G. and Fan, X. (2016) Agronomic nitrogen-use efficiency of rice can be increased by driving *OsNRT2.1* expression with the OsNAR2.1 promoter. *Plant Biotechnology Journal* 14, 1705–1715.

Chun, L., Chen, F., Zhang, F. and Mi, G. (2005) Root growth, nitrogen uptake and yield formation of hybrid maize with different N efficiency. *Plant Nutrition and Fertilizer Science* 11(5), 615–619.

Coque, M. and Gallais, A. (2006) Genomic regions involved in response to grain yield selection at high and low nitrogen fertilization in maize. *Theoretical and Applied Genetics* 112, 1205–1220.

Coque, M., Martin, A., Veyrieras, J.B., Hirel, B. and Gallais, A. (2008) Genetic variation for N-remobilization and post silking N-uptake in a set of maize recombinant inbred lines. 3. QTL detection and coincidences. *Theoretical and Applied Genetics* 117, 729–747.

Dell'Acqua, M., Gatti, D.M., Pea, G., Cattonaro, F., Coppens, F. *et al.* (2015) Genetic properties of the MAGIC maize population: a new platform for high definition QTL mapping in *Zea mays*. *Genome Biology* 16(1), 167.

Edwards, K.D., Fernandez-Pozo, N., Drake-Stowe, K., Humphry, M., Evan, A.D. *et al.* (2017) A reference genome for *Nicotiana tabacum* enables map-based cloning of homeologous loci implicated in nitrogen utilization efficiency. *BMC Genomics* 18(1), 448.

Elings, A., White, J. and Edmeades, G.O. (1996) Modelling tropical maize under drought and low N. In: *Annual Abstracts, Agronomy Meeting, Indianapolis, Indiana, 3–8 November 1996*. American Society of Agronomy, Crop Science Society of America and Soil Science Society of America, Madison, Wisconsin, p. 109.

Fan, X., Tang, Z., Tan, Y., Zhang, Y., Luo, B., Yang, M. and Xu, G. (2016) Overexpression of a pH-sensitive nitrate transporter in rice increases crop yields. *Proceedings of the National Academy of Sciences USA* 113, 7118–7123.

FAO (2019) World Fertilizer Trends and Outlook to 2022. Food and Agriculture Organization of the United Nations, Rome. Available at: http://www.fao.org/3/ca6746en/ca6746en.pdf (accessed 5 April 2020).

Fischer, S., Wagner, A., Kos, A., Aschrafi, A., Handrick, R., Hannemann, J. and Otte, K. (2013) Breaking limitations of complex culture media: functional non-viral miRNA delivery into pharmaceutical production cell lines. *Journal of Biotechnology* 168(4), 589–600.

Gallais, A. and Coque, M. (2005) Genetic variation and selection for nitrogen use efficiency in maize: a synthesis. *Maydica* 50, 531–537.

Gallais, A. and Hirel, B. (2004) An approach to the genetics of nitrogen use efficiency in maize. *Journal of Experimental Botany* 55, 295–306.

Galloway, J.N., Winiwarter, W., Leip, A., Leach, A.M., Bleeker, A. and Erisman, J.W. (2014) Nitrogen footprints: past, present and future. *Environment Research Letters* 9, 115003.
Garnett, T., Conn, V., Plett, D., Conn, S., Zanghellin, J. *et al.* (2013) The response of the maize nitrate transport system to nitrogen demand and supply across the lifecycle. *New Phytologist* 198(1), 82–94.
Garnett, T., Plett, D., Conn, V., Conn, S., Rabie, H. *et al.* (2015) Variation for N uptake system in maize: genotypic response to N supply. *Frontiers in Plant Science* 6, 936.
Good, A.G., Johnson, S.J., De Pauw, M., Carroll, R.T., Savidov, N., *et al.* (2007) Engineering nitrogen use efficiency with alanine aminotransferase. *Canadian Journal of Botany* 85, 252–262.
Gutiérrez, R.A., Lejay, L.V., Dean, A., Chiaromonte, F., Shasha, D.E. and Coruzzi, G.M. (2007) Qualitative network models and genome-wide expression data define carbon/nitrogen-responsive molecular machines in *Arabidopsis*. *Genome Biology* 8(1), R7.
Haegele, J.W., Cook, K.A., Nichols, D.M. and Below, F.E. (2013) Changes in nitrogen use traits associated with genetic improvement for grain yield of maize hybrids released in different decades. *Crop Science* 53, 1256–1268.
Han, J., Wang, L., Zheng, H., Pan, X., Li, H., Chen, F. and Li, X. (2015) ZD958 is a low-nitrogen-efficient maize hybrid at the seedling stage among five maize and two teosinte lines. *Planta* 242(4), 935–949.
He, K., Chang, L., Dong, Y., Cui, T., Qu, J. *et al.* (2018) Identification of quantitative trait loci for agronomic and physiological traits in maize (*Zea mays* L.) under high-nitrogen and low-nitrogen conditions. *Euphytica* 214, 15.
Hirel, B., Bertin, P., Quilleré, I., Bourdoncle, W., Attagnant, C. *et al.* (2001) Towards a better understanding of the genetic and physiological basis for nitrogen use efficiency in maize. *Plant Physiology* 125(3), 1258–1270.
Hirel, B., Le Gouis, J., Ney, B. and Gallais, A. (2007) The challenge of improving nitrogen use efficiency in crop plants: towards a more central role for genetic variability and quantitative genetics within integrated approaches. *Journal of Experimental Botany* 58, 2369–2387.
Hu, B., Wang, W., Ou, S., Tang, J., Li, H. *et al.* (2015) Variation in *NRT1.1B* contributes to nitrate-use divergence between rice subspecies. *Nature Genetics* 47, 834–838.
Huggins, D.R. and Pan, W.L. (2003) Key indicators for assessing nitrogen use efficiency in cereal-based agro-ecosystems. *Journal of Crop Production* 8, 157–185.
Hull, R. and Liu, H. (2005) Turf grass nitrogen: physiology and environmental impact. *International Society of Turf Research Journal* 10, 962–975.
Imsande, J. and Touraine, B. (1994) N demand and the regulation of nitrate uptake. *Plant Physiology* 105(1), 3–7.
Jansen, C., Zhang, Y., Liu, H., Gonzalez-Portilla, P.J., Lauter, N. *et al.* (2015) Genetic and agronomic assessment of cob traits in corn under low and normal nitrogen management conditions. *Theoretical and Applied Genetics* 128(7), 1231–1242.
Jiang, L., Ball, G., Hodgman, C., Coules, A., Zhao, H. and Lu, C. (2018) Analysis of gene regulatory networks of maize in response to nitrogen. *Genes* 9, 151.
Krapp, A. (2015) Plant nitrogen assimilation and its regulation: a complex puzzle with missing pieces. *Current Opinion in Plant Biology* 25, 115–122.
Kuan, K.B., Othman, R., Rahim, K.A. and Shamsuddin, Z.H. (2016) Plant growth-promoting rhizobacteria inoculation to enhance vegetative growth, nitrogen fixation and nitrogen remobilization of maize under greenhouse conditions. *PLoS ONE* 11(3), e0152478.
Kurai, T., Wakayama, M., Abiko, T., Yanagisawa, S., Aoki, N. and Ohsugi, R. (2011) Introduction of the *ZmDof1* gene into rice enhances carbon and nitrogen assimilation under low nitrogen conditions. *Plant Biotechnology Journal* 9(8), 826–837.
Lassaletta, L., Billen, G., Garnier, J., Bouwman, L., Velazquez, E., Mueller, N.D. and Gerber, J.S. (2016) Nitrogen use in the global food system: past trends and future trajectories of agronomic performance, pollution, trade, and dietary demand. *Environment Research Letters* 11(9), 095007.
Léran, S., Varala, K., Boyer, J.C., Chiurazzi, M., Crawford, N. *et al.* (2014) A unified nomenclature of NITRATE TRANSPORTER 1/PEPTIDE TRANSPORTER family members in plants. *Trends in Plant Science* 19(1), 5–9.
Li, P., Chen, F., Cai, H., Liu, J., Pan, Q. *et al.* (2015) A genetic relationship between nitrogen use efficiency and seedling root traits in maize as revealed by QTL analysis. *Journal of Experimental Botany* 66, 3175–3188.
Liao, C., Peng, Y., Ma, W., Liu, R., Li, C. and Li, X. (2012) Proteomic analysis revealed nitrogen-mediated metabolic, developmental, and hormonal regulation of maize (*Zea mays* L.) ear growth. *Journal of Experimental Botany* 63, 5275–5288.

Liu, H., Niu, Y., Gonzalez-Portilla, P.J., Zhou, H., Wang, L. *et al.* (2015) An ultra-high-density map as a community resource for discerning the genetic basis of quantitative traits in maize. *BMC Genomics* 16(1), 1078.

Liu, J., Li, J., Chen, F., Zhang, F., Ren, T., Zhuang, Z. and Mi, G. (2008) Mapping QTLs for root traits under different nitrate levels at the seedling stage in maize (*Zea mays* L.). *Plant and Soil* 305, 253–265.

Luo, B., Tang, H., Liu, H., Shunzong, S., Zhang, S. *et al.* (2015) Mining for low-nitrogen tolerance genes by integrating meta-analysis and large-scale gene expression data from maize. *Euphytica* 206, 117–131.

Majumdar, D. (2003) The blue baby syndrome: nitrite poisoning in humans. *Resonance* 10, 20–30.

Mandolino, C.I., Andrea, K.E.D., Olmos, S.E., Otegui, M.E. and Eyhérabide, G.H. (2018) Maize nitrogen use efficiency: QTL mapping in a US dent × Argentine-Caribbean flint RILs population. *Maydica* 63, 1–17.

Martin, A., Lee, J., Kichey, T., Gerentes, D., Zivy, M. *et al.* (2006) Two cytosolic glutamine synthetase isoforms of maize are specifically involved in the control of grain production. *The Plant Cell* 18, 3252–3274.

Mastrodomenico, A.T., Hendrix, C.C. and Below, F.E. (2018) Nitrogen use efficiency and the genetic variation of maize expired plant variety protection germplasm. *Agriculture* 8(3), 1–17.

McAllister, C.H. and Good, A.G. (2015) Alanine aminotransferase variants conferring diverse NUE phenotypes in *Arabidopsis thaliana*. *PLoS ONE* 10(4), e0121830.

Moll, R.H., Kamprath, E.J. and Jackson, W.A. (1987) Development of nitrogen-efficient prolific hybrids of maize. *Crop Science* 27, 181–186.

Niu, J., Chen, F., Mi, G., Li, C. and Zhang, F. (2007) Transpiration, and nitrogen uptake and flow in two maize (*Zea mays* L.) inbred lines as affected by nitrogen supply. *Annals of Botany* 99(1), 153–160.

Okamoto, M., Vidmar, J.J. and Glass, A.D.M. (2003) Regulation of *NRT1* and *NRT2* gene families of *Arabidopsis thaliana*: responses to nitrate provision. *Plant & Cell Physiology* 44, 304–317.

Pan, X., Hasan, M.M., Li, Y., Liao, C., Zheng, H., Liu, R. and Li, X. (2015) Asymmetric transcriptomic signatures between the cob and florets in the maize ear under optimal- and low-nitrogen conditions at silking, and functional characterization of amino acid transporters ZmAAP4 and ZmVAAT3. *Journal of Experimental Botany* 66, 6149–6166.

Parnell, J.J., Berka, R., Young, H.A., Sturino, J.M., Kang, Y., Barnhart, D.M. and DiLeo, M.V. (2016) From the lab to the farm: an industrial perspective of plant beneficial microorganisms. *Frontiers in Plant Science* 7, 1110.

Peña, P.A., Quach, T., Sato, S., Ge, Z., Nersesian, N. *et al.* (2017) Expression of the maize *Dof1* transcription factor in wheat and sorghum. *Frontiers in Plant Science* 8, 434.

Plett, D., Baumann, U., Schreiber, A.W., Holtham, L., Kalashyan, E. *et al.* (2016) Maize maintains growth in response to decreased nitrate supply through a highly dynamic and developmental stage-specific transcriptional response. *Plant Biotechnology Journal* 14, 342–353.

Pollmer, W.G., Eberhard, D., Klein, D. and Dhillon, B.S. (1979) Genetic control of nitrogen uptake and translocation in maize. *Crop Science* 19(1), 82–86.

Presterl, T., Groh, S., Landbeck, M., Seitz, G., Schmidt, W. and Geiger, H.H. (2002) Nitrogen uptake and utilization efficiency of European maize hybrids developed under conditions of low and high nitrogen input. *Plant Breeding* 121, 480–486.

Raun, W.R. and Johnson, G.V. (1999) Improving nitrogen use efficiency for cereal production. *Agronomy Journal* 91(3), 357–363.

Reetz, H.F. Jr, Heffer, P. and Bruulsema, T.W. (2015) 4R nutrient stewardship: a global framework for sustainable fertilizer management. In: Drechsel, P., Heffer, P., Magen, H., Mikkelsen, R. and Wichelns, D. (eds) *Managing Water and Fertilizer for Sustainable Agricultural Intensification*. International Fertilizer Industry Association, International Water Management Institute, International Plant Nutrition Institute and International Potash Institute, Paris, pp. 65–86.

Schlüter, U., Mascher, M., Colmsee, C., Scholz, U., Bräutigam, A., Fahnenstich, H. and Sonnewald, U. (2012) Maize source leaf adaptation to nitrogen deficiency affects not only nitrogen and carbon metabolism but also control of phosphate homeostasis. *Plant Physiology* 160, 1384–1406.

Seebauer, J.R., Moose, S.P., Fabbri, B.J., Crossland, L.D. and Below, F.E. (2004) Amino acid metabolism in maize earshoots. Implications for assimilate preconditioning and nitrogen signaling. *Plant Physiology* 136(4), 4326–4334.

Shrawat, A.K., Carroll, R.T., DePauw, M., Taylor, G.J. and Good, A.G. (2008) Genetic engineering of improved nitrogen use efficiency in rice by the tissue-specific expression of alanine aminotransferase. *Plant Biotechnology Journal* 6, 722–732.

Sinclair, T.R. and Vadez, V. (2002) Physiological traits for crop yield improvement in low N and P environments. *Plant and Soil* 245(1), 1–15.

Smith, B.E. (2002) Nitrogenase reveals its inner secrets. *Science* 297, 1654–1655.

Soltabayeva, A., Srivastava, S., Kurmanbayeva, A., Bekturova, A., Fluhr, R. and Sagi, M. (2018) Early senescence in older leaves of low nitrate-grown *Atxdh1* uncovers a role for purine catabolism in N supply. *Plant Physiology* 178, 1027–1044.

Sorgonà, A., Lupini, A., Mercati, F., Di Dio, L., Sunseri, F. and Abenavoli, M.R. (2011) Nitrate uptake along the maize primary root: an integrated physiological and molecular approach. *Plant, Cell & Environment* 34(7), 1127–1140.

Su, C., Wang, W., Gong, S., Zuo, J., Li, S. and Xu, S. (2017) High density linkage map construction and mapping of yield trait QTLs in maize (*Zea mays*) using the genotyping-by-sequencing (GBS) technology. *Frontiers in Plant Science* 8, 706.

Sun, H., Qian, Q., Wu, K., Luo, J., Wang, S. *et al.* (2014) Heterotrimeric G proteins regulate nitrogen-use efficiency in rice. *Nature Genetics* 46(6), 652–656.

Tilman, D., Cassman, K.G., Matson, P.A., Naylor, R. and Polasky, S. (2002) Agricultural sustainability and intensive production practices. *Nature* 418(6898), 671–677.

Tollenaar, M., Nissanka, S.P., Rajcan, I. and Bruulsema, T.W. (1997) Yield response of old and new corn hybrids to nitrogen. *Better Crops* 81, 3–5.

Trevisan, S., Nonis, A., Begheldo, M., Manoli, A., Palme, K. *et al.* (2012) Expression and tissue-specific localization of nitrate-responsive miRNAs in roots of maize seedlings. *Plant, Cell & Environment* 35(6), 1137–1155.

Tuberosa, R., Salvi, S., Sanguineti, M.C., Maccaferri, M., Giuliani, S. and Landi, P. (2003) Searching for quantitative trait loci controlling root traits in maize: a critical appraisal. *Plant and Soil* 255(1), 35–54.

Uhart, S.A. and Andrade, F.H. (1995) Nitrogen deficiency in maize I. Effects on crop growth, development, dry matter partitioning and kernel set. *Crop Science* 35, 1376–1783.

Verzeaux, J., Hirel, B., Dubois, F., Lea, P.J. and Tétu, T. (2017) Agricultural practices to improve nitrogen use efficiency through the use of arbuscular mycorrhizae: basic and agronomic aspects. *Plant Science* 264, 48–56.

Vidal, E.A., Araus, V., Lu, C., Parry, G., Green, P.J., Coruzzi, G.M. and Gutiérrez, R.A. (2010) Nitrate-responsive miR393/*AFB3* regulatory module controls root system architecture in *Arabidopsis thaliana*. *Proceedings of the National Academy of Sciences USA* 107(9), 4477–4482.

Wang, H. and Loussaert, D.F. (2015) Functional expression of yeast nitrate transporter (YNT1) and a nitrate reductase in maize. *US Patent No.* US8975474B2.

Ward, M.H., Jones, R.R., Brender, J.D., de Kok, T.M., Weyer, P.J., Nolan, B.T., Villanueva C.M. and van Breda, S.G. (2018) Drinking water nitrate and human health: an updated review. *International Journal of Environmental Research and Public Health* 15(7), 1557.

Wen, Z., Tyerman, S.D., Dechorgnat, J., Ovchinnikova, E., Dhugga, K.S. and Kaiser, B.N. (2017) Maize NPF6 proteins are homologs of *Arabidopsis* CHL1 that are selective for both nitrate and chloride. *The Plant Cell* 29(10), 2581–2596.

Wuebbles, D.J. (2009) Nitrous oxide: no laughing matter. *Science* 326, 56–57.

Yadava, P., Singh, A., Kumar, K., Sapna and Singh, I. (2019) Plant senescence and agriculture. In: Sarwat, M. and Tuteja, N. (eds) *Senescence Signalling and Control in Plants*. Academic Press, London, pp. 283–302.

Yadava, P., Aggarwal, C., Verma, R., Kumar, K. and Singh, I. (2020) Effect of nitrogen-starvation on growth pattern and expression of nitrogen assimilation related genes in maize (*Zea mays* L.). *Indian Journal of Agricultural Sciences* 90(1), 195–200.

Yanagisawa, S. (2004) Dof domain proteins: plant-specific transcription factors associated with diverse phenomena unique to plants. *Plant and Cell Physiology* 45(4), 386–391.

Yesbergenova-Cuny, Z., Dinant, S., Martin-Magniette, M.L., Quilleré, I., Armengaud, P. *et al.* (2016) Genetic variability of the phloem sap metabolite content of maize (*Zea mays* L.) during the kernel-filling period. *Plant Science* 252, 347–357.

Yu, P., Li, X., Yuan, L. and Li, C. (2014) A novel morphological response of maize (*Zea mays*) adult roots to heterogeneous nitrate supply revealed by a split-root experiment. *Physiologia Plantarum* 150, 133–144.

Yuan, L., Loqué, D., Kojima, S., Rauch, S., Ishiyama, K. *et al.* (2007) The organization of high-affinity ammonium uptake in *Arabidopsis* roots depends on the spatial arrangement and biochemical properties of AMT1-type transporters. *The Plant Cell* 19(8), 2636–2652.

# 20 Recent Advancement in Molecular Breeding for Improving Nutrient-Use Efficiency in Maize

**Damar López-Arredondo[1,2]*, Xiaorong Fan[2] and Yinping Jiao[1]**
[1]*Texas Tech University, Lubbock, Texas, USA;* [2]*Nanjing Agricultural University, Nanjing, People's Republic of China*

## 20.1 Introduction

Since the 20th century, the use of chemical fertilizers has contributed to increase food production to feed nearly half of the global population (Roser and Ritchie, 2020). Maize (*Zea mays* L.) is a preferred dietary staple in many developing countries, particularly in southern and eastern Africa, Central America and Mexico. Maize is cultivated in several countries around the world and the USA, China and Brazil are the top three maize-producing countries, accounting for approximately 563 million tonnes of the 717 million tonnes produced per year (Ranum *et al*., 2014). Over the past 50 years, maize breeders have made great progress in producing new varieties of maize with desired agronomic traits, including improved nutritional quality and harvest index, resistance to diseases, insects and weeds (i.e. striga), tolerance to drought, cold and Al, and early-maturing tropical germplasm (Dowswell, 1996). Moreover, genetically modified maize resistant to different herbicides and insects is currently on the market. However, to the authors' best knowledge, no new varieties with significantly improved nutrient uptake or use efficiency, by using molecular or traditional breeding approaches, have been released to date.

A reliable supply of N and P, two essential macronutrients, has greatly improved maize production per unit area of land over the past century, thereby contributing to global economic development, supporting food production for large populations in the Third World nations and preventing the conversion of forests into agricultural lands to meet food demand in Africa and Latin America (Dowswell, 1996; Hallauer *et al*., 2010; Zhang, X. *et al*., 2015). Due the relevance of N and P as two essential macronutrients to sustain maize productivity, this chapter is devoted to summarizing and discussing efforts towards molecular breeding of maize to improve N-use efficiency (NUE) and P-use efficiency (PUE).

First, nutrient-use efficiency is defined and the relevance of N and P as essential macronutrients and the molecular regulation of their metabolism in maize are presented. Then, the efforts towards molecular breeding of maize to improve NUE and PUE are summarized and discussed. Finally, plant phenotyping as one of the main and challenging components of molecular breeding and the potential of genome editing approaches to implement current findings on maize are addressed.

*Corresponding author: damar.lopez-arredondo@ttu.edu

 *Molecular Breeding in Wheat, Maize and Sorghum: Strategies for Improving Abiotic Stress Tolerance and Yield* (eds M.A. Hossain *et al.*)
DOI: 10.1079/9781789245431.0020

## 20.2 What is Nutrient-Use Efficiency?

Several definitions have been coined to describe nutrient-use efficiency over the years, some of them relying on complex mathematical models and formulas. From a physiological perspective, nutrient-use efficiency is contributed by a complex array of physiological, molecular and developmental interactions involving root uptake, root-to-shoot transport, utilization and remobilization, and C and N/P balance in the plant (Xu *et al.*, 2012; Hirel and Lea, 2018; Li, S. *et al.*, 2018) (Fig. 20.1). However, nutrient-use efficiency is also influenced by fertilizer application practices that vary widely from field to field depending upon socio-economic aspects and local practices. Moreover, soil properties (i.e. pH) and the presence of beneficial microorganisms also affect nutrient utilization by plants (Fig. 20.1). Therefore, for practical purposes, nutrient-use efficiency is defined here as the combined effect of two main factors: (i) the efficiency of the plant to take up the nutrient from the soil; and (ii) the efficiency of its utilization for grain yield, or the amount of grain produced per unit of mineral absorbed by the plant (for a detailed review on this topic, see Wang *et al.*, 2018). In agricultural systems, this can be translated into how well applied fertilizer (input) is used by the plant and harvested (output) in the form of grain, fruits, etc. (Fig. 20.1).

## 20.3 Nitrogen and Phosphorus: Two Limiting Nutrients for Maize Productivity

Like all living organisms, plants require P and N as essential nutrients for their growth and reproduction. P participates in numerous essential cellular processes and is a component of different macromolecules such as DNA, RNA and phospholipids; therefore, it cannot be substituted in plant nutrition. The vast majority of organisms acquire and metabolize P in the form of orthophosphate (Pi) and are unable to metabolize other chemical forms of P (López-Arredondo *et al.*, 2014). Similarly, N is a component of many plant structures and takes part in multiple metabolic processes (Wang *et al.*, 2018). N is commonly used as $NO_3^-$ and $NH_4^+$ by plants.

Modern cultivation of maize relies on the application of large amounts of N- and P-based fertilizers to obtain high yields. In 2017, the total world consumption of N fertilizers was approximately 109 million tonnes, of which maize production accounted for 39 million tons, representing 35% of the total (FAO, 2017); the consumption of Pi fertilizers was about 45 million

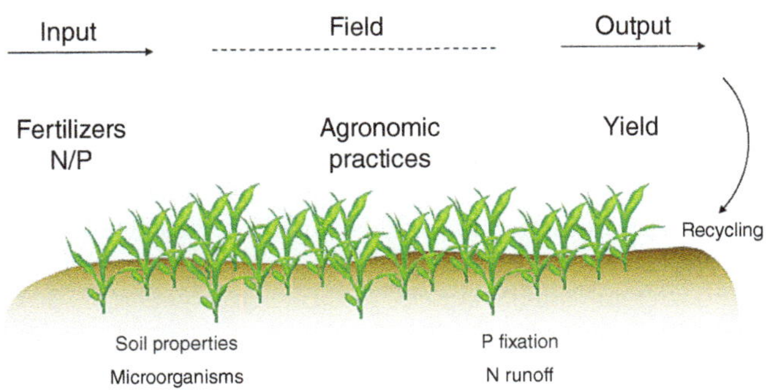


**Fig. 20.1.** General illustration of nutrient-use efficiency in the context of agricultural systems. Achievement of high yield (output) in modern varieties relies on high N and P application rates (input). Under field conditions, multiple factors may influence nutrient-use efficiency, from agronomic practices to soil properties. Therefore, it is necessary to optimize management practices to avoid excessive use of fertilizers and loss of nutrients by soil fixation and runoff.

tonnes in the same year, and its demand is expected to peak by 2033 (FAO, 2017). Being one of the most important cereals, meeting about 60% of global dietary needs, high fertilizer application has been necessary to improve maize yields in intensive agriculture. However, current agricultural practices based on high P and N application rates have led to excessive fertilization which also brings environmental pollution. For example, excessive application of N fertilizers resulted in accumulation of one-third of the N applied to maize-producing soil in the form of $NO_3^-$, which is very soluble and can run off into waterbodies (Fig. 20.1) (Lu *et al.*, 2019).

The case of Pi fertilizers is special because unlike the chemically synthesized N fertilizers, Pi fertilizers are made from phosphate rock, which is a non-renewable resource and expected to become scarce in the near future. Furthermore, most soils are poor in available P content, which has become a major limiting factor for increasing maize production. Of the available P in the soil, less than 0.01% is in the form of water-soluble ions $H_2PO_4^-$ and $HPO_4^{-2}$ that are readily available for plant uptake. This condition directly impacts PUE, which ranges from 20 to 30% in agricultural systems, and thus limits maize yield by 30 to 40% (Li *et al.*, 2019). Hence, there is an urgent need to optimize agronomic management for the rational use of both N and P fertilizers, but more importantly to reduce the requirements for fertilizer to achieve high yields via improving NUE and PUE by implementing integrated molecular breeding strategies in maize.

## 20.4 General Regulation of Nitrogen and Phosphorus Use in Maize

P and N are the main inorganic compounds used to feed crops in agricultural soils, and deficiency of N and P leads to growth retardation and yield losses. However, plants have evolved a series of molecular and adaptive responses to face N and P supply limitation (e.g. altering root system architecture (RSA) and improve biochemical properties of the nitrate transporters). Besides being one of the most important crops cultivated worldwide, maize has served to some extent as a model plant to study these responses, allowing studies on its agronomic performance, physiological responses to different stresses and regulatory networks involved by applying different molecular and genomic approaches. This provides a unique opportunity to expand our knowledge of the regulatory networks controlling N and P use in monocots. In this section, recent findings in this regard are summarized and discussed.

### 20.4.1 Uptake, translocation and assimilation of nitrogen and phosphorus

Molecular and genetics studies have revealed the primary mechanisms of P and N uptake and metabolism and how specific transporters, transcription factors, microRNAs (miRNAs) and enzymes act in concert to regulate root architecture, metabolic adaptations and hormone signalling to define crop performance.

Root cells take up N and P which are then transported and distributed to plant tissues by means of a complex arrangement of specific protein transporters. These transporters, displaying high and/or low affinity for the ion, belong to different gene families which have been identified in numerous plant species (for a detailed review, see López-Arredondo *et al.*, 2014; Wang *et al.*, 2018). $NO_3^-$ transporters belong to the nitrate transporter/peptide transporter (NRT1/PTR) family (NPF), the NRT2 family (or major facilitator superfamily, MFS) or the AMT family of transporters, whereas Pi transporters belong to four different families, PHT1, PHT2, PHT3 and PHT4 (Xu *et al.*, 2012; López-Arredondo *et al.*, 2013). Exhaustive comparative transcriptomic and microarray data analysis of maize in response to low-N and low-P conditions have shown that the expression of most of these transporter genes (i.e. *NRT1.1/NPF6.3*, *NRT2.2*, *NPF7.3*, *NPF6.6*, *AMT1*, *AMT2*, *PHT7*) is upregulated to increase nutrient absorption (Plett *et al.*, 2016; Sawers *et al.*, 2017; Hirel and Lea, 2018; Dechorgnat *et al.*, 2019; Ma *et al.*, 2020). Garnett *et al.* (2013) characterized the transcript profiles of the N-responsive genes in maize. Expression of *ZmNRT2.1* and *ZmNRT2.2* was correlated with two distinct peaks in high-affinity root N-uptake capacity and N availability (Garnett *et al.*, 2013). To date, the functional validation of several maize transporters, homologues of *Arabidopsis*, has been carried out. A good example is the functional characterization of two homologues of AtNRT1.1 (CHL1 or NPF6.3) in maize, ZmNPF6.6 and ZmNPF6.4, which uncovered that ZmNPF6.6 is a pH-dependent high-affinity nitrate-specific transporter whereas ZmNPF6.4 is a low-affinity nitrate transporter with efflux activity in maize. However, $Cl^-$ can switch the affinity of ZmNPF6.6 and ZmNPF6.4; that is, with only $Cl^-$ as substrate, ZmNPF6.6 becomes a low-affinity chloride transporter but ZmNPF6.4 functions as a high-affinity chloride transporter. The finding that ZmNPF6 proteins have different substrate specificity in maize suggests that the competition in soil of $NO_3^-$ and $Cl^-$ drives a new mechanism in maize root on N use, and that controlling their expression is a potential avenue to develop plants in which N nutrition and $Cl^-$ uptake can be manipulated (Wen *et al.*, 2017).

In maize, five *PHT1* genes with differential expression have been identified which are involved in diverse processes, including phosphate uptake from soil and transport at the symbiotic interface with mycorrhizas, phosphate translocation into the shoot and phosphate uptake during pollen tube growth (Nagy *et al.*, 2006).

### 20.4.2 Molecular regulators

Molecular elements potentially important for the global regulation of P and N utilization have been identified in the maize genome based on *Arabidopsis* homologue genes, such as NLPs, miRNAs, PHR1, PHO1, ZAT6 and SXP (Calderón-Vazquez *et al.*, 2011; Van-de-Wiel *et al.*, 2016; Ge *et al.*, 2017; Hirel and Lea, 2018; Jiang *et al.*, 2018; Kumar *et al.*, 2018), and some have been validated, mainly in *Arabidopsis*.

The miRNA399–*ZmPHO2* module has been identified as a key regulator of P uptake in maize (Du *et al.*, 2018). The miRNA *ZmmiR399b* was found to regulate P homeostasis because *ZmMIR399b*-overexpressing lines displayed excessive phosphate content in shoots and typical phosphate toxicity phenotype in leaves. In that study, authors also confirmed that *ZmmiR399b* could target a ubiquitin-conjugating E2 enzyme, ZmPHO2, as demonstrated previously in *Arabidopsis*. From the transcriptome assembly and sequence complementary analysis, the regulation of a long non-coding RNA ncRNA1 (PILNCR1) and *ZmmiR399b* was investigated. The authors demonstrated that PILNCR1 could inhibit *ZmmiR399*-guided cleavage of ZmPHO2 (Du *et al.*, 2018). Interestingly, *ZmmiR399* expression was upregulated under low-P conditions in both P-efficient (CCM454, XI14, HAI9-21) and P-inefficient (31774, FR19, 1538, JI419) lines. However, post-transcriptional repression of *ZmmiR399* was observed only in P-efficient lines, suggesting that this regulatory mechanism increases tolerance to P deficiency in maize (Du *et al.*, 2018). Furthermore, comparative transcriptomic analysis of two genotypes, Qi319 and 99038 (a low-P-tolerant mutant), under normal and P-deficiency conditions, allowed the identification of seven novel miRNA families associated with low-P tolerance, with predicted target genes involved in root development or stress response (Pei *et al.*, 2013). A cDNA clone *pZmCip1*, isolated by differential RNA display, was confirmed to be a primary response gene to cytokinin which could be involved in N-signal transduction mediated by this hormone in maize (Sakakibara *et al.*, 1998). NIN-LIKE PROTEIN (NLP) is a plant-specific transcriptional factor family that plays important roles in $NO_3^-$ signalling and assimilation. To date, nine *ZmNLPs* have been identified in the maize genome by searching the conserved PWP-RK and PB1 domains. *ZmNLP6* and *ZmNLP8a* function was validated by overexpressing them in the *Arabidopsis nlp7-4* mutant background (Cao *et al.*, 2017). The recovered $NO_3^-$ assimilation and induction of $NO_3^-$-responsive genes in the transgenic plants indicated that these two genes play important roles in $NO_3^-$ metabolism regulation. Recently, it was demonstrated that *ZmNLP5* is highly responsive to $NO_3^-$ treatment and its product directly regulates the expression of nitrite reductase 1.1 (*ZmNIR1.1*) by binding to its 5′-untranslated region (5′-UTR) (Ge *et al.*, 2020). These molecular elements represent potential targets to improve NUE and PUE in maize.

### 20.4.3 Nitrogen, phosphorus and carbon trade-off

In plants, coordinated acquisition of the different mineral nutrients is required to achieve optimal development and maximum yield. However, for many years, the general approach has been to study N and P adaptive responses and regulatory mechanisms independently. However, it was shown that during N deprivation, Pi accumulates, and genes involved in Pi-starvation response, especially SPX-encoding genes, are downregulated, suggesting a crosstalk between the signalling pathways that regulate responses to N and Pi starvation (Schlüter *et al.*, 2012). At the same time, genes associated with C metabolism (i.e. *pep3*, *sps2*, *sus2*, *fru*, *pyruvate decarboxylase*) are also strongly regulated by N stress in maize (Liu *et al.*, 2011). In a recent report, it was demonstrated that plants evolved finely tuned mechanisms to coordinate utilization of P and N to face the highly variable environmental nutrient conditions (Poza-Carrión and Paz-Ares, 2019). The novel regulatory module OsNRT1.1B–OsSPX4–OsNLP3 was reported to integrate N and P signalling in rice. OsNRT1.1B can physically interact with OsSPX4, a crucial repressor involved in Pi signalling, which is enhanced in the presence of $NO_3^-$, leading to SPX4 degradation (Hu *et al.*, 2019; Wang *et al.*, 2020). Furthermore, OsNLP3 also interacts with OsSPX4, blocking OsNLP3 capacity to move into the nucleus (Hu *et al.*, 2019). OsNRT1.1B was shown previously to have great potential to improve NUE in rice (Chao and Lin, 2015). Thus the potential of manipulating the complete module to improve PUE and NUE should be evaluated in maize and other plant species.

Rice and wheat varieties produced from the 'green revolution' are characteristically dwarfed and resistant to lodging, and DELLA proteins contribute to this dwarfing. However, as these varieties are generally produced under high fertilization rates, they are not very efficient in N and P uptake and utilization. Recently, a control system was identified that enables rice varieties to have an improved use of nutrients. The transcription factor GROWTH-REGULATING FACTOR 4 (GRF4) was shown to control the expression of N- and C-related genes which is limited by GRF4 interaction with SLR1, a DELLA protein (Li, S. *et al.*, 2018), confirming that C metabolism and N metabolism are strongly interconnected. GRF4 promotes and integrates N assimilation, C fixation and growth, whereas DELLA inhibits these processes. Therefore, rice varieties carrying this allele of *GRF4* show enhanced grain yields and greater NUE, without losing the desirable dwarf characteristic. It remains to be investigated whether a similar mechanism also modulates NUE and PUE in maize and other monocots.

## 20.5 Strategies for Molecular Breeding of Nutrient-Use Efficiency in Maize

As P and N are essential inputs to maize productivity, it is of great significance to dissect the genetic architecture of maize PUE and NUE. Identification of alleles related to PUE and NUE will provide targets for modern maize breeding, to contribute to the development of a more sustainable agriculture. However, maize breeding efforts and public programmes have been mainly focused on abiotic stresses such as drought and waterlogging for example, whereas breeding programmes for improving P utilization in plants are much less frequent.

Given the complexity of quantitative traits like PUE and NUE, along with the strong influence of environmental interactions, the dissection of the genetic architecture in maize has been challenging. Nevertheless, the generation of reference maize genomes, the B73 (Schnable *et al.*, 2009) and the Mexican landrace Palomero (Vielle-Calzada *et al.*, 2009), and recent versions with improved assembly of intergenic regions and centromeres and annotation (Jiao *et al.*, 2017) have provided valuable tools for maize basic and applied research. Similarly, sequencing and comparative analyses of numerous maize inbred lines have provided significant insights on the genetic foundation of maize yield improvement over the years, highlighting an increase of rare alleles in inbred lines as a result of modern breeding techniques (Jiao *et al.*, 2012). Moreover, transcriptomic, proteomic, metabolomic and genomic data have increased substantially over the last 10 years and now more integrative approaches to study P and N use towards NUE and PUE improvement are under application. Important to achieving higher NUE and PUE is the high genetic natural diversity of maize as the basis for crop improvement and breeding.

### 20.5.1 Genetic diversity of maize under low nitrogen and phosphorus availability

Besides being an important food and energy crop and the most cultivated worldwide, maize has proven to be one of the plant species with more diversity and plasticity, adaptable to many different environmental conditions. Maize germplasm represents an important source of rare and favourable alleles related to abiotic stress tolerance and therefore an opportunity exists to exploit these resources for the development of improved varieties.

Heterosis studies, meaning the increased vigour of hybrid plants in comparison to their parental lines, have demonstrated that morphological, physiological and even molecular traits related to NUE and PUE are inherited by the hybrid progeny (Da-Silva *et al.*, 1992; Narang and Altmann, 2001; Wang *et al.*, 2019). Wang *et al.* (2019) compared the difference in NUE between parental hybrids and inbred lines and found that hybrids exhibited a significant heterosis, up to 466% for NUE. Phenotypically, those hybrid lines showed higher leaf appearance rate with higher photosynthetic NUE, rapid reduction in the specific weight of leaf and stalk after flowering, and reduced grain N concentration and enhanced sink strength. Based on these observations, they attributed the heterosis of NUE in maize to three major mechanisms: (i) earlier establishment of pre-anthesis source for N uptake and accumulation; (ii) enhanced leaf N remobilization during grain filling; and (iii) a higher productive efficiency per unit of grain N (Wang *et al.*, 2019). In maize hybrids under low-P stress, reduced above-ground P demand and grain P concentration were observed (Liu *et al.*, 2018). The identification of these traits is extremely useful for the establishment of breeding programmes aiming to develop cultivars more tolerant to low availability of P and more efficient in their use of N.

In a recent study, analysis of 33 three maize genotypes based on 15 variable traits using standard procedures allowed the identification of low-P-tolerant genotypes including inbred and hybrid varieties (i.e. HKI 288-2, HKH 308, HQPM 7, PEHM2, HM 7, HKH 312, HKI 46, Seed Tech, HKI 163, HKI 193, HKH 407, CM 137, CM 138) and low-P-sensitive lines (i.e. EI 116, VIVEK-5, HM 4, NAI 105, PMH 2, CML 172, BIO 9637, CM 212, HKI 193-1, HKI 323, HKI 1105) (Ganie *et al.*, 2015). The most important traits contributing to the low-P tolerance included P uptake, total biomass, root dry weight, PUE, shoot P per leaf area, root P concentration and root length. A comparative metabolomic analysis of PEHM2 and HM 4 genotypes allowed the identification of metabolites that are possibly involved in this mechanism (Ganie *et al.*, 2015). Zhang, H. *et al.* (2015) investigated 826 maize accessions, representing tropical/subtropical and temperate germplasm from the International Maize and Wheat Improvement Center (CIMMYT) and China, to identify traits useful for low-P tolerance and accessions that can be used in breeding programmes. The authors were able to identify 41 low-P-tolerant accessions based on biomass and leaf traits (Zhang, H. *et al.*, 2015).

These studies are just some examples providing evidence of the huge diversity of adaptive mechanisms in maize to use P from the environment, which is a prerequisite for the establishment of selection-based improvement.

### 20.5.2 Quantitative trait locus mapping and genome-wide association studies

While screenings such as those mentioned above are very helpful to dissect the most important phenotypic traits to study, evaluate and select promising varieties, they do not provide direct information about the number and nature of genes involved in NUE or PUE. Thus, to gain insights on the genetic nature and the number of genes responsible for these phenotypes, quantitative trait locus (QTL) mapping and genome-wide association studies (GWAS) are now being implemented (Table 20.1).

So far, most of the studies about the genetic architecture of maize PUE and NUE have deployed the strategy of linkage-based mapping using segregating populations. To perform linkage mapping on maize PUE and NUE, the first step is to identify two parental lines with contrasting performance on PUE and NUE, then segregation populations, normally $F_{2:3}$ families and recombinant inbred lines (RILs), are generated and the related traits evaluated in these segregating populations. The genotyping of the population is characterized by DNA markers depending on the available technologies, such as single-nucleotide polymorphisms (SNPs), simple sequence repeats (SSRs, or microsatellites) and restriction fragment length polymorphisms (RFLPs). Finally, markers that

**Table 20.1.** QTL and GWAS mapping of genes/genomic regions for PUE- and NUE-related traits in maize.

| Trait | Major results | Reference |
|---|---|---|
| **PUE** | | |
| Vegetative growth | Six QTLs: *umc42b*, *umc46*, *umc138*, *umc19*, *umc59*, *umc117* | Reiter *et al.* (1991) |
| Root development | Three QTLs: *umc107a*, *umc110a*, *phi041* | Kaeppler *et al.* (2000) |
| Root hair length and number | Seven QTLs: *phi001/csu3*, *csu164a/phi055*, *nc003/umc36b*, *bn16.16/umc17*, *phi070/umc62*, *bn17.08a/phi121*, *umc131/nc003* | Zhu *et al.* (2005a) |
| Seminal root length | One QTL: *csu093/bn114.28* | Zhu *et al.* (2005b) |
| Seminal root number | Three QTLs: *umc34/bn112.09*, *csu3/umc133b*, *umc34/bn112.09* | Zhu *et al.* (2006) |
| Grain yield | Four QTLs: *umc1757–bnlg1318*, *umc1990–bnlg2323*, *umc1888–bnlg1805*, *umc1773–umc1012* | Cai *et al.* (2012) |
| Leaf area | Six QTLs: *mmc0191–umc1049*, *bnlg1019–umc1655*, *umc1620–umc1194*, *bnlg391–umc2313*, *umc2213–bnlg430*, *umc2003–bnlg1074* | |
| Leaf length | Three QTLs: *phi083–umc2248*, *bnlg2077–bnlg1267*, *umc1006–umc1257* | |
| Leaf width | One QTL: *umc2213–bnlg430* | |
| Chlorophyll level | Four QTLs: *umc1518–bnlg2248*, *umc1757–bnlg1318*, *phi299852–bnlg1740*, *umc2197–phi082* | |
| Flowering time | Three QTLs: *bnlg1556–bnlg1025*, *bnlg2248–phi083*, *umc1164–umc1757* | |
| ASI | One QTL: *umc1164–umc1757* | |
| APA in maize leaf | Major QTL *AP9*: a 546-kb fragment defined by *ac219* at 100,135.78 kb and *ac2096* at 100,681.88 kb on chromosome 9 | Qiu *et al.* (2013) |
| APA in root | One stable QTL: *bnlg1350–bnlg1449* on chromosome 3 | Qiu *et al.* (2014) |
| Kernel number | QTL *qkn*: a ~480-kb region by *SSR15* and *SSR19* on chromosome 10 | Zhang *et al.* (2013) |
| 18 yield-related traits for GWAS | 259 candidate genes mainly involved in four categories: transcriptional regulation, ROS scavenging, hormone regulation and remodelling of cell wall | Xu *et al.* (2018) |
| 22 traits for GWAS | 1062 candidate genes | Luo *et al.* (2019) |
| **NUE** | | |
| Agronomic traits | Five QTLs for leaf $NO_3^-$ content | Hirel *et al.* (2001) |
| Grain yield | One QTL overlapping with the *gln1* gene | Gallais and Hirel (2004) |
| RSA | A total of 184 and 147 QTLs for NUE- and RSA-related traits, respectively, assigned into 64 clusters | Li, P. *et al.* (2015) |
| Root traits | 30 QTLs associated to seven root traits | Pestsova *et al.* (2016) |
| Metabolites in maize leaf tissue for GWAS | 514 candidate genes | Zhang, N. *et al.* (2015) |

APA, acid phosphatase activity; ROS, reactive oxygen species.

are genetically linked to a QTL influencing the trait of interest will segregate more frequently with trait values and thus will inform on the genetic context.

In the 1990s, with the development of molecular DNA marker technology, Reiter *et al.* (1991) conducted the first mapping of PUE QTLs. In that study, they performed QTL mapping using an $F_3$ population from two maize inbred lines: NY821 (low-P tolerance) and H99 (low-P intolerance). A total of 77 RFLP markers were tested on 90 $F_3$ individuals. This effort allowed the identification of six loci related to biomass under low-P stress. Since then, several other efforts have been carried out by studying other plant traits such as leaf area, chlorophyll content, flowering time and grain yield under low-P conditions. Kaeppler *et al.* (2000) first tested the variation among 28 major maize inbred lines in the response to low-P stress. B73 and Mo17 were identified with the most significant effect on root development under low-P stress, which were then used to construct a RIL population for QTL mapping that enabled three QTLs controlling the root growth under low-P stress to be mapped (Kaeppler *et al.*, 2000). Using the same population and similar strategy, other authors (Zhu *et al.*, 2005a,b, 2006) identified additional QTLs related to the tolerance of low P; seven QTLs associated with root-related traits (root length and number), one QTL associated with root hair length and one main-effect QTL associated with seminal root length, and three QTLs associated with seminal root number. By characterizing leaf area, leaf chlorophyll content, flowering time, anthesis–silking interval (ASI) and grain yield at the silking stage under low-P condition, 25 QTLs were detected (Cai *et al.*, 2012). In another study, Qiu *et al.* (2013) conducted QTL mapping and fine mapping of loci associated with maize leaf under low-P stress, narrowing the stable QTL *AP9* at a 546-kb fragment in chromosome 9. With the same cross of maize lines 082 × Ye107, the authors were also able to map one stable QTL affecting root development under low-P condition (Qiu *et al.*, 2014). Kernel number and grain yield have also been used to evaluate the PUE in maize. Zhang *et al.* (2013) identified a QTL named *qKN* in a region of ~480 kb that increases the kernel number under low-P condition. So far, one of the best examples of the successful application of marker-assisted selection is the identification of *Pup1/PSTOL1 (phosphorus starvation tolerance 1)*, a serine/threonine (Ser/Thr) kinase gene identified in rice, which enhances grain yield under low-P conditions in both rice and sorghum (Gamuyao *et al.*, 2012; Hufnagel *et al.*, 2014). In maize, six *PSTOL1* homologues have been predicted, with four members co-localized with P-related QTLs (Azevedo *et al.*, 2015). However, to date, there are no reports of further studies with maize *PSTOL* genes.

Besides the studies mentioned above, other studies integrating meta-analysis and genome mining are an efficient way to integrate the mapping works mentioned above to narrow down positions of QTLs with high confidence. A good example is the work reported by Zhang *et al.* (2014). These authors collected 191 QTLs to perform meta-analysis, which produced 23 consensus QTLs (cQTLs). Gene mining yielded 215 genes, 22 of them located in the cQTL region (Zhang *et al.*, 2014). These genes were found to be homologous of 14 functionally characterized genes encoding enzymes that participate in plant low-P tolerance, including miR399, WRKY75, PHO1, purple acid phosphatases (PAPs), LPR1 and PHT1 phosphate transporters. Four cQTLs, containing QTLs (*cQTL2-1*, *cQTL5-3*, *cQTL6-2*, *cQTL10-2*) with better consistency, were considered to have more possibilities for improving PUE. These studies facilitate the possibility of identifying the true positions of QTLs involved in PUE and thereby provide direct targets for molecular breeding by transgenic and genome editing approaches.

A number of QTL linkage maps on NUE have also been conducted in maize. For example, Hirel *et al.* (2001) developed a quantitative genetic approach by associating metabolic functions and agronomic traits to DNA markers. In that study, leaves were analysed for physiological traits including $NO_3^-$ content and nitrate reductase (NR) and glutamine synthetase (GS) activities. In particular, the authors detected five QTLs that could explain 28% of the leaf $NO_3^-$ content variation. Another interesting finding of the study was that two QTLs for GS activity overlapping with the QTLs for yield components were found (Hirel *et al.*, 2001). Gallais and Hirel (2004) studied relationships between grain yield, NUE and some other specific traits, identifying one QTL that overlapped with the *gln1* gene (encoding cytosolic GS) related to low N input, ten QTLs at high-N condition and five QTLs for leaf $NO_3^-$ content explaining 28% of the phenotypic variations. In another study, 42 QTLs for grain yield and yield components were detected, 23 of them under sufficient-N conditions and 33 under limited-N conditions. Meta-analysis of published maize NUE QTLs revealed 37 cQTLs including 18 under low-N conditions (Liu *et al.*, 2011). Li, P. *et al.* (2015) investigated the relationship between RSA and NUE in maize using a RIL population. A total of 184 and 147 QTLs for NUE- and RSA-related traits were assigned into 64 clusters. Interestingly, the introgression of these QTL clusters into another background increased grain yield by about 15% for the line per se and 16% in the test cross (Li, P. *et al.*, 2015). RSA is crucial for nutrient acquisition and usage efficiencies and thus several efforts have been carried out to study root morphological characteristics in response to $NO_3^-$ availability. By studying seven European maize inbred lines (NUEC1–7), Pestsova *et al.* (2016) were able to identify two lines, NUEC2 and NUEC4, with the most contrasting differences in constitutive root traits, which were used to generate a mapping population; they were able to detect 30 QTLs associated to seven root traits.

The possibility of having (generating and studying) maize populations has remarkably contributed to increase our knowledge on the genetic architecture of NUE and PUE in maize, as they provide a highly diverse genetic and phenotypic repertoire that is the foundation for GWAS. In addition to the linkage mapping, GWAS has also been performed to identify QTLs related to PUE and NUE (Table 20.1). Xu *et al.* (2018) carried out GWAS on two natural populations under low-P stress at two experimental sites representative of different climate and soil types; the genotyping in their study was done by the 55K SNP array. They found 259 candidate genes involved in four categories: transcriptional regulation, scavenging of reactive oxygen species, hormone regulation and remodelling of cell wall. According to transcriptomic data, 98 candidate genes showed differential expression (Xu *et al.*, 2018). The study provided a genetic basis for marker-assisted selection or maize breeding for tolerance to low-P stress. In an interesting approach, Luo *et al.* (2019) combined analyses of metabolite profiles and GWAS to identify candidate genes regulating the response to P deficiency in maize seedlings. Eleven metabolites with significant differences were identified from comparative analysis of metabolite profiles of maize leaves and roots under Pi-sufficient and Pi-deficient conditions (Luo *et al.*, 2019). Moreover, a total of 1062 candidate genes were identified from the GWAS using 338 inbred lines. Within the candidates, five genes were reported to be related directly to Pi-starvation response mechanisms (Luo *et al.*, 2019).

Genome-wide association has also been performed to study N metabolism in maize by use of nested association mapping (NAM) populations. In a GWAS conducted by Zhang, N. *et al.* (2015), 12 metabolites in maize leaf tissue under field conditions were targeted. Taking advantage of a robotized metabolic phenotyping platform, that study investigated more than 100,000 assays across 12,000 samples, with two independent replicates. In total, 514 candidate genes were identified from the test of the association of 1.6 million SNPs from the maize HapMap1. Interestingly, the authors also found that many of the candidate genes were previously known to be involved in C and N metabolism pathways (Zhang, N. *et al.*, 2015).

### 20.5.3 Transgenic approach to improve nitrogen- and phosphorus-use efficiencies

With the purpose of enhancing NUE and PUE, efforts to manipulate some of the molecular elements identified by genetic engineering have increased in the last years. Unfortunately, the difficult genetic transformation of maize has constrained gene validation directly in this species and in most cases, validation is carried out in *Arabidopsis*. Nevertheless, some of these efforts have produced promising results by improving nutrient-use efficiency and open a window of opportunity to improve NUE and PUE in maize as well.

Due the important role of P and N transporters for nutrient acquisition, they represent potential targets for future breeding strategies (Table 20.2, Fig. 20.2). Two of the candidates are ZmNRT1.1 and ZmNRT2.2, that have been found to play a major role in N metabolism by maintaining biomass production in several maize genotypes under both low and high supply of P (Garnett *et al.*, 2013). Using root-specific promoters, *ZmNRT1.1* and *ZmNRT1.3* were overexpressed in transgenic maize which was tested in the field under normal and low-N conditions; a significant increase in yield was observed under normal fertilization conditions (Allen *et al.*, 2014). Indeed, the potential expression of

**Table 20.2.** Genes with potential to improve NUE and PUE in maize.

| Gene | Function | Reference |
|---|---|---|
| *Zmm28* (*Zm00001d022088*) | MADS-box; when overexpressed, induces plant growth, photosynthesis capacity and N utilization | Wu *et al.* (2019) |
| *ZmNPF6.4* | Low-affinity $NO_3^-$ transporter with efflux activity | Wen *et al.* (2017) |
| *ZmNPF6.6* | pH-dependent high-affinity $NO_3^-$-specific transport protein | Wen *et al.* (2017) |
| *ZmNRT2.1* and *ZmNRT2.2* | Correlated with root $NO_3^-$ uptake capacity | Garnett *et al.* (2013) |
| *ZmNLP5* | Directly regulates the expression of nitrite reductase 1.1 (*ZmNIR1.1*) by binding to its 5′-UTR | Ge *et al.* (2020) |
| *ZmCip1* (*Zm00001d001865*) | A primary response gene to cytokinin involved in the N-signal transduction mediated by cytokinin | Sakakibara *et al.* (1998) |
| *ZmNLP6* and *ZmNLP8a* | Similar roles to the master $NO_3^-$ regulatory gene *AtNLP7* in $NO_3^-$ signalling and metabolism | Cao *et al.* (2017) |
| *ZmmiR528* | Reduction of lignin biosynthesis under high-N conditions | Sun *et al.* (2018) |
| *ZmLAC3* and *ZmLAC5* | Targets of miR528 | Sun *et al.* (2018) |
| *ZmDof1* | Promotes N assimilation | Kurai *et al.* (2011) |
| *Ms44* | The mutation of *Ms44* increases N content in ear and kernel number | Fox *et al.* (2017) |
| *ZmPHT1* | Phosphate transporter; contributes 20% more phosphate uptake and 22–26% in PUE | Sawers *et al.* (2017) |
| *ZmPHR1* | Increases phosphate content in shoots | Wang *et al.* (2013) |
| *ZmmiR399b* | PILNCR1 could inhibit *ZmmiR399*-guided cleavage of ZmPHO2 | Du *et al.* (2018) |
| *ZmPHO2* | Ubiquitin-conjugating E2 enzyme that negatively regulates phosphate uptake and translocation | Du *et al.* (2018) |
| *TsVP* | Increases expression of auxin transporters | Pei *et al.* (2012) |
| *ZmATG8* | Key gene for autophagic recycling involved in N remobilization of maize, critical during N stress | Li, F. *et al.* (2015) |
| *ZmGS1* | Overexpression improves yield in maize by increasing kernel production and NUE under high and low N in field conditions | Martin *et al.* (2006) |

nitrate transporters from yeast (i.e. YNT1) has also been proposed as an alternative because when expressed in maize it produced an increased yield (Wang and Loussaert, 2015).

It has been reported that PP2A is a major Ser/Thr protein phosphatase involved in regulating localization of the PIN family of auxin efflux carriers in the cell (Michniewicz *et al.*, 2007) and influencing auxin distribution and RSA in plants. In a recent report, it was demonstrated that *ZmPP2AA1* is upregulated under P deficiency in maize (Wang *et al.*, 2017). Overexpression of *ZmPP2AA1* in the maize inbred line Q319 induced an increased lateral root density and length, whereas suppression of *ZmPP2AA1* by RNAi caused a significant decrease in the same traits (Wang *et al.*, 2017). *ZmPP2AA1*-overexpressing plants showed a 35–37% increase in phosphate accumulation in shoots, development of more tassel branches, larger and higher number of kernels, an increase of up to 45% in ear weight under low-P conditions, and enhanced tolerance to P starvation in both hydroponic and soil pot experiments (Wang *et al.*, 2017). In a recent report, the previously identified QTL *AP9* associated to a 546-kb region on chromosome 9 in maize was fine mapped and correlated with activity of an acid phosphatase, ZmAPRG. The authors cloned and overexpressed *ZmAPRG* in maize and rice (Yu *et al.*, 2019). For the case of *ZmAPRG*-overexpressing maize lines, increases of 15–34% in acid phosphatase activity and 62–102% in phosphate concentration were

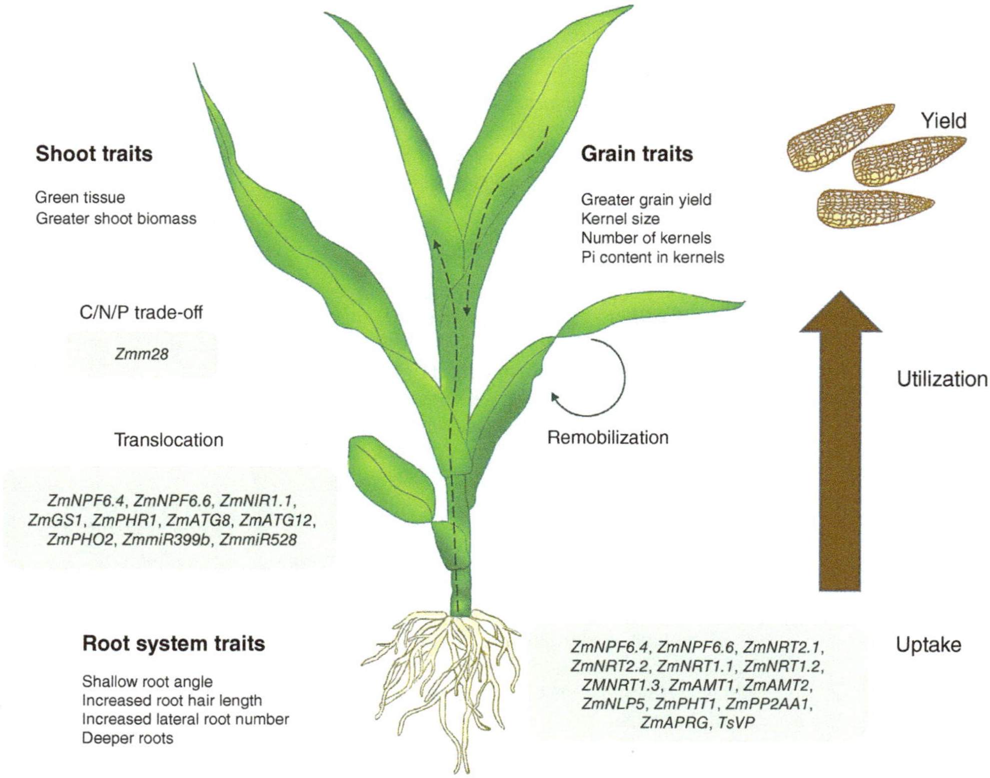


**Fig. 20.2.** Components of nutrient-use efficiency and genes with potential to improve NUE and PUE in maize. Uptake and utilization of N and P are critical components determining nutrient-use efficiency in maize, with the achievement of high grain yields being the final goal. Key traits for nutrient-use efficiency improvement are provided by root system (i.e. shallow root angle/deeper roots), shoot (i.e. biomass) and grain traits (i.e. kernel size), which are controlled by the concerted action of numerous genes, from transporters to miRNAs. Some target genes for NUE and PUE improvement are provided.

observed. A transcriptomic analysis of *ZmAPRG*-overexpressing maize lines showed the upregulation of a number of genes involved in P assimilation, such as those for phosphate transporters, phosphatases and enzymes involved in phospholipid metabolism (Yu *et al.*, 2019).

NR and GS are key enzymes in N metabolism in maize (Gallais and Hirel, 2004; Martin *et al.*, 2006; Liu *et al.*, 2011; Thomsen *et al.*, 2014; Plett *et al.*, 2016). NR catalyses the reduction of nitrate to nitrite, the first reaction in nitrate assimilation, whereas GS incorporates $NH^{+}_{4}$ into glutamine through the GS–GOGAT (glutamine synthetase–glutamate synthase) cycle, a crucial step for converting inorganic N into organic N in plants (Kleinhofs and Warner, 1990). Therefore, manipulation of their activity has represented a relevant factor for improving NUE (Plett *et al.*, 2016). Overexpression of the gene coding for the cytosolic enzyme glutamine synthetase 1 (GS1) has been reported to successfully improve maize yield by increasing kernel production and NUE under high- and low-N conditions in field experiments (Martin *et al.*, 2006). Similarly, the expression of *OsGS1* under control of the *ubiquitin* promoter in rice, and of *SbGln* under control of the *CaMV35S* promoter in sorghum, led also to increased grain yield (Brauer *et al.*, 2011; Urriola and Rathore, 2015).

The function of *ZmPHR1*, a *PHR*-like gene in maize, was validated in *Arabidopsis* (Wang *et al.*, 2013). Transgenic *Arabidopsis* lines overexpressing *ZmPHR1* showed an increase of phosphate content in shoots with upregulation of multiple genes involved in metabolism during P starvation. A plant-specific transcription factor Dof1 was shown to promote N assimilation in *Arabidopsis* (Yanagisawa *et al.*, 2004). ZmDof1 function was validated in rice, enhancing C and N assimilation under low-N conditions (Kurai *et al.*, 2011).

In a recent work, a maize MADS-box transcription factor gene, *Zmm28*, under the control of a moderate-constitutive maize promoter, resulted in maize plants with increased plant growth and higher photosynthesis capacity and N utilization (Wu *et al.*, 2019). The authors generated the transgenic lines by constitutively expressing *Zmm28* instead of expression at late developmental stage. The transgenic lines showed increased plant growth, photosynthesis capacity and N utilization. A mutation in the male sterile gene *Ms44* was found to be able to improve NUE by increasing the N content in ear and to increase kernel number (Fox *et al.*, 2017). The Ms44 maintainer line for fertility restoration was developed to increase hybrid seed production in that study.

Genetic manipulation of miRNAs could also be an alternative method of improving NUE in crops. *OsmiR528*, a monocot-specific miRNA, was reported to enhance N-deficiency tolerance in bentgrass (Yuan *et al.*, 2015). In a recent study, Sun *et al.* (2018) reported that *ZmmiR528* is induced under high-N but reduced under low-N conditions. *ZmmiR528*-overexpressing maize lines showed reduced lignin content and was prone to lodging under high-N conditions (Sun *et al.*, 2018). Manipulation of *ZmmiR528* could be an alternative to improve lodging resistance in maize, that reduces crop yield and grain quality.

Other works have reported the expression of heterologous genes in maize with interesting results. The V-$H^+$-PPase-encoding gene *TsVP* was first cloned from *Thellungiella halophilea* (Gao *et al.*, 2006) and then investigated in maize by generating *TsVP*-overexpressing plants (Pei *et al.*, 2012). Transgenic maize plants overexpressing *TsVP* developed more robust root systems because of increased expression of auxin transporters. The stronger root system enabled the plants to accumulate more P under phosphate limitation.

With the purpose of developing a highly selective P-nutrition system in crops, the expression of the bacterial *ptxD* gene was reported some years ago in tobacco and *Arabidopsis* (López-Arredondo and Herrera-Estrella, 2012) and more recently in cotton (Pandeya *et al.*, 2018). *ptxD* encodes a phosphite oxidoreductase and its expression using the *CaMV35S* promoter confers the plant the capacity to use phosphite, a compound not metabolized by plants and most organisms, as a sole P source. As weeds are not able to metabolize phosphite, the system allowed an effective control of weeds in greenhouse and field experiments (Heuer *et al.*, 2017). Although not reported formally yet, a codon-optimized version of the *ptxD* gene has been expressed in maize under the *ubiquitin* promoter (D.L. López-Arredondo, L. Herrera-Estrella, K. Wang and M.A. Leyva-Gonzalez, 2021, unpublished results). In greenhouse experiments, the transgenic plants were able to grow using phosphite as the only P source, whereas the non-transgenic control and negative lines died after some days (Fig. 20.3). When expressed in tobacco plants, higher biomass and seed yield were observed when fertilized with phosphite in comparison to non-transformed controls fertilized with similar rates of phosphate (López-Arredondo and Herrera-Estrella, 2012). Thus, it could be interesting to evaluate the performance of transgenic maize plants in greenhouse and field tests and determine if an improvement in PUE is revealed.

## 20.6 Phenotyping: An Important Component of Molecular Breeding

Efforts in maize breeding have been mainly focused on phenotyping aerial shoot traits such as ASI, barrenness, stay-green and grain yield. However, as presented above, due the importance of root morphological and physiological traits in nutrient acquisition, in order to select genotypes useful for breeding programmes, root phenotyping when the nutrimental stress is applied is necessary and thus increasing efforts have been devoted to that end. Recently, root traits related to NUE and PUE and associated QTLs important for the uptake of these mineral nutrients have been dissected (Gamuyao *et al.*, 2012).

Technological advances in plant phenotyping now offer now a variety of tools that can be implemented in field and greenhouse tests such a multispectral sensors, aerial imaging, and even automated mobile and fixed platforms, which can now speed up breeding efforts (for a detailed review, see Tracy *et al.*, 2020). However, root maize phenotyping under field conditions is still a critical point to be overcome due to the difficulties of measurements in the field and the extraction of the intact root system.

### 20.6.1 Root phenotyping

RSA of maize is complex and comprises different types of roots that emerge at different physiological stages and display different functional properties (Lynch, 1995). Therefore, variation in root architecture and function will contribute to the efficiency of nutrient uptake. In particular, for P, a non-mobile nutrient that tends to accumulate in the upper layers of the soil, a root architecture favouring topsoil foraging is important. In maize, bean and soybean, shallow root growth angles of axial roots enhance topsoil foraging and, thus, P acquisition; whereas to improve N acquisition the development of deeper root systems can be more important (Lynch and Brown, 2012; Lynch, 2013) (Fig. 20.2). Thus, in order to develop crops with enhanced NUE and PUE, breeding for root traits should prove more effective than conventional breeding practices.

Numerous studies have been conducted at laboratory or greenhouse scale using young plants and implementing high-throughput procedures to measure the root system. Some studies reported the use of hydroponic systems with growth media and agarose or gellan gum (Sanguineti *et al.*, 1998) or the paper roll system (Zhu *et al.*, 2006). More recently, the shovelomics method allowed root phenotyping of field-grown plants (Trachsel *et al.*, 2011). Moreover, innovative software such as WinRhizo, Smart-Root and ARIA (Automatic Root Image Analysis) have been developed to analyse root traits (Lobet *et al.*, 2011; Pace *et al.*, 2014). The shovelomics method is based on the use of shovels to excavate and extract a soil cylinder where the plant is growing, followed by washing out the soil from the roots and then quantification of root traits (Trachsel *et al.*, 2011). While this method offers a real view of the root system under the experimental treatment, it is destructive, labour-intensive, time-consuming and unsuitable for large screenings. Other methods report the use of columns to grow the plants and inert substrates such as sand. The method GroScreen-PaGe accompanied by a high-resolution imaging module allows non-invasive high-throughput root phenotyping and is suitable for different plant species including

**Fig. 20.3.** Maize that metabolizes phosphite. Maize genotype B104 was genetically transformed with the *ptxD* gene under control of the *ubiquitin* promoter, in collaboration with Dr Luis Herrera-Estrella at LANGEBIO-Cinvestav and Dr Kan Wang at Iowa State University. Transgenic lines were screened directly to test their capacity to use phosphite as the P source on sand–vermiculite substrate. Figure shows two putative transgenic lines: a putative positive event (right) and a putative negative event (left). (Photograph courtesy of StelaGenomics, Inc.)

maize; however, only 2- or 3-week-old plantlets can be studied (Gioia *et al.*, 2017). Nevertheless, it is important to recognize that despite the limitations each method presents, substantial advances in maize root phenotyping have been made that have allowed high-throughput evaluation of large populations under nearly natural conditions and identification of important root traits. Like sequencing technologies, technologies for phenotyping are continually expanding and evolving new features, thus exploiting their potential for maize breeding will depend on our capacity to integrate the different technologies into the generated knowledge.

## 20.7 Potential of Genome Editing Approach to Improve Maize Productivity and Performance

Recent advances in genome editing allow the alteration of endogenous genes to improve traits in crops without transferring transgenes across species boundaries (Brandt and Barrangou, 2019; Chen *et al.*, 2019). In particular, the clustered regularly interspaced short palindromic repeats (CRISPR)/CRISPR-associated protein 9 (Cas9) system has emerged as one of the most promising systems to alter in a precise way the genome of any organism. To date, it has been used in numerous crops such as tomato, maize, wheat, potato and rice to improve quantity and quality traits (Waltz, 2016; Li, T. *et al.*, 2018; Nakayasua *et al.*, 2018). The CRISPR systems have been significantly improved over the last years, now allowing the editing of multiple sites of coding and regulatory sequences at the same time, as well as delivery methods of the assembled Cas9 enzyme directly to the target cells without the need for stable integration of transgenes (Brandt and Barrangou, 2019; Chen *et al.*, 2019). In rice, the CRISPR/Cas9 system has been utilized to induce specific mutations in a number of genes with implications in multiple processes such as biotic and abiotic stress tolerance, nutritional improvement and grain yield (Ma *et al.*, 2015; Xu *et al.*, 2016; Huang *et al.*, 2018; Lu *et al.*, 2018). For maize, there are no reports on the use of the CRISPR technology to obtain high NUE or PUE in maize germplasm.

Most of the work proposed to date to use genome editing for crop improvement is based on mutations that inactivate gene function or cause simple nucleotide substitutions. In the case of *OsNRT1.1B*, an SNP in *OsNRT1.1B* between *indica* and *japonica* was reported to induce significant alteration of NUE in rice (Hu *et al.*, 2015). Therefore, this suggests that precise single-base changes can be made via CRIPSR/Cas9 reagents and thus produce elite variants of agronomic traits in crops, which may facilitate NUE improvement. It remains to be investigated whether similar variations for *NRT1.1* are present in different maize varieties and landraces and if they can be manipulated by genome editing approaches to improve NUE. Deep exploration of genetic variation of P and N transporters might be a powerful strategy for the further improvement of PUE and NUE and grain yield in maize. Another potential target for CRISPR/Cas9 application is the N transporters ZmNPF6.6 and ZmNPF6.4. In the case of NPF6.4, for example, a low-affinity $NO_3^-$ transporter that switches to a high-affinity $Cl^-$ transporter in the presence of $Cl^-$, Wen *et al.* (2017) reported that a simple mutation (Y370H) resulted in a saturable high-affinity $NO_3^-$ transport activity and $NO_3^-$ selectivity, whereas another mutation (H362Y) eliminated both $NO_3^-$ and $Cl^-$ transport. By implementing genome editing, these and more complex arrangements of mutations can be achieved and therefore exploited to design fine-tuned strategies for modelling plant nitrate and chloride metabolism.

## 20.8 Challenges and Future Perspectives

Significant efforts have been made in maize towards the dissection and isolation of genes and QTLs involved in nutrient-use efficiency. However, due to complexity of the regulatory networks of PUE and NUE and the challenges in measuring maize phenotypic variations, most of the QTL studies did not narrow down to specific genes. Although transcriptomic, genomic and metabolomic studies have extended our knowledge on use efficiencies of N and P in maize and other non-model plants, more integrative studies are required to be able to develop improved varieties. Unlike herbicide or insect resistance traits that have been conferred on several crops including maize by expressing a single gene, improving nutrient-use efficiency is much more complex. Therefore systems-level approaches are necessary to identify and manipulate all the genes and their interactors involved in determining nutrient-use efficiency, as alteration in morphology of the root system, as well as physiological and metabolic modifications, will be required to achieve significant changes. Furthermore, it is necessary to also incorporate fluxomic studies to connect genomic and metabolic data with the phenotype observed. Fluxomic studies will provide *in vivo* metabolic fluxes in mutants, transgenics and wild-type genotypes exhibiting contrasting PUE and NUE phenotypes. In this regard, to be able to screen and exploit genetic diversity and favourable alleles present in landraces and native maize from different countries, a major commitment and funding is required from the scientific community to execute large-scale trials in different locations with contrasting conditions (i.e. soil characteristics, climate) and in multiple years.

Although several different approaches for plant phenotyping have been developed to date, phenotyping, mainly root phenotyping, remains a major challenge for the improvement of maize and other crop plants. Understanding the foundation of RSA and related traits is crucial to be able to develop varieties with improved NUE and PUE. However, most of the reported systems are difficult to implement in maize or they can be implemented only at seedling stage, constraining phenotyping and evaluations at later stages. Therefore, non-destructive systems dedicated to maize phenotyping under field conditions are needed.

The ability to genetically transform maize is a crucial step for application of gene technologies for maize improvement, and alternative novel and innovate protocols for genetic transformation have been reported recently. For example, the development of transgenic maize overexpressing morphogenic regulators such as *Baby boom* (*ZmBbm*) and *Wuschel* (*ZmWus2*) genes made possible the transformation of numerous previously non-transformable inbred lines with high transformation efficiencies (Lowe *et al.*, 2016). However, these protocols still rely on laborious tissue culture-dependent steps and on the induction and maintenance of embryogenic callus, which is expensive and requires dedicated personnel with specialized skills. There is an urgent need for efforts to develop more efficient, simple and genotype-independent protocols to exploit the increasing knowledge on the genomic basis of maize genetic diversity and adaptive plasticity.

It is important also to consider the study of other mechanisms, such as autophagic recycling and epigenetic changes, for example, which are starting to be understood. Autophagy plays a central role in maize N remobilization during N stress and severely impacts maize NUE. AUTOPHAGY-RELATED8 (ATG8) and ATG12 have been identified as key elements in this process (Li, F. *et al.*, 2015) and their potential to improve NUE remains to be evaluated. Epigenetic changes induced by N and P status have been detected in rice and *Arabidopsis* (Kou *et al.*, 2011; Yong-Villalobos *et al.*, 2015). In rice, 50% of the altered methylation patterns were inherited by the progeny which was reflected in higher tolerance to N starvation (Kou *et al.*, 2011). Therefore, fine-tuned and innovative strategies are required to determine the contribution of epigenetic remodelling to NUE and PUE and to be able to exploit CRISPR/Cas9 technology for epigenome editing towards maize improvement.

## References

Allen, S.M., Guo, M., Loussaert, D.F., Rupe, M. and Wang, H. (2014) Enhanced nitrate uptake and nitrate translocation by over-expressing maize functional low-affinity nitrate transporters in transgenic maize. *US Patent No.* US20160010101A1.

Azevedo, G.C., Cheavegatti-Gianotto, A., Negri, B.F., Hufnagel, B., Silva, L.d.C.e. *et al.* (2015) Multiple interval QTL mapping and searching for *PSTOL1* homologs associated with root morphology, biomass accumulation and phosphorus content in maize seedlings under low-P. *BMC Plant Biology* 15, 172.

Brandt, K. and Barrangou, R. (2019) Applications of CRISPR technologies across the food supply chain. *Annual Review of Food Science and Technology* 10, 133–150.

Brauer, E.K., Rochon, A., Bi, Y.M., Bozzo, G.G., Rothstein, S.J. and Shelp, B.J. (2011) Reappraisal of nitrogen use efficiency in rice overexpressing *glutamine synthetase1*. *Physiologia Plantarum* 141(4), 361–372.

Cai, H., Chu, Q., Yuan, L., Liu, J., Chen, X. *et al.* (2012) Identification of quantitative trait loci for leaf area and chlorophyll content in maize (*Zea mays*) under low nitrogen and low phosphorus supply. *Molecular Breeding* 30(1), 251–266.

Calderón-Vazquez, C., Sawers, R.J.H. and Herrera-Estrella, L. (2011) Phosphate deprivation in maize: genetics and genomics. *Plant Physiology* 156, 1067–1077.

Cao, H., Qi, S., Sun, M., Li, Z., Yang, Y., Crawford, N.M. and Wang, Y. (2017) Overexpression of the maize *ZmNLP6* and *ZmNLP8* can complement the *Arabidopsis* nitrate regulatory mutant *nlp7* by restoring nitrate signaling and assimilation. *Frontiers in Plant Science* 8, 1703.

Chao, D.Y. and Lin, H.X. (2015) Nitrogen-use efficiency: transport solution in rice variations. *Nature Plants* 1, 15096.

Chen, K., Wang, Y., Zhang, R., Zhang, H. and Gao, C. (2019) CRISPR/Cas genome editing and precision plant breeding in agriculture. *Annual Review of Plant Biology* 70, 667–697.

Da-Silva, A.E., Gabelman, W.H. and Coors, J.G. (1992) Inheritance studies of low-phosphorus tolerance in maize (*Zea mays* L.), grown in a sand–alumina culture media. *Plant and Soil* 146, 189–197.

Dechorgnat, J., Francis, K.L., Dhugga, K.S., Rafalski, J.A., Tyerman, S.D. and Kaiser, B.N. (2019) Tissue and nitrogen-linked expression profiles of ammonium and nitrate transporters in maize. *BMC Plant Biology* 19, 206.

Dowswell, C. (1996) *Maize in the Third World.* CRC Press, Boca Raton, Florida.

Du, Q., Wang, K., Zou, C., Xu, C. and Li, W. (2018) The *PILNCR1*–miR399 regulatory module is important for low-phosphate tolerance in maize. *Plant Physiology* 177(4), 1743–1753.

FAO (2017) FAOSTAT. Fertilizers by Nutrient. Food and Agriculture Organization of the United Nations, Rome. Available at: http://www.fao.org/faostat/zh/#data/RFN (accessed 26 January 2020).

Fox, T., Bruin, J.D., Collet, K.H., Trimnell, M., Clapp, J., Leonard, A. *et al.* (2017) A single point mutation in *Ms44* results in dominant male sterility and improves nitrogen use efficiency in maize. *Plant Biotechnology Journal* 15(8), 942–952.

Gallais, A. and Hirel, B. (2004) An approach to the genetics of nitrogen use efficiency in maize. *Journal of Experimental Botany* 55(396), 295–306.

Gamuyao, R., Chin, K.H., Pariasca-Tanaka, J., Pesaresi, P., Catausan, S. *et al.* (2012) The protein kinase Pstol1 from traditional rice confers tolerance of phosphorus deficiency. *Nature* 488(7412), 535–539.

Ganie, A.H., Ahmad, A., Pandey, R., Aref, I.M., Yousuf, P.Y., Ahmad, S. and Iqbal, M. (2015) Metabolite profiling of low-P tolerant and low-P sensitive maize genotypes under phosphorus starvation and restoration conditions. *PLoS ONE* 10(6), e0129520.

Gao, F., Gao, Q., Duan, X., Yue, G., Yang, A. and Zhang, J. (2006) Cloning of an $H^+$-PPase gene from *Thellungiella halophila* and its heterologous expression to improve tobacco salt tolerance. *Journal of Experimental Botany* 57(12), 3259–3270.

Garnett, T., Conn, V., Plett, D., Conn, S., Zanghellini, J. *et al.* (2013) The response of the maize nitrate transport system to nitrogen demand and supply across the lifecycle. *New Phytologist* 198(1), 82–94.

Ge, M., Liu, Y., Jiang, L., Wang, Y., Lv, Y. *et al.* (2017) Genome-wide analysis of maize NLP transcription factor family revealed the roles in nitrogen response. *Plant Growth Regulation* 84(1), 95–105.

Ge, M., Wang, Y., Liu, Y., Jiang, L., He, B. *et al.* (2020) The NIN-like protein 5 (ZmNLP5) transcription factor is involved in modulating the nitrogen response in maize. *The Plant Journal* 102(2), 353–368.

Gioia, T., Galinski, A., Lenz, H., Müller, C., Lentz, J. *et al.* (2017) GrowScreen-PaGe, a non-invasive, high-throughput phenotyping system based on germination paper to quantify crop phenotypic diversity and plasticity of root traits under varying nutrient supply. *Functional Plant Biology* 44(1), 76–93.

Hallauer, A.R., Carena, M.J. and Filho, J.B.M. (2010) Introduction BT-quantitative genetics in maize breeding. In: Carena, M.J., Hallauer, A.R. and Filho, J.B.M. (eds) *Quantitative Genetics in Maize Breeding*. Springer, New York, pp. 1–31.

Heuer, S., Gaxiola, R., Schilling, R., Herrera-Estrella, L., López-Arredondo, D. *et al.* (2017) Improving phosphorus use efficiency: a complex trait with emerging opportunities. *The Plant Journal* 90(5), 868–885.

Hirel, B. and Lea, P.J. (2018) Genomics of nitrogen use efficiency in maize: from basic approaches to agronomic applications. In: Bennetzen, J., Flint-Garcia, S., Hirsch, C. and Tuberosa, R. (eds) *The Maize Genome. Compendium of Plant Genomes*. Springer, Cham, Switzerland, pp. 259–286.

Hirel, B., Bertin, P., Quilleré, I., Bourdoncle, W., Attagnant, C. *et al.* (2001) Towards a better understanding of the genetic and physiological basis for nitrogen use efficiency in maize. *Plant Physiology* 125(3), 1258–1270.

Hu, B., Wang, W., Ou, S., Tang, J., Li, H. *et al.* (2015) Variation in *NRT1.1B* contributes to nitrate-use divergence between rice subspecies. *Nature Genetics* 47, 834–838.

Hu, B., Jiang, Z., Wang, W., Qiu, Y., Zhang, Z. *et al.* (2019) Nitrate–NRT1.1B–SPX4 cascade integrates nitrogen and phosphorus signalling networks in plants. *Nature Plants* 5, 401–413.

Huang, J., Li, J., Zhou, J., Wang, L., Yang, S. *et al.* (2018) Identifying a large number of high-yield genes in rice by pedigree analysis, whole-genome sequencing, and CRISPR-Cas9 gene knockout. *Proceedings of the National Academy of Sciences USA* 115, E7559–E7567.

Hufnagel, B., Sousa, S.M.d., Assis, L., Guimaraes, C.T., Leiser, W. *et al.* (2014) Duplicate and conquer: multiple homologs of *PHOSPHORUS-STARVATION TOLERANCE1* enhance phosphorus acquisition and sorghum performance on low-phosphorus soils. *Plant Physiology* 166(2), 659–677.

Jiang, L., Ball, G., Hodgman, C., Coules, A., Zhao, H. and Lu, C. (2018) Analysis of gene regulatory networks of maize in response to nitrogen. *Genes* 9(3), 151.

Jiao, Y., Zhao, H., Ren, L., Song, W., Zeng, B. *et al.* (2012) Genome-wide genetic changes during modern breeding of maize. *Nature Genetics* 44(7), 812–815.

Jiao, Y., Peluso, P., Shi, J., Liang, T., Stitzer, M.C. *et al.* (2017) Improved maize reference genome with single-molecule technologies. *Nature* 546, 524–527.

Kaeppler, S.M., Parke, J.L., Mueller, S.M., Senior, L., Stuber, C. and Tracy, W.F. (2000) Variation among maize inbred lines and detection of quantitative trait loci for growth at low phosphorus and responsiveness to arbuscular mycorrhizal fungi. *Crop Science* 40(2), 358–364.

Kleinhofs, A. and Warner, R.L. (1990) Advances in nitrate assimilation. In: Miller, B.J. and Lea, P.J. (eds) *Intermediary Nitrogen Metabolism*. Academic Press, New York, pp. 89–120.

Kou, H.P., Li, Y., Song, X.X., Ou, X.F., Xing, S.C. *et al.* (2011) Heritable alteration in DNA methylation induced by nitrogen-deficiency stress accompanies enhanced tolerance by progenies to the stress in rice (*Oryza sativa* L.). *Journal of Plant Physiology* 168, 1685–1693.

Kumar, A., Pandeya, A., Malik, G., Sharma, M., Kumari, H. *et al.* (2018) A web resource for nutrient use efficiency-related genes, quantitative trait loci and microRNAs in important cereals and model plants. *F1000Research* 7, ISCB Comm J-673.

Kurai, T., Wakayama, M., Abiko, T., Yanagisawa, S., Aoki, N. and Ohsugi, R. (2011) Introduction of the *ZmDof1* gene into rice enhances carbon and nitrogen assimilation under low-nitrogen conditions. *Plant Biotechnology Journal* 9(8), 826–837.

Li, D., Wang, M., Kuang, K. and Liu, W. (2019) Genetic study and molecular breeding for high phosphorus use efficiency in maize. *Frontiers of Agricultural Science and Engineering* 6(4), 366–379.

Li, F., Chung, T., Pennington, J.G., Federico, M.L., Kaeppler, H.F. *et al.* (2015a) Autophagic recycling plays a central role in maize nitrogen remobilization. *The Plant Cell* 27(5), 1389–1408.

Li, P., Chen, F., Cai, H., Liu, J., Pan, Q. *et al.* (2015b) A genetic relationship between nitrogen use efficiency and seedling root traits in maize as revealed by QTL analysis. *Journal of Experimental Botany* 66(11), 3175–3188.

Li, S., Tian, Y., Wu, K., Ye, Y., Yu, J. *et al.* (2018a) Modulating plant growth–metabolism coordination for sustainable agriculture. *Nature* 560(7720), 595–600.

Li, T., Yang, X., Yu, Y., Si, X., Zhai, X. *et al.* (2018b) Domestication of wild tomato is accelerated by genome editing. *Nature Biotechnology* 36, 1160–1163.

Liu, R., Zhang, H., Zhao, P., Zhang, Z., Liang, W., Tian, Z. and Zheng, Y. (2011) Mining of candidate maize genes for nitrogen use efficiency by integrating gene expression and QTL data. *Plant Molecular Biology Reporter* 30(2), 297–308.

Liu, Z., Liu, X., Craft, E.J., Yuan, L., Cheng, L., Mi, G. and Chen, F. (2018) Physiological and genetic analysis for maize root characters and yield in response to low phosphorus stress. *Breeding Science* 68, 268–277.

Lobet, G., Pages, L. and Draye, X. (2011) A novel image-analysis toolbox enabling quantitative analysis of root system architecture. *Plant Physiology* 157, 29–39.

López-Arredondo, D. and Herrera-Estrella, L. (2012) Engineering phosphorus metabolism in plants to produce a dual fertilization and weed control system. *Nature Biotechnology* 30(9), 889–893.

López-Arredondo, D., Leyva-González, M.A., Alatorre-Cobos, F. and Herrera-Estrella, L. (2013) Biotechnology of nutrient uptake and assimilation in plants. *The International Journal of Developmental Biology* 57, 595–610.

López-Arredondo, D., Leyva-González, M.A., González-Morales, S.I., López-Bucio, J. and Herrera-Estrella, L. (2014) Phosphate nutrition: improving low-phosphate tolerance in crops. *Annual Review of Plant Biology* 65, 95–123.

Lowe, K., Wu, E., Wang, N., Hoerster, G., Hastings, C. *et al.* (2016) Morphogenic regulators *Baby boom* and *Wuschel* improve monocot transformation. *The Plant Cell* 28, 1998–2015.

Lu, H.P., Luo, T., Fu, H.W., Wang, L., Tan, Y.Y. *et al.* (2018) Resistance of rice to insect pests mediated by suppression of serotonin biosynthesis. *Nature Plants* 4, 338–344.

Lu, J., Bai, Z., Velthofc, G.L., Wu, Z., Chadwick, D. and Ma, L. (2019) Accumulation and leaching of nitrate in soils in wheat–maize production in China. *Agricultural Water Management* 212, 407–415.

Luo, B., Ma, P., Nie, Z., Zhang, X., He, X. *et al.* (2019) Metabolite profiling and genome-wide association studies reveal response mechanisms of phosphorus deficiency in maize seedling. *The Plant Journal* 97(5), 947–969.

Lynch, J.P. (1995) Root architecture and plant productivity. *Plant Physiology* 109, 7–13.

Lynch, J.P. (2013) Steep, cheap and deep: an ideotype to optimize water and N acquisition by maize root systems. *Annals of Botany* 112, 347–357.

Lynch, J.P. and Brown, K.M. (2012) New roots for agriculture: exploiting the root phenome. *Philosophical Transactions of the Royal Society B: Biological Sciences* 367(1595), 1598–1604.

Ma, N., Dong, L., Lü, W., Lü, J., Meng, Q. and Liu, P. (2020) Transcriptome analysis of maize seedling roots in response to nitrogen-, phosphorus-, and potassium deficiency. *Plant and Soil* 447, 637–658.

Ma, X., Zhang, Q., Zhu, Q., Liu, W., Chen, Y. *et al.* (2015) A robust CRISPR/Cas9 system for convenient, high-efficiency multiplex genome editing in monocot and dicot plants. *Molecular Plant* 8, 1274–1284.

Martin, A., Lee, J., Kichey, T., Gerentes, D., Zivy, M. *et al.* (2006) Two cytosolic glutamine synthetase isoforms of maize are specifically involved in the control of grain production. *The Plant Cell* 18(11), 3252–3274.

Michniewicz, M., Zago, M.K., Abas, L., Weijers, D., Schweighofer, A. *et al.* (2007) Antagonistic regulation of PIN phosphorylation by PP2A and PINOID directs auxin flux. *Cell* 130(6), 1044–1056.

Nagy, R., Vasconcelos, M.J., Zhao, S., McElver, J., Bruce, W. *et al.* (2006) Differential regulation of five Pht1 phosphate transporters from maize (*Zea mays* L.). *Plant Biology* 8(2), 186–197.

Nakayasua, M., Akiyamaa, R., Lee, H.-J., Osakabe, K., Osakabe, Y. *et al.* (2018) Generation of α-solanine-free hairy roots of potato by CRISPR/Cas9 mediated genome editing of the *St16DOX* gene. *Plant Physiology and Biochemistry* 131, 70–77.

Narang, R.A. and Altmann, T. (2001) Phosphate acquisition heterosis in *Arabidopsis thaliana*: a morphological and physiological analysis. *Plant and Soil* 234, 91–97.

Pace, J., Lee, N., Naik, H.S., Ganapathysubramanian, B. and Lubberstedt, T. (2014) Analysis of maize (*Zea mays* L.) seedling roots with the high-throughput image analysis tool ARIA (Automatic Root Image Analysis). *PLoS ONE* 9(9), e108255.

Pandeya, D., López-Arredondo, D.L., Janga, M.R., Campbell, L.-A.M., Estrella-Hernández, P. *et al.* (2018) Selective fertilization with phosphite allows unhindered growth of cotton plants expressing the *ptxD* gene while suppressing weeds. *Proceedings of the National Academy of Sciences USA* 29, E6946–E6955.

Pei, L., Wang, J., Li, K., Li, Y., Li, B., Gao, F. and Yang, A. (2012) Overexpression of *Thellungiella halophila* $H^+$-pyrophosphatase gene improves low phosphate tolerance in maize. *PLoS ONE* 7(8), e43501.

Pei, L., Jin, Z., Li, K., Yin, H., Wang, J. and Yang, A. (2013) Identification and comparative analysis of low phosphate tolerance-associated microRNAs in two maize genotypes. *Plant Physiology and Biochemistry* 70, 221–234.

Pestsova, E., Lichtblau, D., Wever, C., Presterl, T., Bolduan, T., Ouzunova, M. and Westhoff, P. (2016) QTL mapping of seedling root traits associated with nitrogen and water use efficiency in maize. *Euphytica* 209(3), 585–602.

Plett, D., Baumann, U., Schreiber, A.W., Holtham, L., Kalashyan, E. *et al.* (2016) Maize maintains growth in response to decreased nitrate supply through a highly dynamic and developmental stage-specific transcriptional response. *Plant Biotechnology Journal* 14(1), 342–353.

Poza-Carrión, C. and Paz-Ares, J. (2019) When nitrate and phosphate sensors meet. *Nature Plants* 5(4), 339–340.

Qiu, H., Mei, X., Liu, C., Wang, J., Wang, G. *et al.* (2013) Fine mapping of quantitative trait loci for acid phosphatase activity in maize leaf under low phosphorus stress. *Molecular Breeding* 32(3), 629–639.

Qiu, H., Liu, C., Yu, T., Mei, X., Wang, G., Wang, J. and Cai, Y. (2014) Identification of QTL for acid phosphatase activity in root and rhizosphere soil of maize under low phosphorus stress. *Euphytica* 197(1), 133–143.

Ranum, P., Pena-Rosas, J. and Garcia-Casal, M.N. (2014) Global maize production, utilization, and consumption. *Annals of the New York Academy of Sciences* 1312, 105–112.

Reiter, R.S., Coors, J.G., Sussman, M.R. and Gabelman, W.H. (1991) Genetic analysis of tolerance to low-phosphorus stress in maize using restriction fragment length polymorphisms. *Theoretical and Applied Genetics* 82(5), 561–568.

Roser, M. and Ritchie, H. (2020) Fertilizers. Our World in Data website. Available at: https://ourworldindata.org/fertilizers (accessed 26 January 2020).

Sakakibara, H., Suzuki, M., Takei, K., Deji, A., Taniguchi, M. and Sugiyama, T. (1998) A response-regulator homologue possibly involved in nitrogen signal transduction mediated by cytokinin in maize. *The Plant Journal* 14(3), 337–344.

Sanguineti, M.C., Giuliani, M.M., Govi, G., Tuberosa, R. and Landi, P. (1998) Root and shoot traits of maize inbred lines grown in the field and in hydroponic culture and their relationships with root lodging. *Maydica* 43, 211–216.

Sawers, R.J.H., Svane, S.F., Quan, C., Grønlund, M. and Wozniak, B. (2017) Phosphorus acquisition efficiency in arbuscular mycorrhizal maize is correlated with the abundance of root-external hyphae and the accumulation of transcripts encoding PHT1 phosphate transporters. *New Phytologist* 214(2), 632–643.

Schlüter, U., Mascher, M., Colmsee, C., Scholz, U., Bräutigam, A., Fahnenstich, H. and Sonnewald, U. (2012) Maize source leaf adaptation to nitrogen deficiency affects not only nitrogen and carbon metabolism but also control of phosphate homeostasis. *Plant Physiology* 160(3), 1384–1406.

Schnable, P.S., Ware, D., Fulton, R.S., Stein, J.C., Wei, F. *et al.* (2009) The B73 maize genome: complexity, diversity, and dynamics. *Science* 326(5956), 1112–1115.

Sun, Q., Liu, X., Yang, J., Liu, W., Du, Q. *et al.* (2018) MicroRNA528 affects lodging resistance of maize by regulating lignin biosynthesis under nitrogen-luxury conditions. *Molecular Plant* 11, 806–814.

Thomsen, H.C., Eriksson, D., Møller, I.S. and Schjoerring, J.K. (2014) Cytosolic glutamine synthetase: a target for improvement of crop nitrogen use efficiency? *Trends in Plant Science* 19(10), 656–663.

Trachsel, S., Kaeppler, S.M., Brown, K.M. and Lynch, J.P. (2011) Shovelomics: high throughput phenotyping of maize (*Zea mays* L.) root architecture in the field. *Plant and Soil* 341, 75–87.

Tracy, S.R., Nagel, K.A., Postma, J.A., Fassbender, H., Wasson, A. and Watt, M. (2020) Crop improvement from phenotyping roots: highlights reveal expanding opportunities. *Trends in Plant Science* 25(1), 105–118.

Urriola, J. and Rathore, K. (2015) Overexpression of a glutamine synthetase gene affects growth and development in sorghum. *Transgenic Research* 24(3), 397–407.

Van-de-Wiel, C.C.M., Van-der-Linden, C.G. and Scholten, O.E. (2016) Improving phosphorus use efficiency in agriculture: opportunities for breeding. *Euphytica* 207, 1–22.

Vielle-Calzada, J.-P., Vega, O.M.d.l., Hernández-Guzmán, G., Ibarra-Laclette, E., Alvarez-Mejía, C. *et al.* (2009) The Palomero genome suggests metal effects on domestication. *Science* 326(5956), 1078.

Waltz, E. (2016) CRISPR-edited crops free to enter market, skip regulation. *Nature Biotechnology* 34, 582.

Wang, H. and Loussaert, D.L. (2015) Functional expression of yeast nitrate transporter (YNT1) and a nitrate reductase in maize. *US Patent No.* US8975474B2.

Wang, J., Pei, L., Jin, Z., Zhang, K. and Zhang, J. (2017) Overexpression of the protein phosphatase 2A regulatory subunit A gene *ZmPP2AA1* improves low phosphate tolerance by remodeling the root system architecture of maize. *PLoS ONE* 12(4), e0176538.

Wang, W., Hu, B., Li, A. and Chu, C. (2020) NRT1.1s in plants: functions beyond nitrate transport. *Journal of Experimental Botany* 71(15), 4373–4379.

Wang, X., Bai, J., Liu, H., Sun, Y., Shi, X. and Ren, Z. (2013) Overexpression of a maize transcription factor *ZmPHR1* improves shoot inorganic phosphate content and growth of *Arabidopsis* under low-phosphate conditions. *Plant Molecular Biology Reporter* 31(3), 665–677.

Wang, Y.-Y., Cheng, Y.-H., Chen, K.-E. and Tsay, Y.-F. (2018) Nitrate transport, signaling, and use efficiency. *Annual Review of Plant Biology* 69, 85–122.

Wang, Z., Ma, B.-L., Yu, X., Gao, J., Sun, J., Su, Z. and Yu, S. (2019) Physiological basis of heterosis for nitrogen use efficiency of maize. *Scientific Reports* 9(1), 18708.

Wen, Z., Tyerman, S.D., Dechorgnat, J., Ovchinnikova, E., Dhugga, K.S. and Kaiser, B.N. (2017) Maize NPF6 proteins are homologs of *Arabidopsis* CHL1 that are selective for both nitrate and chloride. *The Plant Cell* 29(10), 2581–2596.

Wu, J., Lawit, S.J., Weers, B., Sun, J., Mongar, N., Hemert, J.V. *et al.* (2019) Overexpression of *zmm28* increases maize grain yield in the field. *Proceedings of the National Academy of Sciences USA* 116(47), 23850–23858.

Xu, C., Zhang, H., Sun, J., Guo, Z., Zou, C. *et al.* (2018) Genome-wide association study dissects yield components associated with low-phosphorus stress tolerance in maize. *Theoretical and Applied Genetics* 131(8), 1699–1714.

Xu, G., Fan, X. and Miller, A.J. (2012) Plant nitrogen assimilation and use efficiency. *Annual Review of Plant Biology* 63(1), 153–182.

Xu, R., Yang, Y., Qin, R., Li, H., Qiu, C. *et al.* (2016) Rapid improvement of grain weight via highly efficient CRISPR/Cas9-mediated multiplex genome editing in rice. *Journal of Genetics and Genomics* 43, 529–532.

Yanagisawa, S., Akiyama, A., Kisaka, H., Uchimiya, H. and Miwa, T. (2004) Metabolic engineering with Dof1 transcription factor in plants: improved nitrogen assimilation and growth under low-nitrogen conditions. *Proceedings of the National Academy of Sciences USA* 101(20), 7833–7838.

Yong-Villalobos, L., González-Morales, S.I., Wrobel, K., Gutiérrez-Alanis, D., Cervantes-Peréz, S.A. *et al.* (2015) Methylome analysis reveals an important role for epigenetic changes in the regulation of the *Arabidopsis* response to phosphate starvation. *Proceedings of the National Academy of Sciences USA* 112(52), E7293–E7302.

Yu, T., Liu, C., Lu, X., Bai, Y., Zhou, L. and Cai, Y. (2019) *ZmAPRG*, an uncharacterized gene, enhances acid phosphatase activity and Pi concentration in maize leaf during phosphate starvation. *Theoretical and Applied Genetics* 132(4), 1035–1048.

Yuan, S., Li, Z., Li, D., Yuan, N., Hu, Q. and Luo, H. (2015) Constitutive expression of rice microRNA528 alters plant development and enhances tolerance to salinity stress and nitrogen starvation in creeping bentgrass. *Plant Physiology* 169(1), 576–593.

Zhang, G., Wang, X., Wang, B., Tian, Y., Li, M. *et al.* (2013) Fine mapping a major QTL for kernel number per row under different phosphorus regimes in maize (*Zea mays* L.). *Theoretical and Applied Genetics* 126(6), 1545–1553.

Zhang, H., Uddin, M.S., Zou, C., Xie, C., Xu, Y. and Li, W.X. (2014) Meta-analysis and candidate gene mining of low-phosphorus tolerance in maize. *Journal of Integrative Plant Biology* 56, 262–270.

Zhang, H., Xu, R., Xie, C., Huang, C., Liao, H. *et al.* (2015a) Large-scale evaluation of maize germplasm for low-phosphorus tolerance. *PLOS ONE.* DOI: 10.1371/journal.pone.0124212.

Zhang, N., Gibon, Y., Wallace, J.G., Lepak, N., Li, P. *et al.* (2015b) Genome-wide association of carbon and nitrogen metabolism in the maize nested association mapping population. *Plant Physiology* 168(2), 575–583.

Zhang, X., Davidson, E.A., Mauzerall, D.L., Searchinger, T.D., Dumas, P. and Shen, Y. (2015c) Managing nitrogen for sustainable development. *Nature* 528(7580), 51–59.

Zhu, J., Kaeppler, S.M. and Lynch, J.P. (2005a) Mapping of QTL controlling root hair length in maize (*Zea mays* L.) under phosphorus deficiency. *Plant and Soil* 270, 299–310.

Zhu, J., Kaeppler, S.M. and Lynch, J.P. (2005b) Mapping of QTLs for lateral root branching and length in maize (*Zea mays* L.) under differential phosphorus supply. *Theoretical and Applied Genetics* 111(4), 688–695.

Zhu, J., Mickelson, S.M., Kaeppler, S.M. and Lynch, J.P. (2006) Detection of quantitative trait loci for seminal root traits in maize (*Zea mays* L.) seedlings grown under differential phosphorus levels. *Theoretical and Applied Genetics* 113(1), 1–10.

# 21 Molecular Breeding for Increasing Nutrition Quality in Maize: Recent Progress

**Firoz Hossain[1]*, Sujay Rakshit[2], Bhupender Kumar[2], John J. Amalraj[3], Vignesh Muthusamy[1], Bhukya Prakash[4], Rajkumar U. Zunjare[1], Chikkappa Karjagi[2], Rajesh Khulbe[5], Arushi Arora[6], Lydia Pramitha[3], Dharampal Choudhary[2], Savaram V.R. Rao[4], Mantena V.L.N. Raju[4] and Mehar C. Kamboj[7]**

*[1]ICAR–Indian Agricultural Research Institute, New Delhi, India; [2]ICAR–Indian Institute of Maize Research, Ludhiana, India; [3]Tamil Nadu Agricultural University, Coimbatore, India; [4]ICAR–Directorate of Poultry Research, Hyderabad, India; [5]ICAR–Vivekananda Parvatiya Krishi Anusandhan Sansthan, Almora, India; [6]Punjab Agricultural University, Ludhiana, India; [7]CCS–Haryana Agricultural University, Regional Research Station, Uchani, India*

## 21.1 Introduction

Malnutrition caused by consumption of an unbalanced diet has emerged as one of the major health concerns, particularly in the developing and under-developed world (Bouis *et al.*, 2019). Globally, around two billion people suffer from malnutrition, while 820 million people are undernourished (IFPRI, 2018). Nearly 45% of deaths among children under the age of 5 years are linked to malnutrition. Some 150.8 million (22.2%) children (<5 years) are stunted, while 50.5 million (7.5%) do not weigh enough according to height (wasting). In India, 38.4% of the children aged <5 years are stunted, 21.0% are wasted and 35.7% are underweight (IIPS, 2017). Anaemia is a serious health issue in India as well, with 58.4% of children aged 6–59 months, 53.0% of adult women and 22.7% of adult men being affected. A survey revealed that 80% of individuals in most of the states in India are receiving <50% of the suggested dietary intake of essential minerals (ICMR, 2010).

Childhood stunting, anaemia in women of reproductive age and overweight among women are the three most prevalent types of malnutrition observed worldwide. Globally, 88% of countries experience a high level of at least two types of malnutrition, while 29% experience three types of malnutrition (IFPRI, 2018). Malnutrition accounts for a loss in gross domestic product of 11% in Asia and Africa, and it could cost society up to US$3.5 trillion per year. Considering the paramount importance of alleviating malnutrition, world leaders at the United Nations framed the Sustainable Development Goals (SDGs) for meeting the current needs without affecting future generations. Alleviating malnutrition is the most cost-effective step as every $1 invested in a proven nutrition programme offers benefits worth $16 (IFPRI, 2016). Thus, a balanced and nutritious diet for people assumes great significance

*Corresponding author: fh_gpb@yahoo.com



DOI: 10.1079/9781789245431.0021

to mitigate malnutrition (Neeraja *et al.*, 2017; Yadava *et al.*, 2018; Gupta *et al.*, 2019). Various approaches, namely food fortification, medical supplementation and dietary diversification, are generally used for alleviating micronutrient malnutrition (Pfeiffer and McClafferty, 2007). However, the success of these approaches is sometimes limited due of lack of purchasing power, poor infrastructure facilities, crop seasonality, expense and lower bioavailability (Lieshout and Pee, 2005; Hossain *et al.*, 2019a). Biofortification, a strategy of increasing micronutrient density in the edible parts of plants through breeding approaches, is a viable, sustainable and cost-effective means for enhancing required levels of micronutrients in food (Bouis, 2018).

## 21.2 Maize and Its Significance

Maize assumes worldwide significance as a source of food, feed and diverse industrial products (Prasanna *et al.*, 2020). Maize grains are generally consumed as flat bread, porridge, and in boiled and roasted forms (Hossain *et al.*, 2019a). Besides, maize ears are also used for an array of specialty purposes, of which sweet corn and waxy corn assume great significance (Hossain *et al.*, 2015; Devi *et al.*, 2017; Chhabra *et al.*, 2019). Together with rice and wheat, maize provides at least 30% of the food calories to more than 4.5 billion people in 94 developing countries (Shiferaw *et al.*, 2011). It is an important staple cereal food crop for billions of people in South America, Africa and Asia, with an estimated world production of 1148 million tonnes from an area of 197 million hectares distributed in as many as 169 countries (FAO, 2019). In India, maize is an important cereal too, and is grown on an area of 9.2 million hectares with a production of 27.8 million tonnes in 2018/19 (https://www.indiastat.com/data/agriculture/maize-17199/data-year/2019, accessed 6 March 2021). In India 20% of the maize produced is used for human food, while nearly 60% is used for poultry and animal feed (Rakshit and Chikkappa, 2018). The demand for maize will be doubled by 2050 in the developing world (Rosegrant *et al.*, 2009). By 2025, India too would require doubling of the production (50 million tonnes) of maize grain to meet the domestic demand (Yadav *et al.*, 2015). Considering the growing importance of maize as both food and feed, biofortification of maize for enhancement of protein quality, provitamin A, vitamin E and reduction of antinutritional factors like phytic acid assumes immense significance. Further, reducing the glycaemic index (GI) of maize by increasing the proportion of resistant starch in grain will add to its utility. This chapter presents the status of molecular breeding, especially marker-assisted backcross breeding (MABB), followed in each of the nutritional traits. In MABB, the gene(s) of interest is/are selected in the segregating populations through 'foreground' selection', while >90% of the recurrent parent genome (RPG) is recovered through 'background selection' assisted through genome-wide markers (Zunjare *et al.*, 2018a).

## 21.3 Improvement of Protein Quality

Besides serving as a source of carbohydrates and energy, maize is also an important source of protein in the diet, especially in the developing and under-developed world where access to pulses and non-vegetarian protein sources is quite limited. Lack of adequate protein leads to protein–energy undernutrition (PEU), formerly known as protein–energy malnutrition (PEM) (Bouis and Welch, 2010; Morley, 2020). Children under 5 years of age, pregnant women and elderly people are the groups most vulnerable to PEU (Mpofu *et al.*, 2014). Hoseini *et al.* (2015) reported that about 146 million children aged <5 years lack adequate protein in their diet. Presence of all the essential amino acids in the human diet helps combat PEU (Hossain *et al.*, 2019b). The endosperm of traditional maize has all the essential amino acids, except lysine and tryptophan (Mertz *et al.*, 1964). Lysine and tryptophan are necessary for tissue growth through protein synthesis. In addition, tryptophan is essential to avoid niacin deficiency which may lead to pellagra in humans (Gutierrez-Rojas *et al.*, 2008). Further, the deficiency of lysine in humans causes fatigue, poor concentration, irritability, nausea, red eyes, hair loss, anorexia and inhibited growth. Therefore, an adequate quantity of these two amino acids in human diets is essential to avoid health disorders. The recommended daily intake of tryptophan is 4 mg/kg body weight for adults, while it is 8.5 mg/kg body weight for infants. The nutritional requirement for lysine in humans varies from 60 mg/kg body weight in infancy to 30 mg/kg body weight in adulthood (Gupta *et al.*, 2015).

### 21.3.1 Breeding for high lysine and tryptophan

Several maize mutants such as *o1, o2, o5, o6, o7, o9–11, o13, o15, o16, o17, fl1, fl2, fl3, fl4, Mucronate* and *Defective endosperm* that enhance lysine and tryptophan are available (Gibbon and Larkins, 2005). All these mutants have been grouped in three major classes: recessive, semi-dominant and dominant mutations (Wang *et al.*, 2011). The *opaque* mutants, which are recessive, include *o1, o2, o5, o6, o7, o9–11, o13, o15, o16* and *o17*; the *floury* mutations such as *fl1, fl2, fl3* and *fl4* are semi-dominant, whereas *Mucronate* and *Defective endosperm* are dominant mutations. The majority of these alleles affect the storage protein synthesis and endosperm texture. Unfortunately, many of the mutants show adverse pleiotropic effects and thus are not suitable for breeding programmes (Vasal, 2000).

Normal field maize protein possesses a low concentration of lysine (1.5–2.0% w/w) and tryptophan (0.3–0.4% w/w), where lysine is most limiting followed by tryptophan for human metabolism (Yadava *et al.*, 2018). However, the *opaque2* (*o2*) mutant possesses nearly twice the normal amount of lysine (3.5–4.0% w/w) and tryptophan (0.7–0.8% w/w), which makes the protein value equivalent to 90% of milk protein (Bressani, 1990). The nutritional benefit of *o2* was discovered in the 1960s (Mertz *et al.*, 1964). The recessive *o2* gene drastically reduces the level of 22-kDa α-zeins and in turn enhances the lysine-rich non-zein proteins (Habben *et al.*, 1995). Recessive *o2* also significantly reduces transcription of lysine keto-reductase (LKR), the enzyme that degrades lysine in maize endosperm, thereby enhancing the concentration of lysine. Further, *o2* is involved in regulation of various metabolic pathways and causes enhanced synthesis of various lysine-rich proteins and enzymes (Hossain *et al.*, 2019b). The protein quality of *o2* maize is 43% better than that of normal maize and 95% of the value of casein (Mertz, 1992). However, pleiotropic effects of the *o2* mutant, such as soft texture, lower seed density, brittleness, insect susceptibility, as well as breakage of kernels during mechanical threshing, decreased its acceptance among farmers and consumers. These negative features of the mutant were ameliorated by the introduction of *o2*-modifier genes, located throughout the genome, having complex phenotypic effects (Vasal *et al.*, 1980). The modified-*o2* genotypes rich in lysine and tryptophan, resembling normal maize both in kernel phenotype and agronomic performance, were developed by breeders at the International Maize and Wheat Improvement Center (CIMMYT). The hard endosperm-based *o2* genotypes with modifier genes are popularly called 'quality protein maize' (QPM) (Vasal *et al.*, 1980; Bjarnason and Vasal, 1992). CIMMYT played a vital role in disseminating QPM germplasm throughout the world that resulted in the release of QPM varieties/synthetics and hybrids in many countries.

India released its first series of biofortified maize composites, namely Shakti, Rattan and Protina, in 1971 (Gupta *et al.*, 2015). These were rich in protein quality but had soft endosperm, thus could not become popular. In 1997, accumulation of modifiers from CIMMYT germplasm led to the release of the first hard endosperm-based *o2* composite, Shakti 1. The first QPM hybrid in India, Shaktimaan 1 (a white kernel-based three-way cross), was released in 2001, followed by Shaktimaan 2 (a white kernel-based single-cross hybrid) in 2004. The first yellow kernel-based single-cross hybrids, Shaktimaan 3 and Shaktimaan 4, were developed together during 2006, followed by Shaktimaan 5 in 2013. These QPM hybrids are specifically adapted to Bihar. Later, a number of single-cross QPM hybrids, namely HQPM 1, HQPM 4, HQPM 5, HQPM 7 and Pratap QPM Hybrid 1, were released with wider adaptability to different agroecologies of the country. Several experimental QPM hybrids are in different stages of national testing, and IMHQPM 1530 has recently been identified for release through the All India Coordinated Research Project (AICRP) on maize.

### 21.3.2 Marker-assisted selection for high lysine and tryptophan

At the beginning of QPM research around the world, several QPM hybrids and OPVs (open-pollinated varieties) were developed through conventional breeding approaches. However, the conventional procedures are tedious and time-consuming. Rapid advances in genome research and molecular biology have led to the development and use of DNA-based markers to hasten the selection process through marker-assisted selection (MAS) (Ribaut and Hoisington, 1998; Babu *et al.*, 2005). Three simple sequence repeat (SSR) markers, *phi057*, *phi112* and *umc1066*, were found to present within *o2* gene (Danson *et al.*, 2006). The *phi112* marker is dominant, while the other two markers are co-dominant. There are many successful examples in QPM where several normal maize lines have been converted into QPM through MAS. A two-generation marker-based backcross breeding programme was carried out for incorporation of the *o2* gene in the background of an early-maturing normal maize inbred, V 25 (Babu *et al.*, 2005). The authors developed the QPM version of V 25 with tryptophan content of 0.85%. Danson *et al.* (2006) introgressed the *o2* gene into herbicide-tolerant maize inbred lines through MAS. Gupta *et al.* (2013) developed a QPM version of a popular hybrid (Vivek Hybrid 9) through MAS, released as Vivek QPM 9 in 2008. Vivek QPM 9 is the 'first MAS-based maize cultivar' released for commercial cultivation in India. Further, the parental lines of the three maize hybrids HM 4, HM 8 and HM 9 were converted into QPM through MAS. Improved versions of QPM hybrids, namely Pusa HM 4 Improved, Pusa HM 8 Improved and Pusa HM 9 Improved, possessed tryptophan and lysine ranging from 0.68 to 1.06% and from 2.97 to 4.18%, respectively (Hossain *et al.*, 2018).

A recessive *opaque16* (*o16*) (on chromosome 8) was isolated from Robertson's Mutator (Mu) stock (Yang *et al.*, 2005). Recessive *o16* mutants possessed nearly twofold more lysine (0.247%) and tryptophan (0.072%) than *O16O16*-based wild type (0.125% lysine and 0.035% tryptophan) (Sarika *et al.*, 2017). Sarika *et al.* (2018a) reported that *o16* does not influence the endosperm attributes such as grain hardness and vitreousness. The recessive *o16* allele was introgressed into *o2*-based parental inbreds (HKI 161, HKI 193-1, HKI 193-2 and HKI 163) of four QPM hybrids (HQPM 1, HQPM 4, HQPM 5 and HQPM 7) using MAS (Sarika *et al.*, 2018b). Hybrids with *o2o2/o16o16* showed an average increase of 49 and 60% in lysine and tryptophan, over the original hybrids, with the highest enhancement being about 64 and 86%, respectively. Further, *o2* and *o16* genes have been introgressed into the waxy genotype using MAS (Wang *et al.*, 2019).

## 21.4 Improvement of Provitamin A

Vitamin A is required for functioning of the visual system, growth and development, the immune system, maintenance of epithelial cell integrity and reproduction in humans (Sommer and West, 1996). Vitamin A needs to be essentially supplied through diet, as it cannot be synthesized inside the human body. The human body is said to be deficient in vitamin A when retinol reserve is <0.1 μmol/g (Tanumihardjo, 2011). Vitamin A deficiency (VAD) affects at least 190 million preschool children and 19 million pregnant women (WHO, 2009). VAD is prevalent in Africa and South Asia including India, where people depend on cereals for their dietary requirement. The daily requirement for vitamin A in non-pregnant and non-lactating women is 500 μg/g, while it is 275 μg/g per day in the case of children aged 4–6 years (Andersson *et al.*, 2017). Night blindness is the hallmark of VAD; however, it also causes keratomalacia, an inflammation that causes irreversible blindness, besides diarrhoea and respiratory diseases (Sommer and Davidson, 2002; Mayer *et al.*, 2008; Bouis and Saltzman, 2017). VAD may also cause disorders like growth retardation, impaired Fe mobilization, depressed immune response and increased susceptibility to infectious diseases (Sommer and Davidson, 2002; WHO, 2009).

### 21.4.1 Breeding for high provitamin A

Traditional yellow maize, other than white endosperm genotypes, possesses a considerable concentration of carotenoids in the grains compared to rice, wheat and other staple cereals; but it is predominated by non-provitamin A type of carotenoids while provitamin A content is only 0.25–2.50 μg/g (Vignesh *et al.*, 2012; Pixley *et al.*, 2013; Muthusamy *et al.*, 2015a,c). This is well below the target level of 15 μg/g suggested for humans by HarvestPlus (Pixley *et al.*, 2013). Global research efforts have led to the identification of two key genes in the carotenoid biosynthesis pathway in maize, namely *β-carotene hydroxylase* (*crtRB1*) and *lycopene-ε-cyclase* (*lcyE*), mutant alleles of which could significantly increase β-carotene and β-cryptoxanthin (major provitamin A carotenoids) content (Harjes *et al.*, 2008; Yan *et al.*, 2010). However, favourable alleles of these genes causing provitamin A enhancement are rare in the population (3.38% for *lcyE*, 3.90% for *crtRB1* and 1.30% for both alleles) (Muthusamy *et al.*, 2015b). For both the *crtRB1* and *lcyE* genes, PCR-based co-dominant and gene-based markers were developed and validated (Harjes *et al.*, 2008; Yan *et al.*, 2010; Babu *et al.*, 2013). These markers have greatly reduced the intensive large-scale phenotyping of breeding populations required for development of provitamin A-rich biofortified maize (Babu *et al.*, 2013; Muthusamy *et al.*, 2014; Zunjare *et al.*, 2018a,b; Goswami *et al.*, 2019a).

Research efforts at CIMMYT and other institutions worldwide have also led to the development of provitamin A-rich maize genotypes through exploitation of either or both of the two key genes, *crtRB1* and *lcyE* (Prasanna *et al.*, 2020). About 11 provitamin A-rich hybrids and/or OPVs developed by CIMMYT, Mexico, were released in African countries like Malawi, Zambia and Zimbabwe. Around 15 provitamin A-rich OPVs, developed by the International Institute of Tropical Agriculture (IITA) in Ibadan, were released in Nigeria, Ghana and DR Congo (Gupta *et al.*, 2019). Of them, three hybrids (GV662A, GV664A and GV665A) were from Zambia, two hybrids (Ife maize hyb-3 and Ife maize hyb-4) and two synthetics (Sammaz 38 and Sammaz 39) were from Nigeria. One provitamin A-rich synthetic has also been released from Ghana (CSIR-CRI Honampa). So far, more than 40 provitamin A-rich maize cultivars including synthetics, single-cross hybrids and three-way hybrids have been released in African countries such as DR Congo, Ghana, Malawi, Mali, Nigeria, Rwanda, Tanzania, Zambia and Zimbabwe (Andersson *et al.*, 2017). All these hybrids/OPVs are reported to contain 6–8 μg provitamin A/g (Dhliwayo *et al.*, 2014; Simpungwe *et al.*, 2017). Around 460 tonnes of certified seeds of provitamin A-rich cultivars were produced for their cultivation by farmers (Gupta *et al.*, 2019). Additionally, 64 synthetics and 74 provitamin A-enriched hybrids were under extensive testing in 14 African countries (Manjeru *et al.*, 2017).

### 21.4.2 Marker-assisted selection for provitamin A

The favourable allele of the *crtRB1* gene was introgressed into parental inbreds of Vivek QPM 9, Vivek Hybrid 27, HM 4 and HM 8 using MAS; and the improved hybrids showed 5.5- to 10.2-fold increase in provitamin A compared with the original hybrids (Muthusamy *et al.*, 2014). This led to the release of Pusa Vivek QPM 9 Improved in 2017, which is the first provitamin A-rich commercial hybrid in India. Recently, Pusa Vivek Hybrid 27 Improved developed through MAS was also released for commercial cultivation in India during 2020. Goswami *et al.* (2019a) also introgressed the favourable allele of *crtRB1* into HKI 1128Q, a medium-maturing elite QPM inbred and female parent of the hybrid HM 9. Zunjare *et al.* (2018a) introgressed both *lcyE* and *crtRB1* into elite *o2*-based parental inbreds of the four popular QPM hybrids. Pusa HQPM 5 Improved and Pusa HQPM 7 Improved were released across the India and Peninsular Zone, respectively, during 2020. Chandran *et al.* (2019) also combined *o2* and *crtRB1* genes using MAS into the maize inbreds, UMI 1200 and UMI 1230, the parental inbreds of CoMH 6. Further, efforts were also made to diversify and develop new inbreds by crossing promising, locally adapted, recurrent inbreds and *crtRB1*-based donors, followed by selection of homozygous *crtRB1* segregants with good agronomic performance. These locally adapted *crtRB1*-based inbreds with high provitamin A and good agronomic performance have been evaluated across locations and their heterotic potential has been tested (Duo, 2019; Goswami *et al.*, 2019b). In China, the favourable allele of *crtRB1* was introgressed into two promising inbreds (CML 161 and CML 171) using MAS. The provitamin A level was enhanced up to 5.25 μg/g from 1.60 μg/g in CML 161; and up to 8.14 μg/g from 1.80 μg/g in CML 171 (Liu *et al.*, 2015). In 2019, China released a MAS-derived provitamin A-rich maize hybrid, Yunrui-596 (Prasanna *et al.*, 2020). Malawi, Zimbabwe and Zambia also released MAS-derived provitamin A-rich hybrids during 2015–2017.

## 21.5 Improvement of Vitamin E

Vitamin E or tocopherol plays essential biological roles in the human body by protecting against reactive oxygen species (ROS) and free radicals (Bramley *et al.*, 2000). It plays vital roles in scavenging of ROS and free radicals, quenching of singlet oxygen (high-energy oxygen) and providing membrane stability by protecting polyunsaturated fatty acids from lipid peroxidation. Vitamin E helps in preventing Alzheimer's disease, neurological disorders, cancer, cataracts, age-related macular degeneration and inflammatory disease. The recommended dietary allowance (RDA) for vitamin E is 4 mg/day for children aged 0–6 months, 15 mg/day for both males and females, and 19 mg/day for lactating mothers (Institute of Medicine, 2000). Vitamin E deficiency leads to progressive damage to the nervous and cardiovascular systems (Traber *et al.*, 2008).

### 21.5.1 Breeding for high vitamin E

Vitamin E is composed of four isoforms (α, β, δ, γ) and among the various tocopherols, γ-tocopherol constitutes ~80% of the total tocopherol, while α-tocopherol accounts for ~20% of the total pool. However, γ-tocopherol is less absorbed by the body due to lack of affinity with receptors in the body. On the contrary, α-tocopherol is the most favoured fraction and absorbed well by the body. Several researchers (Wong *et al.*, 2003; Chander *et al.*, 2008; Shutu *et al.*, 2012; Feng *et al.*, 2013) have reported various quantitative trait loci (QTLs) for kernel α-tocopherol, although they are limited in number. However, Li *et al.* (2012) identified two insertion/deletions (InDel7 and InDel118) within *vte4* (*γ-tocopherol methyl transferase*) gene which significantly affect the accumulation of α-tocopherol. The favourable allele of *vte4* converts γ-tocopherol more efficiently into α-tocopherol.

Das *et al.* (2020) analysed haplotypic variation for *vte4* among 54 diverse inbreds and observed that the mean α-tocopherol among all three favourable classes (0/0, 7/0 and 0/118) was much higher (15.44 μg/g) compared with 7.63 μg/g in unfavourable genotypes (7/118). Das *et al.* (2018) characterized a set of inbreds with favourable haplotypes of *vte4* using microsatellite markers. Das *et al.* (2019a) identified one single-nucleotide polymorphism (SNP) (G→A) and three insertion/deletions (14 and 27 bp) in the *vte4* that differentiated low and high α-tocopherol accumulating inbreds with favourable haplotype (0/0). These newly identified SNP and insertion/deletions, in addition to the already reported InDel118 and InDel7, can be useful in selection of favourable genotypes with higher α-tocopherol in maize. Das *et al.* (2019b) developed hybrids using inbreds possessing the favourable haplotype of *vte4* and reported higher mean α-tocopherol (mean: 21.37 μg/g) than in the check hybrids (mean: 11.16 μg/g). In some of the hybrids, namely MHVTE 2, MHVTE 18, MHVTE 28, MHVTE 10 and MHVTE 3, α-tocopherol constituted ≥50% of the total tocopherol.

### 21.5.2 Marker-assisted selection for vitamin E

InDel118, located 9 bp upstream of the putative transcription start site of *vte4*, controls α-tocopherol by regulating transcript level, whereas InDel7 affects the translation efficiency of *vte4* transcripts (Li *et al.*, 2012). Robust gene-based markers for InDel7 and InDel118 have been reported which are used for marker-assisted introgression of the favourable allele of *vte4* into elite genetic backgrounds (Das *et al.*, 2020). Feng *et al.* (2015) used MAS for enrichment of vitamin E through introgression of *vte4* in maize. Favourable alleles of InDel7 (deletion) and InDel118 (deletion) at *vte4* locus have recently been transferred from a suitable donor parent (SY999) to four *sh2*-based sweet corn inbreds through MAS (Feng *et al.*, 2015). Average α-tocopherol content of the backcross progenies increased significantly in K140 and K185 (14.58 and 9.06 μg/g) with an average increase of 7.73 and 5.33 μg/g, respectively. Converted recipient lines also showed improved γ-tocopherol and total tocopherol contents. Average γ-tocopherol content of four recipient lines before and after conversion was 29.11 and 51.45 μg/g, respectively, whereas total tocopherol content was 36.63 and 63.34 μg/g, respectively. In India, *vte4* allele (0/0) was introgressed into the parental lines of provitamin A-rich QPM hybrids (APQH 1, APQH 4, APQH 5 and APQH 7) (Das, 2019). Reconstituted *vte4*-based hybrids possessed mean α-tocopherol content of 16.8 μg/g over 8.1 μg/g in the original version (Gowda, 2019; Prasanna *et al.*, 2020).

## 21.6 Reduction of Phytate

Fe and Zn are the most common micronutrients that have been found to be deficient in predominantly cereal-based human diets (Bouis *et al.*, 2011). Fe and Zn are essential for metabolism and their deficiency affects human growth and development (Bouis and Welch, 2010). Although wide variability has been reported for Fe and Zn in maize, strong genotype × environment effects and accumulation of a large number of minor genes in a single genotype pose great challenges to breeders for development of Fe- and Zn-rich maize (Chakraborti *et al.*, 2011; Agrawal *et al.*, 2012; Guleria *et al.*, 2013; Gupta *et al.*, 2015). Phytic acid/phytate is an antinutritional factor that reduces the bioavailability of Fe (5–7%) and Zn (20–30%) in the gut of humans and animals (Andersson *et al.*, 2017). Reduction of phytate in maize assumes great significance for enhancing the bioavailability of Fe and Zn (Pramitha *et al.*, 2019; Abhijith *et al.*, 2020).

Phytic acid (*myo*-inositol hexaphosphate, $InsP_6$, or phytate) is present in most cereal grains, which provides *myo*-inositol and P which is essential for normal seed germination and seedling establishment. Among the cereals, maize seed alone produces approximately 4.8 million tonnes of phytic acid annually around the globe (Corell, 1989; Reddy *et al.*, 1989). Phytic acid is the most abundant P-containing compound in mature seeds, typically representing from 65 to 80% of the mature seed's total P (Raboy, 1997; Chu *et al.*, 2009). Presence of phytic acid in human food reduces the bioavailability of essential mineral cations, such as $Fe^{3+}$ and $Zn^{2+}$. The negatively charged phosphate groups on the phytic acid molecule can form strong chelates with divalent cations. As humans have no phytase activity in the gastrointestinal tract, the phytate–metal complexes pass through the intestine unabsorbed and lead to lower net absorption in the blood (Cosgrove, 1980). Phytic acid also interacts with basic amino acids, seed proteins and enzymes in the digestive tract to form complexes that may reduce amino acid availability, protein digestibility and the activity of digestive enzymes (Raboy, 2000; Hambidge *et al.*, 2004).

Phytic acid in maize grains is also poorly digested by monogastric animals and negatively affects animal nutrition and the environment. On the other hand, the adequately available P content of maize grains gets bound in phytic acid and therefore made not available to monogastric animals that lack sufficient phytase in their digestive tract for optimal P nutrition (Raboy, 2006; Chu *et al.*, 2009). As a result, feed has to be supplemented with inorganic P to meet the requirement for optimal animal growth. The undigested phytic acid excreted from monogastric animals is considered a leading source of P pollution from agriculture (Raboy, 2006). Hence, intake of cereals which are rich in phytic acid is also a reason for nutritional deficiency. To overcome this, in the animal feed industry, treatment of animal feeds with food-grade phytase, an enzyme that can cleave the phosphate groups of hexa- and penta-phosphate forms of phytic acid, is done extensively. In human food, fermentation, malting, soaking and germination are some food preparation methods that can be used to reduce the phytate content of foods (Hotz and Gibson, 2007). However, the acceptability, practicality and sustainability of these methods are limited in various areas.

### 21.6.1 Breeding for low phytate

In maize, three low phytic acid mutants (*lpa1*, *lpa2* and *lpa3*) have been isolated and characterized (Sparvoli and Cominelli, 2015). The *lpa1* is caused by a mutation in the gene *ZmMRP4* responsible for the transmembrane transporter protein. This gene is known to play a key role in the transfer of phytic acid to storage vacuoles in seeds. The *lpa2* mutation is caused due to mutation in the *ZmIPK4* gene. The *lpa2-1* is a result of the sequence rearrangement of *ZmIPK* and *lpa-2-2* is by a transition of cytosine to thymine at the position 158. This eventually causes a stop codon in the open-reading frame of the N-terminal region of *ZmIPK* (Pilu *et al.*, 2008). The *lpa3* is caused by a mutation in the gene coding for myoinositol kinase (MIK). These identified mutant lines serve as an essential genetic resource for developing low phytic acid lines in maize (Sparvoli and Cominelli, 2015). Compared with wild-type kernels, the *lpa1*, *lpa2* and *lpa3* mutations achieved 66, 50 and 50% reduction in phytic acid content, respectively (Raboy *et al.*, 2000). The *lpa2-2* mutation achieved a 30% reduction in phytic acid and a threefold increase in inorganic phosphate (Shi *et al.*, 2005). The mutant lines are of temperate origin and are not adapted to local tropical and subtropical conditions. Therefore, there is a need to have the *lpa* locus introgressed into locally adapted, agronomically superior lines to improve their nutritional benefit.

### 21.6.2 Marker-assisted selection for low phytate

The genetic map of maize shows that there are several SSRs linked with the *lpa2* locus at 1.05 bin location of the short arm of chromosome 1, and *umc2230* is in proximity (0.4 cM) to the *lpa2* allele. Sureshkumar *et al.* (2014) introgressed the *lpa2-2* allele from a low-phytate mutant line, EC 659418, into a well-adapted agronomically superior line, UMI 395. Mutant *lpa2-2* has also been introgressed into an elite normal maize inbred, UMI 285 (Tamilkumar *et al.*, 2014). Based on C (wild) to T (mutant) transition at amino acid position 1432 bp of *lpa1-1* gene, two dominant markers each specific to wild type (*LPA1*) and mutant (*lpa1-1*) allele were developed (Abhijith *et al.*, 2020). Full-length sequence alignment between wild-type (*LPA2*) and mutant (*lpa2-1*) alleles revealed one transition mutation (A→G) and a co-dominant cleaved amplified polymorphic sequences (CAPS) marker was developed. Fourteen $F_2$ populations were successfully genotyped using *lpa1-1* and *lpa2-1* markers. Segregants with *lpa1-1/lpa1-1* (1.77 mg/g) and *lpa2-1/lpa2-1* (1.85 mg/g) possessed significantly lower phytic acid compared with *LPA1/LPA1* (2.58 mg/g) and *LPA2/LPA2* (2.53 mg/g). Overall, homozygous segregants of *lpa1-1* and *lpa2-1* showed 31 and 27% reduction of phytic acid, respectively. Four elite inbreds (PMI PV5, PMI PV6, PMI PV7 and PMI PV7) that are parents of four provitamin A-rich QPM hybrids (APQH 1, APQH 4, APQH 5 and APQH 7) have been targeted for marker-assisted introgression of *lpa1-1* and *lpa2-1* genes from separate donors of exotic origin (Bhatt *et al.*, 2018).

## 21.7 Reduction of Glycaemic Index

Foods having high GI cause a sharp increase in blood sugar that triggers insulin secretion and tissue-specific intracellular uptake of glucose, which in turn lowers blood sugar. Repetition of these hyper- and hypoglycaemic cycles is known to result in insulin resistance and type 2 diabetes (Birt *et al.*, 2013). The solution lies in consumption of coarse cereal crops with low GI. Starch is the major food source for the majority of the human population worldwide. Various studies suggest that the consumption of high-amylose maize starch lowers GI and promotes colon health (Sajilata *et al.*, 2006). Maize starch is one of the best-quality starches having a purity level up to 99.5%. Due to this property, 80% of world starch is procured from maize (Zhang *et al.*, 2013). Starch is composed of a large number of glucose monomers joined together by glycoside linkages and is made up of two homopolymers, amylose and amylopectin. Amylose is a linear molecule of glucose residues linked via α-1,4 bonds, whereas amylopectin consists of both linear and branched glucose chains joined by α-1,6 bonds.

On the basis of digestibility, starch is classified broadly into two types: soluble starch and resistant starch. Soluble starch is readily digested and assimilated by the body and quickly elevates blood sugar level, leading to high GI. On the other hand, the resistant starch escapes from being digested in the small intestine and is fermented in the large intestine. Resistant starch has a lower rate of gastric emptying. Due to this property, it has a lower GI, as it does not contribute to the sudden rise in blood sugar. Based on the rate and extent of digestion and physiological properties, starch is classified into three major classes (Englyst *et al.*, 1992). Rapidly digestible starch (RDS) is digested and converted to glucose molecules within 20 min by enzymatic digestion. Slowly digestible starch (SDS) is converted to glucose after 120 min of enzymatic digestion. Being an inaccessible amorphous starch, it takes a longer time to digest, but is digested completely in the small intestine. Resistant starch (RS) resists digestion and absorption in the small intestine and passes undigested to the large intestine. In the large intestine, it acts as a substrate for microbial fermentation. As it is digested very slowly, it significantly reduces the GI. The increasing importance of resistant starch is due its prominent health benefits and large industrial applications. As a result of the linear structure of amylose, it is tightly packed and is resistant to digestion. Besides, resistant starch does not change the taste or texture, but may improve the sensory properties of products in which it is used as an additive (Sajilata *et al.*, 2006). It acts as a probiotic in increasing the population of beneficial microflora in the human digestive system and decreases the chance of colon cancer. It is also efficient in preventing and curing various bowel-related problems. Its consumption is known to reduce formation of gallstones. Replacement of approximately 5.4% of total carbohydrates in the diet with resistant starch significantly reduces fat accumulation, thus aiding weight control (Higgins *et al.*, 2004). It enhances absorption of various minerals especially Ca and Fe in the body (Morais *et al.*, 1996). High-amylose maize starch has a large number of industrial uses in food, medicines, textiles, paper and environmental protection. High-amylose starches particularly from maize are utilized as an ingredient in gum candies, as an adhesive for corrugated cardboards and in the development of biodegradable thermoplastics (Wu *et al.*, 2009).

### 21.7.1 Breeding for high amylose

Normal maize has approximately 25% amylose and 75% amylopectin (Nelson and Pan, 1995). Natural variation within maize germplasm is found for amylose content in different genetic backgrounds. However, this variation is significantly low and cannot be used as a source of exploitation for the production of high-amylose maize. Mutant *amylose extender1* (*ae1*) is reported to increase amylose in maize starch significantly. Maize inbreds (*ae1ae1*) are available with 50–60% amylase. Campbell *et al.* (2007) released GEMS 0067, as a high-amylose *ae1*-based line with 70% amylose. The recessive *ae1* mutant lacks the major form of the starch-branching enzyme (SBEIIb), which is ideally responsible for the production of branches leading to amylopectin. This mutation creates longer glucan chains of amylose and reduces the branch frequency in the structure compared with normal maize, leading to a high concentration of amylose and lesser amount of the amylopectin component (Liu *et al.*, 2009).

### 21.7.2 Marker-assisted selection for high amylose

Maize *ae1* mutants possess distinguishable characteristics phenotypically. These *ae* mutants can be easily identified by their distinct expression of tarnished glassy seeds, dull endosperm and slight shrinkage. Sequence comparison revealed that *ae1* allele possessed a 4 bp insertion region between the ninth and tenth exon which differentiated it from wild-type *Ae1* allele (Chen *et al.*, 2010). A functional marker was developed which amplified a 474 bp fragment in high-amylose maize cultivars, whereas no band was produced in normal maize genotypes. It was named as *ae-474* and behaved as a dominant marker (Chen *et al.*, 2010). Progenies in $BC_1F_1$, $BC_2F_1$ and $BC_2F_2$ populations of We-4-2 × Chang98 were successfully genotyped using the *ae-474* marker. Marker-positive progenies possessed higher amylose than the marker-negative ones.

Chen *et al.* (2013) further reported that the *ae1* mutant (at ninth exonic region) possessed an 84 bp deletion compared with normal maize. They developed *SbeIIb ae* InDel marker by targeting the sequence flanking this 84 bp deletion. This marker differentiated normal and high-amylose maize by producing a 1576 bp PCR product in normal maize, compared to a 700 bp band in high-amylose genotypes. Sequences of these two fragments were compared which indicated an 11 bp deletion in the intron between the eighth and ninth exons and a 675 bp deletion in the intron between the ninth and tenth exons. The ninth exon, together with the 116 and 675 bp flanking introns, were deleted in a high-amylose line, GEMS 0067 (Campbell *et al.*, 2007). GEMS 0067 is a high-amylose line with *ae1* and has a modifier gene which increases the total amylose content. This modifier has caught attention because it was found to be the contributor of high amylose content. This modifier was named *SbeI* after the enzyme variant to which it is known to cause mutation. The allele responsible for formation of *SbeI* was extensively studied by Chen *et al.* (2013) using mapping data. *SbeI* in GEMS 0067 has 15 SNPs resulting in six amino acid changes compared with normal maize. Out of these differences, they found that an SNP (C→A) present on the sixth exon of *SbeI* allele represents an enzymatic site for a restriction enzyme, *Alu*1. Thus they designed a marker that generates a 906 bp fragment in both normal and *ae* mutant maize genotypes. However, upon digestion with *Alu*1, it produces two fragments (617 bp and 289 bp) in normal maize, whereas the fragment remained undigested in high-amylose maize cultivar GEMS 0067 due to transversion of C→A on the target site of *Alu*1. The markers specific to *SbeIIb* and *SbeI* for tracing *ae* gene and modifier, respectively, are effective in MAS. *SbeIIb* is a co-dominant marker which successfully differentiated *Ae1Ae1*, *Ae1ae1* and *ae1ae1* genotypes. This increases the selection efficiency to transfer high-amylose genes into high-yielding but low-amylose cultivars.

## 21.8 Breeding for Multi-Nutrient Maize

Muthusamy *et al.* (2014) targeted VQL 1 and VQL 2 as parental inbreds for marker-assisted introgression of *crtRB1* allele (Fig. 21.1). Pusa Vivek QPM 9 Improved is the first released variety in India that possesses higher provitamin A (8.15 μg/g), tryptophan (0.74%) and lysine (2.67%). It is also the country's first multi-nutrient-rich maize hybrid. Several researchers have demonstrated the cumulative and positive effects of *crtRB1* and *lcyE* genes for provitamin A accumulation (Babu *et al.*, 2013; Zunjare *et al.*, 2017). Liu *et al.* (2015) also combined *crtRB1* and *o2* genes in CML 161 and CML 171 genetic background. Zunjare *et al.* (2018a) stacked the favourable alleles of *crtRB1*, *lcyE* and *o2* for biofortifying four hybrids for provitamin A, lysine and tryptophan. Four elite QPM parental lines (HKI 161, HKI 163, HKI 193-1 and HKI 193-2), which are the parents for commercial four QPM hybrids (HQPM 1, HQPM 4, HQPM 5 and HQPM 7) with wide popularity in India, were targeted. The mean provitamin A content in HQPM 1-, HQPM 4-, HQPM 5- and HQPM 7-based reconstituted hybrids was 9.95, 10.47, 9.63 and 12.27 μg/g, respectively. These hybrids also possessed high lysine and tryptophan as well.

Similarly, the QPM version of HKI 1128, the elite parental inbred of popular maize hybrids HM 9 (HKI 1105 × HKI 1128), HM 10 (HKI 193-2 × HKI 1128) and HM 11 (HKI 1128 × HKI 163), was targeted for introgression of *crtRB1* (Goswami *et al.*, 2019a). HKI 1128 was earlier converted into QPM through MAS of *o2* allele (Hossain *et al.*, 2018), and other parental lines, namely HKI 1105, HKI 193-2 and HKI 163, had been improved for protein quality and provitamin A in an earlier programme (Hossain *et al.*, 2018; Zunjare *et al.*, 2018a). The *crtRB1*-based progenies of HKI 1128Q possessed higher mean provitamin A (10.75 μg/g) compared with HKI 1128Q (3.38 μg/g). Essential amino acids lysine (mean: 0.303%) and tryptophan (0.080%) were high among the introgressed progenies (Goswami *et al.*, 2019a). Das (2019) and Gowda (2019) stacked *o2*, *crtRB1*, *lcyE* and *vte4* genes using MAS in the genetic background of HQPM 1, HQPM 4, HQPM 5 and HQPM 7 to combine high vitamin E, vitamin A, lysine and tryptophan. Bhatt *et al.* (2018) combined *lpa2-1* with *o2* and *crtRB1*, and *lpa2-2* with *o2* and *crtRB1*, in an elite genetic background to develop low-phytate-based provitamin A-rich QPM hybrids.

## 21.9 Impact of Biofortified Maize in Humans

The beneficial effects of QPM are well demonstrated worldwide (Tessema *et al.*, 2016; Gunaratna *et al.*, 2019). Porridge made from QPM resulted in fewer sick days among children consuming it compared with those who had porridge from normal maize. Higher rates of growth in weight (12%) and height (9%) were observed in infants and young children fed with QPM compared with the group given only normal maize (Gunaratna *et al.*, 2010). A study showed that consumption of 100 g QPM is required for children to maintain adequacy of lysine, a reduction in consumption to the tune of 40% relative to normal maize (Nuss and Tanumihardjo, 2011).

Grains of provitamin A-rich hybrids were analysed for bioaccessibility using the Caco2 cell assay (Dube *et al.*, 2018). The daily consumption of 200 g of provitamin A-rich maize hybrids, Pusa-PV-16-3 and Pusa-APQH8, would contribute 52 and 64%, respectively, of the RDA for adult Indian men after adjusting for cooking losses and conversion factors. Lutein and zeaxanthin content in the maize digesta and micellar fraction was inversely related to the β-carotene micellization and intestinal cell uptake, respectively. These results together suggest that the enrichment of provitamin A carotenoids together with decreasing the oxygenated carotenoid metabolites such as lutein and zeaxanthin will further improve the bioavailability of β-carotene from maize hybrids. Gannon *et al.* (2014) conducted

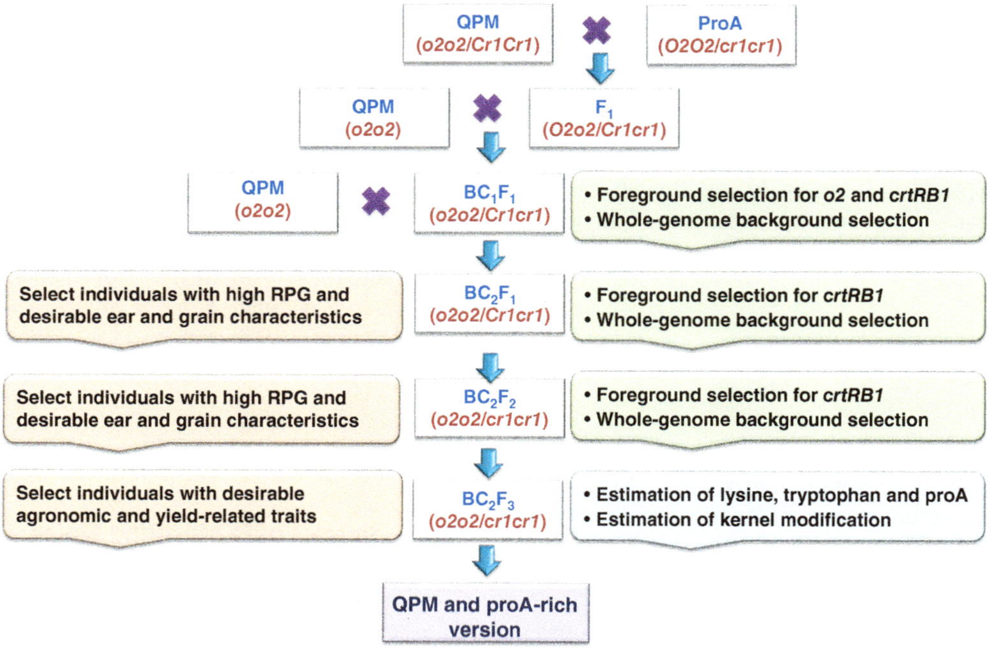


**Fig. 21.1.** Breeding scheme depicting introgression of *crtRB1* gene into *o2*-based QPM genotype. *O2*, unfavourable allele of *opaque2*; *o2*, favourable allele of *opaque2*; *Cr1*, unfavourable allele of *crtRB1*; *cr1*, favourable allele of *crtRB1*; proA, provitamin A.

a study on 133 Zambian children, finding that β-carotene from maize was efficacious when consumed as a staple food in this population and could avoid the potential for hypervitaminosis A. Another feeding trial of rural Zambian children aged 3–5 years was used to determine the impact of provitamin A-rich orange maize intake on serum retinol (Sheftel *et al.*, 2017). The study estimated that maize provided 11% of the recent dietary vitamin A to these children. These results demonstrated that orange maize is efficacious at providing retinol to the vitamin A pool in children through maize provitamin A carotenoids. Zuma *et al.* (2018) also conducted a study in the rural farming district of Mkushi, Zambia, in which children aged 4–8 years were fed with (i) provitamin A-rich orange maize meal or (ii) white maize meal, which were prepared according to standardized recipes. Blood samples before and after the intervention were collected which showed that consumption of provitamin A-rich maize significantly improved the children's serum β-carotene concentrations compared with traditional white maize.

## 21.10 Impact of Biofortified Maize in Poultry and Pigs

Poultry and swine have been recognized as affordable sources of protein worldwide in the forms of egg and meat. The feed alone contributes about 75–80% of the total cost of poultry and swine management under intensive feeding systems. Maize is well recognized as the king of feed ingredients, which contributes about 55–65% of the poultry diet. It is a primary source of supplemental energy and can contribute up to 30% of protein, 60% of energy and 90% of starch in an animal's diet (Dado, 1999). About 70–80% of maize production is used as a feed ingredient for livestock worldwide.

### 21.10.1 Effect of quality protein maize on growth performance of broiler chickens

Chicks weigh 40–45 g on the day of hatch and grow to about 2.2–2.4 kg by 40 days of age, which is mainly due to genetic selection for body weight and better management practices adopted by stakeholders (Rajasekhar *et al.*, 2020). This fast growth is a big challenge to the nutritionist to formulate high-nutrient-density diets that are well balanced in essential nutrients to meet the bird's genetic potential. Maize is a popular energy cereal used in combination with protein meal, such as soybean meal, in the formulation of broiler chicken diets. When QPM replaced normal maize (NM) (weight for weight) in practical broiler diets, higher body weight gain (BWG) (1944 g) was noticed in broilers on the QPM diet (Osei *et al.*, 1998; Panda *et al.*, 2010) compared with the NM diet (1904 g) at 6 weeks of age. Dietary replacement of NM with QPM (weight for weight, without balancing amino acids) improved BWG (2.23 kg in NM versus 2.26 kg in QPM) and feed conversion ratio (FCR) (2.07 versus 2.03) in broilers (Bai, 2002). Amonelo and Roxas (2008) reported higher BWG and better FCR in broilers on a QPM-based diet compared with an NM-based diet. In another study, Onimisi *et al.* (2008) replaced NM at 0, 25, 50, 75 and 100% with QPM in diets for 'Ross' broiler chicks and compared performance against a group fed a control NM diet balanced with synthetic lysine. They reported that 100% dietary replacement of NM by QPM increased BWG and FCR during both starter and finisher phases. Rajasekhar *et al.* (2020) experimented with the 'Cobb 400' breed of chickens and reported that the QPM diet produced higher BWG, breast yield, better FCR, lower abdominal fat and higher ready-to-cook meat over NM. When NM was supplemented with cottonseed meal (CSM) and guar meal (GM) as an alternative source of protein, QPM still improved the performance, slaughter parameters and nutrient utilization over NM with CSM and GM.

### 21.10.2 Effect of feeding quality protein maize-based diets on laying hens

Replacement of NM with QPM in the diet of laying hens increased egg production (EP) (88% in NM versus 91% in QPM) (Osei *et al.*, 1999). Similarly, Zhai (2002) reported higher EP (89.63% in NM versus 90.97% in QPM) and feed consumption due to the complete dietary replacement of NM by QPM, whereas FCR (feed/egg) and egg weight were not affected by feeding a diet based on QPM or NM. Panda *et al.* (2012) studied the utilization of QPM in the diet of 'White Leghorn' layers (28–44 weeks of age) and found that replacement of NM with QPM increased EP and improved feed efficiency. However, feed intake, egg weight, BWG and mortality were not affected among the groups fed NM- or QPM-based diets. Supplementation of lysine to the NM-based diet increased EP, which was comparable to that of the QPM-based diet without supplemental lysine (Tyagi *et al.*, 2011; Panda *et al.*, 2012). It was concluded that synthetic lysine supplementation can be minimized in the QPM-based layer feed.

Replacing NM with QPM, or the NM + lysine diet, had no influence on egg quality parameters such as albumen, yolk or shell, shell weight or shell thickness. However, feeding QPM-based diets improved haugh units and yolk colour index (Zhai, 2002; Panda *et al.*, 2012). Liu *et al.* (2012) reported that provitamin A equivalents increased in the eggs of hens fed maize with high β-cryptoxanthin content. Moreno *et al.* (2016) observed higher accumulation of provitamin A (3.09 ppm) in egg yolks when hens were fed with provitamin A-rich maize, compared with 0.33 ppm in hens fed a commercial yellow maize variety.

### 21.10.3 Effect of feeding quality protein maize-based diets on growth of pigs

The intake of dry matter (DM) was higher in pigs fed QPM compared with those fed NM (1.66 versus 1.60 kg/day) and the DM total-tract digestibility was lower in the QPM-fed pigs compared with the NM-fed pigs (89.0 versus 90.5%). The daily N intake was higher in the pigs fed QPM (24.0 g/day) than in the pigs fed with NM (22.0 g/day). The digestible N intake was higher in the pigs fed QPM. Lysine digestibility was also higher in the pigs fed QPM than in those fed NM. Concerning the growth performance of pigs during the grower phase (20–50 kg), replacement of NM with QPM improved the average daily body weight gain (ADG) and FCR. In the finisher phase (50–80 kg), replacement of NM with QPM in the pig diet remarkably increased ADG. This indicated that feeding QPM-based diets has an advantage over NM-based diets in pigs.

The higher BWG and better FCR might be due to the increase in the lysine content and higher digestibility of critical essential amino acids (Gao, 2002), which might have contributed to better performance in pigs. Similar results were also obtained by Sullivan *et al.* (1989) in grower pigs and Burgoon (1992) in finisher pigs.

## 21.11 Challenges and Future Prospects

The germplasm base for nutritional quality is extremely narrow compared with the traditional maize germplasm developed for high productivity. Thus, widening of the genetic base through systematic crosses and development of diverse inbreds followed by their heterotic grouping should be the priority of breeding programmes for nutritional quality in maize. Further, rainfed ecologies are more prone to several abiotic stresses due to unpredictable weather patterns. There is an urgent need to combine tolerances to drought, heat and waterlogging with nutritional quality. Besides, new diseases like maize lethal necrosis and insect pests such as the fall armyworm have caused significant losses to the crop in recent times. Thus resistance breeding should also be an integral part of biofortification programmes.

There is a strong need to demonstrate nutritionally enriched maize hybrids/OPVs in farmers' fields. Strengthening the seed chain to produce and supply good-quality seeds is an important step for the popularization of biofortified varieties. Low cost and quality seed production remain major issues for enhancing the adoption rate of biofortified hybrids. Farmer, private, cooperative and public-sector participatory seed production programmes need to be evolved in the future for assuring quality seed availability. The 'seed village/district' concept must be adopted to produce quality seeds. Even in commercial production, it is better to move to the 'biofortified maize village' concept to address the issue of contamination by pollen from neighbouring fields having normal maize. Providing subsidized seeds and other inputs to farmers would further contribute to the rapid dissemination of nutritionally improved cultivars among the farmers.

Awareness generation on the health benefits of biofortified crops is also a key factor for the rapid adoption of biofortified varieties by farmers. Educational background of the household head and extent of farmers' participation in demonstration trials and field days are the important factors for the generation of awareness. Further, the decision by households to adopt biofortified crops and the subsequent decision to allocate the nutritious food to young children are also significant factors for impacting children's nutrition and health. Farmers raising poultry and pigs also require sensitization on the beneficial effects of biofortified maize on their animals' growth and development. The apprehension of low yielding potential of biofortified varieties has also been identified as another important factor for slow popularization among farmers. It is now well established that the biofortified varieties are comparable to traditional varieties for their yield potential.

Strong policy support is an important factor for the success of biofortified crops. Segregation of grains of biofortified crops in the market and assurance of remunerative prices through minimum support price and/or premium price in the market would also encourage farmers to grow more biofortified crops. Inclusion of these biofortified cereals in different government-sponsored programmes, especially those related to nutrition for child development and the well-being of pregnant women and school-going children, would help in providing the much-needed balanced food for healthy and disease-free future life.

## References

Abhijith, K.P., Muthusamy, V., Chhabra, R., Dosad, S., Bhatt, V. *et al.* (2020) Development and validation of breeder-friendly gene-based markers for *lpa1-1* and *lpa2-1* genes conferring low phytic acid in maize kernel. *3 Biotech* 10, 121. Available at: https://doi.org/10.1007/s13205-020-2113-x

Agrawal, P.K., Jaiswal, S.K., Prasanna, B.M., Hossain, F., Saha, S., Guleria, S.K. and Gupta H.S. (2012) Genetic variability and stability for kernel iron and zinc concentration in maize (*Zea mays* L.) genotypes. *Indian Journal of Genetics and Plant Breeding* 72, 421–428.

Amonelo, M.O. and Roxas, D.B. (2008) Growth performance of broilers fed a quality protein maize based diet. *Philippine Journal of Veterinary and Animal Sciences* 34, 11–22.

Andersson, M.S., Saltzman, A., Virk, P.S. and Pfeiffer, W.H. (2017) Progress update: crop development of biofortified staple food crops under HarvestPlus. *African Journal of Food, Agriculture, Nutrition and Development* 17, 11905–11935.

Babu, R., Nair, S.K., Kumar, A., Venkatesh, S., Sekhar, J.C. *et al.* (2005) Two generation marker aided backcrossing for rapid conversion of normal maize lines to quality protein maize (QPM). *Theoretical and Applied Genetics* 111, 888–897.

Babu, R., Rojas, N.P., Gao, S., Yan, J. and Pixley, K. (2013) Validation of the effects of molecular marker polymorphisms in *lcyE* and *crtRB1* on provitamin A concentrations for 26 tropical maize populations. *Theoretical and Applied Genetics* 126, 389–399.

Bai, X.F. (2002) Nutritional evaluation of and utilisation of quality protein maize Zhong Dan 9409 in broiler feed. MSc thesis, Chinese Academy of Agricultural Sciences, Beijing.

Bhatt, V., Muthusamy, V., Jha, S., Zunjare, R.U., Baveja, A., Dosad, S. and Hossain, F. (2018) Development of low phytic acid maize through marker assisted introgression of *lpa1-1* and *lpa2-1* genes. In: *Abstracts: 13th Asian Maize Conference on and Expert Consultation on Maize for Food, Feed, Nutrition and Environmental Security, Ludhiana, India, 8–10 October 2018*. CIMMYT, Mexico City, pp. 143–144.

Birt, D.F., Boylston, T., Hendrich, S., Jane, J.L., Hollis, J. *et al.* (2013) Resistant starch: promise for improving human health. *Advances in Nutrition* 4, 587–601.

Bjarnason, M. and Vasal, S.K. (1992) Breeding of quality protein maize. *Plant Breeding Reviews* 9, 181–216.

Bouis, H.E. (2018) Reducing mineral and vitamin deficiencies through biofortification: progress under HarvestPlus. In: Biesalski, H.K. and Birner, R. (eds) *Hidden Hunger: Strategies to Improve Nutrition Quality (special edition). World Review of Nutrition and Dietetics* 118, 112–122. Available at: https://doi.org/10.1159/000484342

Bouis, H.E. and Saltzman, A. (2017) Improving nutrition through biofortification: a review of evidence from HarvestPlus, 2003 through 2016. *Global Food Security* 12, 49–58. Available at: https://doi.org/10.1016/j.gfs.2017.01.009

Bouis, H.E. and Welch, R.M. (2010) Biofortification: a sustainable agricultural strategy for reducing micronutrient malnutrition in the global south. *Crop Science* 50, 20–32.

Bouis, H.E., Hotz, C., McClafferty, B., Meenakshi, J.V. and Pfeiffer, W. (2011) Biofortification: a new tool to reduce micronutrient malnutrition. *Food and Nutrition Bulletin* 32, S31–S40.

Bouis, H.E., Saltzman, A. and Birol, E. (2019) Improving nutrition through biofortification. In: Fan, S., Yosef, S. and Pandya-Lorch, R. (eds) *Agriculture for Improved Nutrition: Seizing the Momentum*. CAB International, Wallingford, UK, pp. 47–57.

Bramley, P.M., Elmadfa, I., Kafatos, A., Kelly, F.J., Manios, Y. *et al.* (2000) Critical reviews produced within the EU concerted action 'Nutritional enhancement of plant-based food in European trade' (NEODIET) – vitamin E. *Journal of the Science of Food and Agriculture* 80, 913–938.

Bressani, R. (1990) Nutritional value of high-lysine maize in humans. In: Mertz, E.T. (ed.) *Quality Protein Maize*. American Association of Cereal Chemists, St Paul, Minnesota, pp. 205–225.

Burgoon, K.G., Hansen, J.A., Knabe, D.A. and Backholt A.J. (1992) Nutritional value of quality protein maize for starter and finisher swine. *Journal of Animal Science* 70, 811–817.

Campbell, M.R., Jane, J.L., Pollak, L., Blanco, M. and Brien, A. (2007) Registration of maize germplasm line GEMS-0067. *Journal of Plant Registration* 1, 60–61.

Chakraborti, M., Prasanna, B.M., Hossain, F. and Singh A.M. (2011) Evaluation of single cross quality protein maize (QPM) hybrids for kernel iron and zinc concentrations. *Indian Journal of Genetics and Plant Breeding* 71, 312–319.

Chander, S., Guo, Y.Q., Yang, X.H., Yan, J.B., Zhang, Y.R., Song, T.M. and Li, J.S. (2008) Genetic dissection of tocopherol content and composition in maize grain using quantitative trait loci analysis and the candidate gene approach. *Molecular Breeding* 22, 353–365.

Chandran, S., Pukalenthy, B., Adhimoolam, K., Manickam, D., Sampathrajan, V. *et al.* (2019) Marker-assisted selection to pyramid the *opaque-2* (*o2*) and β-*carotene* (*crtRB1*) genes in maize. *Frontiers in Genetics* 10, 859. Available at: https://doi.org/10.3389/fgene.2019.00859

Chen, F., Zhu, S.W., Xiang, Y., Jiang, H.Y. and Cheng, B.J. (2010) Molecular marker-assisted selection of the *ae* alleles in maize. *Genetics and Molecular Research* 9, 1074–1084.

Chen, T., Ning, L., Liu, X., Cui, D., Zhang, H. *et al.* (2013) Development of functional molecular markers of *SbeI* and *SbeIIb* for the high amylose maize germplasm line GEMS-006. *Crop Science* 53, 482–490.

Chhabra, R., Hossain, F., Muthusamy, V., Baveja, A., Mehta, B. and Zunjare, R.U. (2019) Mapping and validation of plant *anthocyanin-1* pigmentation gene for its effectiveness in early selection of *shrunken-2* gene influencing kernel sweetness in maize. *Journal of Cereal Science* 87, 258–265.

Chu, G.M., Komori, M., Hattori, R. and Matsui, T. (2009) Dietary phytase increases the true absorption and endogenous fecal excretion of zinc in growing pigs given a corn, soybean meal based diet. *Animal Science Journal* 80, 46–51.

Corell, D.L. (1989) The role of phosphorus in the eutrophication of receiving waters – a review. *Journal of Environmental Quality* 27, 261–266.

Cosgrove, D.J. (1980) *Inositol Phosphates: Their Chemistry, Biochemistry and Physiology*. Elsevier Scientific, Amsterdam, pp. 26–43.

Dado, R.G. (1999) Nutritional benefits of specialty maize grain hybrids in dairy diets. *Journal of Animal Science* 77, 197–207.

Danson, J.W., Mercy, M., Michael, K., Martin, L., Alex, K. and Alpha, D. (2006) Marker assisted introgression of *opaque2* gene into herbicide resistant elite maize inbred lines. *African Journal of Biotechnology* 5, 2417–2422.

Das, A. (2019) Analyses of genetic variability, molecular characterization and marker-assisted enrichment of kernel vitamin E in maize (*Zea mays* L.). PhD thesis, Indian Agricultural Research Institute, New Delhi.

Das, A.K., Jaiswal, S.K., Muthusamy, V., Zunjare, R.U., Chauhan, H.S. *et al.* (2018) Molecular diversity and genetic variability of kernel tocopherols among maize inbreds possessing favourable haplotypes of γ-tocopherol methyl transferase (*γ-VTE4*). *Journal of Plant Biochemistry and Biotechnology* 28, 253–262.

Das, A.K., Muthusamy, V., Zunjare, R.U., Chauhan, H.S., Sharma, P.K. *et al.* (2019a) Genetic variability, genotype × environment interactions and combining ability analyses of kernel tocopherols among maize genotypes possessing novel allele of *γ-tocopherol methyl transferase* (*ZmVTE4*). *Journal of Cereal Science* 86, 1–8.

Das, A.K., Chhabra, R., Muthusamy, V., Chauhan, H.S., Zunjare, R.U. and Hossain, F. (2019b) Identification of novel *SNP* and *InDel* variations in the promoter and 5′ untranslated regions of *γ-tocopherol methyl transferase* (*ZmVTE4*) affecting higher accumulation of α-tocopherol in maize kernel. *The Crop Journal* 7, 469–479.

Das, A.K., Muthusamy, V., Zunjare, R.U., Baveja, A., Chauhan, H.S. *et al.* (2020) Genetic variability for kernel tocopherols and haplotype analysis of *γ-tocopherol methyl transferase* (*vte4*) gene among exotic- and indigenous maize inbreds. *Journal of Food Composition and Analysis* 88, 103446.

Devi, E.L., Hossain, F., Muthusamy, V., Chhabra, R., Zunjare, R. *et al.* (2017) Microsatellite marker-based characterization of waxy maize inbreds for their utilization in hybrid breeding. *3 Biotech* 7, 316.

Dhliwayo, T., Palacios-Rojas, N., Crossa, J. and Pixley, K.V. (2014) Effects of $S_1$ recurrent selection for provitamin a carotenoid content for three open-pollinated maize cultivars. *Crop Science* 54, 2449–2460.

Dube, N., Chandra, M.P., Hossain, F., Pullakhandam, R., Thingnganing, L. and Bharatraj, D.K. (2018) β-Carotene bioaccessibility from biofortified maize (*Zea mays*) is related to its density and is negatively influenced by lutein and zeaxanthin. *Food and Function* 9, 379–388.

Duo, H. (2019) Genetic and biochemical evaluation and molecular marker analysis of locally adapted *crtRB1*-based inbreds for provitamin-A enrichment in maize. MSc thesis, Indian Agricultural Research Institute, New Delhi.

Englyst, H.N., Kingman, S.M. and Cummings, J.H. (1992) Classification and measurement of nutritionally important starch fractions. *European Journal of Clinical Nutrition* 46, S33–S50.

FAO (2019) *FAOSTAT Database. Crops*. Food and Agriculture Organization of the United Nations, Rome. Available at: http://www.fao.org/faostat/en/#data/QC (accessed 8 March October 2021).

Feng, F., Deng, F., Zhou, P., Yan, J., Wang, Q., Yang, R. and Li, X. (2013) QTL mapping for the tocopherols at milk stage of kernel development in sweet corn. *Euphytica* 193, 409–417.

Feng, F., Wang, Q., Liang, C., Yang, R. and Li, X. (2015) Enhancement of tocopherols in sweet corn by marker-assisted backcrossing of *ZmVTE4*. *Euphytica* 206, 513–521.

Gannon, B., Kaliwile, C., Arscott, S.A., Schmaelzle, S., Chileshe, J. *et al.* (2014) Biofortified orange maize is as efficacious as a vitamin a supplement in Zambian children even in the presence of high liver reserves of vitamin A: a community-based, randomized placebo-controlled trial. *American Journal of Clinical Nutrition* 100, 1541–1550.

Gao, J. (2002) Nutritional evaluation and utilization of quality protein maize Zhong Dan 9409 in pig feed. MSc thesis, Chinese Academy of Agricultural Sciences, Beijing.

Gibbon, B. and Larkin, B. (2005) Molecular genetic approaches to developing quality protein maize. *Trends in Genetics* 21, 227–233.

Goswami, R., Zunjare, R.U., Khan, S., Baveja, A., Muthusamy, V. and Hossain, F. (2019a) Marker-assisted introgression of rare allele of *β-carotene hydroxylase* (*crtRB1*) gene into elite quality protein maize inbred for combining high lysine, tryptophan and provitamin A in maize. *Plant Breeding* 138, 174–183.

Goswami, R., Zunjare, R.U., Khan, S., Muthusamy, V., Baveja, A *et al.* (2019b) Genetic variability of kernel provitamin-A in sub-tropically adapted maize hybrids possessing rare allele of β-*carotene hydroxylase*. *Cereal Research Communications* 47, 205–215.

Gowda, M. (2019) *Morphological and biochemical characterization of MAS-derived maize genotypes possessing ZmVTE4, crtRB1, lcyE and opaque2 genes*. MS thesis, Indian Agricultural Research Institute, New Delhi.

Guleria, S.K., Chahota, R.K., Kumar, P., Kumar, A., Prasanna, B.M., Hossain, F. and Gupta, H.S. (2013) Analysis of genetic variability and genotype × year interactions on kernel zinc concentration in selected Indian and exotic maize (*Zea mays* L.) genotypes. *Indian Journal of Agricultural Sciences* 83, 836–841.

Gunaratna, N.S., De Groote, H., Nestel, P., Pixley, K.V. and McCabe, G.P. (2010) A meta-analysis of community-level studies on quality protein maize. *Food Policy* 35, 202–210.

Gunaratna, N.S., Moges, D. and De Groote, H. (2019) Biofortified maize can improve quality protein intakes among young children in southern Ethiopia. *Nutrients* 11, 192.

Gupta, H.S., Raman, B., Agrawal, P.K., Mahajan, V., Hossain, F. and Thirunavukkarasu, N. (2013) Accelerated development of quality protein maize hybrid through marker-assisted introgression of *opaque2* allele. *Plant Breeding* 132, 77–82.

Gupta, H.S., Hossain, F. and Muthusamy, V. (2015) Biofortification of maize: an Indian perspective. *Indian Journal of Genetics and Plant Breeding* 75, 1–22.

Gupta, H.S., Hossain, F., Muthusamy, V. and Zunjare, R.U. (2019) Marker-assisted breeding for enrichment of provitamin-A in maize. In: Qureshi, A., Dar, Z. and Wani, S. (eds) *Quality Breeding in Field Crops*. Springer, Cham, Switzerland, pp. 139–157. Available at: https://doi.org/10.1007/978-3-030-04609-5_6

Gutierrez-Rojas, A., Scott, M.P., Leyva, O.R., Menz, M. and Betrán, J. (2008) Phenotypic characterization of quality protein maize endosperm modification and amino acid contents in a segregating recombinant inbred population. *Crop Science* 48, 1714–1722.

Habben, J.E., Moro, G.L., Hunter, B.G., Hamaker, B.R. and Larkins, B.A. (1995) Elongation factor 1α concentration is highly correlated with the lysine content of maize endosperm. *Proceedings of the National Academy of Sciences USA* 92, 8640–8644.

Hambidge, K.M., Huffer, J.W., Raboy, V., Grunwald, G.K., Westcott, J.L. *et al.* (2004) Zinc absorption from low phytate alleles of maize with and their wild-type iso hybrids. *American Journal of Clinical Nutrition* 79, 1053–1059.

Harjes, C.E., Rocheford, T.R., Bai, L., Brutnell, T.P., Kandianis, C.B. *et al.* (2008) Natural genetic variation in *lycopene epsilon cyclase* tapped for maize biofortification. *Science* 319, 330–333.

Higgins, J.A., Higbee, D.R., Donahoo, W.T., Brown, I.L., Bell, M.L. and Bessesen, D.H. (2004) Resistant starch consumption promotes lipid oxidation. *Nutrition & Metabolism* 1, 8.

Hoseini, B.L., Moghadam, E.Z., Saeidi, M., Askarieh, R.M. and Khademi, G. (2015) Child malnutrition at different world regions in 1990–2013. *International Journal of Pediatrics* 3, 921–932.

Hossain, F., Nepolean, T., Vishwakarma, A.K., Pandey, N., Prasanna, B.M. and Gupta, H.S. (2015) Mapping and validation of microsatellite markers linked to *sugary1* and *shrunken2* genes in maize. *Journal of Plant Biochemistry and Biotechnology* 24, 135–142.

Hossain, F., Muthusamy, V., Pandey, N., Vishwakarma, A.K., Baveja, A. and Zunjare, R.U. (2018) Marker-assisted introgression of *opaque2* allele for rapid conversion of elite hybrids into quality protein maize. *Journal of Genetics* 97, 287–298.

Hossain, F., Muthusamy, V., Zunjare, R.K. and Gupta H.S. (2019a) Biofortification of maize for protein quality and provitamin-A content. In: Jaiwal, P.K., Chhillar, A.K., Chaudhary, D. and Jaiwal, R. (eds) *Nutritional Quality Improvement in Plants*. Concepts and Strategies in Plant Sciences. Springer, Cham, Switzerland, pp. 115–136. Available at: https://doi.org/10.1007/978-3-319-95354-0_5

Hossain, F., Sarika, K., Muthusamy, V., Zunjare R.U. and Gupta H.S. (2019b) Quality protein maize for nutritional security. In: Qureshi, A., Dar, Z. and Wani, S. (eds) *Quality Breeding in Field Crops*. Springer, Cham, Switzerland, pp. 217–237. Available at: https://doi.org/10.1007/978-3-030-04609-5_11

Hotz, C. and Gibson, R.S. (2007) Traditional food-processing and preparation practices to enhance the bioavailability of micronutrients in plant-based diets. *Journal of Nutrition* 137, 1097–1100.

ICMR (2010) *Nutrient Requirements and Recommended Dietary Allowances for Indians: A Report of the Expert Group of the Indian Council of Medical Research*. Indian Council of Medical Research, New Delhi, p. 6.

IFPRI (2016) *2016 Global Nutrition Report*. International Food Policy Research Institute, Washington, DC. Available at: https://globalnutritionreport.org/reports/2016-global-nutrition-report/ (accessed 26 February 2021).

IFPRI (2018) *2018 Global Nutrition Report*. International Food Policy Research Institute, Washington, DC. Available at: https://globalnutritionreport.org/reports/global-nutrition-report-2018/ (accessed 26 February 2021).

IIPS (2017) *National Family Health Survey (NFHS-4), 2015–16*. International Institute for Population Sciences, Mumbai, India.

Institute of Medicine (2000) *Dietary Reference Intakes for Vitamin C, Vitamin E, Selenium and Carotenoids*. National Academies Press, Washington, DC.

Li, Q., Yang, X., Xu, S., Cai, Y., Zhang, D. *et al.* (2012) Genome-wide association studies identified three independent polymorphisms associated with α-tocopherol content in maize kernels. *PLoS One* 7, e36807.

Lieshout, M.V. and Pee, S.D. (2005) Vitamin A equivalency estimates: understanding apparent differences. *American Journal of Clinical Nutrition* 81, 943–945.

Liu, F., Makhmoudova, A., Lee, E., Wait, R., Emes, M. and Tetlow, I. (2009) The amylose extender mutant of maize conditions novel protein–protein interactions between starch biosynthetic enzymes in amyloplasts. *Journal of Experimental Botany* 60, 4423–4440.

Liu, L., Jeffers, D., Zhang, Y., Ding, M., Chen, W., Kang, M.S. and Fan, X. (2015) Introgression of the *crtRB1* gene into quality protein maize inbred lines using molecular markers. *Molecular Breeding* 35, 154.

Liu, Y.Q., Davis, C.R., Schmaelzle, S.T., Rocheford, T., Cook, M.E. and Tanuhardjo, S.A. (2012) β-Cryptoxanthin biofortified maize (*Zea mays*) increases β-cryptoxanthin concentration and enhances the color of chicken egg yolk. *Poultry Science* 91, 432–438.

Manjeru, P., Biljon, A.V. and Labuschagne, M. (2017) The development and release of maize fortified with provitamin A carotenoids in developing countries. *Critical Reviews in Food Science and Nutrition* 59, 1284–1293. Available at: https://doi.org/10.1080/10408398.2017.1402751

Mayer, J.E., Pfeiffer, W.H. and Beyer, P. (2008) Biofortified crops to alleviate micronutrient malnutrition. *Current Opinion in Plant Biology* 11, 166–170.

Mertz, E.T. (1992) Discovery of high lysine, high tryptophan cereals. In: Mertz, E.T. (ed.) *Quality Protein Maize*. American Association of Cereal Chemists, St Paul, Minnesota, pp. 1–8.

Mertz, E.T., Bates, L.S. and Nelson, O.E. (1964) Mutant gene that changes protein composition and increases lysine content of maize endosperm. *Science* 145, 1470–1473.

Morais, M.B., Feste, A., Miller, R.G. and Lifschitz, C.H. (1996) Effect of resistant and digestible starch on intestinal absorption of calcium, iron, and zinc in infant pigs. *Pediatric Research* 39, 872–876.

Moreno, J.A., Díaz-Gómez, J., Nogareda, C., Angulo, E., Sandmann, G. *et al.* (2016) The distribution of carotenoids in hens fed on biofortified maize is influenced by feed composition, absorption, resource allocation and storage. *Scientific Reports* 6, 35346.

Morley, E.J. (2020) Protein–Energy Undernutrition (PEU). Available at: https://www.msdmanuals.com/professional/nutritional-disorders/undernutrition/protein-energy-undernutrition-peu (accessed 8 March 2021).

Mpofu, A., Linnemann, A.R., Nout, M., Zwietering, M.H. and Smid, E.J. (2014) Mutandabota, a food product from Zimbabwe: processing, composition, and socioeconomic aspects. *Ecology of Food and Nutrition* 53, 24–41.

Muthusamy, V., Hossain, F., Thirunavukkarasu, N., Choudhary, M., Saha, S., Bhat, J.S. and Gupta, H.S. (2014) Development of β-carotene rich maize hybrids through marker-assisted introgression of *β-carotene hydroxylase* allele. *PLoS ONE* 9, e113583.

Muthusamy, V., Hossain, F., Nepolean, T., Saha, S., Agrawal, P.K., Guleria, S.K. and Gupta, H.S. (2015a) Genetic variability and inter-relationship of kernel carotenoids among indigenous and exotic maize (*Zea mays* L.) inbreds. *Cereal Research Communications* 43, 567–578.

Muthusamy, V., Hossain, F., Thirunavukkarasu, N., Saha, S. and Gupta, H.S. (2015b) Allelic variations for *lycopene-ε-cyclase* and *β-carotene hydroxylase* genes in maize inbreds and their utilization in β-carotene enrichment programme. *Cogent Food and Agriculture* 1, 1033141.

Muthusamy, V., Hossain, F., Thirunavukkarasu, N., Pandey, N., Vishwakarma, A.K., Saha, S. and Gupta, H.S. (2015c) Molecular characterization of exotic and indigenous maize inbreds for biofortification with kernel carotenoids. *Food Biotechnology* 29, 276–295.

Neeraja, C.N., Ravindra, B.V., Ram, S., Hossain, F., Hariprasanna, K. *et al.* (2017) Biofortification in cereals – progress and prospects. *Current Science* 113, 1050–1057.

Nelson, O. and Pan, D. (1995) Starch synthesis in maize endosperms. *Annual Review of Plant Physiology and Plant Molecular Biology* 46, 475–496.

Nuss, E.T. and Tanumihardjo, S.A. (2011) Quality protein maize for Africa: closing the protein inadequacy gap in vulnerable populations. *Advances in Nutrition* 2, 217–224.

Onimisi, P.A., Dafwang, I.I., Omage, J.J. and Onyibe, J.E. (2008) Apparent digestibility of feed nutrients, total tract and ileal amino acids of broiler chicken fed quality protein maize (obatampa) and normal maize. *International Journal of Poultry Science* 7, 959–963.

Osei, S.A., Atuahene, C.A., Okai, D.B., Donkoh, A. and Tuah, A.K. (1998) The nutritive value of quality protein maize in the diets of broiler chickens. *Ghana Journal of Agricultural Sciences* 31, 1–5.

Osei, S.A., Dei, H.K. and Tuah, A.K. (1999) Evaluation of quality protein maize as a feed ingredient for layer pullet. *Journal of Animal Feed Science* 8, 181–189.

Panda, A.K., Raju, M.V.L.N., Rao, S.V.R., Lavanya, G., Reddy, P.K.E. and Sunder, S.G. (2010) Replacement of normal maize with quality protein maize on performance, immune response and carcass characteristics of broiler chickens. *Asian Australian Journal of Animal Sciences* 23, 1626–1631.

Panda, A.K., Lavanya, G., Reddy, P.K.E., Rao, S.V.R., Raju, M.V.L.N. and Sunder, S.G. (2012) Utilization of quality protein maize in the diet of White Leghorn layers. *Animal Feed Science and Technology* 172, 210–216.

Pfeiffer, W.H. and McClafferty, B. (2007) Harvest plus: breeding crops for better nutrition. *Crop Science* 47, S88–S105.

Pilu, R., Panzeri, D.E., Cassani, F., Badone, C., Landoni, M. and Nielsen, E. (2008) A paramutation phenomenon is involved in the genetics of maize *low phytic acid1-241* (*lpa1-241*) trait. *Heredity* 102, 236–245.

Pixley, K., Rojas, N.P., Babu, R., Mutale, R., Surles, R. and Simpungwe, E. (2013) Biofortification of maize with provitamin A carotenoids. In: Tanumihardjo, S.A. (ed.) *Carotenoids and Human Health*. Nutrition and Health. Humana Press, Totowa, New Jersey, pp. 271–292. Available at: https://doi.org/10.1007/978-1-62703-203-2_17

Pramitha, L., John Joel, A.J., Srinivas, S., Sreeja, R., Hossain, F. and Ravikesavan, R. (2019) Enumerating the phytic acid content in maize germplasm and formulation of reference set to enhance the breeding for low phytic acid. *Physiology and Molecular Biology of Plants* 26, 353–365.

Prasanna, B.M., Palacios-Rojas, N., Hossain, F., Muthusamy, V., Menkir, A. *et al.* (2020) Molecular breeding for nutritionally enriched maize: status and prospects. *Frontiers in Genetics* 10, 1392.

Raboy, V. (1997) Accumulation and storage of phosphate and minerals. In: Larkins, B.A. and Vasil, I.K. (eds) *Cellular and Molecular Biology of Plant Seed Development. Advances in Cellular and Molecular Biology of Plants*, Vol. 4. Springer, Dordrecht, the Netherlands, pp. 441–477. Available at: https://doi.org/10.1007/978-94-015-8909-3_12

Raboy, V. (2006) Seed phosphorus and the development of low-phytate crops.: In: Turner, B., Richardson, A. and Mullaney, E. (eds) *Inositol Phosphates: Linking Agriculture and Environment*. CAB International, Wallingford, UK, pp. 111–132.

Raboy, V., Gerbasi, P., Young, K.A., Stoneberg, S., Pickett, S.G. *et al.* (2000) Origin and seed phenotype of maize *low phytic acid 1-1* and *low phytic acid 2-1*. *Plant Physiology* 124, 355–368.

Rajasekhar, K.V., Prakash, B., Lakshmi, V.K., Rao S.V.R. and Raju, M.V.L.N. (2020) Effect of feeding diet with alternate protein sources and quality protein maize on performance and nutrient utilization in broiler chickens. *Tropical Animal Health and Production* 52, 2297–2302. Available at: https://doi.org/10.1007/s11250-020-02251-4

Rakshit, S. and Chikkappa, G.K. (2018) Perspective of maize scenario in India: way forward. *Maize Journal* 7(2), 49–55.

Reddy, N.R., Pierson, M.D., Sathe, S.K. and Salunkhe, D.K. (1989) Nutritional consequences of phytates. *Phytates in Cereals and Legumes*. CRC Press, Boca Raton, Florida, pp. 81–110.

Ribaut, J.M. and Hoisington, D.A. (1998) Marker assisted selection: new tools and strategies. *Trends in Plant Science* 3, 236–239.

Rosegrant, M.R., Ringler, C., Sulser, T.B., Ewing, M., Palazzo, A. and Zhu, T. (2009) *Agriculture and Food Security Under Global Change: Prospects for 2025/2050*. International Food Policy Research Institute, Washington, DC.

Sajilata, M.G., Singhal, R.S. and Kulkarni, P.R. (2006) Resistant starch – a review. *Comprehensive Reviews in Food Science and Food Safety* 5, 1–17.

Sarika, K., Hossain, F., Muthusamy, V., Baveja, A., Zunjare, R. *et al.* (2017) Exploration of novel *opaque16* mutation as a source for high-lysine and -tryptophan in maize endosperm. *Indian Journal of Genetics and Plant Breeding* 77, 59–64.
Sarika, K., Hossain, F., Muthusamy, V., Zunjare, R., Goswami, R. *et al.* (2018a) *Opaque16*, a high lysine and tryptophan mutant, does not influence the key physico-biochemical characteristics in maize kernel. *PLoS ONE* 13(1), e0190945.
Sarika, K., Hossain, F., Muthusamy, V., Zunjare, R.U., Baveja, A. *et al.* (2018b) Marker-assisted pyramiding of *opaque2* and novel *opaque16* genes for further enrichment of lysine and tryptophan in sub-tropical maize. *Plant Science* 272, 142–152.
Sheftel, J., Gannon, B.M., Davis, C.R. and Tanumihardjo, S.A. (2017) Provitamin A-biofortified maize consumption increases serum xanthophylls and $^{13}C$-natural abundance of retinol in Zambian children. *Experimental Biology and Medicine* 242, 1508–1514. Available at: https://doi.org/10.1177/1535370217728500
Shi, J., Wang, H., Hazebroek, J., Ertl, D. and Harp, T. (2005) The maize low-phytic acid 3 encodes a myo-inositol kinase that plays a role in phytic acid biosynthesis in developing seeds. *The Plant Journal* 42, 708–719.
Shiferaw, B., Prasanna, B.M., Hellin, J. and Banziger, M. (2011) Crops that feed the world, 6. Past successes and future challenges to the role played by maize in global food security. *Food Security* 3, 307–327.
Shutu, X., Dalong, Z., Ye, C., Yi, Z., Shah, T. *et al.* (2012) Dissecting tocopherols content in maize (*Zea mays* L.) using two segregating populations and high-density single nucleotide polymorphism markers. *BMC Plant Biology* 12, 201.
Simpungwe, E., Dhliwayo, T., Palenberg, M., Taleon, V., Birol, E. *et al.* (2017) Orange maize in Zambia: crop development and delivery experience. *African Journal of Food Agriculture, Nutrition and Development* 17, 11973–11999.
Sommer, A. and West, K.P. (1996) *Vitamin A Deficiency: Health Survival and Vision.* Oxford University Press, New York.
Sommer, A. and Davidson, F.R. (2002) Assessment and control of vitamin A deficiency: the Annecy Accords. *Journal of Nutrition* 132, S2845–S2850.
Sparvoli, F. and Cominelli, E. (2015) Seed biofortification and phytic acid reduction: a conflict of interest for the plant? *Plants* 4, 728–755.
Sullivan, J.S., Knabe, D.A., Bockholt, A.J. and Gregg, E.J. (1989) Nutritional value of quality protein maize and food maize for starter and growth pigs. *Journal of Animal Science* 67, 1285–1292.
Sureshkumar, S., Tamilkumar, P., Senthil, N., Nagarajan, P., Thangavelu, A.U. *et al.* (2014) Marker assisted selection of low phytic acid trait in maize (*Zea mays* L.). *Hereditas* 151, 20–27.
Tamilkumar, P., Senthil, N., Sureshkumar, S., Thangavelu, A.U., Nagarajan, P. *et al.* (2014) Introgression of low phytic acid locus (*lpa2-2*) into elite maize (*Zea mays* L.) inbred through marker assisted backcross breeding. *Australian Journal of Crop Science* 8, 1224–1231.
Tanumihardjo, S.A. (2011) Vitamin A: biomarkers of nutrition for development. *American Journal of Clinical Nutrition* 94, 658–665.
Tessema, M., Gunaratna, N.S., Donato, K., Cohen, J.L., Connell, M. *et al.* (2016) Translating the impact of quality protein maize into improved nutritional status for Ethiopian children: study protocol for a randomized controlled trial. *BMC Nutrition* 2, 54.
Traber, M.G., Frei, B. and Beckman, J.S. (2008) Vitamin E revisited: do new data validate benefits for chronic disease prevention? *Current Opinion in Lipidology* 19, 30–38.
Tyagi, P.K., Shrivastav, A.K., Mandal, A.B., Tyagi, P.K. and Elangovan, A.V. (2011) The feeding value of quality protein maize is similar to commercial maize for egg production and quality traits in laying hens. *Indian Journal of Poultry Sciences* 45, 217–219.
Vasal, S.K. (2000) The quality protein maize (QPM) story. *Food and Nutrition Bulletin* 21, 445–450.
Vasal, S.K., Villegas, E., Bajarnason, M., Gelaw, B. and Geirtz, P. (1980) Genetic modifiers and breeding strategies in developing hard endosperm *opaque-2* materials. In: Pollmer, W.G. and Philips, R.H. (eds) *Improvement of Quality Traits for Silage Use*. Martinus Nijhoff, The Hague, the Netherlands, pp. 37–71.
Vignesh, M., Hossain, F., Nepolean, T., Saha, S., Agrawal, P.K. *et al.* (2012) Genetic variability for kernel β-carotene and utilization of *crtRB1* 3′TE gene for biofortification in maize (*Zea mays* L.). *Indian Journal of Genetics and Plant Breeding* 72, 189–194.
Wang, G., Sun, X., Wang, G., Wang, F., Gao, Q. *et al.* (2011) *Opaque7* encodes an acyl-activating enzyme-like protein that affects storage protein synthesis in maize endosperm. *Genetics* 189, 1281–1295.

Wang, W., Niu, S., Dai, Y., Wang, M., Li, Y., Yang, W. and Zhao, D. (2019) The *Zea mays* mutants *opaque2* and *opaque16* disclose lysine change in waxy maize as revealed by RNA-seq. *Scientific Reports* 9, 12265.

WHO (2009) *Global prevalence of vitamin A deficiency in populations in risk 1995–2005. WHO Global Database on Vitamin A Deficiency*. World Health Organization, Geneva, Switzerland. Available at: https://apps.who.int/iris/bitstream/handle/10665/44110/9789241598019_eng.pdf?sequence=1 (accessed 26 February 2021).

Wong, J.C., Lambert, R.J., Tadmor, Y. and Rocheford, T.R. (2003) QTL associated with accumulation of tocopherols in maize. *Crop Science* 43, 2257–2266.

Wu, Y., Campbell, M., Yen, Y., Wicks, Z. III and Ibrahim, A.M.H. (2009) Genetic analysis of high amylose content in maize (*Zea mays* L.) using a triploid endosperm model. *Euphytica* 166, 155–164.

Yadav, O.P., Hossain, F., Karjagi, C.G., Kumar, B., Zaidi, P.H. *et al.* (2015) Genetic improvement of maize in India: retrospect and prospects. *Agricultural Research* 4, 325–338.

Yadava, D.K., Hossain, F. and Mohapatra, T. (2018) Nutritional security through crop biofortification in India: status & future prospects. *Indian Journal of Medical Research* 148, 621–631.

Yan, J., Kandianis, C.B., Harjes, C.E., Bai, L., Kim, E.H. *et al.* (2010) Rare genetic variation at *Zea mays crtRB1* increases β-carotene in maize grain. *Nature Genetics* 42, 322.

Yang, W., Zheng, Y., Zheng, W. and Feng, R. (2005) Molecular genetic mapping of a high-lysine mutant gene (*opaque16*) and the double recessive effect with *opaque2* in maize. *Molecular Breeding* 15, 257–269.

Zhai, S.W. (2002) Nutritional evaluation and utilization of quality protein maize Zhong Dan 9409 in laying hen feed. MSc thesis, North Western Agricultural and Forestry University of Science and Technology, Shaanxi, China.

Zhang, J., Chen, J., Yi, Q., Hu, Y., Liu, H., Liu, Y. and Huang, Y. (2013) Novel role of ZmaNAC36 in co-expression of starch synthetic genes in maize endosperm. *Plant Molecular Biology* 84, 359–369.

Zuma, M.K., Kolanisi, U. and Modi, A.T. (2018) The potential of integrating provitamin A-biofortified maize in smallholder farming systems to reduce malnourishment in South Africa. *International Journal of Environmental Research and Public Health* 15, 805.

Zunjare, R.U., Hossain, F., Muthusamy, V., Baveja, A., Chauhan, H.S. *et al.* (2017) Influence of rare alleles of *β-carotene hydroxylase* (*crtRB1*) and *lycopene epsilon cyclase* (*lcyE*) genes on accumulation of provitamin A carotenoids in maize kernels. *Plant Breeding* 136, 872–880.

Zunjare, R.U., Hossain, F., Muthusamy, V., Baveja, A., Chauhan, H.S. *et al.* (2018a) Development of biofortified maize hybrids through marker-assisted stacking of *β-carotene hydroxylase*, *lycopene-ε-cyclase* and *opaque2* genes. *Frontiers in Plant Science* 9, 178.

Zunjare, R.U., Chhabra, R., Hossain, F., Muthusamy, V., Baveja, A. and Gupta, H.S. (2018b) Development and validation of multiplex-PCR assay for simultaneous detection of rare alleles of *crtRB1* and *lcyE* governing higher accumulation of provitamin A in maize kernel. *Journal of Plant Biochemistry and Biotechnology* 27, 208–214.

# 22 Molecular Breeding for Improving Yield in Maize: Recent Advances and Future Perspectives

**Meiping Zhang[1,2], Yun-Hua Liu[2] and Hong-Bin Zhang[2*]**
[1]*Jilin Agricultural University, Changchun, People's Republic of China;*
[2]*Texas A&M University, College Station, Texas, USA*

## 22.1 Introduction

World food security is being challenged by the rapidly growing global population, emerging global climate change, stagnating crop productivity, and increasing public concern about environmental pollution resulting from intensive food production practices. Population experts are predicting that the world's population will grow to 9.6 billion by 2050. To feed this population, agricultural production must be increased by 45–50% over the next 30 years (Godfray *et al.*, 2010). Climate experts are drawing attention to and showing evidence for rising temperatures, erratic rainfall, declining water tables and increasing drought incidence. Hatfield *et al.* (2011) investigated the potential effects of global climate change on agriculture and found that 'climate disruptions to agricultural production have increased in the past 40 years and are projected to increase over the next 25 years. By mid-century and beyond, these impacts will be increasingly negative on most crops and livestock.' The Intergovernmental Panel on Climate Change (2007) predicted that the current average temperature will increase by 2–3°C over the next 30–50 years. Hatfield and Prueger (2015) showed that warmer temperatures during reproductive stages would reduce maize grain yield by as much as 80–90%. Daryanto *et al.* (2016) analysed historical data of drought effects on crop production and showed that when water was reduced by approximately 40%, wheat and maize yields were reduced by 21 and 39%, respectively. On the other hand, agricultural experts have not been able to raise the yield potential of major cereals and pulses to levels that occurred during the 'green revolution' era. It has been reported that the genetic gain in some major crops has increased only slightly (Ray *et al.*, 2012) or even stagnated (Godfray *et al.*, 2010; Tilman *et al.*, 2011; Schauberger *et al.*, 2018). The level of the reported genetic gain cannot meet the demand for food, feed, clothing and biofuel for the projected global population. Environmentalists are expressing strong concerns about environmental pollution due to excess fertilizer and pesticide use as well as questioning the use of genetically modified crops with foreign genes.

Maize (*Zea mays* L.), commonly known as corn, is the world's leading field crop for human food, animal feed and bioenergy (ethanol). It is a truly worldwide crop that was grown in 169 countries or regions on a harvested area of 236

*Corresponding author: hbz7049@tamu.edu

 *Molecular Breeding in Wheat, Maize and Sorghum: Strategies for Improving Abiotic Stress Tolerance and Yield* (eds M.A. Hossain *et al.*)
DOI: 10.1079/9781789245431.0022

million hectares and produced more than 1404 million tonnes of grains in 2018 (FAO, 2019). The number of countries or regions that grew maize in 2018 was respectively 36 and 43% more than those that grew the two other worldwide staple food crops, wheat and rice. The total harvested area of maize was close to that of wheat, but larger than that of rice by 20%. The total world maize production was much higher than the 865 million tonnes of wheat and the 996 million tonnes of rice (FAO, 2019).

The challenge is how to produce 35–50% more food over the next 40–50 years with less land, less fertilizers, less pesticides and less irrigation water, on the one hand, and erratic rainfall, frequent drought and increasing temperatures, on the other. Enhanced crop improvement through research, development and applications of new technologies (Ronald, 2014; Li *et al.*, 2018) or molecular breeding has been one of the approaches of choice to help feed the world. It is essential to develop and utilize substantially improved crop varieties for enhanced crop production. Therefore, several molecular technologies have been developed and used for enhanced crop improvement, including marker-assisted selection (MAS) (Collard *et al.*, 2008), genetic engineering (GEE) (Datta, 2013), RNA interference (RNAi) (Yogindran and Rajam, 2015), genomic selection (GS) (Meuwissen *et al.*, 2001; Desta and Ortiz, 2014), gene editing (GE) (Abdallah *et al.*, 2015) and gene-based breeding (GBB) (Zhang, M.P. *et al.*, 2020a).

This chapter first clarifies plant breeding and its underlying molecular basis, then reviews the molecular technologies that have been developed thus far for enhanced plant breeding. These are necessary to better understand the applications and perspectives of these molecular technologies for enhanced maize breeding. Next, the chapter updates the recent advances of the molecular technologies for maize grain yield breeding in the past decade. Finally, the chapter compares these molecular technologies and underlines their perspectives for continued maize yield improvement.

## 22.2 Plant Breeding and Its Underlying Molecular Basis

Plant breeding, including pure line variety breeding (Fig. 22.1) and hybrid variety breeding (Fig. 22.2), often includes three steps: (i) parent selection; (ii) cross design; and (iii) progeny selection. In comparison, parent selection is probably the most important to a successful breeding, because it is impossible to

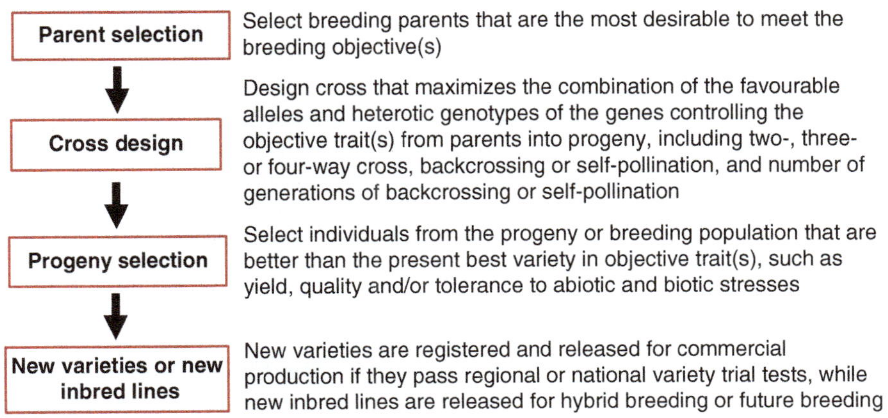


**Fig. 22.1.** A general flowchart of plant breeding for new pure line varieties or new inbred lines. Of these steps of the breeding process, parent selection is the most crucial to success in the development of elite varieties, followed by cross design. It is impossible to develop an elite variety that is competitive for commercial production from a progeny or breeding population without elite individuals. Progeny selection identifies the elite individuals existing in the progeny population that can be potentially developed into elite varieties competitive for commercial production.

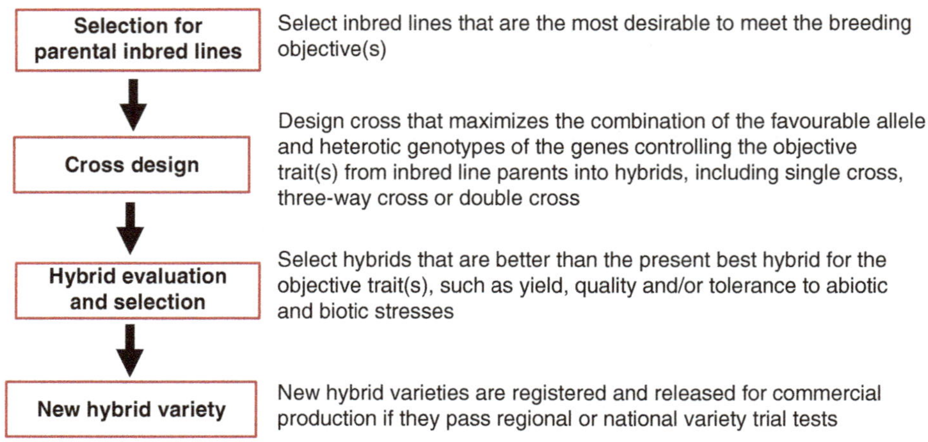


**Fig. 22.2.** A general flowchart of plant breeding for hybrid varieties. Of these steps of the hybrid variety breeding process, parental inbred line selection is the most crucial to success in the development of superior heterotic hybrid varieties, followed by cross design. It is impossible to develop a superior heterotic hybrid variety that is competitive for commercial production from hybrids without a superior hybrid. Hybrid selection identifies the superior hybrid from a number of hybrids developed and under field trial evaluation that can be potentially developed into a superior hybrid variety competitive for commercial production.

develop a superior variety from the progeny derived from parents that do not contain superior individuals. Therefore, it is crucial to the success of plant breeding to identify the parental lines that are the most desirable to approach the breeding objective(s). The next in importance is probably cross design because it is essential to maximize the combination of the favourable alleles and/or heterotic genotypes of genes controlling the objective trait(s) from the selected breeding parents into progeny. Cross design includes, but is not limited to, which of the selected parents is used as the female parent(s) and which is used as the male parent(s); which of the cross methods is used, two-way, three-way or four-way cross; whether backcross is pursued; and for how many generations the backcross or selfing pollination is carried out. In the other words, should all favourable alleles of the genes controlling the objective trait(s) be fixed (homozygous), or the additive or complete dominant alleles of the genes be fixed, but the over-dominant alleles be maintained heterozygous? The least important among the three steps of plant breeding probably is progeny selection, which is aimed at identifying the best individual from the resulting progeny and develop it into a variety that is competitive for commercial production.

Plant breeding overall is a continuous process in which only a progeny line that is significantly better than the best variety that is presently grown for commercial production can be developed into a new variety and released for commercial production. Since most agronomic traits, such as yield, quality and environmental stress tolerances, are controlled by numerous genes, the molecular basis of plant breeding is, in fact, to incorporate continuously and substantially the favourable alleles and/or heterotic genotypes of the genes controlling the objective trait(s), such as yield, from breeding parents into the present best varieties or breeding lines. This molecular basis of plant breeding is applied not only to the conventional phenotype-based breeding (PBB), but also to the modern plant breeding assisted by molecular breeding technologies, such as MAS, GEE, RNAi, GS, GE and GBB.

## 22.3 The Molecular Technologies Developed for Enhanced Plant Breeding

The conventional PBB, in which parent selection, cross design and progeny selection are all based on the visible phenotype of the objective trait(s), has greatly contributed to crop genetic improvement, through which many superior varieties have been developed, thus increasing agriculture production. Nevertheless, it has been a rigorous challenge how to continuously improve crops on the basis of the present crop varieties. It has been the consensus that continued crop improvement will be largely contingent on research, development and application of molecular technologies in plant breeding. Therefore, MAS (Collard *et al.*, 2008), GEE (Datta, 2013), RNAi (Yogindran and Rajam, 2015), GS (Meuwissen *et al.*, 2001; Desta and Ortiz, 2014), GE (Abdallah *et al.*, 2015) and GBB (Zhang, M.P. *et al.*, 2020a) have been developed to improve the ability, efficiency and productivity of plant breeding and/or to accelerate the breeding process.

### 22.3.1 Marker-assisted selection

MAS was the first molecular method initiated for enhanced breeding in the early 1990s (Fig. 22.3). It is an indirect method for the objective trait for parent selection, cross design and progeny selection, and usually used to help improve an existing variety in one or two agronomic traits that are undesirable for commercial production.

Especially, MAS is often aimed at selecting the individual(s) from a breeding progeny that contain(s) the favourable allele of a major gene or quantitative trait locus (QTL) controlling the objective trait using its closest linked flanking DNA markers. MAS can be also used for parent selection and cross design (Table 22.1), but its efficiency is limited for these uses. MAS is often used in the following instances: (i) the objective trait is genetically a simple trait controlled by one or a few major genes, or the objective trait is a quantitative trait with a lower heritability, but the targeted QTL has a major effect on the objective trait; (ii) the objective trait cannot be properly phenotypically selected in early generations of intermating or selfing, and in off-season nurseries or a greenhouse; (iii) backcrossing is used to incorporate the rare, favourable allele of one or a few genes or QTLs from a wild or an exotic germplasm line into an existing variety; or (iv) the breeding objective involves multiple traits that cannot be conducted for progeny selection under a single condition, such as breeding for high yield with disease resistance.

Figure 22.4a summarizes a general flowchart of MAS. Although some of the DNA markers that have been previously mapped and closely linked to the gene or QTL controlling the objective trait can be directly used for MAS for a targeted breeding population, most of them cannot be directly used for plant breeding in a breeding programme. This is because the closely linked DNA markers, such as single-nucleotide polymorphism (SNP), may be no longer polymorphic and/or may have a different linkage phase (repulsion or coupling) with the targeted locus in the targeted breeding population. The effect of the targeted QTL may also vary, due to gene × genetic background interactions and/or genotype × environment (G × E) interactions. Therefore, it is necessary to test and verify the polymorphism and linkage phase of the closest linked DNA markers, and the effect of the targeted QTL on the objective trait, before MAS is carried out. If the closest linked DNA marker is no longer polymorphic in the targeted breeding population, a different marker in the bin of DNA markers containing the closest linked marker or the next closest linked marker is tested and, if polymorphic, used for MAS. If the targeted QTL no longer has a major effect on the objective trait in the targeted breeding population, a different QTL controlling the objective trait may be selected for MAS.

The next concern of MAS is the selection and use of DNA markers for MAS. This influences not only the reliability of MAS, but also

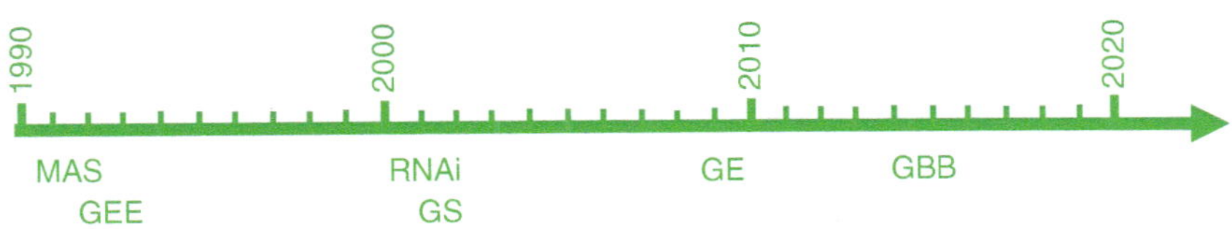


**Fig. 22.3.** History of molecular breeding for crop genetic improvement. MAS, marker-assisted selection; GEE, genetic engineering; RNAi, RNA interference; GS, genomic selection; GE, gene or genome editing; GBB, gene-based breeding.

**Table 22.1.** Comparison of the molecular technologies developed to date for enhanced crop improvement.

| Method | Breeding objective | Potential use | No. of genes | GMO | Trait |
|---|---|---|---|---|---|
| Marker-assisted selection (MAS) | Improve an existing variety or breeding line | • Limited parent selection<br>• Limited cross design<br>• Progeny selection | One or more major genes | No | Simple trait |
| Genetic engineering (GEE) | Improve an existing variety or breeding line | • Gene transformation<br>• Progeny validation | One or a few major genes | Yes | Simple trait |
| RNA interference (RNAi) | Improve or fine-tune an existing variety or breeding line | • RNAi transformation<br>• Progeny validation | One or a few major genes | Yes | Simple trait |
| Gene or genome editing (GE) | Improve or fine-tune an existing variety or breeding line | • gRNA transformation<br>• Progeny validation | One or a few major genes | Possibly not | Simple trait |
| Genomic selection (GS) | Develop new varieties or new breeding lines | • Progeny selection | Genome-wide genes | No | Polygenic trait |
| Gene-based breeding (GBB) | Develop new varieties or new breeding lines | • Parent selection<br>• Cross design<br>• Progeny selection | Genome-wide genes | No | Polygenic trait |

GMO, genetically modified organism; gRNA, guide RNA.

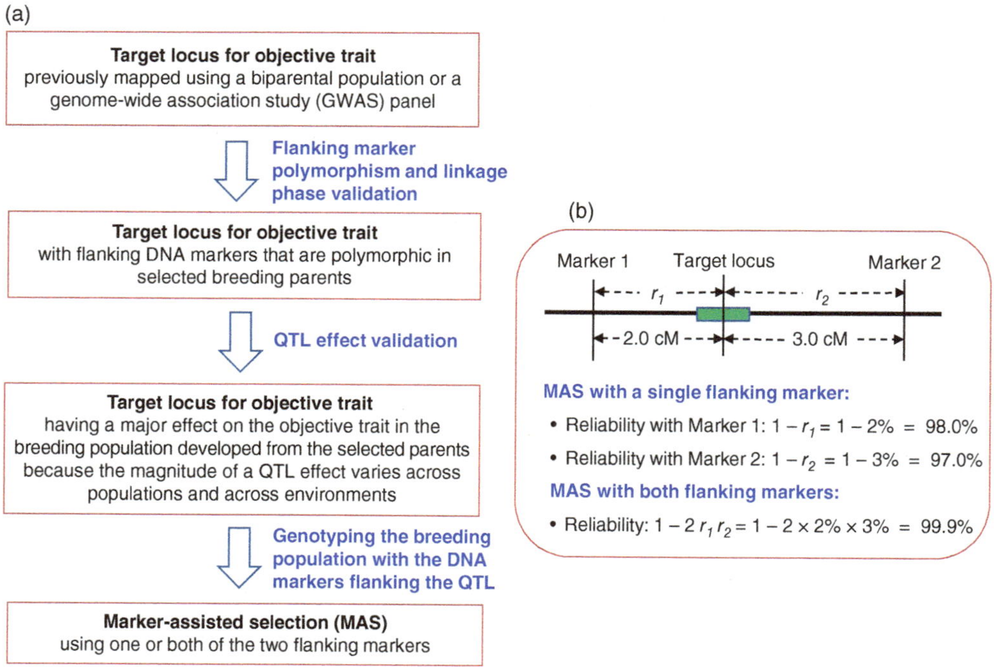


**Fig. 22.4.** Marker-assisted selection (MAS). (a) Flowchart of MAS. (b) The mapped target locus selected for MAS and its selection reliability. The target locus could be a major gene or quantitative trait locus (QTL) controlling the objective trait.

the utility of the plant lines resulting from MAS for subsequent commercial variety development. The reliability of MAS depends on both the genetic distance between the DNA marker and the targeted locus, and the number of markers used for MAS. The closer to the targeted locus or loci the DNA markers selected for MAS, the more reliable the MAS and the more applicable for variety development the plant lines resulting from MAS. It is worthwhile to clarify here that there is usually no correlation between genetic (calculated from genetic recombination frequency) and physical (true) distances. The genetic distance may substantially vary between the population used for targeted locus mapping and the breeding population, due to gene × genetic background interaction, and between earlier and later generations, because of the variation in probability of genetic recombination between the marker and the targeted locus through generations. If both closest linked flanking DNA markers can be used, the reliability of MAS can be increased, relative to use of only one of the two closest linked flanking markers, but the cost of MAS could be largely doubled. The utility of the plant lines resulting from MAS for variety development is also an issue of distance, especially physical distance, between the marker and the targeted locus. The larger the physical distance between the marker and the targeted locus, the larger the chromatic segment containing the targeted locus that would be selected for. Therefore, the likelihood that the chromatic segment also carries undesirable genes would be increased, thus reducing the utility of the plant lines resulting from MAS for variety development.

No genetically modified organism (GMO) is created or involved in MAS, but it is likely that undesirable gene(s) accompany the targeted gene controlling the objective trait. Moreover, since it is often used for progeny selection for one or a few genes controlling the objective trait at a time, MAS is better suited for crop genetic improvement for simple objective traits each controlled by one or a few major genes. MAS is less efficient for genetic improvement for complex or quantitative traits controlled by numerous genes, unless DNA markers associated with most, if not all, of the QTLs controlling the objective trait are applied for MAS.

### 22.3.2 Genetic engineering

GEE is also one of the first developed and used molecular technologies for enhanced crop genetic improvement in the early 1990s (Fig. 22.3). This technology is also often used to improve an existing variety in one or two agronomic traits that are undesirable or absent for commercial production. This is accomplished by genetic transformation of the favourable allele of the gene controlling the objective trait, or of one or two exotic genes controlling a new trait of interest into the variety (Table 22.1).

Figure 22.5a shows a general flowchart of crop improvement through GEE. The genomic DNA of the targeted gene including its regulatory elements, such as promoter and enhancer, or the full-length complementary DNA (cDNA) of the targeted gene is cloned into a plant transformation binary vector, thus creating a so-called gene transformation 'construct'. Nevertheless, when the cDNA of the targeted gene is used to develop the gene transformation construct, it is necessary to use the transcript of the targeted gene that is responsible for the objective trait. Zhang, M.P. *et al.* (2019, 2020b) showed that a number of genes in maize are alternatively spliced into multiple transcripts while they express. Zhang, M.P. *et al.* (2020b) reported that only one to four, but not all, of the transcripts spliced from a maize grain yield gene control grain yield in maize. The construct is then transformed into an existing variety to be improved through either the *Agrobacterium*-mediated or gene gun transformation method. The transformed plants are selected and validated for the presence of the transformed targeted gene or transgene, the performance of the targeted trait and the performance of other agronomic traits that may be affected by the transgene. This is because most of the genes controlling a quantitative trait are pleiotropic and all genes controlling a quantitative trait are functionally correlated, forming an interaction network (Zhang, M.P. *et al.*, 2020b). If the objective trait of the variety has been significantly improved in the transgenic plant, with no other agronomic traits that have been significantly adversely affected, the transgenic plant could be developed into an improved variety or breeding line.

GEE is potentially a straightforward and rapid method for plant genetic improvement through replacing the undesirable alleles of the targeted gene existing in the variety to be improved, or adding a gene controlling a new trait that is absent in the variety, but the variety developed from the transgenic plant is often subjected to the issue of GMO, to the genotypic specificity of genetic transformation and to the limitation of large-scale crop genetic improvement or development of brand-new varieties.

### 22.3.3 RNA interference

RNAi is a molecular tool for enhanced breeding, initiated in the early 2000s (Fire *et al.*, 1998; Kusaba, 2004) (Fig. 22.3), that also involves transgenes and, thus, GMO. Instead of adding the favourable allele of the gene controlling the objective trait or an exotic gene controlling a new trait of interest to an existing variety as does GEE, RNAi aims to improve or fine-tune an existing variety through spatially and/or temporally regulating the expression of the gene controlling the objective trait (Table 22.1). In other words, RNAi can fine-tune the expression of the targeted gene controlling the objective trait in a specific organ, or at a specific plant growth and development stage. It uses a short sequence of the targeted gene contained in the variety to be improved that controls the objective trait to generate small RNAs (microRNA (miRNA) or small interfering RNA (siRNA)) specific to the targeted gene(s). The expression of the targeted gene, therefore, is silenced or downregulated by the small RNAs, thus fine-tuning the phenotype of the objective trait.

Figure 22.5b shows a general flowchart of crop improvement with RNAi. A short sequence (20–800 bp) of the targeted gene contained in an existing variety is cloned in an inverted repeat separated by a short sequence into a plant transformation binary vector. The inverted repeat of the host targeted gene separated by a short sequence will be transcribed into RNA that can form a hairpin or 'stem–leaf' structure, of which the 'stem' part of the hairpin structure is processed into small RNA to regulate the expression of the host targeted gene after transformation into a host plant. The gene RNAi construct is transformed into an existing variety to be improved through either the *Agrobacterium*-mediated or gene gun transformation method as described above for GEE. The transformed plant cells are selected, regenerated into transgenic plants, validated and developed into a transgenic line for variety development.

It is apparent that the RNAi technology is also a type of GEE because both add one or more transgenes to an existing variety. Although it does not transform a foreign gene that is absent in the existing variety to be improved or a favourable allele of a gene that exists in the variety as does GEE, the RNAi technology delivers a short sequence of the targeted gene, 20–800 bp, into the existing variety using the same genetic transformation procedure as GEE. Instead of replacing the favourable allele of the targeted gene or adding a new gene, RNAi regulates or usually silences the expression of the targeted pre-existing gene. As it is a type of GEE, the variety improved by the RNAi technology is also subjected to the issue of GMO, to the genotypic specificity of genetic transformation and to the limitation of large-scale genetic improvement or development into a brand-new variety.

### 22.3.4 Gene or genome editing

Four GE systems have been developed over past decades, including meganucleases (Rouet *et al.*, 1994), zinc-finger nucleases (ZFNs) (Kim *et al.*, 1996), transcription activator-like effector nucleases (TALENs) (Boch *et al.*, 2009) and clustered regularly interspaced short palindromic repeats (CRISPR)/CRISPR-associated protein 9 (Cas9) (Jinek *et al.*, 2012), but the CRISPR/Cas9 system has been most widely used for crop genetic improvement due to its relative ease and high efficiency. GE has been applied as a molecular technology for crop genetic improvement only recently (Fig. 22.3), even though the basic research on its underlying mechanisms has been carried out over decades. It is usually used to modify or improve an existing variety through targeted induction of point mutation(s), such as nucleotide deletion/insertion (InDel) or nucleotide substitution, of the targeted gene controlling the objective trait. As does the RNAi technology, GE modifies endogenous gene(s) controlling the objective trait, but it does not transfer or transform a foreign gene into an existing variety, as does GEE. In comparison with the RNAi technology that modifies the targeted variety

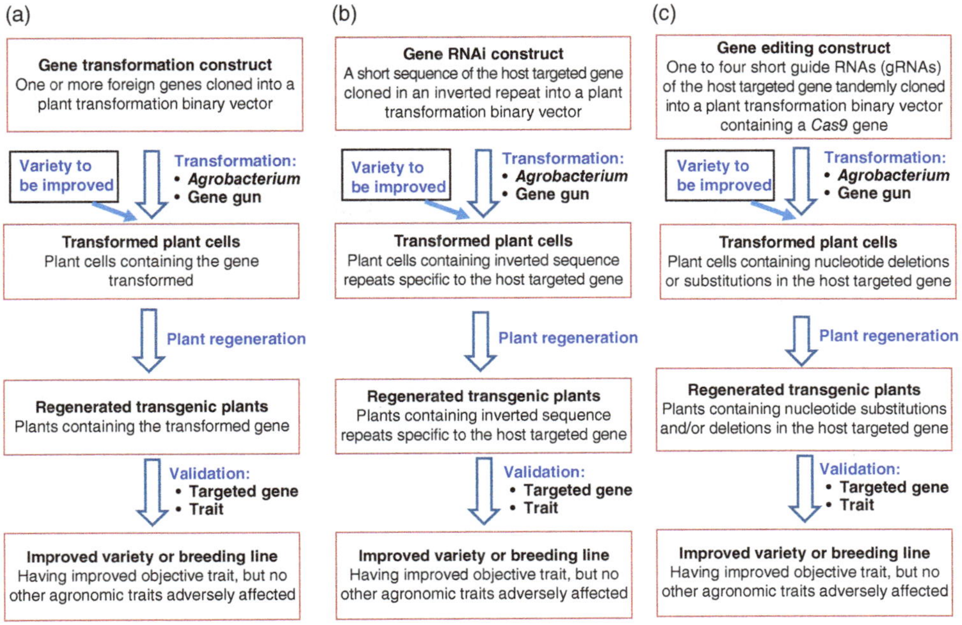


**Fig. 22.5.** General flowcharts of crop improvement through (a) genetic engineering (GEE), (b) RNA interference (RNAi) or (c) gene editing (GE). GEE improves an existing variety by adding one or more favourable foreign genes or alleles to its genome; RNAi improves an existing variety by spatially and/or temporarily silencing or downregulating its one or more existing unfavourable genes; and GE improves an existing variety by targeted knocking-off or mutating its one or more existing unfavourable genes.

through regulating the expression of the gene controlling the objective trait, GE fine-tunes the targeted variety through targeted inducing of point mutations in the gene controlling the objective trait (Table 22.1).

Figure 22.5c shows a general flowchart of crop improvement through GE. A short sequence or a single guide RNA (20 bp) of the host targeted gene is cloned into a plant transformation binary vector containing a Cas9 nickase gene or a Cas9 nickase/adenine or cytidine deaminase fusion. The former vector construct preferentially induces InDels, while the latter vector construct preferentially induces nucleotide substitutions, especially A→G or C→T substitution. The gene GE construct is transformed into an existing variety to be improved through either the *Agrobacterium*-mediated or gene gun transformation method, as is GEE or RNAi construct. The transformed plant cells are selected, regenerated into transgenic plants, validated and developed into a transgenic line for variety development.

Depending on selection of the edited plants, the plants edited with GE may not be transgenic plants, leading to transgene-free plants. If the GE construct is not integrated into the host plant genome, it may still induce point mutations at the targeted sequence of the gene controlling the objective trait through transient expression. If it is integrated into the host plant genome, the transgene could be eliminated from the edited plants through genetic segregation and selection for transgene-free plants. Therefore, no transgene is involved in the edited plants, thus probably being not subjected to GMO. Nevertheless, GE cannot introduce a new gene and thus a new trait into the variety to be improved, as does GEE. Finally, like GEE, RNAi and MAS, GE is restricted to the genotypes that can be transformed and has limited applications for large-scale crop genetic improvement or development of brand-new varieties.

### 22.3.5 Genomic selection

Plant breeding, as indicated above, usually includes three steps: parent selection, cross design and progeny selection. GS was first proposed in 2001 (Meuwissen *et al.*, 2001), but it has not been extensively studied until recently, especially after the advent of the next-generation high-throughput sequencing technology (Fig. 22.3). GS is used to assist progeny selection with genome-wide omics, including genome-wide random DNA (SNP) markers, genome-wide random gene expressions and genome-wide random metabolites. In particular, the phenotype of the objective trait is predicted, instead of phenotyping through field experiments, as does the phenotype-based progeny selection, with genome-wide random omics using a trained prediction model and then the progeny is selected based on the predicted phenotype of the objective trait. Therefore, it is crucial for GS to train a prediction model that enables accurate prediction of the phenotype of the objective trait using genome-wide random omics. Since genome-wide omics is used, GS is well suited for selection of complex objective trait(s) that is/are quantitatively inherited and controlled by numerous genes, i.e. quantitative traits (Table 22.1). Importantly, most agronomic traits, such as crop yield, quality and abiotic and biotic stress tolerances, are quantitative traits, thus making GS attractive to molecular breeding.

Figure 22.6 shows a general flowchart of progeny selection for plant breeding by GS. GS includes two steps: (i) prediction model training and validation; and (ii) genomic prediction of objective trait and genomic selection. To train a prediction model and validate it, a small portion of a targeted breeding population, say from 100 to 1000 plants, is randomly selected, genotyped with genome-wide random omics, such as genome-wide SNPs, and phenotyped through field experiments. This small portion of the breeding population is designated or known as a 'training population or subpopulation'. Since genome-wide omics, such as SNPs, is used for GS, a multiple regression that can be parametric or non-parametric is used as a candidate prediction model. Training a prediction model, in fact, is to calculate the intercept and regression coefficient of the multiple regression model using the genotype of every SNP for every plant of the training subpopulation as the independent variable '*x*' of the regression model, while validating a trained prediction model is to calculate the dependent variable '*y*' (predicted phenotype of the objective trait) using the trained regression model and the genotype of every SNP in a plant of the training subpopulation as the independent variable '*x*' of the regression model, followed by calculating the Pearson's correlation coefficient ($r$) between the predicted ('*y*') and observed phenotypes of the objective trait. The Pearson's correlation coefficient, $r$, is defined as the prediction accuracy or ability, and $R^2$ (squared Pearson's correlation coefficient) as the predictability. The prediction model training and validation is usually accomplished using a cross-validation scheme, such as tenfold cross-validation. Briefly, the training subpopulation is randomly divided into ten groups, with each group having an equal or nearly equal number of plants. These ten groups of plants are then divided into two sets: nine of them as the training set to be used for prediction model training and one of them as the validation set to be used for trained prediction model validation. The phenotype of every plant in the validation set is predicted using the trained model and their genotypes of the DNA markers. This process is repeated until every group has been used once as the validation set and the phenotype of every plant in the training subpopulation is predicted. The prediction accuracy or predictability of the objective trait is calculated using the predicted phenotype of every plant and their phenotypes measured through a field experiment. This is defined as 'one replicate' for some publications. In general, the analysis is repeated for ten or more times and the mean prediction accuracy of the replicates is used to evaluate the accuracy or utility of the trained prediction model for GS. Alternatively, when one of the ten groups is used as the validation set and the phenotype of every plant in the validation set is predicted, the prediction accuracy ($r$) or predictability ($R^2$) is calculated. This is defined as 'one replicate' and when all ten groups are used as the validation sets, the analysis is defined as 'ten replicates' for other publications. Finally, the mean prediction accuracy is calculated to evaluate the accuracy or utility of the trained prediction model for GS. Although the analysis procedure is different between the two methods for calculating prediction accuracy, the final mean prediction accuracy ($r$) or predictability ($R^2$) is essentially the same between the two.

Once validated, with a significant and acceptably high mean value of $r$ or $R^2$, the trained prediction model is used to predict the objective trait phenotype of every plant in the remaining portion of the targeted breeding population that are only genotyped with the genome-wide omic data set. The predicted phenotype of the objective trait for every plant will be used to make selection, namely genomic selection.

One concern about GS is whether the prediction model of an objective trait trained with one breeding population (Population 1, Fig. 22.6) could be used to predict the phenotype of the trait for another population (Population 2, Fig. 22.6) using the genome-wide omic data set. One scenario is to directly use the prediction model trained with Population 1 for genomic prediction of the objective trait in Population 2; and the other scenario is to first validate the prediction model trained with Population 1 using Population 2 and then, if acceptable, the prediction model is used to predict the objective trait phenotype of every plant in Population 2 for GS. The scheme of validating a prediction model trained with one breeding population (Population 1) using an independent population (Population 2) is usually known as 'independent validation'. The former scenario is ideal, but the latter scenario is usually the case that has been widely used for GS. This is because the proper prediction model of the same objective trait may differ for different breeding populations due to gene × genetic background interaction. On the other hand, if the prediction model trained with Population 1 must be validated using Population 2 to evaluate the utility of the prediction model for GS in Population 2, why is a prediction model not directly trained and validated using Population 2, as done for Population 1, because the validation set of Population 2 for validating the

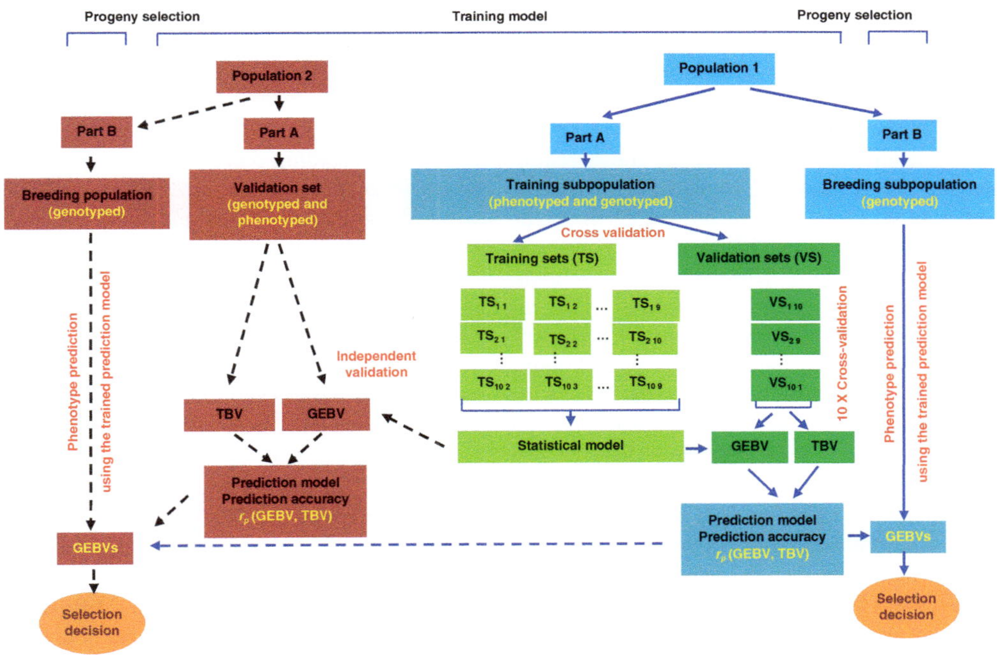


**Fig. 22.6.** Progeny selection of plant breeding by genomic selection (GS) or using the genes controlling the objective trait(s) for gene-based breeding (GBB). The breeding population, including its training subpopulation, is genotyped by genome-wide omic data set(s) for GS or using genome-wide genic data set(s) of the genes controlling the objective trait(s) for GBB. A tenfold cross-validation scheme is used for phenotype prediction of the objective trait for GS or GBB. Since the phenotype of an objective trait is significantly influenced by genetic background × gene interactions, when a prediction model trained and validated for one breeding population (Population 1) is used for progeny selection of another breeding population (Population 2), it must be revalidated using a subset of Population 2 to confirm the applicability of the model for Population 2. For the validation, the validation subset of Population 2 must be both genotyped and phenotyped; therefore, a new model that is desirable for progeny prediction of Population 2 could be trained and predicted as done for Population 1, thus making independent validation nonsense. GEBV, genomic estimated breeding value; TBV, true breeding value. (Modified from Zhang, M.P. *et al*., 2020a)

prediction model trained with Population 1 has been both genotyped and phenotyped anyway?

Unlike MAS, GEE, RNAi and GE, GS is applicable to large-scale crop genetic improvement and development of brand-new varieties because genome-wide omics that are from different loci controlling the objective trait, especially those quantitative traits, are used for prediction of the trait for GS. Unlike GEE, RNAi and GE that create or involve recombinant DNA and/or GMO, GS does not involve recombinant DNA nor GMO. Nevertheless, GS has significant weak points. First, plant breeding, as indicated above, includes parent selection, cross design and progeny selection, while GS is efficient only for progeny selection, but inefficient for parent selection and cross design because genome-wide random omics is used. It is impossible to develop a superior variety from a breeding progeny population without an individual that is superior to the present commercial variety. Second, it is costly to genotype a large breeding population using a large number of genome-wide omic features. For instance, if the objective trait is controlled by 100 genes, of which eight are segregating in the breeding population, at least 65,536 plants are needed in order to select for a plant having all the eight genes that are fixed or homozygous. If the crop has a genome size of 1000 Mb and the genome of its breeding population consists of 1000 recombinant segments, at least 4800 SNPs are needed in order to have a probability of more than 99% that at least one SNP is located at a recombinant segment. If it costs $5.00 to genotype a plant with the 4800 genome-wide omic features, it will cost $327,680 to genotype the 65,536 plants of the breeding population. Third, most quantitative agronomic traits studied thus far had a lower prediction accuracy, with approximately $r = 0.50$, for grain yield in maize, no matter how many omic features, what prediction models and/or which types of omics, such as genome-wide SNPs, gene expressions and metabolites, were used for the prediction (Table 22.2; Zhang, M.P. *et al.*, 2020a).

### 22.3.6 Gene-based breeding

GBB is an innovative molecular breeding technology that was first proposed in 2014 (Liu *et al.*, 2014; Zhang *et al.*, 2014) (Fig. 22.3) and recently demonstrated to be extremely powerful and efficient for enhanced breeding (Liu *et al.*, 2020). GBB develops brand-new varieties or large-scale improved crops by design by making full use of most, if not all, of the genes controlling the objective trait through the entire breeding process, including parent selection, cross design and progeny selection (Table 22.1). In particular, the number of favourable alleles (NFAs), genic SNPs and/or InDels, and expression profiles of the genes controlling the objective trait are used either individually or jointly for parent selection, cross design and progeny selection (Liu *et al.*, 2020). When the genic SNPs and/or InDels and expression profiles of the genes are used for progeny selection, multiple regression prediction models and the working scheme shown in Fig. 22.6 is used for GBB, as it is for GS; while when NFAs are used, a simple linear regression model and a working scheme similar to that shown in Fig. 22.6 are used for GBB, as described below and by Zhang, M.P. *et al.* (2020a). Therefore, GBB allows accurate selection of the breeding parents that are the most desirable to approach the breeding objective(s), wisely designing crosses that maximize the combination of the favourable alleles and/or heterotic genotypes of the genes controlling the objective trait(s) from the breeding parents into breeding progeny, and accurately and rapidly selecting for individuals from the breeding progeny that have the best objective trait(s) and developing them into new superior varieties or breeding lines.

Figure 22.7 shows a flowchart of GBB for superior pure line varieties in comparison with the current PBB assisted by GS or MAS, based on the NFAs of the genes controlling the objective trait(s). As described by Zhang, M.P. *et al.* (2020a), the NFAs is the total number of favourable alleles of the genes controlling the objective trait, calculated by adding up the NFAs of every gene controlling the objective trait. The NFAs of the unfavourable allele homozygote (aa) of a gene is scored as '0'; the NFAs of the favourable allele homozygote (AA) of the gene is scored as '2'; and the NFAs of the heterozygote (Aa) of the gene are scored as '1', if it has an additive effect, '2' if it has a complete dominant effect or '3' if it has an over-dominant effect. Figure 22.7 assumes that the objective trait, such as maize

**Table 22.2.** Examples of applications of molecular breeding for yield improvement in maize.

| Method | Trait | Gene/QTL/marker | Reference |
|---|---|---|---|
| Marker-assisted selection (MAS) | Grain yield | QTL-linked SNPs | Beyene *et al.* (2016a) |
| | Grain yield | QTL-linked SNPs | Beyene *et al.* (2016b) |
| | Grain yield | QTL-linked SNPs | Abdulmalik *et al.* (2017) |
| | Grain yield | QTL-linked SNPs | Bankole *et al.* (2017) |
| | Grain yield | QTL-linked SNPs | Cerrudo *et al.* (2018) |
| | Leaf angle | *UPA2* | Tian *et al.* (2019) |
| Genetic engineering (GEE) | Grain yield | *Mn1* | Li *et al.* (2013) |
| | Grain yield | *ZAR1* | Guo *et al.* (2014) |
| | Grain yield | *ZmDA1*, *ZmDAR1* | Xie *et al.* (2018) |
| | Leaf angle | *ZmRAVL1*, *brd1* | Tian *et al.* (2019) |
| | Grain yield | *zmm28* | Wu *et al.* (2019) |
| | Grain yield | *KNR6* | Jia *et al.* (2020) |
| RNA interference (RNAi) | Grain yield | *CP4 EPSPS* | Yang *et al.* (2018) |
| | Leaf angle | *ZmRAVL1* | Tian *et al.* (2019) |
| | Grain yield | *nac7* | Zhang, J. *et al.* (2019) |
| | Grain yield | *ZmNAC128*, *ZmNAC130* | Zhang, Z. *et al.* (2019) |
| | Grain yield | *KNR6* | Jia *et al.* (2020) |
| Gene or genome editing (GE) | Grain yield under drought stress | *AGROS8* | Shi *et al.* (2017) |
| | Kernel sweet, waxy | *SH2*, *WX* | Dong *et al.* (2019) |
| | Leaf angle | *ZmRAVL1* | Tian *et al.* (2019) |
| | Kernel size | *qKW9* | Huang *et al.* (2020) |
| | Stalk strength | *stiff1* | Zhang, Z. *et al.* (2020) |
| Genomic selection (GS) | Seven biomass- and bioenergy-related traits | 56,110 SNPs, 130 metabolites | Riedelsheimer *et al.* (2012) |
| | Grain yield | 65,995 SNPs | Edriss *et al.* (2017) |
| | Six grain yield-related traits | 100,000 SNPs, 28,769 gene expressions, 748 metabolites | Xu *et al.* (2017) |
| | Grain yield across European environmental conditions | 41,722 SNPs + 1312 SNPs associated with yield QTLs | Millet *et al.* (2019) |
| | Grain yield, days to silking, plant height | SNPs linked to loci with dominance effects; SNPs from the regions enriched in genes, structural features and/or evolutionary features | Ramstein *et al.* (2020) |
| Gene-based breeding (GBB) | Grain yield | *ZmINGY* genes | Zhang, M.P. *et al.* (2020a) |

QTL, quantitative trait locus; SNP, single-nucleotide polymorphism.

grain yield, is controlled by seven genes, defined as A to G, and the potential varieties or lines for breeding have similar phenotypes in the objective trait, because they have the same NFAs for the genes controlling the objective trait. When breeding parents are selected, the parents having six of the seven genes that are complementary could be accurately selected for GBB (Fig. 22.7a), but the parents having any genotypes could be selected for PBB (Fig. 22.7b), including those having the same genotype or all genes that are overlapping, such as AABBCCDDeeffgg, for both parents. Figure 22.7b assumes that the two parents selected for PBB have five of the seven genes that are overlapping and two that are complementary. For GBB, after the selected

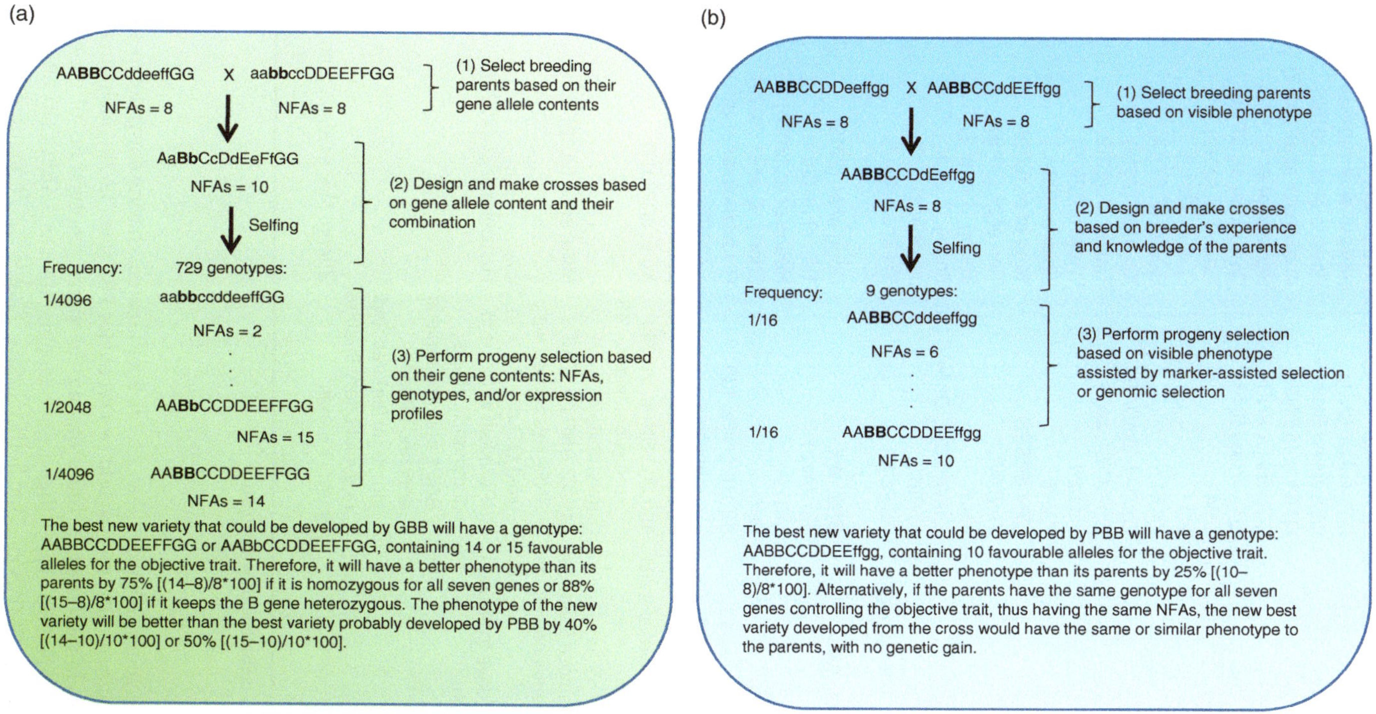


**Fig. 22.7.** (a) Gene-based breeding (GBB) for pure line varieties using the number of favourable alleles (NFAs) of the genes controlling the objective trait(s) versus (b) the current phenotype-based breeding (PPB) assisted by marker-assisted selection or genomic selection (GS) for progeny selection for superior pure line varieties. The NFAs of the genes controlling the objective trait(s) are used for the entire process of breeding: parent selection, cross design and progeny selection. Assume that the objective trait is controlled by seven genes, defined as A to G, with capital letters for favourable alleles and small letters for unfavourable alleles. The effects of all the genes are assumed to be additive, except for that of the B gene that is over-dominant and, thus, heterotic. The NFAs of each gene, such as the A gene, is scored as '0' for aa, '2' for AA; and '1' for Aa. For the B gene, Bb is scored as '3' because it is over-dominant. The total NFAs of the genes contained in a plant or line is the sum of the NFAs of individual genes. Therefore, the cultivars or lines having AABBCCDDeeffgg and AABBCCddEEffgg, or both having AABBCCDDeeffgg or AABBCCddEEffgg, could be selected as the parents for PBB because they are selected based on the phenotype of the objective trait, while those having AABBCCddeeffGG and aabbccDDEEFFGG can be accurately selected as the parents for GBB because they are selected based on gene allele contents, even though they have the same phenotype of the objective trait as those selected for PBB. Nevertheless, if the objective trait is controlled by hundreds or thousands of genes, such as the genes controlling maize inbred grain yield (Zhang, M.P. *et al.*, 2020a), it is impossible or very difficult for breeders to combine and fix most, if not all, of them into a progeny, select it and develop it into a new superior cultivar. GBB can do so because it can accurately select the breeding parents that could best meet the breeding objectives, wisely design crosses that could maximize the combination of the favourable alleles and heterotic genotypes of the genes controlling the objective trait from breeding parents into progeny, and accurately select the progeny that have the largest NFAs.

parents are crossed and the progeny are generated, the progeny plant having all seven genes that are fixed or homozygous, with a total NFAs of 14, could be accurately selected; or the progeny plant having six of the seven genes that are fixed and one of them that is heterozygous and heterotic, with a total NFAs of 15, could be accurately selected. In comparison, for PBB the best progeny plant has a total NFAs of only 10, if it is selected for. Therefore, GBB could develop a new variety that is 75% (all seven genes are fixed) or 88% (six of the seven genes are fixed and the heterotic gene of them is kept heterozygous) higher than the parents, while PBB assisted with GS or MAS could only develop a new variety that is 25% higher than the parents at best. It is apparent that the new variety developed by GBB would be 40 or 50% better than the best variety that is probably developed by PBB. Moreover, if the parents that are selected for PBB have the same genotypes for all seven genes, the new variety developed by PBB would be the same as or similar to its parents, with no genetic gain. Finally, it should be pointed out that many agronomic traits, such as maize grain yield, are in fact controlled by hundreds of or more than a thousand genes (e.g. Zhang, M.P. *et al.*, 2020a). Although GBB could accurately select the breeding parents that are the most desirable to approach the breeding objective, it is difficult, if not impossible, for PBB to accurately identify such parents for breeding.

Figure 22.8 outlines GBB for hybrid varieties in comparison with the current PBB for hybrid varieties. The authors have been conducting GBB research and development for hybrid variety development in maize since 2010 (M. Zhang, Y.-H. Liu, S.C. Murray, W. Xu and H.-B. Zhang, 2021, unpublished results). The breeding project shown in this figure is based on a budget of $100,000. For GBB (Fig. 22.8a), the genes controlling the objective trait, such as maize grain yield, for 1000 parental inbred lines are sequenced using the modern high-throughput sequencing technology. The NFAs, genic SNP/InDel genotypes and/or expression profiles of the genes controlling the objective trait in all 250,000 (test cross) or 499,500 (half-diallel cross) possible $F_1$ hybrids are inferred from those of their inbred line parents, and used for prediction model training and validation, as well as genic selection for superior $F_1$ hybrids. If the top ten hybrids are selected, the selection intensity or rate is 0.002%. These are all completed

(a)

**Budget:** $100,000; **Period:** 1 year

**Testcross:**

- 500 inbred lines developed from a cross
- 500 unrelated inbred lines developed from another cross or diverse inbred lines
- All possible $F_1$ hybrids: 500×500 = 250,000 $F_1$ hybrids
- Neither cross nor field trial is needed

**Half diallel cross:**

- 1000 diverse inbred lines
- All possible $F_1$ hybrids: 1000×(1000–1)/2 = 499,500 $F_1$ hybrids
- Neither cross nor field trial is needed

**Selection rate for top 10 hybrids:** 0.002%, meaning 500-fold more powerful than PBB

(b)

**Budget:** $100,000; **Period:** 2 year

**Testcross:**

- 100 inbred lines developed from a cross
- 10 unrelated inbred lines developed from another cross or diverse inbred lines
- The number of $F_1$ hybrids that can be generated and evaluated by field trial: 100×10 = 1,000 $F_1$ hybrids

**Half diallel cross**

- 45 diverse inbred lines
- Number of $F_1$ hybrids that can be generated and evaluated by field trial: 45×(45–1)/2 = 990 $F_1$ hybrids

**Selection rate for top 10 hybrids:** 1%, meaning 500-fold less powerful than GBB

**Fig. 22.8.** (a) Gene-based breeding (GBB) for hybrid varieties versus (b) the current phenotype-based breeding (PBB) for hybrid varieties. These breeding projects are designed based on the same budget ($100,000), which allows for PBB to produce 1000 $F_1$ hybrids, phenotype the $F_1$ hybrids and make PBB selection among the hybrids through the replicated field trials and for GBB to genotype 1000 parents, predict the phenotypes of all 499,500 possible $F_1$ hybrids and make GBB selection among the hybrids.

*in silico*. The selected top $F_1$ hybrids are finally made and subjected to field trials for hybrid variety test. In comparison, the $100,000 budget allows PBB only to generate approximately 1000 $F_1$ hybrids and conduct field trials to test them for the objective trait(s) (Fig. 22.8b). The superior $F_1$ hybrids are selected based on their performance in the field trials and further subjected to field trials for hybrid variety test. If the top ten $F_1$ hybrids are selected, the selection intensity is 1%. Therefore, GBB is 500-fold more powerful than PBB for hybrid variety breeding.

GBB and GS both enable development of new varieties or large-scale improved breeding lines. However, GBB has significant advantages over other molecular breeding methods, such as MAS, GEE, RNAi, GE and GS:

- GBB, as GS, is efficient for breeding polygenic or quantitative traits, but it is also efficient for breeding simply inherited traits controlled by one or a few major genes, as MAS, GEE, RNAi and GE, but without the creation or involvement of GMOs;
- because GBB is based on the genes controlling the objective trait(s), it is desirable not only for progeny selection, as is GS, but also for parent selection and cross design, while GS is inefficient for parent selection and cross design that are crucial to the success of breeding;
- for progeny selection, GBB is over 60% more accurate and more consistent than GS, due to the absence of or lower probability of genetic recombination between markers and targeted genes, and absence of or lower background noise (Liu *et al.*, 2020);
- GBB is well suited to continuously and accurately integrate, pyramid and fix the favourable alleles and heterotic genotypes of agronomic traits from breeding parents into elite varieties, thus being well suited for substantial and continued crop improvement;
- GBB also efficiently facilitates and enhances breeding assisted by other molecular methods, such as GE, MAS, GEE and RNAi; and
- GBB is much more cost-efficient than GS, because the genes controlling the objective trait can be genotyped by targeted high-throughput sequencing, thus being affordable for plant breeding (Zhang, M.P. *et al.*, 2020a).

Nevertheless, it is prerequisite for GBB to have most, if not all, of the genes controlling the breeding objective trait(s), which would seem to make GBB difficult to deploy in a crop species. Only a very limited number of genes controlling a quantitative agronomic trait have been cloned to date, such as the genes controlling maize grain yield, the genes controlling cotton fibre length and the genes controlling rice grain yield (Zhang, M.P. *et al.*, 2020b). To overcome this barrier (Liu *et al.*, 2020), we have invented an innovative technology, designed the *gExpress* technology, for genome-wide high-throughput cloning of the genes controlling quantitative traits, and cloned over 9000 genes controlling maize grain yield and quality traits, as well as over 10,000 genes controlling cotton fibre lint yield and quality traits (Liu *et al.*, 2020). We have also tested the ability, utility and efficiency of the cloned genes for GBB in maize and cotton using the 1501 cloned maize inbred grain yield genes (Zhang, M.P. *et al.*, 2020a) and the 474 cotton fibre length genes (Liu, 2014; Liu *et al.*, 2020). The *gExpress* technology has been tested and validated using different methods for more than seven years. In comparison to the gene cloning methods that have been developed and used so far, such as map-based cloning, the *gExpress* technology is not only reliable, efficient, simple and rapid, but also more than 1000-fold higher in throughput and widely applicable to genome-wide cloning of genes controlling quantitative traits in any species, regardless of its genome size, genome complexity, ploidy level, and availability of genomic and molecular knowledge and resources. Currently, the tests for the ability, utility and efficiency of the genes cloned using the *gExpress* technology for GBB have been completed. As soon as the results of these tests have been published, the *gExpress* technology will be published and made available to the public. All the maize grain yield and quality genes and all the cotton fibre yield and quality genes cloned so far will be published and made available to the public for GBB.

## 22.4 Recent Advances in Molecular Breeding for Maize Yield

Grain yield is probably one of the most complex traits controlled by numerous genes, some of which may a substantially contribute to grain yield, while others may only have minor effects on grain yield. Zhang, M.P. *et al.* (2020a) showed that at least 1500 genes control maize grain yield. Moreover, it should be pointed out that the effect of a gene on a quantitative trait may vary substantially between populations, due to gene × genetic background interactions, and across environments, due to G × E interactions. Therefore, it is a great challenge to substantially improve maize grain yield using MAS, GEE, RNAi or GE, while GS and GBB have been demonstrated to be well suited for enhanced breeding for grain yield. Summarized below are the advances in molecular breeding for grain yield in maize made in the past decade. Although all these molecular breeding methods have been attempted to improve grain yield in maize, almost all of the studies have been focused on the test of concepts, research and development, function validation of candidate genes and breeding line improvement, with no new high-yielding variety developed and released for commercial production using any of these molecular breeding methods in the past decade.

### 22.4.1 Marker-assisted selection

Studies have been pursued to improve grain yield in maize with MAS (Table 22.2). A vast majority of these studies tested the utility and efficiency of MAS for grain yield improvement in maize using multiple SNP markers that are associated with QTLs controlling grain yield, while MAS was also used to introgress the favourable alleles of genes from teosinte into maize (Table 22.2). Beyene *et al.* (2016a) conducted marker-assisted recurrent selection (MARS) with ten bi-parental tropical maize populations for grain yield using 55–87 SNPs tagging QTLs under drought stress and non-water stress environments in sub-Saharan Africa. MARS was conducted for generations $C_1$, $C_1S_1$ and $C_1S_2$. The results showed that an overall genetic gain of 105 kg/ha per year under non-water stress environments and 51 kg/ha per year under drought stress environments was obtained by MARS. Beyene *et al.* (2016b) also tested the grain yields of the $C_1S_2$ test-cross hybrids derived by MARS from two bi-parental populations relative to those of $S_5$ lines developed by traditional pedigree selection and the founder parents. Consequently, the top ten $C_1S_2$-derived hybrids had 0.5–46.3% and 11.1–55.1% higher mean grain yields than the hybrids developed using the traditional pedigree selection and the commercial checks, respectively, under the drought stress environments. The top ten $C_1S_2$-derived hybrids had 3.4–13.3% and 7.9–36.5% higher mean grain yields than the hybrids developed using the traditional pedigree selection and the commercial checks, respectively, under the non-drought stress environments. Abdulmalik *et al.* (2017) tested the efficiency of MARS for grain yield improvement for generations from $C_0$ to $C_3$ under drought stress and well-watered environments in West Africa with one bi-parental population using 233 SNP markers that were uniformly distributed in the maize genome. The best linear unbiased prediction (BLUP) model was used to predict the genomic estimated breeding value (GEBV) using DNA markers. Significant markers on each chromosome were identified and used for MARS. As a result, genetic gains of 44.9 and 6.0 kg/ha per cycle were obtained for grain yield under drought stress and well-watered environments, respectively. The mean frequency of favourable alleles for grain yield was increased by 10%, from 0.50 at $C_0$ to 0.55 at $C_3$. Similarly, Bankole *et al.* (2017) tested the efficiency of MARS for grain yield improvement for generations $C_1$ and $C_2$ under drought stress, well-watered and rainfed conditions using another bi-parental population. As done by Abdulmalik *et al.* (2017), Bankole *et al.* (2017) predicted the GEBV of each plant using the SNPs significant to grain yield on each chromosome. They found that the test-cross hybrids of $S_1$ lines derived from $C_2$ plants had significantly higher grain yield than those of $S_1$ lines derived from $C_0$ plants under drought stress, well-watered and rainfed conditions, with an average genetic gain of 7, 1 and 3% per cycle, respectively. The frequency of the favourable alleles was increased from 0.510 at $C_0$ to 0.515 at $C_2$. Nevertheless, Cerrudo *et al.* (2018) reported that GS using genome-wide SNPs outperformed MAS using SNPs linked to QTLs for grain yield in a maize doubled haploid population across water treatments. The difference between this result and those obtained by Beyene *et al.* (2016a,b), Abdulmalik *et al.* (2017) and Bankole *et al.* (2017) could be attributed to the fact that only the markers linked to one to three grain yield QTLs were used for MAS under each line or hybrid and environment combination and/or the variation of the QTL effects across environments due to G × E interactions. Tian *et al.* (2019) introgressed the teosinte *UPA2* allele that controls maize upright leaf angle by MAS backcrossing into the two parents of an elite maize hybrid, Nongda108 (HuangC × Xu178). The hybrid containing the teosinte *UPA2* allele had a reduced leaf angle relative to the original hybrid; therefore, it had higher grain yield than the original hybrid because it could be planted at a higher planting density (105,000 plants/ha).

### 22.4.2 Genetic engineering

GEE has continuously been a major molecular technology for function validation and analysis of genes controlling grain yield in maize as well as a molecular breeding method for maize grain yield improvement (Table 22.2). Li *et al.* (2013) investigated the functions of a gene encoding cell wall invertase, particularly maize *Mn1*, *Arabidopsis AtGIF1* (*AtCWINV1*) and rice *OsGIF1* (*OsCIN2*), respectively, transformed by GEE through the *Agrobacterium*-mediated transformation into and constitutively expressed them in an elite maize inbred line, Ye478. They found that the transgenic plants of these genes had grain yield up to 145.3% higher than the wild-type plants. The transgenic plants had enlarged ears with enlarged grain size and grain number, and increased total starch contents, relative to the wild-type plants. Guo *et al.* (2014) overexpressed the maize *ARGOS1* (*auxin-regulated gene involved in organ size*) or *ZAR1* (*Zea mays ARGOS1*) gene in a Hi-II maize or an inbred line (PHWWZ) and found that the *ZAR1* alleles could increase hybrid maize yield. Xie *et al.* (2018) engineered the mutants of the *ZmDA1* and *ZmDAR1* genes, *Zma1* and *Zmdar1*, into a maize elite inbred line (DH4866). Both DA1 and DAR1 are ubiquitin receptors that function as negative regulators of cell proliferation in *Arabidopsis*, thus having a negative effect on seed size. They found that when the mutants of these two genes overexpressed in transgenic plants, the grain yields of the transgenic plants increased by 15% relative to those of the wild-type plants, due to increased grain number, weight and starch contents. Wu *et al.* (2019) engineered *zmm28*, a maize MADS-box transcription factor gene, into maize inbred line, PH17AW, and found that overexpression of the gene in transgenic plants increased maize grain yield by 300–400 kg/ha due to increased plant growth, photosynthesis capacity and nitrogen utilization. Jia *et al.* (2020) studied a serine/threonine protein kinase-encoding gene, *Kernel Number Per Row6* (*KNR6*), that controls pistillate floret number and ear length in a maize inbred line, A188, by GEE. They found that overexpressing the *KNR6* gene in transgenic plants significantly increased grain yield, relative to the wild-type plants, because of their increased ear length and kernel number. Tian *et al.* (2019) transformed *ZmRAVL1*, a gene encoding a positive regulator of maize leaf angle, and *brd1* (*brassinosteroid C-6 oxidase1*), a gene involved in the final step of brassinosteroid synthesis, into a maize inbred line, B73-329, respectively. They found that the overexpression of either the *ZmRAVL1* gene or the *brd1* gene resulted in larger leaf angle than the wild type. The increased leaf angle led to lower grain yield when a higher density of plants per hectare were planted.

### 22.4.3 RNA interference

RNAi is a molecular method for enhanced crop genetic improvement and also has been widely used for function validation and characterization of genes controlling agronomic traits (Table 22.2). Yang *et al.* (2018) created the second-generation Roundup Hybridization System using the RNAi technology for maize hybrid seed production. The expression of the gene encoding CP4 EPSPS, a glyphosate-insensitive 5-enolpyruvylshikimate-3-phosphate synthase in maize pollen, was silenced by maize endogenous male tissue-specific siRNAs, resulting in glyphosate-sensitive male cells, due to lack of the CP4 EPSPS protein. Therefore, male sterility could be induced by glyphosate application at the stages critical to pollen development and the induced male-sterile plants could be then used as the female parent to produce hybrid seeds. Tian *et al.* (2019) downregulated the maize *ZmRAVL1* gene by RNAi and found that the transgenic plants had a smaller leaf angle, thus allowing higher density of plants per hectare for higher yield. Zhang, Z. *et al.* (2019), at Rutgers University, New Jersey, knocked-down the expressions of two endosperm-specific NAC transcription factors, *ZmNAC128* and *ZmNAC130*, with RNAi and showed that the transgenic plants had shrunken kernel phenotypes with significant reduction of starch and protein, thus reducing grain yield. Zhang, J. *et al.* (2019), at Corteva Agriscience, Iowa, downregulated a maize NAC transcription factor-encoding gene, *nac7*, by RNAi and found that the transgenic plants exhibited delayed senescence and increased both biomass and nitrogen accumulation. The test crosses of the RNAi plants with two elite inbred lines resulted in hybrids with prolonged stay-green and increased grain yield by 290 kg/ha. Jia *et al.* (2020) investigated *KNR6*, a serine/threonine protein kinase-encoding gene controlling pistillate floret number and ear length. The RNAi knocked-down lines had significantly shorter ears and fewer kernels per kernel row, thus reducing grain yield.

### 22.4.4 Gene or genome editing

Like GEE and RNAi, GE has largely been used for function validation and analysis of candidate genes that likely control an agronomic trait, while improved breeding lines may be developed from these studies. Shi *et al.* (2017) edited *ARGOS8*, a maize gene encoding a negative regulator of ethylene response, using the CRISPR/Cas system. They inserted the native maize *GOS2* promoter, which confers a moderate level of constitutive expression, into the 5′-untranslated region of the native *ARGOS8* gene or replaced the native promoter of *ARGOS8*. The edited *ARGOS8* plants increased maize grain yield by 109 kg/acre under field drought stress condition at the flowering stage and had no yield loss under the well-watered condition, relative to the wild-type plants. Dong *et al.* (2019) double-edited *SH2* (*SHRUNKEN2*) encoding the large subunit of endosperm ADP-glucose pyrophosphorylase and *WX* (*WAXY*) encoding GRANULE BOUND STARCH SYNTHASE I (GBSS I) that is required for amylose synthesis and determines the amylose content in both endosperm and pollens, using the CRISPR/Cas system. All the edited plants were shown to be InDel-type. The double-gene-edited plants had several-fold higher soluble sugar contents than the wild-type plants, but their grain yields were dramatically reduced. Tian *et al.* (2019) knocked-out the *ZmRAVL1* gene that is involved in determination of maize leaf angle using the CRISPR/Cas9 system. The leaf angle was reduced in the edited plants; therefore, the edited plants produced higher grain yield than the wild-type plants under higher planting densities. Huang *et al.* (2020) validated the function of *qKW9*, a maize kernel weight QTL encoding a DYW motif of a DYW-motif pentatricopeptide repeat protein involved in C→U editing of *ndhB*, a subunit of the chloroplast NADH dehydrogenase-like complex, using the CRISPR/Cas9 system. They analysed two edited plants with a 1 bp deletion and a 4 bp deletion, respectively, and found that the C→U editing of *ndhB* was abolished and photosynthesis was reduced. Consequently, the ear and kernel sizes of the edited plants became significantly smaller. Zhang, Z. *et al.* (2020) validated the function of *stiff1*, a major QTL for maize stalk strength encoding an F-box domain protein, using the CRISPR/Cas9 system. The edited plants had stronger stalks than the wild-type plants, thus validating the function of the *stiff1* gene in stalk strength.

### 22.4.5 Genomic selection

Dozens of studies on the utility and efficiency of GS have been reported, even though no new varieties developed using GS have been reported in the past decade. Several representative studies are reviewed here to update the recent advances of molecular breeding with GS for grain yield improvement in maize. Riedelsheimer *et al.* (2012) reported the genomic and metabolic prediction of complex heterotic traits, including biomass yield, in maize test-cross hybrids. The test-cross hybrids of 285 worldwide diverse *dent* inbred lines crossed with two testers were used to predict their combining abilities for seven biomass- and bioenergy-related traits, including dry matter yield, plant height, dry matter concentration, female flowering, starch content, sugar content and lignin content, using 56,110

SNPs and 130 metabolites, respectively. The ridge regression-best linear unbiased prediction (rr-BLUP) prediction model and a fivefold cross-validation scheme were used for the prediction. Prediction accuracies were obtained for the general combining abilities (GCAs) of the seven traits, ranging from 0.72 to 0.81 for SNPs and from 0.60 to 0.80 for metabolites. The prediction accuracy was the Pearson's correlation coefficients ($r$) between predicted and observed trait GCA values divided by the square roots of the trait heritability and the repeatability of the predictor variables (for SNPs, it is 1 and for metabolites, it is the weighted mean of repeatability of individual metabolites). Edriss *et al.* (2017) performed genomic prediction of 2022 maize diverse breeding lines for grain yield with the BLUP model using 66,000 SNPs. SNP marker analysis clustered the diverse breeding lines into five clusters. Prediction accuracy ranged from 0.20 to 0.36 within clusters and from 0.04 to 0.26 across clusters. The mean genomic prediction accuracy within clusters (0.27) outperformed pedigree-based prediction (0.03). Xu *et al.* (2017) tested the utility and efficiency of different omics, including genomics, transcriptomics and metabolomics, for prediction of six grain yield-related traits, including ear diameter, ear length, ear weight, cob weight, kernel number per row and number of kernel rows. Three hundred and thirty-nine maize inbred lines; 100,000 SNPs, 28,769 gene expressions and 748 metabolites; eight multiple regression prediction models (five parametric and three non-parametric); and a tenfold cross-validation scheme were used for the prediction. Prediction abilities, the correlation coefficients ($r$) between predicted and observed phenotypes, achieved ranged from 0.2041 to 0.5962. The BLUP model overall performed better than the other seven prediction models for different omics data and the genomic prediction was better than the transcriptomic and metabolomic predictions. Millet *et al.* (2019) conducted genomic prediction of maize yield across European environmental conditions with accounting for G × E interactions. Two hundred and forty-six test-cross hybrids plus 56 external hybrids, 41,722 random SNPs plus 1312 yield QTL-associated SNPs, and the BayesR prediction model were used for that study under environment-controlled platforms. For cross-validation, prediction accuracies (i.e. the correlations between observed and predicted grain yields) ranged from 0.20 to 0.85, while for external validation, prediction accuracies ranged from 0.38 to 0.80. They concluded that the strategy with G × E interaction under consideration outperformed prediction of grain yield over a benchmark approach. Ramstein *et al.* (2020) reported improvement in prediction of agronomic traits in hybrid maize with consideration of dominance effects and functional enrichments. A panel of 1106 test-cross hybrids derived from a diverse panel of inbred lines crossed with two testers and 1640 test-cross hybrids of 24 bi-parental populations crossed with a single tester were used for the prediction, with a selection of SNP markers based on their loci dominance effects and their functional features, including genic regions or proximity to genes (gene annotation), structural features (recombination rate and chromatin openness) and evolutionary features (minor allele frequency and evolutionary constraint). Their results showed that although dominance was present for all traits studied, including days to silking, plant height and grain yield, dominance improved genomic prediction only for plant height. However, the genic regions improved genomic prediction for all three traits.

### 22.4.6 Gene-based breeding

Zhang, M.P. *et al.* (2020a) reported the first study worldwide in any species of the utility and efficiency of GBB for enhanced breeding, using maize as the plant material to test its utility and efficiency for progeny selection. One thousand five hundred and one *ZmINGY* (*Zea mays inbred line grain yield*) genes previously genome-wide cloned by these researchers and a bi-parental recombinant inbred line (RIL) population were used for the study. The 1501 *ZmINGY* genes each had an effect on grain yield varying from 14.8 to 61.0%, with 401 increasing grain yield and 1100 decreasing grain yield, when activated or upregulated. Of the 1501 *ZmINGY* genes, 27 contained 63 SNP and/or InDel mutations that increased or decreased grain yield by 14.2 to 54.2%. Moreover, these 1501 *ZmINGY* genes formed a strong co-expression network, of which 15 increased or decreased grain yield by

62–67% when present or absent in the network and 143 had significant impacts on grain yield, when the number of *ZmINGY* genes co-expressing with each of the 143 *ZmINGY* genes varied in the network. Zhang, M.P. *et al.* (2020b) discovered that the expression of at least one of the transcripts alternatively spliced from a gene controlling a quantitative trait, such as maize grain yield, was significantly correlated with the trait performance and the genes controlling a quantitative trait were several times more likely to form a co-expression network than randomly selected unknown genes expressed in a plant organ. The expressions of the 1501 *ZmINGY* genes, the genic SNPs or InDels of the 27 SNP/InDel-containing *ZmINGY* genes as DNA markers, with one SNP or InDel per gene, and the NFAs of the 27 SNP/InDel-containing *ZmINGY* genes, with one pair of alleles corresponding to one SNP or InDel per gene, were used for the genic prediction of grain yield.

For the genic prediction of grain yield with the expressions of *ZmINGY* genes, nine multiple regression prediction models, including five parametric and four non-parametric models, and a tenfold cross-validation scheme were applied, with the maize and rice grain yield genes previously cloned by different researchers as the positive controls and randomly selected unknown maize genes as the negative controls. A series of numbers of *ZmINGY* genes, varying from six to all 1501, were randomly selected from the 1501 *ZmINGY* genes by bootstrap sampling, with ten rounds of the selection for each number of the *ZmINGY* genes, and tested for genic prediction of grain yield. The result showed that both the *ZmINGY* genes and maize and rice grain yield genes previously cloned by others could predict the phenotype of grain yield, while the randomly selected unknown maize genes could not. The *ZmINGY* genes outperformed the maize and rice grain yield genes previously cloned by others in prediction of maize grain yield. All numbers of the genes randomly selected from the 1501 *ZmINGY* genes could predict the maize grain yield and as the number of *ZmINGY* genes increased for the prediction, the prediction accuracy, the correlation coefficient ($r$) between observed and predicted grain yields, increased. When 350 or more of the *ZmINGY* genes were used for the prediction, the prediction accuracy plateaued, approaching 0.77. Furthermore, they compared the prediction accuracy of grain yield using the *ZmINGY* genes with those obtained with GS by different studies using the same prediction model, the same cross-validation scheme and similar types of populations. The *ZmINGY* genes outperformed GS by 3.3–175.9%, with an average of 62.9%. Finally, they identified 150 *ZmINGY* genes from the 1501 *ZmINGY* genes that are key to maize grain yield prediction, approaching a prediction accuracy for grain yield of 0.82.

For the genic prediction of maize grain yield using the 27 genic SNPs or InDels of the 27 *ZmINGY* genes as DNA markers, the same prediction models and cross-validation scheme as above were used. The maize grain yield was predicted at an accuracy of 0.49, which was similar to those of GS using tens of thousands of SNPs. For the genic prediction of maize grain yield using the NFAs of the 27 SNP/InDel-containing *ZmINGY* genes, a simple linear regression model was used for the prediction. The maize grain yield was predicted at an accuracy of 0.52.

Finally, Zhang, M.P. *et al.* (2020a) conducted the genic prediction of grain yield using these three genic data sets and progeny selection based on the predicted grain yields across very diverse environments. They found that the genic prediction accuracy of grain yield remained largely consistent, varying from 0.49–0.82 to 0.43–0.61, across diverse environments. The top ten lines selected by GBB were consistent with those selected based on the observed grain yield determined by replicated field trials at up to 66.7%, when the grain yields predicted with one of the three genic data sets were used for the selection, and at 100%, when the grain yields predicted with two or all three genic data sets were jointly used for the selection. The selection consistencies between GBB and phenotype-based selections across diverse environments was largely the same as those within an environment.

## 22.5 Conclusions and Perspectives

Most of the agronomic traits important to crop production, including those important to crop yield and yield sustainability such as abiotic and biotic stress tolerances, are quantitatively inherited and each controlled by numerous genes.

For instance, Zhang, M.P. *et al.* (2020a) showed that grain yield in maize is controlled by at least 1501 genes. Zhang, M.P. *et al.* (2020b) first revealed that the genes controlling such complex traits are several times more likely to form a network, thus working correlatively to shape the ultimate performance of the traits. Moreover, Zhang *et al.* (2010) and Zhang, H.-B. *et al.* (2020) first discovered that the number of genes, especially those controlling biotic and abiotic stress tolerances that are crucial to yield sustainability, varied by several-fold among varieties and germplasm lines within a species and also among individuals within a bi-parental population, and such variation is significantly associated with plant biology, variation and evolution. All of these make plant breeding, especially continued crop genetic improvement, more complex and more challenging.

Breeding for either pure line varieties or hybrid varieties is usually implemented through three steps: parent selection, crop design and progeny selection. Nevertheless, either the traditional PBB or the modern molecular breeding is a process that continuously incorporates the favourable alleles and/or heterotic genotypes of the genes controlling the breeding objective trait(s) from existing varieties and/or breeding germplasm lines into elite varieties or breeding lines, thus developing new or improved varieties. Liu *et al.* (2019) investigated the genetic potential of the present best crop varieties using 226 *GFL* genes controlling cotton (*Gossypium*) fibre length. They found that the cotton line that has the longest fibres of the best modern cotton varieties contains the favourable alleles of only approximately 50% of the 226 *GFL* genes, suggesting that the present best crop varieties could be significantly improved, if the favourable alleles and heterotic genotypes of all 226 *GFL* genes are incorporated into a new variety. If this result is applicable to maize grain yield and the maize grain yield is controlled by 1501 genes, how could breeders incorporate the favourable alleles and heterotic genotypes of the remaining 750 genes controlling the maize grain yield into the present elite hybrid varieties for maize production?

MAS, GEE and RNAi have been widely used, while GS, GE and GBB are currently under research and development for enhanced crop improvement. MAS, GEE, RNAi and GE are used to improve crops by manipulating individual genes having major effects on the objective trait(s) and are thus more effective for improving or fine-tuning existing varieties. GS and GBB are designed to improve crops by genome-wide manipulation of the genes controlling the objective trait(s) and are thus more efficient for new elite variety development. Nevertheless, GS aims to assist at progeny selection using genome-wide omics, while GBB aims to develop new superior varieties by making full use of the genes controlling the objective trait(s) through the entire process of breeding, including parent selection, cross design and progeny selection. The question is how will these molecular breeding methods help breeders continuously, substantially and quickly incorporate the favourable alleles and heterotic genotypes of the remaining 750 genes controlling maize grain yield into the present best maize varieties?

To this end, the perspectives of molecular breeding for maize grain yield, based on the utility, efficiency and uniqueness of the different molecular methods for enhanced and/or accelerated plant breeding are outlined in Fig. 22.9. It is apparent that GBB will play the most important roles in both maize genetic improvement and breeding, and its roles will be continuously increased as the genes controlling different agronomic traits are genome-wide cloned and used for GBB. GBB is the only molecular breeding technology that enables, by design, accurate and efficient incorporation of the favourable alleles and/or heterotic genotypes of a large number of genes controlling grain yield from different existing varieties or germplasm lines into new varieties or breeding lines through the entire breeding process. Although GS has been demonstrated to be efficient for enhanced progeny selection for polygenic traits, it is inefficient for parent selection and cross design. MAS, GEE, GE and RNAi are not interchangeable, and each has its own unique utility and efficiency for enhanced breeding, but they all have been efficient for the manipulation of a limited number of genes (such as fewer than 100 genes) for enhanced breeding. MAS and GEE are unique for adding new and favourable genes of the breeding objective trait that are absent in an existing variety, while GE and RNAi are unique for fine-tuning the unfavourable alleles or genes of the breeding objective trait already present in an existing variety,

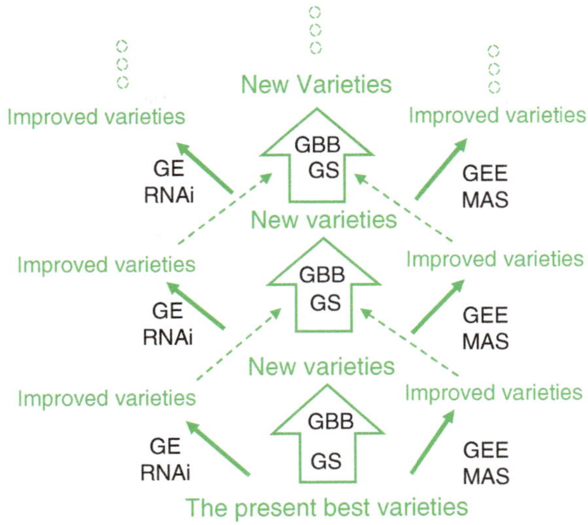


**Fig. 22.9.** Roles and perspectives of molecular breeding for maize grain yield improvement. The font size and positions of the molecular breeding methods indicate their importance and perspectives in maize grain yield genetic improvement. Although marker-assisted selection (MAS), genetic engineering (GEE), gene editing (GE) and RNA interference (RNAi) have been efficient to improve or fine-tune an elite variety, it is difficult to develop a brand-new variety with any of them. Both gene-based breeding (GBB) and genomic selection (GS) develop or help develop brand-new varieties, but GS is efficient only for progeny selection, while GBB is efficient for all steps of breeding, including parent selection, cross design and progeny selection.

with GE permanently fine-tuning the gene(s) in whole individuals and with RNAi fine-tuning the genes temporarily (plant growth and development-specific) and spatially (organ-specific). Nevertheless, GE and RNAi will only be able to assist in maize grain yield breeding. Together and combined with the traditional PBB, these molecular breeding methods are promising to realize the next green revolution, thus promising to help feed the world.

## References


Abdallah, N.A., Prakash, C.S. and McHughen, A.G. (2015) Genome editing for crop improvement: challenges and opportunities. *GM Crops & Food* 6, 183–205. Available at: https://doi.org/10.1080/21645698.2015.1129937

Abdulmalik, R.O., Menkir, A., Mesekam, S.K., Unachukwu, N., Ado, S.G. *et al.* (2017) Genetic gains in grain yield of a maize population improved through marker assisted recurrent selection under stress and non-stress conditions in West Africa. *Frontiers in Plant Science* 8, 841. Available at: https://doi.org/10.3389/fpls.2017.00841

Bankole, F., Menkir, A., Olaoye, G., Crossa, J., Hearne, S., Unachukwu, N. and Gedil, M. (2017) Genetic gains in yield and yield related traits under drought stress and favorable environments in a maize population improved using marker assisted recurrent selection. *Frontiers in Plant Science* 8, 808. Available at: https://doi.org/10.3389/fpls.2017.00808

Beyene, Y., Semagn, K., Crossa, J., Mugo, S., Atlin, G.N. *et al.* (2016a) Improving maize grain yield under drought stress and non-stress environments in Sub-Saharan Africa using marker-assisted recurrent selection. *Crop Science* 56, 344–353. Available at: https://doi.org/10.2135/cropsci2015.02.0135

Beyene, Y., Semagn, K., Mugo, S., Prasanna, B.M., Tarekegne, A. *et al.* (2016b) Performance and grain yield stability of maize populations developed using marker-assisted recurrent selection and pedigree selection procedures. *Euphytica* 208, 285–297. Available at: https://doi.org/10.1007/s10681-015-1590-1

Boch, J., Scholze, H., Schornack, S., Landgraf, A., Hahn, S. *et al.* (2009) Breaking the code of DNA binding specificity of TAL-type III effectors. *Science* 326, 1509–1512. Available at: https://doi.org/10.1126/science.1178811

Cerrudo, D., Cao, S., Yuan, Y., Martinez, C., Suarez, E.A. *et al.* (2018) Genomic selection outperforms marker assisted selection for grain yield and physiological traits in a maize doubled haploid population across water treatments. *Frontiers in Plant Science* 9, 366. Available at: https://doi.org/10.3389/fpls.2018.00366

Collard, B.C.Y., David, J. and Mackill, D.J. (2008) Marker-assisted selection: an approach for precision plant breeding in the twenty-first century. *Philosophical Transactions of the Royal Society B: Biological Sciences* 363, 557–572. Available at: https://doi.org/10.1098/rstb.2007.2170

Daryanto, S., Wang, L. and Jacinthe, P.-A. (2016) Global synthesis of drought effects on maize and wheat production. *PLoS ONE* 11, e0156362. Available at: https://doi.org/10.1371/journal.pone.0156362

Datta, A. (2013) Genetic engineering for improving quality and productivity of crops. *Agriculture & Food Security* 2, 15. Available at: https://doi.org/10.1186/2048-7010-2-15

Desta, Z.A. and Ortiz, R. (2014) Genomic selection: genome-wide prediction in plant improvement. *Trends in Plant Science* 19, 592–601. Available at: https://doi.org/10.1016/j.tplants.2014.05.006

Dong, L., Qi, X., Zhu, J., Liu, C., Zhang, X. *et al.* (2019) Supersweet and waxy: meeting the diverse demands for specialty maize by genome editing. *Plant Biotechnology Journal* 17, 1853–1855. Available at: https://doi.org/10.1111/pbi.13144

Edriss, V., Gao, Y., Zhang, X., Jumbo, M.B., Makumbi, D. *et al.* (2017) Genomic prediction in a large African maize population. *Crop Science* 57, 2361–2371. Available at: https://doi.org/10.2135/cropsci2016.08.0715

FAO (2019) FAOSTAT. Crops. Food and Agricultural Organization of the United Nations, Rome. Available at: http://www.fao.org/faostat/en/#data/QC (accessed 3 March 2021).

Fire, A., Xu, S., Montgomery, M.K., Kostas, S.A., Driver, S.E. and Mello, C.C. (1998) Potent and specific genetic interference by double-stranded RNA in *Caenorhabditis elegans*. *Nature* 391, 806–811. Available at: https://doi.org/10.1038/35888

Godfray, H.C., Beddington, J.R., Crute, I.R., Haddad, L., Lawrence, D. *et al.* (2010) Food security: the challenge of feeding 9 billion people. *Science 327*, 812–818. Available at: https://doi.org/10.1126/science.1185383

Guo, M., Rupe, M.A., Wei, J., Winkler, C., Goncalves-Butruille, M. *et al.* (2014) Maize *ARGOS1* (*ZAR1*) transgenic alleles increase hybrid maize yield. *Journal of Experimental Botany* 65, 249–260. Available at: https://doi.org/10.1093/jxb/ert370

Hatfield, J.L. and Prueger, J.H. (2015) Temperature extremes: effect on plant growth and development. *Weather and Climate Extremes* 10(A), 4–10. Available at: https://doi.org/10.1016/j.wace.2015.08.001

Hatfield, J.L., Boote, K.J., Kimball, B.A., Ziska, L.H., Izaurralde, R.C. *et al.* (2011) Climate impacts on agriculture: implications for crop production. *Agronomy Journal* 103, 351–370. Available at: https://doi.org/10.2134/agronj2010.0303

Huang, J., Lu, G., Liu, L., Raihan, M.S., Xu, J., *et al.* (2020) The kernel size-related quantitative trait locus *qKW9* encodes a pentatricopeptide repeat protein that affects photosynthesis and grain filling. *Plant Physiology* 183, 1696–1709. Available at: https://doi.org/10.1104/pp.20.00374

Intergovernmental Panel on Climate Change (2007) *Climate Change 2007: Impacts, Adaptation and Vulnerability: Contribution of Working Group II to the Fourth Assessment Report of the Intergovernmental Panel on Climate Change*. Cambridge University Press, Cambridge and New York.

Jia, H., Li, M., Li, W., Liu, L., Jian, Y. *et al.* (2020) A serine/threonine protein kinase encoding gene *KERNEL NUMBER PER ROW6* regulates maize grain yield. *Nature Communications* 11, 988. Available at: https://doi.org/10.1038/s41467-020-14746-7

Jinek, M., Chylinski, K., Fonfara, I., Hauer, M., Doudna, J.A. and Charpentier, E. (2012) A programmable dual-RNA–guided DNA endonuclease in adaptive bacterial immunity. *Science* 337, 816–821. Available at: https://doi.org/10.1126/science.1225829

Kim, Y.-G., Cha, J. and Chandrasegaran, S. (1996) Hybrid restriction enzymes: zinc finger fusions to *Fok* I cleavage domain. *Proceedings of the National Academy of Sciences USA* 93, 1156–1160. Available at: https://doi.org/10.1073/pnas.93.3.1156

Kusaba, M. (2004) RNA interference in crop plants. *Current Opinion in Biotechnology* 15, 139–143. Available at: https://doi.org/10.1016/j.copbio.2004.02.004

Li, B., Liu, H., Zhang, Y., Kang, T., Zhang, L. *et al.* (2013) Constitutive expression of cell wall invertase genes increases grain yield and starch content in maize. *Plant Biotechnology Journal* 11, 1080–1091. Available at: https://doi.org/10.1111/pbi.12102

Li, H., Rasheed, A., Hickey, L.T. and He, Z. (2018) Fast-forwarding genetic gain. *Trends in Plant Science* 23, 184–186. Available at: https://doi.org/10.1016/j.tplants.2018.01.007

Liu, Y.-H. (2014) Molecular basis of quantitative genetics revealed by cloning and analysis of 474 genes controlling fiber length in cotton. PhD dissertation, Texas A&M University, College Station, Texas.

Liu, Y.-H., Zhang, M.P., Zhang, Y., Smith, C.W., Hague, S. *et al.* (2014) Large-scale cloning and characterization of genes controlling fiber length for deciphering of the molecular basis of fiber quality and development of a gene-based breeding system in cotton. In: Blake, V.C., Grant, D. and Lazo, G. (eds) *Proceedings of the International Plant & Animal Genome Conference XXII, San Diego, California, 10–15 January 2014*. Scherago International, Jersey City, New Jersey, p. 474.

Liu, Y.-H., Zhang, M.P., Sze, S.-H., Smith, C.W. and Zhang, H.-B. (2019) Continued genetic improvement of cotton through gene-based breeding. In: *Proceedings of the 2019 ASA–CSSA–SSSA International Annual Meeting, San Antonio, Texas, 10–13 November 2019*. American Society of Agronomy, Crop Science Society of America and Soil Science Society of America, Madison, Wisconsin, paper no. 213-10.

Liu, Y.-H., Xu, Y., Zhang, M., Cui, Y., Sze, S.-H., *et al.* (2020) Accurate prediction of a quantitative trait using the genes controlling the trait for gene-based breeding in cotton. *Frontiers in Plant Science* 11, 583277. Available at: https://doi.org/10.3389/fpls.2020.583277

Meuwissen, T.H.E., Hayes, B.J. and Goddard, M.E. (2001) Prediction of total genetic value using genome-wide dense marker maps. *Genetics* 157, 1819–1829.

Millet, E.J., Kruijer, W., Coupel-Ledru, A., Prado, S.A., Cabrera-Bosquet, L. *et al.* (2019) Genomic prediction of maize yield across European environmental conditions. *Nature Genetics* 51, 952–956. Available at: https://doi.org/10.1038/s41588-019-0414-y

Ramstein, G.P., Larsson, S.J., Cook, J.P., Edwards, J.W., Ersoz, E.S. *et al.* (2020) Dominance effects and functional enrichments improve prediction of agronomic traits in hybrid maize. *Genetics* 215, 215–230. Available at: https://doi.org/10.1534/genetics.120.303025

Ray, D.K., Ramankutty, N., Mueller, N.D., West, P.C. and Foley, J.A. (2012) Recent patterns of crop yield growth and stagnation. *Nature Communications* 3, 1293. Available at: https://doi.org/10.1038/ncomms2296

Riedelsheimer, C., Czedik-Eysenberg, A., Grieder, C., Lisec, J., Technow, F. *et al.* (2012) Genomic and metabolic prediction of complex heterotic traits in hybrid maize. *Nature Genetics* 44, 217–220. Available at: https://doi.org/10.1038/ng.1033

Ronald, P.C. (2014) Lab to farm: applying research on plant genetics and genomics to crop improvement. *PLoS Biology* 12, e1001878. Available at: https://doi.org/10.1371/journal.pbio.1001878

Rouet, P., Smih, F. and Jasin, M. (1994) Introduction of double-strand breaks into the genome of mouse cells by expression of a rare-cutting endonuclease. *Molecular and Cellular Biology* 14, 8096–8106. Available at: https://doi.org/10.1128/mcb.14.12.8096

Schauberger, B., Ben-Ari, T., Makowski, D., Kato, T., Kato, H. and Ciais, P. (2018) Yield trends, variability and stagnation analysis of major crops in France over more than a century. *Scientific Reports* 8, 16865. Available at: https://doi.org/10.1038/s41598-018-35351-1

Shi, J., Gao, H., Wang, H., Lafitte, H.R., Archibald, R.L. *et al.* (2017) ARGOS8 variants generated by CRISPR-Cas9 improve maize grain yield under field drought stress conditions. *Plant Biotechnology Journal* 15, 207–216. Available at: https://doi.org/10.1111/pbi.12603

Tian, J., Wang, C., Xia, J., Wu, L., Xu, G. *et al.* (2019) Teosinte ligule allele narrows plant architecture and enhances high-density maize yields. *Science* 365, 658–664. Available at: https://doi.org/10.1126/science.aax5482

Tilman, D., Balzer, C., Hill, J. and Befort, B.L. (2011) Global food demand and the sustainable intensification of agriculture. *Proceedings of the National Academy of Sciences USA* 108, 20260–20264. Available at: https://doi.org/10.1073/pnas.1116437108

Wu, J., Lawit, S.J., Weers, B., Sun, J., Mongar, N. *et al.* (2019) Overexpression of *zmm28* increases maize grain yield in the field. *Proceedings of the National Academy of Sciences USA* 116, 23850–23858. Available at: https://doi.org/10.1073/pnas.1902593116

Xie, G., Li, Z., Ran, Q., Wang, H. and Zhang, J. (2018) Over-expression of mutated *ZmDA1* or *ZmDAR1* gene improves maize kernel yield by enhancing starch synthesis. *Plant Biotechnology Journal* 16, 234–244. Available at: https://doi.org/10.1111/pbi.12763

Xu, Y., Xu, C. and Xu, S. (2017) Prediction and association mapping of agronomic traits in maize using multiple omic data. *Heredity* 119, 174–184. Available at: https://doi.org/10.1038/hdy.2017.27

Yang, H., Qi, Y., Goley, M.E., Huang, J., Ivashuta, S. *et al.* (2018) Endogenous tassel-specific small RNAs-mediated RNA interference enables a novel glyphosate-inducible male sterility system for commercial production of hybrid seed in *Zea mays* L. *PLoS ONE* 13, e0202921. Available at: https://doi.org/10.1371/journal.pone.0202921

Yogindran, S. and Rajam, M.V. (2015) RNAi for crop improvement. In: Bahadur, B., Venkat, M., Rajam, M., Sahijram, L. and Krishnamurth, K. (eds) *Plant Biology and Biotechnology*. Springer, New Delhi, pp. 623–637. Available at: https://doi.org/10.1007/978-81-322-2283-5_31

Zhang, H.-B., Zhang, M.P., Liu, Y.-H., Qi, X. and Su, X. (2020) The DNA 'jigsaw puzzle' structure model as the molecular basis of biology: variation and interaction of genome constituting elements. In: Grant, D., Lazo, G. and Blake, V.C. (eds) *Proceedings of the International Plant & Animal Genome Conference XXVIII, San Diego, California, 11–15 January 2020*. Scherago International, Jersey City, New Jersey, p. W550.

Zhang, J., Fengler, K.A., Van Hemert, J.L., Gupta, R., Mongar, N. *et al.* (2019) Identification and characterization of a novel stay-green QTL that increases yield in maize. *Plant Biotechnology Journal* 17, 2272–2285. Available at: https://doi.org/10.1111/pbi.13139

Zhang, M.P., Wu, Y.-H., Lee, M.-K., Liu, Y.-H., Rong, Y. *et al.* (2010) Numbers of genes in the NBS and RLK families vary by more than four-fold within a plant species and are regulated by multiple factors. *Nucleic Acids Research* 38, 6513–6525. Available at: https://doi.org/10.1093/nar/gkq524

Zhang, M.P., Zhi, H., Chang, F., Zhang, Y., Liu, Y.-H. *et al.* (2014) Large-scale cloning and characterization of genes controlling grain yield for deciphering of the molecular basis of grain yield and development of a gene-based breeding system in maize. In: Blake, V.C., Grant, D. and Lazo, G. (eds) *Proceedings of the International Plant & Animal Genome Conference XXII, San Diego, California, 10–15 January 2014*. Scherago International, Jersey City, New Jersey, p. 875.

Zhang, M.P., Liu, Y.-H., Chang, C.-S., Zhi, H., Wang, S. *et al.* (2019) Quantification of gene expression while taking into account RNA alternative splicing. *Genomics* 111, 1517–1528. Available at: https://doi.org/10.1016/j.ygeno.2018.10.009

Zhang, M.P., Cui, Y., Liu, Y.-H., Xu, W., Sze, S.-H. *et al.* (2020a) Accurate prediction of maize grain yield using its contributing genes for gene-based breeding. *Genomics* 112, 225–236. Available at: https://doi.org/10.1016/j.ygeno.2019.02.001

Zhang, M.P., Liu, Y.-H., Xu, W., Smith, C.W., Murray, S.C. and Zhang, H.-B. (2020b) Analysis of the genes controlling three quantitative traits in three diverse plant species reveals the molecular basis of quantitative traits. *Scientific Reports* 10, 10074. Available at: https://doi.org/10.1038/s41598-020-66271-8

Zhang, Z., Dong, J., Ji, C., Wu, Y. and Messing, J. (2019) NAC-type transcription factors regulate accumulation of starch and protein in maize seeds. *Proceedings of the National Academy of Sciences of the United States of America* 116, 11223–11228. Available at: https://doi.org/10.1073/pnas.1904995116.

Zhang, Z., Zhang, X., Lin, Z., Wang, J., Liu, H. *et al.* (2020) A large transposon insertion in the *stiff1* promoter increases stalk strength in maize. *The Plant Cell* 32, 152–165. Available at: https://doi.org/10.1105/tpc.19.00486

# 23 CRISPR-Mediated Genome Editing in Maize for Improved Abiotic Stress Tolerance

**Ali Razzaq[1], Ghulam Mustafa[1], Muhammad Amjad Ali[2], Muhammad Sarwar Khan[1] and Faiz Ahmad Joyia[1*]**

*[1]Centre of Agricultural Biochemistry and Biotechnology (CABB), University of Agriculture, Faisalabad, Pakistan; [2]Department of Plant Pathology, University of Agriculture, Faisalabad, Pakistan*

## 23.1 Introduction

Food security and climate change are the most pivotal concerns of the 21st century. The global population is growing on an extremely fast track and is predicted to increase to 10 billion people at the end of 2050. This tremendous growth in population will require a 60–100% increase in global crop yield to feed them (Dhankher and Foyer, 2018). It is estimated that around 815 million people are suffering from food shortage and malnutrition, which impedes global food security programmes to accomplish zero hunger by the end of 2030 (Richardson *et al.*, 2018). Despite all the integrated programmes on food security, the large number of hungry and undernourished people is increasing around the globe. According to World Bank estimations, 83 million people are starving in 45 countries. The ratio of undernourished people in developed countries does not go beyond 5%, as compared with 13% of the population in developing states. In addition, an obvious diminishing drift in food security has been observed in many countries of Asia and Africa, where food insecurity stands at 13 and 20%, respectively (Prosekov and Ivanova, 2018). Climate change is linked directly to agricultural production and food security and has a great influence on food security, being the major limiting factor in the reduction of agricultural crop production worldwide. The agriculture sector is adversely affected by climate change, which results in the reduction of crop yield and compromises global food safety and security. In the current scenario, ecosystem resilience and food security are extremely concerning topics internationally.

Climate change is responsible for extreme weather conditions, which ultimately affect crop production globally (Richardson *et al.*, 2018). Climate change poses a great threat to global food security as it adversely affects agricultural production. Biotic and abiotic stresses occur mainly due to climatic changes that are associated with extreme events of rainfall, increasing temperature, variations in $CO_2$ concentration, depletion of the ozone layer and alteration of host–pathogen relationships (Raza *et al.*, 2019). Soil fertility, air pollution, water availability and temperature variations have significant impacts on agricultural production. Continuous variations in climatic conditions increase the chances of crop yield loss due to direct and indirect impacts of abiotic stresses (Noya *et al.*, 2018). Furthermore, extensive usage of fossil fuels and massive deforestation have resulted in the increase of $CO_2$ concentration

*Corresponding email: faizahmad1980@uaf.edu.pk


DOI: 10.1079/9781789245431.0023

to 400 μmol/mol air. It is projected that the level will rise twofold to 800 μmol/mol air over the coming years. The discharge of harmful gases, particularly $CO_2$, is the major cause of global temperature rise and the greenhouse effect (Vaughan et al., 2018). In semi-arid areas, less precipitation and elevated temperature cause negative impacts on crop production and threaten global food security (Nelson et al., 2009).

In coming years, climate change is likely to intensify the spells of abiotic stresses and will hamper agricultural productivity by 20% by the end of 2050 (Vaughan et al., 2018). Agricultural productivity of developing countries is mainly affected because of harsh climatic conditions (Rosenzweig et al., 2014) and plants being sessile suffer predominantly from various abiotic stresses in their natural habitat. The abiotic stresses ultimately impede plant growth and development and lead to losses of up to 50% in yield for major crops, causing food insecurity globally (Benevenuto et al., 2017). Many factors are responsible for the adverse impacts on crop growth and yield due to abiotic stresses, and depend on the crop type, ploidy levels and genotypes (Cramer et al., 2011). Abiotic stresses generally stimulate hyper-accumulation of reactive oxygen species (ROS) that can be fatal for crop plants. In addition, abiotic stresses can disrupt vital enzymatic and cellular metabolic activities, causing damage to DNA, protein, lipid and carbohydrate structures that results in the imbalance of crucial metabolic processes and eventually causes cell death (Gall et al., 2015; Razzaq et al., 2019a). To deal with hostile climatic stresses, plants trigger their cellular, molecular and physiological machinery to produce phytohormones, antioxidant enzymes, secondary metabolites, expression of several stress resistance genes and transcription factors (Razzaq et al., 2019a). In recent years, climate change has forced scientists to develop novel breeding approaches for tackling abiotic stresses to boost agricultural production and guarantee global food security (Rosenzweig et al., 2014).

## 23.2 Significance of Maize and Global Status

Maize (*Zea mays* L.) is ranked third in the major cereal crops after wheat and rice and is one of the most important $C_4$ plants due its greater grain production among the cereals. It has the status of being the most prominent crop in global agriculture and is also known as a miracle crop and the queen of cereals. Maize has a major role in global food security by providing food, feedstock and energy for the expanding human population. Maize left rice and wheat behind in 1996/97 in terms of global production and still its yield is increasing at three times the rate of wheat and twice that of rice production annually (Fischer et al., 2014). About 900 million consumers preferred maize as a staple crop in the developing countries, with 120–140 million of people having shortage of food supply. Developing countries produced 70% of the total maize on an area of almost 100 million hectares. Maize is a highly versatile crop due to the greater adaptability under changing environmental conditions. It is predicted that the global need for maize will double and it will become the most grown crop worldwide by 2050. Maize demand is increasing abruptly due to its multipurpose utilization as food, fodder, feed and in industrial products (Rosegrant et al., 2007).

Maize was cultivated on an area of about 197.1 million hectares with a production of 1134.7 million tonnes in 2017 (FAO, 2019; Fig. 23.1a). Table 23.1 displays the top producers of maize worldwide in 2017. The USA was the largest maize producer, having 50.4% of the total share with a production of 370.9 million tonnes. Asia was the second major maize producer, with 31.4% of global production (Fig. 23.1b). The last decade (2010–2019) has seen remarkable progress in maize productivity as all regions of Asia presented increases in maize yield, including East Asia (31%), South Asia (27.6%) and South-East Asia (11%), with an overall increase of about 28% in Asia (FAO, 2019). China, Indonesia and India have presented striking growth rates of approximately 6% per year in maize production. In 2017, China was the largest producer of maize in Asia and ranked second in global maize production next to the USA (Table 23.1) (FAO, 2019). In China, the total area under maize production was 42.39 million hectares with an annual yield of 259 million tonnes. Maize is also a very important crop for food, feed and bioenergy consumption in Pakistan, where it ranks third after wheat and rice. Maize held 5.93% of total cultivated land and its contribution

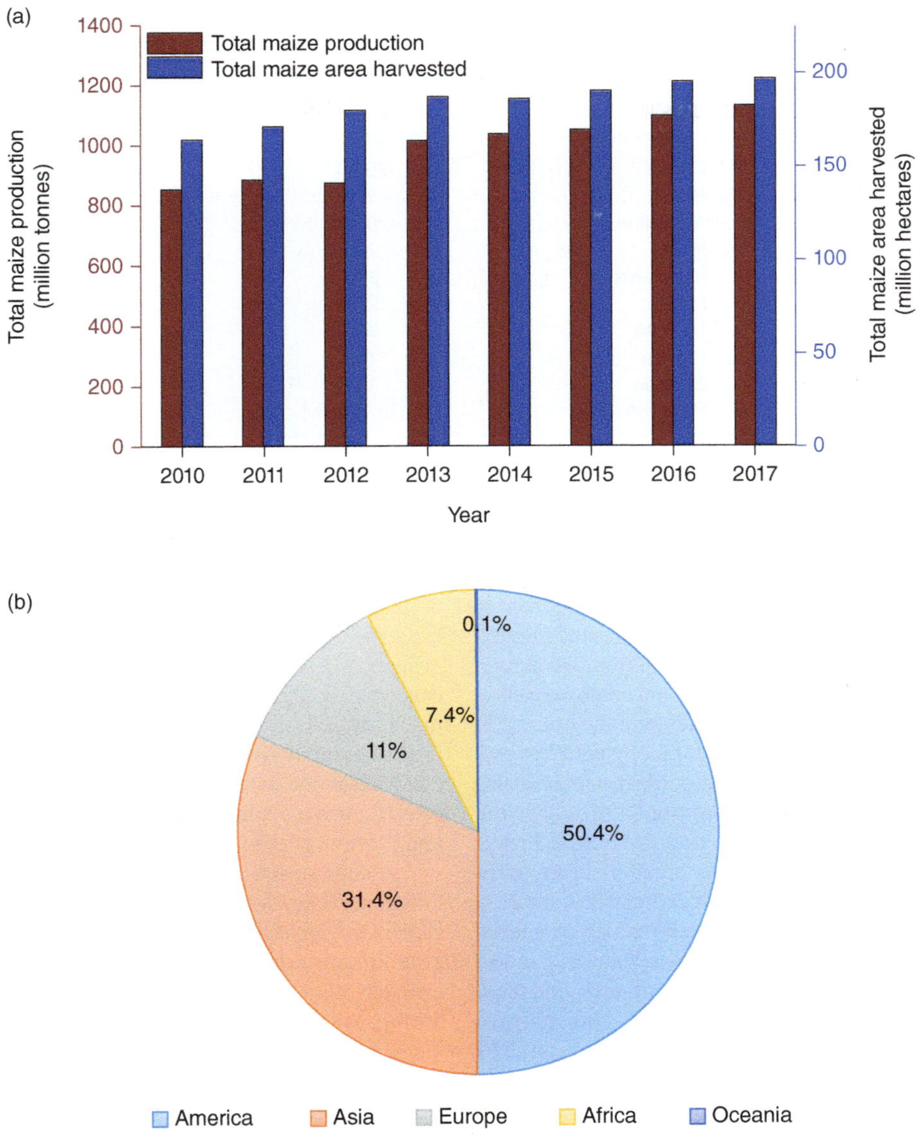


**Fig. 23.1.** (a) Total maize production and total maize area harvested in the world, 2010–2017. (b) The share of global maize production in different regions of the world, 2017. (From FAO, 2019.)

to the country's agricultural production was 4.82% in 2013; Punjab was the major producer with 98% of maize production and contributing 0.4% of gross domestic product (Naqvi and Ashfaq, 2013). In Pakistan, maize covered 1.22 million hectares having 5.70 million tonnes production in 2017 (FAO, 2019). On the other hand, Europe's share of global maize production in 2017 was 10.7% with 110 million tonnes on an area of 17.53 million hectares. Likewise, Africa contributed 7.4% of global production, with a total production of 84 million tonnes covering nearly 40.60 million hectares (FAO, 2019). In most regions of Africa maize is the major staple crop, providing 32% of dietary energy in southern and eastern Africa (Rosegrant *et al.*, 2007).

Maize is beneficial for both humans and animals and can be utilized in a number of ways such as cooking oil, ethanol, starch, sweeteners, syrups, glue, beverages, animal feed, bioenergy

**Table 23.1.** List of important maize-producing countries in the world, 2017. (From FAO, 2019.)

| Ranking | Country | Production (million tonnes) | Cultivated area (million hectares) |
|---|---|---|---|
| 1 | USA | 370.9 | 33.46 |
| 2 | China | 259.0 | 42.39 |
| 3 | Brazil | 97.7 | 17.39 |
| 4 | Argentina | 49.4 | 6.53 |
| 5 | India | 28.7 | 9.21 |
| 6 | Indonesia | 27.9 | 5.37 |
| 7 | Mexico | 27.7 | 7.32 |
| 8 | Ukraine | 24.6 | 4.48 |
| 9 | France | 14.1 | 1.61 |
| 10 | Canada | 14.0 | 1.33 |
| 22 | Pakistan | 5.70 | 1.22 |

and other valuable products. Maize is a nutritious crop and consists of 72% starch, 24% carbohydrates, 4% fats and 10% proteins, offering 365 kcal of energy per 100 g (Ranum *et al.*, 2014). Moreover, it provides numerous minerals such as Fe, Cu, Zn, Mn, Mg and P, some important vitamins such as C, E, thiamin, riboflavin, niacin, pantothenic acid, $B_6$ and folate, as well as fibre (Nuss *et al.*, 2010). Greater daily utilization of maize cornmeal and flour along with less cost of production makes it an excellent staple crop for fortification with micronutrients of which the deficiency poses a serious threat to human health (Ranum *et al.*, 2014). In addition to its agronomic, industrial and nutritional value, maize is an ideal model crop for genetic studies compared with other cereals. Several crucial features of maize like its vast range of nucleotide diversity, large heterochromatic chromosomes and massive mutant pool have given it a significant place in genomic, cytogenetic and genetic studies (Strable and Scanlon, 2009). Apart from being a major staple crop, maize is a very important contributor to exports like animal feed and fuel (USDA, 2018).

## 23.3 Impact of Abiotic Stresses on Maize

To meet the food requirements of the rapidly growing global population, maize production needs to double in order to feed 10 billion people by the end of 2050 (Ray *et al.*, 2013). Maize can be grown in a wide range of agroclimatic conditions, but its production is significantly threatened by harsh climatic stresses (Zuo *et al.*, 2015). Diverse environmental changes have caused intensification of numerous abiotic stresses that hamper maize productivity around the world. The fluctuation in climatic conditions directly affects length of the growing season, precipitation, temperature, the growing cycle and crucial events during crop growth. These varying environmental stresses cause yield reduction and produce poor-quality maize crops for the consumer. Maize is the ideal crop to study potential against the various abiotic stresses due to its extensive cultivation under diverse agro-morphological conditions (Cairns and Prasanna, 2018).

Abiotic stresses such as drought, heat, salinity, waterlogging and cold stress pose serious threats to global agriculture and limit the crop growth and yield (Raza *et al.*, 2019). Maize is generally susceptible to drought and heat stresses during pollination and especially at the grain-filling period that may lead to a substantial reduction in grain production. Many approaches, including early sowing, have been adopted to tackle the climatic stresses (Wijewardana *et al.*, 2015). During 2019, the maize crop in Pakistan, especially that cultivated in the province of Punjab, was severely damaged by a spell of extreme heatwaves before pollination. The immature pollen grains were destroyed by the strong heat and the plants were unable to be pollinated, causing less grain production in the maize cob, which ultimately reduced the crop yield (F.H. Joyia, Faisalabad, 2019, personal communication). There are many regions of the USA in which maize tasselling stage is affected by inconsistent and inadequate rainfall events. Earlier planting under normal growing conditions in

March may induce tasselling at the end of May, where consistent precipitation, plentiful solar radiation and optimum temperature are beneficial for healthy maize growth (Wijewardana *et al.*, 2015). Hence, early sowing would result in greater productivity by escaping harsh climate stresses during summer and could be vital to higher yield generation in the future.

In the Asian tropics, maize is predominantly a rainfed crop that is totally reliant on the prevailing environmental conditions and this is a core reason for its great vulnerability to abiotic stresses. Recently, Cairns and Prasanna (2018) predicted that Asia will observe extreme climate stresses with enhanced diversity beyond the present aptitude. Many environmental modelling reports propose a rapid future increase in temperature, which could negatively influence maize productivity in the tropics (Cairns *et al.*, 2012). In South-East and South Asia, reduction in the days of rainfall has been recognized as one of the key factors of climate change (Manton *et al.*, 2001). This has offered numerous rainfall spells within fewer days, therefore prolonging the dry phases. The unpredictable patterns of monsoon rains cause drought stress at some crop stages and waterlogging at other growing periods. Several tropical regions of Asia are well known as a hotspot for abiotic stresses, with adverse effects on maize production (ADB, 2009). The impact of some important abiotic stresses on maize yield is discussed next.

### 23.3.1 Drought and heat stress

Heat and drought are the two main climatic stresses and frequently happen concurrently due to the unified nature of both stresses. Heat stress can be described as elevation in temperature greater than the threshold position, while drought is water deficit conditions for a longer time that can cause extreme impairment to crop growth. In the future, environmental stress is expected to enhance the threat of excessive temperature, especially heat spells (Horton *et al.*, 2015). Maize is highly prone to heat and drought stress mainly during pollination. About 60% of cultivated land in China is prone to heat and drought stress, which may cause 30% reduction in yield every year (Frey *et al.*, 2015). In Africa, Lobell and colleagues integrated data from 20,000 historical trials of maize in 8 years with climate record data and revealed that for every 1°C rise in temperature above 30°C, maize productivity was decreased by 1.7 and 1% under drought and optimum rainfed conditions, respectively (Lobell *et al.*, 2011). In addition, heat stress damages different plant tissues and depending on the vulnerability of specific metabolic pathways, many processes are triggered during stress responses. Extreme temperatures can cause physiological, anatomical, biochemical and morphological changes during maize growth and development. The reproductive stage of maize is highly vulnerable to heat stress, which can cause sterility, poor light interception and reduced life cycle (Cairns *et al.*, 2012).

In developing countries, maize is particularly susceptible to drought stress because the majority of maize production is dependent upon rainfall. So, drought stress is considered the major limiting factor for maize production across the upland and lowland rainfed areas of developing regions. According to calculated data, approximately 15–20% of maize yield is vanishing every year because of drought stress and this may increase further with the irreversible changes in environmental conditions caused by global climate change (FAO, 2010). In addition, more than 80% of the maize cultivated in rainfed regions of South-East and South Asia produces less than half the yield of irrigated maize. Land coming into rainfed production is about 1.8% annually, which is sixfold greater than that of irrigated land (Edmeades, 2007). The reduction in irrigated land is generally because of shrinking groundwater reserves that bring irrigated maize land under great risk. This scenario is expected to worsen in the future, leading to scarce events of rainfall during the cropping season (ADB, 2009). However, reducing the impacts of drought stress could enhance maize yield in Asia by 35% (Gerpacio and Pingali, 2007).

### 23.3.2 Waterlogging stress

Waterlogging can be defined as the limiting factor hindering crop growth when soil water content is higher than the field capacity. Waterlogging is another main hurdle for maize cropping in numerous agroclimatic zones where rains spells are excessive and unpredictable. It has been estimated that about 16% of fertile land worldwide is prone to waterlogging (Boyer, 1982). In Asia, about 18% of maize-cultivated land is severely damaged by waterlogging, reducing crop yield by 25 to 30% every year (Cairns *et al.*, 2012). The issue of waterlogging in different developmental stages is intensified because of environmental fluctuations in several maize-producing areas in the developing countries (Solomon, 2007). It was reported that maize is relatively more prone to waterlogging stress from seedling phase to tasselling phase and results in weak root development, delayed maturity and stunted growth (Campbell *et al.*, 2015). Excess of water supply in waterlogging conditions is responsible for plant wilting and root decay. High humidity and less stomatal conductance cause limited availability of $O_2$ that results in plant death. Leaching triggers a disorder in the osmotic level of root cortex that stops the radial motion of water from root hairs to xylem. Therefore, the supply of water to various tissues of the plant is restricted and ultimately leads to plant death (Cairns and Prasanna, 2018).

### 23.3.3 Cold stress

Cold stress, which comprises freezing (<0°C) and chilling (<20°C) temperature, severely affects crop growth and considerably reduces crop yield (Chinnusamy *et al.*, 2007). Cold stress is a major limiting factor for maize yield as it is a cold-sensitive crop (Sanghera *et al.*, 2011). Cold stress can disturb maize physiology and development during the whole life cycle, although chilling is particularly damaging at germination stage and initial seedling development (Rodriguez *et al.*, 2014). Hence, early sowing of maize is dangerous, and maturation stage is often affected by chilling stress. Chilling spells in early spring are detrimental for root, shoot and leaf development and also hinder the functional machinery of chloroplasts, thereby affecting photosynthetic activity (Rymen *et al.*, 2007). Chilling stress is broadly understood as having negative effects on maize yield, making cold stress tolerance a crucial feature that requires more investigation for better maize production.

## 23.4 Genome Editing

Due to unpredictable weather conditions and the continuous increase in extreme waves of climatic stresses, current plant biology studies, including the generation of new varieties, require more consideration to develop climate-resilient and stress-tolerant crops. Plant breeders must strive to map novel resistance genes and elucidate several physiological and metabolic pathways that regulate the stress mechanism under various climatic stresses (Raza *et al.*, 2019). Conventional breeding programmes have been carried out to evolve improved crop varieties. However, the exploitation of these approaches for improving resistance against abiotic stresses has decreased with time due to the multidimensional mechanism of inheritance and the complex interactions between a genotype and its environment. Moreover, conventional breeding strategies were practically incapable of tackling the environmental stresses (Razzaq *et al.*, 2019b).

Genetic engineering is a technique that helps to develop improved crop cultivars with increased abiotic stress resistance, more yield and enhanced nutritional value. However, the application of this approach for crop breeding has many drawbacks. Insertion of transgenes or modified genes in plants has been useful to meet the food demands of a growing human population, but these modified plants suffer massively from safety concerns (Marco *et al.*, 2015). The major limitation of genetic engineering is the incorporation of foreign DNA into the genome of a plant without using its own genetic machinery to get the desired quantitative and qualitative results. With time, this approach lost consumer acceptance, and its utilization has been limited because of numerous risks to human health, food safety and the environment. Due to their unstable nature and non-specificity, the consumption of genetically modified crops has been restricted in many countries (Zhang, Y. *et al.*, 2018). Under normal growth conditions, genes' overexpression via promoters can result in growth arrest and produce a small number of seed/grains. This stress-responsive mechanism triggered by overexpressed genes has stunted crop growth, causing reduced yield and few gains in agricultural production (Marco *et al.*, 2015).

To overcome all these hurdles new breeding strategies are required for the development of abiotic stress-tolerant crop varieties to meet the future demands of food security (Razzaq *et al.*, 2019b). Recent progress in plant biotechnology and breeding is enabling scientists to engineer climate-resilient crops; although future breeding approaches must emphasize incorporating multiple genes to offer broad-spectrum tolerance against harsh climatic stresses (Zhang, Y. *et al.*, 2018). Recently, genome editing approaches have emerged as one of the most fascinating tools to tackle abiotic stresses. The accessibility to genome editing approaches has overcome all the bottlenecks of genetic engineering and conventional breeding techniques. In contrast to transgenic strategies, which integrate the phenotype, genome editing techniques have proved to be the most versatile and dynamic tools in plant breeding by the generation of defined mutants (Waltz, 2018). As compared with the transgenic approach, genome editing does not involve incorporation of any foreign gene into a host species and the resulting progeny cannot be distinguished from their parents (Razzaq *et al.*, 2019b). The sequence-specific mediated genome editing can be described as a system of modern molecular approaches that are precise, simple, efficient and specific for targeted mutations at desired genomic loci (Gao, 2015).

In maize, genome editing tools can be efficiently applied to increase the stress tolerance and develop improved maize varieties with stress resilience (Zong *et al.*, 2017; Feng *et al.*, 2018). This strategy needs very little time compared with transgenic or conventional breeding platforms. The accessibility to maize genomic resources like high-throughput genomics data, genomic sequencing, transcriptomics, proteomics and metabolomics techniques has sped up the deciphering of stress-related genes, metabolites and proteins in the maize crop. Traditionally, maize was mutated through irradiation and chemical mutagens and maize genes could be edited via the DNA repair mechanism. Mutation breeding is usually less accurate and results in positive and negative consequences with few switch on genomic regions to be manipulated. Transposon tagging is used to produce mutations in maize genome and helps to identify specific genes (Walbot, 2000). This approach is relatively expensive, time-consuming and can cause random integrations into the genome, and the mutant screening is very cumbersome (Walbot, 2000). The drawback of random genome manipulations triggered the research on sequence-specific genome editing approaches. Modern genome editing approaches have evolved during the last 10 years and have enhanced the fidelity of genome engineering around 1000-fold (Puchta *et al.*, 1993).

## 23.5 Types of Genome Editing Tools

Meganucleases, zinc-finger nucleases (ZFNs), transcription activator-like nucleases (TALENs) and clustered regularly interspaced short palindromic repeats (CRISPR)/CRISPR-associated protein 9 (Cas9) are the four types of genome editing system that have been identified so far (Razzaq *et al.*, 2019b). The basic mechanism of genome editing needs site-specific nucleases to make the double-strand breaks (DSBs) at the desired site on DNA. The DSBs can be repaired via homology-directed repair (HDR) or the non-homologous end joining (NHEJ) process (Jinek *et al.*, 2012; Cong *et al.*, 2013). The utilization of TALENs, CRISPR single guide RNA (sgRNA), sgRNA/Cas9 and CRISPR/Cas9 has been applied in various crops for improving abiotic stress tolerance, including maize (Shi *et al.*, 2017). In addition, genome editing approaches can now widen their application in the era of abiotic stresses to enhance crop yield by producing resistant cultivars. A general mechanism describing the genome editing process is illustrated in Fig. 23.2.

### 23.5.1 Zinc-finger nucleases (ZFNs)

ZFNs (Fig. 23.3) are first-generation genome editing tools and are generally applied to cut a specific DNA segment that is repaired by the NHEJ mechanism, giving site-specific mutagenesis (Urnov *et al.*, 2010). Many types of genomic manipulations including insertion, deletion, duplication, inversion and translocation can be achieved via ZFNs, which gives scientists an excellent technique to introduce genetic mutations. ZFNs are sequence-specific nucleases designed by fusing an array of artificial zinc-finger (ZF) domains, each having a specific DNA-binding domain, to the *Fok*I restriction enzyme having a

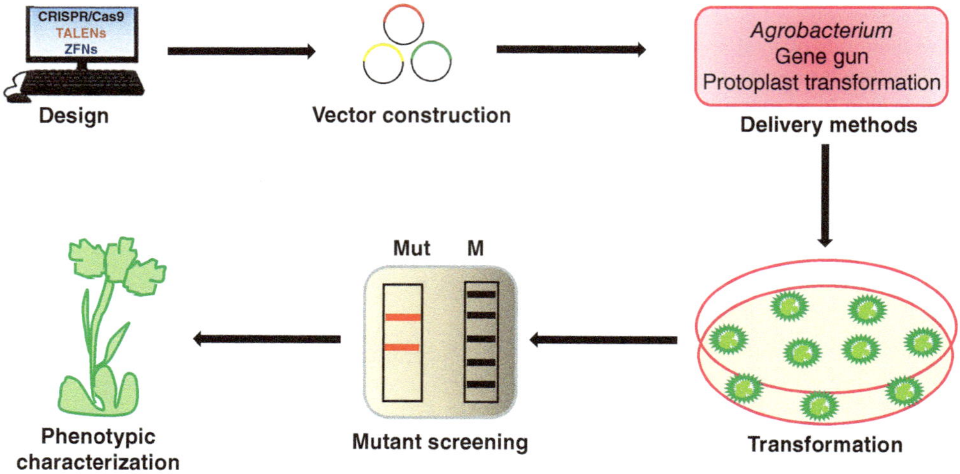


**Fig. 23.2.** Workflow of genome editing mechanism in plants.

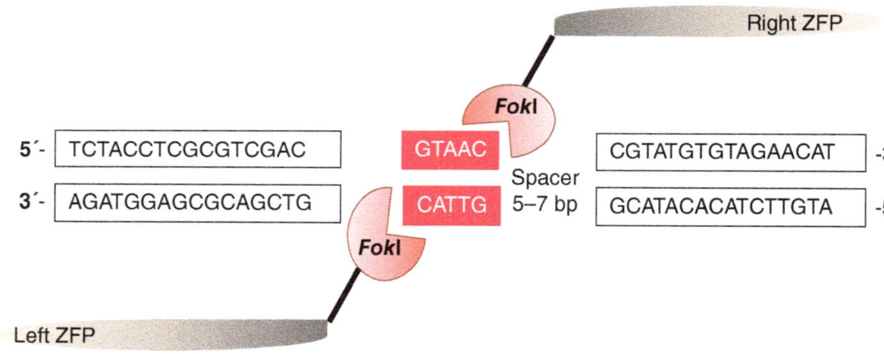


**Fig. 23.3.** Diagrammatic representation of ZFNs. The right and left portions comprise a ZFP and each monomer of ZFN is made up of ZFP that is attached to an 18–24 bp target sequence. The ZFNs are connected with *Fok*I (red) and dimerized to generate DSBs at spacer region of 5–7 bp.

non-specific binding domain. The nuclease domain recognizes the DNA region to be cleaved and each ZF identifies a target sequence of 3 bp (Petolino, 2015). For modular assembly of ZFNs, each ZF should be integrated with the bigger targeting sequence for precise editing of a desired DNA segment. Each ZFN is composed of monomers of two zinc-finger proteins (ZFPs) attached to an 18–24 bp sequence having a spacer region of five to seven nucleotides (Urnov *et al.*, 2010). ZFNs have been successfully applied in maize for genome editing for specific traits in recent years (Shukla *et al.*, 2009). However, the ZFN technology has a large number of disadvantages such as very high cost of protein domain designing, construction complexity, less accessibility to target sites for particular genomic loci, high off-target rate, defective mutations, non-compatible domain interaction and specialized skill required for results interpretation, which have made it less applicable for crop genome editing. Hence, the hunt for novel tools for genome engineering continued and has been directed to the designing of new ways for gene manipulation like TALENs and the CRISPR/Cas9 system (Nemudryi *et al.*, 2014).

### 23.5.2 Transcription activator-like effector nucleases (TALENs)

In contrast to ZFNs, scientists have placed greater attention on TALENs because they can be designed via naturally occurring DNA proteins. TALENs are also a first-generation genome editing tool, developed from *Xanthomonas* bacteria. This genus is a pathogen of some important crops including tomato, pepper and rice, posing a great threat to agriculture. These pathogenic bacteria usually produce transcription activator-like effector (TALE) proteins in the plant cytoplasm and disturb different mechanisms in plant cells that enhances the risk of pathogen damage (Nemudryi *et al.*, 2014). TALENs are applied to generate mutations via NHEJ and HDR processes that use donor DNA to repair the DSBs, introducing gene replacement or gene insertion at the target site. In addition, small insertions or deletions are produced at the joint of new chromosomes (Joung and Sander, 2013). Like ZFNs, TALENs comprise an artificial TALE fused to the cleavage domain of *Fok*I (Fig. 23.4). The sequences of amino acids representing the array of TALEs are quite similar, excluding repeat variable diresidues (RVDs) present at the 12 and 13 positions. Generally, a monomer of TALE in pair form attaches to a sequence of about 50–60 nucleotides having 14–18 nucleotide spacer regions, which is crucial for proper functioning. Taking account of the higher number of RVDs, the designing of TALENs is still a cumbersome task (Joung and Sander, 2013). Although TALEN and ZFN genome editing tools are successful to some level, they have several limitations as well that make their application restricted in many cases. Moreover, high precision is needed to design pairs of TALENs and ZFNs, because both up- and downstream sites of specified regions must be targeted for accurate genome editing events (Kim and Kim, 2014). In addition, both these tools are time-consuming, expensive and difficult to handle, thus cannot be used on a large scale. These approaches rely on proteins and designing of proteins to tailor any DSB, which is a difficult procedure (Choudhary *et al.*, 2017).

### 23.5.3 CRISPR/Cas9 system

Among all the genome editing tools, CRISPR/Cas9 is the most powerful, versatile, simple, popular and modern genome engineering tool for crop improvement. The CRISPR/Cas9 toolbox has greater application than first-generation genome editing tools because it is simpler in design, easy to handle, flexible, robust, precise, accurate and comparatively economical, thus has revolutionized plant genome editing (Razzaq *et al.*, 2019b). This editing tool, isolated from bacterial adaptive immune systems, contains Cas endonuclease and clustered palindromic repeats that guard bacteria from foreign invaders including viruses or plasmids (Barrangou *et al.*, 2007). Three types of CRISPR/Cas (I, II, III) having diverse function are present in archaea and bacteria (Makarova *et al.*, 2011). Type II is the highly efficient system and needs an sgRNA and Cas9 for foreign DNA recognition and cleavage (Cong *et al.*, 2013). Among presently accessible CRISPR/Cas tools, *Streptococcus pyogenes* (SpCas9)-derived Cas9 endonuclease and its variants have been generally applied for extensive applications in crop improvement via site-specific and highly efficient targeted genome engineering (Jinek *et al.*, 2012; Mali *et al.*, 2013), epigenome editing

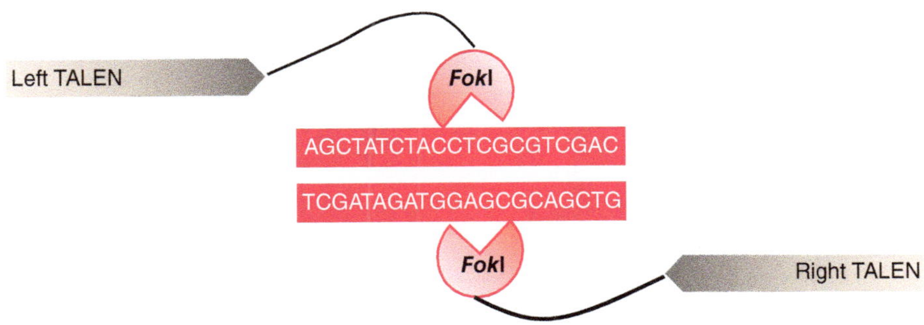


**Fig. 23.4.** Editing mechanism of TALENs. Left and right flanks of TALEN are comprised of 33–35 amino acid repeats. TALEN pairs are flanked by a 12–21 bp spacer region at which TALE molecules attached with *Fok*I cut the target sequence.

(Thakore *et al.*, 2015), precise gene replacement (Zhang *et al.*, 2017), base editing (Zong *et al.*, 2017) and transcriptional reprogramming (Chavez *et al.*, 2015).

The current version of the CRISPR/Cas9 system is derived from type II system and contains Cas9 protein and sgRNA to produce site-specific DSBs at the targeted DNA sequence. In addition, it also consists of a *trans*-encoded CRISPR RNA (tracrRNA), a duplex of CRISPR RNA (crRNA) and a protospacer-adjacent motif (PAM) that flanks the crRNA specific target (Cong *et al.*, 2013). The crRNA recognizes the target site and the tracrRNA assists in crRNA processing as well as DNA cleavage (Jinek *et al.*, 2012). The integration of tracrRNA and crRNA with sgRNA not only made the CRISPR/Cas9 toolbox simple, but also enhanced the cleavage frequency (Mali *et al.*, 2013). An sgRNA is comprised of a 20 bp seed sequence and a hairpin loop-like structure. The sgRNA identifies a specific target sequence and Cas9 produces DSBs adjacent to PAM sequence (Jinek *et al.*, 2012). A specific gene manipulation can be achieved by the repair mechanism via NHEJ, which can generate small insertions/deletions at the cleavage site. On the other hand, DSBs can also be repaired using the HDR mechanism by providing a template having a complementary sequence (Cong *et al.*, 2013). A schematic diagram representing the CRISPR/Cas9 mechanism is shown in Fig. 23.5.

So, remarkably, only 20 bp is enough to give allele character in a single copy site of a genome such as maize because the small size makes designing of the sgRNA simple (Cong *et al.*, 2013). In addition, delivery of the sgRNA/Cas9 expression cassette into the cell can be stable or involve transient maize transformation (Jinek *et al.*, 2012). As compared with other nucleases, the capability of CRISPR/Cas9 to cleave a specific DNA sequence makes for a multidimensional, versatile and excellent tool for genome editing (Bortesi and Fischer, 2015). The targeting efficacy of the CRISPR/Cas9 toolbox is astonishing in contrast to ZFNs and TALENs (Reis *et al.*, 2014). Recently, CRISPR/Cas9-mediated genome editing has been conducted to introduce desired mutations

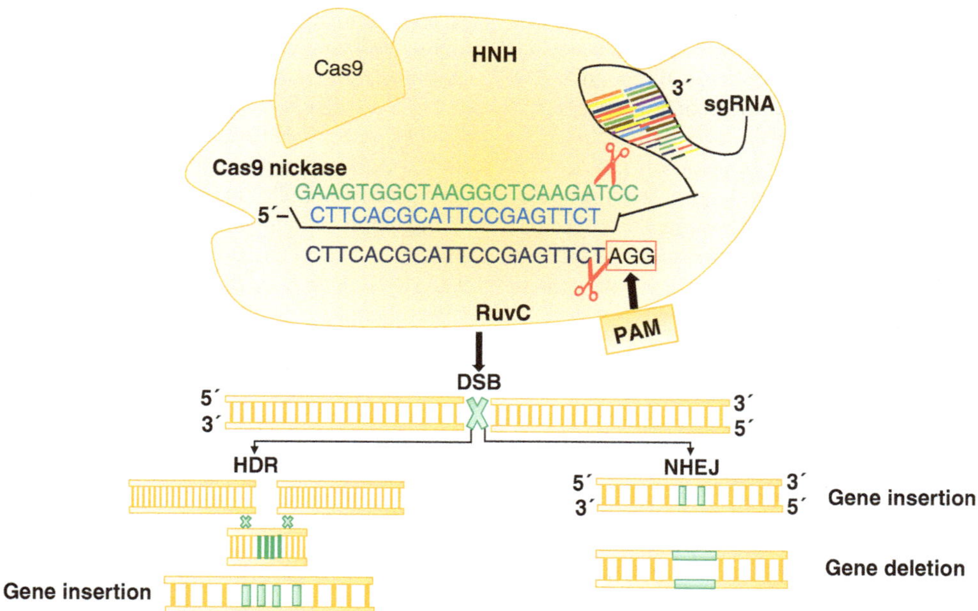


**Fig. 23.5.** The CRISPR/Cas9-mediated genome editing process which consists of Cas9 and sgRNA. The sgRNA/Cas9 complex targets the 20 bp desired sequence and 3 bp upstream of PAM sequence on RuvC site. The DSBs can be filled via the NHEJ or HDR repair pathway, resulting gene insertion and gene deletion. HNH, endonuclease domain named for characteristic histidine and asparagine residues; RuvC, endonuclease domain named for an *Escherichia coli* protein involved in DNA repair.

for maize improvement (Char *et al.*, 2017; Li *et al.*, 2017; Feng *et al.*, 2018). For the transformation via the CRISPR/Cas9 system in maize, different techniques can be adopted such as *Agrobacterium*-mediated transformation (Zhu *et al.*, 2016; Char *et al.*, 2017), particle bombardment or biolistics (Svitashev *et al.*, 2015; Zhu *et al.*, 2016) and protoplast transformation (Xing *et al.*, 2014). The latter is generally used to calculate the efficacy of various sgRNA pools, as there is no standard procedure devised for maize plant regeneration. Biolistics is used when there are regulations limiting the utilization of *Agrobacterium*, which is considered a pathogen in certain countries. However, *Agrobacterium*-mediated transformation of the CRISPR/Cas9 system with removal of the expression cassette backbone via backcrossing makes it the most extensively used technique. DNA-free genome editing can be achieved by using ribonucleoprotein (RNP)-mediated sgRNA/Cas9 delivery to host plants. Furthermore, multiplex genome editing is another exceptional editing technique in which more than one gRNA can be constructed in a single expression cassette to get the desired mutagenesis in maize. Two methods have been established for this purpose: one comprised of tandem repeats having diverse promoters like U3 and U6 regulating the sgRNA, while the other is based on a single promoter controlling multiple sgRNAs (Qi *et al.*, 2016; Char *et al.*, 2017). The DSBs produced by the CRISPR/Cas9 system are generally insertions or deletions of some base pairs and nicks can be filled via the NHEJ or HDR repair pathway. Big deletions like 10 bp have also been achieved, probably repaired by the microhomology-mediated end joining (MMEJ) pathway, but are not common (Feng *et al.*, 2016). Additionally, targeted modification of an allele relay on the repair matrix responsible for desired mutagenesis via HDR and HDR-based promoter exchange has also been reported in maize (Shi *et al.*, 2017).

For example, in 2019 Doll and colleagues carried out CRISPR/Cas9-mediated genome editing in maize and mutated 18 genes to study the 93 mutant alleles by producing single and multiple knockouts. In a single step, double and triple knockouts were produced that are particularly beneficial for functional studies of genes having greater genetic linkage (Doll *et al.*, 2019). In another experiment, the sgRNA/Cas9 expression cassette was co-transferred via biolistics into maize embryos to target five different genes including acetolactate synthase (*ALS1*, *ALS2*), male fertility (*Ms26*, *Ms45*) and ligule-less (*LIG1*). Mendelian segregation was observed in the progeny as a result of targeted gene insertion, edits and mutation, that would be very valuable to increase the maize production globally (Svitashev *et al.*, 2015). Male sterility is a highly beneficial tool to produce hybrid seed and requires male-sterile mutants for the establishment of hybrid seed generation lines. CRISPR/Cas9 tool is the best technique to produce knockout mutants for male sterility in maize. *Agrobacterium*-based transformation was done to mutate the *MS8* gene in maize and results showed the mutations in the $F_1$ and $F_2$ generations. Transgene-free maize plants were produced via $F_2$ screening and male sterility could be incorporated into the best-performing maize cultivars for hybrid seed production (Chen *et al.*, 2018).

## 23.6 Applications of CRISPR/Cas9 for Abiotic Stress Tolerance in Maize

CRISPR/Cas9 has vast potential application in maize genome editing and there are many reported studies in which the maize genome has been successfully mutated to get the desired result. Abiotic stresses are the major concern for agricultural productivity around the world and have severe implications for maize crop yield. In this context the CRISPR/Cas9 system is the most versatile and powerful tool to tackle these abiotic stresses in maize. Although very few studies have been carried out to date, a lot of work has to be done to protect maize from harsh climatic stresses that promises the higher production of maize (Table 23.2). For example, Njuguna and co-workers used the CRISPR/Cas9 system for targeted mutation of *PARP* (*poly(ADP-ribose) polymerase*) and *NUDX* (*ADP-ribose-specific Nudix hydrolase*) genes in maize that play a crucial part in regulating homeostasis under drought stress. The results indicated enhanced tolerance against drought stress in maize cultivars (Njuguna *et al.*, 2017). Previously it was revealed that overexpression of the *ARGOS8* gene, which is a negative regulator of the ethylene response, minimizes the ethylene susceptibility and hence increases yield under drought. To check the

**Table 23.2.** Recent studies on CRISPR/Cas9-based gene editing for abiotic stress tolerance in maize.

| Trait study | Target gene | Delivery method | Editing result | Repair pathway | Reference |
|---|---|---|---|---|---|
| Drought tolerance | *PARP* | *Agrobacterium*-mediated transformation | Knockout | NHEJ | Njuguna *et al.* (2017) |
| Drought tolerance | *ARGOS8* | Particle bombardment | Overexpression | HDR | Shi *et al.* (2017) |
| Heat tolerance | *Dcl5* | *Agrobacterium*-mediated transformation | Knockout | NHEJ | Teng *et al.* (2018) |
| Salinity tolerance | *ZmHKT2* | *Agrobacterium*-mediated transformation | Knockout | NHEJ | Cao *et al.* (2019) |
| Salinity tolerance | *ZmHKT1* | *Agrobacterium*-mediated transformation | Knockout | NHEJ | Zhang, M. *et al.* (2018) |
| Salinity tolerance | *ZmHAK4* | *Agrobacterium*-mediated transformation | Knockout | NHEJ | Zhang *et al.* (2019) |

response of maize to overexpressing *ARGOS8* under drought stress, about 400 inbred maize lines were used and CRISPR/Cas9-mediated targeted mutagenesis was performed to produce novel variants. A native promoter of maize, *GOS2*, was incorporated into *ARGOS8* to swap the native promoter of the gene. Targeted genome manipulation of the maize gene locus was confirmed via real-time quantitative PCR and RNA sequencing. The results showed increased expression of *ARGOS8* gene having *GOS2* promoter in transformed plants compared with wild cultivars and indicated the increase in production of grains under drought stress conditions. This result validated the importance of the CRISPR/Cas9 toolbox in producing new allelic variants for breeding tolerant maize cultivars under drought stress (Shi *et al.*, 2017).

Recently, CRISPR/Cas9-mediated genome editing was executed to knock out *Dcl5* (*Dicer-like 5*) gene to check the temperature sensitivity in maize. *Dcl5* is supposed to control the perfect slicing process generating different 24 nt, phased secondary small interfering RNAs (phasiRNAs) which are produced in meiotic anthers of maize and whose function is still not clear. The results demonstrated that *dcl5* knock-outs have very little or no development of 24 nt phasiRNAs and produce stunted anthers, thus showing that 24 nt phasiRNAs are crucial for heat-susceptible male fertility (Teng *et al.*, 2018). Cao and colleagues selected a recombinant inbred line (RIL) to study the salt tolerance mechanism in maize. In saline soils, $K^+$ homeostasis is regulated by a major quantitative trait locus (QTL) named *ZmHKT2* which encodes a protein of the high-affinity potassium transporter (HKT) family and possibly decreases shoot $K^+$ concentration via removal of $K^+$ ions from xylem sap. In addition, deficiency of *ZmHKT2* enhances $K^+$ concentration in shoot and xylem sap which confers salinity tolerance in maize. CRISPR/Cas9 was used for *ZmHKT2* gene editing that reduced $K^+$ transport delivery in *ZmHKT2* and ultimately assisted maize plants under salt stress. Therefore, the CRISPR/Cas9 system can be vital for improving salinity tolerance in maize (Cao *et al.*, 2019).

Similarly, functional characterization, identification and elucidation of stress tolerance QTLs in maize have been reported. *ZmHKT1* is a gene linked with $Na^+$ concentration in maize and loss of function of that gene via retrotransposons incorporation provides enhanced $Na^+$ concentration in leaves which causes salt hypersensitivity. CRISPR/Cas9-mediated editing of *ZmHKT1* was achieved, with the results showing increased concentration of $Na^+$ in xylem that resulted in enhanced $Na^+$ transfer from root to shoot. Hence, it was concluded that *ZmHKT1* is a vital gene regulating salinity stress and can be targeted to produce elite maize varieties with improved salinity tolerance (Zhang, M. *et al.*, 2018). Recently, the same experiment was carried out to study the salt tolerance in maize by disrupting the *ZmHAK4* gene (Zhang *et al.*, 2019).

## 23.7 Conclusion and Future Prospects

Assuring the food security of the global population of 10 billion people by the end of 2050 is a huge challenge. Food security and safety are the major concerns for policy makers and agricultural scientists. Continuous deterioration of the world's climate increases the chances of abiotic stresses. Maize is the third most important staple crop in the world and largely affected by severe abiotic stresses. To deal with these stresses, modern breeding technology paves the way for the improvement of maize production globally. CRISPR/Cas9 is the most powerful and excellent tool to tackle the abiotic stresses of maize. It has great potential for maize genome manipulation at desired sites. By using CRISPR/Cas9-mediated genome editing, numerous genes can be targeted to produce elite maize cultivars that minimize the challenges of abiotic stresses. In the future, more precise and accurate variants of the CRISPR/Cas9 toolbox are expected to be used for maize yield improvement.

## References

ADB (2009) Food security under threat from climate change. Asian Development Bank, Manila. Available at: https://www.livemint.com/Politics/TedZDXrkLqIEEUszZYa8jJ/Food-security-under-threat-from-climate-change-ADB.html (accessed 11 March 2021).

Barrangou, R., Fremaux, C., Deveau, H., Richards, M., Boyaval, P. *et al.* (2007) CRISPR provides acquired resistance against viruses in prokaryotes. *Science* 315, 1709–1712.

Benevenuto, R.F., Agapito-Tenfen, S.Z., Vilperte, V., Wikmark, O.G., van Rensburg, P.J. and Nodari, R.O. (2017) Molecular responses of genetically modified maize to abiotic stresses as determined through proteomic and metabolomic analyses. *PLoS One* 12, e0173069.

Bortesi, L. and Fischer, R. (2015) The CRISPR/Cas9 system for plant genome editing and beyond. *Biotechnology Advances* 33, 41–52.

Boyer, J.S. (1982) Plant productivity and environment. *Science* 218, 443–448.

Cairns, J.E., Sonder, K., Zaidi, P.H., Verhulst, N., Mahuku, G. *et al.* (2012) Maize production in a changing climate: impacts, adaptation, and mitigation strategies. *Advances in Agronomy* 114, 1–58.

Cairns, J.E. and Prasanna, B.M. (2018) Developing and deploying climate-resilient maize varieties in the developing world. *Current Opinion in Plant Biology* 45, 226–230.

Campbell, M.T., Proctor, C.A., Dou, Y., Schmitz, A.J., Phansak, P. *et al.* (2015) Genetic and molecular characterization of submergence response identifies *Subtol6* as a major submergence tolerance locus in maize. *PLoS One* 10, e0120385.

Cao, Y., Liang, X., Yin, P., Zhang, M. and Jiang, C. (2019) A domestication-associated reduction in $K^+$-preferring HKT transporter activity underlies maize shoot $K^+$ accumulation and salt tolerance. *New Phytologist* 222, 301–317.

Char, S.N., Neelakandan, A.K., Nahampun, H., Frame, B., Main, M. *et al.* (2017) An *Agrobacterium*-delivered CRISPR/Cas9 system for high-frequency targeted mutagenesis in maize. *Plant Biotechnology Journal* 15, 257–268.

Chavez, A., Scheiman, J., Vora, S., Pruitt, B.W., Tuttle, M. *et al.* (2015) Highly efficient Cas9-mediated transcriptional programming. *Nature Methods* 12, 326.

Chen, R., Xu, Q., Liu, Y., Zhang, J., Ren, D., Wang, G. and Liu, Y. (2018) Generation of transgene-free maize male sterile lines using the CRISPR/Cas9 system. *Frontiers in Plant Science* 9, 1180.

Chinnusamy, V., Zhu, J. and Zhu, J.K. (2007) Cold stress regulation of gene expression in plants. *Trends in Plant Science* 12, 444–451.

Choudhary, C., Mushtaq, M., Singh, A.K., Mukhtar, S. and Shah, A.A. (2017) Genome editing using CRISPR/Cas system: new era genetic technology in agriculture to boost crop output. *European Journal of Experimental Biology* 7, 20.

Cong, L., Ran, F.A., Cox, D., Lin, S., Barretto, R. *et al.* (2013) Multiplex genome engineering using CRISPR/Cas systems. *Science* 339(6121), 819–823.

Cramer, G.R., Urano, K., Delrot, S., Pezzotti, M. and Shinozaki, K. (2011) Effects of abiotic stress on plants: a systems biology perspective. *BMC Plant Biology* 11, 163.

Dhankher, O.P. and Foyer, C.H. (2018) Climate resilient crops for improving global food security and safety. *Plant, Cell & Environment* 41, 877–884.

Doll, N.M., Gilles, L.M., Gérentes, M.F., Richard, C., Just, J. *et al.* (2019) Single and multiple gene knockouts by CRISPR–Cas9 in maize. *Plant Cell Reports* 38, 487–501.

Edmeades, G.O. (2007) *Maize in South and South East Asia – A Prospective Study of CIMMYT's Future Role in the Region.* International Maize and Wheat Improvement Center (CIMMYT), Mexico City.

FAO (2010) FAOSTAT Statistical Database. Food and Agriculture Organization of the United Nations, Rome.

FAO (2019) FAOSTAT Statistical Database. Crops. Food and Agricultural Organization of the United Nations, Rome. Available at: http://www.fao.org/faostat/en/#data/QC/visualize (accessed 29 December 2019).

Feng, C., Yuan, J., Wang, R., Liu, Y., Birchler, J.A. and Han, F. (2016) Efficient targeted genome modification in maize using CRISPR/Cas9 system. *Journal of Genetics and Genomics* 43, 37–43.

Feng, C., Su, H., Bai, H., Wang, R., Liu, Y. *et al.* (2018) High-efficiency genome editing using a *dmc1* promoter-controlled CRISPR/Cas9 system in maize. *Plant Biotechnology Journal* 16, 1848–1857.

Fischer, R.A., Byerlee, D. and Edmeades, G.O. (2014) *Crop Yields and Global Food Security: Will Yield Increase Continue to Feed the World?* ACIAR Monograph No. 158. Australian Centre for International Agricultural Research, Canberra.

Frey, F.P., Urbany, C., Hüttel, B., Reinhardt, R. and Stich, B. (2015) Genome-wide expression profiling and phenotypic evaluation of European maize inbreds at seedling stage in response to heat stress. *BMC Genomics* 16, 123.

Gall, H.L., Philippe, F., Domon, J.M., Gillet, F., Pelloux, J. and Rayon, C. (2015) Cell wall metabolism in response to abiotic stress. *Plants* 4, 112–166.

Gao, C. (2015) Genome editing in crops: from bench to field. *National Science Review* 2, 13–15.

Gerpacio, R.V. and Pingali, P.L. (2007) *Tropical and Subtropical Maize in Asia: Production Systems, Constraints, and Research Priorities*. International Maize and Wheat Improvement Center (CIMMYT), Mexico City.

Horton, D.E., Johnson, N.C., Singh, D., Swain, D.L., Rajaratnam, B. and Diffenbaugh, N.S. (2015) Contribution of changes in atmospheric circulation patterns to extreme temperature trends. *Nature* 522, 465–469.

Jinek, M., Chylinski, K., Fonfara, I., Hauer, M., Doudna, J.A. and Charpentier, E. (2012) A programmable dual-RNA–guided DNA endonuclease in adaptive bacterial immunity. *Science* 337, 816–821.

Joung, J.K. and Sander, J.D. (2013) TALENs: a widely applicable technology for targeted genome editing. *Nature Reviews Molecular Cell Biology* 14, 49–55.

Kim, H. and Kim, J.S. (2014) A guide to genome engineering with programmable nucleases. *Nature Reviews Genetics* 15, 321–334.

Li, C., Liu, C., Qi, X., Wu, Y., Fei, X. *et al.* (2017) RNA-guided Cas9 as an *in vivo* desired-target mutator in maize. *Plant Biotechnology Journal* 15, 1566–1576.

Lobell, D.B., Bänziger, M., Magorokosho, C. and Vivek, B. (2011) Nonlinear heat effects on African maize as evidenced by historical yield trials. *Nature Climate Change* 1, 42–45.

Makarova, K.S., Haft, D.H., Barrangou, R., Brouns, S.J., Charpentier, E. *et al.* (2011) Evolution and classification of the CRISPR–Cas systems. *Nature Reviews Microbiology* 9(6), 467–477.

Mali, P., Yang, L., Esvelt, K.M., Aach, J., Guell, M. *et al.* (2013) RNA-guided human genome engineering via Cas9. *Science* 339, 823–826.

Manton, M.J., Della-Marta, P.M., Haylock, M.R., Hennessy, K.J., Nicholls, N. *et al.* (2001) Trends in extreme daily rainfall and temperature in Southeast Asia and the South Pacific: 1961–1998. *International Journal of Climatology* 21, 269–284.

Marco, F., Bitrián, C.P., Venkat Rajam, M., Alcázar, R. and Tiburcio, A.F. (2015) Genetic engineering strategies for abiotic stress tolerance in plants. *Plant Biology Biotechnology* 2, 579–609.

Naqvi, S.A.A. and Ashfaq, M. (2013) Technical efficiency analysis of hybrid maize production using translog model case study in District Chiniot, Punjab (Pakistan). *Agricultural Sciences* 4, 536–540.

Nelson, G.C., Rosegrant, M.W., Koo, J., Robertson, R., Sulser, T. *et al.* (2009) *Climate Change: Impact on Agriculture and Costs of Adaptation*. Food Policy Report. International Food Policy Research Institute, Washington, DC.

Nemudryi, A.A., Valetdinova, K.R., Medvedev, S.P. and Zakian, S.M. (2014) TALEN and CRISPR/Cas genome editing systems: tools of discovery. *Acta Natura* 6, 19–40.

Njuguna, E., Coussens, G., Aesaert, S., Neyt, P., Anami, S. and Van Lijsebettens, M. (2017) Modulation of energy homeostasis in maize and *Arabidopsis* to develop lines tolerant to drought, genotoxic and oxidative stresses. *Afrika Focus* 30, 66–76.

Noya, I., González-García, S., Bacenetti, J., Fiala, M. and Moreira, M.T. (2018) Environmental impacts of the cultivation-phase associated with agricultural crops for feed production. *Journal of Cleaner Production* 172, 3721–3733.

Nuss, E.T. and Tanumihardjo, S.A. (2010) Maize: a paramount staple crop in the context of global nutrition. *Comprehensive Reviews in Food Science and Food Safety* 9, 417–436.

Petolino, J.F. (2015) Genome editing in plants via designed zinc finger nucleases. *In Vitro Cellular & Developmental Biology – Plant* 51, 1–8.
Prosekov, A.Y. and Ivanova, S.A. (2018) Food security: the challenge of the present. *Geoforum* 91, 73–77.
Puchta, H., Dujon, B. and Hohn, B. (1993) Homologous recombination in plant cells is enhanced by *in vivo* induction of double strand breaks into DNA by a site-specific endonuclease. *Nucleic Acids Research* 21, 5034–5040.
Qi, W., Zhu, T., Tian, Z., Li, C., Zhang, W. and Song, R. (2016) High-efficiency CRISPR/Cas9 multiplex gene editing using the glycine tRNA-processing system-based strategy in maize. *BMC Biotechnology* 16, 58.
Ranum, P., Peña-Rosas, J.P. and Garcia-Casal, M.N. (2014) Global maize production, utilization, and consumption. *Annals of the New York Academy of Sciences* 1312, 105–112.
Ray, D.K., Mueller, N.D., West, P.C. and Foley, J.A. (2013) Yield trends are insufficient to double global crop production by 2050. *PLoS One* 8, e66428.
Raza, A., Razzaq, A., Mehmood, S.S., Zou, X., Zhang, X., Lv, Y. and Xu, J. (2019) Impact of climate change on crops adaptation and strategies to tackle its outcome: a review. *Plants* 8, 34.
Razzaq, A., Sadia, B., Raza, A., Khalid Hameed, M. and Saleem, F. (2019a) Metabolomics: a way forward for crop improvement. *Metabolites* 9, 303.
Razzaq, A., Saleem, F., Kanwal, M., Mustafa, G., Yousaf, S. *et al.* (2019b) Modern trends in plant genome editing: an inclusive review of the CRISPR/Cas9 toolbox. *International Journal of Molecular Sciences* 20, 4045.
Reis, A., Hornblower, B., Robb, B. and Tzertzinis, G. (2014) CRISPR/Cas9 and targeted genome editing: a new era in molecular biology. *NEB Expressions* 1, 3–6.
Richardson, K.J., Lewis, K.H., Krishnamurthy, P.K., Kent, C., Wiltshire, A.J. and Hanlon, H.M. (2018) Food security outcomes under a changing climate: impacts of mitigation and adaptation on vulnerability to food insecurity. *Climatic Change* 147, 327–341.
Rodríguez, V.M., Butrón, A., Rady, M.O., Soengas, P. and Revilla, P. (2014) Identification of quantitative trait loci involved in the response to cold stress in maize (*Zea mays* L.). *Molecular Breeding* 33, 363–371.
Rosegrant, M.W., Ringler, C., Msangi, S., Zhu, T., Sulser, T., Valmonte-Santos, R. and Wood, S. (2007) Agriculture and food security in Asia: the role of agricultural research and knowledge in a changing environment. *Journal of Semi-Arid Tropical Agricultural Research* 4, 1–35.
Rosenzweig, C., Elliott, J., Deryng, D., Ruane, A.C., Müller, C. *et al.* (2014) Assessing agricultural risks of climate change in the 21st century in a global gridded crop model intercomparison. *Proceedings of the National Academy of Sciences USA* 111, 3268–3273.
Rymen, B., Fiorani, F., Kartal, F., Vandepoele, K., Inzé, D. and Beemster, G.T. (2007) Cold nights impair leaf growth and cell cycle progression in maize through transcriptional changes of cell cycle genes. *Plant Physiology* 143, 1429–1438.
Sanghera, G.S., Wani, S.H., Hussain, W. and Singh, N.B. (2011) Engineering cold stress tolerance in crop plants. *Current Genomics* 12, 30.
Shi, J., Gao, H., Wang, H., Lafitte, H.R., Archibald, R.L. *et al.* (2017) ARGOS8 variants generated by CRISPR-Cas9 improve maize grain yield under field drought stress conditions. *Plant Biotechnology Journal* 15, 207–216.
Shukla, V.K., Doyon, Y., Miller, J.C., DeKelver, R.C., Moehle, E.A. *et al.* (2009) Precise genome modification in the crop species *Zea mays* using zinc-finger nucleases. *Nature* 459, 437–441.
Solomon, S. (2007) IPCC (2007): Climate change the physical science basis. *Presented at American Geophysical Union, Fall Meeting 2007, San Francisco, California, 10–14 December 2007*, abstract U43D-01.
Strable, J. and Scanlon, M.J. (2009) Maize (*Zea mays*): a model organism for basic and applied research in plant biology. *Cold Spring Harbor Protocols* 10, 132.
Svitashev, S., Young, J.K., Schwartz, C., Gao, H., Falco, S.C. and Cigan, A.M. (2015) Targeted mutagenesis, precise gene editing, and site-specific gene insertion in maize using Cas9 and guide RNA. *Plant Physiology* 169, 931–945.
Teng, C., Zhang, H., Hammond, R., Huang, K., Meyers, B. and Walbot, V. (2018) *Dicer-like 5* deficiency confers temperature-sensitive male sterility in maize. *Nature Communications* 11, 2912.
Thakore, P.I., D'Ippolito, A.M., Song, L., Safi, A., Shivakumar, N.K. *et al.* (2015) Highly specific epigenome editing by CRISPR-Cas9 repressors for silencing of distal regulatory elements. *Nature Methods* 12, 1143–1149.

USDA (2018) Commodities. Grain and Feed. Corn. US Department of Agriculture, Foreign Agricultural Service, Washington, DC. Available at: https://www.fas.usda.gov/commodities/corn (accessed 28 December 2019).

Urnov, F.D., Rebar, E.J., Holmes, M.C., Zhang, H.S. and Gregory, P.D. (2010) Genome editing with engineered zinc finger nucleases. *Nature Reviews Genetics* 11, 636–646.

Vaughan, M.M., Block, A., Christensen, S.A., Allen, L.H. and Schmelz, E.A. (2018) The effects of climate change associated abiotic stresses on maize phytochemical defenses. *Phytochemistry Reviews* 17, 37–49.

Walbot, V. (2000) Saturation mutagenesis using maize transposons. *Current Opinion in Plant Biology* 3, 103–107.

Waltz, E. (2018) With a free pass, CRISPR-edited plants reach market in record time. *Nature Biotechnology* 36, 6–7.

Wijewardana, C., Hock, M., Henry, B. and Reddy, K.R. (2015) Screening corn hybrids for cold tolerance using morphological traits for early-season seeding. *Crop Science* 55, 851–867.

Xing, H.L., Dong, L., Wang, Z.P., Zhang, H.Y., Han, C.Y. *et al.* (2014) A CRISPR/Cas9 toolkit for multiplex genome editing in plants. *BMC Plant Biology* 14, 327.

Zhang, J.P., Li, X.L., Li, G.H., Chen, W., Arakaki, C. *et al.* (2017) Efficient precise knockin with a double cut HDR donor after CRISPR/Cas9-mediated double-stranded DNA cleavage. *Genome Biology* 18, 35.

Zhang, M., Cao, Y., Wang, Z., Wang, Z.Q., Shi, J. *et al.* (2018) A retrotransposon in an HKT1 family sodium transporter causes variation of leaf $Na^+$ exclusion and salt tolerance in maize. *New Phytologist* 217, 1161–1176.

Zhang, Y., Massel, K., Godwin, I.D. and Gao, C. (2018) Applications and potential of genome editing in crop improvement. *Genome Biology* 19, 210.

Zhang, M., Liang, X., Wang, L., Cao, Y., Song, W. *et al.* (2019) A HAK family $Na^+$ transporter confers natural variation of salt tolerance in maize. *Nature Plants* 5, 1297–1308.

Zhu, J., Song, N., Sun, S., Yang, W., Zhao, H., Song, W. and Lai, J. (2016) Efficiency and inheritance of targeted mutagenesis in maize using CRISPR-Cas9. *Journal of Genetics and Genomics* 43, 25–36.

Zong, Y., Wang, Y., Li, C., Zhang, R., Chen, K. *et al.* (2017) Precise base editing in rice, wheat and maize with a Cas9-cytidine deaminase fusion. *Nature Biotechnology* 35, 438.

Zuo, W., Chao, Q., Zhang, N., Ye, J., Tan, G. *et al.* (2015) A maize wall-associated kinase confers quantitative resistance to head smut. *Nature Genetics* 47, 151.

# 24 Molecular Breeding for Combating Salinity Stress in Sorghum: Progress and Prospects

**Nidhi Chakma[1]*, Moutoshi Chakraborty[2], Salma Bhyan[3] and Mobashwer Alam[1],**

[1]*The University of Queensland, Brisbane, Australia;* [2]*Bangabandhu Sheikh Mujibur Rahman Agricultural University, Gazipur, Bangladesh;* [3]*University of the Sunshine Coast, Sippy Downs, Australia*

## 24.1 Introduction

Soil salinity is one of the major constraints that severely limit the growth and development of a crop. Salinity limits plant development not only by reducing the water and nutrient uptake but also by causing metabolic changes (Munns, 2002). It causes serious damage in biological systems and reduces the productivity in crops, especially in arid and semi-arid areas. Due to climate change and global warming, sea level is rising, and saline-affected areas are increasing constantly. Hence, salinity stress is becoming more of a concern and there is now greater demand to develop salinity stress-tolerant varieties. A better understanding of the saline tolerance mechanism in crops will be useful for breeding saline-tolerant crop varieties, which is crucial for sustainable crop production (Mbinda and Kimtai, 2019).

Sorghum (*Sorghum bicolor* L. Moench) is one of the five major cereal crops and is grown all over the world. It was categorized as a moderately salinity-tolerant crop (Igartua *et al.*, 1995) and is known to be comparatively more saline-tolerant than maize, the cereal crop that ranks first in global productivity (Maas *et al.*, 1986). Thus, sorghum has potential as a crop for areas affected by salt (Igartua *et al.*, 1994). Over the years, several breeding methods have been applied for the genetic improvement of this crop. To develop saline-tolerant varieties, researchers are currently exploiting germplasms and employing molecular technologies to understand the saline tolerance mechanism in crops. In recent years, molecular breeding has become very popular to develop biotic and abiotic stress-tolerant crop varieties. Advanced molecular marker technologies, such as the identification of markers closely linked to a quantitative trait locus (QTL) or gene for salt tolerance, would be useful to develop salt-tolerant crop varieties by marker-assisted selection (MAS), thus increasing breeding efficiency (Kumar *et al.*, 2015). In this chapter, current progress and prospects of molecular breeding and strategies for developing better saline-tolerant sorghum varieties are discussed.

## 24.2 Germplasm for Salt Tolerance

Saline-tolerant germplasms are the preliminary sources of further genetic improvement for

*Corresponding author: n.chakma@uq.edu.au


DOI: 10.1079/9781789245431.0024

sustainable crop production (Mbinda and Kimtai, 2019). Although sorghum is a moderately salt-tolerant crop, there is a large genetic variation for saline tolerance (Francois *et al.*, 1984; Igartua *et al.*, 1994). Krishnamurthy *et al.* (2003) have reported 19 highly salt-tolerant sorghum cultivars and elite lines which showed desirable agronomic traits under 250 mM salt concentration. Azooz *et al.* (2004) tested three sorghum cultivars for saline tolerance and cultivar Dorado was found to be the most tolerant cultivar showing a positive correlation with growth characteristics. Krishnamurthy *et al.* (2007) conducted a series of pot experiments for salt tolerance in 100 sorghum lines comprising hybrids from grain and forage, popular and improved varieties, and reported seven highly saline-tolerant genotypes. In another study, 100 sorghum germplasms were tested by Kulhari *et al.* (2008) under saline stress and three genotypes, Raj 27, Raj 30 and Raj 4, were identified as stable salt-tolerant lines. Reddy *et al.* (2010) screened 27 hybrid varieties and 26 restorer lines and concluded that hybrids are superior in the case of establishment and productivity. Sun *et al.* (2014) evaluated ten sorghum varieties in two greenhouse experiments and found that three sorghum varieties, Shallu, Desert Maize and 1790E, were the most salt-tolerant varieties. Three sorghum genotypes, JJ1041, ICSV 745 and S35, were found to be tolerant to saline conditions at the early growth stage by Attia (2016). Kausar *et al.* (2012) recommended that Sandalbar and JS-2002 varieties could be a potential genetic source for sorghum biomass production in saline areas with 20 dS/m. Ding *et al.* (2018) evaluated 300 sweet sorghum germplasms for saline tolerance at the germination stage. Among those, 23 were highly salt-tolerant, 38 were salt-tolerant and 195 were moderately salt-tolerant cultivars. Bafeel (2014) screened seven Saudi local cultivars of sorghum for natural seawater (Red Sea) tolerance and revealed that C3 (mixed white and red seeds) was the most salt-tolerant based on germination ability and shoot development. A list of available salt tolerant cultivars is provided in Table 24.1. These listed germplasms can be used as potential sources for future breeding programmes.

## 24.3 Salinity Response in Growth Stages

In sorghum, the saline response of a variety also depends on the stage of growth. Maas *et al.* (1986) performed research on sorghum saline sensitivity at three growth stages and concluded that sorghum is mostly saline-sensitive at the early growth stage and least sensitive at the maturation stage. However, some researchers reported that the germination stage of sorghum is the most saline-sensitive than other growth stages (Krishnamurthy *et al.*, 2003; Sun *et al.*, 2014). Furthermore, results indicated that the genotypes with higher germination rates had higher biomass and yield potentiality under saline stress conditions (Krishnamurthy *et al.*, 2007), which could be the reason why most of the saline tolerance germplasm screening research has been performed at seedling emergence and seedling growth stage. Tigabu *et al.* (2012) tested 11 sorghum genotypes at seed germination and early seedling stages and found that cultivar ICSV-111 was more salt-tolerant at the germination stage while Teshale and 76T1#23 cultivars showed better salt tolerance at the seedling growth stage. Ali *et al.* (2013) reported that cultivar Mr. Buster was more salt-tolerant at the germination stage but less tolerant at the seedling growth stage compared with the cultivars 'Honey Graze' and 'Extra Sweet'.

## 24.4 Salinity Response in Morpho-Physiological Traits

Evaluation of salinity response in morpho-physiological traits is crucial for accurate screening. Identification of traits associated with saline resistance assists in further molecular investigation and selection of saline-tolerant cultivars. Although numerous studies have been conducted to identify traits for saline tolerance in sorghum, a set of simple and accurate trait indexing methods is still unavailable, which is one of the key reasons for limited success in the genetic improvement programmes (Zeng *et al.*, 2002).

Morphological traits are the primary indicator of salinity tolerance. Investigation in plants usually starts with observations of their surface appearances, and shoots and roots are the most important morphological features for the

**Table 24.1.** List of salt-tolerant sorghum cultivars.

| Cultivars | Criteria (salinity conditions) | Reference |
|---|---|---|
| Northrup King 265 and Asgrow Double TX | Six salinity treatments (osmotic potential of –0.25, –0.35, –0.45, –0.65, –0.85 and –1.05 MPa) | Maas *et al.* (1986) |
| CSV 15, ICSV 766, ICSV 95030, NTJ 2, ICSV 145, S35, ICSV 112, ICSV 300, ICSR 196, SP 40669, SPA2 94029, SP 40672, SPV 1022, SPDM 94006, ICSR 91005, ICSR 89010, SP 40646 and ICSR 93034 | 250 mM, irrigated condition | Krishnamurthy *et al.* (2003) |
| Dorado, Hagen Shandawil and Giza | Five levels (1, 2, 3, 4 and 5 bar) of NaCl | Azooz *et al.* (2004) |
| CSV 15, ICSV 766, NTJ 2, ICSV 95030, S35, ICSV 589 and ICSV 676 | 250 mM NaCl | Krishnamurthy *et al.* (2007) |
| Raj 27, Raj 30 and Raj 4 | 0, 6 and 9 dS/m | Kulhari *et al.* (2008) |
| Soave | 0, 100, 200 and 300 mM NaCl | Almodares *et al.* (2007) |
| Hybrids of ICSA 405 × JJ1041, ICSA 766 × ICSV 96020, ICSA 707 × ICSV 745, ICSA 276 × ICSV 93048 and ICSA 276 × S35; and varieties JJ1041, ICSB 707, SP 47529, CSV 15 and SPV 1022 | 250 mM NaCl | Reddy *et al.* (2010) |
| ICSV-111, Teshale, 76T1#23 | 2, 4, 8 and 16 dS/m | Tigabu *et al.* (2012) |
| CSV-15, HD-19 and HC-171 | 3, 6, 7.2, 10 and 12 dS/m | Rani *et al.* (2012) |
| JS-2002 and Sandalbar | 50, 100, 150 and 200 mM NaCl | Kausar *et al.* (2012) |
| Wad Ahmed | 2, 4, 8 and 16 dS/m | El Naim *et al.* (2012) |
| Mr. Buster, Extra Sweet and Honey Graze | Irrigation with 50, 100 and 150 mM NaCl | Ali *et al.* (2013) |
| Ketian No. 2 and Jintianza | 3.0, 6.0 and 9.0 g NaCl/kg | Zhan *et al.* (2013) |
| C3 | 7 dilutions of Red Seawater (1.6, 3.1, 6.3, 12.5, 25, 50 and 100%) | Bafeel (2014) |
| Shallu, Desert Maize and 1790E | 5, 10 and 17 dS/m | Sun *et al.* (2014) |
| JJ1041, ICSV 745 and S35 | Field irrigation | Attia (2016) |
| Mecca hybrid | 0, 3, 6, 9, 12 and 15 dS/m | Kandil *et al.* (2017) |
| 23 highly salt-tolerant varieties - Jitianzashi (11-5) 7, KORJAJ-2, MN-3086, Jitianza 3, Z1H08902, KORJAJ-1, XTL-4, Tianxuan 171, Nengsi 1, Tianxuan 159, MN-2467, Z1H08912, Bapin 4, Z1H08919, Tianxuan 8, Tianxuan 86, Jintian 09-1, PI196600, BJK156, Tianxuan 192, MN-3808, Tianxuan 94, L-Tian | 0, 100 and 200 mM NaCl | Ding *et al.* (2018) |
| Gadam | 100, 200 and 300 mM NaCl | Mbinda and Kimtai (2019) |
| Hybrid Sorgo, BD703 and BD707 | 0, 6, 12 and 18 dS/m | Sagar *et al.* (2019) |

identification of stress-tolerant plants (Gifford and Foster, 1989; Bai, 2017). In many studies, shoot length and root length have been used as an indicator trait alongside other traits to screen salt-tolerant plants. Saline stress reportedly significantly reduces shoot and root length because soil salinity lowers osmotic potential and increases ion toxicity in plants, thus reducing biomass (Azhar and McNeilly, 1987; Krishnamurthy *et al.*, 2007; Rani *et al.*, 2012; Sun *et al.*, 2012; Tigabu *et al.*, 2012). Germination rate was frequently used to identify saline-tolerant sorghums because seed germination is the first step of performance evaluation. Germination rate severely decreases with increased salinity level due to high accumulation of $Na^+$ and $Cl^-$ and low water potential hindering water absorption (Almodares *et al.*, 2007; Geressu and Gezahagne, 2008; Sun *et al.*, 2012; Tigabu *et al.*, 2012; Zhan *et al.*, 2013). Sun *et al.* (2012) suggested that root length, coleoptiles and leaf weight, and germination rate can be used as key indices to identify salinity-tolerant varieties at an early stage. Plant height, plant shoot and root dry weight, and leaf area per plant were investigated to determine the level of tolerance in these studies.

Physiological responses and stress reactions of plants under salinity are reported to be varied depending on the sorghum variety and growth stage. For instance, Ding *et al.* (2018) evaluated numerous physiological parameters to identify reliable screening traits at the germination stage by using 300 sweet sorghum germplasms. The membership function value (MFV) revealed that germination index (GI) was a better indicator for salt tolerance screening at the germination stage under 200 mM NaCl treatment in sweet sorghum germplasm. Kausar *et al.* (2012) used several physiological parameters such as germination stress tolerance index (GSI), root length stress tolerance index (RLSI), shoot length stress tolerance index (SLSI) and biomass stress tolerance index (BSI) to screen salt-tolerant varieties and suggested that these parameters are efficient in screening a large number of sorghum varieties for saline tolerance. One of the physiological traits to identify salt-tolerant plants could be determination of the content of photosynthetic pigments in plants. It was revealed that a low concentration of NaCl (0.3%) increased the content of chlorophyll a and chlorophyll b in the leaves of sweet sorghum salt-tolerant varieties, thus suggesting that the content of photosynthetic pigment can be used as a marker for salinity stress tolerance (Baiseitova *et al.*, 2018). Tolerance of sorghum cultivars can also be associated with $K^+$ content and $K^+$:$Na^+$ ratio in plants (Azooz *et al.*, 2004). Sun *et al.* (2014) reported that $K^+$ level in leaves increased with increased NaCl concentration. Hence, $K^+$:$Na^+$ ratio can be used as an effective salt tolerance indicator in plants.

## 24.5 Traditional Breeding for Developing Salinity-Tolerant Varieties

Breeding programmes have been performed for thousands of years, as humans continue to use different methods to modify plant traits to achieve desirable characteristics (Sleper and Poehlman, 2006). The presence of large genotypic variability in sorghum provided an excellent opportunity for trait manipulation through a genetic improvement programme (Taylor *et al.*, 1975; Maiti *et al.*, 1994). Success in traditional breeding largely depends on the number of genes controlling the trait (Agrawal, 1998). Usually, qualitative traits are easier to breed than quantitative traits. Qualitative characteristics (controlled by one or few genes) are simple to modify and evaluate in a breeding programme since their expressions generate distinct phenotypes, which can be differentiated by counting and putting them into clear-cut categories (discrete distribution) (Brown and Caligari, 2011). However, quantitative traits are regulated by polygenes (multiple genes), each contributing a small effect to a trait's overall phenotypical expression. The predominant plant characteristics of agronomic and economic significance are quantitative, including crop yield, environmental adaptation and resistance/tolerance to stresses like pests and diseases, drought, salinity, etc. The expression of these traits is often complicated, with phenotypic expression patterns regulated by a complex series of interactions between multiple genes and environmental factors (Kover *et al.*, 2009). It is an ongoing challenge for the plant breeder to dissect the genetic architecture of these complex traits. Many salt-tolerant lines of sorghum have been produced via conventional and molecular breeding approaches. The rest of this chapter provides an update and discussion on the current approaches of developing salt-tolerant sorghum cultivars.

In the conventional method, the development and enhancement of a cultivar can be achieved by manipulating the plant genome within a species' primary gene pool (Acquaah, 2009, 2015). It is the preliminary method of creating new genetic variation by utilizing cross-pollination followed by self-pollination and/or clonal propagation (Briggs and Knowles, 1967). In the last century, conventional breeding predominated, and considerable progress was made in enhancing high yield and stress-tolerant crops. Although many salt-tolerant crop cultivars, such as rice cultivars CSR10, CSR13 and CSR27, have been produced through conventional breeding, very few studies have been conducted for the production of salt-tolerant sorghum varieties (Ashraf and Akram, 2009).

The choice of methodology for breeding depends on the crop's pollination behaviour and the type of cultivar intended for development. Sorghum is an often cross-pollinated crop as cross-pollination varies from 5 to 15% with an average of about 6%, hence is suitable for both self- and cross-pollinated crop breeding practices (Poehlman, 2013). This suggests that sorghum breeders may develop homozygous lines as a cultivar or induce heterosis by hybridization. Although sorghum breeding objectives were mostly centred around yield and yield stability, quality features and tolerance to one or more stresses (such as tolerance to salinity) were also considered (Rakshit and Bellundagi, 2019). For improving sorghum, breeders employed various conventional selection and breeding strategies. These include mass selection, pure-line selection, pedigree method, bulk method, backcross method, single seed descent method (SSD), mutation breeding, etc. Choice of the selection strategy depends on the planned objectives and the handled breeding material. For genetically variable and homozygous populations, mass selection and pure-line selection schemes can be implemented, while in segregating populations other approaches such as pedigree, bulk, backcross, SSD strategy and mutation breeding are implemented.

M-35-1, Yashoda, Sel-3 and Maulee (RSLG-262) are some of the commercial cultivars of sorghum successfully developed through the pure-line selection method; and CSV 17 (SPV 946 × SPV772), CSV 15 (SPV475 × SPV 462) and CSV 20 (SPV946 × Hh89-246) are commercial cultivars of sorghum produced by the pedigree method for yield and quality-related traits. The mutant lines of sorghum – PAHAT for drought tolerance; YT30-39-07, GH-ZB-41-07, B-92 and B-76 for higher yield; ZH30-30-07, ZH30-29-07 and ZH30-35-07 for production of bioethanol; and ZH30-30-07, ZH30-35-07 and ZH30-29-07 for increasing sweetness – were successfully developed via mutation breeding (Ashraf and Akram, 2009; Rakshit and Bellundagi, 2019). These examples illustrate that very little work has been done to develop saline-tolerant sorghum cultivars using the conventional breeding method. Therefore, substantial improvements in the sorghum salt tolerance may not be expected.

## 24.6 Progress and Prospects of Molecular Breeding

Molecular breeding is one of the most important tools for plant breeding. In the last few decades, the advancement of molecular breeding techniques has enabled researchers to speed up the development of different crop varieties. Using DNA-based markers, thousands of QTLs and genes have been mapped in major crops and developed the foundation for advanced MAS strategies (Xu *et al.*, 2017). However, compared with other major crops, the research for salt tolerance in sorghum is still at a preliminary level. More extensive research is required for understanding salt tolerance in sorghum.

An approach to improve selection efficiency and salt tolerance breeding is to detect molecular markers linked with salt tolerance. Depending on the method to identify polymorphism, molecular markers are mainly classified into two categories: hybridization-based and PCR-based markers. Hybridization-based markers, including RFLP (restriction fragment length polymorphism) and VNTR (variable number of tandem repeats), generally use probes and Southern blot analysis. PCR-based markers include RAPD (randomly amplified polymorphic DNA), SSR (simple sequence repeat), AFLP (amplified fragment length polymorphism), ISSR (inter-simple sequence repeat) and SNP (single-nucleotide polymorphism). Several types of molecular markers for salt tolerance have been detected in sorghum. Younis *et al.* (2007) identified 14 RAPD markers and 15 ISSR markers linked to saline stress in sorghum at seedling stage following bulk segregate analysis. Rao *et al.* (2007) found that BADH 1 and Primer RC 15, RFLP and RAPD markers, respectively, can be used in screening salt-tolerant sorghum varieties. Khalil (2013) successfully used 11 ISSR markers to detect salinity stress tolerance in seven sorghum cultivars and detected salt tolerance gene *BADH*. Zhan *et al.* (2013) screened four sweet sorghum varieties using 19 SSR markers and the marker AH24 showed complementary bands between highest salt tolerance and most sensitive varieties. Even though SNP markers are being used for other biotic and abiotic tolerances in sorghum, SNP markers for salt tolerance are yet to be reported.

Over the past two decades, linkage mapping, which is relying on genetic recombination during the construction of mapping populations, has been widely used in different plant species and several QTLs have been cloned or tagged. QTLs are the regions within a genome consisting of genes associated with specific traits (Collard *et al.*, 2005). The identification of QTLs associated with salt tolerance can enhance the speed of marker-assisted breeding. Wang, H. *et al.* (2014) generated a QTL map and reported 12 QTLs for three traits (germination vigour, germination percentage and relative salt-injury rate) at germination stage and 29 QTLs for nine traits (salt injury index, shoot height, root length, shoot fresh weight, root fresh weight, total fresh weight, shoot dry weight, root dry weight and total dry weight) at seedling stage of sorghum using 181 lines derived from a cross of Shihong137 and L-Tian cultivars. One major QTL (*qGP7-1*) was found at the germination stage and six major QTLs (*qSH1*, *qSH4*, *qRL10-2*, *qSFW4*, *qTFW1* and *qTFW4*) were reported at the seedling stage which were assumed to play an important role in salinity tolerance. This result implies that salt tolerance in sorghum is different at the germination and seedling stages and more extensive research is required to determine QTLs at different development stages of sorghum.

As a complement to linkage mapping, association mapping takes advantage of historical recombination events collected over hundreds of generations, resulting in higher precision and higher numbers of alleles (Zhu *et al.*, 2008). With the benefits of high resolution, high allelic richness and lack of need for the tedious construction of a mapping population, natural population-based mapping has become a useful tool for detecting natural variation that underlies complex sorghum traits, especially salinity tolerance. Natural populations composed of many individuals, including inbred lines, landraces, wild relatives and exotic accessions, can be exploited in sorghum breeding for several purposes: (i) to maintain existing genetic diversity by selfing or outcrossing; (ii) to explore phenotypic variability, estimation of broad-sense heritability of desired traits and determination of genotypes of population entries for candidate genes or genome-wide investigation of gene families; (iii) quantification of the magnitude of linkage disequilibrium (LD) of the population selected; (iv) detection of the influence of population structure and kinship; and (v) for utilizing appropriate statistical strategies to test the associations between genotypes and phenotypes. The statistical efficiency of natural population-based mapping depends strongly on the magnitude of LD and the population structure, and also on the sample size and the minor allele frequency (MAF) (Xu *et al.*, 2017). However, until now, there is no report on saline tolerance QTL identification using natural population-based mapping.

Multiple parents having genetic diversity contributes to a population with considerable phenotypic variability, making it ideal for QTL mapping with high resolution. Two experimental designs of multi-parent populations, which include nested association mapping (NAM) and multi-parent advanced generation inter-cross (MAGIC), are becoming increasingly popular (Xu *et al.*, 2017). NAM is a population of the multi-parent design proposed by Yu *et al.* (2008). NAM populations have proved to be extremely powerful resources to dissect the genomic architecture of phenotypes (Yu *et al.*, 2008). The extended statistical capacity and combinatory mapping approaches of NAM populations permit the identification of small-effect loci; however, restricting the false positives common in genome-wide association studies (GWAS) with repeated demonstrations (Kump *et al.*, 2011). Recognizing the distinct benefits of this strategy, researchers of sorghum have also produced NAM populations to dissect essential and distinctive sorghum traits. Currently, there are three conventional NAMs (i.e. SSD from a cross to a recurrent parent) and two backcrossed NAM populations (Boyles *et al.*, 2019). Four out of the five NAM populations concentrate on grain sorghum while a fifth focuses on sorghum for bioenergy. Two of the NAM populations have easily accessible seed and genomic data: the RTx430 population (Bouchet *et al.*, 2017) and the Australian BC-NAM population (Jordan *et al.*, 2011). The populations of BTx623 and RTx430 were part of a collaborative effort between Kansas State University and Texas A&M University. Each utilizes crosses between the two recurrent females (RTx430 and BTx623) with ten different male parents. While these multi-parental populations may be good genetic tools for high-resolution QTL mapping and marker development for salinity tolerance characteristics, there is no study on saline tolerance QTL detection using a NAM population to date.

Recombination is the essential part of all genetic mapping. Conventional linkage analyses with recombinant inbred lines (RILs) and NAMs are impeded by the small number of recombination events and significantly larger size of recombination blocks in their effectiveness. In an effort to overcome the limitations of both approaches, researchers developed inter-cross-generation (MAGIC) populations in order to increase recombination while maintaining an evenly structured population. MAGIC populations depict intermediate to bi-parental crosses and panels of diversity linked to the substructure, diversity of alleles and number of traits to be investigated with high resolution and power (Rakshit *et al.*, 2012). A MAGIC population is typically produced from crossing multiple parental lines across several generations (Ongom and Ejeta, 2018). Compared with other multi-parent populations, MAGIC populations require intermating of multiple inbred founders for multiple generations until inbred lines significantly improve the accuracy of QTL detection. Statistical methods for QTL mapping have become accessible in MAGIC populations, some of them focused on the general linear model (GLM) used in bi-parental populations. The use of MAGIC populations was initially proposed by Threadgill *et al.* (2002) for QTL mapping in mice. In crops, Kover *et al.* (2009) first produced a MAGIC population in *Arabidopsis thaliana* consisting of 527 lines resulting from the intermating of the 19 founders' heterogeneous panel. MAGIC populations have been used in wheat to identify QTL for hectolitre weight and plant height (Huang *et al.*, 2012). Other crop species in which MAGIC populations have been used are tomato (Pascual *et al.*, 2015), barley (Sannemann *et al.*, 2015), maize (Dell'Acqua *et al.*, 2015) and rice (Bandillo *et al.*, 2013). The first MAGIC population of sorghum was reported by Ongom and Ejeta (2018) and was constructed using random mating of 19 different founding lines. MAGIC populations certainly offer great opportunities to dissect complex characteristics, such as salinity tolerance, and to improve sorghum breeding populations. This, however, requires more time and involves higher costs in population creation. To date, no study investigating a MAGIC population design involved in QTL identification for salinity tolerance has been published.

An important step in molecular breeding is to identify candidate genes of desired traits as the genes can provide necessary knowledge for developing an elite variety with desirable attributes. One of the approaches to identify candidate genes is based on expressed sequence tags (ESTs) obtained from different cDNA libraries and clustering the EST sequences, which provides information about genes (Chaduvula *et al.*, 2015). To identify salt stress-related genes, Chaduvula *et al.* (2015) studied 6749 ESTs mined from web resources and clustered and assembled the ESTs into contigs whose biological functions were acquired through gene ontology (GO). A total of 12 candidate genes were detected as salt stress-responsive genes and the candidate genes were validated in nine sorghum cultivars. P5CS ($\Delta_1$-pyrroline-5-carboxylate synthetase) is known as a key regulatory enzyme that plays an important role in proline biosynthesis. Su *et al.* (2011) isolated two closely related *P5CS* genes, *SbP5CS1* and *SbP5CS2*, from sweet sorghum which are located on chromosome 3 and 9. According to expression analysis, *SbP5CS* and *SbP5CS2* transcripts were upregulated after 250 mM NaCl treatment of 10-day-old sweet sorghum seedlings. The peak expression of *SbP5SC1* and *SbP5SC2* under high salt stress was detected at 4 and 8 h, respectively, and the upregulated expression of *SbP5SC1* was higher than that of *SbP5SC2*. The result indicates that *SbP5CS1* might be a good candidate gene in developing stress-tolerant cultivars using genetic engineering.

In general, saline stress includes ion toxicity and osmotic stress. Ion toxicity is mainly caused by increasing $Na^+$ concentration in plants. Genes encoding high-affinity potassium transporters (HKTs) were reported to be involved in salt tolerance in grain crops by excluding $Na^+$ ions from shoot tissues of plants (Waters *et al.*, 2013). Wang, T.T. *et al.* (2014) characterized *SbHKT1;4*, a member of the *HKT* gene family from sorghum, and reported that under $Na^+$ stress *SbHKT1;4* has more upregulated expression in salt-tolerant accessions. Auxin-related genes such as *Gretchen Hagen3* (*GH3*) and *lateral organ boundary domains* (*LBD*) have also shown a relationship with salt resistance properties in sorghum. Wang *et al.* (2010) reported that the *SbGH3* and *SbLBD* genes were highly expressed under salt and drought stresses and low under normal conditions. Also, three genes, *SbIAA1*, *SbGH3-13* and *SbLBD32*, showed higher expression under salt stress. These results provided the information that auxin plays an important role in saline response in sorghum.

Genes related to transcription factors can be useful in improving saline tolerance in sorghum (Kasuga *et al.*, 1999). A common method to determine genes for saline stress is exploiting a comparative study between saline-sensitive and saline-tolerant cultivars (Walia *et al.*, 2005; Wang, H. *et al.*, 2014; Sui *et al.*, 2015). Sui *et al.* (2015) compared transcriptomes between two saline-sensitive and saline-tolerant inbred sweet sorghum lines under 0 and 150 mM NaCl and identified 930 expressed genes in both inbred lines. Most genes were involved in photosystem, carbon fixation and metabolism of sucrose. The genes that regulate photosystem and electron transport were less affected by salt stress in saline-tolerant lines compared with saline-sensitive lines. Furthermore, gene expression encoding $NADP^+$-malate enzyme and sucrose synthetase was upregulated and invertase-encoding genes were downregulated under saline stress in a saline-tolerant sweet sorghum variety. This finding suggests that saline-tolerant sweet sorghum cultivars accumulate more sucrose by escalating sucrose synthesis. DREB2 (dehydration-responsive element-binding protein 2) type transcription factor was characterized in sorghum under saline stress and six putative *SbDREB2* genes were detected in the sorghum genome. All *SbDREB2* genes in root were upregulated under salt stress, while *SbDREB2A* and *SbDREB2C2* were upregulated only in leaves (Akbudak *et al.* 2018). These findings suggest that the *SbDREB2* genes might be useful in developing saline-tolerant sorghum varieties.

Proteomics studies in response to saline stress have been performed in several cereal species including sorghum. Ndimba *et al.* (2010) studied five different proteomes from several parts of plants, including secretome from cell suspension culture. Western blot analysis revealed salt stress-induced expression of HSP70 protein (stress-responsive protein) in the roots of experimental plants. Swami *et al.* (2011) conducted a proteome study to analyse the sorghum response to salt stress in hydroponic culture. According to two-dimensional gel electrophoresis and subsequent mass spectrometric identification, 21 differentially expressed proteins were found and most of them were in the functional categories of signal transduction pathway and inorganic ion transport and metabolism. Similarly, Sekhwal *et al.* (2012) identified five salt tolerance-related proteins in the same functional category of signal transduction mechanisms alongside ribosome maturation and energy production. These categories of proteins suggest a unique mechanism for saline-stress adaptation in sorghum. The expression of proteins was observed at 200 mM NaCl in hydroponic culture after 96 h of saline stress.

Plant breeding for saline tolerance is not very successful due to reproduction barriers and the risk of transferring other undesirable traits. To avoid this problem, genetic engineering is preferable as this technique transfers only specific genes. Plants try to adapt to salinity by inducing various metabolic changes (e.g. production of osmolytes and antioxidative enzymes) and upregulating various genes associated with stress tolerance such as ion transporters, transcriptional factors and various signalling pathway components. These plant responses to salinity have been used by researchers to generate transgenic plants, either by transferring stress-responsive genes into a salt-sensitive elite variety or by altering the expression of genes (Turan *et al.*, 2012). Although numerous research works have been performed in other crops, there is very little information available about improving saline stress tolerance in sorghum using genetic engineering, partly due to slow progress in developing protocols for transformation (Shahbaz and Ashraf, 2013). One successful attempt was reported by Maheswari *et al.* (2010) in developing a transgenic sorghum variety tolerant to water deficit and NaCl stress. The cultivar SPV462 was transformed with the *mtlD* gene encoding for mannitol-1-phosphate dehydrogenase from *Escherichia coli.* Integration and expression of the transgene (*pCAM–mtlD*) were confirmed by PCR, RT-PCR, Southern blot analysis and Western blot analysis. The transgenic plants had 1.7- and 2.8-fold higher shoot and root growth, respectively, compared with the untransformed control under NaCl stress (200 mM), demonstrating that sorghum saline tolerance can be enhanced by engineering the mannitol biosynthetic pathway.

## 24.7 Conclusion

Saline tolerance in sorghum is a complex trait and is controlled by several QTLs or polygenes. It is greatly influenced by environment, genotype and growth stage. Molecular breeding is a powerful technique which can elucidate the saline tolerance mechanism at molecular level in sorghum. Moreover, molecular characterization of germplasms and different genes associated with salt tolerance is important in the utilization and conservation of genetic resources. MAS allows selection of plants with desired traits at seedling stage (or any stage), thus reducing the time frame. Moreover, selecting plants using markers is simpler, more convenient and reliable compared with the conventional type of screening. To meet the growing demand for salt-tolerant varieties, use of molecular markers could hasten the breeding programme. Molecular markers can also be helpful in differentiating genotypes and markers have been used in identifying salt-tolerant varieties. However, most molecular breeding techniques for salt tolerance have been carried out in controlled environments where the plants were not exposed to any variation of the surrounding environment, producing reliable results. Due to the polygenic nature of salt tolerance, the identified QTLs could be false QTLs. Therefore, QTL validation is important in different plant populations and field conditions. Subsequently, marker validation is important before utilizing MAS for screening salt-tolerant plants. Combining molecular breeding with conventional breeding can hasten the development of salt-tolerant sorghum varieties.

## References


Acquaah, G. (2009) *Principles of Plant Genetics and Breeding*. Wiley, Chichester, UK.

Acquaah, G. (2015) Conventional plant breeding principles and techniques. In: Al-Khayri, J., Jain, S. and Johnson, D. (eds) *Advances in Plant Breeding Strategies: Breeding, Biotechnology and Molecular Tools*. Springer, Cham, Switzerland, pp. 115–158. Available at: https://doi.org/10.1007/978-3-319-22521-0_5

Agrawal, R.L. (1998) *Fundamentals of Plant Breeding and Hybrid Seed Production*. Science Publishers, Inc., Enfield, New Hampshire.

Akbudak, M.A., Filiz, E. and Kontbay, K. (2018) DREB2 (dehydration-responsive element-binding protein 2) type transcription factor in sorghum (*Sorghum bicolor*): genome-wide identification, characterization and expression profiles under cadmium and salt stresses. *3 Biotech* 8(10), 426–442.

Ali, Z., Khan, D. and Ahmed, N. (2013) Salt tolerance of three sorghum cultivars during germination and early seedling growth. *International Journal of Biology and Biotechnology* 10(2),193–202.

Almodares, A., Hadi, M.R. and Dosti, B. (2007) Effects of salt stress on germination percentage and seedling growth in sweet sorghum cultivars. *Journal of Biological Sciences* 7(8), 1492–1495.

Ashraf, M. and Akram, N.A. (2009) Improving salinity tolerance of plants through conventional breeding and genetic engineering: an analytical comparison. *Biotechnology Advances* 27(6), 744–752.

Attia, M. (2016) Performance of some sorghum genotypes under salinity conditions. *IOSR Journal of Agriculture and Veterinary Science* 9(4), 8–12.

Azhar, F. and McNeilly, T. (1987) Variability for salt tolerance in *Sorghum bicolor* (L.) Moench. under hydroponic conditions. *Journal of Agronomy and Crop Science* 159(4), 269–277.

Azooz, M.M., Shaddad, M.A. and Abdel-Latef, A.A. (2004) The accumulation and compartmentation of proline in relation to salt tolerance of three sorghum cultivars. *Indian Journal of Plant Physiology* 9(1), 1–8.

Bafeel, S.O. (2014) Physiological parameters of salt tolerance during germination and seedling growth of *Sorghum bicolor* cultivars of the same subtropical origin. *Saudi Journal of Biological Sciences* 21(4), 300–304.

Bai, S.N. (2017) Reconsideration of plant morphological traits: from a structure-based perspective to a function-based evolutionary perspective. *Frontiers in Plant Science* 8, 345.

Baiseitova, G., Sarsenbayev, B., Kirshibayev, E. and Kamunur, M. (2018) Influence of salinity (NaCl) on the photosynthetic pigments content of some sweet sorghum varieties. *BIO Web of Conferences* 11, 00003.

Bandillo, N., Raghavan, C., Muyco, P.A., Sevilla, M.A.L., Lobina, I.T. *et al.* (2013) Multi-parent advanced generation inter-cross (MAGIC) populations in rice: progress and potential for genetics research and breeding. *Rice* 6(1), 11. Available at: https://doi.org/10.1186/1939-8433-6-11

Bouchet, S., Olatoye, M.O., Marla, S.R., Perumal, R., Tesso, T. *et al.* (2017) Increased power to dissect adaptive traits in global sorghum diversity using a nested association mapping population. *Genetics* 206(2), 573–585. Available at: https://doi.org/10.1534/genetics.116.198499

Boyles, R.E., Brenton, Z.W. and Kresovich, S. (2019) Genetic and genomic resources of sorghum to connect genotype with phenotype in contrasting environments. *The Plant Journal* 97(1), 19–39. Available at: https://doi.org/10.1111/tpj.14113

Briggs, F.N. and Knowles, P.F. (1967) *Introduction to Plant Breeding*. Reinhold Publishing Corporation, London.

Brown, J. and Caligari, P. (2011) *An Introduction to Plant Breeding*. Wiley, Chichester, UK.

Chaduvula, P.K., Bhati, J., Rai, A., Gaikwad, K., Marla, S.S., Elangovan, M. and Kumar, S. (2015) *In-silico* expressed sequence tag analysis in identification and characterization of salinity stress responsible genes in *Sorghum bicolor*. *Australian Journal of Crop Science* 9(9), 799–806.

Collard, B.C.Y., Jahufer, M.Z.Z., Brouwer, J.B. and Pang, E.C.K. (2005) An introduction to markers, quantitative trait loci (QTL) mapping and marker-assisted selection for crop improvement: the basic concepts. *Euphytica* 142, 169–196.

Dell'Acqua, M., Gatti, D.M., Pea, G., Cattonaro, F., Coppens, F. *et al.* (2015) Genetic properties of the MAGIC maize population: a new platform for high definition QTL mapping in *Zea mays*. *Genome Biology* 16(1), 167. Available at: https://doi.org/10.1186/s13059-015-0716-z

Ding, T., Yang, Z., Wei, X., Yuan, F., Yin, S. and Wang, B. (2018) Evaluation of salt-tolerant germplasm and screening of salt-tolerance traits of sweet sorghum in the germination stage. *Functional Plant Biology* 45(10), 1073–1081.

El Naim, A.M., Mohammed, K.E., Ibrahim, E.A. and Suleiman, N.N. (2012) Impact of salinity on seed germination and early seedling growth of three sorghum (*Sorghum bicolor* L. Moench) cultivars. *Science and Technology* 2(2), 16–20.

Francois, L.E., Donovan, T. and Maas, E.V. (1984) Salinity effects on seed yield, growth and germination of grain sorghum. *Agronomy Journal* 76, 741–744.

Geressu, K. and Gezahagne, M. (2008) Response of some lowland growing sorghum (*Sorghum bicolor* L. Moench) accessions to salt stress during germination and seedling growth. *African Journal of Agricultural Research* 3(1), 44–48.

Gifford, E.M. and Foster, A.S. (1989) *Morphology and Evolution of Vascular Plants*. W.H. Freeman, New York.

Huang, B.E., George, A.W., Forrest, K.L., Kilian, A., Hayden, M.J., Morell, M.K. and Cavanagh, C.R. (2012) A multiparent advanced generation inter-cross population for genetic analysis in wheat. *Plant Biotechnology Journal* 10(7), 826–839. Available at: https://doi.org/10.1111/j.1467-7652.2012.00702.x

Igartua, E., Gracia, M.P. and Lassa, J.M. (1994) Characterization and genetic control of germination-emergence responses of grain sorghum to salinity. *Euphytica* 76, 185–193.

Igartua, E., Gracia, M.P. and Lasa, J.M. (1995) Field response of grain sorghum to a salinity gradient. *Field Crops Research* 42(1), 15–25.

Jordan, D.R., Mace, E.S., Cruickshank, A.W., Hunt, C.H. and Henzell, R.G. (2011) Exploring and exploiting genetic variation from unadapted sorghum germplasm in a breeding program. *Crop Science* 51, 1444–1457. Available at: https://doi.org/10.2135/cropsci2010.06.0326

Kandil, A.A., Sharief, A.E. and Elbadry, D.E.A. (2017) Germination characters as affected by salinity stress and soaking grain sorghum genotypes in humic acid. *International Journal of Environment, Agriculture and Biotechnology* 2(6), 3268–3278.

Kasuga, M., Liu, Q., Miura, S., Yamaguchi-Shinozaki, K., and Shinozaki, K. (1999) Improving plant drought, salt, and freezing tolerance by gene transfer of a single stress-inducible transcription factor. *Nature Biotechnology* 17(3), 287–291.

Kausar, A., Ashraf, Y.M., Ali, I., Niaz, M. and Abbass, Q. (2012) Evaluation of sorghum varieties/lines for salt tolerance using physiological indices as screening tool. *Pakistan Journal of Botany* 44(1), 47–52.

Khalil, R.M.A. (2013) Molecular and biochemical markers associated with salt tolerance in some sorghum genotypes. *World Applied Sciences Journal* 22(4), 459–469.

Kover, P.X., Valdar, W., Trakalo, J., Scarcelli, N., Ehrenreich, I.M. *et al.* (2009) A multiparent advanced generation inter-cross to fine-map quantitative traits in *Arabidopsis thaliana*. *PLoS Genetics* 5(7), e1000551.

Krishnamurthy, L., Reddy, B.V.S. and Serraj, R. (2003) Screening sorghum germplasm for tolerance to soil salinity. *International Sorghum and Millets Newsletter* 44, 90–92.

Krishnamurthy, L., Serraj, R., Hash, C.T., Dakheel, A.J. and Reddy, B.V.S. (2007) Screening sorghum genotypes for salinity tolerant biomass production. *Euphytica* 156, 15–24.

Kulhari, P.S., Chaudhry, L. and Lakshyadeep (2008) Association studies for salinity tolerance in sorghum [*Sorghum bicolor* (L.) Moench]. *Indian Journal of Plant Genetic Resources* 21(1), 81–84.

Kumar, M., Choi, J.-Y., Kumari, N., Pareek, A. and Kim, S.-R. (2015) Molecular breeding in *Brassica* for salt tolerance: importance of microsatellite (SSR) markers for molecular breeding in *Brassica*. *Frontiers in Plant Science* 6, 688.

Kump, K.L., Bradbury, P.J., Wisser, R.J., Buckler, E.S., Belcher, A.R. *et al.* (2011) Genome-wide association study of quantitative resistance to southern leaf blight in the maize nested association mapping population. *Nature Genetics* 43(2), 163–168.

Maas, E.V., Poss, J.A. and Hoffman, G.J. (1986) Salinity sensitivity of sorghum at three growth stages. *Irrigation Science* 7, 1–11.

Maheswari, M., Varalaxmi, Y., Vijayalakshmi, A., Yadav, S.K., Sharmila, P. *et al.* (2010) Metabolic engineering using *mtlD* gene enhances tolerance to water deficit and salinity in sorghum. *Biologia Plantarum* 54(4), 647–652.

Maiti, R., de la Rosa-Ibarra, M. and Sandoval, N.D. (1994) Genotypic variability in glossy sorghum lines for resistance to drought, salinity and temperature stress at the seedling stage. *Journal of Plant Physiology* 143(2), 241–244.

Mbinda, W. and Kimtai, M. (2019) Evaluation of morphological and biochemical characteristics of sorghum [*Sorghum bicolor* [L.] Moench] varieties in response salinity stress. *Annual Research & Review in Biology* 33(1), 1–9.

Munns, R. (2002) Comparative physiology of salt and water stress. *Plant, Cell & Environment* 25(2), 239–250.

Ndimba, B.K., Thomas, L.A. and Ngara, R. (2010) Sorghum 2-dimensional proteome profiles and analysis of HSP70 expression under salinity stress. *Kasetsart Journal (Natural Science)* 44, 768–775.

Ongom, P.O. and Ejeta, G. (2018) Mating design and genetic structure of a multi-parent advanced generation intercross (MAGIC) population of sorghum (*Sorghum bicolor* (L.) Moench). *G3: Genes, Genomes, Genetics* 8(1), 331-341.

Pascual, L., Desplat, N., Huang, B.E., Desgroux, A., Bruguier, L. *et al.* (2015) Potential of a tomato MAGIC population to decipher the genetic control of quantitative traits and detect causal variants in the resequencing era. *Plant Biotechnology Journal* 13(4), 565–577. Available at: https://doi.org/10.1111/pbi.12282

Poehlman, J.M. (2013) *Breeding Field Crops*. Springer, New York.

Rakshit, S. and Bellundagi, A. (2019) Conventional breeding techniques in sorghum. In: Aruna, C., Visarada, K.B.R.S., Venkatesh Bhat, B. and Tonapi, V.A. (eds) *Breeding Sorghum for Diverse End Uses*. Woodhead Publishing, Cambridge, UK, pp. 77–91.

Rakshit, S., Rakshit, A. and Patil, J.V. (2012) Multiparent intercross populations in analysis of quantitative traits. *Journal of Genetics* 91(1), 111–117.

Rani, C.R., Reema, C., Alka, S. and Singh, P.K. (2012) Salt tolerance of *Sorghum bicolor* cultivars during germination and seedling growth. *Research Journal of Recent Sciences* 1(3), 1–10.

Rao, M.V.S., Kumari, P.K., Manga, V. and Mani, N.S. (2007) Molecular markers for screening salinity response in sorghum. *Indian Journal of Biotechnology* 6, 271–273.

Reddy, B.V.S., Kumar, A.A., Reddy, P.S., Ibrahim, M., Ramaiah, B. *et al.* (2010) Cultivar options for salinity tolerance in sorghum. *Journal of SAT Agricultural Research* 8, 1–5.

Sagar, A., Tajkia, J.E., Haque, M.E., Fakir, M.S.A. and Hossain, A.K.M.Z. (2019) Screening of sorghum genotypes for salt-tolerance based on seed germination and seedling stage. *Fundamental and Applied Agriculture* 4(1), 735–743.

Sannemann, W., Huang, B.E., Mathew, B. and Leon, J. (2015) Multi-parent advanced generation inter-cross in barley: high-resolution quantitative trait locus mapping for flowering time as a proof of concept. *Molecular Breeding* 35(3), 86.

Sekhwal, M.K., Swami, A.K., Sarin, R. and Sharma, V. (2012) Identification of salt treated proteins in sorghum using gene ontology linkage. *Physiology and Molecular Biology of Plants* 18(3), 209–216.

Shahbaz, M. and Ashraf, M. (2013) Improving salinity tolerance in cereals. *Critical Reviews in Plant Sciences* 32(4), 237–249.

Sleper, D.A. and Poehlman, J.M. (2006) *Breeding Field Crops*, 5th edn. Blackwell Publishing, Oxford, UK.

Su, M., Li, X.F., Peng, X.J., Zhao, A.G., Cheng, L.Q., Chen, S.Y. and Liu, G.S. (2011) Cloning two *P5CS* genes from bioenergy sorghum and their expression profiles under abiotic stresses and MeJA treatment. *Plant Science* 181, 652–659.

Sui, N., Yang, Z., Liu, M. and Wang, B. (2015) Identification and transcriptome profiling of genes involved in increasing sugar content during salt stress in sweet sorghum leaves. *BMC Genomics* 16(1), 534.

Sun, L., Zhou, Y.F., Wang, C., Xiao, M.J., Tao, Y., Xu, W.J. and Huang, R.D. (2012) Screening and identification of sorghum cultivars for salinity tolerance during germination. *Scientia Agricultura Sinica* 45(9), 1744–1722.

Sun, Y., Niu, G., Osuna, P., Zhao, L., Ganjegunte, G. *et al.* (2014) Variability in salt tolerance of *Sorghum bicolor* L. *Agricultural Science* 2(1), 9–21.

Swami, A.K., Alam, S.I., Sengupta, N. and Sarin, R. (2011) Differential proteomic analysis of salt stress response in *Sorghum bicolor* leaves. *Environmental and Experimental Botany* 71(2), 321–328.

Taylor, R., Young, E.F. Jr and Rivera, R.L. (1975) Salt tolerance in cultivars of grain sorghum. *Crop Science* 15(5), 734–735.

Threadgill, D.W., Hunter, K.W. and Williams, R.W. (2002) Genetic dissection of complex and quantitative traits: from fantasy to reality via a community effort. *Mammalian Genome* 13(4), 175–178. Available at: https://doi.org/10.1007/s00335-001-4001-y

Tigabu, E., Andargie, M. and Tesfaye, K. (2012) Response of sorghum (*Sorghum bicolor* (L.) Moench) genotypes to NaCl levels at early growth stages. *African Journal of Agricultural Research* 7(43), 5711–5718.

Turan, S., Cornish, K. and Kumar, S. (2012) Salinity tolerance in plants: breeding and genetic engineering. *Australian Journal of Crop Science* 6(9), 1337–1348.

Walia, H., Wilson, C., Condamine, P., Liu, X., Ismail, A.M. *et al.* (2005) Comparative transcriptional profiling of two contrasting rice genotypes under salinity stress during the vegetative growth stage. *Plant Physiology* 139, 822–835.

Wang, H., Chen, G., Zhang, H., Liu, B., Yang, Y. *et al.* (2014a) Identification of QTLs for salt tolerance at germination and seedling stage of *Sorghum bicolor* L. *Moench. Euphytica* 196(1), 117–127.

Wang, S., Bai, Y., Shen, C., Wu, Y., Zhang, S. *et al.* (2010) Auxin-related gene families in abiotic stress response in *Sorghum bicolor*. *Functional & Integrative Genomics* 10(4), 533–546.

Wang, T.T., Ren, Z.J., Liu, Z.Q., Feng, X., Guo, R.Q. *et al.* (2014b) *SbHKT1;4*, a member of the high-affinity potassium transporter gene family from *Sorghum bicolor*, functions to maintain optimal $Na^+$ /$K^+$ balance under $Na^+$ stress. *Journal of Integrative Plant Biology* 56(3), 315–332.

Waters, S., Gilliham, M. and Hrmova, M. (2013) Plant high-affinity potassium (HKT) transporters involved in salinity tolerance: structural insights to probe differences in ion selectivity. *International Journal of Molecular Science* 14(4), 7660–7680.

Xu, Y., Li, P., Yang, Z. and Wu, C. (2017) Genetic mapping of quantitative trait loci in crops. *The Crop Journal* 5(2), 175–184.

Younis, R.A.A., Ahmed, M.F. and El-Menshawi, M.M. (2007) Molecular genetic markers associated with salt tolerance in grain sorghum. *Arab Journal of Biotechnology* 10(2), 249–258.

Yu, J., Holland, J.B., McMullen, M.D. and Buckler, E.S. (2008) Genetic design and statistical power of nested association mapping in maize. *Genetics* 178(1), 539–551.

Zeng, L., Shannon, M.C. and Grieve, C.M. (2002) Evaluation of salt tolerance in rice genotypes by multiple agronomic parameters. *Euphytica* 127(2), 235–245.

Zhan, Q., Shu, C., Li, X., Zhan, M., Li, J. and Lin, P. (2013) Screening of SSR primers and evaluation of salt tolerance in 20 sweet sorghum varieties for silage. In: Michalk, D.L., Millar, G.D., Badgery, W.B. and Broadfoot, K.M. (eds) *Proceedings of the 22nd International Grassland Congress, Sydney, Australia, 15–19 September 2013*. New South Wales Department of Primary Industry, Orange, Australia, pp. 141–142.

Zhu, C., Gore, M., Buckler, E.S. and Yu, J. (2008) Status and prospects of association mapping in plants. *The Plant Genome* 1(1), 5–20.

# 25 Quantitative Trait Locus Mapping and Genetic Improvement to Strengthen Drought Tolerance in Sorghum

**Tariq Shehzad[1] and Kazutoshi Okuno[2*]**
[1]*Syngenta Crop Protection, LLC, Research Triangle Park, North Carolina, USA;*
[2]*University of Tsukuba, Tsukuba, Japan*

## 25.1 Introduction

Conventional plant breeding is relatively slow but has been successful for genetic gain over the past century. The massive increase in population growth and demand for food can only be tackled by combining advanced molecular techniques with conventional approaches. Molecular breeding techniques offer real hope of accelerated progress in increasing yield by utilizing useful genetic variations in several important crops.

Quantitative trait locus (QTL) mapping has been in wide use for nearly two decades, during which molecular markers have become available in conjunction with interval mapping methods (Lander and Botstein, 1986). The goal of QTL mapping is to identify the loci that are responsible for variation in complex quantitative traits. In some situations, the ultimate goals are to determine the number, chromosomal locations and the interaction of these loci. The identification of the actual genes and their functions are often of interest. Multi-environment testing plays a key role in understanding the response of QTLs in different environments or genetic backgrounds that ultimately achieve the development of improved crop varieties. If the gene(s) underlying a QTL is cloned, transgenic approaches can also be used to directly introduce beneficial alleles across wide species boundaries. A schematic representation of QTL mapping and map-based cloning of genes is given in Fig. 25.1. An important development during the last decade in quantitative genetics was the ability to identify genome regions responsible for variation for a trait due to the advent of molecular markers (Paterson *et al.*, 1988). The term 'QTL' has come to refer to a constellation of polygenes underlying a quantitative trait. Numerous studies have been reported on identifying QTLs for various traits in humans, animals and plants.

The prerequisites for QTL mapping are the selection or development of an appropriate mapping population (experimental populations for linkage-based mapping or natural/breeding populations for association mapping), the appropriate method of phenotyping and the selection of molecular markers (Fig. 25.1). The availability of a wide range of molecular markers and powerful statistical methods has significantly facilitated QTL mapping. Linkage analysis and association mapping are the two commonly used tools for dissecting complex traits. The general scheme of linkage disequilibrium (LD)-based association mapping is given in Fig. 25.2. Both QTL mapping methods begin with the collection of genotypic and phenotypic data from either a segregating or a natural population, followed by statistical

*Corresponding author: okuno.kazutoshi2423@gmail.com


DOI: 10.1079/9781789245431.0025

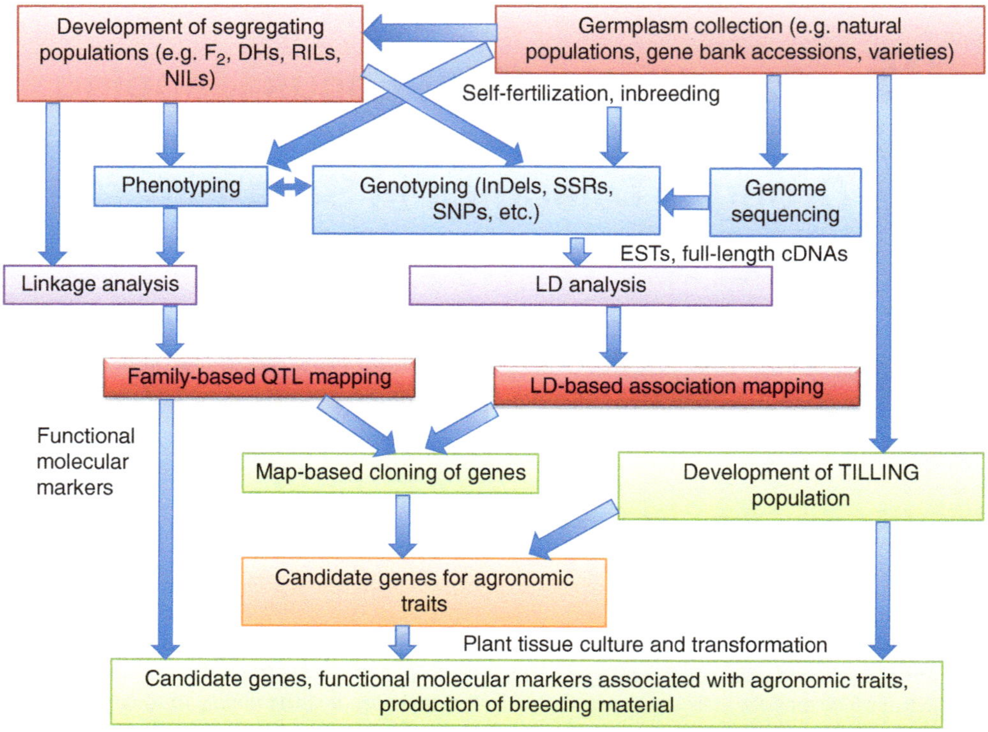


**Fig. 25.1.** Schematic representation of the process of genetic improvement and breeding schemes. DH, doubled haploid; RIL, recombinant inbred line; NIL, near-isogenic line; InDel, insertion/deletion; SSR, simple sequence repeat; SNP, single-nucleotide polymorphism; EST, expressed sequence tag; LD, linkage disequilibrium; QTL, quantitative trait locus; TILLING, targeting induced local lesions in genomes.

analysis to reveal all possible marker loci where allelic variation correlates with the phenotype.

The use of mapping and segregating populations and of diverse germplasm collections to assess the natural variation in different crops including wheat, barley, rice and sorghum for important traits has enhanced breeding improvement in cereals. This chapter overviews the approaches to and application of QTL mapping and positional cloning of genes controlling important traits related to drought tolerance in sorghum, which ultimately yields crop improvement and genetic modification.

## 25.2 Role of Molecular Markers

Molecular markers are essential tools in genomic and genetic research. Molecular markers allow discrimination of DNA sequences among cultivars and breeding lines and thus offer the scientific community singularly powerful tools to monitor, track and exploit sequence variation in germplasm. Many types of DNA markers have been developed and are now an essential part of functional, structural genomics and molecular breeding. Different kinds of molecular markers are being used for the dissection of important traits directly or indirectly related to the tolerance to abiotic stresses in crops. Microsatellites or simple sequence repeat (SSR) markers have proved useful for crop researchers since they offer reproducibility, multiallelic nature, co-dominant inheritance, genome specificity, relative abundance and good genome coverage. Another promising marker is single-nucleotide polymorphism (SNP) or biallelic markers because they form the basis for most genetic variation between individuals; they are widely distributed and amenable to high multiplex detection systems. The identification of SNP markers depends on comparative sequencing of lines or analysis of expressed sequence tags (ESTs). Molecular markers can be derived either from within or from outside the gene

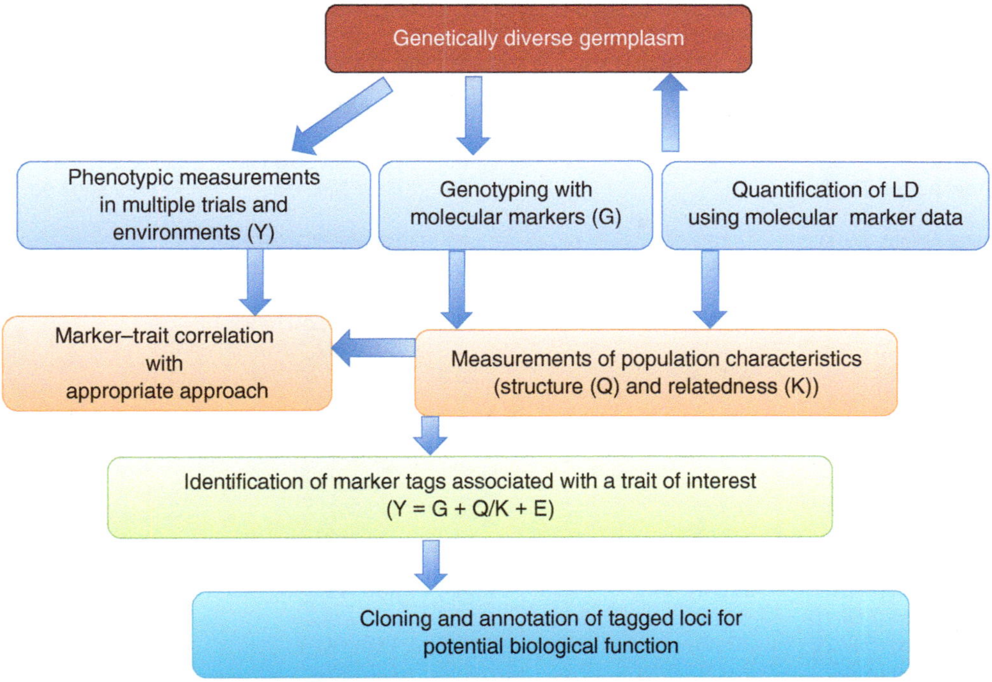


**Fig. 25.2.** Scheme of linkage disequilibrium (LD)-based association mapping for tagging a gene of interest using diverse germplasm.

sequence and the selection of the most suitable marker system depends on the objective and cost. The applicability of marker information across germplasms or cultivar collections would allow the genetically based crop modelling to be performed without recourse to the use of a mapping population. Over the past two decades significant advances in the area of genomics of crop plants have been achieved, in particular the development of high-throughput, cost-effective marker genotyping technologies which have contributed substantially to the identification, mapping and introgression of alien genes from wild species.

## 25.3 Discovery of Quantitative Trait Loci

The advances in functional genomics in delivering bacterial artificial chromosomes (BACs), ESTs, partial gene sequences, full-length cDNA clones, genes and markers have enabled the establishment of molecular maps based on genic markers. The creation of suitable mapping populations and the development of molecular markers have enabled linkage studies in sorghum and many QTLs have been identified for yield under drought and other stress environments (Srinivas *et al.*, 2008, 2009). QTL studies reported in sorghum are presented in Table 25.1.

Linkage studies have shown that QTLs for grain yield reside in several chromosomal regions, and measurements of yield components allow the dissection of complex traits into smaller genetic components more amenable for building knowledge of trait architecture, which informs future strategies for exploitation. Mapping populations have also been developed in sorghum for the study of biotic and abiotic stresses (Subudhi *et al.*, 2000; Xu *et al.*, 2000; Kebede *et al.*, 2001; Felderhoff *et al.*, 2012; Satish *et al.*, 2012; Wang, H. *et al.*, 2014; Wang, T.T. *et al.*, 2014). Despite the available maps, populations and marker technologies, advances in transferring knowledge from QTL studies on yield under drought to breeding remain slow. This is due to three major factors: (i) yield is a quantitative trait with poorly defined genetic architecture of multiple interacting loci; (ii) these loci have a high genotype ×

**Table 25.1.** List of some important studies on mapping of QTLs/genes conferring drought tolerance and its related traits in sorghum identified under drought condition.

| Trait/QTL/Gene | Linkage group/ Chromosome | LOD/$Log_{10}$ /(*P*-value) | $R^2$ (%) | Associated marker/Gene ID | Reference |
|---|---|---|---|---|---|
| Maturity | 6 | 91 | 86 | *pSB189/pSB580* | Lin *et al.* (1995) |
| Plant height (*Dw3*) | 7 | 8 | 29 | *isu123/isu116* | Pereira and Lee (1995) |
| Plant height (*Dw1*) | 10 | 6 | 20 | *isu140/PIO100016* | |
| Yield | E, F, I | – | 12.4 | *UMC109, tM5/75, tC20/58* | Tuinstra *et al.* (1997) |
| Yield stability | B, C, E | – | 35.9 | *UMC22, bC15/95, tH9/82, tD19/170* | |
| Seed weight stability | A, E, N | – | 23.8 | *t375/45, b206/89, tH19/50* | |
| Stay-green | B, F, G, H, I | – | 52.6 | *tF16/68, tM5/75, bH19/170, t229/55, bG6/115* | |
| Stay-green (*SGA*)[a] | A (3)[b] | 6.6 | 28.6 | *TXS307* | Crasta *et al.* (1999) |
| Stay-green (*SGD*) | D2 (2) | 5 | 22.5 | *TXS1537* | |
| Stay-green (*SGG*) | G (1) | 5.8 | 25.8 | *UMC27* | |
| Stay-green (*SGB*) | B (10) | 3 | 14.4 | *pSB115* | |
| Stay-green (*SGI.1*) | I1 (9) | 3.8 | 17.6 | *pSB134.2* | |
| Stay-green (*SGI.2*) | I2 (9) | 2.9 | 13.7 | *TXS1541* | |
| Stay-green (*SGJ*) | J (5) | 2.3 | 11.6 | *TXS713* | |
| Maturity (*DFB*) | B (10) | 12.6 | 47.5 | *TXS1299* | |
| Maturity (*DFG*) | G (1) | 4.4 | 20.3 | *UMC27* | |
| Stay-green (*Stg1*) | 3 | 5 | 20 | *Xtxp442/Xtxp38* | Xu *et al.* (2000) |
| Stay-green (*Stg2*) | 3 | 6 | 30 | *Xtxp2/Xtxp503* | |
| Stay-green (*Stg3*) | 2 | 3 | 16 | *Xtxp430/Xtxp1* | |
| Stay-green (*Stg4*) | 5 | 2 | 11 | *Xtxp225/Xtxp15* | |
| Maturity | 6 | 11 | 36 | *psb521/psb708* | Kebede *et al.* (2001) |
| Seed weight | 4 | 4 | 16 | *txs604/cdo516.1* | Feltus *et al.* (2006) |
| Seed weight | 6 | 7 | 10 | *pSB521a/pSB428a* | |
| Grain yield | 10 | 3 | 15 | *AAG/CTT2* | Ritter *et al.* (2008) |
| Grain yield | 2 | 4 | 18 | *AAG/CAA1* | |
| Plant height (*Dw2*) | 6 | 16 | 27 | *AG/CTG9* | |
| Seed weight | 4 | 5 | 16 | *Xtxp51/txa6257* | Brown *et al.* (2008) |
| Seed weight | 8 | 5 | 12 | *isu145.2/txa558* | |
| Seed weight | 6 | 8 | 15 | *txa2873/txa2067* | Murray *et al.* (2008) |
| Seed weight | 8 | 6 | 11 | *rio65/rio37* | |
| Grain yield | 6 | 5 | 15 | *GlumeT/Xtxp145* | Srinivas *et al.* (2009) |
| Maturity | 1 | 6 | 15 | *txp58/Dsenhsbm63* | |
| Shoot dry weight (*qSDW1*) | 1 | – | – | *Sb01g036220* | Mace *et al.* (2012) |
| Nodal root angle (*qRA1_5*) | 5 | – | – | *Sb05g007450* | |
| Nodal root angle (*qRA1_8*) | 8 | – | – | *Sb08g005781* | |
| Root dry weight (*qRDW1_2*) | 2 | – | – | *Sb02g037700* | |
| Root dry weight (*qRDW1_8*) | 8 | – | – | *Sb08g018270* | |

*Continued*

**Table 25.1.** Continued.

| Trait/QTL/Gene | Linkage group/ Chromosome | LOD/Log$_{10}$ /(*P*-value) | $R^2$ (%) | Associated marker/Gene ID | Reference |
|---|---|---|---|---|---|
| Total leaf area (*qTLA1_8*) | 8 | – | – | *Sb08g017820* | |
| Panicle exertion | 1/10 | 2.7/3.5 | 18.7/61.2 | *Xtxp279/PepC* | Sakhi *et al.* (2013) |
| Panicle length | 2 | 3.4 | 22.5 | *Xtxp315* | |
| Culm diameter | 1 | 3 | 9 | *Xtxp335* | |
| Number of panicles | 10 | 2.5 | 9.6 | *Xtxp270* | |
| Total number of leaves | 3 | 3.2 | 26.8 | *Xtxp228* | |
| Flag leaf length | 1 | 2.6 | 12 | *Xtxp32* | |
| Leaf drying score | 1 | 2.5 | 18.6 | *Xtxp149* | |
| Grain weight (*qGW1*) | 1 | 7 | 22 | *SB00037/SB00219* | Han *et al.* (2015) |
| Grain_yield (*qYLD1.2*) | 1 | – | – | *Sb01g012195* | Sukumaran *et al.* (2016) |
| Grain_yield (*QYLD1.3*) | 1 | – | – | *Sb01g012230* | |
| Chlorophyll content (*qSPAD4.1*) | 4 | – | – | *Sb04g006830* | |
| Grain yield (*qYLD4.1*) | 4 | – | – | *Sb04g019670* | |
| Chlorophyll fluorescence (*qFv/Fm4.1*) | 4 | – | – | *Sb04g034665* | |
| Grain yield (*qYLD2.1*) | 2 | – | – | *Sb02g027900* | |
| Flowering time (*qFT6.1*) | 6 | – | – | *Sb06g001033* | |
| Grain yield (*qYLD6.1*) | 6 | – | – | *Sb06g020970* | |
| Flowering time (*qFT9.1*) | 9 | – | – | *Sb09g004180* | |
| Seed dormancy (*QGI-3*) | 3 | – | – | *Sb03g040510* | Cantoro *et al.* (2016) |
| Seed dormancy (*QGI-4*) | 4 | – | – | *Sb04g027660* | |
| Seed dormancy (*QGI-1*) | 1 | – | – | *Sb01g030510* | |
| Seed dormancy (*qGI-6*) | 6 | – | – | *Sb06g025130* | |
| Seed dormancy (*qGI-7*) | 7 | – | – | *Sb07g024070* | |
| Seed dormancy (*qGI-9*) | 9 | – | – | *Sb09g028980* | |

LOD, logarithm of the odds.
[a]The results of composite interval mapping (Crasta *et al.*, 1999).
[b]Linkage groups correspond to chromosomes as reported by Kim *et al.* (2005).

environment (G × E) interaction component; and (iii) the accurate phenotyping of the traits under study. To enable a study on the interaction of genotype × environment, phenotyping trials have to be carried out on the same population in multiple field sites where environmental covariates are considered in the analysis. To overcome this problem, one would like to grow the same population in multiple environments. This allows the simulation of various gene actions or models in the context of selection and breeding strategies. Despite the complexity of studying yield and its genetic architecture under drought stress, data are now available from multiple studies and from various genetic materials to enable identifying important loci for further work.

The development of new genomic, transcriptomic and metabolomic platforms with bioinformatic analysis has facilitated the identification and fine-mapping of QTLs, which in turn has facilitated the identification of a number of DNA marker–trait associations in various crop species (Abou-Elwafa, 2016).

Adaptation to drought conditions is one of the most complex biological processes. It involves numerous pathways including transcriptional activation/inactivation of specific genes, transient increases in abscisic acid (ABA) levels, accumulation of compatible solutes and protective enzymes, increased levels of antioxidants and suppression of energy-metabolism pathways. High-throughput transcriptome studies and functional genomic approaches have been used to identify genes correlated with the response to water stress in plants. Large numbers of genes have been identified and the diversity of these responsive genes and pathways reflects the complexity of the mechanisms involved in sensing and responding to water stress. These genes have been categorized by Yamaguchi-Shinozaki *et al.* (2006) into two major groups: the first one codes for functional proteins involved in protecting cellular function (water channels, transporters, detoxification, proteases, protection factors of macromolecules, chaperones and osmolyte biosynthesis); the other group codes for regulatory proteins involved in signal transduction (such as transcription factors, protein kinases, protein phosphatases and enzymes of lipid metabolism) (Shinozaki *et al.*, 2003). Genetic modification of the expression of members belonging to these two major groups of genes, further categorized as ABA-dependent or ABA-independent, has resulted in some improvement in traits relating to tolerance under water stress in a variety of plant species. However, only a few genes have been successfully engineered in crops to enhance tolerance to drought under field conditions. An important aspect that is worth highlighting here is that fine-tuning the modulation of the target gene(s) expression to specific cell types may be essential in a successful strategy for modulating the system.

Drought significantly reduces plant productivity by inhibiting growth and photosynthesis (Taiz and Zieger, 1998). In plants at least four independent stress-responsive genetic regulatory pathways are known to exist, forming a highly complex and redundant gene network (Umezawa *et al.*, 2006; Shinozaki and Yamaguchi-Shinozaki, 2007). Two of the pathways are ABA-dependent whereas two are ABA-independent. These pathways are also linked to other biological processes involved in other stress responses, including cold, high temperature and salinity. Many studies on drought tolerance monitored the physiological and biochemical status of stressed plants compared with unstressed plants. Detection of QTLs for drought tolerance is the first step of genetic improvement to stabilize global crop production.

## 25.4 Fine Mapping and Positional Cloning of Quantitative Trait Loci

Primary QTL mapping will identify a QTL within 10–30 cM resolution. To fine-map a locus to a higher resolution, near-isogenic lines specific for the particular QTL under study are required to Mendelize the locus (Alonso-Blanco and Koornneef, 2000). When the QTL region is resolved to a few centimorgans distance, markers closest to the QTL are then used to anchor the genetic map to a physical map and BAC libraries are then used to identify potential candidate sequences. The use of model species, synteny and bioinformatics tools is critical at this stage to enable candidate genes to be identified. While QTL cloning represents a huge undertaking in terms of the technology, time and resources required, the advantages derived from its success are directly applicable to gene cloning. Candidate genes genetically and physically co-segregating with the QTL are then identified and/or selected for evaluation. The increase in mapping resolution required to accomplish QTL positional cloning is substantial, since after primary mapping a QTL is positioned within a chromosome interval of *c.*10–30 cM which usually includes several hundred genes. Eventually, independent proof is required to validate the role of the identified allelic polymorphism on the observed phenotypic effect.

## 25.5 Candidate Genes and Genetic Engineering Approach

High-throughput transcriptome studies and functional genomic approaches have evolved two approaches which could be implemented in the identification of candidate genes behind quantitative traits: candidate gene and genome-wide scanning. With the aid of molecular markers, genome-wide scanning could be applied to locate glancing chromosomal segments, which usually comprise a huge number of candidate genes. Alternatively, the candidate gene approach, which is a cost-effective and more efficient strategy for direct gene identification, has been reported as an extremely powerful approach to dissect the genetic architecture of quantitative traits. However, successful application of the classical candidate gene approach relies on available information about known or presumed biology of the investigated phenotype at the molecular level, which is unfortunately still vague for most biological traits. Therefore, it is essential to develop an alternative approach to break through the limitation of information bottleneck. Rapid development in next-generation sequencing technologies has facilitated sequencing of targeted regions from large genomes including genes (Teer and Mullikin, 2010; Ekblom and Galindo, 2011). Plant molecular breeders are interested in an approach for rapid identification of quantitative traits in crop genomes to empower marker-assisted breeding aimed at overcoming global food insecurity by accelerating breeding programmes.

A bioinformatics pipeline combining both genome-wide scanning and candidate gene approaches was reported as a powerful approach in the cloning of genes underlying quantitative traits (Abou-Elwafa, 2016). The proposed bioinformatics pipeline relies on genomic sequence data and consists of four complementing steps: (i) employing *in silico* mapping of DNA markers located to the QTLs in the genome to determine the physical position of those QTLs; (ii) identification of candidate genes underlying these QTLs by the annotation of the critical chromosomal regions of the QTLs to a known protein database, such as the National Center for Biotechnology Information's (NCBI) Reference Sequence (RefSeq); (iii) based on sequence similarity to orthologues from different plant species as revealed by phylogenetic analysis, conserved functional roles could be presumed; and (iv) genes involved in the response to a given phenotype, such as drought stress, should be differentially expressed in plants possessing or subjected to contrasting phenotypes or environmental conditions.

## 25.6 Use of Core Collections in Genetic Analysis

The complexity of the genetic nature of quantitative traits has complicated the application of classical QTL analysis, which employs molecular markers on bi-parental populations, in accurate dissection of the genetic architecture of quantitative traits. It is important to note that conventional QTL analysis necessitates the development of a genetic linkage map and is identified based on linkage equilibrium that the bi-parental population undergoes in just a few cycles of recombination, which limits the resolution of genetic maps and identified QTLs. Besides, the segregation of major developmental genes in each cross makes it difficult to determine whether the identified loci are robust and transient (Forster *et al.*, 2004). In addition, in a majority of linkage mapping-based QTL analyses, QTLs identified for physiological traits involved in drought tolerance for instance did not co-localize with either the yield or agronomic traits (Comadran *et al.*, 2009). Therefore, association mapping, which has been widely successfully applied to identify QTLs conferring complex traits, is an alternative promising approach. In that context, QTLs are identified in natural populations (no regular genetic structure) using analytical approaches exploiting LD between markers and closely linked QTLs present in a population consisting of large number of accessions (Abou-Elwafa *et al.*, 2019). However, a core collection is a limited set of accessions representing, with a minimum of repetitiveness, the genetic diversity of a crop species and its wild relatives (Frankel, 1984). Core collections have become important and are commonly used to screen for agronomically useful traits. Such collections proved powerful tools to study the genetic mechanisms of important physiochemical processes in crops. Similarly, the extent of either G × E interactions or gene/QTL × environment interactions can be assessed and exploited in a gene pool (Charmet *et al.*, 1993).

Utilization of the natural variation in germplasm to enhance drought tolerance in sorghum has been attempted (Sakhi *et al.*, 2013). An SSR marker-based mini-core collection of 107 sorghum accessions selected from Asian and African regions of the globe has been developed (Shehzad *et al.*, 2009b). The representativeness of the collection can be judged by comparing with the original population. The core set developed in that study retained 99.93% of total alleles present in the whole population. Similarly, no drastic change was observed for genetic diversity attributes like per cent polymorphic loci, similarity coefficient and gene diversity regarding base collection. A wide range of variation can be observed for different important agronomical and physiological traits in the germplasm, including panicle shape ranging from loose to very compact type (Hmon *et al.*, 2013). In the assessments of the 98 SSR markers, a total of 470 alleles were observed and the alleles could uniquely classify the 107 sorghum accessions into three subpopulations by structure analysis. The first population contained 33 sorghum accessions mostly of African origin, while the second population contained 36 accessions from both Africa and Asia regions. The third population contained 38 accessions mostly of East Asian origin (Shehzad *et al.*, 2009b).

In general, the conventional method for QTL analysis in germplasm is linkage mapping. To identify QTLs by linkage mapping one needs to construct one or several segregating populations by crossing between parents (e.g. $F_2$, doubled haploid, backcrossed populations). Linkage analysis depends on recent genetic recombination between two different plant lines (as the result of a genetic cross) to identify general regions of interest, with the advantage of requiring few genetic markers to ensure genome-wide coverage and high statistical power per allele. However, it has the disadvantages of low mapping resolution and low allele richness. An alternative, association mapping based on LD analysis, might be an effective way to identify the function of the gene or targeted high-resolution QTL (Shehzad *et al.*, 2009a; El Mannai *et al.*, 2011, 2012; Hmon *et al.*, 2014; Shehzad and Okuno, 2014). The use of both linkage mapping and association mapping can dissect the mechanism of response of sorghum to drought stress and identify the chromosomal regions controlling such characteristics (Sakhi *et al.*, 2014).

## 25.7 Conclusion and Future Prospects

Conventional breeding programmes for complex agronomic traits have achieved limited success. This lack of success is in part because breeders prefer to evaluate their genetic material in ideal conditions. There is a need to develop higher-throughput phenotyping systems for drought assessments. The statistical power of linkage and association analysis needs to be enhanced by reducing type I and type II errors. Use of advanced molecular techniques will further promote the identification of genes and gene networks for drought stress in crops. There is also a need to establish better comparative systems for genomic studies. To enhance the efficiency of marker-assisted selection (MAS), knowledge of the DNA sequence of the gene enables the design of direct markers which are located within the actual gene, thus eliminating the possibility of recombination between marker and gene (Ellis *et al.*, 2002). With continuous advances in sequencing technologies, genome-based selection is likely to replace the conventional marker-based genotyping approach to provide a powerful tool for high-resolution mapping and large-scale gene discovery.

With increasing threat of climate change, crop productivity is under the danger of severe loss in coming years. A strategy integrating both conventional and molecular approaches would be the best option that could play an important role in understanding the mechanism of drought tolerance. Approaches such as marker-aided backcrossing by pyramiding multiple QTL/genes conferring complex traits like drought stress tolerance into a single cultivar, MAS and genomic selection (GS) are effective. Similarly, functional genomics such as high-throughput sequencing, genotyping and resequencing can help in identifying genes that show response to drought stress. The most accurate and robust methods of phenotyping is the prerequisite for any such improvement. The use of high-throughput phenotyping will help better understand the mechanism involved in response to drought stress by plants. The new paradigm of scientific research should focus on the integration of physiology, genetics, genomics, soil characteristics and breeding to deal with the challenges of food security in the coming years.

## References


Abou-Elwafa, S.F. (2016) Association mapping for yield and yield-contributing traits in barley under drought conditions with genome-based SSR markers. *Comptes Rendus Biologies* 339, 153–162.

Abou-Elwafa, S.F. and Shehzad, T. (2018) Genetic identification and expression profiling of drought responsive genes in sorghum. *Environmental and Experimental Botany* 155, 12–20.

Abou-Elwafa, S.F., Amin, A. and Shehzad, T. (2019) Genetic mapping and transcriptional profiling of phytoremediation and heavy metals responsive genes in sorghum. *Ecotoxicology and Environmental Safety* 173, 366–372.

Alonso-Blanco, C. and Koornneef, M. (2000) Naturally occurring variation in *Arabidopsis*: an under-exploited resource for plant genetics. *Trends in Plant Science* 5, 22–29.

Brown, P.J., Rooney, W.L., Franks, C. and Kresovich, S. (2008) Efficient mapping of plant height quantitative trait loci in a sorghum association population with introgressed dwarfing genes. *Genetics* 180, 629–637.

Cantoro, R., Fernandez, L.G., Cervigni, G.D., Rodriguez, M.V., Gieco, J.O. *et al.* (2016) Seed dormancy QTL identification across a *Sorghum bicolor* segregating population. *Euphytica* 211, 41–56.

Charmet, G., Balfourier, F., Ravel, C. and Denis, J.B. (1993) Genotype × environment interactions in a collection of French perennial ryegrass populations. *Theoretical and Applied Genetics* 86, 731–736.

Comadran, J., Thomas, W.T., Eeuwijk, F.A., Ceccarelli, S., Grando, S. *et al.* (2009) Patterns of genetic diversity and linkage disequilibrium in a highly structured *Hordeum vulgare* association-mapping population for the Mediterranean basin. *Theoretical and Applied Genetics* 119, 175–187.

Crasta, R.R., Xu, W., Rosenow, D.T., Mullet, J.E. and Nguyen, H.T. (1999) Mapping of post-flowering drought resistance traits in grain sorghum: association of QTLs influencing premature senescence and maturity. *Molecular and General Genetics* 262, 579–588.

Ekblom, R. and Galindo, J. (2011) Applications of next generation sequencing in molecular ecology of non-model organisms. *Heredity* 107, 1–15.

Ellis, M.H., Speilmeyer, W., Gale, K.R., Rebetzke, G.J. and Richards, R. (2002) 'Perfect' markers for the *Rht-B1b* and *Rht-D1b* dwarfing genes in wheat. *Theoretical and Applied Genetics* 105, 1038–1042.

Mannai, Y., Shehzad, T. and Okuno, K. (2011) Variation in flowering time in sorghum core collection and mapping of QTLs controlling flowering time by association analysis. *Genetic Resources and Crop Evolution* 58, 983–989.

annai, Y., Shehzad, T. and Okuno, K. (2012) Mapping of QTLs underlying flowering time in sorghum (*Sorghum bicolor* [L.] Moench). *Breeding Science* 62, 151–159.

hoff, T.J., Murray, S.C., Klein, P. and Rooney, W. (2012) QTLs for energy-related traits in a sweet × ain sorghum [*Sorghum bicolor* (L.) Moench] mapping population. *Crop Science* 52, 2040–2049.

F.A., Hart, G.E., Schertz, K.F., Casa, A.M., Kresovich, S. *et al.* (2006) Alignment of genetic maps QTLs between inter- and intra-specific sorghum populations. *Theoretical and Applied Genetics* 1295–1305.

.W., Jones, E.S., Batley, J. and Smith, K.F. (2004) Molecular marker-based genetic ana- f pasture and turf grasses. In: Hopkins, A., Wang, Z.Y., Mian, R., Sledge, M. and Barker, R.E. *Molecular Breeding of Forage and Turf*. Developments in Plant Breeding, Vol. 11. Springer, ht, the Netherlands, pp. 197–238.

(1984) Genetic perspectives of germplasm conservation. In: Arber, W., Llimensee, K., Pea- J. and Starlinger, P. (eds) *Genetic Manipulation: Impact on Man and Society*. Cambridge Press, Cambridge, UK, pp. 161–170.

., Mace, E.S., Liu, Y., Zhu, M. *et al.* (2015) Fine mapping of *qGW1*, a major QTL for grain rghum. *Theoretical and Applied Genetics* 128, 1813–1825.

hehzad, T. and Okuno, K. (2013) Variation in inflorescence architecture associated with ents in a sorghum germplasm. *Plant Genetic Resources: Characterization and Utilization*

Hmon, K.P.W., Shehzad, T. and Okuno, K. (2014) QTL underlying inflorescence architecture in sorghum (*Sorghum bicolor* [L.] Moench) as detected association analysis. *Genetic Resources and Crop Evolution* 61, 1545–1564.

Kim, J.S., Klein, P.E., Klein, R.R., Price, H.J., Mullet, J.E. and Stelly, D.M. (2005) Chromosome identification and nomenclature of *Sorghum bicolor*. *Genetics* 169, 1169–1173.

Kebede, H., Subudhi, P., Rosenow, D. and Nguyen, H. (2001) Quantitative trait loci influencing drought tolerance in grain sorghum (*Sorghum bicolor* [L.] Moench). *Theoretical and Applied Genetics* 103, 266–276.

Lander, E.S. and Botstein, D. (1986) Strategies for studying heterogeneous genetic traits in humans by using a linkage map of restriction fragment length polymorphisms. *Proceedings of the National Academy of Sciences USA* 83, 7353–7357.

Lin, Y.R., Schertz, K.F. and Paterson, A.H. (1995) Comparative analysis of QTLs affecting plant height and maturity across the Poaceae, in reference to an interspecific sorghum population. *Genetics* 141, 391–411.

Mace, E.S., Singh, V., van Oosterom, E.J., Hammer, G.L., Hunt, C.H. and Jordan, D.R. (2012) QTL for nodal root angle in sorghum (*Sorghum bicolor* [L.] Moench) co-locate with QTL for traits associated with drought adaptation. *Theoretical and Applied Genetics* 14, 97–109.

Murray, S.C., Sharma, A., Rooney, W.L., Klein, P.E., Mullet, J.E., Mitchell, S.E. and Kresovich, S. (2008) Genetic improvement of sorghum as a biofuel feedstock: I. QTL for stem sugar and grain non-structural carbohydrates. *Crop Science* 48, 2165–2179.

Paterson, A.H., Lander, E.S., Hewitt, J.D., Peterson, S., Lincoln, S.E. and Tanksley, S.D. (1988) Resolution of quantitative factors by using a complete linkage map of restriction fragment length polymorphisms. *Nature* 335, 721–726.

Pereira, M.G. and Lee, M. (1995) Identification of genomic regions affecting plant height in sorghum and maize. *Theoretical and Applied Genetics* 90, 380–388.

Ritter, K.B., Jordan, D.R., Chapman, S.C., Godwin, I.D., Mace, E.S. and McIntyre, C.L. (2008) Identification of QTL for sugar-related traits in a sweet × grain sorghum (*Sorghum bicolor* [L.] Moench) recombinant inbred population. *Molecular Breeding* 22, 367–384.

Sakhi, S., Shehzad, T., Rehman, S. and Okuno, K. (2013) Mapping the QTLs underlying drought stress developmental stage of sorghum (*Sorghum bicolor* (L.) Moench) by association analysis. *Euphy* 193, 433–450.

Sakhi, S., Rehman, S., Okuno, K., Shazad, A. and Jamil, M. (2014) Evaluation of sorghum (*So bicolor*) core collection for drought tolerance: pollen fertility and mean performance of yield t its components at reproductive stage. *International Journal of Agriculture and Biology* 16,

Satish, K., Gutema, Z., Grenier, C., Rich, P.J., Sharma, A. *et al.* (2012) Molecular tagging and microsatellite sequencing markers linked to the low germination stimulant gene (*lgs*) fo ance in sorghum [*Sorghum bicolor* (L.) Moench]. *Theoretical and Applied Genetics*

Shehzad, T. and Okuno, K. (2014) QTL mapping for yield and yield-contributing traits *ghum bicolor* (L.) Moench) with genome-based SSR markers. *Euphytica* 203, 17-

Shehzad, T., Iwata, H. and Okuno, K. (2009a) Genome-wide association mapping o sorghum (*Sorghum bicolor* (L.) Moench) by using multiple models. *Breeding*

Shehzad, T., Okuizumi, H., Kawase, M. and Okuno, K. (2009b) Development of S *ghum bicolor* (L.) Moench) diversity research set of germplasm and its e traits. *Genetic Resources and Crop Evolution* 56, 809–827.

Shinozaki, K. and Yamaguchi-Shinozaki, K. (2007) Gene networks involve and tolerance. *Journal of Experimental Botany* 58, 221–227.

Shinozaki, K., Yamaguchi-Shinozaki, K. and Seki, M. (2003) Regulatory drought and cold stress responses. *Current Opinion in Plant Bio*

Srinivas, G., Satish, K., Madhusudhana, R. and Seetharama, N. (20 satellite markers from subtracted drought stress ESTs in *So and Applied Genetics* 118, 703–717.

Srinivas, G., Satish, K., Madhusudhana, R., Reddy, R.N., Mu Identification of quantitative trait loci for agronomically i ic-microsatellite markers in sorghum. *Theoretical and*

Subudhi, P., Rosenow, D. and Nguyen, H. (2000) Quanti (*Sorghum bicolor* (L.) Moench): consistency across *and Applied Genetics* 101, 733–741.

Sukumaran, S., Li, X., Li, X., Zhu, C., Bai, G. *et al.* (2016) QTL mapping for grain yield, flowering time, and stay-green traits in sorghum with genotyping-by- sequencing markers. *Crop Science* 56, 1429–1442.
Taiz, L. and Zieger, E. (1998) Stress physiology. *Plant Physiology*, 2nd edn. Sinauer Associates, Inc., Sunderland, Massachusetts, pp. 725–757.
Teer, J.K. and Mullikin, J.C. (2010) Exome sequencing: the sweet spot before whole genomes. *Human Molecular Genetics* 19, 145–151.
Tuinstra, M.R., Grote, E.M., Goldsbrough, P.B. and Ejeta, G. (1997) Genetic analysis of post-flowering drought tolerance and components of grain development in *Sorghum bicolor* (L.) Moench. *Molecular Breeding* 3, 439–448.
Umezawa, T., Fujita, M., Fujita, Y., Yamaguchi-Shinozaki, K. and Shinozaki, K. (2006) Engineering drought tolerance in plants: discovering and tailoring genes unlock the future. *Current Opinion in Biotechnology,*17,113–122.
Wang, H., Chen, G., Zhang, H., Liu, B., Yang, Y. *et al.* (2014a) Identification of QTLs for salt tolerance at germination and seedling stage of *Sorghum bicolor* (L.) Moench. *Euphytica* 196, 117–127.
Wang, T.T., Ren, Z.J., Liu, Z.Q., Feng, X., Guo, R.Q. *et al.* (2014b) *SbHKT1;4*, a member of the high-affinity potassium transporter gene family from *Sorghum bicolor*, functions to maintain optimal $Na^+/K^+$ balance under $Na^+$ stress. *Journal of Integrative Plant Biology* 56, 315–332.
Xu, W., Subudhi, P.K., Crasta, O.R., Rosenow, D.T., Mullet, J.E. and Nguyen, H.T. (2000) Molecular mapping of QTLs conferring stay-green in grain sorghum (*Sorghum bicolor* (L.) Moench). *Genome* 43, 461–469.
Yamaguchi-Shinozaki, K., Sakuma, Y., Ito, Y. and Shinozaki, K. (2006) The DRE/DREB regulon of gene expression in *Arabidopsis* and rice in response to drought and cold stress. In: Ribaut, J.M. (ed.) *Drought Adaptation in Cereals*. The Haworth Press, New York, pp. 583–598.

# 26 Improving Abiotic Stress Tolerance to Adapt Sorghum to Temperate Climatic Regions

**Subhadra Chakrabarty[1], André Schaffasz[1], Harmeet Singh Chawla[1], Maria Laura Federico[2,3], Rod J. Snowdon[1] and Steffen Windpassinger[1]***
[1]*Justus Liebig University, Giessen, Germany;* [2]*CONICET, Buenos Aires, Argentina;* [3]*INTA, Famaillá, Argentina*

## 26.1 Introduction

*Sorghum* is a genus consisting of many species in different levels of domestication, comprising wild (e.g. *Sorghum arundinaceum*), weedy (e.g. *Sorghum halapense*) and domesticated ones (e.g. *Sorghum* × *almum*, *Sorghum sudanense* and *Sorghum bicolor*). *S. bicolor* (L.) Moench is the fifth most important cereal crop globally. It shows a remarkable diversity, including five different subspecies and their intermediates, and several crop types like grain, forage, sweet and broomcorn (Hariprasanna and Patil, 2015). Although it originates in the tropics of Africa, the remarkable scope of genetic diversity among the different subspecies has conferred an extraordinarily broad adaptability and a highly versatile range of end uses (Boyles *et al.*, 2019). Sorghum has a particular advantage over crops like maize in extreme climates, where it achieves superior yields to maize (Farré and Faci, 2006; Staggenborg *et al.*, 2008). Although sorghum is a major subsistence crop worldwide and an important component of industrial agriculture, it is frequently considered an orphan crop (Boyles *et al.*, 2019) and breeding progress has been considerably slower than for most major global crops.

Despite its enormous potential and broad adaptive diversity, breeding to adapt sorghum for agricultural use in temperate Europe has progressed only slowly so far. Until recently, conventional breeding methods were the primary approach for genetic improvement of sorghum. These were mainly based on selection by visual phenotyping accompanied by introgression of desirable traits into elite germplasm. As in the early stages of classical breeding in most crops, these classical approaches present strong challenges to overcome linkage drag and maintain useful diversity in chromosome regions carrying essential adaptation genes under strong selection.

In recent times, modern molecular breeding tools like high-throughput molecular markers and genomic selection (GS) have shown great potential. Both can be combined to great effect with conventional breeding schemes to improve ecogeographical adaptation to new growing environments and further increase genetic gain for essential agronomic traits. For example, the major factors limiting adaptation of sorghum as

*Corresponding author: steffen.m.windpassinger@agrar.uni-giessen.de

 *Molecular Breeding in Wheat, Maize and Sorghum: Strategies for Improving Abiotic Stress Tolerance and Yield* (eds M.A. Hossain *et al.*)
DOI: 10.1079/9781789245431.0026

a cereal or biofuel crop in Northern Europe are its photoperiod sensitivity, along with a strong susceptibility to cool or cold temperatures at sowing and flowering (Windpassinger, 2016). Whereas photoperiod sensitivity is relatively simple to solve by conventional or marker-assisted backcrossing, adaptation to other abiotic stress is much more difficult to deal with in breeding programmes. This chapter reviews germplasm sources for and new developments in the identification and implementation of useful genetic diversity for temperate climate adaptation, along with genomics-based methods for breeding of complex, low-heritability traits like abiotic stress tolerance.

## 26.2 Origin and Genetic Diversity of Sorghum

The origin of early domestication of sorghum is hypothesized to have taken place in sub-Saharan Africa (present Ethiopia and Sudan) around 5000–8000 years ago (Mann *et al.*, 1983). From here, the species spread into different climatic zones of Africa, India, the Middle East and East Asia, initially via anthropogenic migration and later via human trade routes (Morris *et al.*, 2013). The lengthy process of adaptation to these different environments and ongoing selection for different kinds of human agricultural uses (grain, fodder, sugar, fibre) have created a vast genetic and phenotypic diversity among cultivated sorghum forms. Based on panicle and spikelet morphology the cultivated germplasm can be classified into five major races (*bicolor*, *caudatum*, *durra*, *guinea* and *kafir*) and ten intermediate races (Harlan and de Wet, 1972), representing the adaptation into different agroclimatic zones. Phylogenetic studies showed that genetic relatedness in sorghum accessions is predominantly based on geographic origin and race (Morris *et al.*, 2013). Similar to other crops, the identification and implementation of useful genetic diversity remain a paramount goal in sorghum breeding to ensure maintenance of genetic gain and constant replenishment of resistances to important pathogens. Recent studies have underlined the importance of a continuous integration of diverse, exotic material into modern breeding programmes (Jordan *et al.*, 2011), particularly the value represented by the secondary gene pool in the wild species *S. bicolor* subsp. *verticilliflorum* and *Sorghum propinquum* (Dillon *et al.*, 2007; Mace *et al.*, 2013; Muraya, 2014; Venkateswaran *et al.*, 2014).

## 26.3 Existing Molecular Tools to Enhance Abiotic Stress Tolerance

Since the origins of plant breeding as a systematic science, after the rediscovery of Mendel's laws just over a century ago, plant breeders have successfully developed new improved varieties by crossing and selection (Smýkal *et al.*, 2016). In recent decades, integration of new molecular breeding and biotechnological techniques into the process of crop improvement has contributed further to enhanced genetic gain and breeding has increased the productivity and sustainability of major crops (Voss-Fels *et al.*, 2019). The availability of various genomic tools and resources has led to a new revolution in plant breeding, as they facilitate the study of the genotype and its relationship with the phenotype, in particular for complex traits under strong environmental influence. The availability of a reference genome for sorghum (Paterson *et al.*, 2009), along with extensive additional genome sequence resources (e.g. Mace *et al.*, 2013; Morris *et al.*, 2013), greatly facilitates genomic-based breeding approaches. The relatively small diploid genome ($2n = 20$, ~730 Mb) and the high degree of synteny to maize and rice (Paterson *et al.*, 2009) greatly facilitate comparative genomics to take advantage of vast genomics resources from related monocot crops. Progress in this field for sorghum, however, has been relatively slow and limited due to its lower economic importance compared with other cereals.

Reduction in the cost of next-generation sequencing (NGS) technologies has led to mass sequencing of crop genomes and transcriptomes, which has facilitated the discovery of new genes and the development of vast collections of molecular markers like single-nucleotide polymorphisms (SNPs), which in turn are the basis for construction of high-density genetic maps. These facilitate mapping of genes and quantitative trait loci (QTLs) in bi-parental populations or via genome-wide association studies (GWAS), helping to dissect genomic regions associated with complex phenotypes. In addition, reverse genomics approaches including mutagenesis approaches make it possible to screen mutant and germplasm collections for allelic variants of target genes. All these tools and resources facilitate exploration of the genetic diversity and the developed markers can be incorporated into breeding programmes for marker-assisted selection (MAS). Common strategies include marker-assisted backcrossing with trait-linked foreground markers and genome-wide background markers, 'breeding by design' or new strategies like GS. Sorghum was one of the first major crops whose genome sequence was assembled. More recently, large-scale sequencing of transcriptomes (Makita *et al.*, 2015) has opened up new avenues for the global sorghum research community and has drastically simplified SNP marker identification and development. Sorghum genomic databases like SorghumFDB (Tian *et al.*, 2016) have combined various genomic data and functional annotations to create multidimensional biological relationships which could assist in sorghum functional genomics analyses and help in effective crop improvement.

Early genetic linkage maps for sorghum were constructed using labour-intensive or dominant markers such as RFLP (restriction fragment length polymorphism), AFLP (amplified fragment length polymorphism) and RAPD (random amplified polymorphic DNA) (Hulbert *et al.*, 1990; Berhan *et al.*, 1993; Boivin *et al.*, 1999; Peng *et al.*, 1999; Singh and Lohithaswa, 2006; Ejeta and Knoll, 2007). Although these maps played an important role in early sorghum genetic research, they were later superseded by more informative markers like simple sequence repeat (SSR) microsatellite markers, which were commonly used for sorghum gene mapping, genome evolutionary studies, molecular genetics and marker-assisted breeding (Tao *et al.*, 1998; Xu *et al.*, 2001; Yonemaru *et al.*, 2009; Kong *et al.*, 2013).

However, these technologies were limited by their restricted genome coverage and the relatively low number of polymorphic markers. Hamblin *et al.* (2004) first integrated SNP markers in sorghum to study genetic variation. The Diversity Arrays Technology (DArT; Canberra, Australia) was later applied to integrate multiple-component sorghum genetic maps to create a consensus map (Mace *et al.*, 2009). Presently, SNP markers are widely used in sorghum for studies tackling the genetic control of abiotic stress tolerance (see below).

## 26.4 History of Temperate Adaptation in Sorghum

Sorghum is originally a photosensitive short-day plant, conferring the best adaptation to its centre of origin in the semi-arid tropics of the Sahel zone. In these environments, the rainy season ends quite reliably at a latitude-specific time in autumn, with a day length below 12 h, whereas the onset of the new rainy season can vary strongly from year to year. Hence, local landraces are best adapted when flowering starts around 20 days before onset of the dry season, regardless of their sowing time, to allow for sufficient water supply during anthesis and grain-filling period and dry conditions during ripening (Guitton *et al.*, 2015). The relatively simple genetic architecture of photoperiodism in sorghum, which is controlled by four major maturity loci designated *Ma1*, *Ma2*, *Ma3* and *Ma4* (Quinby and Karper, 1945; Quinby, 1966), facilitated the generation of useful mutants. Dominant alleles at these loci induce photosensitivity, with *Ma1* having the largest impact (Klein *et al.*, 2008). A recessive mutation at this locus alone is sufficient to allow flowering under longer days in extra-tropical environments, and corresponding mutations occurred independently in different parts of the world following geographic dispersal of sorghum. By tracing allelic variants of the underlying gene *PRR37*, new insights were obtained into the historical expansion of sorghum into temperate areas of South Africa, China, Europe and the USA (Klein *et al.*, 2015). Sorghum was probably introduced into China as early as AD 400 via trade routes from India. Subsequently, under strong selection pressure for photoperiod insensitivity and early vigor, grain type *kaoliang* and broomcorn diverged there. From China, sorghum was brought to Europe (Klein *et al.*, 2015), where its first description dates back to 1204 in the Piemont region of Italy (Becker-Dillingen, 1927). However, in contrast to new-world crops such as maize and potato, which arrived several centuries later but were rapidly adopted, the importance of sorghum for European agriculture remained limited for a considerable time. It was relatively widespread from the 16th to 18th centuries in Southern and South-Eastern Europe; however, during that time its utilization was confined to broomcorn (Dahlberg *et al.*, 2012), implying low selection pressure for grain yield and further adaptive traits. The introduction of sorghum into North America occurred as broomcorn from Europe during the 1750s (Berenji *et al.*, 2011) and, more importantly, as grain and sweet sorghum arriving from Africa on slave ships during the first half of the 19th century (Sleper and Poehlman, 2006).The number of founder cultivars was low and sorghum cultivation was initially limited to subtropical areas of Texas. However, farmers soon selected early-maturing mutant plants, corresponding to the previously described mutations at the *Ma* loci, with a short stature. Similarly to photoperiodism, plant height in sorghum is determined by four major *dwarf* loci (designated $Dw_{1-4}$) with dominant alleles conferring tallness (Quinby and Karper, 1945; Multani *et al.*, 2003; Hilley *et al.*, 2016, 2017).

Systematic, modern sorghum breeding started in the USA during the first half of the 20th century and the combination of different desirable mutants facilitated the release of early-maturing cultivars that enabled cultivation as far north as Nebraska (Klein *et al.*, 2008). Impressive yield gains were achieved in the 1950s by changing from line to hybrid breeding (see below). Nevertheless, the narrow genetic base of photoperiod-insensitive breeding lines was soon recognized as a bottleneck for further yield gains and improvements in abiotic and biotic stress tolerance. To broaden the genetic diversity for temperate sorghum breeding, the 'Sorghum Conversion Program' was initiated in 1963 by the US Department of Agriculture (USDA). Taking into consideration the inheritance of photosensitivity and plant height in sorghum, a backcrossing programme was conducted to convert genetically diverse tropical accessions to early-maturing lines suitable for combine harvesting. Through this programme, about 850 converted and partially converted lines have been developed (Stephens *et al.*, 1967). Owing to the huge impact of this programme, most temperate sorghum hybrids today have conversion lines in their pedigree (Gabriel, 2005).

In theory, fully converted sorghum genotypes were expected to consist of 97% recurrent tropical parent genome. However, the ability to visualize and characterize genome introgressions using whole-genome sequencing or SNP genotyping revealed that the recovery of the exotic genome in backcrossed progenies containing desirable *dwarf* and *maturity* alleles was not as complete as assumed. Extensive stretches of the donor genome remained in linkage drag, for example on sorghum chromosome Sb06, which harbours crucial adaptive loci (*ma1* and *dw2* genes) (Klein *et al.*, 2008). As a result, little functional diversity in temperate sorghum genotypes has remained on this chromosome, which contains roughly 10% of all sorghum genes. This severely limits the adaptive potential especially for complex traits (Thurber *et al.*, 2013), presenting a key target for genomics-assisted breeding.

## 26.5 Potential of Sorghum in Temperate Climates

Underlining its adaptive potential, a substantial proportion of the world's sorghum harvest is today produced far away from its origin in tropical Africa, with countries like the USA, Mexico, Argentina, China and Australia among the main producers. However, in these countries, sorghum is mainly still grown in hot and dry, predominantly subtropical environments, with little expansion of production into temperate high-latitude areas. In Europe, sorghum has not yet achieved more than minor importance and so far the production is also concentrated in areas with hot summers, like southern France, Italy, Hungary, Romania, Ukraine and southern Russia. Altogether, the sorghum acreage in Europe of around 465,000 ha comprises only 1% of the global acreage. In contrast, maize is planted on 18,000,000 ha in Europe, spanning as far north as southern Scandinavia. This comparison with maize is relevant for two main reasons. First, maize is also a $C_4$ crop derived from the tropics, so its adaptation and expansion into cool temperate areas may serve as a blueprint for sorghum. Second, maize has a highly similar production technique and provides the same range of major uses as sorghum (grain, silage, bioenergy). Presently, maize is the crop of choice for biofuel and feed in temperate zones and the production system is very well established. However, there are several factors favouring an increase in sorghum acreage at the expense of maize in temperate Europe. Most importantly, the continuing expansion of maize monoculture on to vast production areas in Europe presents serious phyto-pathological threats like the western corn rootworm *Diabrotica virgifera* (Wesseler and Fall, 2010), especially in light of European policies on non-use of genetically modified organisms which preclude cultivation of transgenic *Bt* maize as an effective solution to this pest. Sorghum as a non-host (Oyediran *et al.*, 2004) may become one of the most compatible, viable alternatives to maize in quarantine areas. Furthermore, sorghum has a greater tolerance against most types of abiotic stress than maize. In the face of growing environmental and agroecological concerns with regard to climate change and agricultural sustainability, the high nutrient efficiency of sorghum can be a key asset, allowing for significant reduction of N fertilization. Its high water-use efficiency and drought tolerance are becoming more important in the context of climate change, enabling satisfactory yields in recent hot and dry European summers where other crops have frequently failed. Surprisingly, sorghum also tolerates temporary waterlogging considerably better than maize (Promkhambut *et al.*, 2011), ensuring good yield stability even in extreme or fluctuating environments. Despite its potential, as a tropical $C_4$ plant sorghum needs extensive breeding effort for adaptation to temperate climates before it can be broadly established in areas like Central Europe (Windpassinger *et al.*, 2015). In particular, its sensitivity to chilling represents a major constraint. Again, however, the history of North American and European maize breeding demonstrates that successful adaptation of a highly diverse tropical $C_4$ plant into temperate environments is feasible, suggesting similar possibilities for sorghum. However, the prerequisites for maize were arguably more advantageous, with a far longer history of temperate adaptation than sorghum. Already in pre-Columbian times, maize had spread into temperate North America up to what is today southern Canada (Matsuoka *et al.*, 2002), and recent results indicate a divergence between tropical and temperate maize as early as 3400 years ago (Liu *et al.*, 2015). Successful maize introductions to Central Europe from the 16th century onwards consisted of these 'pre-adapted' temperate maize types (*flint* variety group). Subsequently, adapted *flint* landraces developed under a strong selection pressure for early maturity and tolerance to cool spring temperatures. The fast expansion of maize acreage in Central Europe during the second half of the 20th century benefited considerably from (i) the existence of adapted *flint* germplasm and (ii) their heterotic pattern with North American *dents*, which allowed an optimal exploitation of heterosis in well-adapted and high-yielding hybrids. Both of these prerequisites exist only vaguely in sorghum, although Chinese *kaoliang* forms might eventually play a similar role as European *flint* in terms of early ecogeographic adaptation to temperate climatic zones. However, thanks to the modern molecular breeding techniques presented in the following sections, enhancement of quantitative traits like abiotic stress tolerance can potentially be achieved at a significantly faster pace today than during the early days of European maize breeding in the last century, provided sufficient investment is possible. Furthermore, access to extremely diverse tropical sorghum materials, including germplasm with pre-existing cool-temperature adaptation from growth at altitude in the highlands of Central and East Africa, is facilitated by a simple inheritance of photosensitivity. This allows for rapid backcrossing programmes, as successfully proven by the Sorghum Conversion Program in the USA.

## 26.6 Breeding Goals for Sorghum Temperate Adaptation and Their Present State-of-the-Art

Enhancements in abiotic stress tolerances, especially chilling tolerance, are the principal breeding goals required to establish stable sorghum productivity in temperate areas. Obviously, yield and adequate maturity, as the final outcome of genotype × environment interactions, have greater economic importance; however, in temperate sorghum production systems these traits are highly intertwined with chilling tolerance.

### 26.6.1 Juvenile chilling tolerance

Due to its tropical origin, sorghum generally does not tolerate frost and requires temperatures of more than 20°C for optimal growth. Lower temperatures induce different grades of chilling stress (see Fig. 26.1) and are especially problematic during emergence and seedling establishment (Pinthus and Rosenblum, 1961; Peacock, 1982). An improved juvenile chilling tolerance is thus mandatory for a successful adaptation to higher latitudes, since it would allow for earlier sowing, enhancing yield potential and maturity due to a longer growth period. Presently, in temperate areas such as Central Europe, sorghum is still sown several weeks later than maize, implying a loss of growth days which explains most of its present yield penalty in comparison to maize. However, improved early chilling tolerance can also be beneficial for some subtropical regions where sorghum is already well established, since earlier sowing in spring can potentially allow a better utilization of winter moisture (Patane *et al.*, 2006).

Several studies have been undertaken to mine sources of chilling tolerance in sorghum (Singh, 1985; Salas Fernandez *et al.*, 2014). Basically, these studies coincide in the identification of chilling-tolerant germplasm among Chinese *kaoliangs* and tropical highland accessions (e.g. from Ethiopia, Uganda and Yemen). Singh (1985) studied juvenile and pre-flowering cold tolerance in 380 accessions, with a focus on materials from China, Ethiopia, Uganda and the USA. The highest cold tolerance was observed among the Ethiopian and Ugandan accessions Alemaya70, Jewegere 935, Muyra, Mabere, Magune and Nyundo. The Chinese accession PI 610727 was highlighted as highly cold-tolerant by Franks *et al.* (2006) and further used by Burow *et al.* (2011) as the cold-tolerant parent in a bi-parental QTL mapping population. Salas Fernandez *et al.* (2014) tested 38 *kaoliangs* and 18 non-*kaoliangs* and identified new tolerant material from China, Korea and Russia. *Kaoliangs* are often highly cold-tolerant but have poor agronomic characteristics which limit their direct use in breeding programmes (Franks *et al.*, 2006).

In a recent study conducted under Central European field conditions, Schaffasz *et al.* (2019a) revealed that valuable sources for chilling tolerance and early vigour can also be found among existing US sorghum conversion lines. The accessions SC614 and SC1201 performed best for emergence and SC702 best for early shoot biomass. In contrast to *kaoliang*, these conversion lines are more amenable for breeding of grain sorghum. In recent years, increasing research efforts have begun to dissect the genetic architecture of juvenile chilling tolerance in

**Fig. 26.1.** Variation for chilling tolerance and early vigour among different sorghum accessions. (a) Different reaction of sorghum accessions to controlled chilling-stress conditions (13°C day/10°C night during emergence and subsequent growth). Susceptible genotypes show poor emergence and development, and/or complete chlorophyll degradation (white leaves), while tolerant accessions show satisfying emergence and maintenance of photosynthetic apparatus (green leaves). (b) Sorghum genotypes (sown in two-rowed plots) show remarkable variation for establishment and early vigour in a field experiment in Germany.

sorghum by QTL studies in segregating bi-parental populations.

Knoll *et al.* (2008) dissected the early-season cold tolerance in sorghum using a recombinant inbred line (RIL) population derived from a cross between Shan Qui Red (SQR, cold-tolerant) and SRN39 (cold-sensitive). They identified two QTLs for germination under cold stress on linkage group SBI-03a (on chromosome Sb03) and on group SBI-07b (chromosome Sb07), both of which showed significant trait associations under cold temperatures. A region on chromosome Sb01 derived from SQR showed strong associations with seedling emergence and seedling vigour scores under early and late field plantings. One QTL for both early and late emergence and another QTL for early vigour were identified on Sb02 and Sb04, respectively. Shortly after, 14 QTLs associated with different cold tolerance traits were detected on chromosomes Sb01, Sb02, Sb04, Sb07 and Sb09 (Burow *et al.*, 2011). In particular, Sb09 was shown to harbour four QTLs for field emergence that co-localized with QTLs for cold germinability. Bekele *et al.* (2014) identified highly interactive epistatic QTL hotspots, including a previously unknown QTL on Sb06 with a significant effect on prolonged chilling survival, which were found to regulate different physiological mechanisms contributing to maintenance of growth and development even under chilling temperatures.

During the past decade, genetic association studies using various diversity panels have accelerated the identification of genome regions and promising candidate genes highly influencing cold tolerance during emergence. Fiedler *et al.* (2012, 2014, 2016) reported multiple cold tolerance QTLs and identified the gene *Cold-Shock Domain Protein 1* (*CSDP1*) as a potential positional and functional candidate. One QTL region on chromosome Sb06 was identified as a putative hotspot for temperature-mediated seedling emergence and survival. This region was later independently verified (Parra-Londono *et al.*, 2018). Recently, Schaffasz *et al.* (2019a) presented the first study to jointly analyse both agronomical and cold tolerance traits on a broad diversity set under Central European conditions. The findings from these studies show the potential of GWAS to help dissect the genetic complexity of cold temperature susceptibility, an important prerequisite for development of temperature-resilient sorghum cultivars and further characterization of genomic regions responsible for adaptation to thermal stresses (Chopra *et al.*, 2017).

Recently, Marla *et al.* (2019) applied a nested association mapping (NAM) approach to investigate chilling tolerance in a multi-parental population, identifying ten loci explaining 20–41% of the phenotypic variation within these US sorghum accessions. Surprisingly, the results showed the co-inheritance of chilling tolerance loci with wild-type alleles of classical tannin (*Tan1* and *Tan2*) and dwarfing genes (*Dw1* and *Dw3*), four of the five most important genes under selection by US sorghum breeders in the 20th century. The fifth of these, *Maturity1*, did not co-localize with chilling tolerance QTLs. Because there is no clear evidence to suggest that *Tan1*, *Tan2*, *Dw1* or *Dw3* is directly involved in cold tolerance responses, this association seems more likely to be caused by the substantial linkage drag surrounding these gene variants rather than a negative pleiotropic association between the traits. In other words, this result indicates that strong selection for essential recessive adaptation alleles in modern grain sorghum resulted in loss of early-season chilling tolerance by linkage in repulsion. In consequence, the authors suggest revision of the original model of sorghum chilling sensitivity to be only caused by its tropical origin, since African sorghums introduced into the USA harboured basal chilling tolerance which was subsequently lost by breeding. Altogether, the results of the previously discussed mapping studies are concordant and coincide in a strongly quantitative character for all juvenile chilling tolerance-related traits. For practical breeding, these results imply strong limitations regarding the possibilities of MAS. Nevertheless, these studies provide valuable insights into the underlying physiological mechanisms of abiotic stress tolerance. Of special interest are the results of Marla *et al.* (2019), which provide explanations on how the co-inheritance of chilling sensitivity with desired traits has hampered breeding efforts during the last decades. The recently developed Sorghum QTL Atlas platform (Mace *et al.*, 2019), which facilitates identification of candidate genes in sorghum and their comparison across related species like maize and rice, provides important new data about linkage relationships among genome-wide QTLs for important agronomic traits and can provide a starting point (foreground and background markers) for endeavours to identify useful recombinants that disrupt linkage in repulsion between key adaptation and chilling tolerance loci.

### 26.6.2 Reproductive chilling tolerance

Pre-flowering reproductive stage is the second sensitive developmental phase in sorghum affected by temperatures below 15°C (Singh, 1985). Depending on duration and intensity of the stress, male sterility can be induced in sorghum, leading to a reduction or even, in extreme cases, a complete loss of seed yield (see Fig. 26.2) (Downes and Marshall, 1971; Osuna-Ortega *et al.*, 2003). For adaptation of sorghum to temperate climates, reproductive chilling tolerance is at least equally as important as juvenile chilling tolerance. While farmers can choose later sowing dates to reduce juvenile chilling stress (albeit at the expense of yield potential), there is no escape strategy for cold nights during the critical reproductive stage in summer. Tropical high-altitude environments could also benefit from sorghum varieties with improved reproductive chilling tolerance. In contrast to temperate high-latitude environments, where chilling stress at pre-flowering stage occurs rather infrequently followed by intervals of warmer weather, tropical highlands tend to have constantly cool nights, making them suitable selection environments also for temperate breeding programmes. Reproductive chilling tolerance is also considered to be potentially beneficial for sorghum cultivation in the Indian post-rainy season (*rabi*) (Krishnamurthy *et al.*, 2014).

The first scientific description of male sterility in sorghum after a cold treatment (13°C) was provided by Downes and Marshall (1971), who were originally seeking a new crossing method. Brooking (1976) described problems with meiosis in mother-spore cells as a possible reason for this phenomenon. Singh (1985) scored reproductive chilling tolerance (along with juvenile chilling tolerance, see above) in a set of 380 accessions, identifying several tolerance sources. Notable breeding efforts were undertaken to develop sorghum varieties with enhanced

**Fig. 26.2.** Variation for sorghum reproductive cold tolerance in a field experiment in Germany: tolerant line with high pollen viability also at cool temperatures, and hence full seed set (left); susceptible line with cold-induced male sterility, resulting in almost no seed set (right).

reproductive chilling tolerance for the Mexican High Valleys (>2000 m above sea level) (Mendoza, 1988; Osuna-Ortega *et al.*, 2000, 2003; Leon-Velasco *et al.*, 2009; Cisneros-López *et al.*, 2010), using cold-tolerant accessions from Africa and India donated by the International Crops Research Institute for the Semi-Arid Tropics (ICRISAT)–International Maize and Wheat Improvement Center (CIMMYT) as base material (Leon-Velasco *et al.*, 2009). As a result, varieties with a satisfying seed set and yield even at night temperatures of 6°C during the critical stages could be developed (Osuna-Ortega *et al.*, 2003), underlining the feasibility of genetic improvement of this trait. Maulana and Tesso (2013) recommended the accession Shan Qui Red (a Chinese *kaoliang* known for good early chilling tolerance) as a tolerance source also for reproductive chilling tolerance. In contrast to juvenile chilling tolerance, surprisingly little is known about the genetic architecture of reproductive chilling tolerance in sorghum to date. Both Singh (1985) and Schaffasz *et al.* (2019b) described a more or less dominant inheritance. Interestingly, the observed heritability for seed set traits under stress (Schaffasz *et al.*, 2019b) was notably higher than for juvenile chilling tolerance traits (Windpassinger, 2016; Schaffasz *et al.*, 2019a). However, GWAS or QTL mapping for this trait still needs to be performed to unravel its quantitative genetic control.

## 26.7 Breeding Methods

The type of sorghum and the purpose for its production vary widely depending on the region where it is grown. Sorghum is a predominantly self-pollinating crop. The level of cross-pollination depends on panicle architecture and weather conditions and increases under stress (Osuna-Ortega *et al.*, 2003). Thus, breeding procedures applicable to both self- and cross-pollinated crops can be deployed to sorghum improvement (Rakshit and Bellundagi, 2019). However, for both pedigree breeding and hybrid breeding the primary goals of sorghum breeders throughout the world are always grain/biomass yield, adaptation, stress tolerance and product quality.

Heterosis (hybrid vigour) in sorghum was already described by Conner and Karper (1927) but unlike monoecious maize, the perfect flowers of sorghum prevented hybrid seed production on a commercial scale until 1952, when both cytoplasmic-male sterility (CMS) and fertility restorers possessing dominant *Rf* alleles were discovered in the USA (Stephens and Holland, 1954). Commercial CMS-based hybrid seed production began in 1956 and only four years later, the proportion of sorghum production from hybrid cultivar in the USA reached 95%, resulting in a doubling of grain yield compared with 1952 (Quinby, 1974; Smith and Frederiksen, 2000). Presently, sorghum production in regions with an industrialized, commercial agriculture (USA, Latin America, Australia and Europe) relies almost completely on hybrids, while open-pollinated landraces are still predominantly used for subsistence agriculture in Africa.

### 26.7.1 Enhancements of abiotic stress tolerance via heterosis and hybrid breeding

Heterosis in sorghum is not only expressed for grain and biomass yield, but also for maturity (Kirby and Atkins, 1968) and abiotic stress tolerance (see Fig. 26.3), including juvenile (Pinthus and Rosenblum, 1961; Yu and Tuinstra, 2001; Windpassinger, 2016) and reproductive chilling stress (Leon-Velasco *et al.*, 2009; Schaffasz *et al.*, 2019b). Efficient and successful hybrid breeding requires the development of complementary heterotic pools with a sufficiently high genetic distance between them. Well-designed heterotic pools ensure a consistent exploitation of heterosis by increasing the relative contribution of general combining ability (GCA) effects in comparison to effects from specific combining ability (SCA) (Reif *et al.*, 2005; Schnable and Springer, 2013). Sorghum heterotic pools are not yet defined as clearly as they are in maize (Monk *et al.*, 2014). However, the availability of cost-effective molecular markers provides new opportunities to evaluate the phylogenetic and genomic structure of accessions to establish genetically distinct pools for temperate sorghum breeding programmes. Pre-existing heterotic patterns can be easily compromised by arbitrary crosses, a frequent occurrence prior to the implementation of molecular marker

**Fig. 26.3.** Reaction of two different sorghum $F_1$ hybrids and their respective parental lines to prolonged chilling stress (13°C day/10°C night in a climate chamber experiment) induced after emergence at warm temperatures. While heterosis for early biomass production is clearly visible, the ability to survive prolonged chilling (expressed in the maintenance of green leaves) seems to be a rather additive trait.

techniques in sorghum hybrid breeding (Menz *et al.*, 2004). On the other hand, genome-wide markers can also help to rapidly characterize genetic diversity and design genome-assisted cross schemata for separation of heterotic pools.

The development of heterotic pools can potentially make a significant contribution to enhancement of abiotic stress tolerance in sorghum hybrids. An important question for breeders is to what extent hybrid performance can be predicted based on per se line performance. For juvenile chilling tolerance, Windpassinger (2016) showed that line performance per se is in fact a poor predictor of hybrid performance. Hence, an overly strict selection of hybrid parents based on the per se tolerance is in fact counterproductive, whereas GCA tests seem to be a more efficient and useful approach. For emergence and early heterotrophic growth of hybrids, the impact of the female parent is known to be higher (Yu and Tuinstra, 2001; Windpassinger 2016), suggesting priority should be given to improvement of the female pool in a hybrid breeding programme. For reproductive chilling tolerance, correlations observed between per se and hybrid performance were somewhat higher than for juvenile chilling tolerance (Schaffasz *et al.*, 2019b). Nevertheless, GCA tests also appear to be the preferable selection method for this trait. Due to a high GCA:SCA ratio and low GCA × environment interaction, robust enhancements of reproductive chilling tolerance via hybrid breeding seem to be feasible (Schaffasz *et al.*, 2019b) and a more systematic exploitation of heterosis using genomic tools may simultaneously help improve genetic gain for chilling tolerance. In consequence, future association studies and GS approaches for chilling tolerance should focus rather on GCA than per se performance (as done in the past) as a basis for successful hybrid breeding towards more robust and stable plant establishment, pollen fertility and seed set under cool-temperature conditions.

## 26.8 Advancement and Use of Genomics and Bioinformatics Approaches

### 26.8.1 High-throughput genotyping tools

The major prerequisite for application of genomics in genetic analyses or breeding of complex quantitative traits is the availability of suitable platforms and SNP marker panels for rapid, cost-effective, genome-wide marker screening. In sorghum, large population genomics studies have been achieved with sequencing-based marker techniques (see below), but molecular breeding efforts for most other major crops have generally been based on dedicated SNP array genotyping platforms. SNP arrays have a number of advantages for breeding in comparison to sequencing approaches, not least the ability of service providers who can deliver low-cost genotype data sets without the need for breeders to have access to their own sophisticated molecular genetics laboratories or bioinformatics facilities. With an SNP array a fixed marker panel is genotyped for every individual and the customer/breeder is provided with a simple spreadsheet containing genotype calls. This means that considerably less bioinformatics analysis is required than for derivation of SNP variants from genotyping-by-sequencing (GBS) data, for example. In contrast to other major crops like wheat, maize, canola, barley or soybean, however, until recently there has not been a major push to develop a community-driven public SNP genotyping array platform. Bekele *et al.* (2013) developed a small-scale Illumina Infinium 3K SNP genotyping array with 2620 SNP markers and demonstrated its implementation for genetic mapping, diversity analyses and GWAS (Bekele *et al.*, 2014). Later, large-scale genomic resequencing data were used to generate a 90K SNP Affymetrix Axiom genotyping array for GWAS (Parra-Londono *et al.*, 2018); however, to date, no commercial sorghum SNP chip has been made available for public use. In 2019, efforts were initiated to establish a private–public consortium for development of a low-cost Illumina SNP array for sorghum breeding, but to date most large-scale genotyping efforts have been carried out using sequencing-based genotyping technologies.

### 26.8.2 Use of next-generation-sequencing genotyping techniques in sorghum

Incredible progress has been made in modern DNA sequencing technologies and accompanying bioinformatics methods in recent years. NGS technologies can be utilized for identifying the genetic basis of agriculturally important traits and for predicting the breeding value of individuals in a plant breeding population (Varshney *et al.*, 2014). Detailed phenotyping of multiparental NAM populations (Jordan *et al.*, 2012; Marla *et al.*, 2019) has been applied in association with high-resolution sequencing-based sorghum genotype data to dissect different abiotic stress traits.

Sequence-based genotyping platforms that have been applied for sorghum include restriction site-associated DNA sequencing (RAD-seq) (Nelson *et al.*, 2011) and GBS (Morris *et al.*, 2013). These two similar, reduced-representation, genome-wide resequencing methods are capable of identifying, sequencing and genotyping thousands of markers across the genome at low cost in large populations, making them highly suitable for genome-wide analyses of complex traits. In a more targeted approach, Ji *et al.* (2017) implemented genome-wide specific-locus amplified fragments (SLAF) markers, which are highly abundant and evenly distributed across the genome and thus facilitate the scanning of the sorghum genome for gene mining. The main advantage of NGS-based genotyping platforms compared with arrays is their lack of ascertainment bias in the markers assayed, improving their potential for discovery of novel variants of interest for trait improvement. The large numbers of low-cost SNP markers generated by NGS-based genotyping systems make them an economical option for GS in modern breeding programmes. Several studies have shown the potential of GS to enhance abiotic stress tolerance traits like heat and drought tolerance in major cereal crops like rice and maize (Yuan *et al.*, 2018; Bhandari *et al.*, 2019; Trachsel *et al.*, 2019), and recently Velazco *et al.* (2019) demonstrated the efficacy of GS to enhance the stay-green trait in sorghum. To date, however, GS is still at its infancy in sorghum for improvement of abiotic stress traits.

### 26.8.3 Transcriptome analysis

Coupled with precise phenotyping and proper gene annotations, functional genomics can provide crucial information regarding complex biological processes like abiotic stress responses. The impact of transcriptome analysis in sorghum increased rapidly after the completion of the first reference genome and with the advent of next-generation molecular tools. Remarkable progress has been made in regard to transcriptome analysis of traits like drought tolerance, cold stress, heat and salinity in sorghum (Fracasso *et al.*, 2016; Bashir *et al.*, 2019). It has been shown that cold stress induces osmotic stress and the expression of transcription factors for protein kinase genes is altered (Bashir *et al.*, 2019). Kadier *et al.* (2017) identified upregulation of the sorghum NAC-transcription factor family genes *SbNAC17* and *SbNAC73* in leaf tissues under cold stress conditions. The role of NAC transcription factors in general abiotic stress-response regulation has been described in many plant species (see Shao *et al.*, 2015 for a review).

Woldesemayat *et al.* (2018) recently introduced an integrated approach to mine for candidate stress genes across species by combining ontology-based semantic data integration with expression profiling, comparative genomics, phylogenomics, functional gene enrichment and gene enrichment network analysis. As a result, 221 cold stress genes were identified in sorghum and were validated using ontology mapping. In addition, a phylogenetic tree was constructed to infer the evolutionary relationship of the sorghum orthologues.

### 26.8.4 Genetic transformation

To intensify the plant development, genetic transformation has proved to be a powerful tool for gene induction, modulation and expression (Gurel *et al.*, 2009). However, sorghum has been classified as one of the most challenging plant species to perform tissue culture and genetic transformation (Zhu *et al.*, 1998). *Agrobacterium*-mediated and particle bombardment transformation are the two main approaches that have been exploited for the development of transgenic sorghum (Ahmed *et al.*, 2018). Over the years, there have been a few studies where sorghum was genetically transformed to dissect gene complexity of traits like soil salinity tolerance, protein and tannin content (Yellisetty *et al.*, 2015; Kuriyama *et al.*, 2019; Liu *et al.*, 2019) but the efficiency is still behind other major crops like rice, maize and barley (Che *et al.*, 2018). Since traits like juvenile cold tolerance are controlled by a large number of loci and exhibit low heritability (Bekele *et al.*, 2014), it remains challenging to design genetic transformation strategies for improving these traits.

### 26.8.5 TILLING

Although great progress has been made in sorghum genomics, the availability of mutant lines for functional studies via reverse genetics is limited. Chemical mutagenesis of sorghum germplasm, followed by screening for mutants altered in important agronomic traits by targeting induced local lesion in genomes (TILLING), represents a rapid and effective means for studying agronomically important genes (Xin *et al.*, 2008). Jiao *et al.* (2016) identified potential genes involved in drought tolerance using TILLING mutants. Similarly, other abiotic stress traits like cold tolerance and photoperiodism can also be studied. Using EcoTILLING, Bharathi *et al.* (2016) screened naturally occurring mutations in potential candidate genes to study several agronomically important traits in sorghum. Undoubtedly, a large-scale resource of well-characterized mutants and naturally occurring genetic variation would provide an efficient platform for functional validation of genes in sorghum, thereby accelerating sorghum breeding.

## 26.9 Future Prospects

Sorghum is an important failsafe crop which provides food, feed, fuel and fodder in many countries around the globe. It can be used as a model for other $C_4$ crops because of its extensive collection of diverse germplasm, genetic and genomic resources, and breeding information. Recent advancements in NGS, high-throughput phenotyping and bioinformatics tools are helping to accelerate genetic gain in sorghum across different climatic zones. Despite efforts to improve genetic and genomic resources, many such resources are still decentralized and independent. Increased efforts to coordinate cooperation among complementary public research programmes could further help to integrate research platforms available for meta-analysis of complex adaptive traits like chilling tolerance. Using information from related monocot crops can critically increase the power of comparative genomics and help dissect adaptive traits to enhance crop improvement. New molecular breeding methods and tools are already promoting considerable progress in plant breeding, including fast-track adaptation and the genetic dissection and breeding for complex abiotic stress traits (Pérez-de-Castro *et al.*, 2012). The integration of large-scale molecular marker data sets, high-density genetic maps, genome and transcriptome sequences with quantitative genetics can help translate functional genomics knowledge into genome-based improvements in modern breeding populations. This will help to further accumulate base knowledge to develop a framework for implementing GS for sorghum improvement. As a consequence, the historical division between breeding and genomics is becoming increasingly blurred (Deshpande *et al.*, 2017). Crops like sorghum, which to date have been bred by traditional means but for which exceptional genome resources are available, stand to benefit greatly from genomics-based breeding applications.

## References


Ahmed, R.I., Ding, A., Xie, M. and Kong, Y. (2018) Progress in optimization of *Agrobacterium*-mediated transformation in sorghum (*Sorghum bicolor*). *International Journal of Molecular Sciences* 19(10), 2983. Available at: https://doi.org/10.3390/ijms19102983

Bashir, K., Matsui, A., Rasheed, S. and Seki, M. (2019) Recent advances in the characterization of plant transcriptomes in response to drought, salinity, heat, and cold stress. *F1000Research* 8, 658. Available at: https://doi.org/10.12688/f1000research.18424.1

Becker-Dillingen, J. (1927) *Handbuch des Getreidebaues einschließlich Mais, Hirse und Buchweizen*. Paul Parey, Berlin.

Bekele, W.A., Wieckhorst, S., Friedt, W. and Snowdon, R.J. (2013) High-throughput genomics in sorghum: from whole-genome resequencing to a SNP screening array. *Plant Biotechnology Journal* 11(9), 1112–1125. Available at: https://doi.org/10.1111/pbi.12106

Bekele, W.A., Fiedler, K., Shiringani, A., Schnaubelt, D., Windpassinger, S. *et al.* (2014) Unravelling the genetic complexity of sorghum seedling development under low-temperature conditions. *Plant, Cell & Environment* 37(3), 707–723. Available at: https://doi.org/10.1111/pce.12189

Berenji, J., Dahlberg, J., Sikora, V. and Latkovi, D. (2011) Origin, history, morphology, production, improvement, and utilization of broomcorn [*Sorghum bicolor* (L.) Moench] in Serbia. *Economic Botany* 65(2), 190–208. Available at: https://doi.org/10.1007/s12231-0119155-2

Berhan, A.M., Hulbert, S.H., Butler, L.G. and Bennetzen, J.L. (1993) Structure and evolution of the genomes of *Sorghum bicolor* and *Zea mays*. *Theoretical and Applied Genetics* 86(5), 598–604. Available at: https://doi.org/10.1007/BF00838715

Bhandari, A., Bartholomé, J., Cao-Hamadoun, T.-V., Kumari, N., Frouin, J., Kumar, A. and Ahmadi, N. (2019) Selection of trait-specific markers and multi-environment models improve genomic predictive ability in rice. *PLoS ONE* 14(5), e0208871. Available at: https://doi.org/10.1371/journal.pone.0208871

Bharathi, R.R., Agasimani, S., Anusheela, V., Thiruvengadam, V., Chibbar, R.N. and Ganesh R.S. (2016) TILLING and EcoTILLING for discovery of induced and natural variations in sorghum genome. In: Rakshit, S. and Wang, Y.-H. (eds) *The Sorghum Genome*. Compendium of Plant Genomes. Springer, Cham, Switzerland, pp. 257–267.

Boivin, K., Deu, M., Rami, J.-F., Trouche, G. and Hamon, P. (1999) Towards a saturated sorghum map using RFLP and AFLP markers. *Theoretical and Applied Genetics* 98(2), 320–328. Available at: https://doi.org/10.1007/s001220051076

Boyles, R.E., Brenton, Z.W. and Kresovich, S. (2019) Genetic and genomic resources of sorghum to connect genotype with phenotype in contrasting environments. *The Plant Journal* 97(1), 19–39. Available at: https://doi.org/10.1111/tpj.14113

Brooking, I.R. (1976) Male sterility in *Sorghum bicolor* (L.) Moench induced by low night temperature, I. Timing of the stage of sensitivity. *Australian Journal of Plant Physiology* 3(5), 589–596. Available at: https://doi.org/10.1071/pp9760589

Burow, G., Burke, J.J., Xin, Z. and Franks, C.D. (2011) Genetic dissection of early-season cold tolerance in sorghum (*Sorghum bicolor* (L.) Moench). *Molecular Breeding* 28(3), 391–402. Available at: https://doi.org/10.1007/s11032-010-9491-4

Che, P., Anand, A., Wu, E., Sander, J.D., Simon, M.K. *et al.* (2018) Developing a flexible, high-efficiency *Agrobacterium*-mediated sorghum transformation system with broad application. *Plant Biotechnology Journal* 16(7), 1388–1395. Available at: https://doi.org/10.1111/pbi.12879

Chopra, R., Burow, G., Burke, J.J., Gladman, N. and Xin, Z. (2017) Genome-wide association analysis of seedling traits in diverse Sorghum germplasm under thermal stress. *BMC Plant Biology* 17(1), 12. Available at: https://doi.org/10.1186/s12870-016-0966-2

Cisneros-López, M.E., Mendoza-Onofre, L.E., Zavaleta-Mancera, H.A., González-Hernández, V.A., Mora-Aguilera, G., Córdova-Téllez, L. and Hernández-Martínez, M. (2010) Pollen–pistil interaction, pistil histology and seed production in A × B grain sorghum crosses under chilling field temperatures. *The Journal of Agricultural Science* 148(1), 73–82.

Conner, A.B. and Karper, R.E. (1927) *Hybrid Vigor in Sorghum*. Bulletin No. 359. Texas Agricultural Experiment Station, Texas A&M University, College Station, Texas. Available at: http://hdl.handle.net/1969.1/4015 (accessed 5 March 2021).

Dahlberg, J., Berenji, J., Sikora, V. and Latković, D. (2012) Assessing sorghum [*Sorghum bicolor* (L) Moench] germplasm for new traits: food, fuels & unique uses. *Maydica* 56(2). Available at: https://journals-crea.4science.it/index.php/maydica/article/view/688

Deshpande, S., Rakshit, S., Manasa, K.G., Pandey, S. and Gupta, R. (2017) Genomic approaches for abiotic stress tolerance in sorghum. In: Rakshit, S. and Wang, Y.-H. (eds) *The Sorghum Genome*. Compendium of Plant Genomes. Springer, Cham, Switzerland, pp. 169–187.

Dillon, S.L., Shapter, F.M., Henry, R.J., Cordeiro, G., Izquierdo, L. and Lee, L.S. (2007) Domestication to crop improvement: genetic resources for *Sorghum* and *Saccharum* (Andropogoneae). *Annals of Botany* 100(5), 975–989. Available at: https://doi.org/10.1093/aob/mcm192

Downes, R.W. and Marshall, D.R. (1971) Low temperature induced male sterility in *Sorghum bicolor*. *Australian Journal of Experimental Agriculture* 11(50), 352–356.

Ejeta, G. and Knoll, J.E. (2007) Marker-assisted selection in sorghum. In: Varshney, R.K. and Tuberosa, R. (eds) *Genomics-assisted Crop Improvement*. Springer, Dordrecht, the Netherlands, pp. 187–205.

Farré, I. and Faci, J.M. (2006) Comparative response of maize (*Zea mays* L.) and sorghum (*Sorghum bicolor* L. Moench) to deficit irrigation in a Mediterranean environment. *Agricultural Water Management* 83(1–2), 135–143. Available at: https://doi.org/10.1016/j.agwat.2005.11.001

Fiedler, K., Bekele, W.A., Friedt, W., Snowdon, R., Stützel, H., Zacharias, A. and Uptmoor, R. (2012) Genetic dissection of the temperature dependent emergence processes in sorghum using a cumulative emergence model and stability parameters. *Theoretical and Applied Genetics* 125(8), 1647–1661. Available at: https://doi.org/10.1007/s00122-012-1941-4

Fiedler, K., Bekele, W.A., Duensing, R., Gründig, S., Snowdon, R. *et al.* (2014) Genetic dissection of temperature-dependent sorghum growth during juvenile development. *Theoretical and Applied Genetics* 127(9), 1935–1948. Available at: https://doi.org/10.1007/s00122-014-2350-7

Fiedler, K., Bekele, W.A., Matschegewski, C., Snowdon, R., Wieckhorst, S., Zacharias, A. and Uptmoor, R. (2016) Cold tolerance during juvenile development in sorghum: a comparative analysis by genomewide association and linkage mapping. *Plant Breeding* 135(5), 598–606. Available at: https://doi.org/10.1111/pbr.12394

Fracasso, A., Trindade, L.M. and Amaducci, S. (2016) Drought stress tolerance strategies revealed by RNA-Seq in two sorghum genotypes with contrasting WUE. *BMC Plant Biology* 16(1), 115. Available at: https://doi.org/10.1186/s12870-016-0800-x

Franks, C.D., Burow, G.B. and Burke, J.J. (2006) A comparison of US and Chinese sorghum germplasm for early season cold tolerance. *Crop Science* 46(3), 1371–1376.

Gabriel, K. (2005) A study of heterotic relationships in sorghum. PhD dissertation, Texas A&M University, College Station, Texas. Available at https://oaktrust.library.tamu.edu/bitstream/1969.1/3226/1/etd-tamu-2005C-PLBR-Gabriel.pdf (accessed 3 March 2021).

Guitton, B., Vaksman, M., Rami, J.-F., Weltzein, E., Rattunde, F. *et al.* (2015) Enhancing sorghum grain yield and quality for the Sudano-Sahelian zone of West Africa using the backcross nested association mapping (BCNAM) approach. *Presented at XXIIIrd EUCARPIA Maize and Sorghum Conference of the Vth Session – New Insights in the Genetics of Traits (2), Montpellier, France, 10–11 June 2015.*

Gurel, S., Gurel, E., Kaur, R., Wong, J., Meng, L., Tan, H.-Q. and Lemaux, P.G. (2009) Efficient, reproducible *Agrobacterium*-mediated transformation of sorghum using heat treatment of immature embryos. *Plant Cell Reports* 28(3), 429–444. Available at: https://doi.org/10.1007/s00299-008-0655-1

Hamblin, M.T., Mitchell, S.E., White, G.M., Gallego, J., Kukatla, R. *et al.* (2004) Comparative population genetics of the panicoid grasses: sequence polymorphism, linkage disequilibrium and selection in a diverse sample of sorghum bicolor. *Genetics* 167(1), 471–483.

Hariprasanna, K. and Patil, J.V. (2015) Sorghum: origin, classification, biology and improvement. In: Madhusudhana, R., Rajendrakumar, P. and Patil, J.V. (eds) *Sorghum Molecular Breeding*. Springer India, New Delhi, pp. 3–20.

Harlan, J.R. and de Wet, J.M.J. (1972) A simplified classification of cultivated sorghum. *Crop Science* 12(2), 172–176. Available at: https://doi.org/10.2135/cropsci1972.0011183X001200020005x
Hilley, J., Truong, S., Olson, S., Morishige, D. and Mullet, J. (2016) Identification of *Dw1*, a regulator of sorghum stem internode length. *PLoS ONE* 11(3), e0151271. Available at: https://doi.org/10.1371/journal.pone.0151271
Hilley, J.L., Weers, B.D., Truong, S.K., McCormick, R.F., Mattison, A.J. *et al.* (2017) Sorghum *Dw2* encodes a protein kinase regulator of stem internode length. *Scientific Reports* 7(1), 4616. Available at: https://doi.org/10.1038/s41598-017-04609-5
Hulbert, S.H., Richter, T.E., Axtell, J.D. and Bennetzen, J.L. (1990) Genetic mapping and characterization of sorghum and related crops by means of maize DNA probes. *Proceedings of the National Academy of Sciences USA* 87(11), 4251–4255. Available at: https://doi.org/10.1073/pnas.87.11.4251
Ji, G., Zhang, Q., Du, R., Lv, P., Ma, X. *et al.* (2017) Construction of a high-density genetic map using specific-locus amplified fragments in sorghum. *BMC Genomics* 18(1), 51. Available at: https://doi.org/10.1186/s12864-016-3430-7
Jiao, Y., Burke, J., Chopra, R., Burow, G., Chen, J. *et al.* (2016) A sorghum mutant resource as an efficient platform for gene discovery in grasses. *The Plant Cell* 28(7), 1551–1562. Available at: https://doi.org/10.1105/tpc.16.00373
Jordan, D.R., Mace, E.S., Cruickshank, A.W., Hunt, C.H. and Henzell, R.G. (2011) Exploring and exploiting genetic variation from unadapted sorghum germplasm in a breeding program. *Crop Science* 51(4), 1444–1457. Available at: https://doi.org/10.2135/cropsci2010.06.0326
Jordan, D.R., Hunt, C.H., Cruickshank, A.W., Borrell, A.K. and Henzell, R.G. (2012) The relationship between the stay-green trait and grain yield in elite sorghum hybrids grown in a range of environments. *Crop Science* 52(3), 1153–1161. Available at: https://doi.org/10.2135/cropsci2011.06.0326
Kadier, Y., Zu, Y.-Y., Dai, Q.-M., Song, G., Lin, S.-W. *et al.* (2017) Genome-wide identification, classification and expression analysis of NAC family of genes in sorghum [*Sorghum bicolor* (L.) Moench]. *Plant Growth Regulation* 83(2), 301–312. Available at: https://doi.org/10.1007/s10725-017-0295-y
Kirby, J.S. and Atkins, R.E. (1968) Heterotic response for vegetative and mature plant characters in grain sorghum, *Sorghum bicolor* (L.) Moench. *Crop Science* 8(3), 335–339. Available at: https://doi.org/10.2135/cropsci1968.0011183x000800030022x
Klein, R.R., Mullet, J.E., Jordan, D.R., Miller, F.R., Rooney, W.L. *et al.* (2008) The effect of tropical sorghum conversion and inbred development on genome diversity as revealed by high-resolution genotyping. *Crop Science* 48 (S1), S12–S26.
Klein, R.R., Miller, F.R., Dugas, D.V., Brown, P.J., Burrell, A.M. and Klein, P.E. (2015) Allelic variants in the *PRR37* gene and the human-mediated dispersal and diversification of sorghum. *Theoretical and Applied Genetics* 128(9), 1669–1683.
Knoll, J., Gunaratna, N. and Ejeta, G. (2008) QTL analysis of early-season cold tolerance in sorghum. *Theoretical and Applied Genetics* 116(4), 577–587. Available at: https://doi.org/10.1007/s00122-007-0692-0
Kong, W., Jin, H., Franks, C.D., Kim, C., Bandopadhyay, R. *et al.* (2013) Genetic analysis of recombinant inbred lines for *Sorghum bicolor* × *Sorghum propinquum*. *G3: Genes, Genomes, Genetics* 3(1), 101–108. Available at: https://doi.org/10.1534/g3.112.004499
Krishnamurthy, L., Dinakaran, E., Kumar, A.A. and Reddy, B.V.S. (2014) Field technique and traits to assess reproductive stage cold tolerance in sorghum (*Sorghum bicolor* (L.) Moench. *Plant Production Science* 17(3), 218–227. Available at: https://doi.org/10.1626/pps.17.218
Kuriyama, T., Shimada, S. and Matsui, M. (2019) Improvement of agrobacterium-mediated transformation for tannin-producing sorghum. *Plant Biotechnology* 36(1), 43–48. Available at: https://doi.org/10.5511/plantbiotechnology.19.0131a
Leon-Velasco, H., Mendoza-Onofre, L.E., Castillo-Gonzalez, F., Cervantes-Santana, T. and Martínez-Garza, A. (2009) Evaluation of two generations of cold tolerant sorghum hybrids and parental lines. II: Combining ability, heterosis and heterobeltiosis. *Agrociencia* 43(6), 609–623.
Liu, H., Wang, X., Warburton, M.L., Wen, W., Jin, M. *et al.* (2015) Genomic, transcriptomic, and phenomic variation reveals the complex adaptation of modern maize breeding. *Molecular Plant* 8(6), 871–884. Available at: https://doi.org/10.1016/j.molp.2015.01.016
Liu, G., Gilding, E.K., Kerr, E.D., Schulz, B.L., Tabet, B., Hamaker, B.R. and Godwin, I.D. (2019) Increasing protein content and digestibility in sorghum grain with a synthetic biology approach. *Journal of Cereal Science* 85, 27–34. Available at: https://doi.org/10.1016/j.jcs.2018.11.001
Mace, E.S., Rami, J.-F., Bouchet, S., Klein, P.E., Klein, R.R. *et al.* (2009) A consensus genetic map of sorghum that integrates multiple component maps and high-throughput Diversity Array Technology (DArT) markers. *BMC Plant Biology* 9(1), 13. Available at: https://doi.org/10.1186/1471-2229-9-13

Mace, E.S., Tai, S., Gilding, E.K., Li, Y., Prentis, P.J. *et al.* (2013) Whole-genome sequencing reveals untapped genetic potential in Africa's indigenous cereal crop sorghum. *Nature Communications* 4(1), 2320. Available at: https://doi.org/10.1038/ncomms3320

Mace, E., Innes, D., Hunt, C., Wang, X., Tao, Y. *et al.* (2019) The Sorghum QTL Atlas: a powerful tool for trait dissection, comparative genomics and crop improvement. *Theoretical and Applied Genetics* 132(3), 751–766. Available at: https://doi.org/10.1007/s00122-018-3212-5

Makita, Y., Shimada, S., Kawashima, M., Kondou-Kuriyama, T., Toyoda, T. and Matsui, M. (2015) MOROKOSHI: transcriptome database in *Sorghum bicolor*. *Plant & Cell Physiology* 56(1), e6. Available at: https://doi.org/10.1093/pcp/pcu187

Mann, J.A., Kimber, C.T. and Miller, F.R. (1983) *The Origin and Early Cultivation of Sorghums in Africa*. Bulletin No. 1454. Texas Agricultural Experiment Station, Texas A&M University, College Station, Texas. Available at: http://hdl.handle.net/1969.1/128074 (accessed 3 March 2021).

Marla, S.R., Burow, G., Chopra, R., Hayes, C., Olatoye, M.O. *et al.* (2019) Genetic architecture of chilling tolerance in sorghum dissected with a nested association mapping population. *bioRxiv* 130, 622894. Available at: https://doi.org/10.1101/622894

Matsuoka, Y., Vigouroux, Y., Goodman, M.M., Sanchez, G.J., Buckler, E. and Doebley, J. (2002) A single domestication for maize shown by multilocus microsatellite genotyping. *Proceedings of the National Academy of Sciences USA* 99(9), 6080–6084. Available at: https://doi.org/10.1073/pnas.052125199

Maulana, F. and Tesso, T.T. (2013) Cold temperature episode at seedling and flowering stages reduces growth and yield components in sorghum. *Crop Science* 53(2), 564–574.

Mendoza, O.L.E. (1988) Formación de híbridos de sorgo para grano: II. Comportamiento per se de las líneas y su aptitud combinatoria general. *Revista Fitotecnia Mexicana* 11, 39–47.

Menz, M.A., Klein, R.R., Unruh, N.C., Rooney, W.L., Klein, P.E. and Mullet, J.E. (2004) Genetic diversity of public inbreds of sorghum determined by mapped AFLP and SSR markers. *Crop Science* 44(4), 1236–1244. Available at: https://doi.org/10.2135/cropsci2004.1236

Monk, R., Franks, C., Dahlberg, J., Smith, S., Diers, B., Specht, J. and Carver, B. (2014) Sorghum. In: Smith, S., Diers, B., Specht, J. and Carver, B. (eds) *Yield Gains in Major U.S. Field Crops*. CSSA Special Publication, Vol. 33. American Society of Agronomy, Crop Science Society of America and Soil Science Society of America, Madison, Wisconsin, pp. 293–310. Available at: https://doi.org/10.2135/cssaspecpub33.c11

Morris, G.P., Ramu, P., Deshpande, S.P., Hash, C.T., Shah, T. *et al.* (2013) Population genomic and genome-wide association studies of agroclimatic traits in sorghum. *Proceedings of the National Academy of Sciences USA* 110(2), 453–458. Available at: https://doi.org/10.1073/pnas.1215985110

Multani, D.S., Briggs, S.P., Chamberlin, M.A., Blakeslee, J.J., Murphy, A.S. and Johal, G.S. (2003) Loss of an MDR transporter in compact stalks of maize *br2* and sorghum *dw3* mutants. *Science* 302(5642), 81–84. Available at: https://doi.org/10.1126/science.1086072

Muraya, M.M. (2014) Sorghum genetic diversity. In: Wang, Y.-H., Upadhyaya, H.D. and Kole, C. (eds) *Genetics, Genomics and Breeding of Sorghum*. CRC Press, Boca Raton, Florida, pp. 136–162.

Nelson, J.C., Wang, S., Wu, Y., Li, X., Antony, G., White, F.F. and Yu, J. (2011) Single-nucleotide polymorphism discovery by high-throughput sequencing in sorghum. *BMC Genomics* 12, 352. Available at: https://doi.org/10.1186/1471-2164-12-352

Osuna-Ortega, J., Mendoza-Onofre, L.E., González-Hernández, V.A., Castillo-González, F., del Mendoza-Castillo, M.C. and Williams-Alanís, H. (2000) Potential of cold tolerant germplasm in the adaptation and adaptability of sorghum in México: I. High valleys. *Agrociencia* 34(5), 561–572.

Osuna-Ortega, J., Mendoza-Castillo, M.D.C. and Mendoza-Onofre, L. (2003) Sorghum cold tolerance, pollen production. *Maydica* 48, 125–132.

Oyediran, I.O., Hibbard, B.E. and Clark, T.L. (2004) Prairie grasses as hosts of the western corn rootworm (Coleoptera: Chrysomelidae). *Environmental Entomology* 33(3), 740–747. Available at: https://doi.org/10.1603/0046-225X-33.3.740

Parra-Londono, S., Fiedler, K., Kavka, M., Samans, B., Wieckhorst, S., Zacharias, A. and Uptmoor, R. (2018) Genetic dissection of early-season cold tolerance in sorghum: genome-wide association studies for seedling emergence and survival under field and controlled environment conditions. *Theoretical and Applied Genetics* 131(3), 581–595. Available at: https://doi.org/10.1007/s00122-017-3021-2

Patane, C., Cavallaro, V., Avola, G. and D'Agosta, G. (2006) Seed respiration of sorghum [*Sorghum bicolor* (L.) Moench] during germination as affected by temperature and osmoconditioning. *Seed Science Research* 16(4), 251–260.

Paterson, A.H., Bowers, J.E., Bruggmann, R., Dubchak, I., Grimwood, J. *et al.* (2009) The *Sorghum bicolor* genome and the diversification of grasses. *Nature* 457(7229), 551–556. Available at: https://doi.org/10.1038/nature07723

Peacock, J.M. (1982) Response and tolerance of sorghum to temperature stress. In: Mertin, J.V. (ed.) *Sorghum in the Eighties: Proceedings of the International Symposium on Sorghum, ICRISAT Center, Patancheru, India, 2–7 November 1981*. International Crops Research Institute for the Semi-Arid Tropics, Patancheru, India, pp. 143–159.

Peng, Y., Schertz, K.F., Cartinhour, S. and Hart, G.E. (1999) Comparative genome mapping of *Sorghum bicolor* (L.) Moench using an RFLP map constructed in a population of recombinant inbred lines. *Plant Breeding* 118(3), 225–235. Available at: https://doi.org/10.1046/j.1439-0523.1999.118003225.x

Pérez-de-Castro, A.M., Vilanova, S., Cañizares, J., Pascual, L., Blanca, J.M. *et al.* (2012) Application of genomic tools in plant breeding. *Current Genomics* 13(3), 179–195. Available at: https://doi.org/10.2174/138920212800543084

Pinthus, M.J. and Rosenblum, J. (1961) Germination and seedling emergence of sorghum at low temperatures. *Crop Science* 1, 293–296.

Promkhambut, A., Polthanee, A., Akkasaeng, C. and Younger, A. (2011) Growth, yield and aerenchyma formation of sweet and multipurpose sorghum (*Sorghum bicolor* L. Moench) as affected by flooding at different growth stages. *Australian Journal of Crop Science* 5(8), 954–965.

Quinby, J.R. (1966) Fourth maturity gene locus in sorghum. *Crop Science* 6(6), 516–518. Available at: https://doi.org/10.2135/cropsci1966.0011183X000600060005x

Quinby, J.R. (1974) *Sorghum Improvement and the Genetics of Growth*. Texas A&M University Press, College Station, Texas.

Quinby, J.R. and Karper, R.E. (1945) The inheritance of three genes that influence time of floral initiation and maturity date in Milo. *Agronomy Journal* 37(11), 916–936. Available at: https://doi.org/10.2134/agronj1945.00021962003700110006x

Rakshit, S. and Bellundagi, A. (2019) Conventional breeding techniques in sorghum. In: Aruna, C., Visarada, K.B.R.S., Venkatesh Bhat, B. and Tonapi, V.A. (eds) *Breeding Sorghum for Diverse End Uses*. Woodhead Publishing, Cambridge, UK, pp. 77–91.

Reif, J.C., Hallauer, A.R. and Melchinger, A.E. (2005) Heterosis and heterotic patterns in maize. *Maydica* 50(3/4), 215.

Salas Fernandez, M.G., Schoenbaum, G.R. and Goggi, A.S. (2014) Novel germplasm and screening methods for early cold tolerance in sorghum. *Crop Science* 54(6), 2631–2638.

Schaffasz, A., Windpassinger, S., Friedt, W., Snowdon, R. and Wittkop, B. (2019a) Sorghum as a novel crop for central Europe: using a broad diversity set to dissect temperate-adaptation. *Agronomy* 9(9), 535. Available at: https://doi.org/10.3390/agronomy9090535

Schaffasz, A., Windpassinger, S., Snowdon, R. and Wittkop, B. (2019b) Reproductive cold stress tolerance in sorghum $F_1$ hybrids is a heterotic trait. *Agronomy* 9(9), 508. Available at: https://doi.org/10.3390/agronomy9090508

Schnable, P.S. and Springer, N.M. (2013) Progress toward understanding heterosis in crop plants. *Annual Review of Plant Biology* 64, 71–88.

Shao, H., Wang, H. and Tang, X. (2015) NAC transcription factors in plant multiple abiotic stress responses: progress and prospects. *Frontiers in Plant Science* 6, 902. Available at: https://doi.org/10.3389/fpls.2015.00902

Singh, H.P. and Lohithaswa, H.C. (2006) Sorghum. In: Kole, C. (ed.) *Cereals and Millets*. Springer, Berlin/Heidelberg, Germany, pp. 257–302.

Singh, S.P. (1985) Sources of cold tolerance in grain sorghum. *Canadian Journal of Plant Science* 65(2), 251–257. Available at: https://doi.org/10.4141/cjps85-037

Sleper, D.A. and Poehlman, J.M. (2006) *Breeding Field Crops*. Blackwell Publishing, Oxford, UK.

Smith, C.W. and Frederiksen, R.A. (2000) *Sorghum: Origin, History, Technology, and Production*. Wiley, New York.

Smýkal, P., Varshney, R.K., Singh, V.K., Coyne, C.J., Domoney, C., Kejnovský, E. and Warkentin, T., (2016) From Mendel's discovery on pea to today's plant genetics and breeding: commemorating the 150th anniversary of the reading of Mendel's discovery. *Theoretical and Applied Genetics* 129(12), 2267–2280. Available at: https://doi.org/10.1007/s00122-016-2803-2

Staggenborg, S.A., Dhuyvetter, K.C. and Gordon, W.B. (2008) Grain sorghum and corn comparisons: yield, economic, and environmental responses. *Agronomy Journal* 100(6), 1600–1604. Available at: https://doi.org/10.2134/agronj2008.0129

Stephens, J.C. and Holland, R.F. (1954) Cytoplasmic male sterility for hybrid sorghum seed production. *Agronomy Journal* 46(1), 20–23.

Stephens, J.C., Miller, F.R. and Rosenow, D.T. (1967) Conversion of alien sorghums to early combine genotypes. *Crop Science* 7(4), 396. Available at: https://doi.org/10.2135/cropsci1967.0011183x000700040036x

Tao, Y.Z., Jordan, D.R., Henzell, R.G. and McIntyre, C.L. (1998) Construction of a genetic map in a sorghum recombinant inbred line using probes from different sources and its comparison with other sorghum maps. *Australian Journal of Agricultural Research* 49(4), 729–736. Available at: https://doi.org/10.1071/a97112

Thurber, C.S., Ma, J.M., Higgins, R.H. and Brown, P.J. (2013) Retrospective genomic analysis of sorghum adaptation to temperate-zone grain production. *Genome Biology* 14(6), R68. Available at: https://doi.org/10.1186/gb-2013-14-6-r68

Tian, T., You, Q., Zhang, L., Yi, X., Yan, H., Xu, W. and Su, Z. (2016) SorghumFDB: sorghum functional genomics database with multidimensional network analysis. *Database* 2016, baw099. Available at: https://doi.org/10.1093/database/baw099

Trachsel, S., Dhliwayo, T., Gonzalez Perez, L., Mendoza Lugo, J.A. and Trachsel, M. (2019) Estimation of physiological genomic estimated breeding values (PGEBV) combining full hyperspectral and marker data across environments for grain yield under combined heat and drought stress in tropical maize (*Zea mays* L.). *PLoS ONE* 14(3), e0212200. Available at: https://doi.org/10.1371/journal.pone.0212200

Varshney, R.K., Terauchi, R. and McCouch, S.R. (2014) Harvesting the promising fruits of genomics: applying genome sequencing technologies to crop breeding. *PLoS Biology* 12(6), e1001883. Available at: https://doi.org/10.1371/journal.pbio.1001883

Velazco, J.G., Jordan, D.R., Mace, E.S., Hunt, C.H., Malosetti, M. and van Eeuwijk, F.A. (2019) Genomic prediction of grain yield and drought-adaptation capacity in sorghum is enhanced by multi-trait analysis. *Frontiers in Plant Science* 10, 997. Available at: https://doi.org/10.3389/fpls.2019.00997

Venkateswaran, K., Muraya, M., Dwivedi, S.L. and Upadhyaya, H.D. (2014) Wild sorghums – their potential use in crop improvement. In: Wang, Y.-H., Upadhyaya, H.D. and Kole, C. (eds) *Genetics, Genomics and Breeding of Sorghum*. CRC Press, Boca Raton, Florida, pp. 78–111.

Voss-Fels, K.P., Stahl, A., Wittkop, B., Lichthardt, C., Nagler, S. *et al.* (2019) Breeding improves wheat productivity under contrasting agrochemical input levels. *Nature Plants* 5(7), 706–714. Available at: https://doi.org/10.1038/s41477-019-0445-5

Wesseler, J. and Fall, E.H. (2010) Potential damage costs of *Diabrotica virgifera virgifera* infestation in Europe – the 'no control' scenario. *Journal of Applied Entomology* 134(5), 385–394. Available at: https://doi.org/10.1111/j.1439-0418.2010.01510.x

Windpassinger, S. (2016) Breeding strategies for the adaptation of sorghum (*Sorghum bicolor* L. Moench) as a novel crop for temperate Europe. PhD thesis, Justus Liebig University, Giessen, Germany.

Windpassinger, S., Friedt, W., Frauen, M., Snowdon, R. and Wittkop, B. (2015) Designing adapted sorghum silage types with an enhanced energy density for biogas generation in temperate Europe. *Biomass and Bioenergy* 81(3), 496–504. Available at: https://doi.org/10.1016/j.biombioe.2015.08.005

Woldesemayat, A.A., Modise, D.M., Gemeildien, J., Ndimba, B.K. and Christoffels, A. (2018) Cross-species multiple environmental stress responses: an integrated approach to identify candidate genes for multiple stress tolerance in sorghum (*Sorghum bicolor* (L.) Moench) and related model species. *PLoS ONE* 13(3), e0192678. Available at: https://doi.org/10.1371/journal.pone.0192678

Xin, Z., Wang, M.L., Barkley, N.A., Burow, G., Franks, C., Pederson, G. and Burke, J. (2008) Applying genotyping (TILLING) and phenotyping analyses to elucidate gene function in a chemically induced sorghum mutant population. *BMC Plant Biology* 8(1), 103. Available at: https://doi.org/10.1186/1471-2229-8-103

Xu, J.C., Weerasuriya, Y.M. and Bennetzen, J.L. (2001) Construction of genetic map in sorghum and fine mapping of the germination stimulant production gene response to *Striga asiatica*. *Yi Chuan Xue Bao* 28(9), 870–876.

Yellisetty, V., Reddy, L.A. and Mandapaka, M. (2015) *In planta* transformation of sorghum (*Sorghum bicolor* (L.) Moench) using *TPS1* gene for enhancing tolerance to abiotic stresses. *Journal of Genetics* 94(3), 425–434. Available at: https://doi.org/10.1007/s12041-015-0540-y

Yonemaru, J.-I., Ando, T., Mizubayashi, T., Kasuga, S., Matsumoto, T. and Yano, M. (2009) Development of genome-wide simple sequence repeat markers using whole-genome shotgun sequences of sorghum (*Sorghum bicolor* (L.) Moench). *DNA Research* 16(3), 187–193. Available at: https://doi.org/10.1093/dnares/dsp005

Yu, J. and Tuinstra, M.R. (2001) Genetic analysis of seedling growth under cold temperature stress in grain sorghum. *Crop Science* 41(5), 1438–1443. Available at: https://doi.org/10.2135/cropsci2001.4151438x

Yuan, Y., Cairns, J.E., Babu, R., Gowda, M., Makumbi, D. *et al.* (2018) Genome-wide association mapping and genomic prediction analyses reveal the genetic architecture of grain yield and flowering time under drought and heat stress conditions in maize. *Frontiers in Plant Science* 9, 1919. Available at: https://doi.org/10.3389/fpls.2018.01919

Zhu, H., Jeoung, J.M., Liang, G.H., Muthukrishnan, S., Krishnaveni, S. and Wilde, G. (1998) Biolistic transformation of sorghum using a rice chitinase gene [*Sorghum bicolor* (L.) Moench–*Oryza sativa* L.]. *Journal of Genetics & Breeding* 52, 243–252.

# 27 Isolation of Quantitative Trait Loci/Gene(s) Conferring Cadmium Tolerance in Sorghum

**Fırat Kurt[1]* and Ertugrul Filiz[2]***
*[1]Muş Alparslan University, Muş, Turkey; [2]Duzce University, Duzce, Turkey*

## 27.1 Introduction

Sorghum [*Sorghum bicolor* (L.) Moench] is one of the most important crops cultivated in many regions of the world (Billot *et al.*, 2013) and is used as fodder, staple food, and bioenergy crop (Mall *et al.*, 2011). Sorghum is a highly efficient photosynthetic $C_4$ plant (Kumar, 2016) with an extensively diverged genome (Satish *et al.*, 2016), facilitating its cultivation in tropical, subtropical and temperate zones where abiotic stresses such as drought and high temperature restrict the yield potential of many crops (Gill *et al.*, 2014).

Sorghum has 25 species which are clustered into five main sections as follows: *Eusorghum*, *Chaetosorghum*, *Heterosorghum*, *Parasorghum* and *Stiposorghum*. The chromosome numbers of these sorghum species are $2n = 10, 20, 30$ or $40$ with varied genome sizes (Price *et al.*, 2005). *Eusorghum* section involves *Sorghum bicolor* with its subspecies of *arundinacum* and *drummondi* (Price *et al.*, 2005). To make a simpler classification of *S. bicolor* genotypes and accessions, they were classified into a total of 15 races, five basic races and ten intermediate races, in terms of their spikelet and head types. The basic races were *bicolor* (B), *caudatum* (C), *durra* (D), *guinea* (G) and *kafir* (K), whereas the combinations of basic races were named as follows: *guinea–bicolor* (GB), *caudatum–bicolor* (CB), *kafir–bicolor* (KB), *durra–bicolor* (DB), *guinea–caudatum* (GC), *guinea–kafir* (GK), *guinea–durra* (GD), *kafir–caudatum* (KC), *durra–caudatum* (DC) and *kafir–durra* (KD) (Harlan and de Wet, 1972).

Sorghum has a small genome amounting to one-quarter of the maize genome (Kumar, 2016) and because of this relatively small genome size, it is used as a model plant for cereal genomes (Paterson *et al.*, 2009). Due to this property, apart from its agricultural and industrial uses for fibre, fuel, feed and food, sorghum is a very valuable plant for studies aiming to answer questions on the function of genes and the structure of plant genomes in terms of physiology, genetics, evolution, etc. For this aim, DNA marker technology is widely used for the identification of genetic variation among or within genotypes (Satish *et al.*, 2016). Moreover, the recently developed microarray and RNA-seq technologies enable new gene discoveries, identification of co-expressed genes within the sorghum genome and identification of differentially expressed genes (DEGs) in plants exposed to various stressors (Rao *et al.*, 2018). In addition, sorghum has been used in phytoremediation studies to clean

*Corresponding authors: ertugrulfiliz@gmail.com, firat.kurt.usa@gmail.com


DOI: 10.1079/9781789245431.0027

up polluted areas including those affected by Cd. Its high biomass and high level of accumulation render it an attractive plant for such uses (Feng *et al.*, 2018). Therefore, this chapter aims to provide valuable insights into Cd-tolerance genes in sorghum. The term 'sorghum' is used in lieu of *S. bicolor* throughout the chapter.

## 27.2 A Brief Outlook on Sorghum Genome

Sorghum was suggested to be cultivated in 6000 BC, but its domestication (from *Sorghum verticilliflorum* to *S. bicolor*) took place around 5000 BC in central eastern Sudan (Smith *et al.*, 2019). The patterns of repetitive sequences and genes in sorghum are associated with homologous chromosomes despite a considerable number of replacements of repetitive elements (Paterson *et al.*, 2009).

*S. bicolor* has ten haploid chromosomes with about 800 Mbp genome (Price *et al.*, 2005). McCormick *et al.* (2018) reported that alignment of re-sequenced sorghum genotypes to the BTx623 (version 3) reference sorghum genome resulted in identification of 7,375,006 single-nucleotide polymorphisms (SNPs) and 1,876,974 insertion/deletions (InDels) on ten chromosomes and in annotation of 34,211 genes. The positions of nucleosomes in sorghum, like in *Arabidopsis* and maize, is located downstream of the transcription start sites of genes.

In plants, heterochromatic regions of genomes are either centromeric or pericentromeric. The heterochromatic segments are locations where long terminal repeat (LTR) retrotransposons (RTs) accumulate and genetic recombination is low (Tian *et al.*, 2009). The euchromatic region of sorghum genome is observed on distal segments of chromosomes encoding approximately 70% of the sorghum genes, whereas the heterochromatic region contains approximately threefold lower gene density in the pericentromic region. The synteny in sorghum was found high in euchromatic segments with low abundance of retrotransposable elements and low in heterochromatic segments with high abundance of RTs (Kim *et al.*, 2005; Paterson *et al.*, 2009; McCormick *et al.*, 2018). In terms of arrangement of heterochromatic and euchromatic regions on chromosomes, sorghum is similar to tomato (Kim *et al.*, 2005).

Sorghum, maize (*Zea mays*), sugarcane (*Saccharum officinarum*), and millet (*Pennisetum glaucum*) share rice (*Oryza sativa*) as a common ancestor from which they diverged 50 million years ago (Kim *et al.*, 2005). Leiboff and Hake (2019) and Paterson (2008) suggested that the divergence of maize and sorghum happened much later, approximately 12–16 million years ago. Comparison of the genomes of sorghum and rice showed that sorghum has twofold larger genome size with two fewer chromosomes than rice (Price *et al.*, 2005). Sorghum is closer to rice compared with maize in terms of *Copia-* and *Gypsy*-like RTs. The expansion of sorghum genome compared with rice is associated with LTR-RTs. (Paterson *et al.*, 2009). Homologous recombination (UR) and illegitimate (non-homologous) recombination are suggested to have deleted the ~190 Mb of LTR-RTs from the rice genome during evolution (Tian *et al.*, 2009). DNA transposable elements in sorghum, prominent components of genomes providing allelic diversity, were found to be 7.5%, a value intermediate between maize (2.7%) and rice (13.7%) (Paterson *et al.*, 2009). Also, gene duplications recently observed in sorghum are suggested to stem from the ancestral gene family or from a new transposon family found in sorghum. The genome-wide comparison of sorghum, maize and rice SNPs showed that tandem duplications in the euchromatic segment in sorghum and maize induce more SNP accumulation than gene duplications. Thus, the lack of genomic duplication is compensated and genomic innovations for variation are induced (Guo, H. *et al.*, 2019). More SNPs are reported to be found at the end regions of chromosomes, probably due to the repeating sequences near centromeres (Luo *et al.*, 2014). A high quantity of non-synonymous coding SNPs in the sorghum genome indicates probable changes of gene function (Guo, H. *et al.*, 2019).

## 27.3 Cadmium Transport and Tolerance in Plants

Heavy metals or metalloids have five times higher atomic weight and density than water. As it is detrimental to human health, Cd is a toxic substance and harmful to plant life even at very low concentrations, having no known biological function (Hossain *et al.*, 2012). Cd is one of the redox-inactive heavy metals like Hg, Zn, Ni, Al and Pb. It consumes antioxidants and inhibits thiol-containing enzymes and antioxidants (Ercal and Aykin-Burns, 2001). On the other hand, heavy metals such as Fe, Cu, Cr and Co participate in cellular oxidation–reduction reactions. These are redox-active heavy metals whose products are superoxide anion ($O_2^-$) and hydrogen peroxide ($H_2O_2$), as a result of Haber–Weiss and Fenton reactions (Ercal and Aykin-Burns, 2001; Hossain *et al.*, 2012). Cd induces lipoxygenase activity (Hossain *et al.*, 2012) and damages many vital organs, cellular organelles and pathways in plant metabolism (i.e. leaves, chloroplasts, stomata, photosynthesis, lipid peroxidation, N and S metabolism) (Flores-Cáceres *et al.*, 2015; Ibrahim *et al.*, 2017). Cd toxicity, in terms of mechanism of action, induces misfolding of proteins and inactivates enzymes by binding to sulfhydryl groups of structural proteins and enzymes (Hossain *et al.*, 2012). The Fe and Cd toxicity symptoms resemble each other, with chlorosis, necrosis, wilting, red-orange leaves and decline in plant growth (Lindberg and Greger, 2002).

### 27.3.1 Cadmium uptake genes in roots

Cd bioavailability in soil is controlled by several factors including redox potential of the soil, essential trace elements (Sebastian and Prasad, 2014), soil pH and structure, soil organic matter and chemical speciation (Verbruggen *et al.*, 2009a). Cd, like other heavy metals, increases rigidity of cell walls as a result of lignification. Thus root expansion is inhibited, and plants reach less solute and water (Cseh, 2002). The reduced water depending on Cd accumulation in cells induces drought stress and upregulates expression of the gene encoding DREB2A (dehydration-responsive element-binding protein 2A), a protein transcribed under drought and cold stress conditions (Tamás *et al.*, 2008), by bestowing tolerance to the mentioned stressors (Sakuma *et al.*, 2006). The water stress along with abscisic acid (ABA) triggers the expression of two subfamilies of aquaporins, PIP1 and PIP2 (plasma-membrane intrinsic proteins), regulating water passage from plasma membranes (Yaneff *et al.*, 2015). *HvPIP2;1* in barley is one of the aquaporins identified to be involved in water transport to cells, particularly to root cells (Katsuhara *et al.*, 2003).

Cd toxicity, like the toxicities of other heavy metals, induces accumulation of phytochelatins (PCs) in roots and shoots. These non-thiol proteins have been proposed as biomarkers for heavy metal toxicity (Keltjens and van Beusichem, 1998; Filiz *et al.*, 2019). The biosynthesis of PCs is carried out by phytochelatin synthase (PCS) in the presence of reduced glutathione (GSH) and thus phytotoxicities of Cd, Cu, and Zn are detoxified. The expression of *PCS1* in *Arabidopsis* and tobacco induces the transcription of PC, which results in Cd tolerance (Sofo *et al.*, 2013).

The exposure to plants of Cd activates heavy metal signalling molecules (i.e. NO and $Ca^{2+}$) and NADPH enzymes, and thereby increases production of reactive oxygen species (ROS) (Chmielowska-Bąk *et al.*, 2014). Pérez-Chaca *et al.* (2014) reported that NO and $H_2O_2$ are abundant in soybean roots under Cd toxicity. $H_2O_2$, as a main source of NADPH oxidase, induces the production of mitogen-activated protein kinases (MAPKs) (Kalbina and Strid, 2006). The expressions of *MAPK4* and *WRKY25* genes in roots and leaves of *Arabidopsis* is reported under Cd stress (Smeets *et al.*, 2013).

The uptake of Cd in roots occurs via transporter proteins. Like other divalent cations in the soil ($Fe^{2+}$, $Zn^{2+}$, $Ca^{2+}$ and $Mn^{2+}$), the bioavailable $Cd^{2+}$ is transported into roots by a ZIP family protein called iron-regulated transporter 1 (IRT1) (Vatansever *et al.*, 2015). Cd is taken up by AtIRT1 in *Arabidopsis*, and by OsIRT1 and OsIRT2 in rice (Verbruggen *et al.*, 2009a; Uraguchi and Fujiwara, 2012). The natural resistance-associated macrophage protein (Nramp) family is another protein family functioning in Cd transport in roots. AtNramp1, AtNramp3 and AtNramp4 increase $Cd^{2+}$ uptake in *Arabidopsis* and OsNramp1 (Uraguchi and Fujiwara, 2012), OsNramp2, OsNramp5 and OsHMA3 in different rice cultivars (Zhao, J. *et al.*, 2018). Apart from genes involved in Cd uptake, it is also important to note that some plants use an exclusion strategy to protect their cellular homeostasis against Cd. AtABCG36, a member of the ATP-binding cassette (ABC) transporter superfamily of proteins involved in pathogen defence, regulates Cd exclusion from root cells (Kang *et al.*, 2011).

### 27.3.2 Translocation of cadmium from roots to shoots

After Cd is taken up by roots, it is transported to shoots through xylem (Uraguchi and Fujiwara, 2012). Heavy metal translocation in plants is performed through two classes of heavy metal $P_{1B}$-ATPases (HMAs) (Zhiguo *et al.*, 2018). The cations inside cells are pumped out of the membranes by HMAs against an electrochemical gradient using ATP (Verbruggen *et al.*, 2009b; Zhou *et al.*, 2019). One HMA class is responsible for translocation of Cu and Ag, and the other class regulates Zn/Co/Cd/Pb transports (Zhiguo *et al.*, 2018). HMA2 and HMA4 in *Arabidopsis* (AtHMA2/4) are the Cd transporters involved in xylem loading of Cd (Verbruggen *et al.*, 2009a; Feng *et al.*, 2018). *HMA2* gene in rice (*OsHMA2*) is involved in Cd translocation to shoots and distribution of Cd in developing tissues through phloem, whereas *OsHMA3* takes part in the mechanism of Cd sequestration into vacuoles. The highly transcribed *OsHMA3*, at the same time, decreases Cd concentrations in shoot and xylem. Contrary to OsHMA3's function, *OPT3*, expressed in pericycle of *Thlaspi caerulescens* (*TcOPT3*), is reported to transport Cd long distances in plants and to have a role in hyperaccumulation of Cd in shoots (Hu *et al.*, 2012). Likewise, an oligopeptide transporter (OPT) superfamily protein, yellow stripe-like (YSL) in *Solanum nigrum* (SnYSL3), is also involved in Cd translocation. The translocation of Cd by SnYSL3 takes place in the form of Cd–nicotianamine complexes (Feng *et al.*, 2017). Like the YSL3, another peptide transporter in *Arabidopsis*, multidrug resistance protein (AtMRP3), also an ABC superfamily member, is identified as a Cd transporter involved in Cd transport. *AtMRP3*, upregulated in both roots and shoots under Cd stress, is induced just by Cd and no other inducers such as oxidative stress and PCs, and GSH have roles in its induction (Bovet *et al.*, 2003).

Another gene, *cation/calcium exchanger* (*CCX2*), was identified as a putative gene transporting Cd to shoots from roots in rice (Chen, J. *et al.*, 2019). The accumulation of Cd in grains takes place through phloem. Thiol (SH) compounds along with an unknown protein act as substrates for Cd binding. The passage of Cd from xylem to phloem happens via the nodes acting like bridges in rice (Uraguchi and Fujiwara, 2012). Shimo *et al.* (2011) suggested a new soluble vascular protein in rice named low Cd (LCD). *LCD* mutant plants are able to accumulate more Cd in their roots and less in shoots and grains. Likewise, *cadmium tolerance 1* (*CDT1*) gene, isolated from *Digitaria ciliaris*, a plant naturally growing in Cd-contaminated mining sites, increases Cd tolerance in rice (named as *OsCDT1*) and in transgenic *Arabidopsis*. *CDT1* transcriptions are cysteine (Cys)-rich peptides, identified as heavy metal chelators (Matsuda *et al.*, 2009). The constitutive expression of another class of Cys-rich proteins, metallothioneins, is also reported to increase tolerance against Cd in several plants (Yang *et al.*, 2015). From a QTL mapping study, Guo, J. *et al.* (2019) suggested that genes encoding the following proteins are involved in Cd translocation in rice: ZIP4 (ZRT/IRT-like protein), the protein similar to glutathione transferase (GSTs) 16, heat-shock protein Hsp20 domain-containing protein, MAP kinase-like protein and Cd tolerant protein 5.

### 27.3.3 Detoxification and sequestration of cadmium in plants

Plants detoxify Cd by employing several mechanisms including chelation, compartmentalization and ROS scavenging (Matsuda *et al.*, 2009; Chen, H. *et al.*, 2019). As mentioned above, AtHMA2/4, and pleiotropic drug resistance (PDR)-type ABC transporter (AtPDR8) proteins, regulate Cd extrusion from cells and thereby contribute to Cd detoxification strategy (Shimo *et al.*, 2011). The Cd influx from cytosol to organelles in plant root cells is suggested to induce ROS formation in the mitochondrial electron transport chain (Keunen *et al.*, 2011). As a defence mechanism, Cd is transported from mitochondria to cytosol in the form of Cd–GSH and Cd–S complexes by ABCB25 in *Arabidopsis* (Kang *et al.*, 2011). GSH is an important ligand in protecting and balancing cellular redox balance against metals and metalloids (Verbruggen *et al.*, 2009a). Similarly, Cd is excluded from mitochondria by mitochondrial membrane protein (AtATM3). AtATM3 transports Cd in the form of glutamine synthetase-conjugated $Cd^{2+}$ complexes (Kim *et al.*, 2006). Two vacuolar Fe transporters, OsVIT1 and OsVIT2, involved in Fe sequestration, also take part in transport of Cd along with Fe into vacuoles (Eroglu *et al.*, 2019). However, prior to importing Cd into vacuoles, the Cd, like Fe, has to be chelated with nicotianamine (NA), like Fe (Gao *et al.*, 2016).

The vacuolar sequestration of Cd is regulated by several genes such as those encoding HMA3, $H^+$/cation exchanger, and metallothionein-like proteins (MTPs) (Zhang *et al.*, 2019). PC generation in response to Cd toxicity, as mentioned above, is conducive to Cd sequestration into vacuoles; therefore, vacuolar Cd sequestration is not strictly dependent on the Cd. Two ABC member proteins, AtABCC1 and AtABCC2, transporting PCs, are also involved in transport of GSH-conjugated Cd complexes (Mendoza-Cózatl *et al.*, 2011).

Ligands like malate, citrate, metallothioneins and histidine also participate in Cd tolerance by forming metal–ligand complexes (Haydon and Cobbett, 2007; Verbruggen *et al.*, 2009b). Isarankura-Na-Ayudhya *et al.* (2018) suggested that the histidine complexation with Cd is a more efficient way to deal with Cd toxicity compared with prevention of Cd uptake at membrane surfaces. The malate- and citrate-treated rice seedlings under Cd toxicity neutralize the adverse effects of Cd at membrane surfaces by reversing downregulation of *OsNramp1*, *OsIRT1*, *OsHMA3*, and nicotianamine synthase (*OsNAS1*) in root cells. The protective effect of organic acids is due to their chelation with Cd, thereby inhibiting the phytotoxic effect of Cd which induces suppression of mentioned genes (Sebastian and Prasad, 2018).

## 27.4 Methods in Identification of Quantitative Trait Loci

QTLs are genes controlling genotypic or phenotypic traits of interest which are inherited quantitatively. There are two different approaches to identify QTLs in plants: (i) linkage mapping (LM); and (ii) association mapping (AM), also known as linkage disequilibrium (LD) mapping (Álvarez *et al.*, 2014; Verdeprado *et al.*, 2018; Alqudah *et al.*, 2019). AM is conducted with two methods: (i) candidate gene association analyses; and (ii) genome-wide analysis or genome-wide association studies (GWA or GWAS) (Álvarez *et al.*, 2014; Verdeprado *et al.*, 2018). In the candidate gene method, new gene discovery is not the aim. Instead, the focus is on the well-defined loci. A comprehensive prior knowledge about trait(s) of interest, in terms of biological pathway and gene ontology, is a must. On the other hand, the studies on GWA deal with more complex traits requiring a considerable number of markers distributed across whole chromosomes to be identified (Álvarez *et al.*, 2014). The most important parameter of AM is LD, which has to be known before starting a mapping study (Verdeprado *et al.*, 2018). LD is a correlation coefficient showing the strength of association between traits of interest at two or more loci. Gamete frequencies are used for the expression of LD and the population is assumed at equilibrium as long as there are random combinations of alleles of gametes. Therefore, LD shows co-inheritance of allelic variants of gametes since one allele is physically in the vicinity of another (Álvarez *et al.*, 2014). The closer the allelic variants, the higher is the LD. On the other hand, LD should not be interpreted in physical linkage since physically close allele genes may not be co-inherited because of unequal frequencies and recombination rates, and therefore have low LD (Verdeprado *et al.*, 2018).

In AM, the number of markers used in a study is related to which method will be used. GWAS require as many markers as possible to cover all chromosomes of the genome. In this respect, SNPs and InDel markers are used to find out which part of DNA is co-inherited with the trait(s) of interest in GWAS (Álvarez *et al.*, 2014; Verdeprado *et al.*, 2018).

LM and AM approaches have their own advantages and disadvantages (Álvarez *et al.*, 2014; Verdeprado *et al.*, 2018). The main differences between LM and AM can be explained as follows:

- Bi-parental populations such as $F_2$, backcrosses (BCs), recombinant inbred lines (RILs) and near-isogenic lines (NILs) are used in LM, whereas germplasm panels (i.e. populations, genotypes, accessions, etc.) are employed for AM.

- The control of breeders over genetic materials (genotypes or populations) is limited in the AM approach since the recombination events are not generated by the breeders. Consequently, AM involves many phenotypic variations controlled by alleles in populations. However, the recombination is created in progenies by crossing positive alleles versus negative alleles in LM.
- Generally, mapping resolution in the AM approach is higher than for LM. The resolution of LM contains a part of the chromosome, in which there are as many as several hundred genes (Álvarez *et al.*, 2014; Verdeprado *et al.*, 2018; Alqudah *et al.*, 2019).

## 27.5 Recent Quantitative Trait Loci Studies in Sorghum

Molecular markers such as restriction fragment length polymorphisms (RFLPs), amplified fragment length polymorphisms (AFLPs) and simple sequence repeats (SSRs) in the maize genome were used in identifying related gene loci of sorghum, sugarcane, foxtail millet and Johnson grass (Anami *et al.*, 2016). The detection of association between allelic variants and traits in the generation of sorghum QTL maps has become easier by next-generation sequencing (NGS) technology identifying SNPs within the genome (Kong *et al.*, 2018). Genotyping-by-sequencing (GBS), as a genotyping method of NGS, has given extra power for the discovery and accelerated scanning of SNPs in the sorghum genome (Hu *et al.*, 2019). In this respect, Gelli *et al.* (2017) used bi-parental populations, composed of 131 RILs, under normal (100 kg/ha) and low (0 kg/ha) N treatments to construct QTL mapping by GBS in sorghum. They reported that a total of 642 SNPs revealed 38 QTLs for 11 agronomic traits on chromosomes 1, 5, 6, 7 and 9 when the composite interval mapping method was used. These co-localized QTLs were suggested as potential pleiotropic regions associated with flowering time and plant height. The phenotypic variation caused by each QTL was stated to vary from 6.2 to 50.8%. In the study, a total of 726 DEGs was revealed in root tissues of sorghum seedlings as a result of RNA sequencing and 108 genes of these DEGs were associated with chromosomal zones near to QTLs. The DEGs reported were found to be associated with N metabolism (ferredoxin-nitrate reductase), glycolysis (phosphofructo-2-kinase), seed storage proteins, plant hormone metabolism and membrane transport. Moreover, DEGs were suggested by the authors as potential targets for sorghum genome improvement under N fertilization (Gelli *et al.*, 2017). In another study, sorghum RILs (*S. bicolor* × *Sorghum halepense*) and a panel of *S. bicolor* landraces were used to identify grain antioxidant QTL regions in the framework of GWAS (Habyarimana *et al.*, 2019). The newly produced RILs were reported to have higher antioxidant capacity with novel polymorphisms. GWA studies were stated to reveal 55 major QTLs explaining 15 to 31% of the observed variability in antioxidants. In the study, 26 novel markers were discovered. Of these, Chr1_61095994, a novel pleiotropic major-effect marker, was reported to be found in LD with loci of 19 putative *glutathione S-transferase* genes which function in sequestration of anthocyanins into vacuoles (Habyarimana *et al.*, 2019).

If breeding programmes achieve enhancement of bioenergy-related traits such as early-stage chilling tolerance, juvenile development and early maturity, sorghum may be an alternative crop for temperate regions (Schaffasz *et al.*, 2019). Schaffasz *et al.* (2019) scored a sorghum population ($n$ = 338) in terms of mentioned traits and used GWAS to confirm the presence of quantitative traits for cold tolerance and high bioenergy. As a result, they found that sorghum set has high genetic and phenotypic variations and satisfying heritability. Also, the results of GWAS indicated that sorghum genotypes in the set have high quantitative traits for cold tolerance and high biomass accumulation. Nevertheless, these outcomes were evaluated by the authors as major drawbacks for marker-assisted selection in terms of practical breeding (Schaffasz *et al.*, 2019).

To investigate the association of genomic variation with complex traits in *S. bicolor*, many data sets are constructed using GBS to find SNPs. Nevertheless, since reference genome coordinates of sorghum are not the same among data sets, it is not possible to reuse data for mapping studies (Hu *et al.*, 2019). Therefore, 459,304 SNPs were used to construct a data set for 10,323 sorghum genotypes on the version 3.1

reference genome by Hu *et al.* (2019). Also, the authors conducted GWAS to validate SNPs, mapping known genes, and identifying novel associations for traits of interest.

Lopez *et al.* (2017) conducted a study to identify QTLs associated with root depth, growing season length, and stomatal conductance to water vapour ($g_s$) using a high-density bin map. They identified new QTL regions for crown root angle and $g_s$ on chromosome 3 and 7, respectively, in sorghum. Also, they identified new candidate genes depending on physical locations of the QTLs.

In terms of several agronomic traits such as panicle compactness and shape, panicle exertion, etc., and disease resistance like smut resistance, 1425 Ethiopian landrace accessions were identified phenotypically by Girma *et al.* (2019) and 1341 accessions were crossed with the $A_1$ cytoplasmic male-sterile line, ATx623, to investigate fertility or sterility reactions of landraces. The authors reported that 72,190 of 879,407 SNPs, obtained by the GBS method, were identified as reliable SNPs after quality control and the markers thus obtained were used for GWAS using a compressed mixed linear model. The markers showed significant association with traits. These SNPs were stated to explain from 2 to 43% of total phenotypic variation. A *basic helix–loop–helix* (*bHLH*) transcription factor, aborted microspores, was reported to be associated with a gene regulating male fertility. Plant height in sorghum was found to be associated with *RAP2-7* and *clavata1* genes regulating plant growth and suppressing transition to flowering in sorghum, respectively. Similar to these findings, the authors reported that *yellow seed1-like*, an *MYB* transcription factor, and *tannin1* genes were significantly associated with pericarp colour (Girma *et al.*, 2019).

Kante *et al.* (2018) conducted a study to investigate QTLs for fertility restoration (*Rf*) in the $A_1$ cytoplasmic male sterility system in West African sorghum germplasm. For this aim, three $F_2$ populations and parental lines were used by the authors to identify SNPs according to the GBS method. Also, SNPs found within or flanking QTL regions were employed to develop competitive allele-specific PCR (KASP) markers. A total of seven QTLs were reported by the researchers. Two major $A_1$ cytoplasm fertility restoration loci, $Rf_2$ and $Rf_5$, were found on chromosomes SBI-02 and SBI-05, respectively. $Rf_5$ explained 19 and 14% of phenotypic variation in the male parent in any population, whereas 31% of the phenotypic variation in the third population was explained by $Rf_2$ (Kante *et al.*, 2018).

In total 616 GBS-based SNPs were used to construct a genetic map in a RIL population by Kong *et al.* (2018). They used an individual $F_1$ plant, obtained from *S. bicolor* BTx623 and IS3620, to derive $F_7$ and $F_8$ RILs, which constituted a mapping population of 399. As a result, 381 segregated positions were obtained. The authors stated that they were able to separate QTLs for plant height and flowering time and to present evidence opposing to pleiotropy.

A turf grass, seashore paspalum (*Paspalum vaginatum* Sw.), a close relative of sorghum and other cereal crops such as maize and fox tail millet, is able to survive in coastal areas where salt is the main restriction factor for many crops to survive, grow or develop. Therefore, understanding the survival mechanism of seashore paspalum may help to improve cereal crops for saline soils. GBS-based SNPs in a high-density genetic map in seashore paspalum showed that each chromosome of the halophytic plant corresponds to each sorghum chromosome with high collinearity (Qi *et al.*, 2019). Consequently, it is stated that the syntenic relationship between the two crops may also help gain insights into other cereals.

## 27.6 Cadmium-Related Quantitative Trait Locus Studies in Sorghum

Although sorghum is not a hyperaccumulator plant, it can be used to clean up areas which are moderately or partially polluted by heavy metals since it has high biomass production and metal accumulation potential (Zhuang *et al.*, 2009; Al Chami *et al.*, 2015) like maize and alfalfa (*Medicago sativa*) (Jia *et al.*, 2017). Sorghum is able to take up and accumulate Cd, Pb, and Zn in aboveground organs like the stem (Al Chami *et al.*, 2015). The Cd uptake, translocation, and accumulation show a wide range of differences among sorghum genotypes in parallel to its genetic diversity. The accumulator sorghum genotypes may translocate and accumulate four times more Cd than sensitive varieties, suggesting that sorghum has a Cd-tolerance mechanism (Jia *et al.*, 2017).

The genomic diversity of sorghum makes it a very special species providing understanding of the functions of grass genes, although the same principle offers a considerable challenge in the construction of linkage maps and development of a sorghum diversity research set (SDRS). In this regard, some authors have scanned numerous sorghum landraces to identify the best candidates for breeding of Cd accumulators and development of new sorghum accumulator varieties. With this aim, Tsuboi *et al.* (2017) used 38 SSR markers selected randomly from linkage maps to construct a QTL map associated with Cd uptake and to identify Cd accumulation in an SDRS which is a core collection of 107 geographically diverse sorghum landraces. The results of Cd accumulation by AM showed that the marker *SB0753* on chromosome 1 was a marker modulating Cd concentrations after 2 months from planting. Additionally, *SB3412* and *Xtxp104* on chromosome 6 were identified as markers associated with regulation of Cd concentrations in leaf blades, leaf sheaths, and panicles after following head emergence. The landraces used in the study were grouped by their Cd accumulation status and SDRS 33, 47, 48, 55, 95, 106 landraces, and BTx623 cultivar, were found as Cd accumulators. Particularly, SDRS 47 and 48 were suggested for phytoextraction programmes of Cd-polluted areas as they accumulate higher concentrations of Cd in their shoots (Tsuboi *et al.*, 2017). In another similar study, Abou-Elwafa *et al.* (2019) identified QTLs/genes associated with heavy metal tolerance in sorghum SDRS, comprised of 107 accessions, by AM. For this purpose, they used 181 SSR markers. Of these markers, 12 SSRs were associated with five traits such as plant height, leaf length and width, shoot dry and fresh weights, with 14 associations. QTLs found in the study were identified by BLASTP annotation and 19 of the annotated 102 genes (*SbLysMR1*, *SbPPR1*, *SbZFP346*, *SbPPR2*, *SbPPR3*, *SbPPR4*, *SbPPR5*, *SbZc3h18*, *SbLAC9*, *SbZFP8*, *SbMAPKK5*, *SbAVPL1*, *SbWAK2*, *SbPPR6*, *Sb7DGT*, *SbKCS5*, *SbAVP1*, *SbZFP6* and *SbPPR7*) were found to function in phytoremediation and heavy metal tolerance. More specifically, *xtxp315* marker on chromosome 2, *xtxp67* marker on chromosome 9 and *xtxp270* marker on chromosome 10 were associated with genes involved in heavy metal tolerance (Abou-Elwafa *et al.*, 2019).

Element accumulation in seeds is important for seedling establishment and enhancement of plant growth. Shakoor *et al.* (2016) investigated the seed ionome of 407 sorghum accessions genotyped by GBS and conducted a GWA study, to reveal associations between QTLs and traits of interest including Cd accumulation genes in sorghum. *Sobic.002G083000* and *Sobic.002G083100* genes, orthologous genes of *AtHMA*, were identified as candidate genes regulating seed Cd content.

Despite its Cd accumulation capacity, sorghum is not studied well enough to identify its QTLs/genes associated with Cd tolerance. However, Cd-related QTL studies conducted in close relatives of sorghum can be used for sorghum. In this respect, QTL regions found in maize (Soric *et al.*, 2009; Zdunić *et al.*, 2014; Zhao, X. *et al.*, 2018; Asaro, 2019) and rice (Ueno *et al.*, 2010; Sato *et al.*, 2011; Hu *et al.*, 2018; Zhao, J. *et al.*, 2018) can also be employed for sorghum. In doing so, the collinearity of QTLs/genes associated with Cd and robust LDs may facilitate identification or discovery of novel Cd-related genes in sorghum.

## 27.7 Conclusion

Sorghum has a wide range of genetic and germplasm diversity consisting of Cd-sensitive and Cd-tolerant genotypes. Agro-morphological studies have shown that some genotypes are more suitable for higher Cd accumulation in their above-ground organs; that is, phytoextraction. On the other hand, some sorghum genotypes have more desirable traits to reduce Cd mobility by keeping it at the root surface; that is, phytostabilization. Therefore, improvement of the sorghum genome in terms of Cd tolerance can be studied effectively in two different ways by AM studies. The current situation in sorghum has not been offering too much promise about Cd tolerance since there have been few linkage or association studies about the identification of Cd-tolerance genes in sorghum. The inclusion of more sorghum germplasm into Cd tolerance-related studies using NGS methods such as GBS will help gain more knowledge about sorghum genome organization and facilitate improvement of Cd-tolerant genotypes with breeding programmes.

## References

Abou-Elwafa, S.F., Amin, A.E.E.A.Z. and Shehzad, T. (2019) Genetic mapping and transcriptional profiling of phytoremediation and heavy metals responsive genes in sorghum. *Ecotoxicology and Environmental Safety* 173, 366–372. Available at: https://doi.org/10.1016/j.ecoenv.2019.02.022

Al Chami, Z., Amer, N., Al Bitar, L. and Cavoski, I. (2015) Potential use of *Sorghum bicolor* and *Carthamus tinctorius* in phytoremediation of nickel, lead and zinc. *International Journal of Environmental Science and Technology* 12(12), 3957–3970. Available at: https://doi.org/10.1007/s13762-015-0823-0

Alqudah, A.M., Sallam, A., Stephen Baenziger, P. and Börner, A. (2019) GWAS: fast-forwarding gene identification in temperate cereals: barley as a case study – a review. *Journal of Advanced Research* 22, 119–135. Available at: https://doi.org/10.1016/j.jare.2019.10.013

Álvarez, M.F., Mosquera, T. and Blair, M.W. (2014) The use of association genetics approaches in plant breeding. *Plant Breeding Reviews* 38(1), 17–68. Available at: https://doi.org/10.1002/9781118916865.ch02

Anami, S.E., Luo, H., Xia, Y. and Jing, H.-C. (2016) Sorghum genome mapping and its impact generated through public and private efforts. In: Rakshit, S. and Wang, Y.-H. (eds) *The Sorghum Genome*. Compendium of Plant Genomes. Springer, Cham, Switzerland, pp. 95–116. Available at: https://doi.org/10.1007/978-3-319-47789-3_5

Asaro, A. (2019) A combinatorial approach of ionomics, quantitative trait locus mapping and transcriptome analysis to characterize element homeostasis in maize. PhD thesis, Washington University in St. Louis, St. Louis, Missouri.

Billot, C., Ramu, P., Bouchet, S., Chantereau, J., Deu, M. *et al.* (2013) Massive sorghum collection genotyped with SSR markers to enhance use of global genetic resources. *PLoS One* 8(4), e0059714. Available at: https://doi.org/10.1371/journal.pone.0059714

Bovet, L., Eggmann, T., Meylan-Bettex, M., Polier, J., Kammer, P. *et al.* (2003) Transcript levels of *AtMRPs* after cadmium treatment: induction of *AtMRP3*. *Plant, Cell & Environment* 26(3), 371–381. Available at: https://doi.org/10.1046/j.1365-3040.2003.00968.x

Chen, H., Li, Y., Ma, X., Guo, L., He, Y., Ren, Z., Kuang, Z., Zhang, X. and Zhang, Z. (2019) Analysis of potential strategies for cadmium stress tolerance revealed by transcriptome analysis of upland cotton. *Scientific Reports* 9(1), 86. Available at: https://doi.org/10.1038/s41598-018-36228-z

Chen, J., Zou, W., Meng, L., Fan, X., Xu, G. and Ye, G. (2019) Advances in the uptake and transport mechanisms and QTLs mapping of cadmium in rice. *International Journal of Molecular Sciences* 20(14), 3417. Available at: https://doi.org/10.3390/ijms20143417

Chmielowska-Bąk, J., Gzyl, J., Rucińska-Sobkowiak, R., Arasimowicz-Jelonek, M. and Deckert, J. (2014) The new insights into cadmium sensing. *Frontiers in Plant Science* 5, 245. Available at: https://doi.org/10.3389/fpls.2014.00245

Cseh, E. (2002) Metal permeability, transport and efflux in plants. In: Prasad, M.N.V. and Strzałka, K. (eds) *Physiology and Biochemistry of Metal Toxicity and Tolerance in Plants*. Springer, Dordrecht, the Netherlands, pp. 1–36. Available at: https://doi.org/10.1007/978-94-017-2660-3_1

Ercal, N., Aykin-Burns, N. and Gurer-Orhan, H. (2001) Toxic metals and oxidative stress, Part I: mechanisms involved in metal induced oxidative damage. *Current Topics in Medicinal Chemistry* 1(6), 529–539. Available at: https://doi.org/10.2174/1568026013394831

Eroglu, S., Karaca, N., Vogel-Mikus, K., Kavčič, A., Filiz, E. and Tanyolac, B. (2019) The conservation of VIT1-dependent iron distribution in seeds. *Frontiers in Plant Science* 10, 907. Available at: https://doi.org/10.3389/fpls.2019.00907

Feng, J., Jia, W., Lv, S., Bao, H., Miao, F. *et al.* (2018) Comparative transcriptome combined with morpho-physiological analyses revealed key factors for differential cadmium accumulation in two contrasting sweet sorghum genotypes. *Plant Biotechnology Journal* 16(2), 558–571. Available at: https://doi.org/10.1111/pbi.12795

Feng, S., Tan, J., Zhang, Y., Liang, S., Xiang, S., Wang, H. and Chai, T. (2017) Isolation and characterization of a novel cadmium-regulated Yellow Stripe-Like transporter (SnYSL3) in *Solanum nigrum*. *Plant Cell Reports* 36(2), 281–296. Available at: https://doi.org/10.1007/s00299-016-2079-7

Filiz, E., Saracoglu, I.A., Ozyigit, I.I. and Yalcin, B. (2019) Comparative analyses of phytochelatin synthase (*PCS*) genes in higher plants. *Biotechnology & Biotechnological Equipment* 33(1), 178–194. Available at: https://doi.org/10.1080/13102818.2018.1559096

Flores-Cáceres, M.L., Hattab, S., Hattab, S., Boussetta, H., Banni, M. and Hernández, L.E. (2015) Specific mechanisms of tolerance to copper and cadmium are compromised by a limited concentration of glutathione in alfalfa plants. *Plant Science* 233, 165–173. Available at: https://doi.org/10.1016/j.plantsci.2015.01.013

Gao, L., Chang, J., Chen, R., Li, H., Lu, H., Tao, L. and Xiong, J. (2016) Comparison on cellular mechanisms of iron and cadmium accumulation in rice: prospects for cultivating Fe-rich but Cd-free rice. *Rice* 9(1), 39. Available at: https://doi.org/10.1186/s12284-016-0112-7

Gelli, M., Konda, A.R., Liu, K., Zhang, C., Clemente, T.E., Holding, D.R. and Dweikat, I.M. (2017) Validation of QTL mapping and transcriptome profiling for identification of candidate genes associated with nitrogen stress tolerance in sorghum. *BMC Plant Biology* 17(1), 123. Available at: https://doi.org/10.1186/s12870-017-1064-9

Gill, J.R., Burks, P.S., Staggenborg, S.A., Odvody, G.N., Heiniger, R.W. *et al.* (2014) Yield results and stability analysis from the sorghum regional biomass feedstock trial. *BioEnergy Research* 7(3), 1026–1034. Available at: https://doi.org/10.1007/s12155-014-9445-5

Girma, G., Nida, H., Seyoum, A., Mekonen, M., Nega, A. *et al.* (2019) A large-scale genome-wide association analyses of Ethiopian sorghum landrace collection reveal loci associated with important traits. *Frontiers in Plant Science* 10, 691. Available at: https://doi.org/10.3389/fpls.2019.00691

Guo, H., Jiao, Y., Tan, X., Wang, X., Huang, X., Jin, H. and Paterson, A.H. (2019) Gene duplication and genetic innovation in cereal genomes. *Genome Research* 29(2), 261–269. Available at: https://doi.org/10.1101/gr.237511.118

Guo, J., Li, K., Zhang, X., Huang, H., Huang, F. *et al.* (2019) Genetic properties of cadmium translocation from straw to brown rice in low-grain cadmium rice (*Oryza sativa* L.) line. *Ecotoxicology and Environmental Safety* 182, 109422. Available at: https://doi.org/10.1016/j.ecoenv.2019.109422

Habyarimana, E., Dall'Agata, M., De Franceschi, P. and Baloch, F.S. (2019) Genome-wide association mapping of total antioxidant capacity, phenols, tannins and flavonoids in a panel of *Sorghum bicolor* and *S. bicolor* × *S. halepense* populations using multi-locus models. *PLoS One* 14(12), e0225979. Available at: https://doi.org/10.1371/journal.pone.0225979

Harlan, J.R. and de Wet, J.M.J. (1972) A simplified classification of cultivated sorghum. *Crop Science* 12(2), 172–176. Available at: https://doi.org/10.2135/cropsci1972.0011183x001200020005x

Haydon, M.J. and Cobbett, C.S. (2007) Transporters of ligands for essential metal ions in plants. *New Phytologist* 174(3), 499–506. Available at: https://doi.org/10.1111/j.1469-8137.2007.02051.x

Hossain, M.A., Piyatida, P., da Silva, J.A.T. and Fujita, M. (2012) Molecular mechanism of heavy metal toxicity and tolerance in plants: central role of glutathione in detoxification of reactive oxygen species and methylglyoxal and in heavy metal chelation. *Journal of Botany* 2012, 872875. Available at: https://doi.org/10.1155/2012/872875

Hu, D.-W., Sheng, Z.-H., Li, Q.-L., Chen, W., Wei, X.-J. *et al.* (2018) Identification of QTLs associated with cadmium concentration in rice grains. *Journal of Integrative Agriculture* 17(7), 1563–1573. Available at: https://doi.org/10.1016/S2095-3119(17)61847-1

Hu, Y.T., Ming, F., Chen, W.W., Yan, J.Y., Xu, Z.Y. *et al.* (2012) TcOPT3, a member of oligopeptide transporters from the hyperaccumulator *Thlaspi caerulescens*, is a novel Fe/Zn/Cd/Cu transporter. *PLoS ONE* 7(6), e0038535. Available at: https://doi.org/10.1371/journal.pone.0038535

Hu, Z., Olatoye, M.O., Marla, S. and Morris, G.P. (2019) An integrated genotyping-by-sequencing polymorphism map for over 10,000 sorghum genotypes. *Plant Genome* 12(1), 180044. Available at: https://doi.org/10.3835/plantgenome2018.06.0044

Ibrahim, M.H., Kong, Y.C. and Zain, N.A.M. (2017) Effect of cadmium and copper exposure on growth, secondary metabolites and antioxidant activity in the medicinal plant sambung nyawa (*Gynura procumbens* (Lour.) Merr). *Molecules* 22(10), 1623. Available at: https://doi.org/10.3390/molecules22101623

Isarankura-Na-Ayudhya, P., Thippakorn, C., Pannengpetch, S., Roytrakul, S., Isarankura-Na-Ayudhya, C. *et al.* (2018) Metal complexation by histidine-rich peptides confers protective roles against cadmium stress in *Escherichia coli* as revealed by proteomics analysis. *PeerJ* 6, e5245. Available at: https://doi.org/10.7717/peerj.5245

Jia, W., Miao, F., Lv, S., Feng, J., Zhou, S. *et al.* (2017) Identification for the capability of Cd-tolerance, accumulation and translocation of 96 sorghum genotypes. *Ecotoxicology and Environmental Safety* 145, 391–397. Available at: https://doi.org/10.1016/j.ecoenv.2017.07.002

Kalbina, I. and Strid, Å. (2006) The role of NADPH oxidase and MAP kinase phosphatase in UV-B-dependent gene expression in *Arabidopsis*. *Plant, Cell & Environment* 29(9), 1783–1793. Available at: https://doi.org/10.1111/j.1365-3040.2006.01555.x

Kang, J., Park, J., Choi, H., Burla, B., Kretzschmar, T., Lee, Y. and Martinoia, E. (2011) Plant ABC transporters. *The Arabidopsis Book* 9, e0153. Available at: https://doi.org/10.1199/tab.0153

Kante, M., Rattunde, H.F.W., Nébié, B., Weltzien, E., Haussmann, B.I.G. and Leiser, W.L. (2018) QTL mapping and validation of fertility restoration in West African sorghum A1 cytoplasm and identification of a potential causative mutation for *Rf*$_2$. *Theoretical and Applied Genetics* 131(11), 2397–2412. Available at: https://doi.org/10.1007/s00122-018-3161-z.

Katsuhara, M., Koshio, K., Shibasaka, M. and Kasamo, K. (2003) Expression of an aquaporin at night in relation to the growth and root water permeability in barley seedlings. *Soil Science and Plant Nutrition* 49(6), 883–888. Available at: https://doi.org/10.1080/00380768.2003.10410351.

Keltjens, W.G. and van Beusichem, M.L. (1998) Phytochelatins as biomarkers for heavy metal toxicity in maize: single metal effects of copper and cadmium. *Journal of Plant Nutrition* 21(4), 635–648. Available at: https://doi.org/10.1080/01904169809365431.

Keunen, E., Remans, T., Bohler, S., Vangronsveld, J. and Cuypers, A. (2011) Metal-induced oxidative stress and plant mitochondria. *International Journal of Molecular Sciences* 12(10), 6894–6918. Available at: https://doi.org/10.3390/ijms12106894.

Kim, D.Y., Bovet, L., Kushnir, S., Noh, E.W., Martinoia, E. and Lee, Y. (2006) AtATM3 is involved in heavy metal resistance in *Arabidopsis*. *Plant Physiology* 140(3), 922–932. Available at: https://doi.org/10.1104/pp.105.074146.

Kim, J.S., Islam-Faridi, M.N., Klein, P.E., Stelly, D.M., Price, H.J., Klein, R.R. and Mullet, J.E. (2005) Comprehensive molecular cytogenetic analysis of sorghum genome architecture: distribution of euchromatin, heterochromatin, genes and recombination in comparison to rice. *Genetics* 171(4), 1963–1976. Available at: https://doi.org/10.1534/genetics.105.048215.

Kong, W.Q., Kim, C., Zhang, D., Guo, H., Tan, X. *et al.* (2018) Genotyping by sequencing of 393 *Sorghum bicolor* BTx623 × IS3620C recombinant inbred lines improves sensitivity and resolution of QTL detection. *G3: Genes, Genomes, Genetics* 8(8), 2563–2572. Available at: https://doi.org/10.1534/g3.118.200173

Kumar, A.A. (2016) Botany, taxonomy and breeding. In: Rakshit, S. and Wang, Y. (eds) *The Sorghum Genome*. Compendium of Plant Genomes. Springer, Cham, Switzerland, pp. 27–45. Available at: https://doi.org/10.1007/978-3-319-47789-3_2

Leiboff, S. and Hake, S. (2019) Reconstructing the transcriptional ontogeny of maize and sorghum supports an inverse hourglass model of inflorescence development. *Current Biology* 29(20), 3410–3419.e3. Available at: https://doi.org/10.1016/j.cub.2019.08.044

Lindberg, S. and Greger, M. (2002) Plant genotypic differences under metal deficient and enriched conditions. In: Prasad, M.N.V. and Strzałka, K. (eds) *Physiology and Biochemistry of Metal Toxicity and Tolerance in Plants*. Springer, Dordrecht, the Netherlands, pp. 357–393. Available at: https://doi.org/10.1007/978-94-017-2660-3_14

Lopez, J.R., Erickson, J.E., Munoz, P., Saballos, A., Felderhoff, T.J. and Vermerris, W. (2017) QTLs associated with crown root angle, stomatal conductance and maturity in sorghum. *Plant Genome* 10(2), plantgenome2016.04.0038. Available at: https://doi.org/10.3835/plantgenome2016.04.0038

Luo, H., Mocoeur, A.R.J. and Jing, H.-C. (2014) Next-generation sequencing technology for genetics and genomics of sorghum. In: Wang, Y.-H., Upadhyaya, H.D. and Kole, C. (eds) *Genetics, Genomics and Breeding of Sorghum*. CRC Press, Boca Raton, Florida, pp. 226–250.

McCormick, R.F., Truong, S.K., Sreedasyam, A., Jenkins, J., Shu, S. *et al.* (2018) The *Sorghum bicolor* reference genome: improved assembly, gene annotations, a transcriptome atlas and signatures of genome organization. *The Plant Journal* 93(2), 338–354. Available at: https://doi.org/10.1111/tpj.13781

Mall, T.K., Dweikat, I., Sato, S.J., Neresian, N., Xu, K. *et al.* (2011) Expression of the rice CDPK-7 in sorghum: molecular and phenotypic analyses. *Plant Molecular Biology* 75(4–5), 467–479. Available at: https://doi.org/10.1007/s11103-011-9741-9

Matsuda, T., Kuramata, M., Takahashi, Y., Kitagawa, E., Youssefian, S. and Kusano, T. (2009) A novel plant cysteine-rich peptide family conferring cadmium tolerance to yeast and plants. *Plant Signaling and Behavior* 4(5), 419–421. Available at: https://doi.org/10.4161/psb.4.5.8272

Mendoza-Cózatl, D.G., Jobe, T.O., Hauser, F. and Schroeder, J.I. (2011) Long-distance transport, vacuolar sequestration and transcriptional responses induced by cadmium and arsenic. *Current Opinion in Plant Biology* 14(5), 554–562. Available at: https://doi.org/10.1016/j.pbi.2011.07.004

Paterson, A.H. (2008) Genomics of sorghum. In: Xu, Y. (ed.) Genomics of Major Crops and Model Plant Species (special issue). *International Journal of Plant Genomics* 2008, 362451. Available at: https://doi.org/10.1155/2008/362451

Paterson, A.H., Bowers, J.E., Bruggmann, R., Dubchak, I., Grimwood, J. *et al.* (2009) The *Sorghum bicolor* genome and the diversification of grasses. *Nature* 457(7229), 551–556. Available at: https://doi.org/10.1038/nature07723

Pérez-Chaca, M.V., Rodríguez-Serrano, M., Molina, A.S., Pedranzani, H.E., Zirulnik, F., Sandalio, L.M. and Romero-Puertas, M.C. (2014) Cadmium induces two waves of reactive oxygen species in *Glycine max* (L.) roots. *Plant, Cell & Environment* 37(7), 1672–1687. Available at: https://doi.org/10.1111/pce.12280

Price, H.J., Dillon, S.L., Hodnett, G., Rooney, W.L., Ross, L. and Johnston, J.S. (2005) Genome evolution in the genus *Sorghum* (Poaceae). *Annals of Botany* 95(1), 219–227. Available at: https://doi.org/10.1093/aob/mci015

Qi, P., Eudy, D., Schnable, J.C., Schmutz, J., Raymer, P.L. and Devos, K.M. (2019) High density genetic maps of seashore paspalum using genotyping-by-sequencing and their relationship to the *Sorghum bicolor* genome. *Scientific Reports* 9(1), 12183. Available at: https://doi.org/10.1038/s41598-019-48257-3

Rao, M.S., Van Vleet, T.R., Ciurlionis, R., Buck, W.R., Mittelstadt, S.W., Blomme, E.A.G. and Liguori, M.J. (2018) Comparison of RNA-Seq and microarray gene expression platforms for the toxicogenomic evaluation of liver from short-term rat toxicity studies. *Frontiers in Genetics* 9, 636. Available at: https://doi.org/10.3389/fgene.2018.00636

Sakuma, Y., Maruyama, K., Osakabe, Y., Qin, F., Seki, M., Shinozaki, K. and Yamaguchi-Shinozaki, K. (2006) Functional analysis of an *Arabidopsis* transcription factor, DREB2A, involved in drought-responsive gene expression. *The Plant Cell* 18(5), 1292–1309. Available at: https://doi.org/10.1105/tpc.105.035881

Satish, L., Shilpha, J., Pandian, S., Rency, A.S., Rathinapriya, P. *et al.* (2016) Analysis of genetic variation in sorghum (*Sorghum bicolor* (L.) Moench) genotypes with various agronomical traits using SPAR methods. *Gene* 576(1), 581–585. Available at: https://doi.org/10.1016/j.gene.2015.10.056

Sato, H., Shirasawa, S., Maeda, H., Nakagomi, K., Kaji, R. *et al.* (2011) Analysis of QTL for lowering cadmium concentration in rice grains from 'LAC23'. *Breeding Science* 61(2), 196–200. Available at: https://doi.org/10.1270/jsbbs.61.196

Schaffasz, A., Windpassinger, S., Friedt, W., Snowdon, R. and Wittkop, B. (2019) Sorghum as a novel crop for Central Europe: using a broad diversity set to dissect temperate-adaptation. *Agronomy* 9(9), 535. Available at: https://doi.org/10.3390/agronomy9090535

Sebastian, A. and Prasad, M.N.V. (2014) Cadmium minimization in rice. A review. *Agronomy for Sustainable Development* 34(1), 155–173. Available at: https://doi.org/10.1007/s13593-013-0152-y

Sebastian, A. and Prasad, M.N.V. (2018) Exogenous citrate and malate alleviate cadmium stress in *Oryza sativa*, L.: probing role of cadmium localization and iron nutrition. *Ecotoxicology and Environmental Safety* 166, 215–222. Available at: https://doi.org/10.1016/j.ecoenv.2018.09.084

Shakoor, N., Ziegler, G., Dilkes, B.P., Brenton, Z., Boyles, R. *et al.* (2016) Integration of experiments across diverse environments identifies the genetic determinants of variation in *Sorghum bicolor* seed element composition. *Plant Physiology* 170(4), 1989–1998. Available at: https://doi.org/10.1104/pp.15.01971

Shimo, H., Ishimaru, Y., An, G., Yamakawa, T., Nakanishi, H. and Nishizawa, N.K. (2011) *Low cadmium* (*LCD*), a novel gene related to cadmium tolerance and accumulation in rice. *Journal of Experimental Botany* 62(15), 5727–5734. Available at: https://doi.org/10.1093/jxb/err300

Smeets, K., Opdenakker, K., Remans, T., Forzani, C., Hirt, H., Vangronsveld, J. and Cuypers, A. (2013) The role of the kinase OXI1 in cadmium- and copper-induced molecular responses in *Arabidopsis thaliana*. *Plant, Cell & Environment* 36(6), 1228–1238. Available at: https://doi.org/10.1111/pce.12056

Smith, O., Nicholson, W.V., Kistler, L., Mace, E., Clapham, A. *et al.* (2019) A domestication history of dynamic adaptation and genomic deterioration in *Sorghum*. *Nature Plants* 5(4), 369–379. Available at: https://doi.org/10.1038/s41477-019-0397-9

Sofo, A., Vitti, A., Nuzzaci, M., Tataranni, G., Scopa, A. *et al.* (2013) Correlation between hormonal homeostasis and morphogenic responses in *Arabidopsis thaliana* seedlings growing in a Cd/Cu/Zn multi-pollution context. *Physiologia Plantarum* 149(4), 487–498. Available at: https://doi.org/10.1111/ppl.12050

Soric, R., Loncaric, Z., Kovacevic, V., Brkic, I. and Simic, D. (2009) A major gene for leaf cadmium accumulation in maize (*Zea mays*, L.). In: *Proceedings of the International Plant Nutrition Colloquium XVI*. Available at: https://escholarship.org/uc/item/1q48v6cf#author.

Tamás, L., Dudíková, J., Ďurčeková, K., Halušková, L., Huttová, J., Mistrík, I. and Ollé, M. (2008) Alterations of the gene expression, lipid peroxidation, proline and thiol content along the barley root exposed to cadmium. *Journal of Plant Physiology* 165(11), 1193–1203. Available at: https://doi.org/10.1016/j.jplph.2007.08.013

Tian, Z., Rizzon, C., Du, J., Zhu, L., Bennetzen, J.L. *et al.* (2009) Do genetic recombination and gene density shape the pattern of DNA elimination in rice long terminal repeat retrotransposons? *Genome Research* 19(12), 2221–2230. Available at: https://doi.org/10.1101/gr.083899.108

Tsuboi, K., Shehzad, T., Yoneda, J., Uraguchi, S., Ito, Y. *et al.* (2017) Genetic analysis of cadmium accumulation in shoots of sorghum landraces. *Crop Science* 57(1), 22–31. Available at: https://doi.org/10.2135/cropsci2016.01.0069

Ueno, D., Yamaji, N., Kono, I., Huang, C.F. Ando, T., Yano, M. and Ma, J.F. (2010) Gene limiting cadmium accumulation in rice. *Proceedings of the National Academy of Sciences USA* 107(38), 16500–16505. Available at: https://doi.org/10.1073/pnas.1005396107

Uraguchi, S. and Fujiwara, T. (2012) Cadmium transport and tolerance in rice: perspectives for reducing grain cadmium accumulation. *Rice* 5(1), 5. Available at: https://doi.org/10.1186/1939-8433-5-5

Vatansever, R., Filiz, E. and Ozyigit, I.I. (2015) Genome-wide analysis of iron-regulated transporter 1 (*IRT1*) genes in plants. *Horticulture, Environment, and Biotechnology* 56(4), 516–523. Available at: https://doi.org/10.1007/s13580-015-0014-4

Verbruggen, N., Hermans, C. and Schat, H. (2009a) Mechanisms to cope with arsenic or cadmium excess in plants. *Current Opinion in Plant Biology* 12(3), 364–372. Available at: https://doi.org/10.1016/j.pbi.2009.05.001

Verbruggen, N., Hermans, C. and Schat, H. (2009b) Molecular mechanisms of heavy metal hyperaccumulation in plants. *New Phytologist* 181(4), 759–776. Available at: https://doi.org/10.1111/j.1469-8137.2008.02748.x

Verdeprado, H., Kretzschmar, T., Begum, H., Raghavan, C., Joyce, P. *et al.* (2018) Association mapping in rice: basic concepts and perspectives for molecular breeding. *Plant Production Science* 21(3), 159–176. Available at: https://doi.org/10.1080/1343943X.2018.1483205

Yaneff, A., Vitali, V. and Amodeo, G. (2015) PIP1 aquaporins: intrinsic water channels or PIP2 aquaporin modulators? *FEBS Letters* 589(23), 3508–3515. Available at: https://doi.org/10.1016/j.febslet.2015.10.018

Yang, J., Chen, Z., Wu, S., Cui, Y., Zhang, L. *et al.* (2015) Overexpression of the *Tamarix hispida ThMT3* gene increases copper tolerance and adventitious root induction in *Salix matsudana* Koidz. *Plant Cell, Tissue and Organ Culture* 121(2), 469–479. Available at: https://doi.org/10.1007/s11240-015-0717-3

Zdunić, Z., Grljušić, S., Ledenčan, T., Duvnjak, T. and Šimić, D. (2014) Quantitative trait loci mapping of metal concentrations in leaves of the maize IBM population. *Hereditas* 151(2–3), 55–60. Available at: https://doi.org/10.1111/hrd2.00048

Zhang, Z.H., Zhou, T., Tang, T.J., Song, H.X., Guan, C.Y. *et al.* (2019) A multiomics approach reveals the pivotal role of subcellular reallocation in determining rapeseed resistance to cadmium toxicity. *Journal of Experimental Botany* 70(19), 5437–5455. Available at: https://doi.org/10.1093/jxb/erz295

Zhao, J., Yang, W., Zhang, S., Yang, T., Liu, Q. *et al.* (2018) Genome-wide association study and candidate gene analysis of rice cadmium accumulation in grain in a diverse rice collection. *Rice* 11(1), 61. Available at: https://doi.org/10.1186/s12284-018-0254-x

Zhao, X., Luo, L., Cao, Y., Liu, Y., Li, Y. *et al.* (2018) Genome-wide association analysis and QTL mapping reveal the genetic control of cadmium accumulation in maize leaf. *BMC Genomics* 19(1), 91. Available at: https://doi.org/10.1186/s12864-017-4395-x

Zhiguo, E., Tingting, L., Chen, C. and Lei, W. (2018) Genome-wide survey and expression analysis of $P_{1B}$-ATPases in rice, maize and sorghum. *Rice Science* 25(4), 208–217. Available at: https://doi.org/10.1016/j.rsci.2018.06.004

Zhou, M., Zheng, S., Liu, R., Lu, L., Zhang, C. *et al.* (2019) The genome-wide impact of cadmium on microRNA and mRNA expression in contrasting Cd responsive wheat genotypes. *BMC Genomics* 20(1), 615. Available at: https://doi.org/10.1186/s12864-019-5939-z

Zhuang, P., Shu, W., Li, Z., Liao, B., Li, J. and Shao, J. (2009) Removal of metals by sorghum plants from contaminated land. *Journal of Environmental Sciences* 21(10), 1432–1437. Available at: https://doi.org/10.1016/S1001-0742(08)62436-5

# 28 Molecular Breeding for Increasing Micronutrient Content in Sorghum

**K. Hariprasanna* and P. Rajendrakumar**
*ICAR–Indian Institute of Millets Research, Hyderabad, India*

## 28.1 Introduction

Human nutritional requirements must be met from the food that is consumed and the primary source for all nutrients comes from agricultural products. Malnutrition is responsible for more human deaths globally than any other cause of death and all forms of malnutrition including being overweight or obese are associated with various forms of ill health. About 45% of deaths among children below 5 years of age, mainly in low- and middle-income countries, are due to undernutrition alone (Black *et al.*, 2013). An estimated four million deaths (7.1% of all deaths) and 120 million healthy years of life lost (disability-adjusted life years or DALYs) in the global population (4.9% of all DALYs among adults) are due to the health consequences of overweight and obesity (GBD 2015 Obesity Collaborators, 2017), especially in the developed world. Malnutrition also results in huge health-care costs and imposes high human capital costs on individuals, families and nations. It is estimated that malnutrition in all its forms could cost society up to US$3.5 trillion per year (Global Panel, 2016). Dietary diversification holds the key, as the poor people living in the arid and semi-arid tropical regions of the world cannot afford a variety of food items in their diet and suffer from deficiencies of energy and micronutrients.

Among different forms of malnutrition, micronutrient malnutrition, or 'hidden hunger', results primarily from diets poor in bioavailable vitamins and minerals. Hidden hunger is caused by insufficient intake, absorption or utilization of vitamins and minerals. It is a global food-related health problem and affects more than half of the developing world's population, or more than three billion people, especially women and preschool children (Bouis, 2002; Welch and Graham, 2004; Jin *et al.*, 2013). Micronutrient deficiencies can impair the mental and physical development of children and adolescents, and cause blindness and anaemia (even death). Prevalence of stunting and blindness in children, as well as women, is very high in developing countries. Three of the most widespread micronutrient deficiencies are those of Fe, Zn and vitamin A (see Box 28.1), although folate and iodine also remain issues in several countries. Fe deficiency is the most common micronutrient deficiency in the world and the most significant contributor to anaemia. As global data for Fe deficiency do not exist, anaemia is used as an indirect indicator. Approximately 50% of cases of anaemia are considered to be due to Fe deficiency (Stevens *et al.*, 2013). Anaemia affects 800 million women and children globally (HarvestPlus, 2021). An estimated 17.3% of the

*Corresponding author: hari@millets.res.in

 *Molecular Breeding in Wheat, Maize and Sorghum: Strategies for Improving Abiotic Stress Tolerance and Yield* (eds M.A. Hossain *et al.*)
DOI: 10.1079/9781789245431.0028

**Box 28.1.** Micronutrients in Human Nutrition

- Fe is a redox-active constituent of the catalytic site of haem and non-haem Fe proteins. About half of anaemia cases are caused by Fe deficiency. Fe deficiency adversely affects cognitive development, resistance to infection, work capacity, productivity and pregnancy.
- Zn is involved in RNA and DNA synthesis and is a constituent of many Zn-containing enzymes that are critical to cellular growth and differentiation. Zn deficiency leads to impaired growth, immune dysfunction, increased morbidity and mortality, adverse pregnancy outcomes and abnormal neurobehavioural development.
- Vitamin A is a group of fat-soluble nutritionally unsaturated hydrocarbons, which include retinol, retinal, retinoic acid and several provitamin A carotenoids. Preformed vitamin A is found in animal products such as meat, fish, poultry and dairy foods; while provitamin A, the most common being β-carotene, is found in plant-based foods such as fruits and vegetables. Vitamin A is important for growth and development, for the maintenance of the immune system and good vision (both low-light and colour). It also helps in skin and cellular health, as well as maintenance of teeth, skeletal and soft tissues, and mucus membranes.

world's population is at risk of inadequate Zn intake in their diet or Zn deficiency (Wessells and Brown, 2012). The most prone are infants, young children, and pregnant and lactating women due to their elevated requirements for Zn. Vitamin A deficiency is the leading cause of preventable blindness in children, causing blindness in 250,000 to 500,000 children each year and increasing the risk of disease and death from severe infections (http://www.who.int/nutrition/topics/vad/en/, accessed 4 March 2021). One general indicator of micronutrient deficiency in humans is anaemia, as this is caused by the deficiency of many of them, and its effects are aggravated by several diseases.

Interventions like supplementation and food fortification (addition of exogenous nutrients to food products) with micronutrients, which are widely used to address micronutrient malnourishment globally, can directly facilitate easing of hidden hunger to some extent. However, these are short-term approaches, which can address acute cases of malnourishment, and require infrastructure, sophisticated processing techniques, quality control and funding to succeed. Poor purchasing power and lack of access to markets and health-care systems of the poor population further limit the success of such interventions. This warrants new approaches to address the persistent problem of micronutrient malnutrition in a sustained manner, especially to reach the affected population in remote rural areas. Biofortification – the process of enhancing the grain micronutrient content in staple crop cultivars through traditional breeding practices and modern biotechnology – offers a powerful, sustainable and cost-effective tool to overcome micronutrient malnutrition as this is a genetic enhancement that increases the micronutrients in staple crops at their source.

Sorghum, a heat- and drought-tolerant $C_4$ plant with wider adaptation to a range of environments, is the fifth most important crop grown in the world. It is a reliable source of food grain and feed for regions that are regularly affected by drought, where production of other cereals like wheat, maize and barley is challenging (Elkonin *et al.*, 2018). Sorghum is a major staple for nearly 500 million people living in 30 countries of the semi-arid tropical regions (Hariprasanna and Rakshit, 2016). Traditionally, sorghum consumption levels were relatively high among poor people in the semi-arid parts of Africa and India, where more than 70% of the sorghum grain was utilized for human consumption (FAO and ICRISAT, 1996), although consumption has declined sharply in recent years due to a shift in consumer preferences as a result of rapid urbanization and increased economic status (Nagaraj *et al.*, 2013; Hariprasanna and Rakshit, 2016). Sorghum is good source of dietary energy with considerable amounts of micronutrients, K, P, Ca and Na (Hulse *et al.*, 1980; Gopalan *et al.*, 1989; Longvah *et al.*, 2017). Considering the prevalence of micronutrient malnutrition in the predominantly sorghum-producing/consuming regions in the African and Indian subcontinent, the crop holds promise as a key target for further improvement in micronutrient content in the grains to make it more nutrient-dense and thus benefit the micronutrient-malnourished population, lessen the burden of hidden hunger and address the United Nations' Sustainable Development Goals (SDGs) on nutrition. Sorghum grains also have good amounts of phytochemicals like tannins, phenolics and phytate, and have poor protein digestibility, which leads to poor nutritional and processing quality. The Fe and Zn exist predominantly as phytate complexes within the aleurone layer of the grains and because of the presence of phytate, the bioavailability of Fe and Zn in sorghum grains is low (Zhao *et al.*, 2019). Therefore, along with enriching the grains with micronutrients, approaches to enhance the bioavailability are also of prime importance in sorghum. This chapter aims to summarize the limited efforts that have been undertaken to enhance the micronutrient content in sorghum using molecular breeding approaches.

## 28.2 Genetic Resources

### 28.2.1 Germplasm collection – core and mini-core collections

The nutrient content, especially grain Fe and Zn, in sorghum was found to be better compared with cereals like rice (Gopalan *et al.*, 1989; Hariprasanna and Rakshit, 2016). As sorghum is mostly consumed in regions where it is produced unlike many other cereals in India and African countries, enriching the grains with Fe and Zn can alleviate micronutrient deficiency in the sorghum-consuming population living in the semi-arid tropics. Biofortification of sorghum by increasing Fe and Zn contents in grain through conventional breeding approaches is of widespread interest (Zhao, 2008; Ashok Kumar *et al.*, 2009, 2012a, 2013a; Hariprasanna *et al.*, 2012, 2014a). Availability of exploitable genetic variability for target traits is a key requirement for the success of any breeding programme. Upon discovery of genetic variability for grain Fe and Zn content in sorghum, the breeding strategies should focus on combining higher grain yield with more Fe and Zn density in the grains as higher grain micronutrient levels alone will not be sufficient to promote biofortified cultivars.

Globally, a total of 236,617 sorghum accessions have been conserved in different gene banks (98.3% are cultivated and 1.7% wild and weedy relatives) (Upadhyaya *et al.*, 2016). The majority are conserved in Asia (39.18%), the Americas (35.72%) and Africa (16.40%). Major gene banks conserving sorghum germplasm resources are those of: the International Crops Research Institute for the Semi-Arid Tropics (ICRISAT), India (41,816) (http://genebank.icrisat.org/, accessed 4 March 2021); the Plant Genetic Resources Conservation Unit, Southern Regional Plant Introduction Station, University of Georgia, US Department of Agriculture–Agricultural Research Service (39,504) (https://www.genesys-pgr.org/wiews/USA016, accessed 4 March 2021); the National Bureau of Plant Genetic Resources (NBPGR), New Delhi, India (26,118) (http://www.nbpgr.ernet.in, accessed 4 March 2021); and the Institute of Crops Science, Chinese Academy of Agricultural Sciences (ICS-CAAS), China (18,263) (http://ics.caas.cn/en/, accessed 4 March 2021), which together conserve more than 50% of the total global sorghum germplasm. The ICAR–Indian Institute of Millets Research, Hyderabad, India, which is the nodal agency for all millets research and development activities for the country, conserves 27,791 sorghum accessions in medium-term storage as an active and working collection (https://www.millets.res.in, accessed 4 March 2021). The Svalbard Global Seed Vault safeguards >50,000 accessions of sorghum from different gene banks (https://www.seedvault.no/about/the-seeds/, accessed 4 March 2021).

In order to facilitate germplasm maintenance, assessment and utilization, ICRISAT established a core collection that represented 10% of the landraces collection (2247 accessions) from the 22,473 accessions conserved at the ICRISAT gene bank (Grenier *et al.*, 2001a,b). As the core collection was also large in size for meaningful and precise evaluation for important economic traits, a mini-core (10% of core or 1% of entire germplasm accessions) comprising 242 accessions from the existing core collection was developed (Upadhyaya *et al.*, 2009) that is widely used in evaluation for several traits of importance. An initial core collection of 3011 accessions representing 77 different countries out of the sorghum accessions conserved at the US National Plant Germplasm System (NPGS) was developed by random selection of 10% of the accessions from each country (Dahlberg *et al.*, 2004). A refined subset of the core collection representing Sudan, with 352 accessions that represent 13.8% of the total Sudan collection, was made up which was evaluated for relative heat tolerance of sorghum.

### 28.2.2 Identification of donors

Sorghum had only limited research efforts related to biofortification until the early 2000s and thereby had only a limited information base. Initial studies at ICRISAT indicated limited variability for grain Fe and Zn contents (Reddy *et al.*, 2005) in sorghum hybrid parents, advanced breeding lines and germplasm accessions. Only traces of β-carotene content could be detected in the grains of non-yellow endosperm lines, although some yellow endosperm germplasm had β-carotene content ranging from 0.56 ppm (IS 24724) to 1.132 ppm (1S 26886). A similar range of β-carotene content (0.2 to 1.4 ppm) was also reported previously in 20 yellow endosperm sorghum germplasm lines (Kapoor and Naik, 1970). Considering the very low levels of grain β-carotene content, which would not meet recommended daily allowance levels, vitamin A improvement in sorghum was not focused upon subsequently. In the 84 lines tested, the grain Fe content ranged between 20 ppm (ICSR 93031) and 37 ppm (ICSB 472 and 296 B) with an average of 28 ppm, while grain Zn content ranged from 13 ppm (JJ 1041) to 31 ppm (IS 1199) with an average of 19 ppm.

With the support of HarvestPlus (https://www.harvestplus.org, accessed 4 March 2021), sorghum biofortification research gained momentum at ICRISAT. A large number of landraces (2246), hybrid parents (>500 B-lines and 100 R-lines), breeding lines and commercial sorghum cultivars (67) were characterized over the years for assessing grain Fe and Zn concentrations and important agronomic traits. Large genetic variability for grain Fe and Zn concentrations has been reported in these materials (Ashok Kumar *et al.*, 2009, 2010, 2012a; Reddy *et al.*, 2010). Compared with post rainy season-adapted sorghums predominantly grown for food use in India, the rainy season-adapted commercial hybrids possessed better micronutrient contents (up to 44 ppm Fe and 33 ppm Zn). The variability observed in the core collection (2246) along with hybrid parents, breeding lines and commercial cultivars has been put in the public domain in the form of a database (Ashok Kumar *et al.*, 2012b). A number of germplasm accessions with relatively high Fe (>40 mg/kg) were identified for use in the breeding programmes (Table 28.1).

At the ICAR–Indian Institute of Millets Research, characterization of popular Indian cultivars, hybrid parents, breeding lines and some selected germplasm accessions collected from the major sorghum growing states indicated sufficient variability for grain Fe (12–83 mg/kg) and Zn (6–51 mg/kg) contents as well as high heritability (Hariprasanna *et al.*, 2014a, 2018). Significant and positive associations between the grain Fe and Zn concentrations have been reported in most of the sorghum material studied (Ashok Kumar *et al.*, 2009, 2010, 2012a, 2013b; Reddy *et al.*, 2010; Hariprasanna *et al.*, 2014a; Phuke *et al.*, 2017), thus indicating that genetic control of these two minerals is linked or physiological mechanisms for uptake or accumulation in the grains are interconnected, and there is a possibility of simultaneous improvement in Fe and Zn contents (Bhat *et al.*, 2018). Based on the grain Fe and Zn concentrations observed, the base level is estimated to be 30 mg/kg for Fe and 20 mg/kg for Zn. Considering the extent of genetic variability observed among landraces and working collections, and nutrient bioavailability, the target level for genetic improvement has been fixed at >40 mg/kg for grain Fe and >30 mg/kg for grain Zn.

### 28.2.3 Mapping populations

#### *Bi-parental populations*

Genetic improvement of sorghum grains for higher content of micronutrients such as Fe, Zn and vitamin A requires extensive efforts involving resources and funding, encompassing identification of donors, development of mapping populations, QTL mapping and marker-assisted introgression of favourable and effective QTLs for micronutrients. Traditional QTL mapping procedures utilize bi-parental mapping populations such as recombinant inbred lines (RILs), near-isogenic lines (NILs) and doubled haploid (DH) lines generated from controlled crosses involving selected parents contrasting for traits of interest so that the parental alleles will segregate in these mapping populations. Among these mapping populations, RILs are considered as immortal populations since they can be multiplied several times for repeated phenotyping in multiple locations/environments. More importantly, a

**Table 28.1.** Selected sorghum germplasm lines with high grain Fe and Zn contents. (From Ashok Kumar *et al.*, 2009, 2013a.)

| Accession | Race | Origin | Grain yield (t/ha) | Fe (mg/kg) | Zn (mg/kg) |
|---|---|---|---|---|---|
| IS 5427 | *Durra* | India | 2.0 | 61 | 57 |
| IS 5514 | *Guinea–bicolor* | India | 1.4 | 56 | 45 |
| IS 55 | *Durra–caudatum* | USA | 1.3 | 54 | 38 |
| IS 3760 | *Caudatum–bicolor* | USSR | 2.2 | 53 | 37 |
| IS 3283 | *Bicolor* | USA | 1.9 | 50 | 42 |
| IS 17580 | *Caudatum* | Nigeria | 1.6 | 50 | 41 |
| IS 15952 | *Guinea* | Cameroon | 2.5 | 49 | 41 |
| IS 3813 | *Durra* | India | 1.4 | 49 | 38 |
| IS 15266 | *Caudatum* | Cameroon | 2.7 | 49 | 44 |
| IS 2939 | *Kafir* | US | 3.6 | 48 | 37 |
| IS 4159 | *Durra* | India | 1.5 | 48 | 38 |
| IS 3929 | *Kafir–durra* | USA | 2.2 | 48 | 40 |
| IS 3443 | *Guinea–caudatum* | Sudan | 3.3 | 47 | 39 |
| IS 3925 | *Durra–caudatum* | USA | 2.4 | 47 | 39 |
| IS 5460 | *Durra–bicolor* | India | 1.4 | 47 | 46 |
| IS 12452 | *Caudatum–bicolor* | Sudan | 3.2 | 47 | 33 |
| IS 2801 | *Caudatum* | Zimbabwe | 2.3 | 46 | 45 |
| IS 2536 | *Kafir–caudatum* | USA | 2.3 | 45 | 37 |
| IS 5429 | *Durra* | India | 2.8 | 44 | 30 |
| IS 356 | *Durra* | USA | 2.2 | 44 | 33 |
| IS 2265 | *Durra–bicolor* | Sudan | 1.8 | 44 | 41 |
| IS 12695 | *Bicolor* | South Africa | 2.8 | 44 | 39 |
| IS 5538 | *Durra* | India | 1.7 | 44 | 47 |
| IS 5476 | *Durra* | India | 2.1 | 41 | 36 |
| IS 16337 | *Caudatum* | Cameroon | 2.4 | 41 | 34 |
| IS 5853 | *Guinea–durra* | India | 2.4 | 41 | 32 |
| IS 12750 | *Caudatum* | China | 1.9 | 76 | 39 |
| IS 27054 | *Durra–caudatum* | Zimbabwe | 3.6 | 73 | 29 |
| IS 12858 | *Bicolor* | Turkey | 1.8 | 73 | 32 |
| IS 12785 | *Bicolor* | Turkey | 1.7 | 70 | 27 |
| IS 1563 | *Bicolor* | India | 1.3 | 70 | 38 |
| IS 34 | *Bicolor* | USA | 2.4 | 69 | 27 |
| IS 13 | *Bicolor* | USA | 2.9 | 64 | 33 |
| IS 30 | *Bicolor* | USA | 2.1 | 62 | 32 |
| IS 20962 | *Caudatum* | Kenya | 4.7 | 62 | 24 |
| IS 9150 | *Caudatum* | Kenya | 2.9 | 62 | 28 |
| IS 23680 | *Caudatum* | Mozambique | – | 71 | 44 |
| IS 5308 | *Guinea* | India | – | 63 | 45 |
| IS 3790 | *Kafir–bicolor* | Taiwan | – | 58 | 54 |
| IS 3696 | *Guinea–bicolor* | Taiwan | – | 57 | 40 |
| IS 5299 | *Guinea* | India | – | 55 | 40 |
| IS 26962 | *Caudatum* | India | – | 51 | 46 |
| IS 25546 | *Caudatum* | Rwanda | – | 57 | 30 |
| IS 18133 | *Bicolor* | Lebanon | – | 56 | 27 |
| IS 1222 | *Kafir–bicolor* | China | – | 55 | 41 |
| IS 17307 | *Bicolor* | Ethiopia | – | 54 | 38 |
| IS 3106 | *Bicolor* | Kenya | – | 54 | 34 |
| IS 17580 | *Caudatum* | Nigeria | – | 51 | 35 |
| IS 25699 | *Guinea* | Mali | – | 51 | 33 |
| IS 32 | *Bicolor* | USA | – | 50 | 32 |

better map resolution can be achieved using RILs as compared with $F_2$ and DH populations due to the occurrence of several rounds of meiosis and recombination cycles during the development of the RIL population. However, the population size plays a major role in determining the mapping resolution and marker order (Darvasi *et al.*, 1993). For instance, a population size of 100–250 individuals is used for preliminary mapping studies (Collard *et al.*, 2005) while a population of more than 500 individuals is used for the mapping of small-effect QTLs. Bi-parental populations, namely introgression lines (ILs) and NILs, are also used to understand epistatic interactions between multiple loci (Rakshit *et al.*, 2012). Development of bi-parental mapping populations in sorghum for the mapping of genomic regions/QTLs for important traits was reviewed by Rajendrakumar and Rakshit (2015). Use of a bi-parental population for the mapping of QTLs for micronutrient contents has been reported in pearl millet (Sushil Kumar *et al.*, 2016, 2018), maize (Gu *et al.*, 2015), rice (Dixit *et al.*, 2019) and wheat (Liu *et al.*, 2019). Recently, an $F_6$ RIL population comprising 342 lines derived from cross 296B × PVK 801 was employed for the identification of QTLs and candidate genes for high grain Fe and Zn contents in sorghum (Anuradha *et al.*, 2019).

### *Association panels*

The association mapping approach is more widely used nowadays to identify the genomic regions or candidate genes underlying the target traits. Due to its advantage over linkage mapping in terms of higher mapping resolution as it accounts for evolutionary mutations and recombinations, it is becoming a more popular approach for identifying marker–trait associations, particularly with the rapid advancements in next-generation sequencing (NGS) technologies, high-throughput precision phenotyping platforms and improvements in the computational tools for association analysis.

The first association mapping panel in sorghum, comprising 377 accessions representing major cultivated races and important US breeding lines, along with their progenitors, was assembled by Casa *et al.* (2008) and was used for characterizing genotypic and phenotypic diversity. A core collection of 195 sorghum accessions representing the diversity of a core set of basic and intermediate races (210 sorghum genotypes) reported by Deu *et al.* (2006) was utilized as an association mapping panel for the mapping of grain quality and grain yield (de Alencar Figueiredo *et al.*, 2010). A sweet sorghum panel (SSP) of 125 accessions, primarily a mix of diverse old and new sweet sorghum genotypes along with some grain and forage sorghum genotypes, was established by Murray *et al.* (2009). To demonstrate the useful diversity for understanding the key bioenergy traits in sorghum, a bioenergy association panel of 390 accessions comprising 152 sweet types and 238 biomass types that exhibit racial, geographic and phenotypic diversity from major sorghum races (*bicolor*, *caudatum*, *durra*, *guinea* and *kafir*) was established and characterized using 232,303 genetic markers (Brenton *et al.*, 2016). The diversity/association panels constituted worldwide by various scientific groups are reviewed by Boyles *et al.* (2019). Such association mapping panels offer a great opportunity in dissecting the genetic components associated with grain micronutrient contents in sorghum through exploitation of the wide variability present in them for the target traits.

### *Specialized populations*

Genome-wide association studies (GWAS) have been popularly employed in diverse germplasm lines to identify the marker–trait associations for the trait of interest. However, the value of GWAS is restricted by a fundamental trade-off due to the confounding of causative oligogenic variation with polygenic variation leading to false-positive or false-negative associations (Bergelson and Roux, 2010). In addition, it also helps in overcoming the shortcomings of bi-parental populations, such as accounting for limited recombination events and mapping of allelic pairs existing among the two contrasting parents, so that a better map resolution of the QTLs is achieved. Nested association mapping (NAM) populations are specialized populations developed by controlled crossing of a common parent with diverse founder lines in a star design (Huang *et al.*, 2011), which helps in breaking up the population structure thereby enhancing the power of detection of QTLs (Myles *et al.*, 2009). Moreover, the larger population size of NAM populations helps in reducing the overestimation of QTL effects that happens in small populations (Utz *et al.*, 2000). NAM populations have greatly facilitated the dissection of complex traits into genetic components in crops such as maize (Jimenez-Galindo *et al.*, 2019), wheat (Bajgain *et al.*, 2016), rice (Fragoso *et al.*, 2017), etc.

In sorghum, NAM populations have been developed through two different approaches, namely the single-seed descent method and the backcross method, known as traditional NAM and backcrossed NAM (BC-NAM), respectively. A NAM population was developed by crossing ten diverse founder lines with a common parent (RTx430), followed by selfing of the progeny to produce $F_2$ to $F_6$ generations through the single-seed descent method, resulting in a population comprising ten RIL families (Bouchet *et al.*, 2017). Similarly, a different NAM population was developed from a cross between a chilling-sensitive parent (BTx623) and three chilling-tolerant Chinese founder lines (Niu Sheng Zui, Hong KeZi 120 and Kaoliang) (Marla *et al.*, 2019). By employing the backcross method, a NAM population was developed from a large backcross ($BC_1F_1$) population involving an elite line as the recurrent parent comprised of more than 4000 individuals from 100 subpopulations (Jordan *et al.*, 2012). Very recently, a BC-NAM population comprising 1083 $BC_1F_5$ progenies was developed from 13 bi-parental mapping populations derived from crossing the recurrent parent (Lata) with 13 donor parents and selection of $BC_1F_1$ and subsequent generations exhibiting similarity with the recurrent parent for heading date and plant height (Diallo *et al.*, 2019). The drawback of the NAM population is its inability to assess the interaction of target QTLs with genetic background due to the use of a common parent in the development of subpopulations.

Genetic mapping using RIL and NAM populations is affected by limited recombination events and large recombination blocks, while association mapping using diversity/association panels is hindered by more recombination events and higher false positives due to the classification of individuals into groups within diversity panels (Bergelson and Roux, 2010). To overcome these shortcomings, a specialized population known as the multi-parent advanced generation intercross (MAGIC) has been developed to accumulate more recombinations and also maintain the even population structure, thereby helping to achieve high-resolution mapping of quantitative traits. As a result, MAGIC populations are considered intermediate to bi-parental populations and association mapping panels with reference to population structure, allelic diversity, resolution power and number of traits evaluated (Rakshit *et al.*, 2012; Pascual *et al.*, 2016). Designs for developing MAGIC populations involve intercrossing of multiple inbred founder lines several times in a definite fashion to merge the genetic material of these founders in a single line (Cavanagh *et al.*, 2008), resulting in highly diverse lines each containing a unique mixture of founder alleles. MAGIC populations have been developed in crops including rice (Ogawa *et al.*, 2018), maize (Jimenez-Galindo *et al.*, 2019), wheat (Stadlmeier *et al.*, 2018) and sorghum (Ongom and Ejeta, 2018). The first MAGIC population developed in sorghum was comprised of 1000 inbred lines derived from 19 diverse founders through random mating (Ongom and Ejeta, 2018). However, a multi-parent wide diallel population was developed earlier involving 19 founder lines for mapping the heterotic trait locus in sorghum and to identify intra-locus interactions involved in hybrid vigour (Ben-Israel *et al.*, 2012). Very recently, a PP 37 MAGIC population was developed to recombine sorghum accessions possessing diverse putative resistance mechanisms against *Striga hermonthica* (Khangura, 2019). Whole-genome sequences were developed for approximately 1006 individuals of the PP 37 MAGIC population. Similarly, the development of NAM and MAGIC populations involving the donor lines containing high micronutrient contents will help in establishing the role of genomic regions or QTLs associated with grain Fe and Zn contents in sorghum. The drawback of the MAGIC population is that its development is a resource-intensive and time-consuming process. Moreover, the population has limitations in the dissection of complex traits since it is likely to show extensive segregation for plant development traits. NAM and MAGIC populations developed worldwide by various research groups are reviewed by Boyles *et al.* (2019).

## 28.3 Genomic Resources

### 28.3.1 Genome sequence resources

Sorghum, a $C_4$ grass, was the second crop genome sequenced after rice, the model $C_3$ crop plant. Grain, sweet, forage and biomass types are the predominant ones in sorghum and this range of phenotypic diversity allows this crop to serve as a functional model for studying bioenergy and biomass-related traits. Being an interesting crop model for understanding the evolution of cereal genomes, sorghum also offers a unique opportunity to gain insight into weed biology, carbon assimilation at high temperatures as well as forage and biomass traits. The first draft genome of sorghum sequenced was that of a grain sorghum inbred line BTx623 through whole-genome shotgun (WGS) technology (Paterson *et al.*, 2009). By employing gene prediction methods in combination with expressed sequences of sorghum, maize and sugarcane, about 27,640 protein-coding genes were predicted. Among the high-confidence sorghum genes, about 94% have orthologues in rice, *Arabidopsis* and/or poplar; 24% have members only in sorghum and rice; and only 7% were exclusive to sorghum. Recently, the coverage and sequence quality of the reference genome were improved with the advances in sequencing technology and transcriptomics by integrating 29.6 Mbp of additional sequence, which increased annotated genes to 24%, increased average length of gene and reduced the frequency of error by tenfold (i.e. 1 per 100 kbp) (McCormick *et al.*, 2018).

Sweet sorghums differ from grain sorghums in plant height, stem sugar accumulation, grain and biomass production. To get an insight into the genome-wide variations among sweet and grain sorghums, resequencing was performed using two sweet (Keller and E-Tian) and one grain (Ji2731) sorghum genotypes, which revealed the presence of 83,262 single-nucleotide polymorphisms (SNPs) in the coding regions out of 1,057,018 SNPs identified among these sorghum genomes. About 1442 genes revealed genetic variation between sweet and grain sorghums and five of these genes that are involved in the starch and sucrose biosynthesis pathway were present on chromosomes 2, 6 and 9, which are important for sugar production. In addition, the gene for cinnamyl-alcohol dehydrogenase (*Sb06g028240*) present on chromosome 6 plays a crucial role in lignin biosynthesis that is essential for biofuel production (Zheng *et al.*, 2011). Very recently, a sweet sorghum line Rio was sequenced through long-read single-nucleotide sequencing technology (Cooper *et al.*, 2019) and its comparison with the grain sorghum reference discovered the presence of a high rate of non-synonymous loss-of-function mutations with little changes in gene content or genome structure. Gene deletions were more common than gene duplications in Rio and the most fascinating putative deletions were observed in sucrose transporter genes such as *SUT4*, *SWEET3-3* and *SWEET8-2*. Comparing the expression profiles of differentially expressed genes (DEGs) in the internode between Rio and PR22 revealed that the gene *SIP2* (*Sobic.002G075800*) is significantly downregulated in the former during the vegetative stage as compared with the latter, but considerably upregulated at all later stages, confirming its putative function in sugar metabolism and storage. Another candidate gene (*Sobic.009G235700*) that differentiates Rio and BTx623 contains a predicted domain for sugar transport possessing four amino acid substitutions. An important mutation in *NAC* gene underlying the *D* locus was found to be a contributing factor that distinguishes juicy-stalked sorghum from dry-stalked sorghum with an influencing effect on sugar yield (Xia *et al.*, 2018).

A comparison of DNA polymorphisms among BTx623, Tx7000 and BTx642 genotypes, derived from accessions belonging to the races *kafir*, *durra* and *caudatum*, respectively, using the resequence data revealed the presence of >2.8 million SNPs and small insertions/deletions (InDels), of which 1.2 million SNPs and 120,969 InDels discriminated Tx7000 from BTx623, and 1.6 million SNPs and 152,836 InDels discriminated BTx642 from BTx623 (Evans *et al.*, 2013). Genomic DNA spanning the plant height loci, *dw1* (*SBI-09*) and *dw3* (*SBI-07*), revealed identical haplotypes among the three genotypes owing to the selection for dwarf height. The lower density of SNPs in genes located in peri-centromeric regions of the chromosome compared with those located in euchromatic regions is consistent with these low recombination regions. In order to expand the ambit of DNA sequence variations in sorghum, it is essential to resequence a greater number of highly diverse germplasm accessions including landraces and wild relatives of sorghum along with a set of closely related, but phenotypically divergent genotypes. To explore such wide DNA sequence variations, resequencing was performed for a set of extremely diverse cultivated sorghum accessions belonging to major races (18 landraces and 17 improved inbreds), besides its progenitors and *Sorghum propinquum*, which revealed about 4.9 million high-quality SNPs, of which 83 and 4.5% were located in intergenic and coding regions, respectively (Mace *et al.*, 2013). Detected were about 1,982,971 InDels, the majority of them were 1–6 bp in length (86%) and only 2.5% of them were of length >20 bp. Considering only the genome sequence of *S. propinquum*, about 8 million high-quality SNPs and 1.9 million InDels were identified in addition to specific events of gene loss and gain in cultivated sorghum. A greater number of SNPs was found to be specific to wild and weedy sorghum (34%) as compared with landraces (18%) and improved inbreds (8%).

DNA sequence variation from a limited number of diverse sorghum genotypes may not reflect the full landscape of genetic diversity available in a particular plant species. To expand the capture of genetic variation, the pan-genome analysis offers an appropriate platform through the investigation of the entire repertoire of the genome of the species of interest by sequencing multiple individuals of that species. Cheaper sequencing technologies and refined analysis tools have made sequencing of multiple genotypes of a crop species feasible (Golicz *et al.*, 2016) and this can be seen from a large number of pan-genomic studies in crops such as rice (Zhao *et al.*, 2018), maize (Hirsch *et al.*, 2014), wheat (Montenegro *et al.*, 2017), soybean (Li *et al.*, 2014) and a few other crops, which have shed new insights on crop diversity and application of pan-genomic variations in crop improvement. Therefore, pan-genome analysis has great potential in dissecting the complete crop diversity of the sorghum crop for its effective utilization in sorghum improvement programmes.

### 28.3.2 DNA marker resources

Together with genetic resources such as germplasm accessions and mapping populations, genomic resources such as DNA markers are vital for the identification and mapping of the genomic regions/QTLs associated with important target traits for their further use in genetic improvement through marker-assisted selection or molecular breeding. Even though DNA markers such as randomly amplified polymorphic DNA (RAPD), restriction fragment length polymorphism (RFLP), inter-simple sequence repeat (ISSR) and amplified fragment length polymorphism (AFLP) were used initially for genetic diversity assessment and mapping (Ahnert *et al.*, 1996; Uptmoor *et al.*, 2003; Ritter *et al.*, 2007; Knoll *et al.*, 2008; Aruna *et al.*, 2012), simple sequence repeats (SSRs) are the markers of choice for molecular breeders due to their presence in abundance in the genome, hypervariability, co-dominant nature, high reproducibility and also amenability to automation. Prior to the *de novo* sequencing of the draft genome of sorghum, a substantial number of SSR markers were developed, but at a slower pace, through small-insert genome library-based conventional approaches or computational approaches by exploiting the DNA sequence in the public databases. These SSR markers were developed from genomic DNA libraries with or without enrichment, cDNAs/expressed sequence tags (ESTs) and unigenes (Table 28.2). However, the availability of the draft genome sequence of sorghum (Paterson *et al.*, 2009) accelerated and scaled up the development of DNA markers by employing computation tools involving improved algorithms for the analysis of large-scale data. By exploiting the draft genome sequence of sorghum, genome-wide non-redundant SSR markers (5599) were developed by Yonemaru *et al.* (2009) and the physical position of 5012 of these markers was determined by using the electronic PCR-based positions in combination with the predicted locations of gene loci. Experimental validation of 970 of these SSR markers in a set of 12 genotypes comprising 11 sorghum lines and one sudangrass line revealed successful amplifications in 67.8% of markers in sorghum line (BTx623), exhibiting a mean polymorphism rate of 45.1%. With this success rate, ~3400 of these SSR markers could be useful with an expected >1500 polymorphic SSR markers in the sorghum lines used for validation. Microsatellite markers targeting $(GATA)_n$ motifs were identified in sorghum through genome-wide analysis of sorghum genome by Jaikishan *et al.* (2013). Out of the 128 such motifs identified, a set of 110 PCR-based markers was developed and the validation of 50 of these markers in 24 diverse sorghum genotypes belonging to different racial groups resulted in the identification of 38 polymorphic markers with an average polymorphism information content (PIC) value of 0.69. Even though these markers constitute a small set of

**Table 28.2.** DNA markers developed in sorghum by various research groups internationally.

| Type of DNA marker | No. of markers | Reference |
|---|---|---|
| Genomic SSRs | 408 | Brown *et al.* (1996); Taramino *et al.* (1997); Bhattramakki *et al.* (2000); Kong *et al.* (2000) |
| cDNA/EST-derived SSRs | 1,360 | Schloss *et al.* (2002); Arun (2006); Srinivas *et al.* (2008, 2009a); Ramu *et al.* (2009) |
| Unigene-derived SSRs | 1,569 | Srinivas *et al.* (2009b); Nagaraja Reddy *et al.* (2012) |
| Other SSRs | 30 | Mutegi *et al.* (2011); Billot *et al.* (2012) |
| Genome-wide SSRs | 5,599 | Yonemaru *et al.* (2009) |
| Genome-wide $(GATA)_n$ motif-based SSRs | 110 | Jaikishan *et al.* (2013) |
| Genome-wide ILPs | 37,861 | Jaikishan *et al.* (2015) |
| Genome-wide SNPs | 283,000 | Nelson *et al.* (2011) |
| | 1,057,018 | Zheng *et al.* (2011) |
| | 4,946,038 | Mace *et al.* (2013) |
| Genome-wide InDels | 99,948[a] | Zheng *et al.* (2011) |
| | 1,982,971[b] | Mace *et al.* (2013) |

[a]InDels of 1–10 bp length.
[b]InDels of 1–66 bp length.

SSR markers, they could be a good set of markers for genetic diversity and other breeding applications due to their robust amplification along with good allelic diversity.

In addition to SSRs, DNA sequence variations also occur as SNPs and additions or deletions, commonly known as InDels, which can be a potential target for the development of DNA markers. Sequencing of the draft genome of sorghum (Paterson *et al.*, 2009) and simultaneous improvements in NGS technologies as well as computational tools for the analysis of large-scale DNA sequence data have resulted in the rapid genome-wide development of DNA markers for further use in high-throughput genotyping. The first report on genome-wide identification of SNPs was published by Nelson *et al.* (2011), who identified about 283,000 SNPs through the alignment of short-read sequences generated from eight diverse accessions to the reference genome. During the same time, another set of 1,057,018 SNPs was identified through whole-genome resequencing and alignment of sweet sorghum (Keller and E-Tian) and grain sorghum (Ji2731) inbred lines. This investigation also resulted in the identification of 99,948 InDels having a length of 1–10 bp (Zheng *et al.*, 2011). To generate a wide repertoire of SNP data in sorghum for its effective use in SNP genotyping and QTL identification for important target traits, a set of 44 highly diverse genotypes belonging to important races of cultivated sorghum was resequenced along with its progenitors and *S. propinquum*. That study resulted in the identification of genome-wide SNPs (4,946,038) and InDels (1,982,971) of length in the range of 1–66 bp (Mace *et al.*, 2013). DNA sequence variation in the intronic regions of genes in the form of SNPs and InDels can be used as markers; however, intron length polymorphism (ILP) is the most easily identifiable one since it is PCR-based targeting the InDels. A total of 37,861 potential introns were identified. Analysis of 36,139 genes/coding sequences of sorghum resulted in the identification of 37,861 potential introns, which were targeted to develop ILP markers (Jaikishan *et al.*, 2015). The number of ILP markers ranged from 1498 (chromosome 5) to 7290 (chromosome 1) and were distributed across sorghum chromosomes. Experimental validation of 200 ILP markers in a set of 24 diverse sorghum genotypes revealed robust amplification for 172 markers, of which 48 exhibited polymorphisms.

SNPs have been identified in sorghum by employing a genotype-by-sequencing (GBS) approach by different research groups worldwide. Approximately 265,000 SNPs were identified by Morris *et al.* (2013) through GBS for 971 accessions comprised of the mini-core collection, the US Sorghum Association Panel and the reference set developed under the Generation Challenge

Programme. A comparison of resequence data generated from sorghum genotypes Tx7000 and BTx642 with the reference sequence of BTx623 resulted in the identification of 1.2 million SNPs and 120,969 InDels between Tx7000 and BTx623, and 1.6 million SNPs and 152,836 InDels between BTx642 and BTx623 (Evans *et al.*, 2013). Differences in coordinates of the reference genome among these GBS data sets poses a great challenge in the reuse of the large-scale SNP data sets generated by different groups. To overcome this problem, a reference SNP data set was developed for sorghum (Hu *et al.*, 2019) by integrating GBS data generated from various studies involving a large set of sorghum lines (sorghum association panel, bioenergy association panel, NAM population, etc.) and large-scale SNP data, which will assist in the reuse of these data sets. Generation of such large-scale genome-wide SNPs through GBS will accelerate the QTL mapping and precise detection of marker–trait associations. Details of different types of DNA markers developed before and after the draft genome sequencing of sorghum are given in Table 28.2. Together these genomic resources will help the understanding of complex traits and the DNA sequence variations associated with favourable alleles, which will be useful in their deployment in genetic improvement programmes to achieve genetic gains in grain and biomass.

With the availability of sorghum germplasm accessions possessing variability, genome annotations and the advancements in SNP genotyping, SNP–trait associations can be identified through candidate gene-based association analysis, since genes involved in Fe and Zn metabolism are identified in major cereals. Towards this end, about 22 candidate genes reported in major cereals were identified in sorghum encompassing all chromosomes except chromosomes 5, 8 and 9, and their accession IDs were retrieved from MOROKOSHI Sorghum transcriptome database. A total of 143 SNPs were identified in these candidate genes using SorGSD database; the greatest number of SNPs were identified in *zinc-induced facilitator-like* gene (30 SNPs) and the least were found in *FDH*, *IDEF1*, *ID1* and *IDS3* (one SNP each). Primers were designed targeting these 143 SNPs in candidate genes for genotyping using the sorghum association mapping panel through competitive allele-specific PCR (KASP) assay (Rajendrakumar *et al.*, 2018). This set of candidate gene SNP-based primers forms a useful genomic resource that will be helpful for the identification of SNP–trait associations for grain Fe and Zn contents in sorghum.

## 28.4 Biofortification Through Genomics-Assisted Breeding

The presence of extensive genetic variation for the target traits is very important in their genetic improvement for improving the trait of interest. Sorghum germplasm harbours sufficient variability for grain Fe and Zn contents, which could be exploited to achieve their genetic enhancement by a clear understanding of their genetic control. Earlier genetic studies revealed the quantitative inheritance of both Fe and Zn contents in grain with the predominance of additive gene action for Zn content and additive as well as non-additive gene actions for grain Fe content (Ashok Kumar *et al.*, 2013c; Hariprasanna *et al.*, 2014b). Being quantitatively inherited, dissecting the genetic factors governing grain Fe and Zn contents in sorghum and determination of the map positions of these traits along with their linked markers are essential for development of micronutrient-rich cultivars through molecular breeding. A number of QTLs and candidate genes controlling grain Fe and Zn contents have been reported from rice (Anuradha *et al.*, 2012), maize (Jin *et al.*, 2013), wheat (Velu *et al.*, 2018) and pearl millet (Sushil Kumar *et al.*, 2018).

Marker-assisted selection has been applied successfully in hastening the breeding programmes towards the improvement of various yield and quality-related traits, but is in its infancy for QTL identification and validation for grain Fe and Zn contents. Utility of these validated markers in marker-assisted selection for transferring high Fe and Zn QTLs from germplasm to elite high-yielding cultivars is yet to be determined. In pearl millet, using a RIL population comprising 317 lines derived from the cross (ICMS 8511-S1-17-2-1-1-B-P03 × AIMP 92901-S1-183-2-2-B-08) genotyped using SSR and Diversity Arrays Technology (DArT) markers, about 19 QTLs were detected for Fe and Zn contents, of which three were large-effect QTLs that were co-mapped, one on LG1 and the other two on LG7 (Sushil Kumar *et al.*, 2018). In rice, loci associated with grain Fe content including iron homeostasis genes, *nicotianamine synthase* (*OsNAS3*) and *vacuolar iron transporter* (*OsVIT1*), have been identified using a MAGIC Plus population through GWAS (Descalsota *et al.*, 2018). In sorghum, a homology (*in silico*) search was performed on 91 metal homeostasis candidate genes from major cereals and about 77 genes exhibiting homology with the sorghum genome were identified, of which genes associated with grain Fe and Zn concentrations from maize and wheat showed 100% homology on the sorghum genome. Gene identical percentage (similarity of sequences) ranged from 71.9% (*IDS3*) to 95.9% (*HMA*) with an average of 86.4% (Anuradha *et al.*, 2013). Very recently, 11 QTLs (individual environments) and three QTLs (across environments) for grain Fe and Zn contents were identified in sorghum using an $F_6$ RIL population (342 lines) generated from the cross 296B × PVK 801 that was phenotyped across six environments and genotyped using three DNA marker systems, namely SSR, DArT and Diversity Arrays Technology Sequencing (DArTSeq). Two putative candidate genes (*CYP71B34* and *ZFP8*) associated with Fe and Zn metabolism were identified in the QTL interval of *qfe7.1*, *qzn7.1* and *qzn7.2* (across environments) located on chromosome 7 (Anuradha *et al.*, 2019). Use of the markers linked to these QTLs can be made in marker-assisted selection programmes for the development of sorghum cultivars that are rich in grain Fe and Zn.

Marker-assisted selection is considered a promising approach for the introgression of genomic regions/QTLs with major effects. However, the genetic improvement of complex quantitative traits through marker-assisted selection of identified QTLs was found to be unsuccessful in many cases due different genetic backgrounds, QTL × environment interactions and QTLs with minor effects (Bernardo, 2016). As Fe and Zn accumulation in grains is greatly influenced by environmental conditions, detection of QTLs with minor effects is problematic through association mapping due to its low power of detection of QTLs with smaller effects. In this scenario, genomic selection (GS) is promising as it combines the genotypic and phenotypic data to predict the performance of individual genotypes in a population through their genomic estimated breeding values (Crossa *et al.*, 2017). Despite its relatively simple diploid genome and the availability of genomic resources, the implementation of GS in sorghum is in its infancy compared with crops like wheat and maize. The potential of GS for Fe and Zn biofortification was demonstrated in wheat using the HarvestPlus association mapping panel comprising 330 diverse lines, resulting in the genomic prediction ability of 0.331–0.694 for Zn and 0.324–0.734 for Fe across different environments (Velu *et al.*, 2016). GS can be effective in identifying useful sources of genetic variation since high prediction accuracies ($0.67 \leq r \leq 0.83$) were achieved for biomass yield, plant height, stalk number and lodging using GS models (Yu *et al.*, 2016). The first report on the development of genomic prediction models in sorghum for plant height, studied in 151 diverse accessions, was published in 2017 (Watanabe *et al.*, 2017). The predictability of plant height and yield was increased in diverse biomass sorghum lines by using trait-based GS as compared with GS alone (Fernandez *et al.*, 2018). Among related sorghum families, moderate to high prediction was obtained for grain yield across environments and the genomic prediction declined with divergence of individuals between training and testing populations (Hunt *et al.*, 2018). Promising predictions from the above studies in sorghum indicate that there is an immense potential for the implementation of GS models for estimating the breeding value of individuals in the population with a focus on achieving accelerated genetic gains for micronutrient contents.

## 28.5 Biofortification Through Genetic Modification

Transgenic technology is a powerful alternative when an adequate amount of genetic variability for a target trait is not available in the germplasm to follow the genetic improvement through conventional or molecular breeding strategies, or when an inability to exploit the variability available exists due to certain barriers. This is true for micronutrients like Fe and Zn as their variability is either limited or unusable due to certain geographical, social, political/diplomatic or security issues. Development of a successful transgenic requires reliable methods of genetic transformation for introducing gene constructs and efficient methods of plant regeneration from genetically transformed cells. Even though the generation of transgenic plants in sorghum was challenging for a long time, noteworthy progress has been made in the area of microprojectile-mediated transformation and *Agrobacterium*-mediated transformation due to intensive research in recent years (Howe *et al.*, 2006; Liu and Godwin, 2012; Wu *et al.*, 2014; Do *et al.*, 2016; Belide *et al.*, 2017; Che *et al.*, 2018). To overcome the shortcomings of *Agrobacterium*-mediated transformation, an *in planta* method of producing transgenic sorghum was proposed by Elkonin *et al.* (2009) by inoculating panicles with agrobacterial cell suspension containing activated *vir* genes. Modifications to this *in planta* method of genetic transformation in sorghum were reported by Yellisetty *et al.* (2015) and Visarada *et al.* (2016). Advancements in genetic transformation, tissue culture and plant regeneration systems in sorghum were reviewed by Visarada and Sai Kishore (2015).

Transgenics has a greater potential for the development of sorghum rich in essential amino acids (lysine, tryptophan and methionine) and micronutrients due to the occurrence of limited or no variability for some of these nutrients. Specific transgenes can be incorporated in the plant system for redistribution of micronutrients to specific tissues, enhancing the bioavailability of micronutrients by silencing of antinutritional factors and transfer of multiple genes to a single plant (Naqvi *et al.*, 2009; Carvalho and Vasconcelos, 2013). Crops get Fe and Zn from the rhizosphere surrounding them as micronutrients are not synthesized in the plant system (Morrissey and Guerinot, 2009). Genetic engineering has been deployed in various crops to increase mineral content especially Fe and Zn. The transgenic approaches for the development of Fe- and Zn-rich crops are mainly focused on increasing their uptake and use efficiency in plants by regulating the expression of transporters (Kerkeb *et al.*, 2008) and silencing the antinutritional factors such as phytic acid. Considering the poor Fe and Zn bioavailability in sorghum because of high phytate content and also poor protein quality, research efforts in the form of the Africa Biofortified Sorghum (ABS) Project started in 2005 (Zhao, 2008; AHBFI, 2010) to develop transgenic sorghum with increased levels of lysine (80–100% increase), vitamin A (20 μg β-carotene/g dry sorghum), Fe (50% increase) and Zn (35% increase). The ABS Project developed the sorghum transformation system through the technology developed by DuPont Pioneer. Over 250 events of golden sorghum (with yellow/golden endosperm) have been produced and analysed for enhanced carotenoid levels, β-carotene stability and field performance. The Project achieved 50% improvement in β-carotene half-life stability from 3 to 7.5 weeks (AHBFI, 2011).

Genetic manipulation of genes involved in the metabolic pathways, uptake, translocation and sequestration of micronutrients has great potential for improving grain Fe and Zn contents in plants. Increased content of ferritin (non-toxic storage form of Fe) through genetic engineering is one of the possible ways of achieving enhanced accumulation of Fe in edible plant parts. Increase in Fe content and its bioavailability through overexpression of the gene encoding ferritin has been reported in several crops (Lucca *et al.*, 2001; Drakakaki *et al.*, 2005; Aluru *et al.*, 2011; Borg *et al.*, 2012). An increased accumulation of ferritin and enhanced translocation of Fe was achieved in rice by overexpression of *OsYSL2*, the gene coding for Fe(II)-nicotianamine transporter, in rice endosperm (Masuda *et al.*, 2012). The resultant transgenic lines with higher levels of Fe and Zn content indicated the effectiveness of Fe biofortification by the introduction of multiple genes involved in Fe and Zn homeostasis as compared with a single gene. Increase in Fe content up to fourfold in polished rice was achieved by the overexpression of *IRT1*, encoding a divalent metal transporter involved in Fe uptake, along with *PvFER1* in the endosperm (Boonyaves *et al.*, 2017). Overexpression of genes involved in Fe uptake, such as those for nicotianamine synthase, mugeneic acid and Fe deficiency-inducible transcription factor, and analysis of genetically transformed rice plants have shown enhanced accumulation of Fe and Zn contents (Lee *et al.*, 2009; Masuda *et al.*, 2013, 2017).

In spite of large efforts to increase the Fe and Zn contents in seeds or any other edible part, biofortification faces the great challenge of bioavailability. Seeds of sorghum and other cereals store P as phytic acid, which has a strong affinity to chelate divalent cations, especially $Fe^{2+}$ and $Zn^{2+}$, making them unavailable for absorption by intestinal cells. Bioavailability of Fe and Zn can be increased by using genetic engineering to enhance the production of phytase enzyme that is known to hydrolyse phytic acid.

Transgenic plants with enhanced phytase production in their grains have been reported in crops such as maize, rice and wheat (Reddy *et al.*, 2017). Genetic transformation studies concluded that increased expression of phytases from various sources can help in improving the bioavailability of micronutrients (Lucca *et al.*, 2001; Brinch-Pedersen *et al.*, 2002). The genetically modified sorghum developed under the ABS Project had elevated levels of provitamin A (5.7–21 μg/g β-carotene), reduced phytate (35–65%) and an improved protein quality (tryptophan 10–20%, lysine 30–120%, threonine 30–40%) (AHBFI, 2011). Bioaccessibility of provitamin A using an *in vitro* digestion model found that the transgenic event *Homo188-A* contained the greatest bioaccessible β-carotene, with a four- to eightfold increase from null/non-transgenic sorghum (Lipkie *et al.*, 2013). Low-phytate sorghum plants were generated in the line Tx430 through genetic transformation, where the ATP-binding cassette transporter (a protein associated with multidrug resistance) was silenced, resulting in lower phytate content (80–86%) in grains and increased Zn and Fe availability compared with the control plants (Kruger *et al.*, 2013). The lowering of phytic acid content in seeds was achieved in maize by expressing a fungal phytase leading to a threefold enhancement of Fe bioavailability (Drakakaki *et al.*, 2005). Another potential target for enhancing the bioavailability of Fe is cysteine amino acid, which has been proved to enhance the bioavailability of Fe in humans (Glahn and Van Campen, 1997). Bioavailability studies have shown increased Zn absorption of 30–40% and increased Fe absorption of 20–30% (Saltzman *et al.*, 2013) when phytate levels were reduced by ≥30 and ≥80%, respectively, in sorghum. Efforts are underway to increase the levels and stability of vitamin A within the plant and enhance mineral bioavailability through alternative approaches for phytate reduction (Obi *et al.*, 2017).

In contrast to Fe and Zn, genetic engineering strategies for enhancing vitamin and protein contents are quite simple. The incorporation of a limiting step in metabolic pathways of developing seed or modifying the pathway for their increased production are the most promising genetic engineering approaches to develop protein- or vitamin-rich genotypes. The barley gene encoding high-lysine analogue (HT12 protein) of α-hordothionin protein under the control of a maize *γ-zein* promoter and a herbicide-resistant gene (*bar*) under the control of maize *ubiquitin* promoter were co-transformed in sorghum genotypes, P898012 and PHI391, through *Agrobacterium*-mediated genetic transformation and resulted in five independent transgenic events, of which three exhibited high levels of protein in the grain leading to a 40–60% increase in lysine content (Zhao *et al.*, 2003). In order develop sorghum grains rich in β-carotene content, the genetic constructs encoding enzymes implicated in the carotenoid biosynthesis pathway (1-deoxyxylulose 5-phosphate synthase, *Zea mays* phytoene synthase 1 and *Pantoea ananatis* carotene desaturase) were introduced into the genome of sorghum genotype Tx430, resulting in significantly increased levels of β-carotene in mature seeds of transgenic (9.1 μg/g) compared with non-transgenic plants (0.5 μg/g), which degraded during storage due to its oxidation (Che *et al.*, 2016). The stability of provitamin A during seed storage was enhanced by the co-expression of a barley vitamin A biosynthesis gene (*homogentisate geranylgeranyl transferase*) stacked with carotenoid biosynthesis genes, resulting in mitigation of β-carotene oxidation. The success reported in the studies discussed above proves the effectiveness of the transgenic approach in the genetic manipulation of genes involved in plant metabolism as per human needs.

## 28.6 Future Prospects

Increasing the micronutrient content of sorghum grain is of paramount importance for alleviating malnutrition since it will help in overcoming the hidden hunger that is prevalent in millions of women and children in the sorghum-growing/consuming regions across the globe. It is known that biofortification involving crop breeding, genetic modification, and even agronomic augmentation of minerals, is a promising strategy that offers immense promise for addressing the challenges posed by micronutrient malnutrition. An ever-increasing repertoire of genomics and bioinformatics will help in the identification of candidate genes involved in the accumulation of Fe, Zn and provitamin A and study their expression. Moreover, elucidation of the role of mineral ion transporters in nutrient uptake, translocation and storage could assist in accumulation of micronutrients in edible parts of the crop. Identification of SSRs, SNPs and InDels linked to micronutrient contents through biparental mapping, association mapping and GS approaches will help in the development of micronutrient-rich sorghum cultivars in a short time and with limited resources. The genome editing tools, such as clustered regularly interspaced short palindromic repeats (CRISPR)/CRISPR-associated protein 9 (Cas9), zinc-finger nucleases (ZFNs) and transcription activator-like effector nucleases (TALENs), should be exploited to edit sorghum genes involved in the accumulation of micronutrients towards the development of biofortified sorghum. In addition to significant enhancements in micronutrients and vitamins, there is an immense need for improving the bioavailability of micronutrients by reduction of antinutritional factors. Plant scientists should work in tandem with nutritionists and medical practitioners to evaluate the bioavailability of micronutrients through *in vitro* studies or by clinical trials.

## References

AHBFI (2010) *Africa Biofortified Sorghum Project: Five-year Progress Report 2010.* Africa Harvest Biotech Foundation International, Nairobi.

AHBFI (2011) *Africa Harvest Annual Report, 2010.* Africa Harvest Biotech Foundation International, Nairobi.

Ahnert, D., Lee, M., Austin, D., Livini, C., Woodman, W. *et al.* (1996) Genetic diversity among elite sorghum inbred lines assessed with DNA markers and pedigree information. *Crop Science* 36, 1385–1392.

Aluru, M.R., Rodermel, S.R. and Reddy, M.B. (2011) Genetic modification of low phytic acid 1-1 maize to enhance iron content and bioavailability. *Journal of Agricultural and Food Chemistry* 59, 12954–12962.

Anuradha, K., Agarwal, S., Rao, Y.V., Rao, K.V., Viraktamath, B.C. and Sarla, N. (2012) Mapping QTLs and candidate genes for iron and zinc concentrations in unpolished rice of Madhukar × Swarna RILs. *Gene* 508(2), 233–240.

Anuradha, K., Prakash, B., Ramu, P., Shah, T., Ashok Kumar, A. and Deshpande, S.P. (2013) *In silico* identification of candidate genes involved for grain Fe and Zn concentration in sorghum using reported cereals gene homologs. In: Rakshit, S., Das, I.K., Shyamprasad, G. and Mishra, J.S. (eds) *Compendium of Papers and Abstracts: Global Consultation on Millets Promotion for Health & Nutritional Security.* Society for Millets Research, Directorate of Sorghum Research, Hyderabad, India, pp. 10–12.

Anuradha, K., Phuke, R., Hariprasanna, K., Mehtre, S.P., Rathore, A. *et al.* (2019) Identification of QTLs and candidate genes for high grain Fe and Zn concentration in sorghum [*Sorghum bicolor* (L.) Moench]. *Journal of Cereal Science* 90, 102850.

Arun, S.S. (2006) *In silico* EST data mining for elucidation of repeats biology and functional annotation in sorghum [*Sorghum bicolor* (L.) Moench]. MSc thesis, University of Agricultural Sciences, Dharwad, India.

Aruna, C., Priya, A.R., Neeraja, C.N., Patil, J.V. and Visarada, K.B.R.S. (2012) Diversity analysis using ISSR markers for resistance to shoot pests in sorghum. *Crop Protection* 35, 110–117.

Ashok Kumar, A., Reddy, B.V.S., Ramaiah, B., Reddy, P.S., Sahrawat, K.L. and Upadhyaya, H.D. (2009) Genetic variability and plant character association of grain Fe and Zn in selected core collection accessions of sorghum germplasm and breeding lines. *Journal of SAT Agricultural Research* 7, 1–4.

Ashok Kumar, A., Reddy, B.V.S., Sahrawat, K.L. and Ramaiah, B. (2010) Combating micronutrient malnutrition: identification of commercial sorghum cultivars with high grain iron and zinc. *Journal of SAT Agricultural Research* 8, 1–5.

Ashok Kumar, A., Reddy, B.V.S., Ramaiah, B., Sahrawat, K.L. and Pfeiffer, W.H. (2012a) Genetic variability and character association for grain iron and zinc contents in sorghum germplasm accessions and commercial cultivars. *The European Journal of Plant Science and Biotechnology* 6(Sp. Iss. 1), 66–70.

Ashok Kumar, A., Reddy, B.V.S. and Ramaiah, B. (2012b) Database for grain Fe and Zn in sorghum – a proposal. *Journal of SAT Agricultural Research* 10, 1–7.

Ashok Kumar, A., Anuradha, K. and Ramaiah, B. (2013a) Increasing grain Fe and Zn concentration in sorghum: progress and way forward. *Journal of SAT Agricultural Research* 11, 1–5.

Ashok Kumar, A., Reddy, B.V.S. and Ramaiah, B. (2013b) Biofortification for combating micronutrient malnutrition: identification of commercial sorghum cultivars with high grain iron and zinc concentrations. *Indian Journal of Dryland Agricultural Research & Development* 28, 95–100.

Ashok Kumar, A., Reddy, B.V.S., Ramaiah, B., Sahrawat, K.L. and Pfeiffer, W.H. (2013c) Gene effects and heterosis for grain iron and zinc concentration in sorghum [*Sorghum bicolor* (L.) Moench]. *Field Crop Research* 146, 86–95.

Bajgain, P., Rouse, M.N., Tsilo, T.J., Macharia, G.K., Bhavani, S., Jin, Y. and Anderson, J.A. (2016) Nested association mapping of stem rust resistance in wheat using genotyping by sequencing. *PLoS One* 11(5), e0155760.

Belide, S., Vanhercke, T., Petrie, J.R. and Singh, S.P. (2017) Robust genetic transformation of sorghum (*Sorghum bicolor* L.) using differentiating embryogenic callus induced from immature embryos. *Plant Methods* 13, 109.

Ben-Israel, I., Kilian, B., Nida, H. and Fridman, E. (2012) Heterotic trait locus (HTL) mapping identifies intra-locus interactions that underlie reproductive hybrid vigor in *Sorghum bicolor*. *PLoS One* 7, e38993.

Bergelson, J. and Roux, F. (2010) Towards identifying genes underlying ecologically relevant traits in *Arabidopsis thaliana*. *Nature Reviews Genetics* 11, 867–879.

Bernardo, R. (2016) Bandwagons I, too, have known. *Theoretical and Applied Genetics* 129, 2323–2332.

Bhat, J.S., Patil, B.S., Hariprasanna, K., Hossain, F., Muthusamy, V. *et al.* (2018) Genetic enhancement of micronutrient content in cereals. *SABRAO Journal of Breeding and Genetics* 50(3), 373–429.

Bhattramakki, D., Dong, J., Chhabra, A.K. and Hart, G. (2000) An integrated SSR and RFLP linkage map of *Sorghum bicolor* (L.) Moench. *Genome* 43, 988–1002.

Billot, C., Rivallan, R., Sall, M.N., Fonceka, D., Deu, M. *et al.* (2012) A reference microsatellite kit to assess for genetic diversity of *Sorghum bicolor* (Poaceae). *American Journal of Botany* 99, e245–e250.

Black, R.E., Victora, C.G., Walker, S.P., Bhutta, Z.A., Christian, P. *et al.* (2013) Maternal and child undernutrition and overweight in low-income and middle-income countries. *Lancet* 382(9890), 427–451.

Boonyaves, K., Wu, T.Y., Gruissem, W. and Bhullar, N.K. (2017) Enhanced grain iron levels in rice expressing an iron-regulated metal transporter, nicotianamine synthase, and ferritin gene cassette. *Frontiers in Plant Science* 8, 130.

Borg, S., Brinch-Pedersen, H., Tauris, B., Madsen, L.H., Darbani, B., Noeparvar, S. and Holm, P.B. (2012) Wheat ferritins, improving the iron content of the wheat grain. *Journal of Cereal Science* 56, 204–213.

Bouchet, S., Olatoye, M.O., Marla, S.R., Perumal, R., Tesso, T. *et al.* (2017) Increased power to dissect adaptive traits in global sorghum diversity using a nested association mapping population. *Genetics* 206(2), 573–585.

Bouis, H.E. (2002) Plant breeding: a new tool for fighting micronutrient malnutrition. *Journal of Nutrition* 132, 491S–494S.

Boyles, R.E., Brenton, Z.W. and Kresovich, S. (2019) Genetic and genomic resources of sorghum to connect genotype with phenotype in contrasting environments. *The Plant Journal* 97(1), 19–39.

Brenton, Z.W., Cooper, E.A., Myers, M.T., Boyles, R.E., Shakoor, N. *et al.* (2016) A genomic resource for the development, improvement, and exploitation of sorghum for bioenergy. *Genetics* 204(1), 21–33.

Brinch-Pedersen, H., Sorensen, L.D. and Holm, P.B. (2002) Engineering crop plants: getting a handle on phosphate. *Trends in Plant Science* 7, 118–125.

Brown, S.M., Hopkins, M.S., Mitchell, S.E., Senior, M.L., Wang, T.Y. *et al.* (1996) Multiple methods for the identification of polymorphic simple sequence repeats (SSRs) in sorghum [*Sorghum bicolor* (L.) Moench]. *Theoretical and Applied Genetics* 93, 190–198.

Carvalho, S.M.P. and Vasconcelos, M.W. (2013) Producing more with less: strategies and novel technologies for plant-based food biofortification. *Food Research International* 54, 961–971.

Casa, A.M., Pressoir, G., Brown, P.J., Mitchell, S.E., Rooney, W.L. *et al.* (2008) Community resources and strategies for association mapping in sorghum. *Crop Science* 48, 30–40.

Cavanagh, C., Morell, M., Mackay, I. and Powell, W. (2008) From mutations to magic: resources for gene discovery, validation and delivery in crop plants. *Current Opinion in Plant Biology* 11, 215–221.

Che, P., Zhao, Z.Y., Glassman, K., Dolde, D., Hu, T.X. *et al.* (2016) Elevated vitamin E content improves all-*trans* β-carotene accumulation and stability in biofortified sorghum. *Proceedings of the National Academy Sciences USA* 113(39), 11040–11045.

Che, P., Anand, A., Wu, E., Sander, J.D., Simon, M.K. *et al.* (2018) Developing a flexible, high-efficiency *Agrobacterium*-mediated sorghum transformation system with broad application. *Plant Biotechnology Journal* 16(7), 1388–1395.

Collard, B.C.Y., Jahufer, M.Z.Z., Brouwer, J.B. and Pang, E.C.K. (2005) An introduction to markers, quantitative trait loci (QTL) mapping and marker-assisted selection for crop improvement: the basic concepts. *Euphytica* 142(12), 169–196.

Cooper, E.A., Brenton, Z.W., Flinn, B.S., Jenkins, J., Shu, S. *et al.* (2019) A new reference genome for *Sorghum bicolor* reveals high levels of sequence similarity between sweet and grain genotypes: implications for the genetics of sugar metabolism. *BMC Genomics* 20(1), 420.

Crossa, J., Perez-Rodriguez, P., Cuevas, J., Montesinos-Lopez, O., Jarquin, D. *et al.* (2017) Genomic selection in plant breeding: methods, models, and perspectives. *Trends in Plant Science* 22, 961–975.

Dahlberg, J.A., Burke, J.J. and Rosenow, D.T. (2004) Development of a sorghum core collection: refinement and evaluation of a subset from Sudan. *Economic Botany* 58, 556–567.

Darvasi, A., Weinreb, A., Minke, V., Weller, J.I. and Soller, M. (1993) Detecting marker–QTL linkage and estimating QTL gene effect and map location using a saturated genetic map. *Genetics* 3, 943–951.

de Alencar Figueiredo, L., Sine, B., Chantereau, J., Mestres, C., Fliedel, G. *et al.* (2010) Variability of grain quality in sorghum: association with polymorphism in *Sh2*, *Bt2*, *SssI*, *Ae1*, *Wx* and *O2*. *Theoretical and Applied Genetics* 121, 1171–1185.

Descalsota, G.I.L., Swamy, B.P.M., Zaw, H., Inabangan-Asilo, M.A., Amparado, A. *et al.* (2018) Genome-wide association mapping in a rice MAGIC Plus population detects QTLs and genes useful for biofortification. *Frontiers in Plant Science* 9, 1347.

Deu, M., Rattunde, F. and Chantereau, J. (2006) A global view of genetic diversity in cultivated sorghums using a core collection. *Genome* 49, 168–180.

Diallo, C., Rattunde, H.F., Gracen, V., Toure, A., Nebie, B. *et al.* (2019) Genetic diversification and selection strategies for improving sorghum grain yield under phosphorous-deficient conditions in West Africa. *Agronomy* 9(11), 742.

Dixit, S., Singh, U.M., Abbai, R., Ram, T., Singh, V.K. *et al.* (2019) Identification of genomic region(s) responsible for high iron and zinc content in rice. *Scientific Reports* 9, 8136.

Do, P.T., Lee, H., Mookkan, M., Folk, W.R. and Zhang, Z.J. (2016) Rapid and efficient *Agrobacterium*-mediated transformation of sorghum (*Sorghum bicolor*) employing standard binary vectors and *bar* gene as a selectable marker. *Plant Cell Reports* 35(10), 2065–2076.

Drakakaki, G., Marcel, S., Glahn, R.P., Lund, E.K., Pariagh, S. *et al.* (2005) Endosperm-specific co-expression of recombinant soybean ferritin and *Aspergillus* phytase in maize results in significant increases in the levels of bioavailable iron. *Plant Molecular Biology* 59, 869–880.

Elkonin, L.A., Ravin, N.V., Leshko, E.V., Volokhina, I.V., Chumakov, M.I. and Skryabin, K.G. (2009) Agrobacterial transformation of sorghum plants in *in planta* conditions. *Biotechnology* 1, 23–30.

Elkonin, L., Italyanskaya, J. and Panin, V. (2018) Genetic modification of sorghum for improved nutritional value: state of the problem and current approaches. *Journal of Investigative Genomics* 5(1), 39–48.

Evans, J., McCormick, R.F., Morishige, D., Olson, S.N., Weers, B. *et al.* (2013) Extensive variation in the density and distribution of DNA polymorphism in sorghum genomes. *PLoS One* 8(11), e79192.

FAO and ICRISAT (1996) *The World Sorghum and Millet Economies: Facts, Trends and Outlook*. Food and Agricultural Organization of the United Nations, Rome and International Crops Research Institute for the Semi-Arid Tropics, Patancheru, India.

Fernandes, S.B., Dias, K.O.G., Ferreira, D.F. and Brown, P.J. (2018) Efficiency of multi-trait, indirect, and trait-assisted genomic selection for improvement of biomass sorghum. *Theoretical and Applied Genetics* 131(3), 747–755.

Fragoso, C.A., Moreno, M., Wang, Z., Heffelfinger, C., Arbelaez, L.J. *et al.* (2017) Genetic architecture of a rice nested association mapping population. *G3: Genes, Genomes, Genetics* 7, 1913–1926.

GBD (Global Burden of Disease) 2015 Obesity Collaborators (2017) Health effects of overweight and obesity in 195 countries over 25 years. *The New England Journal of Medicine* 377, 13–27.

Glahn, R.P. and Van Campen, D.R. (1997) Iron uptake is enhanced in Caco-2 cell monolayers by cysteine and reduced cysteinyl-glycine. *Journal of Nutrition* 127(4), 642–647.

Global Panel (2016) *The Cost of Malnutrition: Why Policy Action is Urgent*. Global Panel on Agriculture and Food Systems for Nutrition, London. Available at: https://glopan.org/sites/default/files/pictures/CostOfMalnutrition.pdf (accessed 5 March 2021).

Golicz, A.A., Batley, J. and Edwards, D. (2016) Towards plant pangenomics. *Plant Biotechnology Journal* 14, 1099–1105.

Gopalan, C., Rama Sastri, B.V. and Balasubramanian, S.C. (1989) *Nutritive Value of Indian Foods*, edition revised and updated by Narasinga Rao, B.S., Deosthale, Y.G. and Pant, K.C. National Institute of Nutrition, Indian Council of Medical Research, Hyderabad, India.

Grenier, C., Bramel-Cox, P.J. and Hamon, P. (2001a) Core collection of sorghum: I. Stratification based on eco-geographical data. *Crop Science* 41, 379–380.

Grenier, C., Hamon, P. and Bramel-Cox, P.J. (2001b) Core collection of sorghum: II. Comparison of three random sampling strategies. *Crop Science* 41, 241–246.
Gu, R., Chen, F., Liu, B., Wang, X., Liu, J. *et al.* (2015) Comprehensive phenotypic analysis and quantitative trait locus identification for grain mineral concentration, content, and yield in maize (*Zea mays* L.). *Theoretical and Applied Genetics* 128(9), 1777–1789.
Hariprasanna, K. and Rakshit, S. (2016) Economic importance of sorghum. In: Rakshit, S. and Wang, Y.H. (eds) *The Sorghum Genome*. Compendium of Plant Genomes. Springer, Cham, Switzerland, pp. 1–25.
Hariprasanna, K., Agte, V., Prabhakar and Patil, J.V. (2012) Genotype × environment interactions for grain micronutrient contents in sorghum [*Sorghum bicolor* (L.) Moench]. *Indian Journal of Genetics and Plant Breeding* 72, 429–434.
Hariprasanna, K., Agte, V., Elangovan, M. and Patil, J.V. (2014a) Genetic variability for grain iron and zinc content in cultivars, breeding lines and selected germplasm accessions of sorghum [*Sorghum bicolor* (L.) Moench]. *Indian Journal of Genetics and Plant Breeding* 74, 42–49.
Hariprasanna, K., Agte, V. and Patil, J.V. (2014b) Genetic control and heterosis for grain iron and zinc contents in sorghum [*Sorghum bicolor* (L.) Moench]. *Indian Journal of Genetics and Plant Breeding* 74(4 Suppl.), 638–643.
Hariprasanna, K., Venkateswarlu, R., Niharika, G., Manasa, K., Suresh, P. *et al.* (2018) Assessment of genetic variability for grain micronutrient contents in sorghum [*Sorghum bicolor* (L.) Moench] for biofortification. *Presented at Sorghum in the 21st Century – 'Food, Feed and Fuel in a Rapidly Changing World' Conference, Cape Town, South Africa, 9–12 April 2018*, poster 427.
HarvestPlus (2021) Nutrition. HarvestPlus, Washington, DC. Available at: https://www.harvestplus.org/what-we-do/nutrition (accessed 10 March 2021).
Hirsch, C.N., Foerster, J.M., Johnson, J.M., Sekhon, R.S., Muttoni, G. *et al.* (2014) Insights into the maize pan-genome and pan-transcriptome. *The Plant Cell* 26, 121–135.
Howe, A., Sato, S., Dweikat, I., Fromm, M. and Clemente, T. (2006) Rapid and reproducible *Agrobacterium*-mediated transformation of sorghum. *Plant Cell Reports* 25(8), 784–791.
Hu, Z., Olatoye, M.O., Marla, S. and Morris, G.P. (2019) An integrated genotyping-by-sequencing polymorphism map for over 10,000 sorghum genotypes. *Plant Genome* 12, 180044.
Huang, X., Paulo, M.J., Boer, M., Effgen, S., Keizer, P., Koornneef, M. and van Eeuwijk, F.A. (2011) Analysis of natural allelic variation in *Arabidopsis* using a multiparent recombinant inbred line population. *Proceedings of the National Academy of Sciences USA* 108, 4488–4493.
Hulse, J.H., Liang, E.M. and Pearson, O.E. (1980) *Sorghum and The Millets: Their Composition and Nutritive Value*. Academic Press, New York.
Hunt, C.H., van Eeuwijk, F.A., Mace, E.S., Hayes, B.J. and Jordan, D.R. (2018) Development of genomic prediction in sorghum. *Crop Science* 58, 690–700.
Jaikishan, I., Paik, G.R., Madhusudhana, R., Elangovan, M. and Rajendrakumar, P. (2013) Development of microsatellite markers targeting $(GATA)_n$ motifs in sorghum [*Sorghum bicolor* (L.) Moench]. *Molecular Breeding* 31, 223–23.
Jaikishan, I., Rajendrakumar, P., Madhusudhana, R., Elangovan, M. and Patil, J.V. (2015) Development and utility of PCR-based intron polymorphism markers in sorghum [*Sorghum bicolor* (L.) Moench]. *Journal of Crop Science and Biotechnology* 18(5), 309–318.
Jimenez-Galindo, J.C., Malvar, R.A., Butron, A., Santiago, R., Samayoa, L.F., Caicedo, M. and Ordas, B. (2019) Mapping of resistance to corn borers in a MAGIC population of maize. *BMC Plant Biology* 19, 431.
Jin, T., Zhou, J., Chen, J., Zhu, L., Zhao, Y. and Huang, Y. (2013) The genetic architecture of zinc and iron content in maize grains as revealed by QTL mapping and meta analysis. *Breeding Science* 63(3), 317–324.
Jordan, D., Mace, E., Cruikshank, A.W., Hunt, C.H., Hammer, G.L. and Henzell, R.G. (2012) Development and use of a sorghum nested association mapping population. *Poster presented at XX International Plant and Animal Genome Conference, San Diego, California, 14–18 January 2012*. Available at: https://pag.confex.com/pag/xx/webprogram/Paper3995.html (accessed 5 March 2021).
Kapoor, H.C. and Naik, M.S. (1970) Effects of soil and spray applications of urea and storage on the β-carotene content of yellow endosperm sorghum and pearl millet grains. *Indian Journal of Agricultural Sciences* 40, 942–947.
Kerkeb, L., Mukherjee, I., Chatterjee, I., Lahner, B., Salt, D.E. and Connolly, E.L. (2008) Iron-induced turnover of the *Arabidopsis* iron-regulated transporter1 metal transporter requires lysine residues. *Plant Physiology* 146, 1964–1973.

Khangura, M. (2019) A genome-wide association study of the quantitative resistance to *Striga hermonthica* and plant architecture of *Sorghum bicolor* in northwestern Ethiopia. PhD dissertation, Purdue University, West Lafayette, Indiana.

Knoll, J., Gunaratna, N. and Ejeta, G. (2008) QTL analysis of early-season cold tolerance in sorghum. *Theoretical and Applied Genetics* 116, 577–587.

Kong, L., Dong, J. and Hart, G.E. (2000) Characteristics linkage map positions and allelic differentiation of *Sorghum bicolor* (L.) Moench DNA simple sequence repeats (SSRs). *Theoretical and Applied Genetics* 101, 438–448.

Kruger, J., Taylor, J.R.N., Du, X., De Moura, F.F., Lönnerdal, B. and Oelofse, A. (2013) Effect of phytate reduction of sorghum, through genetic modification, on iron and zinc availability as assessed by an *in vitro* dialysability bioaccessibility assay, Caco-2 cell uptake assay, and suckling rat pup absorption model. *Food Chemistry* 141, 1019–1025.

Lee, S., Jeon, U.S., Lee, S.J., Kim, Y.K., Persson, D.P. *et al.* (2009) Iron fortification of rice seeds through activation of the nicotianamine synthase gene. *Proceedings of the National Academy of Sciences USA* 106, 22014–22019.

Li, Y.H., Zhou, G., Ma, J., Jiang, W., Jin, L.G. *et al.* (2014) *De novo* assembly of soybean wild relatives for pan-genome analysis of diversity and agronomic traits. *Nature Biotechnology* 32, 1045–1052.

Lipkie, T.E., De Moura, F.F., Zhao, Z., Albertsen, M.C., Che, P., Glassman, K. and Ferruzzi, M.G. (2013) Bioaccessibility of carotenoids from transgenic provitamin A biofortified sorghum. *Journal of Agricultural and Food Chemistry* 61, 5764–5771.

Liu, G. and Godwin, I.D. (2012) Highly efficient sorghum transformation. *Plant Cell Reports* 31(6), 999–1007.

Liu, J., Wu, B., Singh, R.P. and Velu, G. (2019) QTL mapping for micronutrients concentration and yield component traits in a hexaploid wheat mapping population. *Journal of Cereal Science* 88, 57–64.

Longvah, T., Ananthan, R., Bhaskarachary, K. and Venkaiah, K. (2017) *Indian Food Composition Tables*. National Institute of Nutrition, Indian Council of Medical Research, Hyderabad, India.

Lucca, P., Hurrell, R. and Potrykus, I. (2001) Genetic engineering approaches to improve the bioavailability and the level of iron in rice grains. *Theoretical and Applied Genetics* 102, 392–397.

Mace, E.S., Tai, S., Gilding, E.K., Li, Y., Prentis, P.J. *et al.* (2013) Whole-genome sequencing reveals untapped genetic potential in Africa's indigenous cereal crop sorghum. *Nature Communications* 4, 2320.

Marla, S.R., Burow, G., Chopra, R., Hayes, C., Olatoye, M. *et al.* (2019) Genetic architecture of chilling tolerance in sorghum dissected with a nested association mapping population. *G3: Genes, Genomes, Genetics* 9, 4045–4057.

Masuda, H., Ishimaru, Y., Aung, M.S., Kobayashi, T., Kakei, Y. *et al.* (2012) Iron biofortification in rice by the introduction of multiple genes involved in iron nutrition. *Scientific Reports* 2, 543.

Masuda, H., Kobayashi, T., Ishimaru, Y., Takahashi, M., Aung, M.S. *et al.* (2013) Iron-biofortification in rice by the introduction of three barley genes participated in mugineic acid biosynthesis with soybean ferritin gene. *Frontiers in Plant Science* 4, 132.

Masuda, H., Shimochi, E., Hamada, T., Senoura, T., Kobayashi, T. *et al.* (2017) A new transgenic rice line exhibiting enhanced ferric iron reduction and phytosiderophore production confers tolerance to low iron availability in calcareous soil. *PLoS One* 12(3), e0173441.

McCormick, R.F., Truong, S.K., Sreedasyam, A., Jenkins, J., Shu, S. *et al.* (2018) The *Sorghum bicolor* reference genome: improved assembly, gene annotations, a transcriptome atlas, and signatures of genome organization. *The Plant Journal* 93(2), 338–354.

Montenegro, J.D., Golicz, A.A., Bayer, P.E., Hurgobin, B., Lee, H. *et al.* (2017) The pangenome of hexaploid bread wheat. *The Plant Journal* 90, 1007–1013.

Morris, G.P., Ramu, P., Deshpade, S.P., Hash, C.T., Shah, T. *et al.* (2013) Population genomic and genome-wide association studies of agroclimatic traits in sorghum. *Proceedings of the National Academy of Sciences USA* 110(2), 453–458.

Morrissey, J. and Guerinot, M.L. (2009) Iron uptake and transport in plants: the good, the bad, and the ionome. *Chemical Reviews* 109, 4553–4567.

Murray, S.C., Rooney, W.L., Hamblin, M.T., Mitchell, S.E. and Kresovich, S. (2009) Sweet sorghum genetic diversity and association mapping for brix and height. *The Plant Genome* 2, 48–62.

Mutegi, E., Sagnard, F., Semagn, K., Deu, M., Muraya, M. *et al.* (2011) Genetic structure and relationships within and between cultivated and wild sorghum [*Sorghum bicolor* (L.) Moench] in Kenya as revealed by microsatellite markers. *Theoretical and Applied Genetics* 122, 989–1004.

Myles, S., Peiffer, J., Brown, P.J., Ersoz, E.S., Zhang, Z., Costich, D.E. and Buckler, E.S. (2009) Association mapping: critical considerations shift from genotyping to experimental design. *The Plant Cell* 21, 2194–2202.

Nagaraj, N., Basavaraj, G., Parthasarathy Rao, P., Bantilan, C. and Haldar, S. (2013) Sorghum and pearl millet economy of India – future outlook and options. *Economic & Political Weekly* XLVIII(52), 74–81.

Nagaraja Reddy, R., Madhusudhana, R., Murali Mohan, S., Chakravarthi, D.V.N. and Seetharama, N. (2012) Characterization, development and mapping of unigene-derived microsatellite markers in sorghum [*Sorghum bicolor* (L.) Moench]. *Molecular Breeding* 29, 543–564.

Naqvi, S., Zhu, C., Farre, G., Ramessar, K., Bassie, L. *et al.* (2009) Transgenic multivitamin corn through biofortification of endosperm with three vitamins representing three distinct metabolic pathways. *Proceedings of the National Academy Sciences USA* 106, 7762–7767.

Nelson, J.C., Wang, S., Wu, Y., Li, X., Antony, G., White, F. and Yu, J. (2011) Single-nucleotide polymorphism discovery by high-throughput sequencing in sorghum. *BMC Genomics* 12, 352.

Obi, C.E., Ejiogu, A.O. and Sanou, E.I.R. (2017) The role of transgenic biofortified food in the reduction of hidden hunger in Nigeria. *Paper presented at First International Conference of Food Security and Hidden Hunger held at Federal University Ndufu-Alike, Eboyi State, Nigeria, 8–11 October 2017*. Available at: https://www.researchgate.net/publication/321808550_The_Role_of_Transgenic_Biofortified_Food_in_the_Reduction_of_Hidden_Hunger_in_Nigeria (accessed 10 March 2021).

Ogawa, D., Yamamoto, E., Ohtani, T., Kanno, N., Tsunematsu, H. *et al.* (2018) Haplotype-based allele mining in the Japan-MAGIC rice population. *Scientific Reports* 8, 4379.

Ongom, P.O. and Ejeta, G. (2018) Mating design and genetic structure of a multi-parent advanced generation intercross (MAGIC) population of sorghum [*Sorghum bicolor* (L.) Moench]. *G3: Genes, Genomes, Genetics* 8, 331–341.

Pascual, L., Albert, E., Sauvage, C., Duangjit, J., Bouchet, J.P. *et al.* (2016) Dissecting quantitative trait variation in the resequencing era: complementarity of bi-parental, multi-parental and association panels. *Plant Science* 242, 120–130.

Paterson, A.H., Bowers, J.E., Bruggmann, R., Dubchak, I., Grimwood, J. *et al.* (2009) The *Sorghum bicolor* genome and the diversification of grasses. *Nature* 457, 551–556.

Phuke, R.M., Anuradha, K., Radhika, K., Farzana, J., Anuradha, G. *et al.* (2017) Genetic variability, genotype × environment interaction, correlation, and GGE biplot analysis for grain iron and zinc concentration and other agronomic traits in RIL population of sorghum (*Sorghum bicolor* L. Moench). *Frontiers in Plant Science* 8, 712.

Rajendrakumar, P. and Rakshit, S. (2015) Genomics and bioinformatics resources. In: Madhusudhana, R., Rajendrakumar, P. and Patil, J.V. (eds) *Sorghum Molecular Breeding*. Springer, New Delhi, pp. 117–154.

Rajendrakumar, P., Hariprasanna, K., Saakre, M., Venkateswarlu, R. and Tonapi, V.A. (2018) Identification of candidate genes associated with iron and zinc metabolism along with SNPs in sorghum and designing KASP assay for genotyping. In: *Book of Abstracts, 1st National Genetics Congress on 'Genetics for Sustainable Food, Health and Nutrition Security', ICAR–Indian Agricultural Research Institute, New Delhi, 14–16 December 2018*. Indian Society of Genetics & Plant Breeding, New Delhi, p. 300.

Rakshit, S., Rakshit, A. and Patil, J.V. (2012) Multiparent intercross populations in analysis of quantitative traits. *Journal of Genetics* 91, 111–117.

Ramu, P., Kassahun, B., Senthilvel, S., Kumar, C.A., Jayashree, B. *et al.* (2009) Exploiting rice–sorghum synteny for targeted development of ESTSSRs to enrich the sorghum genetic linkage map. *Theoretical and Applied Genetics* 119, 1193–1204.

Reddy, B.V.S., Ramesh, S. and Longvah, T. (2005) Prospects of breeding for micronutrients and β-carotene-dense sorghums. *International Sorghum and Millets Newsletter* 46, 10–14.

Reddy, C.S., Kim, S.C. and Kaul, T. (2017) Genetically modified phytase crops role in sustainable plant and animal nutrition and ecological development: a review. *3 Biotech* 7(3), 195.

Reddy, S.P., Reddy, B.V.S., Ashok Kumar, A., Ramesh, S., Sahrawat, K.L. and Rao, P.V. (2010) Association of grain Fe and Zn contents with agronomic traits in sorghum. *Indian Journal of Plant Genetic Resources* 23, 280–284.

Ritter, K.B., McIntyre, C.L., Godwin, I.D., Jordan, D.R. and Chapman, S.C. (2007) An assessment of the genetic relationship between sweet and grain sorghums, within *Sorghum bicolor* ssp. *bicolor* (L.) Moench using AFLP markers. *Euphytica* 157, 161–176.

Saltzman, A., Birol, E., Bouis, H.E., Boy, E., De Moura, F.F., Islam, Y. and Pfeiffer, W.H. (2013) Biofortification: progress toward a more nourishing future. *Global Food Security* 2(1), 9–17.

Schloss, S.J., Mitchell, S.E., White, G.M., Kukatla, R., Bowers, J.E., Paterson, A.H. and Kresovich, S. (2002) Characterization of RFLP probe sequences for gene discovery and SSR development in *Sorghum bicolor* (L.) Moench. *Theoretical and Applied Genetics* 105, 912–920.

Srinivas, G., Satish, K., Murali Mohan, S., Nagaraja Reddy, R., Madhusudhana, R. *et al.* (2008) Development of genic microsatellite markers for sorghum staygreen QTL using a comparative genomic approach with rice. *Theoretical and Applied Genetics* 117, 283–296.

Srinivas, G., Satish, K., Madhusudhana, R. and Seetharama, N. (2009a) Exploration and mapping of microsatellite markers from subtracted drought stress ESTs in *Sorghum bicolor* (L.) Moench. *Theoretical and Applied Genetics* 118, 703–717.

Srinivas, G., Satish, K., Madhusudhana, R., Nagaraja Reddy, R., Murali Mohan, S. and Seetharama, N. (2009b) Identification of quantitative trait loci for agronomically important traits and their association with genic microsatellite markers in sorghum. *Theoretical and Applied Genetics* 118, 1439–1454.

Stadlmeier, M., Hartl, L. and Mohler, V. (2018) Usefulness of a multi-parent advanced generation intercross population with a greatly reduced mating design for genetic studies in winter wheat. *Frontiers in Plant Science* 9, 1825.

Stevens, G.A., Finucane, M.M., De-Regil, L.M., Paciorek, C.J., Flaxman, S.R. *et al.* (2013) Global, regional, and national trends in haemoglobin concentration and prevalence of total and severe anaemia in children and pregnant and non-pregnant women for 1995–2011: a systematic analysis of population-representative data. *Lancet Global Health* 1, E16–E25.

Sushil Kumar, Hash, C.T., Nepolean, T., Singh, G., Rajaram, V. *et al.* (2016) Mapping quantitative trait loci controlling high iron and zinc content in self and open pollinated grains of pearl millet [*Pennisetum glaucum* (L.) R. Br.]. *Frontiers in Plant Science* 7, 1636.

Sushil Kumar, Hash, C.T., Nepolean, T., Mahendrakar, M.D., Tara Satyavathi, C. *et al.* (2018) Mapping grain iron and zinc content quantitative trait loci in an Iniadi-derived immortal population of pearl millet. *Genes* 9, 248.

Taramino, G., Tarchini, R., Ferrario, S., Lee, M. and Pe, M.E. (1997) Characterization and mapping of simple sequence repeats (SSRs) in *Sorghum bicolor*. *Theoretical and Applied Genetics* 95, 66–72.

Upadhyaya, H.D., Pundir, R.P.S., Dwivedi, S.L., Gowda, C.L.L., Reddy, V.G. and Singh, S. (2009) Developing a mini core collection of sorghum for diversified utilization of germplasm. *Crop Science* 49, 1769–1780.

Upadhyaya, H.D., Vetriventhan, M. and Deshpande, S. (2016) Sorghum germplasm resources characterization and trait mapping. In: Rakshit, S. and Wang, Y.H. (eds) *The Sorghum Genome*. Compendium of Plant Genomes. Springer, Cham, Switzerland, pp. 77–94.

Uptmoor, R., Wenzel, W., Friedt, W., Donaldson, G., Ayisi, H. and Ordon, F. (2003) Comparative analysis on the genetic relatedness of *Sorghum bicolor* accessions from Southern Africa by RAPDs, AFLPs and SSRs. *Theoretical and Applied Genetics* 106, 1316–1325.

Utz, H.F., Melchinger, A.E. and Schon, C.C. (2000) Bias and sampling error of the estimated proportion of genotypic variance explained by quantitative trait loci determined from experimental data in maize using cross validation and validation with independent samples. *Genetics* 154, 1839–1849.

Velu, G., Crossa, J., Singh, R.P., Hao, Y., Dreisigacker, S. *et al.* (2016) Genomic prediction for grain zinc and iron concentrations in spring wheat. *Theoretical and Applied Genetics* 129(8), 1595–1605.

Velu, G., Singh, R.V., Crespo-Herrera, L., Juliana, P., Dreisigacker, S. *et al.* (2018) Genetic dissection of grain zinc concentration in spring wheat for mainstreaming biofortification in CIMMYT wheat breeding. *Scientific Reports* 8, 13526.

Visarada, K.B.R.S. and Sai Kishore, N. (2015) Advances in genetic transformation. In: Madhusudhana, R., Rajendrakumar, P. and Patil, J.V. (eds) *Sorghum Molecular Breeding*. Springer, New Delhi, pp. 199–215.

Visarada, K.B.R.S., Prasad, G.S. and Royer, M. (2016) Genetic transformation and evaluation of two sweet sorghum genotypes for resistance to spotted stem borer, *Chilo partellus* (Swinhoe). *Plant Biotechnology Reports* 10(5), 277–289.

Watanabe, K., Guo, W., Arai, K., Takanashi, H., Kajiya-Kanegae, H. *et al.* (2017) High-throughput phenotyping of sorghum plant height using an unmanned aerial vehicle and its application to genomic prediction modeling. *Frontiers in Plant Science* 8, 421.

Welch, R.M. and Graham, R.D. (2004) Breeding for micronutrients in staple food crops from a human nutrition perspective. *Journal of Experimental Botany* 55, 353–364.

Wessells, K.R. and Brown, K.H. (2012) Estimating the global prevalence of zinc deficiency: results based on zinc availability in national food supplies and the prevalence of stunting. *PLoS One* 7(11), e50568.

Wu, E., Lenderts, B., Glassman, K., Berezowska-Kaniewska, M., Christensen, H. *et al.* (2014) Optimized *Agrobacterium*-mediated sorghum transformation protocol and molecular data of transgenic sorghum plants. *In Vitro Cellular & Developmental Biology – Plant* 50(1), 9–18.

Xia, J., Zhao, Y., Burks, P., Pauly, M. and Brown, P.J. (2018) A sorghum *NAC* gene is associated with variation in biomass properties and yield potential. *Plant Direct* 2(7), e00070.

Yellisetty, V., Reddy, L.A. and Mandapaka, M. (2015) *In planta* transformation of sorghum [*Sorghum bicolor* (L.) Moench] using *TPS1* gene for enhancing tolerance to abiotic stresses. *Journal of Genetics* 94(3), 425–434.

Yonemaru, J.I., Ando, T., Mizzubayashi, T., Kasuga, S., Matsumoto, T. and Yano, M. (2009) Development of genome-wide simple sequence repeat markers using whole-genome shot-gun sequences of sorghum [*Sorghum bicolor* (L.) Moench]. *DNA Research* 16, 187–193.

Yu, X., Li, X., Guo, T., Zhu, C., Wu, Y. *et al.* (2016) Genomic prediction contributing to a promising global strategy to turbo charge gene banks. *Nature Plants* 2, 16150.

Zhao, Q., Feng, Q., Lu, H., Li, Y., Wang, A. *et al.* (2018) Pan-genome analysis highlights the extent of genomic variation in cultivated and wild rice. *Nature Genetics* 50, 278–284.

Zhao, Z. (2008) The Africa Biofortified Sorghum Project – applying biotechnology to develop nutritionally improved sorghum for Africa. In: Xu, Z., Li, J., Xue, J. and Yang, W. (eds) *Biotechnology and Sustainable Agriculture 2006 and Beyond*. Springer, Dordrecht, the Netherlands. pp. 273–277.

Zhao, Z.Y., Glassman, K., Sewalt, V., Wang, N., Miller, M. *et al.* (2003) Nutritionally improved transgenic sorghum. In: Vasil, I.K. (ed.) *Plant Biotechnology 2002 and Beyond*. Kluwer Academic Publishers, Dordrecht, the Netherlands, pp. 413–416.

Zhao, Z.Y., Che, P., Glassman, K. and Albertsen, M. (2019) Nutritionally enhanced sorghum for the arid and semiarid tropical areas of Africa. In: Zhao, Z.Y. and Dahlberg, J. (eds) *Sorghum*. Methods in Molecular Biology, Vol. 1931. Humana Press, New York, pp. 197–207.

Zheng, L.Y., Guo, X.S., He, B., Sun, L.J., Peng, Y. *et al.* (2011) Genome-wide patterns of genetic variation in sweet and grain sorghum (*Sorghum bicolor*). *Genome Biology* 12, R114.

# 29 Ideotype Breeding for Improving Yield in Sorghum: Recent Advances and Future Perspectives

**P. Sanjana Reddy***
*ICAR–Indian Institute of Millets Research, Hyderabad, India*

## 29.1 Introduction

Among the widely grown cereals across the world, sorghum (*Sorghum bicolor* (L.) Moench) and maize (*Zea mays* L.) are the most suitable crops in the anticipated climate change scenario. Having the $C_4$ carbon fixation pathway, they are able to perform better than $C_3$ plants under increased temperature and light intensity. Sorghum is grown across 87 countries (having more than 1000 hectares under the crop) with India (5.86 million hectares) ranking first, closely followed by Nigeria (5.82 million hectares) and Sudan (5.41 million hectares). Despite having the largest acreage, India ranks fifth, Nigeria second and Sudan ranks sixth in production (FAO, 2019). Sorghum is generally grown in areas that receive 350 to 700 mm rainfall annually. Sorghum yield is determined by the cultivars grown and the agronomic practices followed in the region (Mundia *et al.*, 2019). Sorghum is valued as a dual-purpose crop with both grain and fodder having equal importance in supporting the food and livestock needs in marginal areas. On the African continent, sorghum is predominantly grown for food purposes while in the developed regions of the world such as USA, Australia and South America, it serves the fodder requirements of cattle and the grain is used as feed. It is also an important source of animal feed in some parts of Eastern Africa. Traditional African beers are brewed from sorghum grain. The crop is a driver for economic development in Africa. In India, sorghum is cultivated in both rainy as well as post-rainy/winter seasons. The rainy-season crop is generally grown with hybrids. The crop maturity period is usually coincident with rains and thereby the grain harvested is of poor quality due to grain moulds. The grain from rainy-season crops is used as food or for ethanol production. In contrast, the winter-season crop is mostly grown with landraces that are photoperiod-sensitive and have lustrous grains with superior *roti*/leavened bread-making quality, hence being highly valued for food use. Sorghum with high stalk sugar content (sweet sorghum) is exploited for sugar and ethanol production. The grain is employed for industrial uses such as potable alcohol, malt, beer, liquids, gruels, starch, adhesives, core binders for metal casting, ore refining, grits as packaging material, etc. The starch from the grains is utilized in food, pharmaceutical, textile and paper industries. Malt drinks and malt cocoa-based weaning food and baby food industries are popular in Nigeria. Owing to its multiple uses, suitable cultivars are chosen to meet specific end uses.

*Corresponding author: sanjana@millets.res.in



DOI: 10.1079/9781789245431.0029

## 29.2 Types of Sorghum

Cultivated sorghums are included under *Sorghum bicolor* subsp. *bicolor*. Based on the agronomic importance, they are grouped as grain sorghum, sweet sorghum, sudangrass and broomcorn (Berenji and Dahlberg, 2004). Based on spikelet morphology, grain and head types, the cultivated sorghums are classified into five main races that are *bicolor*, *guinea*, *kafir*, *caudatum* and *durra* and ten intermediate races (*durra–caudatum*, *guinea–bicolor*, etc.) that are due to a combination of the basic races (Harlan and de Wet, 1972; Smith and Frederiksen, 2000). The rainy-season sorghum cultivars of India predominantly belong to *caudatum*, *kafir* and *bicolor* races, while the winter sorghums are mainly of *durra* types. Winter sorghums differ from *kharif* sorghums with respect to adaptation, photoperiod sensitivity, grain and fodder quality. The forage sorghums include cultivars ranging from silage sorghum hybrids to varieties, to *Sorghum sudanense* ($2n = 10$) and sorgo-sudangrass hybrids, to sudangrass varieties ($2n = 10$) and hybrids (Kalton, 1988). Brown midrib (*bmr*) mutants of *S. bicolor* exhibit reddish brown pigment in the midrib that is associated with low lignin and thereby contributes to enhanced digestibility of stover. Such lines were used as forage sources for livestock (Oliver *et al.*, 2004) and are being explored for second-generation ethanol production. Broom sorghum (broomcorn, *Sorghum vulgare*) is also used as a source of animal feed in some regions, although it is less digestible than *S. bicolor* (Nikkhah *et al.*, 2004). Sorghums crossed with sudangrass have high biomass, good regeneration potential, thin stems and high tillering compared with normal sorghums. These hybrids are preferred among the forages and have flooded the market. They are being used as pasture, hay, green chop or silage for livestock. Sweet sorghums are sorghums with high stalk sugar content and Brix value up to 24%, thus have a great potential for ethanol production (Reddy *et al.*, 2004).

## 29.3 Advances in Sorghum Improvement

Sorghum is extensively cultivated in Africa, Asia and North America. In Asia, it is predominantly cultivated in China and India, while in North America, the USA and Mexico are the top producers. In Africa, Nigeria and Sudan are leading producers of sorghum.

### 29.3.1 India

Sorghum is not a very ancient crop in India. According to Indian literature, sorghum cultivation is recorded in the first century AD. The *bicolor* sorghums were probably directly introduced into India from Africa while the *durra* cultivars were introduced by both land and sea routes (de Wet and Huckabay, 1967). There is also a belief that sorghum was introduced into India in the second century AD and through Mongol conquest, spread to China (Kiple, 2000). Asia contributes 15.26% of the world area under sorghum cultivation and 13.44% of the world production. The average yields of Asia are closer to the world average, with India and China contributing the major share of area (84.63%) and production (87.72%). Based on 2018 sorghum crop data, India accounts for 77% of total sorghum area and 60.2% of total sorghum production on the Asian continent. But the yield in India (968 kg/ha) is below the Asian average of 1240 kg/ha (FAO, 2020). The sorghum in India evolved independently and formed a distinct group, and thus the introductions into the USA did not perform well. In India, similar to Egypt and unlike in other countries, sorghum is cultivated in two seasons: rainy season (June/July–September/October) as a rainfed crop and winter (post-rainy) season (October–December/January) under residual soil moisture/limited-irrigated conditions (Sanjana Reddy and Reddy, 2018). Due to seasonal variations and different production constraints experienced, season-specific cultivars are developed and cultivated (Rana *et al.*, 1997). In India during 1965/66, the proportion of area between rainy- and winter-season sorghums in the total sorghum cropped area was 62 and 38%, respectively. Due to the poor grain quality of sorghum grown during the rainy season, these proportions had changed to 37 and 63%, respectively, by 2014/15 (Charyulu *et al.*, 2016) and thereby the area has been diverted to other crops (Charyulu *et al.*, 2013).

Research on sorghum improvement in India is carried out through the All India Coordinated

Research Project on Sorghum administered by the Indian Council of Agricultural Research (ICAR) through a network of 21 centres, spread in 11 states under 17 state agricultural universities. Prior to 1960, variety CSV 1 was released as a result of pure-line selection from IS 3924, a *kafir–durra* from the USA. Later on, pedigree breeding was followed for the development and release of 19 grain sorghum varieties. Recently, a *bmr* sorghum variety, JAICAR Nutrigraze involving a *bmr* source line IS 21891, was released in 2019. With the availability of cytoplasmic-nuclear male sterility (CMS) and temperate germplasm, the period of the 1960s marked the beginning of the development of hybrid sorghums (Reddy and Stenhouse, 1994). The first sorghum hybrid, CSH1 (Coordinated Sorghum Hybrid), bred in India from a cross between CK60A and IS84, was released for commercial cultivation in 1964. The popular hybrids, CSH 5 and CSH 6 in the mid-1970s and CSH 9 in the early 1980s, brought about the wide spread of hybrid technology and boosted productivity. To date, 29 grain sorghum hybrids have been developed. The female parent 296A deserves special mention. It is the best-combining female parental line having an $A_1$-based CMS system and has been popular for four decades. It was developed from a cross between IS 3922 and Karad Local. It is the female parent of five hybrids and has contributed to the pedigree of several female parental lines. Focused breeding on winter sorghum was initiated in the early 1970s with the first variety, CSV 7R, released in 1974 (Reddy *et al.*, 2003). Overall, ten winter sorghum varieties were released. Unlike rainy sorghum hybrids, the hybrid programme is not advanced in winter sorghum, the reasons being lack of sufficient heterosis and poor seed set experienced by hybrids when temperatures fall below 10–12°C. However, seven hybrids have been released for winter-season adaptation although they are not popular with farmers. The variety M 35-1, a selection from local landrace *maldandi* bulk, was released in 1969 and is still popularly grown by farmers. The dual-purpose varieties CSV 15, CSV 20, CSV 23 and CSV 27 could realize good grain and fodder yields in farmers' fields. Apart from this, multi-cut forage sorghum hybrids CSH 20MF and CSH 24MF and forage sorghum varieties CSV 21F and CSV 30F are available for commercial cultivation.

### 29.3.2 China

China is next to India in cultivated area and production but is well ahead of India in terms of yield with 4542 kg/ha (FAO, 2020). In 1918, sorghum was cultivated on a vast area of 14.7 million hectares with a total grain production of 16 million tonnes. The sorghum-growing area began to decline in the 1920s, and in 1952, the cultivation area was 9.4 million ha with a total grain production of 11.1 million tonnes (Gao *et al.*, 2010). From the 1950s onwards, the area started decreasing rapidly, from 6.8 million hectares in 1961 to 5.2 million, 2.6 million, 1.4 million and 0.78 million hectares in 1971, 1981, 1991 and 2001, respectively. The area stabilized around 0.5 million hectares from 2004 onwards until 2018 (FAO, 2020). Prior to the 1980s, sorghum was consumed mainly as a human food in China. Later on, the major portion of the produce was diverted to alternative uses such as liquor and vinegar and only a very small proportion is used for human food. The sorghum lines from China are known by a special name, *kaoliang*. They differ from the lines belonging to India and Africa, which may be due to adaptation to cold climates. Low temperature is an important stress factor in both the seedling and grain-filling stages, especially for north-eastern China. Chinese sorghums or *kaoliangs* have good seedling vigour and high resistance to low temperature, but they lack resistance to important pests and diseases (Gao *et al.*, 2010). The important diseases affecting the commercial value of sorghum in China are anthracnose (*Colletotrichum* spp.), smut (*Sphacelotheca reiliana*), blotch (*Exserohilum turcicum*) and purple blotch (*Cercospora* spp.), while the important pests include European corn borer (*Ostrinia furnacalis*), aphid (*Melanaphis sacchari*) and armyworm (*Mythimna separata*).

Sorghum was widely cultivated during the 1900s and reached a peak during 1918. Breeding was initiated with selection from landraces during the 1920s and involved three steps: (i) pure-line selection from the best local landraces/varieties; (ii) crossing between local varieties and advancement through selection following pedigree breeding; and (iii) with availability of the CMS system, heterosis breeding has been followed since 1965 (Zhen Yang, 1997). Similar to other regions across the world, sorghum breeding in China involved breeding for higher grain and fodder yields, resistance to low temperature and drought among abiotic stresses and resistance to aphids and head smut among biotic stresses, apart from grain quality. Breeding for livestock feed and forage sorghum was also emphasized (Zhen Yang, 1997). As sorghum lost its importance as a human food, sorghum breeding for animal fodder gained popularity after the 1970s. Forage cultivars were diversified for silage and hay. However, research did not receive much focus in forage breeding also. Sweet sorghum and high-biomass grain sorghum hybrids are used for silage and sorghum–sudangrass hybrids are used for making hay. Due to increased interest among the international community on utilization of sweet sorghum for biofuel production, the $A_3$ cytoplasm was exploited to produce sweet sorghum without grains and thereby $A_3$ cytoplasm-based male-sterile lines with high sugar content were identified for development of sweet sorghum hybrids (Zhao *et al.*, 2007). Breeding of sorghum for biofuel is still in its infancy in China (Diao, 2017).

### 29.3.3 Nigeria

Nigeria is the highest sorghum producer in the West African region, accounting for 71% of the total regional output. Sorghum is a very valuable industrial crop for brewing alcoholic and non-alcoholic drinks as well as in the baking and confectionery industries in Nigeria (Sani *et al.*, 2013). Over the years, Nigeria has experienced a tremendous growth and boost in its economy because of the cultivation of sorghum. In 2015–2017, Nigeria rose up to second place in the world regarding sorghum production. Moreso, Nigeria presently accounts for 65–70% of the total sorghum produced in West Africa. Growing industrial demand for sorghum, with about 20% of the total sorghum produced being taken up by industries, rising awareness about sorghum's health benefits and the government's policy of high import prices have increased cultivation of sorghum. Nigeria could increase its sorghum production through cultivation of varieties. In Nigeria, before the 1970s, hybrids directly introduced from the USA and India failed to boost sorghum productivity due to their poor adaptation. Therefore, the exotic seed parents were crossed with local breeding lines to develop male-sterile lines from 1970 onwards. Recently, in 2018, Nigeria's National Committee on Variety Naming, Registration and Release released two medium-maturing sorghum varieties, SAMSORG 47 as ZAUNA-INUWA, SAMSORG 48 as KAURA BORNU, and an early-maturing variety, SAMSORG 49 as CF35:5, in Nigeria. They were developed from germplasm materials from Nigeria and Mali. In Nigeria, the most common landraces of sorghum are *kaura*, *fara-fara* and *guinea*. They are variously tolerant to *Striga* (a parasitic weed) in all the savannah zones (Ogbonna, 2007).

### 29.3.4 Sudan

Sudan is a large producer of sorghum, although production in the country tends to be extensive rather than intensive. A significant part of sorghum cultivation takes place in the eastern region of the country, also known as Sudan's breadbasket. Sorghum yields depend heavily on climatic factors, rainfall in particular as 90% of the area is rainfed. Productivity is, however, constrained by drought and the use of low-yielding varieties and landraces. The area for sorghum cultivation in Sudan is about 6 million hectares and the estimated area infested by *Striga* is 1.6 million hectares. The estimated yield loss is about 1 million tonnes, which is 30% of total sorghum production of the country. Other constraints to the growth of sorghum include weeds other than *Striga*, drought, soil salinity and damages by sorghum midge and stem borer, according to the archives of the *Sudan Journal of Agricultural Research* published by Agricultural Research Corporation, Sudan. Research pertaining to sorghum production focuses on fertilizers, irrigation and tillage, sowing time and effects of rotational crops. While Sudanese agriculture has witnessed the entry of new agribusiness players in recent years, much production is still at the subsistence level carried out by smallholder farmers.

### 29.3.5 The USA

Way back in 1757, Benjamin Franklin wrote about sorghum being used for making brooms.

Thus, sorghum was known in the USA since the 1700s. The 'Sorghum Belt' runs from Texas to South Dakota. Sorghum is grown in about 14 states. Half the sorghum produced in the USA is used for forage and silage and the other half for feed grains. Only a small amount is grown for syrup. With sorghum's use for ethanol production increasing rapidly, 12–40% of sorghum production in the USA goes to ethanol production. With about 70% share in world trade, the USA is the world's largest exporter of grain sorghum. At the beginning of the 19th century, plant breeding in the USA was initiated with 30 primary introductions and sudangrass from Africa. With the discovery of CMS, the first commercial hybrid was produced in 1957 and by 1960, 95% of the grain sorghum crop was grown with hybrids. These mainly involved crosses between *kafir* types of southern Africa and *milo–caudatum* types of central Africa. Similarly, forage sorghum involving sorghum–sudangrass hybrids, like Sudax SX 11, were largely cultivated. Later, O.J. Webster and R.E. Karper introduced Kaura and Korgi germplasms. These were *caudatum* and *durra* yellow endosperm lines. These lines broke through the existing yield plateau and brought in drought tolerance. Due to sorghum diseases such as downy mildew, maize dwarf mosaic virus, small seed malady and sorghum green bug among pests, along with a reduction in public breeding programmes, a second yield plateau was reached during the period 1978–1984 that is still ongoing. The International Sorghum and Millet Collaborative Research Support Program (INTSORMIL CRSP) is a project funded by the US Agency for International Development (USAID), involving seven US universities, meant for sorghum and millet improvement.

Sorghum has now become an important crop in Latin America as well. The crop has gained prominence in Mexico over the past half-century. Research is being focused on processing sweet sorghum in the farmer's field to save transportation costs and the cost of building and maintaining central processing facilities.

## 29.4 Challenges for Genetic Improvement

The yield and quality of sorghum produced worldwide are influenced by a wide array of biotic and abiotic constraints. The significant among the biotic constraints include shoot fly (India, Eastern Africa), stem borer (India, Africa), midge (Eastern Africa, Australia), head bug (India, West and Central Africa), aphid, armyworms and locusts among the insects and the major diseases prevalent in different parts of the world include grain mould (all regions), charcoal rot, downy mildew, anthracnose (West and Central Africa, northern India), rust, leaf blight and viral diseases. *Striga* (*Striga asiatica*, *Striga densiflora*, *Striga hermonthica*) is a parasitic weed seen in many regions of Africa and a few parts of India. Problematic soils such as saline (some parts of India and Middle Eastern countries) and acidic (Latin America), temperature extremities and drought (all regions) are the abiotic factors influencing sorghum productivity. High-yielding and stress-resistant cultivars for each of these have been developed but their dissemination to farmers' fields is affected by their lower percentage of yield advantage compared with the hybrids.

## 29.5 Ideotype Breeding Types

A biological model with defined traits expected to perform in a predictable manner within a defined environment is defined as a crop ideotype (Donald, 1968). In this method of breeding, the breeder selects for the defined target traits rather than grain yield. This was initially developed for cereal crops (Rasmusson, 1987; Peng *et al*., 2008; Hanocq *et al*., 2009). Donald (1968) proposed the ideotype initially for low- or non-stress environments which had limitation of radiation affecting crop yield. As the performance factors differ with the target ecologies, a single trait may not contribute to plant performance in all scenarios (Tardieu, 2012). Depending on the targeted environments, specific ideotypes are designed. Three views of ideotypes were defined by Andrivon *et al*. (2013): (i) the historical, 'genetic' view, as described above; (ii) the 'agronomic' view, where new genotypes are designed for specific cropping systems; and (iii) the 'modelling' view, where the best combinations of traits (usually represented by model parameters) are identified from formal or simulation experiments.

## 29.6 Ideotype Breeding in Cereals

In rice, an ideotype was defined for plant architecture and through breeding for this modified plant architecture, yield potential has been increased in rice (Khush, 2000). The first short-statured variety, IR-8, developed at the International Rice Research Institute (IRRI), had a good canopy architecture contributed by profuse tillering, dark green and erect leaves and sturdy stems. Such an ideotype responded well to nitrogenous fertilizer and had an improved harvest index of 0.45 (Chandler, 1969). It was felt that instead of a greater number of tillers, fewer productive tillers could add to grain yield performance of the plant. Thus, to further increase grain yield, reduced tillering capacity with more productive tillers, each with 200–250 grains per panicle, was added to the ideotype with the very sturdy stems and dark green, thick and erect leaves. A vigorous and deep root system was added to support this new plant type (NPT). This NPT out-yielded IR-72 (Khush, 1995). Later on, the focus in rice breeding shifted towards improving grain quality and incorporating genes for disease and insect resistance that were added to the ideotype. Similar to the NPT breeding programme in India, Chinese scientists proposed an ideotype that included moderate tillering, heavy panicles (5 g/panicle) and slightly taller plant height (about 100 cm). More focus was given on three leaves from the top. The three leaves should be above panicle height and should remain erect until maturity. The leaves should be narrow and V-shaped (2 cm when flattened) and should be thick and dark green. The leaf length of the flag leaf should be 50 cm and with a leaf angle of 5°. The second and third leaves should be 55 cm with leaf angles of 10° and 20°, respectively. Leaf area index of the top three leaves should be 6 and it should have a harvest index of 0.55 (Yuan *et al.*, 1989).

A similar exercise was done by the Wheat Yield Consortium (WYC) in 2011. They formulated high-yielding wheat ideotypes in order to increase wheat yield potential. Traits such as increased capacity and efficiency of photosynthesis, optimal developmental pattern to maximize spike fertility, optimal partitioning to grain, improved grain filling and potential grain size, and lodging resistance were considered. It was found that the increase in efficiency of light conversion, increasing the length of the grain-filling period, resulted in a higher harvest index and optimal phenology, thereby contributing towards higher grain yield (Semenov and Stratonovitch, 2013).

One such study was also made in maize wherein both shoot and root traits were considered. A small leaf angle with faster rate of kernel filling and kernel dehydration among the shoot traits and a deep root system were proposed. The root should have reduced living cortical area and cell number and increased cell size when subjected to drought stress. Drought being an important constraint in several crops, the stress-related genes were suggested to be brought through a transgenic approach along with higher grain yield. The maize ideotype with narrow leaves is supposed to allow the planting of more plants per unit area thus facilitating mechanized harvest, while the deep root system helps to improve water and nutrient uptake especially in resource-poor soils (Gong *et al.*, 2015).

On the whole, designing the ideotype/ideotyping can be split into three steps (Martre *et al.*, 2015):

**1.** Defining the objectives of the breeding programme apart from yield and adaptation (e.g. breeding for biotic and abiotic stresses, end uses, grain and fodder quality, plant architecture amenable to mechanical harvest).
**2.** Identification of morpho-physiological traits, studying the relationship among the traits, deciding the breeding protocol and assembling them in a genotype (e.g. developing early-maturing cultivars with resistance to biotic stress and reduced plant height, developing cultivars with corneous endosperm and grain yield, developing early-maturing cultivars with stay-greenness, developing cultivars with both disease resistance and drought tolerance in a moderate-yield background).
**3.** Evaluation of the suggested ideotypes to prove the agronomic relevance of trait integration in target environments (through simulations or field experiments).

## 29.7 Methodologies for Defining Crop Ideotypes

Plant breeding involves continuous improvement of a crop to meet the challenges emerging from climate change, change in cropping systems, emergence of new pests and diseases, commercialization and postharvest uses, with the traits incorporated in a high-yielding background and adapted to the place where the crop is grown. Some of those objectives can be antagonistic wherein improvement in one trait compromises another desirable trait. It mostly happens in resistance breeding programmes wherein the levels of resistance have to be compromised when placing the resistance genes in a desirable genetic background (Brown, 2002). Such antagonisms between the traits also contribute to the genotype × environment × cropping systems interactions observed in multiple-location varietal trials which are evaluated for performance under decreased chemical inputs, diversification of production techniques and climate change (Van Eeuwijk *et al.*, 2016). When the interactions are very high, evaluation in large areas and identification for broad adaptation should be replaced by more locally adapted varieties presenting contrasted profiles suitable for the region of cultivation. This method is followed for varietal testing in several crops in India wherein the regions are divided based on climatic factors, designated as zones, and after testing the varieties suitable for specific zones or for broad adaptation across all zones are released. As biotic and abiotic stresses vary widely with climate, breeding for specific adaptation helps to release varieties with better resistance to biotic and abiotic stresses that occur in a particular zone. The gene × environment (G × E) interaction can be partitioned and addressed. Based on the local requirements, an ideotype can be defined, a product profile framed, and a breeding programme can be formulated aimed at developing the ideotype-like variety.

In the conventional method of breeding for sorghum improvement, pedigree breeding is often followed. Lines with the trait of interest are crossed and the resultant progenies are advanced with selection until the final product is obtained. It takes several years for the varietal development, more so for hybrid development where parental lines are involved. Due to ever-changing climatic conditions and consequently the emergence and resurgence of pests and diseases, the occurrence of abiotic stresses, changing farmer preferences in lieu of consumer demand and changing cropping pattern, the desired product profile changes with time and also differs from the one under selection. Breeding for the desired product can be speeded up by following selection for the designed ideotype rather than the routine varietal profiles. Such products can be easily adopted by the farming community. In order to breed for a drought-tolerant *rabi* (post-rainy season) sorghum variety, a plant model is fixed with the target traits of seedling emergence rate, seedling growth rate, early maturity, stay-greenness of leaves, plant height and architecture, leaf angle, root angle, root depth, charcoal rot resistance, etc. The genetics of these individual traits are explored in terms of diversity, heritability, genetic advance, correlations to the economic yield and linkage among the traits. Some of these traits can also be obtained through agronomic practices such as impact of growth regulators on plant height, impact of sowing date on the plant phenology, controlling plant architecture through maintaining spacing, etc. Such traits may not be required to be handled through genetic means. The traits that are available within the existing genetic material, those with moderate to high heritability, not efficiently attained through cropping practices, can be attained with breeding. When a number of traits show high correlation, a trait which has high diversity and is easily amenable for genetic improvement can be targeted. An ideotype thus combines several traits that have to be targeted through breeding which do not show high correlations among themselves. Therefore, one must consider the genetic correlation between the possible target traits carefully. This combination of traits into ideotypes is supported by both field trials and crop models. The genotypes contrasting for several traits are evaluated in the target environments and repeated across years for confirmation. Based on the multi-environment trial (MET) data, target traits governing a particular objective are identified. Such MET results involving contrasting genotypes are useful to fill knowledge gaps and confirm some hypotheses about possible traits of interest. However, they generally require expensive phenotypic characterization and are limited by the number of environmental conditions under study. This can be one of the reasons for the reduced literature available on crop ideotypes.

Of late, computer modelling approaches are frequently used. The modelling involves powerful optimization algorithms which can help to decide a target trait or their combinations for reaching a desired objective that can be suitable for a target environment. Thus these techniques can be employed to define ideotypes or identify varieties adapted to particular cropping conditions. However, these models have several limitations. First, they are partial. They do not take biotic and abiotic constraints and weeds into account. For example, plant and canopy architectures are, at least in some pathosystems, powerful levers to limit inoculum production, inoculum dispersion and/or inoculum efficacy, and thus contribute to slow down the progress of an epidemic (Tivoli *et al.*, 2012). Thus, both genetic and agronomic aspects are involved in ideotype development. Second, limited numbers of varieties and cropping conditions were tested and used for calibration of the models. Hence, when new ideotypes with a different product profile from the existing varieties are tested, the models may not give valid results and the inferences need to be drawn cautiously as those results depend on the model used. Several models can thus be tested to improve the robustness of the results before drawing conclusions (Tao *et al.*, 2017). For those model-based approaches to be useful in breeding or recommending varieties, the key parameters of the models (those having an impact on the target goals) must be measurable so that the varieties can be selected on it and the genetic diversity as well as the between-trait correlations must be credible. If the simulations are cheaper than the trials, a major challenge would be to develop a right computing model-based approach. Modelling platforms like RECORD (Bergez *et al.*, 2013) or ISIde (Paleari *et al.*, 2016) give access to models and computing tools to combine them or to design and implement simulation batches for an *in silico* ideotyping that can be worked on by people unfamiliar with the software. Several studies have taken up and utilized the modelling approach for the prediction of crop growth and development in crop management. Simulation models have been used to predict nutrient losses (Solie *et al.*, 2012) and estimate yield responses based on varying combinations of seeding densities and nitrogenous fertilizer rates for a given cropping system (McNunn *et al.*, 2019). The use of such modelling approaches for genotype-to-phenotype prediction is in its infancy (Hammer *et al.*, 2002; Hammer and Jordan, 2007). The Agricultural Production Systems sIMulator (APSIM) is a cropping systems simulation model, designed to combine accurate predictions of economic product for many crop species in response to climate and management conditions, with predictions of the long-term consequences of cropping systems on soil physical and chemical conditions (Keating *et al.*, 2003).

Modelling has been increasingly used over the past two decades to predict cultivar performance in a targeted environment. Prediction of genotypic differences across environments taking G × E interactions into account has led to a focus on certain traits such as leaf elongation, early vigour and flowering time. Thus these process-based models are now increasingly used to define and characterize crop environments, predict an ideotype that may fit into the environment and thereby help breeding programmes to breed for targeted traits using identified genotypes that could lead to a predicted output. Ecophysiological models are also used to assist plant phenotyping and could provide necessary links between controlled-conditions phenotyping and plant performance in the field (Martre *et al.*, 2015). In wheat, two parameters of an ecophysiological model (Vsat and Pbase, representing genotype vernalization requirements and photoperiod sensitivity, respectively) were optimized for 210 genotypes (Bogard *et al.*, 2014). The incorporation of genetic controls in ecophysiological models helps to analyse the genetic control of phenotypic plasticity across a wide range of environments, and the G × E × management interactions are now being explored using efficient algorithms to find ideotypes optimizing many antagonist criteria. This approach derives the best combinations of genetic and agronomic parameters to achieve the predefined objectives. However, not much work has been done in this direction and the relationship between genes and the model parameters is yet to be derived. Considerable efforts are still needed to develop robust links between genetic controls, physiological determinants and traits relevant to breeders (Martre *et al.*, 2015).

## 29.8 Genetics and Breeding for Important Traits in Sorghum

Knowledge on the genetics of targeted traits is required before framing strategies and designing plant ideotypes. Exhaustive reviews on the genetics of various traits may be found in Doggett (1988), Murty and Rao (1997) and Rooney (2000).

Plant height is controlled by four recessive, non-linked and brachytic dwarfing genes, *dw1*, *dw2*, *dw3* and *dw4* (Quinby and Karper, 1954). Based on classical genetics four maturity loci (named *Ma1*–*Ma4*) governing photoperiod sensitivity have been identified, of which dominance at each one of the loci delays maturity under long-day conditions. They are involved in controlling duration of growth and floral initiation (Quinby *et al.*, 1973). Wild sorghums flower during short days with a photoperiod of 12.5 h of daylight or less. Most of the tropical landraces/varieties are dominant at all four loci. The maturity locus *Ma1* has large effect and large allelic series, and thereby a recessive allele at *Ma1* locus will influence photoperiod sensitivity and response to temperature variations greatly. In addition to the four maturity loci, three more (*Ma5*, *Ma6* and *Ma7*) were found (Mullet *et al.*, 2010). High-biomass sorghums were developed by manipulating maturity loci. Two early-flowering lines could be hybridized to produce a late-flowering, high-biomass line (Mullet *et al.*, 2010).

The genetics controlling resistance to various diseases is simple. For example, three races of *Sporisorium sorghi* are known to cause kernel smut, resistance to each is controlled by an incomplete dominant gene (*Ss1*, *Ss2*, *Ss3*). Resistance to head smut is dominant. Milo disease is governed by a single locus, susceptibility being partially dominant. Resistance to rust and anthracnose is governed by a single dominant gene. Susceptibility to leaf blight in sudangrass is inherited as a single dominant trait (House, 1985). The stay-green trait is inherited as dominant with E 36-1 hybrids (Reddy and Stenhouse, 1993). More than two loci are involved in the resistance to downy mildew, preferably three with different interactions (Reddy *et al.*, 1992). The grain mould caused by a complex of fungi has complex genetics (House, 1985). Like in the case of many crops, complex genetics govern insect resistance in sorghum. Non-preference mechanism is the predominant one and it is quantitatively inherited with the predominance of additive gene action (Sharma *et al.*, 1977). Rana *et al.* (1981) reported additive gene action, with $F_1$ being intermediate to the two parents for shoot fly resistance. Under low to moderate shoot fly pressure, resistance was found to be partially dominant. For resistance to stem borer, both tolerance and antibiosis seem to be operating (Jotwani, 1976). Rana and Murty (1971) reported that while additive (A) and A × A interactions explained primary damage, the secondary damage by stem borer was controlled by A and non-additive gene interactions. Resistance to midge was predominantly under the control of additive gene action (Sharma *et al.*, 1996). Ratnadass *et al.* (2002) have reported additive gene action governing head bug resistance.

The genes for yield and plant height were dominant in tropical germplasm while the temperate germplasm harboured dominant alleles for earliness. The tropical × temperate crosses have been shown to produce several high-yielding varieties with desirable plant height (2.0–2.5 m) and maturity (100–110 days). The *guinea* restorer lines contributed to the highest heterosis and grain yield per se in hybrids across the seasons followed by *caudatum* restorer lines. Clasping of glumes was seen in derivatives of *caudatum* and *kafir* crosses. The *guinea* restorers crossed with *caudatum*–*kafir* male-sterile lines can solve the problem of clasping of glumes in hybrids (Reddy and Prasada Rao, 1993). Winter-season adapted landraces cultivated in India belong to *durra* and have excellent grain characteristics for yield and quality, are photoperiod-sensitive and suitable for the prevalent moisture-limiting conditions. The hybrids based on these landraces were found to retain their characteristics with moderate heterosis; that is, around 15% superiority in grain yield over cultivated landraces (ICRISAT, 1995). Therefore, restorers derived from the *rabi* (post-rainy season) landraces were proposed for breaking the yield plateau in the post-rainy season in India (Reddy *et al.*, 2003).

Breeding methodology for simultaneous improvement for grain yield and drought resistance has been established by Reddy (1986). The breeding materials were selected for specific traits such as emergence under crust, seedling vigour, seedling drought recovery and grain yield under drought-prone and yield-potential areas for early-stage drought. For midseason drought, screening under stress-prone areas was suggested. For handling terminal drought, the most frequently occurring mode of drought, screening for drought recovery and grain yield under drought-prone as well as yield-potential areas and breeding for traits such as stay-green, non-lodging and higher grain yield under stress were recommended.

Since sorghum is a crop that is often cross-pollinated; with the CMS system in place, pure lines were developed through pedigree breeding programmes that contributed to hybrids and varieties. Population improvement programmes for specific traits such as high tillering, bold grain and grain mould resistance using *ms3*/*ms7* genetic male sterility was taken up. Thus, in sorghum pedigree selection appears to be more appropriate than population improvement, the latter being mostly used to generate variability. Thus, it is evident that the targeted gene pool approach is appropriate for a programme that aims at a broad geographic mandate (Reddy *et al.*, 2003). A breeding scheme involving simultaneous selection for resistance and grain yield in two programmes was proposed. The maintainer lines which had resistance in the screening trials and good yield in the agronomic trials were converted into male-sterile lines. In this method, male-sterile lines for resistance to pests and diseases were developed in the shortest possible period of four years (Reddy *et al.*, 2003). Considering the independence of antibiosis and the different inheritance patterns of resistance to stem borer at different growth stages (flower, peduncle, head formation), it was proposed that breeding for resistance to stem borer should involve the three traits of foliar and stem damage and percentage of dead hearts (Singh and Rana, 1994). A paired-plot technique with comparisons of infested and non-infested plots was used successfully to identify genotypes with resistance to stem borer (Reddy *et al.*, 2003). Trichomes contributed towards shoot fly resistance in sorghum. Season specificity of trichome development was established based on evaluations of the materials resistant to shoot fly developed during the rainy and post-rainy seasons. On this basis, breeding for season-specific shoot fly resistance breeding materials was proposed (Jayanthi Kamala, 1997). In breeding for grain yield and resistance, it was shown that selection for resistance on a family basis and selecting individual single plants in the selected resistant family for grain yield was most effective (ICRISAT, 1995).

## 29.9 Ideotype Breeding in Sorghum

Depending on the target location, production constraints and end-product utilization, the sorghum improvement programmes across different parts of the world have focused on specific objectives.

### 29.9.1 Grain sorghum

In the 1950s, the US Department of Agriculture (USDA) began assembling, evaluating, characterizing and classifying a base collection of sorghum samples in the USA and India, which continued until the 1960s. Towards this objective, several African and Asian countries contributed their germplasm and support, which resulted in accumulation of 11,000 sorghum lines. Since many of the lines were supposed to be unadapted, a unique 'shuttle-breeding' procedure was followed where breeders grew a first generation of random crossbreeds in the tropics (mainly at Mayagüez, Puerto Rico) where the days are short. The desired lines among them were grown in a temperate zone (Texas) where days are long during the growing season. The lines that had desired traits even in the temperate region were collected. Such lines were photoperiod-insensitive and could grow and produce grain under both tropical and temperate conditions. In the next phase, shorter lines with early maturity were selected from the genetically diverse populations. This was the first ideotype-based breeding method where genotypes broadly adaptable to various day-lengths, shorter height and early maturing were obtained. Thus, out of the immense pool of germplasm with tall stature, slow-growing and photoperiod-sensitive types, suitable only for small farms in the tropics, have come universally useful types for use throughout the world, on any scale. The other traits for which selection was not practised were kept diverse. This helped breeders to fine-tune the lines to meet the specialized demands of specific locations.

The Sorghum Conversion Program is one of the most successful plant breeding programmes and it provides populations that are reservoirs of genes that have become cornerstones for much of the present rise in sorghum production worldwide. Historically, sorghum genetic improvement is related to changes in above-ground biomass production (increased ratio of leaf to stem and higher leaf mass), longer panicle length, decrease in peduncle length and superior root mass (Assefa and Staggenborg, 2011). Sorghum yield improvement is tightly connected to changes in number of panicles per unit land area, increased kernel numbers and increased final total grain weight. Increase in the harvest index is related directly to yield gains in many cereals and has resulted in the 'green revolution'. About 50 years ago, wheat had a harvest index of 32% that has been increased successfully to 48% in some cultivars, which is almost half the weight of above-ground biomass. To bear such heavy weights of grain yield, changes were made in plant stature by reducing height so that they do not get top-heavy and blow over in a summer storm. Similar work has been done in sorghum. Short-statured sorghums can be easily harvested with a combine and this has revolutionized sorghum cultivation in the USA, where all the commercial hybrids are dwarfs. However, in other than developed countries the plant stature is not fashioned in this way for making it amenable for machine harvesting. Initially, sorghums in the USA were tall and had a harvest index of 21 or 22% (about the same as those grown now in West Africa), but careful selection, followed by intensive breeding, has reduced the internode length. Now the harvest index for many improved types used in the USA, Mexico and Argentina is 48–52%, as high as that of wheat. Dwarf sorghums have also been created at research stations in Zambia. This model needs to be pursued in other countries as well.

### *Winter sorghum*

Winter sorghums are photoperiod-sensitive and thermo-insensitive. They are responsive to shorter daylengths. The flowering and maturity coincide with low temperatures and the climatic factors are responsible for good grain quality. They are tolerant to terminal moisture stress and resistant to stalk rot/charcoal rot. The winter sorghums are important from the perspective of both grain and fodder requirements during the season. They usually produce high biomass (grain and stover) and have high lustrous grains with semi-corneous endosperm. The landraces are the native *durras* and dominate the winter sorghum cultivated tract. A landrace selection, M 35-1, developed in 1937, that still dominates the post-rainy season sorghum areas in India, possesses these traits (Sanjana Reddy *et al.*, 2009). As the crop is grown under receding soil moisture conditions, productivity depends on the moisture availability. The area under this crop is divided into three soil depths: shallow, medium and deep. Depending on moisture and soil depth, suitable varieties are available for cultivation (Jirali *et al.*, 2007). Under shallow soils, the cultivars were shorter, flowered and matured early, and in medium–deep soils, mean leaf area, grain number and 1000-grain mass, grain and fodder yields were higher. The ideotype was reflected in the following traits when grown in medium soils under residual moisture conditions: higher grain yield and per-day grain productivity, higher biomass and per-day fodder productivity, harvest index greater than 30, greater earhead exertion (5–10% more than M 35-1), bolder seed (5% bolder than M 35-1), panicle dry matter constituting 50% of total dry matter, and higher relative water content (Sanjana Reddy *et al.*, 2012).

Specific characteristics from among the listed traits can be 'custom-designed' across the world for grain sorghum.

- Grain yield and related traits:
  - higher yield – a balance between grain number and grain size;
  - easy threshing;
  - erect leaves (to increase the amount of sunlight intercepted);
  - greater flag leaf length and width;
  - plant height amenable for mechanical harvesting;
  - resistance to biotic stresses;
  - resistance to diseases such as grain mould, charcoal rot, anthracnose, rust, downy mildew, *Striga* and smuts; and
  - resistance to insects such as shoot fly, stem borer, aphids, greenbug, shoot bug and midges.
- Resistance to abiotic stresses:
  - yield under stresses such as drought, heat, cold, soil acidity and salinity;
  - greater stalk strength;
  - non-senescence or stay-green (to improve grain yield under terminal drought);
  - greater root development (to help plants withstand drought);
  - faster grain filling (to overcome terminal drought); and
  - early flowering.

- Grain quality:
  - grain colour (lighter colours are widely acceptable);
  - better *roti*-making quality when consumed as food and harder grains suitable as feed;
  - increased protein content (greater than 10%);
  - high lysine content;
  - greater digestibility; and
  - suitable for diverse food products.

Combination of traits depends on the area of cultivation. Drought among the abiotic stresses has become an important objective of plant breeding programmes worldwide to maintain the sustainability of the crop in marginal areas. A drought-tolerant sorghum line possessed roots at least 40 cm deeper than a drought-sensitive one, and deeper rooting of stay-green lines under drought conditions was reported (Salih *et al.*, 1999; Vadez *et al.*, 2005). Many quantitative trait loci controlling root traits in sorghum, including root length, number of roots per plant, root volume, root fresh weight and dry weight (Mace *et al.*, 2012), number of brace roots (Li *et al.*, 2014) and nodal root angle (Mace *et al.*, 2012), have been identified. In hot and dry environments with high intensities of solar radiation, smaller leaves are advantageous to reduce transpiration, whereas large leaves with less efficient energy-exchange capacity are advantageous in lower irradiance and cooler and moister environments (Ackerly *et al.*, 2002). Similarly, sorghum diseases and pests are specific to the region and season of cultivation. Shoot fly in India and green bug in the USA are the priority pests to be targeted. Grain mould caused by a complex of fungi has reduced the sorghum area drastically in India, similar to the *Striga* menace in Africa.

Modelling can be used to classify regions and decide the suitable ideotypes for specific adaptation. The APSIM was used to characterize the main water- and heat-stress patterns of the temperate central region of Argentina. Three drought-stress conditions were identified that included a pre-flowering drought stress occurring in 39% of the area, a low terminal-drought stress affecting 38% of the area and a grain-filling drought stress, showing lower frequency of 23%. Flowering heat stress (>33°C) was found at lower latitudes with a frequency of 20–50%. Defined environment can explain observed G × E interactions for yield in an independent data set. This can help to optimize breeding and management strategies across the region of interest. Grouping sites with similar frequency of drought can help to handle the spatial variability and breed for specific adaptation. However, if there is no defined environment, dealing with seasonal variability will be challenging (Carcedo and Gambin, 2019). Genotypes differing in height were found to differ in biomass partitioning among organs and a tall hybrid had significantly increased radiation-use efficiency: a novel finding in sorghum. Introducing the plant height into the model generated simulated phenotypic differences in green leaf area retention during grain filling and thereby enhanced grain yield through managed N dynamics (Hammer *et al.*, 2010).

### 29.9.2 High-biomass sorghum

APSIM is a crop modelling framework that is widely used for predicting the growth and biomass yield of high-biomass sorghum and to identify suitable cultivars and target traits for crop improvement. A modelling study undertaken by Truong *et al.* (2017) showed that crop duration of the high-biomass sorghums had a direct relationship with biomass yield. The long-duration ones had 30% more vegetative growth than short-duration ones under water-limited conditions as they captured water more efficiently. The genotypes with optimized trait for vapour pressure deficit-limited transpiration can be bred for enhancing the productivity and resilience of bioenergy sorghums in water-limited environments, thus bringing more marginal areas under cultivation. Bioconversion efficiency or fermentation efficiency is the term used to measure the degree of conversion of soluble carbohydrates (glucose, fructose, sucrose) and insoluble carbohydrates (cellulose, hemicellulose) into biofuels. Genetic studies for lignocellulosic biomass conversion efficiency trait in sorghum revealed 49 loci associated with enzymatic biomass conversion efficiency and two loci were found to control crystallinity index (Vandenbrink *et al.*, 2013). The process involving breakdown of hydrolytic enzymes in lignocellulosic materials to fermentable sugars is referred to as saccharification, an important step for biofuel production, thereby making less lignin content a preferred trait. The brown midrib mutant (*bmr*) is associated with reduced lignin content in the cell walls and vascular tissues, as well as increased conversion efficiency of sorghum stover to ethanol, and could contribute to increased cellulosic biofuel production (Vermerris *et al.*, 2007).

### 29.9.3 Sweet sorghum

Sweet sorghum is a potential alternative source for ethanol production. It can accumulate juice up to 78% of the total biomass, while the Brix content was found to vary from 14 to 23% (Almodares and Sepahi, 1996; Vinutha *et al.*, 2014). The sugars in the juice extracted from sweet sorghum are easily fermentable. The sugars are composed of 75% sucrose with small amounts (about 2.6%) of fructose and glucose (Kawahigashi *et al.*, 2013). Sweet sorghum has good production potential in tropical climates, while cold temperatures are a major deterrent in temperate regions. The tall and fast-maturing sorghum plants with high Brix content have high potential for breeding as a biofuel crop. However, to reach commercial levels of ethanol for displacement of fossil fuels and also to reduce crop-based biofuels in order to sustain the food requirements, there is a need for breeding and deployment of sweet sorghum ideotypes. The sweet sorghum genotypes should have high biomass, higher juice yields, higher Brix content and overall sugar yield, even when cultivated under marginal conditions, to bring more areas under commercial cultivation. Murray *et al.* (2008) suggested that both traits, high grain yield and soluble sugar content, could be bred into a single sorghum cultivar. Pyramiding genes involved in sucrose metabolism, namely UDP-glucose pyrophosphorylase, sucrose synthase and sucrose phosphate synthase, resulted in higher growth and therefore greater biomass production due to increase in plant height (Coleman *et al.*, 2010). Higher levels of leaf N concentration were correlated with enhanced sugar production of sweet sorghum (Serrão *et al.*, 2012). Agronomic management through application of N fertilizers and breeding for enhanced photosynthetic efficiency can be explored for increasing sugar content in sweet sorghum (Xu *et al.*, 2011). As the photosynthates are not efficiently transported to sink as in the case of grain sorghum, the panicle size is usually small (Sipos *et al.*, 2009) and hence the ideotype should not focus on higher grain yield, while moderate grain yield can be targeted for addressing food sustainability. This can be realized through crossing grain-type seed parents and sweet-type pollen parents (Hunter and Anderson, 1997). Lower expression of two sucrose transporters (*SUT1* and *SUT4*) correlated with higher sugar accumulation in sweet sorghum (Qazi *et al.*, 2012). Greater plant height and higher biomass are generally correlated with late flowering. Per-day biomass production potential can be one component of ideotype (Zhang and Wang, 2015). In addition, cultivars in a wide range of maturity groups can allow for staggered sowing and thereby assure continuous supply of feedstock to industry. The cultivars meant for marginal lands should have drought tolerance traits such as more root biomass in topsoil layers, to absorb the diffusion-limited nutrients from topsoil and reduce runoff on slope terrains, and deeper roots to increase the uptake of water and soluble nutrients (Hirel *et al.*, 2007).

Depending on the climatic zone, the traits required to combine will be different. The cultivars should have the basic traits such as high stem sugar and juice content and high biomass. To achieve the basic traits, the target traits depend on the region of cultivation. In regions with more wind velocity, increase in stem diameter can increase overall sugar yield while resisting the lodging due to windy weather. In other regions, tall plants can be bred for more sugar yield. For northern China and highland Europe, chilling tolerance is desirable for early seed germination, seedling growth and plant establishment. For tropical and subtropical regions, three or four crops can be harvested in a year to increase annual biomass yields. Development of perennial sweet sorghum cultivars can be planned in future breeding programmes that need to be more resilient to extreme environmental conditions (Anami *et al.*, 2015). For arid and semi-arid regions, the ideotype should have enhanced water- and nutrient-use efficiencies apart from terminal drought tolerance.

Thus the sweet sorghum ideotype should contain:

- higher biomass (taller plants/wider stem);
- higher juice yield;

- higher Brix content;
- moderate grain yield;
- higher sugar content (balance between juice yield and sugar content);
- greater percentage of fermentable sugars (sucrose, etc.);
- higher levels of leaf N;
- stay-greenness of leaves;
- higher water- and nutrient-use efficiencies; and
- resistance to biotic and abiotic stresses similar to grain sorghum.

### 29.9.4 Forage sorghum

Development of single-cross, three-way cross, interspecific, single-cut and multi-cut forages is popular in sorghum. During the rainy season, forage hybrids are preferred by farmers compared with the forage varieties grown. Interspecific sorghum–sudangrass hybrids are also being exploited for forage production. Sorghum–sudangrass three-way hybrids are by far the most popular forage hybrids and are based on red-grained sorghum male-sterile lines. The interspecific forage hybrids have not been widely exploited as the variability was low in red-grained sorghum seed parents and sudangrass pollinators. Seed parents are desired to have large-seeded, red- or white-grained, high-density panicles and the sudangrass pollinators should possess resistance to foliar diseases and have high sugar content. Forage quality is as important as forage yield in sorghum for increasing feed intake and improving digestibility. The antinutritional attributes should be low (Smith *et al.*, 1997). Forage quality is determined by: (i) higher contents of crude protein (CP), *in vitro* dry matter digestibility (IVDMD), neutral-detergent fibre (NDF) and acid-detergent fibre (ADF), reflecting degradable and non-degradable proteins, structural and non-structural carbohydrates, lignin and celluloses; and (ii) lower antinutritional attributes such as HCN, oxalic acid, tannins and phenolics. The IVDMD is increased through decreasing cell wall concentration (determined by NDF, *in vitro* fibre digestibility (IVFD)), reducing lignin concentration (measured by ADF), increasing metabolizable energy (measured by water-soluble carbohydrates (WSC)) and/or increasing CP. Bloomless increases digestibility of intact leaves (Cummins and Dobson, 1972). Stay-greenness of leaves also contributes towards improving fodder quality. Brown midrib loci have been reported to improve IVDMD by reducing lignin in sorghum stems by 51% and in leaves by 25% (Porter *et al.*, 1978) and NDF concentration by 13% (Fritz *et al.*, 1981). Tannins are negatively correlated with CP, IVDMD and ADF. Plants with tan plant colour, which is controlled by a recessive gene, have low tannin (8%), while purple plants have 10–18% tannins (Gourley and Lusk, 1978). Phenolics interfere with the digestion of structural carbohydrates and NDF (Raffrenato *et al.*, 2017). HCN is found in sorghum until 45 days after emergence. When absorbed into the blood of animals fed on them, HCN causes cellular asphyxiation and eventual death (Hoveland and Monson, 1980). HCN is under genetic control of a major dominant gene, reinforced by multiple genes with additive effects (Duncan, 1996).

Focused efforts to improve the seed parents and pollinators for forage traits like plant height, biomass and growth index, high tillering, fast growth, stay-green, quick regeneration, thin stems, non-hairy leaves and brown midrib characters, resistance to foliar diseases and stem borer, high stalk sugars, forage intake and digestibility in animals will add further diversity to forage cultivar development.

## 29.10 Conclusions and Outlook

Ideotype breeding involves defining and breeding for the target traits to reach the objectives and differs from classical plant breeding which focuses more on yield. Due to diversification of cropping systems and ever-changing climatic conditions, change in farmers' preferences and several production constraints, breeders need to focus on more traits simultaneously. By studying the genetics and the relationships among the traits, ideotypes can be framed and breeding for such ideotypes makes the objectives easily achievable. Such traits should also be considered in the identification of suitable cultivars in breeding programmes and METs. However, modelling work needs to be more extensively targeted to estimate the impact of climate change with minimum resources. Ideotype approaches are yet to be adopted in commercial breeding programmes.

Sorghum is an important crop of semi-arid regions serving food and fodder needs and has multiple uses, contributing vital shares of the energy, protein, vitamin and mineral requirements for millions of poor people of these regions. Based on the end-use requirements, grain, forage and sweet sorghums with distinct characters of their own are grown. Apart from their basic traits, breeding for resistance to biotic and abiotic stresses, quality traits, value-added traits such as increasing shelf-life and reducing the undesirable attributes like fat content and phenol compounds, HCN content, etc. need to be addressed by crop improvement programmes worldwide. However, the breeding objectives are region/environment-specific and depending on the area grown, they have to be defined. Such a focused breeding programme requires designing of crop ideotypes or product profiles. Focused research is lacking in this area. Based on the background information as discussed, ideotype breeding with respect to the region of cultivation and end use needs to be targeted in the current breeding programmes.

## References


Ackerly, D., Knight, C., Weiss, S., Barton, K. and Starmer, K. (2002) Leaf size, specific leaf area and microhabitat distribution of chaparral woody plants: contrasting patterns in species level and community level analyses. *Oecologia* 130, 449–457. Available at: https://doi.org/10.1007/s004420100805

Almodares, A. and Sepahi, A. (1996) Comparison among sweet sorghum cultivars, lines and hybrids for sugar production. *Annals of Plant Physiology* 10, 50–55.

Anami, S.E., Zhang, L.M., Xia, Y., Zhang, Y.M., Liu, Z.Q. and Jing, H.C. (2015) Sweet sorghum ideotypes: genetic improvement of stress tolerance. *Food and Energy Security* 4, 3–24.

Andrivon, D., Giorgetti, C., Baranger, A., Calonnec, A., Cartolaro, P. *et al.* (2013) Defining and designing plant architectural ideotypes to control epidemics? *European Journal of Plant Pathology* 135, 611–617.

Assefa, Y. and Staggenborg, S.A. (2011) Phenotypic changes in grain sorghum over the last five decades. *Journal of Agronomy and Crop Science* 197, 249–257. Available at: https://doi.org/10.1111/j.1439-037X.2010.00462.x

Berenji, J. and Dahlberg, J. (2004) Perspectives of sorghum in Europe. *Journal of Agronomy and Crop Science* 1905, 332–338.

Bergez, J.E., Chabrier, P., Gary, C., Jeuffroy, M.H., Makowski, D. *et al.* (2013) An open platform to build evaluate and simulate integrated models of farming and agro-ecosystems. *Environmental Modelling & Software* 39, 39–49.

Bogard, M., Ravel, C., Paux, E., Bordes, J., Balfourier, F. *et al.* (2014) Predictions of heading date in bread wheat (*Triticum aestivum* L.) using QTL-based parameters of an ecophysiological model. *Journal of Experimental Botany* 65, 5849–5865. Available at: https://doi.org/10.1093/jxb/eru328

Brown, J.K.M. (2002) Yield penalties of disease resistance in crops. *Plant Biology* 5, 339–344.

Carcedo, A.J.P. and Gambin, B.L. (2019) Sorghum drought and heat stress patterns across the Argentinian temperate central region. *Field Crops Research* 241, 107552. Available at: https://doi.org/10.1016/j.fcr.2019.06.009

Chandler, R.F. Jr (1969) Plant morphological and stand geometry in relation to nitrogen. In: Eastin, J.D., Haskin, F.A., Sullivan, C.Y. and van Bavel, C.H.M. (eds) *Physiological Aspects of Crop Yield*. American Society of Agronomy, Madison, Wisconsin, pp. 265–285.

Charyulu, K.D., Bantilan, M.C.S. and Rajalaxmi, A. (2013) Development and diffusion of sorghum improved cultivars in India: impact on growth and variability in yield. *Presented at 57th AARES Annual Conference, Sydney, New South Wales, 5–8 February 2013*. Australian Agricultural and Resource Economics Society, Vermont, Australia, paper no. 152141.

Charyulu, D.K., Moses Shyam, D., Bantilan, C., Borikar, S.T., Ashok Kumar, A. and Reddy, B.V.S. (2016) *Rainy Season Sorghum Technology Adoption and Impact Study in Maharashtra*. Research Report No. 70. International Crops Research Institute for the Semi-Arid Tropics, Patancheru, India.

Coleman, H.D., Beamish, L., Reid, A., Park, J.Y. and Mansfield, S.D. (2010) Altered sucrose metabolism impacts plant biomass production and flower development. *Transgenic Research* 19, 269–283.

Cummins, D.G. and Dobson, J.W.J. (1972) Digestibility of bloom and bloomless sorghum as determined by a modified *in vitro* technique. *Agronomy Journal* 64, 682–683.

de Wet, J.M.J. and Huckabay, J.P. (1967) The origin of *Sorghum bicolor*. II. Distribution and domestication. *Evolution* 21, 787–802.

Diao, X. (2017) Production and improvement of minor cereals in China. *The Crop Journal* 5, 103–114.

Doggett, H. (1988) *Sorghum*, 2nd edn. Longman Scientific & Technical, Harlow, UK.

Donald, M. (1968) The breeding of crop ideotype. *Euphytica* 17, 385–403.

Duncan, R.R. (1996) Breeding and improvement of forage sorghums for the tropics. *Advances in Agronomy* 57, 161–185.

FAO (2019) FAOSTAT Database. Crops. Food and Agriculture Organization of the United Nations, Rome. Available at: http://www.fao.org/faostat/en/#data/QC (accessed 3 December 2019).

FAO (2020) FAOSTAT Database. Crops. Food and Agriculture Organization of the United Nations, Rome. Available at: http://www.fao.org/faostat/en/#data/QC (accessed 15 March 2021).

Fritz, J.O., Cantrell, R.P., Lechtenberg, V.L., Axtell, J.D. and Hertel, J.M. (1981) Brown midrib mutants in sudangrass and grain sorghum. *Crop Science* 21, 706–709.

Gao, S.J., Wang, Y. and Li, G.Y. (2010) Sorghum breeding and production in China. In: He, Z.H. and Bonjean, A.P.A. (eds) *Cereals in China*. International Maize and Wheat Improvement Center, Mexico City, pp. 97–108.

Gong, F., Wu, X., Zhang, H., Chen, Y. and Wang, W. (2015) Making better maize plants for sustainable grain production in a changing climate. *Frontiers in Plant Science* 6, 835. Available at: https://doi.org/10.3389/fpls.2015.00835

Gourley, L.M. and Lusk, J.W. (1978) Genetic parameters related to sorghum silage quality. *Journal of Dairy Science* 61, 1821–1827.

Hammer, G.L. and Jordan, D.R. (2007) An integrated systems approach to crop improvement. In: Spiertz, J.H.J., Struik, P.C. and van Laar, H.H. (eds) *Scale and Complexity in Plant Systems Research: Gene–Plant–Crop Relations*. Wageningen UR Frontis Series No. 21. Springer, Dordrecht, the Netherlands, pp. 45–61.

Hammer, G.L., Kropff, M.J., Sinclair, T.R. and Porter, J.R. (2002) Future contributions of crop modelling: from heuristics and supporting decision-making to understanding genetic regulation and aiding crop improvement. *European Journal of Agronomy* 18, 15–31.

Hammer, G.L., van Oosterom, E., McLean, G., Chapman, S.C., Broad, I., Harland, P. and Russell, M.C. (2010) Adapting APSIM to model the physiology and genetics of complex adaptive traits in field crops. *Journal of Experimental Botany* 61, 2185–2202.

Hanocq, E., Jeuffroy, M.H., Lejeune-Hénaut, I. and Munier-Jolain, N. (2009) Construire des idéotypes pour des systèmes de culture variés en pois d'hiver. *Innovations Agronomiques* 7, 14–28.

Harlan, J.R. and de Wet, J.M.J. (1972) A simplified classification of cultivated sorghum. *Crop Science* 12, 172–176.

Hirel, B., Le Gouis, J., Ney, B. and Gallais, A. (2007) The challenge of improving nitrogen use efficiency in crop plants: towards a more central role for genetic variability and quantitative genetics within integrated approaches. *Journal of Experimental Botany* 58, 2369–2387.

House, L.R. (1985) *A Guide to Sorghum Breeding*, 2nd edn. International Crops Research Institute for the Semi-Arid Tropics, Patancheru, India.

Hoveland, C.S. and Monson, W.G. (1980) Genetic and environmental effects on forage quality. *Crop Quality, Storage, and Utilization*. American Society of Agronomy and Crop Science Society of America, Madison, Wisconsin, pp. 139–168.

Hunter, E. and Anderson, I. (1997) Sweet sorghum. *Horticultural Research* 21, 73–104.

ICRISAT (1995) Restorers. In: *ICRISAT Asia Region Annual Report 1992*. International Crops Research Institute for the Semi-Arid Tropics, Patancheru, India, p. 75.

Jayanthi Kamala, P.D. (1997) Genetics of shoot fly resistance in sorghum hybrids of cytoplasmic male sterile lines. PhD thesis, Acharya N.G. Ranga Agricultural University, Hyderabad, India.

Jirali, D.I., Biradar, B.D. and Rao, S.S. (2007) Performance of *rabi* sorghum genotypes under receding moisture conditions in different soil types. *Karnataka Journal of Agricultural Sciences* 20, 603–604.

Jotwani, M.G. (1976) Host plant resistance with special reference to sorghum. *Proceedings of the National Academy of Sciences USA* 46, 42–48.

Kalton, R.R. (1988) Overview of forage sorghums. *Proceedings of the Annual Corn Sorghum Research Conference* 43, 1–12.

Kawahigashi, H., Kasuga, S., Okuizumi, H., Hiradate, S. and Yonemaru, J.I. (2013) Evaluation of brix and sugar content in stem juice from sorghum varieties. *Grassland Science* 59, 11–9.
Keating, B.A., Carberry, P.S., Hammer, G.L., Probert, M.E., Robertson, M.J. *et al.* (2003) An overview of APSIM, a model designed for farming systems simulation. *European Journal of Agronomy* 18, 267–288.
Khush, G.S. (1995) Breaking the yield frontier of rice. *GeoJournal* 35, 329–332.
Khush, G.S. (2000) Strategies for increasing yield potential of rice. *Studies in Plant Science* 7, 207–212.
Kiple, K.F. (2000) *The Cambridge World History of Food*. Cambridge University Press, Cambridge, UK.
Li, R., Han, Y., Lv, P., Du, R. and Liu, G. (2014). Molecular mapping of the brace root traits in sorghum (*Sorghum bicolor* L. Moench). *Breeding Science* 64, 193–198. Available at: https://doi.org/10.1270/jsbbs.64.193
Mace, E.S., Singh, V., Van Oosterom, E.J., Hammer, G.L., Hunt, C.H. and Jordan, D.R. (2012) QTL for nodal root angle in sorghum (*Sorghum bicolor* L. Moench) co-locate with QTL for traits associated with drought adaptation. *Theoretical and Applied Genetics* 124, 97–109. Available at: https://doi.org/10.1007/s00122-011-1690-9
McNunn, G., Heaton, E., Archontoulis, S., Licht, M. and VanLoocke, A. (2019) Using a crop modelling framework for precision crop–benefit analysis of variable seeding and nitrogen application rates. *Frontiers in Sustainable Food Systems* 3, 108. Available at: https://doi.org/10.3389/fsufs.2019.00108
Martre, P., Turion, B.Q., Luquet, D., Memmah, M.M.O.S., Chenu, K. and Debaeke, P. (2015) Model-assisted phenotyping and ideotype design: crop physiology: applications for genetic improvement and agronomy. In: Sadras, V. and Calderini, D. (eds) *Crop Physiology: Applications for Genetic Improvement and Agronomy*, 2nd edn. Academic Press, London, pp. 349–373. Available at: https://doi.org/10.1016/B978-0-12-417104-6.00014-5
Mullet, J.E., Rooney, W.L., Klein, P.E., Morishige, D., Murphy, R. and Brady, J.A. (2010) Discovery and utilization of sorghum genes (*Ma5/Ma6*). *US Patent No.* US8309793B2.
Mundia, C.W., Secchi, S., Akamani, K. and Wang, G. (2019) A regional comparison of factors affecting global sorghum production: the case of North America, Asia and Africa's Sahel. *Sustainability* 11, 2135. Available at: https://doi.org/10.3390/su11072135
Murray, S.C., Sharma, A., Rooney, W.L., Klein, P.E., Mullet, J.E. *et al.* (2008) Genetic improvement of sorghum as a biofuel feedstock: I. QTL for stem sugar and grain nonstructural carbohydrates. *Crop Science* 48, 2165–2179.
Murty, U.R. and Rao, N.G.P. (1997) Sorghum. In: Bahl, P.N., Salimath, P.M. and Mandal, A.K. (eds) *Genetics, Cytogenetics and Breeding of Crop Plants*, Vol. 2. *Cereal and Commercial Crops*. IBH Publishing Co. Pvt. Ltd, Oxford, UK, pp. 197–239.
Nikkhah, A., Alikhani, M. and Amanlou, H. (2004) Evaluation of feeding ground or steam-flaked broom sorghum and ground barley on performance of dairy cows in midlactation. *Journal of Dairy Science* 87, 122–130.
Ogbonna, A.C. (2007) Sorghum: an environmentally friendly food and industrial grain in Nigeria. *Nigeria Food Journal* 1, 25.
Oliver, A.L., Grant, R.J., Pedersen, J.F. and O'Rear, J. (2004) Comparison of brown midrib-6 and -18 forage sorghum with conventional sorghum and corn silage in diets of lactating dairy cows. *Journal of Dairy Science* 87, 637–644.
Paleari, L., Bregaglio, S., Cappelli, G., Movedi, E. and Confalonieri, R. (2016) ISIde: a rice modelling platform for *in silico* ideotyping. *Computers and Electronics in Agriculture* 128, 46–49.
Peng, S., Khush, G.S., Virk, P., Tang, Q. and Zou, Y. (2008) Progress in ideotype breeding to increase rice yield potential. *Field Crops Research* 108, 32–38.
Porter, K.S., Axtell, J.D., Lechtenberg, V.L. and Colenbrandu, V.F. (1978) Phenotype fiber composition and *in vitro* dry matter disappearance of chemically induced brown midrib (*bmr*) mutants of sorghum. *Crop Science* 18, 205–208.
Qazi, H.A., Paranjpe, S. and Bhargava, S. (2012) Stem sugar accumulation in sweet sorghum – activity and expression of sucrose metabolizing enzymes and sucrose transporters. *Journal of Plant Physiology* 169, 605–613.
Quinby, J.R. and Karper, R.E. (1954) Inheritance of height in sorghum. *Agronomy Journal* 46, 211–216.
Quinby, J.R., Hesketh, J.D. and Voigt, R.L. (1973) Influence of temperature and photoperiod on floral initiation and leaf number in sorghum. *Crop Science* 13, 243–246.
Raffrenato, E., Fievisohn, R., Cotanch, K.W., Grant, R.J., Chase, L.E. and Van Amburgh, M.E. (2017) Effect of lignin linkages with other plant cell wall components on *in vitro* and *in vivo* neutral detergent

fiber digestibility and rate of digestion of grass forages. *Journal of Dairy Science* 100, 8119–8131. Available at: https://doi.org/10.3168/jds.2016-12364

Rana, B.S. and Murty, B.R. (1971) Genetic analysis of resistance to stem borer in sorghum. *Indian Journal of Genetics* 31, 521–529.

Rana, B.S., Jotwani, M.G. and Rao, N.G.P. (1981) Inheritance of host plant resistance to sorghum shoot fly. *International Journal of Tropical Insect Science* 2, 105–109.

Rana, B.S., Kaul, S. and Rao, M.H. (1997) Impact of genetic improvement on sorghum productivity in India. In: *Proceedings of the International Conference on Genetic Improvement of Sorghum and Pearl Millet, Lubbock, Texas, 22–27 September 1996*. INTSORMIL and ICRISAT, pp. 142–165.

Rasmusson, D.C. (1987) An evaluation of ideotype breeding. *Crop Science* 27, 1140–1146.

Ratnadass, A., Chantereau, J., Coulibaly, M.F. and Cilas, C. (2002) Inheritance of resistance to the panicle-feeding bug (*Eurystylus oldi*) and the sorghum midge (*Stenodiplosis sorghicola*) in sorghum. *Euphytica* 123, 131–138.

Reddy, B.V.S. (1986) Genetic improvement for drought resistance in sorghum: a plant breeder's view point. In: *Genetic Improvement of Drought Resistance, Proceedings of a Discussion Series of the Drought Research Seminar Forums*. International Crops Research Institute for the Semi-Arid Tropics, Patancheru, India pp. 28–32.

Reddy, B.V.S. and Prasada Rao, K.E. (1993) Varietal improvement: genetic diversification in Cereals Program. In: *ICRISAT Annual Report 1992*. International Crops Research Institute for the Semi-Arid Tropics, Patancheru, India, p. 48.

Reddy, B.V.S. and Stenhouse, J.W. (1993) Hybrid Development and Testing. In: *Cereals Program Annual Report 1992*. International Crops Research Institute for the Semi-Arid Tropics, Patancheru, India, p. 45.

Reddy, B.V.S. and Stenhouse, J.W. (1994) Sorghum improvement for semi-arid tropics region: past current and future research thrusts in Asia. *PKV Research Journal* 18, 155–169.

Reddy, B.V.S., Mughogho, L.K. and Jambunathan, R. (1992) Breeding grain mold resistance seed parents and hybrids. In: *Cereals Program Annual Report 1991*. International Crops Research Institute for the Semi-Arid Tropics, Patancheru, India, p. 28.

Reddy, B.V.S., Sanjana, P. and Ramaiah, B. (2003) Strategies for improving post rainy season sorghums: a case study for landrace based hybrid breeding approach. *Paper presented at the Workshop on Heterosis in Guinea Sorghum, Sotuba, Mali*, 10–14 March.

Reddy, B.V.S., Ramesh, S. and Sanjana Reddy, P. (2004) Sorghum breeding research at ICRISAT – goals, strategies, methods and accomplishments. *International Sorghum and Millet Newsletter* 45, 5–12.

Rooney, W.L. (2000) Genetics and cytogenetics. In: Smith, C.W. and Frederiksen, R.A. (eds) *Sorghum: Origin, History, Technology and Production*. Wiley, New York, pp. 261–307.

Salih, A.A., Ali, I.A., Lux, A., Luxova, M., Cohen, Y., Sugimoto, Y. and Inanga, S. (1999) Rooting, water uptake and xylem structure adaptation to drought of two sorghum cultivars. *Crop Science* 39, 168–173. Available at: https://doi.org/10.2135/cropsci1999.0011183X003900010027x

Sani, R.M., Haruna, R. and Sirajo, S. (2013) Economics of sorghum (*Sorghum bicolor* (L) Moench) production in Bauchi local government area of Bauchi State, Nigeria. Invited paper presented at *4th International Conference of the African Association of Agricultural Economists, Hammamet, Tunisia, 22–25 September 2013*. University of Nairobi, Nairobi, pp. 1–12.

Sanjana Reddy, P. and Reddy, B.V.S. (2018) History of sorghum improvement. In: Aruna, C., Visarada, K.B.R.S., Bhat, B.V. and Tonapi, V.A. (eds) *Breeding Sorghum for Diverse End Uses*. Woodhead Publishing, Cambridge, UK, pp. 61–76.

Sanjana Reddy, P., Reddy, B.V.S. and Ashok Kumar, A. (2009) M 35-1 derived sorghum varieties for cultivation during the postrainy season. *E-Journal of SAT Agricultural Research* 7, 1–4.

Sanjana Reddy, P., Patil, J.V., Nirmal, S.V. and Gadakh, S.R. (2012) Improving post-rainy season sorghum productivity in medium soils: does ideotype breeding hold a clue? *Current Science* 102, 904–908.

Semenov, M.A. and Stratonovitch, P. (2013) Designing high-yielding wheat ideotypes for a changing climate. *Food and Energy Security* 2, 185–196.

Serrão, M., Menino, M., Martins, J., Castanheira, N., Lourenço, M. *et al.* (2012) Mineral leaf composition of sweet sorghum in relation to biomass and sugar yields under different nitrogen and salinity conditions. *Communications in Soil Science and Plant Analysis* 43, 2376–2388.

Sharma, G.C., Jotwani, M.G., Rana, B.S. and Rao, N.G.P. (1977) Resistance to sorghum shoot fly [*Atherigona soccata* Rondani] and its genetic analysis. *Journal of Entomological Research* 1, 1–12.

Sharma, G.C., Abraham, C.V., Vidyasagar, P. and Stenhouse, J.W. (1996) Gene action for resistance in *Sorghum bicolor* (L.) Moench to sorghum midge (*Contarinia sorghicola*). *Crop Science* 36, 259–265.

Singh, B.V. and Rana, B.S. (1994) Influence of varietal resistance on disposition and larval development of stalk borer, *Chilo partellus* Swinhoe and its relationship to field tolerance in sorghum. *Insect Science and Its Application* 5, 287–296.

Sipos, B., Reczey, J., Somorai, Z., Kadar, Z., Dienes, D. and Reczey, K. (2009) Sweet sorghum as feedstock for ethanol production: enzymatic hydrolysis of steam-pretreated bagasse. *Applied Biochemistry and Biotechnology* 153, 151–162.

Smith, C.W. and Frederiksen, R.A. (2000) *Sorghum: Origin, History, Technology, and Production*. Wiley, New York.

Smith, K.F., Reed, K.F.M. and Foot, J.Z. (1997) An assessment of the relative importance of specific traits for the genetic improvement of nutritive value in dairy pasture. *Grass and Forage Science* 52, 167–175.

Solie, J.B., Monroe, A.D., Raun, W.R. and Stone, M.L. (2012) Generalized algorithm for variable-rate nitrogen application in cereal grains. *Agronomy Journal* 104, 378–387.

Tao, F., Rotter, R., Palosuo, T., Diaz-Ambrona, C., Minguez, I. *et al.* (2017) Designing future barley ideotypes using a crop model ensemble. *European Journal of Agronomy* 82, 144–162.

Tardieu, F. (2012) Any trait or trait-related allele can confer drought tolerance: just design the right drought scenario. *Journal of Experimental Botany* 63, 25–31.

Tivoli, B., Calonnec, A., Richard, B., Ney, B. and Andrivon, D. (2012) How do plant architectural traits modify the expression and development of epidemics? Consequences for reducing epidemic progress. *European Journal of Plant Pathology* 135, 471–478. Available at: https://doi.org/10.1007/s10658-012-0066-6

Truong, S.K., McCormick, R.F. and Mullet, J.E. (2017) Bioenergy sorghum crop model predicts VPD-limited transpiration traits enhance biomass yield in water-limited environments. *Frontiers of Plant Science* 8, 335. Available at: https://doi.org/10.3389/fpls.2017.00335

Vadez, V., Kashiwagi, J., Krishnamurthy, L., Serraj, R., Sharma, K.K. *et al.* (2005) Recent advances in drought research at ICRISAT: using root traits and rd29a:DREB1A to increase water use and water use efficiency in drought-prone areas. *Poster presented at the Interdrought II Conference, Rome, 24–28 September 2005.*

Van Eeuwijk, F., Bustos-Korts, D. and Malosetti, M. (2016) What should students in plant breeding know about the statistical aspects of genotype × environment interactions? *Crop Science* 56, 2119–2140.

Vandenbrink, J.P., Hilten, R.N., Das, K., Paterson, A.H. and Alex F.F. (2013) Quantitative models of hydrolysis conversion efficiency and biomass crystallinity index for plant breeding. *Plant Breeding* 132, 252–258.

Vermerris, W., Saballos, A., Ejeta, G., Mosier, N.S., Ladisch, M.R. and Carpita, N.C. (2007) Molecular breeding to enhance ethanol production from corn and sorghum stover. *Crop Science* 47, S-142–S-153.

Vinutha, K.S., Rayaprolu, L., Yadagiri, K., Umakanth, A.V., Patil, J.V. and Srinivasa Rao, P. (2014) Sweet sorghum research and development in India: status and prospects. *Sugar Tech* 16, 133–143.

Xu, F., Shi, Y.C., Wu, X., Theerarattananoon, K., Staggenborg, S. and Wang, D. (2011) Sulfuric acid pretreatment and enzymatic hydrolysis of photoperiod sensitive sorghum for ethanol production. *Bioprocess and Biosystems Engineering* 34, 485–492.

Yuan, L.P., Virmani, S.S. and Mao, C.X. (1989) Hybrid rice: achievements and future outlook. In: *Progress in Irrigated Rice Research*. International Rice Research Institute, Manila, pp. 219–235.

Zhang, B. and Wang, Q. (2015) MicroRNA-based biotechnology for plant improvement. *Journal of Cellular Physiology* 230, 1–15.

Zhao, W., Zhang, F., Cheng, Q., Li, M. and Wang, H. (2007) Research and application of sorghum $A_3$ CMS lines. *Crops* 1, 62–64.

Zhen Yang (1997) Sorghum breeding research in China. *International Sorghum and Millets Newsletter* 38, 15–18.

# Index

Note: Page numbers in **bold** type refer to **figures.** Page numbers in *italic* type refer to *tables*